THE LIBRARY
D0215317

EARTH'S EARLIEST BIOSPHERE

EARTH'S EARLIEST BIOSPHERE

ITS ORIGIN AND EVOLUTION

Edited by J. William Schopf

PRINCETON UNIVERSITY PRESS · PRINCETON, NEW JERSEY

Published by Princeton University Press, 41 William Street, Princeton, New Jersey. In the United Kingdom: Princeton University Press, Guildford, Surrey

Library of Congress Cataloging in Publication Data will be found on the last printed page of this book
ISBN 0-691-08323-1
ISBN 0-691-02375-1

This book has been composed in Times Roman

Clothbound editions of Princeton University Press books are printed on acid-free paper, and binding materials are chosen for strength and durability. Paperbacks, while satisfactory for personal collections, are not usually suitable for library rebinding.

Printed in the United States of America by Princeton University Press, Princeton, New Jersey

FRONTISPIECE. Fossil remnants of ancient life: Domical stromatolite (above, ×0.35) and stratiform stromatolitic laminae (below, ×1.43), from strata of the 3.5-Ga-old Warrawoona Group in the North Pole Dome region of northwestern Australia (for discussion, see Chapter 8). Photographs by J. W. Schopf and T. Susuki.

CONTENTS

LIST OF FIGURES

APPENDIX FIGURES

LIST OF PHOTOGRAPHS

LIST OF TABLES

Text Tables

APPENDIX TABLES

LIST OF CONTRIBUTORS

SHERWOOD CHANG
Extraterrestrial Research Division, National Aeronautics and Space Administration, Ames Research Center, Moffett Field, California 94035, U.S.A.

DAVID J. CHAPMAN
Department of Biology, University of California, Los Angeles, California 90024, U.S.A.

PRESTON CLOUD*
Department of Geological Sciences, University of California, Santa Barbara, California 93106, U.S.A.

DAVID DesMARAIS
Extraterrestrial Research Division, National Aeronautics and Space Administration, Ames Research Center, Moffett Field, California 94035, U.S.A.

W. GARY ERNST
Department of Earth and Space Sciences, and Institute of Geophysics and Planetary Physics, University of California, Los Angeles, California 90024, U.S.A.

HOWARD GEST
Department of Biology, Indiana University, Bloomington, Indiana 47405, U.S.A.

JOHN M. HAYES
Departments of Chemistry and Geology, Indiana University, Bloomington, Indiana 47405, U.S.A.

HANS J. HOFMANN
Department of Geology, University of Montreal, Montreal, Quebec H3C 3J7, Canada.

ISAAC R. KAPLAN
Department of Earth and Space Sciences, and Institute of Geophysics and Planetary Physics, University of California, Los Angeles, California 90024, U.S.A.

CORNELIS KLEIN
Department of Geology, Indiana University, Bloomington, Indiana 47405, U.S.A.

RUTH MACK*
Extraterrestrial Research Division, National Aeronautics and Space Administration, Ames Research Center, Moffett Field, California 94035, U.S.A.

UDO MATZIGKEIT
Max-Planck-Institut für Chemie (Otto-Hahn-Institut), Saarstrasse 23, Mainz D-6500, Federal Republic of Germany.

STANLEY L. MILLER
Department of Chemistry, University of California, San Diego, La Jolla, California 92093, U.S.A.

MANFRED SCHIDLOWSKI
Max-Planck-Institut für Chemie (Otto-Hahn-Institut), Saarstrasse 23, Mainz D-6500, Federal Republic of Germany.

J. WILLIAM SCHOPF
Department of Earth and Space Sciences, and Institute of Geophysics and Planetary Physics, University of California, Los Angeles, California 90024, U.S.A.

DAVID J. STEVENSON
Division of Geological and Planetary Sciences, California Institute of Technology, Pasadena, California 91125, U.S.A.

GARY E. STRATHEARN
Department of Earth and Space Sciences, University of California, Los Angeles, California 90024, U.S.A.

JAN VEIZER
Department of Geology, University of Ottawa, Ottawa, Ontario K1N 6N5, Canada.

JAMES C. G. WALKER
Department of Atmospheric and Oceanic Science, The University of Michigan, Ann Arbor, Michigan 48109, U.S.A.

MALCOLM R. WALTER
Baas Becking Geobiological Laboratory, P.O. Box 378, Canberra City, A.C.T. 2601, Australia.

KIM W. WEDEKING
Department of Chemistry, Indiana University, Bloomington, Indiana 47405, U.S.A.

* Invited contributor (non-P.P.R.G. member)

PREFACE

The P.P.R.G.: An Interdisciplinary Experiment

Like so many other fields of natural science, the roots of Precambrian paleobiology extend well into the past century. The pioneering efforts of C. D. Walcott and his contemporaries of the 1880s and 1890s and the historical development of the field are reviewed in depth in the first chapter of this volume by Preston Cloud who over the past quarter-century has himself played a pivotal role in the development of the science. It is only, however, during this past quarter-century that major progress, stimulated particularly by the now classic contributions of Cloud, Martin Glaessner, and Elso Barghoorn, has been made in the study of Precambrian life. As a result of these and related geologic and paleobiologic studies since the mid-1950s it is now recognized that the Precambrian encompasses an enormous sweep of Earth history, characterized by large-scale planetary evolution, by major changes in the Earth's crust, hydrosphere, and atmosphere, and by a mode of biologic evolution based largely on biochemical and intracellular innovations in primitive microbes rather than on the development of new morphologies and innovative body plans that typifies the evolution of higher forms of life. In these and other respects the Precambrian differs, significantly, from later geologic time. It is thus not surprising that as the study of Precambrian life has come of age, the field has taken on a distinctive character, a broad-based multidisciplinary approach that in style and emphasis rather markedly differs from that of the more established subdisciplines of paleontology.

This decidedly multidisciplinary flavor, a hallmark of the science, is both a strength and a weakness. On the one hand, it calls upon disparate, but mutually reinforcing lines of evidence to confront the unsolved problems of the field. Biology, geology, and chemistry, together, seem certain to give a "better answer" than any single discipline alone. Yet such an approach presupposes the ability and inclination of its practioners to master the various disciplines involved—and both to comprehend and to communicate in their specialized, often arcane languages—an ideal not easily achieved. The limitations are not only human, but institutional: the traditional discipline-defined organization of universities, funding agencies, and research institutes is a notoriously effective inhibitor of interdiscipline interaction; students, emulating their teachers, assimilate the all too obvious lesson that specialization is the fast-lane to scientific success; and in an age of dwindling resources and budgetary cutbacks, most institutions are hard-pressed simply to maintain the status quo, much less to strike out in some new, unconventional, multidisciplinary direction. But unlike human institutions, nature is noncompartmentalized. The problems confronting those concerned with the origin and early history of life transcend the normal boundaries of the established sciences; a multidisciplinary approach is thus demanded for their solution.

Given this state of affairs, what could one do about it? Over the years, from talks with numerous colleagues in the field, I had evolved a sort of pipe dream, a notion that if it were possible to set aside the time, and somehow find the means, it would be constructive, productive—and above all, sheer fun—to assemble a multidisciplinary group of specialists at U.C.L.A. to work together on Precambrian problems of mutual interest. For a time, the notion remained just an intriguing possibility, but one that seemed obviously impractical without a source of major funding. Thus, in early February of 1977, when John Hayes and I first broached the subject to Richard Young (currently Executive Director of the American Society for Cell Biology, but at that time controller of the purse strings of N.A.S.A.'s Planetary Biology Program), I was heartened to learn that "in principle" his program "might possibly" be interested in supporting such a project. Thus encouraged, in late March I contacted those who were to be the nucleus of the group (Hayes, Hans Hofmann, Ian Kaplan, David Raup, and Malcolm Walter) to begin the task of putting together a grant proposal. Then, one day in April, less than a month later, it suddenly became apparent that in one form or another the venture would almost certainly come to fruition: I had been selected as recipient of the 1977 Alan T. Waterman Award of the U.S. National Science Foundation, a prize that carried with it \$150,000 in "unrestricted" funds—monies that

I could apply to any project that seemed to me worthwhile; the pipe-dream team was the obvious choice.

In the following months, the project began to take shape: by January 1978 our proposal had been submitted (now requesting only "matching-funds" from N.A.S.A.); in June, by this time calling ourselves the Precambrian Paleobiology Research Group and with our ranks now expanded to some fourteen members, we met at U.C.L.A. for a week-long planning session; in November we were elated to learn that our request to N.A.S.A. had been officially approved; by the following spring, space had been assigned us at U.C.L.A. and our research materials (available rock samples, laboratory equipment, etc.) were being rapidly assembled; in late May, the P.P.R.G. Field Contingent of Hayes, Hofmann, and Walter set off on a four-week trip to Australia, Africa, and Canada to fill in the gaps in our pooled collections of Archean and Proterozoic rock specimens; and in early July, 1979, the entire group met once again, this time to begin work in earnest. In toto, the group included twenty-four scientists (Mario Baur and Don DePaolo of U.C.L.A., George Claypool of the U. S. Geological Survey, Sam Epstein of Cal Tech, David Raup of the University of Chicago, and the 19 other P.P.R.G. members listed as contributors to this volume; Figure P-1), about half of whom were present at U.C.L.A. for the 14-month duration of the project; others were in residence for periods of a few weeks to several months, and the remainder worked solely at their home institutions. During the course of the project, the total group came together every three to four months for quarterly meetings to report research progress, exchange data, and to coordinate ongoing studies. The final meeting, in August, 1980, was held as a three-day symposium, an open forum attended by some 120 additional scientists for the purpose of presenting results of the project to interested members of the scientific community; Preston Cloud delivered the keynote address at the symposium banquet, and Keith Kvenvolden of the U.S. Geological Survey presented a summary and overview of the scientific proceedings.

The P.P.R.G. included two students (Strathearn, Wedeking), one highly talented technician (Matzigkeit), and specialists in the fields of microbiology (Chapman, Gest), paleontology (Raup), Precambrian paleobiology (Hofmann, Schopf, Walter), planetary science (Stevenson), atmospheric physics (Walker), geochronology (DePaolo), mineralogy-petrology (Ernst, Klein), inorganic geochemistry (Veizer), physical chemistry (Baur), synthetic organic chemistry (Chang, Miller), organic geochemistry (Claypool), and light isotope geochemistry (DesMarais, Epstein, Hayes, Kaplan, Schidlowski). Virtually all of the work carried out by the group—from the crushing of the first rock samples to the planning of the August symposium—was a collaborative venture, and this book is no exception: its outline was the result of a group effort; chapters by one or more P.P.R.G. members were reviewed by other members of the group prior to publication; and with the exception of the first author of each chapter (to whom was assigned principal responsibility for the material therein), authorship of the various chapters and appendices is ordered alphabetically. Indeed, one of the most pleasant and rewarding aspects of the venture was the collaborative nature of the entire enterprise and the fact that members of the group developed the knack of "delegating responsibility" to one another. For example, we asked each other to prepare documented "mini-lectures" on specialized topics of mutual interest, reports that were then presented at the quarterly meetings. Our get-togethers were thus spirited and stimulating, and we discovered that interdisciplinary communication is not only possible, but both enjoyable and enlightening as well.

From its inception some four years ago, all of us have viewed this venture as something of an experiment, an exercise in interdisciplinary science that might give some indication of whether this type of format could, in practice, produce tangible results. In fact, a number of my colleagues (as well as several reviewers of our N.A.S.A. proposal!) were initially not at all certain that it would work. In retrospect, however, I think that all of us now believe that on the whole it turned out rather well; there were, of course, disappointments, but the project has yielded a number of interesting discoveries and has resulted in generation of a sizeable body of new data, probably the most extensive such data base now available in the science. Above all, I suspect that as individuals we most profited simply from the broadening of personal horizons that the experience entailed, a matter of each of us being forced to view the entire forest rather than only those few trees with which we were already well familiar. While many questions remain unanswered regarding the origin and evolution of Earth's ear-

FIGURE P-1. The Precambrian Paleobiology Research Group. Front row (left to right): Manfred Schidlowski, Sherwood Chang, J. William Schopf, David DesMarais, Kim W. Wedeking. Middle row: James C. G. Walker, Malcolm R. Walter, John M. Hayes, Isaac R. Kaplan, Howard Gest, David J. Chapman. Back row: Keith A. Kvenvolden (guest speaker at the August 1980 symposium), Gary E. Strathearn, George E. Claypool, David J. Stevenson, W. Gary Ernst, Cornelis Klein, Stanley L. Miller. Not present: Mario Baur, Donald J. DePaolo, Samuel Epstein, Hans J. Hofmann, Udo Matzigkeit, David M. Raup, Ján Veizer. Photograph by L. Weymouth.

liest biosphere, it seems to us that an interdisciplinary effort of the type here attempted provides a promising avenue toward their ultimate solution.

Acknowledgments

A venture of this magnitude simply could not have occurred without the cooperation of a large number of individuals, many of whom went well beyond the bounds of their usual duties in helping the P.P.R.G. accomplish its various tasks. On behalf of my colleagues, therefore, I want to acknowledge the exceptionally generous assistance of the staff of the U.C.L.A. Department of Earth and Space Sciences, especially George Lapins, Diane Hunter, Sue Schuman, and Gail Marshall of the departmental office; Ellen Ledeboer and Hessie

Axelrod of the personnel office; Julie Gunther and Vicki Jones for drafting; Ram Alkaly of the Thin Section Lab; Jerry Stummer of the x-ray Lab; Takeo Susuki and Lowell Weymouth for photographic help; and Julie Knaack who dealt expertly with the problem of finding housing for long-term visitors and their families. In the P.P.R.G. laboratories, the operation was held together by two excellent secretaries, Cheryl Casdorph and, especially, Nancy Nolan as the principal architect of the August symposium with able assistance provided by students Katrin Hafner, Kathryn Hayes, Carl Mendelson, Al Nyberg, and Glen Striegler; together with Yong Leih, Mendelson and Nyberg also provided yeoman service during our quarterly meetings. In Ian Kaplan's Organic Geochemistry Laboratory, help was provided by Pat Breslin, Luis Chu, Tony Idahosa, and, especially, Dave Winter; at Indiana University, stable carbon isotope analyses were carried out by Steve Studley and George Pauly; and at the Max-Planck-Institut für Chemie in Mainz, Germany, carbonate isotope data were gathered by Jan-Dirk Arneth. We are grateful, indeed, to all of these individuals.

The P.P.R.G. Field Contingent was provided generous assistance by numerous individuals: in Australia, by David Groves, John Dunlop, Roger Buick, and L. Bettenay of the University of Western Australia and by David Gregg of the Australian Bureau of Mineral Resources, Geology and Geophysics which also provided a vehicle and other equipment for field work; in southern Africa, by T. R. Mason of the University of Natal and N. J. Grobler of the University of the Orange Free State; and in Canada, by J. B. Henderson of the Geological Survey of Canada and W. A. Padgham of the Department of Indian and Northern Affairs at Yellowknife. In addition, many individuals and institutions provided samples and information to help make our geologic collections and data files as comprehensive as possible. In particular, acknowledgment must be given Janine Sarfati of the Centre Geologique et Geophysique, Montpellier, France, who provided numerous stromatolitic samples for our study; the Australian Bureau of Mineral Resources, Geology and Geophysics, including Marjorie Muir, currently of CRA Exploration Pty. Ltd., Canberra, but at that time a member of the Bureau of Mineral Resources, who made available to us a number of important specimens from her private collections; the Geological Survey of Western Australia, especially the director, A. F. Trendall, who generously made available both specimens and relevant information; the C.S.I.R.O. Division of Mineralogy in Western Australia with special help provided by W. E. Ewers, R. C. Morris, and R. E. Smith; the C.S.I.R.O. Division of Fisheries Research in Western Australia; the Hamersley Exploration Pty. Ltd., Hamersley Iron Pty. Ltd., and Goldsworthy Mining Ltd., all of Western Australia; the Anglo-American Corporation in South Africa with help provided especially by S. G. Buck; the Steeprock Lake Mines Ltd. in Canada; the Geological Survey of Canada; and the N.A.S.A. Ames Research Center in California.

We appreciate also the support of various other institutions that made this venture possible: laboratory and office space was provided by the U.C.L.A. Institute of Geophysics and Planetary Physics; the U.C.L.A. Department of Earth and Space Sciences provided a full range of facilities—from photography and drafting to x-ray diffraction and thin sectioning—and numerous nooks and crannies throughout the Geology Building where we could carry out aspects of our work; the U.C.L.A. Golden Year Program, the U.C.L.A. Divisions of Physical Sciences and of Continuing Education, and Princeton University Press provided funding for our August 1980 symposium; studies at the U.C.L.A. Organic Geochemistry Laboratory were supported by N.A.S.A. Grant NGR 05-007-221 to I. R. Kaplan; studies at Indiana University were supported by N.A.S.A. Grant NGR 15-003-118 to J. M. Hayes and by N.S.F. Grant EAR 76-11740 to C. Klein; studies at Mainz were supported by Deutsche Forshungsgemeinschaft Grant SFB 73 to M. Schidlowski; and support for research carried out by the total group was provided by N.A.S.A. Grant NSG 7489 to the P.P.R.G. and by N.A.S.A. Grant NGR 05-007-407 and N.S.F. Grants DEB 77-22518 (the Alan T. Waterman Award), DEB 77-03888, and DEB 79-21777 to J. W. Schopf.

Acknowledgment for support and assistance in other, more specific aspects of these studies is given at the conclusion of several of the chapters herein. Special thanks should here be extended, however, to the many non-P.P.R.G. scientists who reviewed portions of this volume prior to publication. Among these, B. B. Buchanan of the University of California, Berkeley, P. E. Cloud of the University of California, Santa Barbara, H. D. Holland of Harvard University, J. R. Quayle of the University of Sheffield, M. Schoell of the Federal Institute of Geological Sciences and Natural Resources, Hannover, A. W. Schwartz of the University of Nijmegen, R. Sirevåg of the University of Oslo, H. G. Trüper of the University of Bonn, R. P. Wintsch of Indiana University, and two anonymous reviewers provided by Princeton University Press merit particular mention for their constructive criticism and helpful comments.

Finally, there are several rather more personal acknowledgments that I should like to make. Credit is due Dick Young, without whose encouragement and support this "unorthodox" project would never have eventuated. Thanks are due our wives and families, especially Ingrid, Jane, Janice, and Marilyn, for permitting us to work the long hours and for coping with our sleepless nights that were necessary to get the job done. I am grateful for the facilities and congenial environment provided me by the Beijing Institute of Botany, Academia Sinica, the Nanjing Institute of Geology and Palaeontology, Academia Sinica, and, especially, by Alan W. Schwartz and the Department of Exobiology at the University of Nijmegen, The Netherlands, where I have edited for publication the manuscripts for this book. I much appreciate the help of my wife, Jane, especially her assistance in the onerous task of crosschecking the accuracy of the innumerable references cited in the volume. And above all, I want to acknowledge the friendly cooperation and empathetic understanding of my colleagues in the P.P.R.G. in bringing this pipe dream to fruition; all played a role, but for advice and help I turned most often to John Hayes and Malcolm Walter, and I especially appreciate their contributions to this collaborative venture.

J. William Schopf
Big Bear City, California

EARTH'S EARLIEST BIOSPHERE

How to Read this Book

By J. William Schopf

Purpose of the Book

From its inception, the members of the P.P.R.G. have felt that the ultimate test of this project was the degree to which it resulted in generation of tangible, publishable results. Certainly, as one might expect from the bringing together of such a group of interested specialists, we were convinced that the project would yield substantial scientific spinoff, but in what form, and to what extent, we were initially unsure. In fact, however, in this regard the project seems already to have surpassed our earlier, if rather vague, expectations: to the present, data gathered during the course of the study have contributed to more than a dozen publications (viz., Awramik et al., 1983; Bridgwater et al., 1981; Chameides and Walker, 1981; Gole and Klein, 1981; Schidlowski, 1980a, 1980b, 1981a; Schidlowski and Junge, 1981; Schopf, 1980a, 1980b, 1980c; Schopf and Walter, 1980, 1982; Walker, 1981; Walker et al., 1981; Walter, 1980; Walter et al., 1980; Wedeking and Hayes, 1982). Nevertheless, the majority of these reports, based largely on preliminary findings, have been brief and cursory in scope. What seemed necessary was a more extensive format where we could bring together these and related new findings and place them in the conceptual framework of the field as a whole and at the same time serve to focus our efforts on a major, tangible product. Hence this book, a volume that has four principal goals: (*i*) to report results obtained by the P.P.R.G. during the course of the project; (*ii*) to place these data in a context provided by an in-depth summary of other relevant data now available; (*iii*) to provide an integrated, up-to-date assessment of current evidence relating to the timing and nature of major events in the early history of life; and (*iv*) to highlight unsolved problems and to suggest fruitful lines for future inquiry.

Structure of the Book

The structure of this volume represents something of a compromise, for whereas there was agreement as to the nature of our goals, there was legitimate difference of opinion as to how those goals might here best be met. Of the numerous approaches considered, we opted finally for what might be termed a "vertically model-oriented" presentation—"vertical," because the volume has a strong geochronological component, moving progressively from consideration of older (Archean) to younger (Proterozoic) materials and thus vertically up the conventional geologic column; and "model-oriented," because our overriding aim has been to test, to the extent feasible, certain of the models or working hypotheses that have become widely accepted as organizing principles for the field over the past decade or so. As one might expect, however, by now our original plan of attack has undergone extensive permutation; our final product is thus neither purely "vertical" (many of the chapters being strongly "horizontal" in emphasis as well) nor solely model-dependent (indeed, upon reading each chapter individually, one might be hard-pressed to identify the model-oriented thread that ties them all together).

While it is debatable that this is the best of all conceivable structures for a volume of this type, it does have the advantage of lending itself to the critical analysis of presently accepted scenarios, of permitting, and indeed encouraging the contrast of that which is known with that predicted or demanded by the currently preferred conceptual models. At the same time, however, the deficiencies in the approach are manifest, especially those that derive from its model-oriented character; for if the underlying working hypothesis is fundamentally in error, it may lead us to ask the wrong questions, to seek evidence of a particular process or evolutionary event in materials of the wrong geologic age, or to disregard significant lines of evidence simply because they seem inconsistent with our model-dependent predilections. Throughout the volume we have of course endeavored to guard against these deficiencies, an effort that is particularly explicit in the final chapter of the book where an attempt is made to distinguish between what seems well-established in the field on the basis of "compelling" evidence, what can reasonably be surmised via "presumptive" or "preponderant" evidence, and what can only be speculated from "permissive" or "consistent" evidence.

Because of the length of this book and the diversity of its subject matter, we have designed it so that it can be perused at three differing levels:

(1) An overview of the material can be gained from a cursory reading of the Contents and perusal of the following section of this Introduction, an extended abstract of the volume that highlights, especially, the new findings herein presented.

(2) With the exception of the first chapter of the book, the final summary and conclusions section of each of the chapters ennumerates (à la *Biological Reviews*) the major points discussed therein and makes note of those areas of the field that are in obvious need of additional detailed study; together with the concluding chapter of the volume, these sections are intended to present an up-to-date summary of the current status of the science.

(3) Finally, each of the chapters is designed to be read as a self-contained research paper with research results placed in the context provided by previous studies in that subject area. The individual chapters are cross-referenced within the text so that interested readers can locate relevant ancillary material elsewhere in the volume.

Geologic, geochronologic, and geographic data relating to the several hundred rock specimens analyzed by the P.P.R.G., and details relating to processing procedures, analytical techniques, and various mathematical computations referred to in the text have been brought together in a series of appendices toward the back of the volume. As necessitated by the multidisciplinary nature of the book, an extensive glossary is also included, both of technical terms and of the specialized, and in part new terms herein used to refer to biological energy conversions, electron transport processes, interactions of cellular systems with molecular oxygen, and carbon nutrition. References cited in the various chapters are listed together in a single section at the back of the book.

Overview of the Book

This volume is concerned with the Archean and Early Proterozoic segments of Earth history, the earlier three-quarters of the Precambrian Eon spanning the time range from 4.5 to 1.6 Ga (giga anna, 10^9 years) ago. This relatively unexplored portion of the geologic past, comprising some 65% of the total history of the planet, was a time of momentous geologic events including the formation of the Earth itself, the segregation of its core and mantle, the development of its continents and the onset of plate tectonics, and the formation of its atmosphere and oceans and their evolution from an anoxic to an oxygenic state. Biologically it was no less significant, a time of the origin and earliest evolution of living systems, of the development of such biochemical processes as fermentation, photosynthesis and aerobiosis, and of the evolution of a global oxygen-dependent ecosystem from its more primitive, fully anaerobic precursors. In terms both of geology and biology, the volume thus encompasses the period of transition from a formative, primordial stage in the history of the planet and of life to the establishment of a "normal" geologic environment and a physiologically "modern" biosphere.

In the first chapter of the book (Early biogeologic history: the emergence of a paradigm), Preston Cloud traces the historical development of Precambrian paleobiology from its halting beginnings in the late 1800s to its emergence during recent decades as an established subdiscipline of paleontological science. As pivotal in this development, Cloud emphasizes the 1961 Woodring Conference on Major Biologic Innovations and the Geologic Record, an interdisciplinary workshop organized by Cloud and P. H. Ableson and attended by a score of other workers, several of whom (e.g., E. S. Barghoorn, M. Calvin, S. W. Fox, T. C. Hoering, and H. D. Holland) are now recognized leaders in the field. The chapter includes an evaluation of the more notable contributions of numerous workers worldwide, and a useful synopsis of recent and current activities both in China and in the Soviet Union. Throughout this authoritative, first-person narrative, Cloud seeks to tie together the antecedents of the conceptual model that serves as the underpinning for this volume and much of the other current writing in the field—the now widely embraced "new biogeological paradigm [that] was first advanced in preliminary form in the middle 1960s by Cloud . . . and subsequently modified by him . . . to accomodate new data that accumulated as others probed its weaknesses and challenged its validity on one point or another." The essential elements of this working hypothesis are (*i*) an initially anoxic terrestrial environment; (*ii*) a pervasive interreaction between early biospheric, atmospheric, hydrospheric, and lithospheric evolution; (*iii*) an extensive, Archean-Early Proterozoic interval of fluctuating imbalance between sources and sinks of molecular oxygen that with the development and dispersal of microbial, oxygenic photosynthesizers yielded ultimately to the establishment of an aerobic environment; and (*iv*) a concurrent biospheric evolution from anaerobic, heterotrophic precursors to prokaryotic aerobes (including aerobic photoautotrophs) and, subsequently, to higher, eukaryotic forms of life. In referring to this unifying scenario, Cloud uses the Kuhnian term "paradigm" advisedly, for he notes that these new concepts "depart sharply from once widely held views of substantive uniformity in historical processes," the scenario thus representing "a dramatic shift of previously prevailing views."

The following two chapters (Chapter 2: The nature of the Earth prior to the oldest known rock record: the Hadean Earth, by David J. Stevenson; and Chapter 3: The early Earth and the Archean rock record, by W. Gary Ernst) provide the cosmogonic and geologic setting for subsequent discussion of the origin and earliest development of life.

In Chapter 2, Stevenson outlines the timing and nature of those events thought to have led to the formation of the protoplanet Earth via "homogeneous accretion" some 4.5 Ga ago. A simplified mathematical model of the parameters involved indicates that this process would have resulted in generation of high internal planetary temperatures leading both to the subsolidus convection of silicates and the melting of free iron. According to the scenario here favored, silicate melting would have been confined to a hot, relatively near-surface rind-like zone, whereas the iron would have percolated downward toward the cooler interior of the planet to form a spheroidal liquid shell. To account for the required geologically rapid displacement of this iron-rich layer to form the Earth's geomagnetic core, Stevenson proposes a new model of core formation—that of "catastrophic asymmetry"—one that envisions the development of the core as having occurred simultaneously with planetary accretion, and, thus, over a period of some 10^6 to 10^8 years. As a corollary to this novel scenario, mantle differentiation and associated outgassing of the Earth's interior are for the most part regarded as having similarly occurred "catastrophically," that is, within a geologically short period of time very early in the history of the planet.

In Chapter 3, Ernst reviews the known Archean rock record—the generally highly metamorphosed scraps and pieces of geologic terranes greater than 2.5 Ga old that have survived to the present providing the only direct evidence still available of the Earth's most ancient crustal veneer. Such units are now known from western Greenland, the north-central United States, peninsular India, Western Australia, southern Africa, Canada, Brazil, Fennoscandia, and elsewhere; whether of supracrustal or of plutonic origin, however, they tend to differ, significantly (e.g., in mineralogy, petrology, dominant rock type, style of structural deformation, etc.), from otherwise comparable deposits of younger geologic age. These differences (summarized in Table 3-1), together with analysis of the thermal regime of the early Earth and of other relevant parameters (including the petrotectonic evolution of the other terrestrial planets), lead Ernst to suggest a simplified four-stage model for the plate tectonic evolution of the Earth: (*i*) a Hadean (>3.9 Ga ago) *pre-plate stage*, characterized by a partially molten planet that underwent gravitative separation to form a metallic core, silicate mantle, and a continuously reworked crustal residuum; (*ii*) an Early to Late Archean (3.9 to 2.5 Ga ago) *platelet stage*, with the planet's surface being dominated by hot, thin, unsubductable sialic slabs that aggregated to form protocontinents; (*iii*) an Early, Middle, and possibly Late Proterozoic (2.5 to 1.0 or perhaps 0.6 Ga ago) *supercratonal stage*, characterized by the emergence of large land masses and the development of freeboard, by intracratonal orogeny, and by the drift of supercontinents; and (*iv*) a late Precambrian-Phanerozoic (1.0 or 0.6 Ga ago to the present) Wilson-type cycle of plate tectonics involving rifting, the dispersal of continental fragments, subduction, and the generation of linear mobile belts at or near plate margins.

As a natural sequel to the contributions of Stevenson and Ernst, the following chapter by Sherwood Chang, David DesMarais, Ruth Mack, Stanley L. Miller, and Gary E. Strathearn seeks to place current studies of prebiotic organic syntheses and the origin of life in their appropriate geologic context; attention is focused especially on the geologically determined atmospheric setting in which such syntheses may have occurred. In this geologic emphasis the discussion differs, notably, from much of the other recent writing in this field. The chapter begins with a detailed discussion of the historical development of this subject area—a review of the intellectual co-evolution of studies of prebiotic organic chemistry and of the Earth's early environment—followed by a synopsis of the lines of evidence that support the concept of chemical evolution, including a useful summary of results obtained from laboratory syntheses under various atmospheric conditions. The question then posed is how such materials, once abiotically produced and deposited and preserved in ancient sediments, might be distinguished from similarly preserved biotic products, a matter of crucial import to establishment via direct evidence of the validity of the chemical evolution paradigm. To address this issue, the experimental approach adopted by Chang et al. is to synthesize organic matter by a series of plausibly prebiotic processes, to examine the elemental (viz., N/C and H/C ratios) and isotopic compositions (viz., the $\delta^{13}C$, $\delta^{15}N$ and δD values) of these products both before and after subjecting them to artificial diagenesis, and to compare the measured compositions with those observed in biologic organic matter. The results obtained indicate that neither the elemental nor the isotopic compositions investigated can be used by themselves to distinguish unambiguously between geologically preserved abiogenic and biogenic materials (and, interestingly, that carbonaceous matter produced in spark discharge experiments using CO_2 as a carbon source can be more than 20 permil depleted in ^{13}C relative to the starting material, an isotopic discrimination in the same direction as, and of comparable magnitude to, that observed during carbon fixation via biological photosynthesis).

Finally, as a reprise of topics touched on in Chapters 2 and 3, Chang et al. present an in-depth discussion of current models of the formation of the Earth and its core, and of the implications of these models for the composition of the atmospheric "reaction mixture" from which initially may have derived the components of life. Based on an approach similar to that first introduced by H. D. Holland, results are presented of calculations designed to determine the makeup of seven model atmospheres that would be expected to have been produced via outgassing of various types of magmas held at two different temperatures. All of the atmospheres thus calculated are reducing, with H_2 supplying the bulk of reducing power, and all are anaerobic, having a pO_2 less than 10^{-8} atm. In gaseous composition, however, they vary from a strongly reducing H_2-, CH_4-, CO-, NH_3-rich mixture, to "mildly reducing" atmospheres in which CO_2 accounts for more than 88% of the available carbon, CO is orders of magnitude more abundant than CH_4, and in which nitrogen occurs almost entirely as N_2 rather than as NH_3. As a result of these and other considerations, the authors conclude that any of a rather wide range of anaerobic atmospheric compositions can be plausibly postulated for the prebiotic Earth, compositions that range from hydrogen- and methane-rich (the type of setting in which abiotic syntheses could have proceeded readily) to atmospheres containing mostly N_2 with only minor to trace amounts of H_2, CO, and CO_2 (and a type of environment in which prebiotic syntheses would apparently have been far less facile). Because of this uncertainty and the fundamental importance of atmospheric composition to current concepts of the origin of life, Chang et al. "advocate study of pathways for organic syntheses in all model environments that continue to be consistent with evidence unveiled in the cosmochemical, geologic and biologic records," and urge "a strongly interdisciplinary attack on [the] many correlated issues" that are as yet unresolved regarding both the timing and nature of the beginnings of biologic activity.

Chapter 5 (Precambrian organic geochemistry, preservation of the record, by J. M. Hayes, Isaac R. Kaplan, and Kim W. Wedeking) presents a scholarly discussion both of previous studies of the organic geochemistry of Precambrian sediments and of new data gathered during the course of the P.P.R.G. project. The chapter is encyclopedic, including up-to-date compilations of such relevant information as the quantitative aspects of the modern carbon cycle (Table 5-1); the relationship between the color of the particulate carbonaceous kerogen of ancient sediments and its elemental composition (Table 5-3); correlations between the degree of graphitization of preserved organics and other indicators of their maturity and metamorphic grade (e.g., H/C ratios, coal rank, and mineralogically indicated metamorphic facies; Table 5-4); and products reported from degradative analysis (via pyrolysis, oxidation, and ozonolysis) of Precambrian kerogens (Table 5-6). Hayes et al. also present results of some 320 new measurements of the total organic carbon content and the $\delta^{13}C$ values of organic carbon in about 50 Archean and Proterozoic geologic formations (Table 5-7); a summary of 33 previously reported analyses of the H/C ratios in kerogens from 4 Archean and 17 Proterozoic units (Table 5-5); and results of new analyses of the H/C and N/C ratios, percent carbon content, $\delta^{13}C$, $\delta^{15}N$ and δD values, and the color and x-ray diffraction characteristics of 42 kerogens from 11 Archean and 13 Proterozoic geologic units (Table 5-8).

To a major extent, Chapter 5 focuses on the fate of organic matter in ancient sediments and the critical question of what types of information can be expected to have survived the geochemically harsh conditions to which such material has been subjected. Hayes et al. conclude that (*i*) although the extractable organic components (i.e., the molecular "chemical fossils") of ancient sediments tend to be more readily interpretable biochemically than the associated, insoluble, macromolecular kerogen, the low concentration of such molecules present in Precambrian deposits and their solubility and resulting geologic mobility make them a relatively unreliable source of paleobiologic information; (*ii*) most Precambrian kerogens—including all of those isolated from Archean age deposits—have been extensively dehydrogenated and, thus, severely altered structurally; (*iii*) the degree of such structural alteration, evidenced by a wide variety of parameters, is usually well correlative with the mineralogically indicated thermal history of the kerogen-bearing sediments; (*iv*) carbon isotopic abundances vary far more widely in Precambrian than in Phanerozoic organic matter; and (*v*) although post-depositional effects can result in alteration of such abundances, the observed changes tend to be of relatively small magnitude—the unusual, and in some cases extreme depletion of ^{13}C here reported for numerous Precambrian kerogens (e.g., $\delta^{13}C$ values < -35 ‰ for kerogens in 16 of 19 sedimentary rock samples from the ~2.6-Ga-old Ventersdorp Group and $\delta^{13}C$ values < -40 ‰, including several samples < -50 ‰, for 18 of 21 sedimentary kerogens from the ~2.8 Ga old Fortescue Group) thus seems to be of primary, rather than of secondary and solely inorganic orgin, an issue addressed in more detail in Chapter 12 of the volume.

Of necessity, the search for direct evidence of the origin and evolution of Earth's earliest biosphere must focus on the one repository where such primary evidence may have been sequestered—the preserved rock record. Yet, as is evident from the foregoing chapters of the book, (*i*) rocks are as yet unknown that date from the earliest (Hadean) stage of the development of the planet, a segment of Earth history that may well have encompassed the beginnings of life; (*ii*) rock sequences of Archean age are of rare occurrence, limited areal extent, and are predominantly of high metamorphic grade; and (*iii*) the geochemical remnants of ancient biologic and/or abiotic organic compounds occurring in such sequences are chemically complex and so highly altered that much of the information they originally contained has been all but obliterated. Direct evidence of the earliest history of life is thus far less accessible than one might wish. The picture, however, is not entirely bleak, for in addition to constraints set by the various paleontologic and isotopic data that are now available (discussed in Chapters 7, 8, and 9), it is important to recognize that the geologic record is not the sole repository of potentially useful, significant information. Sequestered within the biochemistry of each extant organism is yet another type of record—albeit one that has been abridged and otherwise modified over the course of time—a molecular record of the past development of its particular evolutionary lineage. Comparative metabolic and biochemical studies of extant organisms can thus provide important insight regarding the nature and occurrence of ancient evolutionary events. The cardinal problem in such studies is to identify those biochemical and molecular biological features that faithfully reflect the history of the evolving lineage. Chapter 6 (Biochemical evolution of anaerobic energy conversion: the transition from fermentation to anoxygenic photosynthesis, by Howard Gest and J. William Schopf) summarizes results of one approach to this problem, with emphasis on the metabolic characteristics of the earliest forms of life and, in particular, on the development of anaerobic photosynthetic prokaryotes from their more primitive fermentative ancestors.

Gest and Schopf begin their analysis by emphasizing the pervasive role played by sugars and sugar metabolism in modern organisms, the ubiquity and importance of which (together with the fact that saccharides can be readily synthesized under plausible "primitive Earth conditions") lead them to suggest that "among the earliest forms of life were anaerobic, heterotrophic prokaryotes capable of fermenting hexose sugar via glycolysis to produce pyruvate and, ultimately, lactic acid." A series of evolutionary steps is then proposed that leads sequentially from these earliest fermenters to forms capable of "accessory oxidant-dependent" fermentation, to anaerobes possessing a primitive "electrophosphorylation module" and, ultimately, to the earliest microbes capable of photophosphorylation, anaerobic photosynthetic bacteria. Evolutionary refinement of the primitive photosystem of such bacteria is postulated to have resulted in the subsequent development of oxygenic (i.e., cyanobacterial) photosynthesis. The timing of these events is a matter that can only be established by referral to the available fossil record, but as the authors note, that they in fact occurred "is perhaps best evidenced by the bioenergetic-biosynthetic coupling so manifest in the metabolism of modern cells"—to paraphrase the well-known nineteenth-century dictum of Charles Lyell, in biology as well as geology it seems evident that "the present is the key to the past."

When did these events occur? What does the geologic record suggest regarding the time of origin of fermentation, electrophosphorylation, autotrophy, and the like? The following chapter (Chapter 7: Isotopic inferences of ancient biochemistries: carbon, sulfur, hydrogen and nitrogen, by Manfred Schidlowski, J. M. Hayes, and Isaac R. Kaplan) summarizes the application of isotope data to questions of this type, focusing particularly on analysis of the available carbon and sulfur isotopic records. With regard to carbon chemistry in particular, the chapter represents a natural extension of the preceding two contributions—based on the isotope data presented in Chapter 5 and the evolutionary framework provided by Chapter 6 it addresses the question of when in Earth history did CO_2-fixing autotrophy first evolve.

To approach this problem, Schidlowski et al. first discuss the major pathways of biological carbon fixation (summarized in Table 7-1), the biochemistry of the carbon isotope fractionation associated with such fixation, and the range of fractionations observed in extant plants and microorganisms (summarized in Figure 7-1). Following consideration of the effects of preservation and metamorphism on the isotope record, they summarize in graphic form (Figure 7-3) the nearly 8,000 measurements of $\delta^{13}C$ now available for carbonate and organic carbon in Phanerozoic and Precambrian deposits. Based on these data they conclude that (*i*) the isotopic composition of sedimentary carbonates has varied within relatively narrow limits over time ($\delta^{13}C_{carbonate} = 0.4 \pm 2.6$ ‰), suggesting that the carbon isotopic content of marine bicarbonate has held more or less constant since early in the Precambrian; (*ii*) the great majority of isotope ratios measured in sedimentary organic matter 3.5 Ga in age or younger fall in the range $\delta^{13}C_{organic} = -27 \pm 7$ ‰, values typical of autotrophically produced organic matter and thus consistent with the existence of autotrophs since the Early Archean; (*iii*) positive deviations of $\delta^{13}C_{organic}$ from this range (reflecting a relative increase in ^{13}C content) commonly are a result of metamorphic alteration; (*iv*) marked negative deviations of such values (e.g., the extreme depletions in ^{13}C reported in Chapter 5 for kerogens 2.6 to 2.8 Ga old) suggest the involvement of methane-utilizing microorganisms in the formation of the kerogen precursors; and (*v*) in comparison with values typical of most materials of younger geologic age, the carbon isotopic values from the oldest sediments now known, those of the Isua Supracrustal Group of western Greenland, are anomalous ($\delta^{13}C_{carbonate} = -2.3 \pm 2.2$ ‰, $n = 187$; $\delta^{13}C_{organic} = -13.0 \pm 4.9$ ‰, $n = 78$). The Isua sediments, however, have been severely metamorphosed and it is thus reasonable to postulate that their initial isotopic composition may have been consistent with the presence of autotrophic forms of life.

In the remainder of the chapter, Schidlowski et al. consider, in turn, available sulfur, hydrogen, and nitrogen isotopic data. Based on arguments analogous to those presented for carbon, they suggest that "isotope patterns of sedimentary sulfides attributable to microbial sulfate reducers can be traced back to about 2.7 Ga ago," but that sufficient data are as yet unavailable regarding the H/D and $^{15}N/^{14}N$ content of ancient organic matter to permit effective use of these isotope ratios in studies of Precambrian paleobiology.

The following two chapters (Chapter 8: Archean stromatolites: evidence of the Earth's earliest benthos, by Malcolm R. Walter; and Chapter 9: Archean microfossils: new evidence of ancient microbes, by J. William Schopf and Malcolm R. Walter) present both a review of previous studies and an impressive body of new data relating to the morphological record of Archean life. Both chapters deal with the accretionary, organosedimentary structures termed "stromatolites"—distinctively laminated megascopic objects produced as a result of the growth and metabolic activities of benthonic, mat-building, microbial communities. Chapter 8 focuses on the stromatolites themselves, and Chapter 9 deals in part with new discoveries of cellularly preserved microfossils in Archean stromatolitic sediments.

In Chapter 8, Walter presents a well-documented discussion of the occurrence and paleobiologic significance of the eight demonstrably stromatolitic units and the three possibly stromatolitic deposits that are now known from sediments of Archean age. Of these eleven occurrences (in the Warrawoona, Fortescue, Hamersley, and Turee Creek Groups of Western Australia; the Onverwacht, Insuzi, and Bulawayan Groups, and the Ventersdorp Supergroup, of southern Africa; and in the Uchi Greenstone Belt, Yellowknife Supergroup, and the Steeprock Group of Canada), most were discovered or first described only during the past few years. This chapter, which presents new observations regarding many of these deposits, summarizes results of the first in-depth study in which all of these units have been considered together and compared. Walter finds that although Archean stromatolites "are not markedly different from younger examples," based on the evidence available it would nevertheless be premature to infer that they are necessarily of cyanobacterial origin (as is commonly assumed for Proterozoic and younger stromatolites) rather than being products of other kinds of prokaryotic communities. Analysis of the microstructure of Early Archean (~3.5 Ga old) stromatolites from the Warrawoona Group leads him to suggest that they were constructed by finely filamentous microbes, an interpretation consistent with the discovery discussed in Chapter 9 of filamentous microfossils in this same geologic unit. Walter finds also that stromatolites had become "an abundant component of the sedimentary environment at least as early as about 2.8 Ga ago"—a notable departure from many earlier studies that had placed their first widespread appearance more than 500 million years later—and that several lines of evidence suggest (but do not prove) that perhaps unlike stromatolites of Early and Middle Archean age, the abundant stromatolitic deposits of the Late Archean may have been produced by biocoenoses dominated by cyanobacterial, oxygenic, photosynthesizers.

During the past half-century, some 43 categories of microfossils and "microfossil-like objects" have been reported from at least 28 Archean units; of the more than 80 publications dealing with these occurrences, well over three-quarters have appeared since 1965. In Chapter 9, Schopf and Walter critically evaluate each of these reported occurrences, arguing that to be acceptable such reports must establish that (*i*) the rock investigated is in fact of Archean age, and that the reported "fossil-like objects" are (*ii*) indigenous to the rock in question, (*iii*) syngenetic with primary lithification, and (*iv*) are assuredly biogenic. As tabulated in an extensive annotated bibliography (Table 9-1), they conclude that "the vast majority of reported 'Archean microfossils' are in fact nonfossil: inorganic pseudofossils, organic pseudofossils, artifacts of sample preparation, or modern contaminants." Indeed, they suggest that although possible microfossils have been reported from several Archean deposits, "authentic, well-established Archean microbiotas are now known from but two geologic units—bedded stromatolitic cherts of the ~3.5-Ga-old Warrawoona Group . . . and cherty portions of stromatolitic limestones of the ~2.8-Ga-old Fortescue Group," both of Western Australia and both discovered only quite recently. In addition to including the first published illustrations of the Fortescue fossils, the chapter contains a well-documented discussion of the several microbial taxa of the Warrawoona assemblage, forms interpreted as probably the oldest microfossils now known, and concludes that this entirely prokaryotic biota probably included both heterotrophic and autotrophic microorganisms of which some, and possibly all, were apparently anaerobic; the major mat-building taxa of the assemblage are inferred to have been filamentous, probably photoresponsive, and possibly photoautotrophic. Although available data do not indicate whether or not the putative photoautotrophs were oxygen-producing (rather, for example, than being anaerobic photosynthetic bacteria), the diversity and complexity of the biocoenose is interpreted to "indicate that the beginnings of life on Earth must have occurred substantially earlier" than deposition of the Warrawoona sediments, some 3.5 Ga ago.

The following two chapters (Chapter 10: Geologic evolution of the Archean-Early Proterozoic Earth, by Ján Veizer; and Chapter 11: Environmental evolution of the Archean-Early Proterozoic Earth, by James C. G. Walker, Cornelis Klein, Manfred Schidlowski, J. William Schopf, David J. Stevenson, and Malcolm R. Walter) set the stage for consideration of one of the most momentous events to have occurred during the history of life, the development of aerobiosis and of the anaerobic-aerobic ecosystem that characterizes the modern Earth.

Building upon concepts introduced in the contributions of Stevenson (Chapter 2) and Ernst (Chapter 3), in Chapter 10 Veizer presents a quantitative, analytical assessment of the Archean-Early Proterozoic geologic evolution of the planet, a stage in Earth history that he views as having been "dominated by the diachronous transformation of tectonically unstable greenstone-granite belts and high-grade gneissic terranes to produce stabilized cratons" beginning about 3.1 Ga ago. The chapter presents a summary of the known distribution of Archean crustal segments (Figure 10-1), a graphical comparison of Archean and post-Archean crustal chemistry (Figure 10-6), and a quantitative consideration of theoretically possible modes of crustal growth (Figure 10-3). The onset of extensive cratonization and associated changes in the internal heat flux of the planet are regarded by Veizer as having had profound effects on the Earth's sedimentary regime, resulting in an important increase in the influence of the continental river flux on the chemistry of sea water, in a marked increase in the total global sedimentary mass, and in major changes in the Earth's sedimentary veneer, a transition from deposition of sequences dominated by volcaniclastics and immature first-cycle sediments to mature shelf and platform assemblages. Chemically, the transition is regarded as being evidenced by secular changes from mafic to felsic compositions in clastic rocks and by elemental and isotopic variations in the composition of chemical sediments. The Archean-Early Proterozoic thus appears to have been a time of major transformation of the Earth's surficial geologic setting, probably the most profound such development to have occurred in all of Earth history.

In the following chapter, Walker et al. consider the environmental changes of potential biologic import that may have accompanied this transition, focusing particularly on day length, average surface temperature, the partial pressures of atmospheric carbon dioxide and of molecular oxygen, and on the availability of required nutrient elements. The chapter is broadly interdisciplinary and summarizes a wealth of information including such relevant geologic data as the known Precambrian distribution of banded iron-formations (Table 11-1, Figures 11-9 and 11-11); of sedimentary sulfate evaporites (Figure 11-2); of glaciogenic sequences (Figure 11-3); and of red beds (and their unoxidized sedimentary equivalents), iron-leached and apparently unleached paleosols, and uraninite-bearing conglomerates (Figure 11-12). Based on such evidence, and on relevant theoretical considerations, the authors conclude that "there is no indication of profound changes in most biologically important parameters since the beginnings of the known rock record." They find that "liquid water oceans and an atmosphere containing carbon dioxide and water vapor (but presumably composed chiefly of nitrogen . . .) have been extant; the chemistry of sea water has remained relatively unchanged; shallow water, open ocean and land surface environments have existed; day lengths have been shorter . . . and average tidal amplitudes have been larger . . . but not markedly so; and the average global climate has apparently been neither decidedly hotter, nor appreciably colder, than that of more recent Earth history." However, three other parameters are singled out as likely to have experienced biologically important change: (*i*) the partial pressure of atmospheric carbon dioxide is inferred possibly to have been greater—perhaps by as much as two orders of magnitude—in the Archean than in subsequent geologic time; (*ii*) the redox state of the Earth's atmosphere and hydrosphere is regarded as having evolved from a reducing to an oxidizing condition during the Early Proterozoic; and (*iii*) the inferred development of a global aerobic environment about 1.7 Ga ago is suggested to have resulted in major modification of the geochemical cycles of a number of biologically important nutrient elements. Thus, a key aspect of these inferred environmental changes is the advent of oxygenic conditions, a topic that forms the basis of the following two chapters of the volume.

It is well established that the photodissociation of water vapor induced by ultraviolet light can result in a net production of uncombined atmospheric oxygen, but that this inorganic mechanism is a weak oxygen source, one capable of producing only trace amounts of free atmospheric oxygen. Because the present atmosphere is highly oxygenic (and because all higher forms of life, and many prokaryotes as well are obligate aerobes), there is obvious need for some other, appreciably more powerful oxygen source. The most plausible such source is oxygenic photosynthesis, a process that first evolved in cyanobacteria (or their immediate precursors) during the Precambrian.

The origin of oxygen-producing photosynthesis—of the *biologically* mediated photodissociation of water—was thus of crucial import in the history of life. The timing of this event, however, is a matter of uncertainty, chiefly because the lines of evidence that can be brought to bear on the problem tend to be qualitative rather than quantitative, permissive rather than compelling. As is discussed in Chapter 11, available geologic and mineralogic evidence, especially the inferred mode and rate of deposition of Early Proterozoic iron-formations, seems to require the presence of biologically produced free oxygen at least as early as 2.5 Ga ago. Analyses of Archean stromatolites in Chapter 8 and of cellularly preserved microfossils in Chapter 9 are consistent with the existence of oxygen-producing cyanobacteria 2.8 Ga ago, but neither they nor the available organic geochemical data per se analyzed in Chapters 5 and 7 rule out the possibility that such organisms may have evolved either far earlier, or much later, in geologic time. In Chapter 12 (Geochemical evidence bearing on the origin of aerobiosis, a speculative hypothesis), J. M. Hayes presents an additional, independent line of argument bearing on this problem, suggesting that aerobic microorganisms (and thus, indirectly, oxygen-producing photosynthesizers) may have been locally abundant as early as 2.8 Ga ago.

The Hayes hypothesis emerges from the need to explain the extreme depletion in ^{13}C observed in many kerogens 2.8 to 2.5 Ga old, as discussed in Chapter 5, and the observation that in modern ecosystems such anomalously "light" isotope values have been recorded only for the methane produced by anaerobic methanogenic bacteria. On numerous grounds, methanogens have been regarded as members of a primitive, early evolving lineage. Their existence in the Precambrian, and that of hydrogen-producing bacteria (other early evolving anaerobes, the presence of which may have been required for generation of the hydrogen needed by methanogens for reduction of CO_2 to CH_4), can thus reasonably be postulated. However, being a gaseous metabolic byproduct, the methane produced by such microbes would not be expected to have become incorporated into kerogen precursors without the intervention of an appropriate methane-fixing process. The scenario therefore calls upon heterotrophic "methylotrophs," a group of somewhat more advanced prokaryotes that derive their carbon solely from fixation of CH_4 or related one-carbon compounds. Significantly, virtually all such methane-oxidizers that are now known are aerobes; their postulated occurrence is thus interpreted as most probably requiring the presence of aerobic conditions and, by extension, the prior origin of oxygen-producing photosynthesis. And the fact that the anomalously "light" kerogens are restricted to local basins and are coeval with isotopically "normal" kerogens occurring in other locales is interpreted as suggesting that the postulated aerobic methane fixers were only locally, rather than globally abundant, a possibility in concert with the view expressed in Chapter 11 that strict aerobes could not have become widespread and abundant prior to about 1.7 Ga ago. In sum, the locally restricted, multicomponent ecosystem envisioned in this chapter seems consistent with all geologic, biologic, and chemical evidence now available; although at this point regarded by its author as only a "speculative hypothesis," it is at the same time a new and useful model that provides a promising basis for further testing and refinement.

Chapter 13, by David J. Chapman and J. William Schopf, is devoted to a consideration of the biological and biochemical effects of the development of an aerobic environment. Of the four principal elements of life—the "biogenetic elements" carbon, hydrogen, oxygen, and nitrogen—oxygen stands out as being a biochemical anomaly: in a combined state, it is a fundamental constituent of all living systems and of the great majority of their biochemical components, yet as free diatomic O_2 it is at the same time both lethal to many types of organisms and an absolute requirement for many others. This seemingly paradoxical behavior is a reflection of the marked reactivity of oxygen and its chemical derivatives, with such oxidation resulting either in the generation of useable chemical energy (as, for example, during aerobic respiration) or in the loss of biologic function (as, for example, results from oxidation of numerous types of enzymes). Thus, the postulated Archean origin of oxygen-producing photosynthesis and the subsequent development of an aerobic environment must have required major adjustments in biologic oxygen relations, adjustments that involved the development of biochemical systems required both for oxygen-protection and for metabolic and biosynthetic oxygen-use.

Chapman and Schopf present a systematic assessment of oxygen interactions with organisms, dividing such relations into two principal categories: (*i*) oxygen metabolism and enzymology (viz., oxygen use in biosynthesis or for production of cellular energy, and interactions resulting in formation of such toxic "side products" like the superoxide and hydroxide radicals, hydrogen peroxide, and singlet oxygen), and (*ii*) oxygen tolerance or toxicity. Such interactions are classed either as "systemic," the oxygen or its derivatives interacting with specific, identifiable biochemical reactions or pathways, or as "organismic," a result of the synergistic or aggregate effect of systemic oxygen interactions on the total organism. Numerous biochemical mechanisms are identified that presumably evolved during the mid-Precambrian to protect intracellular systems from the deleterious effects of the toxic derivatives of molecular oxygen, the most important being those based on superoxide dismutase and on the catalases, peroxidases, and carotenoids. And the authors suggest that "subsequent to the development of biochemical mechanisms involved in the production of and protection from molecular oxygen, capability evolved for the metabolic [e.g., via aerobic respiration] and biosynthetic utilization [e.g., in the synthesis of sterols, polyunsaturated fatty acids, etc.]" of oxygen, in general via "the extension and refinement of pathways already extant in previously evolved anaerobes." Finally, based on the fragmentary data available, they conclude that systemic oxygen-requiring biochemistry could have occurred "once the atmospheric pO_2 had attained a level of about 10^{-4} atm, with the threshold for organismic aerobiosis being about an order of magnitude higher," levels that the fossil record suggests "had become established at least as early as 1.4 Ga ago and . . . that . . . may have been attained much earlier."

In the following chapter of the volume, by Hans J. Hofmann and J. William Schopf, the search for early evidence of oxygenic photosynthesis is carried one step farther, in this case by analysis of the known record of "Early Proterozoic microfossils." During the past 60 years, microfossils and possible microfossils have been reported from at least 40 Early Proterozoic units; of the more than 160 publications dealing with these occurrences, nearly 90 ‰ have appeared since 1965. As tabulated in an extensive annotated bibliography (Table 14-2), Hofmann and Schopf critically evaluate these occurrences, concluding that 23 include authentic microfossils, the remainder being comprised solely of dubiomicrofossils or of mineralic pseudomicrofossils. The major taxa of more than half of the previously reported bona fide microbiotas are here illustrated, as are 22 additional taxa newly detected during the course of the P.P.R.G. project in four of these units (viz., the Windidda Formation and the Duck Creek, Amelia, and Bungle Bungle Dolomites). The chapter also includes a tabulated listing that summarizes the morphology, postulated affinities, and known geologic distribution of the 122 genera (including 23 "taxa" of dubiofossils and pseudofossils) thus far described from Early Proterozoic assemblages (Table 14-3). The authors recognize five principal categories of authentic Early Proterozoic microfossils: (*i*) coccoid "unicell-like" bodies (represented by some 58 genera reported from all but three of the 23 fossiliferous units, the most diverse and abundant components of Early Proterozoic assemblages); (*ii*) septate "bacterium- and/or cyanobacterium-like" filaments (including 18 genera described from about two-thirds of the fossiliferous units); (*iii*) tubular microstructures (cyclindrical "sheaths" of prokaryotic trichomes, represented by nine genera that are known to occur in about half of the described microbiotas); (*iv*) branched filaments (microfossils of uncertain systematic position that are represented in the biotas by only two rarely occurring genera); and (*v*) "bizarre forms" (comprised of 12 genera that are of atypical microbial morphology and of uncertain biologic affinities). From their analysis, Hofmann and Schopf conclude that "the Early Proterozoic biota appears to have been composed largely, and perhaps entirely, of primitive prokaryotic microbes," many of which were probably of non-cyanobacterial affinities; nevertheless, "several lines of evidence suggest that oxygen-producing photosynthetic microorganisms of the cyanobacterial type were also represented," organisms that presumably played a role in the mid-Precambrian development of an oxygenic environment.

The final chapter of the volume (Evolution of Earth's earliest ecosystems: recent progress and unsolved problems, by J. William Schopf, J. M. Hayes, and Malcolm R. Walter), is both a summary and an extension of the foregoing chapters, an attempt to draw together the disparate lines of evidence there presented into a holistic working model that describes the timing and nature of major events in the early history of life. The chapter includes in chart form a rather detailed synopsis of the relevant available data and of their geologic and biologic interpretations (Figure 15-1). Schopf et al. begin their discussion by categorizing the types of evidence potentially available as either (*i*) "compelling" (abundant evidence that permits only one "reasonable" interpretation); (*ii*) "presumptive" (the preponderance of evidence pointing to a single "most likely" interpretation); or (*iii*) "permissive" (the available evidence being consistent with at least two "more-or-less equally tenable" competing interpretations). While recognizing that "such categories are necessarily subjective, the 'compelling' evidence of one investigator perhaps being regarded as only 'presumptive' or even 'permissive' by another," they argue that such classification serves the useful purpose of explicitly identifying the degree of confidence with which various lines of evidence are regarded, a particularly important matter in studies of Precambrian life for which many of the major generalizations have yet to be firmly established.

In contrast with the morphologically based evolution that has characterized the Phanerozoic history of life, Precambrian evolution is viewed as having been largely a function of the development of biochemical and intracellular innovations. In the following discussion, the authors thus identify seven principal metabolic benchmarks in early evolution, the occurrence of each of which is regarded as having played a crucial role in the sequential development of the Earth's earliest ecosystems. Evidence relating to these events is evaluated, in each case by assessing what the authors consider to be the oldest compelling, presumptive and permissive evidence currently available regarding their probable time of occurrence. On the basis of these considerations, they conclude by summarizing their current "best guess scenario" of the timing and nature of major events in Archean-Early Proterozoic evolution, one that dates (*i*) the origin of life as having occurred earlier than 3.5 Ga ago; the development of (*ii*) anaerobic chemoheterotrophy, (*iii*) anaerobic chemoautotrophy, and (*iv*) anaerobic photoautotrophy as probably also having occurred earlier than 3.5 Ga ago; the origin of (*v*) aerobic photoautotrophy and of (*vi*) amphiaerobic metabolism as possibly having occurred about 2.9 Ga ago; and the global appearance of microbes dependent on (*vii*) obligate aerobiosis as having occurred between about 1.7 and 1.5 Ga ago.

Throughout virtually all chapters of the volume, one overriding theme is recurrent—that while recent progress in the field has been encouraging, there is a myriad of basic questions that are as yet unanswered, questions that seem likely to be resolved only through intensive interdisciplinary analysis like that here attempted, and ultimately only by referral to the single known source of direct evidence now available, the Archean-Early Proterozoic geologic record.

CHAPTER 1

Early Biogeologic History: the Emergence of a Paradigm

By Preston Cloud

On all sides they clamor:
Of what interest to us are the sayings or deeds of old?
We are self-taught; our youth has learned for itself.
Our band does not accept the dogmas of the ancients;
We do not burden ourselves in following their utterances.
Rome may cherish the authors of Greece.
I dwell on the Petit-Pont, and am a new authority.
It may have been discovered before; I boast it is mine.

John of Salisbury, The Entheticus, Twelfth century

1-1. Introduction

To study the origin and evolution of Earth's earliest biosphere involves the blending of biology and geology in the broadest sense and at the deepest levels. Here we have the roots of a new interdisciplinary science that asks and attempts to answer questions traditionally seen as outside the main stream of either biology or geology but profoundly relevant to both. The goal is to describe and interpret the record of life-processes in geology, utilizing all relevant materials and means, and extending forward in time from the origin of living systems. The activity is analogous to biochemistry or biophysics. It has been called many things, but the most descriptive is biogeology. It is not merely paleontology extended in time or even paleobiology, although it includes activities appropriately so designated. Rather it extends these subjects and integrates them with the data and ideas of biochemistry, genetics, geochemistry, sedimentology, and atmospheric and aquatic chemistry in a geochronological framework.

Although the concerted effort we now see has emerged only during the last two decades, an interest in biogeological problems is not new, nor is the realization that the solution to such problems involves a great deal more than the analysis of fossils or other biogeological residues. No better sign could be given, however, that the subject has come of age than the interaction of relevant disciplines to create this book. This profoundly interdisciplinary study collates, updates, and extends the results of less extensive, earlier studies that have transformed our views of pre-metazoan Earth history quite as significantly as the plate tectonics paradigm has modified our perceptions of Phanerozoic crustal evolution. In both instances the term paradigm (in the sense of Kuhn, 1970) denotes a dramatic shift of previously prevailing views.

The new biogeological paradigm was first advanced in preliminary form in the middle 1960s by Cloud (1965, 1968a, 1968b) and subsequently modified by him (1972, 1974, 1976a, 1976b, 1980), to accomodate new data that accumulated as others probed its weaknesses and challenged its validity on one point or another. The main elements of this paradigm comprise: (*i*) an initially anoxic atmosphere; (*ii*) a pervasive interaction (feedback) between biospheric, atmospheric, hydrospheric, and lithospheric evolution; (*iii*) a probably long interval of fluctuating imbalance between sources and sinks of oxygen, the eventual main source of free O_2 being photosynthetic; (*iv*) the ultimate accumulation of atmospheric O_2 as a consequence of kinetic lag and the resultant retardation of thermodynamic recombination; and (*v*) a concurrent biospheric evolution from anaerobic, heterotrophic, microbial ancestors to aerobic, photoautotrophic descendants, eucaryotes, and differentiated multicellular organisms, followed by Phanerozoic diversification and evolution. Although numerical ages, durations, and details emphasized vary with the author and year of publication, the main elements of this paradigm continue to recur in some form close to the above. They are both confirmed and importantly refined and modified in the following chapters of this book. An important point here, however, is that they depart sharply from once widely held views of substantive uniformity in historical processes to recognize dramatic changes in Earth states and rates within the bounds of unchanging (or systematically changing) natural laws.

Like plate tectonics also, the new biogeological paradigm has long antecedents in the work of a perceptive minority of forerunners. Its emergence can be described as involving four stages: (*i*) The gestational or embryonic century, 1850–1950; (*ii*) The emergent decade of the 1950s; (*iii*) The breakthrough decade, the 1960s; (*iv*) The take-off decade, the 1970s.

The remainder of this paper surveys the main elements of those four stages, concluding with some remarks about current views and future prospects. To be constructive, such a survey should both encapsulate progress and warn of pitfalls. All who have been truly active in the field have made mistakes and some (e.g., Cloud et al., 1965) have published them. Many false reports, including some of their own, are reviewed by Schopf (1975, 1976), Cloud (1976a), Cloud and Morrison (1979), Schopf and Walter (1980), and by the authors of Chapters 9 (Table 9-1) and 14 (Table 14-2) in this book. I have warned at other places of the ever-present problems of pseudofossils, artifacts of diagenesis and preparation, and post-depositional contamination of several sorts (Cloud, 1973b, 1976a; Cloud and Hagen, 1965; Cloud and Morrison, 1979; Cloud et al., 1980). And I will not hesitate in the following pages to comment frankly on other pitfalls and instances of misinterpretation as seems appropriate. The goal is to eliminate avoidable repetition of error, not to castigate it. Thus, where I have felt that no useful purpose would be served by negative comment, I have not referenced papers that I consider useless or misleading, any more than I have referenced all papers that I consider important and informative. The interested reader will find a comprehensive listing of relevant references in a bibliography by Awramik and Barghoorn (1978).

My aim, then, is to forge a constructive and reasonably balanced assessment of the history and state of the art of pre-Phanerozoic biogeology, with some stress on the biological component. I do not seek to be exhaustive, and I do not claim to be free of bias. I do, however, recognize the scientific imperative to seek objectivity and shall strive to guard against hidden bias. The not-always-gentle reader will be the judge of how well I succeed in this aim.

I have been urged, however, to make this a personal account and I shall. Let me begin by confessing to a few personal biases that I have no intention of hiding and that I have insisted on following in this chapter, contrary to the preferred standards of the editor and usage in the remainder of this volume:

(1) Biogeology, I repeat, is my preferred term for the broad study of life processes in Earth history, including their interactions with the physical world.

(2) The main divisions of Earth history (Hadean, Archean, Proterozoic, and Phanerozoic), in my view, should reflect Earth history. They should be defined, therefore, by historical events which may vary in age from one region to another, not by geochronologic numbers which should be reserved for *calibration* of the historical scale and its often diachronous events as recorded in the rocks.

(3) Thus I see the Archean-Proterozoic transition not as fixed at a specific time, but as related to an historical change in prevailing rock types and structural modes that is grossly diachronous over a time range from ~3.0 to ~2.5 Ga ago.

(4) The event that in my view most appropriately reflects the Proterozoic-Phanerozoic transition is the first appearance of metazoan body fossils and *Lebensspuren* characteristic of the Ediacarian System beginning about 680 Ma ago.

(5) The Ediacarian thus becomes the basal system of the Phanerozoic, and the terms pre- or sub-Cambrian, pre-Ediacarian, and pre-Phanerozoic are used informally to mean precisely what they say. The currently fashionable use of Precambrian as a formal term is abandoned as an impediment to unambiguous discussion of the Proterozoic-Phanerozoic transition.

(6) It is a cardinal principle of systematic biology, essential to nomenclatural stability, that the proper

spelling of a systematic term is that of first use, regardless of logic or etymology. The K sounds in the terms procaryote and eucaryote and derivatives thus are indicated by "c", following the founding paper by Chatton (1938)—as in cosmos and cyanophyte (or "cyanobacteria") and not as in the perhaps more appropriate but inconsistent teutonic K of kosmos or kyanobacteria.

(7) Whatever may be the proper name for the chlorophyll-*a* containing procaryotes under the International Rules of Botanical Nomenclature, it is *not* Cyanobacteria regardless of the long-recognized affinity (e.g., Papenfuss, 1955) between blue-green algae and bacteria. Although Papenfuss uses a different name, he notes (1955, p. 174) that the first legitimately proposed term was Myxophyceae (or Myxophyta), suggested as a restrictive use of an older wastebasket term. That seems to be the nomenclaturally correct name and the one I shall employ when I do not call them blue-green algae. Of course they are blue-green algae in the same ecological sense that mangroves are mangroves or turtle grass, grass. Van Valen and Maiorana (1980) have pointed the way to the cutting of this Gordian knot with their proposal that the chlorophyll-*a* containing (i.e., oxygen-releasing) procaryotes as a group be called *proalgae* (see Glossary I).

1-2. Stage I. The Gestational Century: 1850–1950

We now know, thanks to the impressive achievements of geochronology, that the eon of manifest life on Earth, the Phanerozoic, represents only the last 15% of biogeologic history. The first 85%, the pre-Phanerozoic (or Cryptozoic), had already elapsed before the first Metazoa (and later, vascular plants) appeared. Geology itself, however, was already an old science by human standards before the great length and biologic interest of pre-Phanerozoic history was finally recognized. If the publication in 1669 of Nicolaus Steno's *Prodromus* be taken as the beginning of geology as a distinctive science, it was another 166 years before the realization by Adam Sedgwick in 1835 that there were older, seemingly unfossiliferous rocks beneath the oldest manifestly fossiliferous ones. The current perception that, in fact, the main events in biospheric evolution had already transpired when the Metazoa finally made their debut about 680 Ma ago then had to await Henri Becquerel's discovery of radioactivity in 1896, the perfection of the mass spectrometer, and the present episode of biogeological discovery.

As in so much else, it was Charles Darwin who, in 1859, first grasped the significance of the apparently abrupt appearance of a diversity of metazoan forms of life early in Phanerozoic time. As he and Sedgwick suspected, there was indeed a prior history, of which the main features are only now beginning to be understood.

The first systematic attempts to resolve the dilemma of abrupt appearances were those of Charles Doolittle Walcott, on the instigation of John Wesley Powell. Powell, then director of the U.S. Geological Survey, sent Walcott into the Grand Canyon in 1882 to assess the prospects that the sub-Cambrian sequence there might contain evidence of ancient life. His hunch soon bore fruit, when Walcott, in his very first paper on the region, reported from strata now known to be sub-Cambrian "a small Discinoid shell" (probably the form later called *Chuaria*) and "an obscure Stromatopora-like" form (Walcott 1883) which he later recognized to be similar to some Upper Cambrian structures from New York called *Cryptozoon* by James Hall in 1884 (Walcott 1895). These, of course, were the bounded, laminated, calcareous sedimentary structures later to be called stromatolites (Kalkowsky, 1908). Meanwhile, in 1890, G. F. Matthew described from New Brunswick a stromatolite that he called *Archeozoon acadiense*—the first reasonably authentic pre-Phanerozoic fossil to be named. Matthew's paper was also noteworthy for the then-revolutionary concept expressed in his title: "On the existence of organisms in the pre-Cambrian rocks." Except for "*Cryptozoon*" *acadiense* his evidence has not withstood scrutiny. His work and that of Walcott, however, began the study of Proterozoic stromatolites that was to be continued by the Fentons in the 1930s (e.g., Fenton and Fenton, 1931, 1937, 1939) and accelerated to present levels following intensive investigations by geologists of the Soviet Academy of Sciences, beginning with Maslov in the late 1930s (e.g., Maslov, 1939) and accelerating during the 1950s and beyond.

Because of the fossils mentioned, Walcott at first thought his Chuar strata to be Cambrian. Later he came to realize that they lay unconformably beneath the oldest Cambrian and that his "Stromatopora" was "probably a species of Cryptozoon"

(Walcott, 1895). Walcott also named the first well-documented body-fossil to be recorded from pre-Phanerozoic rocks, in 1899, although a very similar unnamed fossil from the Visingsö beds of southern Sweden was described and illustrated by Wyman in 1894 and was probably also described by Holm as early as 1885 (Wyman, 1894). Walcott's discovery was *Chuaria*, a discoidal form a millimeter or more in diameter from the upper Proterozoic Chuar strata of the Grand Canyon. Although long regarded as a brachiopod, *Chuaria* and similar forms have generally been interpreted as the compressions of large spheroidal planktonic algae since first so identified by Cloud (1968b). The term, in fact, has become a wastebasket for morphologically similar compressions now known worldwide from strata of later Proterozoic ($\sim$1000 Ma ago) to earliest Phanerozoic age (Ford and Breed, 1973; Hofmann, 1977a).

From 1882 to 1916, despite his concurrent Cambrian researches and heavy administrative responsibilities as director of the U.S. Geological Survey, secretary of the Smithsonian Institution, and president both of the National Academy of Sciences and the American Philosophical Society, the self-taught Walcott remained the most productive student of pre-Phanerozoic life. Although his descriptions and illustrations were less than compelling, he was the first to describe and illustrate what now seem to be probably authentic microbial fossils of Proterozoic age. In 1914 and again in 1915, he discussed and illustrated chains of coccoid cells a few microns in diameter from the Belt Supergroup of the Northern Rocky Mountains that he interpreted as bacteria and that are probably procaryotes of proalgal (Van Valen and Maiorana, 1980) affinities. Perhaps even more important than Walcott's pre-Phanerozoic studies was his 1919 paper on the algal flora of the Middle Cambrian Burgess Shale. For it established the presence, on the heels of Proterozoic time, of a diverse algal flora that includes filamentous and probably proalgal microbial forms of the sort now widely known from underlying Proterozoic rocks. In the sense that he planted the seed that was to flower into the activity we know today, Walcott may truly be called the father of pre-Phanerozoic biogeology. Indeed his contribution was fundamental in establishing nearly a century ago, contrary to much then-current folklore, that there is substantial evidence for a pre-Phanerozoic (and pre-Cambrian) record of life.

Yet along with his substantial and prescient achievements, Walcott's work, and that of many who followed him, suffered from a biological naiveté and an over-eagerness for results that continues to plague the search for pre-Phanerozoic life. Many of the objects described as fossils by Walcott and others have turned out to be either not of biologic origin or incorrectly identified as animals. Raymond, who published one of the most perceptive and devastating reviews of the field in 1935, characterized the situation thus: "So anxious are geologists to obtain fossil from these rocks . . . that anything which remotely resembles an organism is carefully saved and studied in the greatest detail." And, he might have added, is likely to be described and given a double-barreled Latin name.

Raymond's warnings, despite some retrospectively unwarranted asperity, were needed but not sufficiently heeded. His comments on the description by Cayeux (1894a) of supposed radiolarians from cherts in the late Proterozoic Brioverian strata of Brittany (first reported by Barrois in 1892) have been considered severe. Yet, although it is now established by the work of Deflandre (1949), Roblot (1964), Chauvel and Schopf (1978), and my own unpublished work that microbial fossils do indeed occur in these rocks (and in equivalent strata from Normandy), those fossils in no way resemble either Radiolaria or the "Foraminifera" and "sponge spicules" reported by Cayeux (1894b, 1895) from the same rocks. Cayeux (1894a), in fact, remarked that it was so difficult to see anything in photomicrographs of his material that he assigned an artist who had never before drawn radiolarians to draw exactly what he thought he saw in the single thin section that contained all the best material (prepared from rock collected at Ville-au-Roi-en Maroué, Lamballe). I studied (courtesy of Mme. Dr. M. Deflandre-Rigaud) the three extant thin sections of this very material (CV51, CV52, CV53), labeled in Cayeux's own hand, at the Museum of Paleontology in Paris in August 1980. Section CV51 is considered to be the slide from which Cayeux's illustrations were drawn. As Raymond hypothesized, Cayeux's artist did indeed have a remarkable imagination, or a good idea about what he was supposed to see. Although tiny, smooth-surfaced, ovoid acritarchs occur in all of these thin sections, and a very few show curvate, bluntly spine-like protuberances, it is hard to believe that a microscopist of Cayeux's legendary skill could have been viewing anything other than the artist's drawings while describing the 20 genera and 10 families of "Radiolaria" he recorded.

Until the 1950s, in any case, the most widely accepted evidence of pre-Phanerozoic life was the indirect evidence of stromatolites. They were abundant among carbonate rocks, easily recognized, and generally agreed to consist prevailingly of microbially mediated sedimentary deposits. More energetically sought and more frequently described, however, were supposed Metazoa. It would be an unprofitable business here to review in detail the extensive published record of nonbiologic objects that have been described by biologically unsophisticated geologists and geologically unsophisticated biologists as authentic metazoan fossils. They include desiccation cracks, mud rolls and flakes, clastic dikes, tectonic wrinkles and gashes, defluidization structures, rill-marks and micro-ripple marks, concretions, crystal rosettes and crystals of diagenetic and metamorphic origin, compaction phenomena, impact marks, gas blisters, drag marks, impressions of vibrating wire ropes, and even man-made objects (Cloud, 1948, 1968b, 1973b). Most misleading of all are the works of living organisms intrusive into authentic Proterozoic sediments (Cloud et al., 1980).

It is bad enough that nature lays traps for us. More vexing to deal with is the fact that fantasy can all too often rule the judgment even of those who at other times practice science. It is amazing that apparently competent physical geologists, who would be unlikely to tolerate a similar lack of rigor in their own fields, are nevertheless willing, without benefit of biological experience or advice, to express far-reaching biological judgments about objects of the most dubious biological nature or ancient provenance.

After many years of intermittent personal inquiry into this bizarre record (1948–1965) I came to realize, in searching for metazoans and trace fossils in the kinds of sedimentary rocks where such things might be expected had they been of Phanerozoic age, that I had been looking for the wrong things in the wrong rock types by the wrong methods.

If stromatolites are omitted, which offer for the most part, no matter how devoutly believed (and I mostly believe), only presumptive and indirect evidence of a biological presence, what have we to work with?

In retrospect, the now productive microbial route could have been seen. Not only Walcott, but also Moore (1918), Gruner (1923), and Ashley (1937) had published illustrations of microstructures of suggestively (though by no means conclusively) microbial aspect from thin sections of cherty rocks in North America and Africa. Had their illustrations been better, these records might have had more impact at the time. Moore's material, from Proterozoic banded iron-formation in the Belcher Islands, preceded by 36 years the epochal discovery of similar forms from cherts of the Gunflint Formation by Tyler and their first description by Tyler and Barghoorn in 1954. In his report, Moore quotes comparisons with modern forms by botanist J. B. Hill and reproduces Hill's camera-lucida drawings. There is little doubt that he had found fossils very similar to those later named *Gunflintia* and *Huroniospora*. But the offhand nature of their mention and the very schematic illustrations carried no conviction. If anyone was paying attention, he or she failed to recognize the significance of the discovery or to act on its possibilities. Even after the startingly seminal work of Wetzel in 1933 on the organically preserved microorganisms in cherts of the Baltic chalk, it still does not seem to have occurred to anyone to search systematically for microbial fossils in the pre-Phanerozoic cherts.

The tide in the gestational history of biogeology was not yet ready for the flood. No one had yet begun to think very hard about the probable biological aspects of atmospheric evolution or about the possibility of geological evidence bearing on the origin and early evolution of living things. Rocks older than Cambrian were still generally thought to be everywhere too highly metamorphosed to preserve much in the way of fossils, let alone delicate microbial forms without hard parts.

Yet the seeds of creative synthesis were being sown. Already in 1927 MacGregor suspected a relation between banded iron formation and atmospheric evolution. Oparin in 1936 presented views and data that for the first time gave a degree of tangibility to questions of the origin and early development of life (Oparin, 2nd ed. 1953). Horowitz outlined a scheme for the evolution of biochemical syntheses that gave the problem a needed biochemical perspective (Horowitz, 1945). And Cloud, although mistakenly denying an interregional stratigraphic utility, recognized the paleoecological potential of stromatolites and their possible use in intrabasinal stratigraphy (Cloud 1942).

A key event was the above mentioned publication of Deflandre's 1949 paper on nannofossils from the late Proterozoic of Brittany. For, while conclusively laying to rest Cayeux's claim to a radiolarian presence, Deflandre unequivocally established the existence of abundant organic walled microbial fossils in a chert matrix of late pre-Phanerozoic age. The problem of micro-contaminants and pseudomicrofossils that continues to plague us (as discussed in Chapter 9 of this book), moreover, would have been greatly reduced had later workers paid more attention to Deflandre's advice to be very cautious about all morphologies not displaying unequivocal para-genetic evidence of a primary origin in thin sections or translucent slivers of cherty rock.

1-3. Stage II. The Emergent Decade: The 1950s

So casually was a foundation laid for the creative events of the postwar decade! The shift of focus from war to science was followed by years of ferment not only in paleomicrobiology but also in studies of biochemical and prebiotic evolution, in concepts of atmospheric and hydrospheric equilibria, in ideas on metazoan origins, in the study of stromatolites, and, above all, in the geochronology that would provide the essential time-frame for studies of biogeologic evolution. Although there was little or no conscious interaction among these several related lines of inquiry during that first postward decade, their parallel development prepared the way for the more consciously integrated biogeological studies that were to follow in the 1960s and 1970s.

In retrospect, the two most important paleo-microbiological events of the 1950s were (*i*) the discovery of a rich and diversified mixed benthonic and planktonic microbiota in stromatolitic cherts of the ~2 Ga-old Gunflint Formation by Tyler and Barghoorn (1954); and (*ii*) the finding by Soviet scientists of abundant, minute, carbonaceous spheroids of biological appearance in palynological macerations of later Proterozoic siltstones and shales.

The first published illustrations of the Gunflint discovery, however, lacked resolution, and it was not until the later more convincingly illustrated paper by Barghoorn and Tyler in 1965 that the full significance of their discovery burst on a previously unimpressed world. Meanwhile, in the U.S.S.R., spore-like bodies described from Uralian and other siltites and shales were generating heated discussion until it was shown that many of them had leaked downward from overlying sediments of Cambrian to Devonian age and were not indigenous to the Proterozoic rocks from where they were reported (Krylov, 1968).

Timofeev did eventually publish a number of records of what appear to be authentic primary acritarchs from fine clastic sediments of later Proterozoic and early Phanerozoic age in east Siberia, the Baltic region, and elsewhere in the U.S.S.R. (1958, 1959, and later). Important results of Timofeev's work were that (*i*) it showed that primary microbiotas composed prevalently of microfossils of spheroidal morphology and probably planktonic habit do occur in fine clastic sediments over a wide area; and (*ii*) it led to his recognition of a significant trend of upward-increasing cell size among younger Proterozoic acritarchs. This trend, later exhaustively documented by Schopf (1977a) as beginning about 1.4 Ga ago and affecting filamentous and spheroidal forms alike, is consistent with the suggested first appearance of the eucaryotic cell at or before that time as proposed by Cloud and others in 1969. Timofeev spoke to me of the observed size increase as early as 1965, although he did not emphasize it in publication until 1973 (Timofeev, 1973a). It seems that Timofeev also was able to avoid a problem that beset most other Soviet students of the so-called "microphytolites" who commonly mistook both contaminants and artifacts for authentic microbial fossils.

The 1950s saw the establishment of studies of biochemical and pre-biotic evolution on a solid foundation. The experiments of Garrison and others, utilizing ionizing radiation to reduce CO_2 in aqueous solutions, suggested ways of making organic compounds from available primitive Earth volatiles (Garrison et al., 1951). Miller went a huge step farther in demonstrating the production of amino acids under similar anoxic conditions and an external source of energy—in this case spark discharge, simulating lightning (Miller, 1953 and later). His work in turn stimulated others to undertake parallel experiments using different starting materials and energy sources, with the result that one could soon visualize making not only amino acids but also other important protein- and nucleotide-building molecules utilizing a range of plausible primitive atmospheres and energy sources, provided free O_2 was excluded and H_2O was present. The publication in 1954 of volume 16 of the now extinct journal *New Biology*, published by Penguin, with papers on the origin of life by J. B. S. Haldane, J. D. Bernal, N. W. Pirie, and J. W. S. Pringle, was also a key event in sorting out ideas about the origin of life and suggesting future approaches.

A review of his own work and that of others on the geochemistry of organic substances by Abelson late in the decade outlined both prospects and pitfalls in biogeochemistry as they seemed at that time (Abelson, 1959). All should have been warned by this work of a too-ready-acceptance of the primary nature of seemingly indigenous organic substances in pre-Phanerozoic rocks, especially of the thermally unstable and metastable amino acids. Yet reports of supposedly ancient pre-Phanerozoic amino acids, including such short-lived and almost certainly contaminant species as serine, threonine, arginine, and tyrosine, still occasionally appear in print. Kvenvolden constructively reviews this problem (Kvenvolden, 1975).

New outlooks on atmospheric and hydrospheric evolution also appeared during the fifties, especially as outlined in two epochal papers by Rubey in 1951 and 1955. His conclusion that both atmosphere and hydrosphere accumulated gradually as a consequence of volcanic outgassing is now yielding to hypotheses that call on catastrophic early outgassing followed by subsequent episodic outgassing at diminishing rates. Yet it is Rubey above all (with his acknowledgments to forerunners Boggild and MacGregor) to whom we owe the now generally accepted basic idea of outgassing as the primary source of Earth's surface volatiles. Rubey's work ignited new lines of thinking about atmospheric and hydrospheric evolution that carry a strong biogeological slant and that are still being actively pursued.

The problem of metazoan origins also came to the fore during the fifties. Cloud reasoned in 1948 that, despite its increasing incompleteness with greater age, we could only deal with the geologic record that we actually knew. We ought to try interpreting this, he said, as if it comprised a reasonably representative sampling of past events, with minimal assumptions as to how it might become enlarged by future findings. He thus proposed a polyphyletic origin to account for the geologically abrupt appearance of a diversity of Metazoa, followed by their eruptive diversification at the beginning of Phanerozoic history, as pioneer types radiated adaptively into a variety of previously unoccupied ecologic niches under low competitive pressure (an example of quantum evolution, now increasingly called "punctuated equilibrium"). He emphasized that we had at that time no records of unequivocal Metazoa in rocks of undoubted pre-Cambrian age. This conclusion, when proposed, was consistent with the then supposedly Early Cambrian age of the Ediacara fauna of South Australia (Sprigg, 1947)—an age assignment that persisted until the late fifties (Sprigg, 1949; Glaessner, 1958a), but which was lowered with increasing confidence to a sub-Cambrian position beginning in 1958 (Glaessner, 1958b, 1966). Stratigraphic assignment, of course, is a matter of definition, and I accept the new definition. But, along with the change in terminology, subsequent findings have borne out Cloud's 1948 suggestion of rapid early Phanerozoic Metazoan evolution (summarized by Stanley, 1976). A theoretical basis for this hypothesis begins with a 1959 paper by Nursall in which he stressed the necessity of appreciable free oxygen as a prerequisite to metazoan evolution.

After a long dormancy the study of stromatolites as potential stratigraphic and biologic indicators reappeared and advanced rapidly during this postwar decade, particularly from the efforts of Soviet investigators. Maslov pioneered this movement from the late thirties through the fifties, demonstrating the variability of stromatolites at different structural levels and through time and proposing a stratigraphic utility for them (e.g., Maslov, 1939, 1960). In the later forties and fifties, others, notably A. G. Vologdin and K. B. Korde pursued the idea proposed by Cloud in 1942 that the key to stromatolite understanding should be sought in any contained microbial remains. More fruitful during this time, however, was a determined and still-active effort by stromatophiles of the Geological Institute of the Soviet Academy of Sciences to discover successional variations in stromatolite morphology and apply them to the biostratigraphic subdivision of Proterozoic strata. This program, in which Soviet science led the world, was inspired and directed by V. V. Menner and B. M. Keller. The actual work was carried out by many now-familiar Soviet researchers beginning with that of I. K. Korolyuk and I. N. Krylov, first published in the late fifties (Korolyuk 1958; Krylov 1959a, 1959b). Korolyuk's ideas about a stromatolite stratigraphy based on branching and columnar morphology, at first distrusted by physical stratigraphers, came to be widely accepted as her suggested sequences were found to be consistent with physical stratigraphy over ever more widely separated regions of the Siberian and Russian platforms.

The *sine qua non* of the biogeological edifice, however, indeed of the whole of modern pre-Phanerozoic history, was an independent method for ordering events in time—an objective, numerical geochronology. And geochronology, with all of the self-checking and overlapping methods and refinements we know today, is primarily a post-World War II development. Of course, it did not in any sense arise independently of the magnificent prewar achievements of the great pioneers—Rutherford, Boltwood, Holmes, Barrell, and Nier. However, the unprecedented numbers of well-trained, well-equipped, well-funded, and gifted researchers in the field and their constructive interactions with a host of able and interested geologists simply generated explosive progress. This progress has continued with the demands of space exploration and the broader awareness of applications arising from new geochronological methods and the possibilities they create of working with ever smaller samples to ever more precise levels. The contributors are so many that to name any may seem invidious. Yet few would disagree that among the leaders have been A. O. Nier, S. S. Goldich, L. T. Aldrich, L. O. Nicolaysen, Clair Patterson, G. J. Wasserburg, G. W. Wetherill, L. T. Silver, G. R. Tilton, S. Moorbath, W. Compston, E. K. Gerling, and A. I. Turgarinov. Their works are too well known to call for citation here.

1-4. Stage III. The Breakthrough Decade: The 1960s

By 1960 my confidence in my earlier conclusion (Cloud, 1948) that the appearance of a metazoan grade of organization was probably a chronologically compressed and polyphyletic early Paleozoic event had reached a high level. I became convinced that efforts to extend the record of life into the more distant past by means of a metazoan record were likely to be unproductive. I saw at last that the discoveries of Walcott in 1914 and 1915, Moore in 1918, Gruner in 1923, Wetzel in 1933, Ashley in 1937, Deflandre in 1949, Tyler and Barghoorn in 1954, and Timofeev and other Soviet paleomicrobiologists of the later fifties were pointing to more fruitful ways to extend the biogeological record in pre-Phanerozoic history. The methods of Walcott, Moore, Gruner, Ashley, Deflandre, and Tyler and Barghoorn involved the study at high magnifications of thin sections of cherty rocks (as it turned out, the more carbonaceous and more stromatolitic the better). Chemical maceration of carbonaceous shales and siltstones following standard palynological practice was the preferred Soviet technique, one that produces mixed results, depending on the care invested in collecting samples and their laboratory preparation and study.

The times seemed right for a new focus on the field and Wendell Woodring, a creative force in paleontology, was about to turn seventy. This occasion provided impetus for the organization of a conference on major biologic innovations and the geologic record, known as the "Woodring Conference" which focussed on major biogeological events from the origin of life to the origin of Metazoa. That conference, sponsored by the U.S. National Academy of Sciences, the Carnegie Institution of Washington, and the National Science Foundation was held in June 1961 at Big Meadows Lodge in the Blue Ridge Mountains of Virginia. It involved a highly interdisciplinary international group of 23, including a number who have since been active in research and training in pre-Phanerozoic biogeology or related microbiological, biochemical, or geochemical studies—P. H. Abelson, E. S. Barghoorn, M. Calvin, P. Cloud, M. Florkin, S. W. Fox, T. C. Hoering, H. D. Holland, R. Stanier, and others (Figure 1-1). We laid out our evidence, discussed interpretations, educated each other, and considered strategies for future research.

It became clear to many of us at the Woodring Conference that there were indeed researchable components to problems of the origin, early development, and biogeochemical aspects of life on the primitive Earth. Perusal of the brief report on that conference (Cloud and Abelson, 1961) reveals that many of the great problems of biogeology that have occupied out attention for the past two decades were there perceived, and that nearly half of the participants have contributed actively to the ensuing research on them. Subsequent work by these people and their students, focussing on the earliest evidence for biological

FIGURE 1-1. Participants at the Woodring Conference on Major Biologic Inovations and the Geologic Record held at Big Meadows Lodge in the Blue Ridge Mountains of Virginia, 16 June 1961. From left to right are M. N. Bramlette, P. Cloud, J. R. Vallentyne (Canada), H. D. Holland Dixie Lee Ray, S. W. Fox, H. L. James, W. H. Bradley, R. Stanier (France), D. Bonner, P. H. Abelson, C. B. Anfinsen, W. P. Woodring, M. K. Hubbert, T. C. Hoering, E. S. Barghoorn, N. W. Pirie (England), H. A. Lowenstam, M. Calvin, K. E. Lohman, and M. Florkin (Belgium); participants not present in the photo are R. M. Garrels and S. B. Hendricks. Country of origin given only for non-U.S. participants. Photograph by K. E. Lohman.

activity and its consequences for sedimentary and atmospheric evolution, marks a quantum advance in understanding life processes and ecosystems on the primitive Earth. For the first time a substantial number of minds of diverse backgrounds came to focus on the evidence for pre-Phanerozoic life under some understood constraints.

I see the Woodring Conference, therefore, as the beginning of the systematic multidisciplinary study of life processes on the primitive Earth and their interactions with the evolving atmosphere, hydrosphere and lithosphere—that is, of biogeology, a subject in which I include the study of the origin of life and the paleoenvironmental circumstances that made that event chemically inevitable.

A few years later, in the spring of 1964 as I recall, Abelson and I organized a very informal sequel at the Carnegie Institution's Geophysical Laboratory in Washington, D.C. to discuss progress. This meeting included some carryovers from the Woodring Conference as well as some newly active participants. J. W. Schopf was one of the latter—the first person to begin at an early age to prepare for a career in the field, and in that sense the first to be adequately prepared for it—brought as a student observer by Elso Barghoorn.

By this time, stimulated by prior inquiry, discussion, and the papers of MacGregor, Rubey, and Nursall, I had formulated ideas about the interrelated evolutions of biosphere, atmosphere, hydrosphere, and lithosphere on the primitive Earth and related them in my imagination to the development of banded iron-formation, its phasing out upward in the geologic column, and the following onset of conspicuous red bed sedimentation. Informal discussion revealed some parallelism between my thinking and that of Lepp and Goldich (1964) about banded iron-formation on the one hand and that of Berkner and Marshall (1964a) about a relation between oxygen and biologic evolution on the other. Thus I delayed publication in order that the more fully developed contrasting views of Lepp and Goldich take precedence and also in anticipation of a symposium that Berkner and I were by then planning for the spring meeting of the National Academy of Sciences in 1965 on the possible relation of atmospheric oxygen to biologic evolution. I also felt that a convincing description and illustration of at least the main elements of the ~2.0-Ga old Gunflint microbiota was an essential prelude to the hypotheses proposed in my then unpublished paper. Because such a documentation of the results of Tyler and Barghoorn was so long delayed, I had at last undertaken my own overview study of the Gunflint microbiota, by then nearly complete.

This was one of two potential sources of misunderstanding that were bruited about and eventually resolved by participants after the 1964 Geophysical Laboratory meeting. The second had to do with the occurrence of the presumed chlorophyll derivatives pristane and phytane in richly carbonaceous Archean deposits of Minnesota—the Soudan Formation—from which I had sent samples to two different geochemical laboratories, both of which were then preparing similar results for publication. The resolution of these issues, involved on the one hand publishing two similar articles on the organic geochemistry of the Soudan end to end in the same 1965 issue of *Science* magazine, and on the other a delay in the publication of my paper (Cloud, 1965) illustrating some conspicuous Gunflint nannofossils and discussing their implications until Barghoorn could complete his part of a descriptive paper with the by-then deceased Tyler (Barghoorn and Tyler, 1965a). That paper by Barghoorn and Tyler, and later works, established the Gunflint microbiota as one of the "golden spikes" of paleomicrobiological history (see, for example, discussion in Chapter 14), but the Soudan pristane and phytane remain suspect. These interesting molecules are in the rock, no doubt, but, being the product of benzene-methanol extracts, they could, like all extractables, be post-depositional contaminants. In fact the common discrepancy between $^{12}C/^{13}C$ ratios of the kerogen and extractable organic compounds from ancient rocks (Hoering, 1961) emphasizes that probability.

Important as the Woodring Conference and that at the Geophysical Laboratory were in bringing a diversity of outlooks and methods to bear on common interests, however, they were but a significant part of a much wider intellectual ferment. Stromatolite studies in the U.S.S.R. continued through the 1960s and beyond with an accelerating vigor that focused attention worldwide, as conveniently summarized by Semikhatov and Preiss in chapters 7.1 and 7.2 of Walter's great compendium (Walter, 1976a). Geochronology continued its explosive growth (partly summarized in Harper, 1973). Significant advances were achieved by Calvin and his colleagues, and by Oro, Abelson, Hoering, and others in organic geochemistry and its applications to pre-Phanerozoic biogeology (summarized in Eglinton and Murphy, 1969). Margulis was focusing her energy and talents on the origin of the eucaryotic cell, culminating in her monographic defense of serial endosymbiosis in 1970.

Thanks also to the space program, leading up to the Apollo 11 landing on the Moon and return of samples in July 1969, unprecedented resources became available to extend the search for evidence of life in time and space. An important contribution to biogeological thinking was the production in 1966 by the National Research Council at the request of N.A.S.A. of two major volumes of collected papers dealing with the nature and origin of life, prebiotic chemistry, biogeochemistry, life under extreme conditions, and the possibility of a broader cosmobiology (Schneour and Ottesen, 1966; Pittendrigh et al., 1966).

Advances in paleomicrobiology also accelerated during the sixties. Schopf published the first reasonably comprehensive paper on a good microbial paleoflora in 1968—that portion of his doctoral thesis under Barghoorn dealing with the ~900-Ma-old Bitter Springs microbiota, a superbly illustrated landmark in descriptive paleomicrobiology. Joining Schopf in the vanguard of emerging western biogeology were Hans Hofmann, M. D. Muir, and Gerald Licari, followed in the 1970s by a flood of students and contributors to the field from related disciplines.

The first papers to apply transmission electron microscopy (TEM) to pre-Phanerozoic paleomicrobiology appeared during the sixties—one by Cloud and Hagen (1965) using TEM to study Gunflint nannofossils and several others that dealt unintentionally with contaminants and formvar artifacts. Toward the end of this decade also, Cloud, Licari, and others (1969) extended the record of probable eucaryotes back in time to the ~1.3-Ga-old Beck Spring Dolomite, an occurrence that became much better documented with the publication of Licari's doctoral thesis in 1978. This remains one of the oldest reasonably well-supported records of a eucaryotic level of organization—a conclusion supported by the presence of the megaspheroidal *Chuaria* in rocks as much as 1.0 Ga old (Ford and Breed, 1973) and by other megascopic algae in ~1.3-Ga-old rocks of the Belt Supergroup (Walter, Oehler, and Oehler, 1976). A contrary view is vigorously defended by Knoll and Barghoorn (1975a) but disputed by Schopf and Oehler (1976) and by Cloud (1976a).

Papers dealing with the mainly spheroidal assemblages, and especially with the Soviet effort to develop a paleomicrobiological zonation of the later Proterozoic, Vendian (i.e., latest pre-Cambrian), and Cambrian strata, continued to appear in the U.S.S.R., authored by Naumova, Timofeev, and associates. Most ambitious, and one of the best, is Timofeev's 1966 treatise, dealing mainly with Vendian and younger Paleozoic forms (although as in most Soviet paleontological publications, the quality of the published illustrations leaves much to be desired). A paper by Naumova and Pavlovsky (1961) was the first to describe an authentic pre-Phanerozoic microbiota from Great Britain. And both Naumova and Timofeev wrote up their views on the possible paleomicrobiological zonation of the later Proterozoic and early Phanerozoic strata for the Twenty-third International Geological Congress at Prague in 1968. In her summary of that work Naumova (1968) claims to recognize "22 zonal assemblages of plant microfossils" in sub-Cambrian and Lower Cambrian strata of the Russian platform—ten in the Lower Cambrian, eleven of Vendian age, but only one in the Riphean (Upper Proterozoic).

The authenticity of many of the microscopical objects and assemblages described from pre-Phanerozoic strata of the U.S.S.R, however, remains moot. Eliminating identified contaminants (e.g., Krylov, 1968) there is much ambiguity about both the biological origin and the provenance of materials described and often inadequately or even badly illustrated. Many of the so-called spores or acritarchs seem to be subspheroidal clusters of fine debris, bubble rings, or other artifacts of preparation, while others resemble (perhaps superficially) the spores of modern mosses. Among forms broadly classed by Soviet workers as microphytolites and catagraphs are many which appear to be the products of sedimentary, diagenetic, or even metamorphic processes that give rise to radiate or otherwise structured ooids (see also Boureau, 1977, 1978, and other non-Soviet authors), clusters of ooidal bodies, crystallites or clusters of them, glomerular bodies and aggregates, patterns of Leisegang diffusion rings, and other configurations of dubious biological significance. Even more doubtfully biogenic are various, perhaps mainly diagenetic acicular structures such as the clusters of gypsum crystallites or pseudomorphs that have sometimes been called "red algae" or otherwise attributed to biological causes and given binomial Linnean names by Soviet (and recently Chinese) authors. In some instances where materials illustrated look as if they might be biogenic (e.g., Vologdin and Drozdova 1969, fig. 1.5) evidence supporting a primary origin is not given or is weak, while illustrations are often simply too poor to permit reasoned judgment.

Still, the case for a succession of flourishing primary microfloras of pre-Phanerozoic age was established during the sixties—both for the mixed benthonic and planktonic biotas now known so widely from thin sections of stromatolitic cherts and, somewhat less convincingly, for the mainly planktonic biotas from palynological macerations of fine clastic sediments. Contributors to this record include not only those named above, but also a more widely international community recording occurrences from Czechoslovakia, Germany, Norway, and Finland as well as from North America, Australia, and the U.S.S.R.

During the sixties too, largely as a result of studies by Glaessner and associates in Australia, Ford and associates in England, a lucky find in a borehole on the Russian platform (Keller, 1963, p. 505, pl. 18, fig. 11), and renewed interest in the definition of the lower limits of Cambrian, Paleozoic, and Phanerozoic (Termier and Termier, 1960; Keller, 1963; Glaessner, 1966; Cowie, 1967; Cloud, 1968b), I began to consider seriously a view first implied by the Termier's—namely that the Ediacarian faunal assemblage probably approximates the beginnings of metazoan evolution and Phanerozoic history, whether it be considered sub-Cambrian or merely a distinctive early Cambrian facies.

Finally, the sixties was the decade when Cloud (1965, 1968a, 1968b) first formulated and proposed his much cussed and discussed (and still evolving) working model of the primitive Earth—linking biospheric, atmospheric, hydrospheric, and crustal evolution and calling for an initially oxygen-free atmosphere and a long interval of fluctuating imbalance between sources and sinks of O_2 before mainly photosynthetic O_2 finally takes up permanent residence in the atmosphere.

1-5. Stage IV. The Takeoff Decade: The 1970s

Whether the growth of pre-Phanerozoic biogeology as a science is measured by the number of active researchers entering the field, the number of papers published, the number of new microbiotas described, or the level of interdisciplinary interaction, the 1970s were a decade of surging advance (see, for example, Figures 9-2 and 14-2 in subsequent chapters of this book). By all of these measures 1970 is near the point in the exponential growth curve at which it begins to ascend toward the vertical. Of the 27 authentic and well-documented microbiotas from cherty microbially laminated sediments listed by Schopf in his 1977 review (Schopf, 1977a), only three were known in 1969. If the microbiotas of the clastic sediments were added the discrepancy would remain marked. Fewer than a dozen active paleomicrobiologists were concerned with the pre-Phanerozoic biotas during the sixties, where now some scores are involved, and there have been parallel increases in related fields. While once we eagerly awaited the appearance of each new relevant paper, it has become an increasingly demanding task to keep current with the growing press of new publication. And, if this book may be taken to suggest the level of interdisciplinary interaction, it is now very high.

It would be a daunting task to extract the essence of all this paleomicrobiological and related biogeological activity; but fortunately a number of helpful review papers and compendia bring it within compass, and the critical developments are all enlarged upon in other chapters of this book. Here I need but mention a few highlights.

The most active scientific contributor of the decade has been, without doubt, the editor of the present volume, J. W. Schopf. In five far-ranging reviews (1970, 1974, 1975, 1977a, 1978) and now in this book, he has kept the biogeological world informed of his evaluations of the evidence and the work of others at an increasing level of sophistication and criticality. I have written four reviews myself over this same interval (Cloud, 1974, 1976a, 1976b, 1980), dealing with a differing but overlapping body of subject matter, and I find his assessments of the paleomicrobiological data, his identification of problems, and his recommendations for future research right on target.

Schopf was a carryover from the vanguard of the 1960s. An even longer-range carryover, who may not have realized at earlier times that he was doing biogeological research, was S. Golubić. Golubić's mastery during prior years of proalgal taxonomy, ecology, and sedimentology prepared him to fulfill the important function during the 1970s of introducing a level of criticality in differentiating between primary and degradative morphology among microbial fossils. He brought to the growing biogeological community a broad understanding of myxophyte ecology and its stromatolitic products.

Students of Barghoorn, Cloud, Glaessner, Golubić, and Schopf, plus others who emerged from training centers elsewhere in North America, Australia, the U.S.S.R., Europe, and southern Africa swelled the ranks. Among them were M. R. Walter, Wolfgang Preiss, S. M. Awramik, Mary Moorman, Dorothy Oehler, John Oehler, Bonnie Bloesser, J. M. Blacic, R. L. McConnell, G. J. B. Germs, David McKirdy, R. J. Horodyski, Brian White, D. G. Darby, A. H. Knoll, T. R. Fairchild, B. J. Javor, S. C. Lo, and others of the same age range (for reference to their papers not here listed, see Awramik and Barghoorn, 1978). The third International Palynological Congress at Novosibirsk in 1971 provided opportunity for a comprehensive review of Soviet work on the biotas of the Proterozoic and early Phanerozoic shales and siltstones and for the publication of an important summary by Timofeev (1974). But the biogeological credibility that finally came to the spheroidal assemblages of the clastic sediments is, as much as anything, a product of the meticulous methods of preparation and study exemplified by the work of Gonzalo Vidal, and in particular his 1976 monograph on the Visingsö beds of Sweden.

During the 1970s, in fact, increasing attention worldwide was directed to the microbial content of the ancient clastic rocks. The utility of these biotas is seriously handicapped by the problem of detecting and eliminating contaminants and artifacts. Yet it seems clear that they will eventually have at least as great a biogeological significance as the cherty biotas and potentially a greater stratigraphic utility. Brief but important recent reports by Horodyski (1980) and by Horodyski and others (1980) support such conclusions, with appropriate references to earlier work.

The authors mentioned and others whose papers are referenced in the 1978 bibliography of Awramik and Barghoorn (e.g., C. W. Allison, M. D. Muir, Magda Konzalova, Chantal Fournier-Vinas, Pierre Debat, V. U. Shenfil, Yu. K. Sovietov, J. A. Donaldson, Marcel Dardenne, and P. K. Maithy) produced an unprecedented flow of papers describing new occurrences in both cherty and clastic facies, but especially the latter. As a result, and in addition to a number of new localities in North

America, Europe, and the U.S.S.R., we now know of several authentic new pre-Phanerozoic microbial biotas from Australia (e.g., Schopf and Fairchild, 1973; Muir, 1974, 1976; Walter et al., 1976; J. H. Oehler, 1977c; D. Z. Oehler, 1978); from central and southern Africa (Maithy, 1975; Binda, 1976), from India (Maithy and Shukla, 1977; Schopf and Prasad, 1978) and Greenland (Vidal, 1979); and of probable nannofossils from Brazil (Fairchild and Dardenne, 1978). However, the only illustrated record that has come to my attention of supposedly sub-Cambrian acritarchs from Antarctica (Iltchenko, 1972) is very unconvincing; at best it requires validation.

China was the last major region to enter the lists of pre-Phanerozoic paleomicrobiology, with a strong slant toward Soviet methods of preparation and interpretation, and with focus on the fine clastic sediments. Among the first and best works from China are those of Sin and Liu (1973, 1976), including material that I judge to include both biological and non-biological components. More recently, in 1979, the Academia Sinica published the *Proceedings of a Symposium on the Stratigraphy, Paleontology, and Geology of Iron Ore*, 1977 (Science Press, Peking) including English abstracts of five papers dealing with the Anshan Group of east Liaoning Province, dated by the potassium-argon method at 2.4 Ga, metamorphosed up to amphibolite and even granulite grade, and containing banded iron-formation. From such rock has been described supposedly biological microstructures and amino acid spectra which I had the privilege of examining in the laboratories of the Academia Sinica in Nanjing and Beijing in June, 1980. The presence of geochemically unstable components in the amino acid spectra implies a contaminant origin. Among the supposed nannofossils shown to me I saw none that I regarded as of convincing biological origin. Nine types of modern iron bacteria, however, are reported from electron microscopical studies of the iron-formation, among which I saw beautifully preserved coccoids and spiral filaments similar to iron bacteria called *Gallionella* and *Leptothix*. The evidence of an association between the bacteria and the iron-formation is convincing, but the bacteria are too delicately "preserved" to be accepted as primary in such highly metamorphosed rocks without more convincing evidence (see also Table 14-2, no. 37c). Their reported stratigraphic restriction may be a function of the fact that they live preferentially on iron-rich substrates. As in another instance where I was shown a dense mat of supposedly fossil filamentous algae that still contained traces of a chlorophyll-like pigment, the origin of these microorganisms might be resolved by an attempt to culture a fresh extract.

Other recent Chinese publications refer to other supposedly sub-Cambrian fossils. Apart from some of the microstructures described by Sin and Liu (1973, 1976), however, and omitting shelly faunas of Tommotian age, the only ones that I have so far found to be biologically convincing were some in strata of the Sinian System (not the Sinian Suberathem) of probable Ediacarian age in western Hubei and Sichuan provinces (i.e., somewhat less than half of the material described by Yin and Li, 1978). As for the Chinese stromatolite work, what I observed seem to be moving in constructive directions, except that, as in other aspects of their pre-Phanerozoic biogeology, Chinese stromatophiles seem to have adopted bad as well as good practices from their Soviet teachers—in this case the interpretation of gypsum and other diagenetic crystallites in stromatolites and associated carbonate rocks as biogenic structures (e.g., Cao and Zhao, 1978). A determined effort is now being made throughout China to bring greater rigor to their pre-Phanerozoic biogeological studies.

Unfortunately the unevenness of the published pre-Phanerozoic biogeological data is not as widely understood as it should be even within the field, let alone by the biologically and geologically unsophisticated. One anecdote that I can relate from recent personal experience may illustrate the problem. At the end of a talk by an eminent biogeochemist, seeking to enlighten a group of planetary scientists about the early biogeological record, he showed a slide, without magnification, illustrating a chain of minute and supposedly very ancient cells. Seeking to be graphic, he described this chain of tiny coccoid cells as having a worm-like configuration. Minutes after his presentation I overheard an attentive listener from that audience excitedly telling a group of colleagues about the new record of 3.5-Ga-old worms!

To return to the central subject, another prime mover of western biogeology in the 1970s has been Walter, whose epochal 1976 compendium, *Stromatolites*, involves a large proportion of the leading workers in the field. It also includes a bibliography of 2,034 titles compiled by Awramik, in which the works of Bertrand-Sarfati, Gebelein, Hofmann,

Krylov, Monty, Preiss, Raaben, Semikhatov, Serebryakov, Walter, and other leading contributors are listed. Suffice it to note here only that the vexing questions of stromatolite morphogenesis and taxonomy are still moot, as well as the matter of stratigraphic application (Monty, 1977; Semikhatov et al., 1979). Most stromatophiles now agree that both ecologic and biologic factors play a role in determining stromatolite form, although not about the part each plays. And most also agree that some of the columnar morphologies and primary microstructural peculiarities also have a stratigraphic value. How much value is another question. A persuasive rationale for perceived evolutionary trends among stromatolites continues to elude us.

If, to the above, we add the growing interest of microbiologists, biochemists, geochemists, and atmospheric chemists in the field, we should probably find a population of perhaps 250 or 300 people worldwide that contribute to the goals of pre-Phanerozoic biogeology. Many of these people are now interacting with one another not only in informal gatherings but also under the aegis of International Geological Correlation Program Project 157 (Early Organic Evolution and Mineral and Energy Resources) and other I.G.C.P. activities, as well as at meetings of the International Symposia on Environmental Biogeochemistry organized by J. Skujins, H. Ehrlich, T. D. Brock, P. A. Trudinger, W. E. Krumbein, R. O. Hallberg, and others.

I have failed so far in this section to bring into perspective the contributions of or interactions with geochemistry, biochemistry, and microbiology. The foci of much pre-Phanerozoic biogeology has had to do with five main problem areas: (*i*) the sources, sinks, levels, and consequences of atmospheric oxygen; (*ii*) the mechanisms for storage, transport, and precipitation of iron in the sedimentary record and its possible relation to biological O_2 production, transfer, and protection; (*iii*) the types of autotrophy and photosynthesis, the enzymatic and other protective responses of microorganisms to UV and free O_2, and the relative abilities of microorganisms to utilize glycolosis and oxidative phosphorylation in energy transfer; (*iv*) the time and mode of origin of life, of O_2-releasing autotrophy, of eucaryotes, and of Metazoa; and (*v*) the ecological feedback throughout the system.

It would not be appropriate for me here to deal in depth with those questions, for they are the subjects of much of the remainder of this book and are probably already familiar to most of its readers. I will, therefore, take only brief notice of some recent reviews and highlights of the relevant subject matter:

(1) Atmospheric and ocean chemistry and evolution are the subjects of two recent summary volumes. One, by Holland (1978), a leading contributor to ideas about atmospheric and hydrospheric evolution beginning in the fifties, focuses on the present state of these systems and promises a second volume on their evolution. A second, by Walker (1977), after reviewing the present state of the atmosphere and its interactions with hydrosphere, lithosphere, and biosphere, deals at depth with aspects of atmospheric evolution and the balances between oxygen sources and sinks as related to the evolving biosphere. The reader is referred to these volumes for an update on present thinking, as well as to Chapter 11 by Walker et al. in this book.

(2) Problems related to the sources, storage, transport, and precipitation of sedimentary iron as related to various aspects of banded iron-formation and possibly causal biological processes were discussed by Trendall, Klein, Goldich, Cloud, Holland, Garrels, Perry, and others in a symposium on pre-Phanerozoic iron-formations (James and Sims, 1973). In that volume Cloud (1973a) and Holland (1973) independently proposed the storage of iron from volcanism or terrestrial sources in anaerobic ocean basins, with upwelling to sites of precipitation where still disputed mechanisms caused the banded structure.

(3) Aspects of biochemistry as related to photosynthesis, O_2, UV, energy transfer, and related matters have been conveniently and clearly summarized in a recent paper by Schopf (1978), who stresses the anaerobic early phases of the most fundamental biochemical processes, the antiquity of UV protective and repair mechanisms, and the late appearance of O_2 as a driving force in biological reactions (see also Chapter 13 in the present volume).

(4) As for the question of origins, Miller and Orgel (1974) give us a comprehensive and balanced survey to mid-decade, setting the stage for a clear and sweeping summary of how it looked to Dickerson in 1978. There are no surprises here. Everything is consistent with present implications from biogeochemical data (Cloud 1974, 1976a, 1980) and evolutionary theory (Cloud 1973c) that life originated soon after the formation of a liquid

hydrosphere, before the advent of persistent atmospheric oxygen, and as simple unicellular or noncellular anaerobic heterotrophs. The biomass attainable by such *Clostridium*-like organisms, Dickerson (1978) rightly stresses, would be small. And, under such conditions, selective pressures would favor the emergence of autotrophic systems, of primary producers that could evade the shortages of abiotic nutrient supply. Anaerobic photosynthesis utilizing H_2S as an electron source probably preceded oxygen-releasing photosynthesis utilizing H_2O and Photosystem II, but Cloud (1976a, 1976b, 1980) would infer that oxygen-producing photosynthesis of some sort was already in operation by the time the oldest banded iron-formation we know was deposited 3.8 Ga ago.

There remains dispute about time of origin of eucaryotes, although all would agree that the minimum atmospheric O_2-level at that time was not less than one percent of the present. As for mode of origin, Margulis (1970) has been the main contender. In her remarkable *tour-de-force* she brought together a formidable mass of information advocating the once minority view that serial endosymbiosis was the process whereby the eucaryotic cell arose from a procaryotic host-ancestor (Margulis, 1970). The vigor with which she presented and defended the case, its appealing simplicity, and the evidence for the independent biochemistry and reproduction of certain eucaryotic organelles persuaded some and turned others to a search for alternatives. But serial endosymbiosis has withstood testing well. A plausible model for the ancestral procaryotic host has been suggested among the archaebacteria (Van Valen and Maiorana 1980), and a reasonable analog has been found in a living thermophilic mycoplasma (Searcy et al., 1978). Although critics balk at a spirochaete ancestor for the $9 + 2$ flagellum and a complicated multiple evolution of sexuality, an endosymbiotic origin seems plausible for mitochondria and convincing for chloroplasts. Even those who take exception to the hard sell accept endosymbiosis as probably at least a part of the story of origins (e.g., Mahler and Raff, 1975).

Going beyond the eucaryotic cell itself, data on the beginnings of metazoan evolution have continued to accrue during the 1970s until they now constitute a substantial body of evidence favoring the first appearance of demonstrably animal life about 680 Ma ago. This evidence implies a circumglobal distribution of a distinctive, diverse, mainly soft-bodied, floating and shallow benthonic community, preceding by 80 to 100 Ma the appearance of the conventional shelly faunas of the Cambrian (Stanley, 1976, summarizing the work of Cloud, Germs, Glaessner, and others; see also Towe, 1970; Sokolov, 1972; and Fedonkin, 1977). The evidence is consistent with the concept of geologically rapid polyphyletic diversification proposed by myself in 1948, but the new availability of geochronologic ages permits us to order our data better. Although in 1948 and later I thought the Ediacarian might be simply a facies of the Cambrian, I now agree that it is sub-Cambrian by any reasonable definition of the latter. Yet, along with a few others, I also judge that any reasonable definition of Paleozoic and Phanerozoic includes the Ediacarian as its basal system. That is why I have abandoned the formal term Precambrian in favor of the informal pre-Phanerozoic (or pre-Paleozoic, or pre-Ediacarian).

Support for such a view may be growing. In a recent vote, members of the "Precambrian-Cambrian Boundary Working Group" (Cowie, 1978b) strongly supported placing the base of the Cambrian "near the evolutionary changes which are signaled . . . by the appearance of diverse fossils with hard parts;" but they split evenly on approving the same stratigraphic level as the "initial point of the Phanerozoic" and rejected this level as the "terminal point of the Proterozoic." I could not agree more wholeheartedly with Cowie (1978a) than I do when he asserts that: "The main aim of the whole project" should be "the investigation and elucidation of the . . . Cryptozoic-Phanerozoic (Proterozoic-Paleozoic) Transition." We diverge, however, where he seems to equate that goal with the "establishment of the Precambrian-Cambrian Boundary," especially considering the implications of the terms transition and boundary.

(5) Although ecological feedback should be at the center of all of our thinking about pre-Phanerozoic biogeology, it rarely comes consciously to the fore. Important in making it overt has been the interest that microbial biologists such as Barghoorn, Brock, Golubić, and Margulis have taken in the field and the interactions they have stimulated with others. As a result we see an increase in numbers and proportions of both biologically trained geologists and geologically trained biologists entering the field and a growing biological sophistication and ecological awareness in

mainstream biogeological practice. This is exemplified in several of the papers in Walter's *Stromatolites* (1976a) as well as in a number of other recent works dealing with biological-ecological problems such as that of cell degradation and its recognition (e.g., Knoll et al., 1975, 1978; Cloud et al., 1975; Hofmann, 1976; Golubić and Hofmann, 1976; Golubić and Barghoorn, 1977; Awramik and Barghoorn, 1977; Schopf, 1977a).

1-6. Current State and Future Prospects

Although it is not part of my charge in this chapter, I want to conclude these remarks with a few comments about what I judge to be now reasonably well understood or accepted and how we might best seek further to test, validate, and extend both theoretical and data bases. In the time frame to which I refer, "now" is just before the writing of the following chapters, which thus become a part of the future to which I allude.

The known record of truly sub-Cambrian life now consists of several score well-documented fossil microbiotas from sedimentary rocks as old as 2 Ga and a few probable older records. A high proportion of these records are from rocks of Ediacarian age—pre-Cambrian, but not pre-metazoan or pre-Phanerozoic. The best preserved and so far most informative of these microbiotas are found in thin sections of primary or early diagenetic cherts. The most numerous are from conventional palynological maceration of fine-clastic rocks, especially toward the younger part of the succession. This sequence of microbiotas also reveals some local peculiarities and some apparent evolutionary trends that already seem to have a crude stratigraphic utility, and it provides evidence favoring the emergence of the eucaryotic cell between about 2.0 and 1.4 Ga ago. There is, then, a reasonable expectation that new discoveries will enhance stratigraphic applications and narrow the time-ranges within which the major biologic innovations are now bracketed.

The presumptive evidence of stromatolites pushes the record of life and probably of photosynthesis back as far as ~3.4 or 3.5 Ga ago (Lowe, 1980a; Walter et al., 1980). And Shidlowski and others (1979) have argued on persuasive grounds that carbon isotope ratios indicate a continuity of biological processes from at least 3.8 Ga ago until the present. Suppose then that one also accepted the reasoning that there may have been a relationship between photosynthetic oxygen and banded iron-formation—likely enough considering the prevalence of iron in the shielding scytonemine pigments that envelop so many myxophytes and the part played by ferrous and other metal ions in the construction of the cytochromes and other vital macrocyclic pigments. One might then conclude with me that life was probably already extant and had evolved to an oxygen-releasing photoautotrophic level by the time the oldest known sediments were deposited in Greenland, Labrador, and the Limpopo fold belt about 3.8 Ga ago. There should, of course be a little biogeopoetry in every biogeologist's life. The following biogeorhapsody is, I believe, consistent with evidence surveyed above:

(1) Life had already arisen by chemical evolution from non-living materials by the time the oldest yet known sediments were deposited about 3.8 Ga ago.

(2) No free O_2 existed at that time as more than evanescent products of the abiotic photolysis of H_2O. The most vital biological processes are reductive ones, and nature is careful to carry out her biological oxidations by the removal of hydrogen rather than by the addition of oxygen. The kinds of chemistry that could have given rise to living systems of the sort we know will not go on in the presence of oxygen. Procaryote phylogeny based on 16*S* ribosomal RNA sequencing shows that the earliest lines of descent are anaerobic and that aerobes arose from anaerobic ancestors (Fox et al., 1980). Even the most vital oxygen-driven biochemical processes have anoxic roots. Thus, life itself is presumptive evidence that free O_2 is a secondary development, and that, in the beginning, there was none (or very little and evanescent).

(3) We may conclude, then, that first life was anaerobic and presumably heterotrophic.

(4) Evidence for the presence of primary sulfate minerals in sediments as old as 3.4 to 3.5 Ga in West Australia (Walter et al., 1980), however, tells us that some oxygen was already coming from somewhere—perhaps from abiotic photolysis of H_2O or the utilization of H from H_2S by photosynthetic green sulfur bacteria to photoreduce nicotinamide adenine dinucleotide (NAD^+) to NADH, releasing sulfate in the process.

(5) The evolution and continuance of an oxidative atmosphere was a highly improbable event. It

could not have happened without a kinetic lag in thermodynamic balances that would even today eliminate all resident O_2 in our atmosphere in a few million years (~3 Ma; Holland, 1978). For no known source of primary O_2 exists and all secondary sources produce thermodynamically unstable co-products that promptly recombine: carbohydrates and O_2 from photosynthesis to form CO_2 and H_2O, hydroxyl or hydrogen and oxygen from photolytic dissociation to produce H_2O.

(6) The geological evidence of thick and extensive detrital deposits of easily oxidized uraninite and pyrite 2.3 Ga old and older, the banded iron-formation of older pre-Phanerozoic deposits, and the conspicuously oxidized continental sediments of younger ones combine with other evidence to indicate important historical changes in O_2 levels. During Archean and early Proterozoic time until ~2.3 Ga ago, O_2 probably fluctuated between about 10^{-4} and one percent of present atmospheric level. Such vanishingly low levels of free O_2 are implied by the detrital uraninites (Grandstaff, 1980) and are sufficient to account for the hematitic stains on some Archean rocks (Shegelski, 1980) and the Lower Proterozoic weathering profiles in South Africa (Button, 1979) and southern Ontario (Gay and Grandstaff, 1980). Indeed, low levels of O_2 may have favored deeper weathering (Siever and Woodford, 1979). Higher levels of O_2 are implied by more typical red beds perhaps as old as 2.3 Ga (Dimroth, 1968, 1970; Chandler, 1980). After ~2.3 Ga ago, therefore, O_2 may have fluctuated near one percent present atmospheric level until the general onset of red beds and general cessation of banded iron-formation around 2 Ga ago. The first common appearance of what seem to be authentic proalgal heterocysts (sites of anaerobic nitrogen fixation) at about the latter time suggests an evolutionary response to a need for shielding sites of biological nitrogen fixation from ambient free O_2. Around 2.0 to 2.3 Ga ago, therefore, seems a likely time for the beginning of O_2 as a significant and more or less permanent atmospheric component at concentrations at or above one percent present atmospheric level.

(7) Nursall (1959) was probably on the right track in relating O_2 levels to the emergence of the Metazoa ~680 Ma ago.

(8) Considering the foregoing, the dates below are suggested for the major biologic innovations indicated: (*i*) Origin of life >3.8 Ga ago; (*ii*) Bacterial photosynthesis >3.4 Ga ago (and perhaps both bacterial photosynthesis and O_2-releasing photosynthesis before 3.8 Ga ago); (*iii*) Eucaryotes 2.0 to 1.3 Ga ago; (*iv*) Metazoa ~680 Ma ago.

How might we improve our data and interpretations in the future? To what degree will the work and conclusions of the past be reflected in new judgments about the early biosphere and its interactions with the physical earth? If I could borrow Aladin's lamp for a day, what might I ask the jinni about pre-Phanerozoic biogeology and how do I think it might respond?

Of course I want better answers to all the questions implied in my biogeorhapsody. I want to know the timing, circumstances, and succession of events leading to (*i*) the origin of life; (*ii*) the first autotrophs; (*iii*) oxygen-producing photoautotrophy; (*iv*) an advanced oxygen-mediating enzyme system; (*v*) oxidative phosphorylation; (*vi*) a eucaryotic level of cellular organization; (*vii*) the origin of meiosis and sexuality; (*viii*) the organization of cells into tissues and organs; and (*ix*) the emergence of the kingdoms and phyla of higher organisms.

It is clear from the foregoing that a key aspect of these questions turns on the relations between processes involved and the presence or absence, levels, and persistence of free oxygen. Such relations constrain the permissible processes and events to such a degree that it becomes important to ask what facts well-informed persons might agree upon. The jinni would tell me, I think, that there was substantial agreement on stanzas two to six of my biogeorhapsody and some concurrence with the other three.

In seeking answers to the things we should like to know about pre-Phanerozoic biogeology, therefore, I see as preeminent at this time, within all the necessary interdisciplinary effort, two main needs. First, we need continued search in the rocks themselves for new, authentic, and well-dated microbial and biogeochemical records that will enlarge our still impoverished knowledge of things past. Second, we need better analyses of the phenomena related to the history of atmospheric oxygen. With reference to oxygen we must ask:

(1) How can we write more believable and better quantified geochemical budgets for oxygen and its co-products, whether carbon, hydrogen, or something else?

(2) What were the early oxygen sinks that delayed accumulation of O_2 in the hydrosphere and atmosphere? Besides ferrous iron, volcanic sulfides

and sulfites, and reducing atmospheric gases such as hydrogen, ammonia, methane, and carbon monoxide, were there other important oxygen sinks on the primitive Earth? What might some yet unidentified oxygen sinks have been? And what are the best estimates we can make of quantities involved at all levels, including quantities transferred to or recycled through the mantle by subduction over geologic history?

(3) What kinds of evidence might tell us when O_2 first appeared in the hydrosphere and atmosphere as more than a trivial and transient gas?

(4) When and why did O_2 start to build up in the atmosphere, and what evidence might there be for earlier episodes of low-level atmospheric oxygenation that were overwhelmed by new oxygen sinks, turning the trend back toward anoxic states? (5) Once O_2 began to accumulate in the atmosphere on a sustained scale, how rapidly (or slowly) did it build up? And what clues might we find to O_2 levels at various times in Earth history? What, for instance, if anything, do the ratios of carbon and sulfur isotopes, ferro-ferric ratios, and the characteristics of uranium and copper ores tell us of O_2 levels and possible swings in abundance with time?

(6) What likely interactions may have taken place between events and processes suggested and living systems, known or yet unknown? The biogeologic evidence of cell diameter and filament branching suggests an age anterior to 1.3 Ga for the origin of the eucaryotes. Oxygen levels probably above the Pasteur Point as far back as 2.0 to 2.3 Ga ago would be consistent with a eucaryotic emergence by then. Can we narrow the permissible range of time for this event more closely than 1.3 to 2.3 Ga ago? Similarly, biostratigraphic, radiometric, and geochemical data suggest the first appearance of Metazoa perhaps 680 Ma ago. Will that date stand the test of time or will older Metazoa be found? What other points or ranges dare we hypothesize or can we establish or bracket on the curves of oxygen growth and evolution?

(7) And, finally, where might we look for evidences of pre-solar-system evolution, either cosmochemical or cosmobiological? The Allende and Murchison meteorites, through the brilliant work of Wasserburg and associates (1979), brought the first hints that there could be, within the solar system, evidence of pre-solar events. What unearthly evidence might there be within the solar system or beyond it that could more brightly illuminate the mystery of the origin of life or of its basic building blocks?

Whatever the height we scale, however, our science is rooted in the rocks. It is in the historical record of this planet from which emerge the only data that tell us unequivocally of its past, that irreducible minimum with which all other aspects of pre-Phanerozoic biogeology must be consistent. However far our minds may wander in theory, they must return to this base-line for testing—for validation or disproof. We deal, to be sure, with the biologic aspects of Earth history—but the subject is geology, *ergo* biogeology.

In closing, then, let me emphasize once more that the admissable subject matter must be validated both biologically and geologically. Whether a body fossil, a sedimentary product of biological activity, or a biogeochemical compound, it must be capable of being shown to be (or to have a high probability of being) of authentic biological origin. It must then be shown to be (or to have a high probability of being) both indigenous to sedimentary rocks where it is found and to have originated at the same time as, or penecontemporaneously with, the original sediments—to be primary in a geological sense, that is. Finally, interpretations of this evidence are valid only in the context of the entire preserved history of the enclosing rock. They are constrained by the limitations and probabilities imposed by the rock's history, by physical laws, and by previously well-established biological sequences and ecological probabilities. Given rigor and consistency in the foregoing matters, the creation of testable hypotheses and models that may eventually become theories or paradigms then becomes a matter of the widening of perspective, of the embracing and integration of all relevant disciplines and subject matter. Future generations of pre-Phanerozoic biogeologists will judge the product.

CHAPTER 2

The Nature of the Earth Prior to the Oldest Known Rock Record: The Hadean Earth*

By David J. Stevenson

2-1. Introduction

Statements about the Earth's history prior to 3.8 Ga ago, the age of the oldest rock sequence now known on Earth, are based on indirect evidence (such as that derived from meteorites and lunar samples) and on plausibility arguments, the persuasiveness of which often depend more on the eloquence of their advocates than on the weight of relevant data. Such statements, however, are of profound importance as they relate to one of the central issues of this volume—the origin of biologic systems (see Chapter 4)—an event that may well have occurred during the pre-3.8-Ga-old "Hadean" segment of Earth history. Yet understanding of the Hadean Earth is far from complete; in particular, and because of the paucity of directly relevant data, theorists dealing with the problem are obliged to work with "models," constructs that can be no better than the input assumptions. New data or interpretations can cause these models to be discarded or recast. The discussion presented in this chapter should be seen in this light: basic physical principles need to be understood, but for the present, detailed scenarios or predictions based upon them are best regarded as "convenient fictions," worthy of discussion but not enshrinement.

* Contribution number 3509 from the Division of Geological and Planetary Sciences, California Institute of Technology, Pasadena, California 91125.

2-2. Origin and Evolution of the Solar System

The most striking feature of the solar system is its high degree of order: the preferred sense of rotation; the coplanarity, circularity and regular spacing of planetary orbits; and the grouping of planets into compositional classes in accordance with their distance from the Sun. Despite the striking dynamical properties of the system, however, it is the physiochemical characteristics of planets and meteorites, together with relevant astrophysical considerations, that provide the best current constraints on the characteristics of solar system formation. The data base is incomplete, and is never likely to be sufficient for a detailed cosmochronological reconstruction, but an examination of our galactic environment suggests plausible scenarios of formation.

In our galaxy, as much mass is contained in the material between stars, as there is within stars, and although the density of this interstellar medium is typically 10 to 100 times too low to initiate the gravitational collapse of gas clouds having solar mass, localized density enhancements (e.g., the material in spiral arms, or concentrations enhanced as a result of a nearby supernova explosion; Cameron and Truran, 1977) may initiate the process. The angular momentum and magnetic field of such collapsing clouds will determine the nature and morphology of the collapse. The role of the magnetic field in this process is poorly understood, but rotation alone is expected to lead to the formation of a diffuse, disc-shaped body of gas and dust. A number of observational characteristics of young stars or protostars suggest the existence of just such discs in the galaxy (e.g., MWC 349, discussed by Thompson et al., 1977). The occurrence of fragmentation or of secondary collapse-centers during the formation of these discs are also suggested by the evidence for low mass companions of sun-like stars (Abt, 1977).

Much research effort is now directed at understanding so-called "accretion discs," flattened gas clouds that consist of a central concentration of matter (the newly formed or forming star) and a comparable or smaller mass of rotating gas and dust distributed in a very oblate nebula (see, for example, Cameron, 1978c). In accordance with cosmic and solar elemental abundances, such discs can be expected to be composed overwhelmingly of hydrogen and helium (the "gases"), leavened with about 1.5 percent by weight of the volatile forms of oxygen, carbon, and nitrogen (primarily H_2O, N_2, NH_3, HCN, CO_2, CO, CH_4) and 0.5 percent by weight of "rock" (principally iron and silicates). The high temperatures generated by the gravitationally induced accretion of material onto the disc are expected to have resulted in vaporization of most pre-existing solid material; the starting material from which terrestrial-type, inner planets form is thus believed to be a fine, particulate condensate from the nebula. Selective condensation due to a radial temperature gradient within the accreting cloud may be part (but is certainly not all) of the explanation for the compositional differences observed between the inner and outer planets of the solar system.

Dust grains formed by condensation can rapidly settle ($\sim 10^3$ years) into the central plane of the nebula. Depending on the relative amounts of gas and dust, either or both components can undergo gravitational instability leading to formation of giant, gaseous protoplanets (Cameron, 1978b) and/or rather loosely assembled planetesimals, rocky bodies of 1 to 10 km in size (Safronov, 1972; Goldreich and Ward, 1973). Of these, planetesimals appear to be the more relevant to an understanding of the origin of terrestrial planets and meteorites (although the disposition of the gaseous component is also of undeniable importance, even for a terrestrial-type planet). The questions of which instabilities will occur, and where they will occur within the disc, are primarily contingent on the mass of the accreting nebula; in the case of the solar system this is a poorly known parameter since most of this material may have subsequently been returned into interstellar space (for a discussion of these and other similarly mysterious events see, for example, Dermott, 1978; or Ringwood, 1979).

As collapse continues, mutual perturbations and collisions between planetesimals lead to the growth of larger bodies and the regularization of orbits, for our solar system on a timescale of 10^6 to 10^8 years (a shorter time scale is dynamically implausible, and a longer time scale is incompatible with isotopic data from meteorites and lunar rocks). An important feature of this accumulation process is the relative infrequency of high velocity impacts, since otherwise, with the rate of fragmentation exceeding that of amalgamation, planets could not form. Nevertheless, the heating that accompanies such impacts is an important source of energy resulting in melting and differentiation of the accreting planetesimals. Internal heating, possibly due to radioactive decay of short-lived ^{26}Al (Lee et al., 1977), may also play a role. In the case of our solar system, as is evidenced by the variety and complexity of the composition and mineralogy observed in planetesimal-like meteorites, extensive physiochemical processing must have taken place between initial condensation and eventual incorporation in the growing planets.

The simplest models for nebular evolution predict a gradual cooling such that the earlier forming condensates and planetesimals are rich in refractories. This is the basis of the *heterogeneous accretion* model for the Earth, a scenario in which most of the Earth's core is thought to have formed by direct accretion, with a surficial silicate layer being subsequently added. However, there are numerous difficulties with this hypothesis (see, for example, Ringwood, 1979), with one such difficulty (not mentioned by Ringwood) being the absence of any mechanism leading to the formation of an adiabatic, liquid outer core as is required for the generation of the Earth's magnetic field.

The most recently proposed models for nebular evolution (Cameron, 1978b) favor a complicated temporal and spatial history of nebular temperatures, a factor that, together with the complicated dynamics required for accumulation of planetesimals, removes the basis for a simplistic heterogeneous accretion. The following discussion of the origin of the Earth thus assumes essentially a *homogeneous accretion* (e.g., beginning with an intimate mixture of particulate metallic iron and silicates), although it is recognized that some degree of heterogeneity may have also occurred (e.g., the possible generation of a late, volatile-rich veneer; for further discussion, see Chapter 4).

2-3. Growth and Primary Differentiation of the Earth

The currently favored accretion scenario is a gravitationally interacting and collisional "swarm" of solid bodies in Keplerian orbits, with the stochastic growth of progressively larger yet fewer bodies occurring at the expense of smaller planetesimals. As the protoearth is assembled, impacts by planetesimals occur with relative velocities that are comparable but slightly larger than the escape velocity of the forming planet. Since most of the bodies incorporated during this accretion will be substantially smaller than the forming protoplanet, such growth and its thermal consequences can be approximately understood by a stochastic model in which the rate at which mass is added is regarded as slowly varying with time.

Among the crucial aspects to understand regarding this process are (*i*) the internal and surface temperatures generated during and immediately after the Earth's accretion; (*ii*) the nature of primary differentiation (viz., of core formation); and (*iii*) the extent of outgassing, crustal formation, and convective recycling that occurred in the Hadean Earth. The following simple model provides an approach to the answers to these problems, none of which is currently well understood.

Consider a growing planet of radius r and mass $M(r)$, assumed to have accreted after the decay of short-lived radiogenic nuclides (such as ^{26}Al) but so rapidly that the important long-lived sources (viz., ^{40}K, ^{235}U, ^{238}U, ^{232}Th) have not yet provided significant heating. In the absence of other external mechanisms (e.g., electromagnetic induction), planetary heating will occur as a result of the release of the gravitational energy of accretion. The kinetic energy per unit mass of impacting planetesimals is approximately $GM(r)/r$ (where G is the gravitational constant) and a portion, h, of this energy is retained as buried heat (the remainder being immediately reradiated into space via particulate ejecta or consumed to do work). Although the value of this parameter, h, is poorly known, it is not very close to either 0 or 1; a plausible estimate is perhaps 0.5 (for discussion, see Kaula, 1980). At low internal temperatures, redistribution of this internal heat is via conduction, a process much slower than the rate of accretion. Thus, except very near the surface, the temperature at radius $r' < r$ within the growing planet is

$$T(r') \simeq \frac{hGM(r')}{cr'} + T_e \tag{2-1}$$

where c is the specific heat and T_e is the ambient planetary temperature. Since $M(r) \propto r^3$, it follows that $T(r') - T_e \propto r'^2$; the highest temperatures are therefore near the surface of the growing planet. The central regions, formed from slowly moving planetesimals, are cold and rigid—a matter of importance in the subsequent discussion of core formation.

Equation 2-1 ceases to be valid in an outer conductive boundary layer of thickness $\sim(K\tau)^{1/2}$, where τ is the characteristic timescale for accreting of the planet and K is the thermal diffusivity. For $K \sim 10^{-2}$ cm^2/s (a value typical of rocks) and $\tau \sim 10^7$–10^8 years (the time scale suggested in Section 2-2, above), this layer has a thickness of 20–50 km. At the solid surface, the temperature T_s is determined by radiation into space and can thus be estimated by equating the accretional energy input with the radiative output. Very approximately (ignoring the blanketing effect of a dense transient atmosphere generated by the partial devolatilization of impacting planetesimals),

$$\frac{GM^2(r)}{r\tau} \simeq 4\pi r^2\sigma(T_s^4 - T_o^4) \tag{2-2}$$

where T_o is the insolation temperature (the surface temperature in the absence of accretion) and σ is the Stefan-Boltzmann constant. For $T_o = 300°$ K, $\tau \sim 10^8$ years and $r \sim R$ (the final, present radius of the Earth), it follows that $T_s \simeq 330°$ K. It can thus be concluded that only a modest increase in average *surface* temperature is likely to have occurred from the accretionary heat (for similar results, see Jakosky and Ahrens, 1979). This conclusion may be substantially modified, however, by a dense opaque atmosphere (Sekiya et al., 1980). Some previous workers (e.g., Ringwood, 1975) have favored shorter accretion times (and thus higher surface temperatures), but these are dynamically implausible (and the resulting higher temperatures are geochemically unnecessary). High *internal* temperatures, however, seem unavoidable since Equation 2-1 with $h \sim 0.5$ predicts that $T(r) \sim 1000^\circ$ K for $r \sim 0.2\ R$, when the accreting Earth had only reached the size of the Moon. Indeed, the clear evidence for early differentiation of the Moon to form anorthositic highlands (see Chapter 3) can be interpreted as being consistent with an accretional scenario with $h \sim 0.5$ (Stevenson, 1980a).

Eventually, as a planet continues to accrete, the internal temperature will increase to exceed that at which the subsolidus convection of silicates, or the melting of free iron, can occur.

Consider, first, the effect regarding subsolidus convection. For simplicity a uniform composition is assumed. The relevant dimensionless number is the Rayleigh number, R_a, defined as

$$R_a = \frac{g\alpha \Delta T d^3}{\nu K} \qquad (2\text{-}3)$$

where g is the gravitational acceleration, α is the coefficient of thermal expansion, ΔT is the temperature drop across a layer of thickness d, and ν and K are the effective viscosity and thermal diffusivity, respectively, of the medium. Recently, Schubert (1979) has reviewed the application of R_a to solid planets. Typical silicate materials have rheologies that can be approximated by

$$\nu \simeq \nu_o \exp(AT_m/T) \qquad (2\text{-}4)$$

where T_m is the solidus temperature and both ν_o and A are constants. Convection will occur once R_a exceeds a critical value, $R_a \sim 10^3$. For plausible parameter values (viz., $g \sim 10^3$ cm/s^2; $\alpha \sim 3 \times 10^{-5}$ K^{-1}; $d \sim 10^7$ cm; $\nu_0 \sim 10^6$ cm^2/s; and $A \sim 25$) criteria for convection are satisfied once $T \gtrsim 0.6\ T_m \sim 1000^\circ$ K. Since the viscosity of silicates is very strongly temperature dependent, the convection will be vigorous if the temperature is substantially higher than this value (yet still subsolidus). Equation 2-1 must therefore have ceased to be valid once the forming Earth was comparable in size to the present Moon since at that size, internal redistribution and outward transport of buried heat became increasingly important.

According to the boundary layer theory of convection (Turcotte and Oxburgh, 1972), the amount of heat that can be transported by convection is approximately $0.1\ k\Delta T(g\alpha\Delta T/\nu K)^{1/3}$ where k is the thermal conductivity. This is about 2,500 $(10^{16}/\nu)^{1/3}$ erg/cm^2-s, with an assumed $\Delta T \sim 10^{3}$°K (a value that gives about the right mantle heat flow for the present Earth if $\nu \sim 10^{21}$ cm^2/s, a value compatible with postglacial rebound studies, is substituted). The accretional energy input per unit area is about 2×10^4 $(10^8$ yrs$/\tau)$ $(r/R)^3$ erg/cm^2-s. For a dynamically plausible accretion time (between 10^7 and 10^8 years; Safronov, 1972), this accretional energy could be transported to the surface, without causing complete melting of the mantle, provided the viscosity of the solid or partially molten transmitting medium was as low as 10^{12} to 10^{14} cm^2/s. If such a low viscosity is possible, then large-scale melting of the primary silicate fraction (e.g., olivine) need not have occurred (i.e., a magma ocean, such as that discussed in Chapter 3, need not have eventuated). Unfortunately, the rheology of partially molten or near-solidus materials is poorly known; the effort to determine the extent to which convection can buffer or prevent melting is thus poorly constrained at present. The accretional energy available must have been sufficient (by a large factor) to melt the Earth completely, but this does not mean that complete melting necessarily occurred, since convection could have transported the heat out as quickly as it was buried! In fact, complete melting seems implausible, since under such circumstances the almost complete explusion of primordial volatiles such as ^{3}He would have occurred. In fact, however, primordial ^{3}He is still escaping from the Earth in substantial quantities (Welham and Craig, 1979). A convective "self regulation" is presumably responsible for preventing complete melting and, as is discussed below, it may have also played an important role in the nature of early differentiation and outgassing. Consideration of pressures within the formative Earth is also essential, since the slope of the adiabatic gradient (appropriate for convection) is less steep than that of the solidus. Thus, the most likely place for melting of silicates to occur within the planet is always near the surface (both in the present Earth and the accreting Earth). Partial melting may therefore have been confined to a shallow, near surface region, ~100 km thick (but in the Moon, where internal pressures are much lower, the depth of early partial melting—and thus of the putative "magma ocean"—was much greater).

Thus, it is likely that the silicate fraction in the early Earth was never completely melted. This was not the case, however, for the free iron component, which has a lower melting point. If accretion were homogeneous, with metallic iron disseminated throughout the primordial Earth in particulate form, then production of an iron melt would already be initiated when the Earth were only the size of the Moon or a little larger. And the presence of alloying constituents, especially sulphur, would increase the ease of melting still further. Since the iron fraction of the total volume was large, liquid iron would have readily percolated downwards through the silicate matrix. However, since the temperatures at deeper levels would have been colder (Equation 2-1), it would not have migrated to the center of the planet. Rather, the iron would have formed a spheroidal layer, the bottom of which would necessarily have been at its freezing point. Below this layer would have existed a zone consisting of an undifferentiated, cold primordial mixture of materials; and the mantle above the iron-rich layer would have been iron depleted and well stirred by thermal convection and Rayleigh-Taylor instabilities initiated by incoming iron-rich planetesimals.

How might the core have formed from this layer? A recent analysis of the dynamics involved (Stevenson, 1980b) suggests difficulties for the popular notion of large drops (hundreds of kilometers in radius) of iron sinking rapidly to the center, as originally proposed by Elsasser (1963). The principal difficulty with this mode of core formation is that the central regions are cold and do not flow viscously in response to the non-uniform load of the heavier iron above. And thermal diffusion is slow such that this central region does not heat up readily, despite the exothermic nature of the core formation process. Another proposed mechanism of core formation is the "sinking layer" model (Vityazev and Mayeva, 1976) in which the iron layer expands downwards by gradual melting of the iron in the primordial mix below, thereby releasing silicate grains that float upward and merge with the mantle above. Like the Elsasser model, this mechanism of core formation is slow, and for a similar reason: the downward transport of heat must be achieved by diffusion, a process that takes $\sim 10^9$ years.

It is here proposed, instead, that the fastest and probably dominant process whereby iron migrated to the center of the planet was by "catastrophic asymmetry": the less dense primordial core is suggested to have undergone a spontaneous, rigid displacement relative to the more dense, liquid iron layer above it (with a characteristic time scale for such displacement on the order of hours). This would have created a pear-shaped planet. However, the resulting non-hydrostatic stresses would have been sufficient to fragment the primordial core. The deep interior of the Earth would thus have turned itself inside-out, with the result being a liquid iron core upon which would have floated "rockbergs" (cold, undifferentiated fragments of the primordial core) overlain, in turn, by the hot, well-stirred, partly fluid mantle. The stirring and rotational energy loss associated with such a process are ample enough to prevent pressure-induced freezing of the iron; the core thus formed would have been adiabatic and entirely fluid. Although the process has been described here as a single "catastrophic" event, in reality the governing time scale would have been that of accretion, since it was the accreting planetesimals that would have provided the free iron from which the core formed. As here postulated, therefore, *core formation would have been contemporaneous with accretion.*

The stirring of the liquid iron in the core in these postulated early stages would have been much more vigorous than it is in the present outer core because the vigorously convecting mantle would have efficiently cooled the core from above. Thus, there would have been no difficulty in generating a magnetic field by the dynamo process. As in the present Earth, the energy source is likely to have been primarily gravitational. The magnetic field of such a forming and newly formed Earth might have been substantially larger than the present field, perhaps as large as 10 Gauss externally (an estimate derived by assuming that Ohmic dissipation would have been proportional to the heat flux driving core convection). Application to the Earth of Hoyle's (1960) model for the transfer of angular momentum via hydromagnetic torques from the primordial Sun suggests that, in principle, the transfer of $\sim 1\%$ of the Earth's mass from the Earth's surface to beyond the Roche limit is achievable. The actual importance of Hoyle's hydromagnetic effects depends crucially on the degree of ionization of the immediate Earth environment, a factor which cannot be determined without detailed consideration of the transient atmosphere and ionosphere. This has not yet been attempted. It is clear, however, that the newly

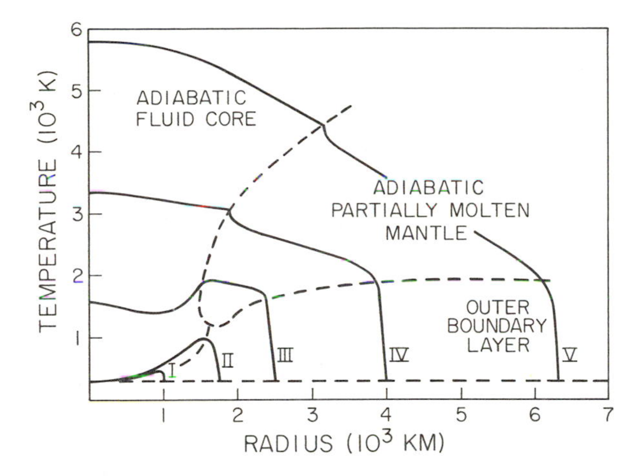

FIGURE 2-1. Schematic representation of temperature profiles in an accreting Earth. The chronological sequence is from I (outer radius $\simeq 1000$ km) to V (outer radius $\simeq$ present Earth radius). The actual time scale represented by I to V is unimportant, but is likely to be $\sim 10^8$ years. In I and II, temperatures are too low for differentiation, but by III, partial separation of core and mantle is achieved; subsequent profiles are qualitatively similar to that of the present Earth. The dashed lines delineate regions of the core, mantle, and outer boundary layer. Quantitative aspects of this figure should only be considered approximate.

formed Earth would have had a large magnetic field, one more than sufficient to screen out charged cosmic rays. The earliest direct evidence for the existence of a geomagnetic field now dates back to about 3.5 Ga (McElhinny and Senanayake, 1980).

The energy released as a result of core formation would have been very large, sufficient to raise the temperature of the whole Earth by about 1500°K, or of the core alone by $\sim 5000°$ K (Shaw, 1978). As above, however, the actual temperature rise was no doubt much lower since such energy would have transported out of the Earth's interior by mantle convection.

Figure 2-1 shows approximately the temperature profile of the Earth as a function of its radius at various stages of accretion (see also Kaula, 1979, 1980).

2-4. Differentiation and Outgassing of the Early Mantle

The general picture, therefore, is that within about 10^8 years after formation of the solar system, the accretion of the Earth was essentially completed. The core and mantle were well separated and each was convecting vigorously. The mantle was primarily solid, but localized high degrees of partial melt were being produced. Accretion rates were rapidly (perhaps exponentially) decreasing, and meteoritic impacts were therefore playing a progressively less important role in the prevailing dynamics. Long-lived radiogenic heat sources had not yet become important, with vigorous mantle convection being sustained by secular cooling alone.

Under these conditions, and letting $T(t)$ be a characteristic mantle temperature and $\theta = T(t)/T(\mathrm{o})$ represent a normalized temperature, consider the following simple model of an Earth with no internal or external heat sources, a planetary body that was convecting only because of loss of its internal heat content (originally emplaced by accretion). Boundary layer theory then leads to a differentiation equation for θ of the form (Stevenson and Turner, 1979)

$$\frac{\partial \theta}{\partial t} = -\frac{\theta^{4/3}}{\tau} \exp\left[-\frac{A}{3}\left(\frac{1}{\theta} - 1\right)\right] \quad (2\text{-}5)$$

where the exponential term arises from the chosen parameterization of mantle viscosity (Equation 2-4). For plausible parameter choices, $\tau \sim 4 \times 10^8$ years, if it is assumed that the mantle starts to cool from a temperature near its melting point (i.e., $T(\mathrm{o}) \simeq T_m$). Initially, at $t = 0$, the heat flux ($\propto -\partial\theta/\partial t$) would have been about 100 times its present value, and the viscosity would have been very low, $\sim 10^6$–10^8 times less than the present value. Using these conditions, Figure 2-2 shows an approximate time history for transient cooling of the Earth. This phase would have lasted for a few hundred million years at most, by which time radiogenic heating would have begun to dominate and the cooling rate would have decreased. During this transient phase, convection would have been extremely vigorous (as illustrated by the estimated convective turnover times indicated on Figure 2-2). The total amount of heat lost in this transient event would have been comparable to the total amount of radiogenic heat lost in the entire subsequent history of the Earth.

The occurrence of this event would have had profound consequences for primordial crustal formation and for Hadean outgassing. The heat flux of the Earth can be thought of as the product of a mass flux (viz., of material brought to the near

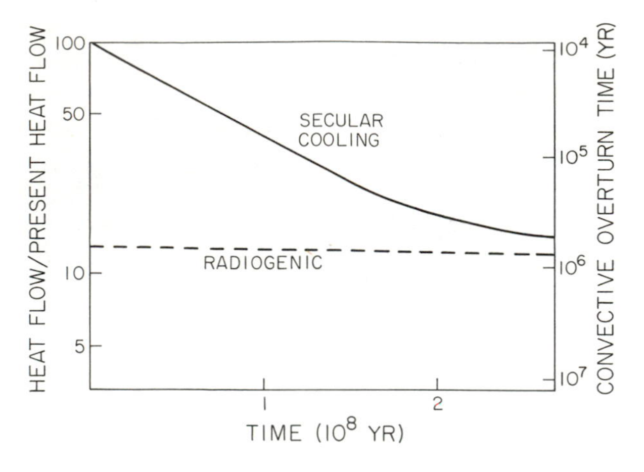

FIGURE 2-2. Transient cooling of a hot primitive Earth showing the surface heat flux (in units of the present heat flow) and mantle convection overturn time, as a function of time elapsed since cessation of accretion. The dashed, approximately horizontal line represents the radiogenic heat. The time scale is approximate (perhaps to within a factor of two) and the convection overturn times, based on boundary layer theory, are even more approximate (probably valid to within an order of magnitude).

surface) multiplied by the excess heat content of such material. The temperature excess driving convection was only ~20% larger then than it is at present. It follows that if the heat flux during this transient event was very much larger than the present heat flux, then the mass flux must have also been very much larger than its present values by a comparable factor. (In the context of modern plate tectonics, this mass flux is the source of new oceanic crust at spreading centers—the dominant form of volcanism on Earth. The style of tectonics on the early Earth may have been different, perhaps much more chaotic and with smaller characteristic length scales, but this does not invalidate the simple argument presented here; for a more extensive discussion, see Chapter 3.)

From the foregoing, it follows that the total mass of mantle cycled to near surface regions, and thus available for differentiation and outgassing during this postulated transient cooling phase, would have been comparable in quantity to the total mantle mass thus cycled during all of subsequent Earth history. This simple consideration suggests that *most of the outgassing and differentiation of the mantle may have occurred "catastrophically" as a transient response to accretional heating* (although it may have occurred after completion of most of the accretion; see also discussion in Chapter 4). It is not possible at present to assess whether "most" means 60% or 90% or 99%; the primitive crust formed at this stage has been recycled since, and there is no evidence of it still remaining. Much of the atmosphere and hydrosphere formed at this early stage may still be present (but, as discussed below, with partial recycling of H_2O and CO_2).

The crust produced during this stage must have been primarily basaltic (though perhaps tending toward an ultramafic composition because of the higher degree of partial melt). Geologic recycling of this crust may have required the pressure-induced transition to eclogite (Ringwood and Green, 1966; Anderson, 1979), so the crustal thickness might have been substantial despite the thin lithosphere (regarded here as the layer cold enough to behave rigidly). Sialic crust was presumably formed via fractional crystallization from extensive magma chambers and by remelted basaltic crust, both in the presence of water. The processes for recycling the sialic crust are not understood (for additional discussion, see Chapter 3).

With regard to the Hadean atmosphere there are four possible sources for its gaseous constituents: the primordial solar nebula; the solar wind; "collisions" of the Earth with asteroids or comets; and outgassing of the Earth's interior (for details, see the recent review by Pollack and Yung, 1980). Of these, outgassing is commonly assumed to have been dominant, much of which may have taken place as a result of impact events and, if during core formation, in the presence of metallic iron. Under these circumstances, the H_2 partial pressure of the gases released would have been substantial and, thus, dominant carbon species would have been CO and CH_4. Nitrogen would have been in the forms N_2 and NH_3, and sulphur would have occurred primarily as H_2S. However, if most of the outgassing responsible for the *present* atmosphere occurred subsequently, during the transient cooling phase discussed above, then the mantle oxidation state would have been similar to its present value; CO_2, N_2, and SO_2 would have been the dominant outgassed constituents. The transient, impact-produced atmosphere discussed by Jakosky and Ahrens (1979) concerns the pre-core formation phase. The behavior and retention of the subsequent atmosphere while the Earth was still accreting are crucial, unresolved issues (see

Chapter 4). At present, on these bases it is not possible to assess the likelihood of a reducing atmosphere for the early Earth; such an atmosphere might equally well have been CO_2-dominated. As noted previously, surface temperatures were low, and recycling of CO_2 and H_2O via weathering reactions with surface rocks is likely.

Finally, there is the question of accretional tail-off and the stabilization of crust toward the end of the Hadean, ~3.9 Ga ago. It is often assumed that these events may have been related (principally, based on analogy with the established cratering chronology of the Moon and those inferred for Mercury and Mars). In fact, however, this inference is tenuous at best. From the second law of thermodynamics, the rate of work that mantle convection could do must be comparable (but smaller) than the heat flux (the Carnot efficiency of mantle convection being actually much higher than most man-made machines). It seems plausible that the stabilization of the Earth's crust might have occurred at the time when the energy flux from accretion became less than the energy output by convection. This hypothesis, when applied to known cratering chronologies (e.g., Neukem and Wise, 1976; Hartmann, 1977) suggests that crustal materials as great as 4.0 Ga in age should have survived to the present. The absence of rocks older than 3.8 Ga thus remains unexplained. Despite its necessity, the "convenient fiction" alluded to in the introduction to this chapter remains unsatisfying; direct evidence of the earliest stages of Earth history are sorely needed.

2-5. Summary and Conclusions

(1) The oldest geologic sequence now known on Earth dates back only to about 3.8 Ga ago; notions regarding pre-rock record geologic history (that of the "Hadean" Earth) must thus be based on indirect evidence (e.g., that derived from meteorites or lunar samples) or on theoretical "models," simplified abstractions of the early Earth which can be no better than their input assumptions.

(2) Physiochemical characteristics of the planets and meteorites, together with relevant astrophysical considerations, seem to indicate that formation of the solar system was initiated ~4.6 Ga ago via the gravitational collapse of interstellar matter to produce an accretion disc composed almost entirely of hydrogen and helium (but containing also small concentrations of water, ammonia and methane ices, and solid iron and silicate particles); and that as collapse continued, condensation of these components and gravitational instabilities within the resulting cloud of gas and dust led to formation of planetesimals (rocky bodies 1 to 10 km in size) that within 10^6 to 10^8 years accreted to form the protoplanet Earth. The model here favored is one of homogeneous accretion (with the protoplanet composed of a homogeneous mixture of particulate metallic iron and silicates); the accretionary processing was completed ~4.5 Ga ago.

(3) During accretion, a portion of the kinetic energy of the impacting planetesimals would have been retained in the protoplanet as buried heat. Consideration of a simplified model of the parameters involved in this process indicates that for a dynamically plausible accretion time (i.e., ~10^8 years), there is only a modest increase in average surface temperature of the forming planet, but that high internal temperatures would have been unavoidable (e.g., ~1000° K when the accreting Earth had reached the size of the Moon).

(4) Eventually, as a function of accretion, internal temperatures would have increased to reach levels at which both subsolidus convection of silicates and the melting of free iron, could have occurred. Consideration of the relevant parameters suggests that although the total buried accretional energy would have been sufficient by a large factor to have completely melted the Earth, this probably did not occur due to the convective transport of such heat to the surface and its subsequent loss to space. Melting of the silicate fraction may thus have been confined to a hot, near surface region, ~100 km thick. But this would not have been the case of the particulate free iron component that, because of its lower melting point, would have become molten in the hot periphery of the planet (beginning when the Earth had reached approximately lunar size) and would thus have percolated downward, toward the colder interior of the planet, to form a spheroidal liquid shell with its lower boundary defined by the iron freezing point.

(5) The mechanisms by which this iron-rich layer ultimately migrated to the center of the planet to form the Earth's geomagnetic core are as yet poorly understood, with most popular

models appearing to require an implausibly long time scale (viz., $\sim 10^9$ years). An alternative model, that of catastrophic asymmetry, is thus here suggested, one involving during accretion the (*i*) spontaneous, rigid displacement of the cold, primordial core relative to the immediately overlying, more dense, liquid iron shell; (*ii*) fragmentation of the displaced rocky core due to nonhydrostatic stresses within the resulting pear-shaped planet; and (*iii*) consequent formation of a liquid iron core encompassed by a shell of rocky primordial fragments which was in turn overlain by a hot, well-mixed mantle. According to this scenario, core formation and the resultant development of the Earth's geomagnetic field would have been contemporaneous with accretion (i.e., occurring over a time scale of $\sim 10^8$ years).

(6) As envisioned, core emplacement and buried accretional heat would have resulted in release of an exceedingly large amount of thermal energy (roughly equivalent to the total amount of radiogenic heat produced during the entire subsequent history of the Earth) over a period of only a few 10^8 years. The volume of material processed near the Earth's surface during this relatively brief transient cooling phase would have been comparable in quantity to the total mantle mass thus cycled over all subsequent Earth history. The model thus suggests that for the most part, mantle differentiation and outgassing of the Earth's interior occurred catastrophically, within a geologically short period of time very early in Earth history.

(7) The composition of the resulting Hadean atmosphere would have been crucially dependent on the timing and nature of heat-generating events on the primitive Earth: if outgassing largely occurred prior to core emplacement (as a result, for example, of energy released via impact events), the resulting atmosphere may have contained substantial concentrations of H_2, with carbon thus occurring as CO and CH_4, nitrogen as N_2 and NH_3, and sulphur principally as H_2S; if, however, outgassing occurred chiefly during the transient cooling phase postulated above, the oxidation state of the mantle would have been comparable to that of the present and the principal outgassed constituents would have been CO_2, N_2 and SO_2.

(8) As noted at the outset of this discussion, understanding of the Hadean Earth is far from complete, the principal difficulty being a function of the paucity of currently available, directly relevant data. Models can be no better than their observational and theoretical underpinnings, and until additional, firmly grounded evidence becomes available, notions regarding the earliest stages of planetary history seem destined to remain just that—simplified, seemingly plausible abstractions of the early Earth that may or may not represent close approximations of reality.

CHAPTER 3

THE EARLY EARTH AND THE ARCHEAN ROCK RECORD

By W. G. Ernst

3-1. Introduction

The initial formation and earliest evolution of the Earth are shrouded in mystery, but fragmentary evidence is contained in preserved tracts of ancient crust. The deep interior of the planet (mantle plus core) has undergone profound thermal, mineralogic, and chemical changes over the course of geologic time; relics of the earliest stages either have been obliterated or—if yet preserved—are virtually unrecognizable (except as they may be reflected in a few rather subtle geochemical parameters) because of the inability to sample and directly measure even the uppermost mantle. The atmosphere and hydrosphere too have evolved with time, but circulation and mixing (extremely rapid on a geologic time scale) have destroyed, many times over, evidence of chemical fractionation. Thus it is that we turn to the rocky outer shell of the Earth for information regarding the earliest stages of planetary development (for more extensive treatments, see Cloud, 1976b; Windley, 1977; Smith, 1979). Moreover, because sea-floor spreading and subduction of oceanic crust result in the geological recycling of the oceanic basement and its underpinnings (at current rates, every 50–200 million years), available evidence regarding the nature of the early Earth resides chiefly in the continents.

The sialic continental crust, however, represents only a very small portion of the differentiated planet. To understand the production of this surficial "slag," we must thus consider at least briefly the chemical maturation of the core and mantle. In turn, the course of differentiation of the Earth as a whole can be placed in a better perspective through consideration of the origin and evolution of terrestrial planets generally. For this reason, let us begin with a brief speculative account of the formation of Earth, Moon, Mars, and Venus (see also discussion in Chapter 2) prior to moving on to a consideration of the Archean rock record. The paleomagnetic evidence of ancient continental motions, and some constraints on past plate tectonic regimes will then be addressed.

At this point, however, a few definitions regarding geologic time are in order. Three principal intervals are recognized: the Archean (from the formation of the planet to 2.5 Ga ago); the Proterozoic (from the close of the Archean to the Precambrian-Cambrian "boundary"); and the Phanerozoic (the time of megascopic life). Time divisions as employed in this book, largely after Harrison and Peterman (1980), are as follows:

PRECAMBRIAN
- Archean, 4-5-2.5 G
  - Hadean, 4.5-3.9 Ga
  - Early Archean, 3.9-2.9 Ga
  - Late Archean, 2.9-2.5 Ga
- Proterozoic, 2.5-0.57 Ga
  - Early Proterozoic, 2.5-1.6 Ga
  - Middle Proterozoic, 1.6-0.9 Ga
  - Late Proterozoic, 0.9-0.57 Ga

PHANEROZOIC
- Cambrain and younger, 0.57-0 Ga

3-2. The Hadean Earth

Isotopic evidence from meteorites and the Moon indicates that the age of the solar system is approximately 4.6 Ga (Patterson, 1956; Papanastassiou and Wasserburg, 1971). Gravitational collapse of a locally high density of interstellar gases resulted in condensation of the solar nebula to form initially a rotating disc. The greatest mass of this "proto-solar system"—from more than half to perhaps as much as 99 percent—became concentrated toward the center of the disc, resulting in ignition of thermonuclear reactions and consequent formation of the Sun, with the accretion of planetesimals in more peripheral regions of the disc giving rise to discrete, planetary bodies orbiting this newly formed central star. Although these condensates may have been rather cold, the burial of heat during successive impacts probably resulted in increasingly hot accretion (Wetherill, 1976; Kaula, 1979). In comparison with the outer planets, the inner, terrestrial planets are relatively refractory because during accretion, volatile elements were differentially condensed and/or subsequently lost; in contrast, planetesimals coalescing at greater distances from the Sun were colder, thus preferentially retained their gassy constituents. This primordial phase of formation of the solar system was probably completed by about 4.5 Ga ago.

Because of the relative abundance of the primordial radioactive isotopes of such elements as uranium, potassium, and thorium in the newly formed Earth, a significant amount of planetary "self-heating" would have occurred directly following the accretionary stage. Moreover, most planetary materials, except for metals, are poor thermal conductors; radioactive and accretionary (impact) heating would thus have promoted incipient melting of the smaller terrestrial bodies and, ultimately, a substantial degree of fusion for the more massive ones. Because iron melts at a lower temperature than do silicates under high confining pressure (iron and silicate liquids being immiscible), a dense iron-rich melt would have formed that would have migrated down the gravity gradient displacing silicates upwards, thus producing a molten metal core (Elsasser, 1963); and, as is discussed in the preceding chapter of this volume, even if the central portion of the Earth were initially colder than the temperature of the iron solidus, asymmetry of metal accumulation at depth would have resulted in a catastrophic downward movement of the iron-rich melt (Stevenson, 1980b). The conversion of kinetic energy to heat during this process would be expected to have liberated additional amounts of thermal energy, which would thereby have contributed to the fusion and separation of metal core and silicate mantle. Slight cooling that has occurred since this earliest stage of planetary differentiation has resulted in formation of the Earth's solid inner core; differential flow in the outer liquid portion of the core is regarded as the source of the dynamo responsible for generation of the Earth's magnetic field.

No terrestrial rocks have survived to the present from this early Hadean stage of planetary evolution; indeed, the apparently most ancient samples now known, from the Isua supracrustal belt of western Greenland, are no older than about 3.8 Ga (Moorbath et al., 1972; Moorbath, 1975). However, evidence of this stage of development is preserved on the Moon: the anorthositic impact breccias of the lunar highlands contain rock fragments at least as old as 4.2 to 4.4 Ga, and the highlands are heavily cratered, attesting to continued, intense meteorite bombardment there—and probably also on Mercury, Venus, Earth, and Mars—to about 3.9 Ga (Wetherill, 1972; Burnett, 1975). Such meteoritic accretion was undoubtedly most intense during the earliest stages of planetary formation, tapering off toward the end of Hadean time. The lunar maria are floored by younger (Early Archean) basalts, indicating that the thermal budget of the Moon was such that at least local production of magma occurred as late as 3.8 to 3.3 Ga ago. The formation of the highlands appears to have resulted from the gradual accumulation of fragments and blocks of plagioclase minerals that floated on a hypothesized lunar "magmatic ocean" 4.2 to 4.4 Ga ago or even earlier.

Because of the far greater mass (and consequent gravitational attraction) of the Earth, higher pressure equilibria would be expected to have dominated terrestrial crystallization other than in the most surficial parts of any analogous, now obliterated terrestrial magmatic ocean. Accordingly, on the Earth, early crystallization of garnet ($\rho = 3.6$ g/cm^3) under attendant high pressures (rather than the corresponding calcic plagioclase of the Moon, $\rho = 2.7$ g/cm^3) would have resulted in the settling of aluminous crystalline phases rather than flotation. In addition, by analogy with present-day lava lakes, the thin refractory crust that would have solidified at the upper cooling surface would overlie less dense molten material and would thus tend to sink on cracking and break-up. During crystal fractionation, therefore, ferromagnesian constituents should have been stabilized in relatively deeper portions of the gradually thickening lithosphere of the primordial Earth, with the more fusible LIL (large ion lithophile) and incompatible elements tending to be concentrated toward the surface in a relatively silicic globe-encompassing rind (Hargraves, 1976). Until temperatures decreased to below the granite solidus, however, this more silicic, alkalic material would have remained largely molten, in contrast to the more refractory but volumetrically less significant anorthosite. Probably during this early, very hot stage, the sialic material would continue to have been reincorporated into the rapidly convecting, rehomogenizing mantle (Jacobsen and Wasserburg, 1981). Outgassing would have been substantial, with the resulting volatiles transferred from the mantle to the atmosphere.

By 3.8 Ga ago (Isua time), water-laid sediments containing silicic, LIL-rich, continental crust-type debris were being deposited. Clearly, the temperature of the surface of the planet was less that a few hundred degrees Celsius, as required for the existence of liquid water (374°C at 221 bars is the critical end point for the boiling of H_2O); in addition, rather fusible sialic crust (which melts at 600–900°C depending on the fugacity of H_2O and considerably lower temperatures would have been required by lower pressures and/or by the existance of life forms, such as bacteria) was also present. Thus no matter how the initial differentiation of the Earth was accomplished, a surficial layer of crust and overlying hydrosphere must have existed at least locally by latest Hadean time. And, because the fusion of mantle-forming peridotites requires higher temperatures than those required for formation of oceanic and continental crust, at least a thin layer of solid mantle basement must also have been present by this time. It seems evident, therefore, that the chief components of the lithosphere—as well as both a hydrosphere and atmosphere—were extant on Earth by 3.8 Ga ago. This condition succeeded and in part no doubt overlapped with the earlier Hadean stage of Earth history, a primordial phase in the development of the planet characterized by intense meteoritic infall during fractional crystallization of the hypothesized terrestrial magmatic ocean, widespread volcanism, abundant release of volatile constituents, sialic recycling (crustal foundering), and meteoritic reworking ("gardening"). It should be noted, however, that the existence of the postulated magmatic ocean on the early Earth is not at all certain; in fact, as is discussed in Chapter 2, although primordial ^{3}He would be expected to have been completely lost during the outgassing accompanying any such episode of worldwide, near-surface melting, this inert gas is still escaping from the mantle.

Phase relations involved for conditions of partial fusion, and inferred geothermal gradients for the present day and for the Early Archean, are illustrated in Figure 3-1. Although calculated temperature/depth relationships are markedly dependent

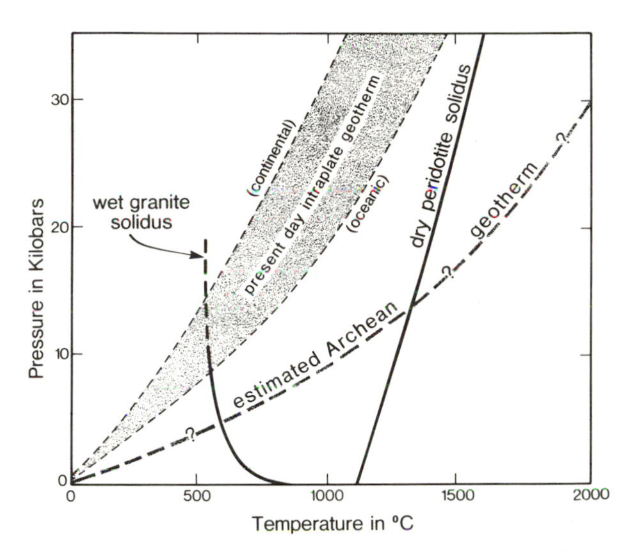

FIGURE 3-1. Phase relations for the initiation of melting of anhydrous, fertile periodotite (after Green and Ringwood, 1967a; Ito and Kennedy, 1967; and O'Hara, 1968), and of ternary granite in the presence of excess H_2O (after Luth et al., 1964). Also shown are computed, modern intraplate geothermal gradients (Clark and Ringwood, 1964), and an estimated 3.6-Ga-old geotherm assuming approximately three times the present-day heat generation and a stratified Earth (Lambert, 1976).

on the assumptions employed, the early Earth unquestionably exhibited a higher average geothermal gradient than currently exists. Therefore, the proportion of this heat lost through a more rapid convective overturn would probably have been far greater during the Archean than at present. In any case, wet sialic crust—and even dry peridotite—would have begun to melt at shallower depths than currently (Figure 3-1). For this reason, Archean continents and their solid mantle underpinnings must have been relatively thin (i.e., both the Mohorovičić Discontinuity and the top of the asthenosphere must have been situated at shallower depths than they are today).

3-3. The Archean Rock Record and Inferred Physical Conditions

3-3-1. MINERALOGY AND PETROLOGY

Apparently the oldest preserved rocks now known on Earth belong to the Isua supracrustal belt of western Greenland (Allaart, 1976); they consist of amphibolites of various types, ultramafic lenses, and metasediments such as metacherts, carbonate-bearing mica schists, and metaconglomerates which contain volcanic clasts. The detrital components suggest the prior existence of continental crust, probably characterized by anorthositic highlands and a fine-grained sedimentary cover. The Isua supracrustals are engulfed by the Aritsoq tonalitic and granodioritic gneisses (Mason, 1975), a complex that was apparently emplaced about 3.7 Ga ago (Moorbath et al., 1975; for a recent summary of geologic events in the Isua area, see Bridgwater et al., 1981). A generalized geologic map of the Early Archean rock units of western Greenland is illustrated in Figure 3-2. Granites and feldspathic gneisses of nearly comparable age crop out in the Minnesota River Valley of the north central United States (Goldich and Hedge, 1974) and in India (Basu et al., 1981).

Archean complexes exposed in South Africa, Rhodesia, Western Australia, peninsular India, Canada, Brazil, and Fennoscandia bear testimony to the abundant occurrence of belts of greenstone and superjacent volcanogenic sedimentary strata (graywackes) as old as 3.0 to 3.6 Ga (Anhaeusser et al., 1969), rock assemblages of the type suggested by Tarney et al. (1976) as representing remnants of ancient marginal basins. In general, these relatively small, linear complexes occur as feebly metamorphosed synclinoria surrounded by higher-grade plagioclase (tonalitic) gneisses, pyroxene-bearing amphibolites, and granulites. The encompassing tonalitic terranes appear to truncate and deform the greenstone-graywacke belts, and are therefore probably of younger age (commonly 3.4 to 2.7 Ga old). However, the so-called "gray gneisses" of such complexes may represent rifted continental blocks between which mafic complexes were later extruded (Goodwin and Ridler, 1970), or they may be remnants of the sialic substrate upon which plateau basalts and associated sediments were initially deposited (this appears to be the situation in Western Australia, Rhodesia, and parts of the Canadian shield, as reported by Glikson, 1976; Nisbet et al., 1977; and Jolly, 1980, respectively). For such gray gneisses, the geochronologic data would merely reflect a time of isotopic homogenization—and a resulting resetting of the "isotopic clock"—rather than their actual geologic age. Such gneisses are antiformal in structure, and their crosscutting relationships to the feebly metamorphosed greenstone keels, as well as their much higher metamorphic grade, are probably indicative of buoyant upward plastic flow from deeper crustal levels. These gray gneisses (and associated granulites) are in turn commonly intruded by migmatites and pink potassic post-orogenic granites, approximately 3.0 to 2.5 Ga in age.

Archean assemblages of this sort thus suggest a model of downward movement of dense volcanogenic (mafic- and ultramafic-rich) supracrustal materials that displaced laterally and in an upward

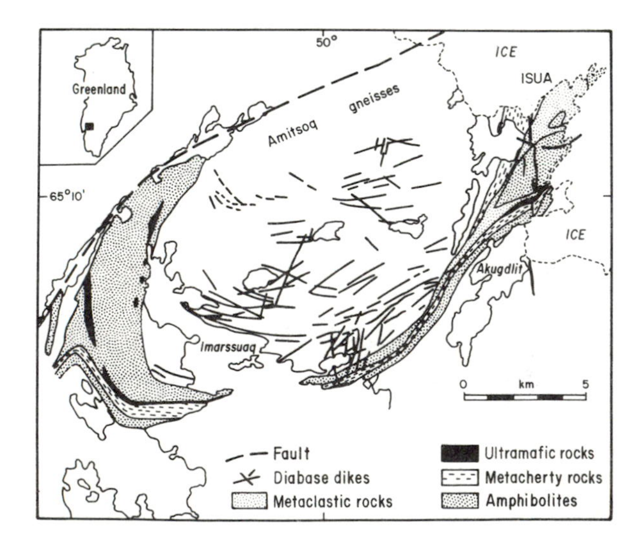

FIGURE 3-2. Map of the Isua supracrustal belt, western Greenland (simplified after Allaart, 1976).

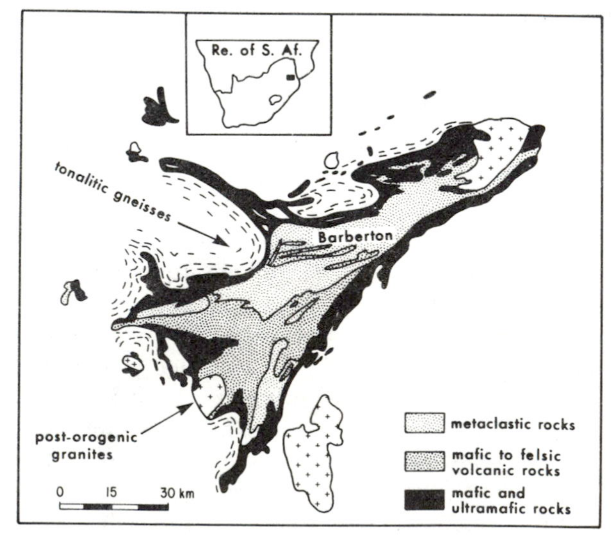

FIGURE 3-3. Map of the Barberton supracrustal belt, South Africa (simplified after Anhaeusser, 1971; and Viljoen and Viljoen, 1971).

direction a complex of thermally softened, relatively buoyant, quartzofeldspathic rocks that were subsequently intruded by post-orogenic granites. Figure 3-3 presents an example of this type from the South African shield. Generally lacking from the geologic section in this area are mature platform-type strata such as orthoquartzites, thick layers of monomineralic limestones or dolomites, and widespread evaporitic sequences; uncommon, too, are peraluminous shales, red beds, impure mixed carbonates and banded iron-formations of great lateral extent (Lowe, 1980b). Structurally, the terranes seem to have been dominated by vertical tectonics. Recumbent folds and low-angle thrust faults are not typical of Archean greenstone-granite terranes (see Figure 3-4).

The igneous rocks so prominent in Archean greenstone belts consist chiefly of tholeiitic, commonly pillowed basalts, indicative of submarine extrusion; these grade upward into andesitic lavas and pyroclastics. Such volcanic cycles, which are generally capped by immature clastic sediments (feldspathic conglomerates and graywackes) or by chert or jaspilite, are commonly repeated stratigraphically upwards.

Thin flows of komatiite and basaltic komatiite are the most distinctive rock types in the lower portions of many of these greenstone belts. These ultramafic lavas are characterized by MgO contents greater than 18 percent by weight, and CaO/Al_2O_3 ratios normally exceeding unity (M. J. Viljoen and R. P. Viljoen, 1969; Nesbitt et al., 1979). That these rocks were completely molten is demonstrated by the occurrence of "spinifex texture" (i.e., intermeshed blade-like crystals of olivine and

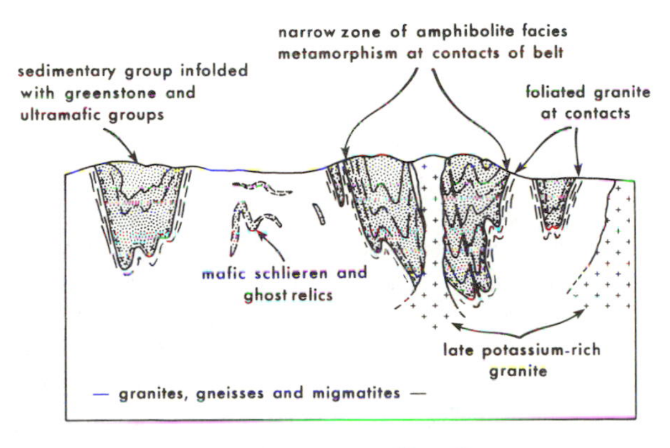

FIGURE 3-4. Diagrammatic cross section of greenstone synclionoria enveloped in a tonalitic gneiss terrane (after Anhaeusser et al., 1969).

orthopyroxene that in some instances exceed a meter in length). These crystals grew perpendicular to the surface of nuclei-free, highly fluid magnesian-rich volcanic flows. The occurrence of komatiite is significant, inasmuch as melts of komatiitic composition can be liquid only at temperatures approaching 1600°C, even at very low pressures. Because this rock type closely approaches pyrolite in composition (i.e., that of an essentially undepleted, fertile mantle; for discussion, see Ringwood, 1975), the occurrence of a komatiitic melt therefore reflects a high degree of mantle partial fusion—probably at considerable depth (Green, 1975). The limited enrichment in REE (rare earth elements) relative to meteorites characteristic of such rocks supports the hypothesis of their derivation through the partial melting of pristine, undepleted mantle material. Thus, although a few, rare examples of komatiites are known also from the Phanerozoic, the virtual restriction of such rocks to Archean terranes seems undoubtedly to be a consequence of the high geothermal gradient of the Archean Earth and the declining average of this gradient over geologic time. Extensive melting of the mantle must therefore have accompanied Hadean and Early Archean crustal differentiation, diminishing as the amounts of undecayed radioactive nuclides decreased in abundance.

Archean metasedimentary associates of the dominantly ultramafic, mafic, and more felsic igneous rocks of greenstone terranes are commonly interpreted as representing poorly sorted, first-cycle clastics, units rich in volcanogenic debris and interlayered siliceous precipitates (Veizer and Jansen, 1979). Local derivation from an emergent volcanic belt seems required. The occurrence of crossbedding, ripple marks, and of other reworking features indicate that many such units were deposited in relatively shallow water (Lowe and Knauth, 1977; Barley et al., 1979).

But what can be said regarding the source of the encompassing gray gneisses and granulites? Their ultimate origin is obscure. Except for charnockites, these rocks are low in potash and have initial $^{87}Sr/^{86}Sr$ ratios of about 0.701, indicating a mantle source. Windley (1979) interprets these largely tonalitic terranes as ancient, deeply eroded analogues of the calc-alkaline batholithic belts that occur at modern convergent continental margins. If this interpretation is correct, the "post-orogenic," Late Archean potassium-rich granites that intrude such high-grade metamorphic terranes would appear to represent a mobilized sialic basement with the required upward buoyant movement presumably being a consequence of a higher geothermal gradient and the partial fusion attending the local concentration of LIL and incompatible elements involved in mantle devolatilization; the reworked, secondary nature of such units seems reflected by their enrichment in alkali elements, their high K_2O/Na_2O ratios, and their elevated $^{87}Sr/^{86}Sr$ values.

Some of the most distinctive characteristics of Archean crustal rocks—both from supracrustal and from plutonic gneissic terranes—are listed in Table 3-1. As indicated there, of equal importance to features exhibited by Archean complexes are those which in comparison with younger geologic units are conspicuous by their absence or, at least, by their relatively lesser abundance.

TABLE 3-1. Distinctive characteristics of Archean crustal rocks.

LOW-GRADE GREENSCHISTIC TERRANES

Synclinoria; small-scale, isoclinal and ptygmatic folding
Near-vertical axial traces
Pillowed greenstone + chert + graywacke
Chemically immature sedimentary rocks (eugeoclinal)
Komatiitic lavas exhibiting spinifex texture
Detrital uraninite, sulfides

HIGH-GRADE GNEISSIC, GRANULITIC TERRANES

Domical, gregarious batholiths, plastic intrusions
Near vertical igneous contacts, flow banding
Tonalitic gneisses, pyroxene-bearing amphibolites and charnockites
Sillimanite, andalusite, cordierite

LESS ABUNDANT OR NEARLY ABSENT

Thick, monomineralic carbonate strata (miogeoclinal)
Orthoquartzites (miogeoclinal)
Peraluminous pelites (miogeoclinal)
Evaporites (miogeoclinal)
Red beds (miogeoclinal)
Tillites
Alkalic igneous rocks (including kimberlites and carbonatites)
Great mafic intrusives, stratiform sheets, dikes, and sills in general
Ophiolitic assemblages
Eclogites and blueschists
Kyanite, in general
Paired metamorphic belts
Overturned folds, decollement thrusts
Major, continuous orogenic belts
Aulacogens

3-3-2. PHYSICAL CONDITIONS

Let us now turn to a consideration of the physical conditions suggested by the mineralogy and petrology of Archean crustal rocks. As is well known, the presence in Archean sediments (see Chapter 11) of water-worn detrital sulfides and, especially, of detrital uraninities—minerals which are resistant to long-term transport and attendant solution weathering only under essentially "anoxic" conditions—is consistent with the presence of an early atmosphere/hydrosphere characterized by low oxygen fugacities. The occurrence of discontinuous, thin, but indisputably sedimentary layers of banded iron-formation in the Archean seems similarly consistent with a low fO_2, apparently required for transport of soluble ferrous iron across sedimentary basins for deposition in regions typified by slightly more oxidizing environments. As is discussed in Chapter 10, the ubiquity of such oxygen fugacity levels (quite low compared to present-day atmospheric values of 0.2 bar) was apparently maintained until about 2.0 Ga ago, permitting the formation of the great banded iron-formations of South Africa, western Australia, Labrador, Minnesota/Ontario, and elsewhere during the Early Proterozoic.

Although $^{18}O/^{16}O$ paleotemperature determinations depend critically on the assumed isotopic composition of sea water, and on the absence of post-depositional exchange, Perry et al. (1978) and Yeh and Epstein (1978) have suggested that ferruginous cherts and related chemically precipitated sediments from the Archean of western Greenland and South Africa crystallized at temperatures of 55° to nearly 100°C, whereas Proterozoic analogues from Canada, western Australia, and California yield apparent temperatures of 22 to 55°C. Exact values may be questioned, but these data suggest that the oceans may have cooled over the course of geologic time (this interpretation, however, cannot as yet be considered firmly established; see Chapter 11).

The virtual restriction of komatiitic lavas to the Archean suggests that, at least in the vicinity of the greenstone belts, early Earth geothermal gradients were higher than at present. The total absence of blueschists, eclogites, and other high-pressure, low-temperature metamorphic rocks (Ernst, 1972) from such terranes also attests to this ancient high heat-flow regime. As noted above, this condition appears to be a consequence of the fact that heat production due to radioactive decay at and before 3.8 Ga was roughly three or more times greater than the present (Lambert, 1976). The Archean abundance of pyroxene-bearing amphibolitic gneisses and sillimanitic granulites likewise suggests relatively high crustal temperatures (although the thickness of the Archean crust is a matter of some debate; see, for example, Lambert, 1976; Grambling, 1981; Wells, 1979; and Baer, 1981). Granulites are proportionately more abundant in Archean terranes than in younger metamorphic sequences, an abundance of high-grade complexes that presumably is a product of an ancient high-gradient thermal regime (Newton, 1978).

Lithologic evidence thus suggests that the early Earth was characterized by high crustal geothermal gradients; a variable, but in part high degree of mantle partial melting; igneous rocks that were predominantly extrusive; an essentially anoxic atmosphere and ocean; and a compositionally immature veneer of locally derived volcanogenic sediments. The geothermal structure for the Archean Earth necessitated both by calculated heat flow and by the occurrence of preserved granulitic and komatiitic terranes requires the presence of thin lithospheric plates (Cook and Turcotte, 1981) and the generation of both oceanic- and continental-type protocrust as relatively near-surface phenomena. The maximum thickness of this early crust is uncertain. Lambert (1976), O'Hara (1977), and Newton (1978) argue that Late Archean continental thicknesses may have approached those of the present-day average depth to the Mohorovičić Discontinuity. In contrast, my view (Ernst, 1972) is that high geothermal gradients within the early Earth would have promoted the partial fusion and upward rise of previously deeply buried wet sialic materials, resulting in the generation of a generally thin continental crust (a suggestion supported by the pressure-temperature relationships illustrated in Figure 3-1).

The dimensions of the Archean plates are poorly constrained, but judging from the fact that extant greenstone belts are relatively small linear features—typically on the order of 20 × 150 km in size—lithospheric plates may have been much smaller than those of today. A plethora of small, rapidly circulating upper mantle convective cells would also be in accord with the high heat loss required by the inferred thermal structure of the early Earth. Because the high-grade metamorphic mineral assemblages preserved in ancient sialic crust bear testament to only slightly elevated geothermal gradients, most of the early heat dissipation must be assigned to rapidly convecting suboceanic asthenospheric flow (see Chapter 10). Thus, the early oceanic lithosphere was thin, hot and probably nearly unsubductable; partial melting of this portion of the outer Earth must somehow account for the observed komatiite-tholeiite-andesite association of the greenstone belts.

3-4. Paleomagnetic Data, Cratonization, and Continental Drift

Students of the Earth's past magnetic field have been obliged to make three important assumptions: (*i*) the magnetic field has always been essentially dipolar; (*ii*) the magnetic poles coincide, approximately, with the terrestrial spin axis; and (*iii*) the remnant magnetism of a rock accurately reflects the external field at the time of its formation, or the time when it cooled below the Curie temperature. It is well known that the Earth's magnetic field reverses sign episodically. Nevertheless, the first assumption appears to be correct, at least on average; the second is a reasonable approximation of actual conditions; and the third is, at least generally, consistent with the available data.

Paleomagnetic measurements on any favorable lithotype provide an indication both of the paleomagnetic latitude (dip of the field) and the azimuth (pole vector) for the time of lithification of the rock, provided the material has not been disturbed by a later heating episode. Thus, for a particular area, if a series of suitable units of different ages is available for investigation, a so-called "apparent polar wander path" can be constructed; one starts by using the current location of the magnetic north (or south) pole and the present geographic location of the area in question as fixed reference points, and proceeds backward in time to construct a global curve coordinated by the paleoazimuth and the paleolatitude of the individual samples. If remnant magnetic measurements for geographically dispersed rocks of the same age-span can be fitted to a single apparent polar wander path, this is considered as evidence that these different areas behaved as parts of a rigid, coherent segment of crust (i.e., a tectonic plate) during the time interval.

In general, available paleomagnetic data suggest that, with the notable exception of a "proto-North Atlantic" rifting/converging couple which produced the Caledonian orogenic belt, a large supercontinent existed from more than 800 million years ago until the initiation during the Late Triassic of the current Wilson cycle of continental drift (McElhinny et al., 1974), when this megacontinent was disrupted by rifting. For rocks lithified during the Early Proterozoic, apparent polar wander curves for South America and most of southern and western Africa approximately coincide (Briden, 1973; Piper et al., 1973); at that time, at least these sialic masses appear to have been joined together as a super continent. A similar relation probably holds for the cratons of northern Africa and western Australia (e.g., see Embleton, 1978). Archean paleomagnetic data, however, are less definitive. Nevertheless, it seems plausible that dispersal of the cratons may be a geologically rather recent development (Engel and Kelm, 1972); during the early portions of the Precambrian, accreting sialic segments may have been rapidly swept against one or more enlarging supercontinents, with other areas of the Earth's surface being typified by open seas. In any case, accreting, presumably relatively rigid plates, as documented by apparent polar wander curves, were apparently extant as early as about 2.6 Ga and perhaps 2.8 Ga ago (McElhinny, 1973; Dunlop, 1981). During the Proterozoic, intracratonic orogenic belts formed across pre-existing "stable" shield areas, as indicated by abundant radiometric data (Hurley, 1968); evidently, their large size helped in some manner to stabilize the supercratons.

Unfortunately, neither paleomagnetic nor geochronologic data provide a basis by which to estimate quantitatively the area ratio of Archean ocean basin to continent. The proportion of the Earth floored by oceanic versus continental crust is a complex function of the amount and thickness of the latter, with the lateral extent and thickness of the continents being determined by such factors as: (*i*) the geothermal gradient and, thus, the depth at which melting of quartzofeldspathic material occurs; (*ii*) the volume of the hydrosphere (and, hence the "freeboard" of the continents, and their potential erosion to wave base); (*iii*) the rate of sweeping together of continental debris via the mechanisms of sea-floor spreading; and (*iv*) the rate of production of sialic material through chemical differentiation of undepleted (fertile) mantle. The Mohorovičić Discontinuity must have been at relatively shallow depths during the early Precambrian, because of the relatively high geothermal gradient and the consequent fusion of deeply buried wet granitic crust (Figure 3-1). The volumes of both hydrosphere and continental crust have apparently increased (at an unknown rate) over the course of geologic time as a result of continued, but waning, mantle evolution. Convective overturn of the hot Archean asthenosphere undoubtedly was much more vigorous than at present. Accordingly, it can be concluded that the sialic crust was initially thinner and that it was rather rapidly accreted into cratonal areas. Thus, perhaps two-thirds, or as much as three-quarters of the total volume of continental crust was produced during the Hadean and Archean (Lambert, 1976), although prior to 3.8 Ga ago some of this material may have been returned to the mantle (De Paolo, 1981).

Judging from the known rock record, the emergence of continental shelves and shallow epeiric seas of great areal extent may not have occurred until about 2.5 Ga; Archean sediments are typically first-cycle (or at least relatively "primitive"), locally derived, volcanogenic, and compositionally immature, whereas Proterozoic strata commonly include reworked, mechanically and chemically recycled units of much greater lateral continuity. It is conceivable that the Archean micro-continental assemblies were largely submerged, as suggested by the apparent lack of platform sediments, with freeboard only being widely attained early in the Proterozoic. Thus, the shallow marine (photic zone) environment suitable for the development of life would have been largely restricted to the greenstone volcanic arcs in Archean times. It is small wonder, then, that evidence of life is so fragmentary in these ancient rocks.

Continentality combined with transpolar drift allows for climatic extremes (e.g., see Crowell and Frakes, 1970) and it is perhaps not a coincidence that the most ancient glaciations date from Proterozoic time (about 2.2 Ga, or 2.7 Ga at the very oldest; for a summary of data regarding such glaciations, see Chapter 11).

Another piece of evidence supporting an increase during the Archean to Proterozoic in the thickness of sialic crustal material involves the nature of emplacement of basaltic magmas. Archean mafic igneous rocks are dominantly extrusive; huge floored intrusives of gabbroic magma appear only in the geologic record in units about 2.7 Ga old or younger. During the earlier Archean, sial may have been repeatedly breached by fluid ferromagnesian melts derived from depth, supporting the hypothesis that the continental crust was somewhat thinner. By the beginning of the Proterozoic, however, locally thick granitic crust apparently provided a gravitatively stable shield that could not be surmounted by dense mafic liquid (i.e., the hydrostatic head on such liquids was insufficient to permit their rise to the continental surface). A factor that may have played a partially compensating role here is the degree of primitiveness of the mantle protolith: given a certain percentage of partial fusion, primordial melts would be expected to have had a high iron/magnesium ratio (reflecting their pyrolite source), whereas later generated magmas, derived from a partially depleted mantle, should be more magnesium-rich, hence less dense and more easily able to rise within the crust. However, it must also be remembered that high Archean geothermal gradients would promote greater degrees of partial melting—thus resulting in more magnesium-rich melts. Nevertheless, the basaltic rock record suggests that thickening of the sialic crust with time was relatively more important than the evolution of more magnesian melts as a consequence of mantle fractionation.

3-5. Plate Tectonics Through Geologic Time

The configuration and present motions of lithospheric plates, a plate tectonic regime initiated during the Early Mesozoic opening of the Atlantic Ocean and the fragmentation of Pangea, provide data regarding the current mechanisms by which oceanic crust is formed and destroyed; they also demonstrate that growth of the continental crust results from the addition of andesites and related calc-alkaline igneous rocks to the stable, nonsubducted side of convergent plate junctions. Effects of this Wilson cycle of plate tectonics can be reliably traced back into Paleozoic terranes where structural evidence of convergent plate boundaries exists in the form of overthrust zones and allochthonous nappes such as those of the Appalachian orogen (e.g., see Cook et al., 1979). Sea-floor spreading as it occurs today produces a distinctive oceanic petrostratigraphic sequence that includes tectonite peridotites, cumulate ultramafic-mafic complexes, sheeted diabase dikes and sills, and an overlying series of massive tholeiitic lavas, pillow basalts, breccias, and pelagic sediments. Ophiolites, as these assemblages are termed (Anonymous, 1972), are typical of Phanerozoic and uppermost Proterozoic terranes (e.g., Dewey and Bird, 1971), but similar suites are not recognized unambiguously from any but the youngest sections of the Precambrian. Paired metamorphic belts (Miyashiro, 1961, 1967), similarly a typical product of Phanerozoic plate tectonics, are also rare or unknown in the Precambrian.

Thus, the typical lithotectonic associations produced by Wilson-type plate motions in the Phanerozoic may be absent from all but the latest Precambrian. Is it possible, therefore, that plate tectonic processes did not operate on the early Earth? Certainly this may have been the case on other terrestrial planets: the preservation of heavily cratered terranes on the Moon, Mercury, and Mars show that these bodies were not subjected to major, crustal reworking via the overturn of lithospheric plates, at least subsequent to the marked decline in intensity of meteritic bombardment some 3.9 Ga ago. On the other hand, all of these planets have considerably smaller masses than the Earth and thus could not be expected to have sustained impact-induced and radioactively induced high internal temperatures, presumably required for the onset of plate tectonic processes, for as extended a period. Preliminary radar imagery-based topographic data for Venus, the "sister planet" of the Earth, suggest that although plate tectonics may have characterized the Archean of that planet, such processes have long ceased (Phillips et al., 1981). Thus, although primordial melting no doubt occurred on other terrestrial planets (as discussed in Section 3-2), the Moon, Mercury, and Mars evidently cooled to the stage at which the thermally driven gravitative instability required for mantle convection ceased prior to termination of the sweep-up of planetesimals. The larger mass of the Earth (and perhaps Venus) allowed its sustained retention of high temperatures and strong thermal gradients, the latter undoubtedly serving to fuel the heat engine that drives the lithospheric plates.

Because this heat engine has existed since the early (Hadean) formation of the Earth's core, it seems evident that at least small, thin platelets must have bounded the Earth's dynamic surface beginning early in Early history, a supposition that is consistent with the preserved Archean rock record and with the phase diagram illustrated in

Figure 3-1. Rapid convective overturn of the mantle probably would have driven such platelets against and beneath one another, but because of the elevated near-surface temperatures and the thinness of the plates, together with the small magnitude of the lithospheric/asthenospheric density inversion and the attendant very minor gravitative instability, the lithosphere would not have been subducted to appreciable depths. This accounts for the lack in the Archean and at least Early Proterozoic of high-pressure metamorphic rocks such as blueschists and eclogites, and the absence of alkalic igneous suites (thought by most petrologists to be derived by extremely small degrees of partial fusion at substantial mantle depths), as well as for the local occurrence in such terranes of the highly refractory komatiites and related lavas. High-pressure modes of generating andesite—and therefore, the tonalitic Archean continental crust (Green and Ringwood, 1967b; Ringwood, 1975)—seem to be precluded; such compositions must therefore have been derived through fractional crystallization of mafic magmas (Bowen, 1928), through the partial melting of hydrated metabasalts (Holloway and Burnham, 1972; Boettcher, 1973) or by partial fusion of upper mantle material (Arculus, 1981). Because of the high rate of outgassing and high heat-flow regime of the primitive Earth, mafic volcanic activity would have been much more voluminous than at present, resulting in the concomitant rapid early development of quartzofeldspathic crust and the hydrosphere/atmosphere. Although Archean gray gneisses were no doubt remobilized during accretion and thickening of the cratons, their mantle-like isotopic and REE geochemistries could well reflect their derivation from just such fertile peridotitic, basaltic and amphibolitic (hydrated metabasaltic) precursors.

According to this scenario, the Late Archean to Early Proterozoic transition would thus have been characterized by a gradual change from small, thin, hot Archean platelets, driven about by numerous rapidly convecting asthenospheric cells, to more "modern," relatively thick, cooler, laterally extensive, coherent plates, the motions of which may have been a function partly of negative buoyancy and partly of asthenospheric flow on a grander scale. By the Early Proterozoic, small sialic masses, produced chiefly during early mantle overturn and chemical segregation, had become assembled into major, thicker continental cratons; as freeboard thus became increasingly important, the processes of mechanical erosion and sedimentation began to play major roles in the production of chemically differentiated sedimentary facies.

3-6. Summary and Conclusions

(1) The known Early, Middle, and Late Archean (viz., 3.9 to 2.5-Ga-old) rock record is markedly incomplete; although spanning a substantially longer duration than the immediately younger (viz., 2.5 to 1.6-Ga-old) Early Proterozoic, far fewer Archean rock units have survived to the present, and more ancient (i.e., Hadean) geologic units are as yet unknown on Earth.

(2) Preserved Archean terranes consist largely of supracustal greenstone synclinoria surrounded by more extensive, domical, plutonic gneiss complexes; the former are predominantly of greens-schist or lower metamorphic grade, whereas the latter are characterized by high-rank amphibolite and granulite/charnockite facies assemblages.

(3) The greenstone synclinoria typically consist of tholeiitic, pillowed metabasalts, but cyclically repeated sequences pass downward into komatiitic lavas and serpentinized ultramafics, and upward into more andesitic, volcanics, pyroclastics, and a sedimentary veneer of ferruginous cherts and locally derived immature clastic sediments (viz., graywackes and conglomerates).

(4) The surrounding, gray tonalitic gneiss complexes commonly crosscut the greenstone keels, but exhibit similar radiometric ages; geological evidence indicates that the high-temperature gneisses and associated granulites were probably emplaced around the dense, low-temperature supracrustals as a result of their buoyant upward plastic flow. The gray gneisses, in turn, have been shouldered aside and partially replaced by remobilized migmatites, and intruded by potassic granites, of younger Archean geologic age.

(5) Consideration of the thermal regime of the early Earth, of the petrotectonic evolution of the other terrestrial planets, and of contrasts between the Archean rock record and that of later geologic time leads to the following tentative scenario for the history of the Earth:

4.6 to 4.5 Ga ago—Condensation of the solar nebula; formation of the Sun and protoplanets.

4.5 to 4.2 Ga ago—Rapid build-up of internal temperature due to energy release from accretionary impacts and radioactive decay; early catastrophic infall and generation of the liquid iron-nickel core; formation of the mantle (perhaps capped by a globe-encompassing magmatic ocean); intense devolatization of the primitive mantle leading to formation of the atmosphere/hydrosphere.

4.2 to 4.0 Ga ago—Crystallization of the (hypothetical) magmatic ocean; flotation and accumulation of minor anorthositic crust and of silicic, alkalic molten residues; settling of dense ultramafic cumulates to deep within the mantle; overturn and rehomogenization of molten sialic material into the mantle; continued intense outgassing.

4.0-3.8 Ga ago—Solidification of thin lithospheric basement (in the upper mantle) surmounted by a ubiquitous veneer of crustal material covered by early seas (condensed volatiles); local erosion resulting in deposition of chemically immature sediments; exuberant volcanism and continued rapid loss of mantle volatiles.

3.9 Ga ago—Fall-off in the rate of planetesimal sweep-up; "gardening" via meteorite impact becomes negligible.

3.8 to 2.7 Ga ago—Initiation of platelet tectonics; accretion of thin, soft, hot surficial slabs of limited area (not profoundly subducted because of their relative buoyancy); accretion of surficial veneer by rapidly convecting asthenospheric cells to form protocontinents; rapid production of sialic crust (with some recycled back into the mantle); derivation of immature, first-cycle sediments draped around volcanic island arcs.

2.7 to 2.5 Ga ago—Continued, but slowing, continental accretion and thickening; formation of first, large-scale, layered intrusives (due to decreased geothermal gradient and resulting increased crustal thickness to exceed the hydrostatic head even of relatively magnesian gabbroic magmas); emplacement of widespread mafic dikes and sills (reflecting cooling, aging and embrittlement of the continental crust); emergence of continents above wave base on a large scale, leading to the formation of epeiric seas and the production of chemically mature, multicycle sediments.

2.5 to 0.8 Ga ago—Continued, very gradual, continental growth; deposition of increasingly abundant multicycle sediments; formation of supercontinents embedded in thickening lithosphere and of orogenic belts along intracratonal zones of weakness; drift of supercontinents (with formation of glacial deposits in polar regions).

1.0-0.8 Ga ago to present—Continued thickening of cooling lithospheric plates, resulting in onset of Wilson-type plate tectonics (characterized by large convective cells, continental fragmentation, formation of paired metamorphic belts, mountain building at plate margins, etc.).

(6) Thus, four more-or-less discrete stages in the plate tectonic evolution of the Earth seem to be recognizeable: (*i*) a Hadean (preplate) stage, characterized by a partly molten planet that was intensely bombarded by meteorites and that underwent profound gravitative separation to form a metallic core, ferromagnesian silicate mantle, and a continuously reworked crustal scum; (*ii*) an Early(?), Middle, and Late Archean (platelet) stage, with the Earth's surface dominated by hot, soft, thin, unsubducted platelets that aggregated sial to form protocontinents; (*iii*) a Proterozoic (supercratonal) stage characterized by the emergence of large land masses, development of freeboard, intracratonal orogeny, and by the drift of supercontinents; and (*iv*) a Late Precambrian-Phanerozoic (Wilson-type) cycle of plate tectonics involving rifting, the dispersal of continental fragments, subduction, and the generation of long, linear mobile belts at plate margins. (A rather similar sequence of events has been postulated by Kroner, 1981).

(7) The above scenario, while apparently consistent with data currently available, is highly speculative—indeed, largely because of the inherent limitations of the ancient rock record (and the fact that studies of the tectonic evolution of the Precambrian Earth are still in their infancy) there are relatively few "firm facts" on which to base such a sweeping overview. Nevertheless, in recent years, progress in the field has been encouraging; what now seems needed are (*i*) additional, detailed, petrotectonic analyses of relatively well-preserved ancient rock sequences, (*ii*) the generation of new, more refined data regarding, especially, the paleomagnetic and geochronologic framework of early crustal evolution and, above all (*iii*) the discovery

and characterization of additional Archean (including Hadean) geologic terranes, units that might well hold important clues by which to unravel the obvious complexities of earliest Earth history.

Acknowledgments

This study has been supported by U.C.L.A. and in part by the National Science Foundation through grant EAR 80-17295. Discussions and preliminary drafts of this summary have been read and criticized by colleagues at U.C.L.A., D. J. DePaolo, E. S. Grew, W. M. Kaula, D. J. Stevenson, and Malcolm Walter. My early interests in the Precambrian rock record were rekindled by the investigations of P. E. Cloud, G. W. Wetherill, and B. F. Windley; what value there is in this synthesis represents an imperfect distillation of their works. In the last analysis, however, J. W. Schopf made me write it!

CHAPTER 4

Prebiotic Organic Syntheses and the Origin of Life

By Sherwood Chang, David DesMarais, Ruth Mack, Stanley L. Miller, and Gary E. Strathearn

4-1. The Chemical Evolution Paradigm

The outline of a modern paradigm for the origins of life on Earth was first formulated by A. I. Oparin and published in Russian in the early 1920s (Oparin, 1924). Similar but less detailed views were expressed a few years later in England by J. B. S. Haldane (1929). A more detailed account of Oparin's ideas appeared in 1938 in English, followed by a series of revised editions in 1957, 1962, 1964, and 1968. (See Farley, 1977, for an historical account of the philosophical, political, and scientific controversies associated with changing conceptions of the origins of life from the time of Descartes to the twentieth century.)

According to the Oparin-Haldane hypothesis, living organisms arose naturally on the primitive Earth through a lengthy process of chemical evolution of organic matter that began in the atmosphere and culminated in the primordial seas. In this process, the flux of energy through the prebiotic environment transformed simple molecules into increasingly complex organic compounds that, through aqueous condensation-polymerization and oxidation-reduction reactions, eventually became the precursors of proteins, nucleic acids, and other biochemicals. Presumably sunlight, electric discharges, thunder shock waves, and other energy sources provided the driving force for the chemical reactions that occurred in the atmosphere and seas and at their boundaries with the land surface.

The expression "chemical evolution" (see Calvin, 1956), has been used in a number of ways, in some cases signifying the chemical development or the chemical processing of the primitive Earth and, in other cases, the organic chemistry of the primitive Earth before the development of biological or Darwinian evolution. In the present context we use the term in reference to prebiotic organic chemistry, with the understanding that inorganic matter also must have participated as reactants and catalysts in the primary synthesis and subsequent transformations of organic matter. But we include the other uses, too, because the organic chemistry of the prebiotic Earth must have been constrained by the global physical-chemical evolution of the planet, which would have determined the temperature, pressure and chemical composition of environments above, below, and at its surface as a function of time. In this context, the chemical evolution of organic matter is an integral aspect of the evolution of a planet, and progress along the chemical pathway to life may be terminated at different, characteristic stages depending on the location and history of a planet. Indeed, the concept of organic chemical (in contrast with "biochemical") evolution need not be restricted to planets; it encompasses all nonbiologic processes that produce organic compounds occurring in any environment in space and time. We shall return to this viewpoint in a later section.

"Chemical evolution" should not be confused with Darwinian evolution with its requirements for reproduction, mutation, and natural selection. These did not occur before the development of the first living organism, and so chemical evolution and Darwinian evolution are quite different processes. However, the term chemical evolution has an appeal because of the continuity it implies for the transformation of lifeless into living matter.

The chemical evolution of organic matter toward life must have been inextricably intertwined with the evolutionary development of Earth itself. A careful study of the geologic record has revealed the presence of diverse types of microscopic organisms as early as 3.5 Ga ago (Chapter 9). But no clear record remains of an earlier prebiotic era, let alone of that geochronologically elusive moment in time when life first emerged from non-living matter. In order to understand the circumstances leading to the origin of life, we are forced to reconstruct an environmental puzzle with pieces from both a later geologic epoch and an earlier cosmochemical epoch, all bounded by the tenuous and changing framework of theoretical models for the formation and early evolution of Earth. Despite these inherent limitations, such an approach places the study of the origin of life on a sound scientific basis.

In the following section of this chapter we review in an historical context the tenuous and changing framework within which the prebiotic environment and the chemical evolutionary processes were formulated by several principal contributors. We consider in Section 4-3 the nature of the evidence, both natural and synthetic, that supports the chemical evolution paradigm. In this section we also report on recent laboratory experiments aimed at assessing the utility of elemental and isotopic criteria that may be applied in the future to distinguish between abiotic and biotic organic matter as preserved in the early geologic record. Within the last ten years, as a result of new knowledge and insight gained by vigorous research in the space sciences, revised models for the origin and early evolution of Earth have emerged. In Section 4-4 we consider the implications of these recent developments for chemical evolution. Finally, in Section 4-5 we address some problems and prospects for future research in the field.

Throughout this chapter, emphasis is placed on the atmospheric setting within which the primary prebiotic synthesis of organic matter may have taken place. As a result, the discussions focus on hydrogen, carbon, nitrogen and oxygen, the elements principally involved in the chemistry of organic matter.

Certainly, when we deal with a problem of natural science that is as fundamental and, yet, is as obscured by the passage of time as is the origin of life, strong differences of opinion may be expected to emerge and controversy may swirl around issues that cannot be resolved readily by direct observation. Nevertheless, consideration must continue to be given to all models for chemical evolution on Earth that are supported by, or not excluded by, available data on the early history of the solar system by the existing record of biological and geological events on Earth.

4-2. The Historical Context

4-2-1. The Oparin Model of the Prebiotic Earth: A Hydrogen-Rich, Strongly Reducing Atmosphere

According to Oparin (1938), a prerequisite for the prebiotic synthesis of organic compounds was a reducing environment lacking uncombined, molecular oxygen (O_2). He postulated that hydrogen, water, ammonia (NH_3), and methane (CH_4) and other simple hydrocarbons were the primary atmospheric ingredients in the prebiotic milieu where organic compounds were initially synthesized. To justify this atmospheric composition, he noted that in stars, comets, meteorites, and in Jupiter and Saturn, carbon occurred in elemental form and bound with nitrogen as cyanide (HCN), and was combined with hydrogen as hydrocarbons. Thus, according to the limited knowledge then available, primordial carbon existed predominantly in reduced valence states. Oparin therefore attributed the carbon dioxide (CO_2) on Earth and Venus to secondary origins, the result of biological activity on Earth and of unknown causes on Venus. The presence of carbon dioxide in Mars' atmosphere was unknown at the time.

Oparin (1938) pointed out that, with few exceptions, earlier writers on the subject of the origin of life had presumed that all the carbon in the prebiotic atmosphere was in the form of carbon dioxide. It is noteworthy that he discussed the ideas of H. Osborn published in 1919 (cited in Oparin, 1938). In Oparin's words, Osborn "pictures the Earth still uninhabited by living creatures, thickly blanketed with an atmosphere containing much water vapor and carbon dioxide. This carbon dioxide served as the source of carbon for the formation of organic substances from which organisms developed by a long process of evolution .... But Osborn fails to explain how such transformation could occur, limiting himself to the rather ambiguous statement about the 'attractive force' between oxygen and hydrogen." Although in principle, Haldane (1929) agreed with Oparin's ideas on the emergence of life from a reservoir of organic matter in a reducing prebiotic atmosphere, he also initially proposed carbon dioxide rather than methane as the primary source of carbon. In his later writings, however, Haldane (1954) adopted Oparin's methane postulate. Osborn provided no insight into the steps involved in the transformation of lifeless to living matter, but it is interesting to note that his brief description of the setting in which the process of chemical evolution began is at least not inconsistent with the evidence now available from the most ancient portions of the known rock record.

Oparin (1938) viewed Earth as having formed by the condensation of super-heated gases torn from the Sun by a catastrophic encounter with another star. He considered the first condensates from the cooling gas to have been heavy metal droplets of iron and its alloys with other elements. Carbon and nitrogen dissolved in this metal to form carbides and nitrides that were incorporated with metal in the droplets that coalesced to form the molten core of the planet. Accumulation of matter on the proto-Earth was thought to have been strongly influenced by gravitational effects and, thus, to have proceeded according to atomic or molecular weight; the heavier rock-forming elements were therefore thought to have accumulated next with their complement of oxygen to form the igneous rocks of the mantle, followed finally by the retention of the light elements (and gaseous water and nitrogen) in a hot atmosphere. Eruptions of deep-seated metal from the core to the planet's surface were suggested to bring metal, carbides, and nitrides into contact with hot water in the atmosphere, whereupon they were supposed to decompose to yield the hydrogen (H_2), hydrocarbons, and ammonia in Oparin's highly reducing atmosphere.

In the next stage of this scenario, Oparin envisioned that reactions in the hot atmosphere between gaseous ammonia and water and unsaturated hydrocarbons produced additional, more complex organic compounds, which eventually "rained" into the primordial sea. There they underwent further transformations and accumulated to form a "broth" of organic matter that ultimately spawned the first living systems. In Oparin's early writings, thermal energy and the chemical free energy of metastable organic compounds provided the driving force for chemical evolution. Not until the 1957 edition of his book did he take into account the importance of sunlight, electrical discharges in the atmosphere, and other natural energy sources.

While many features of Oparin's (1938) scenario are untenable in light of modern theory and knowledge from astrophysics, geophysics, and geochemistry, it does bear some resemblance to the earliest formulation (Eucken, 1944) of the inhomogeneous accretion model for the formation of the Earth (see Section 4-4-3), which it preceded by eight years and which was not resurrected in modern form until more than 30 years later by Turekian and Clark (1969). Later editions of Oparin's book attempt to modify his earlier views to accommodate and integrate pertinent new data. Implicit in his later writings is the acknowledgment that all working models, his own included, would evolve and be modified by new findings.

Throughout his writings, Oparin recognized that the scientific study of the origin of life was an undertaking of such broad scope that knowledge from many disciplines was required. He recognized the value and necessity of deciphering the available record of the past that is preserved in natural materials. Appropriately, astrophysics, the cosmochemistry of other bodies in the solar system, the geology and geochemistry of the Earth, and the biochemistry and molecular biology of past and present life on Earth all contribute to an understanding of the origin of life. Thus, Oparin viewed the origin of life in the larger context of the origin of Earth, Sun, and the solar system; and he drew upon pertinent contemporary ideas from this context to formulate his vision of the prebiotic environment and its change with time. In addition, he outlined the path of chemical evolution leading to life in terms that can be traced to his understanding of the biochemistry of modern organisms and their evolutionary development. Perhaps Oparin's most important contributions to the study of the origin of life were the abundant detail of his concept of chemical evolution together with his proper placement of the problem of life's origins in a universal cosmochemical and planetological framework.

4-2-2. The Urey Model: Support for Oparin's Prebiotic Atmosphere and Forerunner of Modern Models for the Formation of Earth

In 1952, H. C. Urey published his book *The Planets, Their Origin and Development* and thereby ushered in the modern era of scientific thinking about the origin of the bodies in the solar system. Drawing upon then current theories of stellar and solar system evolution, interpretations of new observations of the Moon, and geological and geophysical evidence, Urey presented a model for the origin of the Earth and the Moon containing features that have endured to this day. (A much shorter treatment of the early chemical history of Earth and its relationship to the origin of life appeared in the same year; see Urey, 1952b.)

Urey (1952a) adopted a concept that collapse of a cloud of interstellar dust and gas yielded a rotating disc, the solar nebula, which according to his hypothesis contained hydrogen, methane, ammonia, water, noble gases, and other components in their cosmic abundances. In his scenario the Sun separated from this nebula, and the nebula fragmented into massive protoplanets composed of dust and gas at different distances from the Sun. The proto-Earth could have been 1000 times the present mass of the Earth. The known occurrence of meteorite impacts on Earth, and the Moon's highly cratered surface indicated to Urey an important role for small bodies, planetesimals, in the formation of planets. Thus, planetesimals composed of silicates, iron oxides and sulfides, water, and ammonia or ammonium salts were postulated to have accreted cold from the dust and gas both within the forming protoplanet, and in the solar nebula. These planetesimals contained virtually no carbon because methane was not condensable at the temperatures prevailing at the proto-Earth's distance from the sun. An "Earth-embryo" amounting to perhaps ten times the mass of the Moon formed by accumulation of these planetesimals at the center of gravity of the protoplanet. Supposedly, as a consequence of rapid gravitational contraction of its massive gaseous envelope, the protoplanet passed through a high temperature state in which pressures from 10^{-2} to 10^{2} atm and temperatures from 273° K or lower to 3000° K could have occurred at different times and at different locations.

In the course of this rapid heating, the Earth-embryo and the planetesimals that existed within the protoplanet, as well as those that were gravitationally captured by it, underwent chemical changes to varying degrees depending on their size. At the high temperatures, iron-bearing oxide and silicate phases were reduced to metallic iron and iron-nickel alloys in the presence of a large excess of H_2; silicates and water were vaporized; ammonium and nitride compounds were decomposed to N_2; and methane in the gaseous envelope was converted to elemental carbon and carbides. Small planetesimals vaporized throughout; larger ones were converted to alloys of iron and nickel; while still larger ones acquired metallic regions on their external surface but remained unchanged internally. The Earth-embryo was presumed to be a member of this latter category. At some point, vaporization of silicates and heating of the gaseous envelope reversed the contraction process and led to the loss of most of the protoplanet's mass and nearly all of its gaseous envelope; further development proceeded in the absence of a significant atmosphere. Heat loss was rapid in the absence of a gaseous envelope and the Earth-embryo cooled quickly. Planetary growth continued by accretion of the remaining planetesimals that now resembled stony and iron meteorites.

Although most of these planetesimals (as well as the surface of the Earth-embryo) were depleted in volatiles, vaporization during impacts slowly replenished the atmosphere. (The term "volatiles" usually refers to the halogens F, Cl, Br, and I, noble gases He, Ne, Ar, Kr, and Xe, and compounds of the elements H, N, C, S, but also includes such volatile metals as Cd, In, Hg, Pb, Bi, Tl.) Carbon occurred predominantly in the form of carbides and elemental carbon; nitrogen as nitrides and ammonium chloride; and hydrogen as hydrated minerals. Water, released from chemically bound sites by heating during impacts, reacted with some of these materials (including metallic iron and sulfides) to produce an atmosphere containing H_2, H_2O, CH_4, N_2, and H_2S. Noble gases trapped in the planetesimals were also released. During this major stage of growth, Urey presumed that the temperature of Earth never exceeded 1200° K and was probably considerably lower. As the atmosphere and surface of Earth cooled, N_2 in the presence of H_2 was thought to convert to NH_3, the most stable low temperature form of nitrogen, thereby yielding a highly reduced atmosphere similar to that earlier hypothesized by Oparin (1938). Urey further postulated that as H_2 continued to escape from Earth, atmospheric electric discharges and photochemical processes involving ultraviolet sunlight would have converted methane, ammonia, and water to more oxidized species, notably including numerous types of organic compounds. Some of these compounds would have dissolved in the primitive oceans and undergone further chemical evolution along the paths envisioned by Oparin. Thus, Urey believed that his model for formation of Earth provided a natural means for establishing the prebiotic conditions originally postulated by Oparin (1938) for the origin of life.

According to Urey's scenario, the prebiotic Earth was, more or less, a homogeneous mixture of metal and silicate overlain by the primordial sea and atmosphere; core formation began at a later stage and occurred gradually over geologic time. Urey's model, unlike Oparin's qualitative one, was strengthened by his quantitative approach and by his use of physics and chemistry, especially thermodynamics. He also pointed out the importance of sunlight and electrical discharges as major sources of free energy on the prebiotic Earth. But, of course, like Oparin before him, Urey's vision was limited by the information available to him at the time of writing. Nonetheless, his model was a forerunner of a class of "homogeneous accretion" models for Earth that prevailed during the period of 1950 to 1970 and that continues to be important today (see Section 4-4-2 and Chapter 2).

4-2-3. The Rubey Model: An Alternative to Oparin's Prebiotic Atmosphere and the Rekindling of a Controversy

Just prior to publication of Urey's book, Brown (1949) and Suess (1949) discussed the observation that the ratios of the noble gases to silicon in the total Earth were many orders of magnitude smaller than the corresponding ratios in the Sun. Moreover, the depletion of the individual noble gases with respect to their solar abundances increased dramatically with decreasing atomic weight. These depletions were interpreted as indicating that Earth retained virtually no trace of a primary atmosphere inherited from the solar nebula. Thus, Earth's noble gases and the other volatile elements (like C, N, H, and S) now on its surface must have been contained originally in the solid material that formed the planet, a conclusion consistent with Urey's scenario but not Oparin's.

At about the same time that Urey was preparing his book for publication, W. W. Rubey (1951) reported his investigations into the origin of seawater. He observed that weathering of pre-existing rocks over geologic time could account for the major rock-forming elements (e.g., Al, Ca, Fe, K, Mg, Na, Si) in sedimentary rocks and dissolved in sea water. But the abundance of volatile species and elements (e.g., C as CO_2, H as H_2O, N, Cl, S, Ar) in the atmosphere, hydrosphere, and sediments greatly exceeded the amounts attributable to rock weathering. Moreover, the apparent composition of these "excess volatiles" (predominantly H_2O and CO_2 with smaller amounts of Cl, N, S, and other elements) was remarkably similar to that of magmatic gases released from the interior of Earth by fumaroles, hot springs, and volcanoes. Rubey suggested two possible explanations for the excess volatiles: (*i*) they represented Earth's original endowment of volatiles in the hot dense primitive atmosphere of the molten planet, which upon cooling condensed to form the primitive ocean; or, alternatively, (*ii*) they evolved by degassing of Earth's interior, either rapidly and early in Earth history or gradually throughout geologic time. Rubey found that the geochemical consequences of a dense early atmosphere were not borne out by the known geologic record. He therefore concluded that the excess volatiles must have gradually accumulated over time at a more or less steady rate by outgassing of Earth's interior.

Implicit in Rubey's conclusion was the denial of a prebiotic atmosphere composed predominantly of H_2, CH_4, NH_3, and H_2O. Several years later, Rubey (1955) confronted this issue explicitly. He strengthened his original conclusion with thermodynamic calculations and discussions of the lack of geochemical and atmospheric evidence that might have supported the notion of an early atmosphere containing H_2, CH_4, and NH_3. He also suggested that the present inventory of excess volatiles did not contain enough hydrogen to allow the past existence of an atmosphere rich in H_2, CH_4, and NH_3. In the final analysis, however, Rubey (1955) acknowledged that the abundance of H_2 in the early atmosphere and its rate of loss from Earth would have been a major factor in determining whether CH_4 was present or absent. He concluded on the basis of his best estimates for the rate of escape of H_2 from the Earth's exosphere that CH_4 would have persisted in the atmosphere for no more than 10^5 to 10^8 years. According to Rubey's (1955) view, life originated in an environment resembling the present-day surface of Earth, but lacking free oxygen (O_2). Thus, the pre-Oparin conception of a primordial atmosphere containing CO_2 as the dominant carbon source was given scientific support, a concept that would be subsequently adopted by others.

In their later estimate of the amount of CH_4 in the prebiotic atmosphere, Miller and Urey (1959) took into account the presence of CO_2 in the atmosphere and as carbonate in the oceans. However, their thermodynamic calculations yielded 4×10^3 atm of CH_4 on the basis of current surface pressures for H_2O and CO_2 and an estimated H_2 pressure of 1.5×10^{-3} atm. This amount of CH_4 exceeded estimates of the inventory of carbon in the Earth's atmosphere, crust and upper mantle by more than a factor of ten. Consequently, they suggested that the relevant equilibrium, $CO_2 + 4H_2 \rightleftarrows CH_4 + 2\,H_2O$, was not established near the surface of the planet but, rather, high in the atmosphere as a result of photochemical processes occurring at greatly reduced pressures (viz., $2.5 \times 10^{-9} \times 10^{-9}$ atm), but at the same temperature. In contrast, the relevant equilibrium between nitrogen-containing species, $1/2\,N_2 + 3/2\,H_2 \rightleftarrows NH_3$, was supposed to be established near the Earth's surface in order to convert N_2 to NH_3 and to allow dissolution of the latter in the oceans as ammonium (NH_4^{1+}) ions. The validity of their conclusion regarding the existence of "a reducing atmosphere containing low partial pressures of hydrogen and ammonia and a moderate pressure of methane and nitrogen" rested both on the assumption that thermodynamic equilibria would have been established at low temperatures, and on the oxidation-reduction (redox) state of the atmosphere-crust system of that time.

4-2-4. A Multi-Stage Model for Early Atomospheric Evolution: A Synthesis of Alternatives

The strongly contrasting hypotheses regarding the composition of the primitive atmosphere offered by Oparin and Urey on the one hand, and by Rubey on the other, have become parts of an historical debate that was heralded by the early work of Oparin (1924), that then became more sharply defined scientifically in the 1950s, and that persists to this day. Rubey's arguments in favor of a CO_2–N_2–H_2O prebiotic atmosphere were based on evidence in the known geologic record from which he made extrapolations back to an earlier time. Urey's and Oparin's contentions for a CH_4–NH_3–H_2O atmosphere were based on astrophysical and cosmochemical models for the origin of Earth, for which they found support in the record of early solar system events as evidenced on the Moon, in meteorites, and other solar system bodies. Both points of view involve extrapolations: the former backwards in time, the latter forward in time and from extraterrestrial environments. The viewpoint represented by Urey (1952a, 1952b), though later revised somewhat by Miller and Urey (1959) and by Horowitz and Miller (1962), remained largely unchanged until H. D. Holland (1962) proposed his model for the evolution of Earth's atmosphere.

Holland (1962) pointed out that the evidence for or against a strongly reducing CH_4-rich primitive atmosphere was inconclusive insofar as no direct record was known to remain from the earliest period of Earth's history. He then proceeded to develop a multistage model for the evolution of Earth's atmosphere that could reconcile the two sides of the controversy. His model was based on the assumptions that the Earth formed cold by a process of homogeneous accretion (see Section 4-4-2) and that the primitive atmosphere evolved subsequently during the gradual heating of Earth that led to core formation. The redox state of the magmas generated in the upper mantle during the core formation stage would have been controlled by the presence of metallic iron and of ferrous iron in silicates and oxides (Fe^0–Fe^{2+} pair, e.g., iron-wustite synthetic buffer).

Holland's thermodynamic calculations supposed that such a system at equilibrium at 1500° K would be associated with an oxygen partial pressure of about $10^{-12.5}$ atm. According to equilibrium constants for the relevant reactions, $H_2 + 1/2\,O_2 \rightleftarrows H_2O$ and $CO + 1/2\,O_2 \rightleftarrows CO_2$, this amount of O_2 would permit emission of volcanic gases with the ratios of $H_2O/H_2 = 0.44$ and $CO_2/CO = 0.17$. Methane would be much less abundant than CO_2 if the total water released was less than 10 atm. Therefore, the primordial atmosphere generated while metallic iron remained in the upper mantle would have been dominated by H_2O, H_2, and CO. Similar calculations by Heald et al. (1963) supported these results and showed that when the elemental ratios of H/O/C/S found in Hawaiian volcanic gases were used in the calculations, the ratio of CH_4/CO_2 was less than 10^{-4}.

Holland further assumed that the temperature structure of the pre-core atmosphere was similar to the present-day structure; water condensed to liquid; and re-equilibration of the atmosphere at 300° K occurred rapidly (by unspecified pathways) on a short time scale compared to the time of H_2 exospheric escape. The reactions of interest, $CO + 3\,H_2 \rightleftarrows CH_4 + H_2O$ and $CO_2 + 4\,H_2 \rightleftarrows CH_4 + 2\,H_2O$, proceed far to the right at 300° K; thus, if excess H_2 and kinetically accessible pathways were available, a CH_4-dominated atmosphere would have resulted. By assuming that all of the carbon in Rubey's excess volatiles was in the form of CH_4, Holland estimated a maximum methane pressure of 6.6 atm. Similar considerations led to the conclusion that N_2 was probably a major atmospheric component (0.8 atm), although the presence of small amounts of ammonia could not be excluded. Maintenance of a high abundance of H_2 against loss by exospheric escape would have been critical for achieving re-equilibration of CO to CH_4 (and N_2 to NH_3), and limits were estimated by Holland (1962) for the partial pressure of H_2 to have been 0.27 to 2.7×10^{-5} atm. At the lower limit, CO_2 would have been a significant atmospheric component. In accordance with the then prevailing version of homogeneous accretion models for Earth's origin (see Section 4-4), Holland (1962) presumed that core formation would have proceeded gradually over time, taking up to 500 million years. Holland (1962) did not consider the thermal consequences of core formation for the preservation and subsequent use in chemical evolution of organic matter synthesized during the pre-core stage.

Once the segregation of metallic iron into the core was completed, the composition of volcanic gases emitted from the upper mantle would have been controlled by the ratio of ferrous to ferric iron in silicates and oxides (Fe^{2+}—Fe^{3+} pair, e.g., fayalite-magnetite-quartz synthetic buffer). The oxygen partial pressure in equilibrium with this system was estimated by Holland (1962) to be about $10^{-7.1}$ atm near 1500° K; under these conditions, gases would be emitted with the ratios H_2O/H_2 = 105 and CO_2/CO = 37. Assuming H/O/C/S ratios corresponding to those of Hawaiian volcanic gases, the ratio CH_4/CO_2 would be less than 10^{-8} at 1500° K and approximately 10^{-2} at 400° K, assuming re-equilibration occurred at the low temperature (Heald et al., 1963). With H_2O and most of the CO_2 in the oceans, Holland (1962) estimated that the CH_4 abundance would range from $>10^{-6}$ to $<10^{-4}$ atm as compared to approximately 3×10^{-4} atm for CO_2. Holland (1962) compared the predictions of this second-stage of his model with Hawaiian volcanic gases and found excellent agreement. He went on to suggest that there was no evidence in the geological record that the oxidation state of magmas had changed with time and, therefore, that the hypothesis of "constant oxidation state of volcanic gases emitted during all but the pre-core stage of the Earth's history appears to be tenable."

Holland's (1962) contribution provided a unified model for the early evolution of Earth's atmosphere by accomodating a highly reducing primordial composition at the outset, and establishing a scenario for its conversion to a much less reducing one consistent with the geological record of outgassing. Nonetheless, the validity of the basic assumptions underlying his formulation of the precore stage remained vulnerable.

4-2-5. Other Variations on the Theme of Prebiotic Atmospheres

Since the appearance of Holland's model, other workers have made contributions to the debate. Abelson (1966), drawing on the work of Rubey (1951, 1955) and Holland (1962), estimated the amount of reducing power that could have been carried to the Earth's surface by volcanism. He noted that of the total amount of free oxygen thought to have been produced by photosynthesis over geologic time, only about 30% could be accounted for as free O_2 in the present atmosphere or as fossil O_2 used in the past to oxidize Fe^{2+} to Fe^{3+} and S^{2-} to SO_4^{2-}. Because oxidation of the H_2 and CO that (with CO_2 and H_2O) should have accompanied volcanism over geologic time could account for this apparent oxygen deficit, he concluded that these gases must have been significant sources of atmospheric reducing potential. This and other arguments led Abelson (1966) to propose that with H_2O condensed in oceans and CO_2 dissolved in the oceans and converted to carbonate, the main residual gases in the prebiotic atmosphere would have been CO, H_2 and N_2.

Some constraints on the lifetime and abundance of CO in a prebiotic environment were formulated by Van Trump and Miller (1973). Based on the kinetics of the reaction of CO + OH^- in aqueous solution to give formate, Van Trump and Miller (1973) considered two limiting cases: (*i*) if all the surface carbon on the Earth were released initially as CO, its half life against conversion to formate in an ocean (at pH 8) would range from 17 years at 348° K to 1.2×10^7 years at 273° K; or (*ii*) if it were degassed over 2×10^9 years, the steady state abundance of CO would be 0.06 atm at 273° K and 2.7×10^{-4} atm at 298° K. Other sinks for CO would have made its lifetime in the atmosphere even shorter.

Abelson (1966) also noted that a quantity of NH_3 corresponding to all the nitrogen in the present atmosphere could have been photochemically decomposed to N_2 in about 3×10^5 years. On the other hand, Bada and Miller (1968) pointed out that NH_3 would have dissolved in the primitive oceans to yield ammonium ions (NH_4^+), the maximum concentration (10^{-2} M) of which would have been regulated by the ion-exchange capacity of clays formed by rock weathering. A minimum ammonium ion concentration of 1×10^{-3} M was thus postulated for the oceans on the assumption that preservation of aspartic acid in oceans was required for the origin of life.

The preceding summary shows that atmospheric compositions ranging from highly reducing (H_2, NH_3, CH_4, H_2O) to mildly reducing (H_2O, H_2, N_2, CO, CO_2) to nonreducing (H_2O, N_2, CO_2) have been proposed. The assumptions on which these various compositions were based are discussed further in Section 4-4; there they are considered in light of information obtained since the early 1970s as a result of the study of lunar samples, meteorites, and terrestrial rocks of both ancient and recent origin, together with theoretical modeling studies of atmospheric processes.

4-3. Evidence in Support of the Chemical Evolution Paradigm

The record of biological evolution manifest in the chemistry of living organisms and preserved in the geologic record probably provides the most compelling evidence for a period of chemical evolution early in Earth history. Acceptance of this generalization, however, leaves unsatisfied a scientific appetite for more direct evidence of the process and fuller understanding of the circumstances involved. To satisfy this appetite, extraterrestrial environments have been scrutinized for the presence of organic matter and for the insights they may yield into the nature of chemical evolution. Additionally, putative prebiotic conditions on Earth have been simulated in attempts to test pathways for organic synthesis in model environments. In the first two parts of this section we summarize in cursory fashion evidence obtained by these two approaches. Clearly, the uncovering of unambiguous evidence in the geologic record of an early period of chemical evolution would constitute a major confirmatory discovery. Thus, in the last part of this section we describe experiments designed to assess chemical criteria that may be used to distinguish between organic matter of primary abiotic origin and biogenic matter in ancient sediments.

4-3-1. Extraterrestrial Evidence

The assumption that organic compounds necessary for life would have been formed naturally by non-biological processes serves as the foundation of the chemical evolution paradigm. In our judgment the general validity of this assumption is strongly supported by the observation that organic matter is widespread in the solar system and beyond.

Meteorites of the carbonaceous type were known since the early nineteenth century to contain organic compounds, and most of the organic matter was considered extraterrestrial in origin (see Hayes, 1967; Nagy, 1975). In 1970, unambiguous proof of the abiotic origin of amino acids in the Murchison meteorite was presented (Kvenvolden et al., 1970). This work was especially significant because amino acids play a key role in biochemistry, and they are presumed to be among the essential molecular building blocks in the process of chemical evolution. The presence of a variety of organic compounds in the Murchison meteorite (e.g., amino, hydroxy, mono- and di-carboxylic acids; urea and amides; ketones and aldehydes; hydrocarbons; alcohols; amines; and N- and S-heterocycles; (see Hayes, 1967; Nagy, 1975; Chang, 1979) leaves little doubt that the path of chemical evolution in an extraterrestrial environment had passed the milestone of synthesis of simple organic molecules. Unfortunately, the organic chemical processes involved in these syntheses and where they took place remain poorly understood (Miller et al., 1976; Chang, 1979; Bunch and Chang, 1980). It is noteworthy, however, that many of the objects in the asteroid belt may be carbonaceous meteorite parent bodies (Morrison, 1978).

The simple molecules observed in comets (e.g., CH, C_2, C_3, CO^+, CO_2^+, NH, NH_2) suggest the presence of organic compounds; but hydrogen cyanide (HCN) and methyl cyanide (CH_3CN) are the most complex cometary molecules observed thus far (Delsemme, 1977). Similar sources may have supplied the organic matter in meteorites and comets (Delsemme, 1975; Chang, 1979; Bunch and Chang, 1980), and an interstellar origin for some of it appears very probable (Whittaker et al., 1980; Hayatsu et al., 1980; Ott et al., 1981).

A recent review of interstellar chemistry reports a roster of 52 interstellar molecules (Turner, 1980), of which 36 may be considered organic (i.e., containing C–H bonds); compounds containing up to eleven atoms have been observed [viz., $H(C_2)_4CN$]. Apparently, the environment of "dense" interstellar clouds with its low temperature ($\leq 100°$ K) and gas densities ($\leq 10^6$ particles per cubic cm), as exotic and as seemingly inimical to chemical reactions as it may appear, nonetheless exhibits a rich chemistry that manifests itself in the production of numerous types of organic compounds.

On Jupiter, the presence of simple hydrocarbons in the atmosphere, and the colors of its multihued clouds, are indications of active chemical processes. The hydrocarbons are readily explained as products of solar ultraviolet photochemistry and lightning activity (Prinn and Owen, 1976; Bar-Nun, 1979). Urey (1952) was the first to suggest that complex organic molecules were responsible for the distinct coloration of Jupiter's cloud cover. Colored inorganic matter has also been postulated on the basis of photochemical models, but existing data do not permit the adoption of either viewpoint to the exclusion of the other (Prinn and Owen, 1976; Bar-Nun, 1979). In any case, atmospheric circulation can transport material formed high in the atmosphere to deeper levels where high temperatures and pressures and reaction with gaseous H_2 can reconvert it to the primary ingredients of the atmosphere. Thus, the dynamics of Jupiter's atmosphere limit the potential progress of chemical evolution.

Saturn's moon, Titan, contains methane and small amounts of C_2-hydrocarbons in its atmosphere and exhibits a reddish coloration that may be caused by organic matter. Although the surface of the planet is extremely cold ($\sim 90°$ K), some analogy may be drawn between the processes occurring in Titan's atmosphere and those that may have prevailed on a hypothetical CH_4-rich prebiotic Earth. The organic chemistry on Titan has been reviewed recently by Chang et al. (1979).

The widespread occurrence of organic matter elsewhere in the cosmos, and within our solar system, confirms our expectation that chemical evolution is a natural consequence of the evolution of matter. But the potential pathways for the chemical evolution of organic matter are circumscribed by the evolutionary track peculiar to a specific environment, be it an interstellar cloud, a meteorite parent body, or a planet; progress toward the origin of life will be diverted at different stages depending on the physical-chemical constraints imposed by each environment. Even on planets as similar as Venus, Earth, and Mars, the evolutionary tracks led to very different results. Although all three planets may have experienced a common early history, temperatures on Venus are now too high to support the presence of either organic compounds or life; and the limited sampling taken by the Viking Mission suggests that the surface of Mars may be devoid both of life (Klein, 1977) and organic matter (Biemann et al., 1977). Comparative study of these planets should lead in the future to major insights into the factors that made the origin and evolution of Earth unique (see Pollack and Yung, 1980). In the absence of such insights, however, we are left with the realization that despite the common occurrences of organic matter in extraterrestrial environments and the assurance they provide that chemical evolution occurs naturally, they may have little to tell us of the particular circumstances involved in prebiotic syntheses on the primitive Earth.

4-3-2. Organic Synthesis under Simulated Prebiotic Conditions

Prior to the 1950s there are no reports that Oparin and his colleagues tested his ideas on prebiotic synthesis by studying some of the reactions he postulated under conditions consistent with his model of the prebiotic environment. The possibility of prebiotic synthesis of organic matter by lightning and ultraviolet sunlight was not mentioned in Oparin's writings until 1957 (Oparin, 1957, chap. 5); even though suggestions were made along those lines by F. Allen (1899, cited in Oparin, 1938, p. 52), whose ideas he discussed. Beutner (1938), in his review of Oparin's (1938) book, also suggested that prebiotic organic matter might have been produced by electric discharges. Beutner noted that the eminent chemist F. Haber had passed electric discharges through carbon-containing gases like methane and carbon dioxide in hopes of synthesizing sugars, only to find a complex unseparable mixture of components. According to Beutner (1938, p. 94), these experiments led Haber to conclude that discharges of that sort could produce "any substance known to organic chemistry." But, apparently, Haber did not realize the potential significance of his work; and among Oparin's contemporaries, experimentalists interested in the origin of life were either unaware of his ideas on the prebiotic origins of organic matter or chose to ignore them.

In the first half of the twentieth century, most attempts to simulate the prebiotic synthesis of organic compounds were performed as models for plant photosynthesis. In this context, ultraviolet irradiation of CO_2 and H_2O in the presence or absence of oxygen proved largely unsuccessful (for a summary of such studies, see Rabinowitch, 1945, p. 78–98). Experiments that consistently yielded small amounts of formaldehyde required the presence of hydrogen (H_2) or another reductant (Rabinowitch, 1945, p. 78–98; Oparin, 1957, p. 162–166). As late as 1951, Garrison et al. (1951) reported low yields of formic acid and formaldehyde when aqueous solutions of CO_2 and ferrous sulfate were irradiated by 40 MeV helium ions. Hydrogen (H_2) was not initially present in their experiments, but the presence of dissolved Fe^{2+} ions afforded a reduced reaction medium. While these early experiments tended to substantiate the proposition that a reducing environment was necessary for the initial stages of the chemical evolution of organic matter, they neither denied nor affirmed the importance of ammonia and CH_4 or other hydrocarbons in Oparin's and Urey's models.

It was not until 1953 that a major advance was made in testing atmospheric models for the prebiotic synthesis of organic matter. Miller (1953, 1955, 1957) passed an electric discharge through a mixture of CH_4, NH_3, H_2O, and H_2 and achieved the facile synthesis of a variety of organic compounds, particularly amino acids. The ease with which the syntheses were accomplished under conditions similar to those originally postulated by Oparin (1938) and Urey (1952b) stimulated experimentation by other workers. Below we briefly summarize the results obtained for various model compositions. For more comprehensive reviews of the multitude of experiments performed, the reader is referred to Kenyon and Steinman (1969), Lemmon (1970), Miller and Orgel (1974), Gabel (1977), and Day (1979).

STRONGLY REDUCING ATMOSPHERES

CH_4, NH_3, and H_2O (with or without H_2). The system CH_4, NH_3, and H_2O is strongly reducing either with or without molecular H_2. Passage of an electric discharge through this mixture (containing H_2) resulted in the first laboratory prebiotic synthesis of amino acids (2% of the initial carbon was converted to glycine alone), together with hydroxy acids, short aliphatic acids, and urea (Miller, 1953, 1955, 1957). Many of the compounds produced were of biological importance. The mechanism of synthesis of the amino and hydroxy acids involved solution reactions of smaller molecules produced in the discharge, in particular of hydrogen cyanide and aldehydes:

$$\mathrm{RCHO + HCN + NH_3 \rightleftharpoons RCH(NH_2)CN \xrightarrow{H_2O} RCH(NH_2)\overset{\overset{\displaystyle O}{\|}}{C}NH_2 \xrightarrow{H_2O} RCH(NH_2)COOH}$$

$$\mathrm{RCHO + HCN \rightleftharpoons RCH(OH)CN \xrightarrow{H_2O} RCH(OH)\overset{\overset{\displaystyle O}{\|}}{C}NH_2 \xrightarrow{H_2O} RCH(OH)COOH}$$

Detailed studies of the equilibrium and rate constants of these reactions showed that either amino or hydroxy acids or both, depending on the NH_3 concentration, could have been synthesized at high dilutions of HCN and aldehydes in a hypothetical primitive ocean (Van Trump and Miller, 1980). These molecules can also be produced by the thunder shock waves associated with lightning discharges (Bar-Nun and Shaviv, 1975).

Irradiation of this mixture of gases with short wavelength ultraviolet light <150 nm produces very low yields of amino acids (Groth and Von Weysenhoff, 1960). If the gas mixture is modified by adding H_2S or formaldehyde, amino acids can be obtained at relatively long wavelengths (≤ 230 nm) where considerable energy from the Sun is available (Sagan and Khare, 1971; Becker et al., 1974). Apparently, hydrogen atoms produced by photodissociation of H_2S and formaldehyde carry sufficient excess kinetic energy to initiate reactions with CH_4 and NH_3. In the presence of a titanium dioxide photocatalyst, irradiation of CH_4 and NH_3 in water with near ultraviolet and visible light also yields amino acids (Reiche and Bard, 1979).

Pyrolysis of CH_4 and NH_3 followed by addition of H_2O gives very low yields of amino acids. The pyrolysis temperatures ranged from 1073° to 1473° K with contact times of a second or less (Lawless and Boynton, 1973). However, the pyrolysis of CH_4 and other hydrocarbons gives good yields of benzene, phenylacetylene, and many other hydrocarbons. It can be shown that phenylacetylene can be converted to the amino acids phenylalanine and tyrosine by model pathways possible in the primitive seas (Friedmann and Miller, 1969). Similarly, tryptophan can be synthesized from indole produced by the pyrolysis of the hydrocarbons in the presence of NH_3.

CH_4, N_2 with traces of NH_3 and H_2O. The composition of CH_4, N_2 with traces of NH_3 and H_2O is somewhat less reducing than the mixture above. It is appropriate for an Oparin-Urey atmosphere in which the NH_3 is dissolved in the seas. Passage of an electric discharge through it produces a somewhat lower yield of amino acids than is obtained in the presence of abundant NH_3, but the products are more diverse (Ring et al., 1972; Wolman et al., 1972). Hydroxy acids, short aliphatic acids, and dicarboxylic acids are produced along with the amino acids. Ten of the 20 amino acids that occur commonly in proteins are produced directly in this experiment.

Although it can be shown that HCN and various aldehydes are produced by sparking mixtures of CH_4, N_2, NH_3 and H_2O, the most effective way to study these systems is to analyze the products from sparking mixtures of $CH_4 + NH_3$, $CH_4 + N_2$, and $CH_4 + H_2O$ separately (e.g., Toupance et al., 1975; Allen and Ponnamperuma, 1967). The results show that nitriles are produced in the first two mixtures; aldehydes, ketones, acids, and alcohols in the last; and hydrocarbons in all three. Hydrogen cyanide is probably the most important product since it may have been involved in both amino acid and purine synthesis in the primitive seas. Additional products include cyanoacetylene, which could have been converted subsequently to pyrimidines and aspartic acid; acrolein and propiolaldehyde, which could have been precursors of other biochemicals; and cyanamide and cyanogen, which could have served as dehydrating agents suitable for promoting condensation-polymerization reactions. Ultraviolet irradiation (185 nm) of CH_4, N_2, H_2O mixtures yields hydrocarbons, alcohols, aldehydes, and ketones, but no detectable products containing nitrogen (Ferris and Chen, 1975).

MILDLY REDUCING ATMOSPHERES

CO, N_2, and H_2. Electric discharges acting on the mixture of CO, N_2, and H_2 are not effective in amino acid synthesis, unless the ratio of H_2 to CO is greater than one (Abelson, 1966; Miller and Schlesinger, 1983). Under such conditions, glycine is produced in fair yield, but only small amounts of any higher amino acids are produced. Large amounts of formaldehyde and HCN are obtained, however, compounds that are important in models for the prebiotic synthesis of sugars and purines.

CO_2, N_2, and H_2. The CO_2 is more oxidized than the foregoing system, but the excess H_2 makes this a reduced mixture. As with the $CO + N_2 + H_2$ mixture, amino acid synthesis is quite low with electric discharges in this case, unless H_2/CO_2 is greater than about two (Abelson, 1966; Miller and Schlesinger, 1983); glycine is produced, but again very little of the higher amino acids is formed. When H_2O is included in this mixture, formaldehyde can be produced in models of solar ultraviolet photochemistry (Pinto et al., 1980).

CO and H_2. This mixture of CO and H_2 is used in the Fischer-Tropsch reaction to make hydrocarbons in high yields. The reaction requires a catalyst, usually Fe or Ni supported on silica, a temperature of 673° K, and a short contact time. Depending on reaction conditions, aliphatic hydrocarbons, aromatic hydrocarbons, alcohols, and acids can be produced. If NH_3 is added to the $CO + H_2$ mixture, then amino acids, purines, and pyrimidines can be formed (Studier et al., 1968; Hayatsu et al., 1972). The intermediates in these reactions are not known, but HCN may be involved.

CO and H_2O. Electric discharges are not effective with this mixture of CO and H_2O. But irradiation with ultraviolet light that is absorbed by the water ($\leq$184 nm) produces formaldehyde and other aldehydes, alcohols, and acids in fair yields (Bar-Nun and Hartman, 1978).

NON-REDUCING ("NEUTRAL") ATMOSPHERES

CO_2 and H_2O. Neither electric discharges nor ultraviolet light yield organic compounds with this system of CO_2 and H_2O. Ionizing radiation (e.g., 40 MeV helium ions) gives small yields of

formic acid and formaldehyde (Garrison et al., 1951).

CO_2, H_2O in the presence of Fe^{2+}. The action of γ-rays on an aqueous solution of CO_2 and ferrous ion gives fair yields of formic acid, oxalic acid, and other simple products. Ultraviolet light gives similar results with lower yields. In these reactions (Getoff et al., 1960), the Fe^{2+} is a stoichiometric reducing agent rather than a catalyst. Nitrogen in the form of N_2 does not react; experiments with NH_3 have not been tried.

The experimental results summarized above demonstrate that organic compounds can be synthesized in relatively high abundances and great variety in strongly reducing atmospheres. As the oxidation state of the gases is increased by stepwise substitutions for CH_4 of first CO, then CO_2, and of N_2 for NH_3, both yield and product diversity decrease. However, the molecular composition of the gases is also important. Apparently, solar ultraviolet irradiation cannot produce C—N bonds when nitrogen occurs as N_2 in the atmosphere, regardless of whether carbon is present as CH_4, CO, or CO_2. Moreover, calculations suggest that in the presence of H_2, H_2O, CO, and CO_2, lightning converts N_2 to nitric oxide (NO), which could have been converted to HNO, HNO_2, and HNO_3 and dissolved in primitive seas as nitrite and nitrate (Yung and McElroy, 1979; Chameides and Walker, 1981; Kasting and Walker, 1981). A thermochemical-hydrodynamic model has been used to predict rates of carbon and nitrogen fixation by lightning in the prebiotic atmosphere (Chameides and Walker, 1981). The results suggest that the critical factor in the synthesis of HCN as the major primary product rather than NO would have been the ratio of C/O in the atmosphere and not the amount of hydrogen. According to Chameides and Walker (1981), with water condensed in oceans, passage of lightning through an atmosphere containing N_2 and CH_4 would have yielded predominantly HCN; whereas if CO_2 or CO were present, NO would have been the primary product regardless of whether or not H_2 was present in very much higher abundances than CO_2.

Despite some limitations, model prebiotic syntheses have achieved a notable degree of success. Nearly all the compounds that comprise the monomeric building blocks of proteins and nucleic acids have been synthesized, either directly from such experiments (conducted for the most part in reducing atmospheres) or in secondary experiments starting with the intermediates and products obtained in the former. Considerable progress has also been made toward producing peptides and polynucleotides, but the molecular lengths achieved do not yet approach those of biopolymers. Chemical evolutionary developments at these higher levels of molecular organization will not be addressed here; for references to the early work in this area, the reader is referred to Oparin (1938, 1957) and the reviews cited above; for more recent developments and other citations we suggest Eigen (1971), Orgel and Lohrmann (1974), Miller and Orgel (1974), Usher (1977), Lohrmann et al. (1980), Nelsestuen (1980), and Fox and Dose (1977).

4-3-3. Assessment of Possible Criteria for Identifying Evidence of Chemical Evolution in the Geologic Record

Clearly, the documentation in ancient sediments of the transition between prebiotic organic evolution and the origin of life would demonstrate most convincingly the chemical evolution paradigm. Presumably, prior to life the dominant sources of organic matter would have been prebiotic syntheses and meteoritic and cometary debris (see Section 4-4-3). After life arose, biological organic matter eventually became the dominant source of reduced carbon preserved in sediments. For such a scenario to be verified, one must be able to distinguish between abiotic and biologic organic matter as preserved in ancient rocks. Certainly it is possible to identify fresh biological organic material. It consists of certain distinctive compounds, the amino acids, sugars, nucleotides, and lipids, for example, and many microorganisms have characteristic elemental N/C values due to their high protein content. Moreover, organisms are typically depleted in the ^{13}C isotope, relative to the inorganic carbon in their environment, a depletion caused principally by the CO_2-fixation step of the photosynthetic Calvin cycle (see Chapter 7). Unfortunately it is not possible to examine "fresh" organic matter in rocks of Early Archean and Hadean age, the time in Earth history when the origin of life must have occurred. Organic matter in these rocks has been severely altered by the ravages of time and temperature, and has been commonly subject to post-depositional contamination. As is discussed in Chapter 5, virtually all of the small, soluble, organic molecules initially deposited in sediments of Archean age either have been degraded long ago or they have been diluted with contaminant molecules of a more recent origin. On the other hand, the major component of insoluble, polymeric, organic carbon in these rocks, the so-called "kerogen," is probably authentic Archean material. Thus, it might be possible to discern the elemental and isotopic contents of original Archean organic matter by analyzing the kerogen that derived from it (Tissot and Welte, 1978). The relevant question, then, is the following: Can elemental and isotopic analyses of ancient organic material be used to establish its abiotic or biological origin, despite the chemical changes that have occurred during the material's lengthy storage in the geologic environment?

The approach adopted here to address this issue was to prepare organic matter by several plausibly prebiotic processes, to examine the elemental and isotopic compositions of these products both before and after artificial diagenesis experiments, and to compare the measured compositions with those obtained from biologically synthesized organic matter. Distinctive dissimilarities in the compositions of the abiotic and biotic products were sought as potentially useful criteria by which to distinguish between the two sources.

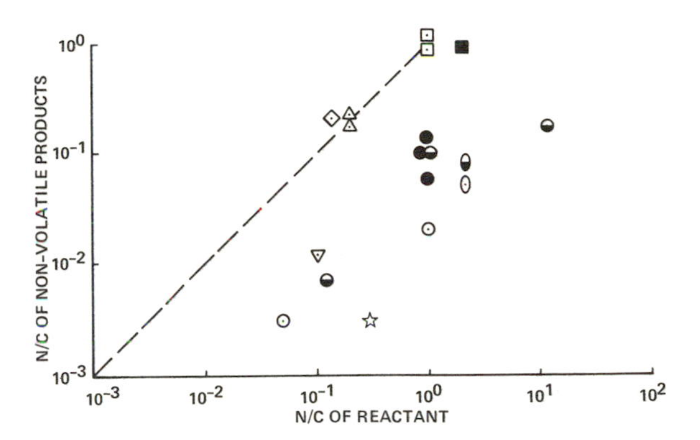

FIGURE 4-1. Plot of N/C values of initial reactant gas mixtures versus the N/C values found for the non-volatile products. The electric discharge initial gas compositions are represented as follows: electrode to water, CH_4–N_2–H_2O (open circles), CH_4–N_2–NH_4Cl–H_2O (partly shaded circles), CH_4–NH_3–H_2O (shaded circles); electrode to electrode, CH_4–N_2–H_2O (open oval), CH_4–N_2–NH_4Cl–H_2O (partly shaded oval). The ammonium cyanide polymer experiment in the present work is depicted by a shaded square. Values obtained from the literature are represented as follows: phenylalanine photopolymer (diamond); Fox microstructures (triangle); Fisher-Tropsch product (inverted triangle); HCN polymer (open square); and Allende acid-insoluble carbon (star). For details and data concerning the electric discharge and ammonium cyanide polymer experiments, see Appendix III, Sections III-1 and III-2 and Tables III-1, III-2, and III-3. For literature values and references, see Appendix III, Table III-6.

CARBON AND NITROGEN ABUNDANCES

Variations in the nitrogen to carbon ratios (N/C) of abiotic products were studied for a number of abiotic synthesis mechanisms, including spark discharge synthesis and HCN polymerization. Additionally, for spark discharge synthesis, the effects of a number of experimental variables upon product N/C values were examined.

In the spark discharge experiments, the effects of the following four parameters on the products were studied: (*i*) the N/C value of the reactant gas mixture containing N_2 and CH_4; (*ii*) the extent of chemical reaction (measured as percentage of CH_4 consumed); (*iii*) the configuration of the electric spark (sparking from one electrode either to the other electrode or to the water surface); and (*iv*) the presence or absence of reduced nitrogen (added as NH_4Cl) in the reactant mixture. Additionally, N/C values were examined for both the total nonvalatile products and the portion of those products that was insoluble after treatment with hot acid.

As is shown in Figure 4-1, the spark discharge products generally had N/C values that were from ten to 100 times smaller than the values in the original reactant gas mixtures; product N/C values ranged from 0.005 to 0.11. The relatively low nitrogen content in the products may in part reflect the difficulty in severing the strong triple bond in the N_2 molecule to allow individual nitrogen atoms to participate in subsequent reactions.

When CH_4, N_2, and H_2O are reactants, a direct correlation is observed between the N/C values of the reactants and those of the products (Figure 4-1). For example, reactant N/C values of 0.06 and 2.0 yielded product N/C values of 0.004 and 0.05, respectively. If NH_4Cl is present as a reactant, the correlation is still observed, but the resulting product N/C values are larger. For example, reactant N/C values of 0.09 and 2.1 yielded product values of 0.017 and 0.08, respectively. The relatively larger N/C values obtained from the NH_4Cl experiments perhaps reflect the greater ease with which reduced nitrogen is incorporated into the products relative to the incorporation of N_2.

A slightly larger product N/C value was obtained in an experiment with CH_4, N_2, H_2O, and NH_4Cl when the spark in the flask passed from one electrode to the water surface, rather than to a second electrode in the headspace (Figure 4-1). This greater N/C value may reflect the enhanced participation of aqueous NH_4Cl in the spark discharge reaction due to the termination of the spark at the solution surface.

In a spark experiment with CH_4, NH_3, and H_2O (Table 4-1), both the total nonvolatile products and the products that remained insoluble after acid treatment were examined. The acid-insoluble products constituted about one-third of the total nonvolatile fraction and yielded N/C values that ranged from 3 to 12 times lower than the values for the total non-volatiles. This observation indicates that most of the non-volatile nitrogen can be converted to soluble products by acid hydrolysis and, therefore, that the more refractory, readily preservable non-volatile material is relatively nitrogen-poor.

Table 4-1 also reveals that the N/C values of the products of this reaction do not vary markedly with the extent of the reaction, with one exception: when over 95% of the methane had been consumed, a very high N/C value was found in the acid-soluble product fraction.

Results of Fischer-Tropsch-type reactions reported in the literature are generally consistent with the spark discharge observations here reported. When CO, H_2, and NH_3 were allowed to react at 573° to 673° K in the presence of iron oxide and aluminum oxide catalysts, a ten-fold depletion in N/C values was observed in the solid products relative to the reactants (see Figure 4-1).

TABLE 4-1. Elemental and carbon isotopic ratios in products formed by spark discharge reaction in CH_4, NH_3, and H_2O (see Appendix III for experimental details).

% CH_4 reacted	*Reaction time (hrs)*	*Total non-volatiles*		*Acid-insoluble non-volatiles*	
		N/C	$\delta^{13}C_{PDB}$*	*N/C*	$\delta^{13}C_{PDB}$*
44 ± 5	6	0.13	−47.4	0.039	−47.0
80 ± 10	20	0.10	−43.2	0.028	−42.7
>95	118	0.48	−39.5	0.039	−36.9

* The carbon isotopic abundances are expressed as $\delta^{13}C_{PDB}$, in permil, which is defined as follows:

$$\delta^{13}C_{PDB} = \left(\frac{R_{Sample} - R_{PDB}}{R_{PDB}}\right) 1000,$$

where R_{Sample} equals the $^{13}C/^{12}C$ value of the sample and R_{PDB} is the $^{13}C/^{12}C$ value in the Pee Dee Belemnite carbonate standard (Craig, 1957). Initial CH_4: −38.6 permil.

Incorporation of nitrogen into products is enhanced in reaction systems in which an appreciable percentage of nitrogen is initially bound to carbon in the reactants. During the polymerization either of amino acid mixtures or of hydrogen cyanide, or during the photodegradation of phenylalanine, the N/C values of the reaction products were very similar to those of the reactants (Figure 4-1). These results suggest that the processes leading to the initial formation of carbon-nitrogen bonds are the limiting reactions principally responsible for the nitrogen depletion observed in the products of the spark discharge and Fischer-Tropsch-type experiments.

For purposes of comparison, the N/C value of organic matter from the Allende meteorite is also plotted on Figure 4-1. If it is assumed that the primitive "reactant" mixture that produced these meteoritic organics indeed contained carbon and nitrogen in their solar system abundance proportions (see Cameron, 1975, for these proportions), then a relative nitrogen depletion of about 100 is observed in the "product" meteoritic organics. Although this depletion is consistent with the magnitudes of nitrogen depletion observed in the abiotic synthesis experiments, it probably reflects the many complex processes, in addition to organic synthesis, that participated in the formation of the meteorite.

To summarize in reaction systems having N_2 or NH_3 as a nitrogen source, carbon is incorporated preferentially relative to nitrogen into the products. Depending upon reaction conditions, N/C values in spark discharge products are highly variable and can exhibit a wide range of ratios (from 0.005 to 0.3, with 0.03 being a typical value). Consequently, *if it is not possible to specify either the initial conditions or the mechanism of an abiotic synthetic process, then one must assume that abiotic organic matter, such as might be preserved in ancient rocks, could have a very wide range of possible N/C values.*

CARBON ISOTOPIC FRACTIONATION

Because the stable isotopes of carbon can react at different rates in kinetically-controlled systems, the isotopic compositions of reaction products can vary markedly, depending upon the mechanisms involved. The isotope fractionation that accompanies various types of abiotic synthesis is reported here to serve as a basis for comparison with isotopic distributions observed in biological systems.

Because bonds containing ^{12}C almost always react more readily than bonds containing ^{13}C, ^{13}C tends to be depleted, relative to ^{12}C, in chemical reaction products. Figure 4-2 illustrates how the carbon isotopic contents of a hypothetical reactant and product can change during the course of a simple reaction where compound A is converted to product B. If the reacting carbon-to-carbon bonds that contain only ^{12}C atoms react 2% faster than do bonds containing a ^{13}C atom, then the first product B that is formed will be approximately 20 parts per thousand (equal to 20 permil, or 20 ‰) depleted in ^{13}C, relative to reactant A. Because ^{13}C reacts more slowly than ^{12}C, the remaining pool of reactant A becomes enriched in ^{13}C as the reaction proceeds. Consequently, molecules of product B that are produced progressively later in the reaction become progressively more enriched in ^{13}C because they are being produced from a relatively ^{13}C-enriched reservoir. Ultimately, when all of A is converted to B, conservation of mass requires that the ^{13}C content of B will be equal to the ^{13}C content of A before the reaction began.

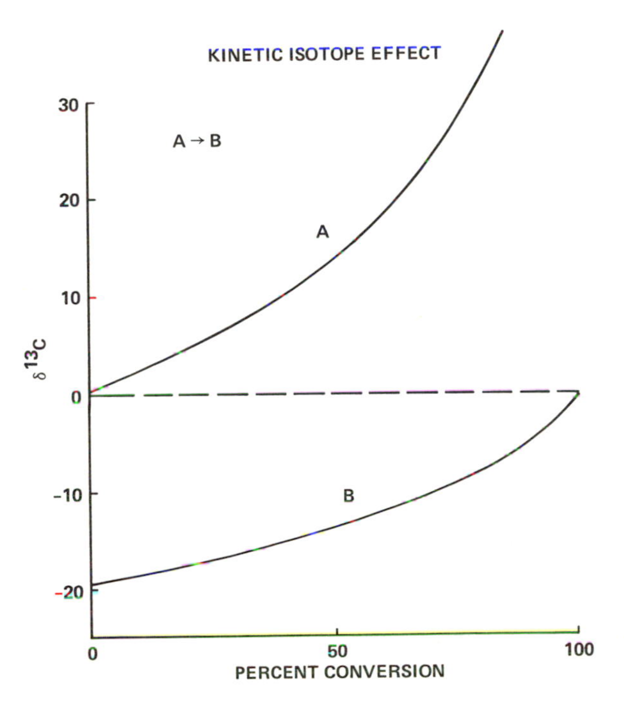

FIGURE 4-2. Carbon isotopic composition of reactant A and product B versus the percentage of conversion of A to B. The curves were obtained using equations described by Melander and Saunders (1980, pp. 95–102), assuming that chemical bonds containing only ^{12}C atoms react two % more rapidly than do bonds containing a ^{13}C atom.

This trend toward increasing ^{13}C enrichment of both reactants and products accompanying the progress of a reaction is a "closed system effect"; this effect is important for interpretation of the experimental results described here. It is possible to obtain a relatively more accurate estimate of the isotopic fractionation associated with the production of a product B by examining its isotopic composition early, rather than late, in the reaction sequence. In the present example, Figure 4-2 shows that an isotope fractionation of approximately −20 ‰ is associated with the conversion of A to B. A useful discussion of this subject is given by Melander and Saunders (1980).

As with the studies of N/C values, most of the isotope work reported here was performed using the spark discharge type of synthesis. In experiments in which methane was the carbon source, the isotopic compositions of individual higher molecular weight hydrocarbons and other products were examined. The carbon isotopic fractionation associated with reactions of carbon sources other than methane was also investigated.

In one experiment, pure gaseous CH_4 was subjected to a spark discharge. The carbon isotopic compositions of individual hydrocarbons in the flask were examined periodically. In these experiments, acetylene (C_2H_2) was the most abundant volatile product. In general, unsaturated and saturated hydrocarbons were of similar abundance. As Figure 4-3 shows, both reactant and products became enriched in ^{13}C as the reaction proceeded, a result consistent with the "closed system effect" explained above. Examination of product $\delta^{13}C$ values early in the reaction reveals that the isotopic fractionation in these reactions ranges from −10 to −25 ‰ and that the degree of fractionation is correlative with the molecular weights of the products formed. It is therefore reasonable to conclude that such spark discharge syntheses are capable of forming less volatile, higher molecular weight hydrocarbons (having more than six carbon atoms) which are depleted by more than 20 ‰ in ^{13}C, relative to the methane carbon source.

Figure 4-4 shows a parallel set of results from reactions involving CH_4, N_2, NH_3, and H_2O. In this case, the carbon isotopic compositions of the nonvolatile reaction products are compared to the carbon isotopic composition of the reactant methane. As above, the ^{13}C content of the products increases with the progress of the reaction (also see Table 4-1), consistent with the "closed system effect," and the $\delta^{13}C$ values obtained for the products early in the reaction suggest that the overall carbon isotopic fractionation is about -10 to -15 ‰. The degree of scatter in these data (Figure 4-4) may be attributable both to experimental uncertainty and to the fact that the reactant mixtures had varying N/C values. For example, the two data points to the lower right of the other points in Figure 4-4 represent experiments where the reactant N/C values were the highest investigated in this work (N/C $\cong$ 12; see Table III-1, Appendix III).

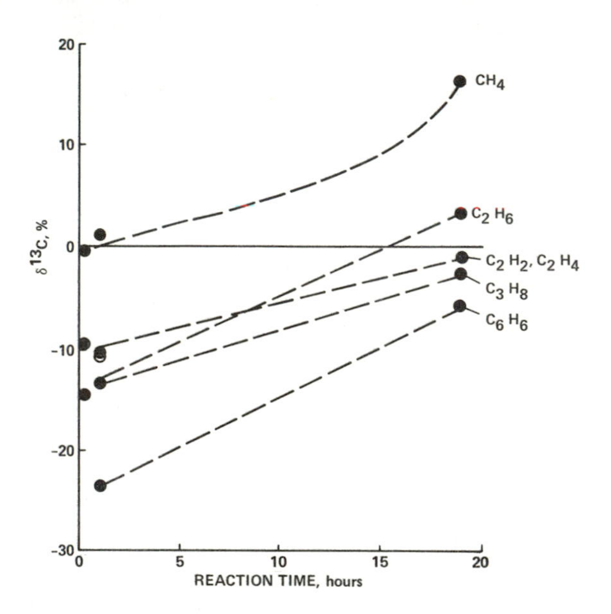

FIGURE 4-3. Carbon isotopic composition of light hydrocarbons generated during spark discharge in methane gas versus time of the reaction. The $\delta^{13}C$ values are expressed relative to a $\delta^{13}C$ value of zero for the methane at the beginning of the reaction. For experimental details and data see Appendix III, Sections III-1-1, III-2-1, and Table III-4.

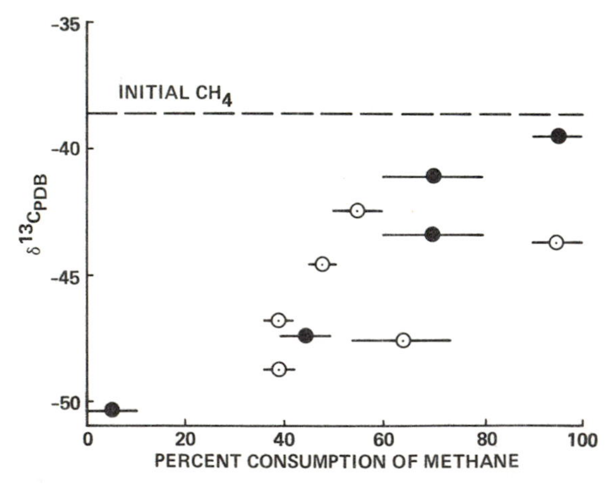

FIGURE 4-4. Carbon isotopic compositions of non-volatile products of spark discharges in CH_4–N_2–H_2O (represented by open circles) and CH_4–NH_3–H_2O (represented by shaded circles). The horizontal error bars reflect the uncertainty in the determination of CH_4 consumed. For details, see Appendix III, Sections III-1-1, III-2-2, and Tables III-1 and III-2.

Another experiment demonstrates both that reduced carbonaceous matter can be obtained abiotically from CO_2 and that this reduced matter can be more than 20 ‰ depleted in ^{13}C. A mixture of carbon dioxide (15%) and hydrogen (85%) was subjected to a spark discharge, and the carbon isotopic compositions of the most abundant product species were examined (see Table 4-2). The principle reaction observed was the conversion of CO_2 and H_2 to CO and H_2O. When the reaction was stopped after 54 hours, the reduction of CO_2 to CO was nearly complete. At that time, traces of hydrocarbons were detected by gas chromatography. Methane and acetylene were the most abundant of these, and constituted approximately 3×10^{-3} and 10^{-5} mole fraction, respectively, of the total carbon present. Presumably the generation of more acetylene with further reaction would have led to the production of non-volatile organic products (Abrahamson, 1977). When only 14% of the reactant CO_2 was converted, the product CO was approximately 11 ‰ depleted in ^{13}C, relative to the starting CO_2. Later in the reaction, when enough acetylene had been generated for isotopic analysis, the acetylene was found to be approximately 23 ‰ depleted in ^{13}C, relative to the starting CO_2.

Figure 4-5 illustrates the isotopic heterogeneity that is found among the various products of a spark discharge synthesis. The compounds CH_4, NH_3, and H_2O were subjected to a spark discharge until half of the methane was consumed,

TABLE 4-2. CO_2–H_2 spark discharge $\delta^{13}C_{PDB}$* values (see Appendix III for experimental details).

%	*$\delta^{13}C_{PDB}$ of Reactant and Products*		
CO_2 consumed	*CO_2*	*CO*	*C_2H_2*
0	−41.5		
14	−39.6	−52.7	
86	−25.8	−44.5	
>95			−64.2

* See Table 4-1 for explanation of $S^{13}C_{PDB}$.

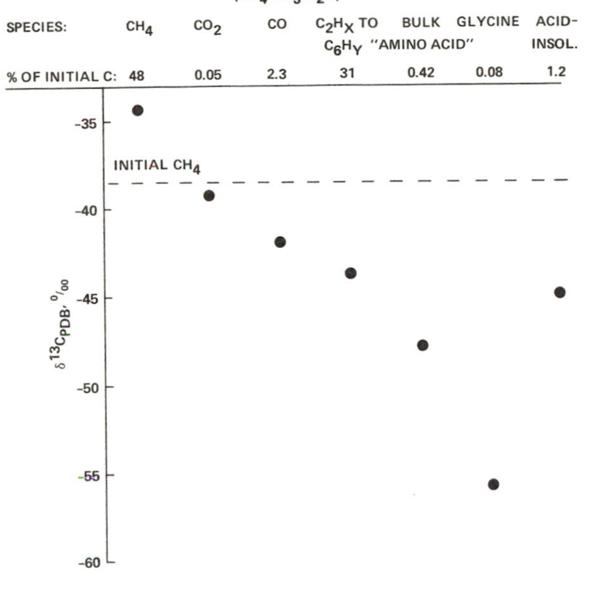

FIGURE 4-5. Carbon isotopic compositions of a variety of products from a single spark discharge experiment (CH_4–NH_3–H_2O). For details, see Appendix III, Section III-1-1, III-2-1, and III-2-2 and Table III-3.

and the products were analyzed. The magnitudes of the ^{13}C depletions in the products, from largest to smallest, were as follows: amino acids, acid-insoluble material, light hydrocarbons, and carbon oxides. The residual CH_4 was enriched, relative to the initial CH_4, due to the "closed system effect." The more complex products, namely those consisting of either four different elements (the amino acids) or higher molecular weight material (the "acid-insolubles") were more depleted in ^{13}C than were the simpler carbon oxides and light hydrocarbons. The more complex molecules are probably products of a longer sequence of chemical reactions; therefore, more reactions were involved wherein ^{12}C reacted preferentially, relative to ^{13}C. Clearly, the severity of the preservation problems for Precambrian organic matter precludes a detailed molecular analysis of this kind for the organics present in ancient sediments. Nevertheless, such observations may ultimately prove useful for the interpretation of organic matter isolated from meteorites.

Data from other modes of abiotic syntheses are generally consistent with the spark discharge results. Figure 4-6 summarizes the isotopic observations from two ultraviolet photolysis experiments

CARBON ISOTOPIC FRACTIONATION IN ABIOTIC SYNTHESES

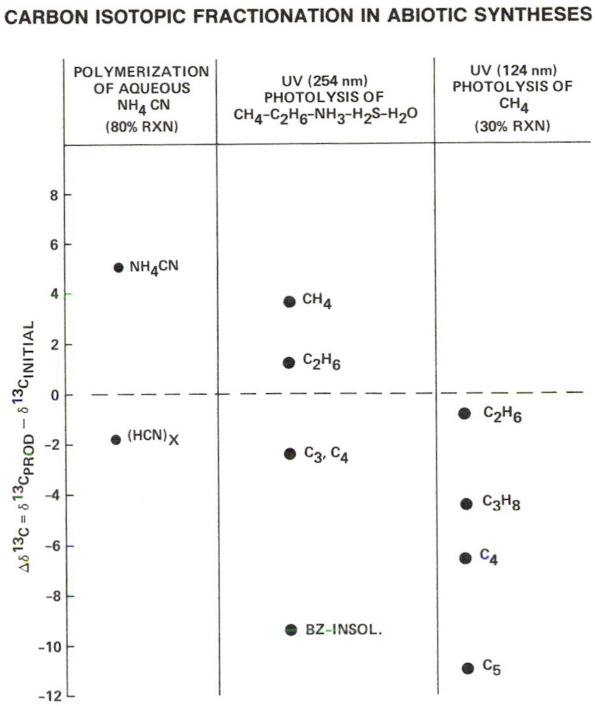

FIGURE 4-6. Carbon isotopic composition of products formed during NH_4CN polymerization and UV photolysis. The $\delta^{13}C$ values are normalized to an assumed initial $\delta^{13}C$ value of zero for the carbon source. In the UV photolysis experiment having CH_4 and C_2H_6 as reactants, the two hydrocarbons had an identical initial $\delta^{13}C$ value. See Appendix III for details concerning the NH_4CN polymerization (Sections III-1-3 and III-2-2 and Table III-3); the photolysis at 254 nm (Sections III-1-2, III-2-1, and III-2-2, and Table III-3); and the photolysis at 124 nm (Sections III-1-2 and III-2-1 and Table III-4).

and from the polymerization of aqueous NH_4CN (see Appendix III for experimental details). The NH_4CN formed an HCN polymer at room temperature that was several permil depleted in ^{13}C. Likewise, the hydrocarbon products in the photolysis experiments are several permil depleted in ^{13}C, and the magnitude of this depletion correlates with the molecular weight of the hydrocarbons produced. In the early stages of Fischer-Tropsch-type syntheses carried out at 400° K, ^{13}C depletions in hydrocarbon products relative to reactant CO have been reported to range from 20 to 60 ‰ (Lancet and Anders, 1970).

Taken together, the significance of these observations is that abiotic reaction mechanisms, such as those which have been studied in prebiotic experiments, do, in fact, commonly discriminate against the incorporation of ^{13}C into their products. The

^{13}C depletions range from a few permil to as high as 60‰. Therefore, *they can be comparable in their magnitude to the degree of* ^{13}C *discrimination observed during the biosynthesis of organic matter via the photosynthetic Calvin cycle* (Degens, 1969; see also Chapters 5 and 7, herein).

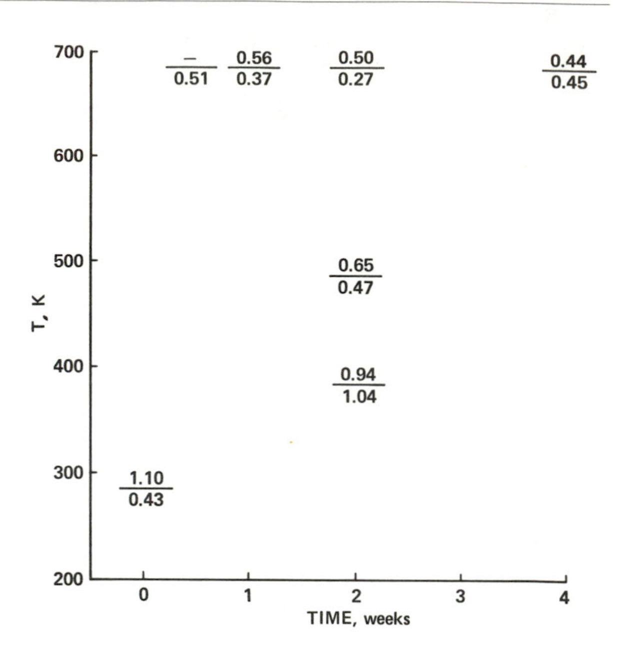

FIGURE 4-7. Plot of H/C values obtained at different times and temperatures for heating of electric discharge and HCN polymerization products. Numbers give H/C values for electric discharge products (numerator) and HCN polymer (denominator). Data are listed in Appendix III, Table III-5.

HEATING EXPERIMENTS

The simulated diagenesis studies consisted of heating the non-volatile products from the electric discharge and HCN polymerization experiments over a range of times and temperatures. Prior to heating, the samples were sealed under vacuum in a glass vessel, a procedure that removed air and most of the water. The time experiments were run at 673° K for 0.5, 1, 2, and 4 weeks. The temperature experiments were run for two weeks at 373° K, 473° K, and 673° K.

Concerning the color of these materials, the HCN polymerization product was initially black, and no visible changes occurred during heating. The electric discharge (ED) material was initially tan in color and became black with increasing time and temperature. In these experiments the higher temperatures appear to have had a more pronounced effect upon the color than did the longer heating times.

The heating also influenced the weight and elemental content of the nonvolatile abiotic products. Solid material in both the ED and HCN polymer experiments was lost with increasing time and temperature (Table III-5, Appendix III). The H/C values of both the ED material and the HCN polymer also decreased with increasing time and temperature, as is illustrated in Figure 4-7. The range of H/C values covered by these materials during heating is identical to the range of values observed during the diagenesis and catagenesis of biogenic kerogen (Tissot and Welte, 1978). Decreasing H/C values are interpreted as reflecting the preferential elimination of hydrogen-rich aliphatic material, relative to the material in which carbon atoms have multiple carbon-to-carbon bonds (Tissot and Welte, 1978). Because *the* H/C *values for biogenic and abiogenic organic matter are so similar, it appears unlikely that these values will be useful in distinguishing between the two types of materials in ancient sediments.*

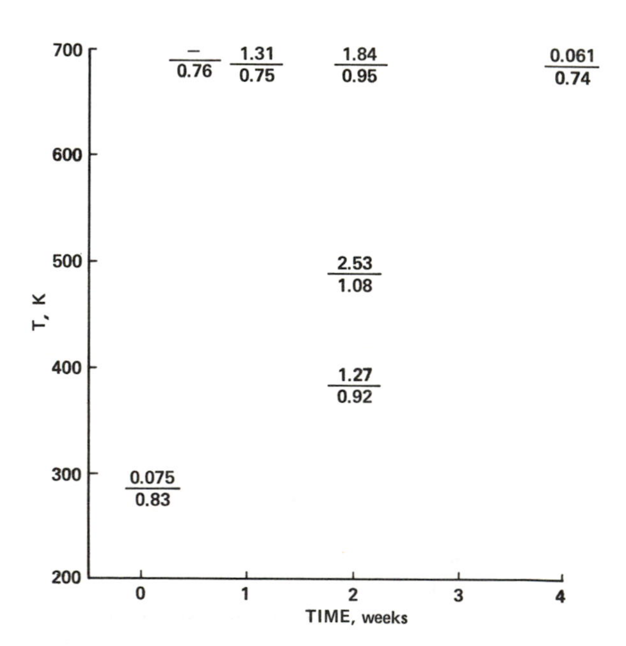

FIGURE 4-8. Plot of N/C values obtained at different times and temperatures for heating of electric discharge and HCN polymerization products. Numbers give N/C values for electric discharge products (numerator) and HCN polymer (denominator). Data are listed in Appendix III, Table III-5.

The measurements of N/C values do not show any clear trends with time or temperature (Figure 4-8). In fact, the results for the heating of the

ED material show considerable scatter, even between replicate measurements (compare, in Table III-5, Appendix III, the two values obtained from heating the material at 373° K for two weeks). Another inconsistency is noted when the N/C values are compared for the unheated electric discharge material and for ED material heated at 373° K for two weeks. The data imply that the N/C value increased from 0.092 to 2.1 during this heating interval, and that 47% of the original material was volatilized. Given a 47% weight loss, the remaining residue would attain its highest N/C value if all the material lost during heating were carbon. But if this had been the case, then it is calculated that material with an initial N/C value of 0.092 would yield, upon heating, a product with an N/C value of only 0.21. Probably, the measured value of 2.1 represents contamination by nitrogen during the analysis. These data are therefore inconclusive.

In contrast, the replicate N/C data from the heating of the HCN polymerization products do appear to be reproducible and valid. These data suggest that the N/C value of this material remained constant with increasing time and temperature. Because this material does sustain a weight loss during heating (Table III-5, Appendix III), these observations indicate that nitrogen is lost at the same rate as carbon. The presence of multiple bonds between the carbon and nitrogen atoms in the HCN polymer material might account for the equal loss rates for carbon and nitrogen by favoring their loss as C–N fragments.

Isotopic measurements of $\delta^{13}C$, $\delta^{15}N$, and δD were made on the heated samples (Table III-5, Appendix III). The $\delta^{13}C$ values varied less than one permil over the entire time and temperature ranges, suggesting that ^{12}C and ^{13}C are lost at an approximately equal rate during heating. This behavior is similar to that observed for coal deposits (Degens, 1969), the carbon isotopic compositions of which do not change appreciably with increasing coal rank. The nitrogen isotope data are sparse, yet they also suggest that no apparent trends occur with increased heating. The δD values show no clear trend with increasing time or temperature, yet there is a weak inverse correlation between H/C values and δD (Table III-5, Appendix III). Taken together, these last two observations suggest that any variation of δD with time or temperature is less than the analytical variability associated with these measurements. The inverse correlation between H/C and δD could be explained by the presence of varying degrees of procedural contamination of the samples by water which was relatively depleted in deuterium.

Both the H/C results and the isotope data from these heating experiments indicate that biologic and abiotic organic matter generally behave similarly during artificial heating experiments, and suggest that such materials would behave similarly during diagenesis in a natural sedimentary environment. The N/C data for the spark discharge heating experiments are inconclusive on this point, perhaps principally because of the questionable quality of the data available. The nitrogen data for the HCN polymerization products do suggest that cyanide polymers might be recognized in ancient sediments because of their notably high N/C values (which are typically 0.7 or higher). Such high values have not been detected in the organic matter of ancient sediments (see Chapter 5), suggesting either that (*i*) these maturation experiments may not be accurate simulations of the maturation of Precambrian organic matter; (*ii*) cyanide polymers may never have been deposited in Precambrian sediments; (*iii*) sediments containing such polymeric material have not yet been found; or that (*iv*) the "cyanide" kerogen in such rocks is masked by larger amounts of non-HCN abiotic polymers and/or biogenic kerogen. *Within the limitations of our experiments, it must be concluded that for bulk samples of kerogen the elemental and isotopic criteria we have examined cannot alone be used to distinguish reliably between ancient terrestrial abiotic organic material and material having a biological origin.* However, isotopic analyses of the individual components of such kerogen may be more fruitful.

The predominant form of organic matter in carbonaceous meteorites ($\geq 75\%$ by weight) is a complex macromolecular phase characterized by its insolubility in solvents and acids (Hayes, 1967). If extraterrestrial organic matter were deposited and preserved in ancient sediments, the similarities in the elemental and carbon isotopic composition of this phase (see Appendix III, Tables III-6 and III-7) with those of terrestrial sedimentary organic matter (Chapters 5 and 7) would make it indistinguishable from the latter using the criteria here studied. The macromolecular phase appears to be the "carrier" of the bulk of the noble gases in carbonaceous meteorites, however (Frick and Chang, 1978; Ott et al., 1980), and its presence in ancient sediments might be detected by its characteristic noble gas signature. Although preliminary analyses of acid-insoluble residues from $\sim$3.2-Ga-old South African cherts gave no clear evidence of a meteoritic contribution (Frick and Chang, 1977), it is conceivable that analyses of older samples could prove more rewarding.

4-4. Accretion of Earth and Formation of Its Core: Modern Models and Implications

4-4-1. Accretion and Early Core Formation

Just as Oparin and Urey placed the origin of the prebiotic atmosphere and of life itself in the context of then current ideas about the formation of the Earth, so it is appropriate now to consider the problem of origins in a more modern framework of such ideas. (For summaries of models and data concerning processes involved in the origin and early development of Earth, the reader is referred to Chapters 2 and 3; Ringwood, 1979; and Smith, 1979.)

Two general types of models for the formation of Earth have been postulated, the *homogeneous accretion* model and the *inhomogeneous accretion* model. For both types of models the collapse and fragmentation of an interstellar cloud of dust and gas to form a solar nebula serves as a prelude. In most homogeneous accretion models, the growth of a planet proceeds by a relatively slow process of accretion, in which planetesimals, formed by the progressive aggregation of dust grains, grow by colliding with and sweeping up smaller bodies in their orbits, ultimately coalescing to form non-gaseous protoplanets. Formation of planets by inhomogeneous accretion may proceed by a planetesimal accumulation mechanism or via giant gaseous protoplanets formed as a result of large-scale gravitational instabilities in the solar nebula; in either case short time scales, $<10^6$ yr, based on model lifetimes for the solar nebula (Cameron, 1978b), are thought to be required for the entire process. Wetherill (1980) has reviewed the assumptions, uncertainties, and complexities of protoplanet and planetesimal accretion mechanisms, placing major emphasis on the latter. Development of the protoplanet mechanism is described by Cameron (1978a, 1978b) and by DeCampli and Cameron (1979).

The major features of the early homogeneous accretion models are: (*i*) Earth accreted as an intimate mixture of metal, silicate particles, and condensed volatiles resembling chondritic meteorites (at the most general level this mixture may be contained within individual planetesimals, or it may represent the average composition of a large number of planetesimals, each containing variable proportions of from 0 to 100% of metal, silicate and volatile components, or both); (*ii*) the chemical and mineralogic composition of this chondritic material was determined by physical-chemical processes operating in the solar nebula prior to accretion into planets; (*iii*) accretion occurred at a slow rate ($\sim 10^8$ yr) and accretional energy was efficiently lost by radiation to space, thus allowing the planet to form cold and unmelted; (*iv*) slow heating by long-lived radioactive elements eventually raised internal temperatures high enough to melt the metal phase, which separated from the less dense silicates and sank to form the core. Evolution of the primordial atmosphere began during the heating process. According to these models, the solar nebula dispersed prior to accretion of Earth, and accretion and core formation were separated by a long time span (up to 1 Ga) when metal and silicate remained in intimate contact. Notable exceptions to these cold accretion models, however, were provided by Ringwood (1966) and Hanks and Anderson (1969). The latter authors developed a model constrained by the requirement that core formation predated the formation of the earliest rocks possessing remanent magnetism. To fulfill this constraint, accretion times of 5×10^5 yr or less were needed, and early core formation within that time was considered a distinct possibility.

In contrast to the homogeneous accretion models, inhomogeneous accretion models (see Ringwood, 1979, for a listing; also Ganapathy and Anders· 1974; Turekian and Clark, 1975) that rely on a planetesimal accumulation mechanism postulate that as condensates formed during cooling of a hot solar nebula, they were rapidly accreted into a growing planet. Thus, as cooling of the nebula proceeded with time so did the composition of the material accreted change with time. The layered state of Earth's core-mantle-atmosphere structure is thought to have resulted directly from the accretion process, with refractory material of high temperature origin concentrated at the outset in the interior of the planet and volatile material at the surface. According to this type of model, accretion began before the nebula dissipated; Earth accreted hot; core formation accompanied accretion; and volatiles were acquired late in the accretion process.

Inhomogeneous accretion by way of a giant, gaseous protoplanet would have occurred in less than 10^6 years. It would have involved melting of solid particles that were circulated into the hot interior by convection, rain-out of the droplets into a molten central mass, followed by loss of the gaseous envelope through tidal stripping by the sun. Volatiles would have been acquired subsequently by accretion of remaining planetesimals.

For our purposes, the homogeneous and inhomogeneous accretion models are distinguished by differences in (*i*) the times required for accretion of Earth; (*ii*) the initial thermal state; (*iii*) the timing of core formation; and (*iv*) the time at which volatiles were accreted. By itself, the core segregation process would have raised the average global temperature of Earth by an estimated 2300° K (Brett, 1976). Since the initial thermal state and the timing and energetics of core formation must have had a profound influence on the prebiotic environment, we now summarize results of some recent investigations relevant to core formation and consider what adjustments of earlier ideas need to be made.

Abundant evidence for lunar differentiation prior to 4.3 Ga years ago (Papanastassiou and Wasserburg, 1975; Wasserburg et al., 1977) argues for early core formation and melting of Earth's surface. Since the larger Earth should have had even more potent heat sources to drive core formation, it is highly likely that the process occurred even earlier on Earth than it did on the Moon. Studies of the isotopic composition of lead in rocks of Archean age (Gancarz and Wasserburg, 1977; Pidgeon, 1978) place the age of Earth at about 4.5 Ga and require core formation either to accompany accretion or to occur within 100 to 200 million years after accretion (also see Vidal and Dosso, 1978). In this time span, heating due to decay of long-lived radioactive elements (K, U, Th) alone cannot provide enough thermal energy to initiate core formation. Decay of the short-lived radioisotope ^{26}Al (Lee, 1979) may have been effective, but its half life (7×10^5 yr) requires that Earth accreted rapidly (in less than 10^6 years) in which case accretional energy would have made a major contribution.

The formation ages of some iron meteorites and basaltic achondrites (4.5 to 4.6 Ga; Wetherill, 1975a) indicate an origin contemporaneous with, or earlier than, Earth. Since partial to complete melting of their parent bodies appears to have been necessary to produce these meteorites (basaltic achondrites: Stolper, 1977; Stolper et al., 1979; irons: Scott, 1979), heat sources that acted effectively on even small bodies (Sonnet and Reynolds, 1979; Mittlefehldt, 1979) must have been available at the earliest stage of solar system history. It would not be unreasonable, therefore, to presume that if Earth formed by accretion of planetesimals some of them were already hot and partially differentiated.

Most models for the formation of Earth from planetesimals yield accretion times of the order of 10^7 to 10^8 yr (see review by Wetherill, 1980), although the possibility of growth of $\sim$4000 km diameter bodies in $\sim 5 \times 10^4$ yr has been suggested recently (Greenberg, 1980). Calculations reported by Hayashi et al. (1979) indicate that if Earth accreted before the solar nebula dissipated, it would have had a primordial atmosphere whose blanketing effect would have reduced the dissipation of accretional energy to space. As a consequence, at 25% of its full size, Earth's surface temperature could have reached 1500° K, sufficiently high to initiate melting and separation of metallic iron. In other model studies the effects of depositing and trapping accretional energy at depth through impacts and burial of large planetesimals from a reasonable size-population have been calculated. The results show that temperatures would rise to the point (> 1500° K) at which melting of iron and core separation would have begun before the Earth was 10% formed (Kaula, 1979, 1980; Safronov, 1978). In any case, the energy released by the gravitational segregation of molten iron and other dense minerals would have augmented the accretional heating of the planet. Steep thermal gradients would have driven vigorous convection within the mantle that would have continued throughout subsequent growth. Furthermore, according to Kaula (1980), the energy "released by differentiation of subsequently infalling material would have gone mainly into the mantle, and would have further contributed to vigorous convection in the mantle." Shaw's (1978) study of the energetics of core formation starting with a fully formed and homogeneously accreted Earth also predicts steep temperature gradients and vigorous convection.

Kaula (1980) pointed out, however, that core formation should have taken place as soon as sufficiently high temperatures were attained on the growing planet. The discussion of physical mechanisms and dynamics of core formation given by Stevenson (1980b, also see Chapter 2) favors a "catastrophic" rather than a gradual process.

Evolution of Earth from a giant gaseous protoplanet (Cameron, 1978a, 1978b; DeCampli and Cameron, 1979) also entails rapid core formation within a very short time after separation of the protoplanet from the solar nebula. By either the gaseous protoplanet or the planetesimal accretion mechanism, Earth would have formed hot, and core formation would have occurred essentially simultaneously with accretion, consistent with the geochronologic constraint imposed by the lead isotope data. This outcome contradicts the common earlier beliefs that homogeneous accretion would have yielded a cold Earth and that core formation would have occurred later and over an extended period of time as a result of radioactive heating. (The reader is referred to Ringwood, 1979, for detailed descriptions of the many geophysical and geochemical constraints imposed on any model for the formation of Earth.)

For our purposes, the important differences remaining between the two types of models are the timing of Earth's acquisition of its component of volatiles and, therefore, its prebiotic atmosphere. According to the homogeneous accretion model, the volatiles were supplied continuously throughout accretion as more or less a constant but small fraction of the infalling material. An inhomogeneously accreted Earth would have acquired its volatiles at a late stage of accretion as a thin veneer. In the following discussion, we examine the implications of each of these models for the range of atmospheric compositions that might have been outgassed initially and involved later in the primary synthesis of organic matter. For each model we consider the factors that would have influenced the evolution of the prebiotic atmosphere, discuss the possible fate of the volatiles in it, and describe relevant theoretical and experimental observations.

4-4-2. Homogeneous Accretion of Earth

A RANGE OF POSSIBLE PREBIOTIC ATMOSPHERIC COMPOSITIONS

In this subsection we delineate the atmospheric compositions that could have been produced within the framework of the homogeneous accretion model. We begin by adopting Kaula's (1980) suggestion for the initial stage of accretion. Accordingly, core formation would have commenced once Earth accreted about 10% of its final mass, and volatiles would have begun outgassing into a primitive atmosphere. Subsequently, impacting planetesimals would (*i*) contribute volatiles to the atmosphere by partially vaporizing during passage through the atmosphere and at impact with the surface; (*ii*) deposit impact energy as heat at depth; (*iii*) contribute to convection in the mantle through impact stirring and release of additional energy associated with their differentiation into metal and silicate fractions; and (*iv*) add volatiles to the upper mantle.

If the upper mantle continued to equilibrate with metallic iron throughout accretion, and vigorous mantle convection provided a mechanism for core-mantle equilibration, the primordial atmosphere generated by outgassing from a hot Earth would have been completely "pre-core" in composition and thus dominated by H_2, H_2O, CH_4, and CO (see below). Brett (1976) and Ringwood (1979) have discussed the evidence for and against core-mantle equilibration, but the issue remains unresolved. Recent measurements of the intrinsic oxygen fugacities of mantle-derived spinels yielded values that were up to four orders of magnitude lower than previous data for basic volcanic rocks (Arculus and Delano, 1980). These results were interpreted by Arculus and Delano (1980) as support for the metal-silicate and core-mantle equilibrium hypothesis. Such equilibration accompanied by release of an atmosphere whose composition was governed by the Fe^0/Fe^{2+} ratio in source magmas remains a distinct possibility.

Once iron sank into the core, however, it might not have been available to control the redox state of the Earth-embryo's upper mantle and, thus, the composition of the evolved gases. Iron in accreting bodies would have continued to differentiate and settle into the core. According to planetesimal accretion theory (e.g., Wetherill, 1976; Greenberg, 1979), the maximum size of the next largest planetesimals in the vicinity of the growing Earth would have been only about 4% of Earth's mass at that time. These bodies would have had a transient existence because of the high probability of their tidal disruption upon near encounters with Earth. Fragments from this disruption would have been swept up. In effect, all accreted planetesimals would have been individually much less than 4% of the mass of the Earth-embryo. Therefore, the mass of metallic iron introduced into the mantle with each impacting body is expected to have been negligibly small compared to the mass of the upper mantle itself; and the redox state of the upper mantle would have been governed primarily by the oxidation states of iron in silicates and oxides.

Under these circumstances, the upper mantle in the growing Earth-embryo would not have attained equilibrium with metallic iron, which is consistent with the occurrence of Fe^{3+} in basalts (e.g., Holland, 1962; Fudali, 1965; Carmichael and Nicholls, 1967; Haggerty, 1978; Bence et al., 1980) and in other mantle-derived rocks (e.g., Harris et al., 1972; Nesbitt, 1972; Frey and Green, 1974; Green et al., 1975; Sobolev, 1977). In this scenario, oxidized iron in the upper mantle could have been introduced as a component of the planetesimals and by the disproportionation of Fe^{2+} to Fe^{3+} and Fe^0 under high pressure in the mantle (Bell and Mao, 1975), with the Fe^0 sinking into the core and the Fe^{3+} remaining in the mantle.

With time the initial pre-core atmosphere released in the presence of metallic iron would have become increasingly diluted by gas of progressively less reducing composition. Such dilution would occur both as a result of the equilibration of previously expelled gas with the hot metallic iron-free surface and the generation of fresh gas from impacting planetesimals. According to this model, the average redox state of the accumulating atmosphere would have become less reducing after core formation had begun, even in the absence of H_2 exospheric escape; and if total outgassing of the atmosphere did not occur during accretion and core formation, the redox state of volatiles outgassed subsequently over geologic time would have been governed by the Fe^{2+}/Fe^{3+} ratio of their source magmas.

In the following paragraphs, we present results of calculations of the compositions of seven model prebiotic atmospheres. In order to place rough compositional limits on these model outgassed atmospheres, it is useful to consider first the possible sources of volatiles in the early solar system. There now seems to be general agreement that Earth's volatiles originated in the planetesimals from which the planet accreted (e.g., Brown, 1949; Fanale, 1971; Anders and Owen, 1977) rather than by direct acquisition from the solar nebula or Sun (see Sekiya et al., 1980, however). Comets and meteorites, and interplanetary dust derived from them remain to bear evidence of the composition of the original planetesimals. Although the age of comets is unknown, the formation ages of most meteorites correspond to the age of the solar system, some pre-dating the age of Earth (Wetherill, 1975a). Some uncertainty exists (Wetherill, 1978; Anders, 1978) regarding the possible derivation of the most volatile-rich meteorites, the carbonaceous meteorites, from comets; but it is highly probable that both types of objects were formed within the same time frame as a result of the collapse of an interstellar cloud and accretion in the solar nebula. Comets and carbonaceous meteorites may represent components of a distribution of planetesimals that, when formed at increasing heliocentric distances, accreted increasing proportions of ice and other volatile-rich phases (Bunch and Chang, 1980). We here focus attention on planetesimals resembling the carbonaceous meteorites and comets because some combination of their volatile contents seems most likely to be able to account for Earth's relative abundances of volatile elements (Anders and Owen, 1977; Bogard and Gibson, 1978; Sill and Wilkening, 1978; Chang, 1979; Ringwood, 1979).

In the discussion that follows, and for the purpose of supplying as highly a reducing prebiotic atmosphere as reasonable (since such an atmosphere would demonstrably have been conducive to prebiotic syntheses), we adopt a maximum value of 100 for the elemental H/C ratio. This value is somewhat larger than the maximum ratio of 80 estimated from the production rates of volatiles released by Comet Kohoutek (Delsemme, 1977). In some calculations we also use a nominal value of 25 for H/C based on estimates of H and C in surface reservoirs that have been involved in geochemical cycles over time. Total abundances for C and N were chosen on the same basis. The elemental abundances used in the calculations are given in Appendix III, Table III-9. Sulfur is not taken into account because its relative abundance falls in the order $C \gg N > S$ (see Walker, 1977, p. 200–201). Qualitatively, the importance of H_2S relative to SO_2 should parallel that of CH_4 relative to CO_2.

To estimate compositions for the model atmospheres we have adopted the approach used by Holland (1962). (Illustrative calculations are provided in Appendix III, Section III-3.) In the calculations, the oxygen fugacity of a magma (equivalent to the oxygen partial pressure for our purposes) plays a key role. Measurements of the oxygen fugacities of a wide sampling of magmatic rocks shows that, to a first approximation and with relatively few exceptions, their trend with temperature (decreasing oxygen fugacity with decreasing temperature) follows or closely parallels that of the synthetic fayalite-magnetite-quartz buffer system over the range of 973° to 1473° K (Carmichael and Nicholls, 1967; Nash and Wilkinson, 1970, Nicholls, 1971; Rosenhauer et al., 1977; Haggerty, 1978). For "typical" values in metallic iron-free systems we therefore use oxygen fugacities taken from this synthetic buffer system (see Buddington and Lindsley, 1964; Haggerty, 1978) rather than from natural samples. For a magma generated in the presence of metallic iron, however, we adopt the oxygen partial pressure originally used by Holland (1962).

For the first five model atmospheres to be considered, we assume that gases were released from magmas at 1473° K. In the first case we assume that the atmosphere was outgassed in equilibrium with metallic iron and that the H/C ratio had the nominal value of 25. We designate this the "*pre-core-nominal-H/C*" atmosphere. In the second case, we assume H/C = 100 and that 10% of Earth's atmosphere was derived by outgassing in equilibrium with metallic iron, silicates, and oxides (pre-core stage). Furthermore, we assume that this 10% did not change in composition by subsequent equilibration with a growing mantle that was continuously depleted in metallic iron (to become less reducing), but simply mixed with the bulk of the atmosphere as it was released (post-core stage). We refer to the resulting gas composition as the "*unequilibrated-maximum-H/C*" atmosphere. For the third case, the "*unequilibrated-nominal-H/C*" atmosphere, we reduce the H/C value to 25. In the fourth case, the "*equilibrated-maximum-H/C*" atmosphere, we take H/C to be 100 and, in effect, allow the pre-core and post-core portions of the atmosphere to equilibrate with metallic iron-free magmas. This is accomplished simply by calculating the equilibrium gas composition compatible with metallic iron-free sources. For the fifth case, the "*equilibrated-nominal-H/C*" atmosphere, we repeat the preceding case with a value of H/C reduced to 25.

Next we calculate two gas compositions at equilibrium with metallic iron-free basalts at 873° K using the maximum H/C value for one and the nominal H/C value for the other. These seven cases represent either atmospheres that outgassed initially at higher temperatures and then came to thermochemical equilibrium with the surface of Earth as it cooled to 873° K or atmospheres that outgassed directly from source magmas at 873° K.

The calculated model compositions are summarized in Table 4-3. All compositions represent reducing atmospheres in which the oxygen partial pressure is less than 10^{-8} atm (see Chapter 11). In all cases H_2 supplies the bulk of the reducing power. However, the composition of the gases varies widely from case to case. As expected, the "pre-core" atmosphere provides the most strongly reduced composition: H_2 is most abundant, CH_4 and CO account for 93% of the carbon, and NH_3 constitutes 13% of the total N. Although an order of magnitude less H_2 and CO occurs in the "equilibrated-maximum-H/C" atmosphere formed at 873° K, relatively high abundances of CH_4 and NH_3 remain. In all other model atmospheres, NH_3 represents 2% or less of the total N, CO_2 supplies 88% or more of the carbon, and CO exceeds CH_4 (except in the "equilibrated-nominal-H/C" atmosphere formed at 873° K). The "equilibrated-nominal-H/C" atmosphere (1473° K) represents the least reducing case. Factors that favor NH_3, CH_4, and CO are high H/C values, equilibration with metallic iron, and outgassing or re-equilibration with magmas at low temperatures. If the primitive atmosphere outgassed at high temperatures in the absence of metallic iron or if it equilibrated with metallic iron-free magmas at high temperatures but failed in either case to re-equilibrate at temperatures below about 973° K, CO_2 would have been the predominant carbon-bearing gas; and CH_4 and NH_3 would have been virtually absent. Because H_2 would have escaped continuously from the Earth's exosphere during and after accretion, the amounts of hydrogen shown in Table 4-3 represent upper limits.

The extent to which the primitive atmosphere would have "equilibrated" at high temperatures can be expected to have depended on the degree to which the atmosphere had been continuously exposed to, and remixed into the hot upper mantle. This equilibration should have been related to the rate at which mantle convection occurred and the general problem of Earth's early thermal evolution. No certain conclusions are warranted as yet regarding the extent of equilibration, but some inferences may be drawn.

TABLE 4-3. Compositions of model prebiotic atmospheres outgassed at 1473° K and 873° K.

Atmosphere type (release temperature)	Pressure (atm)						
	H_2O	H_2	CO_2	CO	CH_4	N_2	NH_3
Pre-core- H/C = 25 (1473° K)	137	319	3	15.0	22.0	1.3	0.4
Unequilibrated- H/C = 100 (1473° K)	1823	174	35	3.0	1.5	1.5	0.06
Unequilibrated- H/C = 25 (1473° K)	456	44	35	4.0	0.1	1.5	0.006
Equilibrated- H/C = 100 (1473° K)	1960	40	38	2.0	$\leq 2 \times 10^{-5}$	1.5	0.02
Equilibrated- H/C = 25 (1473° K)	490	10	38	2.0	$\leq 1 \times 10^{-6}$	1.5	0.003
Equilibrated- H/C = 100 (875° K)	1938	38	28	0.2	12.0	1.3	0.4
Equilibrated- H/C = 25 (875° K)	488	10	39	0.3	0.9	1.5	0.03

During the first 10^8 years, when most of Earth accreted, the high planetesimal impact rate coupled with vigorous upper mantle convection should have prevented the establishment of any significant large-scale lithosphere. According to Stevenson's estimates (Chapter 2), the mantle could have turned over convectively more than 250 times during the first 2.5×10^8 years of its evolution. A much higher level of radioactive heating in the upper mantle then than now, caused by the upward migration and concentration of U, Th, and K, could have prolonged the period of extensive convection for the next several 10^8 years. During the latter period the impact rate would have diminished drastically; it should have, nonetheless, assisted vigorous mantle convection in continuing the fragmentation of a deformable lithosphere and in promoting its recycling into the upper mantle (Solomon, 1980). On the Moon, planetesimal bombardment during its first 5×10^8 years has left virtually no remnant of the protocrust that must have formed when the planet underwent major differentiation at 4.45 Ga (Wasserburg et al., 1977). (With few exceptions, crystallization ages of lunar rock are younger than 4.0 Ga; see Wasserburg et al., 1977. The gap between these crystallization ages and the age of the Moon may be attributed to the obliteration of the geochronologic record by a late period of intense bombardment by planetesimals that occurred at about 4.0 Ga and terminated at about 3.9 Ga; see Tera et al., 1974. It remains unresolved whether the record of this bombardment represents a late "spike" in the accretion process or simply the waning stages of the accretion process, the record of which is preserved for lack of later bombardment, as noted by Wetherill, 1975b, 1977. Bombardment of the Earth would have been about 20 times more intense (Ganapathy et al., 1970). The two processes, planetesimal infall and mantle convection, could have delayed emplacement of a largely continuous, solid lithosphere, thick enough to support material differentiated from the mantle and prevent its rapid recycling, until possibly as late as 4 Ga ago. Impact bombardment has even been suggested to have triggered the formation of differentiated crust (Grieve, 1980).

Geologic data have been interpreted to indicate that early Archean tectonic activity reflected steeper geothermal gradients then than now (Lambert, 1976; chap. 3). Green et al. (1975) have suggested that Archean magmas reached the surface at about 1923° K, as compared to present-day temperatures several hundreds of degrees lower. And Archean rocks appear to have originated largely from the upper mantle rather than by recycling of pre-existing crust (Gunn, 1976; Lambert et al., 1976; Moorbath, 1976). These observations for the Archean support the ideas that the effects of extensive early heating of Earth could have lasted well into the Archean (Solomon, 1980) and that Earth was even more vigorously convective prior to 3.9 Ga (see also Chapter 3).

The hypothesis of an "equilibrated" atmosphere has its vulnerable points, however. Vigorous mantle convection is supposed to have promoted equilibration between the hot silicate-oxide upper mantle and the outgassing atmosphere without simultaneously causing overall core-mantle equilibration. Whether or not such constrained equilibration is reasonable may be clarified by further development of mantle convection models (see Chapter 2 and Stevenson and Turner, 1979).

Production of an "equilibrated" atmosphere also implies that the composition of the major elements and the bulk-averaged, upper mantle magma source regions would have varied little after 10% accretion of Earth. Chemical heterogeneity occurs in the mantle (see Papike and Bence, 1979) and heterogeneities in redox state have been reported (e.g., Arculus and Delano, 1980). Evidence that overall constancy in redox state has been preserved is not clear-out. A stable oxygenic atmosphere appears to have arisen about 2 Ga ago (see Chapter 11). Prior to the earliest firm evidence of life at 3.5 Ga (see Chapters 5, 7, 8, and 9), the Earth's crust was presumably more reducing than now. Over geologic time, exospheric escape of hydrogen would have increased the oxidation state of the crust and, therefore, that of the mantle, as a result of subduction of the crust. The present crust remains relatively reduced according to the Fe^{2+}/Fe^{3+} ratio of its components (Ronov and Yaroshevsky, 1969), and it amounts to only 4% of the upper mantle mass. How much recycling of crust into mantle has occurred over geologic time is not well known. It would be interesting to estimate how much recycling of present-day crust would have been required to yield the ratios of Fe^{2+}/Fe^{3+} observed in rocks derived from the upper mantle assuming that such recycling of surface material was the only source of Fe^{3+}.

The ratios of Fe^{2+}/Fe^{3+} in rocks derived from the upper mantle in the Archean (e.g., Harris et al., 1972; Nesbitt, 1972; Green et al., 1975) fall in essentially the same range as those of more recent volcanic rocks (e.g., Holland, 1962; Fudali, 1965; Frey and Green, 1974; Haggerty, 1978; Bence et al., 1980). Although these data are more or less consistent with constancy in the range of Fe^{2+}/Fe^{3+} over time, it is not clear how much of the Fe^{3+} in early Archean samples was produced at a later time as a result of exposure to a more oxidized surface environment (e.g., see Chapter 11). Furthermore, Sato (1978) has suggested that the redox state of hydrous basaltic magmas may be governed by the diffusive loss of H_2, formed as a result of water reduction via the reaction $2\,FeO + H_2O \rightarrow Fe_2O_3 + H_2$, or by thermal dissociation of H_2O. In this case the occurrence of Fe^{3+} may be a secondary result of the outgassing process rather than a primary determinant of the redox state of magmatic gases. On the other hand, the reverse reaction coupled with others proposed by Sato (1978) (e.g., $2\,Fe_2O_3 + C \rightarrow 4\,FeO + CO_2$ and $4\,Fe_2O_3 + 2\,FeS \rightarrow 10\,FeO + SO_2$) are capable of reducing magmas while at the same time accounting for the gases exhaled during volcanism today. In the context of Sato's (1978) model, magma source regions could have become more reducing as soon as H_2O, CO, CO_2, and SO_2 (rather than H_2, CH_4, and H_2S) became the dominant gases released from Earth's interior. On noting the parallel variation of oxygen fugacity with temperature exhibited by natural systems and by the synthetic fayalite-magnetite-quartz buffer system, Nash and Wilkinson (1970) suggested "that a given mineralogical assemblage crystalizing from a magma is not subject to the whims of an arbitrary gas phase, but rather that the magma is 'internally buffered.' "

Clearly, uncertainties exist about mechanisms of mantle convection, the degree of heterogeneity of redox states in the upper mantle, the extent of recycling of crust into the mantle, and the dominant factors governing the composition of magmatic gases over time. Within the context of these uncertainties, however, Holland's (1962) suggestion that the redox state of source regions for basalts and, therefore, their associated magmatic gases has not changed significantly throughout geologic time remains viable. Although Holland (1962) surmised that the redox state was fixed sometime within 5×10^8 years after Earth accreted, the homogeneous accretion scenario implies that it could have been fixed even before accretion ended.

CATASTROPHIC VERSUS GRADUAL RELEASE OF VOLATILES

The gradual and steady evolution of the atmosphere implicated by Rubey's (1951, 1955) interpretation of excess volatiles appears to be inconsistent with the concept of an Earth that accreted hot and differentiated early. In trying to reconcile Rubey's conclusions with his own statement of a case for catastrophic early degassing, Fanale (1971) suggested that "equilibrium solubility between extensive surface silicate magmas and a massive early atmosphere played an important role in limiting the efficiency with which the major volatiles were outgassed from the interior of the Earth, and that this suggests the early existence of an extensively molten upper mantle and crust." According to this suggestion, resorption and storage of substantial amounts of H_2O and CO_2 in the mantle would have occurred early, and the gases remaining in the atmosphere would have been supplemented by subsequent release over geologic time as crustal differentiation proceeded. This scenario is consistent with present-day volcanism and the considerable evidence for the occurrence of CO_2 and carbonate in the upper mantle (summarized by Irving and Wyllie, 1973, 1975). Other lines of evidence that support Fanale's (1971) suggestion include determinations of the solubilities under mantle conditions of H_2O and CO_2 in mineral assemblages that represent mantle rocks (Mysen, 1977) and petrogenetic models of the mantle that take into account the presence of H_2O and CO_2 (e.g., Boettcher et al., 1975; Mysen, 1977; Wyllie, 1977, 1978).

To varying degrees, models advocating early and sudden degassing of the Earth's atmosphere are based on estimates of the total inventory of primordial noble gas abundances (Fanale, 1971) and on measurements of the isotopic ratios of radiogenic to primordial noble gases in mantle rocks (e.g., Fanale, 1971; Hamano and Ozima, 1978; Thomsen, 1980). In a review of these approaches, Bernatowicz and Podosek (1978) came to the conclusion that the present status of these components could give no definitive answer to the question of whether Earth had outgassed only partially or nearly completely early in its history. They suggested, however, that all the available noble gas evidence is consistent with an early and rapid, but low degree of outgassing from the interior followed by continuing evolution of volatiles over time.

Despite the uncertainties of models, several facts are obvious. The geologic evidence of metamorphosed rocks of sedimentary origin 3.8 Ga in age (Chapter 3) indicates the existence of bodies of water and, therefore, outgassing of an atmosphere prior to that time. Nearly complete outgassing of Earth's interior cannot have occurred early because primordial noble gases (and other volatiles, Section 4-4-2) are still being released today (e.g., Jenkins et al., 1978) and have been found in basalts and other mantle-derived material (e.g., Hennecke and Manuel, 1975; Craig and Lupton, 1976; Kaneoka et al., 1977; Saito et al., 1978; Kaneoka and Takaoka, 1980). It is likely that a significant fraction of Earth's atmosphere existed in the prebiotic epoch, but how large that fraction was remains unclear. Furthermore, as pointed out by Fanale (1971), and more recently by Thomsen (1980), the atmosphere may have largely outgassed during accretion, but Earth itself may still retain the bulk of its volatiles. The rate of outgassing of volatiles since the Hadean and Early Archean probably paralleled the rate of crustal development (see Chapters 3 and 10).

To gain some appreciation of the possible range of atmospheric compositions while taking into account in a crude fashion a high (80%) or low (20%) degree of initial outgassing for the atmosphere, we here consider two extreme redox states and an intermediate one. For these compositions to be relevant for the moderate to low temperatures suitable for the chemical evolution epoch (see subsection below), we assume water and carbon dioxide to have formed seas and carbonates, respectively, and that the partial pressures of these gases were comparable to their present levels. No other atmospheric processes that could have affected the remaining gases are taken into account. Under these conditions, CH_4, CO, N_2, and H_2 would have been the dominant gases in the atmosphere. We also assume that the portion of the atmosphere not outgassed initially would have accumulated at a steady rate over geologic time since 4.5 Ga ago. Within these bounds, the most highly reducing "initial" composition is represented by 80% outgassing of the precore atmosphere (Table 4-3): $P_{H_2} = 255$ atm, $P_{CO} = 12$ atm, $P_{CH_4} = 18$ atm, $P_{N_2} = 1$ atm. The least reducing "initial" composition is provided by 20% outgassing (10% pre-core, 10% post-core) of an "equilibrated-nominal-H/C" atmosphere: $P_{H_2} = 2$ atm, $P_{CO} = 0.4$ atm, $P_{CH_4} \leq 10^{-6}$ atm, $P_{N_2} = 0.3$ atm. An intermediate case is represented by a 20%-outgassed nominal-H/C "unequilibrated" atmosphere: $P_{H_2} = 20$ atm, $P_{CO} = 3.7$ atm, $P_{CH_4} = 0.9$ atm, $P_{N_2} = 0.6$ atm.

EVOLUTION OF THE PREBIOTIC ATMOSPHERE

Whether or not the prebiotic Earth was endowed with an "equilibrated" atmosphere is unknown. It is reasonable to presume, however, that formation of a continuous, hard and cool lithosphere would have inhibited equilibration of the atmosphere with the hot upper mantle. Further development of models for the Earth's early thermal evolution is needed (e.g., Davies, 1980, and Schubert et al., 1980, for recent studies) to place bounds on the rates at which changes in surface and atmospheric temperatures occurred with time. These studies could constrain the period of time available for atmospheric equilibration; they could also yield insight into the length of time required for the atmosphere and surface to cool to temperatures near 373° K were they initially much hotter. We refer to this time as the "pre-chemical evolution" epoch because surface temperature in excess of 373° K is incompatible with the synthesis and preservation of thermally labile organic matter that could participate in subsequent stages of chemical evolution. In some accretion models (Jakosky and Ahrens, 1979), the average surface temperature never exceeds 320° K. If survival of heat-sensitive organic compounds (e.g., amino acids and sugars) over geologic time in a particular locale was required for the early stages of chemical evolution, average local temperatures as low as even 298° K would have been too hot. In general, cool seas with temperatures nearer the freezing point than the boiling point of water may have favored the origin of life (Miller and Orgel, 1974).

Establishment of limits for the pre-chemical evolution time frame would permit an estimate of how much time was available for subsequent chemical evolution, that in turn might provide constraints on the number of plausible chemical pathways leading to the origin of life. Present constraints on the chemical evolution period allow up to a billion years (the time between the origin of the Earth and the earliest firm evidence of life at 3.5 Ga ago).

It is not clear whether a "one-shot" synthesis of organic matter would have been conducive to the origin of life or whether a more or less steady production rate would have been required. If the origin of life required a long period of chemical evolution, perhaps 200 to 800 million years, as seems generally believed, it seems necessary that reservoirs of the required elements H, C, N (and other trace elements) had to have been available over the same time. Therefore, atmospheric processes such as photochemistry and electrical discharges, though capable of producing the organic compounds necessary for chemical evolution, should not have depleted the atmospheric reservoir over the time required for chemical evolution. On the other hand, a recent proposal on the origin of life (White, 1980) suggests the possibility that life emerged in a short period of chemical evolution. (We estimate $\sim 10^6$ years.) The uncertainties associated with the time required for the origin of life underscore the importance of understanding the boundary conditions on the origin and evolution of the atmosphere and its composition.

Factors that could have decreased the reducing power of atmospheres during the pre-chemical evolution period include the photodecomposition of CO and CH_4 and the exospheric escape of H_2. The effect these two processes would have exerted on diminishing the reducing capacity of the prebiotic atmosphere up to the onset of chemical evolution would have depended in large part on the duration of the pre-chemical evolution period, which in turn would have been governed by Earth's thermal evolution. On the basis of work reported by Lasaga et al. (1971), Strobel (1973), and Yung and Pinto (1978), Walker (1980) has pointed out that in a hydrogen-rich atmosphere, photodecomposition of CH_4 to less volatile hydrocarbons could have occurred in the very short time (by geologic standards) of 10^6 to 10^7 years; in a hydrogen-poor atmosphere, oxidation to CO_2 and H_2O could have occurred even faster (Walker, 1977; Logan et al., 1978). Recent calculations of exospheric temperatures in a strongly mixed CH_4–H_2 atmosphere led Shimizu (1976) to suggest that the atmospheric decay time due to gravitational escape of H_2 would have been less than 10^6 years. This is much shorter than earlier estimates by McGovern (1969). If NH_3 existed in the atmosphere during the pre-chemical evolution period at a mixing ratio with other gases of $\leq 10^{-4}$, its irreversible photochemical conversion to N_2 could have taken place in a time as short as 40 years (Kuhn and Atreya, 1979). Although no calculations have been carried out for H_2S, similar minimum lifetimes against photodestruction would not be surprising. Continuous production mechanisms for CH_4 and NH_3 may have been available, however (see below). In the context of the model of homogeneous accretion with simultaneous core formation, outgassing of NH_3, CH_4, and H_2S at $\sim 1500°$ K over geological time is small, and the participation of these gases in chemical evolution would have depended on some fraction of their initially outgassed inventories having survived the above processes during the prechemical evolution period. On the other hand, outgassing at lower temperatures may have provided a continuous supply of these gases.

In addition to photochemical processes and H_2-exospheric escape, the composition of the initially outgassed atmosphere could also have changed by re-equilibration at lower temperatures as the surface and the atmosphere cooled over time. Thermodynamic re-equilibration in the presence of sufficient hydrogen would permit partial conversion of CO_2 and CO to CH_4, and of N_2 to NH_3. Whether there were kinetically feasible pathways by which the atmospheric re-equilibration process could have taken place down to temperatures below 973° K, where the highly reduced forms make up significant fractions of the gas phase (see Table 4-3), is not clear. Sackett and Chung (1979) found no re-equilibration between CH_4 and CO_2 after heating at 773° K in the presence of water and clay catalyst for 10.5 days. This is hardly a test over geologic time, yet their results indicate that strictly thermal processes may have had very slow reaction rates. (Mathematical model studies of the ascent and emplacement of basaltic magmas on the Moon and Earth suggest that 10 days exceeds the transport time for magmas erupted at depths of 0.5 to 20 km to reach the surface; see Wilson and Head, 1980). Photochemical pathways are unlikely because above the critical temperature of water, 646° K, the dominant constituent of the atmospheres listed in Table 4-3 would have been H_2O, whose primary photodissociation yields HO· radical and H atoms. On the average, exospheric escape of H atoms would have left behind the HO· radical that would have acted as an efficient oxidizing agent to convert CH_4 and CO to CO_2, and NH_3 to N_2 (Vander Wood and Thiemens, 1980). The rates of conversion of CO_2 and CO to CH_4 and of N_2 to NH_3 would have had to compete with the rate of H_2 exospheric escape and the combined rates of destruction of reduced species caused by direct photodissociation and oxidation by hydroxyl radicals. An additional constraint on the re-equilibration process would have been the amount of H_2 available. Conversion of CO_2 to CH_4 requires an H_2/CO_2 ratio at least as high as four. In Table 4-3, only for the "unequilibrated-maximum-H/C" and "pre-core" model atmospheres does the ratio exceed four (H_2/CO_2 = 105 and 106, respectively).

A recent study of kinetic models for the low temperature re-equilibration of N_2 and H_2 to NH_3 catalyzed by grain surfaces in the solar nebula was reported by Norris (1980). His calculations indicate that the reaction would have been kinetically inhibited at 400° K and that NH_3 would not have constituted more than 3% of the total nitrogen in the system. If the extrapolations used by Norris in his kinetic model are valid, his results cast doubt on the usefulness of models of the solar nebula that invoke abundant NH_3 and, thus, of models of the primitive Earth atmosphere that rely on a nebular source of NH_3 for its nitrogen endowment. The most important factor in the inhibition was the low hydrogen pressure ($\leq 10^{-3}$ atm). Model kinetic calculations by Lewis and Prinn (1980) led to similar conclusions and suggested, additionally, that the analogous conversion of CO to CH_4 could have been comparably kinetically inhibited in the primitive solar nebula. Although the reports of Norris (1980) and Lewis and Prinn (1980) do not strictly apply to processes on the prebiotic Earth, they serve as a cautionary note that low temperature re-equilibration processes may not have occurred in that environment, too. Facile low-temperature re-equilibration of a prebiotic atmosphere had been assumed in the past (Urey, 1952a; Miller and Urey, 1959; Holland, 1962; Horowitz and Miller, 1962). In light of possible kinetic constraints and uncertainties regarding Earth's thermal evolution, this assumption cannot be accepted at face value, but should now be critically evaluated by experimental and modeling studies (geologic evidence relevant to low temperature re-equilibration process is discussed in subsection below). Indeed, all of the atmospheric processes described above are uncertain in that none of them has been studied in atmospheric models spanning a range of allowable prechemical evolution thermal states. Clearly many opportunities for research and new insight exist in this context (see Hart, 1978). Comparative studies of the origin and evolution of the atmospheres of Venus, Earth, and Mars would be especially enlightening (e.g., Pollack and Yung, 1980; Shimizu, 1979).

The preceding discussion should point out the perils of specifying with any comfortable degree of certainty the composition of the prebiotic atmosphere at the beginning of the chemical evolution epoch. If the chemical evolution of life began when liquid water appeared on the primordial Earth, then weathering of igneous rocks would already have begun. Dissolution of CO_2 in the seas and its precipitation as carbonate would have ensued. The prevailing uncertainties permit a wide range of compositions in the residual atmosphere from ones that were highly reducing, containing variable but very high partial pressures of H_2 and CH_4 along with N_2, to an atmosphere composed almost entirely of N_2 and H_2 with minor to trace levels of CO and CO_2. Attainment of steady state abundances early in the chemical evolution period as a result of balances between volcanic emissions and the exospheric escape of H_2 (Walker, 1978a; Kasting et al., 1979), the production of O_2 as a result of water photodissociation (Kasting et al., 1979) and the conversion of CO to formate in the oceans (Van Trump and Miller, 1973) would have permitted compositions as mildly reducing as the following: $N_2 \simeq 0.3$ atm; $H_2O \simeq 1.5 \times 10^{-2}$ atm; $CO_2 \simeq 3 \times 10^{-4}$ atm; $H_2 \simeq 4 \times 10^{-6}$ to 2×10^{-2} atm; and $CO \simeq 10^{-6}$ to 7×10^{-3} atm. This last composition would have resembled that of the present atmosphere except for its lack of O_2 and its resultant less oxidizing effect on reduced gases.

SOURCES OF CARBON IN EARTH'S INTERIOR

In this subsection we consider the forms in which the key element of life, carbon, occurs in, and is ejected from, Earth's interior. The many observations of carbonate minerals and CO_2 in mantle-derived material have been summarized by Irving and Wyllie (1973, 1975); the origin of diamonds (crystalline carbon) in the upper mantle and the conditions under which they may have formed have been discussed by Deines (1980); and graphite, a polymorph of crystalline carbon, has a stability field consistent with upper mantle conditions (Kennedy and Kennedy, 1976; Deines, 1980) and may occur in highly dispersed form in igneous rocks (Lebedev, 1957; Hoefs, 1965). Atomic carbon may also be dissolved in oxide phases in the mantle (Freund et al., 1980). Apparently, considerable evidence points to the abundant occurrence of carbon in the 4+ and zero valence states in Earth's interior. Thus, the occurrence of CO_2 and elemental carbon in igneous rocks (Craig, 1953; Hoefs, 1965; Pineau et al., 1976; Moore et al., 1977) extruded onto the surface of Earth can be correlated with internal, deep-seated (and presumably primitive) sources. The CO_2 observed in volcanic emissions (Nordlie, 1971; Anderson, 1975; Allard et al., 1979) can be attributed to its exsolution from magmas during ascent, while the CO observed in such emissions is a consequence of high temperature equilibration between CO_2 and its source magma. Some magmatic CO_2 may be produced by decomposition of sedimentary carbonates (e.g., Allard, 1979) as implied by Brett (1976) and Arculus and Delano (1980). However, the carbon isotopic composition of CO_2 in basalts (Pineau et al., 1976; Moore et al., 1977) and in hydrothermal fluids (Craig et al., 1980) associated with deep ocean vents, where most outgassing occurs today, accords with values presumed to be characteristic of primary mantle origin (see also Irwin and Barnes, 1980).

Evidence for upper mantle source regions for the reduced forms of carbon, CO and CH_4, appears to be more indirect and ambiguous. Although there is some uncertainty regarding the existence of a vapor phase in the upper mantle (Hollaway and Eggler, 1976), there seems to be a range of conditions under which the existence and storage of CH_4 and, to a lesser extent, CO are consistent with thermodynamic considerations. Deines (1980) has discussed the variations in the $^{13}C/^{12}C$ ratios of diamonds in terms of the thermodynamic constraints imposed on their formation in the upper mantle (pressures between 30 and 60 kb). He noted that if carbon-bearing gas phases participated in the formation of diamonds, the gases would be predominantly CO_2 and CH_4. Although the inferred participation of CH_4 is consistent with the carbon isotopic composition of some diamonds and the release both of CO_2 and CH_4 when some diamonds are crushed (Melton et al., 1972; Melton and Giardini, 1976), Deines (1980) concluded that an explanation of all the isotopic variations in diamonds may require isotopic heterogeneities in the mantle, or the recycling of reduced organic carbon from the surface, or both. Thus, while the presence of juvenile CH_4 in the mantle is not excluded, recycled organic carbon may account for the observed isotopic variations. Earlier, French (1966) reported thermodynamic constraints on the composition of carbon-bearing gases in the presence of graphite and accessory minerals like magnetite ($Fe_3O_4 \sim FeO\text{-}Fe_2O_3$) under pressure conditions corresponding to shallower depths in the mantle than those considered by Deines (1980). French's calculations showed that equilibrium temperatures above 873° K were sufficient to convert CH_4 to CO_2, H_2O, and H_2 by way of graphite as an intermediate. According to these calculations, it is likely that any primordial CH_4 transported from storage deeper in the Earth (Deines, 1980) would have been converted to graphite, CO_2, H_2, and H_2O when it reached the surface, provided that in near-surface regions re-equilibration to CH_4 at temperature lower than 873° K did not occur.

Nevertheless, CH_4 (and other hydrocarbons) is emitted from geothermal wells and hot springs along with CO_2. Typically the amounts of CH_4 are a small fraction of the CO_2. The sources of this CH_4 may be (*i*) recycled dissolved or sedimentary organic matter; (*ii*) the re-equilibration reaction that favors CH_4 at low temperatures (viz., $CO_2 + 4\,H_2 \rightleftarrows CH_4 + 2\,H_2O$); (*iii*) a primitive mantle source; or (*iv*) combinations of these.

Hulston and McCabe (1962) suggested sources (*i*) and (*ii*) for the CH_4 observed in thermal areas of New Zealand. These workers calculated equilibration temperatures of 523° K to 713° K at depth on the basis of carbon isotopic compositions of coexisting CH_4 and CO_2. In contrast, Gunter and Musgrave (1971) concluded from studies of hydrothermal gases collected at Yellowstone and Lassen National Parks that source (*i*) was responsible for CH_4 (and other observed hydrocarbons) and that the conclusions about equilibration temperatures based on carbon isotopic considerations were invalid. In other studies (e.g., Baskov et al., 1973), source (*i*) was implicated by H- and O-isotopic evidence indicating that much of the associated water had a surface origin. For hydrocarbons of geologic origin, increases in $^{13}C/^{12}C$ ratio with increasing carbon number are indicative of a sedimentary source (Feux, 1977). Where recent data are available from geothermal wells and hot springs in California, the carbon isotopic composition of the C_1 and C_4 hydrocarbons fits the pattern for a sedimentary origin (Des Marais et al., 1980).

Recent observations of CH_4 in hydrothermal fluids associated with deep ocean vents (Welhan and Craig, 1979) may also be attributed to sources (*i*), (*ii*), and/or (*iii*). At this point there are insufficient data to establish the relative importance of these three sources (see Welhan, 1980). The amount of outgassing occurring at the deep sea vents may be as much as 100 times more than that released from land sources. Determination of the source of this CH_4 will critically influence thinking about the composition of the primitive atmosphere. Resolution of this uncertainty will be of major significance.

Russian researchers have also reported many instances in which CH_4 and other hydrocarbons (and H_2 and CO in small amounts) have been extracted from pore spaces and gaseous inclusions in igneous rocks from the Kola Peninsula (reviewed by Galimov, 1973; also Petersilje and Pripachkin, 1977). In some instances the pattern of $^{13}C/^{12}C$ ratios in the hydrocarbons may not be attributable to a sedimentary carbon source, and coexisting CO_2 is either absent or present only in trace amounts. Although the Russian workers found a correlation between the abundances of these hydrocarbons and the alumina content of the host rocks, it is not clear how this correlation may be related to the origin of the gases. Data on the FeO/Fe_2O_3 ratio in the host rocks could shed light on whether the presence of hydrocarbon gases is consistent with the redox state of the host rocks. While available geologic data do not exclude the direct emission of primordial CH_4 from upper mantle source regions now or in prebiotic times, neither do they provide clear support. Additional work on the gases in igneous rocks obtained from the Kola Peninsula seems highly desirable to clarify the origin of the observed hydrocarbons. Along the same lines, in such gases CO has been found to exceed the abundance of CO_2 (White and Waring, 1963; Galimov, 1973); and HCN (Mukhin, 1974) and various types of organic compounds (Markhinin and Podkletnov, 1977; Mukhin, 1976) have been reported in the volcanic gases and dust released from the volcanos in Kamchatka. The fact that organic matter of biological origin is ubiquitous at the surface of Earth and in buried sediments requires that special care be exercised in eliminating such secondary sources of reduced nitrogen and carbon before a primary magmatic source (either directly emitted or resulting from low temperature re-equilibration of primary N_2 and CO_2 or both) can be accepted. Moreover, organic matter can be produced as an artifact during extraction of rocks with organic solvents (Ponomarev et al., 1979).

4-4-3. Inhomogeneous Accretion of Earth and Possible Prebiotic Atmospheric Compositions

Walker (1976) has discussed in detail the implications of a heterogeneous accretion model for the prebiotic atmosphere; here we summarize only briefly his scenario. According to the model, the Earth formed by accretion of solid matter as it condensed during cooling of a hot, massive, solar nebula. Rapid accretion was accompanied by melting and segregation into a molten metallic core and a fluid silicate-oxide mantle. Any volatiles contained in this refractory material were driven into the primitive atmosphere. As temperatures continued to drop, metallic iron particles remaining in the nebula were converted to ferrous iron between 1200° and 500° K, and to ferric iron below 400° K, by reactions with cooling nebular gas. Thus, at late stages of accretion metallic iron was no longer an important constituent in the nebula, and iron accreted in oxidized states to form the material that would become the upper mantle and crust. Since the entire system was embedded in the solar nebula, a mechanism for dissipation of the nebula and the primitive atmosphere was required. Walker presumed that when the sun passed through its T-Tauri stage, the solar wind would have been powerful enough to have blown the remaining nebular gas out of the inner solar system, carrying Earth's primitive atmosphere with it. Alternatively, tidal stripping of the atmospheric envelope of a giant, gaseous, protoplanet by the sun could have occurred early, leaving behind an Earth-sized body of condensed matter (Cameron, 1978a); or the primary atmosphere could have been dissipated by drag effects associated with the mass outflow of H_2 due to heating of the upper atmosphere by a strongly enhanced flux of solar ultraviolet radiation (Sekiya et al., 1980). In any case, residual planetesimals were accumulated by the primitive Earth. These provided volatile-rich material to form the thin surface veneer of the Earth. The volatile-rich component was presumed to be carbonaceous chondritic in composition. Heating of this late-accreted debris either during passage through the atmosphere, during impact with the surface, or while embedded in a hot surface, released the volatiles to form the atmosphere.

According to Walker (1976), H_2O and CO_2 would have dominated the secondary atmosphere; N_2 would have occurred in minor amounts; and H_2 and CO would have been present only in traces. Traces of CH_4 and other hydrocarbons would presumably have been oxidized readily to CO_2 by iron oxides in the hot surface. If the composition of such an atmosphere was determined by the redox state of the silicate crust and upper mantle at ~1500° K, it would have strongly resembled the "equilibrated-nominal-H/C" atmosphere shown in Table 4-3; if it was controlled by the redox state at much lower temperatures, near 873° K, the composition would have been much more strongly reducing (Table 4-3). Once the temperature of the Earth dropped below 646° K (critical temperature of H_2O), water could have condensed to begin formation of the oceans, and weathering of basic igneous rocks by CO_2 would have afforded carbonates. The resulting prebiotic atmosphere would have been subject to the same processes as those described in Section 4-4-2.

A variation on this theme of late accretion of volatiles is also worth considering in that it could have provided both preformed organic compounds and additional reducing capacity to the prebiotic atmosphere. An important outcome of the study of lunar rocks was the discovery that a late period of intense bombardment of the lunar surface ended at about 3.9 Ga ago (Tera et al., 1974). This finding supported the idea that the composition of secondary atmospheres of all the terrestrial planets could have been produced by late-stage impacts of volatile-rich planetesimals. Computer modeling of the late-stage accretion process by Benlow and Meadows (1977) yielded an amount of volatiles derived from vaporization of C1 meteorites that was of the same order of magnitude as the present terrestrial inventory (see also Lange and Ahrens 1980.) Consideration of the dynamical behavior of bodies early in solar system evolution suggests that both comets and meteorites could have been the impacting bodies (Wetherill, 1975b, 1976, 1977; Whipple, 1976).

If accretion of volatiles was delayed until as late as 4.1 to 3.9 Ga ago, Earth's lithosphere might have been sufficiently cool, thick, and continuous to have prevented the rapid recycling of such materials into the crust and upper mantle. Release of volatiles during atmospheric entry and impact then could have been a kinetically controlled process in which their chemical composition would have been governed by the chemistry of the mineral matrices from which the volatiles were released, rather than by equilibration with bulk silicate and oxide melts in the lithosphere. Under these circumstances the composition of the volatiles might have resembled that released by laboratory prolysis of carbonaceous meteorites (Simoneit et al., 1973; Wszolek et al., 1973; Bunch et al., 1979), namely, with H_2O, CO_2, and CO as the dominant species accompanied by minor amounts of hydrocarbons and other volatile organic molecules, and with nitrogen almost entirely in the form of N_2. In such laboratory studies the abundance of CO is at least comparable to (and sometimes higher than) the amount of CO_2 released. Artificial meteorite ablation and impact experiments could provide tests of the chemical consequences of this possible release process.

Let us also consider the possibility that late-stage accretion occurred when liquid water oceans were already present. Such conditions might permit organic compounds to be supplied directly to the primitive environment via the leaching and weathering of carbonaceous meteorites that reached the surface of the Earth intact (as some do today). A simple model-dependent calculation by Chang (1979) suggests that the amounts of amino acids, for example, supplied by this mechanism could have formed a highly dilute (5×10^{-7} molar) solution. At this dilution, it is difficult but possible to formulate geologically plausible scenarios to concentrate the amino acids and continue the course of chemical evolution to more complex molecules (Lahav and Chang, 1976).

If the atmosphere also contained a cometary contribution, comets could have supplied part of the initial inventory of organic matter for chemical evolution, a suggestion made earlier by Oró (1961). Since we have no clear knowledge of the content of organic compounds in comets we cannot make an estimate of the potential contribution from this source as we did for carbonaceous meteorites. Comparison of the scanty data available on production rates of cometary molecules (and estimates of the dust-to-gas mass ratio in comets) with the abundances of organic compounds in meteorites leads one to expect considerably higher abundances of volatiles and extractable organic compounds in comets (see Chang, 1979).

Even if comets did not directly supply organic matter, they may have provided the early atmosphere with some reduced gases, particularly CH_4 and CO, and with HCN and other intermediates that facilitate organic synthesis. However, CH_4 has not yet been detected in comets and the CO observed may be derived in part from decomposition of CO_2 (Delsemme, 1977). Evaporation of volatiles from an icy matrix rapidly and directly into the Earth's atmosphere during entry and impact could free them for prebiotic synthesis. It has been suggested that 50% or more of the Earth's volatile inventory may have been supplied by cometary material (Chang, 1979), but uncertainities regarding the elemental abundances on which this estimate is based make it qualitatively, rather than quantitatively significant.

4-4-4. An Early Earth Regolith, Weathering, and Submarine Environments

Up to this point we have focused attention on the effects that various processes would have exerted on the prebiotic atmosphere. We now briefly consider surface environments, with reference especially to the nature and products of the earliest weathering processes. The late stage of accretion by planetesimal impacts on a cooling surface would have produced a regolith on Earth. Some possible physical properties of an early Earth regolith and their implications for chemical evolution have been discussed by Nussinov and Vekhov (1978). Studies of lunar samples have yielded much information about the regolith formed on a body whose atmosphere (if it ever had one) was very short-lived. Unlike those of Earth, lunar rocks and "soils" show little sign of having been the influenced by water. In contrast, Mars has retained an atmosphere, but although its cratered terrain indicates that its regolith is of great age, the latter's physical, mineralogical, and petrologic characteristics are poorly understood. In addition, Mars' channels and rilles testify to the occurrence of liquid water on its surface during past epochs, but billions of years of surface evolution have passed since then, and the planet's surface properties, as revealed by the Viking Missions (Toulmin et al., 1977), may bear little resemblance to those of the distant past.

Meteorites, however, are relics from the earliest period of solar system history, and many carbonaceous meteorites show evidence of having formed in a near-surface regolith environment of a small planetary body under the influence of water. (Ceres, with a diameter of about 1000 km, is the largest known asteroid exhibiting spectroscopic properties similar to those of aqueously altered carbonaceous meteorites; see Matson et al., 1978.) Meteorite chemistry, mineralogy, and petrology reflect to varying degrees processes in which liquid and gaseous water (or both), either seeping through the regolith or trapped within it (or both), altered a pre-existing assemblage of anhydrous regolith components (Bunch and Chang, 1980). Clays, carbonates, sulfides, and sulfates are among the secondary minerals formed by such aqueous alteration processes and are found coexisting with organic matter. Thermodynamic arguments have been used to suggest the occurrence of a similar alteration assemblage on the Martian surface (Gooding, 1978). The presence of carbonates in these meteorites suggests that CO_2 assisted H_2O in the weathering of anhydrous precursor phases.

Garrels and Mackenzie (1971, p. 290–291) have also speculated on the sequence of geochemical events and the nature of weathering processes and products that may be associated with the transition from a hot to cool planetary atmosphere and surface. One possibility they consider involves a hot, CO_2-rich atmosphere above an ocean at 473° K that was 1 N in HCl. In another, strongly acidic volatiles are envisioned as having reacted quickly with the crust, with water being removed by hydration of silicates before it could condense to liquid so that at temperatures approaching 373° K the surface may have had a CO_2-rich atmosphere but no oceans.

The degree to which the analogy with meteorite parent body processes may apply to the prebiotic Earth is uncertain, but if a surface previously hot and anhydrous became exposed to the weathering effects of gaseous or liquid water at lower temperatures, it is possible that the end products could be similar to those observed in carbonaceous meteorites. By analogy, we suggest that clays, carbonates, sulfates, and sulfides were widespread among the inorganic materials formed on the cooled Earth at the beginning of its chemical evolution period as a consequence of the appearance of liquid water. The detailed chemical composition of those substances would have differed from their meteoritic counterparts, but the presence of these types of materials seems difficult to deny. How they would have interacted with each other and with primordial chemical and energy fluxes remains to be elucidated.

At least clays, with their ion-exchange properties, their catalytic capabilities, their high surface area, and other characteristic physical-chemical properties, are expected to have played diverse and important roles in chemical evolution (summarized by Rao et al., 1980). Clays would have regulated the pH of the seas at the present-day value of ~ 8, and they might have provided (*i*) sites for concentration and separation of various organic compounds (e.g., Bernal, 1951; Lahav and Chang, 1976; Otroschenko and Vasilyeva, 1977; Lawless and Edelson, 1980); (*ii*) catalysts for synthesis and condensation reactions (e.g., Paecht-Horowitz et al., 1970; Anders et al., 1973; Harvey et al., 1975; Schwartz and Chittenden, 1977; Lahav et al., 1978; Paecht-Horowitz, 1978); and (*iii*) templates for the origin of life itself (Carins-Smith, 1965, 1971; Hartman, 1975; Lahav and White, 1980). Properties of primary unaltered minerals, as well as those of other secondary minerals mentioned above, should be explored for their potential contributions to the chemical evolution process (see Section 4-5).

Hargraves (1976) has proposed a model of early Earth history in which a global ocean covered thin, tectonically active lithospheric plates, thus implying the absence of a land surface in the chemical evolution epoch (see, however, Chapter 3). Ingmanson and Dowler (1977) have suggested that Hargrave's model would still permit chemical evolution to occur at plate-spreading centers where environments are reducing, primordial gases are released into overlying water, clays and other minerals are present, and temperatures are moderate. According to Ingmanson and Dowler (1977), chemical evolution may be taking place today in these submarine environments, a proposition that is consistent with, and seems tentatively supported by the reported presence of glycine in hot submarine brines (Ingmanson and Dowler, 1980); this possibility merits additional testing.

4-4-5. Prebiotic Organic Synthesis: Biological Implications and Possible Occurrence in Oxidizing Environments

An integral part of the Oparin-Haldane hypothesis for the origin of life is the concept that the first organisms were chemoheterotrophs capable of utilizing preformed organic matter in their environment as the source both of energy and of cell-building materials. Since even the simplest modern cells exhibit an extremely high degree of physical-chemical complexity, it is difficult to imagine how such complex, intracellular, biochemical machinery, with its sophisticated molecular organization, could have initially arisen without at least having small parts of the assemblage available preformed in the watery primordial stockroom. According to the Oparin-Haldane scenario, chemical evolution supplied a complex chemical environment equipped with both a diverse inventory of organic compounds and the continuous production mechanisms necessary to allow molecular organization to ensue. Presumably, if all the building blocks of primitive proteins and nucleic acids (e.g., amino acids, nucleotides, sugars, or their primitive precursors) were available in the primitive seas, then the major remaining obstacle to the origin of life would have been the formidable one of evolving polymeric structures having useful biochemical functions including the capability for genetic self-reproduction. The apparent ease with which key organic compounds like amino acids could be synthesized from model prebiotic gas compositions containing CH_4 and CO provided strong support for the idea of heterotrophic origins. Horowitz (1945) increased the credibility of this "heterotrophic hypothesis" by describing how the original heterotrophs might have evolved the biosynthetic pathways necessary for utilizing other substances in the environment once the supply of preformed organic matter became limiting. To this day the heterotrophic hypothesis has remained the prevailing viewpoint.

The earliest firm evidence of life, found in rocks 3.5 Ga old, is recorded in stromatolites (Chapter 8), the morphology of microfossils (Chapter 9), and in the isotopic composition of ancient carbon (Chapters 5 and 7). As is discussed in Chapter 9, however, the evidence is consistent with the presence of either heterotrophs or autotrophs or, most probably, both. Recent advances in elucidating phylogenetic relationships among modern microorganisms on the basis of nucleic acid sequences have reached the same pluralistic impasse (Fox et al., 1980); moreover, they have uncovered evidence suggesting that some heterotrophic microbes evolved from autotrophic precursors (a possibility that is by no means implausible inasmuch as numerous lineages of heterotrophic protists have similarly been derived via loss of photosynthetic capability). Within the constraints of the biological and paleobiological records, both heterotrophic and autotrophic models for the first organisms remain viable. Therefore, past (Cairns-Smith, 1965, 1971) and recent (Hartman, 1975; Woese, 1977) statements of alternatives to the prevailing heterotrophic model cannot be excluded and deserve experimental and theoretical assessment.

The assumption that the first organisms could not have had the biochemical sophistication to permit direct use of atmospheric CO_2 as a carbon source has been used as a major argument for heterotrophic origins. Yet the complexity and biochemical sophistication of living heterotrophic microorganisms makes even the utilization of preformed prebiotic organic matter difficult to imagine for the first organisms. The assumption has also been coupled with the apparent difficulty in synthesizing organic compounds from CO_2 to discount the possibility that CO_2 could have been the prebiotic carbon source.

Atmospheric models involving CO_2 and N_2 as the primary sources of carbon and nitrogen, with variable amounts of H_2, remain viable on the basis of cosmochemical, geologic, and biologic evidence. But prebiotic organic syntheses in these atmospheres may pose difficulties for reactions utilizing the three energy sources (lightning, thunder shock waves, and sunlight) generally considered to have been most abundant on the prebiotic Earth. Gabel (1977), Baur (1978), and Walker (1980c) have suggested syntheses in the seas and at the interfaces between the atmosphere and the surfaces of land and seas. (We shall return to these possibilities in Section 4-5.) It is clear from the lack of attention devoted to them in the chemical evolution literature that pathways for prebiotic syntheses based on the presence of H_2, H_2O, CO_2, and N_2 in the atmosphere merit more study. The problem should be viewed as an important challenge to understanding all plausible pathways for chemical evolution on Earth. In the absence of certainty about the composition of the atmosphere in which the processes of chemical evolution occurred, one of the most fruitful approaches to take in validating the initial stage of the paradigm is to demonstrate plausible pathways for organic synthesis in all environments consistent with available constraints.

Walker et al. (Chapter 11) and Kasting et al. (1979) point out that on the prebiotic Earth, the production rates of reduced gases from volcanic sources (H_2 and H_2S), and of reduced species in solution from weathering, were likely to have overwhelmed the production rate of oxygen resulting from photodissociation of atmospheric water followed by hydrogen exospheric escape. This viewpoint is supported by the work of Vander Wood and Thiemens (1980) who investigated the reactions of hydroxyl radicals (HO·) produced by H_2O photodissociation in the Earth's early atmosphere. They found that rapid reaction of HO· with even trace amounts of CO, HCl, SO_2, H_2S, NH_3, and CH_4 would have inhibited the formation of O_2. If the indicated gases were present, then the rate of O_2 production would have been orders of magnitude slower than that calculated simply from H_2O photodissociation rates and exospheric H_2 escape. Until compelling arguments to the contrary appear, we adopt the view that the prebiotic environment had very little free oxygen and, therefore, was conducive to the chemical evolution of organic matter (also see Hart, 1979). Apparently, the emergence and proliferation of oxygen-producing photosynthetic organisms led in time to an oxygen production rate that exceeded the rate of supply of reduced species in the environment (see Chapter 11 for a discussion of the timing of the development of an aerobic atmosphere).

The possibility of simultaneous production of organic matter by nonbiological and biological pathways after the origin of life, but before the advent of an aerobic atmosphere, remains an open question. Evidence in the geologic record (Chapter 11) points to a generally reducing environment 3.5 Ga ago, at which time a diverse population of microorganisms already existed (Chapter 9). However, the isotopic composition of organic carbon in 3.5 Ga-old sediments (Chapter 5 and 7) and the discovery of stromatolites of equal antiquity (Chapter 8) are not inconsistent with the existence of biological (viz., photoautotrophic) sources of O_2 and, therefore, of local environments that were "oxidizing" but anaerobic (see Glossary to this volume and Chapter 11 for a discussion of this distinction). Whether abiotic synthesis of organic matter in such "oxidizing" locales could have occurred would have depended on the level of oxygen that prevailed. Although early attempts at such synthesis were largely unsuccessful (Section 4-3-2), no systematic study of this question has been conducted. However, in experiments designed to simulate Martian surface chemistry, Tseng and Chang (1974) reported evidence that ultraviolet irradiation (254 nm) of silica gel in the presence of CO, CO_2, and O_2 afforded very small amounts of formic acid. It is also noteworthy that the pyrolytic release experiment sent to Mars as part of the Viking Biology experiments yielded results (Horowitz et al., 1977; Hubbard, 1979) that could be interpreted in terms of very low amounts of heterogeneous photoreduction of CO or CO_2 in the presence of O_2.

4-5. Problems and Prospects for Future Studies

Having considered the implications for prebiotic syntheses of the various homogeneous and inhomogeneous accretion models for the Earth, we are left with a range of possible atmospheric compositions and surface environments in which the primary synthesis of organic matter may have begun. Rather than making an arbitrary choice from among these possibilities and regarding it as a well-established "truth," we advocate investigations of the possibilities for chemical evolution in any and all candidate atmospheres and surface environments that continue to be consistent with available cosmochemical, geologic, and biological knowledge.

Although production of the organic compounds necessary for chemical evolution would have proceeded readily in a highly reducing atmosphere, the possibilities in a N_2–H_2O atmosphere containing minor amounts of H_2 and CO_2 and minor to trace amounts of CO remain relatively unexplored. Of the various energy sources on the Earth today, ultraviolet light (>150 nm), electric discharges, and thunder shock waves are generally believed to be the most significant on a global scale; there appears to be no compelling reason to assume a different situation for the primitive Earth. Pinto et al. (1980) have shown in theoretical photochemical models based on a N_2–H_2O atmosphere that CO_2 and CO can be converted to formaldehyde. Bar-Nun and Hartman (1978) have shown experimentally that CO in the presence of H_2O vapor can be photochemically converted to simple alcohols, aldehydes, and organic acids. In Section 4-3-2, above, we demonstrate that with a H_2/CO_2 ratio of about six and no water present, electric discharges can convert CO_2 to CO, and ultimately to hydrocarbons in low yield. Although only in the case of Pinto et al. (1980) were model conditions used that can be viewed as acceptably plausible simulations of a N_2–H_2O atmospheric model, these and the other experiments described in Section 4-3-2 do indicate that organic carbon compounds can be synthesized from CO_2 and CO.

The difficulties in reducing N_2 to organic nitrogen and synthesizing key compounds such as amino acids in a N_2–H_2O atmosphere have been pointed out by Gabel (1977). Even with CH_4 replacing CO_2, Ferris and Chen (1975) were unable to produce amino acids by ultraviolet photochemistry. In our own experience, electric discharges through CO_2–N_2–H_2O mixtures afforded nitrous and nitric acids as the major products rather than organic compounds. Theoretical studies of the fixation of nitrogen by electric discharges (and thunder shock waves) in a N_2–H_2O atmosphere with minor amounts of CO_2 and H_2 indicate that nitric oxide is the dominant product (Yung and McElroy, 1979; Chameides and Walker, 1981). Apparently, only when the elemental ratio C/O in the atmosphere exceeds unity will HCN be formed in greater yields than NO (Chameides and Walker, 1981). Abelson (1966) produced HCN by passing electric discharges through mixtures of H_2, CO, and N_2, and although his experimental conditions may not constitute good models for a N_2–H_2O atmosphere, they show qualitatively that HCN, a key precursor for some model syntheses of amino acids and other nitrogen-containing organic compounds, can be produced.

Clearly, the presence of a reducing gas (H_2, CH_4, or CO) is required if organic synthesis occurred in the atmosphere. Walker (1976) offered the possibility that H_2 produced volcanically through decomposition of H_2O in early tectonic processes and amounting to about 1% of the atmosphere could have persisted for about 0.5 Ga on the early Earth. Whether or not this amount would have been sufficient to permit atmospheric organic synthesis remains to be evaluated. If not, and if other highly reducing species were unavailable, how might the basic chemical building blocks of life have been produced? In solutions containing ferrous iron, CO_2 can be reduced to formaldehyde either by irradiation with ultraviolet light (Getoff et al., 1960) or with high energy particles (Garrison et al., 1951), suggesting that dissolved ferrous iron might have provided an important source of reducing power on the prebiotic Earth. Weathering of igneous rocks should have provided a steady supply of ferrous iron to the primordial seas. Hartman (1975) and Gabel (1977) have proposed other schemes that required reactions in the oceans and on clays. Calculations by Baur (1978) show that spontaneous formation of reduced organic matter, including amino acids, is thermodynamically possible in heterogeneous systems containing N_2 and CO_2 in the presence of ferrous iron-containing minerals and H_2O. Investigations of the potential pathways for organic synthesis in such heterogeneous systems are highly desirable. Recent work by Halmann et al. (1980) has shown that ultraviolet irradiation of suspensions containing clay particles, water, and dissolved CO_2 results in production of methanol and other simple organic compounds. In the context of the Mars Viking biology experiments, Hubbard (1979) has shown that CO_2 can be converted to organic carbon when irradiated with UV light (>220 nm) in the presence of iron oxides and iron-bearing clays.

Formation of reduced nitrogen from N_2 in the absence of highly reducing conditions may also be more feasible than has been widely assumed. An early report by Bahadur et al. (1958) claimed production of amino acids via irradiation of mixtures of paraformaldehyde, water, and colloidal molybdenum oxide in the presence of air. In aqueous solutions, metallo-organic complexes of molybdenum, studied as nitrogenase models, have been found to reduce N_2 to hydrazine (Schrauzer, 1975). Recently, Schrauzer and Guth (1977) reported that in the presence of N_2, near ultraviolet irradiation of powdered titanium dioxide (TiO_2) and iron oxide (α-Fe_2O_3) produces NH_3 and hydrazine, as does irradiation of desert sand in air by sunlight (Schrauzer et al., 1979). The implications of heterogeneous photoreduction of N_2 to NH_3 have been discussed recently by Henderson-Sellers and Schwartz (1980). Apparently, reduction of nitrite to ammonium ion can be achieved readily in aqueous solution at pH 8 with ferrous ion (Moraghan and Buresh, 1977). It is easy to imagine how nitrous acid produced by electric discharges might be dissolved in seas containing ferrous ions and converted to ammonium ions.

In summary, although no clear solutions are yet available for the problem of organic synthesis in a N_2–H_2O atmosphere containing minor amounts of H_2 and CO_2, we anticipate that considerable progress in the understanding of this system will be made in the future as more attention is devoted to it.

An adequate understanding of the course of organic chemical evolution on the primitive Earth, and of the chronology of the major events involved, also requires far more knowledge about the early history of the solar system and Earth than is now available. The course of condensation and accretion of matter in the solar nebula would have been critical in establishing the chemical composition of material that formed the planets and the processes that initiated their geophysical and geochemical evolution. More knowledge of the geophysical and geochemical evolution of early Earth or other planetary bodies (including meteorite parent bodies) would lead to better insight into the thermal history of the mantle-crust-atmosphere system. From this knowledge could come tighter constraints on the origin, composition and early evolution of the atmosphere. Knowledge of its thermal structure (e.g., Morss and Kuhn, 1978) would allow determination of the rate at which H_2 (and, therefore, reducing power) was lost from the top of the atmosphere; it would allow assessment of the magnitude and duration of early greenhouse effects and their effects on the chemical evolution environment (see Chapter 11); and it would also determine, in part, the atmosphere's compositional structure and the kinds of chemical transformations that could occur in its various parts. For similar reasons, it is important that the radiative transmission in the atmosphere, which determines the energy budget associated with incident sunlight, be better defined.

Understanding of transport processes in the early atmosphere is important, too. Numerous questions need to be resolved. For instance, if material were produced by ultraviolet light high in the atmosphere, how long would it take to descend to the surface of Earth? Would the transport time be short enough to prevent either destruction of such material or its conversion into refractory substances incapable of further involvement in chemical evolution (e.g., Pinto et al., 1980)? Similarly, if material were produced on land, how might it have been transported to bodies of water where it could interact with other materials? And if it were produced in oceans, how might it have been concentrated (e.g., Lahav and Chang, 1976; Armstrong et al., 1977)?

Possible production rates for organic compounds, even in the relatively well-studied reducing atmosphere, are not well known. Such syntheses could occur in the atmosphere or in the oceans; at the interfaces between the atmosphere, bodies of water and land; or beneath the seas at sea-floor spreading centers. There are also chemical reactions that act as "sinks" for organic matter. For example, amino acids and sugars react (via the Maillard or so-called "browning" reaction) to form water-insoluble products, which on the primitive Earth would have thus been removed from the reservoir of organic material that would have been available for chemical evolution (e.g., Nissenbaum, 1976). How could material in sinks of this type have been recycled to make it again available (e.g., Henley, 1968)?

What were the environmentally plausible pathways available to convert simple monomeric amino acids and nucleotides to the biopolymers necessary for the origin of life? One pathway which has recently received attention involves the interaction between mineral material (particularly clays) and organic matter (e.g., Lohrmann et al., 1980; Lahav et al., 1978).

There is little doubt that organic chemical evolution must have occurred in a predominantly inorganic realm. An important question is, therefore, how did organic chemistry interact with the prevalent inorganic chemistry and mineralogy (e.g., Cairns-Smith, 1965, 1971; Hartman, 1975; Lahav and White, 1980; Lawless and Edelson, 1980; Rao et al., 1980)? Did the inorganic world provide catalysts for organic reactions? If so, what were they (see Section 4-4-4 and above)? How was phosphate utilized to produce nucleotides and polynucleotides (e.g., Schwartz, 1972; Griffith et al., 1977)? Did inorganic material sequester organic matter, thereby removing it from the realm of chemical evolution?

Environments on Earth are subject to numerous fluctuations—day, night, seasons, tides, etc. In some experiments such environmental fluctuations have been simulated, with changes of temperature and moisture content having been used successfully, for example, to produce peptides from amino acids (e.g., Lahav et al., 1978). How important overall were such fluctuations for chemical evolution? Were they a necessary prerequisite to the origin of life?

At some stage in chemical evolution it would have been necessary to achieve a phase separation between an evolving organic system and the external environment. Thus, the origin of membranes must be considered. Could lipid vesicles (e.g., Deamer and Oró, 1980; Stillwell, 1980), or the organic microstructures produced in a variety of abiotic synthesis experiments serve as models for early membranes (e.g., Nissenbaum et al., 1975; Fox and Dose, 1977)?

Finally, molecular selectivity during chemical evolution must have been important. Carbonaceous meteorites and the products of model prebiotic organic synthesis experiments both contain a rich variety of organic compounds. Even within a single class, for example the amino acids, there is a large number of different molecular structures. How was the limited number of amino acids and other compounds that are now utilized by organisms selected from this larger abiotic set (e.g., Lawless and Levi, 1979; Wong and Bronskill, 1979; Weber and Miller, 1981)? How much of such selectivity arose intracellularly, rather than as a result of external abiotic synthesis? How did the genetic code arise (e.g., Crick, 1968; Orgel, 1968)? And what was the origin of chirality, that is, of the preferential "handedness" characteristic of the amino acids and sugars of all living systems (e.g., Walker, 1979; Blair and Bonner, 1980)?

Partial answers may already exist for some of these questions but clearly, much remains to be learned about organic chemical evolution and the origin of life. As a field of active scientific inquiry it is still in its infancy, and contributions from numerous, seemingly disparate disciplines of the natural sciences—among others, astronomy, astrophysics, atmospheric physics, geophysics, geochemistry, inorganic chemistry, organic chemistry, and biology—are essential for its continued growth. If research efforts continue on many of the issues here mentioned, we can anticipate significant progress in the future.

4-6. Summary and Conclusions

(1) According to the modern paradigm for the origins of life, living systems arose naturally through a lengthy process of chemical evolution that began with the prebiotic synthesis of simple organic compounds in the atmosphere and surface environments of the primitive Earth. Further reactions in these environments transformed the primary products into increasingly complex organic matter until the process culminated in the evolution of polymeric structures having the capabilities for primitive self-replication and other biochemical functions.

(2) The identification of organic compounds in interstellar clouds, meteorites, comets, and atmospheres of the outer planets corroborates the expectation that prebiotic synthesis and chemical evolution of organic matter occur naturally to varying degrees throughout the cosmos. The processes that gave rise to such organic matter and imposed limits on its chemical evolution are characteristic of each extraterrestrial environment, however, and may be unrelated to those on Earth that yielded the organic components necessary for chemical evolution and the origins of life.

(3) Detection of prebiotic organic matter in ancient (Archean) terrestrial rocks would provide direct confirmation of the chemical evolution hypothesis. Experimental assessments suggest, however, that the elemental (e.g., N/C) and isotopic (e.g., $^{13}C/^{12}C$) ratios in kerogens, which appear to be the most suitable criteria from the viewpoint of maximum preservation and minimum contamination, cannot be used by themselves to distinguish unambiguously between abiogenic and biogenic organic matter in ancient sediments.

(4) Models of the energetics and dynamics of Earth's accretion and core formation can provide bounds on the range of physical and chemical conditions that may have prevailed on the prebiotic Earth. Current thinking favors early core formation either during Earth accretion or within about 10^8 years after accretion. Predictions that a 2300° K increase in average global temperature, and that vigorous mantle convention would have resulted, underscore the profound influence of core formation on subsequent planetary evolution. Under such conditions, it is highly unlikely that any organic matter formed prior to core formation could have survived to participate in chemical evolution at a later, cooler time.

(5) Better models are needed for the origin of the atmosphere and of its evolution from the time of core formation to the onset of the organic chemical evolution epoch. These models should take into account how Earth's thermal evolution would have affected the temperature, pressure, and composition of the atmosphere and surface and the processes that operated in them. The conditions explored in these models should be bounded by the environmental conditions reflected in Archean rocks and by the range of initial states that are consistent with constraints imposed by knowledge of Earth accretion and of core formation.

(6) Large uncertainties exist in current understanding of Earth's early thermal history, the time span over which it acquired the volatiles now at and near its surface, the elemental abundances in the inventory of its volatiles, and the degree of outgassing of this inventory prior to the origin of life. Because of these uncertainties, "permissible" prebiotic atmospheric compositions at the time when liquid water first appeared range widely, from a strongly reducing composition, dominated

by high abundances of H_2 and CH_4, to mildly reducing, containing mostly N_2 with minor to trace amounts of H_2, CO, and CO_2.

(7) Geologic evidence from the oldest known rocks, those of the Isua Supracrustal Group of western Greenland, indicates the presence of H_2O and CO_2 in the atmosphere 3.8 Ga ago; it gives no clear evidence regarding the possible presence of other gases such as CO, H_2, and CH_4. At present, outgassing from Earth's interior occurs predominantly at ocean-floor hydrothermal vents rather than at continental volcanoes; in addition to CO_2 and He, H_2, and CH_4 are observed. Although the origin of this CH_4 remains to be established, these observations suggest that similar sources may have supplied reduced gases to the prebiotic environment.

(8) Laboratory experiments indicate that prebiotic synthesis of organic matter would have occurred readily in strongly reducing atmospheres. The possibilities in mildly or non-reducing (i.e., "neutral") atmospheres may have been more limited, but they merit much more study than they have thus far received. The conversion of N_2 to nitrogen-containing organic compounds in any prebiotic atmosphere could have been difficult to achieve by atmospheric photochemical processes; electric discharge reactions may have been more effective.

(9) Prebiotic organic syntheses need not have occurred only in the atmosphere; they could have occurred on land, in the seas, and at the interfaces between atmosphere, seas, and land surfaces. Theoretical and experimental studies suggest that primary synthesis in multi-phase heterogeneous systems may have been facile. In this context, the role of ferrous iron in minerals and dissolved in primordial waters deserves particular investigation.

(10) The involvement of inorganic matter in the origin of life is expected to be a natural consequence of the geological context within which atmospheric and organic chemical evolution must have occurred. Recent research shows that some metal ions and minerals (particularly clays) could have served as reactants, catalysts, and even templates for prebiotic organic synthesis. Further explorations of their role in chemical evolution are needed.

(11) Considerable success has been achieved in producing the monomeric and oligomeric building blocks of present-day proteins and nucleic acids in putative prebiotic syntheses. But the connection between the model environmental conditions implicit in many such syntheses and the geologic and meteorologic realities of the prebiotic Earth may be tenuous and remain to be established clearly. It should also be kept in mind that the primary organic compounds involved in the first living systems may have differed substantially from those utilized by organisms today.

(12) Until tighter constraints can be imposed on the range of possible prebiotic atmospheric compositions and surface environments, and in the absence of natural evidence of terrestrial chemical evolution, we advocate study of pathways for organic synthesis in all model environments that continue to be consistent with evidence unveiled in the cosmochemical, geologic, and biologic records. An important measure of the credibility of reaction schemes that are proposed for the chemical evolution of organic matter is how well the implied environment accords with geologic and meteorologic data and principles.

(13) Considering the scope and magnitude of the problem, it is reasonable to assert that the scientific study of the origins of life is still in its infancy and that we can anticipate many new insights and discoveries in the future as more is learned about the early history of Earth. It is clear, however, that a strongly interdisciplinary attack on many correlated issues holds the greatest promise of yielding a fuller understanding of the circumstances of life's origins.

Acknowledgments

For many beneficial discussions we thank our fellow members of the P.P.R.G. and our colleagues in the Chemical Evolution Group at the N.A.S.A.-Ames Research Center. We are grateful to Thomas Scattergood and to Bishun Khare and Carl Sagan for supplying materials from their photochemical experiments for our elemental and isotopic analyses. Helpful comments on the manuscript from David Stevenson, James Kasting, James Walker, Norman Horowitz, Gary Ernst, and John Wood are appreciated. Work performed at the University of California, San Diego, was supported by N.A.S.A. Grant NAGW-20 to S. L. Miller.

CHAPTER 5

Precambrian Organic Geochemistry, Preservation of the Record

By J. M. Hayes, Isaac R. Kaplan, and Kim W. Wedeking

5-1. Introduction

We present here both a review of earlier studies and a discussion of new results in Precambrian organic geochemistry. The review is intended to orient the reader unfamiliar with organic geochemistry, to provide a framework for the discussion of new results, and to indicate clearly why the particular lines of inquiry utilized in this work were chosen. The review shows that the organic geochemist interested in the Precambrian must accept a double assignment, as summarized in Figure 5-1. On the one hand, there is the conventional task of interpreting the organic geochemical record in terms of the ecosystems that might have generated it. On the other hand, there is the special task of evaluating the quality of the available record prior to undertaking its interpretation. Specifically, the investigator must ask whether original organic materials have survived in a condition that will support meaningful interpretation. Geologists might rephrase this question by asking whether the primary (source-controlled) nature of the organic material had been obscured by secondary (post-depositional) processes and events.

Two lines of evidence, one based on structural organic chemistry, the other based on isotopic analyses, can be developed. Here we present and discuss the results of both structural and isotopic investigations of Precambrian organic matter. Both sets of results have significance in terms of the primary processes that originally produced the organic matter, and both isotopic compositions and structural details can be affected by the secondary processes that rework the material in the sedimentary environment. In this chapter, however, our considerations of the isotopic results will be limited to their relationships with secondary processes. The primary significance of the isotopic results—the constraints they place on ancient carbon cycles and the evidence they provide regarding

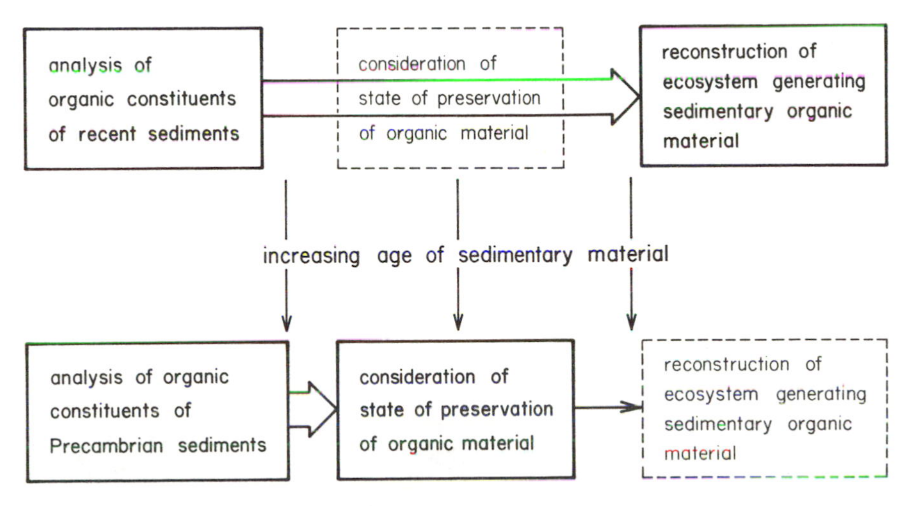

FIGURE 5-1. A schematic representation of logical steps in organic geochemistry.

ancient biochemical pathways—is discussed in Chapters 7 and 12, chapters fully devoted to isotopic considerations.

5-2. Organic Geochemistry—Processes and Products

5-2-1. The Carbon Cycle

Carbon, due to its versatile chemistry and central role in biology, is found abundantly in the atmosphere, hydrosphere, and lithosphere, with its transfer between these reservoirs frequently being biologically mediated. Pathways prominent in the present-day carbon cycle are shown in Figure 5-2, which has been developed from the recent review by Bolin et al., 1979 (this view, which deals only with organic matter, omits important relationships between the cycles of carbon and oxygen and between the carbon cycle and the sedimentary recycling of carbonates; see Garrels et al., 1976).

Organic matter finds its origin in the processes of primary production, presently represented essentially exclusively by the photosynthetic fixation of carbon dioxide, though an interesting contribution of unknown magnitude is made by nonphotosynthetic organisms reducing CO_2 (Karl et al., 1980) and, possibly, oxidizing CH_4 (Welhan and Craig, 1979), such as those near hydrothermal vents on the ocean floor. Although the great majority (see quantitative discussion below) of primary production of organic matter is rapidly recycled to CO_2 by heterotrophic respiration, occurring in the soil or water column where it is produced, a small portion may be buried in sediments where diverse microbial communities diagenetically rework organic material that becomes acces-

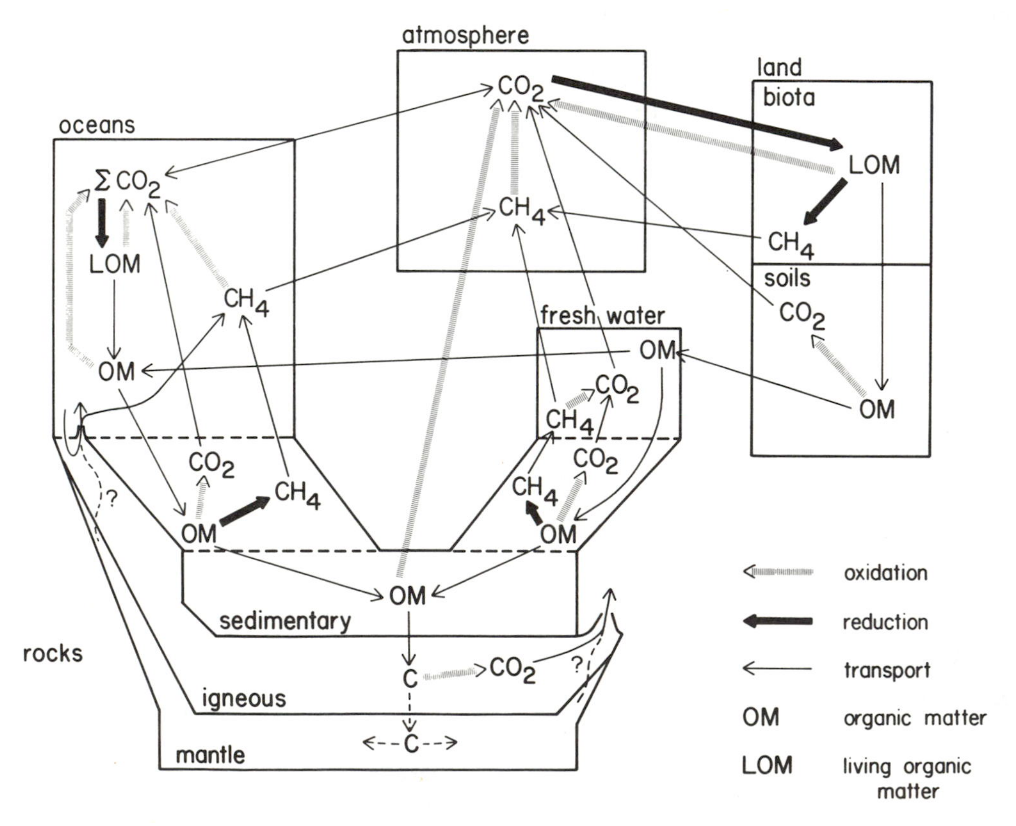

Figure 5-2. An approximate view of the principal pathways of carbon cycling associated with organic matter in the present global ecosystem.

sible to them. It is the residue of this final microbial destruction and transformation that gives rise to the organic material found in sedimentary rocks. Viewed in terms of the efficiency of carbon cycling, this sedimentary carbon component represents a "leak"—it is material which has escaped utilization within the biosphere, and it carries biochemically stored energy which has been lost to direct or immediate use within the ecosystem. Nevertheless, it is this "leak" which, over time, has generated the vast quantities of reduced carbon found in sedimentary rocks and which has, in that way, provided an organic geochemical record of the prior ecosystems of the planet.

Quantitative values can be associated with many of the pools and fluxes shown in Figure 5-2, and present estimates for these values are summarized in Table 5-1. It is believed that more organic material is produced annually on land than in the ocean at the present time, and the synthesis of organic material in freshwater environments is seen to represent much less than 1% of the total, though even it exceeds the amount of preformed organic material reinjected into the cycle by the recycling of organic-rich sediments (for references throughout this discussion, see Table 5-1 and its accompanying notes). All but ~2% of the carbon fixed on land and in freshwater is directly recycled to CO_2. One-third of this unoxidized material is recycled to methane, which is subsequently oxidized in the atmosphere. This methane flux is important, because much of it represents carbon that would otherwise be incorporated in sediments. Two-thirds of the unoxidized residual, according to latest estimates, is transported to the ocean by rivers. If the assessment is correct, this quantity is large enough to make one wonder at the often-repeated maxim (e.g., Lerman, 1979, p. 24) that the land-based carbon cycle is closed, with essentially all of its material being promptly recycled. We will return to this point.

More than 99% of the organic carbon in the ocean is promptly recycled to CO_2. The remainder is incorporated in sediments, where much of it is later oxidized to CO_2 during metamorphism. Two details of this picture deserve further consideration. First, we must note that the role of methane in the recycling of sedimentary organic carbon is unclear. Second, it is interesting to reconsider the fate of organic carbon brought in by rivers. If it is oxidized with the same efficiency as the autochthonous marine organic carbon, then land-derived organic material must constitute a very small fraction of the carbon eventually preserved in sediments. If, on the the other hand, the river-borne material, or some portion of it, is much less efficiently oxidized, then it might make a disproportionate contribution to the organic material incorporated in near-shore sediments. This speculation is not entirely fanciful, since a plausible source can be identified: the recycled sedimentary organic matter (see "Inputs . . . ," Table 5-1), which is conventionally treated as though it were completely recycled to CO_2 during weathering and erosion. There is, in fact, no proof that this oxidation occurs with perfect efficiency and, given the refractory nature of much sedimentary organic matter (see remainder of this chapter), there is every reason to doubt it. Walker (private communication to J. M. H., 1980) has pointed out that this can have significant consequences in the long-term development of the carbon cycle and has called attention to two specific reports (Sackett et al., 1974a; Baxter et al., 1980). For the moment, it is adequate simply to point out that all sediments beyond the first generation might contain not only recent organic material but also ancient organic material that has already passed once (or more) through the rock cycle.

Not immediately evident are the links between the cycling of organic carbon and other environmental and geochemical factors. The burial of large quantities of carbon, for example, can be due to a high sedimentation rate made possible by a high rate of erosion. A mechanism for tectonic control of carbon cycling is, thus, evident, with high relief favoring the preservation of organic material in sediments. Garrels et al. (1976) have discussed some geochemical controls, noting that the availabilities of O_2 and SO_4^{2-} exert control over the rate at which organic carbon can be recycled by oxidation to CO_2. Significant biochemical control would be exerted if the fixation of CO_2 were not linked to nearly stoichiometric O_2 production, as it is in the present ecosystem. The resulting low levels of O_2 would be expected to allow a vastly higher burial rate of organic carbon in sediments. A final example of control arises when we note that the maintenance of any such high burial rate would depend on the continuing availability of CO_2 and of other nutrients required by carbon-fixing organisms. Chief among the latter would be PO_4^{3-} and nitrogen sources.

The complexity and variety of controls on the carbon cycle makes the reconstruction of ancient

TABLE 5-1. Quantitative aspects of the modern* carbon cycle.

Sources, Sinks, and Reservoirs	*Quantities*
	Fluxes (10^{12} *mole C/yr*)
INPUTS OF ORGANIC CARBON	
Net primary productivity:	
on land (Ajtay et al., 1979)	5000
in freshwater (DeVooys, 1979)	48
in the ocean (DeVooys, 1979)	3600
Recycling of sedimentary organic material:	
by uplift and erosion[a]	7
by recovery of fossil fuels (Zimen et al., 1977)	600
FATES OF ALL BUT MARINE ORGANIC CARBON	
Oxidized to CO_2 (directly or via CO):	
in soils, freshwater sediments, and animals[b]	4721
by human destruction of biomass (Hampicke, 1979)	210
by combustion of fossil fuels (Zimen et al., 1977)	600
Reduced to CH_4 prior to oxidation in the atmosphere:	
in land animals (Ehhalt and Schmidt, 1978)	10
in freshwater environments (Ehhalt and Schmidt, 1978)	34
Incorporated in continental sediments[c]:	~0.1
Transported to ocean (Richey et al., 1980):	80
FATES OF MARINE ORGANIC CARBON, INCLUDING INPUT FROM RIVERS	
Directly oxidized to CO_2:	
in the marine water-column[d]	3589
in sediments (Mopper and Degens, 1979)	75
during metamorphism of sediments[e]	4?
Reduced to CH_4 in sediments, then oxidized:	
in the atmosphere (Ehhalt and Schmidt, 1978)	<1
in sediments and in the water column[f]	?
Incorporated in sediments[g]	11
	Reservoirs (10^{12} *mole C*)
OXIDIZED CARBON	
Atmospheric CO_2, 1978 (Bolin et al., 1979):	58,000
Oceanic total carbonate (Bolin et al., 1979):	2,900,000
Carbonate in sediments (Hunt, 1979):	5,400,000,000
Carbonate in metamorphic rocks (Hunt, 1979):	220,000,000
CARBON IN IGNEOUS ROCKS	
(Hunt, 1979):	600,000,000
REDUCED CARBON	
Atmospheric CH_4 (Ehhalt and Schmidt, 1978):	250
Freshwater biomass (Ajtay et al., 1979):	2
Marine biomass (Mopper and Degens, 1979):	250
Terrigenous biomass (Bolin et al., 1979):	49,000
Organic carbon in ocean water (Mopper and Degens, 1979):	83,000
Organic carbon in soils (Atjay et al., 1979):	172,000
Organic carbon in sediments (Hunt, 1979):	1,000,000,000
Organic carbon in metamorphic rocks (Hunt, 1979):	290,000,000

* "Modern" as opposed, for example, to "time-averaged over the Phanerozoic," as discussed by Garrels and Mackenzie (1971, 1972) and Garrels and Perry (1974). Such time-averaging is particularly relevant to models of geologic history; the modern cycle is, however confusedly, observable, and we enumerate its characteristics for that reason only. The present-day sedimentation rate is apparently much higher than the Phanerozoic average (Garrels and Mackenzie, 1971, p. 260; 1972); fluxes associated with sedimentation and rock recycling, therefore, appear substantially higher in this Table than in models intended to represent the average.

[a] Calculated from the present-day sediment flux of 250×10^{14} g/yr (Garrels and Mackenzie, 1971, p. 260). The inorganic material supporting this flux must be derived by erosion of an equal quantity of continental material which is, on average, 63% sedimentary (Garrels and Mackenzie, 1971, p. 249). This sedimentary material contains approximately 0.5% organic carbon (Garrels and Perry, 1974).

[b] Following Bolin et al. (1979) and Lerman (1979, p. 24), this value has been set to indicate complete utilization—apart from the fates identified in separate entries—of organic carbon fixed or recycled from sediments on land or in freshwater. The uncertainty is far greater than that indicated by the number of significant figures.

[c] Kempe (1979) estimates an organic carbon input to continental basins of 1.3×10^{12} mole C/yr. Much of this material is, no doubt, recycled rather than permanently sequestered. The value given here is intended only to indicate the very minor importance of this sink in the present carbon cycle.

[d] This value has been set to balance the fates against the tabulated inputs and, as noted in connection with the major oxidative sink for nonmarine carbon, its uncertainty is far greater than indicated by the number of significant figures. The value is consistent, however, with the estimate of Mopper and Degens (1979) that more than 97% of marine organic carbon is recycled in this way.

[e] The tabulated value is chosen to balance the budget for organic carbon sequestered in sediments.

[f] It is increasingly postulated that the production of methane in marine sediments substantially exceeds the quantities evaded to the atmosphere (e.g., Reeburgh, 1980). Substantial quantities may enter the water column (Martens and Klump, 1980). A substantial portion (perhaps 30%) of the carbon presently thought to be oxidized directly in sediments might follow this path.

[g] Tabulated figure is consistent both with sedimentation of 250×10^{14} g material containing 0.5% C (see note 1) and with an estimate by Mopper and Degens (1979).

cycles difficult. It has, for example, been suggested (Garrels and Perry, 1974) that the average organic carbon content of newly formed sedimentary rocks has not varied greatly over geologic time. If this is correct, it must at some point be reconciled with the evidence that the abundance of O_2 on the early earth was very low (see Chapters 3, 4, and 11). How, when O_2 was extremely scarce, was the formation of carbon-rich sediments prevented? Was primary productivity very much lower, or did some alternative recycling process play a more important role then than now? More specifically, might the absence of O_2 have allowed methanogenic bacteria to occupy a much wider range of environments and to gain prominence as recyclers? Ironically, the intelligence with which that possibility can be discussed is restricted by uncertainties regarding the roles of methanogens in the modern carbon cycle!

5-2-2. Processes Affecting Sedimentary Organic Matter

Petroleum geochemists are interested in understanding all principal "processes affecting sedimentary organic matter," and it is, therefore, not surprising that this field is now quite advanced. Two recent monographs (Tissot and Welte, 1978; Hunt, 1979) summarize present knowledge regarding the transformations of organic materials in sediments so well and in such depth that it is not useful to dwell on those subjects here. The following discussion is largely abstracted from those monographs, to which the reader is referred for more detailed information and primary references.

The input of organic matter to sediments consists of some mixture of dissolved organic compounds, particulate organic matter, and whole cells or larger cellular aggregates. Whatever the distribution among these types, the material is relatively highly organized, containing many unmodified biosynthesized carbon skeletons and some information-rich macromolecules such as proteins and nucleic acids. In principle, at least, an omniscient investigator could analyze and successfully interpret the detailed molecular structures in this sedimentary input, thereby reconstructing many details of the ecosystem that produced it. The sediments, however, have a voice of their own—the benthic organisms and particularly the microbial community characteristic of marine sediments—and it has the last word, generating the message that is most clearly recorded in the organic geochemical medium as it passes from the realm of biochemistry to that of geochemistry.

In the discussion that follows, we use the term "diagenesis" to refer to the earliest natural transformations of sedimentary organic matter. (As is noted in Part 1 of the Glossary to this volume, the term is also used to refer to all post-depositional, but pre-metamorphic changes, exclusive of subsequent weathering, undergone by a sediment and its component minerals, fossils, etc.; the organic geochemical usage is thus equivalent to the geologist's "*early* diagenesis," with "catagenesis" and "metagenesis," as discussed below, being used to refer to progressively later stages of the pre-metamorphic alteration of sedimentary organic material.) As noted in Figure 5-3, such diagenesis is at first microbiologically mediated, with approximately 90% of the organic input being recycled (compare sedimentary entries under "fates of marine organic carbon" in Table 5-1). The metabolic processes involved are initially aerobic, but rapidly become anaerobic after burial to some moderate depth (<1.0 m), with sulfate-reducing and methanogenic microorganisms playing major roles. The expiration of chemically significant biological activity is brought on by an absence of available substrates and nutrients, and diagenesis continues by means of abiological processes as the material is buried at depths up to 1000 m and reaches temperatures of 50° C or less. These conditions are not severe enough to preclude biological activity unless the water content decreases to less than ~20% due to lithification by compaction, cementation, or recrystallization. The attribution of many diagenetic modifications of organic matter to abiological causes (Abelson, 1978) is based on the nature of the products, not on a securely demonstrated absence of biological activity. A time interval of a few million years is typically required for burial to a depth of ~1000 m (Emery and Rittenberg, 1952).

Deeper burial, due to further accumulation of sedimentary overburden, tectonic activity, or both, increases the temperature of the rock and its included organic matter. The term "catagenesis" is applied to transformations occurring between approximately 50° and 150° C. The "principal zone of oil formation" (Vassoevich et al., 1969) lies within this interval. Mobile, hydrogen-rich organic material is produced from highly reduced, paraffinic kerogens around 130° C and from other kerogens at lower temperatures, with the onset of the process

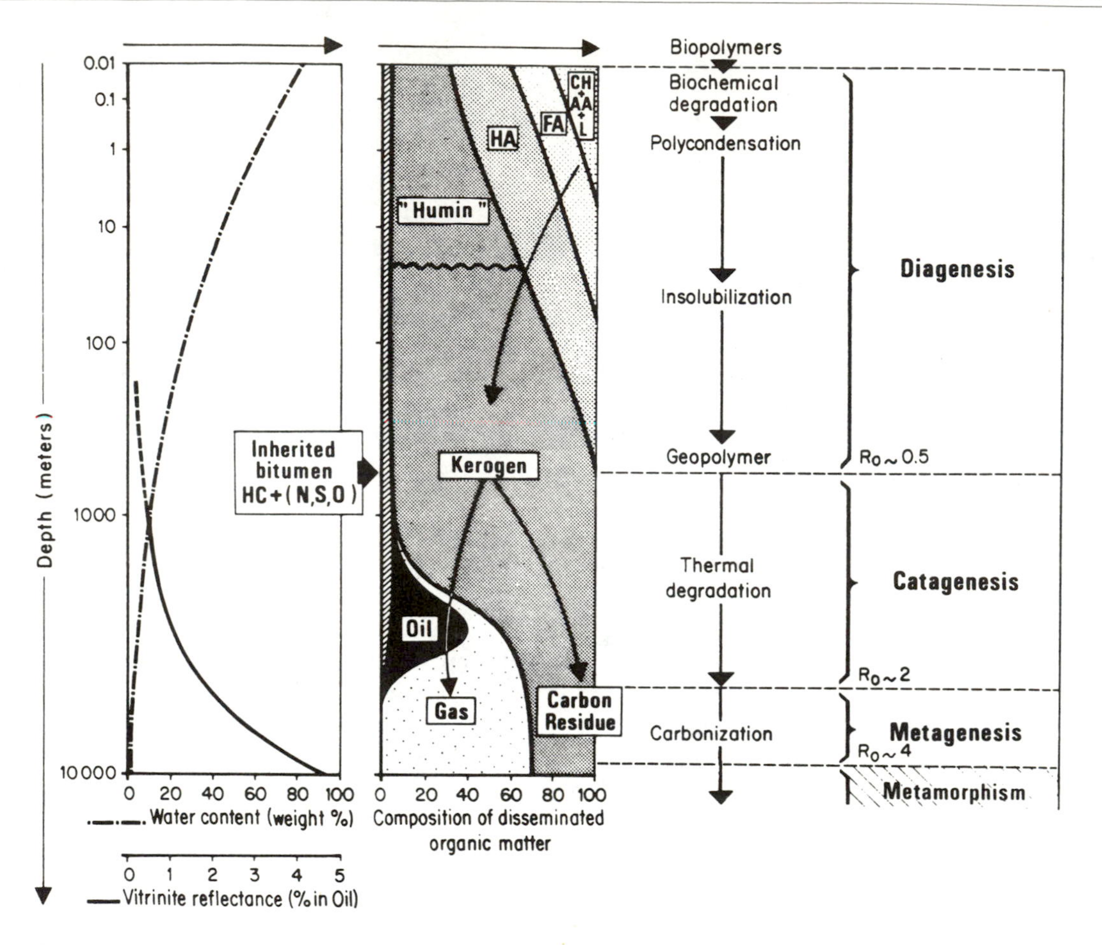

FIGURE 5-3. General scheme of the evolution of organic matter, from the freshly deposited sediment to the metamorphic zone. CH = Carbohydrates. AA = Amino acids. FA = Fulvic acids. HA = Humic acids. L = Lipids. HC = Hydrocarbons. N, S, O = N, S, O compounds (non-hydrocarbons). From Tissot and Welte (1978), with permission.

sometimes occurring as low as 60°. Coexisting inorganic mineral phases are largely unaffected by the conditions occurring during catagenesis, although the smectite to illite transition at 70–100° C is a notable exception.

"Metagenesis" refers to the temperature range between approximately 150° and 250° C, in which further alteration of sedimentary organic matter occurs after all aliphatic materials (other than methane) have been mobilized. The residual kerogen becomes increasingly aromatic, losing hydrogen in part by elimination of methane. This process occurs before extensive thermal alteration of coexisting inorganic minerals can be detected, and, for this reason, metagenesis of organic materials can be resolved from "metamorphism" of organic materials, the latter term being applied to organic transformations occurring under conditions corresponding approximately to the onset of greenschist facies mineralogic metamorphism (~250° C).

5-2-3. Distribution and Types of Organic Matter

The processes described above give rise to a spectacular variation in sedimentary organic matter. Among the different types that can be usefully resolved are gas and oil, other molecular remains with specific structures (e.g., "chemical fossils"), kerogen and coal, and graphite. What is found in a particular sediment depends not only on the initial input of organic matter, but also on time, temperature, and the geological environment, both from the points of view of structural geology and of inorganic mineralogy. We have noted above, for example, that petroleum has its origin in the processes of catagenesis. Time-temperature "trade-offs" in this process have been explored by Connan (1974), who has shown, for example, that the same processes that have occurred in 12 million years at 124° C in the Los Angeles Basin have occurred in 360 million years at 60° C in the Amazon Basin.

Gas, oil, and other molecular species can persist in a given sedimentary environment only if temperatures do not become high enough to degrade them. Substantial accumulation of these mobile species requires, in addition, the geological preservation of whatever rock unit might be trapping them and preventing further migration that would allow them either to be recycled by re-exposure at the surface or to be altered by migration into some high-temperature environment). Evans et al. (1971) have examined thermal alteration of petroleum in Western Canada and have estimated a maximum temperature of preservation of 160–188° C for economic deposits of petroleum. Where this temperature has been exceeded, only methane and pyrobitumen (a solid, largely insoluble hydrocarbon, apparently crosslinked and partly aromatic, usually brown to black) remain (Rogers et al., 1974). Precambrian materials have had a much better chance to be deeply buried than have more recent materials, and, when the time factor is considered in addition, it is not surprising that occurrences of Precambrian petroleum are rare. More specifically, no significant accumulations of crude oil have been found in rocks older than about 10^9 years.

"Chemical fossils," well-preserved, highly characteristic, biosynthetic products, constitute molecular remains of special interest to the paleobiologist. Though nucleic acids, proteins, and specific carbohydrates might constitute some of the most definitive chemical fossils, their early disappearance from the record has already been noted, and, in practice, only especially resistant biomonomers have provided secure organic chemical evidence of the biogenic origin of much sedimentary organic matter. Lipids, or, more explicitly, the defunctionalized and saturated products of their mild diagenetic alterations, have been the most prominent chemical fossils, with amino acids also receiving much attention. An example of the former category is provided by the hopanoids, a subgroup of the pentacyclic triterpanes. The comparative biochemistry of the precursors of these materials is not yet completely resolved, though it is now recognized that they are widely distributed in microorganisms where they probably play much the same role that sterols play in higher organisms (Ourisson et al., 1979). The wide range of structural variation that these compounds offer gives them great potential as chemical fossils amenable to detailed interpretation. Seifert and coworkers (e.g., Seifert and Moldowan, 1979) have provided several examples of investigations of this kind. Chemical fossils can apparently also be preserved by temporary incorporation in the insoluble fraction of sedimentary organic matter. Michaelis and Albrecht (1979) have recently reported that a C_{40} head-to-head polyisoprenoid carbon skeleton characteristic of the cell walls of thermophilic and methanogenic bacteria can be recovered by the application of ether-cleavage reagents to the insoluble fraction of an oil shale, and Moldowan and Seifert (1979) have reported the discovery of this same carbon skeleton (parent and degradation products) in crude oil. The temperature limits for survival of these branched and cyclic carbon skeletons over geologic time intervals are not precisely known, but they must be equal to or lower than those for crude petroleum.

The main chemical product of diagenesis is "kerogen" (Durand, 1980). It is an insoluble macromolecular material with no regular structure (lacking any well defined "monomer," it can not properly be called a "polymer"). The term "kerogen" has about the same level of chemical specificity as the term "metal." "Kerogen" can indicate materials with very different elemental compositions and, equally importantly, it is applied to materials with very different levels of maturity. Kerogens are classified as type I, type II, or type III depending on their ultimate ecological source (Tissot and Welte, 1978, pp. 142–5). Type I kerogens are derived from microbiological debris. Such material has a relatively high lipid content and, consequently, type I kerogens are initially hydrogen-rich, with H/C ratios near 1.7 or more. Type III kerogens are derived chiefly from higher plant debris and, even after bacterial reworking, are relatively hydrogen-poor, having initial H/C ratios near 0.8 and lower. Type II kerogens are intermediate in H/C ratio.

The elemental compositions and chemical structures of all these materials change substantially as they are affected by thermally driven processes occurring in the sedimentary environment. As all kerogens are degraded to graphite under the most rigorous conditions, their compositions eventually become indistinguishable. Notwithstanding this ultimate convergence to a graphitic composition and structure, it is unfortunate that wherever any macromolecular sedimentary organic material might lie along the pathway leading from its initial composition to graphite, the term "kerogen" will still be applied to the substance under discussion.

5-3. The Organic Geochemical Record and Its Elucidation

5-3-1. Chemical Fossils in Precambrian Sediments

The specific identification in Precambrian materials of organic molecules with biosynthetically characteristic carbon skeletons was first reported in 1964 and 1965. Many additional reports have followed, with critical and encyclopedic reviews being provided by McKirdy (1974) and by McKirdy and Hahn (1981). Workers in this field have frequently been their own severest critics, and, for a variety of reasons, the significance of these Precambrian chemical fossils is now unclear.

The development of information regarding Precambrian amino acids is exemplary. Although Abelson carried out and discussed analyses of amino acids in microfossiliferous cherts of the Early Proterozoic Gunflint Formation as early as 1955 (as reported by Barghoorn, 1957), the first detailed report of evidence for the presence of amino acids (at a concentration near 10^{-8} moles total amino acids/g rock) in three different Precambrian sediments was published much later by Schopf et al. (1968). Shortly thereafter, Kvenolden et al. (1969) used newly developed techniques to show that the amino acids in the Fig Tree chert (one of the specimens analyzed in the earlier study), at least, were optically active (all of the L enantiomer). Because it can be shown (e.g., Bada and Schroeder, 1975) that any amino acids stored at or above 25° C for 10^6 yr or more should have been fully racemized, this result required that the amino acids were much younger than the rocks from which they had been isolated, and it was concluded that percolating ground waters must have carried the amino acids to their present locations relatively recently. Investigation of the permeabilities and porosities of even the "tightest" Precambrian sediments has shown that this is quite plausible (Nagy, 1970; Smith et al., 1970; Sanyal et al., 1971). Therefore, if the amino acids are, indeed, chemical fossils, they are representative not of the fossil biocoenose associated with the sediments where they are found but of some relatively recent and quite unrelated ecosystem.

Many other reports of Precambrian chemical fossils have been concerned with hydrocarbons. Although the internal "stereochemical clock" in these materials has not yet been calibrated, the same reasoning that has discredited the amino acids as Precambrian chemical fossils can be applied to them. Moreover, several additional lines of evidence are available. First, a sequential extraction procedure employed by Smith et al. (1970) showed that the extractable hydrocarbons in three different fossiliferous Precambrian cherts were located primarily at grain boundaries and at other readily accessible sites, rather than being included within the silica crystals. Dissolution of these Precambrian cherts by hydrofluoric acid yielded a maximum of 2 parts in 10^9 extractable hydrocarbons from the pre-extracted chert powders, an amount small in comparison to that obtained from the surfaces of the particles prior to dissolution. By contrast, a Phanerozoic control (a Triassic chert rich in plant debris) yielded a substantial fraction of its total extractable hydrocarbons only after dissolution, thus indicating that these materials must be within the silica grains, as is the kerogen, which is quite visible in thin sections. Second, Hoering (1967) has shown that, in five of nine cases examined, the $\delta^{13}C$ content (see Chapter 7) of Precambrian hydrocarbon extracts is more than 7.5 ‰ higher than that of the coexisting kerogen. This contrasts strongly with the relationship more commonly observed in geologic materials, in which the hydrocarbon extract is slightly depleted in ^{13}C relative to the total organic carbon in the host rock. Accordingly, the extractable hydrocarbons are thought to be, in those five cases at least, non-indigenous. Hoering's impeachment of these materials should probably stand, but his suggestion that 2 ‰ should be defined as a maximum acceptable isotopic contrast between kerogen and extract is too extreme in some senses, at least. Extracts more than 2 ‰ heavier than the coexisting kerogen are very likely not indigenous, but extracts more than 2 ‰

lighter might be indigenous in some cases, particularly when the system has not been closed (i.e., when there has been some opportunity for loss of carbon). For example, Gormly and Sackett (1977) have shown that nonpolar extracts of recent marine sediments are usually depleted in carbon-13 by 5 to 7 ‰ relative to the total organic carbon; Baker and Claypool (1970) found that the hydrocarbon extracts in 29 samples of slightly metamorphosed Phanerozoic mudrocks averaged 3.4 ‰ lighter than the coexisting kerogen; and Deines (1980, p. 366) has shown that Phanerozoic petroleums are commonly 2 to 4 ‰ depleted in ^{13}C relative to their presumed source rocks. Finally, J. H. Oehler (1977a) has demonstrated that Precambrian kerogens are capable of selectively and of very tenaciously retaining hydrocarbons to which they might be exposed, thus supporting the possibility that the presently found hydrocarbons are later additions.

The status of chemical fossils in the Precambrian can, thus, be easily summarized: the low concentrations, evident mobilities, and recognized instabilities of many "Precambrian" chemical fossils makes their interpretation problematic and leads to the conclusion that these materials do not provide compelling evidence regarding any details of early life. The crucial question is not whether the many reported occurrences of chemical fossils in Precambrian rocks can be validated; rather it is whether the chemical fossils are demonstrably syngenetic with the associated sediments, whether these (in the words of Tom Hoering) "fancy organic molecules" are in fact as old as the scarred, veteran rocks in which they are found.

McKirdy (1974, p. 105) closes his discussion of chemical fossils with an incisive observation: "no critical examination or systematic assessment of the degree of metamorphic alteration sustained by the Precambrian sediments in question has yet been undertaken. In fact, there is considerable evidence that these and other very old sediments cannot have escaped such alteration." This juxtaposition of topics—chemical fossils, metamorphic histories—brings the problem clearly into focus. It leads naturally to a simple but important question: can the relationship between histories of sediment accumulation and preservation and the survival of chemical fossils be understood? If Phanerozoic organic geochemistry provides any clues regarding productive lines of inquiry—and it certainly must—then the answer to this question can be based on an understanding of the geochemistry of kerogen.

5-3-2. Kerogen Analyses, Methods, and Results

An indispensable step in forming an understanding of kerogen is the development and application of informative analytical techniques. We have already noted that kerogen is insoluble, does not have a regular structure, is the product of a complicated and variable set of diagenetic reactions, and changes its structure and composition over time due to conditions in the sedimentary environment. The resultant complexity presents a challenge far beyond the capabilities of any single technique, and it is evident that a variety of analytical tools must be employed. It is useful to define three groups of such tools: (*i*) "overall" techniques that provide information about the average structure and composition of all the organic matter in a given sediment; (*ii*) structural techniques that allow qualitative recognition or semiquantitative measurement of some of the structural features of the organic matter; and (*iii*) degradative techniques that allow isolation of some more or less well-defined subunits. To apply this sequence of techniques is to know more and more about less and less. While the degradative techniques can provide an exquisitely detailed view of some portions of the kerogen, the material isolated might represent less than one percent of the total organic carbon. In these circumstances, an overall technique has value if it provides information that is representative, however crudely, of all the organic material present.

OVERALL ANALYSES

Elemental Analysis. The organic chemist's standby, elemental analysis, has provided some of the most important information regarding kerogen (Durand and Monin, 1980). Carbon, hydrogen, and nitrogen are usually quantified by analysis of their combustion products. Oxygen is most commonly analyzed indirectly, by difference, because it is almost the only other element present in most pure organic natural products. This approach is dangerous for many practical reasons, however, and Durand and Monin (1980) point out the advantages of direct pyrolytic methods based on the Unterzaucher technique, in which the oxygen present is converted quantitatively into carbon monoxide, which is then quantified in order to finish the analysis.

Because many minerals could contribute exogenous hydrogen and oxygen, it is, in practice, necessary that the elemental analysis be preceded by isolation of the kerogen, a procedure involving the use of hydrochloric and hydrofluoric acids, inorganic ligands, and other reagents to dissolve most of the inorganic materials, leaving an organic-rich residue (Saxby, 1976; Durand and Nicaise, 1980; Appendix IV, this volume). This isolation procedure can itself affect the elemental composition of the kerogen. Although the strong mineral acid solutions will exert a hydrolyzing effect, there is evidence that the net change in elemental composition, even for highly functionalized or immature kerogens, is small (Saxby, 1976; Durand and Nicaise, 1980). Loss of up to 10% of the nitrogen has been observed, though smaller fractions are lost from highly aromatized and cross-linked materials (Saxby, 1970).

Pyrite (FeS_2), an extremely common companion of kerogen in sediments, is unaffected by hydrochloric and hydrofluoric acids and must be oxidized or reduced if its removal is required (amounts of up to 40% can be tolerated without significantly affecting the elemental analysis; see Durand and Monin, 1980). Nitric acid is com-

TABLE 5-2. Elemental compositions of kerogen precursors.

Precursors	*H/C*	*O/C*	*N/C*	*Reference**
COMPOUND CLASSES				
Carbohydrates	1.62	0.85		1
Lignin	0.94	0.38	0.004	1
Proteins	1.57	0.31	0.28	1
Lipids	1.88	0.12		1
LIVING MATERIALS (DRY)				
Trees	1.4	0.6	0.01	2
Grasses	1.5	0.5	0.04	2
Phytoplankton	1.6	0.5	0.1	2
Zooplankton	1.7	0.4	0.15	2
Bacteria	1.6	0.4	0.16	2
MICROBIOLOGICAL DEBRIS				
Laboratory cultures	1.6	0.4	0.2	3
Decomposed algal mat	1.8	0.5	0.08	4
Black anaerobic ooze	1.9	0.5	0.06	4
HUMIC ACIDS, RECENT				
Terrestrial	0.9–1.4	0.4–1.0	0.02–0.09	5, 6
Coastal	1.0–1.4	0.4–0.6	0.01–0.12	5, 7
Open marine	1.1–1.6	0.4–0.9	0.02–0.18	5, 6
HUMIC ACIDS, ANCIENT				
Terrestrial	0.8	0.4–0.5	0.02–0.03	8
Coastal	0.8–1.0	0.3–0.5	0.05	8
Open marine	1.0–1.2	0.2	0.03	8
IMMATURE KEROGENS				
Type I	1.7	0.1	0.05	9
Type II	1.4	0.2	0.03	9
Type III	0.8	0.4	0.04	9

* References: 1. Hunt (1979), p. 85; 2. From elemental compositions of compound classes (above) and compound-class abundances cited by Huc (1980); 3. Philp and Calvin (1976); 4. Philp and Calvin (1977); 5. Nissenbaum and Kaplan (1972); 6. Stuermer et al. (1978); 7. Brown et al. (1972); 8. Huc and Durand (1977); 9. Durand and Monin (1980).

monly employed as an effective oxidant, but it adds significant quantities of nitrogen and oxygen to the kerogen (Lawlor et al., 1963; Saxby, 1976; Durand and Nicaise, 1980). Lithium aluminum hydride ($LiAlH_4$) and sodium borohydride ($NaBH_4$) can be employed as reductants. They are not quantitatively effective, but they cause no significant change in the H/C ratio of the kerogen (Lawlor et al., 1963; Saxby, 1976; Durand and Nicaise, 1980), apparently because the concentration of reducible functional groups is very low and possibly because the elimination of pyrite helps to reduce the contribution of adventitious, mineral-derived water to the analysis.

The elemental compositions of a variety of kerogen precursors are summarized in Table 5-2. Although the various compound classes present in living material differ quite significantly in their elemental compositions (with lignin, in particular, being poor in hydrogen, and carbohydrates being rich in oxygen), the overall elemental compositions of most organisms lie within a rather narrow range, with H/C values generally being within 10% of 1.54 and O/C values being within 20% of 0.5. There is, however, a well-recognized enrichment in nitrogen in marine organisms, which have, on average, an N/C ratio about ten times greater than the average of the terrigenous biomass. Table 5-2 shows that the microbiological debris that presumably contributes most strongly to type I kerogens is, as expected, rich in hydrogen, but that the elemental compositions both of recent and of ancient humic acids cannot be specifically correlated either with their logical precursors or with the kerogens to which they must eventually contribute. Specifically, few if any humic acids are as rich in hydrogen as are the type I kerogens, and most recent land-derived humic acids have an elemental composition which does not match either that of their plant source or the elemental composition of type III kerogens. Two interesting conclusions follow from these observations: (*i*) not all material found in kerogen has passed through an intermediate humic-acid stage, the type I kerogens, in particular, receiving much biological material directly (Tissot and Welte, 1978, p. 87); and (*ii*) the hydrogen-abundance contrasts that are so evident among the kerogen types must develop during diagenesis, and are not due simply to differences in the hydrogen abundances in the biological precursors. Both the environmental conditions usually prevailing during the diagenesis of terrigenous kerogen precursors and the relatively high abundance of oxygen in carbohydrate- and lignin-rich materials are, no doubt, important in the reduction of the hydrogen content of the humic acids and later of type II and type III kerogens.

The evolution of the structure of kerogen during diagenesis, catagenesis, and metagenesis is very clearly reflected by changes in elemental composition. Figure 5-4, a "van Krevelen diagram," plots the relative hydrogen abundance versus the relative oxygen abundance and shows the pathways that the various kerogen types tend to follow as their compositions become more and more graphitic. The network of lines superposed on the graph

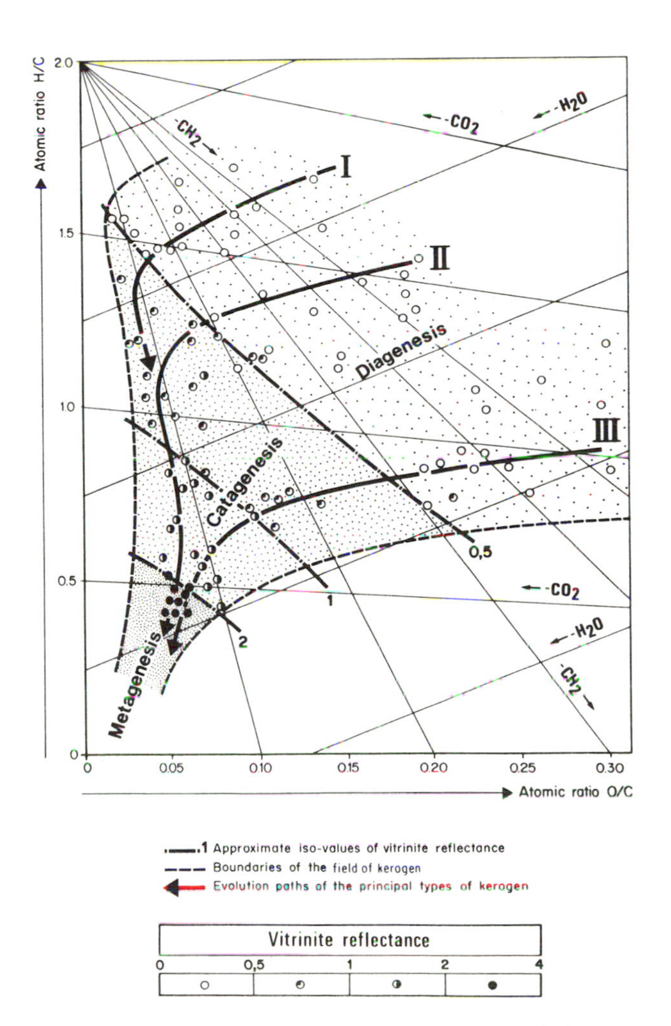

FIGURE 5-4. A van Krevelen diagram showing the evolution of the elemental composition of kerogen during diagenesis, catagenesis, and metagenesis. Approximate values of vitrinite reflectance are shown for comparison. From Tissot and Welte (1978), with permission.

shows the effects of the elimination of water, carbon dioxide, or paraffins from the kerogen structure. The observed parallels between those lines and the kerogen-evolution paths suggest strongly that diagenesis involves primarily the loss of water and carbon dioxide and that the hydrogen loss during catagenesis of type I and type II kerogens can be associated with the production of petroleum. The residue after catagenesis must be highly aromatic (an H/C ratio of 0.5 corresponds, for example, to the aromatic hydrocarbon coronene, $C_{24}H_{12}$—see Figure 5-10), and still more extensive rearrangements of the carbon skeleton must, therefore, accompany the metagenetic evolution toward graphite.

Optical Techniques. Kerogen fractions can also be mounted in epoxy resin and polished for microscopic study. When this is done, it is observed that vitrinite, an easily recognized type of higher plant debris first defined in coals and, hence, named as a coal maceral (other examples: liptinite, inertinite), is widely distributed in sediments, occurring in trace quantities even in many kerogens present in typically marine sediments. It has been shown that the reflectance of vitrinite particles increases smoothly as the coexisting kerogen matures (a detailed review of procedures and supporting evidence is provided by Hunt, 1979, pp. 328–339). Vitrinite reflectance is usually measured using high magnification, oil-immersion microscopy and is assigned the symbol R_o (R_a designates reflectance measured in air). As shown in Figure 5-4, $R_o = 0.5\%$ falls at the boundary between diagenesis and catagenesis, with $R_o = 2\%$ at the boundary between catagenesis and metagenesis. The very large number of measurements available for correlative studies has made vitrinite reflectance an extremely useful parameter in kerogen analysis and has shown that it is one of the most reliable known indicators of the state of maturation of organic matter.

Organic material seen in thin section can display colors in transmitted light ranging from light yellow to black depending on the extent to which it has been thermally modified. Correlations between the color of amorphous organic matter and many other indices of organic maturity have been reported (Tissot and Welte, 1978, pp. 449–471; Hunt, 1979, pp. 320–350; Héroux et al., 1979). Correlations between elemental compositions and colors of a marine kerogen with an initial elemental composition near that shown in Table 5-2 for a type II kerogen have been reported by Peters et al. (1977), who carried out short time (5 to 116 hr.), high temperature (150 to 410°) laboratory heating experiments. These correlations are summarized in Table 5-3. An ability to interpret observed color variations has several advantages: (*i*) it helps to bridge the gap between micropaleontology and organic geochemistry (note, however, that in order to avoid confusion with colored minerals, it is particularly useful to observe the color of the organic matter in strew slides prepared from isolated kerogen fractions or other macerations); (*ii*) it can reveal the presence of otherwise inseparable kerogen mixtures (e.g., the sedimentary combination of a recycled, highly mature, near-black material with an immature, yellow or light brown material); and (*iii*) it provides an alternative to measurements of vitrinite reflectance in materials that contain no vitrinite, an advantage of special importance in Precambrian studies.

Isotopic Analysis. The quantitative oxidation of kerogen to CO_2, H_2O, and N_2 (a step which is, in any case, necessary as part of most elemental analyses), provides an opportunity to obtain average isotope ratios for C, H, and N by analysis of the combustion products. As noted in Chapters 7 and 12, the isotopic composition of the carbon is of particular interest, its value relative to coexisting carbon reservoirs being controlled by isotope effects associated with, for example, biological carbon-fixation processes. Prior to any discussion of ancient biochemistries, however, a preliminary question arises, namely: given the complex chemical reactions involved in the generation and maturation of kerogen, how can it be determined that the isotopic composition of the kerogen residue—the material analyzed—matches that of its biological precursor, a substance that has been stored and reworked, possibly, for aeons? This question can be broken into two parts: (*i*) What is the relationship between the isotopic composition of immature kerogen and that of the primary producers in the ecosystem from which the kerogen is derived? (*ii*) How does the isotopic composition of kerogen evolve during diagenesis, catagenesis, metagenesis, and metamorphism?

A comparison of organic carbon in recent sediments with that in marine organisms can help to answer the first question. Deines (1980a) has considered the results of isotopic analyses of more

TABLE 5-3 Relationship between color and elemental composition of kerogen.

Observed color	*Thermal alteration index*		*H/C*		
	A	B	A	B	C
light yellow	1	1	1.5		
yellow	1+	1+	1.4	1.3	1.24–1.37
orange	2		1.1		
brown yellow		2		1.2	
light brown					1.07–1.22
brown	3	3	0.75	0.7	0.95–1.05
dark brown					0.82–0.91
very dark brown					0.49–0.89
brownish black	4		0.5		
black	5	4 and 5	<0.4	<0.5	<0.5

Columns
A Hunt (1979), pp. 324, 345.
B. Tissot and Welte (1978), p. 467.
C. Peters et al. (1977).

than 1,600 individual samples of Recent marine sediments and has concluded that a representative value lies near $\delta^{13}C_{PDB} = -25.0$ ‰. Marine biological material covers a wide range of isotopic compositions, commonly between −15 and −30 ‰ vs. PDB (Deines, 1980a) [the observed range of values is larger, though of uncertain significance—marine organisms having $\delta^{13}C_{PDB}$ values as high as −9 ‰ have been observed (Craig, 1953; Park and Epstein, 1961), and some pogonophoran worms have recently yielded values as low as −46 ‰ (Rau, 1981, pers. com. to J.M.H.)]. It is not possible to link specifically the average sedimentary value cited above with some properly averaged worldwide isotopic composition of marine biological carbon, but it is certainly true that the relatively heavy isotopic compositions (∼ −20 ‰) of mid- and low-latitude plankton (Sackett et al., 1974) can be correlated with Recent sediment values reported by Degens (1969, p. 326) for those same regions, and that the isotopic compositions of Antarctic plankton (−25 to −30 ‰, Sackett et al., 1974) can also be correlated with those of some high-latitude sediments (Degens, 1969, p. 326; Sackett et al., 1974). The relationship between these observations and a global sedimentary average isotopic composition near −25 ‰ (Deines, 1980a) is unclear. It can be concluded only that the available evidence is consistent with the hypothesis that the isotopic composition of Recent marine sedimentary carbon is controlled by, and is equal to that of, the primary producers in the overlying water column. Alternatively, it appears possible that the organic matter in even the most recent sediments is systematically depleted in ^{13}C by 2–4 ‰ relative to its biological source (Aizenshtat et al., 1973; Sackett et al., 1974; Gormly and Sackett, 1977), possibly by selective preservation of lipid materials that are characteristically depleted in ^{13}C relative to other biosynthetic products.

The second question requires consideration of processes that might cause the isotopic composition of kerogen to change during diagenesis, catagenesis, metagenesis, and metamorphism. Losses of carbon can be significant during these processes, and the isotopic composition of the residue will differ from that of the starting material if the carbon lost has an isotopic composition differing significantly from that of the bulk kerogen. Two kinds of carbon loss can be envisioned: (*i*) the expulsion of a particular subset of carbon atoms without regard to isotopic composition, the loss of carboxyl groups being one possibility, with an isotopic shift resulting if and only if the carboxyl groups have an isotopic composition differing from the residue; and (*ii*) the selective and progressive loss of one isotope by means of a reaction having a kinetic isotope effect, the progressive loss of alkyl substituents to yield gaseous products being an example. In fact, most natural systems probably combine aspects of both of the these models.

A quantitative treatment of the first possibility can be summarized in terms of a "mass balance" equation

$$\begin{bmatrix}\text{moles of } ^{13}\text{C in} \\ \text{kerogen initially}\end{bmatrix} = \begin{bmatrix}\text{moles of } ^{13}\text{C in} \\ \text{material lost}\end{bmatrix} + \begin{bmatrix}\text{moles of } ^{13}\text{C in} \\ \text{residual kerogen}\end{bmatrix} \quad (5\text{-}1)$$

which can be rewritten as

$$n_i F_i = n_l F_l + n_r F_r \quad (5\text{-}2)$$

in which the n terms represent moles of carbon ($^{12}C + ^{13}C$). The subscript, i, denotes the carbon initially present (e.g., at the time of deposition); l, the carbon lost; and r, the carbon in the residual kerogen. The F terms represent fractional abundance of ^{13}C [that is, $F = ^{13}C/(^{13}C + ^{12}C)$], the subscripts having the same significance as noted above. An approximate form of this equation is given by

$$n_i \delta_i = n_l \delta_l + n_r \delta_r \quad (5\text{-}3)$$

where the δ terms represent isotopic abundances relative to some specified standard (see Chapters 4 and 7). The amount lost can be expressed as a fraction of that initially present. This fraction is given by n_l/n_i and can be assigned the symbol f_l. The fractional amount remaining in the kerogen is given by $1 - f_l$. The average isotopic fractionation between the initial material and the material lost can be assigned the symbol Δ_l and is defined by $\Delta_l = \delta_i - \delta_l$; the fractionation between the initial material and the residue being given by $\Delta_r = \delta_i - \delta_r$. Substitution in Equation 5-3 then gives

$$\Delta_r = -[f_l/(1 - f_l)]\Delta_l \quad (5\text{-}4)$$

Note that Δ_r, the shift in the isotopic composition of the kerogen, will be smaller in absolute magnitude than Δ_l whenever less than half of the kerogen is lost, but that relatively large shifts in the isotopic composition of the residue will occur beyond that point.

The isotopic evolution of the residual kerogen during a loss of material by some process having a kinetic isotope effect can be calculated by reference to the Rayleigh distillation equation, an approximate expression for the isotopic composition of the total carbon lost at any value of f_l being given by

$$\delta_l + 10^3 = [1 - (1 - f_l)^{-\alpha}](\delta_i + 10^3)/f_l \quad (5\text{-}5)$$

in which α is the "fractionation factor" (in this case, the ratio of isotopic rate constants for a given chemical process). In the cleavage of carbon-carbon bonds to release hydrocarbons, for example, α would represent $^{12}k/^{13}k$, the latter terms being rate constants for cleavage of bonds to ^{12}C and ^{13}C, respectively. A typical value in this case might be $\alpha = 1.03$, a fractionation factor that a physical chemist would describe as "a three percent kinetic isotope effect." Calculation of the isotopic composition of the residual kerogen requires only the solution of an isotopic mass balance equation (e.g., Equation 5-3)

$$\delta_r = (\delta_i - f_l\delta_l)/(1 - f_l) \quad (5\text{-}6)$$

in which the values of δ_i and f_l are equal to those employed in the calculation of δ_l, that value being obtained using Equation 5-5.

The transition from Recent sediments to lithified, ancient materials appears to cause some preferential loss of ^{13}C. A tabulation by Deines (1980a) of more than 300 analyses of "kerogen and graphite" in "sediments and metasediments" provides an average value of $\delta^{13}C_{PDB} = -26.8$ ‰. The distribution tails significantly on the heavy side, however, an effect possibly due to the inclusion of ^{13}C-enriched metamorphosed materials (see below). If the average is, for that reason, viewed as possibly not representative of immature kerogen, an improved estimate may be provided by the mode of the distribution, which falls at -27.5 ‰ vs. PDB. The latter value is in good agreement with a mean of -27.7 ‰ vs. PDB that can be calculated from a separate tabulation of 222 analyses of unmetamorphosed Phanerozoic sediments (Degens, 1969, p. 327). On the basis of this agreement, $\delta^{13}C_{PDB} = -27.5$‰ can be accepted as representative of immature kerogens just as a value of -25.0 ‰ seems to be representative of total organic carbon in recent sediments, and it can be observed that this difference indicates a 2.5 ‰ depletion in ^{13}C.

The synthesis of immature kerogens from their precursors is marked by a large decrease in the O/C ratio of the organic matter (Table 5-2). Losses of

CO_2 must be important in this transition, since the observed small decreases in the H/C ratio are not compatible with extensive water losses. It is relevant that there is direct evidence that some, though not all, biologically-formed carboxyl (–COOH) groups are relatively enriched in ^{13}C (Abelson and Hoering, 1961; Meinschein et al., 1974; Rinaldi et al., 1974; Vogler and Hayes, 1980). Further, it has been suggested on theoretical grounds that carboxyl groups should generally be enriched in ^{13}C (Galimov, 1973). Peters et al. (1980) have observed that artificially heated "proto-kerogens" display an initial depletion in ^{13}C coupled with an overall carbon loss and a decrease in the O/C ratio, and possibly related indications of early ^{13}C-depletion have been observed in naturally formed kerogens (Brown et al., 1972; Gormly and Sackett, 1977; Redding et al., 1980). Not surprisingly, therefore, it has been suggested (Tissot and Welte, 1978, p. 89; Chung and Sackett, 1979; Peters et al., 1980) that such depletions in ^{13}C are due to the elimination of carboxyl groups, the relatively high ^{13}C content of these groups resulting in a ^{13}C-depletion of the residual kerogen. If it is estimated that the carboxyl groups might be enriched in ^{13}C by 10 ‰ on average and that 10% of the carbon might be lost through decarboxylation reactions, application of Equation 5-4 (f_l = 0.1, Δ_l = 10 ‰) shows that the isotopic composition of the residue might shift by ~1.1 ‰ due to this process. It seems likely, therefore, that some additional mechanism is required to account for the estimated depletion of 2.5 ‰, the preferential accumulation by the kerogen of ^{13}C-depleted lipid carbon skeletons being a likely possibility (Gormly and Sackett, 1977).

Catagenesis and metagenesis involve decreases in the H/C ratio due in part to the production of hydrocarbons, beginning with petroleum (H/C ~ 1.8) and ending with natural gas (H/C ~ 4). The amount of carbon that can be lost through these processes depends on the amount of hydrogen available and on the extent to which hydrocarbons, once formed, are able to migrate away from the reaction zone. The isotopic consequences must depend on the relative importances of the various carbon-loss mechanisms (the production of methane, in theory, at least, being accompanied by the larger isotope effect), on the degree of completion of the reaction, and on details of the structure of the kerogen. This list of genuinely relevant details provides such scope for variation that the isotopic shifts accompanying catagenesis and metagenesis must be recognized as not always predictable and potentially inconsistent.

Large shifts in the isotopic composition of the residue might be expected in some circumstances. For example, Tissot and Welte (1978, p. 455) suggest that the H/C ratio of a type I kerogen at the onset of oil generation is approximately 1.45 and that oil generation continues until an H/C ratio near 0.7 is reached. If the hydrocarbon products have an H/C ratio near 1.8, a mass balance shows that the residual kerogen might account for only 32% of the material initially present. Deines (1980, p. 366) has summarized the isotopic compositions of petroleums relative to their source rocks and has shown that the petroleum is generally 2–3 ‰ depleted in ^{13}C. Application of Equation 5-4 (f_l = 0.68, Δ_l = 2.5 ‰) shows that the residual kerogen could be enriched in ^{13}C by as much as 5.3 ‰. No shift of this magnitude has been documented in nature (a range of 3.8 ‰ in a sequence of type I kerogens analyzed by Redding et al., 1980, may be largely due to primary effects, inasmuch as hydrogen isotopic compositions do not vary within the same sequence, suggesting that hydrocarbon losses have not been extensive) although Peters et al. (1981) have observed an enrichment of 4.5 ‰ in the ^{13}C content of the residue of an artificially heated algal kerogen.

Little or no shift in the isotopic composition of the residual kerogen is far more commonly observed. Degens (1969, p. 316) has noted that coals (type III kerogens) show no evidence of isotopic fractionation during catagenesis and metagenesis, an effect also noted by Redding et al. (1980). Peters et al. (1981) observed a ^{13}C enrichment of only 1 ‰ in an artificially heated type III kerogen. Type II kerogens are not as hydrogen-rich as type I materials and, furthermore, they have fundamentally different structures. Yields of extractable hydrocarbons seldom exceed 15% of the total organic carbon and, consequently, the expected isotopic shift in the residue is less than 0.5 ‰, a prediction that is borne out by many observations of nearly constant isotopic compositions in the H/C range above 0.3 (Baker and Claypool, 1970; McKirdy and Powell, 1974; Hoefs and Frey, 1976; Peters et al., 1978; McKirdy and Hahn, 1981).

During metamorphism still more carbon can be lost or, under the most rigorous conditions, some equilibration or exchange with "heavy" carbon in mobile fluids might take place (Hoefs and Frey,

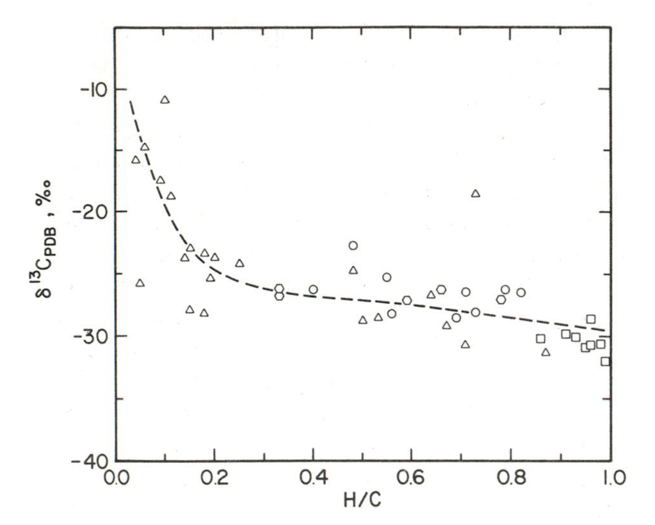

FIGURE 5-5. The relationship between carbon isotopic composition and H/C for reliably analyzed kerogens with ages of 1.6 Ga or less. The broken line is intended to represent the average distribution of the points. Triangles represent Proterozoic and Cambrian materials from Australia (McKirdy and Powell, 1974). Hexagons represent contact-metamorphosed tertiary lignite (Redding et al., 1980). Circles signify Phanerozoic kerogens (Redding et al., 1980). Squares signify Cambrian kerogen (McKirdy and Kantsler, 1980).

1976; Donnelly et al., 1977; Valley and O'Neil, 1981). In either case, substantial shifts in the ^{13}C isotopic abundance in residues can be expected. Figure 5-5 shows this most clearly, but many additional observations of increasing ^{13}C abundances at H/C ratios below 0.3 or at greenschist facies metamorphism and beyond can be cited (Baker and Friedman, 1969; Baker and Claypool, 1970; McKirdy and Powell, 1974; McKirdy et al., 1975; Hoefs and Frey, 1976; Oehler and Smith, 1977; Peters et al., 1978; Schidlowski et al., 1979; Valley and O'Neil, 1981). The general trend of these additional observations is fully consistent with Figure 5-5. Where the maximally shifted kerogen residues (equivalent to graphite) have isotopic compositions within a few permil of coexisting carbonates, the data suggest that exchange (Bottinga, 1969) between residual graphites and mobile fluids having isotopic compositions buffered by abundant sedimentary carbonates, might be taking place. Alternatively, Hoefs and Frey (1976) have suggested that a kinetic isotope effect associated with the oxidation of graphite by H_2O might be responsible. Finally, however, it must be observed that metamorphosed kerogen residues are not inevitably enriched in ^{13}C. In one of the earliest studies of this type, Gavelin (1957) found no apparent enrichments of ^{13}C in graphites from Swedish metamorphic Precambrian sediments, and Hoefs (1980, pers. com. to J. M. H.) has recently found at least one other similar example.

A summary of the isotopic relationships expected is shown in Figure 5-6. Although a substantial range of possibilities must be considered, it should be noted that once the stage of an immature kerogen has been reached, all succeeding shifts are zero or positive. It follows, therefore, that isotopic compositions observed for kerogen residues found in their original depositional environments represent maximum values. Kerogenous precursor materials might have been lighter, but not heavier.

DETERMINATION OF STRUCTURAL CHARACTERISTICS

Many techniques have provided useful views of specific structural aspects of kerogen. Infrared spectrophotometry, for example, has shown which functional groups are most prominent in immature kerogen, and x-ray diffraction techniques have provided a reasonably clear view of the approach of the carbon skeleton to a graphitic final product.

Infrared Spectrophotometry. These techniques and their results have been recently reviewed by Rouxhet et al. (1980). Trends in the distribution of oxygen among different functional groups can be clearly recognized. Ester linkages account for most of the oxygen in an immature type I kerogen, are of approximately equal absolute abundance but account for only one-third of the oxygen in an immature type II kerogen, and are essentially absent in an immature type III kerogen, in spite of the six-fold higher (relative to type I) oxygen content of the latter material. Both absolute and relative abundances of ketone and free-acid functional groups increase in the sequence type I to type III, but the most spectacular increase, accounting finally for half of the oxygen in a type III material, is in noncarbonyl oxygen incorporated in heterocyclic structures and in ether linkages.

The development of hydrocarbon structural units has been particularly studied by Robin et al. (1977), who have documented the expected disappearance of aliphatic C–H bonds at increasing levels of maturity, and who have shown that the disappearance even of aromatic C–H bonds can be recognized as structures approach extreme levels of graphitization. It is apparent that this

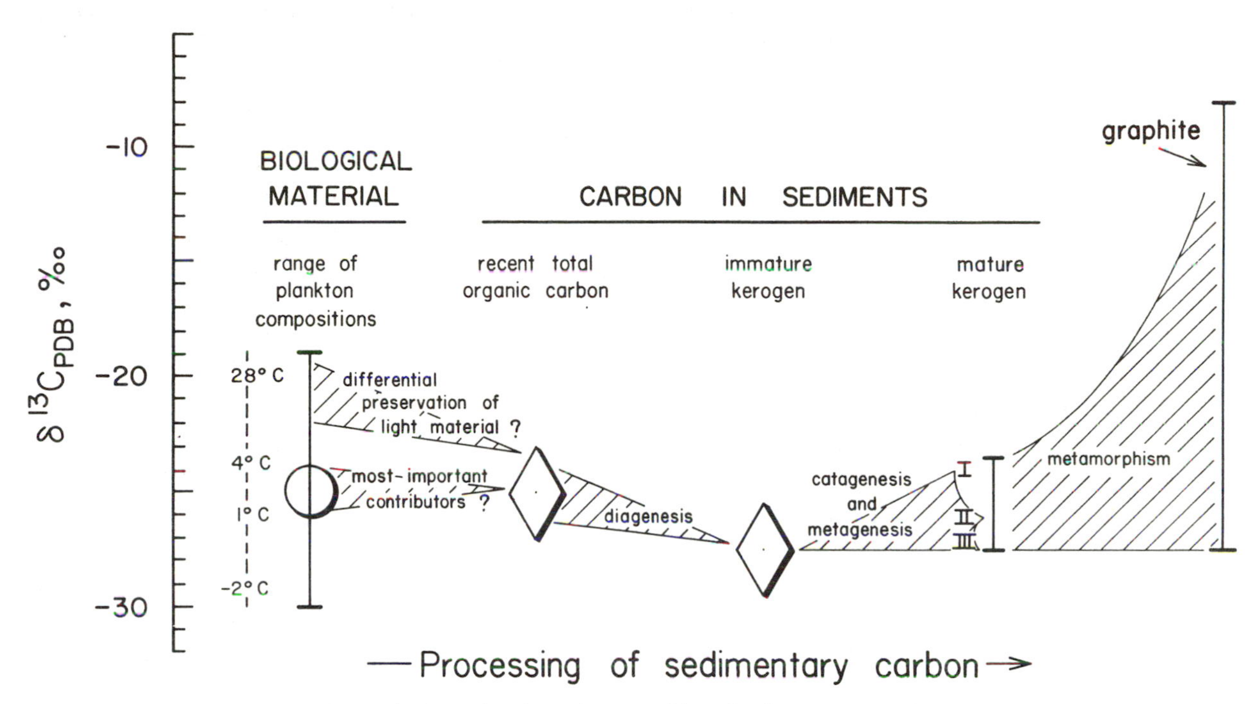

FIGURE 5-6. A summary of the changes in the carbon-isotopic composition of sedimentary organic matter.

final stage of maturation does not differ between the various kerogen types, all of which seem to be structurally equivalent at H/C ratios less than 0.6.

X-Ray Diffraction. Only crystalline materials provide well-defined X-ray diffraction patterns, and it is in the measurement of graphite crystallinity that this technique has been applied in kerogen analysis. The first study of interest in the present context was reported by French (1964), who examined the development of graphite crystallinity at increasing levels of metamorphic alteration of the Early Proterozoic Biwabik Iron Formation. Many later studies are cited in Table 5-4, which summarizes their results in terms of the correlations observed between various x-ray parameters and other indicators of maturation and metamorphism. Although fully ordered graphite displays many peaks in its diffraction pattern, interest has centered on the main (002) reflection because it is the first to appear in the diffraction pattern of kerogen as it develops from an amorphous to an ordered structure. The location of this peak is related to the spacing between the layers of aromatic rings in graphite. The sharpness of the peak is related to (among other things) the regularity of the structure and the size of the crystallites. Highly disordered material displays a broad, ill-defined peak, while fully ordered material, in which all the ring planes are evenly separated by the interlayer distance characteristic of perfectly crystalline graphite, displays a very sharp peak if the crystallites are at least a few thousand angstroms in diameter. In addition to peak location (tabulated in terms of the apparent interlayer spacing) and peak width at half height, Table 5-4 includes in some cases the ratio of peak height to peak width at half height (commonly reported as a dimensionless parameter based on distances measured on the diffractometer output chart). This measure of peak sharpness was introduced by Landis (1971) and is reported here only with reluctance and for purposes of correlation. Values of this parameter depend on experimental conditions, notably the relationship between diffractometer scan rate and recorder chart speed, amplifier output scale factor and recorder sensitivity, and on the distribution and amount of sample material on the supporting slide. Close comparisons between sets of measurements obtained under different conditions are, therefore, difficult, and the danger of misinterpretation is great.

TABLE 5-4. Correlations between graphite crystallinity and other indicators of maturity and metamorphism. (Entries are arranged in order of increasing alternation).

Entry no.	X-ray diffraction parameters: $w^{c}_{1/2}$ °2θ	d_{002} Å	$h/w^{d}_{1/2}$	Correlative parameters: H/C atomic ratio	R_0 %	Metamorphic indicators	Reference[b]
1	No graphitic pattern observed					"Unmetamorphosed" iron-fm.; siderite, ankerite, chamosite	1
2	No graphitic pattern observed				5(max)	Prehnite Zone, prehnite-pumpellyite facies	2
3	No graphitic pattern observed				≤3.2	T < 250°, prehnite-pumpellyite zone	3
4	>10*	>4*	0.15*	1.0			4
5	20*	3.6–3.9		0.7–0.8			5
6	20*	3.6–3.8		0.7[e]	≤1.1[e]	Coal ranks lignite; sub-bituminous A; high-vol. bitum. C, B, A.	6
7	11*	3.5–3.8	<0.5			Zeolite facies, prehnite-pumpellyite facies	7
8	7–12*	3.5–3.7		0.6–0.9			8
9	7*	3.5–3.6	0.1–0.3		3.5	250 < T < 270°, lawsonite zone, lawsonite-albite facies	3
10	8*	3.4–3.6	0.5–1			Lawsonite-epidote-pumpellyite zone, lawsonite-albite-chlorite facies pumpellyite-actinolite facies	7
11	6.5*	3.4*	0.6	0.6			4
12	5–7*	~3.5				Like entry 1 with grunerite and calcite	1
13	~8*	3.5		0.35–0.6[e]	1.1.–4.2[e]	Coal ranks medium and low vol. bitum. semi-anthracite, anthracite	6
14	4–5*	3.4–3.5		0.3–0.5			8
15	3.5–5*	3.5–3.6		0.2–0.3			5
16	5*	3.4–3.5	0.4–1.0		4.0–6.5	270 < T < 320, lawsonite zone, lawsonite-albite facies	5
17	5.6	3.6		0.075		Chlorite zone, greenschist facies	9
18	~5*	3.37–3.52	0.8–2.6	0.11–0.25	5.5–6	Chlorite zone, greenschist facies	10
19	2–4*	3.43				Like entry 12 but with iron pyroxenes	1
20	3.9*	3.4*	1.8	0.6			4
21	2.4*	3.4*	4.6	0.5			4
22	<4*	3.36–3.40	1–20			320 < T < 370, lawsonite-ferroglaucophane zone, lawsonite-albite facies	3
23	"Broad, ill-defined"				6	Chlorite zone, greenschist facies	2

* Value estimated from figure in publication.

[b] References: 1. French (1964). 2. Diessel and Offler (1975). 3. Diessel et al. (1978). 4. Powell et al. (1975). 5. Izawa (1968). 6. Griffin (1967). 7. Landis (1971). 8. Long et al. (1968). 9. Grew (1974). 10. McKirdy et al. (1975).

[c] Indicates width at half height of the 002 reflection of graphite or graphitic material.

[d] Indicates peak height (distance) divided by peak width at half-height (distance). Because the original measurement by Landis (1971) employed a system in which 13.3 mm on the diffractometer chart corresponded to 1° 2θ, all other measurements have been converted to yield values which are, in so far as possible, comparable. For example, for systems with 1° = 10 mm, the originally reported $h/w_{1/2}$ values have been divided by 1.333.

[e] Indicates value estimated from the reported coal rank by reference to the correlations summarized by Stach et al. (1975).

TABLE 5-4. *(continued)*

	X-ray diffraction parameters			*Correlative parameters*			
Entry no.	$w^{c}_{1/2}$ °2θ	d_{002} Å	$h/w^{d}_{1/2}$	*H/C atomic ratio*	R_0 %	*Metamorphic indicators*	*Reference*[b]
24	1.5*	3.37–3.44	3–15			Like entry 10 and in chlorite zone, greenschist facies and blueschist facies	7
25	<1.2	3.39–3.40		0.05–0.1			5
26	0.6–1.3	3.37–3.42		0.05		Chlorite zone, greenschist facies	9
27	<1*	3.36				Iron pyroxenes, grunerite, calcite	1
28	0.7*	3.37–3.39	2.2–6.8			Biotite zone, greenschist facies	2
29	0.5–0.7	3.37–3.38		0.01–0.03		Chlorite-biotite zone, greenschist facies	9
30	1.5*	3.35–3.38	3–15			Biotite zone, greenschist facies; blueschist facies	7
31		3.36–3.40	5–30	0.01–0.11	10.7	Biotite zone and higher	10
32	0.3–0.5	3.36		0.01–0.04		Garnet zone	9
33	0.3	3.36		0.01		Staurolite zone	9
34	0.2*	3.35	18–27			Andalusite-staurolite zone, amphibolite facies	2
35		3.35–3.36	>30			Amphibolite facies	7
36		3.36	15–52			390 < T < 440, epidote zone, amphibolite facies	3
37	0.2–0.3	3.36		0.00–0.01		Sillimanite zone, amphibolite facies	9
38	0.2*	3.36	>75			Sillimanite zone, amphibolite facies	2

Notwithstanding a good deal of scatter and incomplete or inconsistent reporting of correlative parameters, Table 5-4 shows clearly that the crystallinity of graphitic materials can frequently be correlated with the burial history of the host rock. Some development of graphitic structure apparently begins under conditions of very low grade metamorphism, just beyond the stage of organic geochemical catagenesis. The width of the 002 peak is generally less than 5° 2θ in the chlorite zone of the greenschist facies, with the width dropping to and below 1° 2θ by the close of greenschist facies metamorphism. Nearly complete dehydrogenation takes place in this interval, with H/C values dropping from barely postmetagenetic levels to values near zero as the greenschist facies is traversed. Higher levels of metamorphism lead to the development of fully ordered graphite.

CHEMICAL DEGRADATION OF KEROGEN

The isolation of structural subunits by oxidative and reductive techniques has been very well reviewed by Vitorović (1980). Of special importance recently, however, has been the development of practical pyrolytic techniques (reviewed by Tissot and Welte, 1978, pp. 443–47; and by Hunt, 1979, pp. 455–64). The low-temperature (<200°) heating of whole rock samples is used to remove and quantify organic materials not firmly bound to the kerogen (thus yielding "peak 1" in the pyrolytic procedure); a further increase in the temperature (to 550 or 600°) is then utilized to degrade the kerogen present in the sample (thus yielding "peak 2"). A single pyrolysis experiment can thus provide information both on the preformed hydrocarbons already present in a rock and on the potential of the kerogen to generate further hydrocarbons. If peak 1 is small while peak 2 is large, the rock contains an immature kerogen that has not yet produced hydrocarbons to the fullest possible extent. If both peaks are of equal size, the material in question has passed through the principal zone of oil formation and is on the verge of metagenetic production of light hydrocarbon gases. Kerogen with an H/C ratio below 0.5 yields essentially no second peak

(Espitalié et al., 1977), and thus signals its highly aromatic nature and resultant inability to yield further mobile hydrocarbons (this indicator is quite sensitive: kerogen with an H/C ratio of 0.6 yields a hydrocarbon pyrolysate of more than 40 mg hydrocarbons/g organic carbon).

5-3-3. Correlation of Kerogen Analyses with Sedimentary Conditions

Evidence allowing reconstruction of the origin and development of inorganic geochemical materials is provided by index minerals, mineral

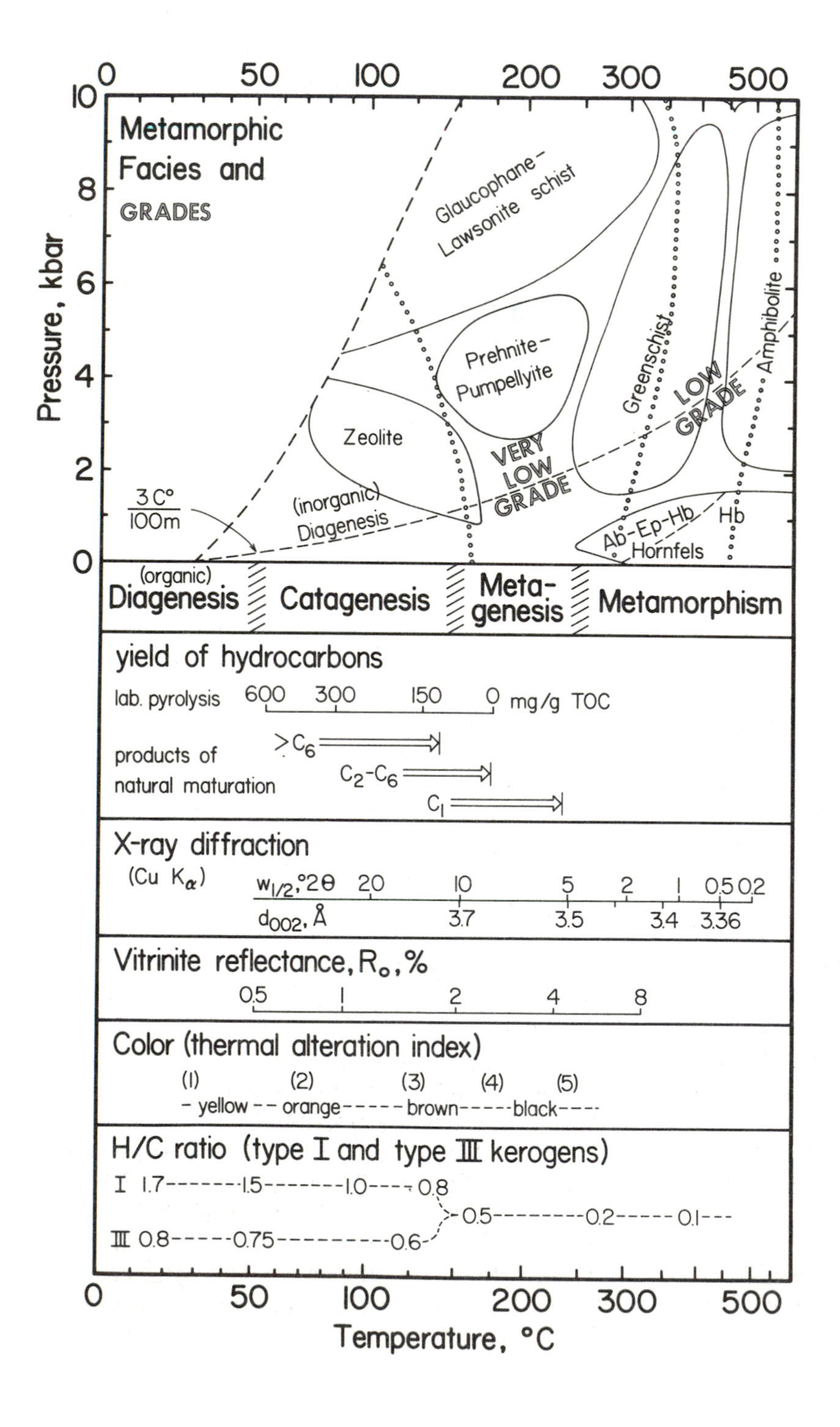

FIGURE 5-7. Summary of correlations between organic and inorganic indicators of thermal alteration of sedimentary materials. The uppermost frame refers to inorganic minerals while all lower frames refer to characteristics of the coexisting organic materials. The indicated geothermal gradient of 3° C/100 m is shown only for purposes of reference. Although this value is representative of many sedimentary basins, both higher and lower gradients are frequently encountered (the gradient appears as a curve because the temperature axis is linear with 1/T). Ab = Albite. Ep = Epidote. Hb = Hornblende. Sources are as follows: metamorphic facies from Turner (1968, p. 366), metamorphic grades from Hurlbut and Klein (1977, p. 468); designation of stages of organic maturation from Tissot and Welte (1978, pp. 69-73); quantitative yield of hydrocarbons in laboratory pyrolysis from Espitalie et al. (1977); qualitative composition of natural products from Hunt (1979, p. 344); X-ray diffraction parameters from Table 5-4; vitrinite reflectance from Tissot and Welte (1978, p. 149), Hunt (1979, p. 345), and McCartney and Teichmuller (1972); color and thermal alteration indices correlated with H/C ratios as indicated in Table 5-3 and with vitrinite reflectance as indicated by Tissot and Welte (1978, p. 467); H/C values greater than 0.5 have been placed by reference to Tissot and Welte (1978, p. 149), while values less than 0.5 have been placed by reference to Table 5-4.

assemblages, compositions of solid solutions, and textural relationships. The interpretation of these features is relatively well developed, and much of the thermal history of most earth materials is understood in some detail. Seeking to reconstruct the depositional history of organic matter, the organic geochemist cannot rely on the same index minerals and petrographic techniques that have provided a firm foundation for the development of inorganic petrology. This problem arises because the processes of primary interest in organic geochemistry occur at low temperatures that affect few of the conventional petrologic indicators. The solution to this problem has, however, been outlined in the preceding sections of this chapter. The structural complexity of kerogen and the range of analytical techniques available provide both a medium capable of displaying subtle variation and an effective means of observing those variations. "Organic petrology" may be relatively undeveloped, but it is not destined for failure.

A range of correlations especially relevant to Precambrian materials is shown in Figure 5-7. The parameters chosen represent a small portion of those potentially available, but do cover the range of techniques utilized in the present and in previous investigations of Precambrian kerogen. Construction of an accurate view of Precambrian organic geochemistry will require very careful correlation and cross-checking of these lines of evidence, but perfect, or, more accurately, simple correlations cannot be expected. The processes involved in the maturation of sedimentary organic matter might be tending toward thermodynamic equilibrium, but they are largely under kinetic control. In these circumstances, marked sensitivities to catalysis (e.g., by mineral surfaces) and to temperature variations can be expected, and the inevitability of evidentiary conflicts which are more apparent than real must be borne in mind.

5-3-4. Analyses of Precambrian Kerogens

elemental analyses

Reports of elemental analyses of Precambrian kerogens are relatively rare, other than from sediments of the Austrialian Proterozoic where McKirdy and coworkers have provided a significant body of results. Table 5-5 summarizes these and all other relevant reports in terms of the observed atomic ratio of hydrogen to carbon. A single specimen has provided an H/C ratio greater than 1.0, and more than half of the reports indicate H/C ratios below 0.5, thus showing that most materials must be heavily altered.

If anything, however, this picture underestimates the general extent of alteration of Precambrian organic materials, since most investigators have not bothered to submit "graphitic" materials for analysis, and since systematic errors (which are probably significant in some cases) will generally shift the H/C results to higher, rather than lower, values. This error arises because samples rich in inorganic material (the "ash" remaining after the organic material has been combusted for analysis) will usually carry excess hydrogen, in the form of water of hydration, bound to the inorganic contaminants. In extreme cases this phenomenon can lead to absurd results—a fortunate happening causing the immediate recognition of the error. Philp et al. (1978), for example, found H/C ratios greater than 4.0 (which would correspond to methane gas) in some recent "kerogens," and Leventhal et al. (1975) found still higher values when attempting to analyze incompletely demineralized organic fractions from Precambrian samples. Less extreme cases are more difficult to recognize, especially because most analyses are reported on a "dry, ash-free" basis that hides the ash content and makes it more difficult to assess the probable reliability of the analytical results. Zumberge et al. (1978) fortunately did not resort to this practice, and it can, therefore, be observed that the carbon contents of their Vaal Reef "kerogen" samples ranged from only 6.1% to a maximum of 32.3% carbon, thus calling for very large ash contents. When it is observed in addition that there is a strong correlation between high ash content and high H/C value within this suite of samples, the case for adventitious hydrogen becomes very strong, and it can be concluded that the reported H/C ratios are systematically in error, perhaps by as much as a factor of two. Details are lacking in the reports cited as references 2 and 4 in Table 5-5, but it can be observed that their results appear high in comparison to related samples.

The significance of low H/C values can best be appreciated by reference to Figures 5-7 and 5-10, which show that these kerogens are more closely related to graphite than to the oil shales where chemical fossils are abundant. The syngenetic nature of Precambrian chemical fossils associated with such kerogens is, thus, highly questionable.

TABLE 5-5. Elemental analyses of Precambrian kerogens.

Sample type	Age Ga	No. of samples	H/C Range	H/C Mean	Reference*
Stromatolite, Wonoka Fm., Austr.	0.6	1		0.47	7, 11
Stromatolite, Etina Fm., Austr.,	0.65	1		0.49	7
Shale, Brachina Fm., Austr.	0.6–0.7	3	0.06–0.15	0.11	8
Stromatolite, Tapley Hill Fm., Austr.	0.7	1		0.24	7, 11
Shale, Tapley Hill Fm., Austr.	0.7	5	0.09–0.25	0.16	8
Shale, Tapley Hill Fm., Austr.	0.7	24	0.01–0.25	0.12	10
Stromatolite, Skillogalee Dolomite, Austr.	0.8	4	0.23–0.33	0.29	7, 11
Stromatolite, Skillogalee Dolomite, Austr.	0.8	2	0.10–0.19	0.15	8, 11
Chert, Skillogalee Dolomite, Austr.	0.8	1		0.20	3
Pertatataka Shale, Austr.	0.8	1		0.15	11
Stromatolite, Bitter Springs Fm., Austr.	0.9	1		0.82	7
Dolomite, Bitter Springs Fm., Austr.	0.9	1		0.73	8
Dolomite, Bitter Springs Fm., Austr.	0.9	2	0.28–0.46	0.37	11
Dolomite, "Proterozoic"		1		0.53	6
Proterozoic cherts and shales	0.7–2.0	6	"<0.1"–"<0.5"		5
Shale, McMinn Fm., Austr.	1.5	1		1.17	8, 11
Shale, McMinn Fm., Austr.	1.5	5	0.26–0.86	0.60	11
Urquhart Shale, Mt. Isa, Austr.	1.5	2	0.20–0.24	0.22	2
Urquhart Shale, Mt. Isa, Austr.	1.5	1		0.13	8
Marimo Slate, Austr.	1.5	1		0.03	8
H. Y. C. shale, McArthur River, Austr.	1.6	1		0.75	2
H. Y. C. shale, McArthur River, Austr.	1.6	1		0.64	8
Shale, Golden Dyke Fm., Austr.	1.8–2.0	1		0.04	8
Thucholites, Witwatersrand, Africa	>2.1	4	0.55–0.67	0.60	2
Vaal Reef carbon seams, Africa	2.5	22	0.47–1.22	0.69	12
Shale, Barreiro Fm., Brazil	~2.5	1		0.05	1
Archean cherts and shales	2.7–3.5	3	"<0.1"–"<0.3"		5
Archean shales, Austr.		2	0.08–0.09	0.09	2
Chert, Fig Tree Group, Africa	3.5	1		0.16	3
Chert, Fig Tree Group, Africa	3.5	2	0.41–0.49	0.45	4
Chert, Onverwacht Group, Africa	3.5	1		0.15	3
Chert, Onverwacht Group, Africa	3.5	11	0.08–1.38	0.59	4
Chert, Onverwacht Group, Africa	3.5	10	0.03–0.19	0.09	9

* 1. Breger et al., 1962; 2. Brooks, 1971; 3. Dungworth and Schwartz, 1972; 4. Dungworth and Schwartz, 1974; 5. Leventhal et al., 1975; 6. McIver, 1967; 7. McKirdy, 1976; 8. McKirdy and Powell, 1974; 9. ref. 8 as corrected by pers. comm. to J. M. H., McKirdy, 1975; 10. McKirdy et al., 1975; 11. McKirdy et al., 1980; 12. Zumberge et al., 1978.

CHEMICAL DEGRADATION OF KEROGEN

Recognition of the uncertainties associated with Precambrian chemical fossils has been a powerful stimulant to the exploration of structural aspects of Precambrian kerogens. Ironically, the literature dealing with the application of sophisticated degradative techniques to Precambrian kerogens is larger than the literature describing rudimentary elemental analyses of these same materials. Reviewing these literatures from the present perspective leads to the painful conclusion that the emphasis has been misplaced.

Table 5-6 summarizes all degradative analyses of Precambrian kerogens reported prior to late 1980. It is evident that pyrolytic analyses have been most widely employed. This choice is logical from several points of view. Minimum sample handling is required, thus providing relatively high immunity from laboratory contamination. Specific functional groups need not be present to facilitate the

TABLE 5-6. Summary of degradative analyses of Precambrian kerogens.

	Sample					
Age, Ga	*identification*	*H/C*[a]	*Technique*[b]	*Yield*[c]	*Principal findings*[d]	*Reference*[e]
0.6	Wonoka Fm.	0.47	Pyrolysis in H_2, 600°/30 sec	?	C_6–C_{20} hc, 20–50% aliphatic (~20% > C_{10}), 10–30% aromatic	7
0.7	Tapley Hill Fm.	0.24	Pyrolysis in H_2 600°/30 sec	?	See Wonoka Fm., above	7
0.8	Skillogalee chert	0.20	Pyrolysis, 580°/12 sec	?	Barely detectable products	2
	(4 samples)	0.33–0.10	Pyrolysis in H_2, 600°/30 sec	?	See Wonoka Fm., above	7
0.8	Pertatataka shale, Amadeus basin Austr.	0.66	Pyrolysis in H_2, 600°/30 sec	?	See Wonoka Fm., above	7
0.9	Bitter Springs carbonates	0.28, 0.46	Pyrolysis in H_2, 600°/30 sec	?	See Wonoka Fm., above	7
1.05	Nonesuch shale	?	HI + P, 200°/16 hr	0.15%	C_{5-8} aliphatic hc	4, 5
		?	KOH fusion	?	Hydroxybenzoic acids	1
		?	Chromic acid oxidation	?	C_{11-22} *n*-alkanoic acids	5
1.2	Muhos shale	?	Pyrolysis *in vacuo*, 250°/4 hr	?	C_{1-7} aliphatic hc	5
1.5	McMinn Fm., Roper River Group	?	Chromic acid oxidation	0.002%	C_{11-21} *n*-alkanoic acids	5
	(6 samples)	1.17–0.36	Pyrolysis in H_2, 600°/30 sec	?	As H/C decreases: *n*-alkanes decrease, C-skeletons > C_{10} decrease, aromatic hc increase	7
1.8	Carbon "sack" from phyllite, Kuusamo, Finland	?	HI + P, 200°/16 hr	0.039%	C_{5-8} aliphatic hc	4, 5
1.8	Anthraxolite, Sudbury, Ontario	?	HI + P, 200°/16 hr	0.11%	C_{5-8} aliphatic hc	4, 5
		?	Pyrolysis, 266°/100 hr	0.009%	C_{1-5} aliphatic hc	5, 6
2.0	Carbon Leader	?	HI + P, 200°/16 hr	0.8%	C_{5-8} aliphatic hc	4, 5
		?	Pyrolysis, 266°/100 hr	0.017%	C_{1-5} aliphatic hc	5, 6
2.2	Transvaal stromatolitic carbonate	?	Ozonolysis	?	C_4-subst. tetrahydrofuran, C_3-subst. tetrahydropyran, heterocyclics	8, 13
		?	Pyrolysis *in vacuo*, 450°/30 min	?	C_{4-11} branched aliphatic hc, C_{6-11} aromatic HC, heterocyclics	8

[a] H/C ratio of organic material analyzed. ? indicates not given.

[b] Use of gas-liquid chromatography for product analysis is assumed. This column refers only to the degradative technique utilized.

[c] Refers to yield of identified products as fraction of total organic carbon.

[d] "hc" indicates hydrocarbon.

[e] References: 1. Brooks and Shaw (1971). 2. Dungworth and Schwartz (1972). 3. Dungworth and Schwartz (1974). 4. Hoering (1964). 5. Hoering (1967b). 6. Hoering and Abelson (1964). 7. McKirdy et al. (1980). 8. Nagy (1976). 9. Nagy and Nagy (1969). 10. Scott et al. (1970). 11. Sklarew, reported by Nagy (1976). 12. Sklarew and Nagy (1979). 13. Zumberge and Nagy (1975). 14. Zumberge et al. (1978).

TABLE 5-6. *(continued)*

Age, Ga	Sample identification	H/C[a]	Technique[b]	Yield[c]	Principal findings[d]	Reference[e]
2.5	Vaal Reef, Witwatersrand	~0.7	Pyrolysis *in vacuo*, 450°/30 min	?	C_{6-13} aromatic hc, C_{2-6} aliphatic hc	14
2.64	Belingwe stromatolite	?	Pyrolysis *in vacuo*, 450°/30 min	?	C_{9-20} aliphatic hc, C_{1-2}-subst. furans	12
2.64	Bulawayan stromatolitic carbonate	?	HI + P, 200°/16 hr	0.05%	C_{5-8} aliphatic hc	4, 5
			Pyrolysis *in vacuo*, 450°/30 min	?	C_{9-16} aliphatic hc, C_{6-11} aromatic hc, heterocyclics	11
3.5	Fig Tree Group	0.16	Pyrolysis, 580°/12 sec	?	C_{9-15} aliphatic hc, possibly some aromatic hc	2
		?	Pyrolysis of rock powder, 500°/5 min	?	C_{1-19} aliphatic hc, some aromatic hc C_{6-12}	10
		?	Pyrolysis *in vacuo*, 250°/12 hr	?	Trace C_6H_6, C_7H_8, no aliphatic hc	5
		?	KOH fusion	?	Hydroxybenzoic acids	1
3.54	Onverwacht Group: Szwartkoppie Fm.	?	Pyrolysis, 580°/12 sec	?	C_{9-15} aliphatic hc, possibly some aromatic hc	3
3.54	Onverwacht Group: Kromberg Fm.	0.15	Pyrolysis 580°/12 sec	?	C_{9-15} aliphatic hc, possibly some aromatic hc	2, 3
		?	Pyrolysis of rock powder, 500°/5 min	?	Mostly aromatic hc, C_6–C_{12}	10
		?	Ozonolysis	?	C_{6-17} dicarboxylic acids, $C_{12,14,16,18,22}$ *n*-alkanoic acids	1
		?	Ozonolysis	?	"Aromatic acids"	9
		?	KOH fusion	?	Hydroxybenzoic acids	1
3.54	Onverwacht Group: Hooggenoeg Fm.	?	Pyrolysis, 580°/12 sec	?	C_{9-15} aliphatic hc, possibly some aromatic hc	3
3.54	Onverwacht Group: Theespruit Fm.	?	Pyrolysis, 580°/12 sec	?	C_{9-15} aliphatic hc, possibly some aromatic hc	3
		?	Pyrolysis of rock powder, 500°/5 min	?	Mostly aromatic hc, C_{6-8}	10

attack of some selective reagent, thus affording general applicability even to refractory materials. Finally, the gas chromatographic techniques usually associated with pyrolytic degradation offer the highest possible return of information from the complex product mixtures sometimes obtained.

A further strength of chromatographic analyses is, however, virtually a weakness in this application. Specifically, the extraordinary sensitivity of chromatographic detection techniques makes possible the identification of very small amounts of material that might represent only an insignificant fraction of the kerogenic precursor material. Most of the pyrolytic investigations have not incorporated a measurement of yield, thus doing nothing to lay this question to rest.

The general failure to determine and consider H/C ratios is of great significance in this connection. The observation of H/C = 0.15 in the Onverwacht chert, for example, is fully consistent with the results of independent elemental analyses subsequently reported by McKirdy and Powell (1974) and is not unrepresentative of Onverwacht kerogens generally (see results of elemental analyses in this work). An H/C ratio of 0.15 would be obtained from a kerogen in which the average structure was an aromatic hydrocarbon ring system containing more than 260 carbon atoms. The idea that aliphatic hydrocarbon pyrolyzates having H/C ratios near 2.0 (see "principal findings" column, Table 5-6) might reasonably be derived from, and representative of, such a structure has been too uncritically accepted. Indeed, it seems most likely that these hydrogen-rich pyrolysis products must be representative, not of the bulk of the kerogen, but of some other material, possibly adventitious hydrocarbons trapped and accumulated as discussed by Oehler (1977).

Given the absence of quantitive data and the presence of unlikely products, it is difficult to conclude that much has been learned thus far from these degradative investigations, in spite of their laudable goals. The same paucity of alternatives that stimulated these investigations in the first place, however, now suggests that they must be followed up in an effort to clarify the situation. Measurements of quantitative yields of pyrolysis products, and of the carbon isotopic compositions of the pyrolysis products relative to those of the kerogens, would be most helpful.

5-4. The Approach Taken in These Investigations

The principal objective of the organic geochemical studies of the P.P.R.G. has been the accumulation of a substantial base of quantitative analyses of Precambrian kerogens. While the completion of a large range of elemental analyses of isolated kerogens is of the greatest interest, the labor involved in the preparation of kerogen fractions is substantial, and it was, therefore, necessary that the samples to be treated in this way be limited in number and carefully chosen. Accordingly, the analytical work was arranged in two levels, with the total organic carbon fraction of nearly all of the available samples first being analyzed both quantitatively and isotopically. The results of these analyses, together with the results of micropaleontological investigations, were utilized in the selection of a smaller set of samples for elemental analysis.

The analyses were performed at two levels of sample preparation: (*i*) cleaned, etched, and crushed whole rock powder; and (*ii*) isolated kerogen fractions. The details of the experimental procedures associated with the preparation of these samples are given in Appendices II and IV. The powder samples were prepared from all specimens in the P.P.R.G. collection—eventually including more than 500 separately cataloged entries ranging in size from approximately 10 g to 10 kg—that were large enough to provide approximately 25 g of whole rock powder as well as petrographic and, where appropriate, micropaleontological thin sections; kerogen fractions were isolated from fifty of these samples which appeared to be of special interest.

The powder samples were analyzed using a procedure that allowed determination of both the total organic carbon content and the isotopic composition of the organic carbon. In this procedure, the samples were treated with hot, concentrated hydrochloric acid in order to remove all carbonates and the residues were burned in pure O_2 (1 atm) at 1400°C in order to produce CO_2 quantitatively from all remaining carbon. Quantitative measurement of the CO_2 provided a total organic carbon value for the sample analyzed and mass spectroscopic analysis of the CO_2 provided the carbon isotope ratio.

Kerogen analyses were undertaken using the procedures outlined in Appendix IV.

5-5. Results and Discussion

5-5-1. Survey Analyses

The results of the survey analyses are summarized in Table 5-7 and in the scatter plots shown in Figure 5-8. It is evident that the carbon abundances and isotopic compositions cover substantial ranges and that they are in some cases correlated.

TABLE 5-7. Quantitative and isotopic analyses of total organic carbon in rock powder samples.

Sampled geologic units are listed in approximate geochronologic order from youngest to oldest. Each entry tabulates: in top line, P.P.R.G. Sample Number (e.g., 455–1; see Appendix I for details regarding sample locality, stratigraphy, etc.) and lithology code (A = carbonate, B = banded iron-formation, C = chert, G = granite, H = schist, I = basalt, N = gneiss, Q = sandstone, and S = shale; combined symbols indicate mixtures); in middle line, total organic carbon (T.O.C.) content (mg C/g rock powder); and in bottom line, $\delta^{13}C_{PDB}$ (‰) value of the total organic carbon. T.O.C. and $\delta^{13}C_{PDB}$ values were determined using the "survey method" of analysis (see Appendix IV).

Burra Group, 0.8 Ga

Sample	Lithology	T.O.C.	$\delta^{13}C_{PDB}$
455–1	C	1.3 ± 0.2	−23.1 ± 0.6

Bitter Springs Formation, 0.9 Ga

Sample	Lithology	T.O.C.	$\delta^{13}C_{PDB}$
133–1	C	0.5 ± 0.1	−21.9 ± 0.6
133–2	C	0.8 ± 0.2	−22.4 ± 0.4

Uinta Mountain Group, 1.0 Ga

Sample	Lithology	T.O.C.	$\delta^{13}C_{PDB}$
135–1	S	1.7 ± 0.3	−17.0 ± 0.4

Oronto Group, 1.0 Ga

Sample	Lithology	T.O.C.	$\delta^{13}C_{PDB}$
497–1	S	2.7 ± 0.5	−33.8 ± 0.2

Allamoore Formation, 1.1 Ga

Sample	Lithology	T.O.C.	$\delta^{13}C_{PDB}$
518–1	C	0.8 ± 0.1	−31.3 ± 0.2

Sibley Group, 1.3 Ga

Sample	Lithology	T.O.C.	$\delta^{13}C_{PDB}$
342–1	A	0.13 ± 0.05	−30.9 ± 1.0
343–1	A	0.3 ± 0.1	−28.4 ± 0.4

Roper Group, 1.4 Ga

Sample	Lithology	T.O.C.	$\delta^{13}C_{PDB}$
114–1	S	6.0 ± 1.1	−32.9 ± 0.2
115–1	S	5.8 ± 1.0	−32.9 ± 0.2
116–1	S	3.0 ± 0.5	CO_2 lost
117–1	S	4.3 ± 0.8	−32.6 ± 0.2
118–1	S	4.3 ± 0.8	−32.7 ± 0.2
119–1	S	3.0 ± 0.5	−33.0 ± 0.2
121–1	S	2.9 ± 0.5	−32.3 ± 0.2
122–2	S	3.5 ± 0.6	−32.1 + 0.2
123–1	S	3.4 ± 0.6	−31.9 ± 0.2
124–1	S	0.3 ± 0.1	−29.2 ± 1.0
125–1	S	0.12 ± 0.05	
126–1 ±	S	0.14 ± 0.05	−28.4 ± 1.8
127–1	S	<0.05	
128–1	S	0.4 ± 0.1	−29.5 ± 1.0
129–1	S	7.4 ± 1.3	−31.0 ± 0.8
130–1	S	4.3 ± 0.8	−32.6 ± 0.4
131–1	S	10.6 ± 1.9	−32.5 ± 0.6
457–1	S	6.7 ± 1.2	−32.9 ± 0.2
458–1	S	10.1 ± 1.8	−32.3 ± 0.2
459–1	S	1.9 ± 0.3	−31.9 ± 0.2
120–1	Q	5.2 ± 0.9	−33.0 ± 0.2

Bungle Bungle Dolomite, 1.6 Ga

Sample	Lithology	T.O.C.	$\delta^{13}C_{PDB}$
140–1	A	0.4 ± 0.1	−30.2 ± 0.4
141–1	A	0.2 ± 0.1	−19.5 ± 1.2
143–1	A	0.2 ± 0.1	−26.3 ± 1.0
147–1	A	0.09 ± 0.04	−27.0 ± 1.4
148–1	A	0.07 ± 0.04	−24.4 ± 2.6
157–1	A	0.2 ± 0.1	−27.8 ± 1.2
158–1	A	0.07 ± 0.04	−22.9 ± 2.4
159–1	C	0.9 ± 0.2	−25.9 ± 0.4
137–1	S	1.8 ± 0.3	−27.7 ± 0.4
144–1	S	0.4 ± 0.1	−29.9 ± 0.2
145–1	S	2.6 ± 0.5	−30.5 ± 0.2
146–1	S	4.1 ± 0.7	−32.4 ± 0.2
149–1	S	0.10 ± 0.05	−23.2 ± 2.4
150–1	S	1.5 ± 0.3	−32.7 ± 0.2
155–1	S	26.6 ± 4.8	−33.6 ± 0.2
156–1	S	11.5 ± 2.1	−32.0 ± 0.2
139–1	AC	0.7 ± 0.1	−28.8 ± 0.2
151–1	AC	0.4 ± 0.1	−30.1 ± 2.0
154–1	AC	0.5 ± 0.1	−28.8 ± 0.4
152–1	ACS	3.6 ± 0.6	−28.4 ± 0.2

McArthur Group, 1.6 Ga

Sample	Lithology	T.O.C.	$\delta^{13}C_{PDB}$
105–1	A	3.9 ± 0.7	−32.3 ± 0.2
103–1	CA	0.7 ± 0.1	−29.4 ± 0.2
106–1	CA	0.4 ± 0.1	−27.2 ± 0.4
107–1	CA	0.2 ± 0.1	−22.1 ± 0.8
108–1	AC	0.8 ± 0.1	−29.1 ± 0.2
109–1	AC	0.9 ± 0.2	−24.7 ± 0.2
110–1	AC	0.9 ± 0.2	−24.4 ± 0.2
111–1	AC	1.7 ± 0.3	−28.6 ± 0.2
112–1	AC	1.3 ± 0.2	−27.8 ± 0.2
452–1	CA	0.10 ± 0.05	−21.6 ± 1.8
113–1	ASC	5.9 ± 1.1	−31.7 ± 0.4

TABLE 5-7. *(continued)*

Vermillion Limestone, 1.7 Ga	332–1	A	0.3 ± 0.1	−23.0 ± 0.6
Earaheedy Group, 1.8 Ga	086–1	A	0.7 ± 0.1	−28.5 ± 0.4
	095–1	A	0.08 ± 0.04	−29.7 ± 2.2
	095–2	A	0.4 ± 0.1	−25.0 ± 0.8
	096–1	A	0.2 ± 0.1	−22.8 ± 1.4
	097–1	A	0.3 ± 0.1	−26.8 ± 0.6
	099–1	A	0.13 ± 0.05	−19.5 ± 1.8
	100–1	A	0.3 ± 0.1	−24.8 ± 0.8
	104–1	A	0.2 ± 0.1	−20.7 ± 2.8
	435–1	A	0.10 ± 0.05	−20.0 ± 2.2
	436–1	A	0.07 ± 0.04	−15.1 ± 3.2
	439–1	A	0.10 ± 0.05	−21.1 ± 2.4
	440–1	A	3.1 ± 0.6	−32.8 ± 1.0
	066–1	C	0.08 ± 0.04	−20.0 ± 4.0
	073–1	C	0.07 ± 0.04	−11.8 ± 5.2
	076–1	C	0.07 ± 0.04	−20.2 ± 3.0
	089–1	C	2.0 ± 0.4	−29.0 ± 0.2
	092–1	C	0.09 ± 0.04	−10.6 ± 4.0
	093–1	C	0.06 ± 0.03	−11.2 ± 5.8
	078–1	CB	0.2 ± 0.1	−14.1 ± 2.8
	080–1	CA	0.4 ± 0.1	−32.2 ± 0.4
	082–1	ACS	0.2 ± 0.1	−23.0 ± 1.4
	091–1	CAB	0.2 ± 0.1	−24.0 ± 1.2
	426–1	AC	0.13 ± 0.05	−23.0 ± 1.6
Rove Formation, 1.8 Ga	498–1	S	8.1 ± 1.5	−32.4 ± 0.2
Chelmsford Formation, 1.9 Ga	450–1	S	5.1 ± 0.9	−31.6 ± 0.2
Onwantin Formation, 1.9 Ga	334–1	S	40 ± 7	−30.7 ± 0.2
	451–1	S	36 ± 6	−30.5 ± 0.2
Kahochella Group, 2.0 Ga	340–1	A	0.09 ± 0.04	−21.8 ± 2.4
Pethei Group, 2.0 Ga	338–1	A	0.13 ± 0.05	−25.1 ± 1.2
	412–1	A	0.11 ± 0.05	−22.8 ± 1.8
Earaheedy Group, 1.8 Ga	054–1	A	0.6 ± 0.1	−23.6 ± 0.4
	055–1	A	0.3 ± 0.1	−19.7 ± 1.0
	056–1	A	0.2 ± 0.1	−19.8 ± 3.0
	057–1	A	0.2 ± 0.1	−24.8 ± 0.8
	050–1	C	0.10 ± 0.05	−23.5 ± 1.8
	059–1	C	0.2 ± 0.1	−25.1 ± 0.8
	060–1	C	0.2 ± 0.1	−29.3 ± 0.8
	049–1	AC	0.2 ± 0.1	−25.9 ± 0.8
	051–1	CA	0.4 ± 0.1	−22.2 ± 0.6
	052–1	CA	0.2 ± 0.1	−19.6 ± 1.4
	053–1	AC	0.14 ± 0.05	−24.6 ± 1.4
	061–1	AC	0.14 ± 0.05	−20.0 ± 1.8
	062–1	CA	0.3 ± 0.1	−22.7 ± 0.8
Gunflint (Biwabik) Iron Formation, 2.0 Ga	335–1	A	9.4 ± 1.7	−34.7 ± 0.2
	134–1	C	0.4 ± 0.1	−30.5 ± 0.6
	134–5	C	0.6 ± 0.3	−32.8 ± 1.0
	336–1	B	0.2 ± 0.1	−27.0 ± 0.5
Krivoi-Rog Iron Formation, 2.0 Ga	401–1	B	0.07 ± 0.03	−21.9 ± 2.6
Upper Albanel Formation, 2.1 Ga	443–1	A	2.9 ± 0.5	−31.4 ± 0.2
Lower Albanel Formation, 2.1 Ga	444–1	A	0.2 ± 0.1	−22.8 ± 0.8
Manitounuk Group, 2.1 Ga	344–1	A	0.14 ± 0.05	−24.0 ± 1.4
	345–1	A	0.11 ± 0.05	−21.1 ± 2.0
Belcher Group, 2.1 Ga	341–1	A	0.2 ± 0.1	−25.6 ± 0.8
	445–1	C	0.4 ± 0.1	−25.3 ± 0.4
	446–1	CA	0.8 ± 0.1	−26.4 ± 1.0
	447–1	CA	0.9 ± 0.2	−22.4 ± 0.2
	449–1	CA	0.13 ± 0.05	−24.5 ± 1.4

TABLE 5-7. *(continued)*

Transvaal Supergroup, 2.2 Ga					
205–1 A 0.14 ± 0.05 −18.7 ± 1.9	240–1 A 0.2 ± 0.1 −29.0 ± 0.8	242–1 A 0.3 ± 0.1 −28.7 ± 0.6	296–1 A 0.8 ± 0.1 −32.8 ± 0.2	297–1 A 0.3 ± 0.1 −30.0 ± 1.2	297–3 A 0.4 ± 0.1 −33.6 ± 0.2
298–1 A 1.2 ± 0.2 −33.1 ± 0.4	298–2 A 1.3 ± 0.2 −33.8 ± 0.2	299–1 A 1.1 ± 0.2 −34.4 ± 0.2	300–1 A 0.7 ± 0.1 −31.1 ± 0.4	302–1 A 0.7 ± 0.1 −33.1 ± 0.4	304–1 A 1.1 ± 0.2 −32.4 ± 0.2
306–1 A 0.6 ± 0.1 −34.2 ± 0.2	307–1 A 0.9 ± 0.2 −33.5 ± 0.2	310–1 A 0.2 ± 0.1 −13.0 ± 1.0	312–1 A 0.9 ± 0.2 −14.9 ± 1.0	314–1 A 0.4 ± 0.1 −30.4 ± 0.4	314–2 A 0.8 ± 0.1 −30.9 ± 0.4
315–1 A 0.7 ± 0.1 CO_2 lost	315–2 A 0.8 ± 0.1 −31.1 ± 0.2	315–3 A 0.8 ± 0.1 −32.3 ± 0.2	316–1 A 0.9 ± 0.2 −34.4 ± 0.4	317–1 A 1.1 ± 0.2 −34.9 ± 0.2	318–1 A 0.6 ± 0.1 −31.2 ± 0.4
320–1 A 1.8 ± 0.3 −13.8 ± 0.2					
Griqualand Group, 2.2 Ga					
295–1 B 0.2 ± 0.1 −24.6 ± 1.4					
Hamersley Group, 2.5 Ga					
370–1 A 0.2 ± 0.1 −29.1 ± 0.8	362–1 C 0.09 ± 0.05 −17.3 ± 3.4	363–1 C 0.08 ± 0.05 −13.6 ± 3.4	364–1 C 0.11 ± 0.05 −16.0 ± 2.2	047–1 S 32 ± 6 −32.7 ± 0.2	048–1 S 41 ± 7 −32.1 ± 0.2
048–2 S 37 ± 7 −32.4 ± 0.2	356–1 S 65 ± 12 −38.5 ± 0.2	357–1 S 2.3 ± 0.4 −34.2 ± 0.2	358–1 S 45 ± 8 −35.5 ± 0.2	359–1 S 18 ± 3 −35.2 ± 0.2	360–1 S 41 ± 8 −29.8 ± 0.2
480–1 S 5.2 ± 0.9 −31.4 ± 0.4	481–1 S 14 ± 3 −31.1 ± 0.2	482–1 S 15 ± 3 −30.9 ± 0.2	485–1 S 14 ± 3 −33.2 ± 0.2	486–1 S 2.0 ± 0.4 −31.1 ± 0.4	487–1 S 1.6 ± 0.3 −30.8 ± 0.2
488–1 S 18 ± 3 −33.2 ± 0.2	490–1 S 0.11 ± 0.05 −21.1 ± 2.2	361–1 B 0.9 ± 0.3 −18.6 ± 3.7	369–1 B 0.3 ± 0.1 −37.8 ± 0.8	372–1 B 0.6 ± 0.1 −39.8 ± 0.2	373–1 B 14 ± 3 −40.0 ± 0.2
374–1 B 0.4 ± 0.1 −34.2 ± 0.2	483–1 B 0.2 ± 0.1 −22.3 ± 2.0	489–1 B 2.6 ± 0.5 −21.4 ± 0.4	491–1 B 0.3 ± 0.1 −18.0 ± 3.6	042–1 CA 0.2 ± 0.1 −9.4 ± 2.2	365–1 CA ≤0.05 −21.7 ± 3.7
366–1 SA 7.3 ± 1.3 −34.2 ± 0.2	367–1 SA 21 ± 4 −32.8 ± 0.2	368–1 SA 2.2 ± 0.4 −43.0 ± 0.2	371–1 SA 3.6 ± 0.6 −32.1 ± 0.2	387–1 SA 14 ± 3 −36.0 ± 0.2	
Turee Creek Group, 2.5 Ga					
063–1 A 0.10 ± 0.05 −28.4 ± 1.4					
Steeprock Group, 2.6 Ga					
321–1 A 0.2 ± 0.1 −25.0 ± 1.0	322–1 A 0.5 ± 0.1 −26.4 ± 0.4	324–1 A 0.6 ± 0.1 −22.1 ± 0.6	323–1 C 2.3 ± 0.4 −30.7 ± 0.4	325–1 C 0.4 ± 0.1 −26.0 ± 0.6	326–1 S 1.7 ± 0.3 −26.9 ± 0.2

Table 5-7. *(continued)*

Group	Sample	Type		
Ventersdorp Supergroup, 2.6 Ga	281–1	A	0.4 ± 0.1	−36.3 ± 0.2
	282–1	A	1.6 ± 0.3	−40.9 ± 0.2
	282–2	A	2.0 ± 0.4	−41.6 ± 0.2
	283–1	A	1.0 ± 0.2	−38.7 ± 0.2
	284–1	A	0.8 ± 0.1	−36.7 ± 0.2
	285–1	A	2.4 ± 0.1	−39.0 ± 0.2
	287–1	A	0.6 ± 0.2	−31.6 ± 0.2
	288–1	A	0.2 ± 0.1	−28.4 ± 1.0
	273–1	C	0.3 ± 0.1	−34.1 ± 0.8
	278–1	C	1.0 ± 0.2	−38.9 ± 0.4
	279–1	C	0.5 ± 0.1	−35.0 ± 0.6
	293–1	C	7.1 ± 1.3	−39.6 ± 0.2
	271–1	G	0.06 ± 0.03	−27.2 ± 2.0
	275–1	I	0.07 ± 0.03	−15.5 ± 3.0
	272–1	AC	0.9 ± 0.2	−42.4 ± 0.2
	274–1	AC	0.6 ± 0.1	−41.8 ± 0.2
	286–1	AC	0.9 ± 0.2	−36.7 ± 0.4
	291–1	AC	1.0 ± 0.2	−37.0 ± 0.2
	292–1	AC	1.6 ± 0.3	−37.2 ± 0.2
	292–2	AC	4.2 ± 0.8	−39.8 ± 0.2
	308–1	AC	0.2 ± 0.1	CO_2 lost
	280–1	AS	2.2 ± 0.4	−41.1 ± 0.2
Yellowknife Supergroup, 2.6 Ga	413–1	A	0.3 ± 0.1	−15.9 ± 0.8
	413–3	A	0.3 ± 0.1	−15.6 ± 1.0
Bulawayan Group, 2.6 Ga	253–1	A	1.0 ± 0.2	−17.2 ± 0.8
	254–1	A	1.0 ± 0.2	−30.6 ± 0.3
	255–1	A	1.8 ± 0.3	−33.1 ± 0.2
Manjeri Formation, 2.6 Ga	218–1	C	0.13 ± 0.05	−28.7 ± 1.1
	219–1	C	0.12 ± 0.05	−29.4 ± 2.8
	221–1	C	0.08 ± 0.04	−13.0 ± 5.6
	224–1	C	0.6 ± 0.1	−8.8 ± 1.0
	225–1	C	0.6 ± 0.1	−32.3 ± 0.4
	226–1	C	1.2 ± 0.2	−8.2 ± 0.6
	227–1	C	2.5 ± 0.5	−29.6 ± 0.4
	228–1	C	0.9 ± 0.2	−12.5 ± 0.4
	223–1	SC	1.0 ± 0.2	−27.1 ± 4.6
"Keewatin" Iron Formation, 2.7 Ga	330–1	C	18 ± 3	−27.1 ± 0.2
Michipicoten Iron Formation, 2.7 Ga	331–1	C	19 ± 3	−16.1 ± 0.2
Lewin Shale, 2.8 Ga	441–1	S	38 ± 7	−43.1 ± 0.2
Fortescue Group, 2.8 Ga	029–1	A	0.2 ± 0.1	−45.8 ± 0.6
	027–1	C	0.7 ± 0.1	−49.9 ± 1.0
	036–1	C	0.4 ± 0.1	−33.0 ± 0.4
	038–1	S	40 ± 7	−41.2 ± 0.2
	039–1	S	68 ± 12	−41.3 ± 0.2
	040–1	S	73 ± 13	−41.4 ± 0.4
	132–2	S	165 ± 30	−40.5 ± 0.6
	375–1	S	31 ± 6	−42.0 ± 0.2
	376–1	S	40 ± 7	−42.4 ± 0.2
	377–1	S	12 ± 2	−38.4 ± 0.2
	035–1	I	<0.05	−3.7 ± 8.0
	024–1	CA	0.08 ± 0.04	−45.4 ± 1.4
	025–1	CA	0.5 ± 0.1	−50.3 ± 0.6
	026–1	CA	0.8 ± 0.1	−50.6 ± 0.4
	028–1	CA	1.8 ± 0.3	−51.7 ± 0.2
	030–1	CA	0.7 ± 0.1	−51.6 ± 0.4
	031–1	CA	0.7 ± 0.1	−51.9 ± 0.4
	032–1	CA	0.4 ± 0.1	−47.6 ± 0.4
	033–1	CA	0.4 ± 0.1	−46.9 ± 0.6
	041–1	CA	0.3 ± 0.1	−44.6 ± 0.6
	037–1	CS	3.9 ± 0.7	−27.9 ± 0.2
Woman Lake Marble, 2.8 Ga	516–1	A	0.9 ± 0.2	−9.6 ± 0.6
"Lower Greenstones," 3.0 Ga	229–1	S	1.69 ± 0.05	−24.1 ± 0.0

TABLE 5-7. *(continued)*

Pongola Supergroup, 3.0 Ga					
258–1 A 0.06 ± 0.03 −13.1 ± 4.0	261–1 A 0.5 ± 0.1 −14.0 ± 0.8	263–2 A 0.3 ± 0.1	264–1 A 0.2 ± 0.1 −16.8 ± 2.0	265–1 A 0.12 ± 0.05 −16.6 ± 2.0	309–1 A 0.4 ± 0.1 −18.3 ± 1.0
269–1 C 0.2 ± 0.1 −14.4 ± 2.6	270–1 C 0.2 ± 0.1 −14.4 ± 3.0	257–1 S 0.3 ± 0.1 −16.3 ± 1.5	260–1 S 0.3 ± 0.1 −24.4 ± 1.4	266–1 S 2.7 ± 0.5 −26.6 ± 0.2	267–1 S 2.6 ± 0.5 −25.6 ± 0.2
256–1 I 0.13 ± 0.05 −18.4 ± 1.8					
Gorge Creek Group, 3.4 Ga					
023–1 C 4.8 ± 2.0 −27.6 ± 0.8	022–1 S 40 ± 7 −30.1 ± 0.2	022–2 S 43 ± 8 −31.5 ± 0.2	022–3 S 39 ± 7 −31.8 ± 0.2	023–2 S 40 ± 10 −27.9 ± 0.4	
Warrawoona Group, 3.5 Ga					
001–1 C 0.2 ± 0.1 −29.8 ± 1.2	002–1 C 0.7 ± 0.1 −34.2 ± 0.2	002–2 C 0.5 ± 0.1 −31.2 ± 0.4	002–3 C 0.5 ± 0.1 −33.4 ± 0.4	003–1 C 0.8 ± 0.1 −37.1 ± 0.2	004–1 C 0.5 ± 0.1 −34.9 ± 0.2
006–1 C 0.11 ± 0.05 −29.4 ± 1.8	007–1 C 0.08 ± 0.04 −25.5 ± 3.0	008–1 C <0.05 −24.0 ± 5.4	009–1 C 0.2 ± 0.1 −29.6 ± 1.0	010–1 C 0.8 ± 0.1 −33.0 ± 0.2	013–1 C 1.4 ± 0.2 −36.8 ± 0.2
016–1 C 0.9 ± 0.2 −34.7 ± 0.2	015–1 I 0.6 ± 0.1 −2.8 ± 1.4	018–1 I 0.13 ± 0.05 −1.3 ± 3.0			
"Swaziland Sequence," 3.5 Ga					
179–1 C 15 ± 3 −15.2 ± 0.2	182–1 C 1.9 ± 0.3 −31.3 ± 0.2	184–1 C 3.3 ± 0.6 −34.2 ± 0.2	186–1 C 1.2 ± 0.2 −32.4 ± 0.2	187–1 C 0.5 ± 0.1 −27.1 ± 0.6	188–1 C 0.5 ± 0.1 −31.5 ± 0.4
189–1 C 2.6 ± 0.5 −28.8 ± 0.2	191–1 C 1.9 ± 0.3 −25.2 ± 0.2	192–1 C 6.0 ± 1.1 −26.3 ± 0.2	193–1 C 1.1 ± 0.2 −37.0 ± 0.2	194–1 C 0.3 ± 0.1 −24.7 ± 0.8	196–1 C 2.0 ± 0.4 −23.4 ± 0.4
197–1 C 0.8 ± 0.1 −25.7 ± 0.4	198–1 C 1.6 ± 0.3 −32.4 ± 0.2	199–1 C 16 ± 3 −26.7 ± 0.2	200–1 C 1.6 ± 0.3 −30.5 ± 0.4	201–1 C 1.8 ± 0.3 −26.9 ± 0.2	204–1 C 3.4 ± 0.6 −28.8 ± 0.2
208–1 C 4.5 ± 0.8 −26.8 ± 0.2	213–1 C 2.2 ± 0.4 −27.5 ± 0.4	215–1 C 1.4 ± 0.2 −31.5 ± 0.2	216–1 C 7.2 ± 1.3 −26.1 ± 0.4	181–1 S 13 ± 2 −15.2 ± 0.4	185–1 S 2.3 ± 0.4 −26.1 ± 0.2
202–1 S 3.0 ± 0.5 −24.3 ± 0.2	203–1 S 62 ± 11 −32.5 ± 0.2	209–1 S 1.0 ± 0.2 −31.1 ± 0.2	212–1 S 64 ± 12 −32.8 ± 0.2	214–1 S 14 ± 2 −30.4 ± 0.2	195–1 H 0.2 ± 0.1 −21.3 ± 1.2
470–1 I 0.4 ± 0.1 −12.7 ± 1.2					
"Isua Supracrustal Sequence," 3.8 Ga					
465–1 B 25 ± 4 −9.8 ± 0.4	492–1 B 0.06 ± 0.03 −12.0 ± 1.6	466–1 H 0.06 ± 0.03 −14.3 ± 2.8	468–1 H 0.06 ± 0.03 −19.3 ± 2.8	494–1 H 0.11 ± 0.05 −28.2 ± 1.8	164–1 N 0.7 ± 0.1 −16.6 ± 0.6
461–1 N 22 ± 4 −16.1 ± 0.2	462–1 N 0.10 ± 0.05 −20.0 ± 3.4	464–1 N 0.09 ± 0.04 −21.6 ± 2.2	469–1 N 0.06 ± 0.03 −16.6 ± 3.4		

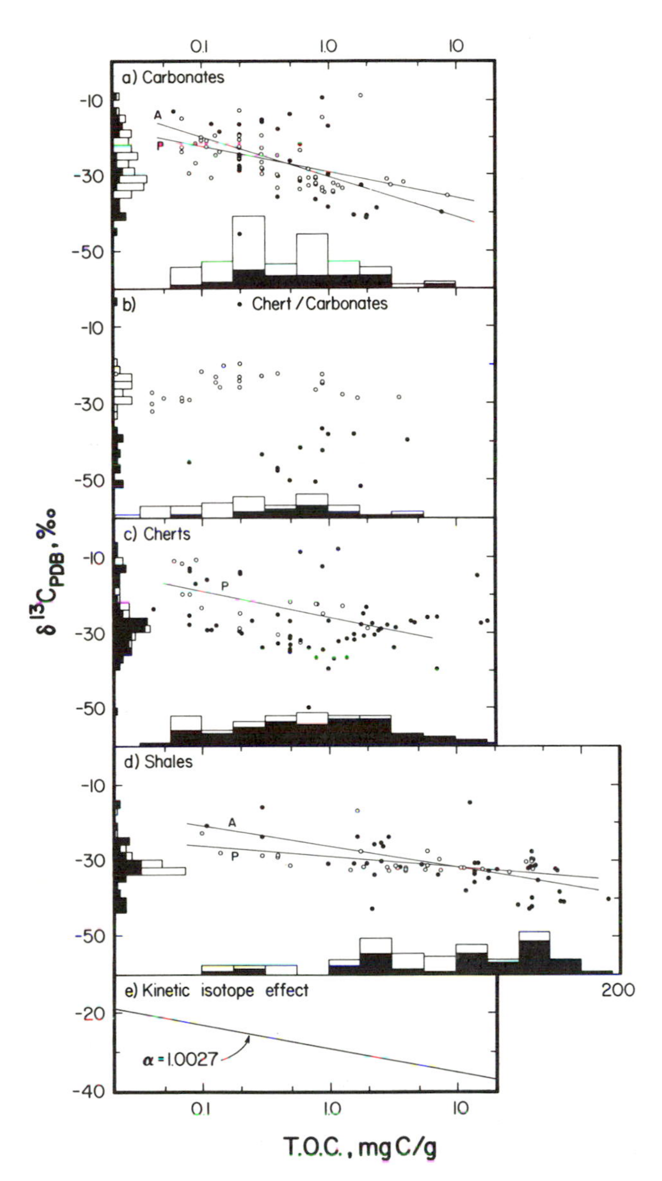

FIGURE 5-8. Scatter plots showing carbon abundances and isotopic compositions sorted by rock type (a–d) and the isotopic fractionation expected from a kinetic isotope effect (e). Open points represent Proterozoic samples, while filled points represent Archean samples. Similarly, the histograms on each axis summarize the distributions of carbon abundances and isotopic compositions, with the filled portions of the bars representing Archean points and the open portions representing Proterozoic points. The lines designate least-squares regression lines for Archean (A) and Proterozoic (P) materials. Where no lines have been drawn, the calculated slope of the regression line did not differ significantly from zero (see Table 5-8). The line shown in (e) was calculated using exact forms of Equations 5-5 and 5-6, the plotted values corresponding to the isotopic composition of the residual carbon in a sample from which carbon is being lost by some process having a 0.27% kinetic isotope effect.

Table 5-8 summarizes the parameters describing these correlations.

The commonly observed increase in ^{13}C-content in carbon-poor sediments evident in the data might be due either to pre- or to post-depositional factors. In the latter case, isotope effects accompanying carbon losses from sediments initially depleted in ^{13}C could cause residues to be enriched in ^{13}C, the isotopic composition of the carbon lost being described by a Rayleigh distillation process (Equation 5-5). Three of the five significant correlations noted in Table 5-8 indicate ^{13}C enrichments of 6 ‰ for each ten-fold decrease in total organic carbon content. Calculation shows that this variation of $\delta^{13}C$ would be expected (and would be accurately linear over at least three orders of magnitude; see Figure 5-8) if the carbon loss process could be modelled by a single reaction having a 0.27% kinetic isotope effect ($\alpha = 1.0027$). In fact, of course, the enrichment of ^{13}C in the organic matter of carbon-poor sediments could have some very different origin related to pre-depositional effects, but the compatibility of the observed variations with even this rudimentary model of post-depositional carbon losses should be recognized.

If all or part of the observed variation in carbon abundances and isotopic compositions is due to post-depositional effects, then several cautionary statements are in order. First, low carbon contents might indicate extensive losses rather than low initial carbon contents. Second, relatively high ^{13}C abundances might represent isotopically fractionated residues rather than unusual ecosystems. It is for these reasons that our reported "representative" isotopic compositions (Chapters 7 and 12) have been derived by mass-weighted averaging, a procedure that reduces the effect of any post-depositional alterations due to carbon losses at the same time it more accurately represents the total carbon inventory in each rock unit.

The average carbon abundances found in most of the rock units are within the range of values previously reported from the Precambrian by other investigators (Reimer et al., 1979; Cameron and Garrels, 1980; Ronov, 1980). The carbon abundances found here for some of the Archean shales are remarkably high, however, and appear to be significantly higher than previously reported average values (Reimer et al., 1979; Cameron and Garrels, 1980). It is tempting to ask whether these results might provide a better view of biological

TABLE 5-8. Analyses of kerogen fractions of Precambrian sediments

P.P.R.G. sample no.	Lithology[b]	Kerogen fraction		X-ray diffraction		Combustion yields (μmole)			Atomic ratios[a]		Isotopic analyses[a]		
		%C found	Color[c]	$w_{1/2}$ °2θ	d_{002} Å	CO_2	H_2	N_2	H/C	N/C	$\delta^{13}C_{PDB}$	δD_{SMOW}	$\delta^{15}N_{air}$
					Burra Group, 0.8 *Ga*								
455–1	C	64	deep orange yellow	4.25	3.52	290	25.1	1.41	0.17	0.010	−23.2	−84.0	5.1
					Bitter Springs Fm., 0.9 *Ga*								
133–2	C	97	strong yellowish brown	3.31	3.48	214	35.9	2.69	0.34	0.025	—[d]	−120.6	7.4
					Roper Group, 1.4 *Ga*								
114–1	S	67	dark orange yellow	3.23[f]	3.55[f]	262	103	2.72	0.79	0.021	−32.4	−95.5	4.5
123–1	S	57	strong yellowish brown	4.06	3.58	183	43.9	1.90	0.48	0.021	−32.0	−104.1	2.8
120–1	Q	85	strong yellowish brown	3.23	3.51	266	—[d]	3.23	—[d]	0.024	−32.1	—[d]	3.1
120–1	Q	—	—	—	—	306	61.6	3.79	0.40	0.025	−32.1	−94.6	3.0
					Bungle Bungle Dolomite 1.6 *Ga*								
159–1	C	65	strong yellowish brown	2.72	3.47	194	30.0	2.31	0.31	0.024	−30.0	−131.4	8.0
159–1	C	—	—	2.84	3.48	—	—	—	—	—	—	—	—
155–1	S	66	dark orange yellow	1.65	3.47	294	68.1	2.23	0.45	0.015	−33.5	−102.6	8.5
138–1	AC	85	strong yellowish brown	2.68	3.45	187	35.8	0.902	0.38	0.010	−30.2	−104.6	9.9
154–1	AC	76	strong yellowish brown	2.90	3.48	259	50.0	2.85	0.39	0.022	−30.7	−97.1	4.8
152–1	ASC	89	deep yellowish brown	—[e]	—[e]	186	39.7	1.58	0.43	0.017	−29.3	−143.1	—
					McArthur Group, 1.6 *Ga*								
113–1	ASC	34	—[d]	—[e]	—[e]	124	24.3	0.780	0.39	0.013	−38.9	−140.5	—[d]
					Earaheedy Group, 1.8 *Ga*								
440–1	A	97	deep yellowish brown	4.45	3.55	432	57.7	1.03	0.27	0.005	−33.2	−94.7	5.3

						Rove Fm. 1.8 Ga							
498–1	S	61	strong yellowish brown	2.44	3.48	320	65.7	1.86	0.41	0.012	−32.1	−131.1	7.1
						Wyloo Group, 2.0 Ga							
054–1	A	76	black	3.50	3.52	255	18.7	1.48	0.15	0.012	−25.2	−110.2	—[d]
054–1	A	—	—	—	—	97.1	8.32	0.562	0.17	0.012	−25.2	−118.0	—[d]
054–1	A	—	—	—	—	263	18.9	1.51	0.14	0.011	−25.3	−94.5	4.8
054–1	A	—	—	—	—	297	20.8	1.72	0.14	0.012	−25.2	−100.2	3.3
054–1	A	—	—	—	—	391	28.1	2.23	0.14	0.011	−25.0	−99.6	2.8
060–1	C	40	strong orange yellow	3.78[f]	3.53[f]	123	19.8	0.303	0.32	0.005	−31.7	−139.9	—[d]
						Gunflint (Biwabik) Iron Fm., 2.0 Ga							
134–4	C	—	—	—	—	114	34.5	1.19	0.61	0.021	−33.5	−77.8	7.2
						Upper Albanel Fm., 2.1 Ga							
443–1	A	71	black	4.29	3.52	320	21.7	0.709	0.14	0.004	−29.8	−95.7	—[d]
						Belcher Group, 2.1 Ga							
447–1	CA	80	strong yellowish brown	4.88	3.57	308	39.3	1.96	0.26	0.013	−28.4	−114.2	4.4
						Transvaal Supergroup, 2.2 Ga							
299–1	A	69	black	4.33	3.49	201	14.0	0.290	0.14	0.003	−35.0	−121.5	—[d]
304–1	A	52	black	4.96	3.56	278	27.5	0.906	0.20	0.007	−32.5	−103.3	—[d]
304–1	A	—	—	—	—	97.3	10.4	0.354	0.22	0.007	−32.5	−117.8	—[d]
						Hamersley Group, 2.5 Ga							
048–2	S	99	black	5.24	3.53	303	17.2	1.18	0.11	0.008	—[d]	−106.2	5.7
357–1	S	53	black	3.70	3.52	345	19.7	1.22	0.11	0.007	−35.3	−111.9	0.8
368–1	SA	52	black	4.96	3.53	219	14.8	0.586	0.14	0.005	−41.8	−93.9	—[d]
						Steeprock Group, 2.6 Ga							
325–1	C	76	black	3.23	3.52	60.8	5.18	0.262	0.17	0.009	−27.2	−99.2	—[d]
325–1	C	—	—	—	—	205	17.7	0.832	0.17	0.008	−27.2	−94.8	—[d]
						Ventersdorp Supergroup, 2.6 Ga							
293–1	C	51	black	4.09	3.53	254	18.0	0.225	0.14	0.002	−38.5	−123.1	—[d]
						Manjeri Fm., 2.6 Ga							
225–1	C	66	black	3.58	3.48	116	8.54	0.262	0.15	0.005	−33.0	−101.7	—[d]
225–1	C	—	—	—	—	227	12.0	0.463	0.11	0.004	−33.2	−106.4	—[d]

TABLE 5-8. (continued)

P.P.R.G. sample no.	Lithology[b]	Kerogen fraction		X-ray diffraction		Combustion yields (μmole)			Atomic ratios[a]		Isotopic analyses[a]		
		%C found	Color[c]	$w_{1/2}$ °2θ	d_{002} Å	CO_2	H_2	N_2	H/C	N/C	$\delta^{13}C_{PDB}$	δD_{SMOW}	$\delta^{15}N_{air}$
Bulawayan Group, 2.6 Ga													
254–1	A	82	black	3.46	3.48	315	15.4	1.24	0.098	0.008	−31.2	−121.1	—
Fortescue Group, 2.8 Ga													
027–1	C	64	black	5.12	3.55	163	20.3	0.146	0.25	0.002	−51.2	−106.8	—[d]
038–1	S	79	black	4.25	3.52	272	19.1	0.488	0.14	0.004	−41.0	−141.2	—[d]
038–1	S	—	—	4.33	3.52	—	—	—	—	—	—	—	—
038–1	S	—	—	4.57	3.52	—	—	—	—	—	—	—	—
038–1	S	—	—	4.49	3.52	—	—	—	—	—	—	—	—
038–1	S	—	—	4.25	3.51	—	—	—	—	—	—	—	—
Mozaan Group, 3.0 Ga													
266–1	S	94	black	1.38	3.39	133	7.01	0.167	0.11	0.003	−27.3	−93.7	—[d]
Gorge Creek Group, 3.4 Ga													
022–1	S	74	black	4.57	3.57	347	16.5	1.89	0.095	0.011	−29.9	−118.1	1.6
022–1	S	—	—	—	—	217	8.50	1.14	0.078	0.010	−30.6	−94.6	3.0
022–1	S	—	—	—	—	545	23.2	—[d]	0.085	—[d]	−29.9	−99.6	—[d]

							Warrawoona Group, 3.5 *Ga*							
002–1	C	64	black		—[e]	—[e]	152	22.4	0.353	0.30	0.005	−34.3	−105.3	—[d]
004–1	C	77	black		4.21	3.58	174	14.7	0.535	0.17	0.006	−35.9	−116.2	—[d]
004–1	C	—		—	—	—	408	33.6	1.34	0.16	0.007	−36.1	−110.8	—
016–1	C	50	black		3.66	3.52	327.1	49.4	0.386	0.30	0.002	−35.2	−109.8	—[d]
							"*Swaziland Sequence*," 3.5 *Ga*							
182–1	C	78	black		3.70	3.52	253	14.7	0.473	0.12	0.004	−31.6	−102.5	—[d]
182–1	C	—		—	—	—	569	34.5	1.04	0.12	0.004	−31.7	−99.3	2.9
189–1	C	82	black		4.02	3.53	230	14.7	0.358	0.13	0.003	−29.9	−93.1	—[d]
198–1	C	64	black		3.74	3.51	347	16.8	0.950	0.096	0.005	−32.0	−103.3	—[d]
198–1	C	—		—	3.39	3.50	—	—	—	—	—	—	—	—
199–1	C	79	black		3.58	3.49	280	11.8	0.903	0.085	0.006	−26.6	−106.9	—
199–1	C	—		—	3.39	3.48	—	—	—	—	—	—	—	—
215–1	C	80	black		3.03	3.48	142	11.3	0.460	0.16	0.006	−31.9	−111.5	—[d]
215–1	C	—		—	—	—	410	33.8	1.28	0.16	0.006	−31.9	−95.6	3.2
181–1	S	98	black		<0.8	3.38	459	4.35	0.123	0.019	0.001	−15.4	−44.1	—[d]
							"*Isua Supracrustal Sequence*," 3.8 *Ga*							
465–1	B	95	black		<0.8	3.39	484	3.80	0.27	0.016	0.001	−10.0	−89.3	—[d]
466–1	H	57	black		<0.8	3.38	455	2.75	0.187	0.012	0.001	−13.2	−103.3	—[d]

[a] Uncertainties (95% confidence limits) associated with tabulated elemental ratios and isotopic compositions are as summarized below, and are discussed in Appendix IV: N/C, ±12%; H/C, ±15%; $\delta^{13}C_{PDB}$, ±0.4 ‰; $\delta^{15}N_{air}$, ±2 ‰; δD_{SMOW}; see discussion of Appendix IV.

[b] Lithology Code: A = carbonate, B = banded iron-formation, C = chert, G = granite, H = schist, I = basalt, N = gneiss, Q = sandstone, S = shale.

[c] Descriptive terms correspond to U. S. National Bureau of Standards nomenclature and observation conditions described in Appendix IV.

[d] Value not determined.

[e] X-ray diffraction pattern poorly defined; parameter not measurable.

[f] X-ray diffraction pattern poorly defined; parameter uncertain.

TABLE 5-9. Total organic carbon: dependence of isotopic composition on abundance. [$\delta^{13}C_{PDB}$(T.O.C.) = $a + b$ log (carbon abundance, mg C/g)].*

Sample type	*n*	*a*	*b*
Proterozoic carbonates	59	−29.4±2.1†	−6.9±3.2†
Archean carbonates	26	−31.0±4.8	−11.4±9.0
Proterozoic cherts	16	−25.8±4.2	−6.2±5.5
Archean cherts	60	−27.7±2.1	(−1.7±3.4)
Proterozoic shales	32	−29.5±1.4	−2.6±1.8
Archean shales	46	−26.7±2.7	−5.3±2.2
Proterozoic chert/carbonates	26	−23.4±6.5	(−0.4±11.8)
Archean chert/carbonates	16	−43.6±6.4	(−4.2±14.8)

* Note that *a* represents the average isotopic composition (‰ vs PDB) of a sample with T.O.C. = 1.0 mg C/g and that *b* represents the isotopic shift (‰) correlated with each 10-fold change in T.O.C. See Fig. 5-8.

† Indicated uncertainties are 95% confidence intervals based on *n* − 2 degrees of freedom. Statistically insignificant values are enclosed in parentheses.

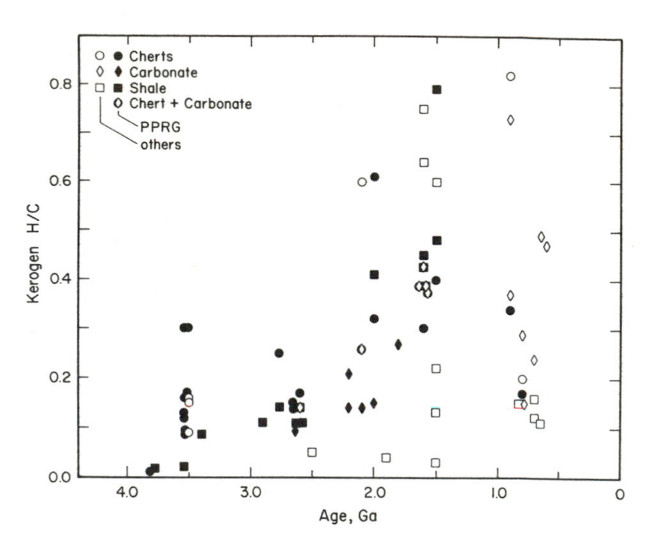

FIGURE 5-9. Elemental compositions of Precambrian kerogens as a function of age. Open symbols represent data from references 3, 8, and 9 of Table 5-5. Filled symbols represent data from Table 5-8.

productivity, rates of sedimentation, or the efficiency of recycling during the Archean, but it appears that no conclusions can be safely drawn because of inadequacies in the sample collection on which the present results are based. The program of sample acquisition in this work has not been designed to support a quantitative inventory of large volumes of sediment. No attempt has been made to collect materials representative of all major rock types in a given unit, and little or no information is available regarding the relative abundances of rock types in each unit. Rather, samples have been acquired specifically for organic geochemical and micropaleontological investigations, and it would appear that the average carbon content in the sample collection is likely to be biased toward atypically carbon-rich materials.

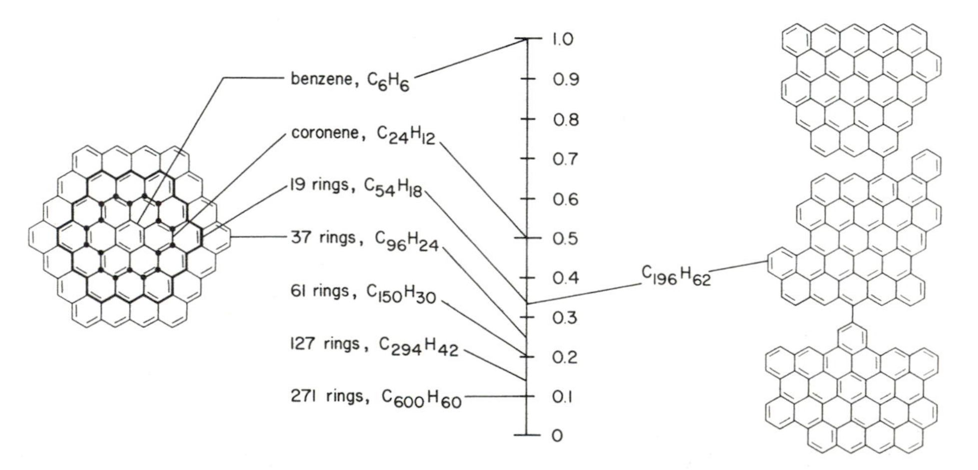

FIGURE 5-10. Representative structures of aromatic hydrocarbons with low H/C ratios. Note that an H/C ratio of 1/n will be obtained from a fully aromatic hexagonal structure with n concentric rings of carbon atoms.

With these limitations in mind, the results of the survey analyses have been utilized chiefly to provide an overview of Precambrian isotopic abundances (as discussed in Chapters 7 and 12) and to aid in the selection of samples for more detailed study. The results of those further investigations will now be discussed.

5-5-2. Analyses of Kerogen Fractions

Table 5-9 summarizes the results of analyses of 42 different kerogen fractions. Procedures used in the collection of these results are described in Appendix IV. Although 50 samples were initially selected for kerogen isolation and analysis, a variety of mishaps prevented completion of satisfactory analyses for 8 of those samples. Kerogen fractions with high ash content were excluded, kerogen fractions from several samples were lost during processing, and a single sample appeared to have been massively contaminated.

ELEMENTAL ANALYSES

Ratios of hydrogen to carbon are plotted in Figure 5-9. It is evident that, while relatively young kerogens have no guarantee of excellent preservation, extremely old kerogens are almost certainly the victims of extensive dehydrogenation. Only three samples of Archean age yielded H/C ratios greater than 0.2, those being the same Warrawoona Group and Fortescue Group samples in which microfossil assemblages were found (see Chapter 9). The structural consequences of such low H/C ratios can be appreciated by referring to Figure 5-10, which dramatically illustrates the very extensive rearrangement that the Archean kerogens must have sustained. Because the same forces that drove the dehydrogenation of the kerogen should also have served either to degrade or to mobilize any labile molecules, it is extremely unlikely that any syngenetic chemical fossils can have survived in these Archean samples.

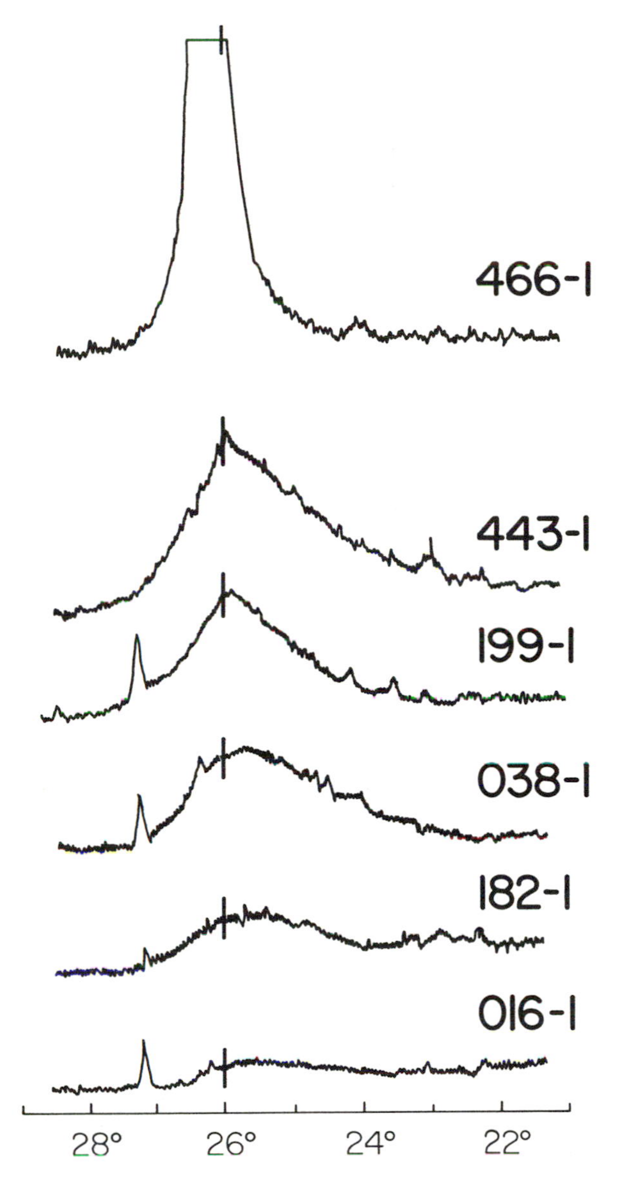

FIGURE 5-11. Representative x-ray diffractograms of kerogen fractions having the P.P.R.G. sample numbers indicated. The vertical tick mark in each recording is at 26° 2θ and is near the maximum corresponding to the 002 reflection of graphite. Procedures employed in the collection of these data are summarized in Appendix IV.

X-RAY DIFFRACTION ANALYSES

Representative x-ray diffractograms are shown in Figure 5-11. Relationships between the d_{002} (computed from the centroid value) and widths at half-height (measured at the centroid) parameters and between the latter value and the H/C ratio are shown in Figure 5-12.

The experimental conditions and procedures employed in these measurements are described in Appendix IV. It is important to note that the use of a glass sample substrate was avoided, thus eliminating any possible confusion between the x-ray diffraction patterns of glass and of disordered, subgraphitic kerogen. Apparent interplanar spacing and width at half-height are well correlated (Figure 5-12A). Although the data collected in this work lie, for the most part, beyond the range

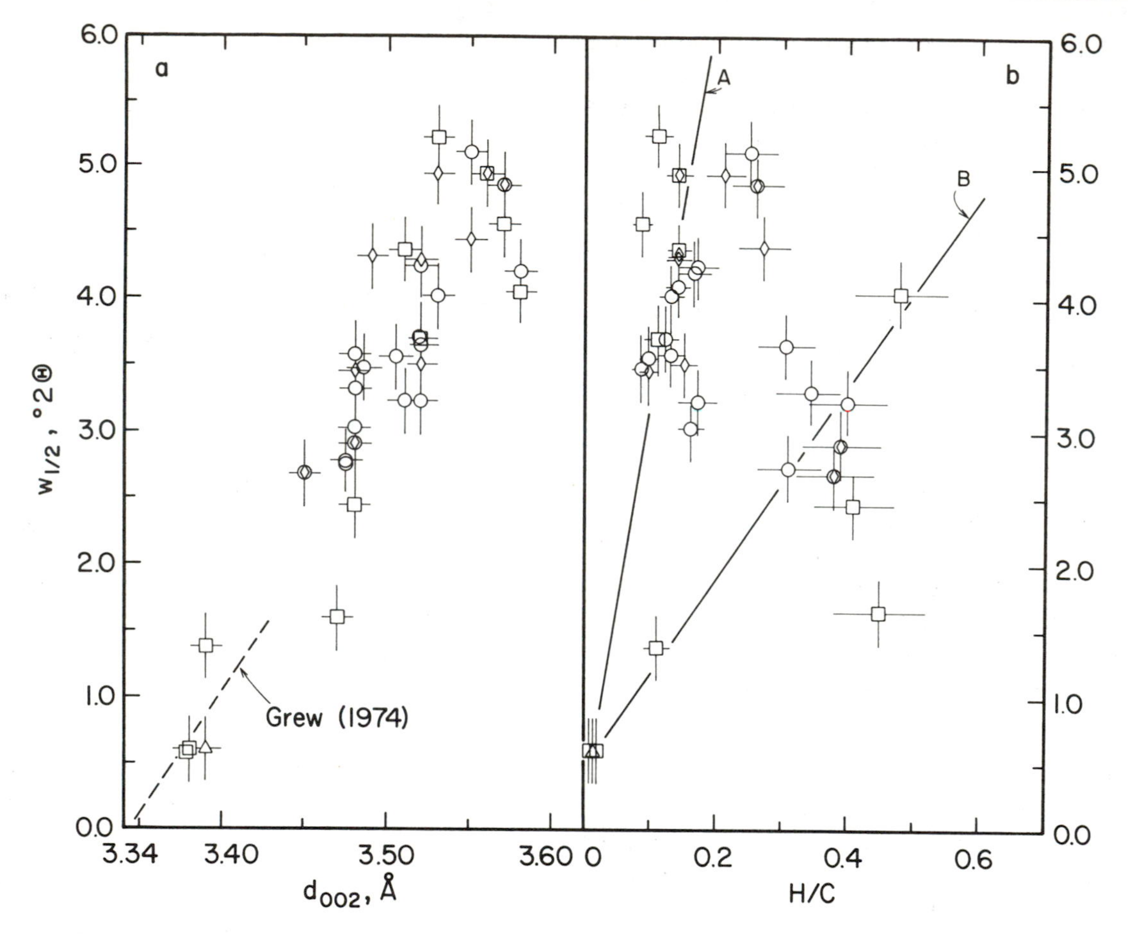

FIGURE 5-12. The relationship between width at half height (measured at the centroid of the 002 reflection) and (a) apparent d_{002} value and (b) elemental composition. Open symbols indicate data from references 3, 8, and 9 of Table 5-5.

investigated by Grew (1974), the present results appear to be fully consistent with his prior observations. This consistency and the strong correlation suggest that these X-ray parameters can be developed as useful tools for the study of subgraphitic kerogens.

It is evident that no regular relationship prevails between the peak width and the H/C ratio (Figure 5-12B), a somewhat surprising result since dehydrogenation and graphitization might naturally be correlated. Possible causes for the rather poor correlation include (*i*) laboratory contamination by water or some other hydrogen source; (*ii*) "natural" contamination of the kerogen by mobile hydrogen-bearing fluids (water or hydrocarbons) under conditions allowing incorporation of the hydrogen by the kerogen; and (*iii*) a mixture of nearly graphitic material and primary biologic material in the original depositional environment (i.e., the incorporation of "pre-metamorphosed" kerogen). It would appear that the first of these alternatives can be safely excluded because the samples with H/C ratios greater than 0.3 are not characterized by any distinctive δD value, a systematic effect which would be expected if the excess hydrogen were due to laboratory contamination.

A choice between the latter possibilities could be offered if it were considered that H/C and $w_{1/2}$ should be related by the process of graphitization. In that case, it might be decided either that line A in Figure 5-12B represents the normal course of graphitization and that the samples lying near line B have somehow accumulated excess hydrogen, or that line B represents the normal course of graphitization and that the samples lying near line A have incorporated hydrogen-poor material. If the data of Table 5-4 are added to Figure 5-12B in order to identify the "normal course of graphitization" by association with either line A or B, two interesting results emerge. First, the gap between lines A and B becomes more densely populated, indicating that no soundly based choice of one or the other as "normal" can be offered at present and, indeed, that a relationship of the type indicated by line A or B may not exist. Second, the gap between the highly ordered material (near $w_{1/2} = 0.5$ and $H/C = 0$) and the rest of the samples does not receive any additional points and becomes a major feature of the results. The persistence of this gap suggests a natural discontinuity analogous to the "jumps" recognized in coalification processes (see Hunt, 1979, pp. 345–346), but does not otherwise identify aspects of the "normal" course of graphitization. In these circumstances, it is not possible to determine whether the postmetamorphic addition of hydrogen-rich materials or the recycling of hydrogen-poor kerogen is responsible for the distribution of results shown in Figure 5-12B.

COLOR

Following the examples cited in the introductory portion of this chapter, the colors of the kerogen samples were carefully assessed (see Appendix IV). This procedure was useful only in distinguishing between kerogen samples that displayed some color and those which were entirely black (opaque). Where some color other than black is noted in Table 5-8, it refers to small (1–10 μm) areas within larger particles that were otherwise black. (Diffraction phenomena were recognizable—the tabulated colors do not refer to optical artifacts.) These results are, thus, not comparable with those reported in Table 5-3, which refers to colors of the lightest (entire) particles within a kerogen sample. Two conclusions can be drawn regarding the use of color measurements for characterization of Precambrian kerogens: (*i*) the range of variation is highly compressed, with most materials falling at or very near the black end of the range and with distinctions between samples being very difficult, and (*ii*) within that highly compressed range, the appearance of some color generally indicates and H/C ratio greater than 0.25.

ISOTOPIC ANALYSES

Elemental and isotopic analyses are plotted together in Figure 5-13, an illustration which, like Figure 5-5 (which includes data from the Phanerozoic), follows the format introduced by McKirdy and Powell (1974). The chart also includes the line drawn through the distribution of points in Figure 5-5. The general trend toward ^{13}C enrichment at low H/C ratios suggests that some metamorphic alteration of ^{13}C abundances has taken place. Because all kerogen types appear to follow similar maturation paths once H/C ratios as low as 0.6 have been reached (Robin et al., 1977), the metamorphic alterations observed here must occur by the same chemical mechanisms responsible for the shifts in $\delta^{13}C$ observed in Phanerozoic materials. It follows that the relationship between H/C (for values less than 0.6) and isotopic shift observed for Phanerozoic materials should also be applicable to these Archean and Proterozoic kerogens, independent of their possibly unusual origins and

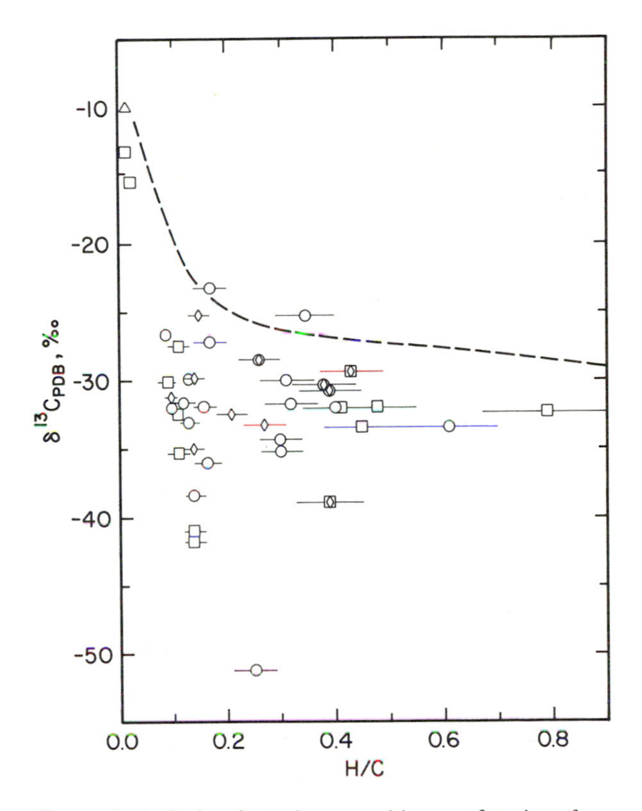

FIGURE 5-13. Carbon isotopic composition as a function of kerogen H/C ratio for the samples analyzed in this work. Open symbols represent data from references 3, 8, and 9 of Table 5-5. Indicated uncertainties in H/C are ±15% and correspond roughly to 95% confidence intervals (see Appendix IV). Uncertainties in $\delta^{13}C$ analyses are smaller than the points. The broken line, shown for reference, duplicates that constructed as representative of Phanerozoic samples and shown in Figure 5-5.

depositional environments. Because the previously observed relationship between $\delta^{13}C$ and H/C indicates that $\delta^{13}C$ will normally shift by less than 10 ‰ in the range between H/C = 0.1 and H/C = 0.5, it can be concluded that not all the observed isotopic variation, which amounts to at least 25 ‰ in the range indicated, can be attributed to metamorphic alteration.

Control of the isotopic composition of the kerogen at its source in the depositional environment furnishes a second possible cause for the observed isotopic abundance variation. Such "source-" or "facies-control" might arise in many different ways, and it is our purpose here only to estimate the possible magnitude of these effects, not the mechanisms of their origin. If the trend line shown near the top of the distribution is displaced to $\delta^{13}C$ values approximately 4 ‰ lower, for example, it is evident that the group of points trending from values near −31 ‰ at H/C = 0.5 to values near −25 ‰ at H/C = 0.15 could be related to a single source with an isotopic composition near −32 ‰ vs. PDB. Interestingly, this group of points comprises most of the Proterozoic rock units investigated, specifically including Skillogallee, Bitter Springs, Roper River, Bungle-Bungle, Wyloo, Biwabik, Rove, Mistassini, Belcher Islands, and Steeprock.

A significant further displacement to values possibly representative of a source with an isotopic composition near −37 ‰ accomodates principally samples from the Warrawoona Group and the Swaziland Supergroup, their different H/C values accounting nicely for the observed difference in carbon isotopic content. No significance can be attributed simply to the correlation, but the same trend line passes through or near (±2 ‰) points associated with several other rock units, specifically Earaheedy, Mistassini, Transvaal (P.P.R.G. Sample No. 304 only), and Pongola. Sources with still lower ^{13}C contents are required to account for points associated with the McArthur Group, the Hamersley Formation, the Ventersdorp Formation, the Belingwe Greenstone Belt, the Bulawayan Limestone, the Fortescue Group, and the Gorge Creek Group.

Application of the H/C vs. $\delta^{13}C$ relationship at H/C values below 0.1 is practically impossible because the curve must be regarded as indeterminate in that region. On the one hand, adequate data are not available. On the other hand, the mechanism responsible for isotopic shifts might change from one of carbon loss to one including the possibility of exchange with other carbon sources (as noted in the preceding review). It is impossible, therefore, to estimate the initial carbon isotopic composition of the carbonaceous material in samples from the Isua Supracrustal Group.

5-6. Summary and Conclusions

(1) The tendency of biosynthesized organic matter to "leak" from the carbon cycle and be incorporated in sediments provides an organic geochemical record that can, in principle, reveal much about ancient ecosystems.

(2) The precision with which the organic geochemical record can be decoded depends very strongly on the level of preservation of the organic material. Problems associated with poor preservation can be expected to be especially acute in materials of Precambrian age. Accordingly, when dealing with these very ancient materials, the accurate appraisal of the level of preservation of the carbonaceous matter becomes a primary objective rather than a preliminary consideration.

(3) Sedimentary organic matter can be classified broadly in two categories: extractable molecules (including chemical fossils) and insoluble, macromolecular material (kerogen). The second category represents a geological end product that eventually incorporates most extractable materials. It is generally true, therefore, that the surviving chemical fossils, which are relatively unaltered materials that have not yet joined the kerogen, can be regarded as better preserved and more likely to be informative.

(4) The very low concentration of chemical fossils in Precambrian materials, together with the inevitable mobility of these relatively small molecules, leads, however, to suspicion that some or all of these interesting and potentially informative species are not as old as the sediments in which they are found, having instead been imported by fluids in the relatively recent geologic past. Stereochemical and carbon-isotopic evidence support this interpretation, though do not prove it universally true.

(5) Elemental and isotopic analyses of Phanerozoic kerogens have been widely reported and interpreted. Maturation pathways describing the development of increasingly altered kerogen structures as sediments reach higher and higher temperatures can be defined in terms of systematically varying elemental abundances. Carbon isotopic abundances can be logically associated with pre-

cursor materials or with post-depositional alterations independently confirmed by mineralogic indicators.

(6) Correlations can be established between inorganic metamorphic indicators and a wide variety of parameters describing aspects of kerogen structure. These correlations are not yet perfectly defined, but they have value as indicators of consistency or convergence of inorganic and organic geochemical lines of evidence.

(7) Elemental and isotopic analyses—particularly in mutually supporting pairs—of Precambrian kerogens have been very rare. Reliable elemental analyses that have been reported, however, generally indicate extensive dehydrogenation and consequent structural alteration of kerogenic carbon skeletons. To the extent that information on metamorphic histories of the analyzed materials has been available, these findings of extensive post-depositional alteration do fit with the inorganic and mineralogic evidence. The apparent persistence of hydrogen-rich and thermally labile chemical fossils has not, on the other hand, been explained.

(8) Forty-two new elemental analyses of kerogen fractions from 23 different Precambrian rock units show that all kerogens of Archean age have been extensively dehydrogenated and, consequently, heavily altered structurally. These results are summarized in Table 5-10, which also shows that higher H/C ratios were found for some Proterozoic kerogens. Even in those cases, however, dehydrogenation has proceeded to an extent that the kerogen must have preserved very few biosynthetic carbon skeletons.

(9) The correlations between structural alteration and metamorphic history shown in Table 5-10 are very similar to those previously developed from the Phanerozoic record and summarized in

TABLE 5-10. Preservation of the Precambrian organic geochemical record.

Rock unit	*Age (Ga)*	*Metamorphism*[a]	*Dehydrogenation, H/C*	*Probable carbon*[b] *isotopic shift, ‰*
Burra	0.8	Lower greenschist	Very extensive, ~0.15	4
Bitter Springs	0.9	Unknown	Extensive, ~0.35	1.2
Roper	1.4	"Little or none"	Mild, ~0.8	0
Bungle Bungle	1.6	Unknown	Moderate, ~0.4	0.8
Earaheedy	1.8	Lower than greenschist	Extensive, ~0.3	1.7
Rove	1.8	"Subgreenschist"	Moderate, ~0.4	0.5
Wyloo	2.0 (054)[c]	Pumpellyite-actinolite	Very extensive, ~0.15	5
	(060)[c]	Pumpellyite-epidote	Extensive, ~0.3	1.2
Gunflint (Biwabik)	2.0	"Subgreenschist"	Moderate, ~0.6	0
Albanel	2.1	Greenschist	Very extensive, ~0.15	5.5
Belcher	2.1	"Subgreenschist"	Extensive, ~0.25	1.6
Transvaal	2.2	"Subgreenschist"	Very extensive, ~0.2	2.5–5
Hamersley	2.5 (368)[c]	Prehnite-pumpellyite	Very extensive, ~0.15	5
	(357)[c]	Pumpellyite-actinolite	Very extensive, ~0.1	8
	(048)[c]	Greenschist	Very extensive, ~0.1	8
Steeprock	2.6	To greenschist	Very extensive, ~0.15	4
Ventersdorp	2.6	Greenschist	Very extensive, ~0.15	5
Manjeri	2.6	Upper greenschist	Very extensive, <0.15	6
Bulawayan	2.6	Greenschist	Nearly complete, ~0.1	9
Fortescue	2.8 (027)[c]	Pumpellyite-epidote	Extensive, ~0.25	1.9
	(038)[c]	Pumpellyite-actinolite	Very extensive, ~0.15	
Mozaan	3.0	Greenschist	Very extensive, 0.11	8
Gorge Creek	3.4	Greenschist	Nearly complete, <0.1	10
Warrawoona	3.5	Lower greenschist	Extensive, <0.3	1.5–4
Swaziland	3.5	Greenschist (excluding 181)[c]	Very extensive, ~0.1	5–10
Isua	3.8	Amphibolite	Complete, ~0.01	?

[a] See Appendix I for details and documentation of metamorphic grades

[b] Crudely estimated by application of the $\delta^{13}C$ vs. H/C relationship shown in Figure 5-5. These values are useful indicators of *possible* or *probable* metamorphic effects, but must be taken *cum granus salus*.

[c] Numbers in parentheses are P.P.R.G. Sample Numbers (see Appendix I).

Figure 5-7. It is evident that organic geochemical investigations of materials subjected to metamorphism at or beyond the upper greenschist facies must amount to a study of graphite and that any level of greenschist facies metamorphism destroys any molecular or structural organic chemical record.

(10) The same physical conditions that lead to dehydrogenation of kerogen must also affect, to a greater or lesser extent, any coexisting isolated molecules (i.e., chemical fossils), causing them either to be mobilized or similarly dehydrogenated. The levels of alteration encountered in all Archean samples investigated in this work would appear to preclude the survival of chemical fossils in those materials. A statement only slightly weaker applies throughout much of the Proterozoic.

(11) Three-hundred-seventeen new isotopic analyses reveal that carbon isotopic abundances in organic matter vary far more widely in Precambrian than in Phanerozoic sediments. Examination of kerogen H/C ratios allows estimation of isotopic shifts due to post-depositional carbon losses. The "probable" isotopic shifts (see Table 5-10) are, in many cases, small enough that substantial errors in their estimation would be of little consequence. It follows that some success in reconstructing the isotopic record of Precambrian biology can be expected. Specifically, some explanation must be offered for the systematic and, in many cases, extreme, depletion of ^{13}C in Precambrian kerogens. As is suggested in Chapter 12, the solution to that problem may contribute significantly to an understanding of the Earth's earliest biosphere.

CHAPTER 6

Biochemical Evolution of Anaerobic Energy Conversion: The Transition from Fermentation to Anoxygenic Photosynthesis

By Howard Gest and J. William Schopf

6-1. Introduction

Of necessity, the search for primary, direct evidence of the origin and evolution of Earth's earliest biosphere must focus on the one repository where such evidence may have been sequestered—the preserved rock record. Yet, as is evident from the foregoing chapters in this book, (*i*) presently surviving rocks are as yet unknown from the earliest Hadean stage of evolution of the planet, a phase in Earth history that may well have encompassed the beginnings of biological activity; (*ii*) known Archean rock sequences are of rare occurrence, of limited areal extent, and quite commonly of high metamorphic grade; and (*iii*) the geochemical remnants of early life and/or abiotic organic compounds occurring in such ancient sequences are chemically complex, highly altered, and difficult to interpret. Although evidence of the early history of life is thus less accessible than one might wish, the picture is not entirely bleak; the occurrence both of microscopic fossils (see Chapter 9) and of the megascopic organo-sedimentary structures (viz., "stromatolites") such organisms produced (as discussed in Chapter 8) set important constraints on the timing and nature of early evolutionary events, and significant information regarding both the biochemistry and physiology of the Archean biota can be extracted from studies of the stable isotopes of other biogenic elements preserved in ancient organic matter (see Chapters 5 and 7). Perhaps most importantly, however, it should be recognized that the geologic record is not the sole source of useful, significant data. Sequestered within the biochemistry and molecular biology of each living organism is another type of record—albeit one that has been abridged and otherwise modified over the course of time—of the history of its evolving lineage. Comparative studies of extant organisms can thus provide important insights into the nature of early evolutionary events, of kind and degree that can never be revealed solely by investigations of the preserved rock record. The problem, of course, is to sort out biochemical and molecular biological features occurring in living systems that accurately reflect their evolutionary history. The present chapter summarizes results of one approach to this problem, with emphasis on the metabolic characteristics of the earliest forms of life and in particular, on the development of anaerobic photosynthetic prokaryotes from their more primitive fermentative ancestors.

The structure and metabolism of all types of living cells are fundamentally based on sugars and their derivatives. This sweeping general assessment can be defended even for prokaryotes that have "unusual" nutrient requirements or that seem to have bizarre biochemical patterns. The enormous number of nutritional and metabolic differences observed among prokaryotes, of course, are clearly of importance for understanding biochemical evolution; they should not be minimized. In fact, it now seems somewhat surprising that many microbiologists and biochemists in the recent past subscribed to such an oversimplified view of the "unity of biochemistry" that S. S. Cohen felt it necessary to conclude his 1963 essay "On biochemical variability and innovation" as follows: "We conclude from this analysis that it is invalid to assume that a biochemical exploration of specialized function will necessarily yield a mere modification of some well-recognized mechanisms. In our estimation the data of biochemistry are not only insufficient to warrant assumptions concerning the nature of substance and mechanism but suggest potentialities and capacity for great complexity in the evolution of function."

Nevertheless, "biochemical unity" at a gross level can be seen in the pervasive occurrence of sugars and their derivatives in cellular structures, and in the widespread use of intermediates of sugar metabolism as biosynthetic precursors of amino acids, purines, pyrimidines, lipids, and numerous other cellular components. With only rare exception, cell walls or envelopes contain hexoses and amino sugars, and reserve materials consist of polymers of sugars or of "C–C" units derived from sugars. Moreover, in all organisms pentoses are integral components of nucleic acids, hydrogen transfer catalysts (pyridine nucleotides and flavins), and the ubiquitous "energy currency" adenosine triphosphate (ATP). The universality of sugars and use of their derivatives for biosynthesis would seem to indicate that the development of sugar-based biochemistry must have been an early evolutionary innovation; indeed, a strong argument can be made that it was only such sugar-based cellular systems that provided the successful starting point for biochemical and cellular evolution.

6-2. Sugars in the Primeval Soup: Early Fermentations

Quayle and Ferenci (1978) have ably summarized current ideas relating to the possible importance of formaldehyde (HCHO) in the prebiotic environment, and of experiments on the abiotic self-condensation of HCHO to form sugars (via the so-called "formose reaction"). They suggest that as the reservoir of abiotic sugars became gradually depleted by anaerobic fermenters in the primitive environment, the shortage of sugar was overcome by net synthesis from HCHO via a biosynthetic pathway similar to the "ribulose monophosphate (RuMP) cycle" now observed in many "methylotrophs" (bacteria that can use reduced one-carbon compounds as their sole source of carbon; for a recent general review of these organisms, see Higgins et al., 1980). It is noteworthy that carbon transformations in the RuMP cycle are strikingly similar to those of the ribulose bisphosphate-based cycle of CO_2 fixation (i.e., the "Calvin cycle"), and it has thus been suggested that the latter evolved from the former (Quayle and Ferenci, 1978; see also Quayle, 1980).

Glucose (the hexose sugar, $C_6H_{12}O_6$) has been described as the universal cellular fuel. Indeed, the capacity for anaerobic fermentative utilization of glucose (or related sugars) undoubtedly is the most ubiquitous of the biological energy-conversion mechanisms known in extant organisms. This is consistent with the view (Oparin, 1938) that the earliest organisms were anaerobic fermenters. And of all known fermentations, it seems most plausible that sugar fermentation in particular was the first energy-conversion process to have been successfully exploited by cells growing in the primitive anaerobic environment; the following are particularly cogent facts in support of this view:

(1) "Thermodynamic stability can be invoked, for example, for the selection of glucose as the universal cellular fuel. Glucose, the most stable of the 16 isomeric hexoses with its bulky substituents in equatorial positions, had the greatest chance to accumulate in the primordial soup" (Bloch, 1979).

(2) Fermentation of glucose, as exemplified by glycolysis to lactic acid, is mechanistically the simplest known type of energy conversion, and requires only one kind of hydrogen (electron) carrier (viz., nicotinamide adenine dinucleotide, NAD).

(3) ATP generation in fermentations occurs via "substrate-level" phosphorylation in completely soluble systems (i.e., membranous cellular structures are not required).

(4) The net yield of ATP from fermentation of sugars is quite low (ordinarily only 1 or 2 ATP per molecule of glucose metabolized as compared with 36 ATP produced via aerobic respiration) as would be expected of a primitive energy-conversion mechanism.

(5) "The view that anaerobic fermentations were the earliest energy-supplying processes is consistent with the finding that some of the components of the fermentation reactions, including ATP and DPN [i.e., NAD], occur in almost every other metabolic process supplying, or depending on, utilisable energy" (Krebs and Kornberg, 1957).

Numerous taxa of modern heterotrophic anaerobes can ferment glucose to obtain energy for growth, and in many instances virtually all of the glucose carbon thus metabolized is converted to extracellular products. For example, *Streptococcus faecalis* can convert glucose almost quantitatively to the three-carbon (C_3) compound lactic acid. This occurs by the Embden-Meyerhof-Parnas (E-M-P) mechanism, which yields two ATP molecules for each molecule of hexose fermented. Subsequent to reactions leading to cleavage of the C_6 sugar to two C_3 fragments, which become 3-phosphoglyceraldehyde (PGAL), the following oxidation-reduction (O/R) reaction occurs:

$$\text{3-phosphoglyceraldehyde} + \text{NAD} + P_i \longrightarrow \text{1,3-diphosphoglyceric acid} + \text{NADH}_2$$

Since NAD is present in cells in only minute ("catalytic") quantities, all of it will be reduced to $NADH_2$ (dihydronicotinamide adenine dinucleotide) after only a very small amount of glucose has been metabolized. For fermentation to continue in quantitatively significant degree, the $NADH_2$ must therefore be continually reoxidized in order that NAD be available for the O/R reaction shown above. This is achieved via transfer of hydrogen atoms from $NADH_2$ to an intermediate of fermentation, pyruvic acid ($CH_3COCOOH$), which is thereby reduced to form lactic acid ($CH_3CHOHCOOH$). In sum:

$$1\ C_6H_{12}O_6 \dashrightarrow 2\ \text{PGAL}$$

$$2\ \text{PGAL} + 2\ \text{NAD} \dashrightarrow 2\ CH_3COCOOH + 2\ NADH_2$$

$$2\ CH_3COCOOH + 2\ NADH_2 \longrightarrow 2\ CH_3CHOHCOOH + 2\ \text{NAD}$$

For the sake of simplicity, the reactions leading from PGAL to pyruvic acid are omitted here, but it should be noted that this sequence generates phosphoryl donors that are used to convert ADP (adenosine diphosphate) to ATP. The focus here is on the redox aspects, especially to emphasize that the simplest energy-yielding glucose fermentation pattern now known requires use of all of the glucose carbon in order to achieve O/R balance. The essence of this pathway, by which "a hexose molecule travels to reach the inert end product, lactic acid" is interestingly described by Lipmann (1946) as follows:

> Initial fission in the middle of the six-carbon chain seems to be prompted by the introduction of phosphate groups at both ends. This phosphorylation is a rather costly investment absorbing just one-half of the gross yield of energy and thus reducing the net yield by about fifty per cent. An initial investment of part of the ultimate energy yield in the operation of the process is a notable feature. It is this need of induction energy which makes the fermentative process autocatalytic. The misleading statement is often made that, in fermentation, one half of the hexose oxidizes the other half, suggesting a dismutative process. What happens, rather, is that a hydrogen donor, after unloading of phosphate bond energy, is transformed into a hydrogen acceptor. This manner of manipulation is expressed in the characteristic shape of the flow lines which, in all fermentations, fold back on themselves. The bending back, to accept a pair of hydrogens released in a previous stage on the molecular flow line, together with the initial expenditure of energy to start the process, may be considered as general and dominant characteristics of anaerobic metabolism.

6-3. More Complex Fermentations

From the standpoint of biochemical evolution, the various fermentation schemes of clostridia are of special interest. Although biological and physiological criteria indicate that the clostridia constitute a closely interrelated group of organisms, they show a remarkably diverse range of fermentation patterns. Glucose fermentation by some species can be represented by the overall equation: $C_6H_{12}O_6 \dashrightarrow 3\ CH_3COOH$ (acetic acid). Fermentation by other species typically yields a number of extracellular products, as summarized in Table 6-1.

Cells of *Clostridium perfringens* grown so as to be iron-deficient ferment glucose mainly by glycolysis, that is, by catabolism to lactic acid via the E-M-P pathway. In iron-sufficient cells, on the other hand, lactate formation is greatly decreased and a large fraction of the glucose fermented is converted to acetic and butyric acids, ethanol, CO_2 and H_2 (rather than to lactic acid). This striking difference primarily reflects alternative mechanisms of metabolism of the key C_3 intermediate, pyruvate. Iron-deficient *C. perfringens* cells achieve redox balance during glucose fermentation mainly by reducing pyruvate to lactate. Cells grown with adequate supplies of iron, however, metabolize

TABLE 6-1. Fermentation of glucose by *Clostridium* species.*

	Moles of product per 100 moles of glucose fermented by:			
		C. perfringens		
Product	*C. butyricum*	*Fe-deficient*	*Fe-sufficient*	*C. butylicum*
Butyric acid	76	9	34	17.2
Acetic acid	42	15	60	17.2
CO_2	188	24	176	203.5
H_2	235	21	214	77.6
Lactic acid	—	160	33	—
Butanol	—	—	—	58.6
Isopropanol	—	—	—	12.1

* Compiled by Wood, 1961. Data for *C. perfringens* (formerly known as *C. welchii*) are from Pappenheimer and Shaskan (1944).

pyruvate largely by the following reactions (in which extracellular end products are underlined):

(a) $CH_3COCOOH + CoASH \longrightarrow CH_3COSCoA + \underline{CO_2} + \underline{H_2}$

(b) $CH_3COSCoA + P_i \longrightarrow CH_3COOPO_3H_2 + CoASH$

(c) $CH_3COOPO_3H_2 + ADP \longrightarrow \underline{CH_3COOH} + ATP$

(d) $2\,CH_3COSCoA \xrightarrow{4H} \underline{CH_3CH_2CH_2COOH} + 2\,CoASH$

(e) $CH_3COSCoA \xrightarrow{4H} \underline{CH_3CH_2OH} + CoASH$

Reaction (a) is of particularly great significance in the normal metabolism of clostridia and numerous other kinds of anaerobes, and can be viewed as a major innovation in the evolution of energy-conversion mechanisms. Rather than being used simply as a terminal electron acceptor, pyruvate is converted to acetyl-CoA (acetyl-coenzyme A, $CH_3COSCoA$) that can serve several functions, namely: (*i*) as a source of additional energy via reactions (b) and (c); (*ii*) as an electron acceptor to aid in facilitating O/R balance [as in reaction (d) and also, to a minor extent, in (e)]; and (*iii*) as a biosynthetic "building block" (e.g., for the synthesis of lipids).

The conversion of pyruvate to acetyl-CoA is an oxidative decarboxylation, and in many anaerobes the electrons are disposed of in the form of H_2 (viz., $2\,e + 2\,H^+ \rightarrow H_2$). This requires the participation of the iron-containing proteins ferredoxin and hydrogenase, and explains the dramatic effect of iron nutrition on the nature of glucose fermentation in *C. perfringens*. The ferredoxin-linked hydrogenase system provides a means of performing certain anaerobic oxidations in the absence of external electron acceptors, and can be interpreted as a "hydrogen valve" that facilitates energy metabolism (Gray and Gest, 1965).

The reaction sequence (a) → (b) → (c), above, in which acetyl-CoA plays such a major role, is thought to be one of the most prominent energy-conversion mechanisms in the metabolism of anaerobes. It is thus noteworthy that acetyl-CoA also serves as the immediate fuel for the molecular oxygen-requiring citric acid cycle, the major energy-yielding process of heterotrophic aerobes. It seems clear that the evolutionary roots of the aerobic citric acid cycle are to be found in the metabolic reactions of anaerobes, and (a), above, is doubtless one such reaction. In contrasting the bioenergetic potential of acetyl-CoA in anaerobes and aerobes, we can envisage progression from a completely anaerobic world where cells could regenerate a maximum of one ATP per molecule of acetyl-CoA produced, to the metabolism of heterotrophic aerobes characterized by regeneration of about 12 ATP molecules for each acetyl-CoA oxidized in the aerobic citric acid cycle.

The electron acceptor function of acetyl-CoA [reaction (d), above, in particular] is reflected by the fermentative production of butyric acid ($CH_3CH_2CH_2COOH$; see Table 6-1). Sugar fermentations in which a number of reduced end products are made in substantial amounts, as in the examples given in Table 6-1, emphasize the "problem" that clostridia and other heterotrophic anaerobes face in obtaining energy for growth. The character of clostridial fermentations is established primarily by the low net yield of ATP (low efficiency) of this type of energy-conversion. Large quantities of substrates are catabolized to energy-rich reduced products that accumulate in the medium, and the net yield of ATP is quite small. Cell yields per unit weight of substrate fermented are correspondingly low. In order to achieve overall O/R balance, the cells are obliged to conduct large-scale electron flow transactions that are superfluous as far as biosynthesis is concerned. Although many clostridia have a "H_2 valve" that facilitates the final balance of electron exchanges, virtually all of the potentially useful metabolites derived from sugar breakdown are preempted for use as the final acceptors of H from $NADH_2$, thus making NAD available for the energy-conserving O/R reaction of the fermentation mechanism. As indicated in Table 6-1, this pattern is especially exaggerated with respect to acetyl-CoA in *C. butyricum*.

6-4. Sugar Fermentation Dependent on an "Accessory Oxidant"

From Table 6-1 it can be seen that acetic acid is a characteristic product of glucose fermentation by saccharolytic clostridia. It is of interest that certain clostridia of this kind, such as *Clostridium lacto-acetophilum* can grow fermentatively on lactic acid as the source of energy only if acetic acid is provided exogenously (Bhat and Barker, 1947). As the bacterium grows under these conditions, acetate disappears from the medium and CO_2, H_2, and butyrate are produced. Presumably, lactate is first oxidized to pyruvate with NAD as the H acceptor, and the pyruvate is subsequently utilized by the reactions (a), (b), and (c) specified in Section 6-3. The significance of the need for added acetate when lactate is the energy source emerges from a comparison of the quantitative aspects of the glucose and lactate fermentations given in Table 6-2.

For reasons still unknown, activity of the "H_2 valve" is considerably lower when lactate is the energy source. This means that correspondingly more electrons resulting from the oxidation of lactate to pyruvate, and from the oxidative decarboxylation of pyruvate, must be disposed of by reduction of acetate to butyrate. Insufficient acetate is produced from lactate, however, to accomodate all of the reducing power ($NADH_2$ or reduced ferredoxin) generated. Thus, an *accessory oxidant*—exogenous acetate—is required to achieve overall O/R balance, even though use of acetic acid in this way must diminish the net energy yield that can be obtained from lactate. [Note that in order for exogenous acetate to be used as an electron acceptor, it must first be converted to acetyl-CoA; this conversion requires reversal of reaction (c), followed by reversal of (b)]. With lactate as the energy source, *C. lacto-acetophilum* is obviously under great bioenergetic stress, primarily due to its limitations in managing O/R balance.

TABLE 6-2. Accessory oxidant-dependent fermentation of lactate by *Clostridium lacto-acetophilum*.*

Compound	*Moles of compound produced from fermentation of 100 moles of:* Glucose	Lactate (+acetate)
CH_3COOH	28	−32 (consumed)
$CH_3CH_2CH_2COOH$	73	65
CO_2	190	100
H_2	182	59

* Data from Bhat and Barker (1947).

The data in Tables 6-1 and 6-2 show that H_2 is a typical end product in the fermentative metabolism of saccharolytic clostridia. Molecular hydrogen is also produced by numerous other microorganisms (Gest, 1954), but the physiological significance of H_2 formation differs depending on the biochemistry involved. For example, photosynthetic bacteria (e.g., *Rhodopseudomonas capsulata*) growing photoheterotrophically under certain conditions produce large quantities of H_2 from organic compounds via an electron transfer process in which the nitrogenase enzyme complex functions as the terminal catalyst (Hillmer and Gest, 1977). In such organisms, energy-dependent H_2-production appears to represent a regulatory device that facilitates "energy-idling" when the energy conversion rate and the supplies of reducing power exceed the capacity of the biosynthetic machinery. In typical clostridia, on the other hand, production of H_2 from organic compounds represents a process central to the energy conversion mechanism itself, the electrons from energy-yielding oxidations being disposed of via their use in the reduction of protons to form H_2 (e.g., pyruvate → acetyl-CoA + CO_2 + $2\,H^+ + 2\,e$; $2\,H^+ + 2\,e \rightarrow H_2$). A reassessment of the physiological significance of H_2 formation by various microorganisms led Gray and Gest (1965) to conclude that:

> From a general standpoint, the formation of molecular hydrogen can be considered a device for disposal of electrons released in metabolic oxidations. We presume that this means of

performing anaerobic oxidations is of ancient origin and that the hydrogen-evolving system of strict anaerobes represents a primitive form of cytochrome oxidase, which in aerobes effects the terminal step of respiration, namely the disposal of electrons by combination with molecular oxygen. We further assume that the original pattern of reactions leading to H_2 production has become modified in various ways (with respect to both mechanisms and functions) during the course of biochemical evolution, and we believe that this point of view suggests profitable approaches for clarifying a number of problems in the intermediary metabolism of microorganisms which produce or utilize H_2. Of special importance in this connection is the basic problem of defining more precisely the fundamental elements in the regulatory control of anaerobic energy metabolism.

Research advances made since 1965 indicate that several interlocking control mechanisms are required to assure regulation of electron flow in complex anaerobic fermentations, but it is still not understood why *C. lacto-acetophilum* must use acetate as an accessory oxidant for lactate fermentation rather than simply produce more H_2. In any event, it seems that the "H_2 formation valve" in clostridia and other anaerobes has an additional regulatory function, namely, it adjusts under different conditions to values that balance the reducing power generated by the fermentative reactions necessary for ATP synthesis with the amount of "H" required for biosynthesis and the formation of extracellular end products. Note should be made of the interesting suggestion that enzyme complexes with nitrogenase activity evolved from hydrogen-producing hydrogenases that originally functioned only as a means of electron discard (Broda and Peschek, 1980).

6-5. Accessory Oxidant-Dependent Fermentation in Photosynthetic Bacteria

During the 1930s several investigators noted the endogenous production of gases and acidic products by resting cells of various types of (photosynthetically grown) purple bacteria incubated in darkness under anaerobic conditions. The acidic products were not identified, and it was claimed that their formation could not be appreciably enhanced by addition of organic substrates. These observations, together with the fact that no one had been able to grow such organisms anaerobically in darkness, led to the view that the fermentative metabolism of photosynthetic bacteria was of a peculiar special nature, one that possibly functioned only for maintenance of the cells during anaerobic dark periods. Later experiments with *Rhodospirillum rubrum* showed, however, that resting cells of this organism could ferment pyruvate anaerobically in darkness by a mechanism similar to that operating in heterotrophic, fermentative "propionic acid bacteria" (Kohlmiller and Gest, 1951). Since the latter organisms *can* obtain the energy and carbon required for growth from pyruvate fermentation, it was concluded (Kohlmiller and Gest, 1951; Gest, 1951) that under appropriate conditions, photosynthetic bacteria should also be able to grow fermentatively in darkness. Experimental trials, however, continued to yield negative results until Uffen and Wolfe in 1970 reported that several photosynthetic bacteria (including *R. rubrum*) could be grown in complex media (containing yeast extract and peptone) in darkness when special procedures were used to establish strict anaerobiosis and a low E_h. Recent studies with *R. rubrum* indicate that the conversion of pyruvate to acetyl-CoA, followed by reactions (b) and (c) (see Section 6-3, above), is the predominant energy-yielding process for the anaerobic fermentative dark growth of this organism (Gorrell and Uffen, 1977).

A number of photosynthetic bacteria can utilize sugars as sole carbon sources for anaerobic *photoheterotrophic* growth, but common experience over many years indicated that sugars could not support anaerobic dark growth of such organisms. It was, consequently, somewhat surprising when Yen and Marrs (1977) demonstrated that *Rhodopseudomonas capsulata* and closely related species can grow anaerobically in darkness in synthetic media with glucose as the sole carbon and energy source provided that dimethyl sulfoxide [$(CH_3)_2SO$, DMSO] is supplied. During growth, the DMSO is reduced to dimethyl sulfide [$(CH_3)_2S$] leading Yen and Marrs to suggest that under the conditions described, ATP is produced either in association with an "anaerobic respiration" in which DMSO functions as the terminal electron acceptor (i.e., electrophosphorylation) or by an unusual type of fermentation (substrate-level phosphorylation) that requires an accessory oxidant. More detailed studies using trimethylamine-N-oxide [$(CH_3)_3NO$, TMAO] as the oxidant in place of DMSO have established that the energy-yielding process involved is indeed a fermentation dependent on an accessory oxidant (Madigan and Gest, 1978; Madigan et al., 1980; Cox et al., 1980).

What is the function of the accessory oxidant? *R. capsulata* apparently contains virtually all of the "classical" fermentation machinery found in many heterotrophic anaerobes, but differs in that it cannot effectively manage O/R balance using only intermediates of fermentation as terminal electron acceptors. For reasons still unknown, during anaerobic growth in the dark the organism is unable to reoxidize the $NADH_2$ generated via the E-M-P pathway by forming highly reduced organic end products (or H_2); an exogenous "accessory" electron acceptor (TMAO or DMSO) must therefore be provided. The essence of this interpretation is summarized in Figure 6-1. As there indicated, NAD is continually supplied to the fermentation apparatus by transfer of H from $NADH_2$ to the TMAO-reduction system. The latter can be viewed simply as a dumping-ground, or "sink," for electrons.

This discovery in a photosynthetic bacterium of a "cryptic fermentation system" that can be activated only by adding an accessory oxidant raises many questions of interest. A plausible interpretation is that this kind of mechanism is in fact an evolutionary relic, one that has survived in a cryptic form from the Hadean or Early Archean when fermentations, perhaps the only energy-conversion devices utilized by the earliest prokaryotes, were undergoing various evolutionary changes leading toward an increase in efficiency in the cellular metabolic economy.

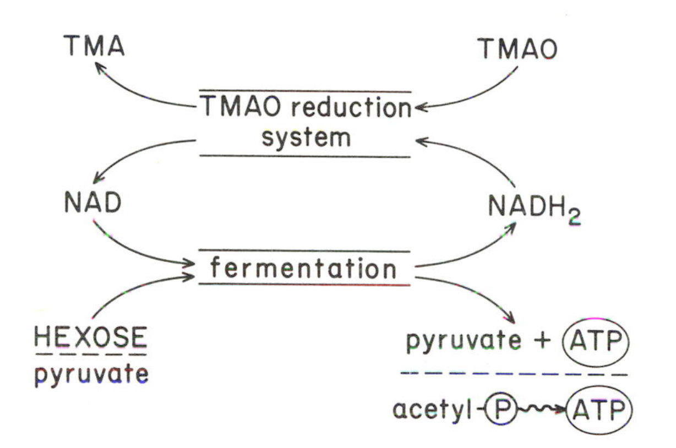

FIGURE 6-1. Schematic representation of the dark fermentative metabolism of *Rhodopseudomonas capsulata*. Energy-yielding catabolism of hexose to pyruvate and the conversion of pyruvate to acetyl phosphate (a phosphoryl donor for ATP synthesis) require NAD as the H (electron) acceptor. In order for fermentation to proceed, NAD must be continuously regenerated from $NADH_2$. This is accomplished by transfer of H to the accessory oxidant trimethylamine-N-oxide (TMAO), which is thereby reduced to trimethylamine (TMA). From Madigan et al., (1980), *Journal of Bacteriology*, vol. 142.

6-6. Other Oxidant-Dependent Fermentations

Reevaluation of the older literature and the results of more recent research indicate that fermentations dependent on accessory oxidants are far more common than has generally been suspected. For example, observations on the anaerobic reduction of nitrate in systems that lack cytochromes led Egami and his colleagues to suggest the concept of "nitrate fermentation", that is, of carbohydrate fermentation in which $NADH_2$ is reoxidized via reaction with soluble nitrate reductase. Figure 6-1 is, in fact, based on a similar diagram published by Egami et al. (1957) to describe this notion (in their scheme, TMAO and TMA [trimethylamine, $(CH_3)_3N$] are replaced, respectively, by nitrate, NO_3^-, and nitrate, NO_2^-). Some experimental support for the catalysis of "nitrate fermentation" by clostridia has been obtained by Hasan and Hall (1977). More recently, Cole and Brown (1980) have provided evidence indicating that anaerobic reduction of nitrite to ammonia (NH_3) during sugar fermentation by several amphiaerobes results from nonphosphorylative oxidation of $NADH_2$ with nitrite as the electron acceptor. Another example is encountered in certain instances of dissimilatory sulfate reduction. Although it has been widely assumed that such sulfate reduction by bacteria signifies energy conservation by electrophosphorylation, it now appears that growth of certain *Desulfotomaculum* species on lactate and sulfate (SO_4^{2-}) is supported by substrate-level phosphorylation (Liu and Peck, 1980)—the fermentation of lactate apparently requires transfer of electrons to sulfate, which thus serves as an accessory oxidant.

6-6-1. CO_2 AS AN ELECTRON SINK

Carbon dioxide, which was abundant in the Archean atmosphere (see Chapters 4 and 11), serves as an electron sink in the metabolism of numerous prokaryotes. In the present context it is of particular interest that CO_2 can function as an accessory oxidant in energy-yielding fermentations carried out by various extant anaerobes and amphiaerobes. As indicated below, two major categories of such processes can be discerned.

USE OF CO_2 AS AN ELECTRON SINK BY "HOMOACETATE-FERMENTING" CLOSTRIDIA

The mechanism of glucose fermentation by *Clostridium thermoaceticum* and *C. formicoaceticum* (see Ljungdahl and Andreesen, 1976) can be summarized by the equations:

(f) glucose $\dashrightarrow$ $2\,CH_3COCOOH + 2\,ATP + 2\,NADH_2\,(=4H)$

(g) $CH_3COCOOH + CoASH \longrightarrow CH_3COSCoA + CO_2 + 2\,H$

(h) $CH_3COSCoA + P_i \longrightarrow CH_3COOPO_3H_2 + CoASH$

(i) $CH_3COOPO_3H_2 + ADP \longrightarrow CH_3COOH + ATP$

(j) $CO_2 + 6\,H + HX \dashrightarrow CH_3\text{–}X + 2\,H_2O$

(k) $CH_3COCOOH + CH_3\text{–}X + H_2O \dashrightarrow 2\,CH_3COOH + HX$

(l) overall: glucose $\dashrightarrow$ $3\,CH_3COOH\ (+\ 3ATP)$

The actual sequence by which CO_2 is reduced to CH_3-X [reaction (j)] is quite complex, and the details are not particularly relevant to the present discussion. In effect, CO_2 serves the function of an electron sink, permitting reoxidation of the electron carriers that are necessarily reduced during formation of the E-M-P pathway intermediates that are required for substrate-level phosphorylations.

Clostridia have been widely regarded as being devoid of cytochromes, but this generalization is now known to be invalid. Indeed, both of the clostridia under discussion have been shown recently to contain a *b*-type cytochrome (Gottwald et al., 1975) that probably functions as an electron carrier in the next reaction mechanism to be considered.

USE OF CO_2 IN BIOSYNTHESIS OF OXIDANTS OF THE SZENT-GYÖRGYI PATHWAY

In a large number of anaerobes and amphiaerobes, carbon dioxide also serves the function of an electron sink in a second and more subtle fashion. This occurs through the condensation of CO_2 with a C_3 intermediate of fermentation [either pyruvate or phosphoenolpyruvate ($CH_2C(OPO_3H_2)COOH$, PEP)] to yield a relatively oxidized C_4 dicarboxylic acid, oxaloacetate ($HOOCCOCH_2COOH$, OAA); OAA and its derivatives can serve as electron sinks for four H atoms. Thus:

CO_2 + "C_3" $\longrightarrow$

$HOOCCOCH_2COOH$ (oxaloacetate)

$\downarrow +2H$

$HOOCCHOHCH_2COOH$ (malate)

$\downarrow -H_2O$

$HOOCCH{=}CHCOOH$ (fumarate)

$\downarrow +2H$

$HOOCCH_2CH_2COOH$ (succinate)

Some bacteria require substantial quantities of exogenous CO_2 in order to ferment sugars and this has been attributed to the role of CO_2 as a precursor of OAA (and thus, fumarate; Anderson and Ordal, 1961; White et al., 1962). In other instances, exogenously added fumarate, by acting as an H acceptor, permits fermentative growth of bacteria on lactate (Quastel et al., 1925) or glycerol ($CH_2OHCHOHCH_2OH$; Singh and Bragg, 1975). Accordingly, fumarate can be considered as another example of an accessory oxidant that facilitates redox balance in fermentative metabolism.

Depending on the organism (or tissue in higher eukaryotes) the "C_3" precursor of OAA can be either pyruvate or PEP. From the standpoint of oxidation-reduction balance, the net result is the same. The special importance of the sequence depicted above is that it represents a mechanism for reoxidation of the $NADH_2$ generated in the E-M-P pathway through utilization of only one of the two C_3 fragments produced from hexose cleavage. Operation of what might be called a "C_4 dicarboxylic acid electron sink sequence" thus has the consequence that the other C_3 fragment is

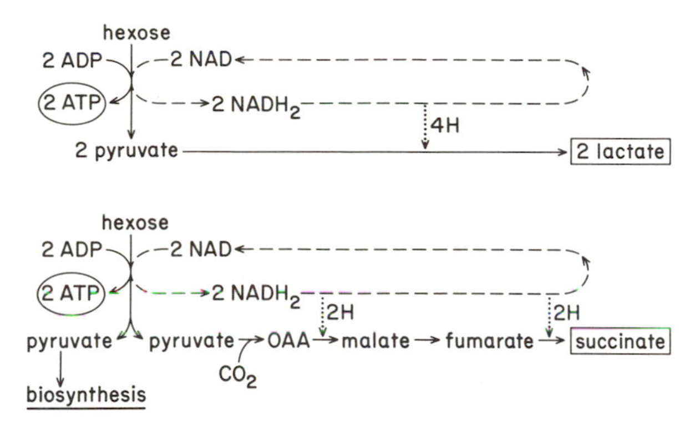

FIGURE 6-2. (Top) Condensed scheme for energy-yielding hexose fermentation by the Embden-Meyerhof-Parnas (E-M-P) pathway. The fermentation is "internally balanced" from the standpoint of oxidation-reduction (O/R); NAD is regenerated from $NADH_2$ by transfer of H atoms to the intermediate, pyruvate, resulting in the formation of lactate. (Bottom) A variation of hexose fermentation in which reoxidation of $NADH_2$ is effected by H transfer to the dicarboxylic acids, oxaloacetate (OAA) and fumarate. The latter is the terminal H acceptor, yielding succinate. Carboxylation of pyruvate or a related C_3 compound gives rise to the relatively oxidized C_4 dicarboxylic acids (in many extant organisms, phosphoenolpyruvate rather than pyruvate is the C_3 moiety that condenses with CO_2 to form oxaloacetate; this has no effect on the redox balance, which is the central point at issue). This pattern of fermentation spares one of the two pyruvates produced from each molecule of hexose for use in biosynthesis. From Gest (1980), © Elsevier/North Holland Biomedical Press, Amsterdam.

potentially available for biosynthesis. The conception is illustrated in condensed form in Figure 6-2. Since pyruvate (or PEP) plays a central role in many biosynthetic mechanisms, it is reasonable to assume that an anaerobic cell with the capacity to exploit CO_2 for the synthesis of C_4 electron acceptors (Figure 6-2, bottom) would have a significant selective advantage over cells constrained to use hexose as an energy source only via the classical glycolytic mechanism (Figure 6-2, top). Thus, it is envisioned "that in its earliest form, fumarate reduction evolved as a means of achieving redox balance and at the same time sparing pyruvate for biosynthetic purposes." (Gest, 1980).

The sequence:

$$\text{hexose} \dashrightarrow \text{"}C_3\text{"} \xrightarrow{+CO_2} \text{OAA} \xrightarrow{+4H} \text{succinate},$$

which has been designated the "Szent-Györgyi pathway" (Ottaway and Mowbray, 1977), is of particular interest for investigators of biochemical evolution. In addition to the electron sink function of the "C_3"$\dashrightarrow$succinate segment of this pathway, the reductive sequence leading from OAA to the formation of succinate can be construed as the evolutionary starting point for the citric acid cycle of aerobes, a central feature of their energy metabolism. In extant aerobes the sequence, of course, runs in the oxidative direction, i.e., succinate → fumarate → malate → OAA. However, prior to the establishment during the Early Proterozoic of an aerobic environment (see Chapter 11), it seems reasonable to suppose that this same sequence operated in the reductive direction, functioning as an electron sink and also as a source of succinate (viz., as succinyl-CoA) needed for biosynthesis. In most aerobes, succinyl-CoA (produced via α-ketoglutarate dehydrogenase activity) is a "branch-point compound" in the sense that it is subsequently either converted to succinate (the next intermediate of the citric acid cycle) or is used as a biosynthetic precursor of metalloporphyrins (i.e., the metal-chelated tetrapyrrole nucleus of cytochromes, chlorophyll, etc.). Anaerobes that produce metalloporphyrins (e.g., as in cytochrome *b*) but lack α-ketoglutarate dehydrogenase apparently produce the required succinyl-CoA via the sequence: OAA → malate → fumarate → succinate → succinyl-CoA.

Ubiquity of a metabolic system argues for its antiquity, and the widespread distribution of the capacity to use fumarate as an electron sink is striking. Numerous and diverse types of modern prokaryotes exhibit this ability (see listings in Thauer et al., 1977; and Kröger, 1977) and, in addition, the Szent-Györgyi pathway plays a prominent role in the metabolism of many invertebrate metazoans that can survive long periods in essentially anaerobic habitats (e.g., see Hochachka and Mustafa, 1972; Hochachka, 1976).

6-7. Fumarate Reductase and Evolutionary Changes in Fermentation Schemes

Effective utilization of fumarate as a terminal electron acceptor depends on the presence of a "reductase" having the appropriate kinetic properties. The existence of an essentially unidirectional "fumarate reductase" was first clearly demonstrated by Peck, Smith, and Gest (1957) in studies of the strict anaerobe *Micrococcus lactilyticus* (now known as *Veillonella alcalescens*). Extracts of this bacterium were shown to contain a soluble, flavin-linked fumarate reductase activity that clearly could not be ascribed to conventional succinate dehydrogenase acting in the reverse direction. In subsequent research with a variety of anaerobes and amphiaerobes it has been shown that fumarate reductase (FR) and other components of the fumarate-reducing enzyme system are almost always membrane-bound (Thauer et al., 1977). As might be expected, the sequence of steps involved in electron transport to fumarate in different organisms is not always identical; the general pattern, however, is as follows:

electron donor ⇢ menaquinone ⇢ cytochrome b ⇢ FR (Fumarate → Succinate)

In a number of instances, $NADH_2$ generated in the course of hexose fermentation serves as an electron donor and the occurrence of electrophosphorylation has been demonstrated (Thauer et al., 1977; Kröger, 1978). These considerations formed part of the basis of the scheme recently advanced by Gest (1980) for the evolutionary development of fermentation mechanisms (Figure 6-3).

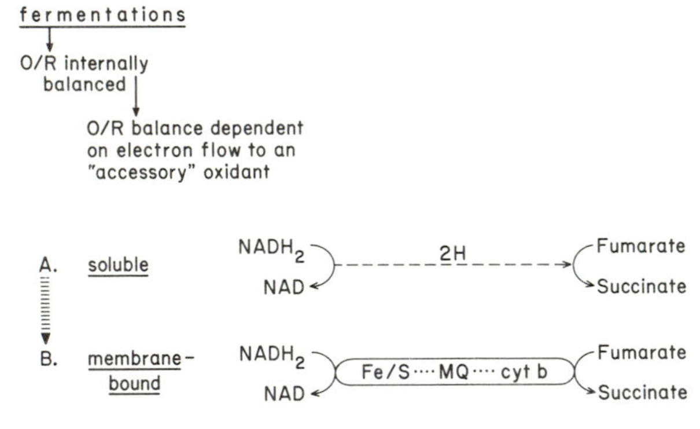

FIGURE 6-3. Proposed sequence of evolutionary change of fermentation schemes. Fermentation in which oxidation-reduction (O/R) is internally balanced (as in Figure 6-2, top) is modified to the pattern in Figure 6-2, bottom, at first as in (A) where reoxidation of $NADH_2$ is effected by a soluble, relatively simple system in which fumarate is the terminal oxidant. Through sequential changes, the accessory oxidant system becomes more complex (B) with iron/sulfur proteins, menaquinone (MQ) and cytochrome *b* (cyt b) participating as catalysts in a membrane-bound system capable of electrophosphorylation. From Gest, (1980), © Elsevier/North Holland Biomedical Press, Amsterdam.

According to this scheme, the earliest (simplest) fermentations are hypothesized to have been internally balanced from an O/R standpoint, as in classical glycolysis (viz., hexose → 2 lactate). At a later stage, fermentations were modified through the acquisition of pathways by which hydrogen was transferred to accessory oxidants such that one of the two pyruvate fragments produced from hexose could be "spared" for use in biosynthesis. Another advantageous aspect of this development was that some pyruvate could then be converted to acetyl-CoA, providing a potential source of additional energy [via reactions (b) and (c) in Section 6-3].

The earliest accessory oxidant systems used for reoxidation of $NADH_2$ presumably were relatively simple and soluble (as indicated in Figure 6-3 "A") and functioned only to ensure redox balance. The ubiquity among modern microbes of the fumarate-reducing system and the obviously close connections of the C_4 intermediates of the Szent- Györgyi pathway to other aspects of metabolism strongly imply that the use of fumarate as a terminal oxidant was an important prototype for evolutionary modification of fermentations. The scheme outlined in Figure 6-3 suggests that further biochemical evolution involved the gradual addition of intermediary electron carriers [iron/sulfur proteins, menaquinone (MQ), cytochrome *b*, etc.] and incorporation of most of the system into the cytoplasmic membrane. The reaction sequence depicted in Figure 6-3 "B" represents the essence of the membrane-bound system as it now exists in numerous organisms, including strict anaerobes—an "electrophosphorylation module" thus occurs in modern organisms that can obtain the bulk of their ATP by conventional fermentation.

The plausibility of the postulated transition from a system in which fumarate acted only as an electron sink (Figure 6-3 "A") to one in which it is the terminal oxidant for electrophosphorylation (Figure 6-3 "B") is reinforced by the recent studies of Singh and Bragg (1975, 1976) of an *Escherichia coli* mutant unable to produce cytochromes unless supplied with 5-aminolevulinic acid ($HOOCCH_2CH_2COCH_2NH_2$). In cytochrome-deficient cells growing anaerobically on glycerol and fumarate, the ATP needed for nutrient transport and for biosynthesis is regenerated by substrate-level phosphorylation; under these circumstances, fumarate appears to act as an electron acceptor for nonphosphorylative reoxidation of $NADH_2$. However, when the mutant is provided with 5-aminolevulinic acid it synthesizes cytochromes and is now capable of electrophosphorylation using the system outlined in Figure 6-3 "B".

It is not difficult to imagine early anaerobic prokaryotes in which sugar was fermented by a combination of ordinary glycolysis and of its modifications as in the Szent-Györgyi pathway. In any event, the evolutionary development of an "electrophosphorylation module" in primitive fermenters can be seen as a major step forward, one that pointed the way toward the later development of even more efficient systems in which enhanced electron flow in membranes was the driving force of energy conversion.

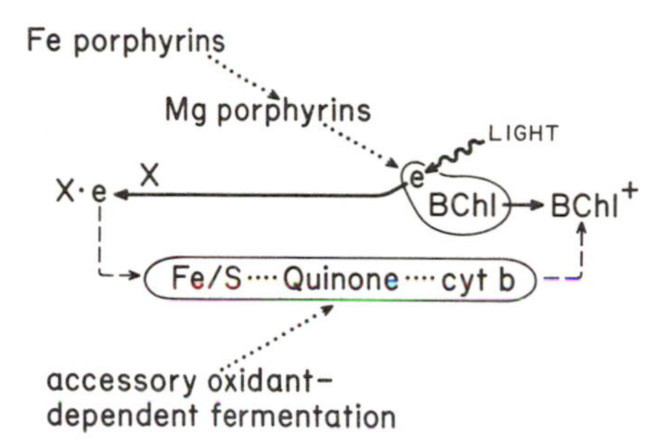

FIGURE 6-4. Proposed scheme for origin of anaerobic photophosphorylation. Synthesis of a form of bacteriochlorophyll (BChl; not necessarily identical in structure to the chlorophylls of extant photosynthetic bacteria) occurred through modification of the pathway for biosynthesis of iron porphyrins. The charge separation induced by the absorption of light became metabolically useful only after the photopigment system was stabilized by fusion with an electrophosphorylation-competent segment of an accessory oxidant-dependent fermentation mechanism. "X" represents the primary electron acceptor of the photosynthetic apparatus. From Gest (1980); © Elsevier/North Holland Biomedical Press, Amsterdam.

6-8. The Origin of Anaerobic Photophosphorylation

In a recent review F. Harold (1978) remarked: "It is quite striking that the constellation of non-heme iron-quinone-cytochrome *b*, however it may have arisen, was strongly conserved thereafter. It is a common feature of the redox chains of both photosynthetic and respiratory organisms, and it will be very interesting to see whether it also occurs in anaerobes such as sulfate reducers, which are thought to be an ancient group." As is discussed above, the "constellation" to which Harold refers is in fact of common occurrence in both anaerobes and amphiaerobes. The close similarities of the "electrophosphorylation module" that on the one hand is linked with the reduction of fumarate and that, on the other hand, is an integral part of the light-dependent energy conversion apparatus of certain extant photosynthetic bacteria suggested, in part, the hypothesis proposed by Gest (1980) for the origin of anaerobic photophosphorylation. As is indicated in Figure 6-4, the essence of this hypothesis is that the earliest photophosphorylation system was established through the fusion in a fermentative anaerobe of an "electrophosphorylation module" with a membrane-associated photopigment (viz., magnesium-porphyrin) complex. The widespread occurrence of the fumarate-reducing module and the close relationships of fumarate and succinate to other aspects of metabolism suggest that this particular system may have been used in the first successful fusion. Subsequently, after photophosphorylative energy conversion became established in this cell line, the continued operation of the more primitive, and relatively inefficient, fermentative pathway based on use of accessory oxidants became unnecessary. It is relevant to note, however, that in extant bacteria fermentative production of propionic acid from sugars or pyruvate occurs through a mechanism that includes the reductive sequence OAA -→ succinate, and that formation of propionate as an end product of *dark* anaerobic metabolism has been observed in a number of photosynthetic bacteria (see Gest, 1980).

Several types of genetic recombination or exchange mechanisms that occur in modern bacteria could have led to a fusion of the kind envisioned in Figure 6-4, but further speculation in this regard is not particularly useful at present. A more relevant question is: How were the extraordinarily rapid photochemical processes initiated by quantum absorption first coupled to the comparatively much slower reactions of cellular chemistry? Gaffron (1965) described the overall result of this coupling as follows: "Students of sensitized photochemical reactions *in vitro* know of course that the greatest obstacle to a practical utilization of light energy (in this manner) is the tendency of the reaction products to recombine either directly or via a short detour of at most one or two intermediate steps .... The living cell has solved the problem of premature and useless back reactions by making the detour longer and longer; in other words, by converting the back reactions into a cycle to which other metabolic processes can be coupled." Stated in another way, an intracellular mechanism was developed that stabilized light-induced separation of charges and facilitated an alternative route of step-wise electron transfer, geared to conservation of energy in forms utilizable for biosynthesis.

Electron transfer catalysts in "electrophosphorylation modules" necessarily have an appropriate vectorial arrangement in membranes (Mitchell, 1979). For a successful mechanistic fusion of such a module with a photopigment complex, the latter would have to have assumed a compatible topological arrangement. One can imagine the occurrence of many trials and errors before this was achieved. The resulting development of a "photosynthetic unit" in which a photoactivatable pigment could replace organic compounds as an electron source for energy conversion was a signal innovation in the early history of life, one that served to link biological systems directly to an inexhaustible (solar) energy source. And despite the obvious complexities that must have been involved in the evolution of this process, it now appears to have originated remarkably early in Earth history, certainly prior to the Proterozoic and probably earlier than 3.5 Ga ago (for discussion, see Chapters 7, 8, and 9). Subsequent evolutionary improvement can be seen in addition of a *c* type cytochrome to the sequence of electrophosphorylation catalysts and, later, coordination of this photosystem with a second photopigment-protein system (the precursor of Photosystem II) that made possible the utilization of water as an electron donor for light-dependent electrophosphorylation and generation of net reducing power (i.e., oxygenic, cyanobacterial photosynthesis). This process, like that of anaerobic (bacterial) photosynthesis also appears to have evolved quite early, with currently available data suggesting the Late Archean (viz., between 2.9 and 2.5 Ga ago) as the most probable time of origin (see discussion in Chapters 12 and 15).

This model for the origin of photosynthetic metabolism suggests new possibilities for tracing biochemical evolution through analysis of the biochemistry and molecular biology of extant organisms. In particular, it would be of special interest to explore the biochemical "fine structure" of accessory oxidant-dependent fermentations in diverse prokaryotes including, especially, all the major types of photosynthetic bacteria.

6-8-1. Comments on the Origin of Aerobic Electrophosphorylation

Broda and Peschek (1979) have recently reconsidered the question: "Did respiration or photosynthesis come first?". In answer, they propose the sequence: "prerespiration" → photosynthesis → respiration. They define "prerespiration" as "single-step electron flow across a membrane with energy conversion through phosphorylation," with the latter presumably being effected by a membrane-bound ATPase. Photosynthetic energy conversion is then imagined as emerging through combination of a "redox photocatalyst" (chlorophyll) "with a proton translocating quinone shuttle evolved from prerespiratory electron transfer." Broda (1975a), in particular, has championed the notion that "aerobic respiratory chains" originated through modification of the transport chains used for photosynthetic electron flow (the so-called "conversion hypothesis"). He has suggested that this might have occurred repeatedly and independently in different evolutionary lineages and, moreover, that all "respirers" were ultimately derived from earlier "photosynthesizers". At present, however, this seemingly implausible scenario is far from established. Indeed, it seems much more likely that photosynthetic and aerobic electrophosphorylation systems developed independently from a common precursor, namely, from an anaerobe with electrophosphorylation capacity.

6-9. Coda

The model presented here for the origin of photosynthetic energy conversion contains elements suggested earlier in general fashion by others. For example, Krasnovskii (1959) envisioned the early evolution of photosynthetic systems as follows: "The system of heterotrophic metabolism in ancient organisms which were not yet capable of light utilization apparently included the steps of catalytic electron transfer and 'anaerobic' oxidoreductive phosphorylation. The development of the pigment system had the result that the active products of the photoprocess were linked to formerly existing biocatalytic systems. The further development of catalytic pigment and systems led to the more rational utilization of light energy."

The main thrust of the more detailed scheme described here stems from the conviction that the remarkable intermeshing of bioenergetics and biosynthesis exhibited in all types of contemporary cells indicates that these two interdependent aspects of cellular biochemistry must have co-evolved in a delicately concerted way. Accordingly, the progression from the earliest anaerobic cells with simple fermentative energy conversion and limited biosynthetic ability, to more advanced cells with electrophosphorylation capacity and the biosynthetic ability to make all cellular constituents from a single organic compound (e.g., a hexose) must have occurred through a lengthy sequence of gradual, interrelated and essentially simultaneous improvements in both biosynthetic and bioenergetic mechanisms.

Previous speculations that attempted analysis of organismal evolution based almost entirely on consideration of bioenergetic mechanisms obviously could only give a one-sided glimpse of the incredibly complex tide of biochemical evolution. In fact, many of the mechanisms of intermediary carbon metabolism elucidated during the past 25 years have as yet to be systematically examined as probable reflections of this evolutionary tide. A more specific indication of biosynthetic-bioenergetic interconnections emerges from even cursory consideration of the highly efficient and complex membrane-associated electrophosphorylation systems typical of modern heterotrophic aerobes. Such systems use $NADH_2$ as the "fuel," with the most prominent and effective fuel source being the citric acid cycle. This cycle, usually thought of mainly as a bioenergetic device, is designed not only for maximal yield of $NADH_2$ from a variety of organic substrates but also as a mechanism for production of intermediates needed for biosynthesis of the monomeric units of larger biomolecules (e.g., OAA $\rightarrow$ aspartate $\rightarrow$ proteins; α-ketoglutarate $\rightarrow$ glutamate $\rightarrow$ proteins; succinyl-CoA $\rightarrow$ porphyrins). Thus, tracing the threads of the evolution of bioenergetic mechanisms will inevitably lead us into an interwoven mosaic in which biosynthesis is a major part of the fabric; since the origin early in Earth history of the first living systems, the evolution of each of these phases of metabolism has played a pervasive role in the co-evolution of the other.

6-10. Summary and Conclusions

(1) Although direct evidence of the origin and earliest evolution of life on Earth can in principle be provided only by studies of the preserved rock record, the biochemical and molecular biological characteristics of all living systems contain within them evidence of their heritage. Comparative metabolic and biochemical studies of extant organisms can thus provide important insight into the nature of early evolutionary developments, insight of a kind and detail that are beyond the scope of geologically based investigations.

(2) In modern organisms, sugars and their derivatives are pervasive constituents of cellular structures; intermediates of sugar metabolism are widely used as biosynthetic precursors of numerous biologically important compounds; and the hexose sugar glucose has been aptly described as the "universal cellular fuel." The ubiquity and importance of sugars and sugar metabolism in extant forms of life, together with the apparent ease by which sugars may have been abiotically synthesized (via the formose reaction) in the essentially anaerobic early environment suggest that among the earliest forms of life were anaerobic, heterotrophic prokaryotes capable of fermenting hexose sugar via glycolysis to produce pyruvate and, utimately, lactic acid.

(3) Modern bacteria exhibit a remarkably broad range of fermentation patterns. Consideration of these patterns leads to the hypothesis that the earliest fermenters, capable of catabolic breakdown of hexose to produce two pyruvate (C_3) molecules via the Embden-Meyerhof-Parnas pathway, were the evolutionary antecedants of "accessory oxidant-dependent" fermenters, organisms that condensed one of the two hexose-derived pyruvate molecules with CO_2 to form oxaloacetate for use as an electron acceptor, thus sparing the remaining pyruvate molecule as a precursor for biosynthesis.

(4) Through a series of sequential evolutionary changes (addition of intermediary electron carriers such as iron/sulfur proteins, menaquinone, cytochrome *b*, etc.), and the incorporation of electron transport catalysts into the cytoplasmic membrane, such accessory oxidant-dependent fermenters are envisioned as having given rise to forms possessing an "electrophosphorylation module" of the type occurring in many modern organisms.

(5) Close similarities exist between the "electrophosphorylation module" linked with the reduction

of fumarate in accessory oxidant-dependent anaerobic fermenters, and the light-dependent energy conversion apparatus of certain extant photosynthetic bacteria. Accessory oxidant-dependent dark fermentation also occurs as a cryptic metabolic pathway in such bacteria. These and other considerations suggest that the earliest photophosphorylation system may have been established in a primitive fermentative anaerobe via the fusion of an "electrophosphorylation module" with a membrane-associated photopigment (i.e., bacteriochlorophyll-like) complex. Geologic evidence suggests that this event, resulting in the origin of anaerobic photosynthetic bacteria, probably occurred earlier than 3.5 Ga ago.

(6) Subsequent evolutionary refinement of this photosystem, involving the addition of a *c* type cytochrome and, later, coordination of the system with a second photopigment-protein complex (a "proto-Photosystem II"), is postulated to have resulted in the origin of oxygenic (i.e., cyanobacterial) photosynthesis, an event that apparently also occurred quite early in Earth history (viz., perhaps between 2.9 and 2.5 Ga ago).

(7) Thus, the bioenergetic and biosynthetic pathways of primitive, Archean anaerobes are viewed as having necessarily co-evolved—developments in either of these two interrelated aspects of cellular biochemistry are seen as having had a profound impact on the development of the other. The timing of these early evolutionary events is a matter that can only be established by referral to the known geologic record; their occurrence, however, is perhaps best evidenced by the bioenergetic-biosynthetic coupling so manifest in the metabolism of modern cells.

Acknowledgments

H. G. is the recipient of a fellowship from the John Simon Guggenheim Memorial Foundation; his research on photosynthetic bacteria is supported by N.S.F. Grant PCM 79-10747.

CHAPTER 7

Isotopic Inferences of Ancient Biochemistries: Carbon, Sulfur, Hydrogen, and Nitrogen

By Manfred Schidlowski, J. M. Hayes, and Isaac R. Kaplan

7-1. Introduction

Most elements—including the key elements of life with the exception of phosphorus—are mixtures of isotopes. When subjected to physicochemical processes, such elements undergo isotopic fractionation governed by thermodynamic and kinetic controls that are based largely on the masses and quantum characteristics of the atoms involved. The Earth's upper mantle and crust are believed to reflect broadly the isotopic distribution patterns of chondritic meteorites (Kaplan, 1975). In the case of the four elements dealt with in this chapter (carbon, sulfur, hydrogen, and nitrogen), the average terrestrial ratios of the dominant isotope to that next in abundance are $^{12}C/^{13}C = 89.4$, $^{32}S/^{34}S = 22.22$, $^{1}H/^{2}D = 6410$, and $^{14}N/^{15}N = 277$.

From the moment primordial mantle material entered geochemical and biogeochemical cycles, the initial isotope mixtures were redistributed, with different isotope ratios of the various elements accumulating in differing reservoirs. During such processes, fractionation between any two phases A and B can be characterized by a fractionation factor, K, defined in the case of carbon as:

$$K = \frac{(^{12}C/^{13}C)_B}{(^{12}C/^{13}C)_A}.$$

In a thermodynamically controlled exchange reaction, K represents the (temperature-dependent) equilibrium constant which can be calculated from the ratio of the partition functions of the isotopically different substances. In a unidirectional process, however, K gives a quantitative expression of the kinetic isotope effect (KIE) equal to the ratio k_1/k_2 of the rate constants for the reactions of the different isotopes. In the case of carbon, for instance, such rate constants are defined by the following relationships:

$$^{12}C_A \xrightarrow{k_1} {}^{12}C_B, \text{ and}$$
$$^{13}C_A \xrightarrow{k_2} {}^{13}C_B.$$

When $k_1 \neq k_2$, the $^{12}C/^{13}C$ ratio in the product B necessarily differs from that in the reactant A.

It is now well known that biological incorporation and subsequent biochemical processing of the key elements of life entail sizable isotope effects as a result of both thermodynamic and kinetic fractionations that occur during metabolic and biosynthetic reactions. With biochemical reactions being largely enzyme-controlled, such fractionations are for the most part due to kinetic isotope effects inherent in enzymatic activity. An outstanding example is the bias in favor of ^{12}C exercised by the ribulose-1.5-bisphosphate (RuBP) carboxylase reaction of Calvin cycle (C_3) photosynthesis that discriminates against "heavy" carbon (^{13}C) by about -20 to -35 ‰, while a quantitatively less important carbon-fixing pathway utilizing phosphoenolpyruvate (PEP) carboxylase entails fractionations of a few permil only (see Section 7-2-1). Fractionations of similar magnitude (and variability) have been reported for both dissimilatory and assimilatory sulfur metabolism, as well as for the principal biologic reactions involving hydrogen and nitrogen. As the isotopically distinctive products of such fractionations can be preserved in sediments with but minor alteration, these biochemical processes have exerted an impact on the long-term biogeochemical cycles of the respective elements that can be traced through geologic time.

In this chapter, both a review of earlier work and a synthesis of recent P.P.R.G. studies will be given on isotope fractionations in the biogeochemical cycles of carbon, sulfur, hydrogen, and nitrogen. The ultimate aim of these efforts is to track down the oldest signals in the geologic record of biologically mediated isotope effects, the detection of which might impose temporal constraints on the inferred time of origin of the underlying biochemical processes.

7-2. Carbon

7-2-1. Biochemistry of Carbon Isotope Fractionation

PRINCIPAL PATHWAYS OF BIOLOGICAL CARBON FIXATION

Biological carbon fixation is, in essence, the assimilation of carbon dioxide (CO_2), bicarbonate ion (HCO_3^{1-}) or carbon monoxide (CO) by autotrophic organisms, or of methane (CH_4) or related one-carbon compounds by heterotrophic "methylotrophs" (see Part 2 of the Glossary to this volume), that proceeds by a limited number of biochemical pathways in which all the cellular components are synthesized from these simple carbon precursors (see Table 7-1). Biological fixation of CO is probably negligible from a quantitative point of view and the quantitative significance of biological CH_4-oxidation is not well known. Moreover, biological utilization of both CO and CH_4 is contingent on the presence of an oxidant (viz., either O_2 or SO_4^{2-}) which suggests that the respective assimilation sequences are unlikely to have been among the initially evolving pathways of biological carbon fixation. With the bulk of primary production in the present biological carbon cycle due to either the chemically or solar-powered conversion of carbon dioxide to organic matter, biological carbon fixation is largely synonymous with fixation of CO_2 by green plants and autotrophic protists (e.g., phytoplankton) and by prokaryotes (e.g., photosynthetic bacteria and cyanobacteria). In fact, autotrophy has often been biochemically defined by the presence of ribulose-1.5-bisphosphate (RuBP) carboxylase, the key enzyme of the reductive pentose phosphate cycle of CO_2-fixation (the so-called "Calvin cycle"). As solar energy provides the most abundant and ubiquitous energy source for this conversion, low-temperature reduction of oxidized carbon primarily proceeds as photosynthetic fixation of CO_2 by the Calvin cycle.

ISOTOPE FRACTIONATIONS IN BIOLOGICAL CARBON FIXATION

It is known that the principal pathways of autotrophic carbon assimilation are accompanied by marked fractionations of the stable carbon isotopes ^{12}C and ^{13}C. These fractionations are mainly due to kinetic isotope effects inherent in the carbon-fixing enzymatic reactions and result in preferential metabolization of the lighter isotope, ^{12}C. Isotopic compositions are generally reported in terms of $\delta^{13}C$ values, with

$$\delta^{13}C = \left[\frac{(^{13}C/^{12}C)_{sa}}{(^{13}C/^{12}C)_{st}} - 1\right] \times 10^3 \ [‰, \text{PDB}],$$

by which the ^{13}C content of a sample (sa) is expressed as a numerical permil difference relative to a standard (st), commonly the Peedee belemnite standard (PDB) in which $^{12}C/^{13}C = 88.99$ and $\delta^{13}C$ is defined as equaling 0.00 ‰. As biological carbon fixation discriminates against ^{13}C, the $\delta^{13}C$ values of biosynthesized cell material are more negative than those of the carbon substrate (generally either atmospheric carbon dioxide, with $\delta^{13}C = -7$ ‰, or marine bicarbonate with $\delta^{13}C = -0.5 \pm 1$ ‰). The difference in isotopic content between the cells and substrate,

$$\Delta\delta = \delta^{13}C_{cells} - \delta^{13}C_{substrate},$$

gives a convenient expression of the magnitude of ^{13}C discrimination inherent in a specific assimilatory pathway.

As may be inferred from Table 7-1, data on carbon isotope fractionations are available for numerous, but by no means for all, assimilatory sequences. In particular, data on fractionations in CO- and CH_4-utilizing pathways are scant. On the other hand, the discrimination properties of the most common CO_2-fixing pathways, the processes responsible for the bulk of carbon transfer from the inorganic to the organic reservoir, are reasonably well understood. Basically, these involve the operation of the reductive pentose phosphate (Calvin) cycle that constitutes the principal mechanism for effecting a large-scale reduction of oxidized carbon to the carbohydrate level.

TABLE 7-1. Pathways of biological carbon fixation relevant to the geochemical carbon cycle. From the quantitative point of view, CO_2-fixing pathways (Nos. 1–4), most notably the RuBP carboxylase reaction of the Calvin cycle or "C3"-photosynthesis (No. 1), are by far more important than assimilatory sequences utilizing CO and CH_4 (Nos. 5–8). Reduction of CO_2 primarily leads to the formation of organic compounds possessing C3 (phosphoglycerate, pyruvate), C4 (oxaloacetate), and C2 skeletons (acetate, acetyl CoA). Principal sources: Buchanan (1979), Colby et al. (1979), Quayle and Ferenci (1978), Thauer and Fuchs (1979) and Walker (1979).

	Chemical process and representative organisms	*Principal electron donors [D] or acceptors [A]*
(1)	CO_2 + ribulose-1,5-bisphosphate ⟶ phosphoglycerate	
	*C3 plants, *algae, *cyanobacteria	[D]: H_2O
	*Purple photosynthetic bacteria	[D]: H_2, H_2S, S, $S_2O_3^{2-}$, organic molecules
	Chemoautotrophic bacteria	[D]: H_2, H_2S, S, $S_2O_3^{2-}$, NH_3, NO_2^-, Fe^{2+}
(2)	CO_2/HCO_3^- + pyruvate/phosphoenolpyruvate ⟶ oxaloacetate	
	*C4 plants, *CAM plants	[D]: H_2O via (1)
	Anaerobic and facultatively anaerobic bacteria	[D]: Organic molecules
(3)	CO_2 + acetyl coenzyme A ⟶ pyruvate	
	*Green photosynthetic bacteria[a]	[D]: H_2, H_2S, S, $S_2O_3^{2-}$
	Clostridium kluyveri, autotrophic forms of sulfate reducing bacteria, *methanogenic bacteria	[D]: H_2
(4)	CO_2 + CO_2 ⟶ acetate/acetyl coenzyme A	
	*Green photosynthetic bacteria[b]	[D]: H_2, H_2S, S, $S_2O_3^{2-}$
	Anaerobic bacteria (*Acetobacterium woodii, Clostridium acidiurici*, *methanogens (?)[c]	[D]: H_2
(5)	CO ⟶ intracellular CO_2	[A]: O_2
	[CO_2 subsequently processed as in (1)]	
	Higher plants	[D]: H_2O
	Carboxydobacteria	[D]: H_2, H_2S (?), H_2O
(6)	CO ⟶ intracellular CO_2	[A]: SO_4^{2-}, H_2O, S
	[CO_2 ⟶ organic molecules by varying pathways]	
	Anaerobic bacteria (Thermoproteus)	[D]: H_2, organic molecules, H_2O
(7)	CH_4 ⟶ formaldehyde (HCHO)	[A]: O_2
	HCHO + ribulose monophosphate ⟶ hexulose monophosphate	
	Type I methanotrophs	No redox reaction involved
(8)	CH_4 ⟶ formaldehyde (HCHO)	[A]: O_2
	HCHO + glycine ⟶ serine	
	Type II methanotrophs	No redox reaction involved

* Groups of organisms for which data on carbon isotope fractionation are available.

[a] This reaction is preceeded by (4) and followed by (2)—"reductive carboxylic acid cycle."

[b] Primary CO_2 fixation by succinyl CoA and α-ketoglutarate.

[c] Primary CO_2 fixation by C1 acceptors.

The biochemical background of carbon isotope fractionation during CO_2-uptake and metabolism by plants and autotrophic microorganisms has been dealt with by numerous investigators (e.g., Park and Epstein, 1960; Smith and Epstein, 1971; Wong et al., 1975, 1979; Sirevåg et al., 1977; Benedict, 1978; Fuchs et al., 1979), an exhaustive review of the field having been recently given by O'Leary (1981). It is generally believed that biological processes are largely dominated by enzymatically controlled kinetic fractionations rather than by equilibrium effects, although the latter have been claimed to be important in inter- and intramolecular isotope exchange among, and within, individual metabolites (Galimov, 1980).

The principal isotope-discriminating steps in biological carbon fixation are (*i*) the uptake and intracellular diffusion of external CO_2, and (*ii*) the first CO_2-fixing (carboxylation) reaction. Hence, the essentials of biological carbon isotope fractionation may be described, with adequate approximation, by the two-step model first proposed by Park and Epstein (1960),

$$CO_2(e) \underset{k_2}{\overset{k_1}{\rightleftharpoons}} CO_2(i) \xrightarrow{k_3} R\text{–}COOH, \quad (7\text{-}1)$$

in which $CO_2(e)$ and $CO_2(i)$ represent, respectively, external and internal CO_2, and k_1, k_2, and k_3 are the rate constants of the respective reactions. It would follow from this simplified scheme that interconversion of $CO_2(e)$ and $CO_2(i)$ is a reversible process while the enzymatic carboxylation of $CO_2(i)$ to yield R–COOH is irreversible. However, the "ideal" model conditions symbolized by Equation 7-1 are probably complicated by superimposed factors such as the diversion of part of the CO_2 to other, unrelated pathways.

Total fractionation is, accordingly, mainly dependent on an interplay of the processes indicated in Equation 7-1 and, notably, on which of the two steps there indicated becomes dominant or rate-controlling in a specific instance. Following O'Leary (1981), we may assume that kinetic fractionations associated with k_1 and k_2 are roughly -4 ‰ (-4.4 ‰ being the value for CO_2 diffusion in air, a value which within broad limits is independent of temperature, pressure, and CO_2-concentration). In specific cases (e.g., in aquatic plants), additional fractionations caused by dissolution, liquid diffusion, and hydration of CO_2 may be superimposed on the above value, but there is general consensus that the overall fractionation during this first step of carbon incorporation is relatively small.

On the other hand, fractionations inherent in the second step, the irreversible enzymatic carboxylation reaction, are considerably larger though variable in detail; hence, the kinetic isotope effect associated with k_3 must be assigned a range rather than a discrete value. The quantitatively most important carboxylation reaction is catalyzed by ribulose-1.5-bisphosphate (RuBP) carboxylase by which CO_2 is fed into the Calvin cycle, yielding a C3 compound (phosphoglycerate, PGA) as the first product of carbon fixation (Table 7-1, No. 1). Photoautotrophs exhibiting this process are thus commonly referred to as "C3 plants." Values reported for ^{13}C discrimination in the process range from -17 ‰ (Park and Epstein, 1960) to -40 ‰ and even lower, a large number of experimental determinations having yielded fractionations within the range -27 to about -38 ‰ (see Wong et al., 1979; O'Leary, 1981; and others). Hence, this reaction discriminates very effectively against ^{13}C, but the magnitude seems to vary over a wide range, a plausible average lying between -25 and -35 ‰. According to O'Leary (1981), such a range is not surprising since isotope fractionations resulting from enzymatic reactions may be expected to vary with pH, temperature, metal ion availability, and other variables.

In contrast, carboxylation by phosphoenolpyruvate (PEP) carboxylase, a reaction characteristic of "C4 plants" yielding a C4 dicarboxylic acid (oxaloacetate, OAA) as the first product of carbon fixation (Table 7-1, No. 2), entails a discrimination against ^{13}C of -2 to -3 ‰ only relative to bicarbonate ion which is the "active" carbon species in this reaction (Reibach and Benedict, 1977). This is one of the reasons for the small gross fractionation observed in the C4 dicarboxylic acid pathway of carbon assimilation. Since subsequent processing of the resulting OAA (inter alia, decarboxylation to pyruvate and recarboxylation of CO_2 as phosphoglycerate in the Calvin cycle) takes place within the closed system of the "Kranz"-type leaf anatomy typical of C4 plants, no further fractionation is apt to occur during these later stages. From the average $\delta^{13}C$ yielded by this group as a whole (about -13 ‰, see Figure 7-1) it is safe to say that the discrimination of about -4 ‰ in the initial diffusion step of Equation 7-1 becomes a major contribution toward total isotope fractionation in C4 plants, while this step is of negligible effect in the C3 pathway. Preliminary studies of green sulfur bacteria operating the reductive carboxylic acid cycle unique to photosynthetic prokaryotes (Table 7-1, Nos. 3 and 4) have yielded fractionations of about -12 ‰ for this pathway (Sirevåg et al., 1977).

Apart from those fractionations occurring during assimilation, photorespiration and related dissimilatory processes may also contribute to the overall isotopic economy of an autotroph. Respiratory loss of carbon would imply the preferential removal of ^{12}C, and discrepancies between predicted and observed fractionations have therefore often been ascribed to respiration (O'Leary, 1981). Other differences may be due to the relative amounts of cell components present in the material analyzed. For example, Park and Epstein (1960) showed that lipids display lower $\delta^{13}C$ values than coexisting carbohydrates and proteins, the difference usually ranging between 4 and 9 ‰ (see also DeNiro and Epstein, 1977). It is also known that the isotopic composition of such metabolites as amino acids may vary systematically (Abelson and Hoering, 1961).

7-2-2. Carbon Isotope Fractionation in Extant Plants and Microorganisms

From the foregoing discussion it is evident that the total isotopic fractionation that occurs during transformation of inorganic to organic carbon depends on the enzymatic carbon-acceptor involved (RuBP or PEP), the initial substrate utilized (principally CO_2 or HCO_3^{1-}), and on numerous other parameters, and that it can therefore be expected to vary according to the organism studied and the nature of its environment. Such variability is well illustrated by the following compilation of the ranges of $\delta^{13}C$ values exhibited by some major groups of higher plants, algae, and autotrophic prokaryotes. A graphic summary of these data (Figure 7-1) indicates that the terrestrial biomass as a whole is enriched, on the average, by 25 ± 5 ‰ in ^{12}C as compared to oceanic bicarbonate carbon that represents the bulk of the inorganic carbon stored in the exogenic exchange reservoir of the surficial environment (see Figure 7-2).

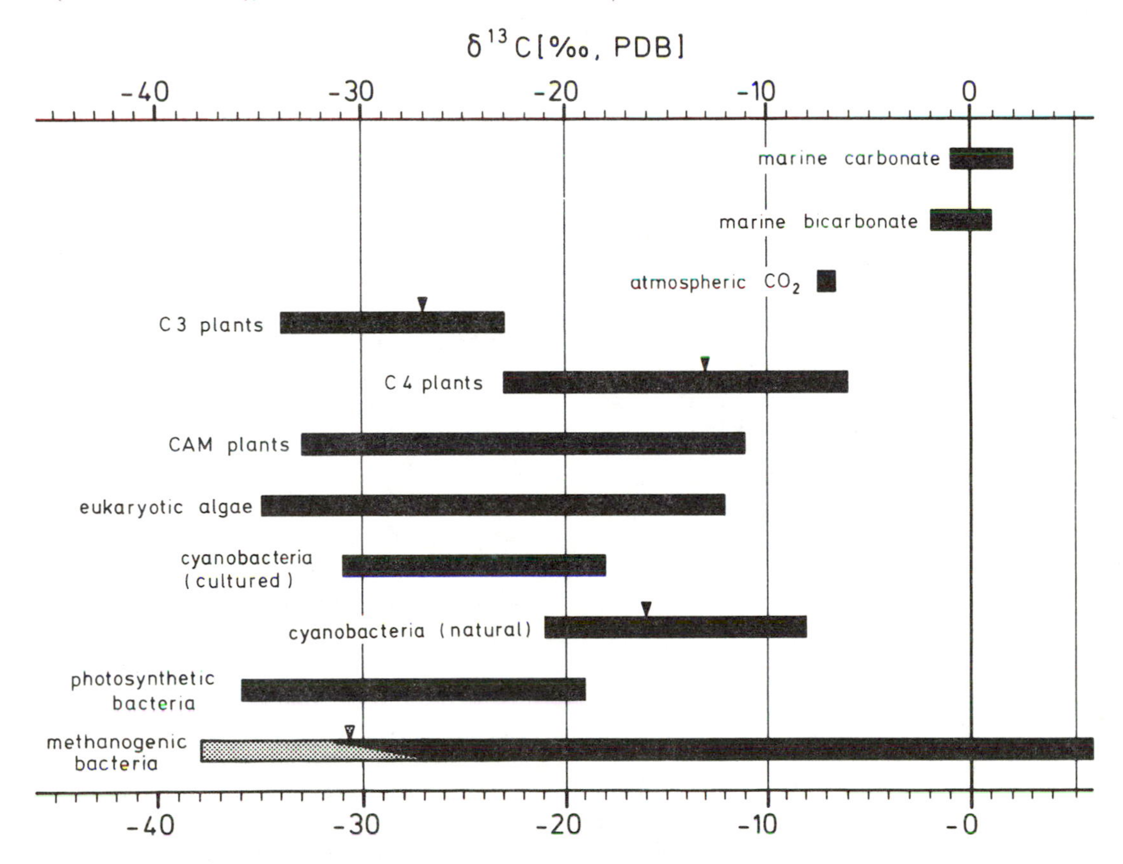

Figure 7-1. Carbon isotope composition of extant higher plants, algae, and autotrophic prokaryotes compared with that of sedimentary carbonate and the environmental reservoirs of oceanic bicarbonate and atmospheric carbon dioxide (for sources of data, see text). Means for some groups are indicated by triangles. Fractionation ranges given for cultures are recalculated for a CO_2 source with $\delta^{13}C = -7‰$. Note the virtual absence of overlap in the ranges of C3 and C4 plants. The higher fractionations measured in cultured cyanobacteria as compared with naturally occurring communities were obtained at $P_{CO_2} > 0.5\%$. Two sets of data are plotted for cultured methanogens: (*i*) those for *Methanogenium thermoautotrophicum*, *M.* strain *ivanovii*, and *M. barkeri* (stippled range; from M. J. DeNiro, pers. comm. to M. S., 1980), and (*ii*) those for *M. thermoautotrophicum* (black range; Fuchs et al., 1979). The relatively heavy values for the latter organism were obtained at very low gassing rates of CO_2 when competition between CO_2 reduction (consuming > 90% of the available substrate with an extreme preference for ^{12}C) and CO_2 assimilation is assumed to give rise to unusually heavy cell material.

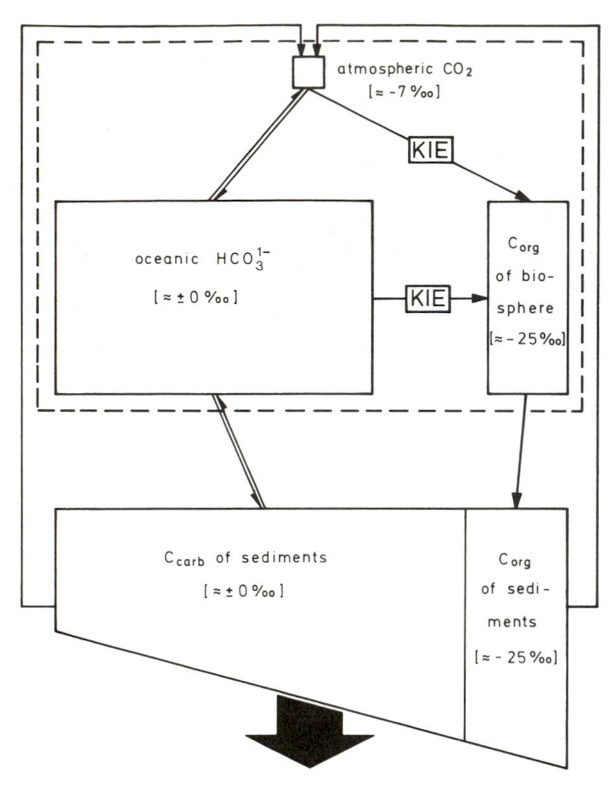

FIGURE 7-2. Synopsis of the terrestrial carbon cycle. The areas representing components of the exogenic exchange reservoir (dashed box) are drawn approximately to scale; at this scale, the box representing the sedimentary components of the cycle (viz., C_{carb} and C_{org} of sediments) would have to be extended to a length of 39.2 m (areas representing the various components have all been calculated in terms of the mass of carbon, not that of the respective carbon-bearing compounds). Oceanic bicarbonate is the dominant carbon species in the exogenic exchange reservoir, in which atmospheric carbon dioxide accounts for only about 1%. Enzymatically controlled kinetic isotope effects (KIE) are imposed on the two pathways leading from CO_2 and HCO_3^{1-} to living matter (C_{org} of biosphere), the resulting fractionation being retained in subsequent geochemical cycling of this carbon species. Isotope fractionations in the CO_2—HCO_3^{1-}—CO_3^{2-} system are governed by equilibrium effects (double arrows) and are thus reversed in back reactions (a fractionation of ~1‰ between bicarbonate and sedimentary carbonate has been omitted for simplicity). Note that the totality of carbon stored in the exogenic compartment (~10^{19} g) is miniscule in comparison with that in the carbon reservoir of the sedimentary shell (~10^{22} g).

C3 PLANTS

As illustrated in Figure 7-1, observed $\delta^{13}C_{PDB}$ values for C3 plants range from about −23 ‰ to −34 ‰, with an average of −27 ‰ (Smith and Epstein, 1971; Benedict, 1978). Since carbon fixation in C3 plants proceeds exclusively via the Calvin cycle, the sizable fractionation found in this group is usually ascribed to the large isotope discrimination typical of the RuBP carboxylase step. However, because the range of fractionation observed in this step is larger ($\Delta\delta$ between −20 and −40 ‰) than the average total fractionation exhibited by C3 photosynthesis ($\Delta\delta \approx -20$ ‰), it can be inferred that the initial diffusion step of Equation 7-1 becomes, in part, rate-limiting, thus substantially reducing the isotope effect of the enzymatic reaction. With photorespiration being very pronounced in C3 plants, dissimilatory decarboxylation processes probably have the largest share in counterbalancing the intrinsic fractionations of the RuBP carboxylase reaction.

C4 PLANTS

The $\delta^{13}C_{PDB}$ values of these plants range from −6 to about −23 ‰ (Smith and Epstein, 1971; Benedict, 1978), with an average between −12 to −14 ‰. In the C4 pathway, bicarbonate is the "active" carbon species for the initial carboxylation by PEP carboxylase (Cooper et al., 1968), a step discriminating to only a limited extent (−2 to −3 ‰) against ^{13}C. If the CO_2–HCO_3^{1-} interchange within such plants is kept close to equilibrium, then PEP would react with a bicarbonate substrate having an isotopic composition close to zero permil (since hydrated CO_2 is enriched by about 7 to 8 ‰ in ^{13}C as compared to gaseous CO_2; Wendt, 1968). The fact that the observed $\delta^{13}C$ values of C4 plants do not approach the expected value of −2 ‰ arrived at by this reasoning suggests that control of isotope fractionation in C4 photosynthesis occurs in the initial diffusion step (see Equation 7-1), the fractionation of −4 ‰ inherent in this process yielding a value of −11 ‰ that closely approximates the average of −12 to −14 ‰ observed for the group as a whole (Figure 7-1). The wide range of values found for C4 plants seems to indicate that, in the specific instance, neither diffusion nor carboxylation is entirely limiting. As noted above, the sizable fractionation inherent in the second carboxylation step of this pathway is suppressed due to the closed system conditions that result from compartmentation of the RuBP carboxylase reaction within the "Kranz anatomy" of such plants.

CAM PLANTS

In succulent plants utilizing Crassulacean acid metabolism (CAM), the first and second carboxylations of the process are not separated in space (as in the different compartments of the "Kranz anatomy" of C4 plants), but in time. During the night, fixation of CO_2 proceeds via PEP carboxylase to produce OAA that is subsequently transformed into malate. This latter compound is decarboxylated during the next light period, with the CO_2 thus generated being immediately recarboxylated in the Calvin cycle. While the sizable isotope discrimination of this second carboxylation reaction (by RuBP carboxylase) is not expressed in diffusion-limited C4 photosynthesis, it does occur in some CAM species that circumvent the PEP carboxylase step and, in the light, carry out direct CO_2 fixation via the Calvin cycle. Depending on whether CAM plants principally operate in the C3 or C4 mode, their isotopic compositions will reflect the range typical either of C3 or C4 plants. Hence, the wide spread of $\delta^{13}C$ values reported for CAM plants (−11 to −33 ‰) suggests the extensive utilization by this group of both of the carboxylation options (Benedict, 1978; O'Leary, 1981).

ALGAE

The $\delta^{13}C_{PDB}$ values reported for marine and freshwater algae cover the range from −12 to −35 ‰, with the bulk of values for marine phytoplankton lying between −20 and −35 ‰ (Smith and Epstein, 1971; Seckbach and Kaplan, 1973; Wong and Sackett, 1978). Cultured species have yielded fractionations of −13 to −27 ‰ between CO_2-substrate and cell material (Seckbach and Kaplan, 1973; Pardue et al., 1976; Sirevåg et al., 1977; Wong and Sackett, 1978) corresponding to $\delta^{13}C$ values of −20 to −34 ‰ in the natural environment (in which $\delta^{13}C_{CO_2} = -7$ ‰). With carbon fixation by algae assumed to proceed through the Calvin cycle, the range of isotope values for this group as whole could be expected to resemble that of higher plants of the C3 type; as shown in Figure 7-1, however, it often approaches the spread characteristic of C4 plants. It has been argued that the small fractionations typical of many such algae may be due either to (*i*) partial fixation of CO_2 by PEP carboxylase (Wong and Sackett, 1978), or to (*ii*) slow diffusion of CO_2 in water, the CO_2 supply thus becoming rate-limiting and the intrinsic fractionation of the RuBP carboxylase reaction being therefore largely suppressed (O'Leary, 1981). This second alternative should, however, apply to all aquatic algae, which is evidently not the case.

CYANOBACTERIA

Reported $\delta^{13}{}_{PDB}$ values for marine and freshwater cyanophytes ("cyanobacteria") and their mat communities vary between −8 and −21 ‰, with an average close to −16 ‰ (Behrens and Frishman, 1971; Smith and Epstein, 1971; Calder and Parker, 1973; Barghoorn et al., 1977). Fractionations obtained with cultured taxa are typically in excess of those found in natural environments, the largest differences between source carbon dioxide and cell material having been observed at high concentrations of CO_2 (>0.5%) in cultures with low cell densities (Calder and Parker, 1973; Pardue et al., 1976). Maximum fractionations recorded in such culture experiments amount to −23.9 ‰, corresponding to a value of −30.9 ‰ for the whole organism if grown on atmospheric CO_2.

PHOTOSYNTHETIC BACTERIA

Carbon isotope data for photosynthetic bacteria have hitherto accrued from laboratory experiments only. Recalculation of the reported results to give values that would have been obtained had atmospheric carbon dioxide ($\delta^{13}C = -7$ ‰) been used as a standard substrate, yields values between −19 and −36 ‰ for the cell material synthesized. A comprehensive study by Wong et al. (1975) has shown autotrophically grown cultures of *Chromatium* sp. to range between −30 and −36 ‰ (recalculated assuming an atmospheric CO_2 substrate), whereas the comparable (recalculated) value reported by Sirevåg et al. (1977) for this bacterium is −29.5 ‰. Such fractionations would lend support to previous proposals (Fuller et al., 1971) that these purple sulfur bacteria operate the Calvin cycle as do autotrophically growing purple nonsulfur bacteria that exhibit similar isotope effects (e.g., *Rhodospirillum* sp. with $\delta^{13}C = -27.5$ ‰; Sirevåg et al., 1977). In contrast, the less negative values yielded by *Chlorobium thiosulfatophilum* (−19.2 and −20.8 ‰: Sirevåg et al., 1977; Barghoorn et al., 1977) may indicate that green sulfur bacteria utilize the reductive carboxylic acid cycle, a pathway of carbon assimilation unique to photosynthetic bacteria (Table 7-1, Nos. 3 and 4).

CHEMOAUTOTROPHIC BACTERIA

Available data on carbon isotope fractionation by this group are restricted to methanogens. According to Fuchs et al. (1979), *Methanogenium thermoautotrophicum* (a strictly anaerobic chemoautotroph growing on CO_2 and H_2 as sole carbon and energy sources) displays a discrimination ranging from $+13$ to -34 ‰ relative to the ^{13}C content of the substrate, in response to the concentration of available CO_2 (i.e., at low CO_2 concentrations ^{13}C is actually enriched in the growing cells). Recalculation of these values (using atmospheric carbon dioxide as the reference medium) yields a $\delta^{13}C$ range $+6$ to -41 ‰ for the bulk organic matter synthesized by this chemoautotroph. Comparable recalculation of observed fractionations between -19.8 and -30.8 ‰ found by M. J. DeNiro (pers. comm. to M. S., 1980) for *M. thermoautotrophicum*, *M. barkeri*, and *M.* strain *ivanovii* yields a range between -26.8 and -37.8 ‰, with an average of -30.6 ‰ for the group as a whole. (With regard to such chemoautotrophs, it should be noted that the methane generated by methanogenic bacteria is usually extremely enriched in ^{12}C, with $\delta^{13}C$ values often reaching -60 ‰ or lower; see Rosenfeld and Silverman, 1959; Games et al., 1978.)

7-2-3. Isotope Fractionations in the Terrestrial Carbon Cycle

The total amount of carbon stored in the atmosphere-ocean-sediment system in both oxidized and reduced form is estimated to be $\sim 6 \times 10^{22}$ g, with the bulk residing in the sedimentary shell and a very minor fraction, about 0.1% (i.e., $\sim 10^{19}$ g), occurring in the various compartments of the exogenic exchange reservoir (Ronov, 1968; Hunt, 1972; Junge et al., 1975). Hence, the exogenic reservoir (the ocean, atmosphere, and biosphere) actually constitutes only a miniscule appendix to the total carbon inventory of the crust (see Figure 7-2). Oxidized carbon exists primarily as *sedimentary carbonate* (viz., as limestone or dolomite), while the prevalent form of reduced carbon is kerogen, the polycondensed, acid- and organic solvent-insoluble end product of diagenetic alteration of sedimentary organic matter derived from living systems and the products of their metabolism (for further discussion, see Chapter 5 of the present volume).

The sedimentary reservoirs of carbonate carbon (C_{carb}) and organic carbon (C_{org}) have been built up, and are still being fed, by carbon from the exogenic compartments of the biogeochemical carbon cycle (Figure 7-2). Additions to the crustal C_{org}-reservoir are provided by a small "leak" in the biological portion of the cycle by which a fraction (between 10^{-2} and 10^{-3}) of the annual primary production of some 10^{16} g C_{org} a^{-1} (Whittaker and Likens, 1973; Woodwell et al., 1978) escapes biological recycling (e.g., via respiration) and is diverted into newly formed sediments. Increments to the C_{carb}-reservoir are added by deposition of carbonate minerals in appropriate sedimentary environments. With the Earth's sedimentary shell in a quasi-stationary state (Garrels and Mackenzie, 1971, p. 255 ff.), the above additions must be balanced by a corresponding subtraction of the two carbon species from the sedimentary reservoir, largely as a result of weathering.

Input into the exogenic exchange reservoir, whether of primordial or of recycled origin, is mostly in the form of oxidized carbon (largely CO_2). Ever since the origin of autotrophic forms of life, any such input has been subject to an isotopic disproportionation into a light and a heavy fraction due to the sizable fractionation effect inherent in biological carbon fixation (Figure 7-2). Preferential incorporation of ^{12}C into organic matter will necessarily produce a relative increase of ^{13}C in the remaining portion of the reservoir, thus resulting in displacement of the $\delta^{13}C$ values of the residual carbon in a positive direction. Assuming a primary carbon influx with $\delta^{13}C_{prim} \approx -5$ ‰ and a fractionation of 25 ‰ between inorganic and organic carbon, a $\delta^{13}C$ value close to zero ‰ for marine bicarbonate would indicate that about 20% of the original input had been converted to C_{org}, and that a ratio of $C_{org}/C_{carb} \approx 20/80$ had become established in the exogenic compartment as a whole. These relationships follow from the constraints of the isotope mass balance

$$\delta^{13}C_{prim} = R\,\delta^{13}C_{org} + (1 - R)\,\delta^{13}C_{carb}, \quad (7\text{-}2)$$

yielding $R = C_{org}/(C_{org} + C_{carb}) = 0.2$ for the above assumptions. If, as at present, the marine bicarbonate reservoir is flushed at a relatively rapid rate (viz., having a current residence time of about 10^5 yr; Holland, 1978, p. 156), the stationary state characterized by the above ratio would become established within a geologically short time. Since HCO_3^{1-} is precipitated as CO_3^{2-} with but minor fractionation, the state of the system should be faithfully monitored by marine carbonates (Broecker, 1970; Schidlowski et al., 1975).

7-2-4. Effects of Diagenesis and Metamorphism on the Isotopic Composition of Sedimentary Carbon

During sedimentation, both the oxidized and the reduced carbon species are incorporated into the crust with approximately the same isotopic compositions imparted to them in terrestrial near-surface environments; the isotopic signature indicating the occurrence of the biological fractionation effect is thus transmitted from the exchange reservoir into the preservable rock record. The wide range of $\delta^{13}C$ values characteristic of the reduced organic matter of living systems (+6 to −38 ‰; Figure 7-1) tends to be averaged out during burial in sediments, though in some sedimentary facies the isotopic characteristics of specific ecosystems may be retained.

In general, diagenetically stabilized carbonate rocks tend to reflect rather faithfully the isotopic composition of their parental carbonate muds, staying well within 1 ‰ of the original $\delta^{13}C$ values (Veizer et al., 1980). On the other hand, the isotope shifts occurring during the burial and diagenesis of organic matter are more pronounced and may total a few or several permil over the maturation pathway of kerogenous materials. These effects are primarily linked to (*i*) preferential mobilization and consequent loss of isotopically light lipids and hydrocarbons during progressive dehydrogenation of kerogen; (*ii*) preferential loss of thermally and biologically labile functional groups; and (*iii*) preferential rupturing under thermal stress of ^{12}C–^{12}C bonds with concomitant release of light carbon (Sackett et al., 1968; Peters et al., 1981; Galimov, 1980). In total, these secondary processes tend to make the residual kerogen somewhat heavier while accumulating ^{12}C-enriched compounds in the mobilized fractions (e.g., methane and crude oil). Although such secondary effects are of relatively small magnitude, and are thus insufficient to totally obscure the biological pedigree of the precursor material, they may markedly blur the isotopic signature of the original ecosystem. Some of the scatter displayed by the $\delta^{13}C_{org}$ values assembled in Figure 7-3 probably results from these post-depositional processes.

During metamorphism, changes in the isotopic composition of both the oxidized and the reduced sedimentary carbon species proportionately increase. A remarkable exception are major beds of pure marbles that retain the $^{13}C/^{12}C$ ratios of their sediment precursors virtually unchanged through successive stages of metamorphism (Shieh and Taylor, 1969; Sheppard and Schwarcz, 1970; Schidlowski et al., 1975, p. 23 ff.). On the other hand, metamorphic isotope effects are common in impure (notably siliceous) carbonate rocks. Such carbonates are susceptible to various metamorphic decarbonation reactions in which the primary carbonate minerals react with silica to produce Ca-Mg-silicates (tremolite, actinolite, etc.) with a concomitant release of CO_2. Decarbonation processes of this type start in the greenschist facies (i.e., 300° to 450° C) and become increasingly important with increasing metamorphic grade. They give rise to evident shifts in the negative direction (up to −4 ‰) of the $\delta^{13}C$ values of the residual carbonate minerals, the magnitude of the effect being ostensibly related to the modal abundance of newly formed calc-silicates (Sheppard and Schwarcz, 1970). The observed retention of light carbon in the residual carbonates is in agreement with calculated fractionation factors for the system CO_2–$CaCO_3$ at metamorphic temperatures (Bottinga, 1969).

As in the case of carbonate carbon, major accumulations of organic carbon (e.g., coal deposits) do not seem to suffer large-scale changes in their bulk isotopic composition as a result of increasing metamorphic rank (Colombo et al., 1970), whereas the response of the more common forms of sedimentary organic matter (e.g., particulate kerogen) to either contact or regional metamorphism is usually pronounced. Metamorphic alteration of kerogen has been noted to shift $\delta^{13}C_{org}$ from original values of -25 ± 5 ‰ to −10 ‰ and less (McKirdy and Powell, 1974; Eichmann and Schidlowski, 1975; Hoefs and Frey, 1976; Peters et al., 1978). The enrichment of heavy carbon in highly metamorphosed kerogens is, in part, due to excessive dehydrogenation under conditions of increasing thermocatalytic stress, with preferential cleavage of ^{12}C–^{12}C bonds and subsequent loss of the light fraction. However, the most important process responsible for the formation of ^{13}C-enriched kerogens seems to be isotope exchange with coexisting carbonates which starts at about 350° C, somewhat below the greenschist/amphibolite facies boundary (Valley and O'Neil, 1981). Increasing $\delta^{13}C_{org}$ values are generally correlated with decreasing H/C ratios, an elemental abundance relationship which is often used as a principal rank indicator of kerogen maturity (see Chapter 5). The observed spread for $\delta^{13}C_{org}$ values in high-rank kerogens is −25 to −8 ‰ with

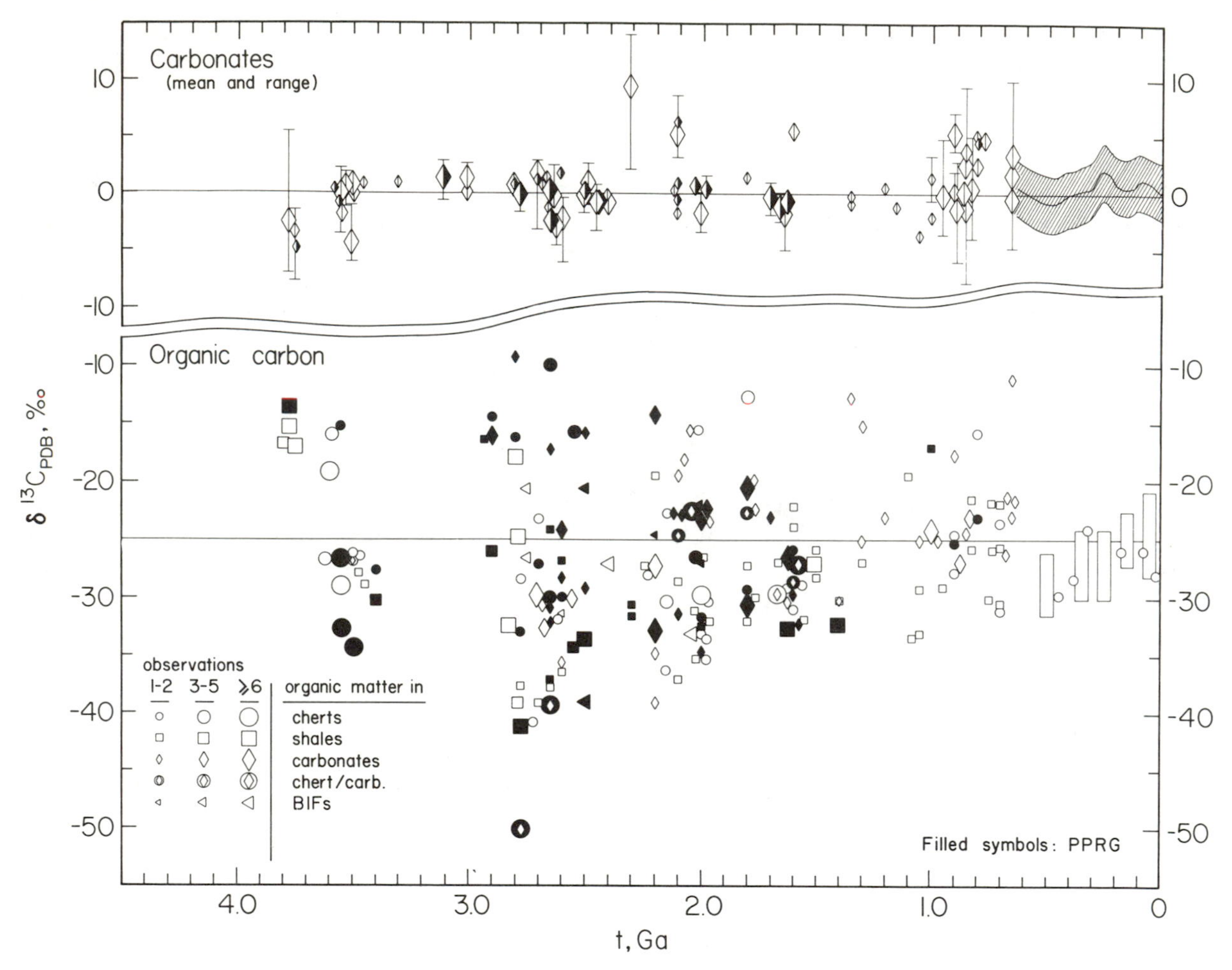

FIGURE 7-3. Isotopic composition of sedimentary carbon as a function of time (filled symbols represent P.P.R.G. data as listed in Tables 5-7, 5-8, and 7-2 of the present volume). The approximate constancy of the carbonate record is in marked contrast with the scatter of data points for organic carbon (kerogen) that reflects both primary variations and the later effects of diagenesis and metamorphism. The gross average for sedimentary organics of all ages is approximately $-25‰$. The age trend for Phanerozoic carbonate values (mean and standard deviation) is according to Veizer et al. (1980), and that for kerogens of the same age is according to Galimov et al. (1975) and Welte et al. (1975). Bars imposed on the Phanerozoic kerogen record are standard deviations for selected age groups as reported by Degens (1969). Sources of non-P.P.R.G. Precambrian data points are: Barghoorn et al. (1977); Donnelly et al. 1977); Eichmann and Schidlowski (1975); Hoefs and Schidlowski (1967); Hoering (1967a); Jackson et al. (1978); Leventhal et al. (1975), McKirdy and Powell (1974); Oehler et al. (1972); Oehler and Smith (1977); Perry et al. (1973); Perry and Ahmad (1977); Schidlowski et al. (1975, 1976a, 1976b, 1979); Schopf et al. (1971); and Smith et al. (1970).

an average in the -14 to -17 ‰ range. As discussed in Chapter 5, $\delta^{13}C_{org}$ values $\gtrsim -20$ ‰ are commonly coupled with H/C ratios $\lesssim 0.1$. Dehydrogenation of kerogen culminates in the formation of graphite, the hydrogen-free end member of the alteration series.

In summary, it can be stated that diagenetic and metamorphic alteration processes are capable of imposing measurable overprints on the isotopic compositions of both oxidized and reduced sedimentary carbon, most notably on particulate organic matter. Although the original isotopic composition may be altered during storage in the lithosphere, the "noise level" imposed will, as a rule, not seriously obscure the isotopic signature of the primary carbon source. Possible borderline cases are represented by highly graphitized kerogens from metamorphic terranes with $\delta^{13}C > -10$ ‰.

7-2-5. Isotopic Composition of Sedimentary Carbon through Time

As is discussed above, the processes and isotope effects involved in the generation, deposition, and preservation in sediments of both C_{carb} and C_{org} are reasonably understood. With this background, we can now attempt to trace the fractionation between the two carbon species back into the geologic past. As such fractionation is biologically mediated, this approach holds the potential of identifying the time when autotrophic ecosystems first became a dominant component of the terrestrial carbon cycle.

Figure 7-3 presents a synopsis of the $\delta^{13}C_{carb}$ and $\delta^{13}C_{org}$ records from the Recent to Isua time (3.8 Ga ago) based on some 8000 measurements. Data gathered during the course of the P.P.R.G. project, represented in Figure 7-3 by filled symbols, are listed in Table 7-2 ($\delta^{13}C_{carb}$ values) and in Tables 5-7 and 5-8 ($\delta^{13}C_{org}$ values) of the present volume. The salient features of the carbonate record are exemplified by its Phanerozoic section, an updated summary of which (based on some 3016 values) has been recently compiled by Veizer et al. (1980). It is obvious from this compilation that $\delta^{13}C_{carb}$ has displayed distinct secular variations over a magnitude of 2.8 ‰ during the last 800 Ma, with a minimum (−0.8 ‰) in the Ordovician and a maximum (+2.0 ‰) in the Permian. Even if the extreme values on this curve have been accentuated by an overrepresentation in the respective age groups of facies-dependent δ-values (which probably applies to the Permian with its abundant hypersaline basins being characterized by isotopically heavy carbonates), the long-term consistency of the trend and its correlation with (*i*) a corresponding trend for $\delta^{13}C_{org}$ (discussed below), and (*ii*) the $\delta^{34}S$ curve of marine sulfate evaporites and attendant corollaries for the oxygen budget (Schidlowski and Junge, 1981), lend credence to its reality.

As can be inferred from Figure 7-3, the moderate oscillations of the Phanerozoic $\delta^{13}C_{carb}$ curve apparently continue into the Precambrian, but both a lack of sufficient data points and the limited geochronologic resolution that can be applied to such points preclude the definition of a well-defined trend line similar in quality to that available for the Phanerozoic. Conspicuous positive extremes represent carbonates from closed or semi-closed basins (i.e., reflecting the so-called "Lomagundi effect" reported by Schidlowski et al., 1976b). Negative excursions from the mean indicate participation of biogenic CO_2 in the process of carbonate formation, a phenomenon apparently typical of banded iron-formation (Becker and Clayton, 1972). The negative shift displayed by the 3.8-Ga-old Isua carbonates (the only markedly metamorphosed carbonates included in the data plotted) is almost certainly due to metamorphism.

The isotope record for organic carbon shows a very much greater degree of scatter. For Phanerozoic kerogens, both Galimov et al. (1975) and Welte et al. (1975) have proposed an age curve largely paralleling that found for $\delta^{13}C_{carb}$ (with a minimum close to −29 ‰ during Ordovician-Silurian and a maximum around −24 ‰ in the Carboniferous). If two constraints are assumed—namely, (*i*) that $C_{org} + C_{carb} =$ constant, and (*ii*) that $\delta^{13}C_{carb} - \delta^{13}C_{org} \approx 25$ ‰—these long-term variations of $\delta^{13}C_{org}$ (and their covariance with $\delta^{13}C_{carb}$) would be a necessary consequence of changes in the relative proportions of the partial reservoirs of organic and carbonate carbon over time, changes which seem likely to have resulted from the considerable chemical inertia of the atmosphere-ocean-crust system.

Calculated deviations of $\delta^{13}C_{org}$ for various Phanerozoic formations have been shown to lie within the confines of the −21 and −32 permil lines (Degens, 1969). The average for all Precambrian kerogens represented in Figure 7-3 falls into this same range, but the total spread of measured values is apparently larger than that for the Phanerozoic, ranging from about −10 to −50 ‰. Isotopically heavy kerogens are, as a rule, in various stages of graphitization and thus characterized by extensive dehydrogenation (exhibiting H/C ratios mostly <0.1). Accordingly, as discussed in Chapter 5, there is little doubt that the bulk of these heavy organics were subjected to metamorphism (McKirdy and Powell, 1974; Eichmann and Schidlowski, 1975; Hoefs and Frey, 1976). Coupling between positive $\delta^{13}C_{org}$ and low H/C values is particularly pronounced in the case of the Isua organics (see Chapter 5).

More enigmatic than these positive excursions are negative extremes in the −35 to −50 ‰ range encountered in Late Archean kerogens approximately 2.8 to 2.5 Ga old, most notably those from the Australian Fortescue and South African Ventersdorp Groups (Figure 7-3). The most negative of these values are among the lightest ever reported for terrestrial nonvolatile carbon. In spite of the

TABLE 7-2. Isotopic analyses of carbonates in the P.P.R.G. collections.

P.P.R.G. sample no.[a]	$\delta^{13}C_{PDB}$[b,c] (‰)	$\delta^{18}O_{SMOW}$[c] (‰)
Burra Group, 0.8 Ga		
455	+4.62	+21.65
Bitter Springs Formation, 0.9 Ga		
133–2	+3.39	+22.50
Bungle Bungle Dolomite, 1.6 Ga		
138	−0.87	+24.78
139	−0.94	+22.77±0.7[d]
143	−0.76	+23.35
152	−0.50	+19.85
153	−0.30	+21.51±0.7[d]
154	−0.25	+22.47
159	−1.00	+20.81
average (n=7):	*−0.66*	*+22.22*
McArthur Group, 1.6 Ga		
103	−0.09	+20.53
105	+0.15	+23.09
106	−2.41	+17.71
107	−1.58	+20.74
108	−1.22	+19.49
109	−1.10	+19.24
average (n=6):	*−1.04*	*+20.13*
Earaheedy Group, 1.8 Ga		
080	−0.89	+18.15
086	−1.79	+19.28
095–1	−1.11	+21.30
096	+1.00	+23.39
097	+0.58	+17.90
435	−1.19	+13.82
439	+0.01	+18.52
440	−2.41	n.d.*
average:	*−0.73* (n=8)	*+18.91* (n=7)
Kahochella Group, 2.0 Ga		
340	−0.18	+16.55
407	−0.73	+17.05
average (n=2):	*−0.45*	*+16.80*
Pethei Group, 2.0 Ga		
338	+1.62	+18.35
412	+1.25	+15.70±0.7[d]
average (n=2):	*+1.43*	*+17.03*
Wyloo Group, 2.0 Ga		
054	+0.99	+16.54
055	+0.65	+15.97
056	+0.54	+15.33
057	+0.35	+15.13
060	+0.79	+18.36
average (n=5):	*+0.66*	*+16.27*
Upper Albanel Formation, 2.1 Ga		
443	+0.91	+23.76
Lower Albanel Formation, 2.1 Ga		
444[e]	+6.38	+21.85
444[e]	+6.33	n.d.
average:	*+6.36* (n=2)	*+21.85* (n=1)
Belcher Group, 2.1 Ga		
447	−0.36	+18.98
449	−0.64	+23.82
average (n = 2):	*−0.50*	*+21.40*
Transvaal Supergroup, 2.2 Ga		
205[e]	−1.44	+21.64
205[e]	−1.28	n.d.
296	−1.32	+21.62
297	−0.52	+22.40
298	−0.52	+21.31
299	−0.71	+21.99
300	−0.83	+22.52
302	−0.93	+21.90
304	−0.11	+23.43
312	−0.91	+21.45
314	−0.61	n.d.
316	−0.99	+21.59
317	−0.82	+21.61
318	−0.77	+24.86
320	−0.89	+19.98
average:	*−0.82* (n=16)	*+22.02* (n=14)
Hamersley Group, 2.5 Ga		
361	−12.00	+20.92
366	+0.51	+22.26
367	−0.07	+19.68
368	+0.36	+20.52
369	−7.35	n.d.
370	+0.02	+16.95
371	−0.08	+16.94
483	−7.91	n.d.
489	−9.35	+20.47
491	−8.52	+20.12±0.7[d]
average:	*−4.44* (n=10)	*+19.73* (n=8)
Steeprock Group, 2.6 Ga		
322	+1.43	n.d.
324	+1.09	n.d.
325	+2.08	+22.93
average:	*+1.53* (n=3)	*+22.93* (n=1)
Ventersdorp Supergroup, 2.6 Ga		
280	−3.63	n.d.
281	−1.88	n.d.
282–1	−3.55	n.d.
282–2	−1.84	n.d.
283	−2.82	n.d.
286	−1.91	n.d.
287	−2.32	+11.03

TABLE 7-2. *(continued)*

288	−2.22	n.d.
291–1	−5.37	n.d.
293	−6.21	n.d.
average:	*−3.18* (n=10)	*+11.03* (n=1)
	Yellowknife Supergroup, 2.6 Ga	
413–3	+1.82	n.d.
	Manjeri Formation, 2.6 Ga	
224	−5.73	+18.47
225	−7.17	+15.46
228	−7.08	+16.73
average (n=3):	*−6.66*	*+16.89*
	Bulawayan Group, 2.6 Ga	
206	+0.51	n.d.
247	−0.7	n.d.
248	0.0	n.d.
249	−0.8	n.d.
251	+0.8	n.d.
254	+1.2	n.d.
average:	*+0.17* (n=6)	*n.d.*
	Fortescue Group, 2.8 Ga	
024	−1.66	n.d.
025–1	−0.49	n.d.
025–2	−1.32	n.d.
026	−0.22	n.d.
027	+0.54	+11.20
028	+0.24	n.d.
029	+0.79	n.d.
030	+0.70	n.d.
032	+0.65	n.d.
033	+0.61	n.d.
041	−1.04	n.d.
042	+0.15	+21.63±0.7[d]
average:	*−0.09* (n=12)	*+16.42* (n=2)
	Pongola Supergroup, 3.0 Ga	
258[e]	+1.75	+14.85
258[e]	+1.39	n.d.
261[e]	−0.14	+10.63
261[e]	−0.61	n.d.
262	+2.05	n.d.
263–1	+1.81	n.d.
263–2	+2.08	+15.11
264	+1.06	n.d.
265	+2.87	n.d.
average:	*+1.36* (n=9)	+13.53 (n=3)
	Warrawoona Group, 3.5 Ga	
013	+1.99	n.d.
016	−0.85	+12.22
average:	*+0.57* (n=2)	*+12.22* (n=1)
	Onverwacht Group, 3.5 Ga	
183	−0.17	n.d.
190	+0.89	n.d.
average:	*+0.36* (n=2)	*n.d.*
	"Isua Supracrustal Sequence," 3.8 Ga	
465	−4.71	+11.80±0.7[d]
492	−5.26	+15.28±0.7[d]
average (n=2):	*−4.99*	*+13.54*

[a] See Appendix I for details regarding sample localities and radiometric ages.

[b] Entries listing $\delta^{13}C_{PDB}$ values only (e.g., P.P.R.G. Sample No. 440) record measurements performed at the Max-Planck-Institut für Chemie, Mainz, Germany.

[c] Entries listing both $\delta^{13}C_{PDB}$ and $\delta^{18}O_{SMOW}$ values (e.g., P.P.R.G. Sample No. 455) record measurements performed at Indiana University, Bloomington, Indiana, U.S.A.

[d] ± values indicate uncertain correction for mineralogy.

[e] Measurements of this sample (e.g., P.P.R.G. Sample No. 444) were performed both at the Max-Planck-Institut für Chemie ($\delta^{13}C_{PDB}$ value only) and at Indiana University (both $\delta^{13}C_{PDB}$ and $\delta^{18}O_{SMOW}$ values).

* n.d. = not determined.

limits imposed by insufficient geochronologic control, it is obvious from Figure 7-3 that these isotopically anomalous kerogens are essentially coeval with occurrences of "normal" kerogen. Thus, the anomalies must reflect relatively local, rather than global events, a conclusion consistent with the apparent absence of isotopic disturbances in the coeval carbonate record. It should be noted that extremely light kerogens have also been reported from younger sediments (Kaplan and Nissenbaum, 1966).

Any attempt to explain these anomalously light Late Archean $\delta^{13}C_{org}$ values has to take into consideration that values of this range have hitherto only been reported in connection with methane. It is, therefore, reasonable conjecture that bacteria of the methane cycle (methanogens and methylotrophs) were either directly or indirectly involved in the formation of these kerogens. As is discussed in Chapter 12 of the present volume, feasible mechanisms might include multi-step fractionations (with carbon passing through more than one biologically mediated step, e.g., $C_{org} \rightarrow CH_4 \rightarrow CO_2 \rightarrow C_{org}$), or possibly direct utilization by methylotrophs of isotopically light methane (see Table 7-1, Nos. 7 and 8). The requirement of an external oxidant for either of these pathways may be a major obstacle to the postulation of their operation during the Archean, but there are recent reports that sulfate (rather than molecular oxygen) may serve as the electron acceptor in an anaerobic conversion of methane to carbon dioxide (Panganiban et al., 1979). Assimilation of ^{13}C-deficient

CO_2 or CH_4 will necessarily give rise to isotopically light cell material and, consequently, to light kerogen. Alternatively, one might postulate the occurrence of intrastratal diffusion of isotopically light methane and its subsequent polymerization in situ, the resulting mixture of the polymerisate and indigenous "normal" organics yielding intermediate $\delta^{13}C_{org}$ values in the range -35 to -50 ‰. Optical microscopic study of the kerogens in question, however, both in petrographic thin sections and freed from the rocks by acid maceration, has yielded no evidence in support of this possibility (see Chapter 5).

7-2-6. Implications of the Sedimentary Carbon Isotope Record

The implications of the sedimentary carbon isotope record seem relatively straightforward. The available $\delta^{13}C_{carb}$ values (about $+0.4 \pm 2.6$ ‰), inherited almost unchanged from an original bicarbonate precursor, suggest that the isotopic composition of oceanic bicarbonate has varied little through time, having been always rather closely tethered to a mean value between 0 and -1 ‰. As carbonates monitor the state of the exogenic carbon compartment (see Section 7-2-3), the $\delta^{13}C_{carb}$ values observed would constrain the isotopic composition of coevally produced organics to an average close to -25 ‰ (assuming $\delta^{13}C_{carb} - \delta^{13}C_{org} \approx 25$ ‰). The observed $\delta^{13}C_{org}$ record over time basically confirms this mean, the majority of data points falling into the range -27 ± 7 ‰. As noted above, however, the total spread of $\delta^{13}C_{org}$ exceeds this range considerably, extending from about -10 to -50 ‰. This scatter reflects both primary variations in the isotopic composition of the parental organic matter (see the spread of biological fractionations summarized in Figure 7-1) and post-depositional alterations that have imposed a noise level on the primary patterns. Biological fractionations are likely to account for the negative extremes of the record (suggesting the involvement of light, bacteriogenic, methane), while the positive extremes seem due to secondary processes, notably metamorphism.

The approximately constant average fractionation between C_{org} and C_{carb} over the presently known record compels us to assume that the main responsibility for the ^{12}C enrichment observed in ancient kerogens has always rested with the process that gave rise to the biological precursor materials, specifically with the step of enzymatic carboxylation of CO_2 in the Calvin cycle of autotrophic organisms. Accordingly, the sedimentary carbon record can be reasonably interpreted as evidencing a rather remarkably consistent isotopic signal of biological (viz., autotrophic) activity on Earth beginning at least 3.5 Ga ago. Furthermore, the uniformity of this signal suggests an extreme degree of evolutionary conservatism for the basic biochemical mechanisms of carbon fixation. It should be noted that this interpretation of the carbon isotope geochemistry of Precambrian sediments is fully consistent with the morphological fossil record of apparently autotrophic life (including both cellular microfossils and the stromatolites they built, now known to extend to at least 3.5 Ga ago; see Lowe, 1980a; Walter et al., 1980; Awramik et al., 1983; and Chapters 8 and 9, herein). Since the isotope shifts recorded for both the oxidized and reduced sedimentary carbon of the Early Archean Isua Supracrustal Belt can be attributed to severe (amphibolite facies) metamorphism, the continuity of the original signal may have possibly extended to at least 3.8 Ga ago.

For this interpretation to be invalidated, one would have to postulate the occurrence of an inorganic chemical process that had mimicked fractionations in biological carbon fixation with a striking degree of precision. Moreover, the quantitative capacity of such process would have to have been comparable to that of biological autotrophy in order to bring the bicarbonate values in the exogenic exchange reservoir up to about zero permil from a value around -5 ‰ assumed for the primordial carbon input (see Equation 7-2). Although there are two known processes that might be called upon, neither seems likely to meet these requirements completely.

Fractionations in Fischer-Tropsch reactions generally range from 50 to 100 ‰ at 400° K (Lancet and Anders, 1970) and increase with decreasing temperatures (i.e., when approaching the temperature range characteristic of sedimentary environments). Such reactions thus seem unlikely to have been involved in the production of sedimentary kerogenous organic matter. Miller-Urey spark discharge syntheses, on the other hand, are accompanied by relatively small isotope effects (usually <10 ‰), although fractionations reported for individual reaction products (e.g., amino acids) may approach 20 ‰ and thus overlap the biological range (see Section 4-3-3, herein). Here, the borderline between biogenic and abiogenic fractionations may indeed appear blurred, but with the isotopically light fractions comprising well less than 1 % of the total yield from such reactions, its quantitative importance is negligible. Maximum fractionations between 9 and 12 ‰ obtained, at low reactant consumption, for the gross reaction products of spark discharge syntheses encroach just marginally on the range attributed to biological fractionations (Figure 7-1) and their fossil manifestations (Figure 7-3).

It is reasonable to expect that the relatively heavy organic matter comprising the bulk product of spark discharge reactions would lend itself to quick detection in the geologic record, at least in unmetamorphosed sediments. The virtual absence of such an isotopic signature in the sedimentary carbon record (Figure 7-3) suggests that large-scale abiotic production of organic matter by this process can be precluded for the last 3.5 Ga. Also, if it is assumed that the products of such syntheses were potentially metabolizable by primitive microorganisms, then the sedimentary contribution from such sources should be largely confined to sediments older than those containing the oldest known morphological fossils, that is, a time earlier than 3.5 Ga. ago. We are, moreover, extremely hesitant to invoke an abiotic pedigree for even the very ancient Isua organics since (*i*) the isotope shifts apparently evidenced in the Isua rocks are consistent with the amphibolite-grade metamorphism of the suite, and (*ii*) the occurrence of sedimentary carbonates within the series seems clearly indicative of the presence of CO_2 in the Isua environment, thus rendering any postulated CO- or CH_4-based abiogenic production of organic matter relatively implausible.

Summing up, the continuity of the sedimentary $\delta^{13}C_{org}$ and $\delta^{13}C_{carb}$ records compels us to assume that carbon fixation by autotrophic life forms had gained control of the terrestrial carbon cycle as early as 3.5, if not 3.8 Ga, ago. The data do not permit a distinction between photoautotrophic and chemoautotrophic pathways, differentiation between which would require application of additional criteria. Conspicuous negative excursions of $\delta^{13}C_{org}$ suggest, moreover, a possible involvement of methylotrophs in the formation of several Archean kerogens. In spite of a limited overlap between biological fractionations and those obtained in Miller-Urey type spark discharge reactions, substantial contributions to the sedimentary C_{org} budget by abiogenic sources of this type should not go undetected in the unmetamorphosed portion of the record. The most conspicuous single discontinuity in the carbon record is between post-Isua and Isua sediments, the latter of which display C_{org}–C_{carb} differences approaching those observed between reaction products and parent CH_4 in Miller-Urey type abiogenic syntheses. However, with the circumstantial evidence discussed above, and the well-established occurrence of high grade metamorphism of the Isua sediments, we cannot seriously entertain the possibility that the bulk of the Isua graphitics were derived from abiogenically produced precursor materials.

From the above considerations, both the potential and the limits of carbon isotope analyses in efforts to decipher the early history of life should be apparent. In the absence of morphological fossils, studies of carbon isotopes have proved to be a valuable predictive tool allowing first conjectures about a possible involvement of autotrophy in the early terrestrial carbon cycle (e.g., Junge et al., 1975; Schidlowski et al., 1979). On the other hand, isotope data typically provide "low resolution" information, yielding only partial answers to the questions asked (e.g., such data are unable to differentiate between organic matter produced by photoautotrophs and chemoautotrophs, by anaerobic and aerobic photoautotrophs, or between some products of spark discharge syntheses and biogenic carbon that has been severely altered by metamorphism). Accordingly, interpretations based on isotopic data should be constrained by the introduction of additional parameters capable of eliminating other degrees of freedom from the system. Integrated in a web of independent, partially circumstantial evidence, carbon isotope data may ultimately provide important clues for the elucidation of the earliest stages of organic evolution.

7-3. Sulfur

Sulfur is another key element of life constituting, on average, between 0.5 and 1.5 ‰ (dry weight) of plant and animal matter. It occurs mainly in proteins that typically display a C/S ratio of about 50/1. Primary synthesis of sulfur-containing cell constituents relies on assimilatory reduction of inorganic sulfate involving a valence change in the element from +6 to −2. Dissimilatory sulfate reduction, on the other hand, is an energy-yielding metabolic process involving the release of hydrogen sulfide that is characteristic of a relatively restricted group of bacteria occurring in reducing environments. Turnover rates of sulfur in this reaction surpass those during assimilation by several orders of magnitude. Accordingly, the decisive biological control of the terrestrial sulfur cycle is exercised by dissimilatory sulfate reducers that are ultimately responsible for a large-scale interconversion of sulfur between the oxidized and reduced reservoirs. Both assimilatory and dissimilatory sulfate reduction are associated with isotope effects of varying magnitudes which are briefly summarized in the following section. As in the case of carbon, isotopic compositions are expressed in terms of the conventional δ-notation,

$$\delta^{34}S = \left[\frac{(^{34}S/^{32}S)_{sa}}{(^{34}S/^{32}S)_{st}} - 1\right] \times 10^3 \ [‰, \text{CDM}],$$

by which the ^{34}S content of the sample (sa) is given as permil difference from a standard (st) which is troilite from the Canyon Diablo Meteorite (CDM) with $^{32}S/^{34}S = 22.22$ and $\delta^{34}S = 0.00$ ‰. As in the case of carbon, biological sulfur-involving reactions usually tend to discriminate against the heavier isotope.

7-3-1. Biochemistry of Sulfur Isotope Fractionation

FRACTIONATIONS IN ASSIMILATORY SULFUR METABOLISM

Uptake of sulfur by plants and microorganisms occurs, for the most part, via the assimilatory reduction of inorganic sulfate (Figure 7-4). In this process, sulfate is first phosphorylated to give "activated" sulfate species like adenosine-phosphosulfate (APS) and phosphoadenylyl sulfate (PAPS) that, in turn, are reduced via sulfite and other intermediates to the sulfide level (Trudinger, 1969; Roy and Trudinger, 1970; Trüper, 1982). The resulting sulfide reacts with serine to give cysteine; cysteine-based transformations then yield other S-containing amino acids (viz., cystine and methionine). With few exceptions (e.g., *Clostridium pasteurianum* and *Saccharomyces cerevisiae*), no hydrogen sulfide is released by organisms during sulfate assimilation. Isotope discriminations in assimilatory processes have been shown to be minor (Figure 7-5), observed differences between cell sulfur and the SO_4^{2-} substrate mostly ranging between $+0.5$ and -4.5 ‰ in the case of marine and freshwater plants (Kaplan et al., 1963; Mekhtieva, 1971), and between -0.9 and -2.8 ‰ for microorganisms (Kaplan and Rittenberg, 1964).

FRACTIONATIONS IN DISSIMILATORY SULFATE REDUCTION

In this energy-yielding process the reduction of sulfate is coupled with the oxidation of organic substrates, commonly lactate, pyruvate or ethanol, for example

$$2 \text{ lactate} + SO_4^{2-} \longrightarrow 2 \text{ acetate} + 2\,H_2O + 2\,CO_2 + S^{2-}.$$

With SO_4^{2-} rather than free oxygen serving as the electron acceptor, dissimilatory sulfate reduction

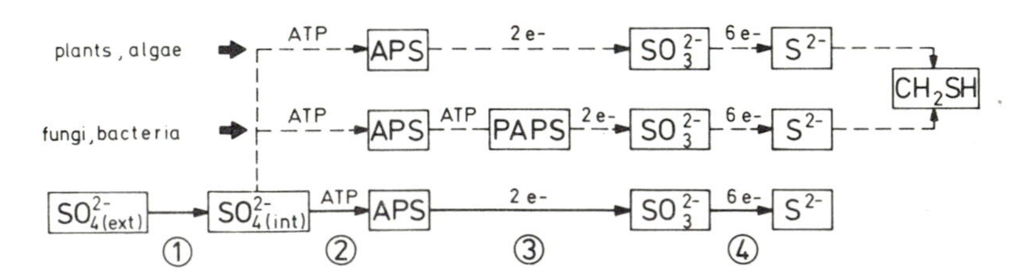

FIGURE 7-4. Synopsis of the assimilatory (dashed lines) and dissimilatory (solid lines) pathways of biological sulfate reduction. Following its uptake from external sources, SO_4^{2-} is phosphorylated to give adenylyl sulfate ("adenosine-5′-phosphosulfate," APS) and, in the case of assimilatory bacteria and fungi, also phosphoadenylyl sulfate (PAPS); these "activated" sulfate species are subsequently reduced to sulfite (SO_3^{2-}) and sulfide (S^{2-}). In either of the assimilatory pathways, sulfur ends up in the sulfhydryl (CH_2SH) group of cysteine; in dissimilatory reduction it is released to the environment as H_2S. The initiation of these reactions is contingent on an investment of energy in the form of ATP; eight electrons are transferred to sulfur from an external electron donor during the process.

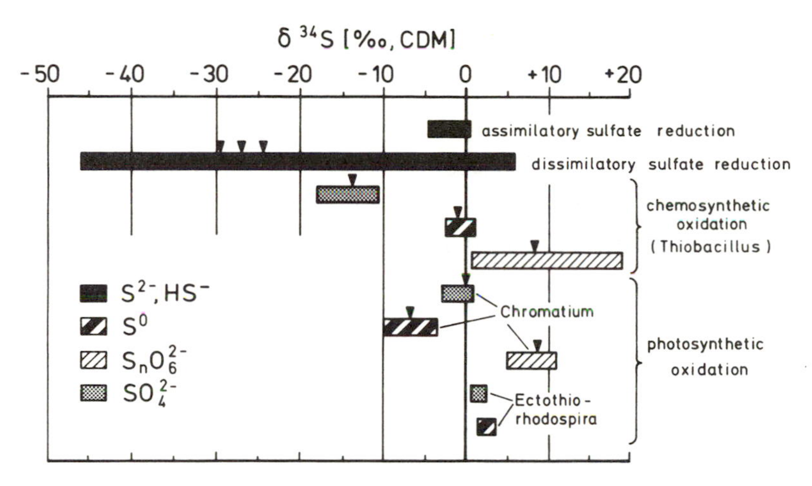

FIGURE 7-5. Isotope fractionations in assimilatory and dissimilatory sulfur metabolism, measured against a standard substrate with $\delta^{34}S = 0.0$ ‰ (for sources of data, see text). Approximate means for some groups are indicated by triangles. The range shown for dissimilatory sulfate reduction (+6 to −46‰) is that obtained in culture experiments (for the range measured in natural environments, see Figure 7-6). The products of the biologically mediated oxidation processes tend to encroach upon the positive part of the δ-scale; this is particularly evident for the polythionate ($S_nO_6^{2-}$) intermediates of these reactions.

has come to be regarded as a form of anaerobic respiration. Although carried out by members of only a few extant bacterial genera (viz., *Desulfovibrio*, *Desulfotomaculum*, and *Desulfomonas*) that are confined to strictly anaerobic habitats, this process is of paramount geochemical relevance as the principal means of bringing about a large-scale low temperature reduction of sulfate to sulfide. This conversion constitutes an important link in the terrestrial sulfur cycle, but has not been demonstrated to proceed at temperatures less than about 200° C unless it is biologically mediated.

The pathway of dissimilatory sulfate reduction is composed of four principal steps (Trudinger, 1969; Roy and Trudinger, 1970; Rees, 1973) which are summarized in Figure 7-4. Following the uptake of external SO_4^{2-}, sulfate is "activated" by ATP sulfurylase to yield APS that is subsequently reduced enzymatically (by APS reductase), producing sulfite. Sulfite then is reduced by sulfite reductase to the sulfide level. The sizable isotope shifts in the negative direction associated with this reaction series both in culture experiments (Figure 7-5) and in natural environments (Figure 7-6) are, accordingly, composites of the discriminations inherent in the four numbered reaction steps depicted in Figure 7-4 (see Kaplan and Rittenberg, 1964; Kemp and Thode, 1968; Rees, 1973; Chambers and Trudinger, 1979). Attempts to account for the observed fractionations in terms of the biochemistry of the reduction pathway must, in principle, consider four boundary conditions with one of the component reactions becoming rate-controlling under specific circumstances. Unfortunately, kinetic isotope effects hitherto assigned to each of the single reactions are rather poorly constrained. There is, however, widespread agreement that fractionations associated with steps (1) and (2) in Figure 7-4 are on the order of a few permil only, the value for (1) even becoming positive (+3 to +6 ‰) at SO_4^{2-} concentrations ≤0.01 mM (Harrison and Thode, 1958; McCready and Krouse, 1980). Values proposed for step (3) vary between −25 ‰ (Rees, 1973) and about −10 to −15 ‰ (Chambers and Trudinger, 1979), whereas the effect linked to sulfite reduction (step 4) has been shown to range between 0 and −33 ‰. Chambers and Trudinger (1979) have suggested −14 to −19 ‰ as a possible range for the fractionation associated with rupture of the S–O bond, the salient reaction in steps (3) and (4).

Apart from these uncertainties it is, however, obvious from experiments performed with bacterial cells under controlled conditions (using organic electron donors and unlimited sulfate), that the largest fractionations (−20 to −46 ‰) are obtained

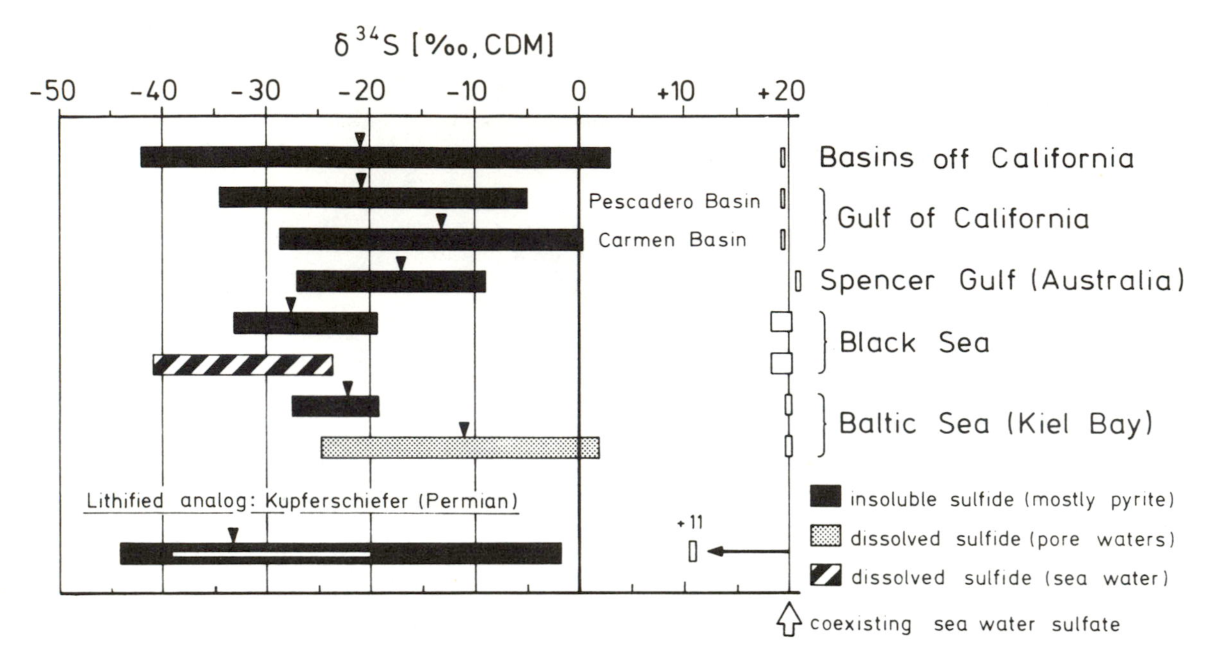

FIGURE 7-6. Approximate ranges of isotope fractionation produced via bacterial sulfate reduction in various present-day marine environments (the mean value for each group is indicated by a triangle). The substrate for these reductions is seawater sulfate with $\delta^{34}S \approx +20‰$ (viz. California: +19.7; Spencer Gulf: +21.0; Black Sea: +18.2 to 20.2; Baltic Sea: +20.0). Values for dissolved hydrogen sulfide in Black Sea water are from depths ≥200 m. The isotope distribution pattern of the Permian Kupferschiefer exemplifies a lithified facies (main range indicated by white bar; $\delta^{34}S$ for Permian sea water ≈ +11‰). Data plotted are from Chambers (1979); Goldhaber and Kaplan (1980); Hartmann and Nielsen (1969); Kaplan et al. (1963); Marowsky (1969); Sweeney and Kaplan (1980a); and Vinogradov et al. (1962).

at low rates of sulfate reduction, whereas the minimum effect under such conditions (at high reduction rates) ranges close to −10‰. At low sulfate concentrations (<1 mM), the magnitude of the fractionation decreases and the direction of the effect is apparently inverted at $[SO_4^{2-}] \leq 0.01$ mM (or 1 mg/liter), a concentration equivalent to that of the solubility of barite ($BaSO_4$).

FRACTIONATIONS IN BIOLOGICAL OXIDATION OF REDUCED SULFUR COMPOUNDS

Biological oxidation of inorganic sulfur compounds is linked to the electron supply for either photosynthetic or chemosynthetic activity by prokaryotic microorganisms. With bacterial (anoxygenic) photosynthesis requiring electron donors other than water (for discussion, see Chapter 6), sulfur compounds at oxidation levels lower than sulfate (e.g., hydrogen sulfide, elemental sulfur, thiosulfate) constitute convenient sources of reducing power for purple and green sulfur bacteria (Chromatiaceae and Chlorobiaceae), for example,

$$2\,H_2S + CO_2 \xrightarrow{h\nu} CH_2O + H_2O + 2\,S^0$$

$$2\,S^0 + 3\,CO_2 + 5\,H_2O \xrightarrow{h\nu} 3\,CH_2O + 2\,H_2SO_4$$

$$2\,Na_2S_2O_3 + 4\,CO_2 + 6\,H_2O \xrightarrow{h\nu} 4\,CH_2O + 2\,Na_2SO_4 + 2\,H_2SO_4.$$

Chemosynthetic sulfur bacteria (the "thiobacilli") utilize the energy released by reactions of sulfide with free oxygen, that is,

$$H_2S + CO_2 + O_2 + H_2O \longrightarrow CH_2O + H_2SO_4.$$

Hence, thiobacilli are aerobes (with the exception of the nitrate respirer *Thiobacillus denitrificans*), while photosynthetic sulfur bacteria are strict anaerobes. It is also known that some species of cyanobacteria can facultatively utilize H_2S rather than H_2O as an electron donor in photosynthesis (Cohen et al., 1975).

Isotope data for the oxidation equivalents resulting from these processes (S^0, SO_4^{2-}) are as yet scanty, but seem to indicate a difference between chemoautotrophs and photoautotrophs (Figure 7-5). Chemosynthetic oxidation of H_2S by *Thiobacillus concretivorus* yields fractionations between -10.5 and -18.0 ‰ for the resulting sulfate and between $+1.2$ and -2.5 ‰ for elemental sulfur, while the respective ranges for photosynthetic oxidation by *Chromatium* sp. are $+0.9$ to -2.9 and -3.6 to -10.0 ‰, respectively (Kaplan and Rittenberg, 1964). Hence, SO_4^{2-} production by bacterial photosynthesis does not seem to discriminate effectively between ^{34}S and ^{32}S. Polythionate ($S_nO_6^{2-}$) intermediates accrued by the two taxa in these experiments were found to be enriched consistently in the heavier isotope (with $\delta^{34}S = +0.6$ to $+19.0$ ‰).

Accumulation of heavy sulfur in the oxidized phases during photosynthetic oxidation of H_2S has also been indicated, albeit in a qualitative way, in culture experiments conducted by Mekhtieva and Kondratieva (1966) with the purple sulfur bacterium *Rhodopseudomonas* sp. (now *Ectothiorhodospira shaposhnikovii*) and by Kondratieva and Mekhtieva (1966) with *Chloropseudomonas ethylicum*. In the case of *E. shaposhnikovii* (see Figure 7-5), quantitative data have been added by Ivanov et al. (1976) who recorded a consistent enrichment in ^{34}S, notably in elemental sulfur ($+1.4$ to $+3.8$ ‰), but also in sulfate ($+0.4$ to $+2.3$ ‰), the $\delta^{34}S$ value of the H_2S medium decreasing from 0 to -1.6 ‰ during the experiments.

As pointed out by Chambers and Trudinger (1979), attempts to interpret these data in terms of the biochemistry of sulfur oxidation must await the charting of the enzymatic pathways in both photosynthetic and chemosynthetic sulfur bacteria which are largely unknown at present (for a tentative approach see Trüper, 1982). It might be noted that discrimination against the lighter isotope as in the case of oxidative sulfur metabolism occurs very rarely in biochemical reactions.

7-3-2. Biological Fractionation of Sulfur Isotopes in the Natural (Sedimentary) Environment

Of all biologically mediated sulfur isotope fractionations, those associated with dissimilatory sulfate reduction are quantitatively the most important. As stated above, bacterial sulfate reduction is the principal agent responsible for a large-scale release of hydrogen sulfide to terrestrial near-surface environments. With part of this H_2S escaping biological (and other) reoxidation and ending up as sedimentary sulfide (mostly as pyrite), this process ultimately accounts for the fact that the sulfur flux into the Earth's crust dichotomizes to form a sulfide and a sulfate branch (see Figure 7-7). Hence, from the geochemical standpoint, biological fractionation of sulfur isotopes is, in essence, a fractionation due to dissimilatory sulfate reducers (i.e., "desulfobacteria").

Isotope discriminations occurring during bacterial sulfate reduction in natural environments are decidedly more difficult to assess than those proceeding under the controlled conditions of culture experiments that have yielded a wide range of $\delta^{34}S$ values (viz., $+6$ to -46 ‰; Figure 7-5). Under natural conditions, additional parameters are superimposed on biological processes, the most salient of these being linked to the specific habitat of sulfate reducers. Being obligate anaerobes, the domains of desulfobacteria are preferably anoxic ("euxinic") sedimentary basins characterized by restricted circulation that thus often approximate closed systems.

The prototype of such "closed system" habitats are interstitial waters of newly formed marine sediments that constitute the principal sites of biological sulfate reduction. Here, reduction of sulfate is usually coupled with sulfate depletion. Assuming a constant kinetic fractionation with preferential removal of light sulfur, the $\delta^{34}S$ values of both residual sulfate and late-formed sulfide will increase with increasing extent of the reaction. It has been shown that these relationships can often be described, with fair approximation, in terms of a Rayleigh (fractional distillation) process, thus allowing calculation of an apparent fractionation factor $K = (^{32}S/^{34}S)_{sulfide}/(^{32}S/^{34}S)_{sulfate}$ on the assumption that "closed system" conditions are maintained for pore water solutions during diagenesis (Goldhaber and Kaplan, 1974; Sweeney and Kaplan, 1980b). Utilizing this and related approaches (see Chambers and Trudinger, 1979), fractionation factors generally between 1.015 and 1.045 have been obtained for sulfate reduction in recent sediments, corresponding to $\delta^{34}S$ values between $+5$ to -25 ‰ for bacteriogenic sulfide derived from present sea water sulfate with $\delta^{34}S \approx 20$ ‰ (Figure 7-6). Fractionation is often found to decrease from about 1.040 to 1.020 with increasing

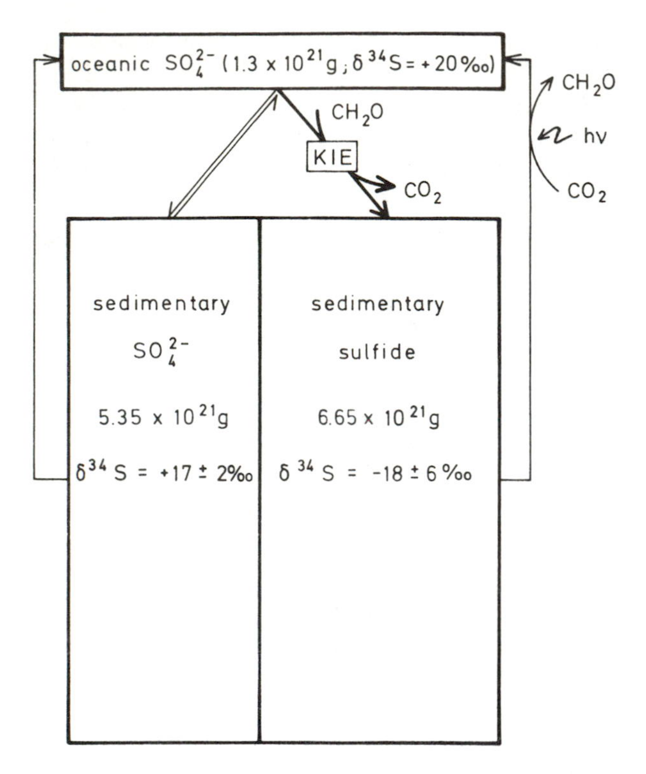

FIGURE 7-7. Sulfur reservoirs of the ocean-sediment system, with relative proportions drawn approximately to scale. The sum of the three boxes represents 1.33×10^{22} g sulfur of which about one-tenth (1.3×10^{21} g) is stored in the exogenic exchange reservoir as seawater sulfate (the amounts of other sulfur species present in the exogenic reservoir are smaller by at least four orders of magnitude). The sedimentary reservoir (sulfate + sulfide) represents 1.2×10^{22} g sulfur, equivalent to 0.5% (Holser and Kaplan, 1966) of the total sedimentary mass (2.4×10^{24} g; Garrels and Mackenzie, 1971). According to the isotope mass balance of Equation 7-3 in the text, the total sulfur stored in the system should be about equally partitioned between the oxidized (sulfate) and reduced (sulfide) species. With 20% of the sulfate residing in the ocean, the amount of SO_4^{2-} in the sedimentary shell is proportionately reduced. Isotope exchange between marine and sedimentary sulfate is governed by an equilibrium effect leading to a slight enrichment (+1‰) of ^{34}S in the sedimentary phase; the pathway from SO_4^{2-} to sulfide is dominated by the kinetic isotope effect (KIE) inherent in bacterial sulfate reduction. Recycling of sulfide is contingent on either direct (photosynthetic) or indirect (by biogenic oxygen) biological oxidation, thus being powered ultimately by solar energy ($h\nu$). Note that the difference in δ-values between the oceanic and sedimentary sulfate reservoirs does not reflect the equilibrium fractionation between $SO^{2-}_{4(solid)}$ and $SO^{2-}_{4(aq)}$; rather, it results from the fact that the sedimentary mean is the average of evaporites ranging from +11 to +32‰ (Claypool et al., 1980).

depth in the sediments, the lower discrimination occurring in deeper layers and probably reflecting the progressive depletion in both sulfate and nutrients in the interstitial substrate. An "open system" model, which makes allowance for continuous diffusion of SO_4^{2-} into pore waters (assisted, in part, by the activities of burrowing benthos), has been suggested by Goldhaber and Kaplan (1980) to yield fractionation factors as large as 1.060.

With bacteriogenic H_2S being largely preserved in the form of sulfide minerals, the $\delta^{34}S$ patterns of sedimentary sulfides (Figure 7-6) reflect the range of sulfur isotope fractionation in contemporaneous marine environments. It is obvious from the examples discussed above that biogenic sulfide patterns are characterized (*i*) by a general shift toward negative (light) $\delta^{34}S$ values relative to the parent sulfate, and (*ii*) by a large spread of these $\delta^{34}S$ values. For comparison, the values for a "fossil" euxinic facies plotted in Figure 7-6 (Permian Kupferschiefer) demonstrate the wide range of isotope distributions that might be expected from ancient sedimentary basins.

7-3-3. Isotope Fractionations in the Terrestrial Sulfur Cycle

The geochemical cycle of sulfur (Figure 7-7) resembles the carbon cycle in many respects. As in the case of primordial carbon, primordial sulfur appears to be partitioned between an isotopically heavy, *oxidized*, and an isotopically light, *reduced*, sulfur species. And like carbon, the sedimentary depositories of both sulfur species derive from an oxidized precursor (marine sulfate) in the exogenic exchange reservoir, the precipitation of which as evaporite sulfate entails an equilibrium fractionation of negligible magnitude, whereas the pathway from marine SO_4^{2-} to sedimentary sulfide is governed by the sizable kinetic effect inherent in bacterial sulfate reduction. With the $\delta^{34}S$ value of primordial sulfur close to zero per mil (Shima et al., 1963; Schneider, 1970; Nielsen, 1978), it is reasonable to assume that preferential accumulation of light sulfur in bacteriogenic sulfide is ultimately responsible for the isotopic disproportionation of terrestrial sulfur into a "light" and a "heavy" fraction (-18 ± 6 ‰ for sulfide versus $+17 \pm 2$ ‰ for sulfate; Figure 7-7). Hence, the crustal sulfur cycle appears to be largely governed by the biosphere, with few genera of microbial sulfate respirers controlling the geochemical distribution and the isotopic compositions of the principal forms of sulfur in the Earth's sedimentary shell.

The fractionation between sulfide and sulfate originates in the exogenic reservoir, notably in marine near-surface sediments, the main habitat of sulfate reducers. With bacteriogenic H_2S continuously reacting with the trace amounts of iron (and other metals) present in such sediments to yield sulfide minerals, the biological effect is preserved and subsequently propagated into the crust. The isotopic compositions of the resulting sulfide phases reflect, as a rule, the Rayleigh fractionation in closed (or semi-closed) pore water environments, giving rise to the large spreads in "light" $\delta^{34}S$ values depicted in Figure 7-6. With light sulfur preferentially concentrated in sulfide, the residual reservoir of marine sulfate will necessarily be enriched in the heavy species.

Due to the mobility of dissolved SO_4^{2-} ions and the rapid mixing of the marine reservoir, the $\delta^{34}S$ values of the residual sulfate span an extremely narrow range (Figure 7-6), thus contrasting markedly with the wide spread of such values in sedimentary sulfides. Hence, the oceanic sulfate mean can be expected to integrate over the exogenic reservoir as a whole, thus probably best reflecting the state of the system at any given time (as does bicarbonate in the case of carbon; Section 7-2-3). With the residence time of oceanic sulfate being on the order of 20 Ma (Holser and Kaplan, 1966), the stationary state characterized by a specific $\delta^{34}S$ mean should be established rapidly in terms of geological time. Since the isotopic composition of dissolved sulfate is preserved in sulfate evaporites with but minor change (viz., between 1–2 ‰ enrichment in ^{34}S in the solid phase; Thode and Monster, 1965), the sedimentary evaporite record will monitor the state of the sulfur cycle through geologic history. (This, of course, is also true for sedimentary sulfides, but with considerably less precision due to the enormous scatter in the observed values.)

The isotopic composition of the Earth's primordial sulfur (S_{prim}) originally fed into the exogenic system is believed to be best reflected by sulfides from ultramafic and mafic rocks of probable mantle affiliation that have consistently yielded $\delta^{34}S$ values close to the CDM standard (see, inter alia, Schneider, 1970; Schidlowski, 1973; and Nielsen, 1978). With the averages for the sedimentary sulfide and sulfate reservoirs being about equally different from this value in opposite directions (−18 and +17 ‰, respectively, corresponding to an overall fractionation $\delta^{34}S_{sulfide} - \delta^{34}S_{sulfate} = -35$ ‰), the isotope mass balance

$$\delta^{34}S_{prim} = R\delta^{34}S_{sulfide} + (1 - R)\delta^{34}S_{sulfate} \quad (7\text{-}3)$$

would yield $R = S_{sulfide}/(S_{sulfide} + S_{sulfate}) \approx 0.5$; that is, sulfur passed through the exogenic cycle should have been almost equally partitioned between the reduced and the oxidized form. This is indicated for the 1.33×10^{22} g sulfur of the ocean-sediment system represented in the box model of Figure 7-7.

7-3-4. Sulfur Isotope Fractionation in the Sedimentary Cycle through Time

$\delta^{34}S$ as an index of biogenicity

As is discussed above in the case of carbon (Section 7-2-6), we can now attempt to trace the characteristic biological fractionation between the two sedimentary sulfur species into the geologic past. Accepting that a difference in $\delta^{34}S$ of −30 to −50 ‰ between bacteriogenic sulfide and marine sulfate has been firmly established in present-day sedimentary environments (Figure 7-6), similar fractionations in ancient sedimentary rocks may be interpreted as evidence for a contemporaneous activity of microbial sulfate reducers. The presence or absence in sedimentary sulfur of such fractionations may thus constrain the time of emergence of the underlying biological process, namely, dissimilatory sulfate reduction or "sulfate respiration." Ideally, the isotopic evidence should be based on measurements of coeval (or, better yet, coexisting) sulfide and sulfate; this requirement, however, is difficult to meet for the major part of the record because of preferential recycling of soluble sulfate evaporites. Accordingly, sulfates tend to disappear from the rock record with increasing geologic age. Sulfur thus occurs mainly as sulfide (mostly pyrite) in Precambrian sedimentary rocks. Almost all Precambrian sulfide occurrences compiled in Figure 7-8 have not been subjected to metamorphism beyond the greenschist facies; noteworthy exceptions are those from Isua and the Adirondack deposits (Figure 7-8, Nos. 1 and 18). As has been demonstrated in several studies, however, bacteriogenic sulfur isotope patterns are unlikely to be completely obliterated during progressive metamorphism, surviving even in fairly high-grade terranes (Buddington et al., 1969; Brown, 1973; Ohmoto and Rye, 1979).

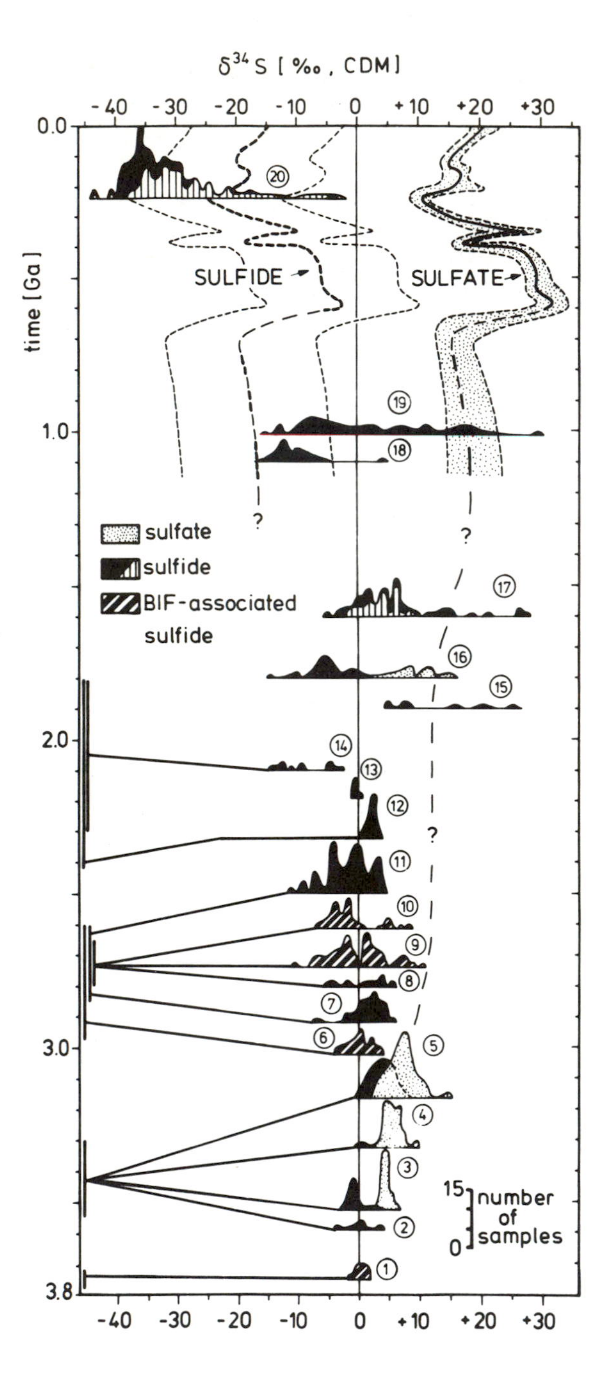

FIGURE 7-8. Isotopic composition of sedimentary sulfide and sulfate through time. The isotope age curve of Phanerozoic and Late Proterozoic sulfates (younger than about 1.2 Ga) is the better documented part of the record, the stippled area being the estimated uncertainty according to Claypool et al. (1980). Prior to 1.2 Ga ago, the record of sulfate evaporites is extremely sparse, restricted almost exclusively to sedimentary barite ($BaSO_4$) that is more readily preserved than calcium sulfates (gypsum, anhydrite). The values for Early Archean (>3.3 Ga old) barites tend to approach the zero permil line. The estimated isotope age curve for Phanerozoic and Late Proterozoic sulfides has been constructed assuming a difference of 35‰ between sulfide and sulfate, a value that approximates the average fractionation observed in modern sedimentary environments (see Figure 7-6). Considering the large spread in $\delta^{34}S$ values of sedimentary sulfides, the bulk of the actually measured values should fall into a range ± 12‰ of this average (as indicated by the dashed lines on either side of the estimated curve). The $\delta^{34}S$ pattern of the Permian Kupferschiefer (No. 20) illustrates a euxinic facies characterized by fractionations clearly in excess of the average value (note that $\delta^{34}S$ of Permian seawater sulfate is about $+11$‰). Although data become scanty with increasing geologic age, there is an apparent tendency for sulfide values to cluster around zero permil in rocks older than 2.8 Ga. Numbered occurrences are as follows. (1) Banded iron-formation from Isua, West Greenland (Monster et al., 1979). (2) Swartkoppie Formation, Onverwacht Group, South Africa (P.P.R.G. data). (3) Sulfide and bedded barite from the Warrawoona Group, Pilbara Block, Western Australia (Lambert et al., 1978). (4) Sulfide and bedded barite from the Fig Tree Group, South Africa (Perry et al., 1971; Vinogradov et al., 1976; Lambert et al., 1978). (5) Sulfide, barite, and anhydrite from the Iengra Series, Aldan Shield, Siberia (Vinogradov et al., 1976). (6) Banded iron-formation from Rhodesian schist belts, mostly the Sebakwian Group (Fripp et al., 1979). (7) Black shales from greenstone belts of the Yilgarn Block, Western Australia (Donnelly et al., 1977). (8) Fortescue Group, Hamersley Basin, Western Australia (P.P.R.G. data). (9) and (10) Michipicoten and Woman River Iron Formations, Superior Province, Canadian Shield (Goodwin et al., 1976. (11) Syngenetic "barren" sulfide deposits of the Birch-Uchi Greenstone Belt, Superior Province, Canadian Shield (Seccombe, 1977). (12) Bedded sulfides in the Cahill Formation of the Pine Creek Geosyncline, Northern Territory, Australia (Donnelly and Ferguson, 1980). (13) Frood Series, Sudbury District, Canadian Shield (Thode et al., 1962). (14) Black shales from Outokumpu, Finland (Mäkelä, 1974, and pers. comm. to M. S., 1978). (15) Onwatin Slate, Sudbury Basin, Canadian Shield (Thode et al., 1962). (16) Stratiform pyrite-barite deposits at Åsen, Skellefte District, Sweden (Rickard et al., 1979). (17) Shales, siltstones and carbonates from the McArthur Basin, Australia (black: pyrite; hatched: galena and sphalerite; Smith and Croxford, 1973, 1975). (18) Adirondack sedimentary sulfides, Grenville Province, Canadian Shield (Buddington et al., 1969). (19) Nonesuch Shale at White Pine Mine, Michigan, U.S.A. (Burnie et al., 1972). (20) Permian Kupferschiefer, Central Europe (black: pyrite; hatched: other sulfides; Marowsky, 1969).

THE SEDIMENTARY SULFUR ISOTOPE RECORD

Isotope trends in the sedimentary records of sulfide and sulfate have been investigated to varying extents, depending on the sulfur species and time interval in question. Accordingly, the confidence level of the different parts of the record depicted in Figure 7-8 varies considerably, generally rather markedly declining with increasing geologic age.

Sedimentary Sulfate. Starting with the pioneering work by Thode and Monster (1965), Nielsen (1965), and Holser and Kaplan (1966), an impressive body of sulfur isotope data has accumulated for sulfate evaporites of Phanerozoic age, the latest updated isotope age curve having been compiled by Claypool et al. (1980). These surveys have documented sizable variations in the $\delta^{34}S$ values of marine sulfates over the last 600 Ma (as indicated in Figure 7-8), with a maximum of +32 ‰ in the Cambrian (~550 Ma ago) and a minimum of +11 ‰ during the Permian (~230 Ma ago). The Phanerozoic record continues, albeit strongly reduced in quantity of available data, into the Late and Middle Proterozoic, where the oldest known anhydrite beds occur in the Upper Roan Group of Zambia and in the Grenville Series of North America (between 1.0 and 1.2 Ga in age). The virtual absence of gypsum and anhydrite from older strata is primarily due to the preferential recycling of evaporites (see Chapter 11, herein) whose mass half-age is about 3 times smaller than that of sandstones and shales (Garrels and Mackenzie, 1971, p. 272). The original presence of salt-bearing strata in older rocks (which would cogently follow from the constraints of the geochemical mass balance) is attested to by abundant indirect evidence, most notably by the occurrence of halite casts, of barite pseudomorphs of gypsum, and of chert and carbonate beds after primary gypsum and anhydrite (see Button, 1976; Crick and Muir, 1980; Lowe and Knauth, 1977; Lambert et al., 1978). As discussed in Chapter 11 of the present volume, the oldest sedimentary sulfates yet detected are barites from Early Archean terranes of South Africa (Barberton Greenstone Belt), Australia (Pilbara Block), and Siberia (Aldan Shield). In the case of the South African and Australian occurrences, convincing evidence has been presented that these barites were deposited as either chemical sediments (Heinrichs and Reimer, 1977) or formed as diagenetic replacement of an evaporitic calcium sulfate precursor by Ba^{2+}-bearing intrastratal solutions without a significant change in the bulk isotope geochemistry of the sulfates (Lowe and Knauth, 1977; Dunlop, 1978; Lambert et al., 1978). Altogether, the Precambrian sulfate record is deplorably scanty (Figure 7-8; see also Figure 11-2, herein).

With $\delta^{34}S$ of marine sulfate monitoring the state of sulfur in the exogenic compartment (Section 7-3-3), secular oscillations of the isotope age curve shown in Figure 7-8 necessarily reflect imbalances of the sulfur cycle, such imbalances constrasting sharply with the balanced condition of the corresponding "index" curve for the carbon cycle represented by marine carbonates (Figure 7-3). In principle, these imbalances may be due to two causes, namely, (*i*) secular changes in the isotopic composition of the sulfur input to the ocean by weathering solutions, and (*ii*) changes in the relative proportions of sulfide and sulfate in the sulfur transferred from the marine exchange reservoir to the sedimentary shell. The isotope curve should reflect the interplay of these two processes. However, major periodic dilutions of the oceanic SO_4^{2-} pool with either light (sulfide) or heavy (sulfate) sulfur seem to be rather unlikely events in view of the fact that weathering processes integrate over the whole surface to the globe (thus being apt to always mobilize an average sample of crustal sulfur per unit time) and of the long residence time of marine sulfate (~20 Ma). Hence, the $\delta^{34}S$ value of the sulfur input can be expected to be closely tethered to a mean value that is possibly skewed in a positive direction (compared to that of primordial sulfur) due to the relatively higher recycling rates of sedimentary sulfates compared with sulfides. Findings by Smith and Batts (1974) on the isotopic composition of low-sulfur coals lend support to such a concept of a relative constancy of the isotopic composition of freshwater sulfate through time.

Thus, a reasonable case can be made for the proposition that the observed oscillations over time of the isotope curve reflect, for the major part, changes in the relative intensities of the two sinks, competing for the removal of sulfur from the marine exchange reservoir. In the case of an ocean at steady state, the sulfur influx with weathering solutions (F_{si}) will be balanced by the rate of bacterial sulfate reduction (F_{ss}) and the rate of evaporite formation (F_{se}), that is $F_{si} = F_{ss} + F_{se}$. With one of these sinks (F_{ss}) entailing a strong isotope fractionation and the other not, the isotopic composition of seawater sulfate will be determined ultimately by the fraction $R = F_{ss}/F_{si}$ of the sulfur input ending up as sedimentary sulfide in accordance with relationship expressed by Equation 7-3. Therefore, the isotopic composition of sulfate evaporites can be expected to provide a measure of the intensity of bacterial sulfate reduction within the total rate of sulfate removal from the ocean (see Rees, 1970; Schidlowski et al., 1977), with increasing $\delta^{34}S$ values implying increasing sulfide/sulfate ratios in the sulfur flux from the marine reservoir into sediments. It is obvious from the later geological record that this relationship holds at least qualitatively, since times of extensive evaporite formation (e.g., the Permian) are generally characterized by low $\delta^{34}S$ values, while periods with an overrepresentation of pyritic shales (like the Early Paleozoic) coincide with maxima of the isotope curve. It seems most likely that the secular oscillations of the curve during the Phanerozoic had counterparts in the Precambrian that, however, are not revealed by the scanty available record.

With the isotope curve of marine sulfate evaporites thus probably indicating sizable fluctuations in the proportions of oxidized and reduced sulfur transferred to the sedimentary reservoir, the sulfur cycle must have had severe repercussions on the operation of the oxygen cycle. In terms of Equation 7-3, periods with low $\delta^{34}S$ values were times of excess evaporite formation and, accordingly, of excess oxygen storage in sedimentary sulfates, the fluctuations of sulfate-bound oxygen being amenable to calculation utilizing known geochemical parameters (Schidlowski et al., 1977; Schidlowski and Junge, 1981). As the oxygen cycle is, in turn, tied to the carbon cycle, changes in the sulfide/sulfate ratio can also be expected to be reflected in the carbon isotope record in that positive $\delta^{13}C_{carb}$ values (indicating increased burial of organic carbon with concomitant release of excess oxygen) should be coupled with negative trends in $\delta^{34}S$ of sulfates (signaling excess oxygen demands due to increased formation of evaporites). Such negative correlation between $\delta^{13}{}_{carb}$ and $\delta^{34}S$ has indeed been reported recently for the Phanerozoic and Late Proterozoic (Veizer et al., 1980).

Sedimentary Sulfide. As is obvious from the $\delta^{34}S$ patterns displayed by bacteriogenic sulfides in contemporary marine sediments (Figure 7-6), the large spread of sulfide values contrasts markedly with the narrow range of the parental sulfate pool. Nevertheless, the data allow us to define an approximate average fractionation between the two sulfur species that lies in the range of -35 to -40 ‰. Accepting the lower limit of -35 ‰, and assuming that marine sulfate faithfully monitors the state of sulfur in the exogenic system, we can construct an isotope curve for sulfide such as depicted in Figure 7-8. The majority of sulfide values measured for each time period can be expected to fall within a field of about ± 12 ‰ from this curve. Although systematic investigations of sulfide minerals from common sediments are rare, shale-hosted sulfides of Phanerozoic and Proterozoic age usually display such distribution patterns. Fractionations observed in typically euxinic sediments are often larger than the average represented by the curve (see, for example, the sulfide pattern of the Permian Kupferschiefer, shown in Figure 7-8, No. 20). Total conversion into sulfide of the limited sulfate reservoir of secluded or semi-secluded basins may sometimes give rise to a marked encroachment of the sulfide values on the positive field as in the case of the sediments of the McArthur Basin of Australia or parts of the Nonesuch Shale of northern Michigan, U.S.A. (Figure 7-8, Nos. 17 and 19).

With few exceptions, isotope data hitherto accrued for Precambrian sulfides are confined to banded iron-formation and to stratiform sulfide deposits and their associated sediments. It is obvious from the compilation presented in Figure 7-8 that isotope patterns suggestive of bacterial origin (characterized by wide spreads and a preponderance of negative values) can be traced to about 2.7 Ga ago. $\delta^{34}S$ values of all known older occurrences closely cluster around zero permil. This is particularly pronounced for a suite of 13 sulfide samples from banded iron-formation of the Isua Supracrustal Belt whose mean of $+0.5 \pm 0.9$ ‰ (Monster et al., 1979) represents the oldest data point in the sulfur isotope record.

THE SULFUR ISOTOPE RECORD AND THE ANTIQUITY OF DISSIMILATORY SULFATE REDUCTION

Based on the record at hand, an attempt to determine, or bracket, the time of emergence of dissimilatory sulfate reduction seems to be called for.

Since the advent of sulfate respiration constituted an important quantum step in the evolution of bioenergetic processes, the assignment of reasonable time limits to this event would provide a benchmark in the temporal framework of the early evolution of life. The crucial question in this context is, how far can isotope patterns attributable to the activity of bacterial sulfate reducers be traced into the geologic past.

There is ample evidence that sulfate reducing bacteria were extant during the whole time span covered by the quasi-continuous sulfate record that extends from the Recent to about 1.2 Ga ago. During this period, the characteristic bacteriogenic fractionation between sulfide and sulfate is always apparent, with sulfate staying well within the positive range (+10 to +30 ‰) and the corresponding sulfide patterns shifted, on average, by some 30 to 50 ‰ to the negative side of the scale. In deposits older than about 1.2 Ga the record is extremely patchy, but it is important to note that the isotope geochemistry of the 1.8-Ga-old pyrite-barite occurrence from Åsen (Figure 7-8, No. 16) basically resembles geologically younger patterns (here, one of the rare isotope surveys of coexisting Precambrian sulfide and sulfate has been carried out by Rickard et al., 1979). Between 1.8 and 3.8 Ga, the sulfate record is virtually nonexistent, but sedimentary sulfides often display what might be accepted as bacteriogenic patterns (e.g., Figure 7-8, Nos. 9, 10, 11, 14, 15). The $\delta^{34}S$ distributions displayed by the 2.7-Ga-old Michipicoten and Woman River Iron Formations (Nos. 9 and 10) have come to be widely accepted as the oldest presumptive evidence of bacterial sulfate reduction (Goodwin et al., 1976; Lambert, 1978; Schidlowski, 1979). This conclusion seems to be corroborated by the occurrence of a coeval and basically analogous pattern yielded by syngenetic sulfides from the Birch-Uchi Greenstone Belt (No. 11). The most convincing of these patterns is that of the Woman River sulfides where, in addition to a fairly large spread, the bulk of the values are distinctly shifted to the negative side.

All older sulfide occurrences thus far investigated lack bacteriogenic characteristics, displaying instead relatively narrow spreads around the CDM standard suggestive of a deep-seated (magmatic) sulfur source. Taken in conjunction with the minor fractionation of −3 to −5 ‰ observed between sulfide and sulfate (barite) in Early Archean sediments (Figure 7-8, Nos. 3, 4, 5), this can be interpreted as constraining the advent of sulfate respirers to the interval between 2.8 Ga ago and the time of formation of these ancient barite deposits (Schidlowski, 1979). While the oldest isotope patterns presumably attributable to microbial sulfate reducers thus date the beginnings of bacterial sulfate reduction at about 2.7 Ga ago, there is independent evidence (Heinrichs and Reimer, 1977; Lowe and Knauth, 1977; Dunlop, 1978) indicating that sulfate may have been abundantly present in the oceans already 3.5 Ga ago. This could suggest that the first appearance of presumably bacteriogenic sulfide patterns may give only a minimum age for the underlying biological event.

The substantially reduced fractionations between bedded barite and sulfide from Early Archean sediments seem to be typical of rocks of this age since isotopic differences of such magnitude have been reported from several ancient shields (e.g., those of South Africa, Australia, Siberia). Since we may assume that the first large-scale introduction of sulfate into the environment was due to the activity of photosynthetic sulfur bacteria (Broda, 1975, p. 77), it is a reasonable conjecture that the bulk of the SO_4^{2-} ions incorporated in the oldest sedimentary sulfates may have stemmed from this source. The enrichment of ^{34}S in the ancient barites by just a few permil as compared to sulfides from the same sequence would be consistent with photosynthetic oxidation of reduced sulfur compounds yielding similar fractionations (see Section 7-3-1). Inorganic oxidation of volcanic H_2S or SO_2 seems an unlikely mechanism for the formation of these sulfates for want of a suitable oxidant. It is also safe to say that the isotopic difference of 3 to 5 ‰ between oxidized and reduced sulfur in these deposits does not reflect an equilibrium fractionation that, in low-temperature environments (~25° C), should give rise to a 74 ‰ enrichment of ^{34}S in sulfate relative to sulfide (Tudge and Thode, 1950; Sakai, 1957; Ohmoto, 1972). If the observed fractionations were to reflect isotopic equilibrium, they would indicate temperatures >700° C that are precluded by geologic evidence for the environment of mineralization. Respective fractionations in hydrothermal systems (at temperatures <400° C) would yield minimum values around 15 ‰ that are considerably in excess of those found in the deposits.

7-4. Hydrogen

7-4-1. The D/H Ratio in Naturally Occurring Materials

The average ratio of hydrogen to deuterium in crustal material is 6410. With this large predominance of hydrogen (also referred to as protium), one might expect only slight isotope variations. However, because of the large mass difference between the two isotopes, isotopic effects are strongly magnified.

THE δD OF METEORIC WATERS

As water is the dominant reservoir for hydrogen on Earth, the isotopic variations of sedimentary, igneous, and metamorphic rocks are thought to reflect the isotopic ratios of the waters with which they were in contact during their formation. The distribution of the D/H ratio (expressed as δD) in various geologic samples is shown in Figure 7-9 plotted relative to SMOW (Standard Mean Ocean Water). It is quickly evident that sea water is on the "heavy" side of most terrestrial samples whereas snow can display high concentrations of hydrogen over deuterium.

Such differences in the D/H ratio are brought about by a step-wise distillation during which hydrogen moves more rapidly than deuterium, a process that has been described by a modified Rayleigh equation (Dansgaard, 1964). For this reason, precipitation (rain or snow) falling farthest from the source of origin will be isotopically enriched in hydrogen. Bodies of water that have undergone evaporation, however, will be enriched in deuterium relative to the original ratio of the water body. Following this principle, natural waters in high latitudes and altitudes are isotopically light, whereas those in the subtropics with low rainfall are frequently heavier than sea water. This distribution has given rise to a well-documented curve first published by Craig (1961) that describes a relationship between δD and δ^{18}O for various waters (Figure 7-10).

THE δD OF PLANT MATERIAL

Because of the relationship of δD and δ^{18}O to geographic location, and hence by inference to climatic conditions, measurements of these isotopes from ice in Antarctica, Greenland, the Arctic, and in mountain glaciers have been used to interpret historic fluctuations in terrestrial temperatures

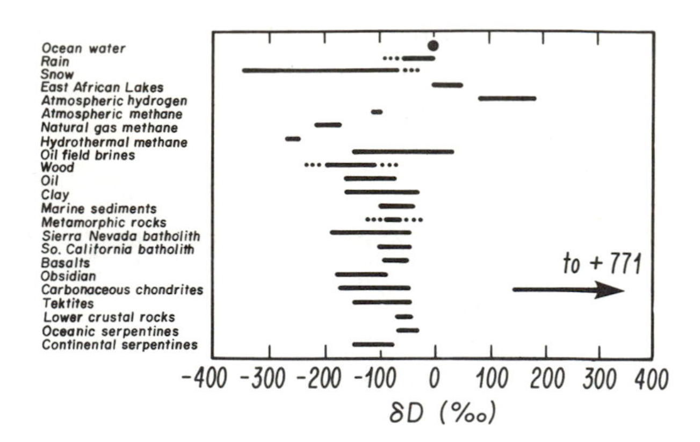

FIGURE 7-9. Range of relative deuterium concentrations in various terrestrial and meteoritic materials (relative to SMOW, Standard Mean Ocean Water). From Friedman and O'Neil (1972).

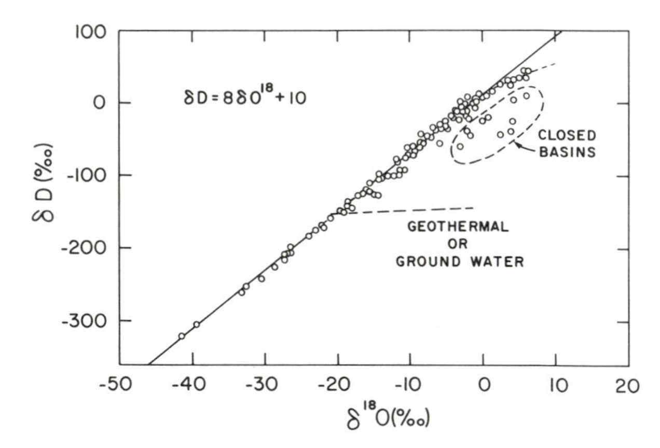

FIGURE 7-10. δD and δ^{18}O values for meteoric waters. From Craig (1961).

(Friedman et al., 1961; Friedman and Smith, 1970; Dansgaard et al., 1969; Epstein et al., 1970). Results of these studies have encouraged an investigation of various plants, particularly of trees that grow slowly and reach ages of several centuries. The approach in such studies is to compare the δD and δ^{18}O in organic material from growth rings that represent various ages (and stages) in the development of the plant. A depletion in deuterium and ^{18}O in growth rings from numerous specimens of the same plant species growing in the same location could be an indication of lower temperature, whereas an increase in concentration of these isotopes may be interpreted as reflecting a higher overall temperature and interaction with isotopically less fractionated rain or ground water. This method has also been applied to measurements of fossil (but undecomposed) wood of Pleistocene age

(Schiegl, 1972; Libby and Pandolfi, 1974; Epstein et al., 1976; Yapp and Epstein, 1977; Epstein et al., 1977).

Schiegl and Vogel (1970) first showed that the δD in wood of growing trees varies as a function of taxon and location. By comparing similar species from South Africa, The Netherlands, and southern Germany, they measured a δD range (vs. SMOW) from -34 to -112 ‰ and concluded that, on the average, whole wood is about 30 ‰ depleted in D relative to the average D/H ratio of the precipitation at the location of growth (Table 7-3). Smith and Epstein (1970), in a study of salt marsh aquatic plants from southern California, found an average difference of 55 ‰ between whole plants and the environmental water in which they grew. However, they demonstrated that the lipid fraction of the plants were depleted in D by about 90 ‰ relative to the total plant (Table 7-4).

TABLE 7-3. Comparison between the deuterium content of wood from different localities and the average deuterium content in summer precipitation (ΔD = difference in the deuterium content between precipitation and wood).*

	Deuterium content δD_{SMOW} (‰)		
Locality	*Wood*	*Precipitation*	ΔD (‰)
Pretoria (South Africa)	− 49	−10	39
Groningen (The Netherlands)	− 73	−40	33
Nussdorf (southern Germany)	−109	−80	29

* From Schiegl and Vogel (1970).

TABLE 7-4. Hydrogen and carbon isotope abundance ratios of different plant parts of *Limonium commune* (Plumbaginaceae, "Marsh-Rosemary").*

	δD_{SMOW} (‰)		$\delta^{13}C_{PDB}$ (‰)	
Plant parts	*Whole organ*	*Lipid*	*Whole organ*	*Lipid*
Green leaves	−67.0	−149.0	−23.2	−28.3
Dead leaves	−62.0	−181.0	−23.1	−27.9
Roots	−67.0	−152.0	−24.1	−28.5
Average	−65.0	−161.0	−23.5	−28.2

* From Smith and Epstein (1970).

Because of intracellular heterogeneity in the δD distribution within plant organic matter, Epstein et al. (1976) suggested that those studies that attempt to determine climatic changes based upon analyses of whole wood samples may be erroneous, due to the presence of an untested mixture of cellulose, starch, lignins, and lipids within the wood tissues, and that D/H ratios of only single, specified components should be compared. They chose cellulose as that component of wood that could be isolated in the purest form, by first replacing the exchangeable OH groups of the polymer with nitrate. In so doing, they demonstrated that the average difference between δD of plant cellulose and environmental water was ~ -22 ‰ (Figure 7-11) and that this difference ($\Delta\delta D = \delta D_{plant} - \delta D_{H_2O}$) was dependent upon the temperature of plant growth and could therefore be used as an environmental indicator (Epstein et al., 1976).

THE δD OF PHYTOPLANKTON AND CYANOBACTERIA

In contrast to vascular plants, algae and cyanobacteria generally appear to yield δD values that are lighter than the average for higher plants. In a detailed study of natural phytoplankton populations from various geographical locations, Estep and Hoering (1980) concluded that the average $\Delta\delta D$ ($=\delta D_{plankton} - \delta D_{H_2O}$) was approximately -100 ‰. In several strains of cyanobacteria and diatoms cultured in the laboratory, they obtained

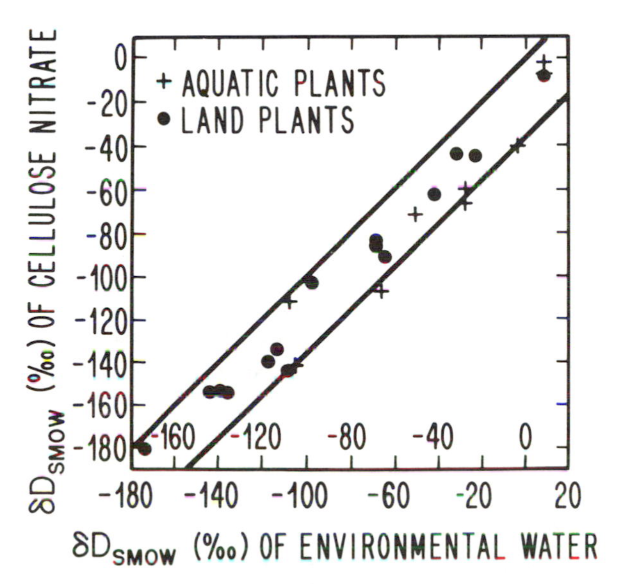

FIGURE 7-11. The relationship between the δD values of cellulose nitrate extracted from plant material and that of meteoric waters associated with the plants. From Epstein et al. (1976).

$\Delta\delta D$ values of -107 to -148 ‰. It is therefore evident that sedimentary organic matter derived from sapropelic algal debris could be isotopically lighter than material derived from higher plant debris. Nissenbaum (1974a,b) showed this to be the case. The average δD value for humic acids isolated from near-shore marine sediments from the northeast Pacific was shown to be -105 ± 5 ‰, whereas the δD of soil humic acids from a variety of locations covered a range from -57 to -97 ‰. As the near-shore samples probably also contained some undetermined proportion of land-derived humic material (Nissenbaum and Kaplan, 1972; Peters et al., 1978), the actual δD values of the marine end-member organic matter is probably lower than the measured average.

Hoering (1977) and Estep and Hoering (1980) confirmed the previous measurements of Smith and Epstein (1970) in which the δD values of the lipid fractions were found to be about 100 ‰ lighter than the δD value for the whole cell. However, the non-saponifiable fraction of such lipids (including the hydrocarbon fraction) was shown to be about 80 ‰ lighter than the saponifiable (fatty acid) fraction of the total lipid extract.

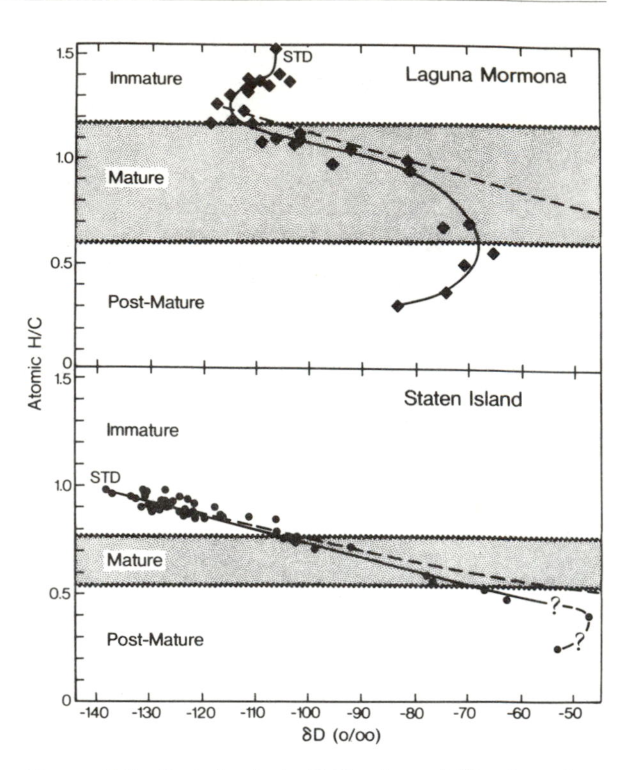

FIGURE 7-12. Variation in the H/C ratios and δD values of organic matter of varying maturity produced by laboratory heating of modern cyanobacterial mat sediments from Laguna Mormona, Mexico, and of modern peat from Staten Island near Sacramento, California, U.S.A. H/C and δD values of kerogen from starting material is indicated as "STD." From Peters et al. (1981).

7-4-2. The D/H Ratio of Sedimentary Organic Matter

Effects of diagenesis and maturation on the δD of organic matter

Schliegl and Vogel (1970) and Schliegl (1972) have suggested that during early diagenesis, compounds enriched in deuterium are preferentially lost leaving the residual peat with a δD depletion of about -34 ‰. This would suggest that during mild biological alteration, oxygen-rich carbohydrate material is first lost, probably as metabolic CO_2 and H_2O, leaving behind lipid-enriched residues.

During thermal maturation, this process is reversed and hydrogen is lost relative to deuterium. Peters et al. (1981) have shown that peat derived from vascular plants behaves similarly to sapropelic-rich organic matter derived from algal mats, in which enrichments from 60 to 100 ‰ in D can be obtained in laboratory heating experiments (Figure 7-12).

The δD of Phanerozoic organic matter

The distribution of δD in Phanerozoic organic material has been measured in kerogen, coal, oil, and natural gas. Schiegl and Vogel (1970) indicated from analysis of coal from one deposit in South Africa and one in northern Europe a δD range of -90 to -150 ‰. Redding (1978) reported results for a series of North American coals, of differing origins and maturation histories, having a range in δD from -65 to -154 ‰. Smith et al. (1981), for a much larger series of Australian coals, have shown a δD spread of from -70 to -160 ‰. There are regional differences that probably reflect the isotopic content of the water of growth, the plant type, and the history of maturation.

Values for four Middle East crude oils were reported by Schiegl and Vogel (1970) to give an average δD value of -89 ‰. More recently, D/H measurements on fifteen Paleozoic oils from Michigan, U.S.A., yielded a δD range of -90 to -130 ‰ (Brand et al., 1980). In this latter study it was determined that the paraffin fraction was approximately 10 ‰ lighter than the aromatic fraction.

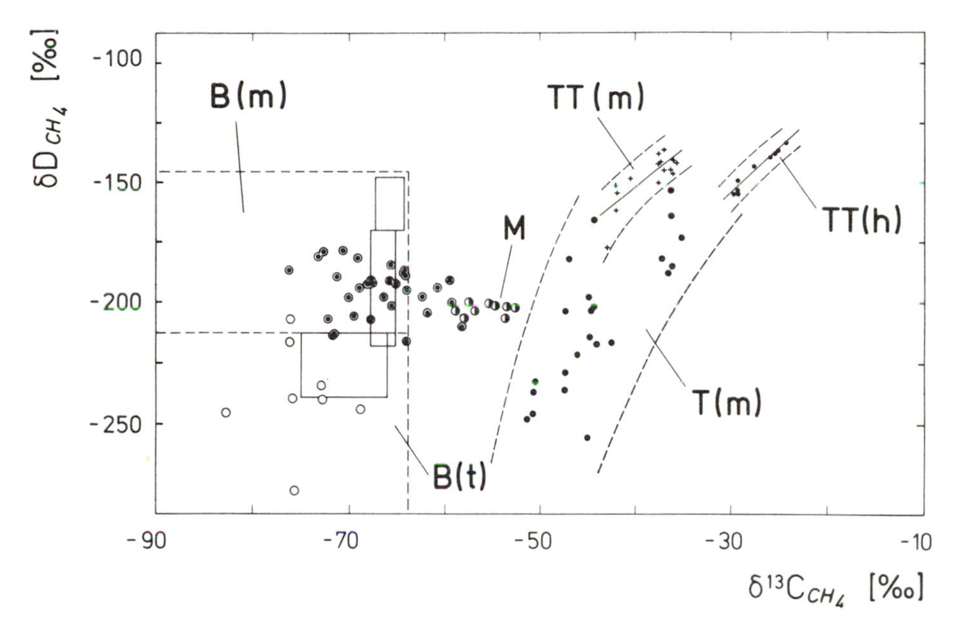

FIGURE 7-13. An isotopic comparison of methane in natural gases according to Schoell (1980). B(m) = Biogenic marine. B(t) = biogenic terrigenous. T(m) = Wet gas associated with crude oil. TT(m) = dry gas, thermogenic, marine source. TT(h) = dry gas, thermogenic. humic source. M = mixed origins.

The δD distribution for natural gases shows a range of values depending upon their source. Schoell (1980) has recently summarized the results of many hundreds of measurements and concludes that the origin of the gas can be deduced from a δ^{13}C vs. δD plot. This plot (reproduced in Figure 7-13) shows that biogenic methane can be differentiated from thermogenic methane or wet gas. It shows also that biogenic methane from marine sources seems to be isotopically enriched in deuterium by about 75 ‰ relative to that derived from terrigenous sources. Schoell interprets this as a response to the D/H ratio of the environmental water. Smith et al. (1981) have found that the δD and also δ^{13}C values in methane from Australian coals appear to be isotopically lighter than values measured by Schoell in European coal seams.

THE δD OF HYDROTHERMAL GASES

Values for δD in methane and hydrogen have also been measured in hydrothermal gases at Yellowstone National Park, U.S.A. (Gunter and Musgrave, 1971), at Broadlands, New Zealand (Lyon, 1974), and more recently from submarine vents at Cerro Prieto, Mexico (Welhan, 1981). The δD values for methane range from an average of -180 ‰ in Broadlands to -240 ‰ in Yellowstone and Cerro Prieto. Values for hydrogen gas are -475 ‰ in Broadlands and -661 ‰ in Yellowstone.

Values for δD and δ^{13}C in CH_4 were substantially heavier (-100 ‰ and -17 ‰, respectively) in samples from the East Pacific Rise (Welhan, 1981). It therefore appears that geothermal methane has a D/H ratio similar to methane derived from biogenic sources. However, values for δ^{13}C of geothermal gases (-27.5 ‰ for Broadlands and -23.5 ‰ for Yellowstone) tend to be somewhat higher than those most frequently encountered in biogenic or fossil fuel sources.

THE δD OF PRECAMBRIAN ORGANIC MATTER

Hoering (1977) compared the δD of extractable bitumen from four Precambrian shales (Table 7-5) with various fractions from Tertiary and Devonian crude oils. With the exception of extracts from the Nonesuch Shale, which were significantly heavier than the other samples, Hoering concluded that there does not appear to be any time-related trend in δD. He further suggested that the deuterium-enriched extracts from the Nonesuch Shale may be due to exchange of hydrogen with ground water or hydrothermal water.

TABLE 7-5. Hydrogen isotope content of petroleums and organic solvent rock extracts.*

Sample	δD total*	δD hexane eluate	δD benzene eluate	δD methanol eluate	Location	Age
798	−153.3	−165.0	−159.4	−159.0	Midway-Sunset Field, California	U. Pliocene
799	−150.5	−163.7	−151.5	−158.5	Midway-Sunset Field, California	U. Miocene
4159	−121.1	−128.0	−113.2	−121.2	Santa Fe Springs, California	U. Miocene
801	−142.5	−147.4	−132.7	−148.5	Midway-Sunset Field, California	L. Miocene
3944	−105.9	−109.5	−89.5	−112.6	Lab-E-Safid, Iran	Miocene
Green River Shale	−187.3	−234.6	−215.6	−150.0	Parachute Creek, DeBeque, Colorado	Eocene
4190	−109.4	−112.1	−112.5	−103.9	Deep River, Michigan	Devonian
Nonesuch Shale	−69.2	−68.8	−64.7	−65.0	White Pine Copper Mine, Michigan	1100 Ma
Muhos Shale	−117.2	n.d.**	n.d.	n.d.	Finland	1300 Ma
McMinn Shale	−111.5	−115.8	−116.9	−100.7	N. Territory, Australia	1400 Ma
Barney Creek Shale	−97.1	−102.4	−94.1	−94.1	McArthur River, N. Territory, Australia	1600 Ma

* From Hoering (1977).
** n.d. = not determined.

7-5. Nitrogen

Although nitrogen is a highly-depleted element on Earth, it plays a tremendously important role in the biological world. Its abundances, biogeochemical pathways and fluxes are discussed briefly in the following sections.

7-5-1. Pathways of the Biogeochemical Cycle of Nitrogen

Nitrogen is redistributed and recombined continuously by biochemical, in addition to physical and geochemical, mechanisms. The basic biochemical reactions are outlined below.

NITROGEN FIXATION

It is well known that the bacteria *Rhizobium* spp., occurring as symbionts in the nodules of the roots of plants in the legume family, can convert molecular nitrogen into organic nitrogen compounds. In addition to these major nitrogen-fixers, many other prokaryotes (e.g., some cyanobacteria) on land as well as in the ocean can fix atmospheric nitrogen, both autotrophically and heterotrophically, but their overall quantitative significance is not known. An oversimplified equation describing the reaction involved can be written as follows:

$$N_{2(g)} + 3\,H_2O_{(g)} \xrightarrow{\text{nitrogenase}} 2\,NH_{3(g)} + 3/2\,O_{2(g)}. \quad (7\text{-}4)$$

The high energy needed to break the triple bond of the diatomic nitrogen molecule makes nitrogen fixation a very inefficient process. However, because nitrogen gas represents the major form of nitrogen in the Earth's crust and near-surface environment, those organisms capable of accomplishing this process have a distinct ecological advantage over their competitors.

NITRIFICATION

The conversion of ammonia into nitrate is carried out by nitrifying organisms following a two-step process:

$$NH_3 + 3/2\,O_2 \longrightarrow HNO_2 + H_2O, \quad (7\text{-}5)$$

$$HNO_2 + 1/2\,O_2 \longrightarrow HNO_3. \quad (7\text{-}6)$$

Chemosynthetic bacteria of the genus *Nitrosomonas* can accomplish the first oxidation and *Nitrobacter* the second.

Apart from phosphorus, inorganic nitrogen compounds are the most limiting nutrients for primary production of organic substances. Under normal oceanic conditions, nitrate is the most stable, and therefore the most common form of combined nitrogen. Any cyanobacterium that assimilates nitrate can also use ammonium or nitrite. In soils, ammonium ion is the dominant form of inorganic combined nitrogen and normally is retained on soil colloid surfaces. After oxidation by nitrifying bacteria, inorganic nitrogen is mobilized as nitrate and can be utilized by vascular plants.

DENITRIFICATION

In the degradation of organic matter by aerobic heterotrophic bacteria, molecular oxygen is used as the electron acceptor. After the oxygen is almost exhausted, nitrate can serve as the electron acceptor. For example, the oxidation of glucose with nitrate can be expressed by the following reaction:

$$C_6H_{12}O_6 + 24/5\ NO_3^- + 24/5\ H^+ \longrightarrow 6\ CO_2 + 42/5\ H_2O + 12/5\ N_2. \quad (7\text{-}7)$$

Denitrification results in the conversion of nitrate to nitrogen gas, and is thought to be the only quantitatively significant mechanism occurring in nature that is capable of converting combined nitrogen into nitrogen gas. On the basis of known rates of biological nitrogen fixation, atmospheric nitrogen would be exhausted in less than 100 Ma if there were no denitrification.

Denitrification takes place in poorly aerated soil, such as in rice paddies, and in the stratified or stagnant water masses in the ocean. One well-known species of denitrifying organisms is *Pseudomonas denitrificans*. It has been demonstrated that nitrous oxide (N_2O) is also a product of denitrification by soil bacteria (Alexander, 1961). Its production by oceanic microorganisms is still not well understood, but it now appears that dissolved N_2O in seawater is formed during the oxidation of nitrogen compounds rather than during their reduction (Cohen and Gordon, 1979).

MINERALIZATION OF ORGANIC NITROGEN

Organic nitrogen is degraded principally by heterotrophic bacteria to simple compounds of nitrogen during mineralization, leading to ammonia as the final product. Most animals excrete the waste nitrogen as urea or uric acid; some marine organisms excrete ammonia.

7-5-2. Global Inventory of, and Fluxes between, Nitrogen Reservoirs

The Earth's ecosystem can be divided into four major subsystems: the atmosphere, the land, the ocean, and the lithosphere. The land subsystem includes the biological products formed on or near the surface of the continents plus soils and in groundwaters. The Earth's lithosphere includes the sediment column, the crust, and the upper mantle; among these, the sediment column is much more involved in biogeochemical processes than are the other two. Table 7-6 shows estimates of the reservoir quantities of nitrogen as compiled from a series of sources (Hutchinson, 1944; Emery et al., 1955; Delwiche, 1970; Wlotzka, 1972; Leventhal, 1976; McElroy et al., 1976; Soederlund and Svensson, 1975). As is evident, many of these estimates are inconsistent with each other. A set of values has therefore been chosen as representing the most reasonable estimates based upon the most recent information available (Table 7-6, No. 8).

More than 99% of the known nitrogen on or near the Earth's surface is present as N_2, either in the atmosphere or dissolved in sea water. Only a minor amount is combined with other elements, mainly with C, O, or H. Within the atmosphere, the troposphere contains most of the mass, whereas the stratosphere contains the ozone layer that absorbs the UV light present in the incident solar radiation. It is now widely recognized that oxides of nitrogen may play a role in controlling the quantitative aspects of the ozone layer. It has been estimated that between 9 and perhaps 20 megatons of nitrogen per year (Crutzen, 1976; Soederlund and Svensson, 1975) diffuse as N_2O across the tropopause into the stratosphere to catalyze decomposition of ozone. Within the land and the ocean subsystems, nitrogenous nutrients are recycled at rates of 2000 and 35000 megatons per year of nitrogen, respectively. In Figure 7-14, the fluxes of nitrogen exchanged between the various subsystems have been tentatively estimated, based upon many sources of data (Soederlund and Svensson, 1975; Garrels et al., 1975; Emery et al., 1955; Holland, 1970; Delwiche, 1970; McElroy, 1975), so as to fulfill the requirements of mass balance for a steady-state condition.

TABLE 7-6. The global inventory of nitrogen as estimated by various authors* (in gram units).

		1	2	3	4	5	6	7	8
Atmosphere	N_2	3.8×10^{21}	3.8×10^{21}	3.8×10^{21}	3.8×10^{21}	3.8×10^{21}	4×10^{21}	3.87×10^{21}	3.8×10^{21}
	N_2O	—	—	—	—	—	—	1.3×10^{15}	1.3×10^{15}
	Other Combined N	—	—	—	—	—	3×10^{12}	1.5×10^{12}	1.5×10^{12}
Land	Organic (including coal)	—	—	8×10^{17}	2×10^{17}	—	7×10^{16}	$\sim 4\text{–}9 \times 10^{17}$	8×10^{17}
	Inorganic	—	—	1.4×10^{17}	—	—	1×10^{17}	1×10^{17}	1.4×10^{17}
Ocean	N_2	—	2.2×10^{19}	2.2×10^{19}	2.2×10^{19}	—	—	2.2×10^{19}	2.2×10^{19}
	Organic	—	3.4×10^{17}	9×10^{17}	—	—	2×10^{17}	$3.4\text{–}9.0 \times 10^{17}$	3.4×10^{17}
	Inorganic	—	5.8×10^{17}	1×10^{17}	—	—	6×10^{17}	$5.7\text{–}7.0 \times 10^{17}$	5.8×10^{17}
Sedimentary rocks		4×10^{20}	7.7×10^{19}	4×10^{21}	7.5×10^{20}	10×10^{20}	—	—	7.5×10^{20}
Crust		1.2×10^{21}	—	1.4×10^{22}	6.5×10^{20}	6×10^{21}	—	—	6.5×10^{20}
Mantle		2.0×10^{23}	—	—	5.6×10^{22}	8×10^{22}	—	—	5.6×10^{22}

* From Sweeney et al. (1978).

Columns 1. From Hutchinson (1944). 2. From Emery et al. (1955). 3. From Delwiche (1970). 4. From Wlotzka (1972). 5. From Leventhal (1976). 6. From McElroy et al. (1976). 7. From Soederlund and Svensson (1975). 8. Adopted values for this paper.

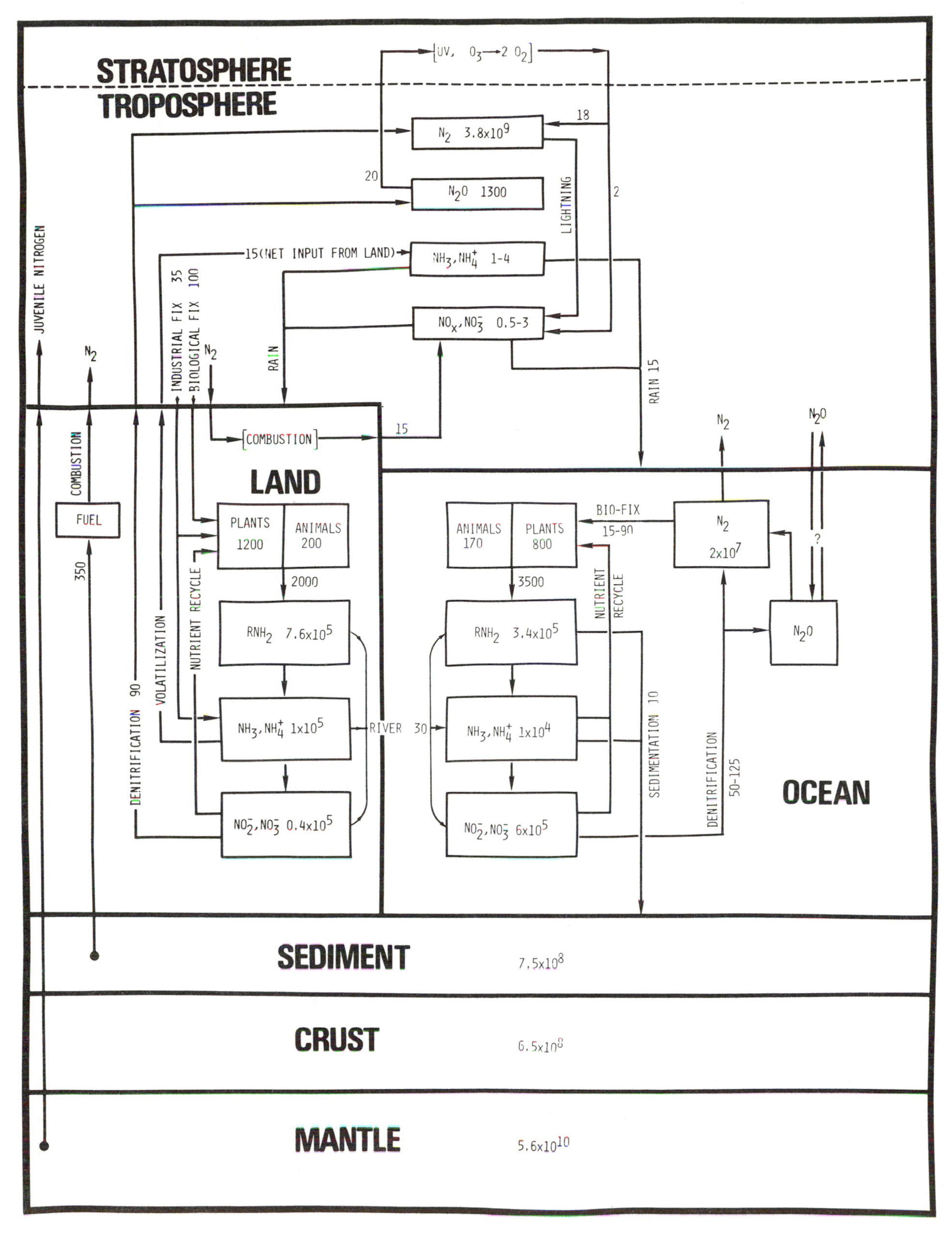

FIGURE 7-14. Inventory of the quantity of nitrogen (in units of 10^{12} g or megatons) in the major reservoirs and the flux of exchange (megaton/yr) between reservoirs. Part of the figure is revised after Delwiche (1970). The estimates of reservoir size are taken from Table 7-6; estimates of fluxes are taken from Soederlund and Svensson (1975) and Garrels et al. (1975). From Sweeney et al. (1978).

7-5-3. Biogeochemical Fractionation of Nitrogen Isotopes

The fractionation factors for the three main processes involved in the biogenic utilization of nitrogen are listed in Table 7-7. The isotope effect during nitrogen fixation ($\alpha = 1.000$ to 1.004) is small relative to the effects for bacterial nitrification and denitrification, (viz., $\alpha = 1.02$). The results of Miyake and Wada (1971) indicate that little overall isotope fractionation occurs during bacterial mineralization of organic nitrogen.

Two important equilibrium processes that commonly occur in nature are ammonia volatilization from ammonium ion, and the solution of nitrogen gas. Table 7-7 shows that the former process has a significant isotope effect with $\alpha = 1.034$, whereas the latter process has an $\alpha = 1.00085$.

Atmospheric nitrogen is not only the largest reservoir of nitrogen on Earth, but it is also isotopically the most homogeneous. The $^{15}N/^{14}N$ ratio in this reservoir, as determined by Junk and Svec (1958), is 1/272. Neither Dole et al. (1954) nor Junk and Svec (1958) could detect any significant isotope differences for atmospheric nitrogen collected at different altitudes or geographic locations. This has been recently confirmed by a series of world-wide measurements (Sweeney et al., 1978).

Table 7-7. Observed isotopic fractionation factors for the major nitrogen-involving processes occurring in nature.*

Isotope effects	*Observed fractionation factors*	*Reference*
KINETIC ISOTOPE EFFECTS		
Nitrogen Fixation:		
Atm. $N_2 \rightarrow$ fixed nitrogen	$\alpha = 1.000$	1
	$\alpha = 1.004$	2
Nitrification: $NH_4^+ \rightarrow NO_2^-$	$\alpha = 1.02$	3
Denitrification:		
$NO_3^- \rightarrow N_2$	$\alpha = 1.02$	2, 3, 4
EQUILIBRIUM ISOTOPE EFFECTS		
Ammonia Volatilization:		
$NH^+_{4(aq)} \rightleftarrows NH_{3(g)}$ 25° C	$\alpha = 1.034$	5
Solution of Nitrogen Gas		
$N_{2(soln.)} \rightleftarrows N_{2(g)}$ 0° C	$\alpha = 1.00085$	6

* From Sweeney et al. (1978).

References: 1. Hoering and Ford (1960); 2. Delwiche and Steyn (1970); 3. Miyake and Wada (1971); 4. Wellman et al. (1968); 5. Kirschenbaum et al. (1947); 6. Klots and Benson (1963).

Figure 7-15 shows the general distribution of $\delta^{15}N$ values for various natural substances (Miyake

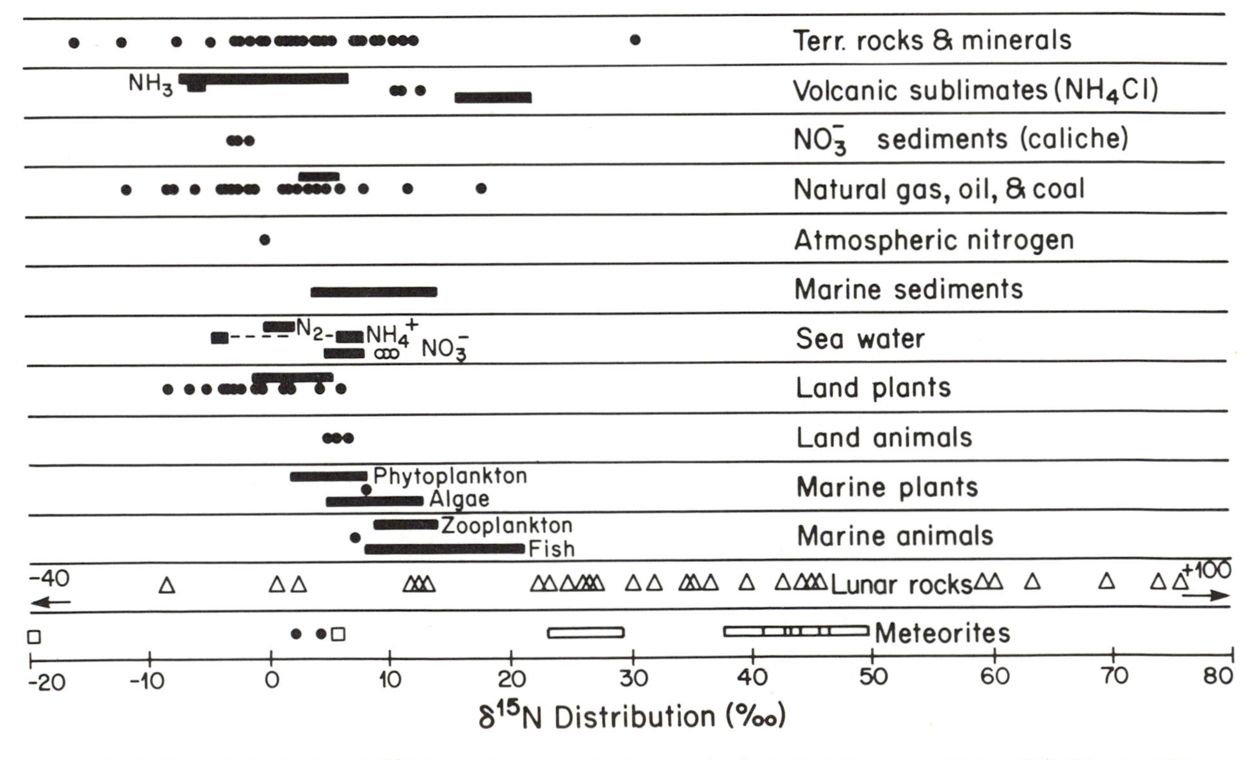

Figure 7-15. General distribution of $\delta^{15}N$ for various natural substances (revised after Miyake and Wada, 1967; Wlotzka, 1972; and Kaplan, 1975). From Sweeney et al. (1978).

and Wada, 1967; Wlotzka, 1972), relative to atmospheric nitrogen. The range of values measured extends from −40 to +100 ‰, both extremes occurring in the nitrogen of lunar rocks. For terrestrial substances alone, the range is from −17 to +30 ‰, both extremes occurring in inorganic rocks and minerals.

The major source of nutrient nitrogen for land plants is molecular nitrogen in the atmosphere; fixation of this nitrogen into a biologically utilizable form is performed by microorganisms in the soil. Considering the fractionation factor for nitrogen fixation (Table 7-7), newly formed nitrogen compounds in the soil should have a $\delta^{15}N$ between zero and −4 ‰. Table 7-8 lists reported values of $\delta^{15}N$ measured in natural soils; the mean $\delta^{15}N$ value of the 223 measurements there included is +5.0 ‰. Mixture of the nitrogen compounds formed in the soil and isotopically altered by fractionations during biogenic cycling with the nitrogen released by rock weathering results in the rather larger observed range of $\delta^{15}N$ values than expected.

Soils are dynamic systems in which the various forms of nitrogen are being continually turned over and complexed. At any given instant, organic compounds of all degrees of refractivity are present. Table 7-9 (after Cheng et al., 1964) shows that the isotope ratio of "non-hydrolyzable" nitrogen (−3 to +1 ‰) is similar to the air input, whereas the hydrolyzable forms of nitrogen may be quite different. The more refractory, non-hydrolyzable nitrogen would be expected to be preferentially retained in the soil profile during transport by rivers to lakes or the oceans. Table 7-10 lists the $\delta^{15}N$ values for measurements made on humic

TABLE 7-8. Summary of published $\delta^{15}N$ values for soil nitrogen.*

	Total nitrogen		
References	*No. of samples*	*Range $\delta^{15}N$ (‰)*	*Mean $\delta^{15}N$ (‰)*
1	28	−1 to +17	+6.3
2	40	−3 to +18	+6.2
3	39	+2 to +11	+5.4
4	69	−7 to + 6	+2.6
5	16	−4 to + 3	−0.2
6	10		+11.7
7	21		+6.2

* The mean of the 223 measurement listed is +5.0 ‰; from Sweeney et al., (1978).

References: 1. Chang et al. (1964); 2. Bremner et al. (1966); 3. Delwiche and Steyn (1970); 4. Riga et al. (1971); 5. Bremner and Tabatabai (1973); 6. Edwards (1973); 7. Wada et al. (1975).

TABLE 7-10. $\delta^{15}N$ values of "non-soil" terrestrial organic matter.*

Organic matter	
PEAT AND COAL	
7 measurements[a]	−2.8 to +1.9
21 measurements[b]	Mean value: +0.8 (standard deviation: 1.6)
HUMIC ACIDS	
Laguna-Mormona, Baja California (cyano-bacterial mat)	+1.7
Staten peat, Sacramento, California	+2.7
Hula Lake peat (Israel)	+1.2
Lake Biwa sediment (Japan)	+0.6

* From Sweeney et al. (1978).
[a] From Hoering (1955).
[b] From Wada et al. (1975).

TABLE 7-9. $\delta^{15}N$ values of various forms of nitrogen occurring in soil.*

	Soil type [values in $\delta^{15}N$ (‰)]				
Nitrogen form	*Grundy*	*Hayden*	*Austin*	*Clarion*	*Glencoe*
Total nitrogen	+16	+7	+5	+3	+2
Hydrolyzable	+18	+10	+7	+5	+4
Ammonium	+7	+7	+3	+6	+5
Hexosamine	+25	+8	0	+2	−2
Amino acid	+16	+14	+12	+5	+8
Hydroxyamine acid	+19	+11	+8	+7	+3
Non-hydrolyzable	−3	−2	−1	0	+1
N-mineralized	+6	+2	+1	+1	+1
Fixed Ammonium	+6	+6	+4	+2	0

* From Cheng et al. (1964) as presented by Sweeney et al. (1978).

acids isolated from several lacustrine or swamp environments. The $\delta^{15}N$ range of these highly complexed organic compounds is, in general, zero to +3 ‰. It is therefore evident that although soil nitrogen may have a large range in $\delta^{15}N$ values, the residual organic matter of vascular plant origin has an isotopic composition corresponding nearly to that of atmospheric nitrogen.

DENTRIFICATION AND NITROGEN BALANCE IN THE OCEAN

Denitrification is a crucial control-mechanism maintaining the isotope balance as well as the mass balance of nitrogen in the ocean. Estimates have been made both of the fractionation factor produced during denitrification in the ocean and of the mass of nitrate reduced (Cline and Kaplan, 1975).

Due to the transient nature of nitrate in the ocean and the large isotope fractionation for denitrification, significant $\delta^{15}N$ variation for dissolved nitrate exists. The $\delta^{15}N$ of nitrate from the eastern tropical North Pacific Ocean has been shown by Cline and Kaplan (1975) to range from +6.5 ‰ (for the Antarctic intermediate water mass) to +18.8 ‰ (in the active denitrification zone). Similarly, in the western Caribbean Sea, a range from +6.4 to +12 ‰ has been observed (Cline, unpublished data). In areas where denitrification is not important, the range in $\delta^{15}N$ for oceanic nitrate is much less. An average $\delta^{15}N$ of +6.6 ‰ for dissolved nitrate was measured by Miyake and Wada (1967) for the western North Pacific Ocean, and an average of +5.8 ‰ was measured by Cline and Kaplan (1975) for dissolved nitrate in the eastern North Pacific Ocean near Hawaii. Measurements of $\delta^{15}N$ for nitrate in the ocean off southern California range between +8 and +9 ‰ (Liu, 1979).

In the temperate regions of the world's oceans, where the abundance of nitrogen-fixing cyanobacteria is limited, nitrate assimilation is the dominant source of nutrient nitrogen utilized by organisms. Off the coast of Japan, the $\delta^{15}N$ of dissolved nitrate, phytoplankton, and sea weeds were all, on the average, +7 ‰ (Miyake and Wada, 1967). Plankton samples collected off the west coast of California (Table 7-11) have a mean $\delta^{15}N$ value of +9 ‰, similar to the value for dissolved nitrate (+8 to +9 ‰).

At present, measurements of the $\delta^{15}N$ values for organisms living in tropical oceans are limited.

TABLE 7-11. $\delta^{15}N$ of plankton collected off the coast of California, U.S.A.*

Location	$\delta^{15}N$ (‰)
120°W, 35°N	+12.1
119°W, 34°N	+11.1
119°W, 34°N	+9.0
120°W, 33°N	+8.5
120°W, 33°N	+8.2
119°W, 34°N	+6.9
118°W, 30°N	+6.5

* From Sweeney and Kaplan. (1980).

Nitrogen fixation by cyanobacteria in such oceans appears to be an important process in decreasing the mean $\delta^{15}N$ of organic matter synthesized in these regions (Wada and Hattori, 1975). The organic nitrogen so produced is isotopically similar to dissolved N_2.

Based on the assumption that organic nitrogen produced in plankton has a $\delta^{15}N \approx 10$ ‰ and terrigenous organic nitrogen is ≈ 0 ‰, Sweeney and Kaplan (1980a) used a simple mixing model to predict the relative proportion of marine and non-marine derived nitrogenous organic matter in a reducing coastal basin off California. They were able to show using 5 m-long marine cores that there were fluctuations in the sedimentation rates and, hence, the relative proportions of marine and non-marine detritus reaching the basin floor.

Applying a similar model, Peters et al. (1978) used $\delta^{13}C$ and $\delta^{15}N$ to determine relative sources of organic matter in coastal waters (Figure 7-16). The marine end member has a $\delta^{13}C = -22$ ‰ whereas the terrigenous end member is −27 ‰. This graph may prove useful in evaluating relative proportions of mixed sources.

$^{15}N/^{14}N$ IN KEROGEN AND CRUDE OIL

Few data for $\delta^{15}N$ in kerogen from shales have as yet been published. Those analyses performed by Scalan (1959) on whole rocks appear to indicate that the mode for $\delta^{15}N$ in a variety of different rocks lies between +2 to +6 ‰.

In crude oil, $\delta^{15}N$ covers the range between zero and +10 ‰ with few measurements falling outside this range (Hoering and Moore, 1958; Grizzle et al., 1979). In the Phanerozoic, most marine kerogens and crude oils appear to have mixed terrigenous and non-terrigenous sources.

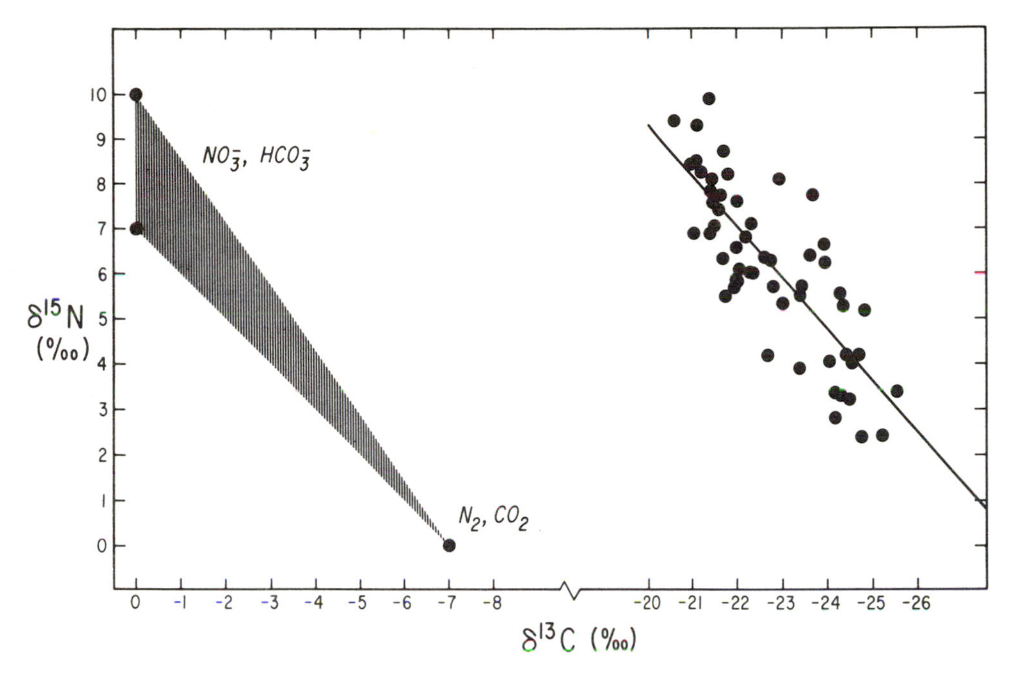

FIGURE 7-16. Carbon versus nitrogen stable isotope ratios in sedimentary organic matter from continental shelf sediments of the northeastern Pacific Ocean. Average slope of a hypothetical mixing line between inorganic end members in this area is about 1.2. Slope of correlation line for organic matter is about 1.1. From Peters et al. (1978).

In the Precambrian, organic nitrogen should be presumed to be of algal and microbial origin. Where cyanobacteria predominated, the dominant process of cellular nitrogen incorporation could conceivably have been N_2 fixation. If so, $\delta^{15}N$ in the ensuing kerogen should be ≈ 0 ‰. Where such values become isotopically heavier and approach +10 ‰, it would indicate that a fractionation process had occurred. This could be indicative of nitrogen cycling with the formation and selective reduction of nitrate. There is presently insufficient information to determine whether nitrate reduction could have occurred on the abiotic Earth. If biological nitrate reducers are necessary to recycle oxidized nitrogen, then the occurrence of ^{15}N-enriched organic matter in Precambrian kerogen could be regarded as indicative of the existence of autotrophic nitrogen bacteria. Data obtained during the course of the P.P.R.G. project on nitrogen in Precambrian kerogens are summarized in Table 5-8 of this volume. The spread in $\delta^{15}N$ values there listed is similar to that observed in Phanerozoic sediments (having an average of +4.2 ‰, with all of the 23 values reported falling in the range between +0.8 and +9.9 ‰) and is thus suggestive of the existence of an active biological nitrogen cycle apparently as early as ~2.5 Ga ago (evidenced presumably by kerogen from P.P.R.G. Sample No. 048-2 from the Hamersley Group of Western Australia with $\delta^{15}N$ = +5.7 ‰).

7-6. Summary and Conclusions

(1) Incorporation into living systems and/or biochemical processing of the principal elements on which life is based (with the exception of phosphorus) entail sizable isotope fractionations. Evidence of such fractionations is retained when organic substances enter the rock cycle as constituents of fossil organic matter ("kerogen"). Decoding of the isotopic information (blurred, in part, by subsequent diagenetic or metamorphic alterations) may furnish, therefore, temporal constraints for the emergence of some basic biochemical processes, most notably (*i*) biological carbon fixation and (*ii*) dissimilatory sulfate reduction.

(2) With the majority of $\delta^{13}C_{PDB}$ values of sedimentary organic matter falling into the range

$-27 \pm 7\%$ (Figure 7-3), the isotope record of organic carbon has basically preserved the composition of the biogenic prescursor materials. Hence, the record as a whole can best be interpreted as giving a continuous isotopic signal of biological (viz., autotrophic) activity on Earth beginning at least 3.5 (if not 3.8) Ga ago.

(3) The total spread of Precambrian $\delta^{13}C_{org}$ values seems to be inflated as compared with that of the younger (Phanerozoic) record, primarily as a result of extremely negative spikes in the range -35 to -50 ‰ which are imposed on the mainstream of the record around 2.7 Ga ago. These spikes are suggestive of the extensive involvement of methane-utilizing biochemical pathways in the formation of the respective kerogen presursors.

(4) Positive excursions in $\delta^{13}C_{org}$ are primarily a characteristic of high-rank (metamorphosed) kerogens. The principal mechanism producing such isotopically heavy kerogens is isotope exchange with carbonate minerals which commences in the uppermost greenschist facies and becomes especially pronounced in amphibolite grade rocks. Accordingly, isotope shifts towards positive values displayed by metamorphosed kerogens cannot be used to infer an abiogenic pedigree for these organics and their graphitic derivatives.

(5) The isotope age curve for sedimentary carbonates varies within relatively narrow limits ($\delta^{13}C_{PDB} = 0.4 \pm 2.6$ ‰), suggesting that the isotopic composition of marine bicarbonate has not changed very much through time. Assuming an average fractionation of -25 to -27 ‰ between organic and inorganic carbon, the near constancy of sedimentary $\delta^{13}C_{carb}$ values at about zero permil would be a necessary corollary to a biological (autotrophic) interpretation of the isotope record for reduced carbon.

(6) An extrapolation of the quasi-continuous record of these carbon isotope relations beyond the termination of the known unmetamorphosed rock record (around 3.5 Ga ago) is permissible if it is assumed that the isotope shifts observed for both of the carbon species in the 3.8 Ga old Isua sediments ($\delta^{13}C_{org} = -13.0 \pm 4.9$ ‰; $\delta^{13}C_{carb} = -2.3 \pm 2.2$ ‰) are due to the amphibolite-grade metamorphism of these rocks. Such interpretation is consistent with the thermodynamics of isotope exchange between reduced and oxidized carbon, and with data obtained from studies of geologically younger metamorphic terranes.

(7) At present, the isotope geochemistry of the terrestrial sulfur cycle is primarily controlled by the process of bacterial sulfate reduction that results in a partitioning of crustal sulfur between "light" sulfide and "heavy" sulfate. Isotope patterns of sedimentary sulfides presumably attributable to microbial sulfate reduction can be traced back to about 2.7 Ga ago. This, however, could be a minimum age since other evidence suggests that sulfate was present in the oceans at least as early as 3.5 Ga ago.

(8) The considerable enrichment of hydrogen over deuterium in plant matter as compared to the D/H ratios of environmental waters is largely retained in sedimentary organics. Based on the limited data now available, there does not appear to be any time-related trend in the δD values of fossil organic matter.

(9) From the small number of $\delta^{15}N$ values thus far accrued for kerogenous materials and oils we may infer the presence of biologically mediated nitrogen isotope fractionation in the geologic past. It is, however, difficult to unequivocally assign the fractionations observed to one or more of the specific processes known to occur during biological nitrogen utilization. The similarity of the spread of Precambrian $\delta^{15}N$ values to that obtained for Phanerozoic sediments is, however, suggestive of the very early establishment of a biological nitrogen cycle.

(10) In sum, studies in recent years of the geochemistry of the stable isotopes of C, S, H, and N have provided a fruitful basis for inference and speculation regarding the antiquity of a number of biochemically important processes. As is true of almost all other aspects of paleobiology as well, however, such studies of the Precambrian record have lagged well behind those of the Phanerozoic. In particular, very few data are as yet available regarding the $\delta^{15}N$ and δD values of Precambrian organic matter, and only a small number of isotopic measurements have been performed on Precambrian sulfur. Even in the case of carbon—by far the most studied of such systems—the available Precambrian data base is by no means quantitatively comparable to that available for younger sediments. Clearly, much additional data will be required before such studies can provide the considerable degree of insight regarding ancient biochemistries of which they seem potentially capable.

CHAPTER 8

Archean Stromatolites: Evidence of the Earth's Earliest Benthos

By Malcolm R. Walter

8-1. Introduction

In terms of the major interests of the P.P.R.G., and the principal thrust of the present volume, stromatolites are of importance primarily as a source of biological information. Indeed, when the P.P.R.G. project began, stromatolites were regarded as "the only undisputed fossils in Archean rocks" (Walter, 1977). That has changed (see, for example, Chapter 9), but they still constitute a prime source of data for Archean paleobiology even though such data are limited: as yet only eleven such occurrences have been documented (counted by major stratigraphic units); of these, three cannot yet be interpreted unequivocally as stromatolites (Table 8-1).* Of the eleven occurrences, seven were discovered or first described within the last decade.

The first Archean stromatolites to be described (although their Archean age was not then known) were those of the Steeprock Group in Canada; initially they were interpreted as stromatoporoids (Rothpletz, 1916). Even earlier, Walcott (1912) had drawn attention to a stromatolite-like structure from the same rock unit, but his specimen is now considered to be non-stromatolitic (Hofmann, 1971). As is discussed below, this problem of recognizing stromatolites, of differentiating them from "stromatolite-like" but non-fossil megascopic structures, still besets workers in this field. Subsequently, a paper that generated considerable interest in Archean stromatolites was that by Macgregor (1941), in which he described Late Archean stromatolites from strata of what was then known as Southern Rhodesia. The field, however, was slow to develop and few additional occurrences (Joliffe, 1955; Winter, 1963) were described until the widespread application of isotopic dating techniques, during the middle to late 1950s, generated new interest in Archean Earth history. Many new isotopic dates are now available, to the extent that at least a temporal scaffolding (if not a complete framework) of the occurrence of Archean stromatolites can be constructed; the age of most of the eleven occurrences are known probably to

TABLE 8-1. Ages and paleoenvironmental settings of known Archean stromatolite occurrences. See text and Appendix I for details.

ARCHEAN STROMATOLITE RECORD

AGE (Ga)	FRESH-WATER LAKES(?), VOLCANO-SEDIMENTARY ENVIRONMENT "BASIN & RANGE"	MARGINAL MARINE	
		EPEIRIC SEAS	VOLCANO-SEDIMENTARY ENVIRONMENT GREENSTONE SEQUENCES
2.5		TUREE CREEK HAMERSLEY 2.490±0.030	
2.6	VENTERSDORP 2.64±0.08		STEEPROCK, BELINGWE, BULAWAYAN } 2.64 ±0.14
2.7	FORTESCUE 2.768±0.014		YELLOWKNIFE ~2.650
2.8			WOMAN LAKE 2.794±0.006
2.9			
3.0		INSUZI 3.09±0.09	
3.1			
3.2			
3.3			
3.4			
3.5			ONVERWACHT 3.540 ± 0.030
3.6			WARRAWOONA 3.556 ± 0.032

* While this paper was in press, a fourth example of possible, but as yet unproven Archean stromatolites was reported by Orpen and Wilson (1981), and *Conophyton*-like stromatolites were described by Grey (1981) from the Late Archean of Western Australia.

within 100 million years or so, a degree of resolution that may seem appallingly imprecise to anyone not familiar with such studies, but that nonetheless represents a great advance over age estimates of only a decade ago.

Since Schopf et al. (1971) published the first modern paleobiological analysis of an Archean stromatolite, much has been learned about extant stromatolites and the microorganisms that construct them. These new data are gradually allowing the formulation of a theoretical "model" or framework for the interpretation of stromatolites, part of which can be summarized as follows: The construction of a stromatolite with a uniform fabric requires a consistent and repeatable set of behavioral responses from a particular, relatively invariant, microbial community. That set of responses can be referred to as a *syndrome*. Examples of such responses include filament orientation, gliding, positive phototaxis, mucus production, and sheath pigmentation. All these factors have recognizable results in the stromatolite fabric, even though no cells might be preserved. Each single response is a product of one or more biochemical systems and the total behavioral syndrome may be a consequence of several interrelated systems. Thus, a stromatolite fabric can be regarded as recording the former presence of a set of interrelated biochemical systems. Because of the biochemical complexities that must underlie each syndrome, it seems reasonable to deduce that the recognition of homeomorphy between an ancient and a modern fabric indicates the occurrence of analogous syndromes. The microbial community that constructed the ancient stromatolite fabric was probably in some fundamental respect physiologically and biochemically comparable with that of the modern example. The more complex the fabric, the more confidently can be established this conclusion. Even when the construction of a modern microbial fabric is not understood in detail (which is true of most modern stromatolites), this approach makes it possible to draw reasonable, if tentative, inferences about the nature of the microbial communities that built ancient stromatolites.

It is also possible to make reasonable inferences about the shape and size of the constructing organisms, based only on geometric features of a stromatolite fabric, although this is a poor substitute for direct observation of the microfossils themselves. Many stromatolite microbiotas are now known from the Proterozoic (see, for example, Awramik et al., 1976; Schopf, 1977; and Chapter 14 of this volume), but an overwhelming majority of stromatolites lack microfossils. However, many stromatolites contain faint remnants of microbial cells and filaments, in the form of streaks and clots of kerogen, lines and clusters of pyrite grains (e.g., Gutstadt and Schopf, 1969), and diffuse structures pigmented by submicroscopic grains of iron oxide. These remnants are very commonly observed by those who study stromatolites and microfossils, and all gradations are known from well-preserved cells to barely recognizable cellular remnants (e.g., Hofmann, 1975). Even though these remnants are faint and would not normally be regarded as microfossils, they can be a significant source of biological information when they are considered along with other fabric characteristics. It is useful to draw attention to fabrics characterized by such remnants by introducing the term *palimpsest microstructure* that can be defined as a microstructure in which the distribution of kerogen, iron oxide, pyrite or some other pigmenting material indicates the former distribution of microbial cells. These microstructures rank with those termed *skeletal* (Riding, 1977) as direct sources of biological information. They differ from skeletal microstructures in that they are not distinguishable from their surrounding matrix by distinct grain-size differences (of the type that probably resulted from some form of biogenic mineralization). Thus, it would be appropriate, for instance, to refer to a "palimpsest pustular microstructure," or a "palimpsest palisade microstructure". As yet, no comprehensive terminology has been proposed for the biogenic microstructures and fabrics of stromatolites, and so it has proven convenient here to use "palimpsest" as defined above and to modify the term, for instance, to "sinuous linear palimpsest microstructure".

True microfossils are known from stratiform stromatolites of the Early Archean Warrawoona Group (and possibly also from the Onverwacht Group of the same age) and from variously structured stromatolites of the Late Archean Fortescue Group (see 8-2, and Chapter 9). The outstanding Early Proterozoic example is the stromatolite microbiota of the Gunflint Iron Formation; other examples of this age occur in the Frere Formation, various units of the McArthur and Mount Isa Groups, the Bungle Bungle Dolomite, Belcher Supergroup, and other units discussed in Chapter 14. As a result of selective, differential preservation, a suite of cellularly preserved microfossils may be a very distorted sample of the original stromatolite microbiota. Palimpsest microstructures may give a more accurate indication of the species composition of the constructing community.

The relative rarity of stromatolitic carbonates and cherts in the Archean contrasts sharply with their abundance in the Proterozoic. Armstrong's suggestion (1960) that many Archean stromatolites may have been obliterated by metamorphism is likely to be correct, but the contrast remains evident in the relative abundances of carbonate rocks. The great pulse of crustal growth that apparently occurred during the Late Archean (Chapters 3 and 10) generated the platform environments that characterize most Proterozoic carbonate sequences, and provides a ready explanation for the increased abundance of stromatolites, as such structures are found mostly in platform sequences. Such a profound crustal event, with its attendant geographic and ecological consequences, undoubtedly stimulated evolutionary changes in benthonic microbiotas. It is likely, therefore, that superimposed on the effects of an increased distribution of carbonate-depositing environments are the results of a microbial invasion, and perhaps more effective biologic utilization, of such environments (Gebelein, 1976).

Most of the descriptions of stromatolites included in this chapter (Section 8-4) are based on field work undertaken, and specimens collected, especially for this project by H. J. Hofmann, J. M. Hayes, and myself; some material, however, was collected earlier by Hofmann and myself individually, and in a few examples descriptions are based on specimens supplied by colleagues from localities we have not visited. In this way, material has been studied from almost every Archean stromatolitic unit known by early 1979 to members of the P.P.R.G. (and the inventory of Archean stromatolites has been extended by establishing the origin of structures from the Warrawoona Group, in a collaborative effort with R. Buick and J. S. R. Dunlop of the University of Western Australia). The only Archean stromatolitic units not here studied are several units in Zimbabwe which we chose not to visit and from which we were unsuccessful in attempts to obtain specimens. Many Early Proterozoic stromatolites were included in the geochemical studies carried out during the course of the project (Chapters 5 and 7) but are not described here. Brief reference is made to Early Proterozoic stromatolites later in this chapter (Section 8-5-4).

The approach used in evaluating the biogenicity of ancient stromatolites is outlined below. By definition (Walter, 1976a), stromatolites are biogenic: they are "organosedimentary structures produced by sediment trapping, binding, and/or precipitation as a result of the growth and metabolic activity of microorganisms, principally [cyanobacteria]." The term "cyanobacteria" has now largely replaced "blue-green algae" and "cyanophytes." The noun "stromatolite" can be modified by adjectives such as "prokaryotic," "algal" (with the understanding that all algae are eukaryotes), and "fungal"; or, more specifically, "chlorobacterial," "cyanobacterial," and so forth.

8-2. Distinguishing Stromatolites from Homeomorphic Abiogenic Structures

Most non-stromatolitic "stromatolite-like" structures fall into two genetic groups: (*i*) deposits resulting from subaerial evaporation, particularly of splashing water, and (*ii*) deposits formed by the evaporation of water in the vadose zone of the regolith (i.e., in the soil). The former have been termed stiriolites (Walter, 1976b); the latter are known as duricrusts (Goudie, 1973). Holocene examples of stiriolites include the coniatolites of the intertidal zone of the Persian Gulf (Purser and Loreau, 1973), cave deposits or speleothems (Thrailkill, 1976), and geyserites (Walter, 1976b). Duricrusts include calcrete (Esteban, 1976; Hay and Reeder, 1978; Read, 1976), phoscrete, silcrete, laterite, bauxite and manganese oxide deposits (Goudie, 1973; Maignien, 1966). Some stiriolites and particularly duricrusts might be partly biogenic (e.g., Klappa, 1979). In addition to these partly or wholly abiogenic structures, there are more or less flat-laminated sedimentary rocks that can easily be mistaken for stratiform stromatolites (e.g., the "stylolaminites" of Logan and Semeniuk, 1976). These are not discussed here because the few Archean stratiform stromatolites examined were interpreted as biogenic from their association with structured (e.g., domical) stromatolites.

In hand specimens, stiriolites and duricrusts can be indistinguishable from true stromatolites. Observations needed to allow these various deposits to be distinguished can be considered under the headings of relationship to bedding, facies relationships, macromorphology, micromorphology, and chemical composition. A stromatolite must:

(1) Be oriented in relation to sedimentary bedding in such a way as demonstrably to have formed synchronously with the bed. Examples of stromatolite-like structures that fail this test are laminated, columnar-branching, manganese oxide structures developed within rocks along joints (viz. "three-dimensional dendrites").

(2) Occur within a sedimentary facies in which the stromatolite is explicable as a primary sedimentary structure. Well-exposed speleothems would frequently fail this test, as they form structures that trangress bedding planes.

(3) Have a macromorphology consistent with a stromatolitic origin. A stalactite, with the apices of its acutely conical laminae directed downward, would be eliminated by this test, as would be the fitted pisolites of some calcretes (which might otherwise be confused with small, spheroidal, unattached stromatolites of the oncolite type).

(4) Have a micromorphology (lamina shape and microstructure) consistent with a microbial origin (e.g., see the detailed descriptions of Hardie and Ginsburg, 1977).

(5) Have a chemical composition consistent with an origin as a stromatolite. A possible example of a deposit that may fail this test would be that of a calcrete developed on a marine carbonate in which the carbon isotopic composition of the calcrete carbonate is markedly different from that of the contiguous marine carbonate.

Although biogenesis can be inferred from these features, some stromatolites exhibit other distinctly biogenic characteristics as well. Examples include (*i*) an ordered arrangement and distribution of cellularly preserved microfossils (Hofmann, 1975; Awramik, 1976) or of remnants of cells or their sheaths (Monty, 1976; Bertrand-Sarfati, 1976); (*ii*) evidence of lamina flexibility resulting from cohesion produced by microbial filaments and mucilage (e.g., penecontemporaneous rip-up and fold structures as described by Davies, 1970, fig. 12); (*iii*) distinctive desiccation cracks reflecting the mechanical properties of microbial laminae (Walter et al., 1973; Hardie and Garret, 1977); (*iv*) evidence of microbial "stickiness", such as the occurrence of detrital laminae which formed at greater than the angle of repose of unattached free grains (Golubić, 1973; Hardie and Ginsburg, 1977, p. 57); (*v*) evidence of selective trapping of only the finer-size fractions of available detritus (Black, 1933; Gebelein, 1969; Hardie and Ginsburg, 1977, p. 110); and (*vi*) chemical evidence of the former presence of abundant microorganisms, for instance a regular alternation of dolomite and calcite laminae (Gebelein and Hoffman, 1973).

In addition to their lack of distinctly biogenic characteristics, stiriolites and duricrusts have their own characteristic features. Duricrusts frequently occur within recognizable weathering profiles and may occur on karst surfaces (Goudie, 1973; Maignien, 1966; Read, 1976); they may contain fitted polygonal pisolites (Dunham, 1969); and in situ brecciation of pisolites is a common feature (Walls et al., 1975). The grains within calcretes are coated with laminated carbonate; the lamination of which is due to a variable content of iron and manganese oxide, not to the distribution of kerogen as it is in many stromatolites (Walls et al., 1975). Grain envelopes may thicken upward or downward. Silcretes frequently contain anomalously high quantities of TiO_2 (Goudie, 1973; Smale, 1973), and calcretes have carbon isotopic compositions characteristic of deposition from groundwater.

These features have been considered in determining the origin as stromatolites of the structures described in this chapter (and have been discussed elsewhere for the Warrawoona Group stromatolites; Walter et al., 1980). Structures for which there is insufficient evidence for confident interpretation are so indicated. Spheroidal structures collected from the Carawine Dolomite of the Hamersley Basin and initially regarded as oncolites (P.P.R.G. Sample No. 043) were later recognized as calcrete pisolites and are not here described.

8-3. Some Aspects of the Paleobiology of Stromatolites

8-3-1. Evidence of Autotrophy

Throughout the Precambrian there are many examples of lenticular stromatolitic dolomites or limestones enclosed by siliciclastic rocks, a relationship that has been interpreted as evidence that the carbonate is authigenic and biogenic, precipitated in place by some biological mechanism in an environment predominantly of clastic, non-carbonate sedimentation. Several such examples are known from the Archean: the stromatolitic dolomites of the Insuzi Group and Yellowknife Supergroup, and the stromatolitic limestones of the Fortescue Group and Ventersdorp Supergroup (see Section 8.4). Since it is known that some algae can precipitate carbonate, as a result of the photosynthetic fixation of carbon dioxide that serves to shift carbonate equilibria in the direction of precipitation (Krumbein, 1979a), it has been argued by analogy that the presence of lenticular stromatolitic carbonates in the Archean is evidence of autotrophic CO_2-fixation and, probably, of photoautotrophy (e.g., Walter, 1978). This interpretation can be challenged on two grounds. Firstly, the precipitation of carbonate and the growth of stromatolite-building microbes could both be consequences of a common cause rather than being directly linked. For example, both may result from the formation of a standing water body, a pond or lake, in an otherwise subaerial environment; the carbonate could be an evaporitic precipitate, and thus authigenic but not biogenic. And secondly, some bacteria do precipitate carbonate, but not solely as a result of photoautotrophy or other mechanisms of CO_2-fixation. However, if it can be established that the lenticular carbonates in question are not evaporitic, then it is reasonable to infer that they are biogenic. Thus, a lenticular carbonate unit that (*i*) occurs within a siliciclastic sequence, (*ii*) lacks desiccation features and sulfate or halite deposits or pseudomorphs, and (*iii*) appears to be marine on geochemical grounds (e.g., $\delta^{13}C_{PDB} \approx 0$ ‰; Peedee belemnite standard) and occurs within a marine facies, can reasonably be interpreted as biogenic. Unfortunately, it is a remarkable feature of a large fraction of known Archean stromatolitic carbonates that they contain evidence of evaporative conditions (including, specifically, those of the Belingwe Greenstone Belt, Bulawayan Greenstone Belt, Steeprock Group, Ventersdorp Supergroup, and the Warrawoona Group). Carbonates of the Insuzi Group meet most of the criteria for biogenicity, although there is geochemical evidence that they may not be marine (viz., very high Sr contents; J. Veizer, pers. comm., 1980). Similarly, the Fortescue carbonates may qualify, but facies analyses suggest that they are lacustrine rather than marine, and that they possibly formed in playa lakes; it seems to be impossible to establish that these lacustrine carbonates are not evaporitic. Finally, the Yellowknife Supergroup carbonates seem to qualify, but they are not well known. Henderson's (1975a, 1975b, 1977) interpretations suggest deposition in a marine environment on the flanks of emergent volcanoes; as reported in Chapter 7, carbon isotopic measurement of these carbonates (viz., P.P.R.G. Sample No. 413) gives $\delta^{13}C_{PDB} = +1.82$ ‰. For illustrative purposes, then, the following discussion is posited on the interpretation of the Yellowknife carbonates as biogenic.

A consideration of carbonate deposition in Solar Lake, on the Sinai Peninsula, is especially instructive in this context (Krumbein et al., 1977). Solar Lake is a hypersaline lagoon in an arid environment, yet it is apparent from the distribution of aragonite within the lagoon that its deposition is biogenic, not evaporative. Furthermore, carbonate deposition is not a direct result of photosynthesis in the thick cyanobacterial mats that coat the lagoon floor. Maximum photosynthesis occurs within the upper 3 mm of the mat, whereas the maximum concentration of aragonite occurs at depths of 30 to 80 cm. Clusters of aragonite crystals with distinctive forms occur on and within non-photosynthetic bacterial micro-colonies deep within the mat. Numerous strains of bacteria cultured from the mat have been shown in the laboratory to precipitate aragonite. As has been noted by Krumbein et al. (1977, p. 653–654),

> "the astonishing similarity between the carbonates produced in the laboratory experiments and those found in the mats supports the assumption that carbonate precipitation within the mats is governed by the degradation of organic carbon compounds by bacteria. The carbonate aggregates that replace the organic matter are closely related to the bacterial morphology and distribution within the mats during the initial stages of precipitation. . . . The bacteria involved in these precipitation processes are aerobic ammonia-producing bacteria in the aerated zone, *Desulfovibrio* species, facultative anaerobic proteolytic bacteria, and many denitrifying bacteria in the anaerobic zone. . . We observed . . . that carbonate particles form initially on the cell sufaces of the bacteria and that the precipitation of carbonate aggregates is accelerated in decaying and lysing bacterial aggregates when the cytoplasm is released into the environment."

As Krumbein (1979a, p. 49) points out, not only photoautotrophs fix CO_2; even heterotrophs can fix a small amount. In short, the mere presence of carbonate, even if demonstrably biogenic and stromatolitic, proves nothing about the biochemistry of the stromatolite-building biocoenose.

The texture of the preserved carbonate may record original crystal shapes and orientations, in principle thus allowing interpretation by analogy with Holocene bacterial carbonates (Monty, 1976; Krumbein, 1979b; Krumbein and Giele, 1979; Krumbein and Potts, 1979; Golubić, 1973). Such fine textural details have not yet been recognized in any fossil stromatolites, but this is an approach with promise, especially if it is linked with a similarly detailed isotopic study of the carbonate phases.

If the carbonate is demonstrably authigenic, then carbon isotopic differences between the carbonate and degraded bacterial and/or cyanobacterial cells (kerogen) can be a very significant source of biochemical information about the local microbial ecosystem (see Chapters 5 and 7). The following Archean stromatolitic carbonates are sufficiently well preserved and well known to be confidently interpreted as authigenic: those of the Fortescue Group, Ventersdorp Supergroup, Insuzi Group, and Warrawoona Group. Carbon isotopic compositions of the carbonate and reduced carbon (kerogen) in samples from these sequences are summarized in Table 8-2. In each example the $\Delta\delta$ (viz., the difference in isotopic composition between coexisting carbonate and kerogen carbon) is large (from 16 to 49 ‰). These results are interpreted in Chapters 5 and 7 to indicate that the carbon cycle at these locations was controlled by CO_2-fixing organisms using the Calvin Cycle (viz., autotrophs, but not necessarily photoautotrophs). As both the carbonate and the kerogen are components of the stromatolites, it can reasonably be concluded that autotrophic bacteria and/or cyanobacteria were part of the stromatolite ecosystem. If such prokaryotes were chemoautotrophs, then suitable electron donors must have been present; these could have been, for example, H_2, H_2S, or NH_3. As Archean stromatolites are all to some extent associated with volcanics, these gases may have been available. Or, they may have been provided by a contemporary biological source as well. Pyrite is abundant in the Fortescue stromatolites, indicating an abundant sulfur source despite a probably nonmarine environment (and hence no ready inorganic source of sulfate). More problematic is the requirement for electron acceptors, which in extant chemoautotrophs are O_2 or NO_3^-, chemical species that might not have been readily available during the Archean. However, some methanogens use CO_2 as an electron acceptor (see Chapter 7).

It is apparent from the arguments adduced above that Archean stromatolite ecosystems probably included autotrophs (in which CO_2 served as the carbon source), but that such arguments do not allow the distinction between the presence of chemoautotrophs (which derive energy from the oxidation of an inorganic electron donor with an inorganic electron acceptor) and photoautotrophs (which use the energy of light).

TABLE 8-2. The carbon isotopic compositions of coexisting kerogen and carbonate minerals from cherts and carbonate rocks of four stromatolitic Archean rock units in which the carbonate most probably is authigenic. The numbers in parentheses are the numbers of analyses averaged (see Chapter 5 for additional discussion).

Rock unit	$\delta^{13}C_{PDB}$*kerogen* (‰)	$\delta^{13}C_{PDB}$*carbonate* (‰)
Ventersdorp Supergroup	−37.5 (18)	−2.7 (9)
Fortescue Group	−48.8 (11)	0.0 (9)
Insuzi Group	−14.3 (5)	+1.2 (3)
Warrawoona Group	−31.0 (12)	−0.9 (1)

8-3-2. Evidence of Photic Responses

If we can establish that the microbes that constructed a stromatolite responded to light, then we have another source of biochemical information. It has been argued that production of a laminated fabric indicates that the constructing microorganisms were phototactic and, by inference, phototrophic (Schopf et al., 1971), but it is now known that many different mechanisms can induce a microbial community to construct a laminated stromatolite (Monty, 1976). For example, Doemel and Brock (1974) have described a laminated stratiform bacterial mat in which the lamination is a consequence of positive aerotaxis. Nevertheless, there are some fabric features that can be related to phototaxis and phototropism. At present, this approach has promise, but has yielded few positive results largely because very little is known about fabric construction in extant stromatolites.

It has been shown that, in the modern environment, construction of the distinctive, conically-laminated stromatolite *Conophyton* is a result of a behavioral syndrome that includes phototaxis and microaerophilia in filamentous cyanobacteria (Weller et al., 1975; Walter et al., 1976; Walter, 1977); such may also have been the case in the Precambrian (Schopf and Sovietov, 1976). Construction of a modern *Conophyton* column begins with the clumping of filaments, achieved by gliding motility. Clumping promotes a microaerophilic microenvironment that stimulates CO_2 fixation. Clumps enlarge to form cones as a result of growth and the positive phototaxis of the filaments. The fine filaments cohere to form a firm, laminated structure. The thinnest laminae are only as thick as single filaments, about 1 μm wide. The top of each cone has on it a tuft of erect filaments, expressed as a lamina thickening on buried cones. In environments with little mineral precipitation, the cones are soft, and often flop over. Where abundant opaline silica or calcium carbonate is precipitated, the cones are stiffened. On the margins of the cones it is possible infrequently to observe enrolled trichomes, that formed balls as they glided up the cone flanks. Thus, the combination of acutely conical laminae with crestal complexities representing former tufts appears to be firm evidence of construction by phototactic filaments. In fossil *Conophyton*, the occurrence of filaments oriented parallel to the laminae and of abandoned sheaths (Schopf and Sovietov, 1976), or of remnants of enrolled filaments, would strengthen the interpretation. On the basis of the one available model from studies of modern stromatolites it is reasonable to consider microaerophilia as the probable cause of clumping.

Other potential evidence of photic responses seems much less convincing. In the modern environment, some finely filamentous oscillatoriacean cyanobacteria form laminated mats with alternately erect and prostrate filaments. It has been demonstrated that the filaments are erect during the day and prostrate at night, with laminae of erect filaments being as much as 1 mm thick (Monty, 1967; Gebelein, 1969; Walter et al., 1976; Golubić and Focke, 1978). From these observations it has been deduced (but not experimentally confirmed) that the trichomes are positively phototactic, and that during the day they move up through any deposited sediment. On this basis, it can be suggested that fossil stromatolites with relatively thick laminae containing preserved alternately erect and prostrate fine filaments were similarly built by phototactic microorganisms. However, there are at least two problems with this interpretation: (*i*) Walter (1976c) reports the occurrence of similar laminae that are probably seasonal, not diurnal; and (*ii*) Doemel and Brock (1974) describe an example where the same filament arrangement is diurnal, but is due to positive aerotaxis, not phototaxis. Evidence of a high sedimentation rate, such as abundant detritus in the laminae, can be used to argue against the possibility of seasonal lamination, but the results of positive aerotaxis and positive phototaxis may be indistinguishable (although it seems likely that trichomes reacting through positive aerotaxis would take the shortest route to a mat surface, and therefore would be radially arranged in a convexity, whereas those reacting through positive phototaxis would tend to move toward the source of light and therefore, generally, would tend to be vertical).

Stromatolite fabrics that preserve evidence of erect filaments are suggestive of phototropic filament orientation or phototactic movement of filaments. These include palisade fabrics (e.g., Davies, 1970; Truswell and Eriksson, 1975a), tufts of filaments (e.g., Golubić, 1973; Horodyski, 1977), walled columnar stromatolites with erect filaments in the walls (Walter et al., 1976), and stromatolites with skeletal or palimpsest microstructures recording erect filaments. The interpretation of these depends on the assumption that the filament orientation is a response to light.

Finally, there is one other indirect approach. Evidence of rapid accretion of stromatolites (e.g., the presence within laminae of abundant detritus) suggests that the microbiota possessed a stratagem to cope with this. In present-day environments that stratagem includes phototaxis.

8-3-3. Significance of the Recognition of a Response to Light

Recognition of a response to light has no value as an end in itself (except that it shows that a stromatolite formed in the photic zone). It is a means by which we may gain some insight into the biochemistry of the constructing microorganisms. This approach is limited by the present level of understanding of the biochemistry of photic responses, but the following remarks (based on Presti and Delbrück, 1978) at least draw attention to an interesting field of research. For many photoresponsive prokaryotes, light acts as a source of metabolic energy for photoheterotrophy and photoautotrophy. It can also damage genetic material (DNA). One of the DNA repair-mechanisms, photoreactivation, utilizes a "photo-reactivating enzyme." This may be the oldest of all physiological photoreceptors, and may have enabled survival in environments with a relatively high UV flux. It is possible that even before bacteria used light as a source of energy for metabolism, they may have used it to repair damaged DNA and conceivably as a stimulus for negative phototaxis so they could avoid damaging radiation.

Light-receiving mechanisms (photoreceptors) in prokaryotes are not limited to organisms that use light as a metabolic energy source. Besides photo-reactivation, for instance, light stimulates carotenoid synthesis in some heterotrophic bacteria.

It seems that we cannot unfailingly use the recognition of phototaxis or phototropism in bacteria to infer phototrophy, although generally that inference would be correct. Such responses do, however, indicate the presence of photoreceptors; the molecules that act as photoreceptors in prokaryotes are bacteriorhodopsin, various porphyrins (hemes and chlorophylls), and riboflavin. Riboflavin, acting as a blue light receptor, is especially implicated in phototropism. Nonetheless, as Presti and Delbrück (1978) have noted, "the phototactic responses of photosynthetic bacteria . . . seem to be intimately coupled to the general photosynthetic metabolism of these organisms."

8-3-4. Significance of the Recognition of Gliding

According to Burchard (1980), "gliding motility is a form of surface-associated, active, translocational movement by bacteria [and cyanobacteria]. No obvious changes of cell shape, other than occasional bending or flexing, are involved. Furthermore, gliding [prokaryotes] bear no resolvable, external organelles that are obviously involved in locomotion." This form of locomotion occurs in some cyanobacteria, chloroflexaceans, myxobacteria, members of the cytophagales, and mycoplasmas. It is the only form of motility known to occur in stromatolite-building microorganisms. Gliding bacteria are variously aerobic or anaerobic, heterotrophic or autotrophic, photoautotrophic or chemoautotrophic. Most gliders are rod-shaped or filamentous, although there are some gliding, spheroidal, unicellular, cyanobacteria (Burchard, 1980). Within the cyanobacteria, gliding motility is a feature of taxonomic significance apparently linked with genetic characteristics (Rippka et al., 1979). Its role in stromatolite morphogenesis has been discussed by Golubić (1976a, 1976b). Except for the mycoplasmas, all gliding bacteria and cyanobacteria are enclosed in a multilayered envelope. In addition, "gliding bacteria generally produce slime (or mucilage) during growth and movement," in what appears to be part of the mechanism of motility (Burchard, 1980). It may be that the slime provides an adhesive interface between these prokaryotes and their substratum. In some rapidly gliding cyanobacteria, a relatively large proportion of the carbon they fix must be excreted and then abandoned as slime or empty sheaths. This process, apparently expensive of carbon and using energy also (only the latter has been measured; Halfen and Castenholz, 1976) would not persist if it did not confer on such gliding prokaryotes a substantial selective advantage. That advantage is that in chemotaxis, it allows them to move to a source of food, away from a toxic chemical, or to a needed chemical (e.g., oxygen, as in the positive aerobic chemotaxis of the stromatolite-building chloroflexacean *Chloroflexus aurantiacus*, Doemel and Brock, 1977); via phototaxis, such prokaryotes are able to maintain themselves in a locale with an optimum light intensity for photoautotrophy or photoheterotrophy.

The excreted slime has a fibrillar structure, suggesting the presence in the cell envelope of pore-like structures, and indeed these have been observed in some gliders. Contraction of these fibrils is one of the suggested mechanisms of motility. "Rotary assemblies," resembling in general aspect the basal bodies of flagella, have been observed in the cell envelopes of some gliders, and it has been suggested that the spinning of these could be the mechanism of motility. According to Burchard (1980), "gliding motility is not powered directly by ATP. Rather, an electrochemical gradient imposed across the cell membrane, established by the extrusion of protons from the cell, results in the production of a yet-to-be-identified intermediate of oxidative phosphorylation. It is this proton motive force that powers gliding. Motility of flagellated bacteria is similarly energized."

Thus, and quite unfortunately from our point of view, gliding motility is still a biological enigma. We can at least conclude that the recognition of gliding probably indicates that the cells involved had multilayered envelopes, excreted slime or mucilage, were receptive to chemical or light stimuli, and may have had "proton pumps" powered by oxidative phosphorylation.

8-3-5. Ultraviolet Protection

Organisms exposed to sunlight are at risk from DNA damage and death due to ultraviolet irradiation, if no atmospheric UV screen is present. This problem has been discussed recently by Margulis et al. (1976) and Rambler et al. (1977). These authors have demonstrated that within cell clumps the outermost dead or moribund cells form an effective shield for healthy cells within each clump, and that some salts within the growth media can also strongly absorb UV radiation. In addition, at least some anaerobic and aerobic bacteria have mechanisms that during hours of darkness repair UV-induced damage done during daylight (Rambler and Margulis, 1980). It could be concluded from these experiments that stromatolite-building microbial mats could have lived exposed to the atmosphere even in the absence of an atmospheric UV screen. However, as Gebelein (1976) has observed, continued accretion of stromatolites is a result of active microbial movement and growth at the sediment surface. No experiments have yet shown whether stromatolite accretion will occur in the presence of a high UV flux.

Walter et al. (1980) have presented evidence suggesting that the 3.5-Ga-old Warrawoona Group stromatolites were at least intermittently exposed to the atmosphere. Stromatolites of the 2.64-Ga-old Ventersdorp Group are cut by desiccation cracks indicative of subaerial exposure. Similarly, desiccation cracks and intraclasts indicative of desiccation occur in sediments interlaminated with the 2.77-Ga-old Fortescue stromatolites. Also within the stromatolitic carbonates of the Fortescue Group are minute tuft-like structures interpreted as remnants of free standing mucus-enclosed trichomes; sediment types associated with these structures, and the desiccation features mentioned above, indicate that these tufts may well have been exposed to the atmosphere (like extant tufted mats of *Lyngbya*) and that they are unlikely ever to have been submerged under more than a meter of water. It thus seems that exposure to the atmosphere did not prevent stromatolite accretion or the growth of microbial communities during the Archean. However, at present there is insufficient experimental information to allow this observation to be interpreted in terms of the UV flux during the Archean, and hence the presence or absence of a biologically effective ozone screen.

8-3-6. Cell and Filament Morphology

Some authors have made the assumption that laminated stromatolites were constructed by filamentous organisms. It seems to be true that all known examples of extant stromatolites with thin and even laminations (i.e., those with microstructures that would be described as banded, streaky, or striated) were constructed by filamentous organisms (e.g., Monty, 1976), and that the laminae within stromatolites constructed by colonies of coccoid prokaryotes (e.g., *Entophysalis major*) are thick and irregular (e.g., Logan et al., 1974).

There are examples of laminated fossil stromatolites where the constructing organisms apparently were predominantly coccoid rather than filamentous, to judge from the contained microfossils: for example, the Eocene Green River stromatolites that contain abundant spheroidal microfossils (Bradley, 1929) and the Middle Proterozoic stratiform stromatolites of the Balbirini Dolomite (D. Z. Oehler, 1978; Chapter 14, herein). Some laminated stromatolites of the Early Proterozoic Belcher Supergroup were constructed by alternating communities of coccoid and filamentous microorganisms (Hofmann, 1975). In each of these examples the lamination is either diffuse or irregular. Radke (1980) has observed in his study of Cambrian carbonates of the Georgina Basin that stromatolites are well laminated where the remains of an associated skeletal metazoan infauna are rare, but that they grade progressively into more and more poorly laminated stromatolites (including thrombolites) as the skeletal remains become more abundant. It is apparent that in at least this example the presence or absence of lamination has more to do with the activities of the associated burrowing infauna than with the morphology of the stromatolite-constructing microorganisms.

Both coccoid and filamentous microorganisms seem to have built, and certainly now build, laminated stromatolites. At least in the presently known modern and fossil examples, the results are distinguishable, the laminae constructed by the spheroidal forms being wrinkled and irregularly lenticular, whereas those built by filamentous organisms tend to be of more uniform thickness and smoothly curved or wavy rather than wrinkled.

Prior to compaction, the thickness of a lamina must be at least as thick as the constructing cells and filaments. This relationship will persist where the laminae are cemented early in diagenesis or where the inorganic detrital component of the laminae greatly exceeds in volume the organic component. In these circumstances, lamina thickness can be used as a crude indicator of maximum cell dimensions (although pressure solution frequently reduces lamina thickness, and care must therefore be exercised when applying this reasoning to samples containing concordant stylolites).

8-4. Archean Stromatolites

The eleven Archean stromatolite occurrences now known are listed in Table 8-1 and their geographic distribution is shown in Figure 8-1. Geological and geographic data relating to occurrences discussed below are assembled in Appendix I. Many datings are imprecise (Table 8-1; Appendix I); the Hamersley Basin stromatolites, for instance, may be earliest Proterozoic rather than Archean.

8-4-1. Warrawoona Group, Pilbara Block, Australia

Stromatolites were discovered in the Warrawoona Group by J. S. R. Dunlop in 1977. Dunlop et al. (1978) reported "a texture which resembles a stromatolitic edgewise conglomerate with pre- or syn-depositional oncolitic overgrowth," and Lambert et al. (1978) illustrated a "possibly stromatolitic" laminated chert, but the first reports convincingly to demonstrate the occurrence of stromatolites in this unit are those of Lowe (1980a) and Walter et al. (1980). An earlier report by Hickman (1973) illustrated a "large bulbous swelling" in layered barite and "a reniform structure in interbedded barite and chert"; as Walter (1978) noted, these structures resemble nodular stromatolites about 1 m wide. Such structures are abundant in the chert-barite units of the Warrawoona Group and are currently being studied by J. S. R. Dunlop and R. Buick of the University of Western Australia. These and other more recently discovered examples are all in units attributed to the Towers Formation of the Warrawoona Group (Hickman, 1980).

At a relatively low stratigraphic level in the Towers Formation, bulbous and nodular stromat-

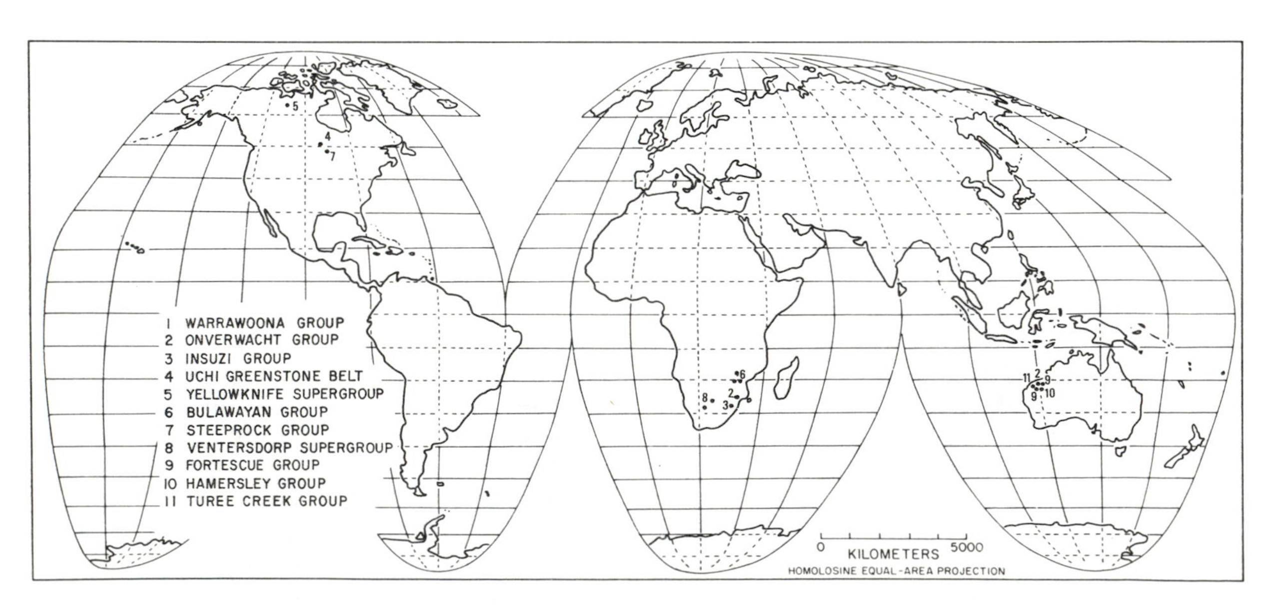

FIGURE 8-1. Map summarizing the distribution of known Archean stromatolite localities.

olites occur in a 40 m-thick, possibly lenticular unit of chert and barite that can be traced discontinuously for at least 30 km. Much of the barite has a distribution consistent with precipitation originally as gypsum in standing water and within soft sediment, and at several localities barite crystals and silica crystallites have been shown to have interfacial angles characteristic of gypsum (Dunlop, 1978; Barley et al., 1979). The cherts contain intraclast grainstones that are probably indicative of desiccation and later reworking in turbulent water (Dunlop, 1978; Barley et al., 1979; Walter et al., 1980). Additional indications of turbulence include the presence of small scour features, ripple cross-lamination, cross-bedding, and imbricate conglomerates. The original composition of the sediments is not clear: some may have been primarily siliceous, but other cherts, which contain stylolites and scattered small rhombs of dolomite, may have been calcareous (Dunlop, 1978). The observed features are all consistent with deposition in a very shallow, hypersaline, water body. Whether this body consisted of a series of salt lakes or a marine barred basin is not clear, but the students of this sequence favor a marine, peritidal environment (Dunlop and Buick, 1980). Similar stromatolitic cherts at a higher stratigraphic level have many of the same sedimentary features, as well as desiccation cracks, and are interpreted by Lowe (1980a) as having been carbonates originally, and having been deposited in a marine subtidal to intertidal environment. Laterally equivalent rocks include explosive felsic volcanics within which volcanic centers have been recognized. The volcanism was both subaerial and subaqueous (Barley, 1978; Barley et al., 1979).

PSEUDOCOLUMNAR STROMATOLITES

The stromatolites described by Lowe (1980a) apparently form very extensive tabular biostromes, up to 60 cm thick. They are pseudocolumnar, composed of erect, parallel, unbranched columnar units about 2 to 6 cm wide. Those illustrated by Lowe are mostly elliptical in transverse section, with the long axes of the ellipses being parallel. Within the columnar units, laminae are conical, with apical angles averaging 70 to 80°, but in some examples as low as 30°. The lamination is described as being "fine and continuous," but no details are available and a slab from the illustrated specimen that is now deposited with the Geological Survey of Western Australia, is very poorly preserved (Walter, personal observation). Lowe (1980a) compares these stromatolites to *Conophyton*, but they differ from members of that group in having less acutely conical laminae, and a diagnostic feature of *Conophyton*, the complex crestal zone, appears not to be present. It has yet to be demonstrated that the conical lamina profile and elliptical transverse sections are not the products of deformation of stromatolites with originally convex laminae.

NODULAR AND STRATIFORM STROMATOLITES

The stromatolites described by Walter et al. (1980) are partially linked, spaced, nodular, and stratiform varieties; the largest nodular example (Frontispiece and Photo 8-1A) is about 30 cm high by 20 cm wide, with at least 8 cm of growth relief. The nodules protrude upwards from a plinthe of wavy-laminated stratiform stromatolites (see Frontispiece). The nodular forms are known with confidence from only one large specimen (P.P.R.G. Sample No. 349; 60 × 60 × 40 cm in size), but other nodular structures of more uncertain origin are abundant in the Warrawoona cherts and are currently being studied by J. S. R. Dunlop and R. Buick. Some may prove to be stromatolites. Similarly, the underlying plinthe can be interpreted as a stratiform stromatolite because of its relationship to the nodular structures, but similar wavy-laminated cherts are widespread in the sequence yet are less readily interpretable because of their lack of distinctive features.

The modular stromatolites have wavy and wrinkled laminae, frequently with domical structures up to 3 cm wide (Photo 8-1B); many of the domical structures persist vertically for more than a centimeter, producing pseudocolumnar structures. Laminae of many of the domes are flat-topped, with sharply deflexed margins; junctions between contiguous columns are V-shaped. As noted by Walter et al. (1980), small discrete columns occur rarely within the large stromatolite nodule.

In these stromatolites the microstructure, where it is well preserved, is distinct and banded to streaky, with laminae predominantly 50 to 200 μm thick, although many are as thin as 20 μm (Photo 8-1C, D). The lamination has been extensively disrupted, apparently by the growth of barite (or gypsum) crystals, and probably also by desiccation. The stromatolite now consists of chert with a grain size of about 50 μm; barite as crystals up to 5 mm long; scattered but abundant rhombs of carbonate up to 1 mm wide; and iron oxide. Tiny mineral inclusions in some laminae outline abundant, slightly sinuous, filamentous structures arranged approximately perpendicular to the laminae; these are about 20 μm wide.

MICROFOSSILIFEROUS STROMATOLITES

The microfossiliferous stromatolites (see Chapter 9) are stratiform, and were not recognized as stromatolites in the field. They come from two localities, and at both they are preserved in flat-laminated black cherts. At the locality from which P.P.R.G. Sample No. 002 was obtained, stratiform stromatolites 1 to 10 mm thick are interbedded with thin beds of peloid grainstone. The laminae are lenticular, which gives the lamination a wavy and wrinkled form without distinct pseudocolumnar structures. At this locality the microstructure is striated, but in samples from the other microfossiliferous locality (P.P.R.G. Sample No. 517) it ranges from striated to banded; these latter samples are extensively weathered and contain abundant chalcedony which may be of relatively recent origin and the presence of which obscures the original fabric. The following description is therefore based only on P.P.R.G. Sample No. 002 (Photo 8-2A, B).

In this sample, lamination is distinct with the laminae ranging in thickness from 10 to 200 μm; the thicker examples locally consist of merged thin laminae. Both light and dark laminae are distinctly lenticular. Dark laminae range from distinct and homogeneous to diffuse and speckled in appearance, depending on the distribution of kerogen and finely crystalline pyrite (which is now mostly oxidised to goethite). Color variation of laminae from black to brown appears to depend on the relative proportions of black kerogen and brown iron oxide.

8-4-2. Onverwacht Group, Barberton Mountain Land, Kaapvaal Craton, South Africa

No macroscopic stromatolites are known from the Onverwacht Group (nor indeed, from the entire Swaziland Supergroup), but some black cherts from the Kromberg Formation appear to be stratiform stromatolites. These carbonaceous cherts were deposited in low energy, shallow water environments on a platform-type setting with subaerial felsic volcanic vents (Lowe and Knauth, 1977; Lowe, 1980b).

The example described by Lowe and Knauth (1977, their fig. 12B) is not known to contain microfossils, but as those authors have also remarked, the fabric appears to be stromatolitic. Of particular interest is the remarkable resemblance of this fabric to that of the microfossiliferous stratiform stromatolites from the Warrawoona Group (Photo 8-2). The fabric interpreted as stromatolitic by Muir and Grant (1976) is quite different in that no lamination is apparent. However, at high magnification it can be seen that carbonaceous spheroids form groups that in some areas of one thin section appear to outline remnants of laminae, some of which contain small convexities (see the following chapter, Photo 9-2). Both of these examples of possible stromatolites from the Kromberg Formation must be studied further, with additional specimens, before they can be interpreted with confidence.

8-4-3. Insuzi Group, Wit Mfolozi Inlier, Kaapvaal Craton, South Africa

Stromatolites were first reported from this unit by Mason and von Brunn (1977) and by von Brunn and Mason (1977). These authors considered that the 30 m-thick stromatolitic dolomite and the enclosing siliciclastics were deposited in intertidal and shallow subtidal conditions at the margin of an epeiric sea, although J. Veizer (pers. comm. 1980) has determined that the carbonates are unusually rich in strontium and therefore possibly nonmarine. Dr. Mason guided H. Hofmann and me to the stromatolite localities previously reported, and also to several others known to him but previously undescribed; all of these localities are in the valley of the Wit Mfolozi River (see Appendix I).

Three forms of stromatolites can be recognized:

(1) Isolated bulbous forms occur that range in size from 1 × 1 cm to 30 × 30 cm (P.P.R.G. Sample No. 309). The larger individuals consist of columnar-layered stromatolites with bulbous, irregular columns up to 24 cm wide. These have been found only in large boulders (Photo 8-3A) in the talus slope 200 to 300 m upstream from "Cloud Cliff." Laminae range from flat to steeply convex, but within columns they are consistently hemispheroidal, locally coating column margins to form patchy walls (Photo 8-3B). They are wavy and wrinkled, with abundant bulbous and nodular protrusions 1 to 10 mm wide with up to 2 mm of relief (Photo 8-3C). The waviness and wrinkling is caused by the juxtaposition of hemispheroidal lenses, producing a microstructure similar to the

"tussock microstructure" as defined by Bertrand-Sarfati (1976). Each of the hemispheres ("tussocks") is laminated but, in contrast to the examples described by Bertrand-Sarfati, a radial fabric is only infrequently present. Much of the fine structure has been destroyed by the growth of sparry dolomite, but where thin laminae are well preserved, a fine, linear, radial palimpsest microstructure is present. In these areas the laminae are banded and are 15 to 50 μm thick; the light laminae are thickest. The radial structures are 5 to 30 μm wide.

Angular detrital quartz grains 30 to 150 μm wide are abundant in the interspace sediment and occur frequently within the stromatolite. They occur even in vertical and overhanging laminae.

No microfossils have been found within these stromatolites, but the radial linear palimpsest microstructure appears to record the former presence of filaments. The finest of these features are 5 to 10 μm wide, indicating that the filaments were this size or smaller.

(2) Stratiform stromatolites are present together with variously contiguous to widely spaced nodular and bulbous individuals from about 1 to 15 cm wide, with up to 5 cm of growth relief (Photo 8-4A). These occur as what appear to be tabular biostromes, in one example 30 cm thick, interbedded in intraclast and ooid grainstone, some of which is cross-laminated. These are the stromatolites described by Mason and von Brunn (1977). The bulbous and nodular forms contain abundant columnar-layered units with columns 1 to 5 mm wide (Photo 8-4B). These have irregular shapes and are composed of laminae ranging in shape from acutely conical to gently convex; most are steeply convex. Column axes are not parallel; there is a crude radial arrangement within the host bulbous and nodular individuals. Some columns with conical laminae have crestal zones where the laminae are thicker and contorted (Photo 8-5A), but because of the small column size it is not clear whether crestal zones are consistently developed. Locally there are hemispherical and subspherical thickenings in the laminae, 100 to 300 μm wide (Photo 8-5B). Acute flexures in many laminae indicate deformation while the stromatolite was soft; the orientation of the flexures suggests that deformation was due to gravity settling.

The microstructure varies from distinct banded to diffuse streaky, with laminae 15 to 500 μm thick; most are less than 50 μm thick. The stromatolites consist predominantly of xenotopic dolomite with a grain size of about 10 μm, and there are laminae of chert of about 20 μm grain size.

(3) Stratiform and columnar-layered forms, with abundant inclined small conical individuals, occur in beds several metres thick. T. R. Mason, H. J. Hofmann (pers. comm. 1980), and I consider that the conical form is in part a result of tectonic deformation, and that before deformation these stromatolites were like those described in item 2, above.

8-4-4. Uchi Greenstone Belt, Superior Province, Canada

The Woman Lake Marble occurs at the top of a volcanic sequence in this greenstone belt (Thurston and Jackson, 1978). It has been metamorphosed to the amphibolite facies. A large sample (P.P.R.G. Sample No. 516) provided by Dr. Thurston contains inclined, spaced, pseudocolumnar structures 1 to 2 cm wide, which persist "upwards" (the original orientation being unknown) for 5 to 10 cm; these are formed from laminae 1 to 2 mm thick that are convex within the columnar units. These structures are characterized by a fenestral fabric and may be true stromatolites, but their origin cannot be established based on the material available for our study (viz., GSC type 65651).

8-4-5. Yellowknife Supergroup, Slave Province, Canada

In this sequence, "stromatolites occur in a discontinuous dolomite unit at or near the contact between the dominantly felsic volcanic sequence and the greywacke-mudstone turbidites" (Henderson, 1975a). The dolomite is up to 40 m thick and discontinuous, but has been traced for 6 km. Where best exposed it occurs within black, pyritic, carbonaceous mudstone that grades up into tuff, sandstone, and conglomerate; elsewhere it occurs within the volcanic sequence. The dolomite contains 10 to 15 cm-thick beds of intraclast grainstone and oncoid grainstone. The felsic volcanics were probably emergent, providing shallow water environments for the stromatolites. The intraclasts and oncoids are considered to be indicative of a very shallow, intermittently exposed, turbulent environment (Henderon, 1975a, 1975b, 1977).

Henderson (1975a) described two localities; at one the stromatolites are stratiform, with some spaced pseudocolumnar structures variously having gently convex or cuspate laminae. The samples here described are from this locality. At the other locality there are linked bulbous forms up to 30 cm wide by 20 cm high, with an internal pseudocolumnar structure of stacked gently convex laminae.

P.P.R.G. Sample No. 413 (GSC type 65650) contains columnar-layered forms with columns 1 to 15 mm wide, with α-β parallel branching and gently convex laminae (Photo 8-5C). Also present is a bed about 10 cm thick containing tabular to spheroidal oncoids up to 5 cm wide. Within the columns, laminae are gently convex and smoothly curved. The columns are naked (i.e., lacking distinct outer walls). The stromatolites have a distinct banded to streaky microstructure with dark grey laminae 50 to 500 μm thick alternating with light laminae 50 to 1000 μm thick (Photo 8-6). In places the lamination is diffuse. A diffuse linear palimpsest microstructure is almost ubiquitous; its light and dark components are 20 to 50 μm wide, slightly sinuous, and more or less perpendicular to the laminae. These appear to be remnants of mat-building filaments. In addition, there appear to be remnants of acicular crystallites that are straight rather than sinuous. Very fine (about 10 μm wide) wisps of kerogen are not consistently oriented in relation to the laminae and probably represent kerogen remobilized along a fracture cleavage. The oncoids and the stratiform stromatolites have the same fabric.

8-4-6. Bulawayan Group, Bulawayan Greenstone Belt, Zimbabwe

The stromatolites of the Bulawayan Group were discovered in 1935, and first described by Macgregor (1941). They have since been restudied by Cloud and Semikhatov (1969) and by Schopf et al. (1971). Their age and stratigraphic setting is discussed by Bond et al. (1973) and Wilson et al. (1978), and aspects of their geochemistry have been considered by numerous authors (Bell et al., 1974; Eichmann and Schidlowski, 1975; Perry and Tan, 1972; Schidlowski et al., 1975; Veizer and Compston, 1976; Veizer and Hoefs, 1976; and by other authors listed by Bond and Falcon, 1973). Microfossil-like objects previously reported from the stromatolites are now considered to be modern contaminants or abiogenic artifacts (see Chapter 9).

Stromatolites from the Belingwe Greenstone Belt in Zimbabwe (Bickle et al., 1975) have recently been described in some detail (Martin et al., 1980). Stromatolites are also known from the Midlands Greenstone Belt (Cheshire, 1978) and the Salisbury-Enterprise Gold Belt (Clay, 1978).

Only stromatolites from the Huntsman quarries in the Bulawayan Greenstone Belt were included in our study. Photographs of some of the Belingwe stromatolites (Photos 8-7, 8-8) were kindly provided by Dr. A. Martin, and the paleobiological significance of those stromatolites is briefly considered below (Section 8-5-3).

Macgregor (1941) reports that the stromatolitic limestone in the Huntsman quarries "forms part of a generally dolomitic calcareous bed, which can be traced along the strike for several miles associated with banded ironstone, arkoses and conglomerate." Bond et al. (1973) show that the limestone is interbedded with quartz-sericite schist, banded ironstone and greenstones. From Macgregor's illustrations of stromatolites subsequently destroyed by quarrying it can be seen that they were linked nodular forms with a growth relief of 20 to 40 cm and perhaps more in some examples. Macgregor also records structures which "are comparable with . . . oncolites." An additional observation of paleoenvironmental significance is that the carbonate has a $\delta^{13}C_{PDB}$ value of close to zero per mil (Schopf et al., 1971). No other sedimentological information is available, but these few facts are consistent with deposition in a marginal marine environment in water of unknown depth, but probably subtidal as indicated by the stromatolite morphology.

The stromatolites are linked nodular forms about 1 m wide by 60 cm thick; according to Macgregor (1941), "it is probable that the domes contain structures of different kinds, as the dentate form seems to occur as forming part of a large dome." The illustrated "dentate" structures appear to be pseudocolumnar (his pl. 4) to columnar-layered (his pl. 3, fig. 2) stromatolites with columnar units about 0.5 to 2.5 cm wide. The laminae are steeply parabolic to conical, and elliptical in transverse sections. This form has not subsequently been restudied.

Macgregor's (1941) figure 2 indicates that the other columnar and pseudocolumnar forms, at least in one example, occur within nodular stromatolites, as well as forming "continuous beds." This appears to be the form studied by Schopf et al.

(1971), although they considered that the form was previously unknown. The specimens available to us are those studied by Schopf et al. (1971) and are all pseudocolumnar, although the columns have been adpressed by stylolitic solution. The column units are mostly 0.8 to 1.5 cm wide. The same form has been illustrated by Cloud and Semikhatov (1969).

In these specimens, laminae predominantly are gently convex, although locally they are steeply deflexed near column margins (Photo 8-9A). They are smoothly curved but frequently have flexures in areas with lenses of sparry calcite, here interpreted as secondary flexuring due to diagenetic growth of an acicular mineral. Similarly, lamina crinkling that is clearly developed in areas with the most distinct remnant acicular fabric is considered to be secondary. Where best preserved, the laminae have a diffuse banded microstructure and are 10 to 30 μm thick (Photo 8-9B,C). A columnar palisade structure is present in one sample.

A remnant acicular fabric is pervasive and marked by aligned coarse grains of prominently twinned calcite. They are aligned more or less parallel to column axes and pass through laminae and "fenestrae" (i.e., lenses of sparry calcite). Each "fiber" of the acicular fabric is a composite of several large, differently oriented grains. These have very irregular margins resulting from the later growth of finer-grained (mostly 20 to 50 μm) equidimensional calcite grains. The prominently twinned calcite is especially coarse-grained (up to about 500 μm) in many of the fenestrae. Anastomosing linear patches of clear carbonate are interpreted as crack fillings in an incipient cataclastic fabric. The twin planes of the coarse crystals are bent and kinked. The fine grains are considered to be the recrystallization products of stressed coarse grains.

Because the original acicular crystals passed through the fenestrae it is probable that the fenestrae represent laminae forced apart by the growth of the crystals (e.g., of gypsum early in diagenesis). They seem unlikely to represent former gas holes, contrary to the suggestion by Schopf et al. (1971).

The laminae are pigmented by black material interpreted as kerogen. Frequently the kerogen is arranged in linear streaks about 2 to 5 μm wide that parallel the formerly acicular fabric and so vary from being perpendicular to the laminae (where the laminae are horizontal) to being oriented at about 60° to the laminae (on the column flanks); that is, the streaks are parallel to the column axes. These may be filament remnants, or kerogen redistributed along the margins of the acicular minerals.

8-4-7. Steeprock Group, Quetico Belt, Superior Province, Canada

Early work on the stromatolites of the Steeprock Group is reviewed by Hofmann (1971, pp. 58–59). Stromatolites occur within a thick, extensive limestone, said to be a "normal sedimentary carbonate formation" (Joliffe, 1955). No detailed sedimentological analysis is available. The limestone is interbedded in a sedimentary sequence with conglomerate, quartzite, banded iron-formation, lavas, and volcaniclastics. Some of the observations here presented contribute to a paleoenvironmental interpretation. The pseudofossil "*Atikokania*" (Walcott, 1912) is common in the limestone and is considered here to be composed of carbonate pseudomorphs after clusters of radially arranged selenite (gypsum) crystals. Its mineralic, crystalline nature has been noted by Hofmann (1971a) and of the two most plausible original mineral phases that may have been involved, aragonite and gypsum, the latter is the more likely because of the large size of the pseudomorphs (up to 25 cm long) and their close resembalance to better preserved examples of gypsum pseudomorphs (e.g., Walker et al., 1977; Debrenne and Lafuste, 1979; Glaessner, 1980). The fabric of some of the stromatolites indicates construction from detrital particles of carbonate. The stromatolites are elongate large nodular forms (up to 17 m wide according to Joliffe, 1955). These features are all consistent with deposition in a shallow subtidal marine environment.

Joliffe (1955) mentions nodular stromatolites ("gentle domes") up to 50 feet (17 m) wide. Hofmann (1971a) describes decimeter-sized bulbous and centimeter-sized pseudocolumnar forms. Those we sampled (Photo 8-10) are elongate, contiguous, linked nodular forms 2 to 3 m wide by 4 to 5 m long, with about 50 cm of relief (P.P.R.G. Sample No. 324). Laminae within them conform to the outer surface, and some laminae have on them polygonal structures resembling desiccation cracks. "*Atikokania*"-like structures occur within the domes. Lamination is discontinuous and locally only crudely developed. It is marked by wispy layers of kerogen and by closely spaced lenses of sparry calcite (Photo 8-11A, B). The lenses are up to 6 mm long by 4 mm thick; in addition, there are abundant, irregular, angular patches of sparry calcite of mostly 1 to 2 mm width. Remnants of coarse acicular crystals (now calcite) intersect the laminae at high angles; these were up to 5 mm wide by more than 5 cm in length. This fabric closely resembles that of the Bulawayan stromatolites, and in this example also the lenses of sparry carbonate are here interpreted as products of the growth of an acicular mineral, perhaps originally gypsum.

A pseudocolumnar to columnar layered form (P.P.R.G. Sample No. 328) has columns about 3 cm wide with laminae varying from gently convex to rectangular (Photo 8-12A, B). They are similar to those previously described except that detrital particles (peloids) are abundant. Many particles are elongate and these lie in the plane of the laminae. The particles are mostly 0.1 to 1.0 mm wide. Abundant patches of sparry carbonate locally produce a "chicken-wire" texture suggestive of the disruptive growth of a sulfate mineral, although the smaller patches also resemble fenestrae. P.P.R.G. Sample No. 327 is columnar layered; this has an especially well-preserved fabric with a film microstructure (Photo 8-13A, B). Brown-pigmented laminae 100 to 150 μm thick alternate with colorless or weakly pigmented laminae 100 to 1000 μm in thickness. A linear palimpsest microstructure locally is prominent: the linear structures are slightly sinuous, 50 to 150 μm wide, and perpendicular to the laminae. Very fine (10 μm wide) streaks of black material (possibly kerogen), more or less perpendicular to the laminae, are abundant. These streaks resemble highly degraded remnants of filaments, but may be kerogen redistributed along grain boundaries during recrystallization. Pigmented laminae consist of xenotopic calcite with a grain size of about 20 μm. Calcite in the unpigmented laminae is as coarse as 1 mm. Some laminae are composed of chert.

8-4-8. Ventersdorp Supergroup, South Africa

Sediments and volcanics of the Ventersdorp Supergroup were deposited in fault-bounded intermontane basins, in a region of basin-and-range topography (de la R. Winter, 1963, 1976; Visser et al., 1976; Grobler and Emslie, 1976a, 1976b; Buck, 1980). According to Buck (1980),

In the lowermost Klippan Formation, sediments accumulated within numerous small but deep intermontane graben basins formed by the block faulting of an older andesitic lava terrain. During the early stages of deposition, coarse clastic debris accumulated along the scarped margins of horsts forming talus scree and debris flow alluvial fan deposits, while within basins fine grained terrigenous and chemical sediments, including stromatolites and ooids, accumulated under lacustrine conditions. Later fluvial processes predominated, resulting in the widespread prograding of alluvial fans across basins.

The sediments of the succeeding Bothaville Formation accumulated on a regionally extensive alluvial plain dominated by braided rivers. Stromatolites occur intermittently throughout these fluvial sediments, and are interpreted as having developed within pools remaining upon the alluvial plain during intervals between major fluvial discharges.

Buck's (1980) interpretation of deposition in nonmarine environments is supported by carbon isotopic measurements of carbonates from the localities he describes: as reported in Chapters 5 and 7, they range from $\delta^{13}C_{PDB}$ -5.37 to -1.84 ‰, and in the eight samples measured average -2.76 ‰. Marine carbonates give carbon isotopic values vs. PDB close to zero. Grobler and Emslie (1976a) interpret the T'Kuip Hills sequence as marine, primarily, it seems, because of the presence of elongate stromatolites comparable with elongate intertidal forms. No geochemical data that relate to this interpretation are available.

T'KUIP HILLS STROMATOLITES

Most Ventersdorp stromatolites are known only from drill cores so their gross morphology is unknown. However, those from the T'Kuip Hills occur in outcrop (Grobler and Emslie, 1976a). There are stratiform, pseudocolumnar, and low, linked nodular forms (Photos 8-14, 8-15).

(1) The stratiform stromatolites have wavy, irregular lamination with a distinct streaky to banded microstructure (Photo 8-16), with light laminae of variable thickness up to about 250 μm and dark laminae of more uniform thickness, generally 20–30 μm (P.P.R.G. Sample No. 278). Some specimens contain abundant pseudocolumnar structures from 100 μm to 5 mm wide, individual units of which persist upward for up to 5 mm

(as in P.P.R.G. Samples Nos. 272 and 274). The columnar units are frequently widely spaced, and range from cylindrical to turbinate (expanding upward) in form. Within them, laminae are gently convex to acutely conical (although the rock is cleaved, and the conical form may therefore be an artifact of tectonic deformation). The columnar units are lighter colored than their matrix because they contain many fewer detrital grains and, in addition, because the light laminae are thicker (150 to 200 μm). Within them the microstructure is banded and diffuse to distinct. They are preserved in calcitic black chert with grain size of 10 to 20 μm. They resemble the stromatolite *Alcheringa narrina* Walter, but poor preservation makes confident identification impossible.

(2) Low, linked, nodular stromatolites like those noted by Grobler and Emslie (1976a, figs. 4, 6) were sampled from a 30 cm-thick domal biostrome. The sampled specimen (P.P.R.G. Sample No. 273) contains very poorly developed small pseudocolumnar structures like those described above, but otherwise the laminae are wavy with a very low degree of inheritance. The microstructure is distinct and coarsely striated, with macrolaminae 150 to 1000 μm thick, within which are locally developed diffuse banded laminae 15 to 20 μm thick. The stromatolite is predominantly chert with a grain size of about 10 μm, but there is abundant detrital mica, quartz and felspar.

(3) Also occurring in this sequence are 40 to 50 cm thick tabular biostromes of erect pseudocolumnar stromatolites; the columnar units are 5 to 10 cm wide and elongate in plan view. The fabric of these is identical to that of the stratiform stromatolites, except that there are larger pseudocolumnar units as well as numerous small forms. They are composed of calcitic chert with detrital mica, quartz and felspar.

WELKOM GOLDFIELD STROMATOLITES

The stromatolites of the Welkom Goldfield, known only from bore cores, are stratiform, pseudocolumnar, and take small columnar forms (Buck, 1980). Those available to us for study were stratiform and linked bulbous and nodular forms.

(1) The linked bulbous and nodular forms (P.P.R.G. Samples Nos. 291 through 293) are contiguous and 1 to 2 cm wide (Photo 8-17). Laminae within them are convex and slightly wavy. The microstructure is striated with laminae 10 to 300 μm thick. The thicker examples in some cases can be seen to be macrolaminae; in well-preserved areas, laminae 10 to 50 μm thick are predominant. The stromatolites are preserved in calcitic cherts, in which the quartz has a grain size of 10 to 20 μm and in which the calcite is coarse-grained and appears to postdate the quartz. The calcite in P.P.R.G. Sample No. 291 also occurs in lenses up to 9 mm wide by 4 mm thick, against which laminae abut or over which they are domed. Within these, calcite is very coarse grained (up to 2 mm). Some are rimmed with an irregular layer of finer-grained calcite. These structures are calcite-filled coarse lenticular fenestrae. Pyrite is abundant in some laminae.

Large areas of some of these stromatolites are brecciated and infilled with fine intraclast grainstone, a feature that is evidently a result of desiccation.

(2) The stratiform stromatolites (P.P.R.G. Samples Nos. 281 and 285 through 290) are wavy-laminated, with a very low degree of inheritance. Their microstructure is of the filmy to streaky type, with dark laminae 15 to 30 μm thick (less frequently up to 300 μm thick) separated by greatly varying thicknesses of light, detritus-rich laminae (Photo 8-18). Brecciation is a common feature, and sheet and prism cracks are numerous. Some prism cracks are overgrown by stromatolitic laminae. Calcite-filled fenestrae are very abundant and diverse, including fine and medium laminoid and irregular forms (in the terminology of Logan, 1974). Some of these calcite-filled structures have a regular spheroidal form 1 to 2 mm in diameter suggesting a concretionary origin; these may be replacements of nodular anhydrite, which is known to form comparable spheroidal nodules. Laminae are generally preserved in xenotopic calcite with a grain size of almost 50 μm.

Stromatolites were first described from the Ventersdorp Supergroup by de la R. Winter (1963), from bore cores from the Wesselsbron District of the Orange Free State. He illustrates small bulbous and laterally linked nodular forms, as well as stratiform, pseudocolumnar, columnar and oncolitic forms. The specimen available to us for study was that illustrated by Walter (1972, pl. 16, figs. 4, 5) and referred by him to *Gruneria* f. nov.; it is a slab from the core illustrated by de la R. Winter (1963) in his plate 1, figure 1 (Photo 8-19 herein). The oncoids are spheroidal to ovoidal and 2 to 4 mm wide. Pseudocolumnar, columnar-layered and columnar stromatolites intergrade and have irregular, tuberous forms, with frequent parallel to markedly divergent branching. The columns are 1 to 7 mm wide. Laminae are slightly wavy and wrinkled and range from gently convex, often with sharply deflexed margins, to steeply convex; in the columnar forms, laminae locally coat column margins to form walls. Throughout the various stromatolite forms the microstructure is uniform and is of the diffuse, streaky type, with laminae 15 to 60 μm thick where best preserved, where they consist of chert with a grain size of about 20 μm and calcite with a grain size of 20 to 40 μm. Detrital grains are abundant in the interspace sediment but are rare within the stromatolites. Very coarse calcite, interpreted as open-space filling, fills the spaces between some stromatolite columns. It also forms spheroidal to biconvex lenses within some of the stromatolites; the lenses range from 100 μm to 4 mm wide, and the laminae dome over them (Photo 8-19B). Some lenses have a rind about 20 to 150 μm thick composed of fine grained carbonate. It is not clear whether the lenses are former gas holes or perhaps sulfate concretions. The smoothly rounded form of the lenses is not what would be expected for gas holes, where gas pressure would split apart laminae and produce a hole with tapered edges. In addition, it is doubtful that gas pressure could maintain a spherical hole long enough to enable that shape to be fixed by carbonate precipitation. Thus, it seems more likely that these structures are former nodules, perhaps of anhydrite, that were dissolved during diagenesis leaving voids that were later filled with calcite. Such an interpretation is supported by the abundant occurrence within some laminae of tabular and lenticular crystal pseudomorphs 15 to 30 μm wide by up to 150 μm long (Photo 8-19C); these are now carbonate but the crystal form resembles anhydrite.

8-4-9. Fortescue Group, Hamersley Basin, Australia

The stromatolites of the Fortescue Group were found in 1959 by geologists of the Geological Survey of Western Australia; studies prior to 1970 are reviewed by Walter (1972). All known stromatolite occurrences in the Fortescue Group are listed by Grey (1979).

There has been no detailed sedimentological study of the Fortescue Group, but what little is known supports comparison with the Ventersdorp Supergroup and, thus, interpretation of this sequence as having been deposited in intermontane basins in a basin-and-range environment. Sedimentation seems to have been localized over Archean greenstone belts, in valleys between Archean granite batholiths (see Trendall, 1975, 1976, and earlier work quoted therein; for a different view, see Horwitz and Smith, 1978). The group is dominated by volcanics, ranging from basalts to rhyolites. Detailed mapping by Smith (1979) has revealed the location within the Fortescue Group Maddina Volcanics of former volcanic vents, fumaroles, and hot springs. Around the northern margin of the basin the various volcanic units are interpreted as predominantly subaerial, but pillow structures, indicative of subaqueous extrusion, appear to become more abundant to the south, toward the center of the basin (Trendall, 1976). Interbedded with the volcanics are conglomerates, sandstones, shales, and stromatolitic limestones. Button (1976) has interpreted the siliciclastics as deposits of braided stream and alluvial fan environments. A geochemical study of a black shale interbedded in one of these sandstones showed boron, gallium, and rubidium contents consistent with deposition in fresh water (Hickman and de Laeter, 1977).

At "Knossos," the principal collecting locality within the Fortescue Group for this project (see Appendix I), cherty stromatolitic limestones are interbedded and interlaminated with shard-bearing tuffaceous sandstones (Photo 8-20). Evidence of extremely shallow water deposition is abundant, and includes evidence of desiccation such as prism cracks and intraclast grainstones; ripple marks are abundant in the sandstones. Growth relief of the stromatolites ranges from a few millimetres up to about 20 cm, in stromatolites that range from minute cuspate forms to broad, linked nodular and bulbous forms. This evidence is all consistent with deposition of the stromatolitic limestones within ephemeral ponds and "permanent" lakes in intermontane basins. However, as reported in Chapters 5 and 7, the carbon isotopic compositions of carbonates from the eight analyzed samples of stromatolitic limestone (from three localities in the Fortescue Group) range from -1.64 to $+0.79$ ‰, and average -0.1 ‰ $\delta^{13}C_{PDB}$, values more consistent with deposition in a marine environment. Possibly the valleys were at about sea level and close to a coast, and the groundwater chemistry was dominated by that of the nearby ocean.

A distinctive stromatolite from the Fortescue Group is presently known only from two large specimens (P.P.R.G. Sample Nos. 024-1 and 024-2) supplied by Dr. A. Glikson from a single locality. The nature of their field occurrence is unknown. The specimens appear to be parts of bulbous or nodular stromatolites with a columnar-layered to pseudocolumnar internal structure (Photo 8-21). The columnar units are 1 to 5 cm wide, are bulbous and irregular, and have frequent markedly divergent branching. Laminae are gently convex and wavy and wrinkled. They have a diffuse streaky to filmy microstructure, with dark laminae 15 to 130 μm thick and light laminae of greatly varying thickness but predominantly about the same thickness as the dark laminae. Coarse-grained (up to 1 mm) sparry calcite is prominent as concordant patches that are interpreted as a coarse laminoid fenestral fabric. There may also be some neomorphic calcite spar. The fenestrae are frequently 1 to 2 mm thick. Spheroidal patches of spar in the interspace filling of peloid grainstone and wackestone (Photo 8-21C) resemble those in the Ventersdorp Supergroup stromatolites; the very coarse grain size and spheroidal form is suggestive of carbonate replacement of anhydrite nodules. Many light laminae contain abundant detrital grains and have a faint, diffuse grumulous texture with clots about 20 to 50 μm wide. The clots probably are carbonate peloids such as are common in the interspace sediment.

At "Knossos," both nodular and stratiform stromatolites occur forming tabular and domical biostromes.

NODULAR STROMATOLITES

At the top of the outcropping section there is a 2 m thick domical biostrome of linked nodular stromatolites 1 to 2 m wide, equidimensional in plan, and with about 20 cm of growth relief (Photo 8-20A). The internal lamination (P.P.R.G. Sample Nos. 032 and 033) is wavy, generally with a very low degree of inheritance, but there are some pseudocolumnar structures up to 1 cm wide and high (Photo 8-22). The microstructure ranges from diffuse and banded, through streaky, to unlaminated diffuse palimpsest. As described below, filament remnants are abundant but generally very indistinct. Where the microstructure is banded, the light and dark laminae are about the same thickness and are 15 to 60 μm thick. Non-carbonate detrital grains about 30 to 60 μm wide are common, and a clotted or mottled appearance of the laminae may indicate that the laminae consist mostly of carbonate peloids of about the same size. In more coarsely banded to streaky laminae (about 150 μm thick), the detrital particles are coarser (up to 150 μm) and siliciclastic grains are more abundant. Finally, there are laminae up to 3 mm thick where the distribution of kerogen outlines the former distribution of coarse microbial filaments. Between the filament remnants there is sparry carbonate interpreted as an open-space fill, and abundant detrital particles up to about 100 μm wide.

In the palimpsest microstructure the filament remnants are 60 to 350 μm wide by up to 3 mm long (Photo 8-22B-E). They are mostly perpendicular to the laminae (even where the laminae are inclined) but are sinous, and in places laminae containing subhorizontal and subvertical filament remnants alternate. The larger structures appear to have been tufts of numerous subparallel filaments, each about 60 μm wide; some of these have along their axes a darker, structureless thread about 10 μm wide, interpreted as a remnant of the original trichome surrounded by the remnants of a very thick sheath. In some of the banded to streaky laminae, 15 to 20 μm wide, faint filament remnants are abundant and are perpendicular to the laminae.

Fine laminoid fenestrae filled with sparry calcite occur uncommonly. The stromatolites consist predominantly of xenotopic calcite with a grain size of 50 to 100 μm. Tuffaceous siliciclastic detritus is abundant, and includes volcanic shards.

STRATIFORM STROMATOLITES

The stratiform stromatolites in this section are much the same as the nodular stromatolites described above, except that the banded microstructure is much more prominent and there are many millimetric and centimetric nodular and bulbous individuals (e.g., P.P.R.G. Samples Nos. 025, 026, 028, 029 and 041; Photo 8-23A). These individuals have been described previously as *Alcheringa narrina* Walter (1972), and need not here be redescribed, but some features of biological significance can be highlighted. In the new specimens (and those obtained previously from elsewhere) occur every stage in a gradation from erect tufts formed of a single filament 30 to 40 μm wide (Photo 8-23B,C), which closely resemble the erect growth habit of *Frutexites microstroma* Walter and Awramik (1979) from the Gunflint Iron Formation, through centimetric nodular and bulbous individuals with a banded microstructure but no filament remnants (Photo 8-24A), to similar individuals with a prominent radial fabric (Photo 8-24B,C); to larger irregular nodular and bulbous forms with a thick streaky microstructure; and to similar forms with a sinuous linear palimpsest microstructure (Photo 8-24D). The banded laminae are frequently 10 to 80 μm thick. The radial fabric is interpreted as a straight linear palimpsest fabric, and is marked by the arrangement of kerogen in 15 to 30 μm-wide lines that radiate but tend to remain near vertical, and that on the flanks of individual stromatolites are thus at an acute angle to the lamination. The sinuous linear palimpsest microstructure has radially arranged kerogen streaks 15 to 30 μm wide that are generally perpendicular to the lamination. One example was observed (Photo 8-25) of a palimpsest-film microstructure with remnants of sinuous filaments 2 to 15 μm wide that are alternately parallel and perpendicular to the laminae.

Siliciclastic grains about 10 μm wide are very abundant in the finely banded stromatolites. Clastic grains occur much more rarely in stromatolites with a sinuous linear palimpsest microstructure. The erect tufts were buried by tuffaceous sand.

8-4-10. Hamersley Group, Hamersley Basin, Australia

Stratiform and discoidal stromatolites were described from the banded iron-formations of this group by Edgell (1964), but in subsequent studies (Trendall and Blockley, 1970; Walter, 1970b) these structures were reinterpreted as being of diagenetic origin; to date, no definite stromatolites are known from the Hamersley Group iron-formations.

The Carawine Dolomite of the Hamersley Group was deposited on a platform flanking the eastern end of the Hamersley Basin. Although it is stromatolitic, its correlative in the basin is the generally non-stromatolitic Wittenoom Dolomite; this latter unit is known to contain stromatolites (Grey, 1979) only in the eastern part of the basin (and the unit is thus notable, since such little metamorphosed sedimentary carbonate formations lacking stromatolites are extremely rare in the Archean and Proterozoic). It is likely that the Wittenoom Dolomite was deposited mostly below the photic zone (resulting in its general lack of stromatolites) and that the Carawine Dolomite is a shallower water equivalent. According to Goode (1981) stromatolites previously attributed to the Carawine Dolomite in fact occur in younger units. Those we have examined (P.P.R.G. Sample No. 042) occur at a locality to which I was directed by Dr. Goode and are definitely in the Carawine Dolomite; they form a tabular biostrome 60 cm thick where there are contiguous decimeter to meter-sized nodular and bulbous stromatolites and erect centimetric pseudocolumnar forms (Photo 8-26A). The stratiform to gently convex laminae have a distinct banded to diffuse streaky microstructure (Photo 8-26B), with laminae 15 to 1000 μm thick (most dark laminae being at the thin end of this range). Locally a sinuous, linear, palimpsest microstructure is well developed (Photo 8-26C). The dark units are 30 to 150 μm wide and vertical, even where the laminae are inclined; they are up to 700 μm long.

8-4-11. Turee Creek Group, Hamersley Basin, Australia

Stromatolites were discovered in this unit by geologists of Hamersley Exploration Ltd. A sample was provided for study by Dr. M. D. Muir. No paleoenvironmental information relating to the sample is currently available. The sample contains pseudocolumnar structures with columnar units about 5 cm wide; however, there is insufficient information to allow these to confidently be interpreted as stromatolites. The structures have laminae with an obtuse conical shape with apical angles of 80 to 120°; there are no apical complexities. Cleavage in the rock has produced a crinkliness in the lamination. The microstructure is diffuse and banded to streaky, with laminae 50 to 300 μm thick. The rock consists of xenotopic carbonate (apparently calcite) with a grain size of 20 to 150 μm, and contains abundant concordant stylolites.

8-5. Biological Interpretation

Very few of the interpretations that follow can be considered compelling. Far too little is known about stromatolite morphogenesis, stromatolite microbiology, and the causes of cell clumping, cell orientation, and movement among extant stromatolite-building organisms, to allow confident conclusions to be drawn by analogy. The problems encountered here are quite different from those associated with interpreting microfossils (as discussed, for example, in Chapter 9). It is very likely that an improved understanding of the morphogenesis of extant stromatolites would lead to new interpretations of Archean stromatolites. At present, the field is most severely limited not by the availability of material for study, or by inherent limitations of the material (as with microfossils), but by current ignorance of morphogenetic processes. This section thus consists mainly of a discussion of reasonable possibilities rather than of well-established interpretations.

8-5-1. Early Archean

The interpretation of the ~3.5-Ga-old Warrawoona Group stromatolites has been discussed by Lowe (1980a) and by Walter et al. (1980), and is the subject of continuing research by R. Buick. Lowe (1980a) expressed the conventional view that because extant stromatolites are constructed mostly by cyanobacteria, those of the Warrawoona Group probably were also. However, as other kinds of prokaryotes, and algae, are also known to build stromatolites (Walter, 1972; Walter et al., 1972; Walter, 1978; Golubić, 1976a), the cyanobacterial origin of the Warrawoona Group stromatolites must be established independently. It is assumed here that algae (eukaryotes) did not evolve until some time in the Proterozoic, probably the mid-Proterozoic; the problem thus becomes to identify what kind(s) of prokaryotes were the constructing organisms of the Warrawoona structures. Lowe (1980a) suggested that the stromatolites he described resemble *Conophyton*—a suggestion that would appear to carry rather specific implications concerning the microorganisms that built them (Walter et al., 1976)—but, as noted above, the obtusely conical lamina profile and the lack of any crestal zone in the specimens he studied preclude their identification as *Conophyton*. Neither microfossils nor kerogen is preserved within the stromatolites described by Walter et al. (1980), nor are they reported by Lowe (1980a) from the specimens he examined. Stratiform stromatolites from the same sequence contain both (Awramik et al., 1982); in these, the carbon isotopic evidence suggests that the microbiota included autotrophs, and there are abundant narrow filaments (see Chapter 9). However, it is not possible to interpret the large bulbous and nodular stromatolites by analogy with the stratiform examples because the microstructures are quite different. Nevertheless, the regular, even lamination of the bulbous and nodular stromatolites is comparable with that demonstrably proproduced by filamentous organisms in other stromatolites, and is unlike that known to be produced by coccoid organisms. Thus, it seems likely that the major mat-formers were filamentous and, as many of the laminae are as thin as 20 μm, such filaments must have been no coarser than that (Walter et al., 1980).

The nodular, bulbous, and columnar stromatolites of the Warrawoona Group seem to have developed in environments where sulfate minerals were precipitating and which were intermittently desiccated (Lowe, 1980a; Walter et al., 1980). Thus, and although the microfossiliferous stratiform stromatolites are not known to be closely associated with features indicative of hypersalinity or desiccation, it seems that at least some of the Warrawoona mat-forming organisms were able to survive the stresses of high salinity and low water potential. In addition, they must have had some stratagem for coping with high light intensity and the danger of burial by sediment. In extant cyanobacterial mats this stratagem involves gliding motility and, most probably, sheath production to combat desiccation. It seems very likely therefore that the various filamentous organisms presumed to have built the structured stromatolites of the Warrawoona Group were capable of gliding motility in response to gradients in light intensity.

It is worth noting that even if it could be concluded that these or other ancient stromatolites were constructured by cyanobacteria, it does not automatically follow that they would have been exclusively oxygenic. More and more cyanobacteria are being found to be facultatively anaerobic photosynthesisers that can use H_2S rather than H_2O as an electron donor (Garlick et al., 1977); some of these form benthonic mats (Krumbein and Cohen, 1977; Hirsch, 1978).

If the fabrics of the comparably aged Onverwacht Group are accepted as stromatolitic, then the example described by Muir and Grant (1976; and discussed and illustrated in Chapter 9 of the present volume) can be compared to "pustular mat" structures described, for instance, from Shark Bay (Logan et al. 1974). These are built by the unicellular cyanobacterium *Entophysalis major* Ercegović (Golubić, 1976a, p. 117). This organism is non-motile, and the stromatolites accrete as a result of the mat growing over sediment deposited in irregularities on its surface.

8-5-2. Middle Archean

Two distinctly different fabrics occur in the ~3.1-Ga-old Insuzi Group stromatolites: that consisting of juxtaposed hemispheroidal lenses some of which have a fine, linear, radial palimpsest microstructure; and that consisting of banded to streaky laminae with convex and conical forms. The former appears to have been constructed by a filamentous organism with filaments (or groups of filaments) at least 5 μm wide arranged radially in tussocks. Hemispheroidal laminae within the tussocks are interpreted as marking successive intermittent growth episodes. Somewhat similar structures are formed by the oscillatoriacean cyanobacterium *Phormidium hendersonii* (Golubić and Focke, 1978; Monty, 1976) and by the rivulariacean cyanobacterium *Rivularia haematites* (Monty, 1976), although the radial fabric is much less evident in the fossil structures.

The Insuzi stromatolites with conical laminae can be interpreted by comparison with the extant stromatolite *Conophyton weedii*, which is constructed by the finely filamentous oscillatoriacean cyanobacterium *Phormidium tenue* (Walter et al., 1976). Points of similarity between these ancient and modern stromatolites suggesting comparable morphogenesis are their acutely conical lamina profile, the marked thickening of some laminae in their crestal areas, the hemispherical and subspherical thickenings within laminae away from such crestal areas, and their banded to streaky microstructure. A difference between the two is the irregular orientation of column axes in the ancient example, whereas in *C. weedii* the columns are parallel. The points of similarity all suggest that the Insuzi stromatolites, like their extant counterparts, were built by finely filamentous microorganisms that were positively phototactic. However, it still needs to be demonstrated for the Insuzi stromatolites that the conical lamina form and the crestal complexities are consistently developed since until now only a small number of specimens has been studied. In *C. weedii*, filament clumping is apparently a result of microaerophilia (Weller et al., 1975), but as very little is known about causes of cell clumping in cyanobacteria and other prokaryotes it would be premature to ascribe this same characteristic to the microorganisms that built the Insuzi stromatolites.

8-5-3. Late Archean

Belingwe, Bulawayan, Steeprock, Carawine, and Yellowknife Stromatolites

The Steeprock Group and Bulawayan Group stromatolites here studied are very similar in both age and morphology and can be considered together. Little can be inferred from the large linked nodular forms; the fabric, where best preserved, is banded, and was therefore most probably constructed by filamentous organisms. A columnar-layered stromatolite from the Steeprock Group has a combined film-linear palimpsest microstructure. Microstructures of this kind are produced in the modern environment either by (*i*) motile, filamentous cyanobacteria (Monty, 1976, figs. 2A, 4B; Bertrand-Sarfati, 1976a) or filamentous chlorobacteria (Doemel and Brock, 1974) occurring in mats where the organisms have an alternating erect and prostrate growth habit; or by (*ii*) a mat where one species forms the prostrate films and a second grows within the mat to form a framework of erect filaments (Monty, 1976). These two possibilities cannot be distinguished in the Steeprock stromatolite. As noted in Chapter 5, all of the Bulawayan and Steeprock stromatolites here analyzed contain light kerogen ($\delta^{13}C_{PDB}$ lighter than -20 ‰) and are composed of carbonate with $\delta^{13}C_{PDB}$ values near zero per mil (see also Schopf et al., 1971), which support the interpretation that they were built by autotrophic organisms. P.P.R.G. Sample No. 328 from the Steeprock Group contains abundant carbonate peloids and appears to have been constructed by the trapping of these detrital grains. The laminae have a film microstructure suggestive of construction by finely filamentous organisms. This is the oldest known example of a stromatolite apparently constructed via sediment trapping by filamentous organisms (the older Archean stromatolites may have formed this way from finer detritus, but no detrital texture is preserved). As is discussed in Section 8-3-3, above, the abundant presence of detritus may be indirect evidence that the constructing organisms were phototactic.

Martin et al. (1980) have described a diverse assemblage of stromatolites from the Belingwe Greenstone Belt in Zimbabwe. A geochemical and microstructural study of these could provide significant biological information. The authors draw attention to the similarity of these stromatolites to younger forms and infer by analogy that they were constructed by oxygenic, photosynthetic "algae" (i.e., cyanobacteria). As for all other Archean stromatolites, this interpretation would need to be supported by independent evidence. Among the stromatolites they tentatively identify is one that is "*Conophyton*-like." If this identification can be confirmed, these would be the oldest known examples of *Conophyton*, and could be interpreted by analogy with extant examples (Walter et al., 1976) as having been constructed by filamentous, positively phototactic organisms that may have been microaerophilic.*

The Yellowknife stromatolites have a sinuous linear palimpsest microstructure suggestive of construction by filamentous organisms with filaments (or clusters of filaments) more than 20 μm wide. Particular caution is necessary with this interpretation because in one specimen the linear structures are straight and could be pseudomorphs after an acicular mineral. Additional caution is needed because the contained kerogen in the two analyzed samples (Chapters 5 and 7) is relatively heavy ($\delta^{13}C_{PDB} = -16$ ‰). The banded to streaky microstructure of the Carawine Dolomite stromatolites similarly indicates construction by filamentous organisms. The same or other filamentous organisms within the constructing mats produced a linear palimpsest microstructure with units that are vertical even where laminae are inclined, suggesting that they were oriented in response to light. Kerogen in these stromatolites is heavy ($\delta^{13}C_{PDB} = -9$ ‰); the significance of this unusual isotopic composition is at present unknown.

While no one set of these Late Archean stromatolites is especially informative biologically, taken together the Belingwe, Bulawayan, Steeprock, Carawine, and Yellowknife stromatolites present a picture of a diverse benthonic microbiota in marine environments, apparently dominated by filamentous microorganisms. Evidence of the erect orientation of filaments and of construction by sediment trapping, suggests that some of the organisms oriented themselves and moved in response to light; coupled with the carbon isotopic composition of the kerogen (which apparently indicates the presence of autotrophic organisms), this suggests that the organisms were photosynthetic.

* While this paper was in press, Grey (1981) described *Conophyton*-like stromatolites from the Late Archean of Western Australia.

FORTESCUE AND VENTERSDORP STROMATOLITES

The Fortescue and Ventersdorp stromatolites are here considered separately from those of the other Late Archean units because of their distinctiveness, similarity to each other, and occurence in probable lacustrine and fluviatile facies rather than marine facies. The various filmy, banded, streaky, and striated microstructures of the Ventersdorp stromatolites are all comparable with those produced by filamentous microorganisms, some of which were no thicker than 10 μm (the thickness of some of the finest laminae). The abundance of sand- and silt-sized detritus in some of the stromatolites suggests that the formative organisms were able to cope with rapid sedimentation (i.e., they were probably phototactic). The pseudocolumnar structures from the T'Kuip Hills locality are considered to have formed through carbonate precipitation in situ within a microbial thallus since they contain much less siliciclastic detritus than does their matrix (although the difficulties in drawing biological inferences from this observation are discussed in Section 8-3-1).

The Ventersdorp and Fortescue stromatolites contain the oldest demonstrably fenestrate structures now known. These are thought to form in extant stromatolites as a result of the oxidation of organic matter (Logan, 1974), desiccation-induced shrinkage, or gas production (Shinn, 1968), but the details of such processes are unknown. Among the Ventersdorp fenestrate structures are relatively large, irregularly bulbous examples that are unlikely to have formed as a result of desiccation because the shapes of the sides of the fenestrae do not match each other; the possibility that these are indicative of aerobic decomposition of organic matter is intriguing but cannot be pursued until more is learned about how fenestrae form (and, of course, they might alternatively be products of solely anaerobic decay).

The Fortescue stromatolites are interpreted as the products of at least three distinct microbial communities. One of these cannot be readily interpreted, because of a paucity of specimens and inadequate preservation (viz., that evidenced in P.P.R.G. Thin Sections Nos. 024-1 and 024-2).

However, the thin sections evidencing this community have a microstructure that is comparable with that of Early Proterozoic stromatolites demonstrably constructed (Hofmann, 1975) by alternating communities of coccoid entophysalidacean and filamentous, possibly nostocalean, cyanobacteria. There is one example of a palimpsest-film microstructure composed of 2 to 15 μm wide remnants of sinuous filaments that are alternatively parallel and perpendicular to the laminae; a few filament remnants can be followed from where they are erect through where they curve over to become prostrate. This is very similar to fabrics produced in modern stromatolites by motile, finely filamentous oscillatoriacean cyanobacteria such as *Phormidium hendersonii* (Monty, 1976; Golubić and Focke, 1978) and *P. truncatum* (Walter et al., 1976, figs. 2–5) and by the motile, finely filamentous chlorobacterium *Chloroflexus aurantiacus* (Doemel and Brock, 1974). In the fossil stromatolite the filament remnants are vertical although the laminae are inclined, suggesting that the orientation is a result of positive phototaxis as in the oscillatoriaceans, rather than of positive aerotaxis as in *Chloroflexus*. The same filament orientation occurs in the non-motile, filamentous rivulariacean cyanobacterium *Calothrix coriacea*, where the lamination results from changing, possibly seasonal, growth conditions (Walter, 1976c, fig. 8). The laminae in the *Calothrix* mats represent relatively long time periods (perhaps seasons), whereas those of the *Phormidium* mats commonly form daily. Sedimentological indications are that the fossil example occurs in rapidly deposited sediments; rapid mat accretion involving filament motility is thus likely. It seems reasonable to conclude that the fossil filamentous organisms were positively phototactic, and may have been oscillatoriaceans, but since only one example of this fabric has been recognized thus far from the Fortescue Group, this conclusion must be regarded as tentative.

The intergrading tufted-nodular-bulbous stromatolites of the Fortescue Group are remarkably well preserved and can be interpreted with some confidence. Here they are all considered to be products of a single, essentially monospecific, microbial community, the constructing organism consisting of trichomes at least 10 μm wide (as shown by the width of degraded trichome remnants within filament tufts) that were possibly 15 to 20 μm wide originally (the width of many faint remnants). Where there is abundant coarse detritus in and around the stromatolites, filament remnants are distinct and separate, in many examples being arranged as tufts that are enclosed by detritus. In these areas, the remnants range from single filaments 30 to 40 μm wide (each of which is interpreted as a single trichome enclosed by a thick sheath), through tufts of such filaments, to filament clusters arranged to form millimetric and centimetric nodular stromatolites. Where there is less coarse detritus, the filament remnants are contiguous and the stromatolites have a prominent banded microstructure. The banding is identical to that in the tufts where it is considered to be the layering of a former gelatinous sheath, and thus in the nodular and bulbous stromatolites it is interpreted as growth layering in a former common gel that enclosed the trichomes of the thallus. In some examples where no coarse detritus is apparent, the banding is the most prominent feature of the stromatolite. These stromatolites were able to establish themselves and grow on rippled tuffaceous sand, and to survive frequent burial by 1 to 2 mm of sand (as shown by the presence of sandy layers within many of the stromatolites). The filament remnants are vertical, or in nodular stromatolites are oriented like the stalks in a sheaf (radially but remaining near-vertical). Only in one example, from the edge of large nodular stromatolite, were the filament remnants observed to be consistently near-horizontal. It is likely, then, that this microorganism was phototropic or phototactic, and was able to cope with rapid sedimentation. The carbon isotopic data for the Fortescue Group are considered to indicate the former presence of autotrophs (Chapters 5, 7). The carbon data cannot be attributed to any particular stromatolite type, but the forms interpreted here are the most abundant so the additional interpretation is made that the stromatolite-constructing organism was an autotroph.

Very similar stromatolites are constructed now by oscillatorialean and nostocalean cyanobacteria. Coarsely filamentous examples that form tufts are *Lyngbya*, among the Oscillatoriaceae, and *Scytonema*, among the Scytonemataceae. *Rivularia* makes a gelatinous thallus within which trichomes are radially arranged, similar to that of some of the small, nodular, Fortescue stromatolites. A single example of a 10 μm wide *Lyngbya*-like fossil trichome is known from the Fortescue stromatolites (Chapter 9). The Fortescue tufts closely resemble the fossil scytonematacean *Frutexites microstroma* from the early Proterozoic Gunflint Iron Formation (Walter and Awramik, 1979), but only in its erect habit. The hard calcareous stromatolites constructed in lakes by *Rivularia haematites* (Monty, 1976) are very similar to the Fortescue stromatolites.

It seems reasonable to postulate, therefore, that the intergrading series of Fortescue stromatolites was built by a *Lyngbya*-like oscillatoriacean that was different from extant species in that it sometimes produced a common gelatinous thallus enclosing many trichomes. This interpretation is strongly influenced by the single occurrence of a *Lyngbya*-like microfossil of a size consistent with it having been a member of the stromatolite-building community. The Fortescue stromatolites are extraordinarily well-preserved, so further work can be expected to confirm or deny this hypothesis.

8-5-4. Early Proterozoic Stromatolites

Stromatolites are abundant and diverse in basins of the Early Proterozoic, and some have been studied in detail by other investigators. Many such stromatolites were included among the samples subjected to geochemical analysis by P.P.R.G., but they are not morphologically described here. However, to allow a comparison with Archean stromatolites, some of the pertinent observations are assembled below.

Early Proterozoic stromatolites apparently formed in many different environments, both marine and nonmarine, and in some places were responsible for constructing very extensive carbonate shelves and reefs: those of the Coronation Geosyncline and Athapuscow Aulacogen in Canada (e.g. Hoffman, 1976) and the Transvaal Basin in South Africa (e.g., Truswell and Eriksson, 1972, 1973, 1975a, 1975b; Eriksson et al., 1976) are the best known. It is apparent that stromatolites formed throughout the photic zone in near-shore oceanic environments, and that there are many similarities between such forms and the stromatolites now forming in the intertidal zone of Shark Bay, Western Australia (as noted by the authors listed above). It may be significant that no assuredly deep-water fossil (or modern) stromatolites have been recognized, as this leads to the inference that the constructing organisms were dependent on light as a source of energy. The poorly laminated stromatolites of the McLean Formation in the Athapuscow Aulacogen (Hoffman, 1976) are interbedded with turbidite greywackes and formed in a relatively deep basin flanking a carbonate shelf. Though it is not possible to estimate the former water depth, it would have been several tens of meters at least. The poor lamination may be indicative of growth at a low light intensity, and it may be that they formed near the base of the photic zone. In the Late Archean to Early Proterozoic Hamersley Group of the Hamersley Basin, Australia, there is a further indication of the light-dependence of the stromatolite-building microbiota. The stromatolitic Carawine Dolomite occurs on what was a relatively shallow shelf on the eastern margin of the basin; the same stratigraphic interval within the basin is occupied by the Wittenoom Dolomite, one of the very few known examples of an essentially non-stromatolitic Precambrian dolomite. The iron-formations that occur above and below the Wittenoom Dolomite were deposited under quiet-water conditions, below wave base (Trendall, 1976), and it is very likely that the Wittenoom Dolomite lacks stromatolites because it was deposited below the photic zone.

Prior to the past few years it seemed that stromatolites may have been less diverse in Early Proterozoic time than later in the Proterozoic (Awramik, 1971), but recent interest in the Early Proterozoic has led to the description of many new forms ("species"), such that the known diversity of named stromatolites of this age (Bertrand-Sarfati and Walter, 1981) is now comparable with that from the Late Proterozoic. Gebelein (1974) considered that at least by the end of the Early Proterozoic, all major groups of stromatolite-building cyanobacteria were present, as evidenced by microfossils and stromatolite microstructures. Following a study of the 2.2-Ga-old columnar stromatolites of the Transvaal Basin (Bertrand-Sarfati and Eriksson, 1977), it became apparent that the microstructures of these ancient stromatolites have many features comparable to those of extant cyanobacterial stromatolites (Bertrand-Sarfati, 1980). The oldest firmly identified *Conophyton* stromatolites are also Early Proterozoic,* from the Athapuscow Aulacogen and the Coronation Geosyncline (Semikhatov, 1978); these have been considered to be cyanobacterial in origin (Walter et al., 1976).

As is reviewed in Chapter 14, some Early Proterozoic stromatolites are microfossiliferous. The

* See footnote p. 187

affinities of such microfossils are there discussed; some additional observations that strengthen the interpretation that the stromatolite-constructing organisms of this age were in fact cyanobacteria are set out below. Particular stromatolites of the Gunflint Iron Formation were built both by *Gunflintia minuta* Barghoorn and *Huroniospora* spp. (Awramik and Semikhatov, 1979). Difficulties of determining the affinities of *G. minuta* have been discussed by several authors (see Chapter 14). It is notable, though, that in the columnar stromatolites with walls (i.e., with laminae that coat the column margins), *G. minuta* filaments in the walls are oriented parallel to the laminae, as are the filaments of the cyanobacterium *Phormidium truncatum* in the walls of an extant columnar stromatolite *Vacerilla walcotti* (Walter et al., 1976). This similar, and probably phototactic, response is consistent with the view that *G. minuta* was a cyanobacterium (a conclusion that is similarly consistent with its morphology and stromatolite-building habit).

The coccoid microfossil *Eoentophysalis belcherensis* seems to have been the dominant component of a community that constructed stratiform stromatolites in the Early Proterozoic (Chapter 14). Its very close resemblance to the extant entophysalidacean cyanobacterium *Entophysalis major* extends not only to its cellular morphology and pattern of gel production, but in addition to the detailed morphology of the microbial mats it formed (as controlled by patterns of cell division) and the environment where it constructed stromatolites (viz., the marine intertidal zone). Thus, the interpretation that *Eoentophysalis belcherensis* was a cyanobacterium seems firmly established.

8-6. Summary and Conclusions

(1) Counted by major stratigraphic units, eleven occurrences of stromatolites (and possible stromatolites) are known from Archean sequences dating back to 3.5 Ga ago. Three of these are too poorly known to be interpreted as definitely stromatolitic. More than half of the eleven occurrences were discovered or first described during the last decade, and with the increasing attention being given to Archean sediments it can be expected that many more occurrences will be discovered. Certain of the Late Archean occurrences are very imprecisely dated, and may prove to be earliest Proterozoic. Nonetheless, it is apparent that stromatolites were an abundant component of the sedimentary environment at least as early as about 2.8 Ga ago.

(2) Archean stromatolites are not markedly different from younger examples. Although most have some distinctive features, all can be interpreted by reference to Holocene analogs. What most limits the interpretation of Archean stromatolites is the current profound ignorance of the morphogenesis of Holocene stromatolites. Thus, it would be premature always to equate the presence of stromatolites with the former presence of mats of cyanobacteria, rather than other kinds of prokaryotes. Furthermore, even if stromatolites are interpreted as cyanobacterial, it must be recalled that use of water as an electron donor is not obligatory for all cyanobacteria; many can use H_2S and thus not release oxygen to their environment.

(3) Early Archean stromatolites were constructed by filamentous prokaryotes (e.g., those of the Warrawoona Group) and possibly also by unicellular prokaryotes (e.g., those of the Onverwacht Group), among which were autotrophs. It is probable that the principal stromatolite-builders were phototactic, photoautotrophic, and produced mucus sheaths. Such microbes were able to cope with high salinities, desiccation, and high light intensities, indicating that the appropriate protective mechanisms were already well developed. In the Early Archean, the occurrence of at least two, or possibly three, different microbial communities seems indicated by the range of stromatolite microstructures now known.

(4) The products of two distinct microbial communities can be recognized in the only Middle Archean stromatolites yet described, those from the Insuzi Group of Natal. Both apparently included filamentous prokaryotes, which in one example were radially arranged to form hemispherical tussocks, and which in the other formed conical tufts. The conical tufts are similar in several respects to *Conophyton*, suggesting that the constructing organisms were probably phototactic and perhaps microaerophilic. Carbon isotopic analyses suggest that both of these communities contained autotrophs, but the kerogen is less enriched in ^{12}C than that of most other Precambrian stromatolites.

(5) Late Archean stromatolites are known from what appear to be both lacustrine and marine paleoenvironments. They are morphologically and microstructurally diverse, suggesting a similar degree of diversity of microbial communities. The kerogen they contain is enriched in ^{12}C relative to

coexisting carbonates, apparently indicating the influence of autotrophs. Some are closely comparable to extant stromatolites, suggesting that they may well have been constructed by cyanobacteria. Several lines of evidence strongly suggest that at least some of the constructing organisms were photoautotrophic. All appear to have been built by filamentous organisms, some of which had thick sheaths or were embedded in a common gel. There is one example of a fabric possibly constructed by alternating communities of filamentous and coccoid bacteria. Some of the fabrics in the 2.77-Ga-old Fortescue Group stromatolites are very similar to those now constructed by oscillatoriacean cyanobacteria. The oldest known fenestrate fabrics occur in Late Archean stromatolites; these may be indicative of heterotrophic decomposition, possibly by aerobes.

(6) During the Early Proterozoic, benthonic microbial communities constructed stromatolites in peritidal and relatively deep subtidal environments in the oceans, probably down to the base of the photic zone (about 100 m). Such communities were so abundant that it is rare to find an Early Proterozoic sedimentary carbonate rock unit that is not at least partly stromatolitic. Many of these stromatolites seem to have been built by microbial communities that were dominated by cyanobacteria.

(7) Studies of early Precambrian stromatolites, especially those of the Archean, are still in their infancy; much remains to be learned before questions such as those relating to their potential biostratigraphic usefulness (if any) can be fruitfully examined. Nevertheless, progress during the past decade has been impressive. It seems reasonable to expect that the now rapidly increasing interest and activity in such studies will in the near future yield significant new insight regarding the distribution and mode of formation of these ancient microbial deposits and the evidence they provide of the Earth's earliest benthos.

CHAPTER 9

Archean Microfossils: New Evidence of Ancient Microbes

By J. William Schopf and Malcolm R. Walter

9-1. Previous Studies

As is discussed in the preceding chapter, it now seems well established that microbially laminated stromatolites became both abundant and relatively widespread at least as early as 2.7 Ga ago, during the Late Archean, and that such structures remained ubiquitous components of carbonate facies environments throughout the whole of the Proterozoic, decreasing in abundance only near the close of the Precambrian (apparently as a result of competitive exclusion of stromatolite-building microbes by eukaryotic, calcareous and non-calcified algae, and of the disruption of stromatolitic habitats by burrowing and grazing metazoans; see Garrett, 1970; Monty, 1973; and Awramik, 1971). Although preservation of the mat-building microorganisms involved in stromatolite formation requires a rather special set of circumstances (in general, the early diagenetic replacement of stromatolitic carbonate by chemically precipitated silica), and for this reason cellularly preserved microbes are actually of only rare occurrence in stromatolitic units, the known microbial fossil record more or less parallels that established for the megascopic stromatolites, as is shown in Figure 9-1; however, two principal differences are apparent. First, and although in the Phanerozoic microbial fossils have been reported from numerous stromatolites and stromatolitic deposits (e.g., oncolites, *Girvanella*-containing intraclasts, etc.), they are also known from various non-stromatolitic facies as well (e.g., petrified peats, lignites and brown coals, boghead coals, Carboniferous "coal balls," etc.); thus, while stromatolites decreased in abundance, and apparently diversity, during the Phanerozoic, the microbial fossil record of this Era (albeit unimpressive in comparison with that of

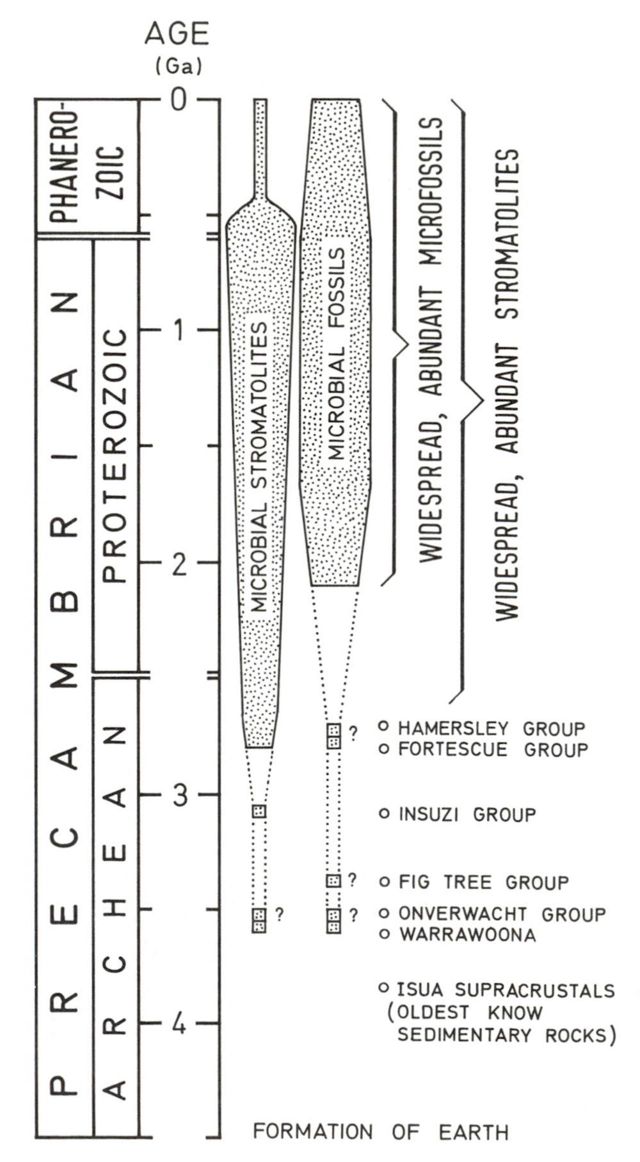

Figure 9-1. Schematic diagram summarizing known geologic distribution of stromatolites and microbial microfossils.

coexisting plants and animals) remained relatively diverse (for recent reviews of the cyanobacterial portion of this record, see Schopf, 1974; and Schopf and Walter, 1981). And second—in marked contrast with the known distribution of stromatolitic sediments—microbial fossils are as yet all but unknown from the Archean and early portion of the Proterozoic (Figure 9-1). Indeed, it has only been during the past year or so that bona fide filamentous microbes, fossils comparable in form to the abundant, well-known, mat-building taxa of the Proterozoic (see Chapter 14), have been discovered in sediments older than about 2.0 Ga and, to date, only two such occurrences have been reported (Schopf and Walter, 1980; Awramik, 1981; Awramik et al., 1983; see Section 9-5; for a revised assessment of the filamentous "microfossils" described by L. Nagy, 1974, from the ~2.3-Ga-old Transvaal Dolomite, see Table 14-2, No. 23). Thus, as now known, there exists a major disparity between the microbial fossil records of the Archean and Proterozoic.

Comparison of the data summarized in the present chapter (Table 9-1) with those presented in Chapter 14 (Tables 14-2 and 14-3) well illustrates the magnitude of the disparity. To date, some 86 categories of microfossils and "microfossil-like objects" have been reported from 40 geologic units of Early Proterozoic age; of these units, 23 are regarded as containing established, bona fide microbiotas (an average of about one such assemblage per 40 Ma over the 900 Ma-long duration of the time segment). In contrast, of the 28 geologic units (containing 43 categories of microfossils and "microfossil-like objects") thus far investigated from the decidedly longer Archean rock record (Figure 9-1), only two are known to include unequivocal microfossils (an average of one firmly established microbiota per 650 Ma over the 1,300 Ma-long duration of the known Archean record). In other words, although the Archean rock record encompasses a time span nearly one and one-half times the duration of the Early Proterozoic, it includes only about 10% as many established microfossiliferous units (and only about 5% as many described genera of authentic microfossils). Why is there this disparity? Might it reflect fundamental differences between the Archean and

TABLE 9-1. Archean microfossils and microfossil-like objects.

Lithology	*Reported objects*	*Interpretation*
	Isua Supracrustal Belt *Godthabsfjord region, southwestern Greenland* 3.770 ± 0.042 *Ga*	
(1) Greenish-white metaquartzite	(1) *Isuasphaera isua* Pflug, "yeast-like microfossils" (Pflug, 1978a, 1978b, 1978c; Pflug & Jaeschke-Boyer, 1979; Pflug et al., 1979).	(1) NONFOSSIL: metamorphically produced multiphase inclusions (Schopf & Walter, 1980; Bridgwater et al., 1981), possibly containing organic fluids (compare with structures described by Mueller, 1972), see Photo 9-1.
(2) Micaceous metaquartzite	(2) Globular graphite microstructures (B. Nagy et al., 1975, 1976; B. Nagy & L. Nagy, 1976).	(2) NONFOSSIL: graphite particles (B. Nagy et al., 1976).
	Warrawoona Group *North Pole Dome, Western Australia* 3.556 ± 0.032 *Ga*	
(3) Carbonaceous stromatolitic chert	(3) "Filamentous fossil bacteria" (Awramik et al., 1983; see also Awramik, 1981).	(3) MICROFOSSILS: filamentous Archean microorganisms (Awramik et al., 1983); see Photos 9-4, 9-5, 9-7A-E, and 9-8.
(4) Carbonaceous stromatolitic chert	(4) Schopf, "unicell-like organic spheroids" and "colony-like aggregates of radiating filaments." (Awramik et al., 1983).	(4) DUBIOMICROFOSSILS: possibly biogenic organic spheroids and filaments (Awramik et al., 1983); see Photos 9-6 and 9-7F-I.
(5) Chert	(5) "Spheroidal carbonaceous microfossils" (Dunlop et al., 1978).	(5) DUBIOMICROFOSSILS: carbonaceous probably nonbiogenic globules (Schopf & Walter, 1980).

TABLE 9-1. (*continued*)

Lithology	*Reported objects*	*Interpretation*
	Onverwacht Group *Swaziland System, eastern Transvaal, South Africa* 3.540 ± 0.030 *Ga*	
(6) Carbonaceous cherts and shales of the Swartkoppie, Kromberg, and Theespruit Formations	(6) *Archaeosphaeroides barbertonensis* Schopf & Barghoorn, and similar "alga-like" or "unicell-like" carbonaceous spheroids [see Schopf, 1975, p. 235–238 regarding stratigraphic source of *A. barbertonensis*] (Pflug, 1966, 1967; Schopf & Barghoorn, 1967; Engel et al., 1968; B. Nagy & L. Nagy, 1969, 1976; Brooks, 1970; Barghoorn, 1971; Brooks & Muir, 1971; L. Nagy, 1971, 1972; Dungworth & Schwartz, 1972; Brooks & Shaw, 1973; Brooks et al., 1973; Brooks & Muir, 1974; Muir & Hall, 1974; Lopuchin, 1976; Muir & Grant, 1976; B. Nagy et al., 1976; Muir et al., 1977; Knoll & Barghoorn, 1977; Strother & Barghoorn, 1980).	(6) DUBIOMICROFOSSILS: although smaller spheroids of this category (e.g., Muir & Grant, 1976; Knoll & Barghoorn, 1977) may be biogenic, larger spheroids (viz., >55 μm; Schopf, 1977a) are probably nonbiogenic floccule-like globules (see Schopf & Prasad, 1978); see Photo 9-2.
(7) Carbonaceous chert	(7) *Eobacterium isolatum* Barghoorn & Schopf, filamentous strands, and other "ultra-microfossils" detected by transmission electron microscopy (Barghoorn & Schopf, 1966; Barghoorn, 1971; Prashnowsky & Oberlies, 1972).	(7) NONFOSSIL: bubbles and similar artifacts, and modern bacterial contaminants (Barghoorn & Schopf, 1966, Figure 5?), introduced during sample preparation (Schopf, 1975, p. 235; Cloud, 1976a, p. 355–359).
(8) Chert	(8) *Ramsaysphaera ramses* Pflug, "yeast-like microfossil" (Pflug, 1976, 1978a, 1978b, 1979; Pflug & von Klopotek, 1978; Pflug et al., 1979).	(8) NONFOSSIL: mineralic dendrite (Schopf & Walter, 1980; compare Pflug, 1976, figs. 1, 3, 4 with Saratovkin, 1959, figs. 3, 24, 25, 59–62).
(9) Carbonaceous chert and shale	(9) "Blue-green alga-like filaments" (Engel et al., 1968; Brooks & Shaw, 1973; Brooks et al., 1973; Brooks & Muir, 1974; Muir & & Hall, 1974; Muir & Grant, 1976).	(9) NONFOSSIL: modern contaminants (e.g. Brooks et al., 1973, fig. h) or linear aggregates of amorphous carbonaceous matter (Schopf, 1975, p. 236).
	Gorge Creek Group *Mt. Goldsworthy region, Western Australia* ~3.4 *Ga*	
(10) Iron-formation (jaspilite)	(10) Mineralic spheroidal "microfossils" (LaBerge, 1967, 1973; Hamilton, 1976); in addition to the units here listed, LaBerge (1973) has reported comparable "microfossils" from the Timagami (Ontario, Canada), Kaministikwia (Ontario, Canada), Wilgie Mia (Western Australia) and Wilgenia (South Australia) Archean iron-formations.	(10) NONFOSSIL: mineralic (largely hematitic) pseudofossils (Muir, 1978b, p. 16) similar to colloid-derived spheroids of the Proterozoic Sokoman Iron Formation (Klein & Fink, 1976, p. 471, 474) and to laboratory-produced microstructures (Merek, 1973; J. Oehler, 1976b).
(11) Iron-formation (jaspilite)	(11) Mineralic filamentous "microfossils" (Hamilton, 1976).	(11) NONFOSSIL: filamentous mineral crystallites (compare with microstructures described by Hawley, 1926, and Merek, 1973).
	Fig. Tree Group *Swaziland System, eastern Transvaal, South Africa* <3.5, >3.1 *Ga*	
(12) Carbonaceous chert and shale	(12) "Unicell-like" carbonaceous spheroids (Pflug, 1966, 1967; Pflug et al., 1969; Dungworth & Schwartz, 1972; L. Nagy, 1972; Brooks & Shaw, 1973; Brooks & Muir 1974; Lopuchin, 1976; B. Nagy et al., 1976).	(12) DUBIOMICROFOSSILS: see 6, above (in part, probably *nonbiogenic*; in part possibly *biogenic*).

TABLE 9-1. *(continued)*

Lithology	Reported objects	Interpretation
(13) Carbonaceous chert	(13) "spore-like microfossils" detected by transmission electron microscopy (Prashnowsky & Oberlies, 1972).	(13) NONFOSSIL: artifacts of preparation (cf. 7, above).
(14) Carbonaceous chert	(14) "*Kakabekia umbellata*-like" microstructures studied by scanning electron microscopy (Dungworth & Schwartz, 1972).	(14) NONFOSSIL: artifacts of preparation; desiccated, clumped organic matter.
(15) Carbonaceous chert and shale	(15) "Blue-green alga-like filaments" (Pflug, 1966, 1967).	(15) NONFOSSIL: linear aggregates of amorphous carbonaceous matter (see 9, above).
	Uitkyk Formation *Pietersburg Schist Belt, northern Transvaal, South Africa* <3.1? *Ga*	
(16) Conglomerate	(16) Globular carbonaceous bodies ("fly-speck carbon") regarded as "spores" of primitive "differentiated" plants (Saager & Muff, 1980).	(16) NONFOSSIL: spheroidal aggregates of carbonaceous matter.
	Madras Gneiss *"Predcharvarian Complex," South India* >3.0? *Ga*	
(17) Graphitic gneiss	(17) *Menneria*, *Microtaenia*, *Archaeosphaeroides*, and irregularly shaped "microfossils" (Lopuchin & Moralev, 1973; Lopuchin, 1975, 1976).	(17) NONFOSSIL: artifacts of preparation (see 18, below) and irregular graphitic aggregates (Schopf & Prasad, 1978, p. 349).
	Yengrian Series *Aldan Shield, Siberia, U.S.S.R.* >3.0? *Ga*	
(18) Calciphyre (marble)	(18) *Menneria* spp., spheroidal, carbonaceous "fossil phytoplankton" (Lopuchin, 1975, 1976).	(18) NONFOSSIL: carbonaceous artifacts, in part resulting from isolation technique (carbonaceous floccules produced during "water separation and thin chemical preparation"), in part due to clumping on microscope slides (e.g., Lopuchin, 1975, figs. 1d, e), and in part being modern contaminants (e.g., Lopuchin, 1975, figs. 2a, b).
(19) Calciphyre (marble)	(19) *Microtaenia* sp. carbonaceous "ribbon-like scraps" (Lopuchin, 1975).	(19) NONFOSSIL: artifacts (see 18, above).
(20) Calciphyre (marble)	(20) Narrow, branching, "filamentous alga" (Lopuchin, 1975).	(20) NONFOSSIL: artifacts (see 18, above).
	Fortescue Group *Tumbiana Formation, Western Australia* 2.768 ± 0.024 *Ga*	
(21) Cherts in stromatolitic carbonates	(21) "Filamentous fossil bacteria" (Schopf & Walter, 1980, 1982; Schopf, 1980a, 1980b).	(21) MICROFOSSILS: filamentous Archean microorganisms (Schopf & Walter, 1980, 1982); see Photos 9-9 and 9-10.
	Southern Cross Quartzite *Yilgarn, Western Australia* >2.7 *Ga*	
(22) Opal veinlets in quartzite	(22) "Hematitic and goethitic fossilized microorganisms" (Marshall et al., 1964).	(22) NONFOSSIL: Tertiary crystallites, products of weathering (Cloud, 1976, p. 356; Muir, 1978b, p. 16; Cloud & Morrison, 1979a, p. 89).

TABLE 9-1. (*continued*)

Lithology	*Reported objects*	*Interpretation*
	Soudan Iron Formation northeastern Minnesota, U.S.A. >2.7 *Ga*	
(23) Iron-formation (jaspilite)	(23) Mineralic spheroidal "microfossils" (Cloud & Licari, 1968; LaBerge, 1973).	(23) NONFOSSIL: mineralic pseudofossils (see 10, above).
(24) Iron-formation (banded black jasper)	(24) Sinuous, mineralic(?), fossilized "blue-green algae" (Gruner, 1924).	(24) NONFOSSIL: apparently filamentous crystallites (see 11, above).
(25) Pyrite nodules in "coal-like" rock	(25) "Blister-like objects" compared with "bacteria or blue-green algae," detected by transmission electron microscopy (Cloud et al., 1965).	(25) NONFOSSIL: artifacts of preparation (see 7, above).
	Woman River Iron Formation Ontario, Canada ~2.7 *Ga*	
(26) Iron-Formation (jaspilite)	(26) Spheroidal, mineralic "microfossils" (La Berge, 1973).	(26) NONFOSSIL: mineralic pseudofossils (see 10, above).
	Witwatersrand Orange Free State, South Africa <2.82, >2.64 *Ga*	
(27) Carbonaceous gold-bearing conglomerate	(27) *Thuchomyces lichenoides* Hallbauer & Jahns, columnar "lichen-like fossils" composed of filamentous "hyphae," and *Witwateromyces conidiophorus* Hallbauer, Jahns & van Warmelo, "filamentous bacteria or fungi" (Hallbauer & van Warmelo, 1974; Hallbauer, 1975, 1978; Hallbauer et al., 1977).	(27) NONFOSSIL: artifacts of preparation (Cloud, 1976a, p. 357–359) and, possibly, modern contaminants (e.g., Hallbauer et al., 1977, fig. 22?).
(28) Shale and chert	(28) Ellipsoidal carbonaceous "micro-fossils," some containing acicular crystals (Pflug et al., 1969).	(28) NONFOSSIL: irregularly aggregated carbonaceous matter.
(29) Carbonaceous conglomerate	(29) "Filamentous and coccoid micro-fossils" detected by scanning electron micro-scopy (B. Nagy et al., 1976; B. Nagy & L. Nagy, 1976).	(29) NONFOSSIL: "unconvincing" (Muir, 1978b, p. 16), apparently artifacts of prepa-ration (B. Nagy et al., 1976, fig. 1d) and irregular scraps of carbonaceous matter (B. Nagy et al., 1976, fig. 1g).
(30) Carbonaceous conglomerate	(30) Blister-like "ultra-microfossils" detected by transmission electron microscopy (Schidlowski, 1970a).	(30) NONFOSSIL: artifacts of sample prepara-tion (see 7, above).
(31) Pyrite	(31) "Bacterium-like" possible microfossils (Schidlowski, 1965, 1966; Saager, 1970).	(31) NONFOSSIL: inorganic microstructures.
	(32) Petrographic texture similar to that of boghead (algal) coal (Snyman, 1965).	(32) DUBIOFOSSILS: textures may be non-biogenic, typical of remobilized(?) fibrous thucholite (Schidlowski, 1969).
	Bulawayan Group Rhodesia, Africa 2.64 ± 0.14 *Ga*	
(33) Stromatolitic limestone	(33) Lens-shape and coccoid "microfossils" (B. Nagy & L. Nagy, 1976; B. Nagy et al., 1976; L. Nagy & Zumberge, 1976).	(33) NONFOSSIL: mineralic (carbonate) pseudofossils (cf. Muir, 1978b, p. 16).
(34) Stromatolitic dolomite	(34) Spiny "fossil microspores" detected by transmission electron microscopy (Oberlies & Prashnowsky, 1968).	(34) NONFOSSIL: modern fungal spores (Schopf, 1975, p. 235; Cloud, 1976a, p. 357; Cloud & Morrison, 1979a, p. 89)

TABLE 9-1. *(continued)*

(35) Stromatolitic dolomite	(35) Fibrous "ultra-microfossils" detected by transmission electron microscopy (Oberlies & Prashnowsky, 1968).	(35) NONFOSSIL: artifacts of preparation (see 7, above).
	Keewatin Chert *southern Ontario, Canada* <2.8, >2.6 *Ga*	
(36) Gray chert	(36) "Seven species of fossil protozoans"—flagellates, euglenoids, heliozoans, etc. (Madison, 1958).	(36) NONFOSSIL: mineralic pseudofossils (based on restudy of material from the same outcrop by J. W. S.; see also Hofman 1971a, p. 54–55).
	Guddadarangayanhalli Formation *(or G. R. Formation or Guddadarangappanahalli Beds)* *Dharwar System, Chitaldurga Schist Belt, southern India* >2.6? *Ga*	
(37) Impure clastic sediments ("shale" and "sandstone")	(37) "Acritarchs," spheroidal, carbonaceous "microfossils" (Gowda & Sreenivasa, 1969).	(37) NONFOSSIL: modern contaminants and artifacts of weathering and preparation (Cloud, 1976a, p. 355, 359, 375, 382).
	Archean Marble *southern Finland* >2.5 *Ga*	
(38) Marble	(38) Elongate "ultra-microfossils" detected by transmission electron microscopy (Oberlies & Prashnowsky, 1968).	(38) NONFOSSIL: artifacts of preparation (Schopf, 1975, p. 235; Cloud & Morrison, 1979a, p. 89).
	Archean Baltic Shield *Kola Peninsula* >2.5 *Ga*	
(39) Lithology not specified	(39) *Menneria* sp., spheroidal "fossil phytoplankton" (Lopuchin, 1975, 1976).	(39) NONFOSSIL: modern contaminants and artifacts of preparation (see 18, above).
	Ogishke Conglomerate *northeastern Minnesota, U.S.A.* >2.5 *Ga*	
(40) Gray chert pebbles	(40) "*Microcoleus*- and *Inactis*-like blue-green algae" (Gruner, 1923; =*Palaeomicrocoleus gruneri* Korde, in Vologdin & Korde, 1965).	(40) NONFOSSIL: trails of ambient pyrite grains (Tyler & Barghoorn, 1963).
	"Suppositional Archaean" *North Tien Shann, U.S.S.R.* >2.5? *Ga*	
(41) Lithology not specified	(41) *Menneria* sp., spheroidal "fossil phytoplankton" (Lopuchin, 1975).	(41) NONFOSSIL: modern contaminants and artifacts of preparation (see 18, above).
	Hamersley Group *Brockman and Marra Mamba Formations, Western Australia* ~2.5 *Ga*	
(42) Iron-formation (jaspilite)	(42) Spheroidal, mineralic "microfossils" (La Berge, 1967; Muir & Plumb, 1976).	(42) NONFOSSIL: mineralic pseudofossils (see 10, above).
(43) Iron-formation (jaspilite)	(43) Spheroidal "carbonaceous microfossils" (LaBerge, 1967; see also Karkhanis, 1976).	(43) DUBIOMICROFOSSILS: spheroidal to subhedral microstructures; see Photo 9-3.

Early Proterozoic biospheres (e.g., in abundance of organisms, biological diversity, size of biomass, etc.), or is it more likely to be some type of artifact, a disparity that is more apparent than real?

As is illustrated in Figure 9-2, the occurrence of Archean "microfossils" was first reported more than half a century ago (Gruner, 1923, 1924a). These earliest reports, however, met with skepticism (e.g., Hawley, 1926), and although there remained considerable intellectual interest in the early history of life (e.g., Seward, 1933), active research in the field languished for the following several decades. In fact, impetus for renewed activity in the field did not arise until the early 1950s, spurred initially by the discovery by Stanley Tyler in 1953 of the now-famous ~2.0-Ga-old microbiota of the Gunflint cherts (see Chapter 1), and vigorous activity did not commence until more than a decade later (Figure 9-2). Thus, to some extent, the current disparity between the Archean and Proterozoic fossil records can be viewed as a product of "historical accident"—the fact that the first unequivocal Precambrian fossils to be reported (Tyler and Barghoorn, 1954) and reasonably well documented (those of the Gunflint Formation; Barghoorn and Tyler, 1965a; Cloud, 1965) happened to occur in Proterozoic, rather than in Archean-age deposits; that the second such microbiota to be discovered (Barghoorn and Schopf, 1965), and the first Precambrian assemblage to receive detailed monographic treatment (that of the Bitter Springs Formation; Schopf, 1968; Schopf and Blacic, 1971), was similarly of Proterozoic age; and that in subsequent years, because of its proven paleontologic potential, workers in the field have tended to concentrate on the Proterozoic rather than the Archean. The early successes in the Proterozoic bred subsequent similar successes and, to some extent, a concomitant neglect of the earlier, and presumably less fruitful, Archean rock record.

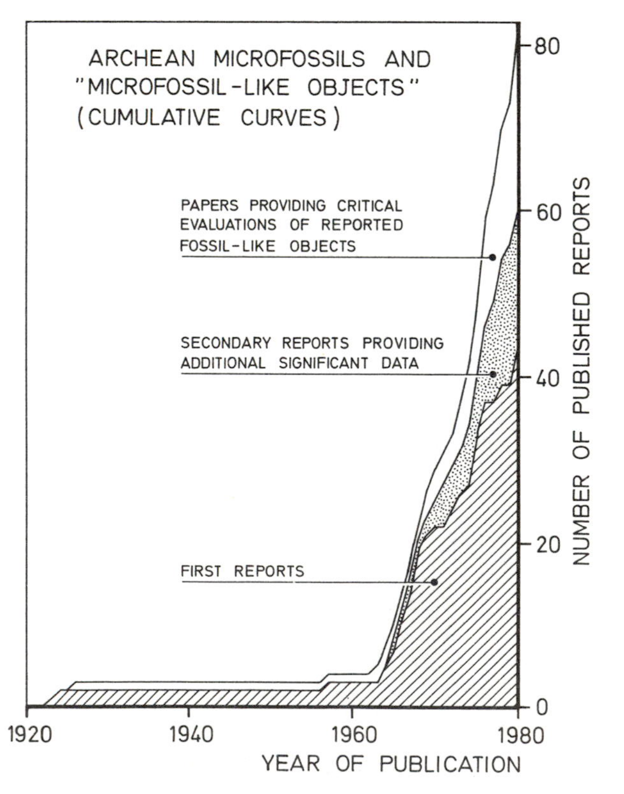

FIGURE 9-2. Published reports from 1920 to the present describing Archean microfossils and "microfossil-like" objects.

Were this more or less benign neglect the only cause for the Archean-Proterozoic disparity, the situation might be easily remedied by simply extending the search for early fossils into progressively older and older terranes. But while such a change in emphasis seems certainly desirable—especially in light of recent developments in the field (Lowe, 1980a; Walter et al., 1980; Schopf and Walter, 1980; Awramik, 1981; Awramik et al., 1983)—it is unlikely that the problem will be so easily solved. In the first place, and despite the Proterozoic emphasis of most such studies to date, in recent years considerable effort has been expended in the search for Archean microfossils; to the present, more than 80 primary and derivative reports have appeared in the literature, well over three-quarters of which have been published since 1965 (Figure 9-2). And although results of these studies have been relatively disappointing (Section 9-3), it is nonetheless true that significant, albeit limited progress has been made (Sections 9-4, 9-5). Moreover, and perhaps of equal importance, the field has recently entered a more mature phase of development, one of rigorous, critical evaluation that has resulted in the winnowing out of quite a number of earlier, erroneous reports (Table 9-1) and a useful (if nevertheless disheartening) exercise that should ultimately lead to well-based understanding of the early history of life.

By far the most serious constraints facing Archean paleobiology, however, are those derived from the nature of the early rock record itself. As is discussed in Chapters 3 and 10 of this volume, the Archean appears to have been characterized by a tectonic regime significantly different from that of the Proterozoic. This difference, inferred to have resulted in the general absence of widespread cratonal sedimentation prior to the Late Archean, together with the occurrence of normal geologic processes (sedimentation, burial, uplift, erosion, etc.) and the resultant recycling of sedimentary rock units, has served to limit, markedly, the abundance of preserved Archean sediments. In comparison with the Proterozoic, there are simply fewer available rock units where cellular fossils might be sought. Moreover, and of special relevance to the possible detection of "organically preserved" (i.e., permineralized, kerogenous) Archean microfossils—a mode of preservation exhibited by numerous Proterozoic microfossil assemblages—available organic geochemical data (see Chapter 5) indicate all too clearly that well-preserved Archean kerogen is of rare occurrence: with but few exceptions, moderate- to high-grade metamorphism and accompanying catagenesis are pervasive (and paleontologically discouraging) aspects of the Archean rock record.

To date, therefore, as a result both of the historical development of the field and of numerous, chiefly geologically imposed, limitations inherent in its study, the search for cellularly preserved remnants of Archean life has met with only limited success. Such difficulties, however, are not wholly insurmountable: in recent years, useful progress has been made both in terms of critical evaluation of the early fossil (and pseudofossil) record (Section 9-3), and of the discovery of a significant, if still quite limited body of evidence that has provided new insight regarding the composition and evolutionary status of the Archean biota (Section 9-5). And among the most constructive developments to have occurred during the brief history of this still fledgling field has been the establishment of a set of criteria by which the authenticity of putative Archean microfossils can be effectively evaluated, a matter to be considered in the following section.

9-2. Criteria for Establishing the Authenticity of Archean Microfossils

Because so few hard data are as yet available regarding the earliest biosphere, the discovery of virtually any new evidence of Archean life will represent an important contribution to available knowledge and, as such, be more than likely to have significant impact on previously accepted generalizations. Thus, reports of the occurrence of microfossils in Archean rocks merit close scrutiny. Although such scrutiny is of course also required of reports of putative microfossils from the Proterozoic as well, the problem is particularly acute in the case of Archean-age materials for which critical evaluation is absolutely required.

For reports of Archean microfossils to be considered valid, four principal criteria must be met. Considered in turn, it must be unequivocally established: (*i*) that the rock unit containing the putative microfossils is in fact of Archean age; (*ii*) that the fossil-like objects are demonstrably indigenous to the Archean rock in question (rather, for example, than being modern contaminants); (*iii*) that these indigenous microfossils were in fact emplaced in the rock at the time of its deposition (i.e., that they are assuredly syngenetic with sedimentation, rather than being of secondary, post-depositional origin); and (*iv*) that the fossil-like microstructures are actually biogenic (rather than being pseudofossils, laboratory-produced artifacts, or some other type of nonfossil).

Among these four criteria, establishment of biogenicity has commonly proved to be the most difficult (Section 9-4), but in recent years problems have arisen with regard to each of these specific factors—all four of the criteria must be satisfied before a putative microfossil can be considered authentic evidence of Archean life.

9-2-1. Geologic Age

As a result of the marked advances that have occurred in recent years in Archean geology generally and, especially, the vastly improved geochronologic control that has been provided by the now widespread application of radiometric dating techniques, far fewer reports of Archean microfossils and fossil-like objects have failed to meet this criterion than any of the other three (or, indeed, than has been the case in studies of reputed Proterozoic fossils; see Cloud, 1968b). Basically, the problem here is one of establishing the geologic provenance of the materials studied, and although numerous difficulties could be envisioned (e.g., the detection of supposed Archean microfossils in lithified soils such as the Tertiary silcretes that immediately overlie Archean exposures in parts of Western Australia), geologically based misinterpretations have been relatively rare. It is worth emphasizing, however, that no bona fide fossiliferous Archean sediment has been dated directly—that all such assigned ages are based on analyses of igneous, commonly volcanic rocks having a known stratigraphic position relative to the fossiliferous unit in question, and that aside from the various analytically derived uncertainties inherent in any radiometrically determined "absolute age," such inferred stratigraphic relationships represent a potentially significant source of error (especially if they are based on correlations over relatively long distances, or if they involve extrapolations within a structurally complex region, situations that are common in Archean terranes).

9-2-2. Indigenousness

Lack of critical application of this criterion—the matter of establishing that the fossil-like microstructures are actually indigenous to (i.e., that they are physically embedded within) the rock matrix—has led to numerous misinterpretations of supposed Archean microfossils. Two principal categories of such nonfossils can be identified. In the first category are those fossil-like objects (studied chiefly by optical microscopy) that have been introduced during, detached from rock surfaces as a result of, or even formed by the process of acid maceration. As is well known to students of Phanerozoic spores, pollen, and micropaleontologia generally, palynological macerations (acid-resistant organic residues produced by digestion of rock specimens in hydrochloric, hydrofluoric and other mineral acids) are notoriously susceptible to contamination by modern organisms or plant parts, either windborne (e.g., pollen, spores, and particles in laboratory dust), water- or acid-carried (especially aquatic fungi, bacteria, diatoms, and other microalgae), or introduced mechanically during sample preparation (e.g., cellulose fibers derived from tissue used to clean microscope slides). For example, the name *Rifenites* has been applied to reddish-amber, multilayered, roughly actinomorphic bodies that have been described in the Russian literature (e.g., Timofeev et al., 1976, pl. 43, figs. 1-7, 9-14) as "Precambrian microfossils" but which in fact are brachysclerids ("stone cells") derived from wooden stirring rods that are commonly used in the Soviet Union to agitate macerating rock chips. Included in this category also are numerous cell-like artifacts (e.g., clumps of coagulated organic detritus and bubbles coated with particulate matter) that can be formed during the process of maceration and slide preparation, and various types of modern contaminants that are present on the surface of (e.g., lichens, fungi, soil algae, and bacteria), or are sequestered in cracks or microcavities within (viz., chasmolithic and endolithic microorganisms) the macerating rock chips, and that as a result of acid treatment become freed from their associated rock matrix.

Members of the second category of non-indigenous nonfossils have the common property of having been detected by electron (transmission or scanning) microscopy. Two types of such objects have been reported: modern bacterial, fungal, and other ultramicroscopic contaminants, commonly air-borne, that settle onto microscope grids or exposed rock surfaces prior to electron microscopic study, and an array of nonbiogenic, but surprisingly life-like artifacts (spheroids, elongate rods, filamentous strands, tissue-like aggregates, etc.) produced as a result of the various manipulations involved in sample preparation.

In Table 9-1 are given specific examples of reported Archean microfossils falling into these two categories. Without belaboring the matter, the cardinal point is that convincing evidence has not been presented to establish that the structures in question are actually indigenous to the rock samples investigated. Indeed, in the case of acid macerations, the very evidence required to solve the problem—evidence demonstrating that the reported microstructures were in fact thoroughly embedded in the rock matrix—is destroyed during digestion of the macerating rock chips. This is not to suggest that the use of macerations is without merit in the study of Archean microfossils nor, in fact, should it be assumed that electron microscopic studies are incapable (at least in principle) of providing convincing evidence of the indigenous nature of objects detected. To date, however, neither of these approaches has provided effective means for solving this problem. At present, the technique of choice—indeed, the only technique that has thus far provided unequivocal evidence of the indigenous nature of Archean microfossils—is optical microscopic study of petrographic thin sections, a technique of special value since it provides evidence at a microscopic scale of the relations of the objects in question to their associated rock matrix, data that have bearing not only on the question of indigenousness but on the problem of determining the possible syngenicity of the objects as well.

9-2-3. Syngenicity

Are the indigenous microstructures demonstrably contemporaneous (syngenetic) with sedimentation, or could they have become enclosed in the rock as a result of some post-depositional process? As above, the answer to this question hinges on the physical relationship between the microstructures and their encompassing matrix, but in this case evidence must be adduced to establish that the structures were present in the sediment at the time of primary lithification (usually by demonstrating that they are embedded within primary, rather than secondary, mineral phases). And, as above, the most effective means by which to demonstrate syngenicity is by optical microscopic study of petrographic thin sections. Thus, reports of "Precambrian microfossils" enclosed in metastable mineral phases (e.g., amorphous opaline silica; see Cloud and Morrison, 1979a, p. 89) should be immediately suspect, as should be reports of such "fossils" occurring in second-generation, void-filling minerals (e.g., botryoidal chalcedony; see Schopf, 1975, p. 223–224). Purely physical relationships are also of import: reports of "fossils" from planar, mineral-infilled veinlets (Marshall et al., 1964), of their concentration at mineral grain boundaries rather than along primary depositional laminae (see Bridgwater et al., 1981), or of cross-cutting relationships in which the "fossils" have disrupted primary features of the lithified rock fabric (see Tyler and Barghoorn, 1963) are similarly subject to question. And in at least one example (see Table 14-2, No. 23), an important, initial clue to the non-syngenicity of putative Precambrian microfossils was provided by their color which differs significantly from that of the particulate, finely laminated kerogenous components of the remainder of the rock matrix, a color difference suggesting that the reddish-amber "fossils" were geochemically less mature (see Chapter 5), and in fact much younger, than the associated, dark gray, syngenetic kerogen.

In actual practice, however, and in addition to demonstrating the absence of relationships that are inconsistent with a primary origin, syngenicity is most commonly reflected by various positive factors as well, aspects of the assemblage indicating that the fossils were emplaced prior to, or concurrent with, lithification. Perhaps the most telling such evidence is that exhibited in stromatolitic microbiotas (see Section 9-5) in which the orientation and distribution of fossil filaments (or colonial unicells) provide evidence indicative of their role in the formation of stromatolitic laminae, and in which the laminae, in turn, are primary components of megascopic stromatolites, organosedimentary structures that can be demonstrably related to the original depositional history of the sedimentary rock unit.

9-2-4. Biogenicity

Among the four principal factors, this criterion—the matter of establishing that the microstructures in question are in fact bona fide remnants of fossil microorganisms—has commonly proved the most difficult to satisfy, largely because (*i*) a rather wide variety of non-biogenic objects are known that effectively mimic "biologic morphologies" (e.g., mineral dendrites, crystallites, etc.); and (*ii*) in the absence of a well-established fossil record with which to compare the objects detected (as, for example, is the case in the Phanerozoic), it has been virtually impossible to know, a priori, just what types of organisms might reasonably have been expected to occur in the Archean. This matter of "reasonability," or lack thereof, deserves special emphasis. It is well to recall, for example, that not long ago (Tyler and Barghoorn, 1954; Barghoorn and Tyler, 1965a), certain of the Gunflint fossils—forms now generally regarded as being solely of prokaryotic affinities (Schopf, 1970, p. 334; Awramik and Barghoorn, 1977)—were variously compared with such relatively advanced eukaryotes as fungi, discoasters, green algae, free-swimming dinoflagellates, and even hydrozoan coelenterates, interpretations that became increasingly implausible as additional data became available (and, especially, as the discovery of other fossil assemblages made it possible to place the Gunflint organisms in a realistic evolutionary context). In this light, and in view of the paucity of "hard data" as yet available regarding pre-Proterozoic evolution, it is perhaps not surprising that a similar array of cellularly complex, implausibly advanced microorganisms (e.g., flagellates, euglenoids, heliozoans, yeasts, lichens, etc.) has been described from the Archean as well.

Unfortunately, hard and fast rules for determining the biogenicity of Archean microstructures—applicable in principal to all such objects in all conceivable modes of preservation—are nearly impossible to define. Thus, what follows is in essence a set of guidelines, a listing of what might be termed "reasonably expectable attributes of unquestionable Archean microfossils" that when considered together provide an effective framework for dealing with the problem. Deviation from these guidelines, while perhaps not necessarily indicative of non-biogenicity, requires special pleading and the occurrence of a relatively unusual set of biological and/or preservational circumstances. In general, therefore, for reported Archean microstructures to be considered bona fide microfossils, they should:

(1) Be of relatively abundant occurrence (if conditions are favorable for preservation of one microorganism, others should also occur) and, most commonly, be members of a multi-component biologic assemblage (in nature, monospecific communities, even as preserved in the fossil record, are of rare occurrence).

(2) Be of carbonaceous composition or, if mineralic, be a result of biologically mediated mineral encrustation (as in iron- or manganese-precipitating bacteria, or in calcified bacterial colonies or cyanobacterial sheaths) or a product of minneral replacement (microorganisms exhibiting siliceous or calcareous tests are unknown from the Precambrian).

(3) Exhibit "biological morphology"—be characterized by a range of (statistically demonstrable) variability (rather than uniformity), including life-cycle variants, comparable to that exhibited by morphologically similar modern and/or fossil microorganisms.

(4) Occur in a geologically plausible context (e.g., in relatively unmetamorphosed sedimentary rocks, associated with stromatolites, etc.).

(5) To the extent feasible (depending on available data), fit within a well-established evolutionary context (i.e., exhibit a level of organization consistent with that of bona fide fossils of comparable geologic age, be part of a demonstrable evolutionary continuum, etc.).

(6) Be dissimilar from potentially coexisting abiological organic bodies (e.g., proteinoid microspheres, carbonaceous "organized elements," and products of abiotic syntheses).

As is dicussed below, the few examples of assured Archean microfossils now known satisfy all of these guidelines (Section 9-5), whereas those forms currently regarded as only possible microfossils (i.e., dubiomicrofossils, Section 9-4) are so considered largely because they cannot be distinguished with certainty from morphologically similar nonbiogenic microstructures (see item 6, above).

9-3. Nonfossils: Pseudofossils, Artifacts, and Contaminants

As is summarized in Table 9-1, over the past half-century some 43 classes of microfossils and microfossil-like objects have been reported from at least 28 Archean rock units; for the reasons there specified, however, the vast majority of these objects are now considered to be nonfossils of one of the following four types:

(1) Inorganic pseudofossils, including such objects as mineralic dendrites (viz., *Ramsaysphaera ramses*; compare Pflug, 1976, figs. 1, 3, and 4 with Saratovkin, 1959, figs. 3, 24, 25, 59–62); metamorphically produced microstructures (viz., *Isuasphaera isua*, described by Pflug, 1978c; see Bridgwater et al., 1981 and Photo 9-1 herein); spheroidal, filamentous, decussate or actinomorphic microcrystallites, commonly of hematitic or goethitic composition (e.g., Gruner, 1924a; Marshall et al., 1964; LaBerge, 1967, 1973; Cloud and Licari, 1968, Figs. 12 and 13); filamentous trails produced by "ambient pyrite grains" (Gruner, 1923); and various other mineralogic pseudomicrofossils (e.g., Madison, 1958; Schidlowski, 1965; L. Nagy and Zumberge, 1976).

(2) Organic pseudofossils, including such objects as globular, apparently solid, carbonaceous bodies (e.g., Saager and Muff, 1980; and, possibly, those described by Dunlop et al., 1978); irregular clumps of graphitic particles (e.g., Dungworth and Schwartz, 1972, fig. 4; Lopuchin and Moralev, 1973, figs. 2–6; B. Nagy et al., 1976, figs. 2a, c); and carbonaceous spheroidal floccules (e.g., Pflug, 1966, pl. 1, figs. 8 and 10; Lopuchin and Moralev, 1973, fig. 1) or linear "filament-like" aggregates (e.g., Pflug, 1966, pl. 3, fig. 14; Lopuchin, 1975, fig. 1e; Muir and Grant, 1976, figs. 15–23).

(3) Artifacts of sample preparation, such structures as blisters, bubbles, and various types of fibrous or filamentous artifacts produced during

preparation of samples for transmission electron microscopy (e.g., Cloud et al., 1965; Barghoorn and Schopf, 1966; Oberlies and Prashnowsky, 1968; Schidlowski, 1970; Prashnowsky and Oberlies, 1972) and the anastomosing, filamentous, metallic artifacts produced during studies of Archean thucholites (viz., *Thuchomyces lichenoides* and *Witwateromyces conidiophorus* described by Hallbauer et al., 1977; see Cloud, 1976a, p. 357–359).

(4) Modern contaminants, including fungal spores (Oberlies and Prashnowsky, 1968, fig.2), bacilliform bacteria (Barghoorn and Schopf, 1966, fig. 5?), spheroidal algae and pollen grains (e.g., Gowda and Sreenivasa, 1969, pl. 9 fig. 5; Lopuchin, 1975, figs. 2a, b), and filamentous microorganisms (e.g., Brooks et al., 1973, fig. 1h; Hallbauer et al., 1977, fig. 22?).

9-4. Possible Archean Microfossils

During the past dozen or so years, some 25 publications have appeared reporting the occurrence of "alga-like" or "unicell-like" spheroids from Archean sediments of the Warrawoona (Table 9-1, Nos. 4 and 5), Onverwacht (Table 9-1, No. 6), Fig Tree (Table 9-1, No. 12), and Hamersley Groups (Table 9-1, No. 43); to date, size data for more than 1,300 such bodies have been reported in the literature (see Schopf, 1976, 1977b). In general morphology, the spheroids are of microscopic size (all <200 μm, most <50 μm); are hollow, rather than solid (except, possibly, those forms shown in L. Nagy, 1971, fig. 1a, and in Dunlop et al., 1978, fig. 3); and are delimited by a rough, irregularly textured, thin (~ 1 μm thick), wall-like layer of carbonaceous material (or, in one case, apparently composed of "siderite"; L. Nagy, 1971). In these and other characteristics (e.g., the occurrence of paired spheroids, interpreted as representing products of cell division; Muir and Grant, 1976; Knoll and Barghoorn, 1977), many of these spheroids rather closely resemble unicellular, microbial, and algal microfossils of a type that are now well known from sediments of the Proterozoic (see Chapter 13). However, the biogenicity of the Archean objects has been questioned (Schopf, 1975, 1976; Cloud, 1976; Cloud and Morrison, 1979; Schopf and Walter, 1980), largely because (*i*) they are similar in size, form, and composition to simple organic spheroids that can be produced by solely non-biological processes—namely, floccule-like carbonaceous globules known from Proterozoic cherts (Schopf and Prasad, 1978, figs, 5e, f); kerogenous spheroids occurring in carbonaceous, chondritic meteorites (Rossignol-Strick and Barghoorn, 1971); and organic microspheres formed during various types of abiotic syntheses (Fox and Yuyama, 1963; Folsome et al., 1975)—because (*ii*) some of these forms exhibit a pattern of size distribution that is similar to that of such non-biologic spheroids, but that apparently differs from that typical of authentic, comparably preserved, Proterozoic unicells (Schopf, 1976); and because (*iii*) several of the Archean assemblages include exceptionally large individuals (microstructures well over 100 μm in diameter; Engel et al., 1968, fig. 3; B. Nagy and L. Nagy, 1969, fig. 3), forms that are many tens of microns larger than both the largest known modern prokaryotes (Schopf, 1976; Schopf and Oehler, 1976) and than any unicells known elsewhere from the fossil record prior to the Middle Proterozoic (Schopf, 1976; Cloud and Morrison, 1979). The point here is not that these Archean objects have been proven to be non-biogenic; it is, rather, that sufficient grounds for doubt exist regarding the matter as to make uncertain their interpretation. As Knoll and Barghoorn (1977, p. 398) have noted in evaluating their report of one of the more "biologic-like" such assemblages yet described, "the simple morphology of these Archean microstructures makes their 100 percent unequivocal identification as biological entities virtually impossible" or, as Cloud and Morrison (1979, p. 89) have averred in assessing this same report: "permissive as this evidence may be . . . we cannot accept it as compelling or even presumptive." Such objects are thus here regarded as only possible rather than as bona fide Archean microfossils. And in view of the rather extensive literature already available regarding their analysis, the following discussion is devoted solely to a series of previously unpublished observations.

9-4-1. Possible Microfossils from the Warrawoona Group

Three categories of Archean dubiomicrofossils have been detected in the $\sim$3.5-Ga-old Warrawoona Group, all from cherts of the North Pole Dome region of northwestern Western Australia. The earliest of these reports (in retrospect, an especially important contribution since its publication prompted additional studies that led to the discovery of unquestionable microfossils in this same sequence of Archean rocks some two years years later), by Dunlop, Muir, Milne, and Groves, appeared in 1978 (Table 9-1, No. 5). "Five morphologies of carbonaceous spheroids," gray to black in color and ranging from about 1 to 12 μm in diameter, were reported from these cherts, as were "pale-coloured ... silica-filled ... abiogenic spheroids" 3 to about 17 μm in size. Two principal problems exist in accepting a biogenic interpretation of the carbonaceous objects: first, and although it is indicated in the publication that four of the morphologies were observed also in petrographic thin sections, the report is based almost exclusively on studies of palynological macerations, a technique fraught with difficulties (see Section 9-2-2); indeed, all of the objects illustrated in the paper (Dunlop et al., 1978, fig. 3) are shown in acid-resistant residues. And second—and unlike bona fide fossil unicells of younger geologic age—the objects have a surprisingly smooth, globule-like outline (viz., Dunlop et al., 1978, figs. 3A-D) and appear to lack a distinct, differentiated cell wall. Despite published descriptions to the contrary (e.g., "the spheroids ... are hollow," and "a broken specimen ... was clearly hollow"), the published photomicrographs give the impression of showing intact or partially disrupted solid, rather than hollow objects. Studies of these bodies in thin section being carried out by R. Buick of the University of Western Australia should provide new data relevant to their interpretation. For the present, however, they seem most likely to be nonfossils, solid carbonaceous globules of apparently non-biologic origin.

The second category of Warrawoona microstructures that are of possible, but as yet unproven biogenicity (Table 9-1, No. 4) are small (commonly 4.0 to about 7.5 μm-diameter), brownish, apparently carbonaceous, thin-walled (and demonstrably hollow), solitary or paired unicell-like spheroids of the type shown in Photo 9-7F-I and described by Awramik et al. (1983) as only "possible microfossils." Of all such Archean spheroids now known, these forms appear to be the most fossil-like in terms both of their morphology and of their context of occurrence: unlike many other reported such microstructures, the spheroidal form of these bodies is clearly and distinctly defined (presumably as a result of the unusually low grade of metamorphic alteration exhibited by the enclosing cherts; see Appendix I, P.P.R.G. Sample No. 517); they occur not only as solitary spheroids, but also in pairs that exhibit flat-sided adjacent walls (Photo 9-7F and G), a morphology typical of paired products of cell division as known from numerous Proterozoic assemblages (e.g., Schopf, 1968; Nyberg and Schopf, 1981; Mendelson and Schopf, 1982); like the unicellular microfossils of many younger Precambrian deposits, they are components of a diverse, filament-dominated, stromatolitic microbiota (see Section 9-5-1); and, as evidenced by their occurrence in petrographic thin sections (Photo 9-7F-I), they are demonstrably indigenous to the cherts investigated. Indeed, had these specific objects been detected in Proterozoic stromatolitic cherts—in which similar spheroidal unicells are commonly abundant (e.g., *Huroniospora* spp. of the Gunflint cherts; Barghoorn and Tyler, 1965)—there would be little hesitancy in regarding them as biogenic. To date, however, relatively few such spheroids have been detected in the Warrawoona deposit and, because of their similarity to known non-biogenic microstructures, in the absence of additional, more telling data (e.g., demonstrating their occurrence in ordered, mucilage-embedded colonies) it would be erroneous, or at least premature to regard them as compelling evidence of Archean life.

The third category of Warrawoona microstructures that are of possible, but as yet unproven biogenicity (Table 9-1, No. 4) are rosette-like aggregates of fine radiating filaments also first described by Awramik et al. (1983). As is shown in Photo 9-6A-G, these colony-like aggregates are composed of fine unbranched filaments, 0.5 to 1.0 μm in diameter and more than 100 μm in length, that radiate irregularly from a "holdfast-like" base, a morphology and organization comparable to those exhibited by such extant filamentous microbes as *Thiothrix* and *Leucothrix* (Table 9-2). Moreover, as is shown in Photo 9-6A, the radiating fossil-like filaments are of essentially constant diameter and range commonly from gently curved to highly sinuous, features decidedly biologic-like in appearance. Nevertheless, other of the filaments are straight rather than curved (Photo 9-6B, D-F) and virtually all are disjointed, composed of short rod-shaped segments of variable length; none of the filaments contains structures that can be interpreted confidently as preserved cells. In addition, and unlike each of the four types of undoubted

TABLE 9-2. Comparison of detected Warrawoona morphotypes and their modern morphological analogues.

Detected morphotypes	Dimensions	Modern morphological analogues*	
		Modern taxa	*Physiological characteristics*
	Diameter (μm)		
Long, sinuous, apparently nonseptate narrow threads (Photo 9-4A-I)	0.3–0.7		
		PIGMENTED GLIDING BACTERIA	Aerobes-chemoheterotrophy
	~0.5	*Flexibacter aggregans*	
	~0.5	*Flexithrix dorotheae*†	
	~0.5	*Herpetosiphon geysericola*†	
		COLORLESS GLIDING BACTERIA	Amphiaerobes-chemoautotrophy
	0.6–0.7	*Pelonema subtilissimum*†	
		GREEN SULFUR BACTERIA	Anaerobes-photoautotrophy and amphiaerobes-photoheterotropy (in dark, aerobes-chemoheterotrophy)
	0.5–1.0	*Chloroflexus aurantiacus*†	
		METHANOGENIC BACTERIA	Anaerobes-chemoautotrophy
	0.5–0.7	*Methanospirillum hungatii*†	
		CYANOBACTERIA	Amphiaerobes and aerobes-photoautotrophy
	~0.6	*Oscillatoria angustissima*†	
	0.6–0.8	*Phormidium angustissimum*†	
Intermediate-diameter possibly septate filaments (Photo 9-5A-C)	0.8–1.1		
		PIGMENTED GLIDING BACTERIA	Aerobes-chemoheterotrophy
	0.7–1.0	*Herpetosiphon persicus*	
		COLORLESS GLIDING BACTERIA	Amphiaerobes-chemoautotrophy
	0.7–1.2	*Archroonema angustum*	
		COLORLESS SULFUR GLIDING BACTERIA	Amphiaerobes and aerobes-chemoheterotrophy and chemoautotrophy
	0.5–1.0	*Beggiatoa minima*	
		GREEN SULFUR BACTERIA	Anaerobes-photoautotrophy and amphiaerobes-photoheterotrophy (in dark, aerobes-chemoheterotrophy)
	0.7–1.1	*Chloroflexus aurantiacus* var. *mesophilus*	
		SHEATHED BACTERIA	Aerobes-chemoheterotrophy
	0.8–1.0	*Leptothrix ochraceae*	
		CYANOBACTERIA	Amphiaerobes and aerobes-photoautotrophy
	0.8–1.2	*Oscillatoria angusta*	
	0.8–1.0	*Phormidium bigranulatum*	

* Data from Buchanan and Gibbons, 1974.
† Septate forms in which component cells would be difficult to discern in the fossil state.

TABLE 9-2. (*continued*)

Detected morphotypes	*Dimensions*	*Modern morphological analogues**	
		Modern taxa	*Physiological characteristics*
Large-diameter tubular sheaths (Photos 9-5D and 9-7A-E)	3.0–9.5		
		COLORLESS GLIDING BACTERIA	Amphiaerobes-chemoautotrophy
	6.0–8.0	*Desmanthos thiocrenophilum* (common sheath)	
		COLORLESS SULFUR GLIDING BACTERIA	Amphiaerobes and aerobes-chemoheterotrophy and chemoautotrophy
	1.5–30.0	*Thioploca minima* (common sheath)	
		GREEN SULFUR BACTERIA	Anaerobes-photoautrophy and amphiaerobes-photoheterotrophy (in dark, aerobes-chemoheterotrophy)
	4.5–6.0	*Oscillochloris chrysea* (sheath)	
		SHEATHED BACTERIA	Aerobes-chemoheterotrophy
	3.0–6.0	*Phragmidiothrix multiseptata* (sheath)	
	3.0–7.0	*Clonothrix fusca* (sheath)	
		CYANOBACTERIA	Amphiaerobes and aerobes-photoautotrophy
	4.0–9.0	*Lyngbya cryptovaginata* (sheath)	
	5.0–8.0	*Symploca muscorum* (sheath)	
	4.5–7.5	*Phormidium autumnale* (sheath)	
	Cellular dimensions *Width* (*μm*) × *Length* (*μm*)		
Large-diameter septate filaments (Photo 9-8)	4.0–6.0 × 4.5–7.0		
		COLORLESS SULFUR GLIDING BACTERIA	Amphiaerobes and aerobes-chemoheterotrophy and chemoautotrophy
	5.0–9.0 × 5.0–8.0	*Thioploca schmidlei*	
	5.0–14.0 × 5.0–7.0	*Beggiatoa arachnoidea*	
		GREEN SULFUR BACTERIA	Anaerobes-photoautotrophy and amphiaerobes-photoheterotrophy (in dark, aerobes-chemoheterotrophy)
	4.5–5.5 × 3.5–7.0	*Oscillochloris chrysea*	
		CYANOBACTERIA	Amphiaerobes and aerobes-photoautotrophy
	4.0–7.0 × 2.0–7.0	*Phormidium autumnale*	
	4.7–6.5 × 2.5–8.0	*Phormidium subincrustatum*	

	Filament dimensions			
	Width (μm)	*Length (μm)*		
Rosette-like aggregates of fine "fossil-like" filaments (Photo 9-6A-G)	0.5–1.0	>100		
			ROSETTE-FORMING GLIDING BACTERIA	Aerobes-chemoautotrophy and chemoheterotrophy
	0.5–2.5	>100	*Thiothrix nivea*	
	2.0–5.0	>100	*Leucothrix mucor*	
	Diameter (μm)			
Solitary or paired "cell-like" spheroids (Photo 9-7F-I)	3.0–7.5			
			SPHEROIDAL GLIDING BACTERIA	Amphiaerobes-chemoautotrophy
	5.0–40.0		*Achromatium volutans*	
			PURPLE NONSULFUR BACTERIA	Anaerobes-photoautotrophy and photoheterotrophy (in dark, aerobes-chemoheterotrophy)
	0.7–4.0		*Rhodopseudomonas sphaeroides*	
			PURPLE SULFUR BACTERIA	Anaerobes-photoautotrophy and photoheterotrophy
	3.0–4.5		*Chromatium buderi*	
			METHANOGENIC BACTERIA	Anaerobes-chemoautotrophy
	0.5–4.0		*Methanococcus vannielii*	
			CYANOBACTERIA	Amphiaerobes and aerobes-photoautotrophy
	5.0–6.0		*Synechocystis aquatilis*	

fossil filaments described below from cherts of the North Pole region (Section 9-5), distinctive radiating aggregates of this type have not previously been described from other Precambrian fossil biocoenoses. At present, therefore, it seems most appropriate to regard these rosette-like aggregates as only possible fossils; they may represent a previously undescribed variety of Precambrian microbe, or they may be non-biogenic, aggregated microstructures of solely non-biologic origin.

9-4-2. Possible Microfossils from the Onverwacht Group

Of the numerous published reports describing putative microfossils from cherts and shales of the ~3.5-Ga-old Onverwacht Group of South Africa (Table 9-1, Nos. 6–9), only one will be considered here, that dealing with microstructures detected in thin sections of black, carbonaceous cherts of the Kromberg Formation (from the upper third of the group) co-authored by Muir and Grant in 1976. Among all such published reports, this is a particularly interesting and potentially significant contribution since it describes in some detail an array of purportedly biogenic microstructures that is unusually diverse in comparison with that noted in most other reports from the Archean—included are objects interpreted as solitary and paired "unicells," "multicellular tissue," biogenic "filaments," and small-scale "domal stromatolites." Unfortunately, the reproductions of photomicrographs included in this publication are of poor quality, and it is thus virtually impossible to evaluate this report properly without studying firsthand the original material. We therefore much appreciate the generosity of Dr. Muir, who upon request provided for our study the thin sections containing the originally described and figured specimens. Thus, to the extent feasible (i.e., for those previously described microstructures that it proved possible for us to relocate), the observations here reported are based on the same specimens as those described by Muir and Grant (1976).

Restudy of the Kromberg thin sections has confirmed many of Muir and Grant's original observations. With regard to the unicell-like objects, for example, as is shown in Photo 9-2 they are of small size, commonly 2 to about 5 μm in diameter (according to Muir and Grant, and based on measurement by them of more than 650 individuals, such bodies range in size from 1.3 to 12.5 μm, and most commonly are about 3.0 μm in diameter). And, as previously reported, they are composed of dark brown to brownish-black, apparently carbonaceous material; they are relatively abundant (especially in the thin section labeled "no. 41447, Komati, S. Africa 1A," the section referred to as "K1A" in Muir and Grant's figure descriptions); they occur, rather uncommonly, in pairs that bear some resemblance to "algae in the process of dividing" (Photo 9-2B, F, J, P, and Q); and some such spheroids are aggregated in arcuate, apparently layered groupings (Photo 9-2A-F) that are similar in several respects to stromatolitic laminae. But Muir and Grant (1976) are in error in reporting that the "small domed arcs" of "cells following the laminations of domal stromatolites" are "domed upwards"; in fact, the "doming" is perpendicular to the bedding laminations of the chert (and the bedding is thus perpendicular, rather than "parallel," to the long side of the photomicrographs shown in figures 1-3 of their paper). In addition, globular, unstructured clusters of such objects (not mentioned by Muir and Grant) also occur in these cherts (Photo 9-2G, P, and Q). Finally, although it should be noted that mineralogic relationships are rather difficult to interpret in certain of these relatively thick petrographic thin sections, at least the majority of the spheroids do not appear to represent carbonaceous coatings of mineral grains, nor do they appear to be structures resulting from metamorphic recrystallization of the chert matrix.

With regard to the other purportedly biogenic objects described from these thin sections, our restudy yielded no evidence of the multicellular tissue reported by Muir and Grant (1976, figs. 12–14), and the putative biogenic filaments of the deposit (of which we were able to relocate only three of the eight objects figured in the paper) were seen to be linear aggregates of amorphous carbonaceous matter concentrated along bedding laminae, structures that exhibit no evidence of cellularity or of other obvious biological organization.

Thus, of the various "microfossils" described by Muir and Grant, it would appear that those most likely to be biogenic are the small, cell-like spheroids. Indeed, there seems no reason to doubt that these objects are both indigenous to, and syngenetic with the deposition of, a sediment of well-established Archean age. With the exception of their similarity to spheroidal microstructures of known non-biological origin, they appear to satisfy nearly all of the guidelines suggested above for the establishment of biogenicity. In fact, if it could be shown that these spheroids were actually the formative agents of stromatolitic laminae (see Chapter 8), or if the globular clusters of such objects (Photo 9-2G) could be clearly demonstrated to represent degraded microbial colonies (rather than simply random aggregates of spheroids), their biogenic interpretation would be well established. As is evident in Photo 9-2 however, the objects are in fact of quite variable and irregular form and, if actually biogenic, they must be incompletely, and rather poorly preserved. Thus, and in the absence of additional, more convincing evidence (as might be provided in this case by the discovery of appreciably better preserved specimens), the origin of these objects—whether biological or solely abiotic—remains uncertain.

9-4-3. Possible Microfossils from the Hamersley Group

The final example of "dubiomicrofossils" here considered are spheroidal, apparently carbonaceous bodies of the type originally described by LaBerge (1967) from cherty (jaspilitic) iron-formation of the ~2.5-Ga-old Hamersley Group of Western Australia (Table 9-1, No. 43). LaBerge's specimens (his "Type C" and "Type D" structures, shown in pl. 3, figs. 1–6 and 8 of LaBerge, 1967) are apparently of rare occurrence, having been detected by him "only in two slides from one piece of drill core" (LaBerge, 1967, p. 336) of the Brockman Iron Formation. Although we have been unable to detect structures of this type in our samples of the Brockman sediments, we were given by Dr. A. F. Trendall a sample from the stratigraphically closely related Marra Mamba Iron Formation, also of the Hamersley Group, that contains similar fossil-like forms (see Appendix I, P.P.R.G. Sample No. 520); thus, the specimens here illustrated (Photo 9-3) are of slightly differing geologic provenance from those reported by LaBerge, but their similarity to his material in all other salient characteristics suggests that they are otherwise essentially identical.

Unlike the objects considered here, the majority of spheroidal "microfossil-like objects" reported by LaBerge (1967, 1973) from Precambrian iron-formations are of ferruginous composition and are now regarded as mineralic, non-biogenic microcrystallites (Table 9-1, Nos. 10, 23, 26, and 42 and Table 14-2). In addition to these nonfossils, however, the Brockman sediments are reported to contain other microscopic spheroids, 12 to about 25 μm in diameter, that are described as being thin- or thick-walled, "internally chambered," and "preserved by black carbonaceous material" (LaBerge, 1967, p. 336). It is because of this reported organic composition that they are of particular paleontologic interest, and although LaBerge does not specify the basis on which the "carbonaceous material" was so identified (other than to indicate elsewhere in the paper, and with regard to a different set of objects, that "the best-preserved structures are usually pigmented by a black carbonaceous material amorphous to X-rays"; LaBerge, 1967, p. 334), it is nevertheless true that, like the microstructures here shown from the Marra Mamba Iron Formation (Photo 9-3), in published photomicrographs the Brockman objects have at least the appearance of poorly preserved carbonaceous bodies (and that they are distinctly different in this respect from hematite and stilpnomelane spheroids figured in the same publication).

What can be said regarding the possible biogenicity of indigenous fossil-like objects of this type? If it is assumed that they are in fact of carbonaceous composition, they would appear to satisfy several of the biogenic guidelines: they are exceptionally abundant (Photo 9-3A; LaBerge, 1967, pl. 3, fig. 8); they occur in a sedimentary, geologically plausible context; and their size and general form, that of simple spheroidal unicells, seems not unreasonable in terms of phylogenetic considerations. Their detailed morphology, however, is suspect: although "good spheroids" occur in the assemblages (Photo 9-3B-D; LaBerge, 1967, pl. 3, figs. 1 and 6), other specimens are more angular (Photo 9-3G-I; LaBerge, 1967, pl. 3, fig. 3) or even subhedral in shape (Photo 9-3F, J)—characteristics that if not solely a result of metamorphic alteration seem suggestive of a mineralogic, rather than a biologic origin—and adjacent occurring spheroids are not uncommonly interconnected (Photo 9-3E-G; LaBerge, 1967, pl. 3, fig. 2) to form short, uniseriate to irregularly branched groupings (Photo 9-3H-K; LaBerge, 1967, pl. 3, fig. 8), a decidedly non-biologic type of organization. Moreover, the occurrence of angular "partitions" within these bodies (LaBerge, 1967, pl. 3, fig. 5), reported by LaBerge to outline as many as eight "internal chambers," and the diffuse, "lacey" nature of their outer walls, evident both in LaBerge's material (p.

336; pl. 3, fig. 4) and in the Marra Mamba specimens (Photo 9-3B-F), indicate that if these objects are in fact biogenic, their original morphology has been rather markedly altered. In sum, then, the biogenicity of these microstructures is far from established: indeed, like the ferruginous microcrystallites that are apparently of common occurrence in iron-formations generally (LaBerge, 1967, 1973), it seems most likely that these forms are solely of non-biologic origin (possibly formed via mechanisms similar to those demonstrated experimentally by J. Oehler, 1976b); were it not for their possible carbonaceous composition, suggesting that they might therefore represent degraded biologic remnants, they would probably be regarded as "non-biogenic nonfossils."

9-5. Bona Fide Archean Microfossils

As noted above, of the 43 categories of putative microfossils thus far reported from the Archean rock record, only two—those of filamentous fossil bacteria from bedded stromatolitic cherts of the ~3.5-Ga-old Warrawoona Group (Table 9-1, No. 3) and from cherty portions of stromatolitic carbonates of the ~2.8-Ga-old Fortescue Group (Table 9-1, No. 21)—appear to fully satisfy all criteria (viz., geologic age, indigenousness, syngenicity, and biogenicity) discussed in Section 9-2 for establishing the authenticity of Archean microfossils. In both cases, the permineralized, cellularly preserved fossils are relatively complex filamentous microbes that appear to have been the principal components of diverse, stromatolite-building, microbial communities; and in both cases, the fossiliferous rock units occur in the Pilbara Block-Hamersley Basin region of northwestern, Western Australia, some 1,000 km north-northeast of Perth, in geologically and geochronologically relatively well-studied Archean to Early Proterozoic terranes (known especially because of the presence of iron ore, barite, and other economically important mineral deposits). But in both cases, the microfossils of these sediments have been discovered only quite recently; their study is thus in an early stage. The Warrawoona assemblage has been described in a single, short report (Awramik et al., 1983) augmented by discussion of their possible evolutionary significance (Awramik, 1981; Schopf and Walter, 1982), and the occurrence of the Fortescue fossils has simply been noted, without extensive illustration, or detailed discussion (Schopf and Walter, 1980, 1982; Schopf, 1980a, 1980b). What here follows, therefore, is perhaps best regarded as an extended preliminary report, a set of observations that amplify those that have previously appeared in the literature, but that fall well short of the in-depth monographic treatment that will ultimately be required for each of these newly discovered biocoenoses. Particular emphasis is here placed on the Warrawoona assemblage, because of its great age the more obviously significant (and also, to date, the relatively better studied) of the two assemblages.

9-5-1. The Warrawoona Microbiota

HISTORY OF THE DISCOVERY

Discovery of fossil filaments in the Warrawoona cherts occurred only quite recently. While the details are still fresh, it thus seems appropriate to recount, briefly, the circumstances leading to this new find, a matter of some interest in illustrating the multifaceted and interdisciplinary character of such discoveries generally.

As is discussed below, the Warrawoona microbiota occurs in black, fine-grained, bedded cherts from the North Pole Dome region of the Archean-age Pilbara Block of Western Australia, a locale said to have been christened "North Pole" by early miners because of its remote location and desolate surroundings. As is typical of many recent paleontological discoveries in the Precambrian, however, the search for fossils in these North Pole rocks occurred only after geochronologic studies had previously provided firm indications regarding the age of the Warrawoona sequence, and after a well-founded geologic framework for the area had already been established, one based on careful geologic mapping at both a regional and a local scale. In this case, the framework was provided largely through studies by the Geological Survey of Western Australia, particularly those of Arthur Hickman, Steven Lipple, and subsequently by students of the Department of Geology at the University of Western Australia, particularly John Dunlop and Roger Buick who have carried out geologic investigations at North Pole as part of their doctoral thesis research. It was Dunlop, in fact, who in 1977 was first to discover megascopic, domical stromatolites in these units (see Frontispiece and Chapter 8), structures that were collected during the P.P.R.G. field work of June 1979 (by Hayes, Hofmann, and Walter, together with D. J. Gregg of the Australian Bureau of Mineral Resources) and that were subsequently described in *Nature* in April 1980 (Walter et al., 1980) adjacent to a second report by Donald Lowe of Louisiana State University that described his independent discovery of stromatolites at a different locality in this same Archean terrane (Lowe, 1980a).

In February 1980, a few weeks preceding publication of these two reports, we were contacted at U.C.L.A. by Stanley Awramik (of the Department of Geology at the University of California, Santa Barbara) who informed us that he had detected microfossils in cherts collected by him in 1977 from rock strata cropping out in the North Pole region that he thought very likely were of Warrawoona age. At Awramik's invitation, we joined him in the study of this material (listed in Appendix I as P.P.R.G. Sample No. 517) and within a matter of days were able to confirm his identification of fossil filaments in these thin sections and to detect numerous additional fossils that included other filamentous morphotypes as well. At about the same time, we found other filaments—the rosette-like possible fossils shown in Photo 9-6—in cherts from a separate North Pole locality (Appendix I, P.P.R.G. Sample No. 002), material obtained from the Warrawoona Group in June 1979 during the same field work that was involved in collection of the domical stromatolites. Subsequently, studies of the x-ray diffraction characteristics and carbon isotopic composition of kerogen isolated from this sample (carried out by Hayes, Kaplan, Wedeking, and Matzigkeit; see Chapter 5) provided data apparently indicative both of its relatively low grade of metamorphic alteration and of its biogenicity (viz., a product of autotrophic CO_2 fixation). The provenance of this material (P.P.R.G. Sample No. 002) was known with certainty. However, there still remained some question regarding the source of Awramik's 1977 collection, the principal concern being whether these fossiliferous cherts were in fact of Warrawoona age or whether they might be derived from the Fortescue Group, a Late Archean sequence that is particularly well developed in the nearby Hamersley Basin. Although exact field relations of the cherts collected by Awramik still remain uncertain, the location of the outcrop area sampled by Awramik was established as being in strata mapped as within the Warrawoona, rather than the Fortescue Group, as a result of three additional visits to the North Pole area, one by Walter (in May 1980), a second by Awramik, Dunlop, and Buick (in June 1980; P.P.R.G. Sample Nos. 534 through 542), and a third by several members of the P.P.R.G. (viz., Chapman, Hayes, Klein, Schopf, Walker, and Walter) who were accompanied by Hickman and David Blight, both of the Geological Survey of Western Australia, in June 1982. Laboratory study of material collected during the latter of these visits is ongoing.

Thus, discovery and initial study of the Warrawoona microfossils was a decidedly collaborative venture, one that involved direct participation of nearly a dozen individuals and integration of the data and techniques of a diverse set of specialties (spanning the range from field geology and geochronology to micropaleontology, petrology, and stable isotope and organic geochemistry). Collaboration of this type is becoming increasingly typical of Precambrian paleobiologic studies generally, a trend necessitated by the multifaceted nature of the problems investigated in this field of science.

AGE RELATIONS AND GEOLOGIC SETTING

The Warrawoona Group, the lowest unit of the Pilbara Supergroup, is a 14 km-thick sequence of dominantly ultramafic to mafic metavolcanic rocks containing regionally extensive, interstratified, cherty sediments <50 m thick (Hickman and Lipple, 1978; Barley et al., 1979). Portions of the group, apparently including both the microfossiliferous cherts (Awramik et al., 1983) and the stromatolite-containing chert-barites of the sequence (Lowe, 1980a; Walter et al., 1980), were deposited in shallow subaqueous to intermittently exposed settings. As is reported in Chapter 5, x-ray diffraction studies indicate that carbonaceous matter preserved within the fossiliferous cherts is nearly amorphous (viz., kerogen isolated from P.P.R.G. Sample No. 002 is within the "graphite-d_3" class of Landis, 1971), a degree of preservation consistent both with the occurrence of cellular, carbonaceous fossils within the rocks and with metamorphism only to approximately the prehnite-pumpellyite facies (and a somewhat lower grade of alteration than the greenschist facies metamorphism exhibited elsewhere in the sequence; Hickman and Lipple, 1978, record the effects of prehnite-pumpellyite facies and greenschist facies metamorphism in the North Pole region).

The age of the Warrawoona Group of about 3.5 Ga is well established: the North Star Basalt, a volcanic unit in the lower third of the sequence, has been dated by the Sm-Nd method at 3,556 ± 32 Ma (Hamilton et al., 1980); extrusive volcanics of the Duffer Formation, occurring near the middle of the sequence, have yielded a model lead age on galena of about 3,500 Ma (Sangster and Brooks, 1977) and a U-Pb date on zircons of 3,452 ± 16 Ma (Pidgeon, 1978a); galena in vein barite cutting the fossiliferous sequence has yielded a model lead age of about 3,400 Ma (J. Richards, quoted in Walter et al., 1980); and the Pilbara Supergroup as a whole was deformed by a post-depositional event about 3,100 to 2,900 Ma ago (DeLaeter and Blockley, 1972; Oversby, 1976).

Within this well-dated and relatively well-preserved Archean terrane, the cherts of interest come from two localities about 4.7 km apart (Awramik et al., 1983, fig. 1), and although the exact collecting site at one of these localities is unknown (viz., that at which Awramik collected fossiliferous cherts in 1977), the strata at both of the outcrop areas are well established to be from the Towers Formation of the Warrawoona Group at a stratigraphic level termed by Dunlop the "second stratiform chert horizon" and they outcrop within a region that is well mapped (1:250,000 geologic map by Hickman and Lipple, 1972; North Shaw 1:100,000 topographic map, 1975) and where the stratigraphy has been investigated in considerable detail (both by Hickman and Lipple, 1978, and by Dunlop and Buick of the University of Western Australia).

INDIGENOUSNESS AND SYNGENICITY

Like the permineralized microbiotas of many other Precambrian stromatolitic cherts (those of the Gunflint and Bitter Springs Formations, for example), the Warrawoona microfossils are three dimensionally preserved in a physically relatively unaltered state, embedded within an interlocking mosaic of fine-grained quartz. Their indigenousness is thus readily demonstrated by studies of petrographic thin sections; although macerations have proved useful for the isolation of kerogen from these cherts for isotopic and organic geochemical analyses, all of the photomicrographs here shown (Photos 9-4 through 9-8) illustrate the microfossils in thin section (commonly by means of photomontages, necessitated by the three-dimensional, sinuous nature of the preserved filaments). Moreover, numerous lines of evidence appear to establish that these fossil microorganisms were emplaced in the cherty sediments prior to, or concurrent with, primary lithification: (*i*) like the mat-building fossil microbes of younger Precambrian stromatolitic assemblages, the Warrawoona filaments are commonly oriented in growth position parallel to, and in part comprising, well-defined stromatolitic laminae (Photo 9-5E-G and Photo 8-2A, B); (*ii*) the fossils are absent from later generation mineral phases (e.g., vein-filling coarse-grained quartz, and void-filling fibrous chalcedony) and in places are disrupted by penetrative, quartz-filled, secondary veinlets (e.g., Photo 9-5C, upper left); (*iii*) petrologically, the fossiliferous cherts exhibit the same characteristic fabric (including the preservation of sharply defined, unweathered carbonate rhombs) as that of other primary or early diagenetic carbonate-containing cherts of the sequence; (*iv*) within individual chert samples, the carbonaceous microfossils and associated finely particulate kerogen are comparably preserved, exhibiting the same range of dark brown to black color; (*v*) the x-ray diffraction pattern of kerogen isolated from the cherts is consistent with the known geologic history of the region (viz., the occurrence of relatively mild metamorphic alteration ca. 3,000 Ma ago); (*vi*) the $\delta^{13}C_{PDB}$ (PDB is Peedee belemnite standard) values of this kerogen are notably light [viz., as reported in Chapter 5 for P.P.R.G. Sample No. 002, -31.2 ‰, -33.4 ‰, -34.2 ‰ (survey analyses) and -34.3 ‰ (kerogen analysis), an average of -33.3 ‰, $n = 4$] and are thus consistent with values typical of Archean organic matter (see Chapter 5) but are distinctly different from values that would be exhibited by modern or Phanerozoic contaminants (ca. -25 ‰); (*vii*) the geologically inferred shallow water to intermittently exposed paleoenvironment of the bedded cherts is in agreement with the finely laminated, stromatolitic organization of the preserved microbial community; and (*viii*) the inferred prokaryotic affinities of the microfossils (discussed below) are wholly compatible both with the stromatolitic organization of the biota and with the remainder of the Precambrian fossil record as now known.

It appears, therefore, that both the indigenousness and the syngenicity of the Warrawoona fossils have been amply demonstrated. This latter aspect, however—the matter of the syngenicity of these fossils—merits further comment. Secondarily emplaced dikes, veins, and veinlets of carbonaceous chert are common in the North Pole region. One such dike occurs within 5 m of P.P.R.G. Locality Number 002. The possibility has thus been alluded to (Buick et al., 1980, p. 162; Groves et al., 1980, p. 73) that the laminar fabric of the cherts from this locality may have resulted from sheet crack formation (Ramsay, 1980) during the intrusion of this dike, rather than being microbially produced stromatolitic laminae, and that the kerogen and the rosette-like possible microfossils may have been emplaced during this intrusive episode. By analogy, this possibility might also be suggested for the cherts from P.P.R.G. Locality Number 517, but at that locality the detailed field relations of the fossiliferous cherts are unknowns. In addition, in the cherts from Locality 517 several generations of silica can be recognized in petrographic thin sections, including that in secondarily infilled veinlets. The possibility might thus be raised that at this locality the microfossils and kerogen may have been emplaced during weathering rather than being syngenetic with primary sedimentation.

Neither of these possibilities is supported by available evidence. In addition to being inconsistent with the evidence cited in items (*i*), (*ii*), and (*iii*), above, the dike-sheet crack hypothesis is incongruous with the micromorphology of the laminae of the fossiliferous cherts that are indistinguishable in all salient characteristics from laminae of fossil and modern stratiform stromatolites (including their intimate interlamination with intraclast grainstone; see Section 8-4-1 and Photo 8-2, herein) but that differ significantly from the sheet cracks recently described by Ramsay (e.g., they lack any sign of disruption or displacement such as the occurrence of fractured grains with serrated margins). Moreover, to the best of our knowledge it has never been demonstrated that microfossils can be introduced into sedimentary rocks by this or any other comparable mechanism. Similarly, the possibility that at either locality microfossils may have been emplaced during weathering is incompatible with numerous lines of evidence, especially those cited in items (*ii*), (*iii*) and (*iv*), above, and for those samples on which the relevant organic geochemical analyses have been carried out (all from Locality 002), such a possibility is inconsistent as well with the evidence cited in items (*v*) and (*vi*).

Thus, the syngenicity of the Warrawoona fossils seems well demonstrated. Indeed, in view of the evidence noted above it seems to us quite implausible that these microfossils were emplaced in the cherty Warrawoona sediments other than prior to, or concurrent with, primary lithification.

BIOGENICITY AND BIOLOGICAL AFFINITIES

In addition to the spheroidal, unicell-like, possible microfossils, and the rosette-like aggregates of fine fossil-like filaments discussed above (Section 9-4-1; Table 9-1, No. 4; Photos 9-6A-G and 9-7F-I), four types of filamentous fossil microbes have been detected in the stromatolitic Warrawoona cherts (Table 9-1, No. 3): long, sinuous, narrow threads, 0.3 to 0.7 μm in diameter and up to 180 μm in length (Photo 9-4); intermediate-diameter, possibly septate filaments, about 1 μm broad and up to 340 μm long (Photo 9-5A-C); large-diameter tubular sheaths, 3.0 to about 9.5 μm broad and up to 600 μm long (Photos 9-5D and 9-7A-E); and large-diameter septate filaments, up to 120 μm long, composed of a uniseriate row of approximately equant cells 4 to 7 μm in size (Photo 9-8).

To a considerable extent, this assemblage is reminiscent of those known from numerous Proterozoic-age permineralized biocoenoses: in addition to having been similarly preserved (as carbonaceous, in part mineral-encrusted, residues) in an ecologic setting apparently comparable to that of many such younger communities, like the majority of Proterozoic filaments the Warrawoona fossils are all of microbial dimensions and organization; all are simple, unbranched, and relatively undifferentiated (e.g., they lack heterocysts, akinetes, and similarly advanced cell types or reproductive structures); and all bear resemblance to specific, modern, filamentous prokaryotes (Table 9-2). Moreover, in terms of the abundance of its specific components (ranging for the various taxa from fewer than 60 to more than 600 individuals/cm^3 of chert), the assemblage is not dissimilar from those of the younger Precambrian (e.g., Schopf and Sovietov, 1976; Schopf and Prasad, 1978). There is some indication, however, that the Warrawoona microbiota (at least as now known) may be somewhat more primitive than its younger counterparts—although biologically diverse (including at least four filamentous morphotypes), it seems relatively impoverished in comparison with well-known Proterozoic communities (those of the Gunflint, Belcher and Bitter Springs deposits, for example); it lacks the diverse set of solitary and irregularly to well-ordered colonial unicells of the type typical of many Proterozoic microbiotes (e.g., Schopf and Sovietov, 1976; Muir, 1976; Chapter 14); and the dominant microbes of the assemblage, long, sinuous, thread-like forms of the type shown in Photo 9-4, of which several hundred specimens have been observed, are morphologically less complex (viz., apparently being nonseptate, elongate rod-shaped bacteria of decidedly small diameter) and, thus, perhaps phylogenetically less advanced than the principal taxa of such younger communities.

In addition to providing some indication of the possible evolutionary status of the Warrawoona microbiota, all of the above observations are of course fully consistent with a biogenic interpretation of the assemblage. And in this regard, it perhaps need only here be noted in addition that (*i*) the fossils also exhibit numerous other characteristics indicative of their biogenicity—that is, evidence of the occurrence of cell division, indicated by the alignment of cells in uniseriate, septate filaments (Photo 9-8); of life cycle variants, illustrated by the presence of both adult (Photo 9-4A, B, and F-I) and juvenile forms (Photo 9-4C-E) of the narrow, thread-like morphotype; of coloniality, indicated by the presence of septate filaments; and, possibly, of the occurrence of gliding motility, suggested by the presence of tubular, empty, sheaths (Photos 9-5D, and 9-7A-E) and of naked, evidently unsheathed trichomes (Photo 9-8)—; and (*ii*) they are morphologically unlike known pseudomicrofossils or other microstructures of non-biologic origin.

What can be said regarding the biological affinities of these microfossils? As is summarized in some detail in Table 9-2, each of the morphotypes detected is morphologically comparable to specific modern microbes. Indeed, and with but one exception, each is comparable to a diverse, and rather impressive array of such microorganisms—the intermediate-diameter, possibly septate filaments, for example, have modern morphological analogues among the pigmented gliding bacteria (Cytophagaceae), the colorless gliding bacteria (Pelonemataceae), the colorless sulfur gliding bacteria (Beggiatoaceae), the green sulfur bacteria (Chloroflexaceae), the "sheathed bacteria," and the cyanobacteria (Oscillatoriaceae). Even the large-diameter septate filaments—for which such comparisons can be based on measurements of cell size rather than solely on filament diameter and length—have numerous modern counterparts including beggiatoaceans (e.g., *Thioploca schmidlei* and *Beggiatoa arachnoidea*), chloroflexaceans (viz., *Oscillochloris chrysea*), and oscillatoriacean cyanobacteria (e.g., *Phormidium autumnale* and *P. subincrustatum*). Thus, other than as an indicator of the primitive, evidently entirely prokaryotic composition of the assemblage, the simple, generalized morphology of these microfossils is of little aid in determining their phylogenetic affinities. Moreover, as is also summarized in Table 9-2, the modern morphological analogues of the fossil taxa generally exhibit a broad spectrum of physiological characteristics, for a single morphotype commonly spanning the range from anaerobe to amphiaerobe to aerobe and including both photoautotrophs and -heterotrophs and chemoautotrophs and -heterotrophs (see Glossary, Part 2). For this reason (and even if it is assumed that the fossils do not represent members of extinct lineages, forms that may have differed significantly in physiology from their modern morphological counterparts), morphology alone can provide only the most limited of insight into the probable metabolic characteristics of the assemblage, a matter of particular import in assessing the level of ecologic organization represented by the Warrawoona biocoenose.

INFERRED METABOLIC CHARACTERISTICS

In addition to morphology, however, both the organic geochemistry of the Warrawoona kerogen and the mode of occurrence of the preserved microorganisms (their distribution, physical organization, etc.) provide grounds for metabolic inference. As is noted above, cherts from both of the localities of interest are distinctly and finely laminated, lamination that grades from relatively planar and regularly banded (Photo 9-5F and G; P.P.R.G. Sample No. 517) to an undulatory and lenticular stromatolitic microstructure (Photo 9-5E and 8-2A, B; Sample Nos. 517 and 002) like that previously described from the Proterozoic as being of the distinct, wavy, striated type (Walter, 1972a). Thus, the filaments are preserved in stromatolitic microbial mats, finely layered accretionary deposits that formed at the sediment-water interface and an organization that in modern sediments is known to be produced partly, and perhaps largely as a result of the light-dependent growth of filamentous microbes (Monty, 1967; Gebelein, 1969; Walter et al., 1976). By analogy, therefore, it seems reasonable to infer that at least some of the filamentous Warrawoona microorganisms were probably photo-responsive. This possibility is consistent also with the morphologic similarity of several of the detected morphotypes (Table 9-2) to extant, photo-responsive, chloroflexaceans and oscillatoriaceans (forms that are capable of movement via "gliding motility"); with the occurrence in the assemblage both of naked (sheath-lacking) cellular trichomes (Photo 9-8) and of empty, tubular sheaths (Photos 9-5D and 9-7A-E), including sheaths of sufficient size to have originally enclosed such trichomes; and with the common preservation of these filaments in growth position, parallel to the stromatolitic laminae—features that all suggest (if they are not merely a result of the vagaries of preservation) that some of the filaments may have been capable of gliding, possibly phototactic, motility. In addition, and by further analogy with modern mat-building microbial communities where aerobes and amphiaerobes are dominant only within the upper several millimeters of the biocoenose, it seems likely that at least some of the Warrawoona organisms were anaerobes. Moreover, it is worth noting that if the sediment-water interface at this time was essentially anoxic, a possibility that seems consistent with available geochemical data (Chapter 11), it is at least conceivable that the Warrawoona biota may have been entirely anaerobic—a possibility that has the interesting corollary of requiring that the major mat-building organisms of the assemblage would have to have been strict anaerobes, forms that if photosynthetic did not produce oxygen as a by-product.

Additional metabolic inferences can be derived from studies of the Warrawoona kerogen. As is discussed in Chapters 5 and 7, carbon isotopic abundances measured in organic matter from the fossiliferous cherts (especially when viewed in the context provided by the rather extensive set of data now available from similar measurements on numerous other Archean and Proterozoic kerogens and carbonates) suggest that the Warrawoona assemblage probably included autotrophs, organisms that used CO_2 as their immediate source of cellular carbon, an inference consistent also both with the paleoecologic setting of the microbiota and with the physiological characteristics of the modern morphological analogues of several of its components (Table 9-2). And the fact that the total organic carbon content of these sediments (viz., for P.P.R.G. Sample No. 002, about 0.6 mg/g of chert) is similar to that of stromatolitic cherts of younger geologic age (e.g., for cherts from the Gunflint Formation, P.P.R.G. Sample No. 134, about 0.5 mg/g of rock, and for those from the Bitter Springs Formation, P.P.R.G. Sample No. 133, about 0.7 mg/g of rock) suggests that organic carbon was probably being recycled within the Warrawoona microbial mats by heterotrophic microorganisms.

IMPLICATIONS

Thus, from the initial studies of the Warrawoona microbiota summarized above, together with data available from analyses of the megascopic stromatolites occurring in this same rock sequence (Lowe, 1980a; Walter et al., 1980; Chapter 8), it seems reasonable to conclude that: (*i*) shallow water and intermittently exposed environments (and possibly also, land surfaces and open oceanic waters; Awramik et al., 1983) were habitable by prokaryotic microorganisms at least as early as 3.5 Ga ago; (*ii*) such organisms comprised finely laminated, multi-component, stromatolitic communities at the sediment-water interface, biocoenoses where the principal surficial mat-building taxa were probably filamentous and photo-responsive, forms that may have been capable of phototactic, gliding motility; and (*iii*) such communities probably included anaerobes and both autotrophic and heterotrophic microorganisms.

With regard to such communities, however, it can only be speculated as to: (*iv*) whether aerobes (including amphiaerobes) were represented; (*v*) whether the presumed phototactic forms were in fact photoautotrophs (rather, for example, than being photoheterotrophs); and (*vi*) if photoautotrophs were represented, whether they were anaerbic, amphiaerobic, or aerobic photosynthesizers.

9-5-2. The Fortescue Microfossils

The Fortescue Group, the lowest stratigraphic unit of the well-studied Western Australian Hamersley Basin (Trendall, 1975, 1976), is a 3 km-thick sequence of dominantly volcanic rocks (ranging from basalts to rhyolitic tuffs) with interstratified and locally extensive clastic and carbonate sediments. As is discussed in Appendix I, available geochronologic data suggest an age for the sequence of about 2.8 Ga (based on U-Pb analyses of zircons from contemporaneous volcanics that have yielded an age of 2,768 ± 24 Ma; R. T. Pidgeon, pers. comm. to M. R. W., 1979), although for reasons there noted the actual age of the group may be somewhat younger.

The extraordinarily well-preserved cherty stromatolitic limestones from the "Knossos" locality of the Tumbiana Formation (from near the base of the group at this locality), where they occur interbedded and interlaminated with tuffaceous sandstones, are described in Chapter 8. Both domical and stratiform carbonate stromatolites occur at this locality, and both types contain abundant microstructural evidence of microbial activity (e.g., erect and radiating filament molds; fine, kerogenous, filament remnants; and calcite-infilled open space structures that may initially have been gas-filled pockets resulting from bacterial decay). In addition, rare, rather poorly preserved, cellular filaments have been detected in fine-grained, carbonaceous, cherty portions of these same stromatolitic specimens (Appendix I, P.P.R.G. Sample Nos. 025 and 029). Two filament types occur: narrow, ~1 μm-diameter, unbranched threads composed of a uniseriate row of short, disc-shaped cells (Photo 9-9); and (*ii*) broad, ~10 μm-diameter, unbranched trichomes composed of a series of short, barrel-shaped cells that are capped by a well-defined, subconical terminal cell and that are enclosed by a prominent, apparently multilamellated, encompassing sheath (Photo 9-10). Additional details regarding these microfossils, and of their possible formative role in the deposition of one or another of the stromatolite types of the Fortescue sediments, are not as yet available. However, their similarity in cell shape and general morphology to certain extant cyanobacteria (viz., for the smaller form, *Oscillatoria* spp., and for the larger, *Lyngbya* spp.) is of more than passing interest since such an affinity would be consistent with, and apparently supportive of, the notion that oxygen-producing photoautotrophs may have first appeared at about the time of deposition of the Fortescue sediments (Chapter 12).

9-6. Summary and Conclusions

(1) During the past half-century, some 43 categories of microfossils and microfossil-like objects have been reported from at least 28 geologic units of Archean age; of the more than 80 publications dealing with these reported occurrences, well over three-quarters have appeared since 1965 (Figure 9-2).

(2) To be acceptable, such reports must provide evidence establishing (*i*) that the reputedly fossiliferous rock investigated is actually of Archean age; (*ii*) that the fossil-like objects are indigenous to the Archean sediment in question; (*iii*) that the putative microfossils were emplaced in the deposit prior to, or concurrent with primary lithification; and (*iv*) that the fossil-like microstructures are in fact biogenic.

(3) Many such reports, however, have failed to satisfy one or more of these criteria; indeed, it now seems evident that the vast majority of reported "Archean microfossils" are in fact non-fossil: inorganic pseudofossils, organic pseudofossils, artifacts of sample preparation, or modern contaminants (Table 9-1).

(4) Thus, and in contrast both with the relatively well-established microbial fossil record of the Proterozoic and the known distribution of microbially laminated Precambrian stromatolites, cellularly preserved fossil microbes are as yet only poorly known from the Archean rock record (Figure 9-1).

(5) Carbonaceous, spheroidal microstructures of possible, but unproven biogenicity ("dubiomicrofossils") have been reported from four Archean units (sediments of the Warrawoona, Onverwacht, Fig Tree, and Hamersley Groups); although considerable information is available regarding these objects (including size data for more than 1,300 such bodies as reported in some 25 publications), their simple morphology and similarity to organic microstructures of known non-biologic origin "makes their 100 percent unequivocal identification as biological entities virtually impossible" (Knoll and Barghoorn, 1977, p. 398).

(6) Authentic, well-established Archean microbiotas are now known from but two geologic units: bedded stromatolitic cherts of the ~3.5-Ga-old Warrawoona Group (from the North Pole Dome region of the Pilbara Block of Western Australia) and cherty portions of stromatolitic limestones of the ~2.8-Ga-old Fortescue Group (the lowest stratigraphic unit of the northwestern Australian Hamersley Basin). Both of these microbiotas were discovered in 1980 and only one, that of the Warrawoona cherts, has as yet been analyzed in some detail (Section 9-5-1).

(7) Studies of the Warrawoona biocoenose, an entirely prokaryotic community that formed laminar microbial mats at the sediment-water interface in a shallow water to intermittently exposed setting, suggest that it included both heterotrophs and autotrophs, that at least some of the microorganisms were anaerobes, and that the major mat-building taxa of the community were filamentous, phototactic, and presumably photoautotrophic (although there are no data indicating whether the presumed photoautotrophs were oxygen-producing).

(8) The Warrawoona fossils thus establish that surficial sedimentary environments were habitable by microorganisms at least as early as 3.5 Ga ago, and the diversity and cellular complexity of the preserved microbes indicate that the beginnings of life on Earth must have occurred substantially earlier.

(9) In sum, while the Archean fossil record remains extremely poorly known, recent progress in the field has been encouraging; the task now at hand is to increase understanding of the few bona fide Archean microbiotas now known, to discover additional such assemblages with which to bridge the paleontologic gap that now exists between the Warrawoona fossils and the relatively well-known microbiotas of the later Precambrian and, especially, to develop new techniques and uncover new, more telling evidence that will better define both the physiology and phylogeny of Archean life.

CHAPTER 10

Geologic Evolution of the Archean-Early Proterozoic Earth

By Ján Veizer

10.1. Introduction

The general evolutionary features of the early Earth are summarized in Chapters 2 and 3 of this volume. The present chapter represents an attempt to quantify, where possible, these broad generalizations and to discern their impact on the development of the Earth's surficial environment (viz., on the atmosphere, hydrosphere, biosphere, and the sedimentary layer of the Earth's lithosphere). Although this exogenic system is, at present, essentially self-contained and self-perpetuating in the outer three terrestrial "spheres"—and even the sedimentary cycle is largely cannibalistic (see Garrels and Mackenzie, 1971)—the direct interaction with the surficial expressions of endogenic mantle convection cannot be entirely disregarded. Such endogenic-exogenic coupling must have been particularly important in the early, formative stages of Earth's history. At the outset of this discussion, it will therefore be essential to set some limits on possible modes of growth and differentiation of the continental and oceanic crusts over geologic time. The limits thus derived will in turn have direct bearing on understanding the evolution of sedimentary environments, the nature and chemistry of sediments, the configuration of sedimentary basins, the distribution and chemistry of water bodies, and, ultimately, the biologic history of the planet. For details regarding many of the points that are considered only briefly here, the reader is referred to Veizer (1984), a discussion more detailed, as well as complementary, to that of the present chapter.

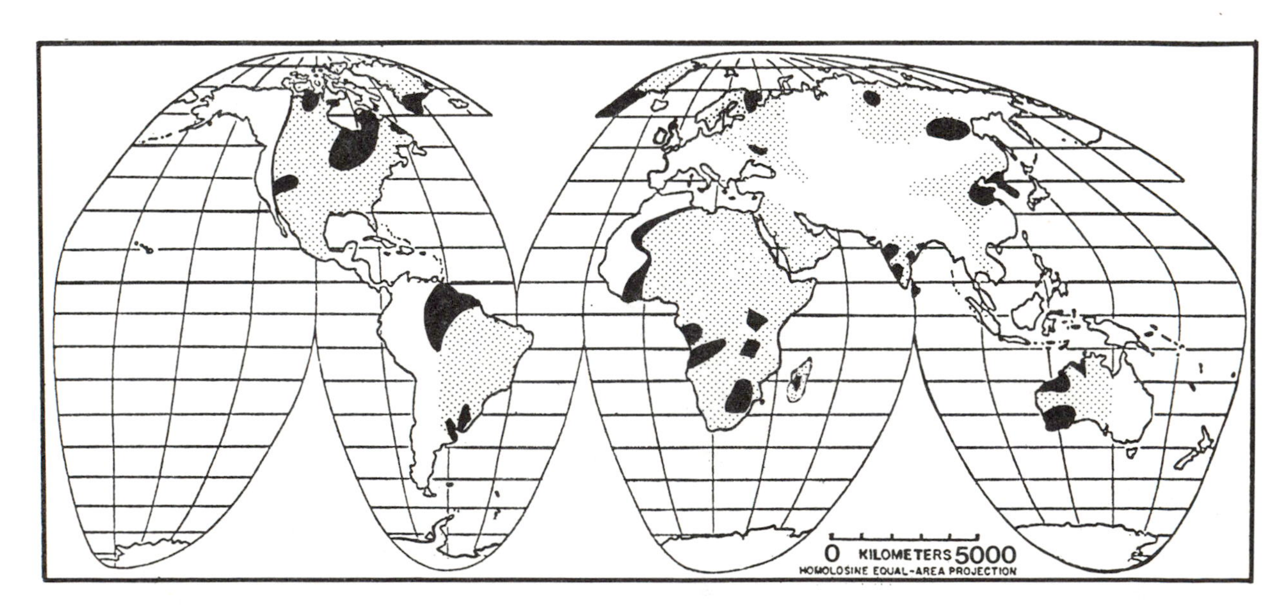

FIGURE 10-1. Map summarizing the approximate distribution of known Archean (dark), Phanerozoic and undifferentiated (light), and Proterozoic and undifferentiated Precambrian (stippled) crustal segments within the confines of present-day continents. Modified after Windley (1977, fig. 1). A considerable portion of the greenstone belts in the Guyana Shield and in western Africa, now considered to be Archean, may be of Early Proterozoic age.

10-2. Growth and Recycling of the Continental Crust

The present-day distribution of Archean and Proterozoic rock units within the confines of the existing continents is illustrated schematically in Figure 10-1 as based on the compilation of Windley (1977). It should be noted that the distribution shown there is a minimal estimate, particularly for the Archean, since some old crustal segments were not yet identified when Windley's compilation was carried out, whereas parts of others were no doubt incorporated (i.e., recycled) into younger rocks, their ancient identity being difficult to discern. As is shown in Figure 10-2, a cumulative plot of known crustal segments (tectonic provinces) as a function of time shows that their area, per time unit, increases toward the present. Theoretically, the observed power law (or exponential) pattern of distribution can be generated in three principal ways (Figure 10-3):

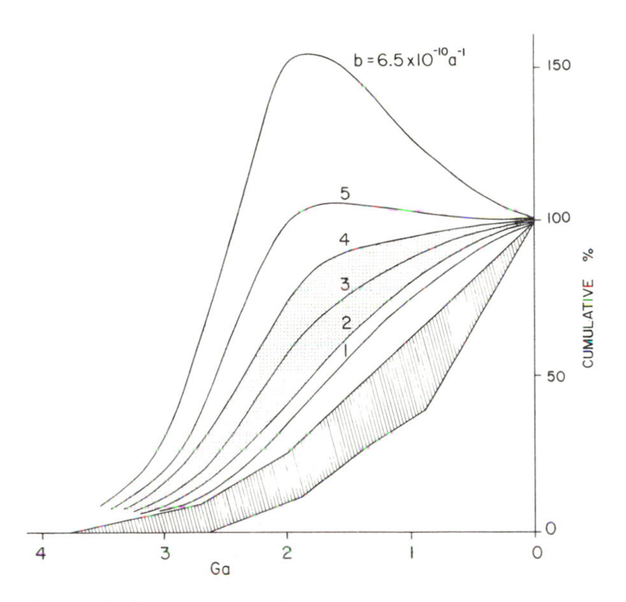

FIGURE 10-2. Present-day distribution of preserved areas of continental basement and platforms as a function of their age. Hachured area (bottom) encompasses all measured curves. Note the power law (exponential) increase of areal increments with decreasing geologic age. Smooth curves (upper part of the figure) represent calculated crustal growth patterns corrected for the effect of geologic recycling, assuming an average recycling constant $b = 1$ to $6.5 \times 10^{-10}\ a^{-1}$ and taking the upper boundary of the hachured area as the observed curve. The preferred range of growth patterns is indicated in the stippled region ($b = 2$ to $4 \times 10^{-10}\ a^{-1}$). Modified from Veizer and Jansen (1979).

(1) It could reflect an exponentially increasing rate of fractionation of continental crust from the mantle in progressively younger geologic periods (see Hurley and Rand, 1969; Figure 10-3A, Case IV, herein). This possibility, however, is at variance with expected consequences of a high geothermal

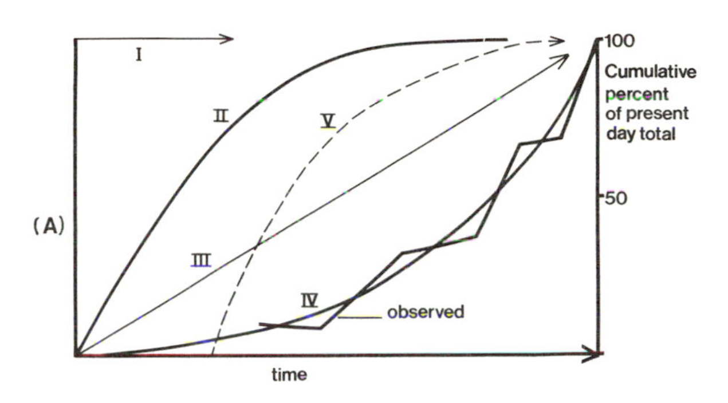

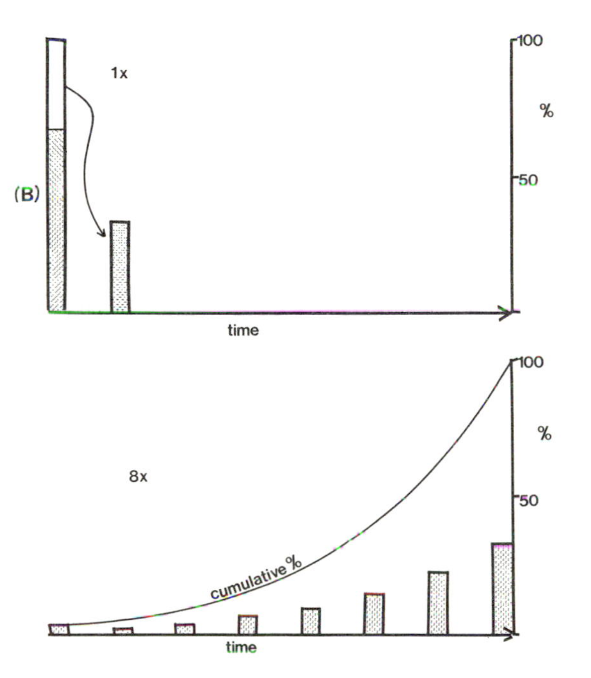

FIGURE 10-3A (upper figure). Schematic representation of some theoretically possible crustal growth models. Case I = Instantaneous growth to present-day total. Case II = Exponentially declining growth rate. Case III = Growth at a constant rate. Case IV = Exponentially increasing growth rate; Case V = As II above, but starting at a later stage of geologic evolution.

FIGURE 10-3B (lower figures). Schematic histograms illustrating the generation of the observed exponential (power law) present-day distribution of crustal segments while maintaining a steady state cumulative area (Case I above). Except for the "cumulative %" curve (bottom figure), the percentages indicated in the histograms are not cumulative.

gradient early in Earth history (as discussed in Chapter 3), which would demand that mantle convection and crustal generation were relatively more rapid during the youthful stages of terrestrial evolution than in later stages.

(2) The exponential pattern might have been produced via the continuous recycling and reworking of a continental crust of present-day size (Figure 10-3A, Case I). According to this alternative, the zonal, age-dependent arrangement of continental tectonic provinces (with ancient nuclei at continental centers and younger segments toward continental margins) would be regarded as a consequence of progressively younger orogenies tending to be either of decreasing areal extent, of relatively greater concentration toward continental margins, or both. Thus, the observed zonal age pattern would be viewed as reflecting the times when particular continental segments were last involved in geologic events that reset radiogenic clocks, rather than the times of fractionation of such segments from the mantle.

(3) The observed exponential pattern of distribution might have been generated through a multitude of combinations of both crustal growth and recycling (e.g. Figure 10-3A, Cases II, III, and V).

Assuming a constant size of continental crust over geologic time (alternative 2, above), it is relatively easy to calculate the average rate of recycling (and, thus, the rate of resetting of radiogenic clocks) necessary to duplicate the observed power law distribution pattern (Figure 10-3B). Similarly, average recycling rates can be calculated for the other possible growth patterns. It is evident that the required rate of recycling will be maximal for a constant size continental crust (Case I in Figure 10-3A) and that it will diminish to zero for an exponentially increasing rate of fractionation of crust from mantle (Case IV in Figure 10-3A). Computer simulation of these recycling processes (Veizer and Jansen, 1979) shows that the theoretically possible average recycling constant b is $\leq 6.5 \times 10^{-10}$ per year. Utilizing such values of b, and correcting the actually observed distribution for the effect of geologic recycling, it is possible to estimate the shape of the original crustal growth curves (Figure 10-2, lines for $b = 1$ to $6.5 \times 10^{-10}\ a^{-1}$). The growth pattern obtained with $b = 6.5 \times 10^{-10}\ a^{-1}$ (the uppermost line in Figure 10-2) can be excluded from further consideration since it would require a net destruction of continents since about 2.0 Ga ago, an unlikely possibility in view of the high buoyancy of continental crustal materials. The overall picture of continental growth (involving fractionation from the mantle followed after some 10^8 years by cratonization) emerging from this simulation process is one resulting in generation of small continental nuclei in the Early and Middle Archean (Figure 10-2). However, the available data base does not as yet provide high enough resolution for selection of a specific b from the $\leq 5 \times 10^{-10}\ a^{-1}$ range, although the best fit calculations suggest that a value for b of 2 to 4×10^{-10} per year (viz., Figure 10-2, lines 2 and 4) is preferable. It is important to note, however, that this calculated growth curve could also be a consequence of (*i*) gross underestimate of the extent of the oldest crustal segments by the prevailing Rb-Sr and U/Th-Pb dating techniques, or (*ii*) the occurrence of recycling rates prior to 2.0 Ga ago that were considerably higher than the values of b used in the calculations.

The relatively limited yield of an intensive search for old previously unrecognized crustal segments during the past decade does not support alternative (*i*). It therefore appears necessary to evaluate the role of recycling in preferential destruction of older crustal segments as demanded by alternative (*ii*).

10-2-1. Limits on Recycling of the Upper Continental Crust

Results obtained by application of the recently developed Sm-Nd systematics—a system which usually dates the time of initial magma generation and is not as easily subsequently reset as are the K-Ar, Rb-Sr, and U/Th-Pb isotope pairs—broadly support previous datings of Archean crustal segments (O'Nions et al., 1979). This consistency indicates that the absence of large remnants of early continents is not an artefact of dating or of their large-scale recycling.

Further support for this interpretation is provided by analyses of isotopic initial ratios. Rb-Sr and Sm-Nd systematics, because of the considerable parent/daughter fractionation that occurs in these element pairs during the process of magma generation, are sensitive to contamination from pre-existing older continental crust. It has been observed that Rb/Sr ratios are higher, and Sm/Nd ratios are lower, in calc-alkaline magmas than in the mantle or basalt precursors of such magmas.

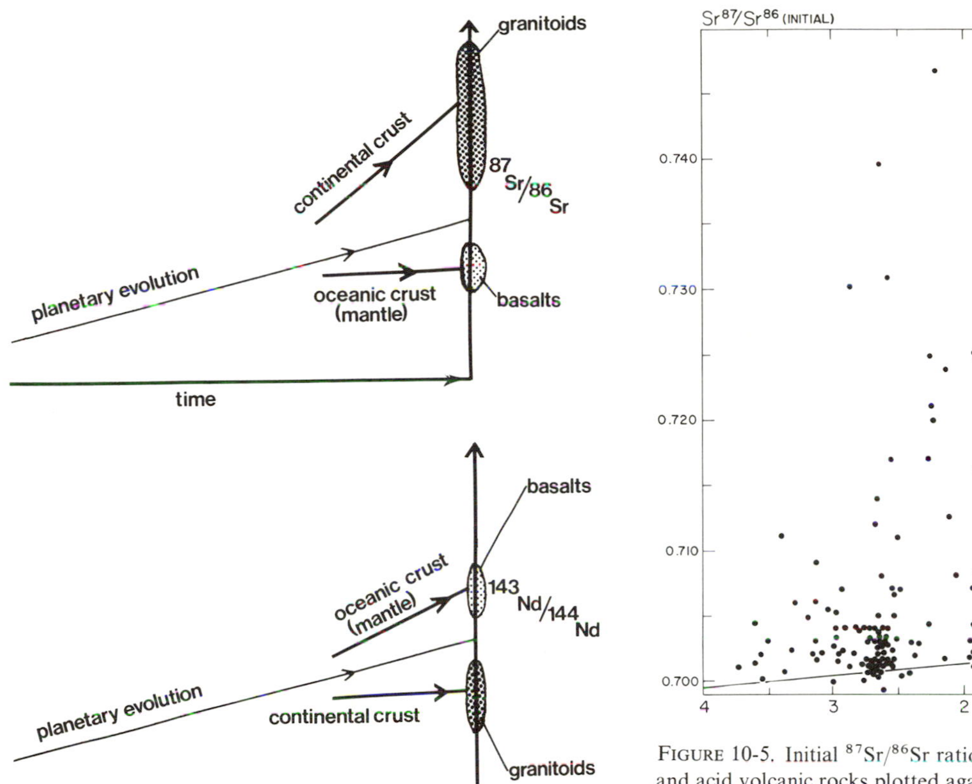

FIGURE 10-4. Schematic representation of the evolution of $^{87}Sr/^{86}Sr$ and $^{143}Nd/^{144}Nd$ ratios during geologic history in the continental and oceanic crust (and mantle). Modified from Allègre (1980).

FIGURE 10-5. Initial $^{87}Sr/^{86}Sr$ ratios of granitoids, gneisses and acid volcanic rocks plotted against their Rb-Sr isochron ages. Data are for the North Atlantic Craton, the Canadian Shield, Australia and southern Africa. Modified from Glikson (1979). See Allègre and Ben Othman (1980) for a similar interpretation based on Sm-Nd systematics.

Evidently, over time such magmas acquire a greater concentration of radiogenic ^{87}Sr, but concentrate less ^{143}Nd, than remains in the residual mantle (Figure 10-4). Therefore, calc-alkaline volcanics or granitoid plutons newly generated at the edges of continental margins should acquire $^{87}Sr/^{86}Sr$ and $^{143}Nd/^{144}Nd$ ratios of the contemporaneous mantle if derived from the mantle source (or its young products) but should acquire crustal ratios if derived from remelting of older continental rocks. Studies of Phanerozoic plutons (DePaolo, 1980a; Hart and Allègre, 1980; Hamilton et al. 1980; Allègre and Ben Othman, 1980) indicate the presence of a considerable crustal component, thus supporting the role of continental recycling in the generation of many Phanerozoic granites, a conclusion that is reinforced also by available oxygen isotope data (Taylor, 1978). However, for progressively older (and particularly for Archean) granitoids, the proportion of plutons with continental crust-like $^{87}Sr/^{86}Sr$ and $^{143}Nd/^{144}Nd$ initial ratios rapidly diminishes, as confirmed by the tight clustering around mantle values for the most ancient examples (see Figure 10-5). These isotopic constraints preclude interpretation of the absence of surviving large early continents as being a consequence of large-scale intracrustal recycling (see, Moorbath, 1977; Hart and Allègre, 1980; DePaolo, 1979, 1980b; Allègre, 1980). On the other hand, the data do not preclude the possibility of large-scale recycling of early continental crust back into the mantle, regardless of whether such recycling is considered in a relatively crude way or in a more sophisticated fashion (e.g., Armstrong, 1981). At the same time, the previously noted buoyancy of continental crustal materials poses difficulties for at least the crude, physical (subduction) models of the process. From the point of view of the evolution of stable continents, however, it is immaterial whether their absence early in the Earth's history was due to limited crustal differentiation and a continuous intermixing of the early crust with the mantle, to a slow rate of formation without such intermixing, or, more likely, to a combination of both of these processes.

In summary, geometric and isotopic constraints favor a growing, rather than a steady-state, continental crust. The reasons for its delayed formation are related to a decrease in the geothermal gradient with the passage of time (Davies, 1980; Turcotte, 1980; Chapter 3 herein), although the actual triggering causes are at present disputed (see, for example, the scenario of Condie, 1980). Strontium initial ratios suggest that $\sim$20 to $\lesssim$50% of granitoids contain a relatively large component of older crustal material (see Chapter 5 in Faure and Powell, 1972). The value of b here preferred of 2-4 $\times$ 10^{-10} a^{-1} yields approximately this amount of recycling. The pattern of continental accretion suggested by these constraints is one of slow growth during the Early Archean, relatively fast accretion during the period 3.0 to 2.0 Ga ago, and slow growth during subsequent geologic time (Figure 10-2), although the actual pattern of fractionation was no doubt episodic rather than of a continuous nature. This picture is identical to that derived by Allègre (1982) from a careful evaluation of all isotopic dating pairs for mantle and crust. Furthermore, the isotopic data indicate that the early evolution of the Earth was characterized by the multi-stage generation of new continental crust from the mantle with little intracrustal recycling and, possibly, with some intermixing with the mantle. With advancing growth and cratonization of the continental crust, the role of intracrustal recycling became progressively more prominent.

10-2-2. Composition of the Continental Crust

The major physical and lithological differences between the granite-greenstone and high-grade gneissic terranes of the Archean and those of the Proterozoic and younger crustal segments have been summarized in Table 3-1 of Chapter 3. A review of this table indicates that the Archean assemblages are relatively less differentiated and more mafic (as evidenced, for example, by the prevalence of komatiites, tholeiitic volcanics, tonalitic gneisses, and greywackes). Although the observed differences are sometimes qualitative (e.g., the apparent absence or rarity of alkalic igneous

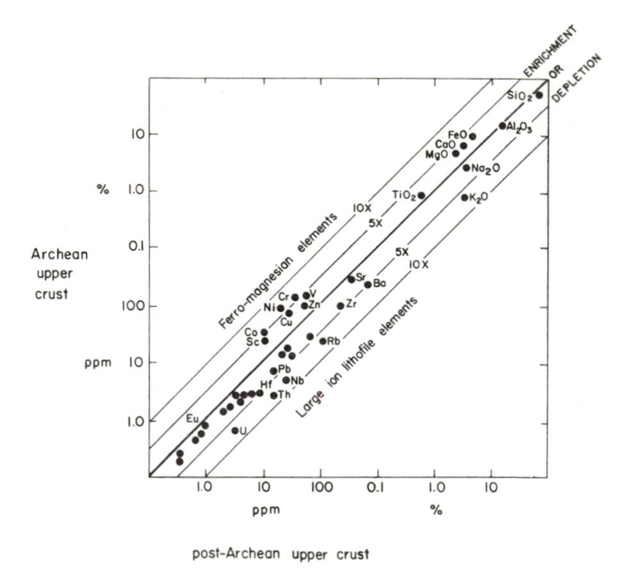

FIGURE 10-6. Chemical composition of the Archean compared with the post-Archean upper continental crust. Unlabelled dots are REE (rare earth elements). Modified from Taylor and McLennan (1981).

rocks in the Archean), for the most part they are of quantitative type. The overall more mafic nature of the Archean upper continental crust is also reflected in the commonly andesitic chemistry of Archean clastic sediments, in contrast with the more typically granodioritic nature of such post-Archean rocks (Figure 10-6; Taylor and McLennan, 1981), and in the relatively high frequency of occurrence of ore deposits typical of ultramafic to mafic associations [e.g., Ni, Cr, (Au)] in the older crustal segments (see, Veizer and Jansen, 1979). The overall andesitic composition of the Archean upper continental crust is due to the presence of a higher proportion of mafic lithologies generally, rather than to the more mafic nature of particular rock types, since compositional differences between Archean and Phanerozoic basalts, for example, may or may not be reflected with regard to trace elements only.

Acceleration of the rate of production of continental crust in the Late Archean and Early Proterozoic, followed by an increasing level of tectonic stabilization (i.e., cratonization) and by a more pronounced degree of chemical differentiation and stratification, was perhaps the most important benchmark in continental evolution. This transition was reflected not only in a marked change of tectonic style (from the type typified by the frequently ensimatic Archean greenstone belts to the style characteristic of the mostly ensialic Proterozoic mobile belts; Kröner, 1981), but also in the abundance, lithology and chemistry of contemporaneously deposited sediments. The diachronous stabilization process, apparently first evidenced in units about 3.1 Ga old in the Kaapvaal Craton of southern Africa (the Pongola Supergroup being the oldest known cratonic sequence; Cloud, 1976b), was largely—but not entirely—completed by about 2.0 Ga ago.

10-3. The Early Oceanic Crust

Due to the transient nature of the oceanic crust (resulting from the processes of plate tectonics), direct vestiges of its early counterparts have not survived to the present. Indirect evidence, based on the nature of surviving petrotectonic assemblages, and the possible implications of such evidence, are summarized in Chapter 3 of the present volume. The following discussion deals with several features not considered in that chapter that have direct bearing on the nature of early exogenic terrestrial environments and cycles.

It seems likely that the high internal heat generation in the early Earth (resulting in higher geotherms) must have been convectively dissipated mainly through the oceanic crust, since early continental geotherms do not appear to have much differed from those of the present day (Condie, 1973; Davies, 1979). This implies a broader areal front of heat loss than is represented by present-day oceanic ridges and hot-spots, regardless of whether the physical expression of such heat loss was a more extensive ridge system (as, for example, might have accompanied smaller tectonic plates) or a broader distribution of hot-spot volcanic centers. This also implies that, on the average, the oceanic crust was generated more rapidly ($\gtrsim$4–7 times faster; Allègre, 1982) and consequently was hotter and of lower density than that of the present.

The depth below sea level ($\sim$4–5 km at present) and the size of the abyssal plain of the oceans is a function of the thickness and the area of the continental crust (Wise, 1974). For a globe of present-day radius, the area of the oceans diminishes as the continents grow in size. With a thinner continental crust, as has been suggested for the Archean (see Chapter 3), the requirements of isostatic compensation demand a higher-standing abyssal plain and, hence, shallower oceans. The lower density of the early oceanic crust implied above will further accentuate this requirement, as clearly demonstrated by the topography of present-day hot ridge systems which stand some 3 km above the surrounding abyssal plain (see, Parsons and Sclater, 1977). In theory, from such considerations one could calculate the depth of the early oceans (Wise, 1974), but the calculation is beset by uncertainties in the assumed variables (viz., the densities of the early oceanic and continental crusts, the continental thickness, and the volume of the coeval hydrosphere) and has therefore no unique solution. However, it is conceivable that the average depth of the early Archean oceans could have been as much as 2 to 3 km less than that of today. If so, the global response to a decrease in the geothermal gradient over time would have been the growth, and (at least in the early stages) thickening, of the continental crust accompanied by the shrinking and deepening of the oceans. This, in turn, would have resulted in a progressively decreasing interchange between global endogenic and exogenic cycles, a development of considerable implication for the evolution of the terrestrial hydrosphere and atmosphere as discussed below.

10-4. The Sedimentary System

The most recent direct estimate of the present-day dimensions of the global sedimentary layer (including interlayered volcanics) is 1.1×10^9 km^3 representing some 2.74×10^{24} g (Ronov, 1980, tab. 10). The bulk (up to 93%) of this sedimentary mass is associated with the continental crust and, to a lesser extent, with the continental shelves and slopes (Ronov, 1980). Since clastic rocks account for two-thirds of all sediments (Ronov, 1980), this association points out that the generation of sedimentary mass is a function of relief and, ultimately, of the development of the buoyant continental crust with its concomitant freeboard. Despite the continental derivation of the bulk of the present-day global sedimentary mass, its chemical composition is distinctively different from that of its presumed ultimate source (upper continental crust). Compared to this apparent source, the sediments (Ronov, 1968, 1980): (*i*) are more mafic in their overall composition, approximating a 70:30% mixture of the "granitic" layer and basalt (Sibley and Wilband, 1977; Veizer,

1979a, 1984); they (*ii*) have higher concentrations of such components as H_2O, CO_2, C, Cl, F, B, and so forth (i.e., the so-called "excess volatiles" of Rubey, 1951); and they (*iii*) have a higher oxidation state.

As is outlined in the following sections of this chapter, these features are a consequence of the evolution of, and the interrelations among, the terrestrial lithosphere, atmosphere, hydrosphere, and biosphere.

10-4-1. Growth and Recycling of the Sedimentary Mass

At present, Early Proterozoic and Archean sediments crop out only on a negligible part of the continental surface (viz., the area indicated in Figure 10-7); they do not at all occur in the transient oceanic environments. Volumetrically, such sediments—covered mostly by younger strata—represent about one-eighth of the global sedimentary mass, with the Archean component accounting for perhaps 5% of the global total (Figure 10-7). By analogy with the distribution discussed above of basement age-provinces, the temporal distribution patterns shown in Figure 10-7 seem certain to be caused by recycling with the rates of such recycling being functions of the assumed growth geometries for sedimentary area, thickness, and mass.

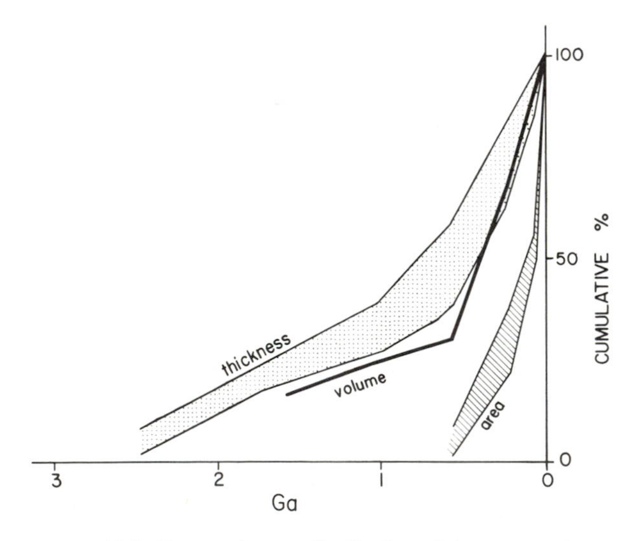

FIGURE 10-7. Present-day age distribution of the preserved cumulative area, thickness and mass (volume) of sedimentary rocks. See Veizer and Jansen (1979) for sources of area and thickness data. The mass (volume) distribution curve (from Ronov, 1980, tab. 7), which includes post-1.6-Ga old sediments only, accounts for 1.57×10^{24} g; the difference between this value and the estimated global continental sedimentary mass of 1.89×10^{24} g (Ronov, 1980, tab. 10) was taken as the mass of pre-1.6-Ga-sediments (viz., 0.32×10^{24} g).

With regard to the growth geometry, Veizer and Jansen (1979) have argued that the continental sedimentary mass should have been growing at a rate similar to that of its underlying basement. If so, the cumulative pre-1.75 Ga mass could have been originally equal to about three-quarters (and that of the Archean to about one-third) of the present-day global sedimentary mass (Figure 10-2). To this amount should be added the mass of transient oceanic sediments, which at present account for about 7% of the total global value (Ronov, 1980). This value, however, could have been higher in the early stages of the Earth's history because of a higher ratio of ocean to continent area and, perhaps also because of a smaller difference in density (and relief) between continental and oceanic crust. Thus, the cumulative global sedimentary mass during the Archean was probably as much as one-half of, and at ~1.75 Ga ago almost equal to, the present-day total. Accepting such a generalized growth geometry, computer simulation suggests $\sim 21 \times 10^{-10}\,a^{-1}$ as the probable average recycling constant (Veizer and Jansen, unpublished). With this values of *b*, the total mass of sediments recycled on continents during all of geologic history amounts to about 3.9 to 5.6 times the present continental sedimentary mass.

The above conclusions give some appreciation of the cyclic nature of sedimentary processes. To what extent this cycle is closed, with no net input or loss (for example to igneous and metamorphic rocks), is difficult to determine. Veizer and Jansen (1979), utilizing McCulloch and Wasserburg's (1978) Sm-Nd data on sediments, argued that the sedimentary cycle should be about two-thirds cannibalistic for a steady-state global sedimentary mass. If, however, as suggested above, the sedimentary mass was in fact growing, less cannibalism would be required to explain the observed Sm-Nd data.

10-4-2. Evolution of Sedimentary Facies Assemblages

The tectonic evolution described above resulted in development of the following sedimentological

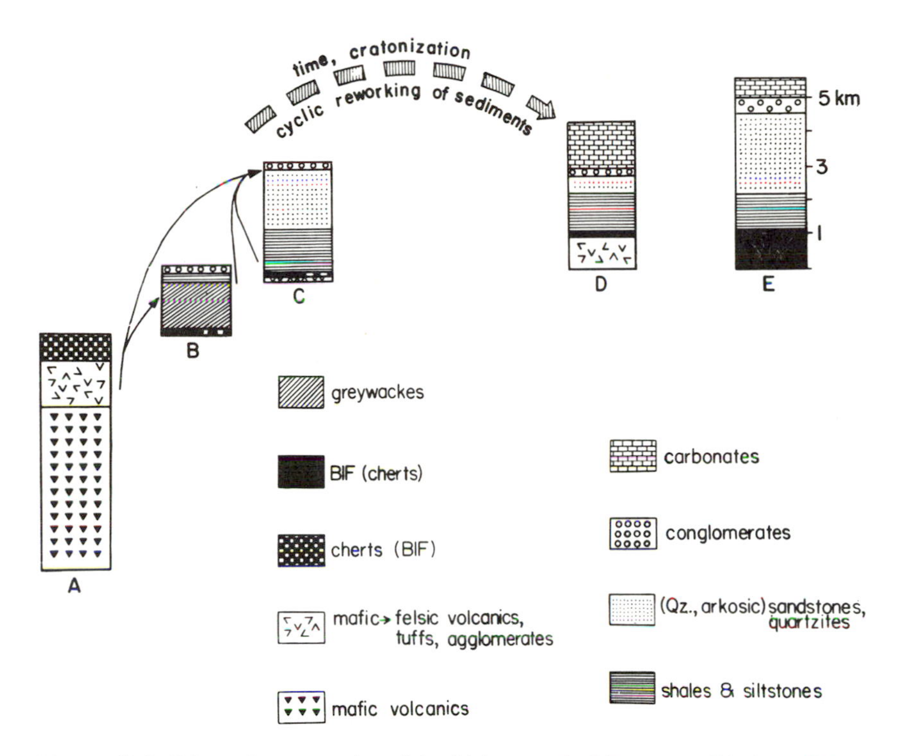

FIGURE 10-8. Schematic presentation of the thickness and of the average lithology of the following sedimentary successions. (A) Upper Onverwacht Group (Lowe and Knauth, 1977, fig. 2). (B) and (C) Fig Tree and Moodies Groups, respectively, as developed in the Eureka-Ulundi syncline (Eriksson, 1980, fig. 2). (D) Transvaal Supergroup in the north-eastern part of its depositional basin (Button, 1976; Haughton, 1969). (E) Earaheedy section of the Nabberu Basin (Hall and Goode, 1978, fig. 6). Sections A–C are from Kaapvaal craton of South Africa, ~3.3-Ga-old; sections D (from Kaapvaal craton) and E (from Western Australia) are of Early Proterozoic age.

facies assemblages:

(1) Cyclic, mafic to felsic, volcanogenic and volcaniclastic accumulations that were capped, particularly at higher stratigraphic levels, by lenses of chemical sediments such as cherts, silicified volcaniclastics, Algoma-type iron-formations, carbonates (in part of exhalative origin), and barites (see Figure 10-8A; and Lowe and Knauth, 1977). The lower parts of these volcanogenic sequences represent shallow submarine lava flows while the overlying volcaniclastics signify a subsequent buildup of subaerial felsic volcanic edifices with large relief (see Lowe, 1980b, and the volume edited by Dimroth et al., 1980). Slumping produced debris flows and lahars that deposited their load on slopes either in terrestrial milieus or as submarine fans (Figures 10-9, 10-10, and 10-11). Erosion of these volcanic cones ultimately produced submarine platforms where chemical sediments were deposited. This assemblage is particularly frequent in the older (> 3.0 Ga) greenstone belts (Lowe, 1980b).

(2) In larger, coalescing volcanic islands, stratified conglomerates and sandstones with minor siltstones and argillites represent braided alluvial floodplains (Figure 10-12), and perhaps local eolian dunes (i.e., the nonmarine facies association of Hyde, 1980, and of Turner and Walker, 1973). Facies assemblages (1) and particularly (2) pass vertically (Figure 10-8A to B) and laterally (Figures 10-10, 10-11, and 10-12) into:

(3) Graded graywacke-shale couplets and resedimented deep to shallow water turbidites (viz., submarine fans and flysch) with occasional paraconglomerates and with a background basinal sedimentation of chemical cherts and/or iron ores (Figure 10-8B; the re-sedimented facies association of Turner and Walker, 1973, and of Hyde, 1980).

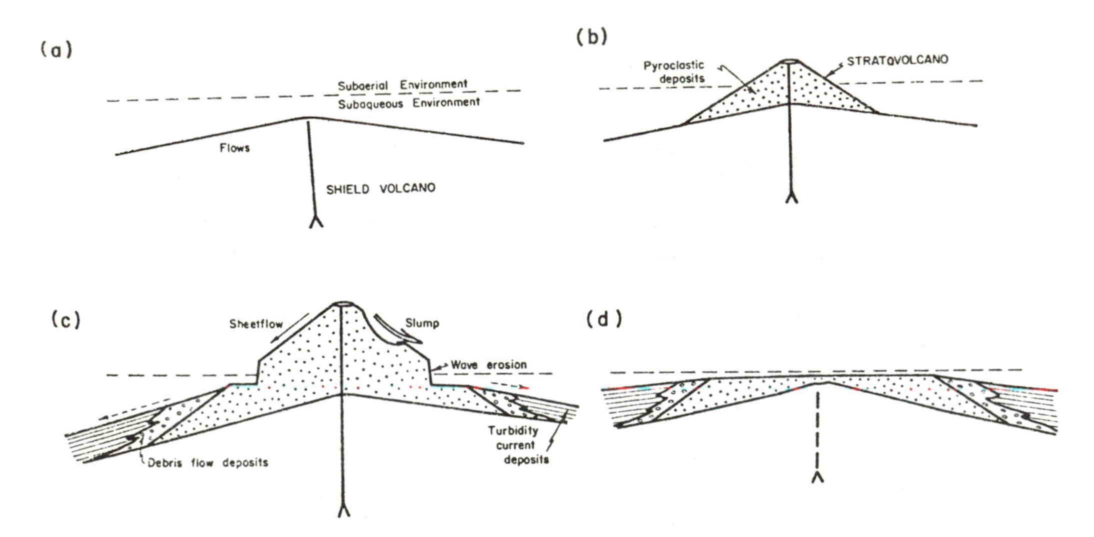

FIGURE 10-9. Diagrammatic sketches showing major stages in the development of stratovolcano on top of a previously formed shield volcano. (A) Build-up of shield volcano into shallow watër. (B) Initial emergence of stratovolcano. (C) Initial erosion of stratovolcano. (D) Erosion of stratovolcano to sea level. From Bailes (1980), after Ayres (1978).

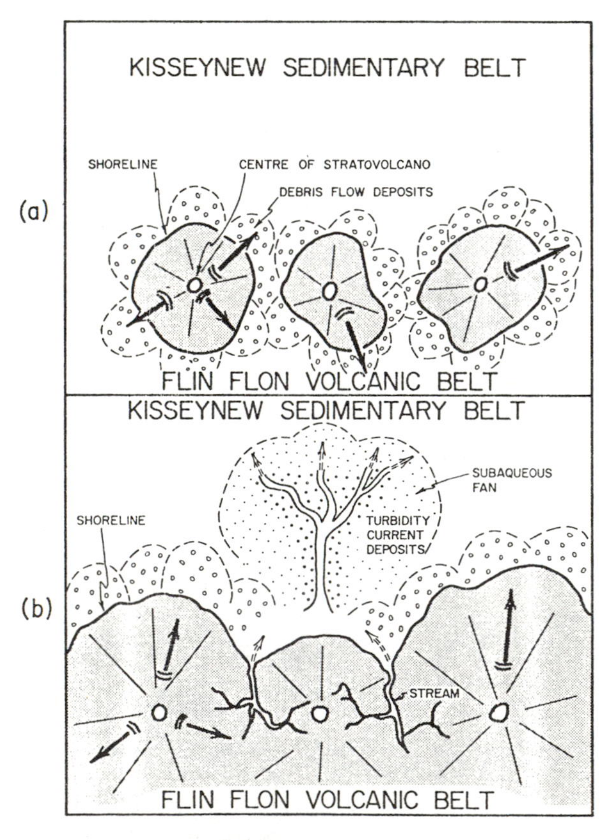

FIGURE 10-10. Plan views of inferred patterns of sedimentation in the Flin Flon and Kisseynew Belts, (A) as stratovolcanoes initially emerge, and (B) as stratovolcanoes coalesce to form large land areas. From Bailes (1980).

(4) The above re-sedimented clastic assemblage passes laterally and vertically into (and in some cases, for example in India, perhaps evolves from) alluvial to marginal marine, cross-bedded sandstones, siltstones and clast-supported conglomerates (Figure 10-8C) representing such coastal environments as deltaic plains, tidal flats, channels, shoals, barrier islands, and shallow marine shelves (Lowe, 1980b; Eriksson, 1980c). This assemblage marks a transition to the mature sediments (assemblage 5) discussed below.

The juxtaposition of alluvial fan/braided river deposits (2) and the re-sedimented conglomerates/turbidites (3), with no or only a narrow intervening shelf, is typical of recent volcanic island settings (see Hyde, 1980; Figure 10-12). In contrast, the development of the coastal (type 4) assemblage suggests the existence of a stable hinterland (Eriksson, 1980c; Lowe, 1980b).

The tectonic development of a greenstone belt (Lowe, 1980) starts with an anorogenic, tectonically inactive phase with abundant volcanism (viz., the stratovolcanoes of type 1, above). This is followed by a tectonic phase characterized by waning volcanism, plutonism, and subsequent uplift of greenstone margins (resulting in deposition of the type 2 nonmarine association) and by development of the adjacent subsiding sedimentary basins (viz., the re-sedimented association of type 3). This

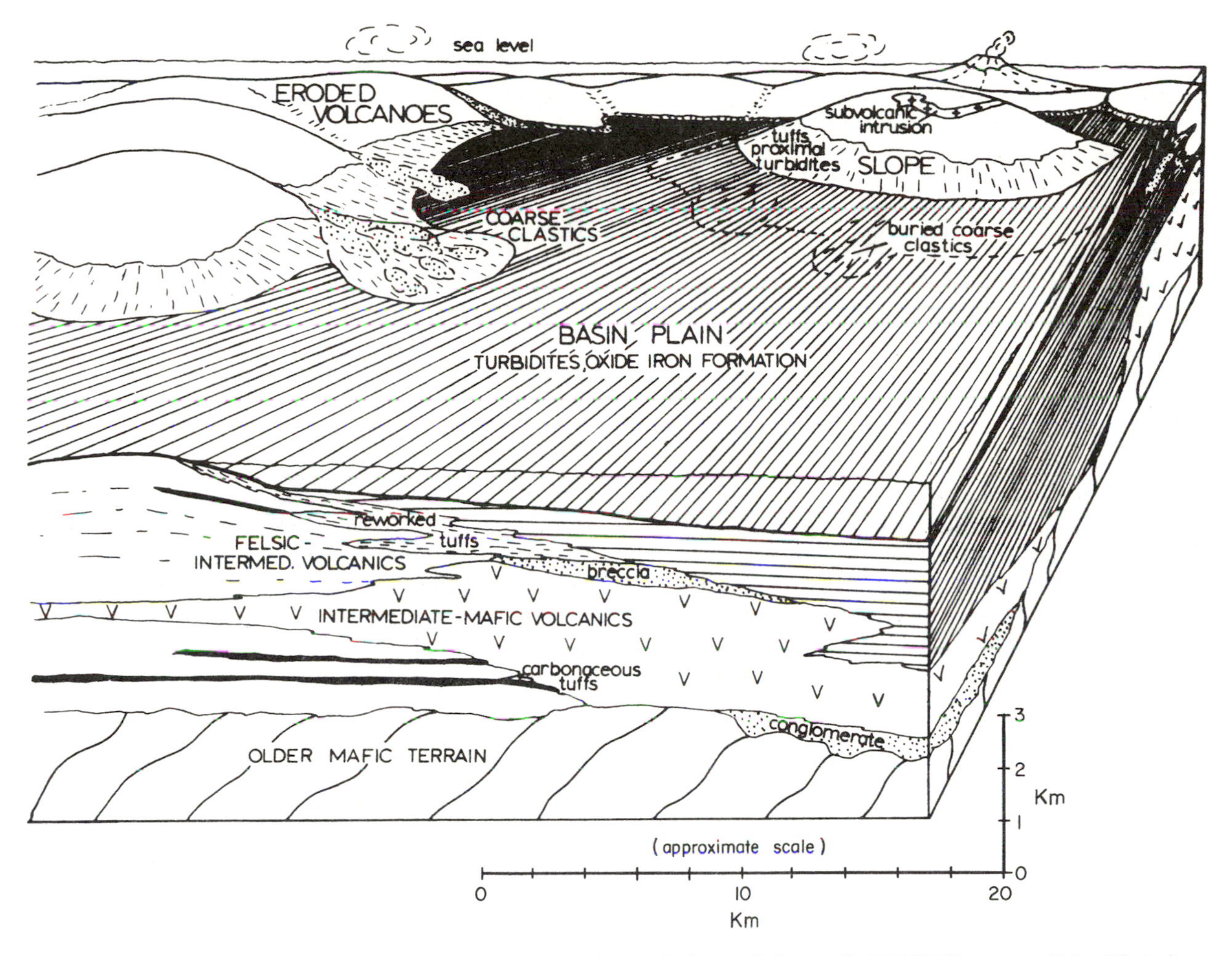

FIGURE 10-11. Three-dimensional paleotopographic reconstruction of the Savant Lake terrain (Abitibi Greenstone Belt of Ontario, Canada). From Shegelski (1978).

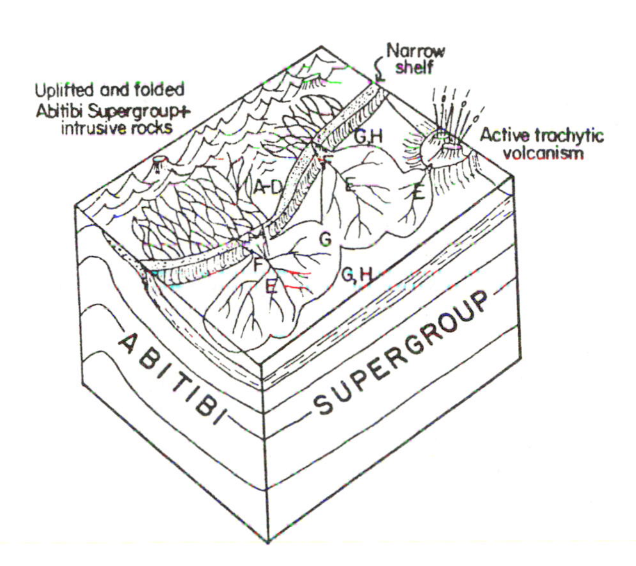

FIGURE 10-12. Model of deposition of the Timiskaming Basin. Mountainous volcanic island is drained by braided river network with deposition of facies A–D (non-marine facies association). A narrow shelf, possibly in places absent, is cut by submarine canyons resulting in development of submarine fans and deposition of facies E–H (re-sedimented facies association). Volcanism is contemporaneous with sedimentation. From Hyde (1980).

evolution culminates in cratonization through intrusion and deformation of the belts.

The resulting post-cratonization marginal basins have wide belts of shallow marine sediments on their broad shelves. Their facies assemblage is characterized by:

(5) Mature orthoconglomerate/quartizitic (and arkosic) sandstone/shale/carbonate/iron-formation (Superior-type) assemblages (see Figure 10-8D and E).

Although—and disregarding long-term evolutionary differences (see Ronov, 1964; Veizer, 1973, 1976, 1984 for details)—a somewhat comparable change in the relative proportion of facies can be observed in Phanerozoic eugeoclinal versus miogeoclinal and platformal areas, a strictly uniformitarian approach is not warranted. Shallow water lithologies, in some instances of a mature character, are known in ancient greenstone belts (see for example, Dunlop, 1978; Heinrichs, 1980; and Naqvi et al., 1978). Similar lithologies were also reported from high-grade gneissic terranes (Chapter 1 of Windley, 1977), although some of these "quartzites" and "carbonates" are in reality metamorphosed cherts and exhalative carbonates. However, no clear-cut example of a mature (type 5) facies assemblage is known prior to about 3.1 Ga ago (viz., the Pongola Supergroup of South Africa). A similar pattern of advancing cratonization and facies development was duplicated toward the end of the Archean on what are now the continents of North America, Australia, and India (Stockwell et al., 1970; Glikson, 1979; Naqvi et al., 1978). Nevertheless, as discussed below, not all variations in facies and in composition of sedimentary rocks can be related to this first order tectonic factor.

10-4-3. Clastic Sediments and Their Chemistry

The Proterozoic increase in the relative proportion of mature orthoconglomerates, quartzites, sandstones and shales, at the expense of immature graywackes and siltstones, is a consequence of the expanding high energy epicontinental seas made possible by advancing cratonization of the precursor greenstone belts. This enabled cyclic reworking and, hence, chemical and mineralogical disproportionation of the older as well as coeval immature assemblages. The above mechanical disproportionation in second-cycle sediments was concomitant with the growth of the continental crust and with the change in its chemistry from andesitic to granodioritic (see Section 10-2-2, above), resulting in formation of a large mass of first-cycle sediments with granodioritic composition. In particular, the unroofing of late orogenic potassium-rich granites (e.g. Brown, 1981; Barker et al., 1981) provided K-Na feldspar-rich detritus for deposition of immature arkoses and arkosic sandstones, both of which are relatively underrepresented in Archean assemblages (Figure 10-8; Ronov, 1964; Schwab, 1978).

The cumulative effect of these Late Archean and Early Proterozoic developments was (*i*) a large increase in the global sedimentary mass (due to the formation and erosion of extensive buoyant continents); (*ii*) an increase in the relative proportion of mature sedimentary deposits (due to the evolution of large epicontinental seas); and (*iii*) within facies, a gradual shift toward more felsic chemical compositions. In the case of shales, which account for ~ 70 to 80% of all present-day clastics (Garrels and Mackenzie, 1971 p. 247; Ronov, 1980), the oldest, >3.0 Ga old, sequences frequently contain high Ni and Cr concentrations inherited from their komatiite-rich source regions (Condie et al., 1970; Naqvi, 1976). Younger assemblages (3.0 to 2.5 Ga old) are less mafic, but still have lower K/Na ratio (Engel et al., 1974), less enrichment in light REE (rare earth elements) (Taylor, 1979), higher Fe^{2+} concentrations (Ronov and Migdisov, 1971), and a less felsic overall composition (Figure 10-6; Cameron and Garrels, 1980) than their post-Archean counterparts. Where adequate sampling density and time resolution exist (specifically, with regard to the K/Na and REE data), it is evident that the major jump in chemical composition coincides with the Archean-Proterozoic transition at approximately 2.5 Ga ago.

The evolution after 2.0 Ga ago of the global sedimentary mass is characterized by a large degree of cannibalistic recycling. As a consequence, the Archean portion of this mass, with its overall (basaltic) andesite chemistry, was not entirely dispersed during subsequent geologic history. This, complemented by addition of new first-cycle sediments from island arcs, is the reason why the global sedimentary mass is more mafic than its presumed source, the upper continental crust of granodioritic composition (see Veizer and Jansen, 1979; Veizer, 1984).

10-4-4. (BIO)CHEMICAL SEDIMENTATION

The greenstone belt sedimentary successions (facies assemblages 1 through 3, above) commonly contain cherts and lenticular Algoma-type iron-formations in their oxide, carbonate, or sulfide facies as well as minor carbonates and barites. The relatively rare carbonates are mostly limestones (see Gimmel'farb, 1975), whereas ankerites and iron-rich dolostones are more typically products of secondary carbonation, or of exhalation and secretion from volcanic piles (Veizer, unpublished). The observed Archean gradations resemble the present-day spatial facies distributions in the vicinity of oceanic spreading centers and island arcs. These gradations range from proximal zones of volcanic silicification and carbonation (with gold deposits) and "sedimentary" barites to Algoma-type iron-formations (with Au) and stratiform base metal (Cu, Zn) deposits to marine carbonates and cherts (e.g., the Abitibi and Barberton Greenstone Belts discussed by Stockwell et al., 1970 and Heinrichs and Reimer, 1977). Such (bio)chemical sediments and ores are somewhat analogous to the products of exhalative activity in the modern oceans. The origin of some barites, which are apparently pseudomorphs after evaporites, is discussed in Chapters 8 and 11 of the present volume.

In contrast to the greenstone belt sedimentary successions, the Early Proterozoic (and, in Africa, Late Archean) mature shelf assemblages (type 5 above) contain well-developed carbonate and banded, Superior-type, iron ore facies. The latter are discussed in detail in Chapter 11 and will be considered only briefly here. The shelf carbonates are predominantly early diagenetic (and possibly primary) Fe- and Mn-rich dolostones (Ronov, 1964; Ronov and Migdisov, 1971; Veizer, 1978b). The reason for the predominance of dolostones over limestones in the Early Proterozoic is not clear, but it may be at least partially related to the lagoonal to subtidal, commonly evaporitic, settings of their depositional environments, since such settings increase the propensity for dolomitization even in the Phanerozoic (Veizer, 1978b).

10-5. The Archean-Early Proterozoic Oceans

Since the mixing time for the present oceans is about 10^3 years (Broecker, 1963), all water in the oceans has been recycled many times during geologic history and may or may not have retained any features indicative of its composition before 1.8 Ga ago. The latter has to be deciphered from (*i*) plausible theoretical scenarios, from (*ii*) consideration of stabilities of chemically precipitated mineralogical phases, or from (*iii*) elemental and isotopic tracer studies in such minerals. The latter two approaches are less dependent on initial assumptions than the former, and are directly testable. The mineralogical approach (Holland, 1978) yields broad limits and should preferably be complemented by tracer studies. The (bio)chemical sea water precipitates dating from this early period are carbonates, cherts, banded iron-formations, and some of the barites. Except for carbonates, relatively little is known about the distribution of elemental and isotopic tracers in these phases. The only clear-cut features are the ^{18}O-depleted composition of ancient cherts (Perry et al., 1978; Knauth and Lowe, 1978) and the non-radiogenic, mantle-like, initial Sr isotopic ratios in Archean barites (as summarized in Veizer and Compston, 1976).

The overall advantage of carbonate rocks—compared with cherts, iron-formations, and barites—is that, in relative terms, they are: (*i*) abundant; (*ii*) mineralogically simple (i.e., mono- to bi-mineralic); (*iii*) able to accommodate a considerably larger number of tracers (e.g., Mn^{2+}, Sr^{2+}, Mn^{2+}, Pb^{2+}, Fe^{2+}, Na^{+}, Ba^{2+}, Zn^{2+}, $^{18}O/^{16}O$, $^{13}C/^{12}C$, $^{87}Sr/^{86}Sr$, etc.) in their lattices at reliably measurable levels; (*iv*) relatively better understood in terms of the theoretical aspects of trace element partitioning and isotope fractionation; and (*v*) better investigated with regard to their potential diagenetic history.

From previous studies of the chemical composition of carbonates over geologic time, the following broad chemical trends can be identified:

(1) Carbonates display an increase in the dolomite/calcite ratio with increasing geologic age. However, there appears to be a reversal of this trend for the Archean (Ronov and Migdisov, 1971; Gimmel'farb, 1975; Veizer, 1978b).

(2) Carbonate-bound Fe^{2+}, Mn^{2+}, Ba^{2+}, and perhaps also SiO_2 concentrations increase with age in both limestones and dolostones (Ronov and Migdisov, 1971; Veizer, 1978b; Veizer et al., 1982).

(3) Concentrations of carbonate-bound Sr^{2+}, and perhaps also of Na^{+}, decrease with age, with

a likely reversal of the trend for strontium in the Archean (Veizer, 1977; Veizer et al., 1982).

(4) The ratio of $^{18}O/^{16}O$ becomes lighter, and the $^{87}Sr/^{86}Sr$ ratio becomes less radiogenic, with increasing geologic age (Schidlowski et al., 1975; Veizer and Hoefs, 1976; Veizer and Compston, 1976; Veizer, et al., 1982); for strontium, there is a major drop in the concentration of its radiogenic ^{87}Sr isotope at about 2.5 Ga ago, concomitant with many other tectonic, lithological and chemical changes.

(5) The $^{13}C/^{12}C$ ratio is comparable to that of Phanerozoic carbonates (Schidlowski et al., 1975; Veizer and Hoefs, 1976; Veizer, et al., 1982; Chapters 7 and 12, and Figure 7-2, herein).

Although the overall trends are known, the details of the secular variations in these parameters are, except for Sr isotopes, not yet resolved. However, it is my belief that when sufficient resolution is attained, these trends will show again a major shift at about 2.5 Ga ago. The observed trends are a consequence of a combination of (*i*) post-depositional alteration phenomena, with older rocks having statistically a higher chance of being more severely altered, and of (*ii*) evolution of the chemical composition of ocean water, and the hydrosphere as a whole, during geologic history. In order to decipher the evolutionary component, it is necessary to correct for the occurrence of post-depositional alteration. The covariance of tracer patterns in a given sample population can be utilized for this purpose, a technique that is briefly outlined below.

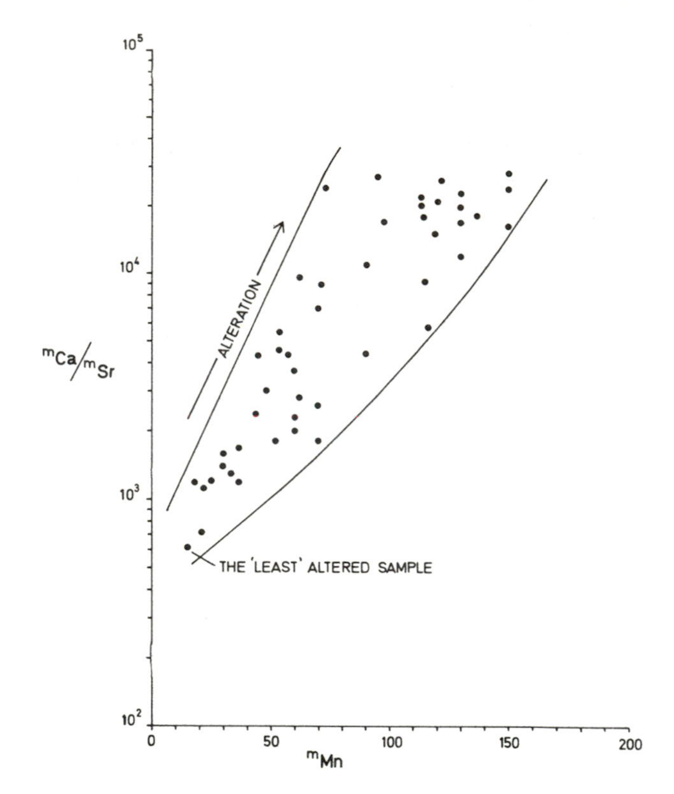

FIGURE 10-13. Graph showing $^mCa/^mSr$ vs. ^mMn concentrations in the ~2.6-Ga-old limestones on the Steeprock Group of Ontario, Canada. Note that strontium is plotted as 1/Sr ($^mCa/^mSr$), thus giving a positive correlation with manganese (from Veizer et al., 1982). A detailed account of the technique and diagenesis is given in Veizer (1983).

10-5-1. Selection of Data

A $CaCO_3$ mineral phase (whether aragonite or a high-Mg calcite) precipitated originally in equilibrium with sea water will, upon exposure to meteoric water, tend to approach equilibrium with the latter. It will dissolve and reprecipitate as a stable low-Mg calcite or dolomite with repartitioned tracers. Due to a number of factors (Brand and Veizer, 1980, 1981), this will lead to an increase in the concentrations of Mn^{2+}, Fe^{2+}, Zn^{2+} and in the $^{87}Sr/^{86}Sr$ ratio, to a decrease in the concentrations of Sr^{2+}, Ba^{2+}, Na^+ and in the $^{18}O/^{16}O$ ratio, and, in some circumstances, also to decrease in the concentration of Mg^{2+} and in the $^{13}C/^{12}C$ ratio. Such alteration trends will be particularly prominent for tracers with partition coefficients or fractionation factors greatly departing from unity. For example, $k_{Sr}^{calcite}$ is ~0.1 (Kinsman and Holland, 1969) and $k_{Mn}^{calcite}$ is ~17 (Bodine et al., 1965). Consequently, a negative correlation would be expected for a population of samples that were originally of comparable mineralogy and chemistry but that underwent differing degrees of post-depositional alteration (Figure 10-13; see also Brand and Veizer, 1980, 1981, regarding the underlying theory and the application of this concept). The two-dimensional approach can be extended into three dimensions (Figure 10-14) permitting the evaluation of all measured parameters. This latter approach enables selection of the "least altered" sample from a given sequence; such a sample can then be compared with others similarly selected from contemporaneous sequences (Figure 10-15) or from their younger counterparts. The obtained values do not necessarily represent concentrations for calcite (or for dolomite) in equilibrium with contemporaneous sea water (and neither do any other data for the remaining three

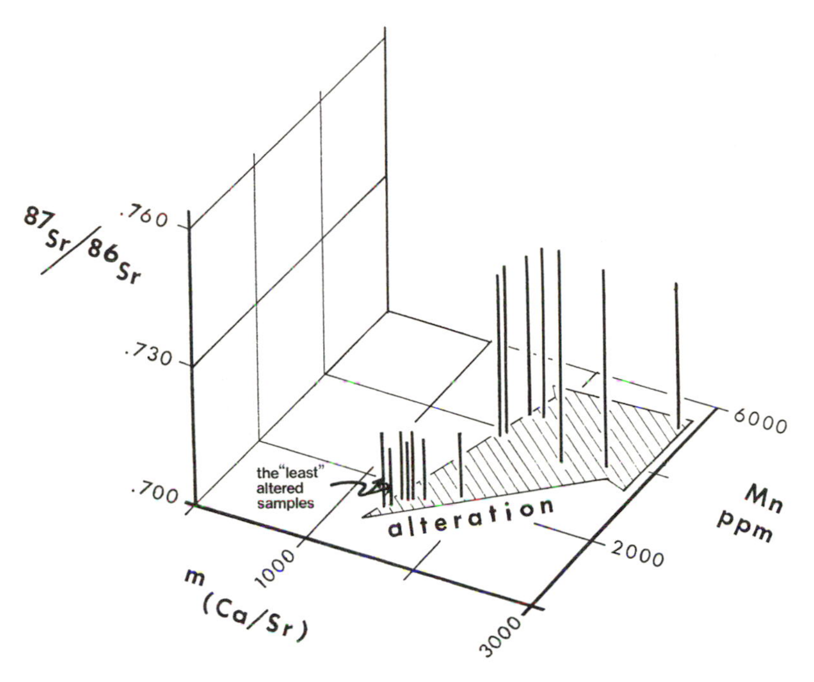

FIGURE 10-14. Plot of $^{m}Ca/^{m}Sr$, Mn concentration, and $^{87}Sr/^{86}Sr$ in dolostones of the ~3.1-Ga-old Pongola Supergroup of South Africa. Note the simultaneous increase in all three variables (from Veizer et al., 1982).

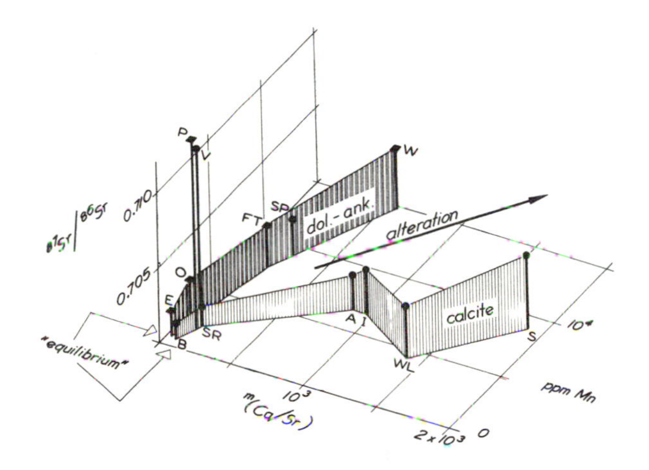

FIGURE 10-15. Three-dimensional projection of $^{87}Sr/^{86}Sr$, $^{m}Ca/^{m}Sr$ and Mn composition of the "least" altered sample for each of the following Archean carbonate sequences (from Veizer et al., 1982). From Zimbabwe: (B) Bulawayan (~2.6 Ga); (S) Sebakwian (~3.3 Ga). From South Africa: (O) Onverwacht (~3.4 Ga); (FT) Fig Tree, (~3.3 Ga); (P) Pongola (~3.1 Ga); (V) Ventersdorp (~2.6 Ga). From Canada: (SP) Yellowknife (~2.6 Ga); (SR) Steeprock (~2.6 Ga); (A) Abitibi Belt (~2.7 Ga); (E) exhalative carbonates of the Abitibi Belt (~2.7 Ga); (W) siderites of the Michipicoten (Wawa) region (~2.7 Ga); (WL) Woman Lake of the Uchi Belt (~2.7 Ga). From India: (I) Sargur marbles ($\gtrsim$ 3.0 Ga). Circles indicate limestones; squares indicate dolostones, ankerites, and siderites. The approximately 3.5-Ga-old "best" ankeritic samples of the Warrawoona Group (Pilbara Block of Western Australia) are not included here due to their probable secondary nature, although on the basis of their $^{87}Sr/^{86}Sr$ (.7032–.7052), strontium ($^{m}Ca/^{m}Sr$ = 1200–2200) and manganese (2200–2500 ppm) contents they would plot in the lower left corner of the diagram. The exceptionally high $^{87}Sr/^{86}Sr$ of the Ventersdorp (V) and Pongola (P) samples are interpreted as being a consequence of their non-marine origin (Veizer, 1979b). Sedimentological criteria support this interpretation for the Ventersdorp (Buck, 1980) but not necessarily for the Pongola (Mason and von Brunn, 1977) sequence.

types of chemical sediments). However, this technique enables selection of relatively well preserved and comparably altered samples, thus self-correcting for the effect of post-depositional alteration. Observed chemical differences can, therefore, be considered to reflect real differences inherited from the originally precipitated mineralogical phases and, ultimately, from sea water or its early diagenetic derivatives.

10-5-2. Results and Consequences

Comparison of the "best" Archean calcites with similarly preserved Cenozoic counterparts (Table 10-1) shows that the former are (*i*) enriched in Mn^{2+}, Fe^{2+}, Sr^{2+}, and Ba^{2+}; are (*ii*) depleted in ^{18}O (and, possibly, in Na^{+}); and (*iii*) exhibit mantle-like $^{87}Sr/^{86}Sr$ ratios (Figure 10-15) and $\delta^{34}S$ in their associated sulfur-bearing phases (see Figure 10-19). Although alternative explanations for some of these observations do exist, most of them are also consistent (at least qualitatively) with a single assumption, namely, that the exchange between sea water and oceanic crust (viz., the "mantle" flux) was—in response to the higher heat flow and thus the higher convective heat dissipation from the early mantle—exceptionally high early in the Earth's history. By analogy with comparable

Table 10-1. Summary of trace element concentrations and of isotopic data for the "best" (least altered) Archean calcites and their Cenozoic (Phanerozoic) counterparts (from Veizer et al., 1982). Listed elemental concentrations are for carbonate-bound (i.e., leachable) values, not whole rocks. The data are presented on 100% carbonate (insoluble residue-free) basis. Note that the comparison is with the Cenozoic counterparts, which are already mineralogically, texturally, and chemically stabilized. This is because early in their depositional history both (Archean and Cenozoic) facies passed through a stage of diagenetic stabilization, which caused some, but comparable and systematic, shift in the preserved chemical signal. Such a comparison therefore largely self-corrects for this diagenetic shift and the observed chemical differences, particularly if large, reflect differences inherited from the precursor carbonate minerals (see Veizer, 1974, 1983; Brand and Veizer, 1980, 1981). It is frequently argued (e.g., Garrels and Mackenzie, 1971) that carbonate rocks, if compared for example to clastics or cherts, have a tendency for faster destruction and recycling and for more pervasive perpetual dissolutions-reprecipitations. In other words, the older the sequence the smaller the abundance and the greater the degree of alteration of carbonates. This, however, is an exception rather than a norm. Geological (Veizer, 1984), textural, and chemical criteria (Bathurst, 1975, p. 425; Veizer, 1977) show that carbonate rocks, once stabilized, are no less stable than other types of sedimentary (and many igneous) rocks. In addition, diagenetic stabilization is a property of all sediments and the rocks can be considered as "stable" only after a passage through this stage. In the case of carbonates, this is achieved mostly in near-surface conditions over a relatively short time-span ($\leq 10^6$ years; see Bathurst, 1975), while, for example, for clastics and cherts this stabilization is protracted over long time-spans ($\geq 10^7$ years) and depths of burial.

Elemental or isotopic species	*Concentration in "Best" Archean Samples*[a]	*Concentrations in stabilized Cenozoic (Phanerozoic) samples*	*Degree of enrichment*	*Degree of depletion*
Mn^{2+}	200 to 900 ppm	20 to 30 ppm[b]	~10 to 30x	
Fe^{2+}	(<20) to 250 ppm	20 to 70 ppm[b]	~3x	
Sr^{2+}	1400 ppm	150 to 600 ppm[c]	~3 to 6x	
Ba^{2+}	130 to 250 ppm	10 to 30 ppm[d]	~10x	
Na^{1+}	40 to 75 ppm	50 to 250 ppm[e]		<2 to 3x
Mg^{2+}	variable	variable	——	——
$^{87}Sr/^{86}Sr$	0.7012 to 0.7025	0.707 to 0.709[f]		0.005 to 0.008
$\delta^{18}O_{PDB}$	-7 to -10 ‰	~0 to -3 ‰[g]		~7 ‰
$\delta^{13}C_{PDB}$	-3 to $+2$ ‰	$+0.2 \pm 1.5$ ‰[g,h]	——	——
Sulfate $\delta^{34}S_{CDM}$	$+1$ to $+4$ ‰	$+15$ to $+20$ ‰[i]		~15 to 20 ‰
Sulfide $\delta^{34}S_{CDM}$	-1 to $+3$ ‰	-15 to -20 ‰[i]	~15 to 20 ‰	

[a] Unpublished data of Veizer, Compston, Hoefs, and Nielsen (see Section 10-5-1 for discussion of "best" used here).
[b] Veizer (1974, and unpublished data)
[c] Veizer (1978a)
[d] Puchelt (1972)
[e] Veizer et al. (1978) and Brand and Veizer (1980)
[f] Peterman et al. (1970)
[g] Veizer and Hoefs (1976)
[h] Veizer et al. (1980)
[i] Claypool et al. (1980)

present-day settings (Boström et al. 1969; Bonatti, 1978), such a flux could account for the high supply of Ba, Mn and Fe (and also Sr?), for the mantle-like composition of Sr and S isotopes, for the high frequency of occurrence of barites (Bonatti, 1978) and of base metal deposits (Degens and Ross, 1969; Edmond et al., 1979b), for the widespread presence of carbonation and silicification alteration phenomena and, ultimately, could also serve as a source for iron and silica in the coeval Algoma-type iron-formations. Low temperature ($<250°$ C) alteration of basalts acts as a sink for ^{18}O (Muehlenbachs and Clayton, 1976; Perry et al., 1978; Gieskes and Lawrence, 1981; Lawrence and Gieskes, 1981); this may be responsible for the $\delta^{18}O$ of Archean oceanic waters to have been more depleted in ^{18}O than they are today. The alternative explanation invoking hot oceans (Knauth and Epstein, 1976; Knauth and Lowe, 1978) is not in accord with other geological observations (see Chapter 11, and Veizer et al., 1982). Furthermore, such a process could conceivably add Ca to, and subtract Mg (and perhaps also Na) from, sea water (Edmond et al., 1979a), thus resulting in the predominance of sedimentary limestones over dolostones (Holland, 1976) in the Archean sedimentary record. However, Archean carbonates are of relatively infrequent occurrence (Cameron and Baumann, 1972; Veizer, 1978b) and are related mostly to volcanic settings, perhaps representing a facies deposited on the submarine platforms of eroded volcanoes (see Section 10-4-2). Such an association is supported by oxygen and carbon isotope systematics (Veizer and Hoefs, unpublished). Some carbonate sequences have "normal" sedimentary Archean values of about -7 to -10 ‰ $\delta^{18}O_{PDB}$ (PDB is Peedee belemnite standard) and near zero or slightly positive $\delta^{13}C_{PDB}$ (e.g., carbonates of the Steeprock Group). Others have values that have been reset thermally or via recrystallization with a trend toward light $\delta^{18}O$ but with little-affected $\delta^{13}C$ (e.g., those of the Bulawayan Group: Schidlowski et al., 1975, and Veizer and Hoefs, 1976, and those of the Fig Tree, Onverwacht, and Sargur sequences). Still others (carbonates of the Yellowknife and Woman Lake Groups, for example), are spread between sedimentary and exhalative end-members, the latter with $\delta^{18}O_{PDB}$ of about -17 ‰ and $\delta^{13}C_{PDB}$ usually between -1 to -5 ‰. Typical examples of exhalative carbonates are those of the extensive (10^2 km in total length) Larder Lake and correlative gold-bearing zones in the Abitibi Greenstone Belt of the Canadian Superior Province (Ridler, 1976). Carbonates of the Isua Supracrustal Group of Greenland appear to coincide isotopically (see Schidlowski et al., 1979, and Chapter 7 herein) with those of the exhalative type.

In view of the foregoing observations, one may propose that the composition of the early ocean was influenced strongly by the (exhalative, hydrothermal and low-temperature submarine weathering) "mantle" flux (Perry et al., 1978; Fryer et al., 1979; Veizer, 1979b, 1984; Veizer et al., 1982). The waning internal heat generation and the coupled coeval growth of continents during the Late Archean-Early Proterozoic led to a subsequent progressive increase in the relative importance of the continental river flux, the present-day ocean chemistry being controlled by the kinetics of the two fluxes and by sedimentary output (see Holland, 1978). This transition in the oceanic buffering system is likely to have been the single most important factor causing the development of the previously discussed secular trends and of the relatively rapid changes in these trends that occurred 2.5 Ga ago.

10-6. Oxidation State of the Early Atmosphere-Hydrosphere System

The modern atmosphere-hydrosphere system is dynamic, the concentration of gaseous or dissolved free oxygen in any particular reservoir being a consequence of its aggregate rate of supply and consumption (as is discussed in some detail in Chapter 11). This situation must also have been the case during the Archean and Early Proterozoic. Blanket statements about coeval oxygenic or anoxygenic oceans (but not necessarily about the well-mixed lower levels of the atmosphere) thus have little value.

The wind mixed, uppermost, 200 m-thick surficial oceanic layer, which interacts with the present-day atmosphere with its abundant free oxygen, is well oxygenated. However, in the present oceans the concentration of molecular oxygen decreases rapidly with depth (Figure 10-16) due to its consumption via the oxidation of organic matter. In a static, unmixed ocean this would lead to a layered structure, with the depth at which Eh = 0 being controlled by the relative rates of surficial productivity of organic matter and of its subsequent oxidation. Due to the dynamic nature of the

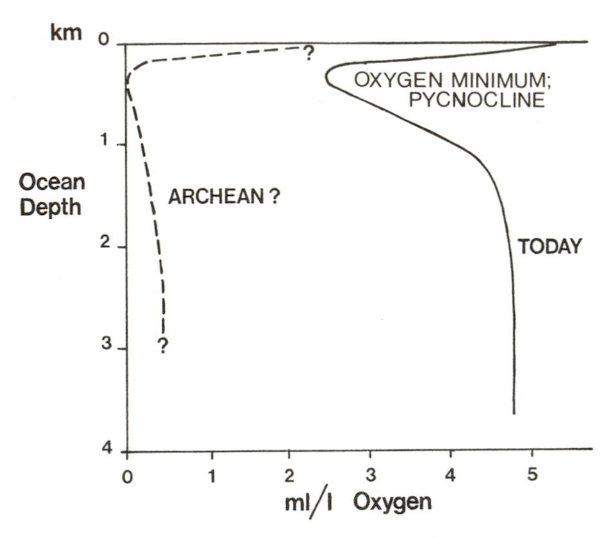

FIGURE 10-16. Schematic illustration of the distribution of dissolved oxygen as a function of depth in present-day (modified from Holland, 1978, p. 220) and Archean oceans.

modern ocean, its deeper parts are ventilated chiefly by sinking of dense, cold, high latitude-derived waters (Wilde and Berry, 1982; Figure 10-17) that transport large quantities of oxygen, causing the observed increase in pO_2 below the pycnocline (Figure 10-16). Those areas where the oxygen minima coincide with high surficial organic productivity and simultaneously abut continental shelf and slope have the strongest potential to be anoxygenic (Wilde and Berry, 1982). During periods of equable climate, a norm for most of the geologic past (including probably the Archean), waters of the deep ocean are derived chiefly from high salinity mid-latitude waters that are less dense and contain less dissolved oxygen than cold, high-latitude waters. This situation results in a deepening of the pycnocline (and, thus, of the oxygen minimum layer) and in an enlargement of the equatorial oxygen deficient bulge (Figure 10-17). Consequently, the oceans have had a tendency to be relatively more anoxygenic in non-glacial, rather than in glacial periods (for detailed discussion see Wilde and Berry, 1982).

Discussion of the possible atmospheric pO_2 during the Archean has led to the conclusion that it may have been about 10^{-2} to 10^{-6} times less than the present atmospheric level (see Chapter 11 herein, and Grandstaff, 1980). The presence of some free oxygen, although at low levels, is indicated not only by the existence of banded iron-formations and of other oxidized mineral phases (see Chapter 11) but also by the Fe^{2+} and Mn^{2+} concentrations present in the "best," least altered, Archean calcites (Table 10-1) that are about two orders of magnitude lower than would be expected for calcites in equilibrium with an anoxygenic water body (Veizer, 1978b). An appeal to post-depositional alteration as a cause of this discrepancy will only compound the difficulty, since such processes would only tend

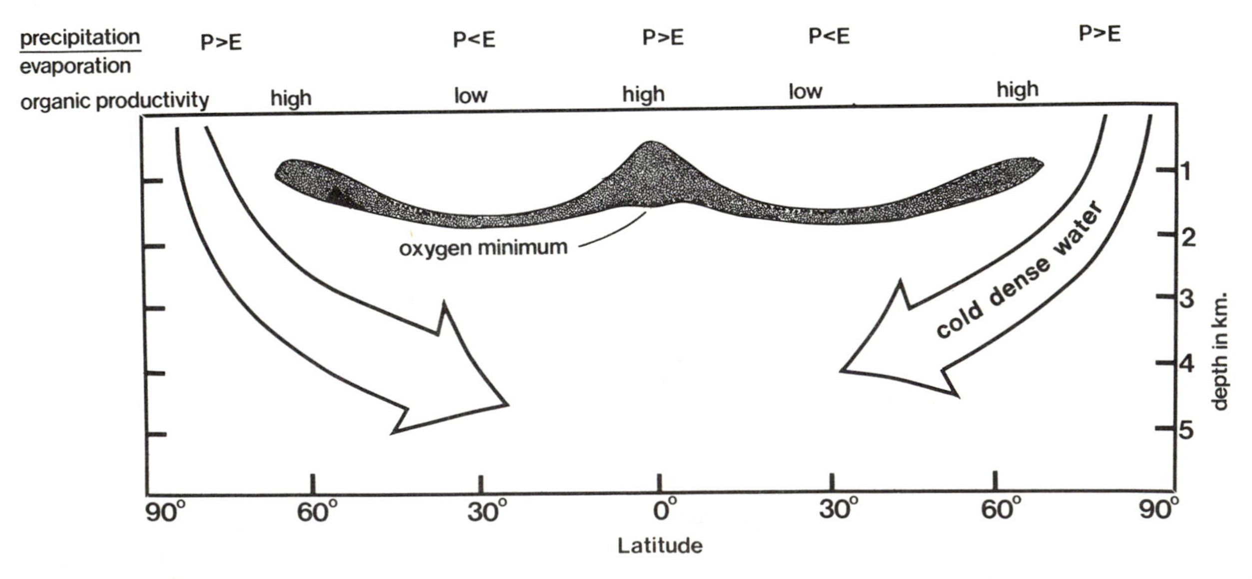

FIGURE 10-17. Diagrammatic illustration of the latitude-dependent distribution of major water masses in the present-day ocean. Modified from Wilde and Berry (1982).

to increase the Fe^{2+} and Mn^{2+} concentrations in calcites (see Section 10-5-1). Consequently, the original values must have been either as listed in Table 10-1 or less. With the above low atmospheric pO_2, there would be, as today, a thin surficial wind mixed layer with some free oxygen (Figure 10-16). However, the sinking denser water would become rapidly depleted in free oxygen, thus vastly expanding the volume of oxygen-poor water masses and perhaps even leading to the development of an essentially anoxygenic ocean at and below the pycnocline (Holland, 1973; Drever, 1974). The apparent absence of Archean glaciations (Chapter 11) and, thus, the scarcity of cold, dense, high-latitude waters would only accentuate this trend. Furthermore, it is likely that the hotter Archean oceanic crust would have created shallower oceans (see Section 10-3, above) and generated numerous ridge systems and/or volcanic hot-spot centers leading to the formation of many small, poorly interconnected basins, a situation that would have retarded circulation and ocean ventilation. Such an ocean could have contained significant quantities of dissolved (and/or suspended fine particulate) iron and manganese (Holland, 1973; Cloud, 1973a; Drever, 1974). However, it must be again emphasized that this was no doubt a dynamic system with constant recharge of iron and manganese, principally derived from submarine sources (rather than from subaerial weathering as has been assumed by the authors cited above), and the deposition of these metals as local banded iron-formations, as disseminated oxidized minerals in forming sediments (see Section 10-4-3) and, perhaps, in alteration of ocean-floor basalts.

10-6-1. Possible Alternative to Biogenic Control of pO_2 Evolution

In present-day discussions of the terrestrial exogenic system it is commonly suggested that the apparent increase of pO_2 at about 2.0 ± 0.2 Ga ago may have been a consequence of a massive increase of oxygen-producing photosynthetic biomass at about this time (for discussion of this and other alternatives, see Chapter 11). While this may be so, and the ultimate source of molecular oxygen seems likely to have been biologic, it is not obligatory that the biota controlled long-term atmospheric pO_2 levels during geologic history. As currently understood, the known carbon isotope record (see Chapter 7) can be interpreted in essentially steady-state terms since about 3.5 (or even 3.8) Ga ago, although this is certainly not the only possible interpretation (see, for example, Chapter 12). However, if this were the case (or if oxygen-producing photosynthesis had become quantitatively important prior to about 2.0 Ga ago), the problem would not be that of insufficient oxygen production but, rather, the explanation for its lack of accumulation in the atmosphere. If an oxygen source were plentiful, it must have been a sink that controlled the pO_2 (see Chapter 11). Such a sink could have been provided in the early oceans by the prolific "mantle" flux of divalent iron and manganese (in basalts as well as emitted in dissolved and particulate form) and of reduced volcanic gases discussed above (Section 10-5-2). The capacity of this oxygen sink has been dramatically demonstrated by the recent discovery of "black smokers," ocean ridge vents discharging H_2S although their recharge is from sea water with its abundant SO_4^{2-} (see Edmond et al., 1979a, b). With decreasing geotherms resulting in a declining rate of sea water/ocean-floor exchange, the efficiency of this sink would have gradually diminished and oceanic pO_2 could have risen to the level permitted by the next steady state buffer system—such buffering at present being provided apparently by biologically mediated redox coupling within and between the carbon and sulfur cycles (Garrels and Perry, 1974; Veizer et al., 1980; Lerman, 1982), with the iron cycle and internal outgassing being of lesser importance (see Chapter 6 in Holland, 1978). The evolution of this system would have been concomitant with the development of extensive continental shelves, and oxidation of iron in surficial oceanic layers would have led at first to local (Algoma-type) and later—through upwelling of deep waters onto the oxygenated shelves—to massive (Superior-type) iron ore deposition (Figure 10-18; Button et al., 1982). It is an intriguing coincidence that the development of large continents, of continental glaciation, and of Superior-type iron-formations was broadly coeval (for relevant data, see Chapter 11). Continentality may be a prerequisite for global glaciation, and cooling, in turn, may have aided ocean ventilation and upwelling.

The above outlined mantle-dependent scenario provides a unifying and consistent explanation for the evolution of the oceanic and continental crusts, of tectonic patterns, and of sedimentary rock types and their chemistry, as well as for ore genesis and

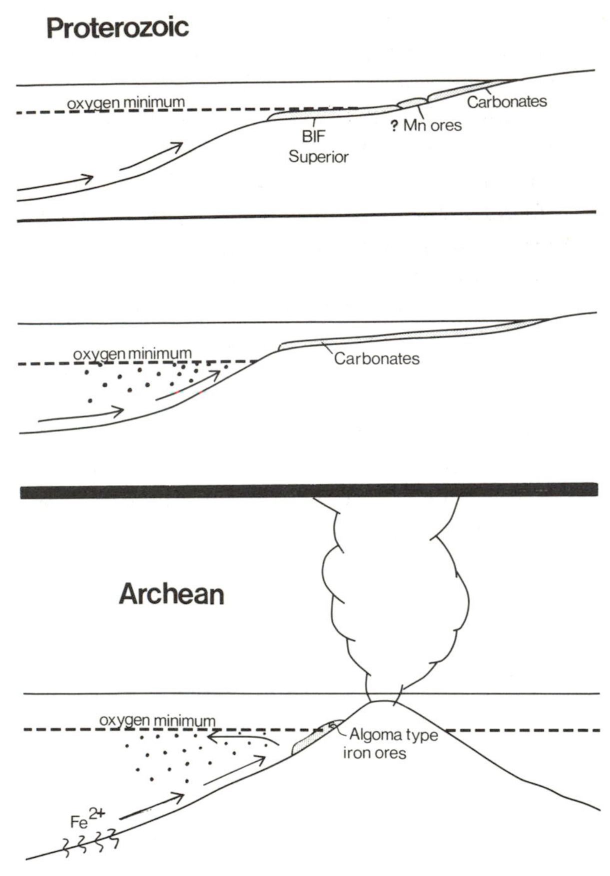

FIGURE 10-18. Schematic representation of the depositional environments of Archean (below) and Early Proterozoic (above) iron-formations. Note that the existence of the Superior-type banded iron-formations on the shelf (as opposed to purely carbonate shelves) may be a function of the depth of the oxygen minimum layer in relation to the depth of the sill.

for the composition of the oceans and perhaps of the atmosphere. If this scenario withstands the trials of qualitative and quantitative testing, living systems—instead of being catalysts—would be viewed largely as simply responding to external factors and opportunistically adjusting to new, geologically produced circumstances. This interpretation, however, is not exclusive and can accommodate variable degrees of significant biological roles (Veizer et al., 1982).

10-6-2. CONSEQUENCES FOR THE SULFUR CYCLE

A rapid turnover of ocean water through the oceanic crust would not only be a net sink for oxygen, but would have important effects on the sulfur cycle as well. Specifically, such turnover could lead to a quantitative removal of SO_4^{2-} by precipitation of metal sulfides and by oxidation of Fe^{2+}, as in present-day ridge systems (Edmond et al., 1979a; Wolery and Sleep, 1976). Such a net drain on exogenic sulfur would have to be compensated by an endogenic input with a mantle-like isotopic composition (i.e., 0‰ $\delta^{34}S_{CDM}$; Wolery and Sleep, 1976). Even a simple recycling of sea water sulfate and its inorganic reduction to sulfide in subsurface brine will result in a considerably lighter $\delta^{34}S$ (probably through nonequilibrium isotope fractionation) at the vent, as is clearly documented by Red Sea deposits where the average sulfide $\delta^{34}S_{CDM}$ is +5.4‰ although the sulfur was likely derived from marine sulfate with $\delta^{34}S_{CDM}$ of about +20‰ (Kaplan et al., 1969; Shanks and Bischoff, 1980). Furthermore, if the concentration of dissolved Fe^{2+} in the oceans were to exceed that of sulfur—as could have been the case in the Archean—the latter would be essentially stripped off as iron sulfide below the pycnocline. However, low levels of sulfate would persist in the oxygenated surficial layer devoid of soluble iron (Drever, 1974). Thus, marine sulfate and sulfide, if present, would be dominated by mantle-like isotope values.

As is summarized in Chapter 7, the presently known $\delta^{34}S$ composition of sulfates and sulfides more than 2.7 Ga old appears to cluster within the 0 to +10‰ range with little fractionation between oxidized and reduced sulfur phases. Since most of these samples represent mineralized specimens, the sulfur being derived mostly from hydrothermal sources (Williams et al., 1982), the tight grouping of these values may only be a reflection of a uniform source (Sangster, 1980). Nevertheless, measurements on disseminated sulfur phases in Archean carbonates (Figure 10-19) appear to support the overall picture. The apparent appearance of patterns with a large spread of isotopic values beginning at about 2.7 Ga ago is interpreted as indicating the existence, and perhaps the emergence, of dissimilatory bacterial sulfate reduction at about this time (see Chapter 7 herein for details). This is a reasonable interpretation, but inorganic alternatives cannot yet be discounted. With the above outlined consequences of rapid pumping of sea water through oceanic crust during the Archean, the isotopic signature of dissimilatory sulfate reduction (regardless of whether sulfide of such origin was present or not) would be largely swamped by environmental sulfur with its mantle-like isotopic composition (Veizer et al., 1982).

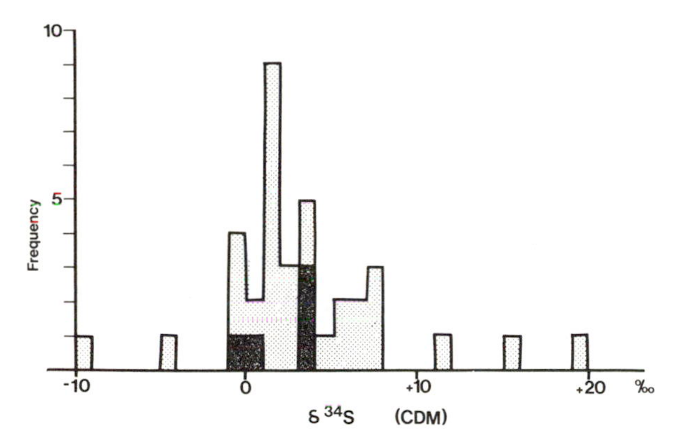

FIGURE 10-19. Histogram summarizing the $\delta^{34}S_{CDM}$ distribution in sulfur-bearing phases disseminated in Archean carbonates (Veizer and Nielsen, unpublished). Dark field shows data for exhalative carbonates of the Abitibi belt (see Section 10-5-1). Samples falling outside the major field in the centre of the diagram are those from the Ventersdorp (light $\delta^{34}S$) and those of the Steeprock and Michipicoten (Wawa) sequences (heavy $\delta^{34}S$). Utilizing the criteria outlined in Section 10-5-1, elemental sulfur content frequently increases, and $\delta^{34}S$ becomes heavier, with increasing degrees of alteration; this indicates that at least a portion of the heavy $\delta^{34}S$ may be of post-depositional, secondary origin.

10-7. Summary and Conclusions

(1) The geologic evolution of the Archean-Early Proterozoic Earth was dominated by the diachronous transformation of tectonically unstable greenstone-granite belts and high-grade gneissic terranes to produce stabilized cratons. This development was apparently a response of the outer layers of the Earth to an exponential decline of internal heat generation and to the pattern of convective dissipation of this heat from the mantle. However, the actual triggering mechanism for the initiation of cratonization at about 3.1 Ga ago is, at present, a matter of speculation only.

(2) Because the existence of a large sedimentary mass is a byproduct of the presence of an extensive, buoyant, continental crust, Late Archean-Early Proterozoic time was likely also a period of major generation of the global sedimentary mass and of its transfer onto the emerging continents and their shelves.

(3) The decrease in internal heat flux, combined with the emergence of continents, led to a reduction in the role of oceanic crust/sea water exchange (i.e., the "mantle" flux) and to an increase in the importance of continental river flux as a controlling factor of sea water chemistry.

(4) As a consequence of these developments, Early and Middle Archean sedimentary successions were dominated by volcaniclastics and first-cycle immature graywackes, and were supplemented by chemical sediments of direct or indirect (i.e., products of degassing, leaching or secretion from volcanic piles) exhalative origin (e.g., cherts and lenticular iron-formations). In contrast, the Late Archean and Early Proterozoic interval was characterized not only by the formation of less mafic first-cycle sediments, but also by the massive cyclic reprocessing of immature facies into mature shelf and platformal assemblages. The subsequent, evolution of the planet after 2.0 Ga ago was dominated by cannibalistic recycling, with comparatively subordinate generation of first-cycle sediments from young volcanic sources.

(5) Chemically, this evolutionary sequence is reflected in secular mafic to felsic compositional trends in clastic rocks and in secular variations in the elemental (viz., Sr^{2+}, Ba^{2+}, Fe^{2+}, Mn^{2+}, Ca^{2+}, Mg^{2+} and possibly Na^{+}) and isotopic (viz., $^{87}Sr/^{86}Sr$, $^{18}O/^{16}O$ and $^{34}S/^{32}S$) composition of (bio)chemical precipitates.

(6) It seems possible that photosynthetically produced free oxygen may have been available in the early oceans since at least 3.5 Ga ago (see Chapters 7, 9 and 12). If so, it may be that the development of an oxygen-rich atmosphere-hydrosphere system was delayed until the Early Proterozoic as a consequence of the diminishing access of ocean water to reductants (viz., the waning "mantle" flux) rather than being a result of an increase in the oxygen-producing photosynthetic biomass. A similar inorganic cause may be responsible for the absence of "biogenic" patterns in $\delta^{34}S$ distributions before 2.7 Ga ago.

Acknowledgments

Research leading to the results here summarized has been supported by the National Sciences and Engineering Research Council of Canada, the University of Ottawa, the RSES-Australian National University, the A. von Humboldt Foundation, and the Geochemical Institute of Göttingen University. Publication 05-83 of the Ottawa-Carleton Centre for Geoscience Studies.

CHAPTER 11

Environmental Evolution of the Archean-Early Proterozoic Earth

By James C. G. Walker, Cornelis Klein, Manfred Schidlowski, J. William Schopf, David J. Stevenson, and Malcolm R. Walter

11-1. Introduction

Although, with regard to most parameters, the rock record shows little evidence of an environment on the early Earth vastly different from that of today, there are several biologically important aspects of the environment that are likely to have changed. These include:

(1) The length of the day. Throughout the Phanerozoic, day-length is known to have gradually increased as a result of dissipation of tidal energy; in the Precambrian the length of the day was probably quite different from that of the present.

(2) The average surface temperature of the planet and the precipitation and weather that it controls. These are likely to have changed as a result of changing solar luminosity and atmospheric composition.

(3) The partial pressure of carbon dioxide in the atmosphere. As a result of man's activities, this parameter is changing today; it seems quite likely to have changed in the past in response to biological and geological evolution.

(4) The partial pressure of oxygen in the atmosphere. At present, the level of atmospheric oxygen is sustained by oxygen-producing photosynthetic organisms; prior to their origin, the oxygen content of the atmosphere should have been much lower.

(5) The availability to organisms of important nutrient elements, including biologically useable nitrogen and dissolved sulfur species. The solubility of many such nutrients depends on their oxidation state; their availability may thus have varied with the partial pressure of atmospheric oxygen. Nitrogen in a biologically useable form is chiefly produced today either via lightning-mediated oxidation reactions in the atmosphere or by nitrogen-fixing microorganisms; it thus may have been in short supply prior to the origin of nitrogen-fixing forms of life.

This chapter discusses the changes that may have occurred in these biologically important aspects of the environment. To the extent possible the discussion is based on geological evidence, but because this evidence is commonly indirect and incomplete, it must be interpreted and extrapolated with the help of theory. Much of the theory that we shall invoke is either a gross simplification of the real world, or extremely speculative, or both. It is not yet possible to reach many firm conclusions concerning the environmental evolution of the Archean-Early Proterozoic Earth. Our tentative conclusions should, therefore, be accepted with caution (and with regard to the history of some biologically important environmental parameters, not even tentative conclusions have yet been reached).

11-2. Length of the Day

In 1695, the English astronomer, Halley, discovered that ancient and modern eclipse records could not be reconciled under the assumption of uniform apparent lunar motion; he thus postulated the need for changes in the previously assumed "constants" of motion for the Earth-Moon system. The evidence is now overwhelming that these changes are predominantly secular (i.e., persistent over geologic lengths of time), caused mainly by tidal friction that transfers angular momentum from the Earth's rotational motion to the orbital motion of the Moon. By this mechanism, the rotation of the Earth has gradually slowed—and the length of day has thus increased—over the course of geologic time. The shorter length of day in the

Precambrian had important physical and biological consequences because of a concomitant increase in the Coriolis effect (resulting in a modified climate due to changes in oceanic and atmospheric currents and in normal mode frequencies), modified diurnal temperature variations, and increased tidal amplitudes. More speculatively, if it occurred, a "Gerstenkorn event" (violent heating and tides accompanying a close approach of the Moon to the Earth) could have profoundly altered the Archean environment.

11-2-1. Review of the Relevant Data

Although astronomical data provide the most accurate measure of present changes in the Earth's rotation and lunar motion (for recent reviews, see Brosche and Sünderman, 1978), there is no a priori basis for extrapolating these results to geologic timescales, since in principal such changes could be a result of solely non-secular effects (e.g., core-mantle coupling). On a geologic timescale, the only relevant data now available are those that have been derived from studies of periodic growth features preserved in fossil organisms.

The potential value of such paleontological evidence was first recognized by Wells (1963) who used daily growth increments of Devonian corals to estimate the number of days per Devonian year. Subsequent studies (reviewed by Scrutton, 1978) have extended this pioneering work to other fossil invertebrates (primarily bivalves, but also cephalopods and brachiopods), to other geologic periods (from the Cambrian to the Tertiary), and to estimates of the number of days per month and months per year. Attempts have also been made to use the laminae of Precambrian stromatolites as a basis for such studies, but the resulting data are conflicting and their interpretation ambiguous (see Scrutton, 1978). Thus, attention is here focussed on data from the Phanerozoic, derived principally from studies of fossil corals and bivalves.

Even the data for this most recent ~15% of Earth history, however, are subject to uncertainties and some degree of subjective interpretation. The techniques used for obtaining the data are simple, basically a matter of counting the daily growth increments on specimens from a particular locale, measuring the thicknesses of such increments, and identifying higher order (semi-monthly, monthly, or seasonal) features. The time series data thereby generated are invariably difficult to interpret and incomplete, partly due to inherent (genetically determined) biologic variability and partly due to various physical factors (e.g., incomplete preservation, dissolution and resorption of previously deposited growth increments, etc.). Since both the incremental growth and noise-generating processes are incompletely understood, the statistical inferences based on such data can only be judged empirically and pragmatically. However, and although there undoubtedly exists a subjective element in these studies—the danger that observers will see what they want to see—the extent of agreement between coral and bivalve results, in each case involving several groups of workers, provides some measure of confidence in the interpretation.

The data generated provide estimates of N_y (the number of solar days per year), N_m (the number of solar days per synodic month) and, hence, the ratio $N_y:N_m$ (the number of synodic months per year). Thus,

$$N_y \equiv \frac{\Omega}{\omega_\odot} - 1$$

$$N_m \equiv \frac{\Omega - \omega_\odot}{\omega_m - \omega_\odot}$$

where Ω is the rotational angular velocity of the Earth, $\omega_\odot$ is the orbital angular velocity of the Earth about the Sun, and ω_m is the orbital angular velocity of the Moon about the Earth. As there are two independent observables (N_y and N_m), but three variables (Ω, $\omega_\odot$ and ω_m), the time variation of Ω (for example) can only be deduced by making an additional assumption, namely that $\omega_\odot$ is constant, a reasonable assumption inasmuch as the solar tidal effect is small, the mass of the Sun is essentially constant (mass outflow or inflow being negligible for the present Sun or similar stars on the main sequence), and the gravitional constant, G, has varied by less than about one part in 10^{11} per year on a geologic timescale (McElhinny et al., 1978). For $\omega_\odot$ assumed to be constant, statistical analysis of Phanerozoic coral and bivalve data (Lambeck, 1978) leads to a time rate of change for Ω, or equivalently N_y:

$$\frac{dN_y}{dt} = -(95 \pm 9) \text{ days/Ga [coral data]}$$

$$= -(100 \pm 12) \text{ days/Ga [bivalve data]}$$

valid for the last 400 million years, and corresponding to a Devonian year approximately 400 to 410 days in length (i.e., one day being equivalent to about 21 to 22 hours). The data for N_m have more scatter (because both the number of days per month and the change in this value over time are proportionately smaller), but the calculated Devonian N_m is between 30 and 30.5 days. A parameter of interest is $(d\Omega/dt)/(d\omega_m/dt)$, the rate of change of the rotational angular velocity of the Earth relative to the rate of change of the orbital angular velocity of the Moon about the Earth. For the coral data, this ratio is 42 ± 11, for the bivalve data it is 46 ± 8, while for modern astronomical data it is 51 ± 5 (thereby indicating that non-secular perturbations of the system have been relatively small). One set of data for growth rhythms in Recent *Nautilus* (Kahn and Pompea, 1978) is not consistent with the above results, and although this particular report is apparently not relevant to interpretations of the history of the Earth-Moon system (Runcorn, 1979), it serves to illustrate that the paleontological data must be treated with caution.

11-2-2. Extrapolation to the Precambrian

No extrapolation is meaningful without an adequate theoretical basis. Application of Kepler's third law and the conservation of angular momentum yields

$$\Omega = \Omega_o \left\{ 1 + 4.9 \left[1 - \left(\frac{a}{a_o} \right)^{1/2} \right] \right\}$$

where a is the Earth-Moon distance and subscript o refers to the present day. The present-day value of $(d\Omega/dt)/(d\omega_m/dt)$ is predicted to be about 48, a value consistent with the ratios noted above for both the data from fossils and from modern astronomical observations. The inferred energy dissipation by tides is also consistent with the observed change in rotational plus orbital energy of the Earth-Moon system (Lambeck, 1977). This provides strong support for the predominance of the tidal mechanism in producing changes in day-length over time and the relative unimportance of other effects such as changes in the Earth's moment of inertia (Lambeck, 1979).

The tidal torque is proportional to $a^{-6}(\Omega - \omega_m)\delta$, where δ is the effective phase lag (about 6° at present). If $k(t)$ is defined as the ratio of the phase angle at time t to its present value (i.e., k is the relative rate of specific tidal dissipation), then:

$$a/a_o \simeq \left[1 - \frac{(t_o - t)}{\tau(t)} \right]^{2/13}$$

$$\tau(t) \equiv 1.4 \text{ Ga}/\bar{k}(t)$$

$$\bar{k}(t) \equiv \frac{1}{(t_o - t)} \int_t^{t_o} k(t')\, dt'$$

where t_o is the present day. This simplified result (see Lambeck, 1979) neglects changes in orbital inclination and eccentricity and omits the tidal effects of the Sun (for a more detailed analysis, see Goldreich, 1966). For $k(t) = 1$ (i.e., with the specific tidal dissipation rate assumed to be the same in the past as it is at present), it follows that the Moon must have been close ($a \ll a_o$) to the Earth only $\sim$1.4 Ga ago. This is an unacceptable solution, if only because of the identification of 3.5-Ga-old tidal sediments in South Africa (Eriksson, 1977; see also Chapter 8). Because there is no known mechanism for keeping the Moon close to the Earth for a long period of time (the tidal torque being very large for a small Earth-Moon distance) and because capture of the Moon by the Earth is dynamically implausible, especially for a solar system close to its present configuration, it seems most reasonable to assume that the Moon has not been close to the Earth since at least 4 Ga ago. This implies that the specific tidal dissipation rate must have been lower in the past. The present mechanism involved in tidal dissipation is poorly understood, although it is clearly dominated by effects of oceanic tides (Lambeck, 1977). There are, however, a number of plausible ways by which such dissipation could have been less in the past, including a different distribution of oceans and shallow seas (Sünderman and Brosche, 1978). The evidence presented in Chapters 3 and 10 for continental growth with time is relevant in this regard because tidal dissipation results largely from interference, by land masses, with the propagation of oceanic tides.

Figure 11-1 summarizes the results of several model calculations for variations in the length of day through geologic time using the above simple formalism. Except for the unacceptable Model I (which assumes a constant rate of specific tidal dissipation), the models assume that the Moon was close to the Earth at 4.5 Ga ago, consistent with (but not limited to) scenarios in which the Moon

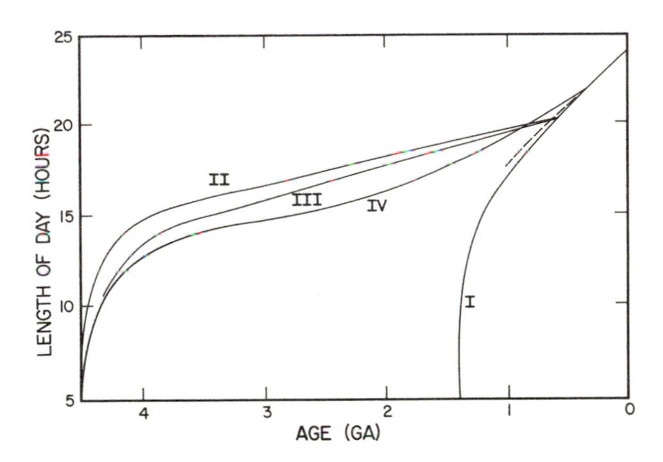

FIGURE 11-1. Models for the evolution of the Earth's rotation through geologic time. The dashed line is an extrapolated linear trend of the fossil data (see text). Model I (which assumes $k = 1$ at all t) leads to an unacceptably recent (~ 1.4 Ga) approach of the Earth by the Moon. The other three models all assume that the close approach was at 4.5 Ga. Model II has $k = 1$ back to 0.6 Ga and $k = 0.2$ in the Precambrian. Model III has $k = 1$ back to 0.6 Ga, $k = 0.25$ between 0.6 Ga and 3.8 Ga and $k = 0.03$ earlier (corresponding to a Hadean Earth with no oceans). Model IV has $k = \exp[-(t_o - t)/1.46 \text{ Ga}]$.

formed in near-Earth orbit (see Ringwood, 1979). In these scenarios, neither the tidal heating of the Moon (Peale and Cassen, 1978) nor the tidal heating of the Earth would have left a discernable trace. Cloud (1968) has proposed that a Gerstenkorn event occurred about 3.8 Ga ago, but there is no compelling case for associating such a postulated close approach of the Moon with the beginning of the terrestrial rock record. The three acceptable models shown in Figure 11-1 make substantially different assumptions for $k(t)$ but typically predict about 15 hours for the length of the day some 3.5 Ga ago, the age of the oldest fossils now known (Chapter 9).

11-2-3. Biological Implications

Reduced day length increases the role of the Coriolis effect in atmospheric and oceanic motions. The atmospheric circulation models of Hunt (1979) suggest that more rapid rotation of the Earth would lead to a warmer tropical region and colder polar regions because of the reduced eddy transport of latent and sensible heat; reduced wind speeds and wind stress would lead to a shallower oceanic mixed layer with greater capacity for seasonal change; and smaller scales of atmospheric motion would cause a reduction in continental precipitation with tropical circulation and associated subtropical arid zones being latitudinally restricted. Other mechanisms may have also affected the Precambrian climate to make it substantially different from that of the present day (e.g., variations in solar luminosity, amount of "greenhouse"-producing atmospheric constituents, and the sizes and locations of the continents), but the effects of a shorter day, if substantiated, would have had important biological implications.

The other important effect of change in the rate of rotation of the Earth is with regard to average tidal amplitude, a factor which scales as a^{-3}, and would thus have been about 50% larger for a day-length of 15 hours. Because variations of this magnitude or more could arise solely from a redistribution of the ocean basins (or even from local shoreline geography as in the narrow channel at the Bay of Fundy, Nova Scotia, where resonant amplification gives rise to a tidal range of as much as 15 m), it is not clear that this increase would have had a substantial effect on biological evolution. In some cases the form and amplitude of stromatolite laminae can provide an indication of tidal amplitudes (Cloud, 1968; Walter, 1970a; Pannella, 1976), but a systematic examination of Precambrian stromatolites in this regard (considered, especially, in the light of current understanding of controls on stromatolite morphogenesis; see Chapter 8), has yet to be conducted.

11-3. Ambient Surface Temperature

11-3-1. Theoretical Considerations

The average surface temperature of a planet like Earth depends on three factors (Goody and Walker, 1972), all of which may have varied with time. The first of these is the radiant energy emitted by the Sun, the "solar luminosity." The second is the fraction of the solar energy incident on the planet that is reflected back into space, a fraction termed the "albedo." The energy that is not thus reflected is absorbed by the planet, sustaining the temperature of the atmosphere and ground against the cooling that results from emission of thermal infrared radiation to space. The third factor is the "greenhouse effect" of the atmosphere: some of the infrared radiation emitted by the planetary surface is absorbed in, and then reradiated by, the atmosphere; the portion of this reradiated energy that returns to the surface provides an additional source of heat that raises surface temperature above the value it would have in the absence of an atmosphere.

The implications of an early cool Sun (i.e., of an increase in solar luminosity over geologic time) for the evolution of the terrestrial atmosphere were first pointed out by Sagan and Mullen (1972). Theories of stellar structure and evolution indicate that the Sun has increased in luminosity by perhaps 25 to 30% during the history of the solar system. The precise magnitude of the increase is sensitive to assumptions made in the calculations, but the approximate rate and direction of change depend only on quite general arguments relating to the physics of gravitationally bound fluid spheres (Newman and Rood, 1977). A solar luminosity some tens of percent lower in the Archean (well after the Earth had cooled from the hot primordial stages envisioned in Chapters 2 and 3), would imply an average surface temperature for the Earth well below the freezing point of water, unless one of the other factors that influences surface temperature was also different. The sedimentary rock record provides evidence of abundant liquid water as far back in time as 3.8 Ga ago (see discussion in Chapter 3), and stromatolite-forming microorganisms, presumably unlikely to have inhabited an ice-covered Earth, were present at least 3.5 Ga ago (see Chapter 9). Evidently the effect of the cool Sun was compensated during the Archean-Early Proterozoic by a different albedo or different atmospheric greenhouse.

There are several reasons why this compensation is not likely to have been solely due to a lower albedo. For one thing, the albedo of bare land is much larger than that of land covered by vegetation (Sellers, 1965; Barrett and Curtis, 1976), so any exposed land surfaces would have contributed more to albedo than they do today (Walter, 1979). On the other hand, the total land mass may have been substantially smaller early in Earth history (see Chapter 3). Planetary albedo would decrease by about 6% if there were no land (and if cloudiness and atmospheric turbidity were at their present values), because the Earth's oceans have a lower albedo than do its land surfaces (Sellers, 1965). These factors are minor, however, compared with the very high albedo of ice and snow, the degree of reflectivity of which can produce an interesting positive feedback effect on average surface temperature: a decrease in solar luminosity could result in a decrease in the average surface temperature and, thus, an increase in the fraction of the globe covered in ice; this increased ice cover, however, would also result in an increased planetary albedo—the planet would thus absorb even less solar radiation and the temperature would fall further still.

According to theory (Sellers, 1969, 1973; Budyko, 1977), if the Earth's average surface temperature were to be more than a few degrees lower than it is today, the ice-albedo feedback effect would cause a catastrophic fall in temperature ending with a completely ice-covered planet. A postulated ice-covered planet with a narrow tropical zone of liquid water in which sedimentary rocks could be deposited and stromatolites could form appears to be climatically impossible.

If the early cool Sun did not lead to the ice-albedo catastrophe, the atmospheric greenhouse effect during the Archean must have been larger than it is today. Sagan and Mullen (1972) demonstrated that water vapor alone could not have provided the necessary enhanced greenhouse, because of the escape of terrestrial radiation through windows in the absorption spectrum of water vapor in the current atmosphere, which is already essentially saturated with regard to this component. An additional greenhouse gas is thus needed—one that absorbs in the water vapor windows. They suggested ammonia as a candidate for the additional absorber. Ammonia is unattractive because of its photochemical instability and the absence of any obvious abiotic sources (Kuhn and Atreya, 1979; however, see Henderson-Sellers and Schwartz, 1980). The much more plausible alternative of carbon dioxide has recently been put forward by Owen et al. (1979). These authors have shown that the greenhouse effect of carbon dioxide and water together is great enough to solve the cool Sun problem, provided the partial pressure of carbon dioxide in the primitive atmosphere was 100 to 1000 times larger than it is today. Because carbon dioxide is released to the atmosphere by volcanoes and because the early Earth was presumably more active tectonically than it is today (for discussion, see Chapter 3), it seems reasonable that carbon dioxide partial pressures may have been higher in the Archean. With this theoretical background we can turn now to the geologic record. What constraints on Precambrian climate does it impose?

11-3-2. Observational Limits

The presence of enough liquid water at the surface to have sustained such normal geologic processes as weathering, erosion, and sedimentation has already been invoked as an observational constraint on the history of surface temperature. Since 3.8 Ga ago there have been areas on the globe where the temperature has been below the boiling point and above the freezing point of water. Because of the postulated ice-albedo feedback effect, the lower limit can probably be somewhat raised: it seems implausible that the average surface temperature of the planet has been as low as the freezing point of water since at least the beginning of the known rock record. The existence of life at least since 3.5 Ga ago does not strictly allow the upper limit to be lowered because thermophilic microbes can survive in boiling water (Brock, 1978); however, if photosynthetic microbes were extant at this time, a likely possibility (see Chapter 9), analogy with modern prokaryotic photoautotrophs would suggest an upper limit (for at least those locales where the fossils occur) of about 70° C (Brock, 1978).

There is another argument based on the properties of living organisms, however, that is less direct but that may set tighter constraints on the highest temperatures under which life first evolved (D. J. Chapman, private communication). Thermophiles, microbes that can live at high temperatures (e.g., in hot springs, etc.), are characterized by enzymes that carry out normal metabolic processes but that are resistant to destruction at high temperatures. Many of these enzymes function at low temperatures as well. The enzymes that perform these functions in microbes adapted to more normal surficial environments, however, are inactivated by high temperatures. If early life had been subjected to selection by high temperatures, the argument goes, temperature-resistant enzymes would be far more widespread than they are. Moreover, if the earliest forms of life evolved in a high temperature regime, one might expect to see a phylogenetic correlation between degree of "primitiveness" and thermophily among extant microbes, but such is not the case—forms commonly regarded as primitive (see Chapter 6), such as *Clostridium* spp., are typically mesothermic (with optimum laboratory growth observed between 25 and 37° C) and although thermophiles occur among the so-called "archaebacteria" (e.g., *Methanobacterium thermoautrophicum*, with an optimum growth temperature of 65 to 70° C), most such forms (e.g., other methanogens, halophiles, etc.) are not notably heat tolerant. Arguments of this sort are of course subject to the criticism that because of evolution and subsequent adaption the capabilities of Archean microbes may not be faithfully represented in their modern descendants. Nevertheless, the tentative picture that emerges from these observational limits (such as they are) seems to be one of an Archean average surface temperature not very different from that of the more recent geologic past.

EVAPORITE DEPOSITS

This picture of unremarkable temperatures receives qualitative support from the Precambrian record of sulfate evaporite deposits. The distribution of such deposits in time and space is summarized in Figure 11-2. The record is uneven, particularly in the Archean (apparently because of imperfect preservation), but the figure gives a general impression of uniformity in time and space. There are no marked discontinuities. It thus appears that evaporites have formed at a more or less steady rate on most of the continents since at least as early as about 3.5 Ga ago. What does this record tell us about climates?

The notion that evaporite formation necessarily implies high temperatures (see, for example, Frakes, 1979) is not correct. Evaporite minerals form today in frigid regions as well as in warm ones (Dort and Dort, 1970). For evaporites to form, obviously, conditions must be arid—evaporation must exceed precipitation—but this condition can be met at low as well as high temperatures. Indeed, as far as global averages are concerned (and assuming that other relevant factors—cloud cover, patterns of wind circulation, etc.—are held constant), high temperatures must imply high rates of precipitation because warm saturated air contains more (potentially precipitable) water vapor than cold saturated air. As a very broad generalization, we might thus expect that warm periods in Earth history would have been relatively wet, and cold periods (e.g., glacial epochs) relatively dry.

On the other hand, the rate of accumulation of an evaporite deposit does depend on temperature, because warm water evaporates into dry air more rapidly than does cold water. This circumstance cannot be used to limit temperatures on the Precambrian Earth, however, because the rates of

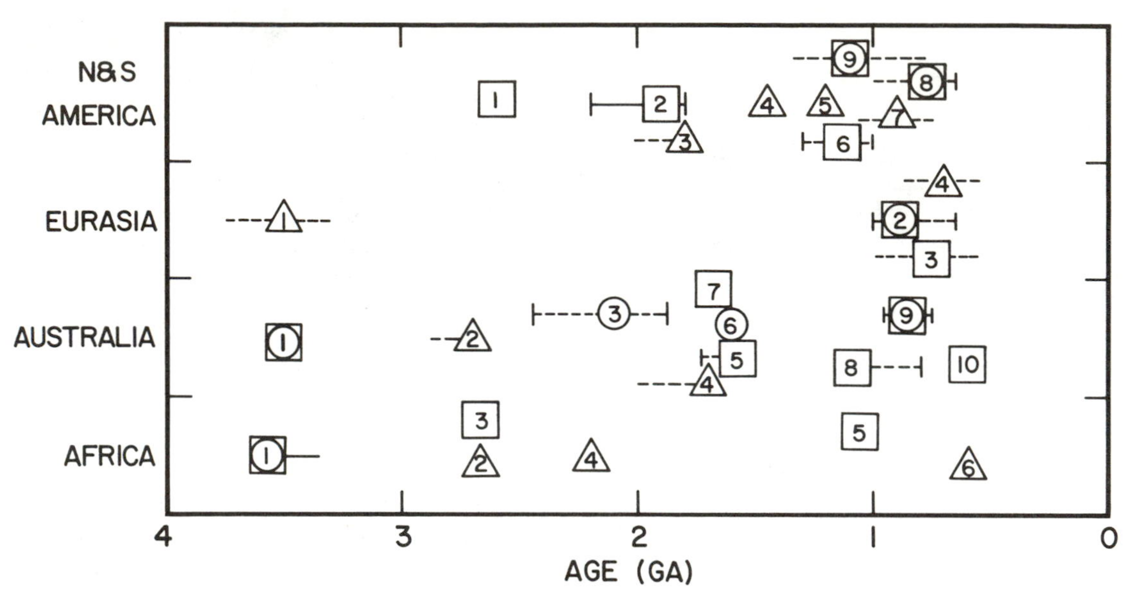

FIGURE 11-2. The temporal distribution of Precambrian sedimentary sulfate evaporites. Deposits are ranked in three categories: (*i*) bedded or massive sulfates, (circle inside squares), in several cases including units tens of meters thick and covering thousands of square kilometers and including a subcategory, (circles), of such deposits later replaced by carbonates; (*ii*) formations with abundant evidence of the presence or former presence of sulfates (squares); and (*iii*) units from which sulfate evaporites or their pseudomorphs are known, (triangles), many examples being only scattered occurrences or thin and isolated beds of pseudomorphs after sulfate minerals. The listing for the Proterozoic undoubtedly is far from comprehensive, but serves to demonstrate the continuity of the record. Broken lines indicate age uncertainties; solid lines indicate age ranges. From Africa: (1) Onverwacht and Fig Tree Groups (Heinrichs and Reimer, 1977; Lowe and Knauth, 1977; Dunlop and Groves, 1978); (2) Ventersdorp Group (see Appendix I); (3) Belingwe Greenstone Belt (Martin et al., in press); (4) Transvaal Group (Bertrand-Sarfati, 1976b; Truswell and Eriksson, 1973); (5) Kitwe Formation (Clemmey, 1978); (6) Nama Group (Debrenne and Lafuste, 1979; Glaessner, 1980). From Australia: (1) Warrawoona Group (Hickman, 1973; Dunlop, 1978; Barley et al., 1979); (2) Black Flag Beds (Golding and Walter, 1979); (3) Batchelor Group (Crick and Muir, 1979); (4) Frere Formation (Goode, in press); (5) Bungle Bungle Dolomite (M. D. Muir, personal communication to M.R.W., 1979); (6) McArthur Group (Walker et al., 1977; Muir, 1979); (7) Mount Isa Group (M.R.W., personal observation); (8) Callanna Beds (Preiss and Forbes, 1980); (9) Bitter Springs Formation (Stewart, 1979); (10) Elkera Formation (Walter, in press). From Eurasia: (1) Aldan Shield (Vinogradov et al., 1977); (2) Qingbaikou System (Chen et al., 1979; see also Glaessner, 1980); (3) Porsanger Dolomite (Siedlecka, 1976; Tucker, 1976); (4) Dalradian (Anderton, 1975). From North and South America: (1) Steeprock Group (Hofmann, 1971a; Appendix I); (2) Great Slave Supergroup (Hoffman, 1968, 1978; Badham and Stanworth, 1977); (3) Belcher Group (Bell and Jackson, 1974); (4) Belt Supergroup (Horodyski, 1976; Vaughan et al., 1977; White, 1977); (5) Dismal Lakes Group (Donaldson, 1976); (6) Grenville (Brown, 1973); (7) Bambui Group (Fairchild and Dardenne, 1978); (8) Shaler Group (Thorsteinsson and Tozer, 1962); (9) Little Dal Formation (Aitken, 1977).

accumulation of Precambrian evaporites can be estimated only very approximately.

All things considered, the record of Precambrian evaporites seems supportive of the notion that climate on the early Earth was not very different from climate during the Phanerozoic. The interpretation of the evaporite record can, however, be made more precise (Dunlop and Buick, 1980; H. D. Holland, pers. comm.). Most of the Precambrian evaporites, including the oldest known (Figure 11-2), show morphological and mineralogical evidence that they originally contained gypsum ($CaSO_4 \cdot 2\,H_2O$), in some cases subsequently altered to pseudomorphic barite ($BaSO_4$). However, laboratory studies under reversible, equilibrium conditions (Hardie, 1967), supported by observations of modern evaporites on the Trucial Coast (Holland, 1978), show that gypsum is converted to anhydrous calcium sulfate (anhydrite, $CaSO_4$) at temperatures above about $58 \pm 2°$ C in pure water, and at lower temperatures in saline water. Since gypsum (monoclinic) and anhydrite (orthorhombic) belong to different crystal classes, the two minerals can be differentiated, even in pseudomorphs. The identification of evaporitic gypsum throughout Precambrian time, therefore, implies that temperatures in evaporitic basins (and presumably globally) have probably not been much above 58° C since at least 3.5 Ga ago.

Even tighter temperature constraints seem suggested by the reported coexistence of gypsum and halite (NaCl) in the ~2.0-Ga-old Great Slave Supergroup of Canada (Badham and Stanworth, 1977), and the ~1.6 Ga old McArthur Group of Australia (Walker et al., 1977). If these minerals were co-precipitated, then the observations on modern evaporites noted above would indicate that the temperature under which they formed did not exceed 18° C (Holland, 1978).

GLACIAL DEPOSITS

The known record of Precambrian glaciations (as shown in Figure 11-3) is strikingly different from that of sulfate evaporites. In the whole span of the Precambrian, there are only two time segments in which firm evidence of glacial activity has been recognized: during the Early Proterozoic, between 2.5 and 2.0 Ga ago (from which are known glaciogenic units in Australia, North America, and South Africa), and during the Late Proterozoic, between about 1.0 and 0.57 Ga ago (in which occur glacial deposits on every continent but Antarctica). Although preservation of such units in Precambrian terranes is no doubt incomplete, because of the widespread distribution of continental-type glaciation the potential for preservation of such glaciogenic deposits should be decidedly greater than that of soluble evaporites which commonly are of rather limited areal distribution. Thus, in spite of the cool Sun, it appears that it was only during these two rather broadly defined time segments that continental glaciation was at all widespread. Taken together, the records of glaciogenic rocks and sulfate evaporites suggest climatic conditions for most of the Precambrian that were neither markedly cooler nor markedly warmer than those of the Phanerozoic.

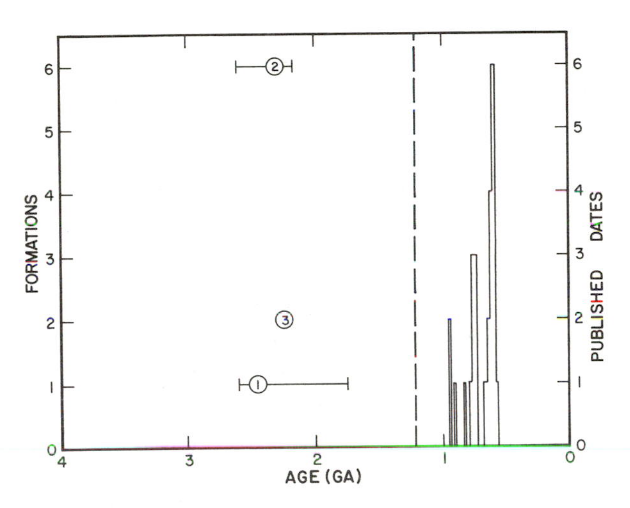

FIGURE 11-3. The temporal distribution of Archean and Proterozoic glaciogenic rocks, based on the reviews of Young (1973a), Williams (1975), Chumakov (1978), and Frakes (1979). The histogram of published dates (to the right of the dashed vertical line) is from Williams (1975); many more dates are now available (Chumakov, 1978), but the distribution remains essentially the same. Both Chumakov (1978) and Frakes (1979) list other possible Archean and Early Proterozoic glaciogenic deposits, but only the well-documented occurrences are plotted here (to the left of the dashed vertical line): (1) Kungarra Formation, Australia (Trendall, 1979). (2) Huronian Supergroup occurrences and their stratigraphic correlatives in North America (Young, 1973a). (3) Transvaal Supergroup, South Africa (Visser, 1971). Horizontal lines indicate the range of uncertainties for the ages of the three Archean-Early Proterozoic occurrences (see Appendix I).

Why this clement climatic regime was twice replaced by episodes of glaciation is likely to be debated for some time to come, but the better known geologic record of the Phanerozoic also contains evidence of intermittent glacial episodes (Frakes, 1979). The causes of these later episodes are known only in speculative outline (Beaty, 1978); the causes of the Precambrian glaciations may have been similar. However, it can only be guessed as to what extent the Precambrian and Phanerozoic episodes may be analogous; geochronologic control (such as that provided in the Phanerozoic by biostratigraphy) is largely lacking for the Precambrian glacial deposits and it is thus unknown, for example, whether the Early Proterozoic glaciogenics are a product of a single widespread episode (bracketed in age between 2.5 and 2.0 Ga) or whether they are a result of at least three such episodes (in Australia, North America, and South Africa) of perhaps significantly different geologic ages. Moreover, the Late Proterozoic glaciation, at least, appears to have been unusual with respect to the latitudes where glaciation occurred. Paleomagnetic data (McWilliams and McElhinny, 1980) reveal that many if not all of these glaciogenic rocks formed in low paleolatitudes, thus suggesting (if the rocks are largely or entirely coetaneous) either unusually severe (quasi-global) glaciation at that time, or an episode of exclusively equatorial glaciation. An interesting feature of the Early Proterozoic glacial deposits, at least in the Huronian sequence of Canada, is that they are underlain by sedimentary rocks that appear to have been deposited under reducing conditions and are overlain by rocks that appear to have been deposited under oxidizing conditions (Roscoe, 1969). There is the possibility, therefore, that Early Proterozoic glaciation was in some way associated with the transition, discussed below, from an "anoxic" to an oxidizing atmosphere.

11-3-3. Conclusions

The climatic data, although sparse, suggest that clement conditions, much like those of later geologic time, prevailed throughout most of the Precambrian. How such conditions were maintained for several Ga in the face of steadily increasing solar luminosity is a question of the greatest interest. There may well have been a feedback mechanism that coupled the atmospheric greenhouse to surface temperature in such a way as to maintain equable conditions. A possible mechanism of this type is described below. The evidence seems reasonably convincing that average surface temperatures during most of the Precambrian were neither above 60°C nor below a few degrees C. Within this broad range—and despite recent attempts to solve the problem based on studies of the hydrogen and oxygen isotopic compositions of ancient cherty deposits (e.g., Knauth and Epstein, 1976; Knauth and Lowe, 1978)—it seems to us that it is not yet possible to say whether temperatures were generally lower or higher than they are today. No evidence has yet been identified for thermal stress as being a significant factor in the early evolution of life. The related question as to whether climate might have been generally wetter or drier early in Earth history is also not answerable at present, although it seems most plausible that the very broad ranges in such conditions that characterize the Phanerozoic were also extant during the Precambrian.

The mechanism of stabilization of the Precambrian climate broke down on at least two occasions, one during the Early Proterozoic and the other in the Late Proterozoic. Episodes of glaciation were the result. During the latter glacial phase, conditions were sufficiently severe to have permitted glaciogenic rocks to form even near the equator.

11-4. Carbon Dioxide Partial Pressure

11-4-1. Theoretical Considerations

There is as yet no widespread agreement concerning the processes that control the partial pressure of atmospheric carbon dioxide. One point of view holds that the pH of sea water is controlled by conditions of thermodynamic equilibrium among clay minerals on the sea floor (Sillén, 1961) that thus control the carbon dioxide partial pressure because of its link to pH by a series of quantitatively known equilibrium reactions involving carbonate and bicarbonate ions (Broecker, 1974). However, this point of view seems to devote too little attention to the problem of maintaining a balanced oceanic budget of both carbon and cations in the face of inputs that might have fluctuated widely through geological time. An alternative kinetic view, originally proposed by Broecker (1971), is discussed in the following paragraphs.

As a basis for this view it is noted that carbon dioxide is released from the solid Earth by volcanoes and other hydrothermal processes at a high rate, one estimated to be large enough to double the total amount of carbon in the atmosphere and oceans in only 400,000 years (Holland, 1978). For the carbon content of the system to remain more or less constant over geologically significant periods of time carbon dioxide must be returned to the solid Earth at an equal rate. The mechanism of this recycling is well known (Siever, 1968; Garrels and Mackenzie, 1971). It is the weathering of silicate minerals followed by the deposition of carbonate minerals in sedimentary rocks as summarized in the following highly schematic reactions:

$$\text{weathering: } CaSiO_3 + 2\,CO_2 + 3\,H_2O \longrightarrow Ca^{2+} + 2\,HCO_3^{1-} + H_4SiO_4$$

$$\text{deposition: } Ca^{2+} + 2\,HCO_3^{1-} \longrightarrow CaCO_3 + CO_2 + H_2O$$

Because the rate of the weathering reactions should thus increase with increasing partial pressure of carbon dioxide, it is reasonable to suppose that the concentration of atmospheric carbon dioxide is buffered by the requirement of a long-term balance between the rate of removal of carbon dioxide by silicate weathering and the rate of addition of carbon dioxide by geologic processes (Figure 11-4).

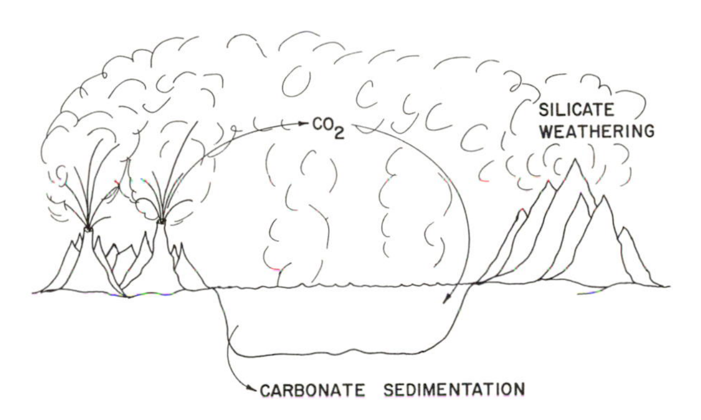

FIGURE 11-4. Processes that control the partial pressure of carbon dioxide. The rate of release of CO_2 by volcanoes is balanced by the rate of its removal as a result of the weathering of silicate minerals and the deposition of carbonate minerals.

The biological removal of carbon dioxide from the atmosphere and ocean system (e.g., via photosynthesis) does not alter this conclusion because, in the long term, such removal must be balanced by equal flow of carbon dioxide (as a product of aerobic respiration and weathering) back into the atmosphere and oceans (Holland, 1973; Walker, 1974) in order to maintain the oxygen content of the atmosphere (see also Chapter 5). In the long term, weathering of carbonate minerals, e.g.,

$$CaCO_3 + CO_2 + H_2O \longrightarrow Ca^{2+} + 2HCO_3^{1-} \quad (11\text{-}2)$$

is also not a significant sink for carbon dioxide because such weathering must ultimately be balanced by the precipitation of carbonate minerals, for example,

$$Ca^{2+} + HCO_3^{1-} \longrightarrow CaCO_3 + CO_2 + H_2O \quad (11\text{-}3)$$

in order to maintain a balanced oceanic budget of cations and carbon.

According to the kinetic view, therefore, the carbon dioxide partial pressure adjusts to a value that will yield a rate of CO_2 use during weathering just equal to its rate of geologic input into the system and the resulting partial pressure, in turn, determines the pH of the ocean. If other factors are held constant, a change in atmospheric carbon dioxide content by a factor of 100 would yield a pH change in the Earth's oceans of less than 1; geological evidence of a more or less constant oceanic pH over time thus imposes little constraint on possible changes in the pCO_2 of the atmosphere (Holland, 1972, 1976).

The kinetic view of the processes that control atmospheric carbon dioxide offers wide opportunities for interesting speculation concerning Earth history. One possibility (Walker et al., 1981) concerns a feedback mechanism that may have sustained equable surface temperatures throughout the Precambrian in the face of steadily increasing solar luminosity. This feedback mechanism depends on the idea that the partial pressure of carbon dioxide adjusts to provide equality between the rate of consumption of carbon dioxide by silicate weathering reactions and its rate of input from geologic sources. It may be assumed that the rate of silicate weathering at a given carbon dioxide partial pressure is a strong function of ambient temperature. According to the kinetic model, the equilibrium partial pressure of carbon dioxide would thus also be a function of temperature. But temperature depends on carbon dioxide partial pressure through the greenhouse effect; carbon dioxide and temperature would therefore be closely coupled in a feedback system.

The assumption that weathering rate increases with temperature is consistent with, and apparently supported by the observation that tropical soils today typically are more chemically weathered (i.e., are more deficient in all but those minerals most resistant to chemical attack) than are the soils of higher latitudes. At the other extreme, glacial deposits contain feldspars, which are among the silicate minerals most susceptible to chemical weathering. Two factors presumably contribute to an increased rate of chemical weathering at higher temperatures. Firstly, weathering reactions may be kinetically favored by increased temperature; and secondly, higher temperatures permit higher humidity, higher rainfall, and therefore higher rates of solution weathering and an increase in the availability of the water needed to carry off the resultant products.

Thus, a feedback mechanism that might have stabilized Earth's surface temperature is this: atmospheric carbon dioxide is held at a sufficient concentration to produce, through the greenhouse effect, a high enough surface temperature to yield a rate of carbon dioxide consumption in the weathering of silicate minerals equal to the rate of release of carbon dioxide by geologic processes. By way of illustration: if the solar luminosity were suddenly to decrease, the surface temperature would fall causing a decrease in the rate of silicate weathering; volcanically derived carbon dioxide would therefore accumulate in the atmosphere (and ocean) causing an increase in surface temperature as a result of the greenhouse effect; and equilibrium would be restored with the perturbation in temperature that occurred during the cycle having been effectively damped, reduced to a level smaller than that which would have occurred in the absence of feedback from the temperature dependence of silicate weathering.

In essence, what is proposed is the converse of the familiar and well established idea that climate is influenced by atmospheric carbon dioxide. But perhaps, as well, atmospheric carbon dioxide is influenced by climate through the dependence of weathering rate on temperature. There is, therefore, the possibility of a closely coupled carbon dioxide-climate system that has regulated both average temperature and atmospheric carbon dioxide throughout geological history. Could this system have maintained equable surface temperatures in the Archean-Early Proterozoic at a time of lower solar luminosity?

To answer this question it would be necessary to know how the global rate of silicate weathering depends on the average surface temperature and on carbon dioxide partial pressure. These dependences are not known. Walker et al. (1981) suppose that the pressure dependence is weak and that the temperature dependence is contained mainly in the dependence of precipitation on temperature. To demonstrate the plausibility of the feedback mechanism described above they have applied a simple greenhouse model to calculate a history of surface temperature and carbon dioxide partial pressure under the assumption that the only external factor that has changed significantly with time is solar luminosity (Figure 11-5). Their results indicate an average surface temperature early in Earth history of some 6° C cooler than that of the present and a pCO_2 some 100 times larger. If, as seems likely, volcanism was appreciably more widespread during the Archean than it is today (see Chapter 3), both the average surface temperature and the partial pressure of carbon dioxide would have been higher than these calculated values. Changes in terrestrial albedo resulting from differing distributions of cloud or snow cover, land masses, or of surficial vegetation, would have perturbed not only temperature but also carbon dioxide levels.

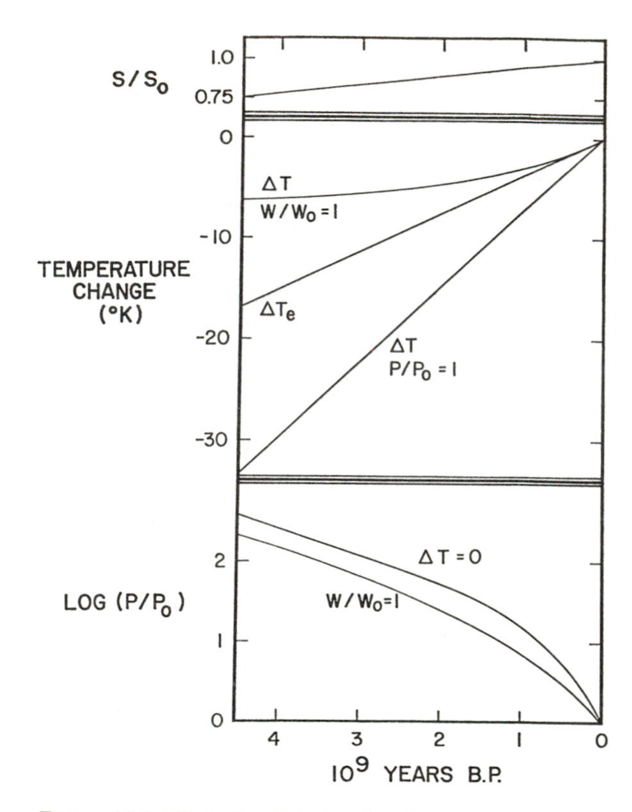

FIGURE 11-5. Illustrative histories of surface temperature and carbon dioxide partial pressure. It is assumed that solar luminosity (S) increases linearly with time as shown in the top panel, causing the increase in effective temperature (T_e) shown in the middle panel. The middle panel compares surface temperature changes (ΔT) that result if weathering rate (W) is held fixed, and if weathering feedback is ignored and the carbon dioxide partial pressure (P) is held fixed. The bottom panel compares partial pressures for constant weathering rate and for the case of no change with time in average surface temperature. This case requires an early weathering rate (volcanic source) four times larger than that of the present day. Zero subscripts refer to the present day (Walker, et al., 1981).

These results are, of course, extremely speculative, but at least one direct consequence of the kinetic view of the processes controlling atmospheric carbon dioxide that underlies the proposed feedback mechanism similarly points to a considerably higher carbon dioxide partial pressure in the Precambrian atmosphere. Most ground water today exhibits partial pressures of carbon dioxide that are enhanced by factors of 10 to 100 over the atmospheric value (Holland, 1978). This enhancement is a consequence of a release of carbon dioxide due largely to the respiration and decay of the roots of vascular plants. Thus, the waters responsible for the solution and chemical weathering of silicate minerals in the soil profile are substantially more corrosive than waters in equilibrium with the atmosphere. In the Precambrian, before the origin of higher plants, such enhancement must have been minimal, carried out solely by soil microorganisms. In order, therefore, for the rate of silicate weathering to have balanced even the present-day rate of release of carbon dioxide from volcanic and other geologic sources, a considerably larger atmospheric partial pressure would have been needed.

11-4-2. Observational Limits

The constraints imposed by the geological record on possible excursions in the composition and pH of sea water have been examined by Holland (1972, 1974). He finds the record of marine evaporites to be most useful in this regard, concluding that the concentrations of the major constituents of sea water have not changed by more than a factor of two since the late Precambrian. Greater excursions would have yielded changes in the mineral associations of evaporitic deposits that would be likely to have been reported. The older Precambrian evaporites (Figure 11-2) were less well known at the time of Holland's work and have not been critically examined from this point of view. How far back in time his arguments can be extended is, therefore, not known.

Permissible excursions in oceanic pH, according to Holland, are about one unit since the Late Proterozoic. The corresponding excursion in the partial pressure of atmospheric carbon dioxide is about a factor of 100 in either direction. There is no immediately obvious conflict between the sedimentary rock record and the suggestion that equable surface temperatures were sustained on the Precambrian Earth due to the greenhouse effect of a larger partial pressure of carbon dioxide, but the requirements of the carbon dioxide greenhouse are sufficiently close to the observational limits that more careful interpretation of the older evaporite deposits in the light of Holland's criteria would be very worthwhile. Moreover, it seems likely that very high partial pressures of carbon dioxide could not have occurred on the primitive Earth at least since the mid-Precambrian development of aerobic life—algal respiration, for example, is inhibited when the carbon dioxide partial pressure is increased to about 0.3 atmospheres (Holland, 1978, p. 51), a value roughly 1000 times the present atmospheric level.

From another point of view, the existence of photoautotrophs, organisms that can capture light energy and can use carbon dioxide as their immediate carbon source, imposes a lower limit on possible excursions in carbon dioxide partial pressure. The growth of such organisms must cease once the concentration of available carbon dioxide becomes lowered to the "compensation point," the level at which the consumption of CO_2 via photosynthesis is just balanced by the production of CO_2 due to light-driven photorespiration. Among extant plants this lower limit is rather variable, generally about 0.3 to 0.1 of the present atmospheric value (Zelitch, 1971; Holland, 1978), and in primitive prokaryotes it can be even lower—the compensation point for the cyanobacterium *Anabaena cylindrica*, for example, is a carbon dioxide level of about 10 ppm (Lex et al., 1972), approximately 0.03 of the present atmospheric value. Evidence presented in Chapters 5 and 9 indicates that primitive prokaryotic photosynthesizers probably originated more than 3.5 Ga ago.

11-4-3. Biological Impact

Judging from extant plants, increased partial pressures of carbon dioxide up to some maximum level (apparently somewhat less than about 100 times the present value; Holland, 1978) would have facilitated biological carbon fixation by autotrophs, whether photosynthetic or chemosynthetic. Below this maximum level, most plants respond to increases in carbon dioxide by increased rates of photosynthesis, provided their growth is not limited by an inadequate supply of water, light, space, or nutrient elements. Why is the photosynthetic apparatus not optimized to the present-day partial pressure of carbon dioxide? Is it possible that this apparatus (viz., ribulose bisphosphate carboxylase/oxygenase) adapted to the higher partial pressures of carbon dioxide (and the potentially deleterious effects of even low levels of ambient free oxygen) that may have existed at the time when photosynthesis first evolved?

Equally speculative is the suggestion that changing carbon dioxide partial pressure and changing oceanic pH might have affected the solubility of calcium carbonate in sea water and, thus, inhibited Precambrian microorganisms from precipitating calcium carbonate tests (or late Precambrian metazoans from forming carbonate shells). This possibility, which seems at least generally consistent with the paucity of ancient (Archean) carbonates (see Chapter 3), has received no serious examination.

11-5. Oxygen Partial Pressure

11-5-1. Theoretical Considerations

The levels of molecular oxygen in the ancient atmosphere have been studied more extensively than those of any other atmospheric constituent (Berkner and Marshall, 1964b, 1965, 1966, 1967; Brinkmann, 1969; Margulis et al., 1976; Walker, 1977, 1978a, 1978b; Kasting et al., 1979; Vander Wood and Thiemens, 1980; Kasting and Walker, 1981). In part this is because of fairly clear (though still debated) sedimentological evidence for a change in the oxidation state of the atmosphere, from neutral (or reducing) to oxidizing, during the Early Proterozoic (MacGregor, 1927; Cloud, 1968a, 1972). In part, also, it is because of the importance of the oxidation state of the Earth's atmosphere to the origin and evolution of life (viz., the evolution of bioenergetic processes). There is little doubt that prebiotic chemical evolution, the series of abiotic processes that culminated in the origin of life, would have been impossible if the atmosphere had contained much free oxygen (Miller and Orgel, 1974; see also Chapter 4). On the other hand, the multi-celled and macroscopic eukaryotes of today are sustained by energy derived from aerobic (oxygen-requiring) respiration. The biota could not have evolved to its present level of diversity and complexity in the absence of molecular oxygen.

It is not hard to see why the early Earth was anaerobic. The Earth, like other bodies of the solar system—and, indeed, the universe in general—is reducing in overall composition (see Chapter 2). Free oxygen is abundant in the atmosphere today only because it is produced as a metabolic byproduct of "green plant-type" (including cyanobacterial) photosynthesis (Van Valen, 1971; Holland, 1973; Walker, 1974, 1980a). The only other potentially significant source of atmospheric oxygen that has yet been identified comes from photochemical reactions that have the net effect of dissociating water vapor and allowing the resultant hydrogen to escape to space. The rate of such oxygen production is very low today, and most theoretical arguments indicate that it would have been even lower in the Archean. Before the origin of oxygen-producing photosynthesis, the free oxygen produced via this inorganic mechanism was likely to have been overwhelmed by an abundance of unoxidized gases from volcanic sources and unoxidized species in solution produced by weathering; the oxygen would thus have been rapidly consumed by the oxidation of available substrates.

It would seem, therefore, reasonable to assume that the biosphere was wholly anaerobic until oxygen-producing (cyanobacterium-like) photosynthesizers began to produce oxygen fast enough to overwhelm the rate of supply of reduced species to the environment. However, such was probably not the case: as is discussed in Chapter 13, intracellular molecular oxygen is toxic to strict anaerobes, even at low concentrations, and it is therefore difficult to envision how the capacity for photosynthetic (necessarily intracellular) oxygen-production could have evolved in microorganisms that were not pre-adapted for at least some degree of oxygen-tolerance. Such tolerance apparently did not evolve within the immediate evolutionary precursors of the cyanobacteria, the non-oxygen-producing photosynthetic bacteria, since in such extant microbes photosynthesis is obligately anaerobic (and the biosynthesis of the light-capturing molecules involved in the process, the bacteriochlorophylls, is inhibited by free oxygen). It seems most probable, therefore, that—perhaps locally, or in times (e.g., of diminished volcanism) when the supply of reduced substrates to the environment was relatively low—photochemically produced oxygen levels built up to (low) concentrations such that oxygen-tolerance was of selective value. In such environments, oxygen-tolerant amphiaerobes, including proto-cyanobacteria, evolved. Over time, their photosynthetic activity ultimately led to the generation of a stable oxygenic atmosphere (and the evolution and dispersal of strict aerobes). The problem is to determine when and how this transition occurred.

There are some rather general ideas that are helpful in a consideration of this problem. The first is the concept of a reservoir. A reservoir is a body of matter that, for purposes of a particular study, may be taken to be fairly homogeneous in composition and to exchange material relatively slowly with other, interconnected reservoirs. In a study of the global balance of oxygen, appropriate reservoirs would include the atmosphere, the mixed layer of the ocean, and the deep ocean. Second is the idea of fluxes of material into or out of a reservoir either from interconnected reservoirs or from outside the system under immediate consideration. Sources supply material to a reservoir; volcanic activity, for example, is a source of atmospheric carbon dioxide. Sinks remove material from a reservoir; the process of aerobic respiration, for example, serves as a sink for oxygen (but as a source for carbon dioxide). Third is the idea of redox titration reactions. There are many chemical reactions that take place within a reservoir essentially instantaneously in terms of geological time, or even on the timescale associated with the transfer of material between reservoirs. Photochemical reactions in the atmosphere provide an example: at present, hydrogen is converted to water vapor in the atmosphere by photochemical reactions with oxygen that proceed as fast as the hydrogen becomes available for reaction; the characteristic time for the process is only two years. If the rate at which the hydrogen became available were to exceed that for the oxygen source, however, free oxygen would be utilized equally fast. Such titration reactions also occur in the hydrosphere, some via entirely inorganic mechanisms and others being metabolically mediated. (For example, free oxygen is removed by amphiaerobic organisms from the upper reaches of anaerobic environments as fast as it is supplied; that is why such environments are anaerobic. And dissolved ferrous iron, when exposed to oxygen, is rapidly oxidized even without the intervention of organisms.)

In considering the oxidation state of a particular reservoir, therefore, it is necessary to examine the relative rates of supply both of free oxygen and of unoxidized (or only partially oxidized) compounds to the reservoir. Of the two, if the rate of supply of oxygen is the greater it is probable that titration reactions will hold the reduced compounds to low levels while excess oxygen accumulates and becomes available for export to other reservoirs. Conversely, we can expect virtually no free oxygen to accumulate in a reservoir that receives reduced compounds more rapidly than oxygen is supplied.

The fluxes and reservoirs that are likely to be important in a first order interpretation of the geological record of free oxygen are illustrated in Figure 11-6. At some rate, R_1, the atmosphere receives potentially oxidizable gases from subaerial volcanic emanations and possibly also by export from the mixed layer of the ocean or as a result of biotic or abiotic processes at the land surface. At the same time, the atmosphere also receives oxygen, at rate O_1, from the abiotic photochemical source (produced at a rate corresponding to the rate of escape of hydrogen to space) and if oxygen-producing photoautotrophs are present, also as a result of photosynthesis in the mixed layer (essentially the photic zone), of the ocean and, possibly, as a result of photosynthesis of cyanobacteria or other microalgae on the Precambrian land surface. As a result of weathering reactions and the transport of material from land surfaces, the mixed layer receives inorganic reduced compounds in solution (and possibly organic compounds of biotic and/or abiotic origin) at rate R_2; this reservoir might also exchange reduced gases with the atmosphere. At the same time, the mixed layer receives oxygen, at rate O_2, from in situ photosynthesis and possibly also from exchange with the atmosphere. Finally, at rate R_3, the deep ocean receives reduced compounds from the overlying mixed layer and potentially oxidizable mineral matter as a result of sea floor weathering, submarine volcanism, and, if present (see Chapter 3), the geothermal sources associated with mid-ocean, plate tectonic spreading centers. However, the deep ocean receives oxygen (at rate O_3) only from the mixed layer.

Estimates of modern values of the relevant fluxes suggest that the rate of input of reduced

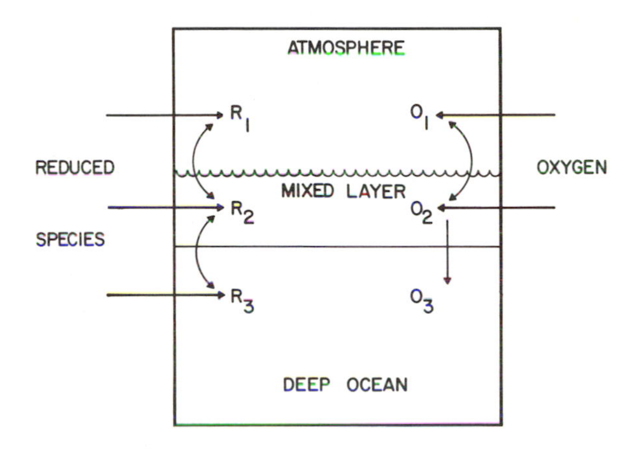

FIGURE 11-6. Reservoirs and fluxes. As discussed in the text, whether a given reservoir is oxidizing or reducing depends on the relative magnitudes of the fluxes of reduced species and oxygen.

material into all three reservoirs would exceed the rate of oxygen supply, were it not for oxygen-producing photosynthesis (Garrels and Mackenzie, 1971; Holland, 1978). Specifically, the current volcanic source of reduced gases is about five times larger than the present-day abiotic source of oxygen, and the weathering source of reduced inorganic compounds in solution is about 500 times larger than such oxygen production. The rate of supply of reduced compounds to the deep ocean has not been estimated, but even without this estimate there is every reason to suppose that reduced compounds were far more abundant than abiotically produced oxygen in the prebiological atmosphere and ocean. Free oxygen must thus have been transient, and held at low levels of concentration, with the excess, unreacted, reduced material being removed from the atmosphere via photochemical reactions and the escape of hydrogen to space, and being removed from the ocean by the deposition of reduced sedimentary minerals and organic matter. Accumulation times in the various reservoirs are too short to permit significant imbalance between rates of supply and rates of removal to persist over geologic time spans.

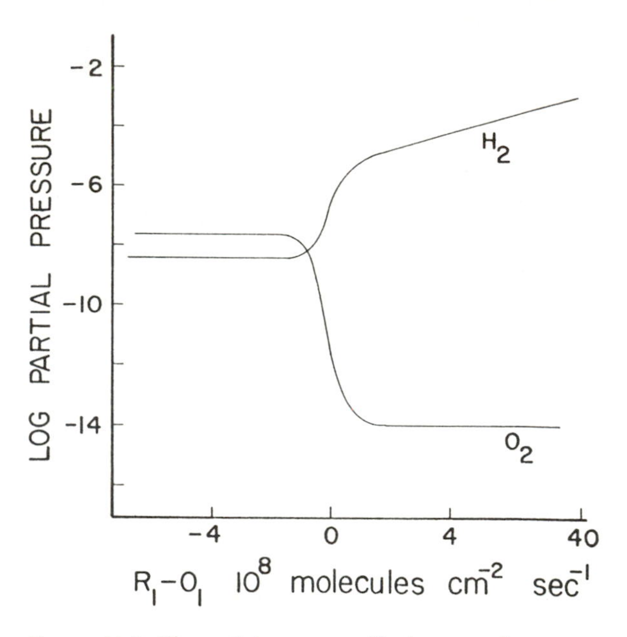

FIGURE 11-7. The partial pressures of hydrogen and oxygen at ground level depend strongly on the relative magnitudes of the rates of supply of oxygen and reduced gases to the atmosphere. The figure shows results of photochemical calculations by Kasting and Walker (1981).

The redox balance of the prebiological atmosphere has been explored by means of detailed photochemical calculations by Kasting et al. (1979) and by Kasting and Walker (1981). Depending on whether the (principally volcanic) source of reduced gases, R_1, is greater than or less than the abiotic source of oxygen, O_1, two quite different situations can be envisioned (Figure 11-7). Using estimates of the modern values of these fluxes (in which $R_1 > O_1$), they calculate that the partial pressures at ground level would be 10^{-14} bar for oxygen and 10^{-5} bar for hydrogen. However, a marked change in the results is obtained if it is assumed that the volcanic source is less than the oxygen source (i.e., $R_1 < O_1$). Assuming no flux of reduced gases to the atmosphere at all, they calculate an oxygen partial pressure of 10^{-8} bar and a hydrogen partial pressure of $10^{-8.5}$ bar with the hydrogen in this case being entirely the product of photochemical reactions, and the excess oxygen being removed from the atmosphere by transfer to the mixed layer of the ocean, assumed to be reducing (i.e., $R_1 = 0, R_2 > O_1$).

In both of these simplified models the atmospheres contain very little oxygen and are firmly anaerobic (unable to support aerobic metabolism): but the second one, interestingly, provides an oxidizing environment at the ocean-atmosphere interface. In the absence of abundant atmospheric oxygen and ozone, solar ultraviolet radiation penetrates deep into the troposphere, producing abundant $HO\cdot$ and $H\cdot$ by the photolysis of water vapor. These free radicals are much more reactive than molecular oxygen and hydrogen, and can be expected to react on essentially every encounter with a suitable (i.e., unreacted) substrate. In the second of the atmospheres described above, the abundance of oxidizing free radicals (viz., $O\cdot$, $HO\cdot$, $HO_2\cdot$) and H_2O_2 in the near surface environment is much greater than the abundance of $H\cdot$, the only reducing free radical present in significant quantities. This atmosphere would have oxidized any iron, for example, with which it came into contact. In the first atmosphere described, the abundance of $H\cdot$ at ground level is much greater than the abundance of oxidizing free radicals. The first atmosphere thus is not only anaerobic, but also definitely "reducing" in the sense that weathering under this atmosphere would not have yielded oxidized minerals.

As a further illustration of conditions that might possibly have characterized the prebiological Earth, it is at least conceivable that both R_1 and

R_2 were zero (or nearly so) and that O_1 was less than R_3 (Figure 11-8). Perhaps the only source of oxygen was photolysis of water vapor followed by escape of hydrogen; perhaps this source was less than the source of reduced minerals in the deep sea as generated by submarine weathering, volcanism, and the like; and perhaps both the subaerial volcanic source of reduced gases and the weathering source of reduced materials from the land surface were negligibly small. Because the rate of transfer of oxygen between mixed layer and deep ocean is about 1000 times slower than the rate of transfer between the mixed layer and the overlying atmosphere (Broecker, 1974), we can expect that the atmosphere and the mixed layer would have achieved equilibrium. Under these conditions, the oxygen partial pressure at the surface can thus be calculated by a simple extension of the arguments of Kasting and Walker. It would have been about 10^{-5} bar, while the hydrogen partial pressure would have been less than 10^{-8} bar; as in the second atmosphere described above, therefore, the near-surface environment would have been anaerobic but definitely oxidizing.

Let us now consider a world shortly after the advent of oxygen-producing photosynthesizers (with such microorganisms not yet inhabiting land surfaces) where the rate of in situ photosynthetic

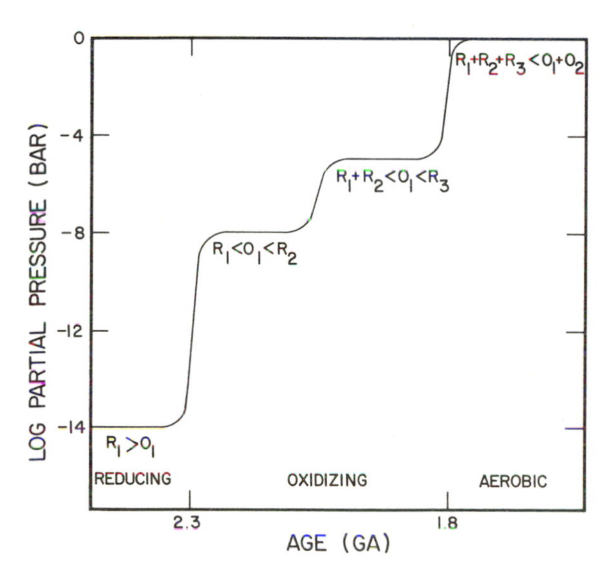

FIGURE 11-8. Schematic view of the increase in oxygen partial pressure as the rate of supply of oxygen to the atmosphere and oceans rises above the rate of supply of reduced gases and reduced species in solution (for notation, see Figure 11-6).

oxygen-production, O_2, was initially less than R_2, the rate at which weathering supplied reduced materials to the mixed layer of the ocean. Under these conditions, the photosynthetic oxygen would have reacted with the excess of reduced matter available in solution, and the mixed layer would thus have remained predominately reducing, although less strongly so than prior to the origin of such organisms or on the prebiological Earth. Since virtually all of the oxygen would have been consumed within the mixed layer, little would have been transferred to the deep ocean or to the atmosphere. The deep ocean would thus presumably have remained reducing and the atmosphere would have remained anaerobic, but whether it was reducing or oxidizing would have depended on the relative magnitudes of the sources of reduced gases (R_1) and oxygen (principally photochemical in origin) that entered this reservoir.

Alternatively, one could suppose that the photosynthetic source of oxygen exceeded the rate of supply of reduced material to the mixed layer but was still less than the combination of such matter supplied by weathering together with that entering the deep ocean. Under these conditions, excess oxygen would have escaped from the mixed layer into the atmosphere and the deep ocean; the deep ocean would have remained anoxic (due to consumption via inorganic reactions of the entering oxygen) while the atmosphere and the mixed layer achieved higher oxygen levels than before. Finally, one could suppose that the rate of photosynthetic oxygen production exceeded the rate of supply of reduced matter to the atmosphere and the oceans combined. Under these conditions, the atmosphere and oceans would have been swept free of reactive reduced compounds, and free oxygen would have accumulated until it achieved a level high enough to support aerobic respiration, the enzyme-mediated, intracellular oxidation of organic matter that serves as the principal energy-yielding metabolic pathway of aerobic forms of life.

The oxidation state of the Earth's habitable environments, therefore, depends on the relative magnitudes of the sources of oxygen and of reduced compounds, and on the reservoirs that these sources supply. As the models illustrate it is possible to imagine circumstances on the early Earth that would have yielded a relatively oxidized atmosphere over a relatively reduced ocean, or a relatively oxidized mixed layer between a less oxidized atmosphere and an essentially anoxic deep ocean. Such circumstances may well have appeared and disappeared as the rates of tectonic activity varied, a matter not considered in the models here outlined. Moreover, the models are, of course, highly simplified since they neglect, among other factors, local variability in the sources and reservoirs concerned (e.g., products of subaerial weathering would have been far more abundant in near shore, river-fed environments than in the offshore ocean; products of volcanism would have been locally concentrated; oxygen-producing photoautotrophs were probably far more abundant per unit area in near shore benthonic communities than in the open ocean, an environment that they may not have invaded until Early Proterozoic, etc.). But in no circumstances could an oxygen level approaching that of the present have been achieved on a global scale until the oxygen source had risen above the total rate of supply of volcanic reduced gases and of reduced minerals and other materials in solution.

So large an oxygen source almost certainly was not possible without 'green plant-type" photosynthesis. Once the rate of photosynthetic oxygen-production exceeded the source of reduced matter, the accumulation of oxygen in the atmosphere and oceans could have been halted only by the onset of aerobic respiration. However, the flux of photosynthetically produced oxygen entering the environment is dependent to a considerable degree on the photosynthetic biomass, a value which on the anaerobic Earth was probably substantially smaller than it is at present (Holland, 1978). Moreover, the time of origin of aerobic respiration is uncertain; since all oxygen-producing photosynthesizers (including cyanobacteria) are also capable of aerobiosis, it is possible that the two processes evolved together but, as noted above, it is also conceivable that some amphiaerobes—microorganisms that can carry out aerobic respiration if oxygen is available but that are otherwise anaerobes—may have evolved even prior to the appearance of aerobic photoautotrophs. The relative rates of oxygen-production and oxygen-consumption in the evolving biota are therefore essentially unknown, and the amounts and fluxes of potentially oxidizable, unreacted materials in the various reservoirs can be estimated only very approximately. It is thus not possible to state with certainty the rate of transition from anaerobic to aerobic conditions (viz., the time required for the rate of oxygen-production to have exceeded the source of reduced matter). As a first approximation, however, it seems likely that within the photic zone the transition would have occurred relatively rapidly in terms of a geologic timescale. Other zones of the environment (particularly the deep sea) could have lagged behind, especially if they contained large amounts of unreacted substrates. The establishment of highly oxygenic conditions, like those of the modern Earth, could have occurred over a long segment of geologic time.

For illustration of these theoretical concepts, we turn now to the known geologic record.

11-5-2. The Geologic Record

BANDED IRON-FORMATIONS

Potentially the most useful sedimentary indicators of redox conditions on the ancient Earth are the banded iron-formations, a rock type that is of rather common occurrence in Archean and Early Proterozoic terranes. As defined by James (1954, p. 239), a banded iron-formation is a "chemical sediment, typically thin-bedded or laminated, containing 15 percent or more iron of sedimentary origin, commonly but not necessarily containing layers of chert."

Composition and Mineralogy. The best preserved, least altered (not secondarily oxidized), and least metamorphosed (i.e., of late diagenetic or, at most, very low grade metamorphic facies) Precambrian banded iron-formations now known, be they Archean or Early Proterozoic in age, are all very similar in both their bulk compositions and their mineral assemblages (Gole and Klein, 1981). Major components (those greater than 1 percent by weight from analyses recalculated on a water- and carbon dioxide-free basis) are SiO_2, FeO, Fe_2O_3, MgO, and CaO. MnO may range from zero to a maximum of over 10 percent by weight (in manganiferous iron-formations); Al_2O_3 ranges from zero to a maximum of about 3.9 percent by weight (Klein, 1978); Na_2O is generally less than 1 percent by weight (but may be as much as several weight percent in Na-rich mineral assemblages); and K_2O, TiO_2, and P_2O_5 generally exhibit concentrations well below 0.5 percent by weight.

An important parameter in understanding the significance of these rocks relative to the evolution of the Earth's environment is their $Fe^{3+}/(Fe^{2+} + Fe^{3+})$ ratio. Values of this parameter as compiled by Gole and Klein (1981) for Archean and Proterozoic banded iron-formations (including data from several highly metamorphosed units) range from 0.31 to 0.58. These relatively low values reflect the abundance of magnetite (a ferro-ferric mineral) as well as such Fe^{2+}-minerals as the iron-carbonates and -silicates. Had some of the very Fe_2O_3-rich Proterozoic iron-formations also been included in this tabulation, the range of ratios would have shown a larger maximum.

The mineral assemblages in unaltered, relatively unmetamorphased banded iron-formations consist of various combinations of the following principal components: quartz [SiO_2], magnetite [$Fe^{2+}Fe_2^{3+}O_4$], hematite [$Fe_2^{3+}O_3$], siderite [$Fe^{2+}CO_3$], members of the dolomite-ankerite series [$CaMg(CO_3)_2$-$CaFe^{2+}(CO_3)_2$], calcite [$CaCO_3$], greenalite [$(Fe^{2+}, Mg)_6Si_4O_{10}(OH)_8$], stilpnomelane

$$[K_{0.6}(Mg, Fe^{2+}, Fe^{3+})_6Si_8Al(O, OH)_{27}.2\text{-}4H_2O],$$

minnesotaite [$(Fe^{2+}, Mg)_3Si_4O_{10}(OH)_2$], riebeckite [$Na_2Fe_3^{2+}Fe_2^{3+}(Si_8O_{22})(OH)_2$], pyrite [$FeS_2$], pyrrhotite [$Fe_{1-x}S$] and, locally, carbonaceous material (see, for example, data summarized in Chapter 5). Minor amounts of chamosite

$$[Fe_{3.6}^{2+}Al_{1.6}(Mg, Fe^{3+})_{0.8}(Si_{2.6}Al_{1.4})O_{10}(OH)_8],$$

ripidolite [an Fe^{2+}-rich chlorite], and talc [$(Mg, Fe^{2+})_6Si_8O_{20}(OH)_4$] also occur. Of the above listed minerals, ferric iron (Fe^{3+}) is an essential constituent only of hematite, magnetite, riebeckite, and some stilpnomelanes; all other iron-containing minerals of such rocks are rich in ferrous iron (Fe^{2+}). Most of the above minerals occur in "microbands" (millimeter-thick alternating layers of quartz and iron-rich minerals), in fine laminations, or in oolitic or granular textures.

All late diagenetic to very low grade metamorphic (i.e., greenschist) facies banded iron-formations that have been described from the Archean and Early Proterozoic (viz., spanning a time span from about 2.7 to 1.8 Ga ago) exhibit the bulk compositions and mineral assemblages described above. Younger iron-bearing units (often referred to as "ironstones" if of Phanerozoic age; Brandt et al. 1972; James, 1966), which commonly have oolitic or pisolitic textures, generally consist of chamosite (rarely greenalite) and ferric oxides and hydroxides with or without kaolinite, carbonate, magnetite, and/or detrital quartz (e.g., Cochrane and Edwards, 1960; Edwards, 1958; O'Rourke, 1961; Schweigart, 1965; James, 1966). Thus, in bulk composition these younger rocks tend to differ distinctly from most Precambrian banded iron-formations: they are considerably more Al_2O_3- and P_2O_5-rich (Lepp and Goldich, 1964) as well as more Fe_2O_3-rich (Govett, 1966) than the Precambrian units. Thus, and although it has been suggested that certain Phanerozoic ironstones may be analogous to Precambrian iron-formations (O'Rourke, 1961), unequivocal banded iron-formations appear to be unique to the Precambrian. Moreover, it now appears that in terms of mineralogy, bulk chemistry, textures, and lithological associations, the Late Proterozoic (viz., $\sim$0.8 Ga; Young et al., 1979) iron-formations described from the Rapitan Group of the Northwest Territories, Canada (Young, 1976), differ from those of the earlier Precambrian (Gross, 1973). Thus, banded iron-formations appear to be restricted in occurrence to terranes of Archean or Early Proterozoic age.

Distribution in Time. The temporal distribution of such banded iron-formations has been suggested by Cloud (1972, 1973a) to be closely tied to the evolution of oxygen-producing, photosynthetic microbes. In 1973, on the basis of then available geochronologic data, Goldich (1973) concluded that the bulk of Precambrian iron-formations was deposited between 2.2 and 1.8 Ga ago; this apparently time-restricted, "major episode" of banded iron-formation deposition has played an important role in models of Earth history. However, more recent data suggest that formations of the Hamersley Basin, perhaps the largest repository of iron-rich rocks now known, are on the order of 2.5 Ga in age (Compston et al., in prep.), and that the Kuruman and Penge Iron Formations of the Transvaal Supergroup of South Africa may well be of similar age (A. F. Trendall, pers. comm. to M. R. W.; Cornell, 1978). Thus, it now appears that there is a considerable age spread among Proterozoic iron-formations, far greater than was thought to be the case a decade ago. If this assessment is combined with the observation that Archean iron-formations are considerably more abundant than was previously imagined (see Table 11-1 for a tabulation of Precambrian iron-formations as a function of geologic age; also see Gole and Klein, 1981), the earlier accepted picture

TABLE 11-1. Geochronologic tabulation of Precambrian banded iron-formations, and the geologic units in which they occur. Only units for which ages (or bracketed ages) are available in the literature are tabulated. For a considerable number of entries the age bracketing is rather wide, and it is impossible, based on data available, to estimate the real age within such brackets. The entries have been arbitrarily ordered on the basis of the oldest age reported. Metamorphic ages, as available, are noted in a separate column. The age of individual banded iron-formations in Archean terranes is commonly not known; in such cases, the general age of the relevant Archean shield is here listed, but this form of entry does not imply that the iron-formations within a particular shield are necessarily coeval.

Approximate age of deposition (Ma)	*Approximate age of metamorphic alteration (Ma)*	*Geologic unit and location*	*References*
3770 ± 42	2900–2800; ~2000	Isua Supracrustal Belt, Greenland	Moorbath et al., 1973; Bridgwater et al., 1974
3600–3300	?3300	Sebakwian Group, Rhodesia	Bliss & Stidolph, 1969
3500–3100	2800	Konsky Series (I), Ukraine, U.S.S.R.	Semenenko et al., 1972; Semenenko, 1973
?3400–3000	2150	Imataca Complex, Venezuela	Hurley et al., 1968; Gaudette et al., 1972
3400–3000		Swaziland Supergroup, South Africa	Eriksson, 1980b
3300–3000	2750; 1600	Tobacco Root and Ruby Ranges, Montana, U.S.A.	James & Hedge, 1980
3200–2750	2700	Beartooth Range, Montana, U.S.A.	Nunes & Tilton, 1971; Gast et al., 1958
3200–2700	2500–2000; 1140–900	Iron Ore Group, Singhbhum Region, India	Sarkar et al., 1969
3100		Pilbara Block, Western Australia	Arriens, 1971
3100–2750		Olecmian Series, Siberia, U.S.S.R.	Vorona et al., 1973
3100–2700	2700; 2100; 500	Liberian Shield, Liberia, Sierra Leone	Hurley et al., 1971; Dodge et al. 1978
3060–2870		Pongola Supergroup, Natal	Davies et al., 1970 Hunter, 1974
3000–2800		Archean Greenstone Belt, Finland	Gáal et al., 1978
3000–2500		Subganian Series, Siberia, U.S.S.R.	Vorona et al., 1973
2900–2700	2800	Konsky Series (II), Ukraine, U.S.S.R.	Semenenko et al., 1972 Semenenko, 1973;
2800–2100	1800–1700	Kola and Tundra Series, Kola Peninsula, U.S.S.R.	Salop, 1968; Gerling et al., 1968
2800–1900		Muya Series, Baikal Region, U.S.S.R.	Salop, 1968
2750		Deoss-Leglierian Series, Siberia, U.S.S.R.	Vorona et al., 1973
2750–2700		Vermilion District (Soudan Formation), Minnesota, U.S.A.	Sims & Morey, 1972
2750–2600		Archean Greenstone Belts (including e.g., Michipicoten), Canada	Goodwin, 1977; Peterman, 1979
>2700	1300	Nova Lima Group, Brazil	Aldrich et al., 1964
2700	1800–1700	Tiris Group, Mauritania	Bronner & Chauvel, 1979
2700–2600		Yilgarn Block, Australia	Arriens, 1971
2700–2300		Witwatersrand Supergroup, South Africa	Hunter, 1974; McElhinney, 1966; Allsopp, 1964
2700–2000		Bazavluksky Zone, Ukraine, U.S.S.R.	Semenenko et al., 1972
	1300	Itabira Group, Minas Gerais, Brazil	Dorr, 1969, 1973
		Transvaal Supergroup, South Africa	Cornell, 1978; Davies et al., 1970; Allsopp, 1964
2520–1800	1800	Homestake Formation, South Dakota, U.S.A.	Zartman & Stern, 1967; Bayley & James, 1973

TABLE 11-1. (*continued*)

Approximate age of deposition (Ma)	*Approximate age of metamorphic alteration (Ma)*	*Geologic unit and location*	*References*
2500		Hamersley Group, Western Australia	Compston et al. (1981); Trendall, 1973
2500–1900		Kursk Magnetic Anomaly, U.S.S.R.	Salop, 1968
2500–1600		Ketilidian Supracrustal Belt, Greenland	Neuman-Redlin & Zitzman, 1976
2345		Dharwar Group, Karnataka, India	Srinavasan & Sreenivas, 1976
2300		Karelian Formation, Finland	Kouvo & Tilton, 1966
2300–1750		Shushong Group, Botswana	Key, 1977
2200		Nimba Range, Liberia	Hurley et al., 1971
2200–1900	1850	Lake Superior Region, U.S.A.	Goldich, 1973
2200–1700	1800–1700	Ijil Group, Mauritania	Bronner & Chauvel, 1979
2180–2020	1490; 1175	Paramaca System, Guyana	Spooner et al., 1971
2000–1800		Krivoy Rog Series, Ukraine, U.S.S.R.	Semenenko et al., 1972
>2000		Serra Dos Carajás, Brazil	Dorr, 1973
1900–1600; 2100–1900; 2600–1900		Karsakpay Series, Kazakhstan, U.S.S.R.	Salop, 1968; Novokhatsky, 1973
1900–1800		Kipalu Iron Formation, Belcher Islands, Canada	Fryer, 1972
1870		Labrador Trough (and Extensions), Canada	Fryer, 1972
1865–1735		El Graara System, Morocco	Charlot, 1976
1820–1790		Kimberley Group, Western Australia	Compston & Arriens, 1968
1820–1770		Yavapai Series, Arizona, U.S.A.	Anderson et al., 1971
>1780	1550	Middleback Group, South Australia	Compston & Arriens, 1968
1750–1500		McMinn and Mullera Formations, Northern Territory and Queensland, Australia	Compston & Arriens, 1968; Trendall, 1973
1635–1600		Porphyry Group, Kiruna, Sweden	Frietsch, 1973, 1976
~1600		Nabberu Basin, Western Australia	Gee, 1979
850–750		Braemar and Holowilena Iron Formations, South Australia	Compston & Arriens, 1968; Trendall, 1973
~800		Rapitan Group, Northwest Territories, Canada	Young et al., 1979; Crittenden et al., 1972
?800–?350		Maly Khingan Series, U.S.S.R.	Egorov & Timofeieva, 1973
<800		Urucum-Mutun Region, Brazil-Bolivia	Dorr, 1973

of the temporal distribution of banded iron-formations is substantially modified. As is summarized in Figure 11-9, currently available data suggest that iron-formation deposition did not reach a unique maximum around 2.0 Ga ago, but that this "Major episode" took place over a large span of the Late Archean-Early Proterozoic. The oldest known banded iron-formation (3.8 Ga old; Appel, 1980) occurs in the Isua Supracrustal Group of West Greenland. The occurrence of chemically and mineralogically very similar banded iron-formations from 3.8 to 1.7 Ga ago presumably reflects the occurrence of very similar chemical sedimentary conditions in iron-formation basins throughout this very large span of time.

Interpretation. Most students of Precambrian banded iron-formations would consider the mineral assemblages and textures observed in relatively unmetamorphosed units to represent diagenetic

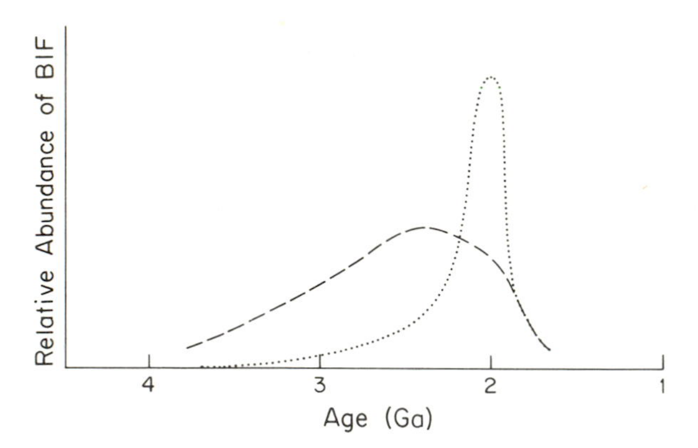

FIGURE 11-9. Highly schematic diagram showing the relative abundance of Precambrian banded iron-formations against time. The tall curve represents banded iron-formation based on the concept that the major basins containing Proterozoic iron-formations may have been approximately coeval. The broad, lower curve takes account of (*i*) recent isotopic age dates that indicate that the major iron-formation basins were not coeval, and (*ii*) the number and extent of banded iron-formations in the Archean (after Gole and Klein, 1981).

products of earlier sedimentary materials (e.g., Gruner, 1946; James, 1954; Trendall and Blockley, 1970; Eugster and Chou, 1973; Mel'nik, 1973; Klein, 1974; Klein and Bricker, 1977; and many others), although some would interpret such features as having resulted from replacement of earlier aragonitic deposits (Dimroth, 1977; Kimberley, 1979). The enormous amounts of iron that entered such depositional basins—almost certainly in the form of soluble Fe^{2+} that was distributed within essentially anoxic waters over huge basinal areas (e.g., 100,000 km^2 or more; Trendall and Blockley, 1970), a set of circumstances that could not have occurred were the iron to have been in the form of insoluble ferric compounds—may have originated from weathering, from volcanic activity, from upwelling of deep ocean waters, or from a combination of such mechanisms. And there is general agreement that the mechanism involved in the deposition of most, and perhaps all such banded iron-formations was reaction of the ferrous iron with low concentrations of ambient free oxygen, presumably in the upper reaches of the water column, and the resulting formation of flocculent, insoluble, ferric and ferro-ferric compounds that became deposited in fine layers (possibly reflecting a seasonality to the process) at the sediment-water interface. The literature on these subjects is very extensive (including, for example, Trendall and Blockley, 1970; Mel'nik, 1973; Holland, 1973; Drever, 1974; Ewers, in press; and many others).

According to the preferred model outlined above, there can be little doubt that banded iron-formations represent very large "oxygen sinks," but the question as to what partial pressure of free oxygen might have prevailed in the atmosphere during their deposition is complex. Values of pO_2 in equilibrium with the stability fields of mineral assemblages on Eh-pH diagrams at 25°C and 1 atmosphere total pressure can be calculated (Garrels and Christ, 1965). However, in the case of the mineral assemblages now present in essentially unmetamorphosed (late diagenetic) iron-formations, the question arises as to what the precursors of the now crystalline minerals may have been. In answer to this question, it is very probable that $Fe(OH)_3$ was a precursor to hematite (Fe_2O_3). The precursor to magnetite (Fe_3O_4) is less well-defined; it may have been a mixture of $Fe(OH)_2$ and $Fe(OH)_3$, a hydromagnetite ($Fe_3O_4 \cdot nH_2O$), or perhaps even a very magnetite-like material. The choice of precursor material will alter, often significantly, its Eh-pH stability field (see Frost, 1978; Klein and Bricker, 1978). Figure 11-10A and B shows Eh-pH diagrams for several of the common mineral assemblages in banded iron-formations, at 25°C and 1 atmosphere pressure (from Klein and Bricker, 1977); superimposed on the diagrams are calculated values of pO_2 (see Garrels and Christ, 1965). Klein and

FIGURE 11-10. Eh-pH diagrams at 25°C and one atmosphere total pressure, depicting stability fields and relations for minerals (or their precursors) that are common in banded iron-formations. Boundaries between aqueous species and solids at $A_{[aqueous\ species]} = 10^{-6}$. Contours show partial pressure of oxygen. The fields of hematite and $Fe(OH_3)$ have been highlighted to show the variability of the Eh-pH conditions of these phases. (A) Minerals or their precursors in the system Fe—H_2O—O_2—SiO_2—dissolved carbon; a stilpnomelane-like composition with $Fe^{2+}/Fe^{3+} = 2:1$ is shown (after Klein and Bricker, 1977). (B) Minerals or their precursors in the system Fe—H_2O—O_2—S; two activities of total dissolved sulfur are shown; a stilpnomelane-like composition is not considered (after Klein and Bricker, 1977). (C) Stability fields of hematite and magnetite in water (after Garrels and Christ, 1965). (D) Stability fields of ferrous and ferric hydroxides in water (after Garrels and Christ, 1965).

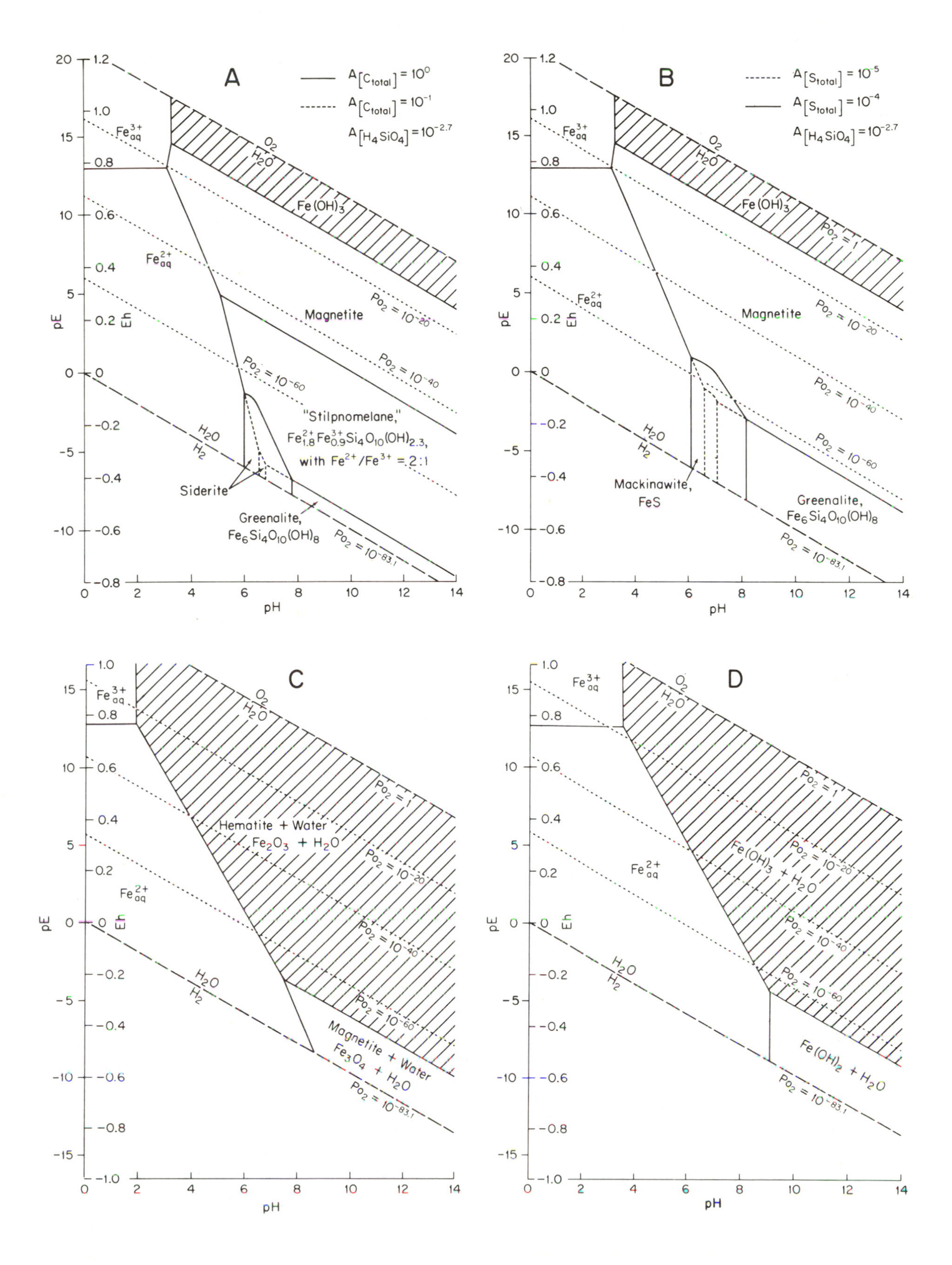
A
B
C
D
$A[C_{total}] = 10^0$
$A[C_{total}] = 10^{-1}$
$A[H_4SiO_4] = 10^{-2.7}$
$A[S_{total}] = 10^{-5}$
$A[S_{total}] = 10^{-4}$
Fe^{3+}_{aq}
Fe^{2+}_{aq}
O_2
H_2O
$Fe(OH)_3$
Magnetite
$P_{O_2} = 1$
$P_{O_2} = 10^{-20}$
$P_{O_2} = 10^{-40}$
$P_{O_2} = 10^{-60}$
$P_{O_2} = 10^{-83.1}$
"Stilpnomelane,"
$Fe^{2+}_{1.8}Fe^{3+}_{0.9}Si_4O_{10}(OH)_{2.3}$,
with $Fe^{2+}/Fe^{3+} = 2:1$
Siderite
Greenalite,
$Fe_6Si_4O_{10}(OH)_8$
Mackinawite,
FeS
Hematite + Water
$Fe_2O_3 + H_2O$
Magnetite + Water
$Fe_3O_4 + H_2O$
$Fe(OH)_3 + H_2O$
$Fe(OH)_2 + H_2O$
pE
Eh
pH

Bricker (1977) chose Fe_3O_4 to represent the magnetite field, but $Fe(OH)_3$ as the precursor to hematite. This choice of oxides allows for the Eh-pH delineation of very common associations such as magnetite-siderite, magnetite-greenalite and magnetite-stilpnomelane (but makes impossible the coexistence of hematite and greenalite, for example, a mineral pair that in fact is extremely rare, or absent from most banded iron-formations). In Figure 11-10A and C are shown the variations in stability fields of crystalline hematite, magnetite, $Fe(OH)_3$ and $Fe(OH)_2$. As is illustrated in Figure 11-10A and B, the hematite precursor field (i.e., that of $Fe(OH)_3$) is very small and exists only at relatively high pO_2 values, but in Figure 11-10C and D the hematite and $Fe(OH)_3$ fields are much larger and extend to very low pO_2 values. It is clear from such calculated equilibrium diagrams that hematite (or its precursor) can be formed over a large range of pO_2 values. The magnetite field is stable only at substantially lower values of pO_2 and the sulfide-rich and carbonate-silicate-rich mineral assemblages observed commonly in banded iron-formations reflect even lower pO_2 values. In short, the data indicate that under equilibrium conditions a major part of iron-formation mineral assemblages reflects anaerobic, essentially anoxic conditions of deposition. It is important to note, however, that such equilibrium diagrams must be applied with caution to an evaluation of environmental conditions during the Precambrian, since it is common to find that the first formed phases of diagenetic mineral assemblages are metastable species, and that it is only with the passage of time that these assemblages adjust through dehydration and crystallization to yield thermodynamically more stable systems.

In light of the theoretical analysis presented above (Section 11-5-1) and the foregoing remarks regarding banded iron-formations, it seems clear that these Archean and Early Proterozoic rock units were deposited at a time when the hydrosphere was predominantly reducing. Evidence of relatively shallow water deposition of some of these iron-formations suggests that pO_2 might have been low in the mixed layer as well as in the deep sea, but this condition may not have existed everywhere and all the time.

Reducing waters on the early Earth are not unexpected. The difficulty with the iron-formations is that they represent a very substantial sink of oxygen. For the rates of accumulation estimated by Trendall and Blockley (1970), Walker (1978a) has shown that the abiotic oxygen source represented by photolysis of water vapor (followed by escape of hydrogen to space) was not large enough to have provided the oxygen incorporated in the ~2.5-Ga-old iron-formations of the Hamersley Basin in Western Australia. The question, as posed by Cloud (1973a, 1974) and reiterated by others, is whether such units always require the presence of oxygen-producing photosynthetic organisms. Because a banded iron-formation is present in the 3.8-Ga-old Isua supracrustals, a positive answer to this question would imply the origin and substantial metabolic evolution of life before the beginning of the known geological record (Walker, 1980b). However, there are at least two reasons why such a conclusion would be premature. First, Cairns-Smith (1978) has raised the possibility that the oxygen in iron-formations might have been extracted from sea water by purely photochemical reactions within the water (Walker, 1980c). Until this possibility has been adequately dealt with, the possible relationship between photosynthesis and iron-formations must remain obscure. And second, there is an absence of information on the rate of accumulation of most iron-formations. It is entirely possible that the 2.5-Ga-old iron-formations in the Hamersley Basin were deposited with photosynthetic oxygen, while the older units, including that at Isua, were deposited much more slowly using oxygen produced abiotically in the atmosphere. Thus we must accept, for the time being, that the mere presence of banded iron-formation does not necessarily require the existence of oxygenic photosynthesizers (nor, apparently, the necessary occurrence of any type of biological activity).

The information represented by the temporal distribution of banded iron-formations is nonetheless striking. Figure 11-11 is derived from Table 11-1 and represents the number of entries in the table per unit of time, with due allowance for uncertainties in the geologic ages of most of the entries (but with no attempt to distinguish between entries that represent single and multiple geologic units, or between those that represent large volumes or only minor occurrences of iron). What is potentially important about this distribution is the sudden drop in the rate of occurrence of reported banded iron-formations about 1.7 Ga ago. If allowance could be made for the smoothing of the distribution that results from substantial uncertainty in ages, it is likely that the drop would be

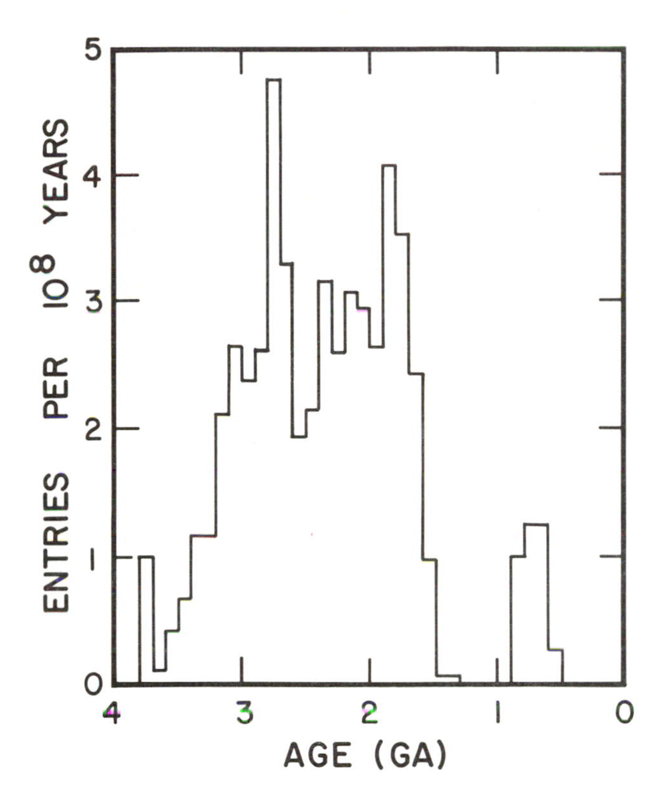

FIGURE 11-11. Distribution of named Precambrian iron-formations as a function of time. The histogram shows the number of entries in Table 11-1 per unit of time; the contribution of each entry is distributed over the age range indicated in the table.

even more precipitous. In spite of all the uncertainties, there are no iron-formations attributed to the period between 1.3 and 0.9 Ga ago (and very few from the Late Proterozoic, which, as noted above, are different in mineralogy, chemistry, and probably also in mode of formation, from those of the earlier Precambrian). It seems clear that production of banded iron-formations stopped relatively suddenly, about 1.7 Ga ago. As proposed by Cloud (1972, 1973a), this may have been the time when the photosynthetic oxygen source finally exceeded the rate of supply of reduced matter to atmosphere and oceans combined, the oceans were swept free of reduced compounds, and a stable aerobic environment became established.

RED BEDS

Red beds have been defined by van Houten (1973) as being detrital sedimentary rocks with reddish-brown ferric oxide pigments coating mineral grains, filling pores, or dispersed in a clay matrix. They generally form in fluvial or alluvial environments on land. The reddening process under desert conditions has been studied in detail by Folk (1976). In general, in these and other red beds, the production of ferric oxides (if of primary origin) is regarded as resulting from subaerial oxidation; their presence has thus been interpreted as requiring the occurrence of free oxygen in the overlying atmosphere. The distinctive coloration of such units, however, commonly results from the presence of only small amounts of iron oxides (usually one percent by weight or less); they thus differ in this respect from the far more iron-rich banded iron-formations (and, if molecular oxygen has served as the oxidant, their formation thus requires a correspondingly smaller oxygen flux). In many red beds it is difficult to assess whether the red oxide pigmentation is syngenetic or of post-depositional origin (van Houten, 1973). As such, red beds appear to be even more difficult to interpret as possible indicators of original environmental conditions than are banded iron-formations.

Despite these problems, it may be significant that red beds are rare or unknown in the Archean (Figure 11-12). A possible example is the Archean Rouyn-Noranda formation described by Dimroth and Kimberley (1976), but earlier accounts of red beds in the approximately 3.1-Ga-old Moodies Group of South Africa appear to be definitely in error (Eriksson, 1980b). Moreover, there can be no doubt that red beds are far more abundant in the Proterozoic than they are in older deposits, and that they are generally absent from the Archean despite the presence of alluvial and fluvial rock sequences of the type that are commonly red in younger terranes. The transition from drab colored rocks toward the base of the approximately 2.3-Ga-old Huronian Supergroup of Canada to red beds higher in that sequence has been interpreted as indicating the time of oxygen accession to the atmosphere (Roscoe, 1969, 1973). Many of these drab colored units are only mildly metamorphosed (viz., prehnite-pumpellyite to lower greenschist facies), so that red coloration due to hematite, had it been originally present, would be expected to have been preserved.

It is not unreasonable to suppose that Archean red beds are rare or absent because the Archean atmosphere was oxygen deficient, in the sense that hematitic deposits did not readily form as a result of subaerial weathering. According to the analysis

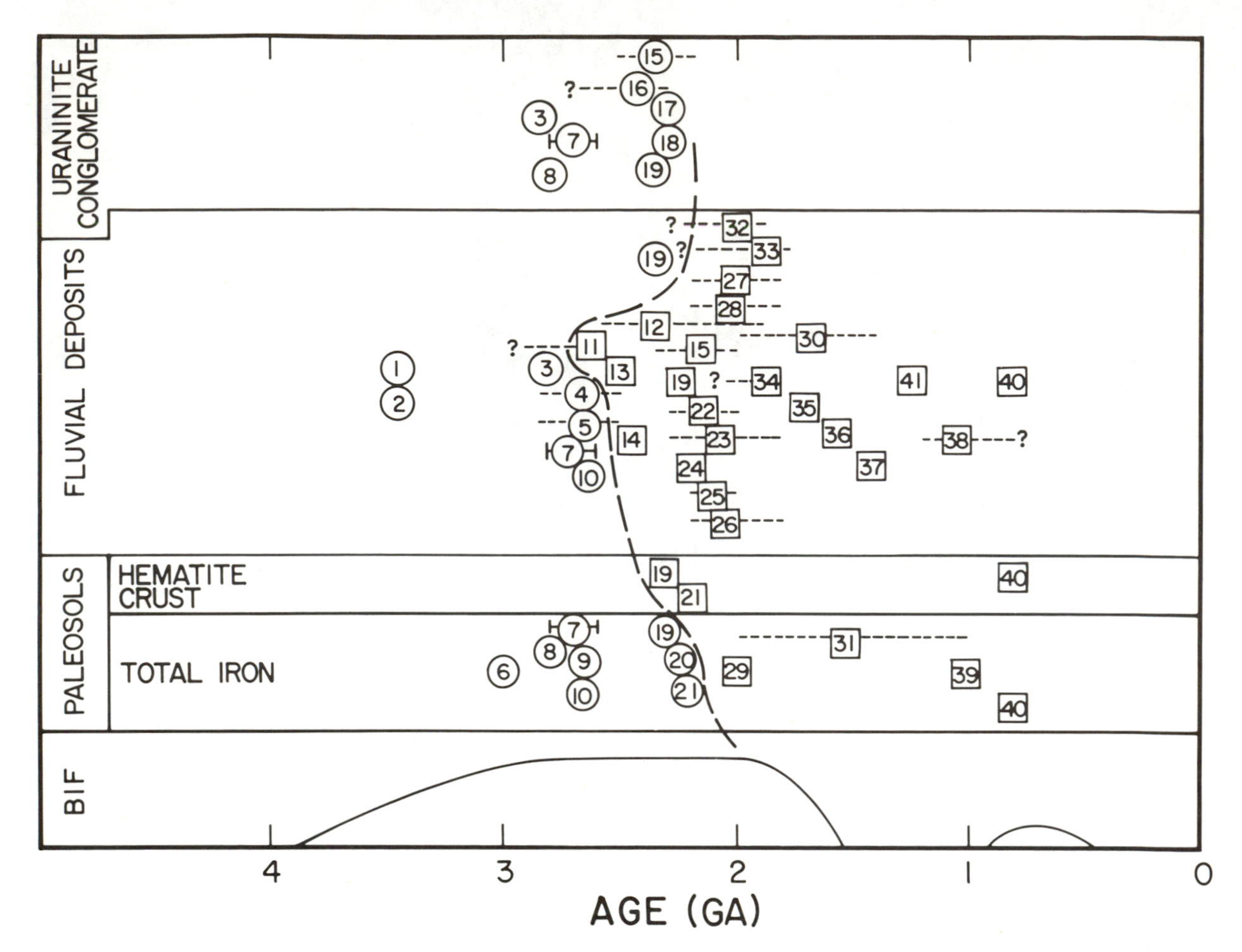

FIGURE 11-12. Sedimentary indicators of the reducing (circle) and oxidizing (square) state of the Precambrian atmosphere and hydrosphere, based in part on the tabulation by Knoll (1979). Error bars, horizontal dashed lines, and question marks indicate uncertainties in ages, but many of these formations are poorly dated and the uncertainties in age may be considerably larger. The heavy dashed line separates circles from squares as an aid to the eye. For *uraninite-bearing conglomerates*, only circles are shown; virtually all sedimentologically comparable facies of younger age are devoid of detrital uraninite. For *fluvial deposits*, squares denote red beds; circles denote deposits sedimentologically equivalent to red beds but lacking their characteristic ferric oxide pigmentation. For *paleosols*, circles denote formations showing evidence of leaching of total iron (and manganese) in the weathering profile; squares denote paleosols not clearly deficient in iron (hematite crust refers to an enhanced ratio of ferric to ferrous iron at the top of the soil profile). The distribution of *banded iron-formations* (BIF) through time is a qualitative estimate of volumes. (1) Gorge Creek Group, Australia (Ericksson, 1980a). (2) Moodies Group, South Africa (Eriksson, 1979, 1980b). (3) Fortescue Group, Australia (Button, 1976; Goode, in press). (4) Late Archean, Canada (Turner and Walker, 1973). (5) Late Archean, Western Australia (Donaldson and Platt, 1975; Marston and Travis, 1976). (6) Pongola Supergroup, South Africa (Button and Tyler, 1979). (7) Witwatersrand Supergroup, South Africa (Pretorius, 1976; Button and Tyler, 1979). (8) Dominion Reef Group, South Africa (Button and Tyler, 1979; Ramdohr, 1958). (9) Belingwe Greenstone Belt, Zimbabwe (Button and Tyler, 1979). (10) Ventersdorp Supergroup, South Africa (Button and Tyler, 1979; Buck, 1980). (11) Archean, Rouyn-Noranda, Canada (Dimroth and Kimberley, 1976). (12) Udokan and Muya Groups, U.S.S.R. (Salop, 1977). (13) Muruwa Formation, Guyana (Cannon, 1965). (14) Burzyan Group, U.S.S.R. (Salop, 1977). (15) Karelian Supergroup, Finland and U.S.S.R. (Salop, 1977). (16) Serra de Jacobina Group, Brazil (Cox, 1967). (17) Montgomery Lake Beds, Canada (Bell, 1970). (18) Chibougamau and Otish Mountain Groups, Canada (Young, 1973b). (19) Huronian Supergroup, Canada (Roscoe, 1969, 1973; Frarey and Roscoe, 1970; Gay and Grandstaff, 1980). (20) Transvaal Supergroup, South Africa (Button and Tyler, 1979). (21) Pretoria Group, South Africa (Button, 1979). (22) Lomagundi Group, Zimbabwe (Swift, 1961). (23) Hurwitz Group, Canada (Bell, 1970). (24) Waterberg Group, Botswana (Haughton, 1969; Crockett and Jones, 1979; Eriksson and Vos, 1979). (25) Charadokan to Kebetka Formations, U.S.S.R. (Salop, 1977). (26) Mistassini Group, Canada (Chown and Caty, 1973). (27) Coronation Geosyncline, Canada (Hoffman et al., 1970). (28) Pine Creek Geosyncline, Australia (Walpole et al., 1968). (29) Bothnian Schists, Finland (Rankama, 1955). (30) Dharwar Group, India (Gowda and Sreenivasa, 1969). (31) Grand Canyon Series, U.S.A. (Sharp, 1940). (32) Labrador Trough, Canada (Dimroth et al., 1970; Dimroth and Kimberley, 1976; Fryer, 1972). (33) Belcher Group, Canada (Bell and Jackson, 1974). (34) Kimberley Group, Australia (Canavan and Edwards, 1938). (35) Roraima Formation, Guyana (Cannon, 1965; Grabert, 1969). (36) Siberian Shield, U.S.S.R. (Davidson, 1965). (37) Ukrainian Shield, U.S.S.R. (Davidson, 1965). (38) Vindhyan Supergroup, India (Akhtar, 1976). (39) Keweenawan, Canada (Kalliokoski, 1975). (40) Torridon Group, Scotland (Williams, 1968; Stewart, 1975). (41) Balkan Shield, Sweden, and Finland (Davidson, 1965).

of Kasting and Walker (1981), a reducing atmosphere in this sense would have resulted from a rate of supply of reduced gases to the atmosphere that exceeded the rate of supply of free oxygen, precisely the conditions that would be expected prior to the time when the rate of photosynthetic production of oxygen rose above the rate of supply of reduced materials to the mixed layer of the ocean (assuming that such oxygen production was absent or minimal on land). It is quite possible, however, that the volcanic source of reduced gases fell to low values upon occasion, even before the origin of life. Under these conditions, the lowest levels of the atmosphere would have become oxidizing because of the abundant presence of oxidizing free radicals. As is apparently the case on Mars today, iron-bearing minerals exposed to such an atmosphere could have acquired the thin coating of hematite that characterizes red beds in spite of a very low oxygen partial pressure. The distinction between an oxidizing atmosphere, which may be essentially devoid of free oxygen, and an aerobic atmosphere, one rich enough in oxygen to support aerobic metabolism (about 0.2 percent), is important. An atmosphere under which red beds can form need not yield a large enough flux of oxygen to have produced banded iron-formations; and neither red beds nor banded iron-formations need necessarily reflect an aerobic environment.

The geologic record of red beds suggests that oxidizing atmospheres arose infrequently if at all during the Archean, but became increasingly common in the middle to later portion of the Early Proterozoic.

PALEOSOLS

Fossil soil profiles should be a fruitful source of information on the redox state of ancient atmospheres. Such paleosols appear to be reasonably abundant in the geological record, even in the Precambrian, but have not yet received extensive study.

A number of Archean and Early Proterozoic paleosols (see Figure 11-12) exhibit a deficiency of iron near the top of the profile that can be attributed to leaching. In modern, aerobic soils iron is concentrated upwards in the soil profile due to the insolubility of ferric oxides. The apparently quite different profile of the Archean paleosols is consistent with weathering under anaerobic conditions. Unfortunately, profiles of total iron concentration have not been measured in enough younger, Proterozoic, paleosols to provide any firm indication of when the downward leaching of iron came to an end. Moreover, the interpretation of fossil soils is complicated commonly by various secondary effects (post-depositional leaching by ground water, metamorphism, etc.), and although studies of such units hold great promise for elucidation of past environmental conditions, available data are too few to permit firm conclusions to be drawn regarding the history of atmospheric oxygen.

Nevertheless, it is possible to imagine an oxidizing atmosphere that would have delivered enough oxidizing free radicals to the surface to have caused a surficial enrichment of hematite, but not enough oxygen to have immobilized all of the iron in the soil as insoluble ferric compounds. Evidence of such conditions about 2.3 Ga ago may be provided by studies by Gay and Grandstaff (1980) and by Button (1979). These authors describe paleosols that are leached in total iron but exhibit a surficial enhancement in the $Fe_2^{3+}O_3/Fe^{2+}O$ ratio. Such a hematitic crust might form under conditions like those resulting in formation of red beds. A larger flux of oxidant into the soil would be required to bring an end to the downward leaching of iron.

URANINITE-BEARING CONGLOMERATES

Although long a subject of controversy (e.g., Ramdohr, 1955; Leibenberg, 1955; Davidson, 1957), the presence in several Precambrian formations (Figure 11-12) of detrital uraninite (UO_2), like that illustrated in Figure 11-13, and sulfides (mostly pyrite, FeS_2), is now well established (Roscoe, 1969; Schidlowski, 1970b, 1980b; Feather and Koen, 1975; Grandstaff, 1980).

The absence of any uraniferous (or pyritiferous) conglomerates of economic grade dating from later than the mid-Early Proterozoic is equally firmly established (Cloud, 1976a). Minor occurrences of detrital uraninite or sulfides in recent sediments are plainly mineralogical curiosities (Steacy, 1953), in no way quantitatively comparable to the large-scale deposits of the Precambrian. Although modern occurrences of detrital uraninite have also been reported at National River, British Columbia (Steacy, 1953) and from Cornwall, England (Davidson and Cosgrove, 1955), apparently the only well-documented such example occurs in the sediments of the Indus River, Pakistan, in the foothills of the Himalayas (Zeschke, 1961; Simpson and Bowles, 1977). As McLennon (in

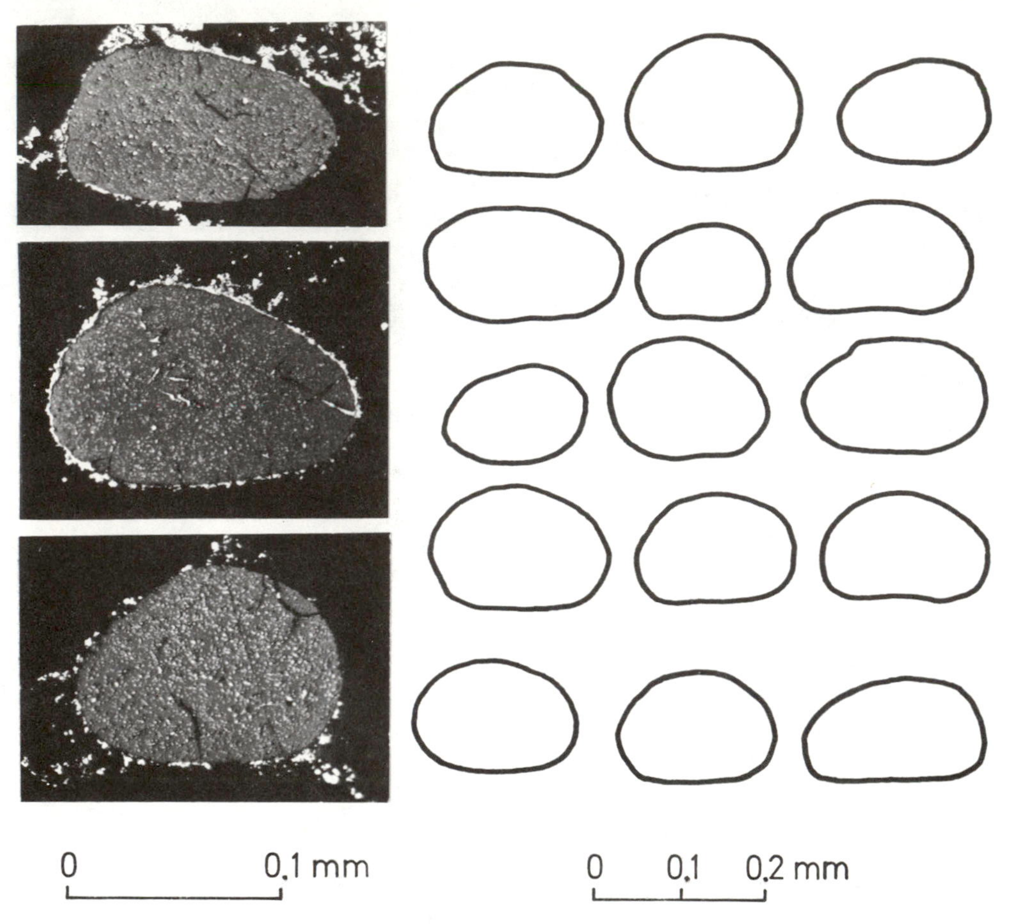

FIGURE 11-13. *Left:* Detrital grains of uraninite from the Basal Reef, Loraine Gold Mines, Ltd., Orange Free State, South Africa. The grains are dusted with myriads of tiny inclusions of galena (PbS) showing up as white specks; part of the galena has been "sweated out" and forms delicate PbS-aureoles on the grain surface. *Right:* Outlines of detrital monazite grains from a West Australian beach placer. Note the resemblance in shape between abrasion forms of uraninite on the left and these detrital "muffin-shaped" monazite grains.

press) has pointed out with regard to this occurrence, however, these uraniferous sediments are derived from a cold, tectonically active area undergoing rapid erosion and deposition with minimal chemical weathering. At least some of the uraniferous conglomerates of the Archean and Early Proterozoic, in contrast, formed in stable areas of low relief.

The occurrence of these Precambrian deposits is relevant to redox conditions at the surface of the Earth because both uraninite and pyrite are unstable under oxidizing conditions, the uranium passing readily from the insoluble tetravalent form ($U^{4+}O_2$) to the hexavalent form (U^{6+}) in which it enters solution as the highly soluble uranyl complex $[UO_2]^{2+}$, and the pyrite being readily oxidized to sulfate and oxides of trivalent iron (Fe^{3+}). The formation of uraniferous conglomerates of economic grade on the Archean and Early Proterozoic Earth indicates that weathering did not lead to the total oxidation and dissolution of the uranium. Calculations by Grandstaff (1976, 1980), based on the dissolution kinetics of uraninite, indicate that oxygen partial pressures must have been below 10^{-2} to 10^{-6} times the present atmospheric level for the survival of detrital uraninite under environmental and physiographic conditions like those that apparently prevailed during deposition of the uraniferous conglomerates of the Witwatersrand. This conclusion seems consistent with the data from red beds discussed above, since a concentration of oxidizing free radicals sufficient to generate (during the course of weathering, erosion, and deposition) a micron-thick coating of hematite in a red bed need not be large enough to oxidize a grain of detrital uraninite with a diameter of perhaps 0.1 mm.

11-5-3. Integration

The evidence from red beds, paleosols, and detrital uraninite (Figure 11-12) is consistent with the concept that the atmosphere changed from a non-oxidizing to an oxidizing condition early in the Proterozoic. Theory indicates that this transition would have occurred when the rate of supply of oxygen to the atmosphere rose above the rate of supply of reactive reduced gases. An oxidizing atmosphere need not necessarily have been one that was rich in oxygen: according to Kasting and Walker (1981), an atmosphere with only 10^{-8} bar partial pressure of oxygen could still have yielded an environment at ground level where exposed minerals would have been oxidized by reactive free radicals.

The distribution of banded iron-formations suggests that the hydrosphere remained generally reducing until about 1.7 Ga ago. The implication is that the photosynthetic oxygen source was less than the rate of supply of reduced materials to the ocean until about that time.

The geologic record, therefore, suggests two transitions of potential biological importance. The first was when the atmosphere changed from reducing to oxidizing, but remained anaerobic. The second was when both atmosphere and ocean changed from anaerobic to aerobic. The first transition may have been triggered by a declining volcanic source of reduced gases (Walker, 1978b) with the oxygen being derived via photochemical reactions in the atmosphere or by leakage of photosynthetic oxygen from the mixed layer of the ocean. Because of its neglect of horizontal heterogeneity, the simple theoretical analysis we have applied has not dealt with the possibility of restricted areas of oxygen-rich surface waters. The second transition may have been triggered by increasing rates of photosynthesis (Knoll, 1979), by declining rates of tectonic activity, or by a combination of the two. More analysis will be needed to gain a clearer idea of why these transitions occurred and what factors determined their timing.

Consideration of the history of oxygen partial pressure has provided no new information concerning such biologically important questions as the time of the origin of life, the onset of bacterial photosynthesis, or the onset of oxygenic photosynthesis, but it does suggest constraints regarding the time of development of widespread aerobic respiration. On a global scale, aerobic respiration could not have developed prior to about 1.7 Ga ago, when the atmosphere and oceans apparently first achieved oxygen levels high enough to support it.

As noted above, oxygen-tolerance must have existed much earlier (e.g., in oxygen-producing photosynthesizers and in the microorganisms closely associated with such forms in cyanobacterial bioceonoses), and it seems quite likely that aerobiosis originated in such forms, giving rise to a diverse suite of amphiaerobes. During daylight hours, local oxygen-production would have permitted such microbes to respire aerobically; at night, however, as oxygen concentrations dropped, such forms would have reverted to more primitive anaerobic metabolism. The widespread occurrence of the amphiaerobic mode of life among modern prokaryotes (but, significantly, not among later evolving eukaryotes), and the preference of most such organisms (cyanobacteria included) for ecologic settings having low oxygen levels (so-called "microaerophily"), presumably harks back to this earlier stage in the evolution of the Earth's environment. Strict aerobes, however, could not have become widespread until a stable aerobic environment had become established, apparently no earlier than about 1.7 Ga ago. And while the biota need not have responded immediately to this development, there would be no need for organisms evolving within the aerobic zone to retain the capability of anaerobic metabolism. Thus, it seems likely that strict aerobes diversified at about this time (a development that may be reflected in the apparently first appearance of abundant, widespread fossil phytoplankton in sediments of about this age; see Chapter 13). The development of widespread aerobes was of considerable environmental importance for there is no other process that could have limited the accumulation of free oxygen once the rate of photosynthesis exceeded the rate of supply of reduced material in the atmosphere and oceans.

11-5-4. Implications

The layer of ozone (O_3) in the terrestrial atmosphere, which today shields the ground from biologically deleterious solar ultraviolet radiation, would have been absent from a primitive atmosphere lacking oxygen. The matter has been extensively studied (Berkner and Marshall, 1964; Ratner and Walker, 1972; Hesstvedt et al., 1974; Carver, 1974; Blake and Carver, 1977; Katsumori, 1979; Levine et al., 1979; Kasting and Donahue, 1980), largely via the application to paleoatmospheres (having varying oxygen contents) of modern photochemical theories regarding atmospheric ozone. The results of these studies are rather disparate, but the most recent work, that by Kasting and Donahue (1980), is probably the most reliable. They find that enough ozone to provide a biologically effective ultraviolet shield does not accumulate unless the oxygen partial pressure is at least 0.02 bar, one-tenth of the present level (and roughly ten times that required by amphiaerobes for aerobic metabolism). It seems most unlikely that such an oxygen level could have arisen before the anaerobic-aerobic transition about 1.7 Ga ago. After this transition there might, in principle, have been enough oxygen to support aerobic respiration but not enough to sustain an ozone screen, but no plausible explanation has been offered of why oxygen should not have achieved values similar to those of the present relatively soon after the disappearance of inorganic reduced species from the atmosphere and oceans.

Thus, it seems possible that solar ultraviolet radiation may have had a strong influence on biological evolution during the Archean and Early Proterozoic, but has been less of an environmental factor since then. On the other hand, experimental studies have recently shown that at least some obligately anaerobic bacteria exhibit effective mechanisms to repair ultraviolet-induced damage (viz., photoreactivation) and that such organisms have a high level of intrinsic resistance to the effects of ultraviolet irradiation relative to aerobes and amphiaerobes (Rambler and Margulis, 1980). Moreover, Margulis et al. (1976) and Rambler et al. (1977) point out that the "matting habit" of stromatolite-forming microorganisms provides a natural defense against ultraviolet radiation. Thus, it may be that the most important effect of the establishment of an effective ultraviolet shield was to make possible a planktonic mode of life (Walker, 1978b), a development that might possibly have first occurred during the mid-Early Proterozoic (see Chapters 9 and 14).

It is, of course, possible that during the Archean some gas other than ozone shielded the surface from solar ultraviolet radiation. The only likely candidate that has been suggested so far is sulfur dioxide (Sagan, 1973). Sulfur dioxide is relatively abundant in volcanic gases (White and Waring, 1963), and might also have been a product of photochemical oxidation of hydrogen sulfide (the photochemistry of the reactions involved in the primitive atmosphere has not yet been studied, but free oxygen would probably not be needed to oxidize H_2S in an atmosphere rich in free radicals like $HO\cdot$).

The photochemical fate of gaseous sulfur compounds is perhaps the most urgent question in paleoatmospheric theory. We need to know the most abundant species because of the possibility of an ultraviolet shield of SO_2, and we also need to know whether sulfur was washed out of the atmosphere as sulfide, sulfite, or sulfate.

The latter question bears on the availability of sulfur to primitive aquatic organisms. In addition to the atmosphere, sources of dissolved sulfur might have included weathering and submarine volcanism (MacGeehan and MacLean, 1980). Relative geochemical abundances suggest that dissolved ferrous iron would have been supplied to the ocean more rapidly than dissolved sulfur compounds. In this case, a titration reaction to produce insoluble ferrous sulfide (pyrite) would have held the concentration of dissolved sulfide to very low values. Dissolved sulfite could have reacted inorganically to produce either sulfate or sulfide, depending on the Eh and pH of sea water. Whether sulfate was stable would have depended also on these ambient conditions. The geologic record of sulfate-containing evaporites (Figure 11-2) suggests that sulfate was stable. Thus, early life may have evolved in an environment deficient in reduced sulfur species but rich in sulfate. This hypothesis needs to be tested against the geological record of sulfur minerals and their isotopic compositions (see Chapter 7).

Other nutrient elements that might have had quite different concentrations in the anaerobic or reducing Archean ocean include manganese (the oxidized form of which is insoluble) and molybdenum (the reduced form being insoluble). The latter element plays a key role in the enzymes involved in nitrogen fixation. Early biological nitrogen fixation may, therefore, have been inhibited, although this seems unlikely since many nitrogen-fixers of the modern biota are obligate anaerobes (the nitrogenase enzyme complex being irreversibly inactivated in the presence of even small concentrations of molecular oxygen) that have no obvious difficulty in extracting sufficient molybdenum from their immediate surroundings. In any case, the study of nutrient availability under differing conditions of environmental redox state could be a fruitful area of research (Ochiai, 1978).

Nitrogen is fixed abiotically by lightning today, and the same source would have been present, although somewhat diminished in size, in the anaerobic atmosphere (Yung and McElroy, 1979; Kasting and Walker, 1981). At the high temperatures of a lightning bolt the oxygen required to convert N_2 to NO can be extracted from H_2O or CO_2 as well as from O_2. The subsequent photochemical fate of fixed nitrogen is not yet certain, however. Yung and McElroy argued in favor of substantial rainout of nitrate (produced photochemically from lightning-generated NO) as a significant source of fixed nitrogen in solution, but Kasting and Walker find that key reactions neglected by Yung and McElroy destroy most of the fixed nitrogen in the atmosphere. Since photochemical reactions would have limited the availability of ammonia on the primitive Earth, if Kasting and Walker are correct a shortage of biologically-useable nitrogen could have limited early biotic evolution prior to the origin of biological nitrogen fixation.

The abiotic fixation of carbon in anaerobic atmospheres of nitrogen, carbon dioxide, water, and hydrogen has also been studied theoretically. Chameides and Walker (1981) find that fixation by lightning would have produced small amounts of hydrogen cyanide (HCN), while Pinto et al. (1980) indicate that photochemical reactions would be expected to serve as a larger, but still not abundant source of formaldehyde (H_2CO). More work on such systems is certainly needed, but there is no reason to believe that reactions in atmospheres of this type would ever have provided an abundant source of fixed carbon (for further discussion, see Chapter 4).

11-6. Summary and Conclusions

(1) Geological evidence of environmental conditions on the Archean-Early Proterozoic Earth is commonly indirect and incomplete. Such evidence must therefore be interpreted and extrapolated with the help of theory, but the theory that can be invoked is generally either a gross approximation of the real world, or highly speculative, or both. Thus, it is not yet possible to reach many firm conclusions regarding environmental evolution during the Archean and Early Proterozoic.

(2) However, within the limitations of the evidence available, there is no indication of profound changes in most biologically important parameters since the beginning of the known rock record some 3.8 Ga ago; habitable environments, in many respects like those of the present, appear to have existed since the Early Archean.

(3) In particular, since the beginning of the known rock record: liquid water oceans and an atmosphere containing carbon dioxide and water vapor (but presumably composed chiefly of nitrogen; Walker, 1977) have been extant; the chemistry of sea water has remained relatively unchanged; shallow water, open ocean and land surface environments have existed; day lengths have been shorter (to a minimum of about 15 hours) and average tidal amplitudes have been larger (to perhaps 1.5 times greater than present values), but not markedly so; and the average global climate has apparently been neither decidedly hotter, nor appreciably colder, than that of more recent Earth history.

(4) Against this broadly uniformitarian background, however, a number of parameters of biologic significance appear likely to have changed during Archean-Early Proterozoic time, of most interest being (*i*) the concentration of atmospheric (and dissolved) carbon dioxide; (*ii*) the redox state of the atmosphere and ocean (and the related concentration of molecular oxygen); and (*iii*) factors that may have served to limit the distribution of early life such as the availability to organisms of mineral nutrients and the ambient ultraviolet flux at the Earth's surface.

(5) Theoretical interpretation of preserved climatic indicators, combined with the model-dependent prediction of an early cool Sun, suggest that carbon dioxide—a minor atmospheric gas at present—may have been 100 times more abundant in the Archean. This suggestion, while apparently not inconsistent with relevant geological, geochemical, and biological data, is as yet only speculative. Confirmation or refutation of the theory will require improved understanding of the global geochemical cycles of carbon dioxide and the processes that control them and, especially, careful analysis of the geochemistry of the early ocean as based particularly on studies of the chemistry and

mineralogy of the chemical sediments deposited therein.

(6) The geologic distribution and mineralogic composition of banded iron-formations, together with an understanding of the conditions required for their deposition, suggest strongly that the Earth's hydrosphere was predominantly reducing prior to about 1.7 Ga ago. Sedimentological evidence of subaerial weathering processes (viz., from red beds, paleosols, and detrital uraninite-containing conglomerates) suggests that the atmosphere changed from a predominantly reducing to a more oxidizing condition somewhat earlier in the Early Proterozoic.

(7) Simplified theoretical considerations of reservoirs, fluxes, and titration reactions suggest how the atmospheric transition could have preceded that in the ocean. As a result of interaction between solar ultraviolet radiation and water vapor in the troposphere, the anaerobic atmosphere exhibited a high density of reactive free radicals at ground level that were capable of oxidizing exposed minerals; although thus "oxidizing," the atmosphere was nonetheless anaerobic and oxygen-deficient (with the partial pressure of oxygen being less than 10^{-2} to 10^{-6} of the present atmospheric level and possibly being as low as 10^{-8} bar)—thus, the occurrence of red beds, for example, need not evidence the presence either of oxygen-producing microorganisms or of a "modern" oxygenic environment. On the other hand, rates of deposition inferred on the basis of geologic evidence for Early Proterozoic banded iron-formations (viz., those of the Hamersley Basin of Western Australia) seem to indicate that these units could not have been deposited without a far greater oxygen source than that provided by solely photochemical reactions; the most plausible such source is oxygen-producing photosynthesis, the origin of which must thus have occurred earlier than about 2.5 Ga ago (the age of the Hamersley Basin iron-formations) and the presence of which by about 1.7 Ga ago had provided a large enough oxygen source to overwhelm the sources of reactive, reduced materials in solution, leading to the accumulation of excess free oxygen and the onset of aerobic conditions in the ocean and atmosphere.

(8) Consideration of the biologic effects of this major change in the redox state of the Earth's surficial environment about 1.7 Ga ago suggests that although oxygen-tolerant, and oxygen-utilizing (amphiaerobic) microorganisms had evidently evolved considerably earlier, strict aerobes probably first became widespread at about this time. The diversification of such aerobes would have resulted in major modification of the geochemical cycles of a number of biologically important elements; and the development of an aerobic atmosphere would have resulted in establishment of a biologically effective ultraviolet-absorbing ozone layer that may have led to increased exploitation of the open ocean and of exposed land surfaces by microorganisms.

(9) One of the principal lessons to be learned from this exercise is that we are a long way from understanding the environmental evolution of the Archean-Early Proterozoic Earth. The problem area is replete with important, but as yet unanswered questions: What were the components (and what were their partial pressures) in the evolving early atmosphere? Was gaseous hydrogen (or methane, or ammonia) ever quantitatively important in the Archean-Early Proterozoic environment? If not, how—under what conditions—did the abiotic synthesis occur that led to the origin of life? Do any identifiable vestiges remain of this hypothesized "primordial soup," or were such syntheses limited to the earlier, Hadean segment of Earth history? When, in fact, did life originate? The first photoautotrophs? Oxygen-producing photosynthesizers? Aerobic respirers? Could an anaerobic biosphere have persisted indefinitely? Is it true, as argued here, that average global surface temperatures have varied only minimally over the geologic past? If so, are changes in concentration of atmospheric carbon dioxide responsible for offsetting the effects of the early cool Sun? And, finally, following the origin of oxygen-producing photoautotrophy, what limited the fluxes of oxygen and reduced constituents into the reservoirs of interest. What caused changes in these fluxes that allowed transitions in redox conditions to occur? Have the banded iron-formations and the detrital uraninites—the keystones of current concepts regarding the buildup of atmospheric oxygen—been interpreted correctly?

(10) Progress in understanding the environmental evolution of the primitive Earth calls for the closest association of theory and observation. Theory can serve as a guide in the interpretation of the geologic record and suggest new questions to ask of the record, but in the final analysis the theories can be validated only by reference to the rocks and the evidence that they contain.

PHOTOGRAPHS

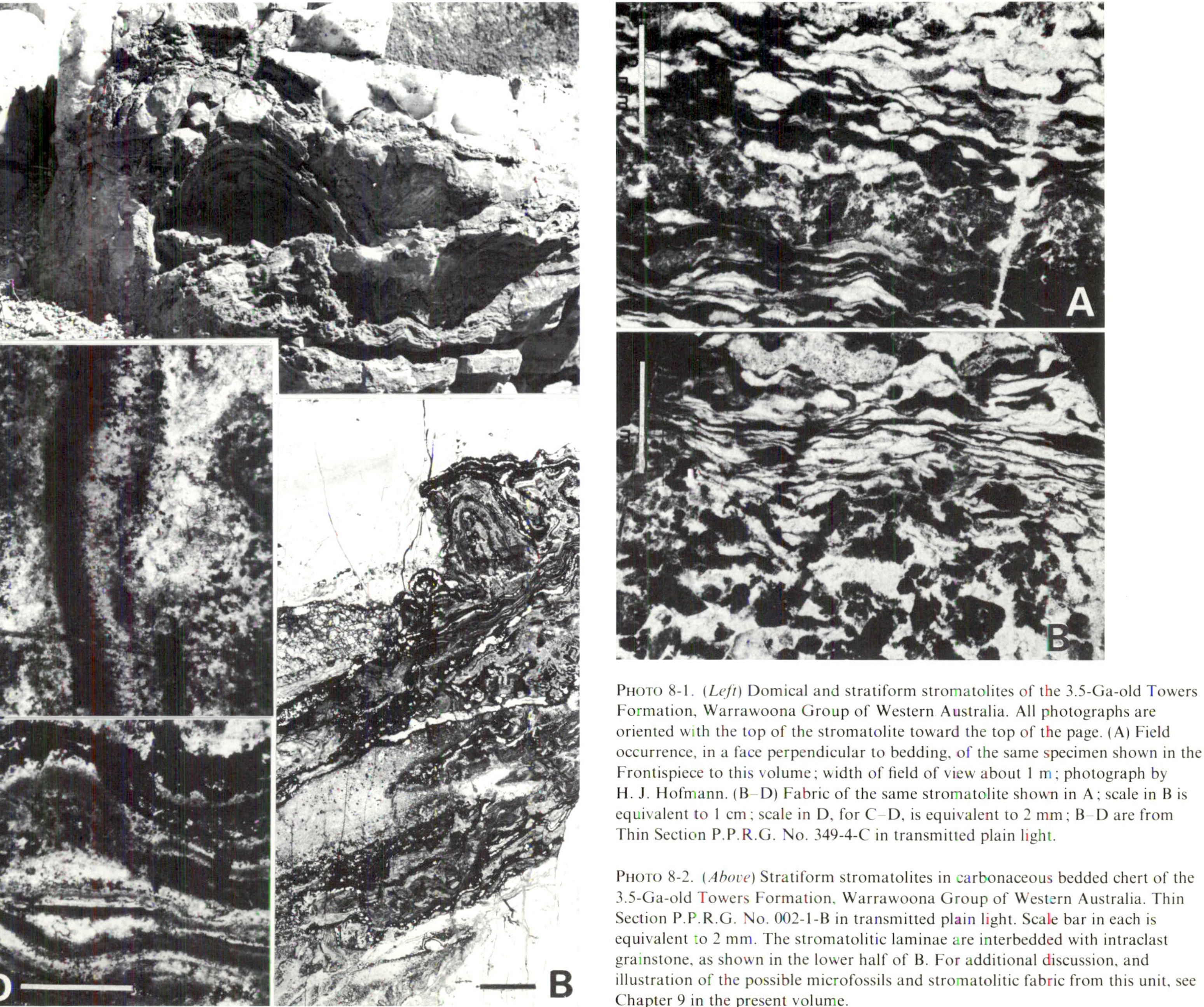

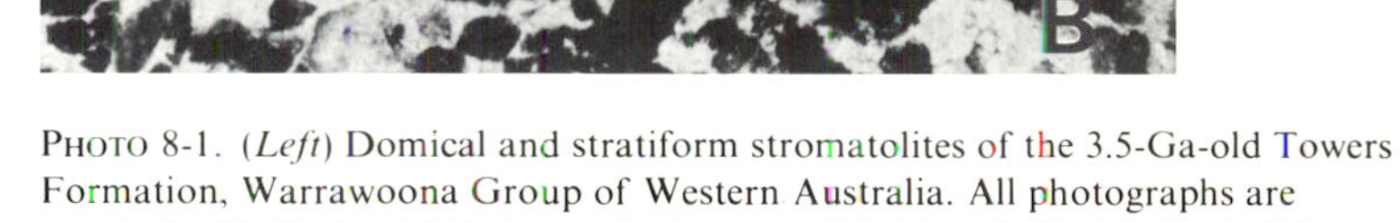

PHOTO 8-1. (*Left*) Domical and stratiform stromatolites of the 3.5-Ga-old Towers Formation, Warrawoona Group of Western Australia. All photographs are oriented with the top of the stromatolite toward the top of the page. (A) Field occurrence, in a face perpendicular to bedding, of the same specimen shown in the Frontispiece to this volume; width of field of view about 1 m; photograph by H. J. Hofmann. (B–D) Fabric of the same stromatolite shown in A; scale in B is equivalent to 1 cm; scale in D, for C–D, is equivalent to 2 mm; B–D are from Thin Section P.P.R.G. No. 349-4-C in transmitted plain light.

PHOTO 8-2. (*Above*) Stratiform stromatolites in carbonaceous bedded chert of the 3.5-Ga-old Towers Formation, Warrawoona Group of Western Australia. Thin Section P.P.R.G. No. 002-1-B in transmitted plain light. Scale bar in each is equivalent to 2 mm. The stromatolitic laminae are interbedded with intraclast grainstone, as shown in the lower half of B. For additional discussion, and illustration of the possible microfossils and stromatolitic fabric from this unit, see Chapter 9 in the present volume.

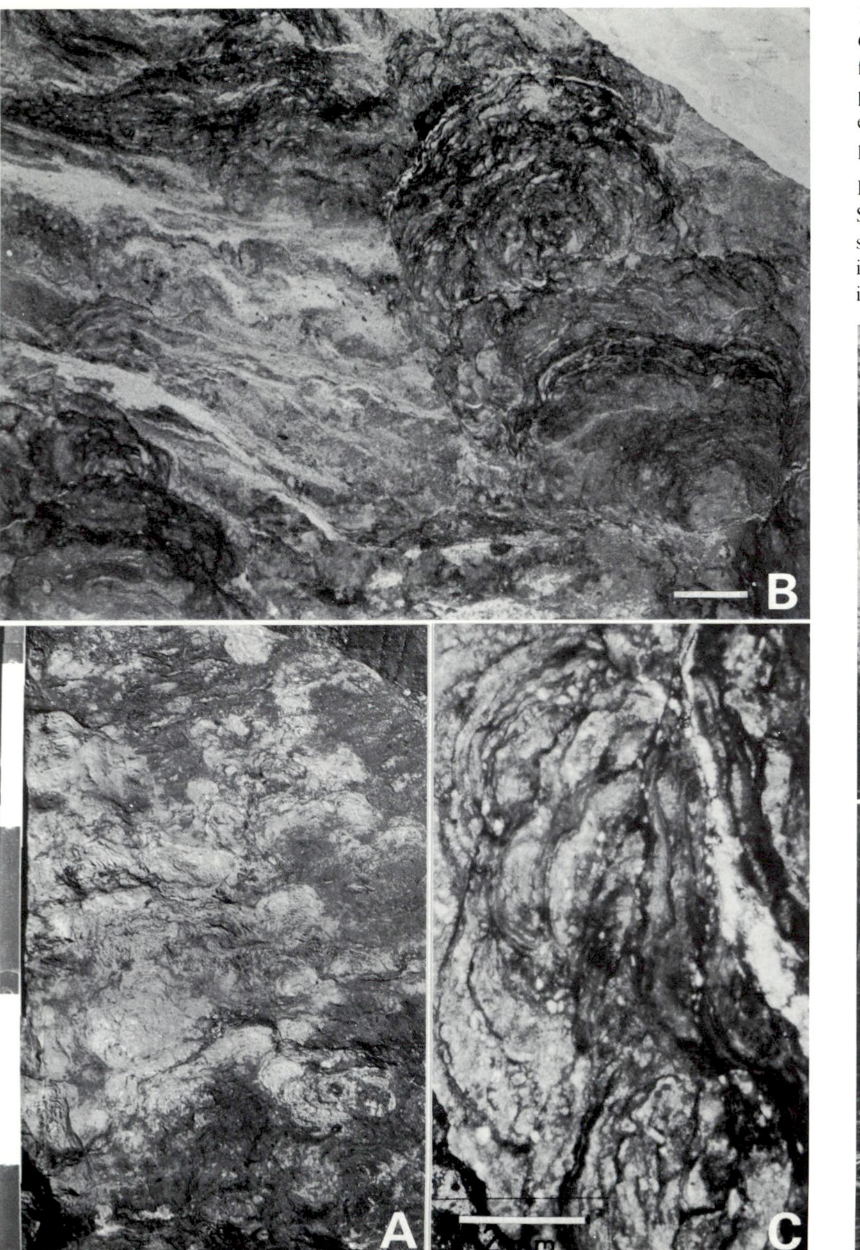

PHOTO 8-3. (*Left*) Bulbous and columnar-layered stromatolites of the 3.1-Ga-old Insuzi Group of South Africa. (A) Field occurrence, in a large boulder, showing a stromatolite face apparently perpendicular to bedding; divisions in scale bar represent 10 cm; photograph by H. J. Hofmann. (B, C) Fabric of the stromatolites shown in A; scale in B equivalent to 1 cm and in C to 2 mm; Thin Section P.P.R.G. No. 309-1-C in transmitted plain light.

PHOTO 8-4. (*Below*) Nodular and bulbous stromatolites of the 3.1-Ga-old Insuzi Group of South Africa. (A) Field occurrence, a bedding plane view showing the tops of individual stromatolites; scale bar equivalent to 10 cm. (B) Thin section perpendicular to bedding an individual like those in A; scale bar equivalent to 1 cm; Thin Section P.P.R.G. No. 263-2-C in transmitted plain light. A and B are enlargements of part of the same field of view.

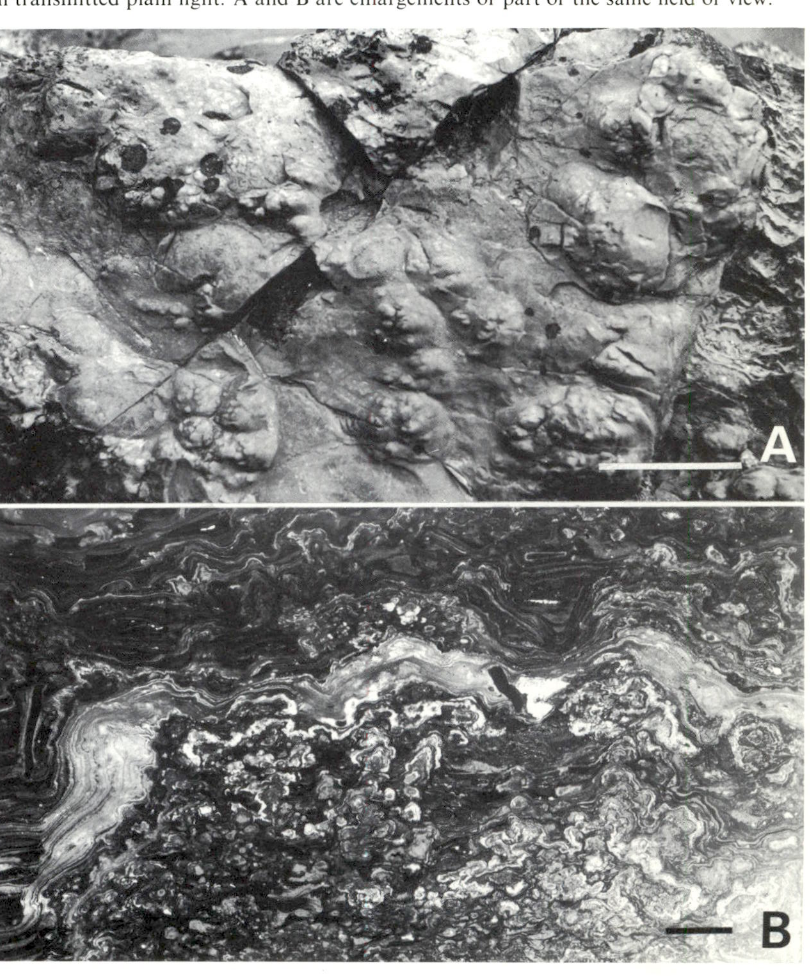

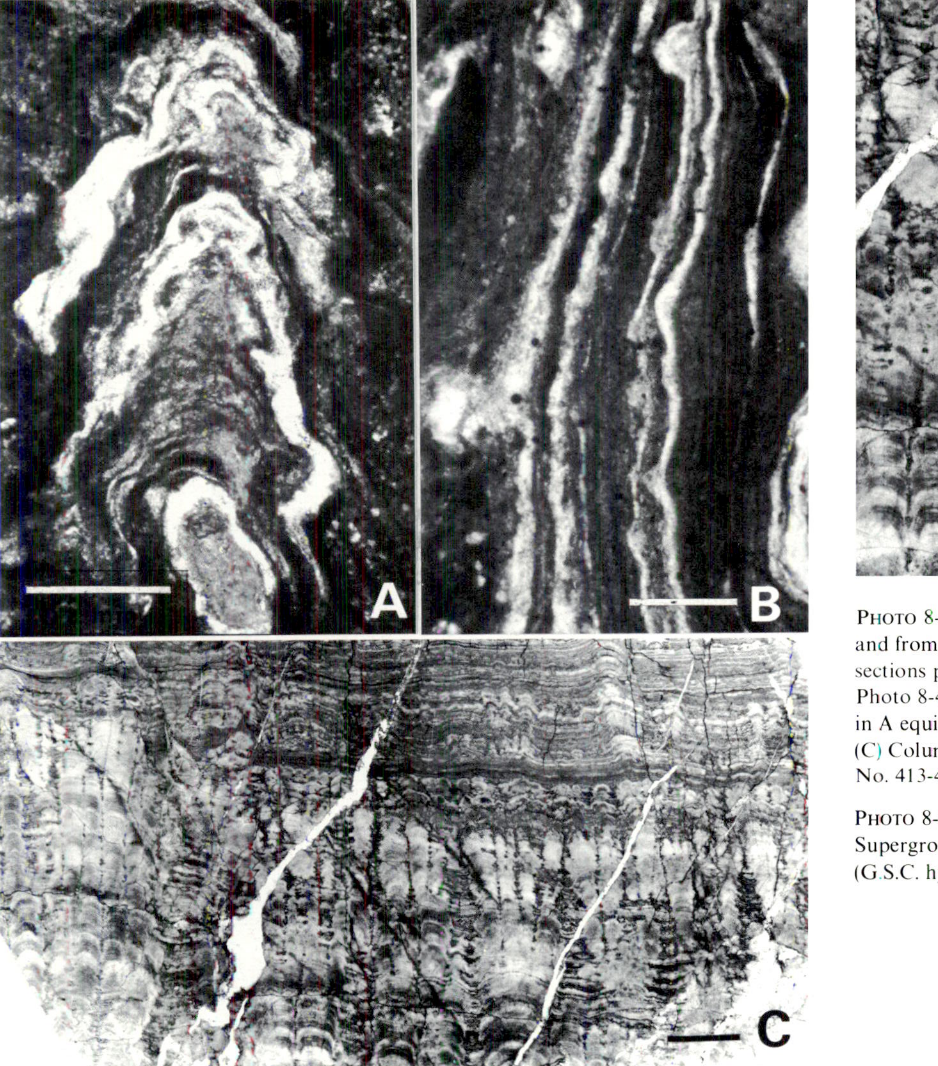

PHOTO 8-5. (*Left*) Stromatolites from the 3.1-Ga-old Insuzi Group (A, B) of South Africa, and from the 2.6-Ga-old Yellowknife Supergroup (C) of Canada. All parts show thin sections photographed in transmitted plain light. (A, B) Fabric of stromatolites shown in Photo 8-4, oriented with the top of the stromatolites toward the top of the page; scale bar in A equivalent to 2 mm; in B equivalent to 1 mm; Thin Section P.P.R.G. No. 263-2-C. (C) Columnar-layered stromatolites; scale bar equivalent to 1 cm; Thin Section P.P.R.G. No. 413-4-C (G.S.C. hypotype 65650). Photograph by H. J. Hofmann.

PHOTO 8-6. (*Above*) Columnar-layered stromatolites from the 2.6-Ga-old Yellowknife Supergroup of Canada. Scale bar is equivalent to 1 cm. Thin Section P.P.R.G. 413-4-C (G.S.C. hypotype 65650) in transmitted plain light. Photograph by H. J. Hofmann.

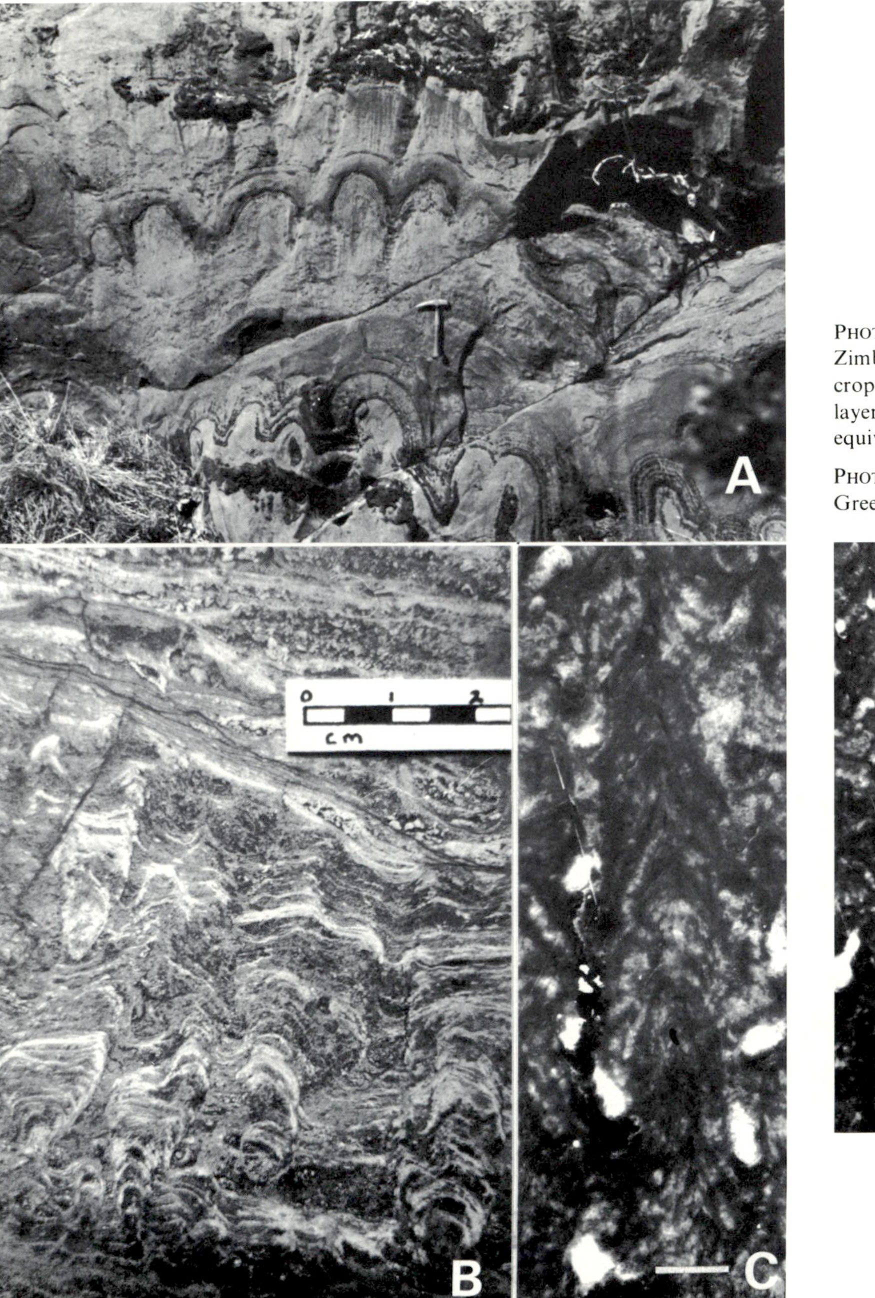

Photo 8-7. (*Left*) Stromatolites from the 2.6-Ga-old Belingwe Greenstone Belt of Zimbabwe. Photographs by A. Martin. (A) Linked nodular and bulbous forms, in outcrop; scale shown by geologic hammer in center of photograph. (B) Small columnar-layered form on a cut face. (C) *Conophyton*-like form on a cut face; scale bar equivalent to 5mm.

Photo 8-8. (*Below*) *Conophyton*-like stromatolites from the 2.6-Ga-old Belingwe Greenstone Belt of Zimbabwe. Scale bar is equivalent to 1 cm. Photograph by A. Martin.

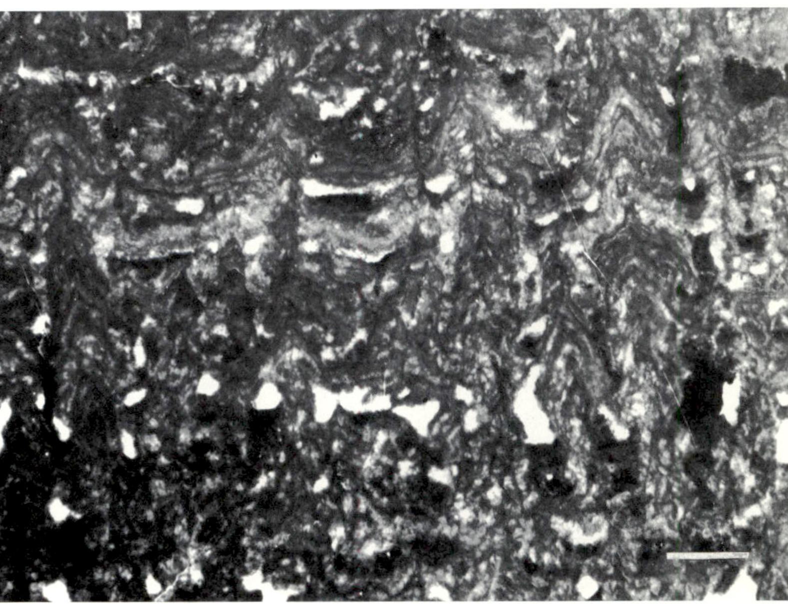

PHOTO 8-9. (*Left*) Pseudocolumnar stromatolites from the 2.6-Ga-old Bulawayan Greenstone Belt of Zimbabwe. All parts show thin sections in transmitted light, of specimens from the Huntsman Quarries. (A) Scale bar equivalent to 5 mm; Thin Section Bul. 4 (P.P.R.G Sample No. 250). (B) Scale bar equivalent to 2 mm; Thin Section Bul. 7 (P.P.R.G. Sample No. 253). (C) Scale bar equivalent to 2 mm; Thin Section P.P.R.G. No. 254-1-B.

PHOTO 8-10. (*Above*) Linked nodular stromatolites from the 2.6-Ga-old Steeprock Group in near-vertical limestone beds in an open cut pit of the Steeprock Mine of Canada. Scale represents 1m. Photograph by H. J. Hofmann.

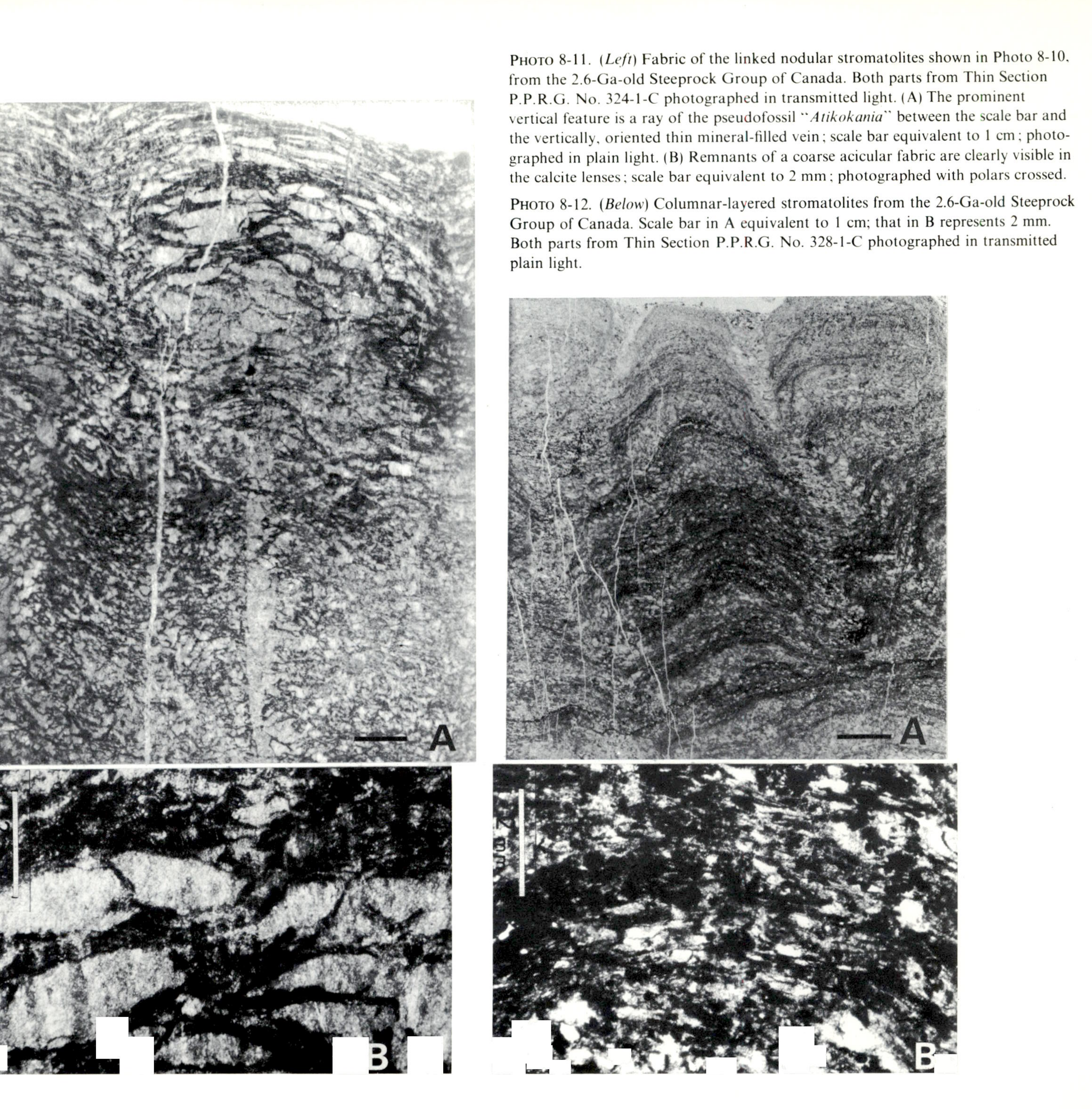

PHOTO 8-11. (*Left*) Fabric of the linked nodular stromatolites shown in Photo 8-10, from the 2.6-Ga-old Steeprock Group of Canada. Both parts from Thin Section P.P.R.G. No. 324-1-C photographed in transmitted light. (A) The prominent vertical feature is a ray of the pseudofossil "*Atikokania*" between the scale bar and the vertically, oriented thin mineral-filled vein; scale bar equivalent to 1 cm; photographed in plain light. (B) Remnants of a coarse acicular fabric are clearly visible in the calcite lenses; scale bar equivalent to 2 mm; photographed with polars crossed.

PHOTO 8-12. (*Below*) Columnar-layered stromatolites from the 2.6-Ga-old Steeprock Group of Canada. Scale bar in A equivalent to 1 cm; that in B represents 2 mm. Both parts from Thin Section P.P.R.G. No. 328-1-C photographed in transmitted plain light.

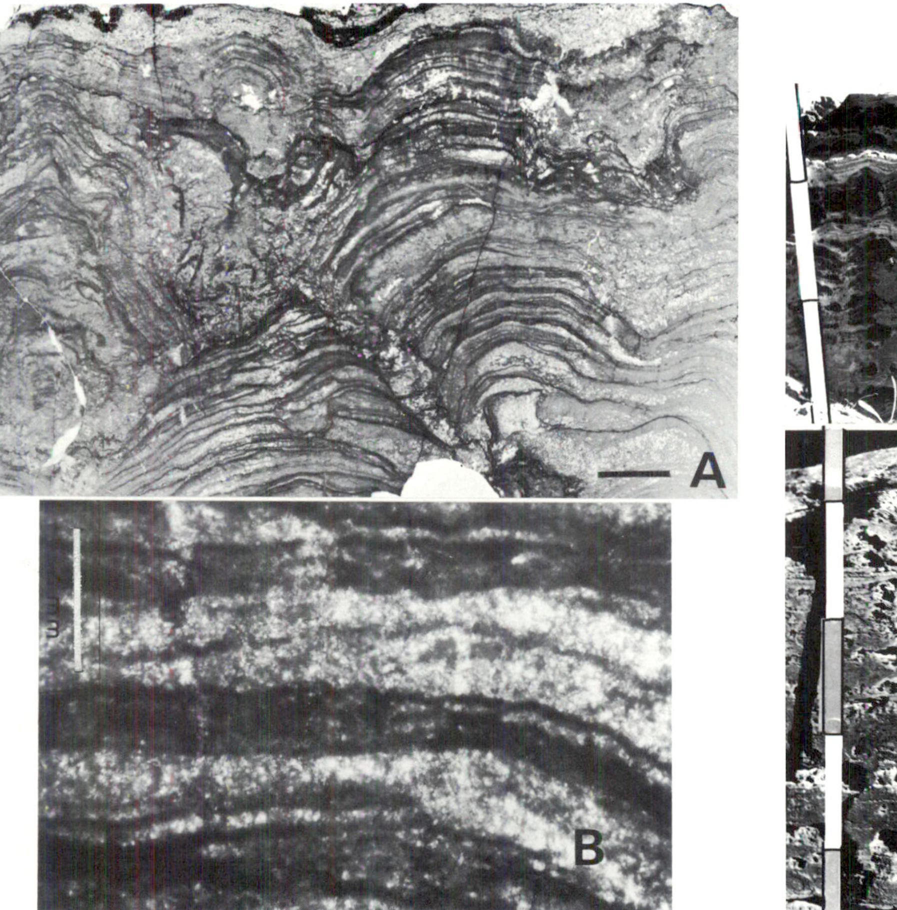

PHOTO 8-13. Columnar-layered stromatolites from the 2.6-Ga-old Steeprock Group of Canada. Scale bar in A equivalent to 1 cm; that in B represents 2 mm. Both parts from Thin Section P.P.R.G. No. 327-1-C photographed in transmitted plain light.

PHOTO 8-14. Stratiform stromatolites in outcrops of the 2.6-Ga-old Ventersdorp Supergroup, at the T'Kuip Hills locality of South Africa. Divisions in the scale bars represent 10 cm. Photographs by H. J. Hofmann.

PHOTO 8-15. (*Below*) Stromatolites in outcrops of the 2.6-Ga-old Ventersdorp Supergroup, T'Kuip Hills of South Africa. In A, divisions in scale bar represent 10 cm; in B, scale is equivalent to 10 cm. Photographs by H. J. Hofmann.

PHOTO 8-16. (*Right*) Stromatolites from the T'Kuip Hills section of the 2.6-Ga-old Ventersdorp Supergroup of South Africa. Both parts show thin sections photographed in transmitted plain light. (A) Scale bar is equivalent to 5 mm; Thin Section P.P.R.G. No. 274-1-C. (B) Scale bar represents 2 mm; Thin Section P.P.R.G. No. 278-1-B.

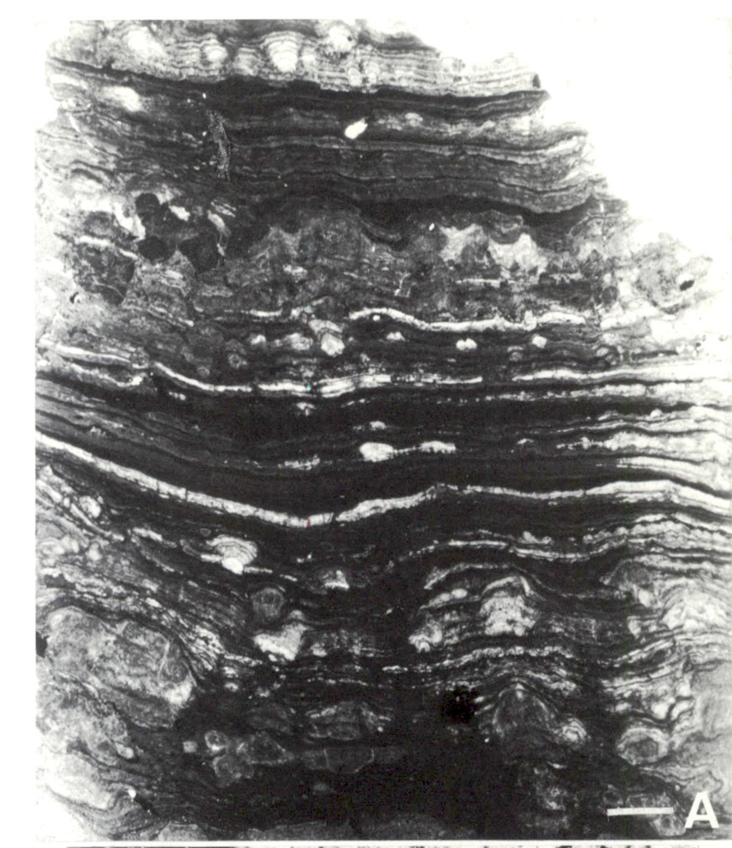

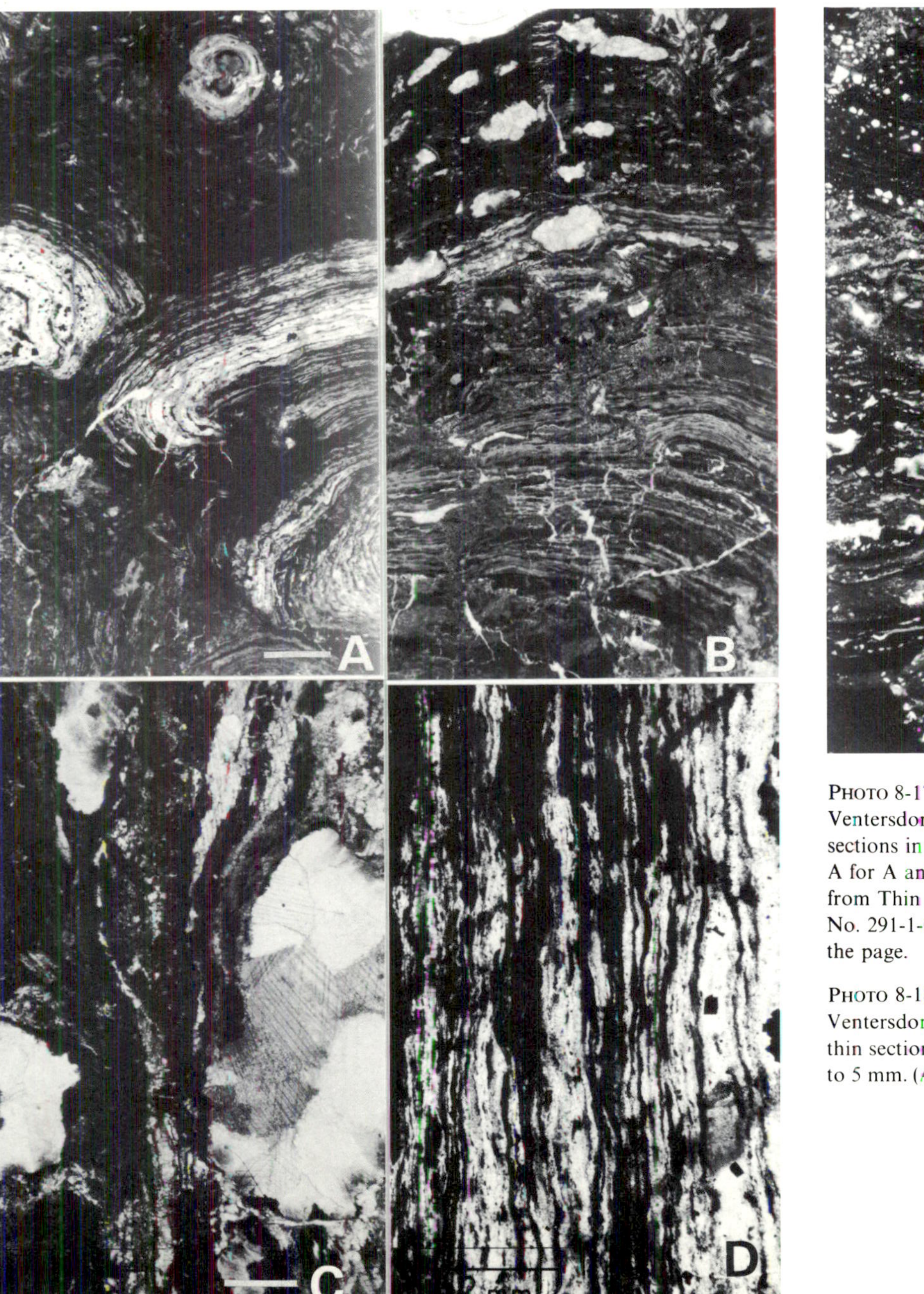

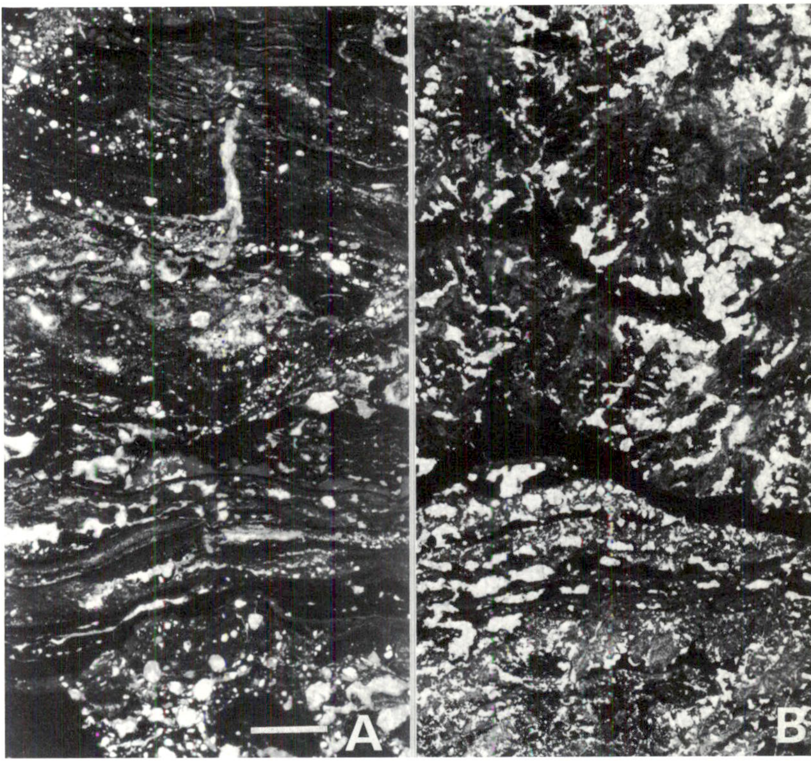

PHOTO 8-17. (*Left*) Stromatolites in diamond drill cores from the 2.6-Ga-old Ventersdorp Supergroup in the Welkom Goldfield, South Africa. All parts show thin sections in transmitted light; A and D, plain light; B and C, crossed polars. Scale bar in A for A and B is equivalent to 5 mm; that in C and D represents 1 mm. A and D from Thin Section P.P.R.G. No. 292-1-C; B and C, from Thin Section P.P.R.G. No. 291-1-C. Photographs C and D are oriented with bedding parallel to the width of the page.

PHOTO 8-18. (*Above*) Stromatolites in diamond drill cores from the 2.6-Ga-old Ventersdorp Supergroup in the Welkom Goldfield of South Africa. Both parts show thin sections in transmitted light with polars crossed. Scale bar in A and B is equivalent to 5 mm. (A) Thin Section P.P.R.G. No. 287-1-C; (B) Thin Section P.P.R.G. No. 281-1-C.

PHOTO 8-19. (*Below*) Stromatolites in diamond drill cores from the 2.6-Ga-old Ventersdorp Supergroup in the Wesselsbron District of South Africa. Scale bar in A is equivalent to 1 cm; in B, to 2 mm; and in C, to 1 mm. All parts show photographs of Thin Section S535 (P.P.R.G. No. 294) in transmitted plain light.

PHOTO 8-20. (*Right*) Cherty stromatolitic limestones in outcrops of the 2.8-Ga-old Fortescue Group at the "Knossos" locality of Western Australia. (A) Linked nodular stromatolites at the top of the cliff section; scale represents 1 m. (B) Stratiform and nodular stromatolites, intraclast breccias, and ripple marks. (C) Climbing ripples. Scales in B and C are equivalent to 15 cm. Photographs by H. J. Hofmann.

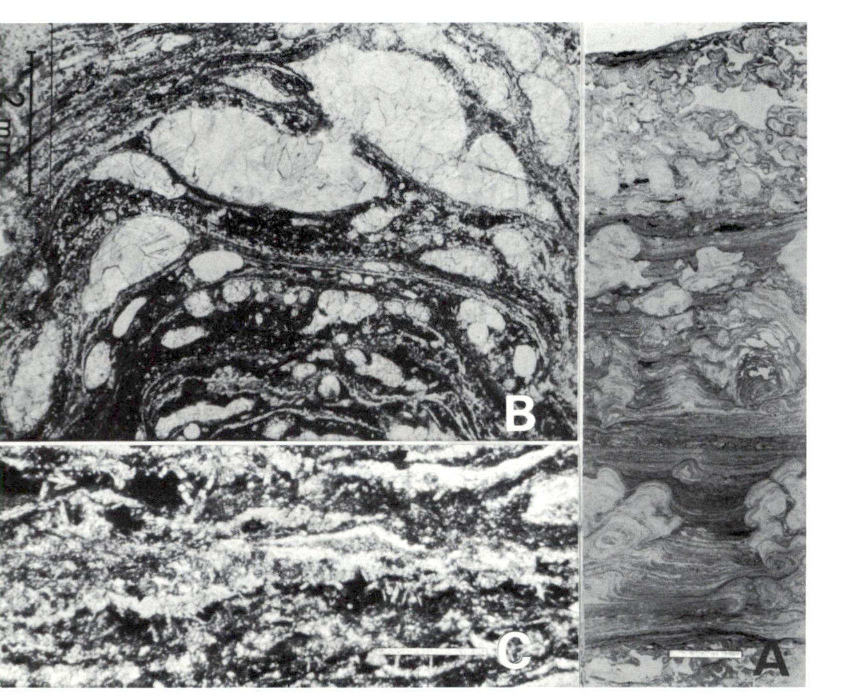

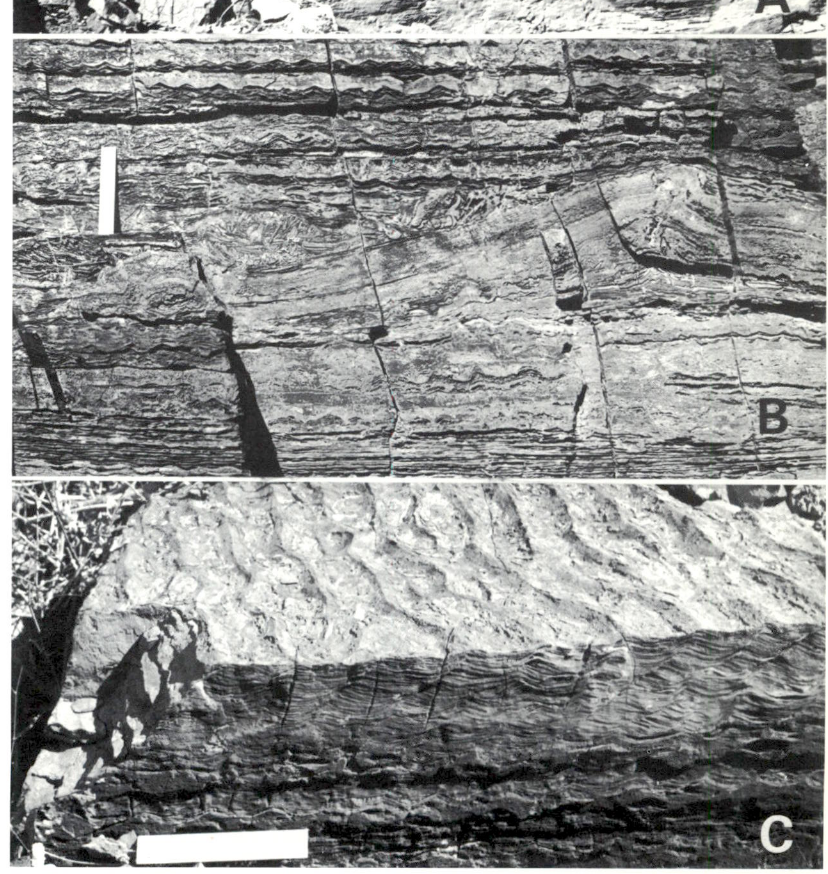

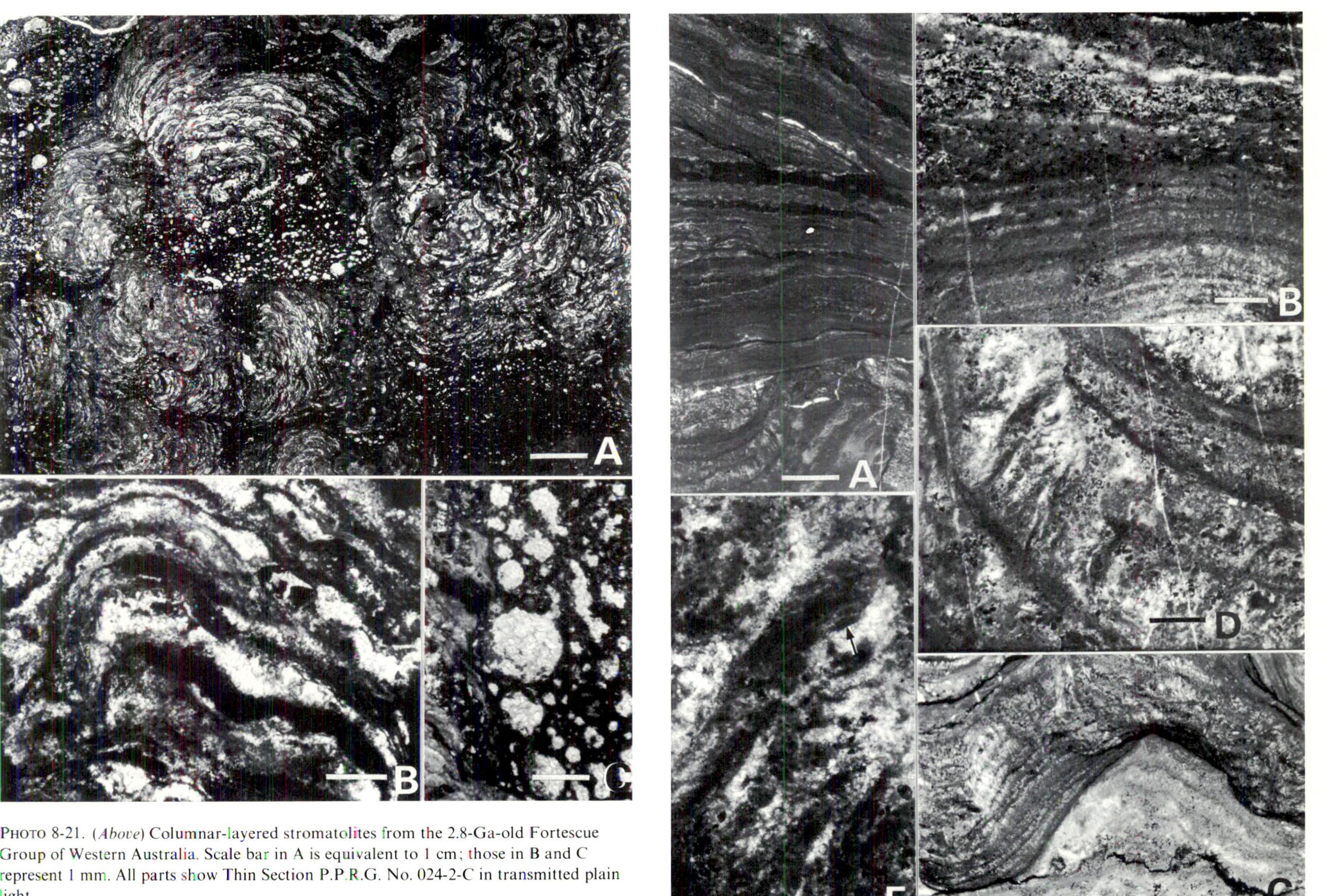

Photo 8-21. (*Above*) Columnar-layered stromatolites from the 2.8-Ga-old Fortescue Group of Western Australia. Scale bar in A is equivalent to 1 cm; those in B and C represent 1 mm. All parts show Thin Section P.P.R.G. No. 024-2-C in transmitted plain light.

Photo 8-22. (*Right*) Stromatolites from the "Knossos" locality of the 2.8-Ga-old Fortescue Group of Western Australia. All parts show variously developed linear palimpsest microstructures. E is an enlargement of a portion of D showing especially prominent filament remnants (e.g., at arrow). Scale bar in A is equivalent to 5 mm; those in B–E represent 1 mm. All parts show Thin Section P.P.R.G. No. 033-1-B in transmitted plain light.

PHOTO 8-23. (*Below*) Stromatolites from the 2.8-Ga-old Fortescue Group of Western Australia. All parts show thin sections photographed in transmitted plain light. (A) *Alcheringa narrina* Walter (1972); scale bar equivalent to 1 cm; Thin Section S95c from the Mt. Herbert locality of Walter (1972). (B, C) Single microbial tufts in rippled tuffaceous sand (now cherty limestone); C is an enlargement of a portion of the area shown in B; scale bar in B equivalent to 2 mm; in C to 1 mm; from Thin Section P.P.R.G. No. 029-1-A from the "Knossos" locality.

PHOTO 8-24. (*Right*) Stromatolites from the "Knossos" locality of the 2.8-Ga-old Fortescue Group of Western Australia. All parts show thin sections photographed in transmitted plain light. Scale bar for A–D is equivalent to 1 mm. (A) Thin Section P.P.R.G. No. 025-2-C. (B) Thin Section P.P.R.G. No. 025-1-B. (C) Thin Section P.P.R.G. No. 029-1-A. (D) From the side of a large bulbous stromatolite, oriented with the top of the stromatolite toward the top of the photograph; Thin Section P.P.R.G. No. 025-2-2-B.

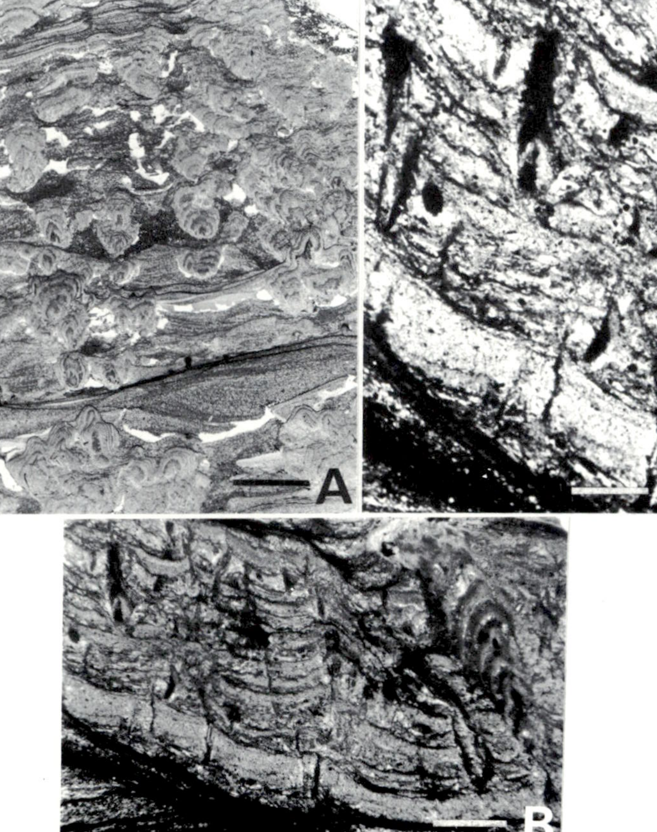

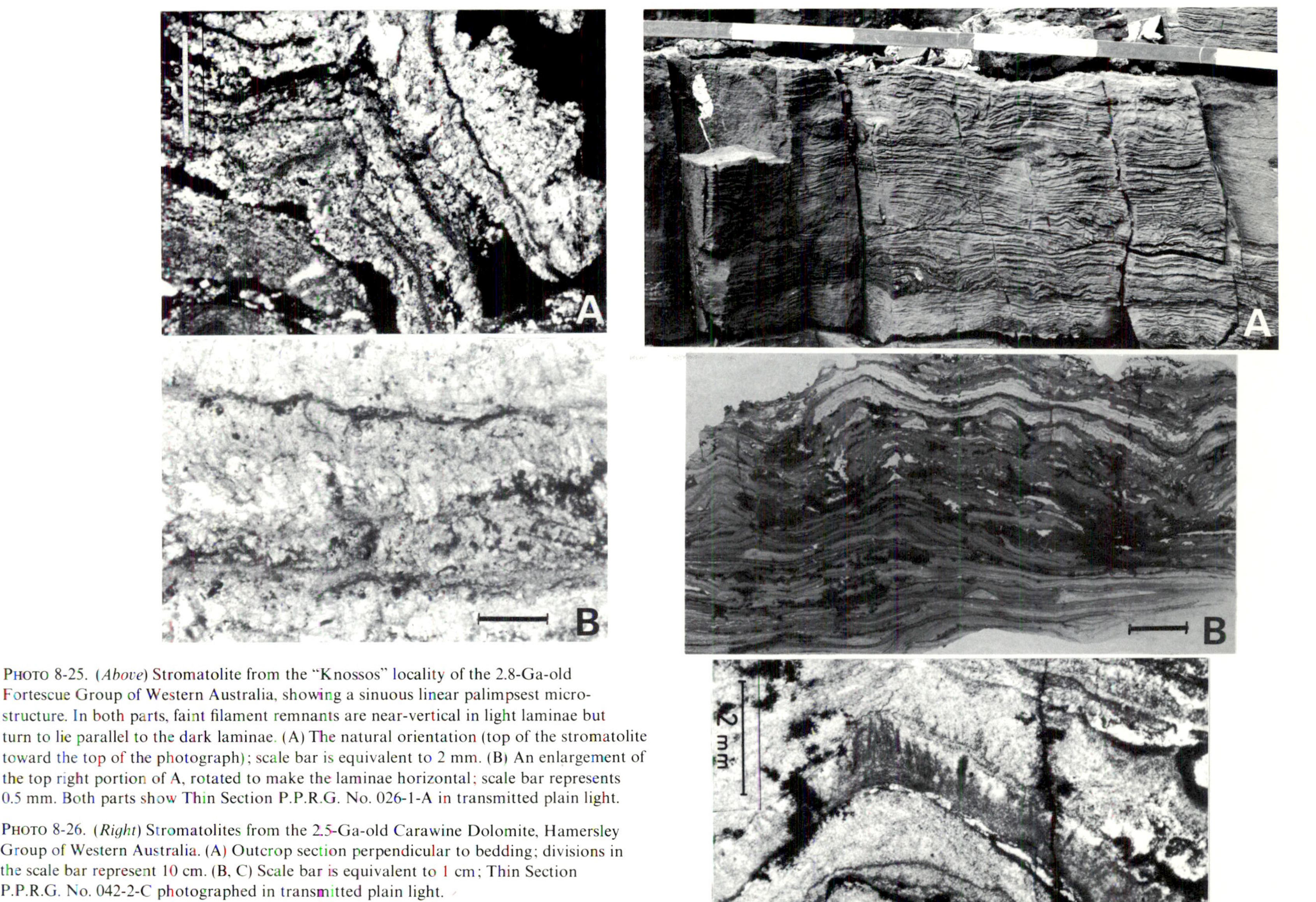

PHOTO 8-25. (*Above*) Stromatolite from the "Knossos" locality of the 2.8-Ga-old Fortescue Group of Western Australia, showing a sinuous linear palimpsest microstructure. In both parts, faint filament remnants are near-vertical in light laminae but turn to lie parallel to the dark laminae. (A) The natural orientation (top of the stromatolite toward the top of the photograph); scale bar is equivalent to 2 mm. (B) An enlargement of the top right portion of A, rotated to make the laminae horizontal; scale bar represents 0.5 mm. Both parts show Thin Section P.P.R.G. No. 026-1-A in transmitted plain light.

PHOTO 8-26. (*Right*) Stromatolites from the 2.5-Ga-old Carawine Dolomite, Hamersley Group of Western Australia. (A) Outcrop section perpendicular to bedding; divisions in the scale bar represent 10 cm. (B, C) Scale bar is equivalent to 1 cm; Thin Section P.P.R.G. No. 042-2-C photographed in transmitted plain light.

PHOTO 9-1 (*Right*) Nonfossils: metamorphically produced, non-biogenic mineralic microstructures—objects originally described as "*Isuasphaera isua*" and interpreted as "yeast-like microfossils" (Pflug, 1978c)—in a petrographic thin section of greenish-white metaquartzite from the 3.8-Ga-old Isua Supracrustal Group of southwestern Greenland (Table 9-1, No. 1). The pseudomicrofossils exhibit a continuum of forms from highly angular (and obviously crystalline) bodies (D and E), to subhedral ellipsoids (I), to more rounded ellipsoids (F and G), to elongate forms with angular ends (C), to globular, teardrop-shaped and elongate structures of somewhat more "life-like" morphology (A, B, and H), and they tend to be concentrated at quartz grain boundaries (F, H, and I) rather than in sedimentary layers as would be expected of true microfossils. All specimens from Rock Sample No. 2377 (the same sample from which the putative microfossils were first described; see Bridgwater et al., 1981); stage coordinates cited are for a Leitz microscope with the thin section (No. 2377-1-B, P.P.R.G. Sample No. 171) oriented with the slide label to the right. Scale in B is for A–C; scale in D is for D–H; scale in I is for I. (A, B) Hollow (arrows), elongate, pseudomicrofossil at two different focal depths; 47.1/107.2. (C) Photomontage showing elongate microstructure; note the non-biologic angularity of the mineral-stained (limonitic?) subhedral ends; 53.2/106.6. (D) Rhombohedral mineral-stained microstructure; 42.2/105.3. (E) Angular, rhombohedral pseudomicrofossil; 33.1/104.8. (F) Ellipsoidal mineral-stained microstructure at grain boundary; 29.8/103.3. (G) Ellipsoidal opaque specimen; 31.7/105.4. (H) Teardrop-shaped pseudomicrofossil, at grain boundary, together with clear, bubble-like, fluid inclusions (lower left and upper right); 40.7/99.1. (I) Large, ellipsoidal to subhedral specimen at grain boundary; 35.7/108.7.

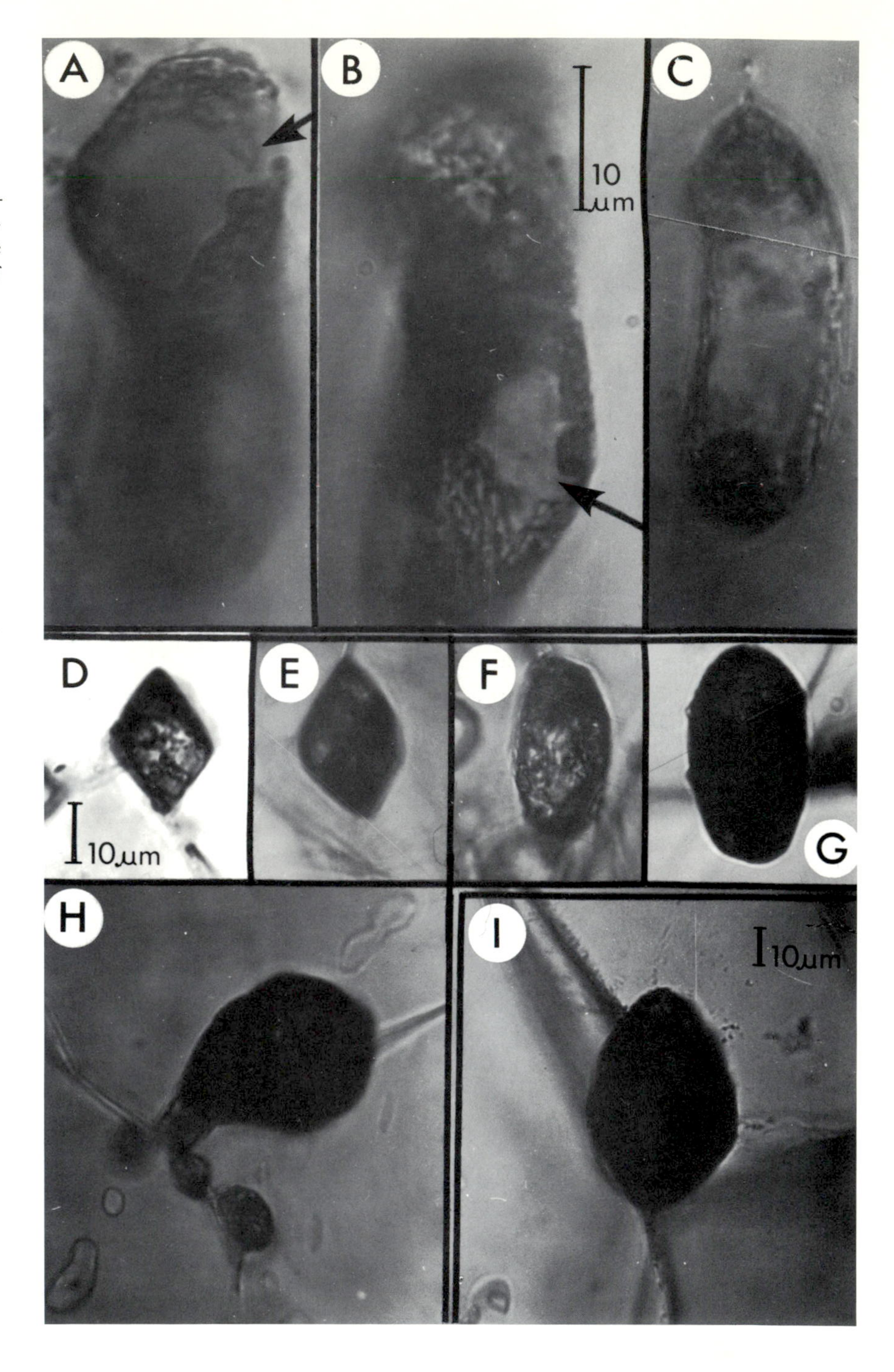

PHOTO 9-2. (*Facing page*) Dubiomicrofossils: carbonaceous microfossil-like spheroids (arrows) in a petrographic thin section of black chert from the 3.5-Ga-old Onverwacht Group (Kromberg Formation) of South Africa (Table 9-1, No. 6). Stage coordinates cited are for a Leitz microscope with thin section no. 41447 Komati S. Africa 1A (the section referred to as "K1A" in Muir and Grant, 1976) oriented with the slide label to the right. Scale in G is for A and G; magnification of all other parts indicated by scale in Part E. In B, F, J, P, and Q, "DIV" refers to apparently paired spheroids of the type described by Muir and Grant (1976) as resembling "algae in the process of dividing." (A) "Small domed arc" of spheroids, oriented perpendicular to the bedding laminations of the chert (see text); boxes indicate areas shown at higher magnification in B–F; 33.3/101.1. (B–F) Carbonaceous microfossil-like spheroids (arrows); 33.3/101.1. (G) Globular, unstructured cluster (irregular colony?) of spheroids; boxes indicate areas shown at higher magnification in P and Q; 32.8/106.9. (H) Small carbonaceous spheroid; 32.8/99.6. (I) Small carbonaceous spheroid; 32.8/99.7. (J) Small carbonaceous spheroid; 40.6/107.0. (K) Small carbonaceous spheroid; 32.7/99.6. (L, M) Relatively large spheroid, shown at two different focal depths; 32.8/99.6. (N, O) Relatively large spheroid, shown at two different focal depths; 33.2/107.2. (P, Q) Carbonaceous, microfossil-like spheroids (arrows); 32.8/106.9.

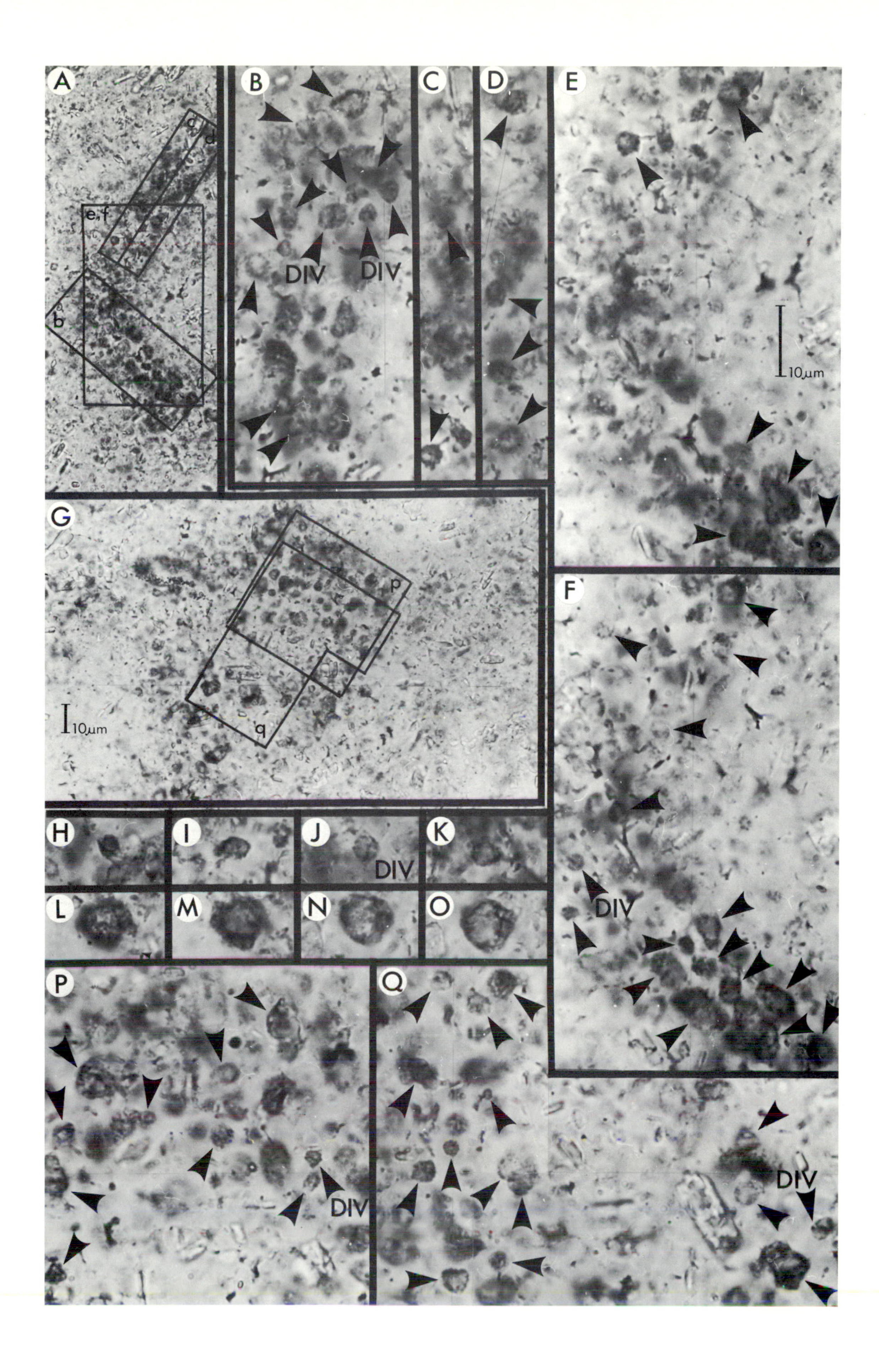
A
B
C
D
E
c
d
e,f
b
DIV
DIV
10µm
G
p
q
10µm
F
H
I
J
K
DIV
L
M
N
O
DIV
P
Q
DIV
DIV

PHOTO 9-3. Dubiomicrofossils: carbonaceous(?) microfossil-like bodies in a petrographic thin section of ferruginous chert (jaspilite) from the 2.5-Ga-old Hamersley Group (Marra Mamba Iron Formation) of Western Australia (Table 9-1, No. 28). Stage coordinates cited are for a Leitz microscope with Thin Section P.P.R.G. No. 520-1-2-A oriented with the slide label to the right. Scale in A is for A; magnification of all other parts indicated by the scale in F. (A) Area containing numerous microfossil-like bodies; 32.9/113.5. (B) Spheroidal, microfossil-like body; 39.4/112.4. (C) Spheroidal, lacey, microfossil-like body; 39.4/114.4. (D) Spheroidal, microfossil-like body; 39.4/112.3. (E) Interconnected spheroids; 39.4/112.4. (F) Interconnected spheroids and angular microfossil-like body (arrow); 32.9/113.8. (G) Interconnected spheroids; 40.0/114.4. (H) Short, pseudofilamentous grouping of spheroids; 39.9/114.5. (I) Branching, pseudofilamentous grouping of spheroids; 39.4/114.4. (J) Pseudofilamentous grouping of spheroidal and subhedral (arrow) microfossil-like bodies; 40.1/114.5. (K) Interconnected pseudofilamentous grouping of spheroids; 39.5/114.0.

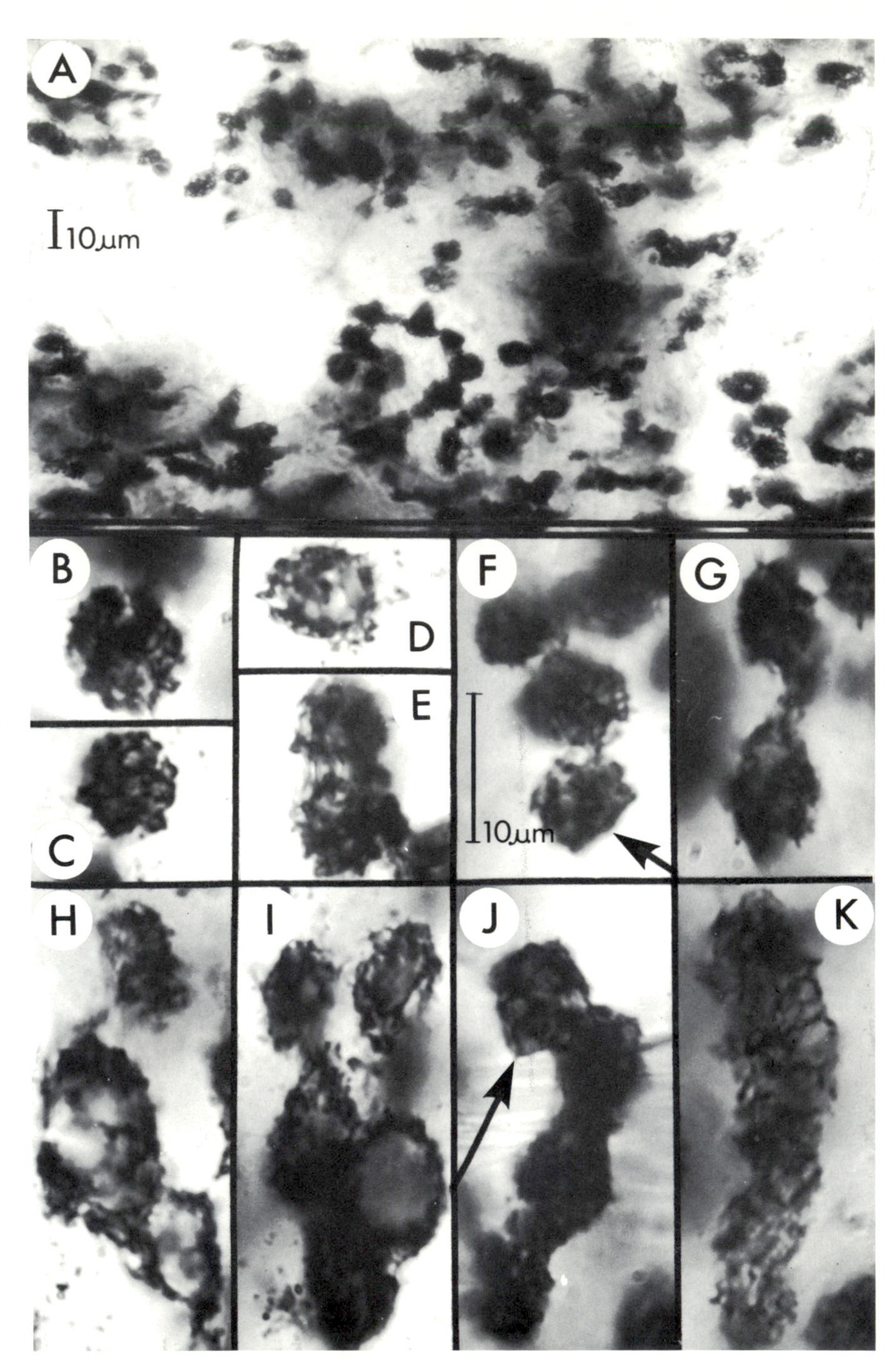

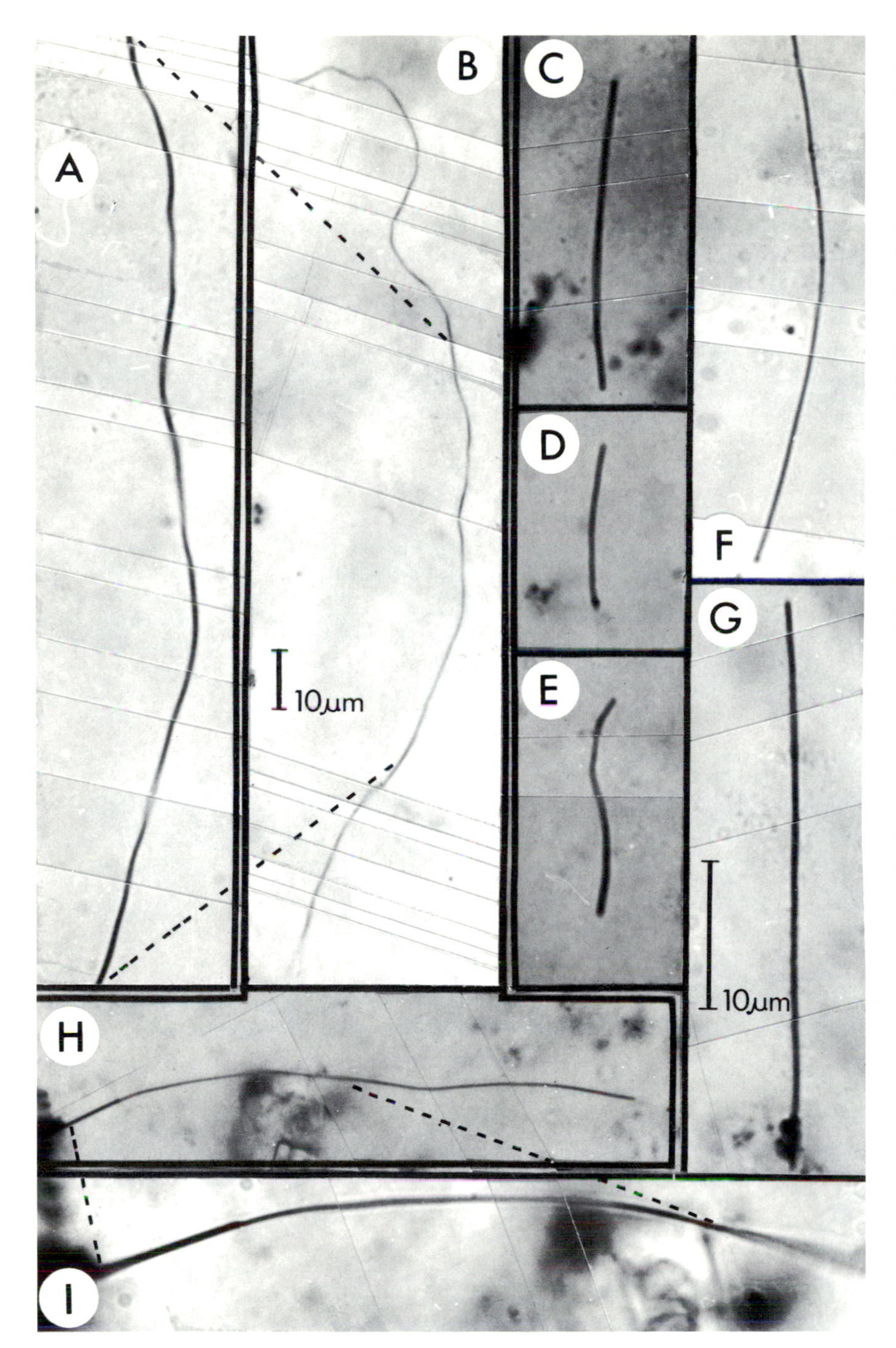

PHOTO 9-4. Elongate rod-shaped, apparently nonseptate fossil bacteria shown in a petrographic thin section of stromatolitic black chert from the 3.5-Ga-old Warrawoona Group (Pilbara Supergroup) of Western Australia (Table 9-1, No. 3). Stage coordinates cited are for a Leitz microscope with thin section no. SMA 299 oriented with the slide label to the left (coordinates for an "x" marked with a diamond scribe on the left front corner of the microscope slide are 66.1/111.6; thin section is deposited in the Commonwealth Paleontological Collections of the Bureau of Mineral Resources, Canberra, Australia; specimen figured in A and B is CPC no. 20207. Scale in B is for B and H; magnification of all other parts indicated by scale in G. (A) Narrow filamentous fossil bacterium, a portion of the specimen shown (as indicated by the dashed lines) at lower magnification in B; 51.7/100.9. (B) Sinuous specimen; 51.7/100.9. (C) Short rod-shaped threadlike filament, apparently a juvenile form (see D and E); 49.0/108.0. (D) Juvenile form; 50.5/104.3. (E) Juvenile form; 51.7/110.6. (F) Narrow, gently curved specimen; 52.3/111.8. (G) Elongate rod-shaped specimen; 48.2/105.7. (H) Specimen shown (as indicated by the dashed lines) at higher magnification in I; 45.9/105.5. (I) Gently curved specimen; 45.9/105.5.

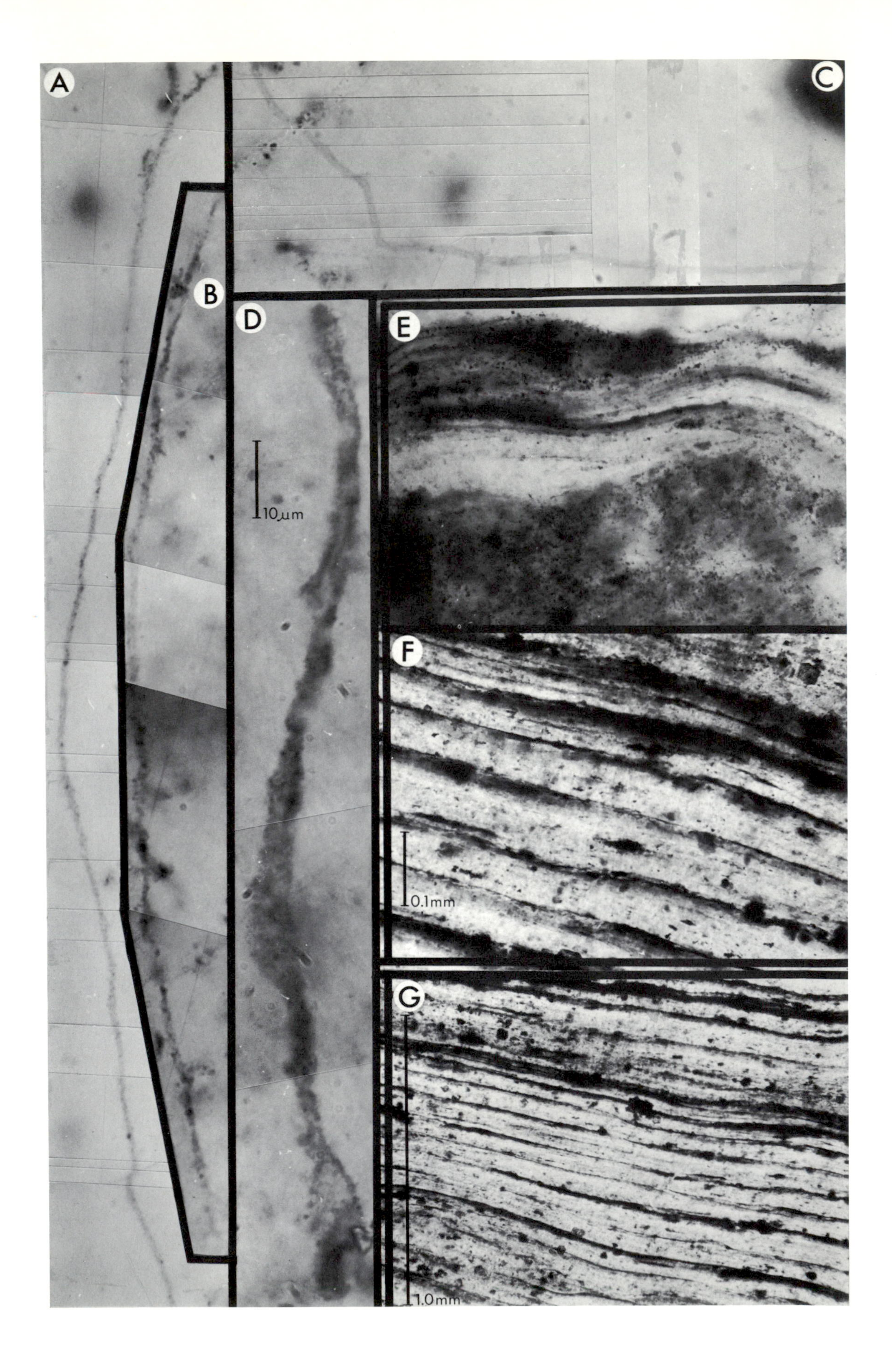
A
B
C
D
E
F
G
10µm
0.1mm
1.0mm

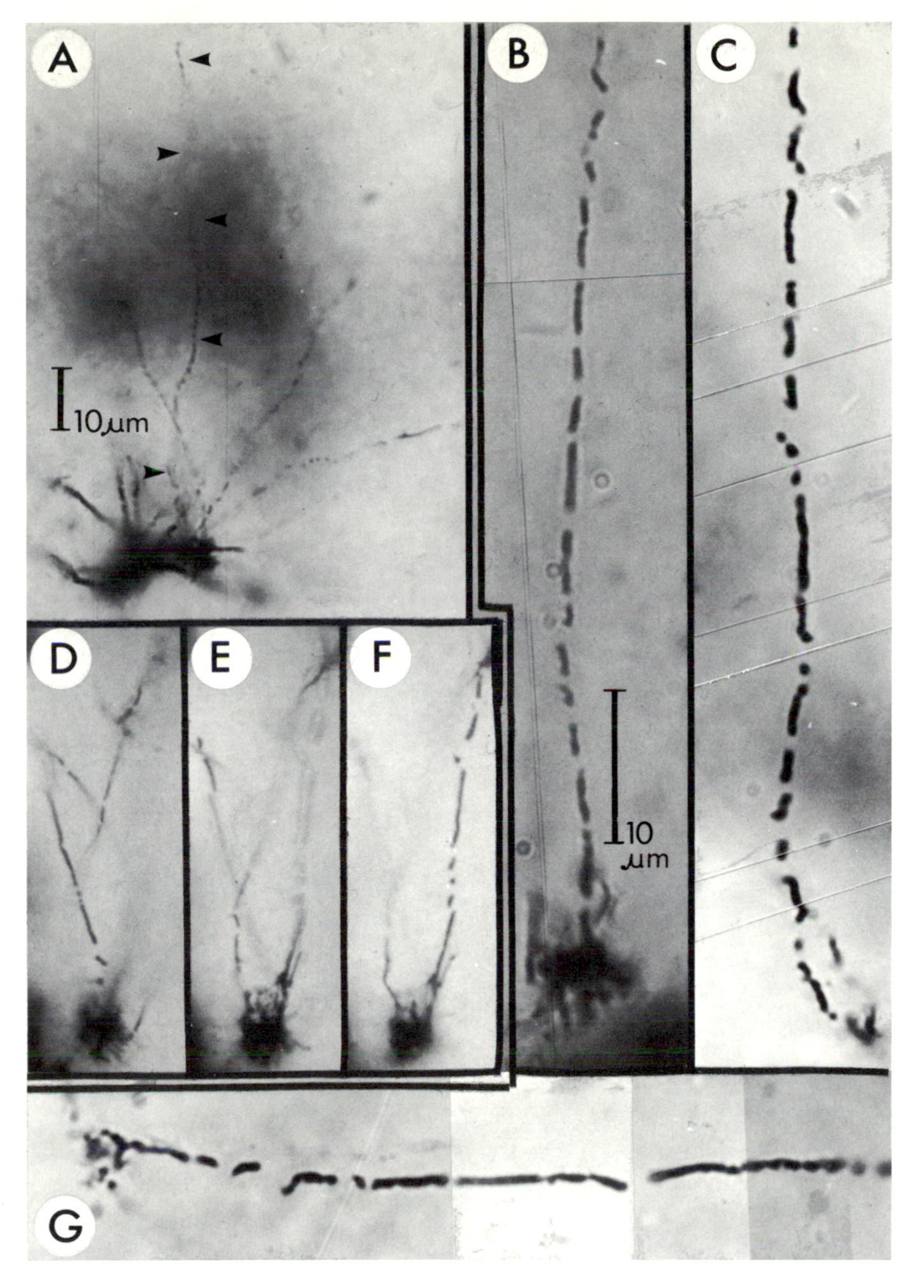

PHOTO 9-6. (*Left*) Dubiomicrofossils: unbranched filaments occurring singly (Parts C and G) or radiating from a holdfast-like base to form irregular hemispheroidal colony-like aggregates (Parts A, D–F) shown in petrographic thin sections of stromatolitic black chert from the 3.5-Ga-old Warrawoona Group (Pilbara Supergroup) of Western Australia (Table 9-1, No. 4). Stage coordinates cited are for a Leitz microscope with thin sections oriented with the slide label to the right (with slide labels oriented to the left, coordinates for an "x" marked with a diamond scribe on the left front corner of each of the microscope slides are for Thin Section P.P.R.G. No. 002-1-A, coordinates 74.3/113.8 and for P.P.R.G. No. 002-1-B, 73.3/113.7 and coordinates for the specimen shown in A are 18.9/100.5 and for the specimen shown in C, 48.4/99.4; thin sections are deposited in the Commonwealth Paleontological Collections of the Bureau of Mineral Resources, Canberra, Australia; specimen shown in A, is CPC no. 20205; specimen in C is CPC no. 20206). Scale in A is for A, D-F; scale in B is for B, C, and G. (A) Filaments radiating from a holdfast-like base (arrows point to a sinuous, vertically oriented, continuous filament); Thin Section P.P.R.G. No. 002-1-A, coordinates 58.0/105.7. (B) One of two filaments radiating from a holdfast-like base (at bottom); P.P.R.G. No. 002-1-A; 57.3/104.3. (C) Long solitary filament; P.P.R.G. No. 002-1-B, 27.8/106.8. (D, E, F) Small aggregate shown at three different focal depths; P.P.R.G. No. 002-1-A, 55.7/103.3. (G) Long solitary filament; P.P.R.G. No. 002-1-B, 28.1/112.5.

PHOTO 9-5. (*Facing page*) Carbonaceous filamentous microfossils (intermediate-diameter filaments, Parts A–C, and tubular morphotype, Part D) and stromatolitic laminae (Parts E–G) shown in petrographic thin sections of black chert from the 3.5-Ga-old Warrawoona Group (Pilbara Supergroup) of Western Australia (Table 9-1, No. 3). Stage coordinates cited are for a Leitz microscope with thin sections oriented with the slide label to the left (coordinates for an "x" marked with a diamond scribe on the left front corner of each microscope slide are as follows: Thin Section No. SMA 298, coordinates 67.0/111.1; No. SMA 299, 66.1/111.6; P.P.R.G. No. 517-1-A, 71.9/134.6; P.P.R.G. No. 002-1-A, 74.3/113.8. Thin sections are deposited in the Commonwealth Paleontological Collections of the Bureau of Mineral Resources, Canberra, Australia; specimen shown in C is CPC no. 20204; specimen in E is CPC no. 20209; specimen in F and G is CPC no. 20208). Scale in D is for A–D; scale in F is for E and F; scale in G is for G. (A) Long, degraded, gently curved specimen; Thin Section No. SMA 299, coordinates 48.1/98.1. (B) Long poorly preserved specimen; SMA 299, 45.8/105.1 (C) Relatively well-preserved, possibly septate specimen; SMA 299, 51.1/103.7; (D) Elongate, partially flattened, bacterial or cyanobacterial sheath; SMA 298, 37.6/95.5. (E) Stromatolitic laminae of the "distinct, wavy, striated" type; P.P.R.G. No. 002-1-A, 34.6/104.8. (F) Relatively planar and regularly banded stromatolitic laminae in the same area as shown at lower magnification in G; P.P.R.G. No. 517-1-A, 51.2/103.2. (G) Regularly banded stromatolitic laminae; P.P.R.G. No. 517-1-A, 51.2/103.2.

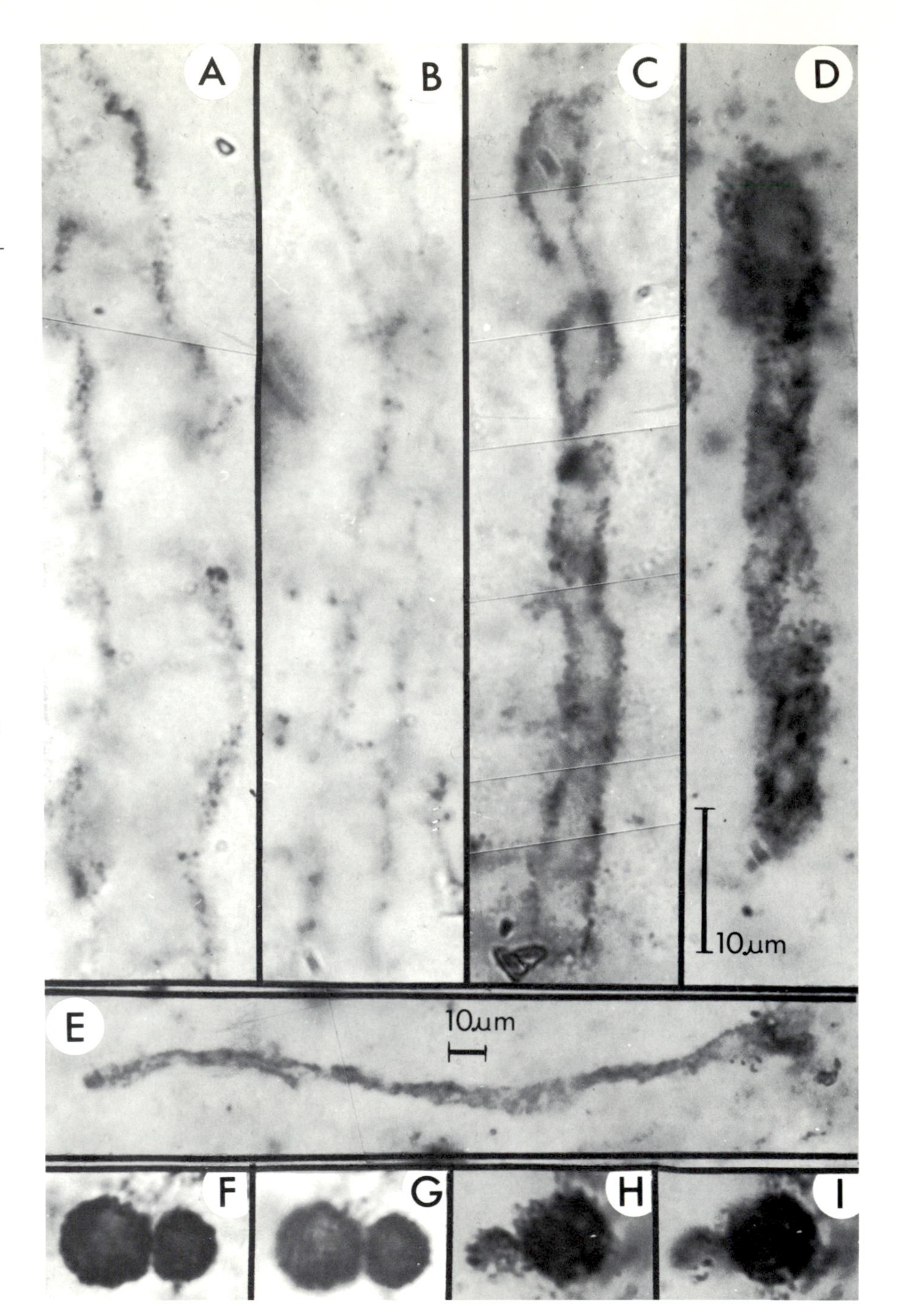

PHOTO 9-7. Tubular or partially flattened, bacterial or cyanobacterial sheaths (A–E), and spheroidal, apparently carbonaceous, dubiomicrofossils, shown in petrographic thin sections of stromatolitic black chert from the 3.5-Ga-old Warrawoona Group (Pilbara Supergroup) of Western Australia (Table 9-1, Nos. 3 and 4). Stage coordinates cited are for a Leitz microscope with thin sections oriented with the slide label to the left (coordinates for an "x" marked with a diamond scribe on the left front corner of each of the microscope slides are for Thin Section No. SMA 298, coordinates 67.0/111.1, and for No. SMA 299, 66.1/111.6; thin sections are deposited in the Commonwealth Paleontological Collections of the Bureau of Mineral Resources, Canberra, Australia; specimen shown in C is CPC no. 20202; specimens shown in F and G are CPC no. 20203). Scale in D is for A–D, F–I; scale in E is for E. (A) Hollow, particularly broad specimen; Thin Section No. SMA 299, coordinates 43.7/101.5. (B) Long, gently curved hollow sheath; No. SMA 299, 44.3/99.9. (C) Well-preserved tubular sheath; No. SMA 299, 46.7/103.0. (D) Relatively poorly preserved, degraded sheath; No. SMA 299, 52.0/112.0. (E) Elongate, partially flattened sheath (the same specimen as shown at higher magnification in Photo 9-6 D; No. SMA 298, 37.6/95.5. (F, G) Well-preserved pair of unicell-like spheroids shown at two different focal depths; No. SMA 299, 48.5/96.9. (H, I) Solitary unicell-like spheroid at two different focal depths; No. SMA 299, 48.5/96.8.

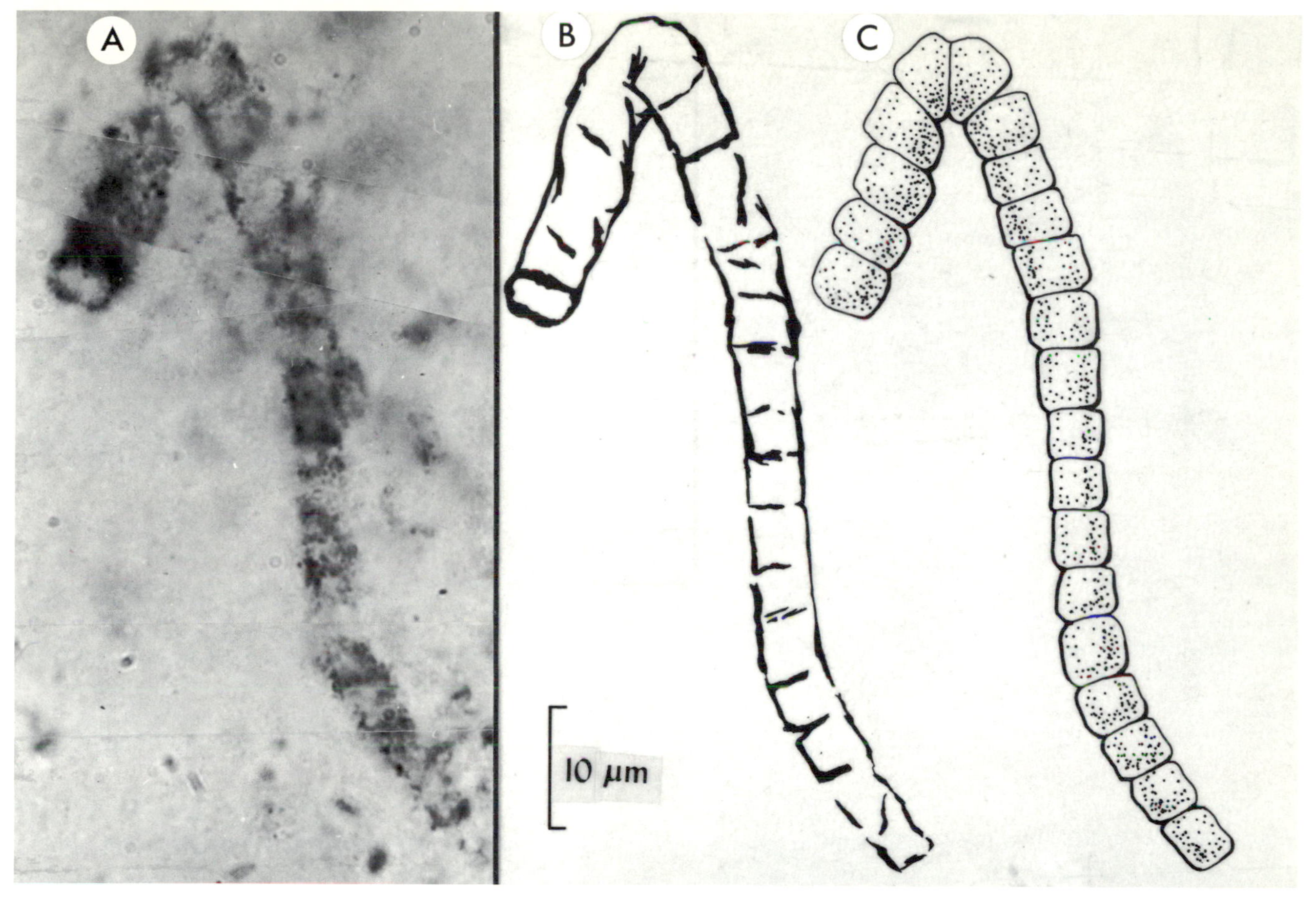

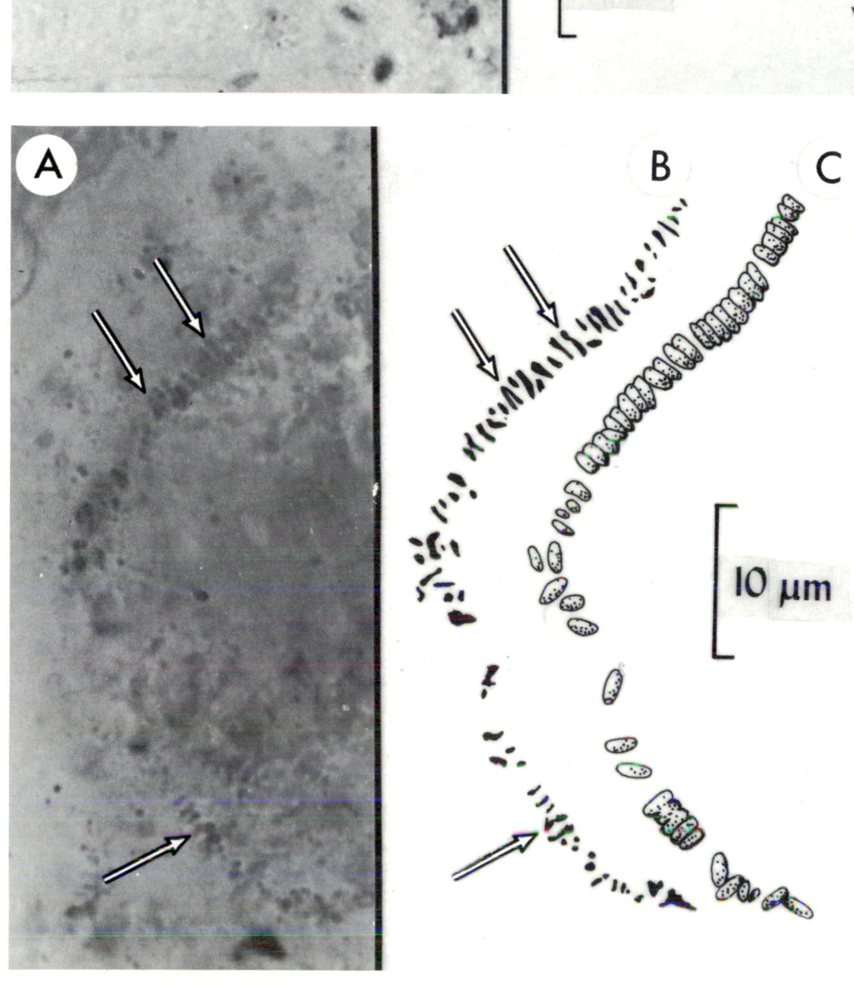

PHOTO 9-8 (*Above*) Unbranched, septate, apparently somewhat tapering filamentous fossil prokaryote shown in petrographic thin section (A) and in schematic reconstructions (B and C) from stromatolitic black chert of the 3.5-Ga-old Warrawoona Group (Pilbara Supergroup) of Western Australia (Table 9-1, No. 3). Stage coordinates for this specimen, with Thin Section No. SMA 299 oriented with the slide label to the left, are 45.1/98.4 (coordinates for an "x" marked with a diamond scribe on the left front corner of the slide are 66.1/111.6; the thin section is deposited in the Commonwealth Paleontological Collections of the Bureau of Mineral Resources, Canberra Australia; CPC No. 20201). Scale for all parts is shown in B.

PHOTO 9-9. (*Left*) Narrow oscillatoriacean-like filamentous fossil prokaryote shown in petrographic thin section (A) and in schematic reconstructions (B and C) from carbonaceous cherty portions of carbonate stromatolites of the 2.8-Ga-old Fortescue Group of Western Australia (Table 9-1, No. 21). Stage coordinates for this specimen, with Thin Section P.P.R.G. No. 025-2-C oriented with the slide label to the left, are 22.2/109.1. Scale for all parts is shown in C. Arrows in B point to cellular remnants visible (at arrows) in A.

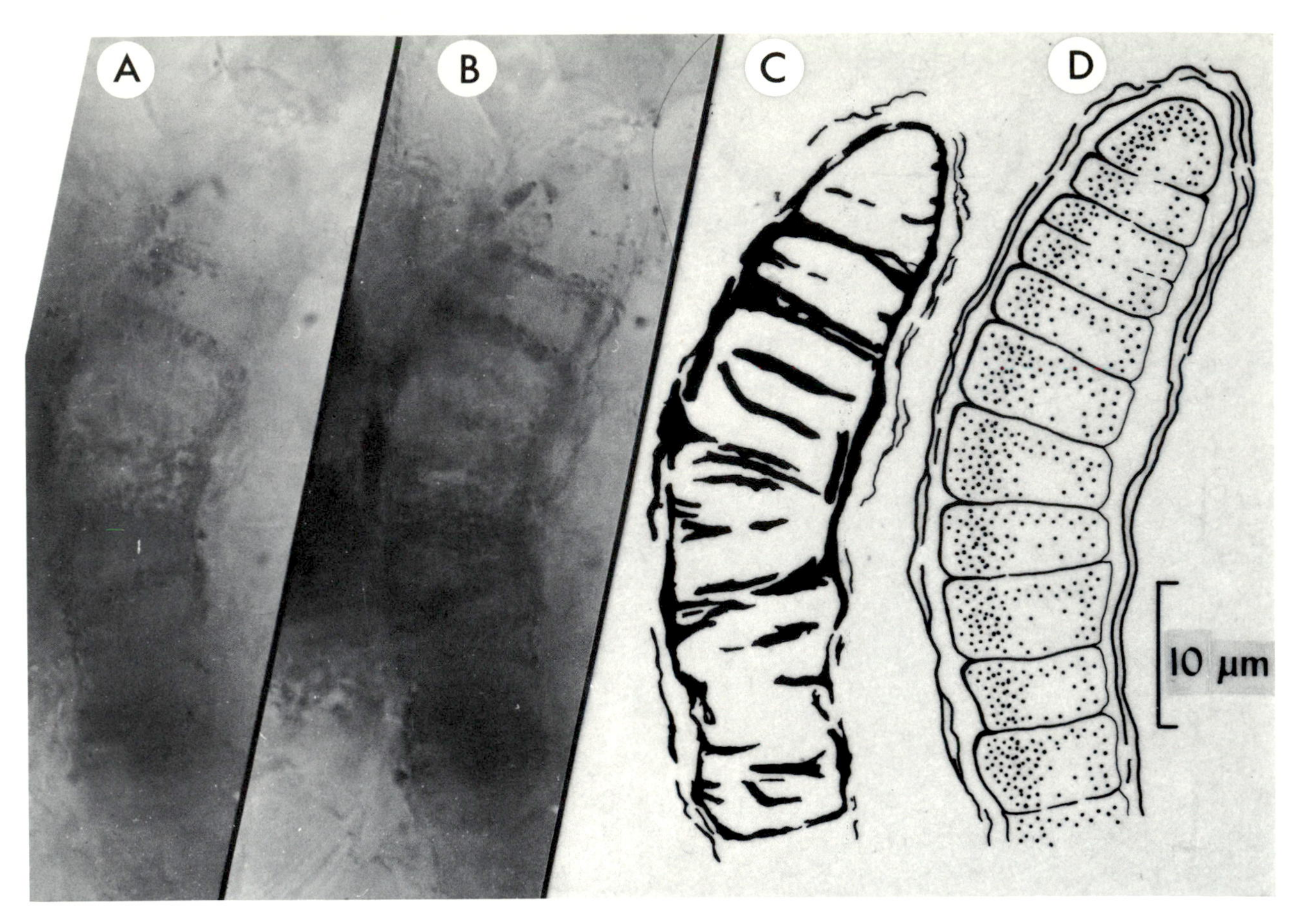

PHOTO 9-10. Broad, sheath-enclosed, *Lyngbya*-like filamentous fossil prokaryote shown in petrographic thin section (at two different focal depths in A and B) and in schematic reconstructions (C and D) from carbonaceous cherty portions of carbonate stromatolites from the 2.8-Ga-old Fortescue Group of Western Australia (Table 9-1, No. 21). Stage coordinates for this specimen, with Thin Section P.P.R.G. No. 029-1-A oriented with the slide label to the left, are 45.8/115.8. Scale for all parts is shown in D.

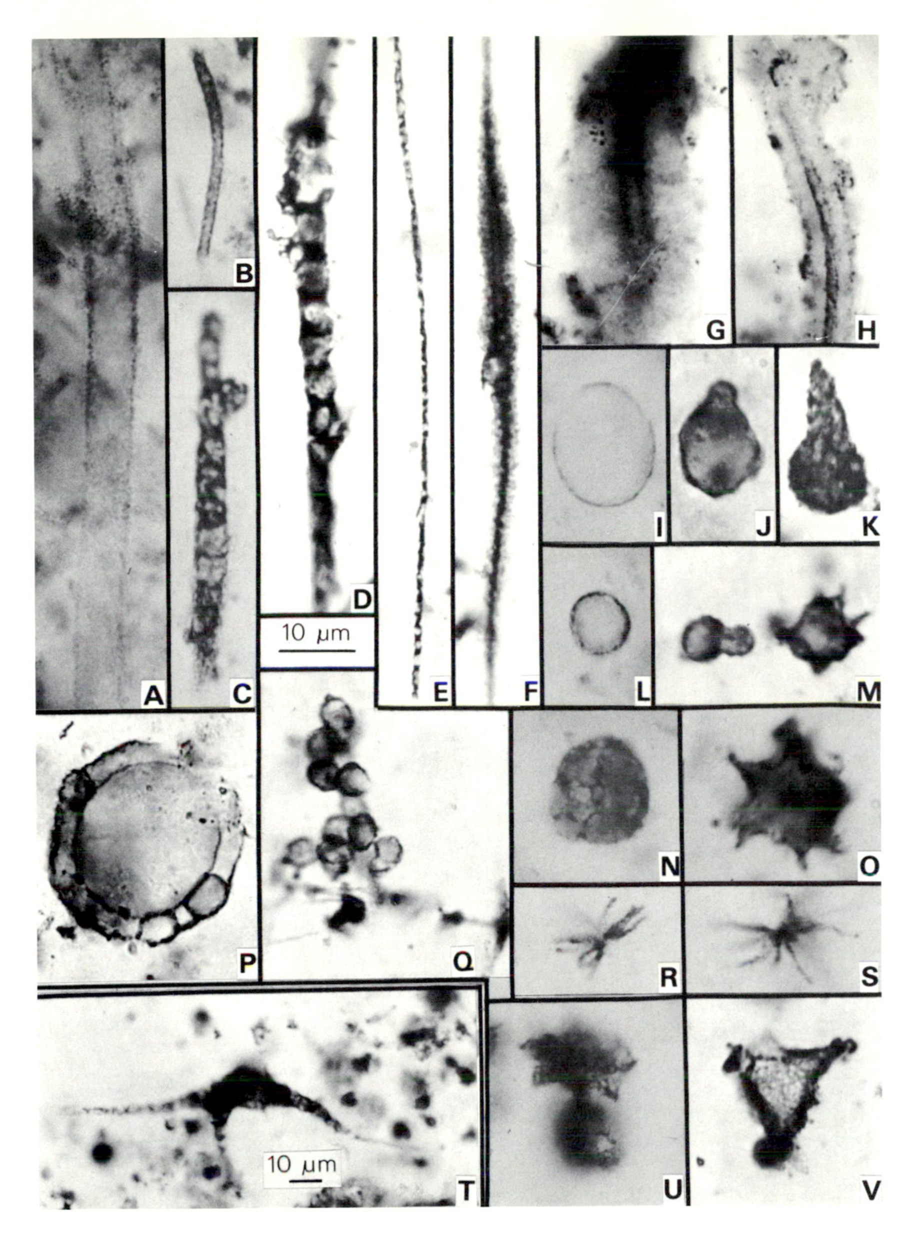

PHOTO 14-1. Microfossils from the Gunflint Formation. Previously illustrated and new material. All magnifications identical, except Part T. Coordinates cited for new material refer to distances, in mm, from distal margin (y = 0) and right margin (x = 0) of slide (with label to right of observer) Photos from Hofmann, 1971a and courtesy of Geological Survey of Canada. (A) *Animikiea septata* (from Hofmann 1971a, pl. 15, fig. 1). (B) *Gunflintia minuta* (from Hofmann 1971a, pl. 15, fig. 3). (C) *Gunflintia grandis* (from Hofmann 1971a, pl. 15, fig. 4). (D) *Gunflintia grandis*, showing enlarged and ruptured cells resulting from diagenetic alteration; new; GSC type 24380c, 24.0x – 5.5y. (E) *Gunflintia minuta* (from Hofmann 1971a, pl. 15, fig. 2), degraded trichome with spiraliform pattern. (F) *Gunflintia minuta*?, degraded filament encompassed by particulate organic matter or remnants of a degraded sheath; new; GSC type 24380c, 22.0x – 11.3y. (G) Degraded filament and associated sheath-like material; new; GSC type 24380c, 33.0x – 15.7y. (H) *Gunflintia minuta*?, surrounded by sheath-like material; new; GSC type 24380c, 13.9x – 16.5y. (I) *Huroniospora psilata* (from Hofmann 1971a, pl. 15, fig. 12). (J) *Huroniospora* sp., showing bud-like protuberance; compare with Darby (1974, fig. 1); new; GSC type 24380b, 21.5x – 11.0y. (K) *Huroniospora* sp., suggestive of "germination"; compare with Darby (1974, fig. 1, 4:33); new; GSC type 24380b, 18.7x – 12.2y. (L) *Huroniospora microreticulata* (from Hofmann 1971a, pl. 15, fig. 10). (M) *Huroniospora microreticulata* (left), with "bud-like" protuberance and *Eomicrhystridium barghoorni*? (right); new; GSC type 24380c, 21.7x – 17.8y. (N) *Huroniospora macroreticulata* (from Hofmann 1971a, pl. 15, fig. 11). (O) ?*Eomicrhystridium barghoorni* (from Hofmann 1971a, pl. 25, fig. 6), if actually *Eomicrhystridium*, the oldest acanthomorph acritarch now known. (P) *Eosphaera tyleri*; new; GSC type 24550, 26.1x – 18.5y. (Q) *Huroniospora* sp., clustered along a degraded filament; compare with Tyler and Barghoorn (1954, fig. 3); new; GSC type 24380c, 24.8x – 19.1y. (R) *Eoastrion simplex* (from Hofmann 1971a, pl. 15, fig. 14). (S) *Eoastrion simplex* (from Hofmann 1971a, pl. 15, fig. 13). (T) *Veryhachium* (?) sp. (from Hofmann 1971b, pl. 67); if actually *Veryhachium*, the oldest polygonomorph acritarch now known. (U) *Kakabekia umbellata* (from Hofmann 1971a, pl. 15, fig. 16). (V) *Exochobrachium triangulum*; new, one of only two specimens of this taxon yet illustrated; compare with Part T; GSC type 24380a, 34.3x – 16.9y.

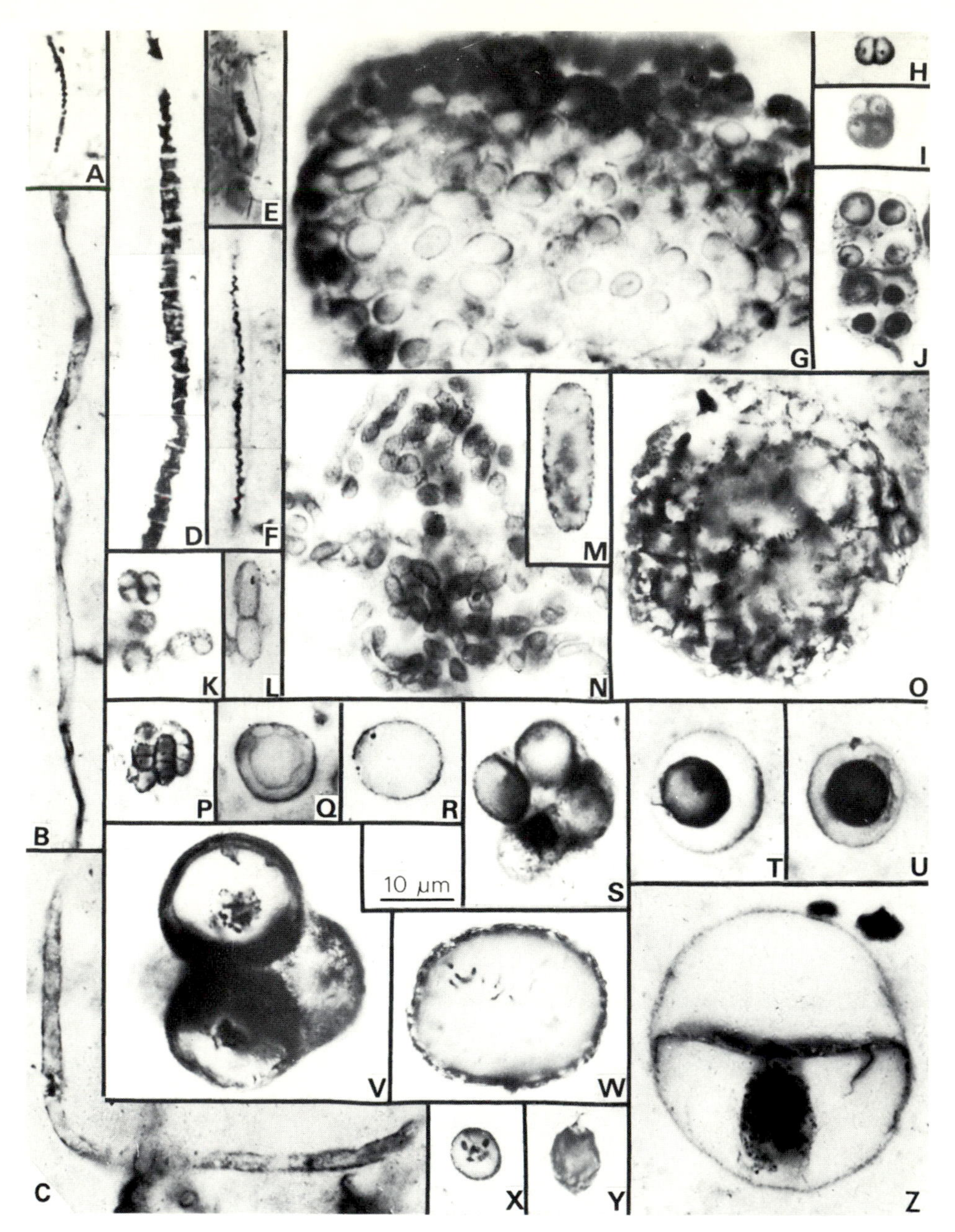

PHOTO 14-2. Microfossils from the Kasegalik (A–L, N, P, Q) and McLeary Formations (M, O, R–Z). All magnifications identical. Previously illustrated and new material; coordinates cited for new material refer to distances, in mm, from distal margin (y = 0) and right margin (x = 0) of slide (with label to right of observer). (A) *Biocatenoides sphaerula* (from Hofmann 1976, pl. 1, fig. 8). (B) *Eomycetopsis* sp. (non *Archaeotrichion*; from Hofmann 1976; pl. 1, fig. 1). (C) *Eomycetopsis filiformis*, new; GSC type 43589, 31.0x – 10.3y. (D) *Halythrix* sp. (from Hofmann 1976, pl. 1, fig. 19). (E) *Rhicnonema antiquum*, catenoid degradational form (from Hofmann (1976, pl. 1, fig. 9). (F) *R. antiquum*, spiraliform degradational form (from Hofmann 1976, pl. 1, fig. 13). (G) *Eoentophysalis belcherensis*, capsulate and capsulopunctate degradational forms; GSC type 42770, 26.6x – 15.0y. (H) *E. belcherensis*, punctate degradational form (from Hofmann 1976, pl. 6, fig. 7). (I) *E. belcherensis*, planar tetrad, punctate degradational form (from Hofmann 1976, pl. 6, fig. 9). (J) *E. belcherensis*, planar tetrads, capsulopunctate degradational form (from Hofmann 1976, pl. 6, fig. 13). (K) *Sphaerophycus parvum* (from Hofmann 1976, pl. 3, fig. 1). (L) *Eosynechococcus medius* (from Hofmann 1976, pl. 2, fig. 9). (M) *E. grandis* (from Hofmann 1976, pl. 2, fig. 12). (N) *E. moorei* (from Hofmann 1976, pl. 2, fig. 4). (O) *Palaeoanacystis vulgaris* (from Hofmann 1976, pl. 5, fig. 2). (P) *Pleurocapsa*? sp. (from Hofmann 1976, pl. 3, fig. 7). (Q) cf. *Eosphaera* (from Hofmann and Jackson 1969, fig. 19); compare with Photos 14-1P and 14-9F. (R) *Leptoteichos golubicii*, new; H-73-11-875b, 28.9x-10.1y. (S) *Myxococcoides minor* (from Hofmann 1976, pl. 7, fig. 7a). (T) *Caryosphaeroides* sp. (from Hofmann 1976, pl. 9, fig. 7). (U) *Globophycus* sp. (from Hofmann 1976, pl. 9, fig. 4). (V) *Melasmatosphaera media* (from Hofmann 1976, pl. 8, fig. 9a). (W) *M. magna* (from Hofmann 1976, pl. 8, fig. 1a). (X) *M. parva* (from Hofmann 1976, pl. 8, fig. 7). (Y) *Kakabekia*? sp. (from Hofmann 1976, pl. 7, fig. 10). (Z) *Zosterosphaera* sp. (from Hofmann 1976, pl. 9, fig. 1).

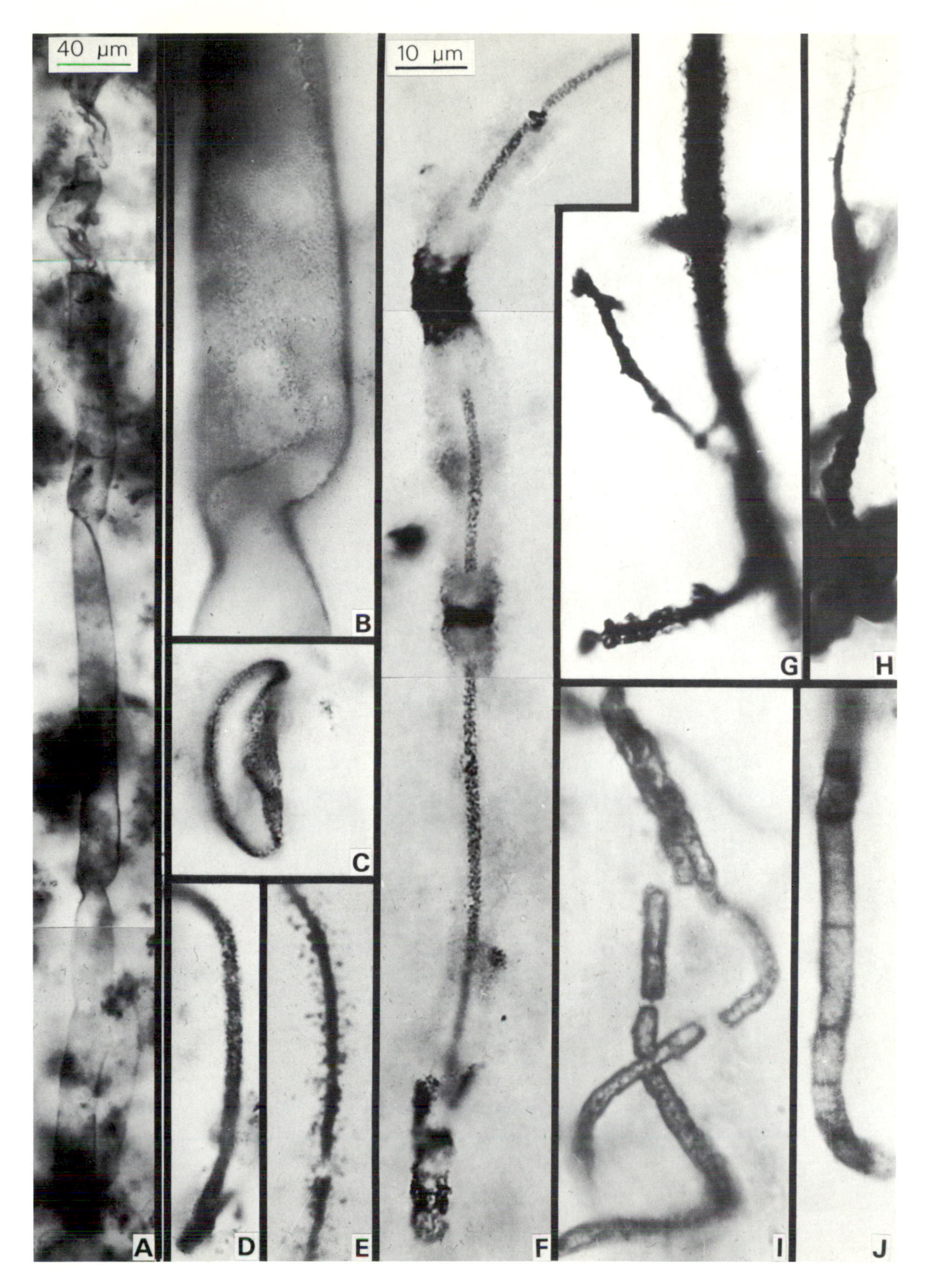

PHOTO 14-3. Microfossils from the Duck Creek Dolomite. All new material. All magnifications identical except Part A. Coordinates cited in captions refer to distances, in mm, from distal margin (y = 0) and right margin (x = 0) of slide (with label to right of observer). (A) *Siphonophycus* sp.; this organically preserved specimen has the largest diameter of any filament known from 2-Ga-old rocks (also see Awramik & Barghoorn, 1977, fig. 7c, d); Thin Section P.P.R.G. 062-1-B, 18.6x – 22.0y. (B) *Siphonophycus* sp., portion of specimen in Part A, showing diagenetically produced granulose surface texture. (C) *Siphonophycus* sp., cross section of deformed specimen; Thin Section P.P.R.G. 062-1-A, 34.5x – 16.6y. (D) *Gunflintia* sp., organically preserved, degraded filament; Thin Section P.P.R.G. 062-1-A, 33.5x – 14.5y. (E) *Gunflintia*? sp., highly degraded filament; Thin Section P.P.R.G. 049-1-B, 21.4x – 32.1y. (F) *Rhicnonema* sp., variably degraded trichome and sheath, preserved as dark brown organic material; Thin Section P.P.R.G. 059-1-A, 31.0x – 13.3y. (G) *Gunflintia*? sp., degraded filament preserved by extremely small iron oxide crystals; Thin Section P.P.R.G. 049-1-A, 14.4x – 20.2y. (H) *Ferrimonilis* sp., preserved by pyrite (wide part with large opaque crystals) and hematite (narrow part at top with tiny crystals); Thin Section P.P.R.G. 049-1-A, 16.2x – 20.0y. (I) *Gunflintia*? sp., preserved by hematite and limonite; Thin Section P.P.R.G. 049-1-B, 23.6x – 32.7y. (J) *Gunflintia*? sp., septate filament with unusually elongate cells, preserved as dark brown organic matter; Thin Section P.P.R.G. 062-1-B, 24.0x – 18.1y.

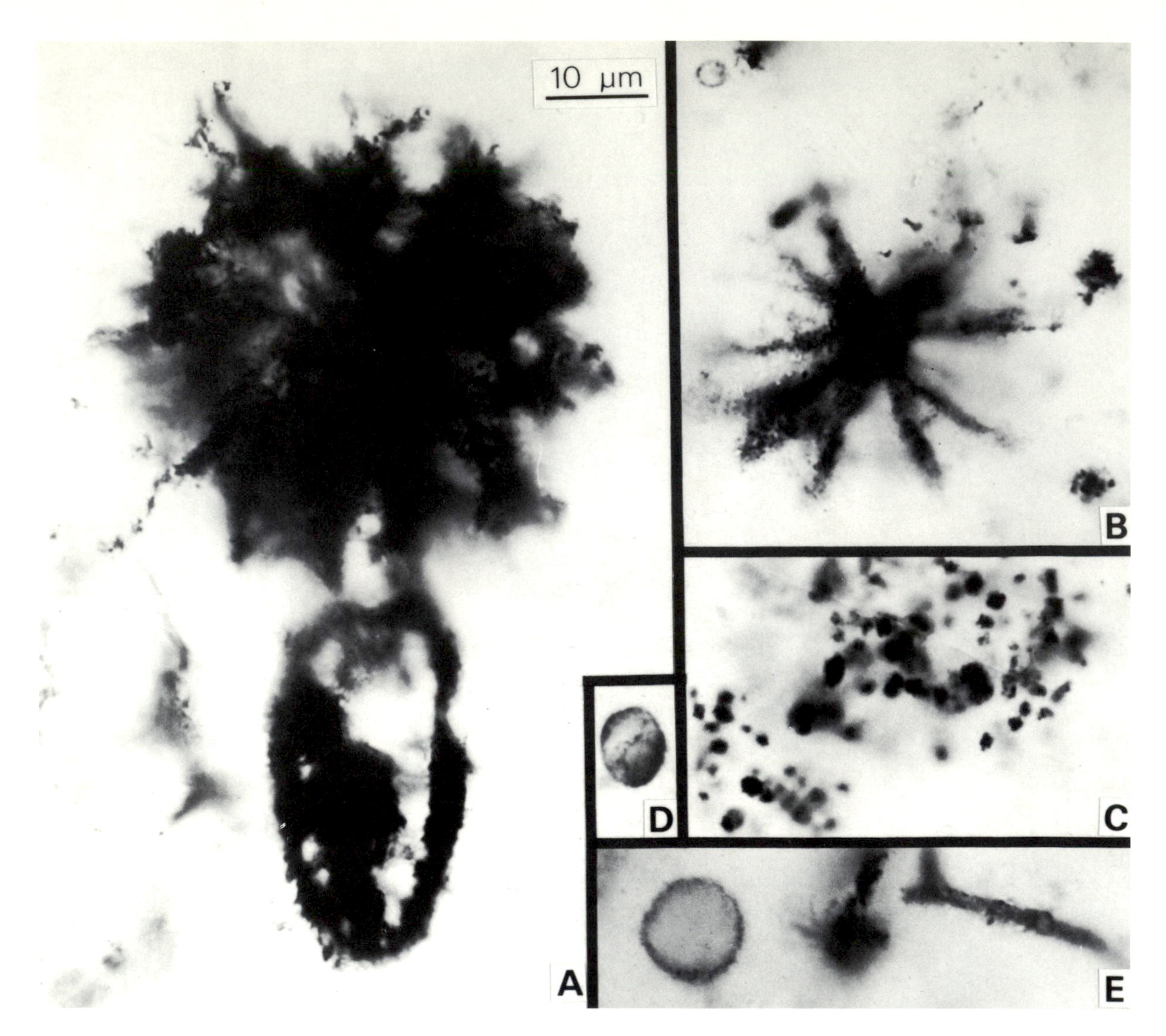

PHOTO 14-4. Microfossils from the Duck Creek Dolomite. All new material. All magnifications identical. Coordinates cited in captions refer to distances, in mm, from distal margin (y = 0) and right margin (x = 0) of slide (with label to right of observer). (A) *Kakabekia* sp., "giant" specimen, about three times larger than previously reported *K. umbellata* (see Barghoorn & Tyler, 1965a, fig. 7.); Thin Section P.P.R.G. 062-1-A, 31.1x – 11.5y. (B) *Eoastrion* sp., unusually large specimen; Thin Section P.P.R.G. 062-1-A, 35.4x – 12.6y. (C) Coccoid bacteria?; compare with Hofmann 1971a, pl. 15, fig. 20; Thin Section P.P.R.G. 060-1-A, 36.5x – 5.4y. (D) *Huroniospora* sp.; Thin Section P.P.R.G. 062-1-A, 33.9x – 10.5y. (E) *Huroniospora* sp., left, and *Gunflintia* sp. with enlarged cells; Thin Section P.P.R.G. 062-1-B, 18.9x – 20.0y.

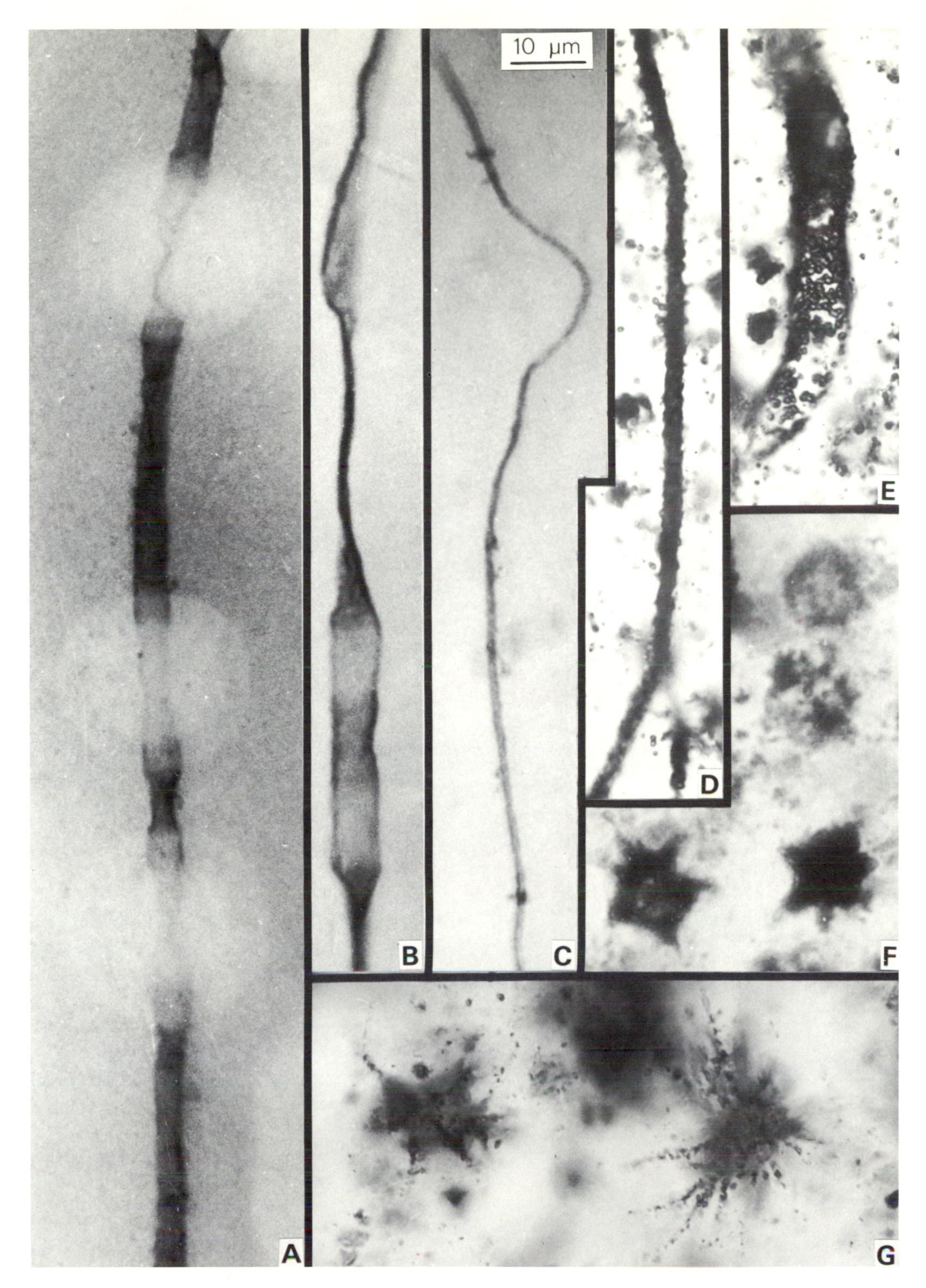

PHOTO 14-5. Microfossils from the Windidda (A–C) and Frere Formations (D–G). All magnifications identical. All new material. Coordinates cited in captions refer to distances, in mm, from distal margin (y = 0) and right margin (x = 0) of slide (with label to right of observer). (A) *Eomycetopsis* sp., discoloration (oxidation?) spheres along organically preserved filament; Thin Section P.P.R.G. 091-2-B, 56.5x – 9.5y, (B) *Eomycetopsis* sp., partially collapsed, organically preserved specimen; Thin Section P.P.R.G. 091-2-B, 49.4x – 12.0y. (C) *Archaeotrichion* sp., Thin Section P.P.R.G. 091-2-B, 36.0x – 15.3y. (D) *Gunflintia* sp., preserved in hematite; Thin Section P.P.R.G. 076-1-A, 11.2x – 15.2y. (E) *Animikiea* sp., preserved in hematite; new; Thin Section P.P.R.G. 076-1-A, 11.2x – 15.0y. (F) *Huroniospora* sp. (at top) and two specimens of ?*Eomicrhystridium barghoorni*, preserved in hematite; Thin Section P.P.R.G. 080-1-B, 24.5x – 39.0y. (G) *Eoastrion* sp., two specimens preserved in hematite; Thin Section P.P.R.G. 080-1-A, 25.5x – 11.9y.

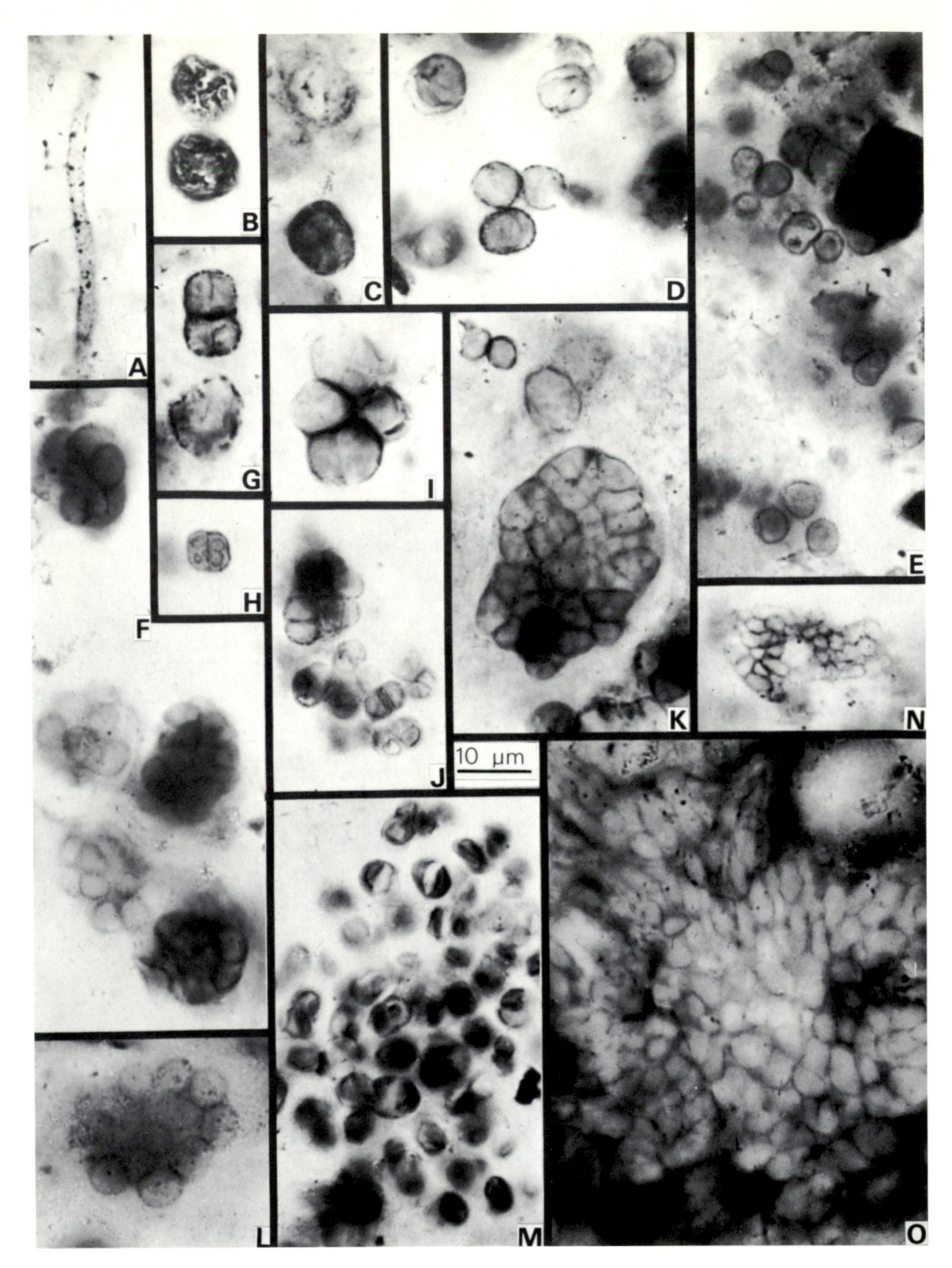

PHOTO 14-6. Microfossils from the Amelia Dolomite. All new material. All magnifications identical. Coordinates cited in captions refer to distances, in mm, from distal margin (y = 0) and right margin (x = 0) of slide (with label to right of observer). (A) *Eomycetopsis filiformis*, degraded specimen; Thin Section P.P.R.G. 111-1-A, 25.5x – 8.6y. (B) *Huroniospora ornata*; Thin Section P.P.R.G. 111-1-A, 18.0x – 11.5y. (C) *Tetraphycus* sp., Thin Section P.P.R.G. 111-1-A, 25.5x – 8.6y. (D) *Huroniospora* sp. or *Myxococcoides* sp. (cf. "*Corymbococcus*"); Thin Section P.P.R.G. 111-1-A, 23.7x – 12.3y. (E) Plasmolysed coccoid cells; Thin Section P.P.R.G. 111-1-A, 13.0x – 11.1y. (F) *Myxococcoides kingii* (cf. *Eoentophysalis belcherensis*); Thin Section P.P.R.G. 111-1-A, 13.9x – 11.0y. (G) *Tetraphycus* sp.; Thin Section P.P.R.G. 111-1-A, 20.6x – 9.0y. (H) *Tetraphycus* sp., capsulopunctate form of planar tetrad; Thin Section P.P.R.G. 109-1-A, 17.6x – 17.8y. (I) *Tetraphycus* sp., cluster of faintly preserved tetrads; Thin Section P.P.R.G. 111-1-A, 16.8x – 10.6y. (J) *Paleoanacystis plumbii* (compare with Muir, 1976, fig. 6k); Thin Section P.P.R.G. 109-1-A, 14.4x – 18.3y. (K) *Paleoanacystis* sp., Thin Section P.P.R.G. 111-1-A, 21.5x – 12.7y. (L) cf. *Eoentophysalis belcherensis*, punctate degradational form; Thin Section P.P.R.G. 111-1-A, 14.7x – 11.1y. (M) *Eoentophysalis belcherensis*, capsulate degradational form; Thin Section P.P.R.G. 111-1-A, 26.7x – 13.2y. (N) *Eosynechococcus moorei*; Thin Section P.P.R.G. 111-1-A, 20.6x – 9.9y. (O) *Myxomorphia janecekii*; Thin Section P.P.R.G. 109-1-A, 17.6x – 17.3y.

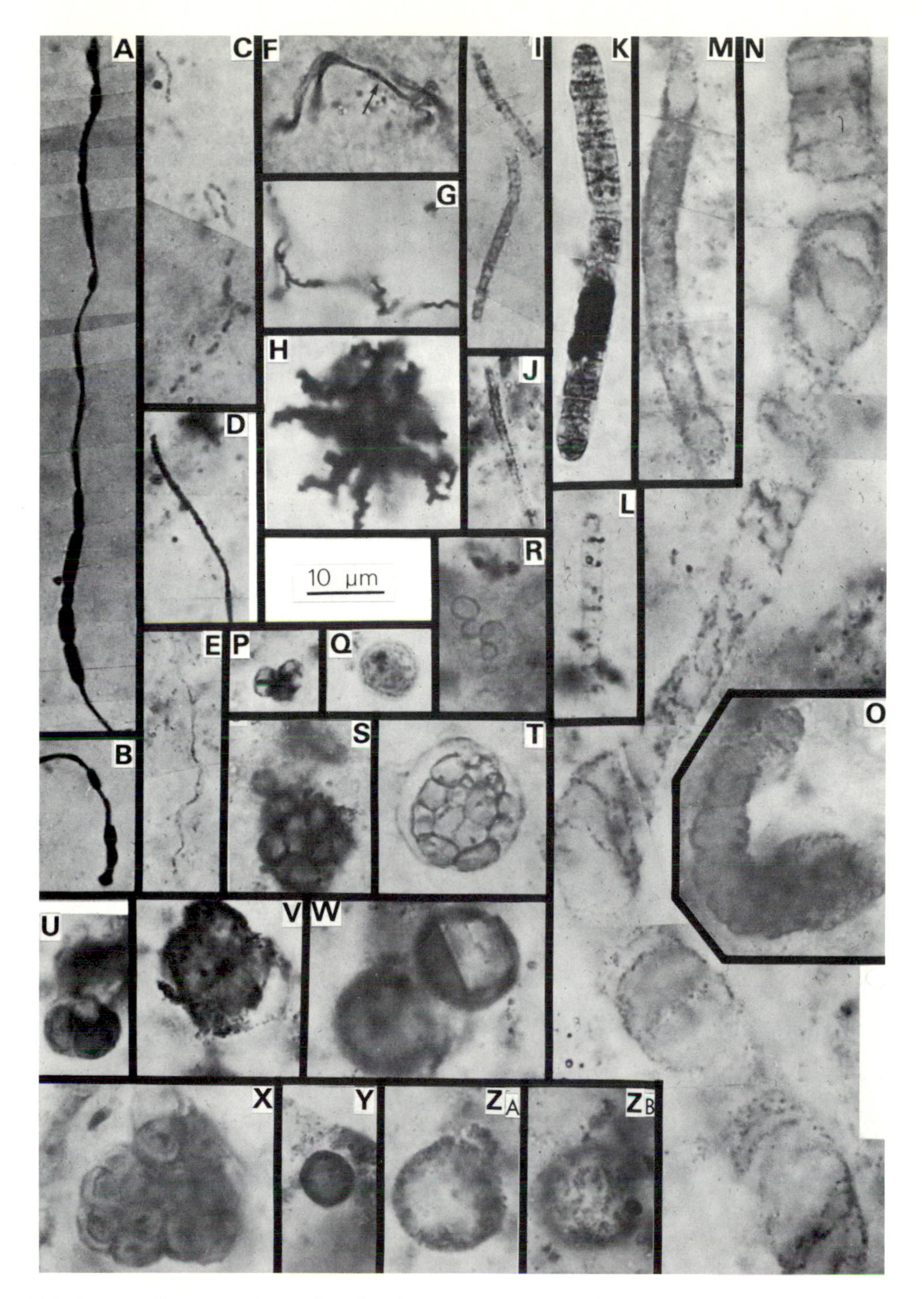

PHOTO 14-7. Organically preserved and mineral-replaced (A, B, D, G?, H) microfossils from the H.Y.C. Pyritic Shale (A–Y) and Cooley Dolomite (Z). All photographs courtesy of J. H. Oehler. (A) *Ferrimonilis variabile* (from Oehler 1977c, fig. 7a). (B) *F. variabile* (from Oehler 1977c, fig. 7c). (C) *Biocatenoides rhabdos* (from Oehler 1977c, fig. 6j). (D) *B. incrustata* (from Oehler 1977c, fig. 5g). (E) *B. pertenuis* (from Oehler 1977c, fig. 7q). (F) *Coleobacter primus* (from Oehler 1977c, fig. 6k). (G) *Ramacia carpentariana* (from Oehler 1977c, fig. 6f). (H) *Eoastrion* sp. (from Oehler 1977c, fig. 8d). (I) *Cyanonema inflatum* (from Oehler 1977c, fig. 13i). (J) *Gunflintia septata* (from Oehler 1977c, fig. 13n). (K) *Oscillatoriopsis schopfii* (from Oehler 1977c, fig. 13a). (L) *Cyanonema inflatum* (from Oehler 1977c, fig. 13d). (M) Tubular structure; cf. *Eomycetopsis* (from Oehler 1977c, fig. 14h). (N) Unnamed trichome; cf. Photo 14-9, G, and *Obconicophycus* (from Oehler 1977, fig. 12m). (O) Sheath, cf. *Animikiea* (from Oehler 1977c, fig. 14e). (P) *Sphaerophycus parvum* (from Oehler 1977c, fig. 12h). (Q) *Globophycus minor* (from Oehler 1977c, fig. 12k). (R) *Nanococcus vulgaris* (from Oehler 1977c, fig. 10k). (S) *Myxococcoides kingii* (from Oehler 1977c, fig. 10u). (T) *Clonophycus elegans* (from Oehler 1977c, fig. 11a). (U) *Bisacculoides vacua* (from Oehler 1977c, fig. 12f). (V) *Bigeminococcus mucidus* (from Oehler 1977c, fig. 12d). (W) *Bisacculoides grandis* (from Oehler 1977c, fig. 12b). (X) *Bisacculoides tabeoviscus* (from Oehler 1977c, fig. 11h). (Y) *Huroniospora* sp. (from Oehler 1977c, fig. 11p). (Za, b) *Huroniospora* sp., two different focal views (from Oehler 1977c, fig. 11L).

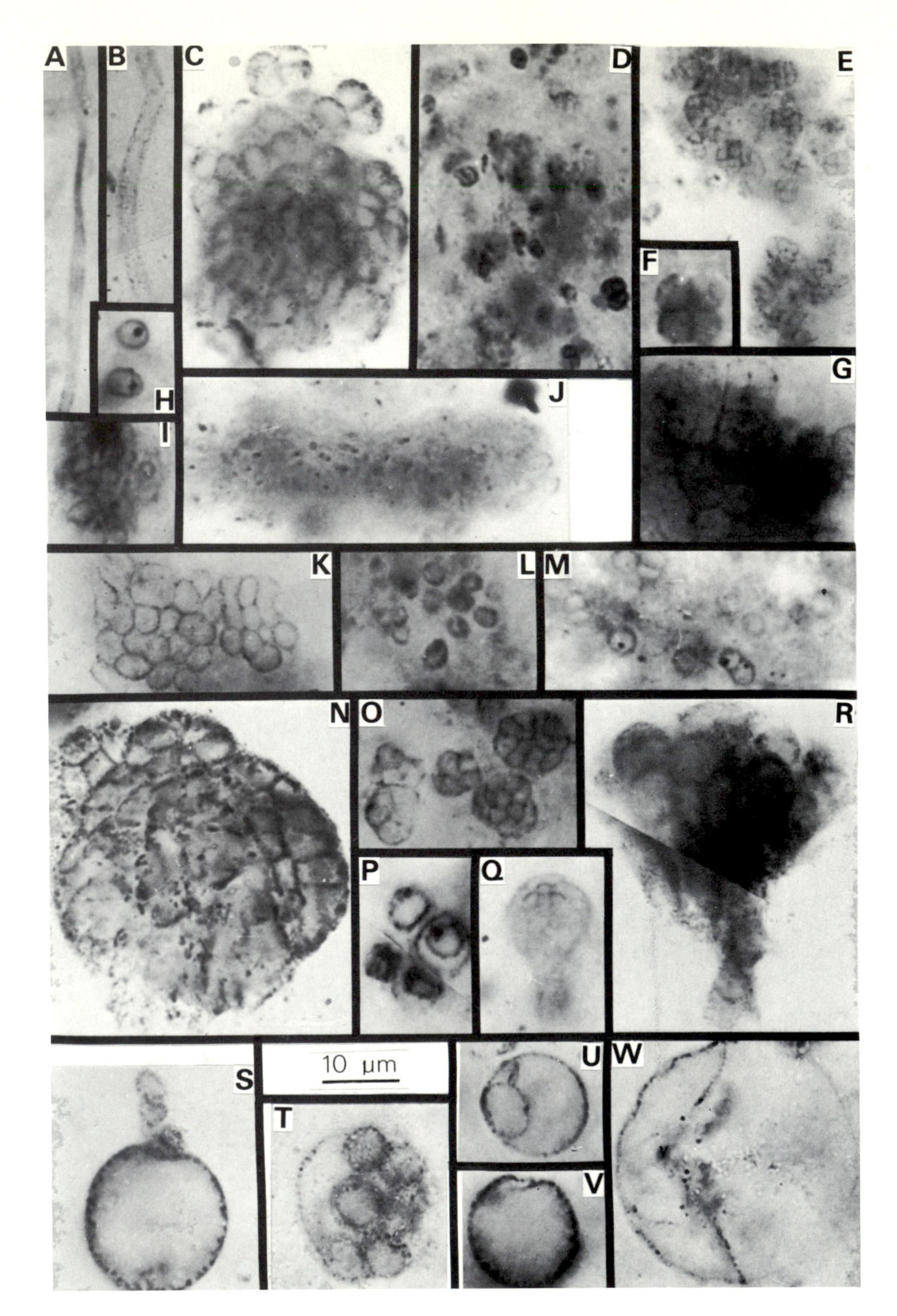

PHOTO 14-8. Microfossils from the Balbirini Dolomite. All photographs courtesy of D. Z. Oehler. (A) Unnamed filament, cf. *Eomycetopsis* sp. (from D. Oehler 1978, fig. 12a). (B) *Eomycetopsis* sp. (from Oehler 1978, fig. 12g). (C) *Eoentophysalis belcherensis* (from D. Oehler 1978, fig. 11e). (D) *Tetraphycus acinulus* (from D. Oehler 1978, fig. 10y). (E) *T. diminutivus* (from D. Oehler, 1978, fig. 9a). (F) *T. major* (from D. Oehler 1978, fig. 10q). (G). *T. gregalis* (from D. Oehler 1978, fig. 9k). (H) *Pilavia maculata* (from D. Oehler 1978, fig. 10a). (I) *Myxococcoides minuta* (from D. Oehler 1978, fig. 8q). (J) *Sphaerophycus reticulatum* (from D. Oehler 1978, fig. 7a). (K) *Myxococcoides cracens* (from D. Oehler 1978, fig. 8d). (L) *Nanococcus vulgaris* (from D. Oehler 1978, fig. 8u). (M) *N. vulgaris* (from D. Oehler 1978, fig. 10j). (N) *Paleoanacystis vulgaris* (from D. Oehler 1978, fig. 7h). (O) *P. plumbii* (from D. Oehler 1978, fig. 9c). (P) *Sphaerophycus parvum* (from D. Oehler 1978, fig. 10r). (Q) *Myriasporella pyriformis* (from D. Oehler 1978, fig. 4v). (R) *Myxomorpha janecekii* (from D. Oehler 1978, fig. 7i). (S) *Clonophycus ostiolum* (from D. Oehler 1978, fig. 4t). (T) *C. biattina* (from D. Oehler, 1978, fig. 4k). (U) *C. vulgaris* (from D. Oehler 1978, fig. 4o). (V) *C.* sp., (from Oehler D. 1978, fig. 4b). (W) *C. refringens*, cf. *Zosterosphaera* (from D. Oehler 1978, fig. 4g).

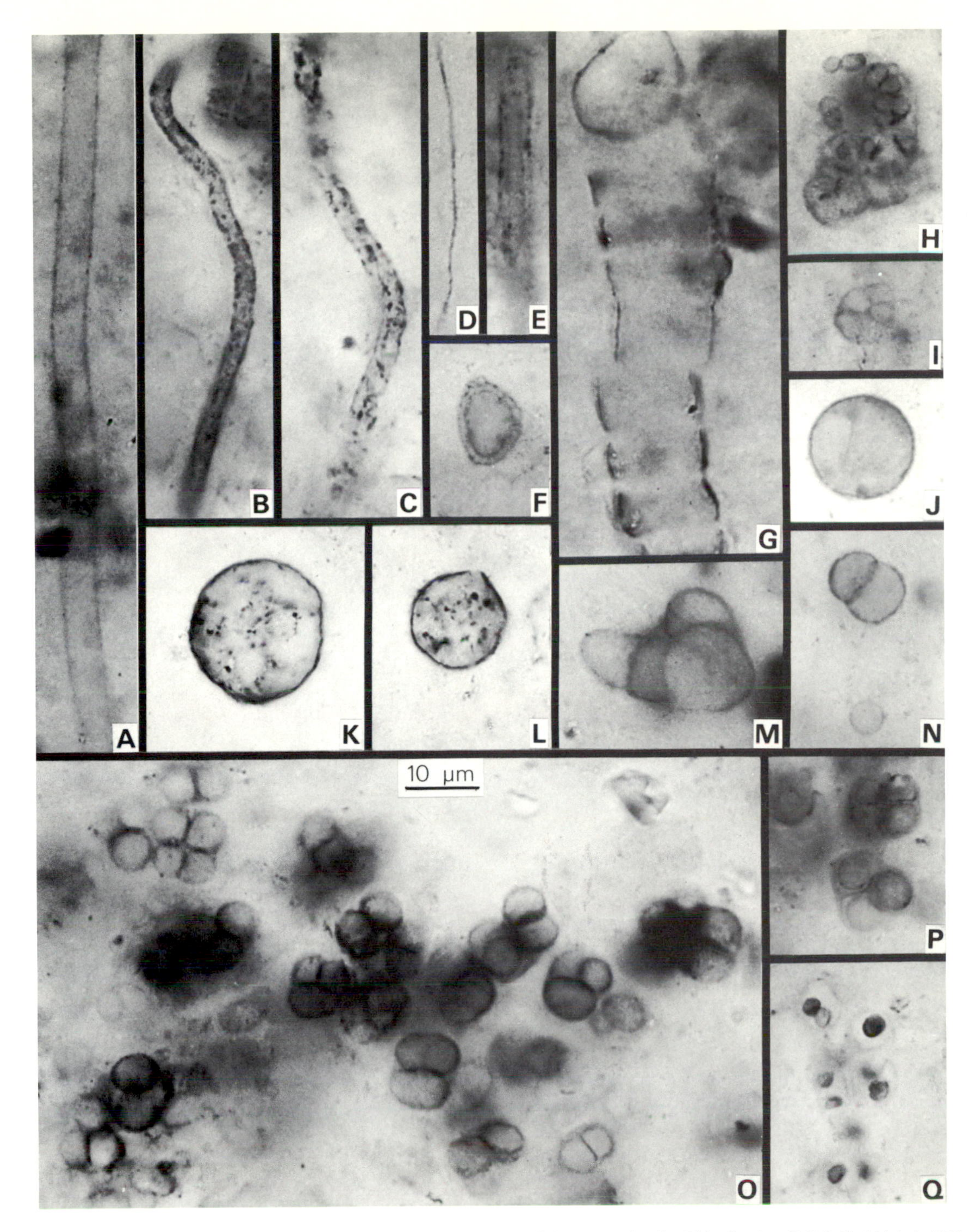

PHOTO 14-9. Microfossils from stromatolitic chert of the Bungle Bungle Dolomite. All new material. All magnifications identical. Coordinates cited in captions refer to distances, in mm, from distal margin (y = 0) and right margin (x = 0) of slide (with label to right of observer). (A) *Eomycetopsis* sp.; Thin Section P.P.R.G. 154-1-A, 31.2x – 9.3y. (B) *Eomycetopsis filiformis*; Thin Section P.P.R.G. 159-1-B, 40.4x – 22.6y. (C) *Rhicnonema antiquum*; Thin Section P.P.R.G. 159-1-B, 38.0x – 3.7y. (D) *Archaeotrichion* sp. or *Biocatenoides* sp.; Thin Section P.P.R.G. 153-1-B, 45.5x – 17.3y. (E) Oscillatoriacean sheath containing faint trichome; Thin Section P.P.R.G. 159-1-B, 38.1x – 3.4y. (F) cf. *Eosphaera*, compare with Photo 14-1P and 14-2Q; Thin Section P.P.R.G. 153-1-B, 47.5x – 15.8y. (G) Large oscillatoriacean, cf. *Obconicophycus*, containing very faint remnants of disc-shaped cells; Thin Section P.P.R.G. 154-1-A, 27.8x – 2.5y. (H) *Sphaerophycus parvum*; Thin Section P.P.R.G. 159-1-A, 21.0x – 10.5y. (I) *S. parvum*; Thin Section P.P.R.G. 153-1-A, 21.2x – 12.6y. (J) *Leptoteichos golubicii*; Thin Section P.P.R.G. 153-1-B, 41.9x – 13.5y. (K) *Melasmatosphaera media*; Thin Section P.P.R.G. 153-1-A, 19.8x – 11.3y. (L) *M. media*; Thin Section P.P.R.G. 153-1-A, 19.4x – 11.2y. (M) *Myxococcoides minor*; Thin Section P.P.R.G. 159-1-A, 19.3x – 18.8y. (N) *M. minor*; Thin Section P.P.R.G. 153-1-B, 48.5x – 17.0y. (O) *Eoentophysalis belcherensis*; Thin Section P.P.R.G. 159-1-B, 36.0x – 17.7y. (P) *E. belcherensis*, capsulate degradational form; Thin Section P.P.R.G. 159-1-B, 35.9x – 16.9y. (Q) *E. belcherensis*, capsulopunctate degradational form; Thin Section P.P.R.G. 159-1-B, 41.8x – 19.6y.

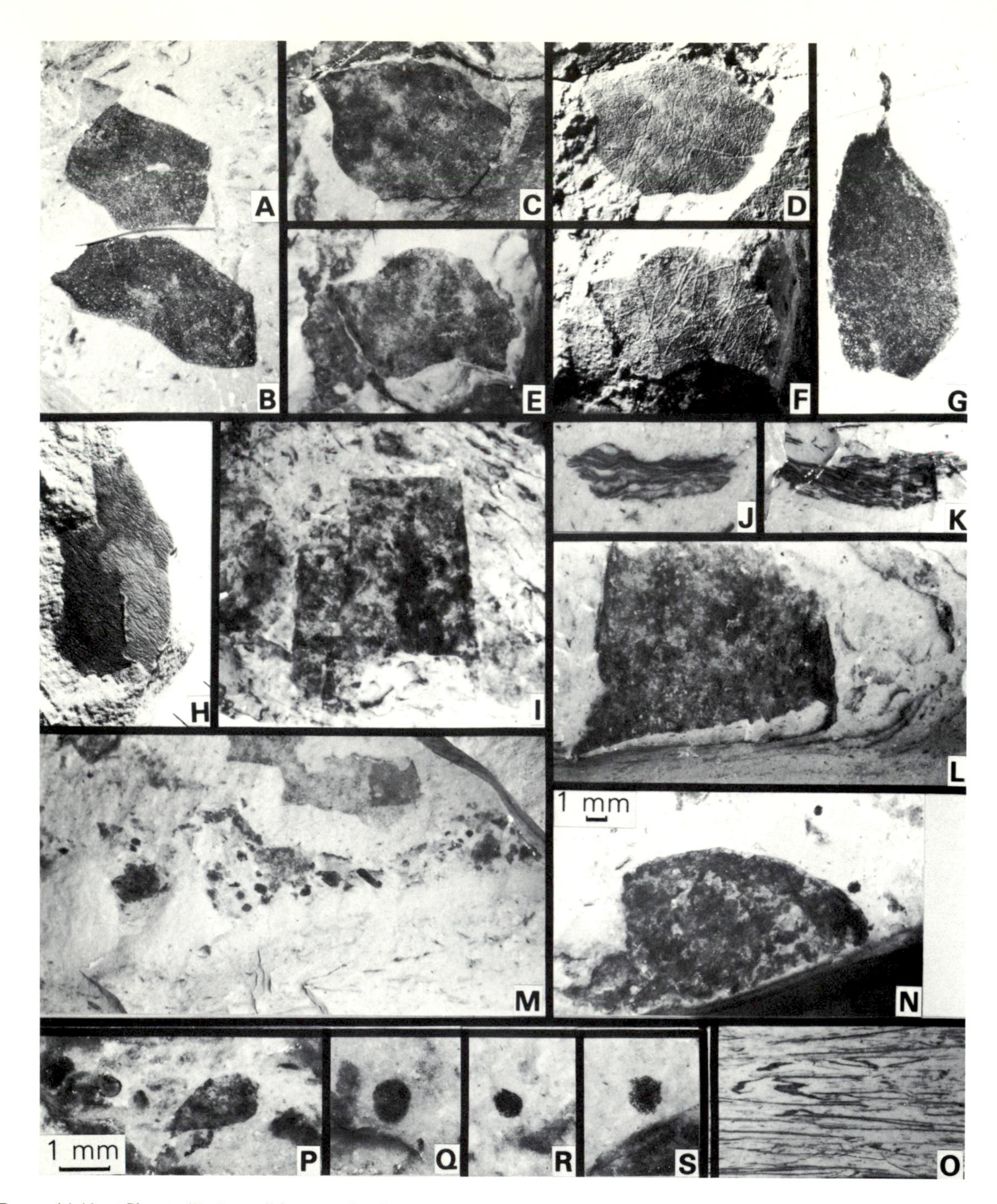

PHOTO 14-10. "*Chuaria*-like" possible macrofossils (N, P–S) and "*Beltina-like*" carbonaceous films (A–O) from pale green "shale" (dolosiltite) for the Bungle Bungle Dolomite. All new material. Scale in (N) is for (A)–(O); magnification for (P)–(S) shown in (P). (A), (B) Angulate carbonaceous films without folds, photographed in air; Rock Specimen P.P.R.G. 144–(7). (C), (D) Angulate film with folds, photographed under water (C) and in air under low oblique illumination (D); Rock Specimen P.P.R.G. 144–(2). (E), (F) Angulate film with folds, photographed under water (E) and in air under low oblique illumination (F); Rock Specimen P.P.R.G. 144–(6). (G) Rounded film photographed under water; Rock Specimen P.P.R.G. P.P.R.G. 144-(10). (H) Angulate film with folds, photographed in air under low oblique illumination (same specimen shown in Part M, top; Rock Specimen P.P.R.G. 144-(1). (I) Overlapping angulate films, photographed under water; Rock Specimen P.P.R.G. 144-(3). (J) Crenulated film, photographed under water; Rock Specimen P.P.R.G. 144-(8). (K) Crenulated film, photographed under water; Rock Specimen P.P.R.G. 144-(9). (L) Tightly and obliquely folded film, photographed under water (plan view upper 3/4 of photo, oblique section lower $\frac{1}{4}$ of photo); Rock Specimen P.P.R.G. 144-(5). (M) Angulate film (top) and small discs (middle), photographed under water; Rock Specimen P.P.R.G. 144-(1). (N) Rounded film and two discs (to right), photographed under water; Rock Specimen P.P.R.G. 144-(1). (O) Photograph of polished slab showing films, highly deformed parallel to bedding laminae; Rock Specimen P.P.R.G. 144-(11). (P) "*Chuaria*-like" discs and irregular films, photographed under water; Rock Specimen P.P.R.G. 144-(4a). (Q) "*Chuaria*-like" disc, photographed under water; Rock Specimen P.P.R.G. 144-(4). (R), (S) Higher magnification photograph of the two "*Chuaria*-like" discs shown in (N), photographed under water; Rock Specimen P.P.R.G. 144-(1).

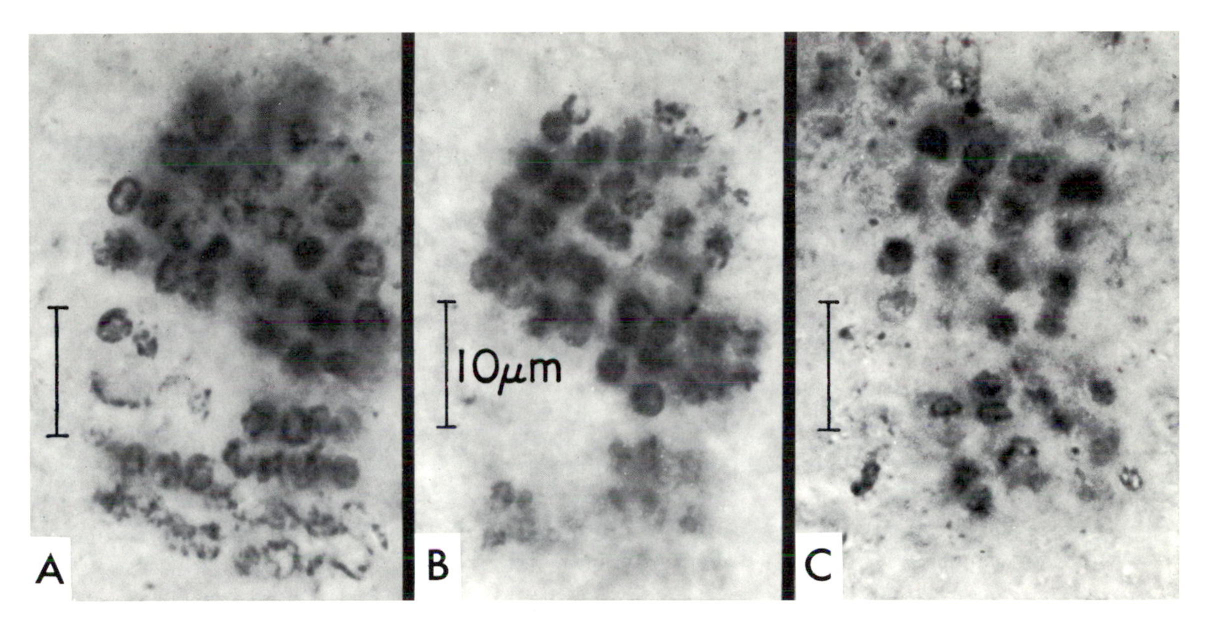

PHOTO 14-11. Microfossils from the Paradise Creek Formation (Licari et al., 1969). Serial optical sections (Parts A, B, C) of cubic aggregates of cells attributed to *Eucapsis*(?) sp. from chert blebs found associated with the columnar branching stromatolite *Eucapsiphora paradisa*. Photos courtesy of G. Licari.

CHAPTER 12

Geochemical Evidence Bearing on the Origin of Aerobiosis, A Speculative Hypothesis

By J. M. Hayes

12-1. Introduction

The atmosphere and hydrosphere of the early Archean earth were essentially anoxic; the crust was at that time relatively rich in unoxidized inorganic materials (Chapter 11). An environment without oxygen, the Earth was then a different planet, from the perspective of either the biochemist or geochemist. The paleontological record (Chapters 8 and 9) shows, nevertheless, that life existed on that different planet, and it is widely held that the advent of oxygenic photosynthesis was the singular event that led eventually to our modern environment. The significance and scope of this transition can hardly be overstated, but evidence regarding details of the process and its timing has been surprisingly difficult to find. To the organic geochemist, it has been particularly puzzling that an event of such significance in the development of the biogeochemical cycle of carbon should somehow have passed unrecorded.

It is now evident that the record of carbon isotopic abundances in sedimentary organic matter (Figure 7-3) displays marked variations about 2.8 Ga before the present. This isotopic record, being virtually the first line of evidence to break an otherwise monotonous sequence of suggestions that the Precambrian carbon cycle differed little from its Phanerozoic counterpart, demands interpretation. What is presented here is an estimate, a plausible sequence of events, linking features in the carbon isotopic record to the origin of oxygenic photosynthesis. The conclusions reached are not perfectly secure, but it is hoped that this exploration will be useful and stimulating.

12-2. The Carbon Isotopic Record

The isotopic analyses summarized in Figure 7-3 (which presents not only results obtained by the P.P.R.G., but by many previous investigators as well) show that the relative abundance of carbon-13 in sedimentary organic material varies especially widely for about 10^9 years beginning about 2.8 Ga ago. To explore possible explanations for this isotopic fluctuation, it is necessary first to consider the mechanisms by which isotopic abundances are controlled during the biogeochemical cycling of carbon.

The distribution of carbon isotopes between inorganic and organic carbon pools during geochemical cycling is discussed in Chapter 7 (Figure 7-2 and accompanying text). Figure 12-1 summarizes this process schematically. The total pool of oxidized carbon in the atmosphere and hydrosphere (ΣCO_2) receives an input from geological recycling. The isotopic composition of that input must, on the average, be equal to the isotopic composition of total crustal carbon, δ_Σ. A portion of the oxidized carbon is biologically reduced in order to form, first, living organic material ("biomass") and, eventually, dead organic matter. Carbon from these pools is subject to burial in sediments but also can be reoxidized by biological processes associated with the recovery of stored chemical energy. In the present environment, biological recycling is so efficient that, on the average, the rate at which organic carbon is immobilized in sediments is about 1000-fold smaller than the rate at which biomass is produced by fixation of CO_2 (see Table 5-1).

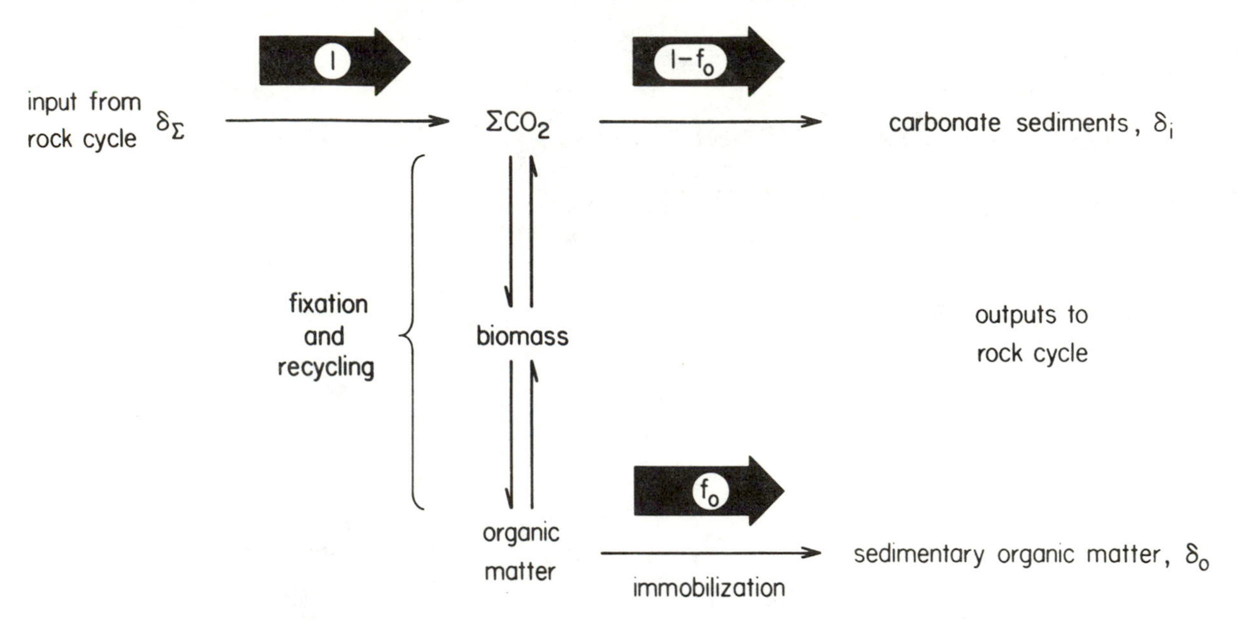

FIGURE 12-1. Schematic representation of the system in which isotopic fractionation occurs during carbon cycling. A unit quantity of carbon enters the system and is partitioned between oxidized (ΣCO_2) and reduced (biomass and organic matter) forms. As a result, the carbon flow is divided between two outputs, the oxidized form (carbonate sediments) receiving a fractional amount $1 - f_o$ while the reduced form (sedimentary organic matter) receives a fractional amount f_o.

The efficiency of Precambrian carbon cycles and the roles that abiotic processes might have played in them is unknown. Independent of the mechanism and efficiencies of carbon fixation and recycling, however, the amount and isotopic composition of the carbon entering the system must, on the average, be balanced by the outputs. The requirement can be expressed in mathematical form:

$$\delta_\Sigma = f_o \delta_o + (1 - f_o)\delta_i, \tag{12-1}$$

where f_o is the fraction of the input that leaves as organic carbon, δ_o is the isotopic composition of the organic carbon, and δ_i is the isotopic composition of the inorganic carbon. Note that because δ_Σ is a constant, f_o can be calculated whenever δ_i and δ_o can be measured. Provided that evidence can be obtained showing that the measured values of δ_i and δ_o are globally representative, it would appear that this means of estimating the rate of sequestration of organic material in sediments is superior to approaches based on estimates of total organic carbon abundances in sediments.

It is important to point out that neither f_o nor sedimentary total organic carbon abundances (T.O.C.) can be simply related to biomass or to the rate of production of biomass from CO_2 (often termed "primary productivity"). The value of f_o or level of T.O.C. is set by the rate at which organic material is immobilized in sediments (see Figure 12-1). That rate need not be related to primary productivity unless the latter process fails to produce enough organic material to supply the demands of the immobilization pathway. For example, a single value of f_o might be related either to an ecosystem with very high primary productivity and very active recycling or to an ecosystem with very low primary productivity and very slow recycling. Similarly, the division of organic carbon between biomass and dead material could vary significantly, making f_o a very unreliable indicator of biomass. In short, f_o provides information about the quantities of organic material available for sedimentation, but tells little about the rate of formation of that material or about the division of organic carbon between living and dead material.

Observed isotopic abundances can also furnish information about the mechanisms responsible for partitioning ^{13}C between the inorganic and organic carbon pools. As noted in Chapter 7, different forms of carbon fixation tend to be characterized by different isotope effects that result in different levels of fractionation of ^{13}C between the organic and inorganic carbon pools. In this connection, it is useful to point out that the level of isotopic fractionation can be expressed simply in terms of the difference between the isotopic abundances:

$$\Delta\delta = \delta_i - \delta_o, \qquad (12\text{-}2)$$

and that Equation 12-1 can then be rewritten as:

$$\delta_i = \delta_\Sigma + f_o\Delta\delta, \qquad (12\text{-}3)$$

or

$$\delta_o = \delta_\Sigma - (1 - f_o)\Delta\delta. \qquad (12\text{-}4)$$

The utility of these independent formulations is illustrated below.

12-2-1. The Phanerozoic

Throughout the Phanerozoic, the inorganic and organic carbon pools have remained near 0 and −27 ‰ vs. PDB (PDB is Peedee belemnite standard), respectively (Figure 7-3). The essentially constant difference ($\Delta\delta = 27$ ‰) is most simply interpreted in terms of consistent control of isotopic abundances by the kinetic isotope effect associated with ribulose-bisphosphate carboxylase, the enzyme responsible for carbon fixation in almost all primary production (Chapter 7). As indicated in Figure 12-2A, a system displaying a constant fractionation of 27 ‰ could operate at any value of f_o between zero and unity. In spite of that range of possible values, however, the isotopic records of both carbonates and sedimentary organic carbon indicate that f_o has been approximately constant near a value of 0.19. The constancy of f_o may appear notable and surprising, if it is asked why the biosphere has not "taken over" at some point, capturing a larger fraction of the carbon and driving f_o to higher values. Redfield (1958) has provided an answer by observing that, given the existence of nitrogen fixation, the fertility of the present global ecosystem is most rigorously limited by the crustal abundance of phosphorus. A global increase in amounts of organic matter would thus require an increase in available phosphorus levels or a decrease in the average P/C ratio of biologically formed organic material. However, a sustained increase in available levels of phosphorus would be impossible without somehow adding to the

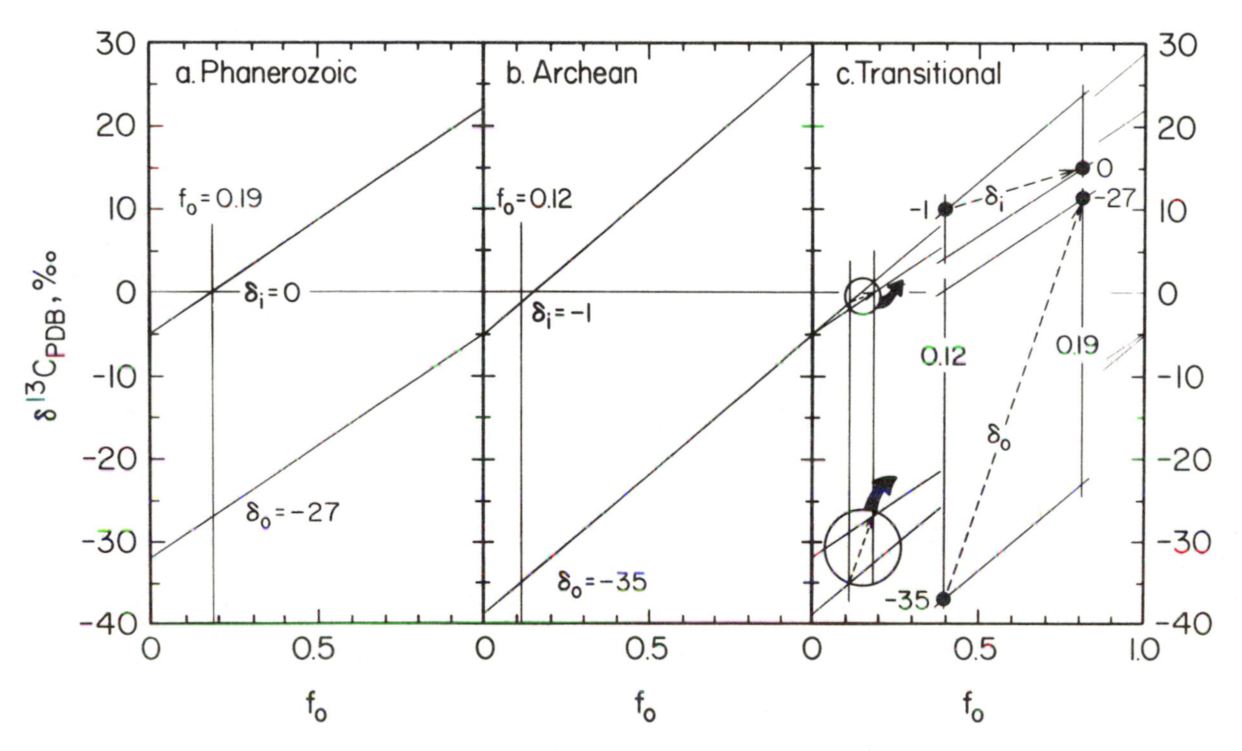

Figure 12-2. Fractionation of carbon isotopes in the global carbon cycle (a) during the Phanerozoic; (b) during the Early and Middle Archean; and (c) during the intervening transition. The expanded portions of (c) indicate how it happens that a 1 ‰ change in δ_i can be correlated with an 8 ‰ change in δ_o.

total amount of phosphorus circulating in the rock cycle. It can be suggested, therefore, that, at least after the origin of N_2 fixation, the constancy of f_o reflects the constancy of available phosphorus levels and that the particular value of f_o observed is characteristic of the average P/C ratio in organic materials.

It is important to bear in mind that the sedimentary record must be related not only to the mechanism of primary production, but to all steps between carbon fixation and the final immobilization of organic carbon in sediments. In the present ecosystem, this includes aerobic food chains and a complicated set of degradative processes in anaerobic sediments. In the latter, nitrate- and sulfate-reducing bacteria commonly attack the organic matter before the final stages of degradation usually associated with methanogenesis (Fenchel and Blackburn, 1979). In short, although it may appear to record isotopically only the mechanism of primary production, sedimentary organic material is, in fact, the product of a complex and highly stratified ecosystem with at present distinct aerobic and (several different) anaerobic zones. The isotopic composition of sedimentary organic matter is potentially sensitive to changes at any point in this pathway.

12-2-2. The Early and Middle Archean

As in the Phanerozoic, the isotopic composition of Archean sedimentary organic material must encode information about the ecosystem from which it was derived. Any inferences, however, must be very cautiously drawn. For one thing, the record is extremely fragmentary; it is not possible to be sure that observed isotope ratios are globally representative or that they represent compositions that prevailed over long periods of time. For another, the number of additional constraints is minimal; it can be said only that the biosphere was largely, if not completely, anaerobic and involved nothing but prokaryotes. It is true, however, that where samples are available, the preservation of isotopically distinct inorganic and organic carbon pools is consistent with the occurrence of isotopic fractionation in an open system like that depicted in Figure 12-1. The observation of equal ^{13}C abundances at localities separated in time and space (e.g., the "Swaziland Sequence" and the Warrawoona and Gorge Creek Groups) suggests consistent control of isotopic abundances in organic matter. The evidence favors consideration of the problem in terms of a dynamic system with some definable input, a fractionation process or processes, and inorganic and organic outputs; that is, a carbon cycle.

With some allowance for post-depositional alteration of isotopic abundances (Table 5-10), the observations summarized in Figure 7-3 suggest a constant difference in ^{13}C contents between carbonates near -1 and organic matter near -35 ‰ vs. PDB. In the modern ecosystem, a similar control of isotopic abundances is associated with a consistent and dominant form of autotrophic carbon fixation, and it is logical to adopt that same interpretation for the Archean observations, thus placing the origin of autotrophic carbon fixation at or before ~3.5 Ga before the present. The fractionation diagram corresponding to $\Delta\delta =$ 34 ‰ (with $\delta_\Sigma = -5$ ‰ vs. PDB) is shown in Figure 12-2B. It can be observed that a somewhat lower value of f_o is called for. If phosphorus were limiting, this finding would indicate a higher P/C ratio in organic matter at that time, an observation fully consistent with the generally higher abundance of phosphorus in prokaryotes than in eukaryotes (Bowen, 1979, pp. 83–98; Lerman, 1979, p. 23). The reasonableness of this finding suggests that nitrogen—on a global scale, at least—must not have been a limiting nutrient at this time.

The observed isotopic fractionation might be consistent with operation of the Calvin cycle at higher CO_2 pressures than at present (Farquhar, 1980; O'Leary, 1981), or it might be associated with an entirely different form of autotrophy. DeNiro et al. (1981), for example, have shown that the biomass of methogens is characteristically depleted by approximately 30 ‰ relative to the CO_2 used as a carbon source.

Whatever the case, it is known that the ecosystem of the Early and Middle Archean must not have depended on oxygen as the principal electron donor and acceptor in carbon cycling. Some oxidizable and reducible element in addition to carbon must have been turning over rapidly, and there may have been H_2-rich and H_2-poor zones (or there may have been environments without sulfate and environments rich in that potential electron acceptor) but, save for a few possible circumstances in which traces of O_2 produced photolytically in the atmosphere might have survived briefly before being scavenged by inorganic material at the surface, there were no environments in which appreciable concentrations of free O_2 persisted. On that anoxic world, obligate anaerobes could have lived freely in close communication with the mobile carbon reservoirs in the atmosphere and hydrosphere.

A carbon cycle in which primary production was due principally to bacterial photosynthesis, with hydrogen or reduced sulfur compounds acting as the electron donors, seems most likely. Recycling of carbon could be due to sulfate-reducing bacteria and (or) anaerobic fermentation, the latter process providing abundant H_2 for possible bacterial consumers. Modern analogs for such cycles based on sulfur do exist (Fenchel and Balckburn, 1979, p. 140) and the existence of some sulfate/sulfide deposits in Archean sediments (Figure 7-6) is probably significant. Alternatively, CO_2-reducing methanogens might have acted as primary producers, with recycling being due simply to anaerobic fermentation and to oxidation of CH_4 by hydroxyl radicals photolytically produced in the atmosphere (J. C. G. Walker, pers. comm. 1981).

12-2-3 The Late Archean

AN OUTLINE OF THE PROBLEM

At least by 2.8 Ga before the present, if not before, something in the carbon cycle began to allow the incorporation in sediments of organic material extraordinarily depleted in ^{13}C. As noted in Chapter 5 and graphically summarized in Figure 7-3, the phenomenon was widespread. Schoell and Wellmer (1981) have reported an additional instance of isotopic compositions below -40 ‰ vs. PDB in sediments from the Superior Province in the Canadian Shield, and further investigations of related materials (M. Schoell, pers. comm. 1981) have revealed still more negative values, approaching those here observed in the Fortescue Group of Western Australia (Tables 5-7 and 5-8, Figure 7-3). These kerogens strongly depleted in ^{13}C are found together with carbonates near 0‰ vs. PDB, and it is generally true that the wide variations in the organic carbon isotopic record are not paralleled by related variations in the isotopic record of carbonate carbon (the few outliers in the population of carbonates represent special circumstances and have been interpreted as not representative of oceanic total carbonate; see Becker and Clayton, 1972; Schidlowski et al., 1976a). This contrast is remarkable, and it should be clear that any reconstruction of the late Archean carbon cycle must account not only for wide variations in the abundance of ^{13}C in organic matter but also for simultaneous isotopic constancy in carbonates.

The range of all possible lines representing the global-average isotopic composition of carbonates (that is, the δ_i-lines of Figure 12-2) is graphically indicated in Figure 12-3. It should be evident that constancy of the isotopic composition of carbonates requires either that both f_o and $\Delta\delta$ have been constant or that both have changed in a specific and coordinated way that happens to maintain δ_i constant. If it is hypothesized that Figure 12-2B represents isotopic partitioning in the carbon cycle that preceded the Late Archean, the isotopic compositions of Late Archean carbonates and organic

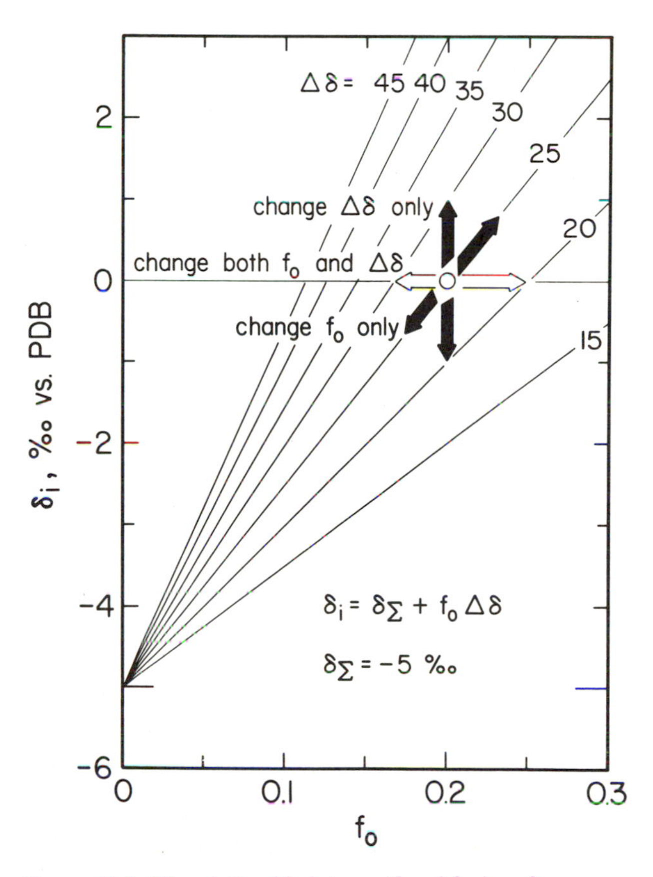

FIGURE 12-3. The relationship between f_o and δ_i given $\delta_\Sigma = -5$ ‰ vs. PDB and various values of $\Delta\delta$. It can be seen that changing only $\Delta\delta$ or f_o will have the result of changing δ_i. Given any change in either f_o or $\Delta\delta$, constancy of δ_i can result only if the remaining variable is also changed in a specific way. For example, for a system initially producing $\delta_i = 0$ ‰ while $\Delta\delta = 25$ ‰ and $f_o = 0.20$, maintenance of $\delta_i = 0$ ‰ if $\Delta\delta$ is changed to 30 ‰ would be possible only if f_o were simultaneously changed to 0.16.

matter can be addressed separately in the following terms:

(*i*) The isotopic composition of Late Archean carbonates is unchanged from that prevailing in the Early and Middle Archean because f_o and $\Delta\delta$ are unchanged (where $\Delta\delta$ refers, in this case, to the weighted average difference between inorganic- and organic-carbon isotopic compositions). This suggestion is introduced because it is easier to believe that neither parameter has changed than to suggest that both happen to have changed in a way that maintained δ_i constant. While this argument loses some of its power because there are apparently some natural variations in δ_i, it is conceptually sound and is, at least, a useful guide for speculation.

(*ii*) The isotopic composition of Late Archean organic matter is variable because of changes in the carbon cycle "downstream" from primary production. This suggestion follows directly from the hypothesized constancy of $\Delta\delta$.

Together, these statements describe a carbon cycle in which the dominant means of primary production (which affects $\Delta\delta$ globally) is unchanged from earlier times but in which the isotopic record (in organic material) of that constancy has been scrambled by reprocessing of the primarily produced organic matter. It is useful to explore ways in which that reprocessing might have occurred in order to examine the plausibility of this hypothesis.

A MODEL FOR THE DEVELOPMENT OF ^{13}C-DEPLETED KEROGENS

Among natural carbon pools, biogenic methane is unique in terms of its extraordinary levels of ^{13}C depletion (see Section 7-2-2 and Deines, 1980a). This same material is prominently involved in the reprocessing of organic material in the modern environment (see Section 5-2-1). Because it has been suggested (above) that the isotopic record of the Late Archean carbon cycle is best interpreted in terms of a development in recycling mechanisms, and because that record features extreme levels of ^{13}C depletion, it is logical to suggest that the ^{13}C-depleted kerogens are linked in some way to the production and utilization (i.e., to make cellular material which is incorporated in kerogen) of biogenic methane.

Attribution of the ^{13}C-depleted kerogens to methane utilization has far-reaching consequences. First, it is required that methanogenic and hydrogen-producing bacteria must certainly have developed by the late Archean. Second, the biological utilization of methane requires the presence of an adequate oxidant and the development of some form of methylotrophy. Both of these requirements could be related to the introduction of O_2 to the environment. The only well-characterized bacteria capable of directly incorporating methane carbon into biomass require O_2 for methane utilization (Hanson, 1980). While there is abundant evidence that methane is presently being consumed in some anaerobic sedimentary environments, apparently with the use of sulfate as an oxidant (e.g., Martens and Berner, 1977; Reeburgh, 1980), the possibility that a similar process would have been responsible for the late Archean production of ^{13}C-depleted organic matter can be virtually excluded. When sulfate acts as the oxidant, the oxidation of methane to CO_2 is not exergonic at standard conditions ($\Delta G^{o\prime} = 3.1$ kcal/mole; Thauer et al., 1977). Not surprisingly in these circumstances, the only bacterial cultures thus far studied that are capable of anaerobic methane oxidation do not incorporate the methane-derived carbon in biomass (Hanson, 1980). While it might be suggested that the methane-derived CO_2 could be segregated and selectively "refixed" in order to produce ^{13}C-depleted organic matter, the maintenance of an isotopically distinct CO_2 pool should have yielded some ^{13}C-depleted carbonates, none of which are observed. No absolutely firm conclusion regarding anaerobic utilization of methane can, however, be drawn on the basis of available evidence.

An additional line of evidence suggests that O_2 may have become available at about this time. The banded iron-formations are at once indicators of a generally anoxic environment that allowed the transport of Fe^{2+} to localized sedimentary basins and of the occurrence of a great deal of oxidation of materials within those sedimentary basins (Chapter 11). Oxygen could have served as the required oxidant; sulfate could not.

A sedimentary carbon cycle—one occurring in a specific environment, not necessarily globally—involving the production and consumption of methane is shown in Figure 12-4. Carbon flows at steady-state are indicated in Figure 12-4A, in which g (moles C/time) designates the input from the environment and the output to geochemical cycling, and in which s (moles C/time) indicates the rate of synthesis of organic matter. A fraction, p, of the synthetic activity is due to primary production of organic matter (OM) from oxidized carbon (ΣCO_2). The remainder of the organic

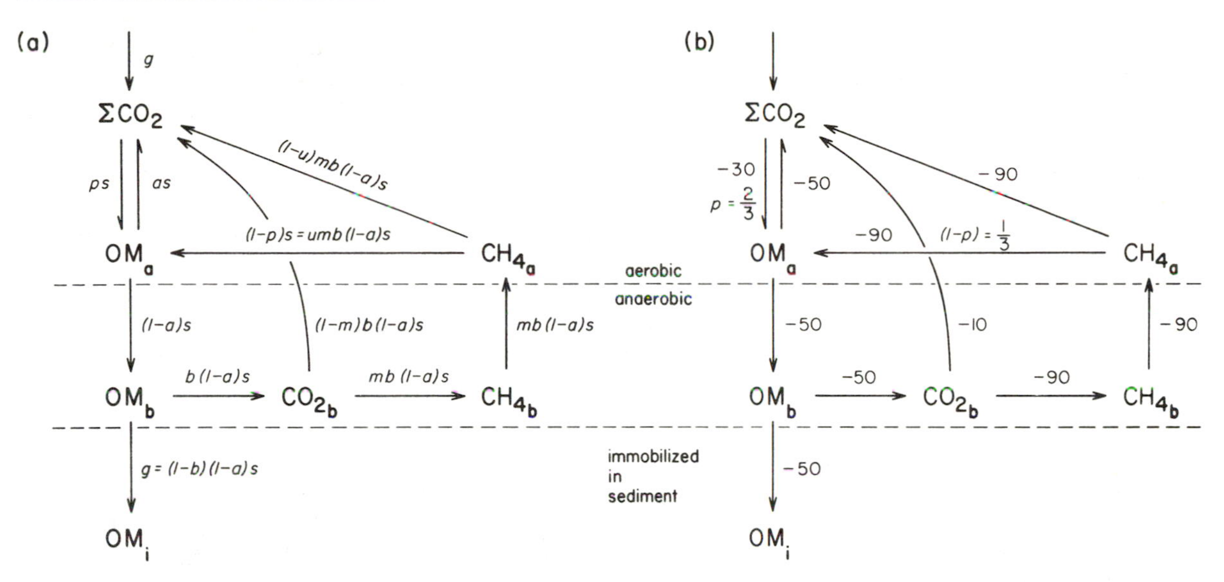

FIGURE 12-4. A reaction network possibly leading to the production of kerogen depleted in ^{13}C; (a) designation of carbon flows; (b) isotopic abundances leading to $\delta_o = -50$ ‰. In (a), g represents the carbon input and output (steady state is assumed) and s represents the rate of synthesis of organic matter. Both of these quantities have dimensions of moles of carbon per unit of time. The coefficients assigned to s are dimensionless and represent the following fractions: p = fraction of synthesis due to primary production; a = fraction of OM_a that is aerobically degraded; b = fraction of OM_b that is anaerobically degraded; m = fraction of CO_{2_b} that is reduced to methane; and u = fraction of CH_4 that is utilized to produce biomass within the system. In (b) it is assumed that the $\Delta\delta$ characteristic of primary production is 30 ‰, that the isotopic fractionation between CO_2 and methane is 80 ‰, and that no isotopic fraction occurs in any other steps; as noted, it is then required that $p = 2/3$.

matter synthesized results from the activities of methane-utilizing bacteria, which produce a fraction $1 - p$ of the organic material synthesized. A fraction, a, of the organic material is aerobically recycled while the remainder (fraction $= 1 - a$) eventually reaches the anaerobic zone. A fraction, b, of that organic material is recycled anaerobically while the remainder (fraction $= 1 - b$) is eventually immobilized in the lithified sediment. Not all of the anaerobically recycled carbon passes through CO_2 (see Games and Hayes, 1976), but carbon flows can be adequately represented by indicating that a fraction, m, is transformed into methane while the remainder (fraction $= 1 - m$) is released as CO_2. A fraction, u, of the methane can be utilized to produce new organic matter while the remainder (fraction $= 1 - u$) is reoxidized to CO_2 (in an open system, some may also escape as unutilized methane, a carbon flow which can be combined with methane-derived CO_2 in the present calculations). As shown in Figure 12-4A, the carbon flows in the various pathways must be related by the equation

$$(1 - p) = umb(1 - a). \qquad (12\text{-}5)$$

This relationship can be used to predict possible isotopic compositions of the organic carbon under various conditions of recycling.

The isotopic compositions that must prevail within the reaction network in order to develop organic matter with $\delta_o = -50$ ‰ vs. PDB are summarized in Figure 12-4B. In this model it is assumed that the isotopic fractionation characteristic of primary production is $\Delta\delta = 30$ ‰. In the process of methane production, it is assumed that the carbon flow is divided evenly between CH_4 and CO_2 (that is, that $m = 1 - m = 0.5$, equivalent to assuming that the average elemental composition of the organic matter is CH_2O and that reducing power is neither imported nor lost during anaerobic recycling) and that the characteristic isotopic fractionation between CO_2 and CH_4 is 80 ‰ (Deines, 1980a, p. 358). In the process of methane utilization, it is assumed that no isotopic fractionation occurs because all of the CH_4 is either oxidized to CO_2 or is utilized to produce new organic matter (this assumption is further discussed below). Finally, it is assumed that, on the average, no isotopic fractionation occurs in reactions in which organic matter is the reactant.

Under this circumstances, the isotopic composition of the organic matter synthesized by primary producers will be -30 ‰ when $\delta_i = 0$ ‰. At steady state when $\delta_o = -50$ ‰, the methane being produced will be strongly depleted in ^{13}C ($\delta = -90$ ‰). In that case, it will be necessary for methane utilization to contribute only one-third of the total production of organic matter; that is, $1 - p = 1/3$.

The relative importances of various processes within the ecosystem can be considered as follows. Substitution of values derived above ($1 - p = 1/3$ and $m = 1/2$) in Equation 12-5 yields:

$$2/3 = ub(1 - a). \qquad (12\text{-}6)$$

Because u, b, and a must all be less than 1.0, it follows that u, b, and $(1 - a)$ must all have minimum values of 2/3. It is certainly possible that less than one-third of the organic matter is aerobically recycled ($1 - a > 2/3$, $1 > a \geq 0$), and that more than two-thirds is anaerobically recycled ($b > 2/3$), but the requirement that more than two-thirds of the methane be utilized in the production of organic matter ($u > 2/3$) is particularly severe. Reviewing 14 different reports of experimental yields of cellular material produced by methane utilizers, Harder and van Dijken (1976) found an average efficiency of utilization corresponding only to approximately $u = 0.64$. While some yields corresponded to values of u as high as 0.8, it appears very unlikely that such efficient utilization of methane would be obtained under natural conditions in an open system. Indeed, there is evidence that $u = 1/3$ is more representative of natural conditions (Fenchel and Blackburn, 1979, p. 96), and it is likely either that excess CH_4 must be imported to the system or that some of the methane-derived CO_2 produced by methylotrophs must be refixed in order to produce additional organic matter depleted in ^{13}C. Because the latter process should be linked to the precipitation of some ^{13}C-depleted carbonates ($\delta_i < 0$), none of which is observed, it is concluded that biogenic methane was probably imported to the microbial communities producing kerogens near -50 ‰ vs. PDB.

It was assumed in the preceding discussion that no excess reducing power was available during methane production; that is, that m could not be greater than 0.5. The question now arises, could the production of more methane (supported, perhaps, by the importation of H_2) within the system change the conclusion reached above? Quantitative consideration of isotopic partitioning within the system described in Figure 12-4 shows that increasing m actually leads to the development of organic material with higher ^{13}C contents because increasing m necessarily increases the ^{13}C content of the CH_4. The importation of reducing power, therefore, does not solve the problem; it exacerbates it.

It was assumed that the cellular carbon produced by the methylotrophs matched in isotopic composition the methane substrate. It is known that methylotrophs tend to produced CO_2 depleted in ^{13}C when supplied with excess CH_4 (Coleman et al., 1981; Barker and Fritz, 1981), and it can be suggested that the cellular material should also be depleted in ^{13}C. There is evidence that this does occur and that, in fact, the isotopic composition of the biomass is equal to that of the respired CO_2 (Hayes and Harder, unpublished observations). A substantial partitioning of ^{13}C can occur, however, only when the bulk of the CH_4 is unutilized ($u \ll 1/3$). Accordingly, this mechanism cannot lead to the generation of organic matter more strongly depleted in ^{13}C.

CONCLUSIONS REGARDING THE LATE ARCHEAN CARBON CYCLE

It is essential that a mechanism be found for the simultaneous deposition of carbonates with $\delta_i = 0$ ‰ and, in some locations at least, or organic material with $\delta_o = -50$ ‰ vs. PDB. Because it furnishes material strongly depleted in ^{13}C and need not be associated with the formation of isotopically exotic carbonates, it is very probable that biogenic methane played an important role in this mechanism.

Model calculations suggest strongly that a significant portion of the methane utilized in the production in a given area of organic material depleted in ^{13}C must have been imported to that area. That is, the methane must have been derived from the recycling of organic material produced elsewhere. Neither the mechanism of methane transport nor aspects of the timing of the steps in the process are clear. The methane might have reached the site of utilization by atmospheric transport or by bubbling or diffusing through sediments or bodies of water. The organic material recycled to produce the methane might have been synthesized approximately simultaneously with, or long before the time of its utilization.

The likelihood that the methane was transported to the oxidant rather than vice versa does not help to clarify the nature of the oxidant. It is, however, highly plausible that the oxidant was O_2 and that, therefore, this disturbance in the carbon isotopic record represents a minimum age for the origin of oxygenic photosynthesis. This suggestion is supported by many convergent lines of evidence: (*i*) well-characterized present-day methylotrophs require O_2 for the incorporation of methane-derived carbon in biomass; (*ii*) the postulated onset of biological production of O_2 correlates well with the global increase in the rate of deposition of iron-formations at about this time; (*iii*) the magnitude of the disturbance in the carbon isotopic record (Figure 7-3) is far greater than that in the sulfur isotopic record (Figure 7-8), and it is logical to interpret the latter as recording the development of increased quantities of sulfate as O_2 concentrations gradually increased; and (*iv*) the origin of oxygenic photosynthesis at this time can be integrated with a coherent view of further aspects of the carbon isotopic record (following sections of this chapter). In contrast, "delaying" the advent of oxygenic photosynthesis to some later time carries with it the necessity of reconciling that event with subsequent aspects of the carbon isotopic record, a difficult task.

The origin of oxygenic photosynthesis would represent the first step in a sequence of events leading to the eventual differentiation of aerobic and anaerobic environments. An ecosystem that was initially homogenously anaerobic would, at that time, begin the journey toward oxygen-dependent stratification. The post-Archean carbon isotopic record can be interpreted in that context.

12-3. The Transition in the Global Ecosystem

Two kinds of environments can be imagined at the beginning of this transition. Free oxygen should have been present wherever oxygenic photosynthesis took place and there was not any reduced inorganic material present in order to scavenge it immediately. In those basins, aerobiosis could develop, methane could be utilized, and sedimentary organic material depleted in ^{13}C could accumulate. In contrast, free oxygen should have been absent wherever abundant reduced inorganic material immediately scavenged any O_2 that was imported or locally produced. In those basins, methanogenesis could continue and, by exporting

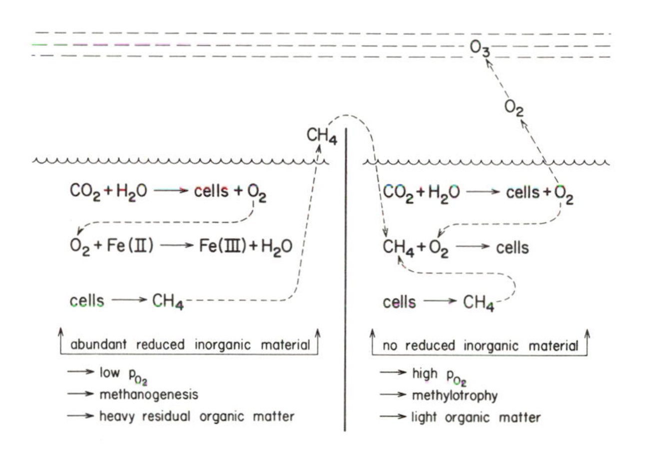

FIGURE 12-5. The carbon cycle during the transitional period when both aerobic and anaerobic sedimentary basins were in in existence.

large quantities of ^{13}C-depleted carbon, may sometimes have noticably enriched the ^{13}C content of the residual sedimentary organic matter. These oxic and anoxic environments could, at that time, have been separated geographically but not, as in later sedimentary basins, vertically. In that case, distinct sediments could develop, sometimes enriched, and sometimes depleted in ^{13}C. The situation is schematically summarized in Figure 12-5 and its isotopic consequences may well be evident in Figure 7-3.

12-3-1. THE EARLY PROTEROZOIC

As more and more oxygen was produced and supplies of reduced inorganic material were consumed, the atmosphere and hydrosphere would become more uniformly and consistently oxygenated and the range of surface environments more constricted. That inevitable trend may be correlated with the gradual convergence of the range of isotopic compositions to values near the Phanerozoic average. If so, the time scale for that convergence might be plausibly related to the rate at which the rock cycle brought reduced inorganic material to the surface. Garrels and Mackenzie (1971, p. 270) have suggested 600 Ma as the mass half-age for the sedimentary cycle, a value that is of the same magnitude as the time constant for the convergence of the isotopic record (Figure 7-3). The facts are, thus, consistent with, but do not compel acceptance of, the postulated interaction between the rock cycle and the development of aerobic and anaerobic environments.

12-3-2. The Middle and Late Proterozoic

It appears possible that the isotopic composition of carbonates through much of the Precambrian has been nearer −1 than 0 ‰ vs. PDB, reaching the latter value relatively recently, and that the average value of $\Delta\delta$ has drifted slowly from 34 ‰ in the Archean to 27 ‰ in the Phanerozoic (Figure 7-3). These coordinated shifts can be interpreted as shown in Figure 12-2C. That sketch depicts a gradual change of the fractionation lines and f_o values over time; the decrease in isotopic fractionation being associated with decreasing CO_2 pressures and, accordingly, decreased expression of the ribulose-bisphosphate carboxylase kinetic isotope effect (Farquhar, 1980; O'Leary, 1981); and the increase in f_o being associated with a decrease in the average phosphorus requirements of the biosphere as multicellular, higher life forms (with lower P/C ratios than microorganisms) became more prevalent.

12-4. Summary and Conclusions

(1) The isotopic record of carbon over time shows structure that must be interpretable in terms developments in the carbon cycle.

(2) As a starting point, it is useful to note that the ratio of organic to inorganic carbon in sediments is controlled at present by the requirements of the biosphere for phosphorus. If biosynthesized organic material contained less phosphorus, more organic material could be synthesized and the ratio of organic to inorganic carbon in sediments would increase.

(3) While the isotopic fractionation between the organic and inorganic carbon pools appears, in the present global ecosystem, to be associated mainly with the mechanism of carbon fixation by primary producers, it is important to recognize that many steps intervene between primary production and immobilization of organic material in lithified sediments, and that isotope effects associated with those steps could cause net isotopic fractionation if carbon flow patterns were to change. In principle and in fact, therefore, the carbon isotopic record encodes information about more than just the mode of primary production.

(4) The isotopic record of the Early and Middle Archean suggests that some form of autotrophic carbon fixation had developed at that time. The Calvin cycle (yielding a higher isotopic fractionation because of higher prevailing CO_2 pressures) and the fixation process associated with CO_2-reducing methanogenic bacteria are at least two of the possibilities.

(5) If the observed isotopic compositions of Early and Middle Archean organic and inorganic carbon are globally representative, then there must at that time have been a lower ratio of organic to inorganic carbon in sediments. Though some other factor might have been limiting, this lower ratio can be interpreted in terms of a higher phosphorus content in organic material produced by an all-prokaryotic Archean biosphere. The plausible persistence of phosphorus as the globally-limiting nutrient suggests that nitrogen was not a limiting nutrient even at that time, and therefore that nitrogen fixation must already have arisen.

(6) Some unidentified redox couple must have been cycling in partnership with carbon. Hydrogen and sulfur both furnish possibilities; there is some geochemical evidence for the latter.

(7) At 2.9 to 2.7 Ga before the present, the observation of extreme ^{13}C depletions in sedimentary organic matter coupled with no recognizable change in the isotopic record in carbonates suggests a change in parts of the carbon cycle other than primary production. Some change allowing utilization of recycling methane would be consistent with the observed isotopic fluctuation and is particularly likely. Both methanogenesis and methylotrophy can be inferred, therefore, to have developed by this time.

(8) The most probable pathway of methane utilization requires O_2. The inorganic-geochemical record furnishes further evidence of oxygenation; it is therefore suggested that oxygenic photosynthesis developed by 2.9 to 2.7 Ga

(9) The simultaneous preservation of ^{13}C-rich and ^{13}C-poor organic matter in sediments can be explained in terms of coexisting, but separate, aerobic and anaerobic sedimentary basins. The presence or absence of reduced inorganic material would determine into which category any given basin fell. The gradual "titration" of inorganic material in the crust would be controlled by the rate at which the sedimentary cycle delivered inorganic material to the reaction zone. As the supplies of that material decreased, the oxygenation of the hydrosphere and atmosphere would have become more uniform and consistent; the range of aerobic and anaerobic environments would have

contracted; and the range of carbon isotopic compositions would have collapsed to presently observed values.

(10) A gradual shift in the overall isotopic fractionation associated with the global carbon cycle can be defined in terms of (*i*) a decrease in the magnitude of the isotopic fractionation, and (*ii*) an increase in the ratio of organic to inorganic carbon in sedimentation pathways. The former change could be due to decreasing CO_2 pressures and the latter to a decreasing average biospheric phosphorus requirement.

CHAPTER 13

Biological and Biochemical Effects of the Development of an Aerobic Environment*

By David J. Chapman and J. William Schopf

13-1. Introduction

13-1-1. Oxygen: A Geochemical and Biochemical Anomaly

Of the four principal elemental components of life—the "biogenetic elements" carbon, hydrogen, oxygen, and nitrogen—oxygen stands out as being a geochemical and biochemical anomaly. In a combined state it is an abundant constituent of all of the inner planets and their satellites, yet among these only the Earth has uncombined, molecular oxygen as a major atmospheric component, one for which there is no known primary (i.e., geochemical) source. And in a combined state oxygen is a fundamental constituent of all living systems and of the great majority of their diverse biochemical components, yet as free diatomic O_2 it is at the same time both lethal to many types of organisms and an absolute requirement for many others.

All of the four biogenetic elements share the property of being of small atomic size; in fact, they are the four smallest elements of the periodic table that can make covalently bonded compounds by sharing one (H−), two (O=), three (N≡), or four (C≡) electrons with combining atoms. All form small, diffusible compounds that occur as gases in the pressure-temperature regime that characterizes the Earth's habitable surficial environment but that are soluble in cytoplasm and can thus play a role in biologic metabolism (e.g., H_2, N_2, O_2, CO, CO_2, NH_3, CH_4, etc.); and all four are of markedly abundant occurrence—indeed, of all potentially reactive (i.e., non-noble gas) elements, these four have the highest cosmic abundances.

* National Research Council of Canada contribution number 20471.

Yet of the four, only oxygen is abundant in the crust of the Earth. In fact, oxygen is the most abundant element of average crustal rocks (followed in rank order by silicon, aluminum, and iron) with the remaining biogenetic elements lagging far behind (viz., H being 10th most abundant, comparable in abundance to titanium and phosphorus; C ranking 16th, similar in abundance to sulfur and zirconium; and N being about 28th, with an abundance comparable to that of scandium, lithium, and niobium). Interestingly, the notably high crustal abundance of oxygen is a reflection of the same factor that is responsible for the seemingly anomalous behavior of the element in organisms; namely, oxygen is so abundant in the crust simply because it reacts readily with silicon, aluminum, iron, and other elements to form nonvolatile oxides, the common rock-forming minerals of geologic materials (the composition of an "average" igneous rock, for example, being about 59% SiO_2; 15% Al_2O_3; 7% Fe-oxides; 5% CaO; and about 3% each of Na_2O, MgO, and K_2O). It is this marked reactivity, both at the temperatures of magma formation (500 to 1200°C) and at the ambient temperature of the Earth's surficial environment that accounts for the potential of molecular oxygen to be both highly useful, and decidedly deleterious, to life, for the biochemistry of living systems is basically that of reduced (relatively H-rich) compounds, biochemicals that are thus potentially oxidizable by O_2 or its derivatives (it is for this reason, for example, that with the addition of molecular oxygen, sugars can be metabolically oxidized to yield CO_2, H_2O, and energy via the process of aerobic respiration). Moreover, although molecular oxygen is required in biosyntheses vital to higher forms of life (e.g., of sterols, phenols, polyunsaturated fatty acids, and various carotenoids, cytochromes, amino acids, etc.), without exception such use occurs only in relatively advanced organisms of a given lineage, and there only in distal, later evolving portions of the pathway. Furthermore, all successful experiments designed to synthesize organic compounds under plausible "primitive Earth conditions" have been carried out under conditions that exclude molecular oxygen (see discussion in Chapter 4); the one possible exception to this generalization, that of relatively O_2-insensitive photocatalytic synthesis on mineral surfaces (Hubbard et al., 1973), is inhibited by liquid water and is thus regarded as being of "questionable importance on the primitive Earth" (Hubbard et al., 1975). And, in fact, virtually all organic compounds tend to be rather easily oxidized and are thus intrinsically unstable in the presence of molecular oxygen (Miller and Orgel, 1974, p. 119). Thus, it is not at all surprising that molecular oxygen is a highly effective biologic toxin, a substance that because of its diffusivity and reactivity (and, more specifically, because its occurrence leads to the formation of highly reactive derivatives) is lethal even in small concentrations to those organisms that lack the biochemical (or behavioral) mechanisms that are required to protect effectively their oxygen-sensitive intracellular biochemistry.

13-1-2. The Evolutionary Context of Aerobic Biology

In terms of the subject matter of this book, two principal questions about biologic oxygen relations are of particular interest. First, what can be said regarding the origin of "aerobic biology" (i.e., of those biochemical processes involved in intracellular protection from, and the photosynthetic production and metabolic and biosynthetic use of, molecular oxygen)? And second, can quantitative aspects of the aerobic biochemistry of modern organisms provide useful insight regarding the pO_2 of the Earth's early environment? More specifically, what concentrations of O_2 are required for its metabolic and/or biosynthetic use, and at what point in the geologic past can the occurrence of such concentrations be reasonably inferred from the known fossil record?

The latter of these aspects is considered in Section 13-4. With regard to the former, for the reasons noted above (and discussed in detail together with other lines of evidence elsewhere; see, for example, Holland, 1978; Schopf, 1978; and Chapters 4 and 11 in the present volume) there can be little doubt that (*i*) the Earth's earliest environment was essentially anoxic, with only minor, trace amounts of molecular oxygen being provided via ultraviolet-induced photolysis of water vapor in the primitive atmosphere; that (*ii*) virtually all (i.e., $\geq 95\%$) of the molecular oxygen that has been produced during Earth history (most of which is now sequestered in various inorganic "sinks" as a result of its consumption during oxidative weathering reactions) is a product of "green plant-type," including cyanobacterial or "blue-green algal," photosynthesis; and that (*iii*) life originated in an anaerobic milieu with the Earth's biota becoming adapted to an anaerobic-aerobic ecologic organization over the course of geologic time.

Thus, in brief outline, the earliest forms of life appear to have been simple, anaerobic prokaryotes (presumably chemoheterotrophic fermenters, as is discussed in Chapter 6) from which evolved other anaerobes including chemoheterotrophs, chemoautotrophs (e.g., methanogens), photoheterotrophs, and photoautotrophs; these earliest photoautotrophs—obligately anaerobic photosynthesizers similar to some extant photosynthetic bacteria—gave rise to aerotolerant, oxygen-producing microorganisms (viz., "proto-cyanobacteria" and their derivatives, the cyanobacteria), the cumulative photosynthetic activity of which ultimately resulted in the transformation of the Earth's surficial environment from an anaerobic to an aerobic condition; and this transformation, in turn, permitted the subsequent development of obligately aerobic forms of life. Biologic aerotolerance (resulting from the presence of biochemical systems capable of effectively scavenging intracellular O_2 or its immediate derivatives with an appropriate reductant) must therefore have arisen prior to, or concurrent with, the development of oxygen-producing photosynthesis (with its possibly earlier origin perhaps being a response to the presence of molecular oxygen of photolytic, non-biologic origin). Amphiaerobic prokaryotes must either have been lacking, or of only limited ecologic significance, prior to the appearance and global dispersal of oxygen-producing photoautotrophs. And obligate aerobes could not have become widespread and well established prior to the establishment of a photosynthetically produced, stable oxygenic environment (see Chapter 11).

However, with regard to the time(s) and mode(s) of origin(s) of obligate aerobiosis and the resulting development of a "modern" anaerobic-aerobic ecosystem, two principal possibilities must be considered:

(1) The advent of such an ecosystem may have been a direct response to, occurring essentially concurrently with, the establishment of a global aerobic environment. The biochemistry of obligate aerobiosis would thus be interpreted as having originated in prokaryotes (whether only once, or on multiple occasions) during the transition from anoxic to oxygenic conditions, a transition indicated by geologic and paleontologic evidence (as discussed in Chapters 11 and 14) to have occurred during the Early Proterozoic, apparently over a time segment some several hundreds of millions of years in duration.

(2) The origin of oxygenic photosynthesis and the Early Proterozoic establishment of a global aerobic environment may have been necessary, but not necessarily sufficient pre-conditions for the advent of a fully developed anaerobic-aerobic ecosystem; obligate aerobiosis could not have developed earlier, but may have appeared much later, possibly with both aerobes and amphiaerobes originating repeatedly as one or another anaerobic lineage gave rise to forms capable first of coping with, and then of utilizing available free oxygen. According to this view, the development of aerobic biology would thus be interpreted as a result of adaptation in differing groups at differing times to similar local conditions (rather than as a direct response to a mid-Precambrian change in the global environment) and the oxygenic biochemistry of extant organisms would be viewed as a result of parallel, polyphyletic, evolution.

Both of these possibilities are at least broadly consistent with available data, for the biochemical characteristics of living systems provide only limited evidence as to the time of their development. Indeed, if it is in fact true that "aerobic respiration has arisen many times" (Fox et al., 1980, p. 461), it is at least conceivable that the process may have evolved relatively early in some lineages and much later in others. Nevertheless, because of the fundamentally aerobic nature of all eukaryotes (e.g., Schopf, 1978) and the variability in oxygen relations exhibited by their more primitive prokaryotic precursors, it seems evident that aerobic biology was a solely prokaryotic "invention"—that if the biochemistry required for intracellular oxygen-protection, -production and -use arose on multiple occasions, it did so only among prokaryotes—and the striking biochemical commonalities of the complex process of aerobic respiration as exhibited by all known aerobes and amphiaerobes seem to us far more likely to reflect a single, rather than a multiple origin. Moreover, there are numerous lines of evidence indicating that both aerobiosis and an anaerobic-aerobic ecosystem were well established at least as early as the Middle Proterozoic. Among these, perhaps the most telling are: (*i*) Since all eukaryotes are basically aerobic, the origin of obligate aerobiosis must pre-date the advent of this cell type, apparently about 1.4 Ga ago (Schopf and Oehler, 1976; Schopf, 1978). (*ii*) High concentrations of molecular oxygen are biologically deleterious, even to "normal" aerobes of the present world; following the advent of oxygenic photosynthesis and initial consumption of the oxygen produced via reaction with oxidizable substrates, aerobic respiration is the only process known that could have restricted oxygen buildup to within biologically tolerable limits (see Chapter 11). And (*iii*) as is discussed in detail elsewhere in this volume (see, especially, Chapters 5, 7, 8, 11, and 14), all salient geological, geochemical, and palentological data are consistent with the mid-Precambrian occurrence of a fully developed anaerobic-aerobic ecology.

Thus, in the discussion that follows we accept the premise that aerobic biology originated in prokaryotes early in Earth history, and that to a major extent the marked variability in oxygen relations exhibited among extant prokaryotes—a variability that is notably lacking among later-evolving eukaryotes—is an evolutionary relic that reflects the diversification and adaptation of the grouping during the Early Proterozoic transition of the Earth's global environment from an essentially anoxic to an aerobic condition.

13-1-3. Definitions and Terminology

The literature on biologic oxygen relations has virtually its own lexicon of terms and definitions, many of which were established before the underlying biochemistry was fully understood. These terms have been recently re-examined by Chapman and Gest who have provided in Part 2 of the Glossary of this volume a list of recommended definitions for oxygen responses and utilization, as well as for carbon nutrition and biological energy conservation. As in the remainder of this book, it is these definitions that are here followed. Similarly, for consistency with other chapters in this book, the oxygen-producing photosynthetic prokaryotes, described variously as "blue-green algae," Cyanobacteria (Stanier et al. 1978), Myxophyceae (Papenfuss, 1955), or Cyanophyceae (Geitler, 1979; Lewin, 1979), are here referred to as Cyanobacteria.

The term "concept organism" is here used to refer to the simplest, most primitive, ancestral cellular prokaryote. The term thus refers to a hypothetical microorganism, but it can be argued (Chapman and Ragan, 1980) that comparable organisms (e.g., *Clostridium* spp.) exist today. It is possible to make an educated guess about the basic energy metabolism, biosynthetic pathways and probable carbon nutrition of such an organism, thereby providing a basis for discussion of subsequent biochemical evolution.

13-1-4. Oxygen Interactions with Organisms

Oxygen interactions with organisms can be conveniently divided into two principal categories: (*i*) oxygen metabolism and enzymology; and (*ii*) oxygen tolerance or toxicity. A considerable amount of literature and information is available with regard to both of these categories. Oxygen enzymology has been reviewed recently by Keevil and Mason (1978), Gunsalus and Sligar (1978), and by White and Coon (1980), and is discussed in numerous recent papers on individual enzymes (see, for example, volumes edited by Boyer, 1975, 1976). The topic of oxygen tolerance or toxicity (a subject often regarded as essentially synonymous with the biochemistry of superoxide dismutase) similarly has been reviewed recently—by Fridovich (1975, 1976, 1978), Halliwell (1978) and by Hill (1978)—and is the subject of volumes edited by Hayaishi and Asada (1977) and Michelson et al. (1977).

Such studies have focused largely on the oxygen relations of discrete ("systemic") systems investigated under controlled laboratory conditions. In nature, however, these systems are in fact not isolated; aerobes in the environment are simultaneously confronted both with the requirement for oxygen utilization and the need for oxygen tolerance. This is the "organismic" aspect of biological oxygen relations, the result of a synergistic or aggregate effect of the individual (systemic) biochemical reactions or pathways that that are present, and a subject on which considerably less work has been undertaken (see reviews by Wimpenny, 1969; Cole, 1976; and Morris, 1975, 1976).

13-1-5. "Concept Organisms" and Anaerobic Biochemistry

The Early Archean atmosphere, whether reducing (H-rich) or "neutral" (e.g., with carbon occurring as CO and CO_2, rather than as CH_4), was essentially anoxic. Although the partial pressure of O_2 may have reached a level of 10^{-6} atmospheres or somewhat higher (for discussion, see Chapter 11), from a biological viewpoint the atmosphere was strictly anaerobic. The biochemistry of the concept organism, accordingly, would have been that of an obligate anaerobe. Such metabolic pathways that may have existed are shown within the dotted lines in Figures 13-1 through 13-4. (Pathways in these figures outside the dotted lines and not designated with "O_2" are still anaerobic, but appear to have been later evolutionary developments, as discussed by Chapman and Ragan, 1980, although probably appearing prior to establishment of global aerobic conditions). One can assume that in such concept organisms biological energy conversion occurred via substrate level phosphorylation. Heterotrophy may have been the method of carbon nutrition (see Chapter 6), but a primitive type of autotrophy, perhaps not necessarily involving CO_2 fixation via ribulose 1, 5 bisphosphate carboxylase, cannot be ruled out. Metalloproteins and other metallo-organics would have been typically Fe, Mn, Zn, Co, Ni, and Mo complexes; as is discussed below, copper may have been essentially unavailable to the concept organisms (and it is thus perhaps of some significance that Cu-compounds are apparently absent from extant anaerobes). The basic metabolism of the postulated concept organism is summarized in Table 13-1. Although simplicity does not necessarily equate with primitiveness, it is difficult to imagine an organism with simpler metabolic characteristics. In numerous respects the concept organism may not have been radically different from modern prokaryotes of the genera *Clostridium* and *Acetobacterium* (for a discussion

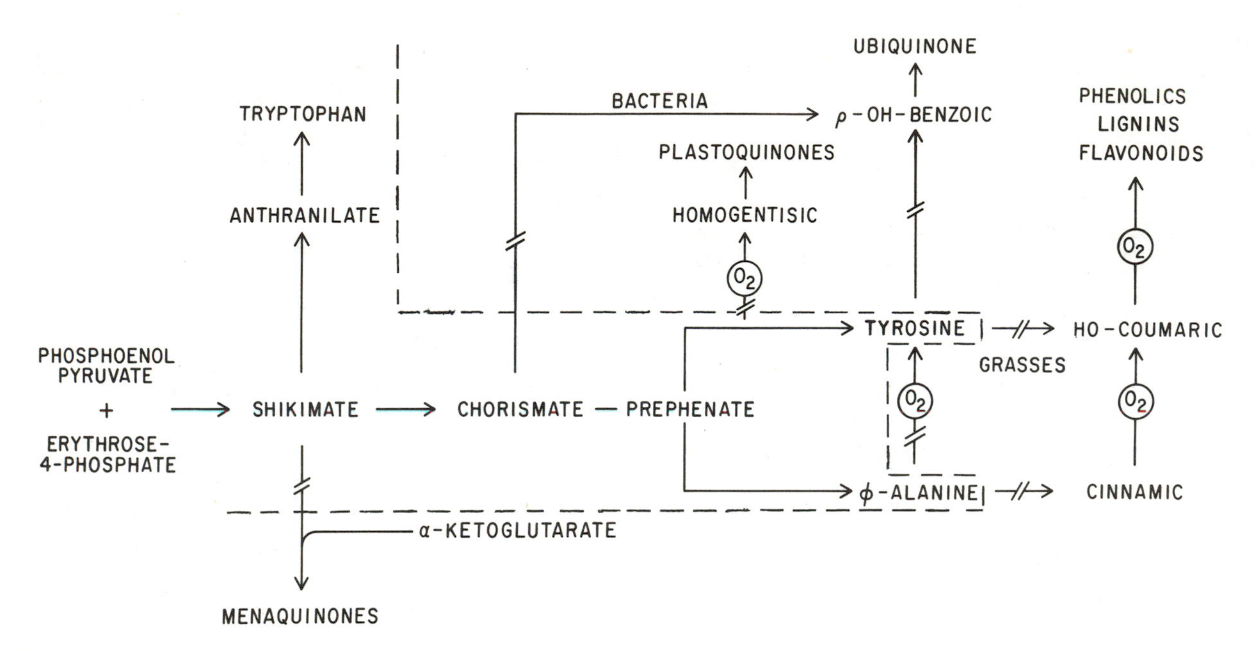

FIGURE 13-1. Basic outline of the shikimate biosynthetic pathway.

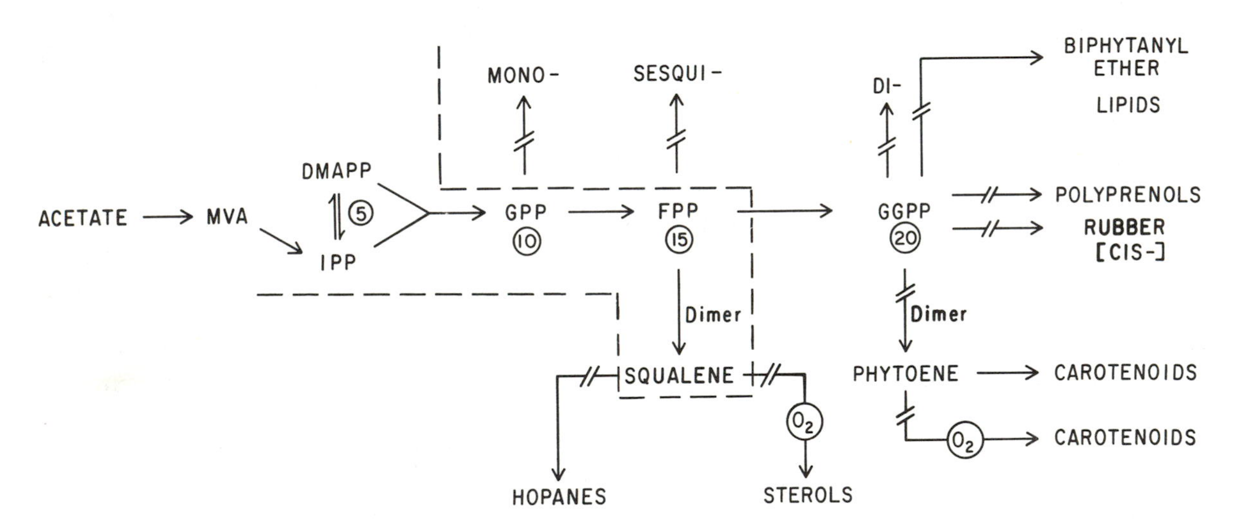

FIGURE 13-2. Basic outline of mevalonate biosynthetic pathway.

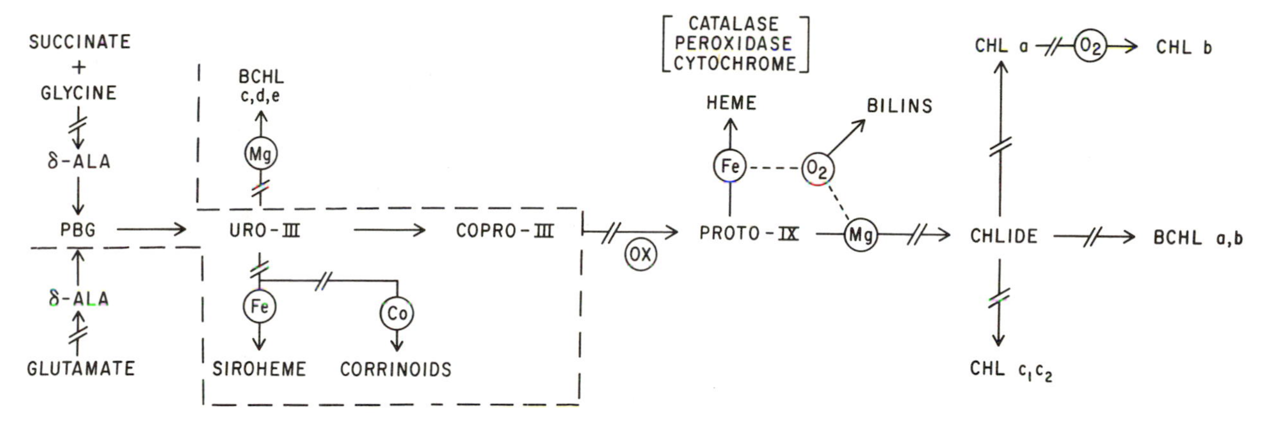

FIGURE 13-3. Basic outline of the δ-aminolevulinate biosynthetic pathway.

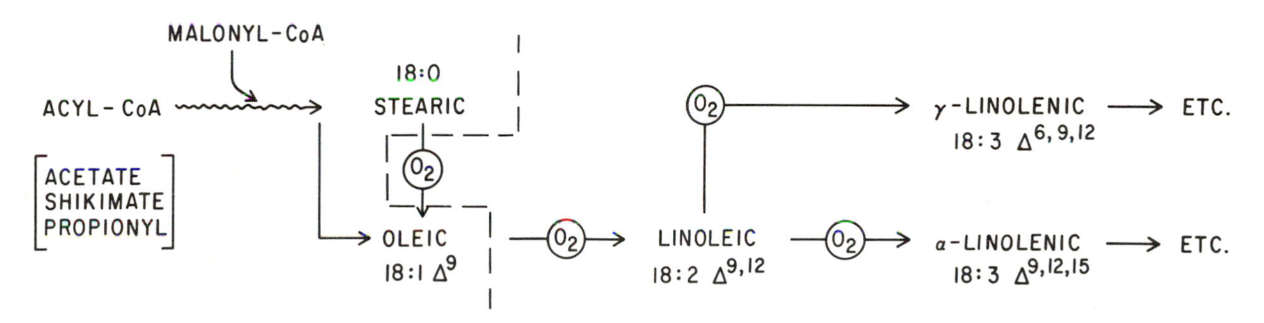

FIGURE 13-4. Basic outline of the fatty acid biosynthetic pathway.

O_2 (circled)	Oxygen-requiring step.
//	Step representing major evolutionary development.
- - - - - -	Part of pathway found in primitive, anaerobic "concept organism" (see text and Table 13-1).
MVA	Mevalonic acid.
IPP	Isopentenyl pyrophosphate.
DMAPP	Dimethyl allyl pyrophosphate.
GPP (10)	Geranyl pyrophosphate, C_{10}.
FPP (15)	Farnesyl pyrophosphate, C_{15}.
GGPP (20)	Geranyl geranyl pyrophosphate, C_{20}.
MONO	Monoterpenes.
SESQUI	Sesquiterpenes.
DI	Diterpenes.

δ-ALA	δ-aminolevulinic acid.
PBG	Porphobilinogen.
URO III	Uroporphyrinogen III.
COPRO III	Coproporphyrinogen III.
PROTO IX	Protoporphyrin IX.
BCHL	Bacteriochlorophyll.
CHLIDE	Chlorophyllide.
CHL	Chlorophyll.
O_X	Special oxidative decarboxylation step as discussed in text.

Glutamate branch to δ-ALA has been identified in cyanobacteria, some photosynthetic purple non-sulfur bacteria, and green plants.

TABLE 13-1. Postulated metabolic characteristics of a clostridial type "concept organism."

Fatty acids biosynthesized:
Saturated and monounsaturated fatty acids
Mevalonic acid pathway:
Pathway to squalene
δ-aminolevulinic acid pathway:
Pathway to coproporphyrinogen and sirohemes
Shikimate biosynthetic pathway:
Pathway to tryptophan, tyrosine and phenylalanine

of methanogens as potential concept organisms, see Chapman and Ragan, 1980).

13-2. Oxygen and Metabolism

13-2-1. Origin of Oxygen-Producing Photosynthesis

Inasmuch as this topic is discussed in some detail in Chapter 6 of the present volume, it is sufficient here to note simply that perhaps the most significant evolutionary change providing the capability for photosynthetic oxygen production was the biosynthetic switch from production of chlorophyllide-bacteriochlorophyll to formation of chlorophyllide-chlorophyll. In terms of formal chemistry the difference between these two compounds, bacteriochlorophyll-*a* and chlorophyll-*a*, quite simple involves only the substitution of a $-CO-CH_3$ group for a $-CH{=}CH_2$ moiety on one pyrrole ring. In addition to the change this necessitated in chlorophyll biosynthesis, the development of oxygenic photosynthesis required the assemblage of the chemical-structural complex of Photosystem II—the photoreaction center capable of mediating photolysis of water—and the subsequent linkage of this system via electron flow to the previously established system like Photosystem I. The origin of oxygenic photosynthesizers (viz., "proto-cyanobacteria") from their anaerobic, photosynthetic, bacterial precursors may thus have been a relatively small evolutionary step (for further discussion, see Section 6-8).

13-2-2. Oxygen and Metabolism

As Williams and Da Silva (1978) have discussed, the metabolic consequences of oxygen utilization fall into one of three categories: (*i*) biosynthesis of metabolites; (*ii*) energy capture (e.g., aerobic respiration); and (*iii*) the production of reactive "side products," substances that are invariably toxic (e.g., superoxide, peroxide, hydroxyl radicals and singlet oxygen) making it necessary for all oxygen-producing and/or -utilizing organisms to exhibit effective intracellular protective devices. All three categories can be regarded as "direct" effects and all are systemic, involving discrete, identifiable, biochemical reactions or systems. Oxygen effects regarded as "indirect" are those correlative with the switch from an anaerobic redox range [viz., $-0.42V$ (H_2/H^+) to 0.0V (H_2S/S)] to an aerobic oxygen-based range [extending up to about $+0.8V$ (H_2O/O_2)]. Williams and Da Silva point out that the development of aerobic oxygen-based metabolism was "almost concomitant with the evolution of new catalysts which, naturally enough for redox chemistry, are based upon transition metals." Thus, the appearance of an aerobic environment brought about, and required, the evolution of higher redox-potential biochemistry.

There is no evidence that the metabolic effects of molecular oxygen have ever differed in significant degree from those exhibited by its interaction with existing metabolic pathways. Thus, the origin of aerobic metabolism apparently involved only the development of new catalysts that acted on preexistent non-oxygen-requiring metabolism or the use of O_2 as a substrate for incorporation into biosynthetically produced metabolites. Admittedly, the oxygen-requiring alkaloid biosyntheses and the pathways of polyketide synthesis (Figure 13-5) are found only in aerobic organisms. However, the latter pathways and many alkaloid pathways do not apparently require the direct involvement of O_2; in terms of present considerations they thus appear to be of subsidiary importance only, apparently arising as a result of terrestrial plant-animal co-evolution.

Processes involved in the evolution of oxygen relations, (as opposed to present-day "oxygen interactions" with organisms) can be divided into two categories: (*i*) direct oxygen intervention, and (*ii*) indirect oxygen intervention. The former involves both types of oxygen interactions (viz., metabolism/enzymology and tolerance/toxicity) and includes oxygen enzymology and the functions of monoxygenases, dioxygenases, and oxidases. The latter category, that of "indirect interventions," includes those effects brought about by an increase

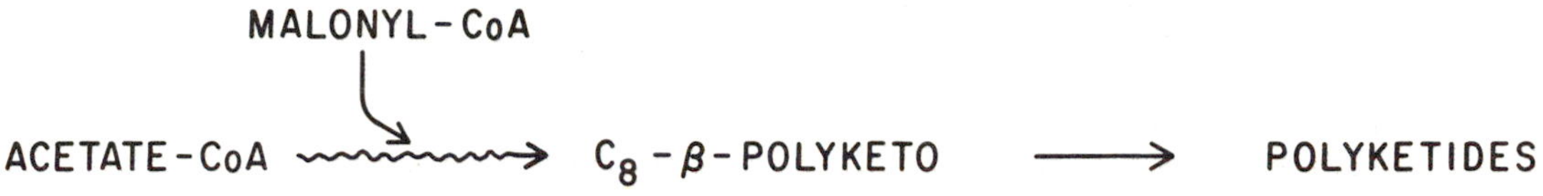

MALONYL-CoA

ACETATE-CoA ⟿ C_8-β-POLYKETO ⟶ POLYKETIDES

FIGURE 13-5. Basic outline of the polyketide biosynthetic pathway (not identified in prokaryotes).

in the E_h of the evolving environment and the resulting changes directly traceable to oxygen-involving reactions.

For convenience, the toxic and inhibitory effects of oxygen or its immediate derivatives (and the biochemical alterations or additions that resulted from the evolutionary pressure of this toxicity) are discussed below as a separate topic. It should be remembered, however, that in terms of the biochemistry involved these toxic effects are examples of "direct intervention." The operative concept is that of toxicity with the deleterious effect deriving from a direct interaction that produces a toxic or inhibitory metabolite or derivative. This is in contrast to other direct effects that are advantageous (or at least non-deleterious).

13-2-3. DIRECT OXYGEN INTERVENTION

Keevil and Mason (1978) have provided a comprehensive list of monooxygenases, dioxygenases, and oxidases, classified according to the presence of one of four specific prosthetic groups: copper, non-heme iron, heme iron, and flavin. Of these enzymes, only a few are of concern in the present context; those of particular evolutionary consequence are listed in Table 13-2.

Monooxygenases mediate the following general reactions (Nozaki, 1979):

$$R + O_2 + H_2X \longrightarrow RO + H_2O + X$$
$$RH_2 + O_2 \qquad RO + H_2O$$

in which R represents an organic substrate and X represents the electron donor (usually NADH or NADPH). The effect of dioxygenases can be summarized as follows:

$$R + O_2 \longrightarrow RO_2$$
$$RH_2 + O_2 \longrightarrow HO{-}R{-}OH$$

and,

$$R + R' + O_2 \longrightarrow RO + R'O$$

in which R′ represents α-ketoglutarate and R′O is succinate. And the effect of oxidases, written empirically as

$$2\,RH_2 + O_2 \longrightarrow 2\,R + 2\,H_2O$$
$$RH_2 + O_2 \longrightarrow R + H_2O_2$$

differs from that of the oxygenases in that molecular oxygen is not incorporated into an organic substrate.

TABLE 13-2. Enzymes of O_2 metabolism of precambrian evolutionary significance.*

Enzyme type	*Prosthetic group***
MONOXYGENASES	
p-Coumaric-3-hydroxylase	Cu
Tyrosinase	Cu
Ribulose bisphosphate carboxylase-oxygenase	—
Fatty acid ω-hydroxylase	HFe
Cytochrome P-450 hydroxylation with Heme α-methenyl oxygenase and Cinnamate-4-hydroxylase	HFe
Squalene epoxidase	—
Fatty acid α-hydroxylase	—
Sterol-C-4-demethylase	—
Carotene (C-3) hydroxylase	—
Zeaxanthin epoxidase	—
Kynurenine-3-hydroxlyase	FAD
DIOXYGENASES	
3-OH anthranilate 3,4, dioxygenase	NHFe
Prolyl hydroxylase	NHFe
p-OH phenylpyruvate oxidase	NHFe
Tryptophan 2,3, dioxygenase	HFe
OXIDASES	
Fatty acyl-CoA desaturase	NHFe
Cytochrome oxidase	HFe, Cu
Glycollate oxidase	FMN

* After Keevil and Mason (1976).

** Cu = Copper; NHFe = Non-heme iron; HFe = Heme iron; FAD = Flavin adenine dinucleotide; FMN = mononucleotide.

TABLE 13.3. Key metabolites derived from O_2 metabolic enzymes.

Enzymes	*Metabolites produced*
DIRECT O_2 ENZYMES	
Fatty acid ω-hydroxylase	Hydroxy fatty acids
Fatty acid α-hydroxylase	Hydroxy fatty acids
Heme α-methenyl oxygenase	Bilins-Biliproteins
Squalene epoxidase	Sterols
Sterol-C-4-demethylase	
Carotene hydroxylase (C-3)	Carotenoids
Zeaxanthin epoxidase	
Kynurenene-3-hydroxylase	NAD Synthesis
3-OH anthranilate 3,4,dioxygenase	
Tryptophan 2,3,dioxygenase	
Prolyl hydroxylase	Hydroxyproline
p-OH phenylpyruvate oxidase	Homogentisic acid
Glycollate oxidase	Glyoxalate, peroxide
Cinnamate-4-hydroxylase	Coumarate
p-Coumaric-3-hydroxylase	Phenols
Tyrosinase	Phenols
Cytochrome oxidase	Respiration
Fatty acyl-CoA desaturase	Unsaturated fatty acids
COPPER ENZYMES	
(Evolutionarily dependent on O_2)	
Cu/Zn Superoxide dismutase	O_2 detoxification
Tyrosinase	Phenols
Cytochrome oxidase	Respiration
Ascorbate oxidase	?
Galactose oxidase	C-6 oxidation
Amine oxidase	Oxidative deamination
Laccase	Phenol oxidation: free radicals
Plastocyanin	Photosynthetic electron transport

TABLE 13-4. Significant structures or functions dependent on O_2 enzymology (Derived from Tables 13-2 and 13-3).

Products of O_2-enzymes/Structures or Functions
Unsaturated fatty acids/Membranes
Sterols/Membranes
Carotenoids/Photosynthesis
NAD/Oxidation-reduction
Coumaric acid, Phenols/Phenols, lignins, flavonoids
Phenol oxidation: free radical/Polymerization wound healing
Hydroxy fatty acids/Plant cutins, suberins (plant cuticles)
Homogentistic acid/Plastoquinone ($\bar{e}$ transport)
Plastocyanin/Photosynthesis
Cytochrome oxidase/Terminal O_2 respiration
Bilins-Biliproteins/Photosynthesis, Phytochrome

Oxygenases and oxidases catalyze a multitude of metabolic processes, producing in turn a vast array of metabolites. The fewer than 20 enzymes listed in Table 13-2, for example, are responsible for the synthesis of numerous key metabolities (as summarized in Table 13-3), compounds that in turn are integral components of a wide range of important structural or functional cellular components (Table 13-4). It is obvious that aerobic biochemistry has played a major role in the evolution of life.

13-2-4. INDIRECT OXYGEN INTERVENTION

For those effects included in this category, oxygen is not involved at the enzymological level; rather, its influence is exerted indirectly, usually as a result of an increase in extracellular E_h (and, via diffusion, presumably also of intracellular E_h). In the absence of molecular oxygen, and in the absence of an appropriate redox reagent [e.g., $Fe(CN)_6^{3-}$], the E_h of the environment (whether in nature or in culture) will be low—especially if reductants (e.g., S^{2-}, H_2) are present. It has been estimated that the E_h of the primitive ocean may have been about −350mV (Osterberg, 1974). If so, biologically important transition elements would have been present in their reduced states (i.e., Fe^{2+}, Cu^{1+}, Co^{2+}, Mn^{2+}, Ni^{2+}, Zn^{2+}), and the presence of sulfide and of alkaline pH under such conditions would have caused the precipitation of some such elements as insoluble sulfides. This may have been of particular importance with regard to copper. Osterberg (1974) first pointed out that under these conditions the very low solubility of Cu_2S (Table 13-5) would probably have ensured that copper was unavailable to organisms. Ochiai (1978) extended these discussions and has calculated that the Cu^{1+} concentration under such conditions would have been on the order of only 10^{-17} M. The other transition metals, although also forming insoluble sulfides or hydroxides, were apparently present in sufficient concentrations to be biologically useful (Ochiai, 1978, Egami, 1975). It is worthy of note that the metalloproteins of "primitive" anaerobes are predominantly metal-sulfur compounds (e.g., ferredoxins, molybdosulfur nitrogenases, etc.) and that copper metalloproteins are unknown in strict anaerobes (i.e., in those prokaryotes that may be considered to be similar to the anaerobes of the Archean). To a considerable degree, the biochemistry of the transition elements in anaerobes appears to be a reflection of metal availability during their early evolution (see McClendon, 1976); the comparative biochemistry of metalloproteins and metallosulfur proteins can in the future thus be expected to shed some light on evolutionary events of the Archean and Early Proterozoic.

TABLE 13-5. Solubility products of metal sulfides and cationic concentrations in the primitive oceans.

Metal sulfides	*Solubility products*	*Ionic concentrations**	
Cu_2S	2.0×10^{-47}	Cu^{1+}	1×10^{-17}M
CuS	8.5×10^{-45}		—
NiS	1.4×10^{-24}	Ni^{2+}	1×10^{-11}M
CoS	3.0×10^{-26}	Co^{2+}	3×10^{-11}M
ZnS	1.2×10^{-23}	Zn^{2+}	1×10^{-11}M
FeS	3.7×10^{-19}	Fe^{2+}	2×10^{-9}M
MnS	1.4×10^{-15}	Mn^{2+}	1.5×10^{-5}M

* After Ochiai. (1978).

A second indirect oxygen effect of particular interest involves the so-called "nitrate problem." Broda (1975b) has argued on thermodynamic grounds that nitrate would have been essentially absent under the anaerobic conditions of the Archean Earth. Williams and Da Silva similarly suggested that nitrate would have been synthesized only under oxidizing conditions, with or without biological catalysts, while Ochiai (1978) estimated the time of available nitrate appearance as being between 1.5 and 0.65 Ga ago, during the Middle and Late Proterozoic. However, Egami (1973, 1976), based principally on biological arguments, has suggested that nitrate respirers preceeded oxygen respirers and that nitrate would have been locally present in a global anaerobic environment (see Rambler and Margulis, 1976). The use of nitrate as a terminal electron acceptor does not involve any oxygen-dependent biosyntheses. Nevertheless, those prokaryotes that do use nitrate can also use oxygen (as, for example, in amphiaerobes), the use of which requires the presence of cytochrome oxidase (viz., heme-protein with copper). While it may not be possible to prove or disprove whether nitrate was present in the early anaerobic environment, it would appear that nitrate probably became quantitatively available to organisms simultaneously with, or just subsequent to, the appearance of oxygenic conditions during the Early Proterozoic (Chapman and Ragan, 1980).

13-2-5. Toxicity Effects

The toxicity of uncombined oxygen is well established. Molecular oxygen, per se, is not the toxic agent, but rather its reactive derivatives, the superoxide anion radical, $O_2^{1\bar{}}$; hydrogen peroxide, H_2O_2; the hydroxyl radical, HO·; and singlet oxygen, $^1O_2$ (Fridovich, 1975, 1976, 1978; Hill, 1978; Halliwell, 1974, 1978.)

Superoxide, perhaps the best known of these toxic derivatives, is a product of the reaction of molecular oxygen with numerous intracellular sources (as summarized in Table 13-6). Hydrogen peroxide can be derived directly from superoxide via the intervention of superoxide dismutase, or

TABLE 13-6. Biological sources of the superoxide radical.

Xanthine oxidase
Aldehyde oxidase
Dihydro-orotic dehydrogenase
Flavoprotein dehydrogenase
Galactose oxidase
Tryptophan dioxygenase
Ferredoxin-NADPH reductase
NADP-Cytochrome *c* reductase
Microsomal hydroxylases
Illuminated chloroplasts

TABLE 13-7. Oxidation reduction potentials of oxygen and derivative species.*

Reactions	*Redox potentials*
$HO\cdot + {}^{1}\bar{e} + H^{1+} \longrightarrow H_2O$	+2.33 V
$1/2\, H_2O_2 + {}^{1}\bar{e} + H^{1+} \longrightarrow H_2O$	+1.35 V
$O_2^{1\,\bar{\cdot}} + {}^{1}\bar{e} + H^{1+} \longrightarrow H_2O_2$	+0.87 V
$1/4\, O_2 + {}^{1}\bar{e} + H^{1+} \longrightarrow 1/2\, H_2O$	+0.81 V
$H_2O_2 + {}^{1}\bar{e} + H^{1+} \longrightarrow HO\cdot + H_2O$	+0.38 V
$1/2\, O_2 + {}^{1}\bar{e} + H^{1+} \longrightarrow 1/2\, H_2O_2$	+0.27 V
$O_2 + {}^{1}\bar{e} \longrightarrow O_2^{1\,\bar{\cdot}}$	−0.33 V

* After Fee and Valentine (1977).

from a two-electron reduction of oxygen via various oxidases. The hydroxyl radical is a product of the reaction of superoxide and hydrogen peroxide.

Thus, peroxide can be formed by the following two reactions:

$$O_2^{1\,\bar{\cdot}} + O_2^{1\,\bar{\cdot}} + 2\,H^{1+} \xrightarrow{\text{(superoxide dismutase)}} O_2 + H_2O_2$$

$$RH_2 + O_2 \xrightarrow{\text{(oxidase, e.g., glycolate oxidase)}} R + H_2O_2.$$

And the reactive hydroxyl radical can be produced as follows:

$$O_2^{1\,\bar{\cdot}} + H_2O_2 \longrightarrow O_2 + OH^{1-} + HO\cdot.$$

Singlet oxygen, $^{1}O_2$, is a product of the interaction of molecular (ground state triplet) oxygen with a "singlet sensitizer" (e.g., a porphyrin, metalloporphyrin, chlorophyll, or flavin). The possible production of singlet oxygen directly from the superoxide radical now appears to have been discounted. Singlet oxygen is thus produced intracellularly via the following three step reaction:

$$[\text{Sensitizer}] \xrightarrow{h\nu} {}^{1}[\text{sensitizer}]$$
$$^{1}[\text{Sensitizer}] \longrightarrow {}^{3}[\text{Sensitizer}]$$
$$^{3}[\text{Sensitizer}] + O_2 \longrightarrow {}^{1}O_2 + [\text{Sensitizer}].$$

The toxic effect of each of these reactive oxygen derivatives results from their strong oxidizing power (Table 13-7), while the superoxide radical can act also as a reductant. Obviously, therefore, the appearance of oxygen-producing photosynthesis and a resulting oxygenic environment necessitated the development of a series of intracellular protective devices and scavengers, particularly in those organisms producing oxygen and in those nonmobile organisms that were unable to use behavioral mechanisms to escape the effects of this newly abundant reactive gas.

The best known of such protective agents is superoxide dismutase (Fridovich, 1975; Malmström et al., 1975), an enzyme system that, as noted above, dismutates superoxide according to the reaction:

$$O_2^{1\,\bar{\cdot}} + O_2^{1\,\bar{\cdot}} + 2\,H^{1+} \longrightarrow O_2 + H_2O_2$$

The potentially deleterious peroxide thus produced is then removed by a second enzyme, hemeprotein catalase, as follows:

$$2\,H_2O_2 \longrightarrow 2\,H_2O + O_2$$

Superoxide dismutase (SOD) poses some interesting evolutionary problems. The three forms of this enzyme, Fe-SOD, Mn-SOD, and Cu/Zn-SOD, exhibit markedly differing phylogenetic and cellular distributions (Table 13-8). The distribution of the Cu/Zn form, absent from the most primitive anaerobic prokaryotes, is consistent with the above discussed postulate regarding the evolution of copper proteins. However, the reported occurrence of Fe-SOD enzymes in a number of obligate anaerobes (Hewitt and Morris, 1975; Hatchikian and Henry, 1977; Lumsden et al., 1977) seems unusual. Why should an obligate anaerobe need superoxide dismutase? Was this metalloprotein originally present in such organisms (perhaps performing some

TABLE 13-8. Phylogenetic and cellular distribution of superoxide dismutases.

Mn-Superoxide dismutase: Intracellular
Purple non-sulfur photosynthetic bacteria
Cyanobacteria
Aerobic bacteria
Mitochondria: yeast, animals
Algae
Fungi
Fe-Superoxide dismutase: Periplasmic
Green sulfur photosynthetic bacteria
Purple sulfur photosynthetic bacteria
Cyanobacteria
Sulfur-reducing anaerobic bacteria
Aerobic bacteria
Anaerobic bacteria
Algae
Cu/Zn Superoxide dismutase: Cytosol
Higher land plants
Green algae (*Nitella**, *Spirogyra**)
Fungi
Animals
Photobacterium leiognathii (also Fe-SOD)

* Green algae on the evolutionary line to higher plants; Cu/Zn-SOD is apparently absent from green algae not on the evolutionary line to higher plants. See Hewitt and Morris. (1975); Lumsden et al. (1976, 1977); Henry and Hall (1977); Hatchikian and Henry (1977); Asada and Kanematsu (1978).

non-oxygen related function) before the advent of an oxygenic environment, or might its development have been a response to short-term exposure to molecular oxygen that was derived from non-biologic (i.e., photolytic) sources? Lumsden et al., (1977) have considered both of these possibilities, in addition to others. Certainly, the periplasmic distribution of the enzyme (Table 13-8) suggests a response to an extracellular agent (the Mn-SOD, in contrast, being more intracellularly located). It would thus appear that Fe-SOD arose in a response to external, perhaps locally produced, toxic oxygen derivatives. It has been shown (Walker, 1978; VanderWood and Thiemens, 1980) that hydroxyl radicals could have been produced photolytically from H_2O in an anaerobic environment (for discussion, see Chapter 11). Was the development of Fe-SOD in obligate anaerobes an evolutionary response to the occurence of this photolytically produced radical? Possibly.

Singlet oxygen is quenched by carotenoids and tocopherols (Krinsky, 1978, 1979). The evolution of carotenoid biosynthesis, however, did not occur as a direct response to the onset of oxygenic conditions. Triplet sensitizers (e.g., triplet chlorophylls) are themselves deleterious and it appears well established, based on comparative biochemical studies (Chapman and Ragan, 1980), that the capability for carotenoid biosynthesis arose concommitantly with that for bacteriochlorophyll biosynthesis, which is an obligately anaerobic process (as is, in fact, bacterial photosynthesis itself). However oxygenation of the environment did enable carotenoid hydroxylation to occur at the C-3 position, an aerobic step, leading to production of such compounds as zeaxanthin in the cyanobacteria (for a summary of this biosynthetic sequence, see Schopf, 1978).

The occurrence of peroxidases and catalases in organisms, although themselves not directly involved with oxygen intervention, seems undoubtedly a result of the establishment of an oxygenic environment. Hydrogen peroxide is an intracellular toxin and it must therefore be scavenged or utilized. As noted above, formation of the peroxide derives either from the activity of superoxide dismutase or of an oxidase. Catalases scavenge this H_2O_2 without the involvement of other substrates and are an essential adjunct to the effectiveness of superoxide dismutase. Peroxidases also scavenge H_2O_2, in this case by mediating an oxidative reaction with an organic substrate:

$$H_2O_2 + RH_2 \xrightarrow{\text{(peroxidase)}} 2\,H_2O + R.$$

13-3. Evolutionary Consequences

13-3-1. Evolutionary Developments After the Establishment of an Aerobic Environment

It is evident that the appearance of an oxygenic environment resulted in a considerable expansion of metabolic versatility compared with that exhibited among pre-existing anaerobes. Some of this expansion and its consequences have been listed in Table 13-3. One hesitates to rank events in terms of their perceived evolutionary "importance". Nevertheless, and allowing for some measure of subjectivity, one can select a few major events that, because of their subsequent impact in biologic evolution appear to have been of particular import.

The appearance of molecular oxygen-ultilizing biosynthetic pathways leading to formation of sterols (and their derivatives) and of polyunsaturated fatty acids must undoubtedly rank high in any such consideration. The formation of these compounds was a significant step in the evolution of membranes, and it is tempting to suggest that without these structural components the extensive membrane structures of eukaryotes might never have evolved. Indeed, molecular oxygen plays such a pervasive, fundamental role in the biochemistry of eukaryotes that it is virtually impossible to envision that this cell type could have arisen without the prior establishment of oxygenic conditions.

The effect of O_2 on transition element biochemistry has been largely ignored in the past, but it seems quite likely that the appearance of an oxygen-rich environment was an essential feature for development of the biochemistry of copper. Indeed, the answer to Williams and Da Silva's (1978) question "Could it be that copper was very largely introduced into biology with and in response to the introduction of oxygen" appears to be yes. The possibility that redox conditions prior to the development of an oxygenic environment were in essence neutral (rather than reducing) and perhaps conducive to a beginning of copper biochemistry cannot be excluded. However the appearance of abundant molecular oxygen would have greatly facilitated the biological use of copper, a development that in turn would have facilitated the increased utilization of oxygen. The presence of cytochrome oxidase, for example, a Cu-containing enzyme, permits the utilization of O_2 as the terminal electron acceptor. The evolution of this protein made possible fully "modern" aerobic respiration and aerobic electrophosphorylation.

The significance of this development is most easily demonstrated by comparing the energy yields of fermentation and aerobic respiration (see also Chapter 6). Anaerobic fermentation, for example of glucose to form two molecules of lactic acid ($\Delta G = -47$ Kcal), or of glucose to form ethanol and carbon dioxide ($\Delta G = -57$ Kcal), results in production of only two molecules of adenosine triphosphate (ATP), containing a total of only about 15 Kcal of cellularly available energy for each molecule of glucose metabolized. In contrast, aerobic respiration, the oxidation of glucose with six molecules of O_2 to yield 6 CO_2 and 6 H_2O ($\Delta G = -686$ Kcal), results in a net production of 36 ATP, representing about 263 Kcal of useable energy for each molecule of glucose oxidized.

Respiration is thus a rich energy-yielding metabolic process, providing perhaps at least a partial explanation for the dependence of eukaryotic cell division (viz., mitosis) on the presence of molecular oxygen (Amoore, 1961). Oxygenation of the environment may also have been responsible for the interconnection of Photosystems I and II (a linkage that we regard as having occurred subsequent to the first appearance of an oxygen-producing Photosystem II). In view of the fact that Photosystem II is a high redox potential system, it seems that the environment must have been neutral, or only very weakly reducing, at the time of development of oxygenic photosynthesis. The evolutionary relationship between Photosystem I and II on the one hand, and Photosystem I and the more primitive bacterial photosystem on the other, is somewhat unclear. However, it seems that although Photosystem I probably developed from the photosynthetic bacterial system, Photosystem II may have arisen separately. Among the redox compounds acting as electron carriers between Photosystem I and II is plastoquinone, the biosynthesis of which is apparently oxygen dependent (since it is synthesized from homogentisic acid the formation of which from hydroxyphenylpyruvic acid is presumably O_2 requiring; see Whistance and Threlfall, 1970). Another electron carrier, the copper protein plastocyanin, presumably appeared via an evolutionary development similar to that outlined above for Cu-containing metalloproteins and thus just after, or concurrent with, the appearance of Photosystem II. These events would require that Photosystems I and II became fully connected through the photosynthetic electron carrier chain (which includes plastoquinone) after the development of Photosystem II and O_2 production. There thus is no a priori requirement that initially Photosystems I and II had to be interconnected.

13-3-2. Biochemistry on the Pre-Oxygenic Earth

The versatility of "post-oxygenic" metabolism does not necessarily imply that pre-existing anaerobic prokaryotes were locked into a metabolic straight jacket. To be sure, there are numerous oxygen-dependent metabolic and biosynthetic processes, the end products of which lack anaerobic analogues or alternative anaerobic modes of biosynthesis. But anaerobes are not metabolic cripples. Indeed, there are a number of examples known in which a particular metabolite, produced by an aerobic pathway in aerobes can also be produced by anaerobes. For example, nicotinamide is synthesized by differing pathways in aerobes and anaerobes. The former use an O_2-requiring pathway (viz., that from tryptophan to kynurenine to 3-OH anthranilate to quinolinate to nicotinamide adenine dinucleotide) involving the use of three separate oxygenases (trypotophan 2,3-dioxygenase, kynurenine 3-monooxygenase, and 3-hydroxyanthranilate 3,4-dioxygenase). Anaerobes, in contrast, use an entirely O_2- independent pathway (starting with aspartate, and formate or dihydroxyacetone phosphate, to yield quinolinic acid and, subsequently, nicotinamide).

The biosynthesis of ubiquinone appears to be another such example. All bacteria (presumably including those anaerobes synthesizing the compound) apparently synthesize ubiquinone directly from chorismic acid. In anaerobes, at least, this must be an O_2-independent pathway. However, in the amphiaerobe *E. coli*, the methoxyl oxygen atoms of ubiquinone are derived from molecular oxygen (Mann, 1978). The oxidation of *p*-hydroxybenzoic acid in bacteria can thus apparently proceed by two separate routes, one aerobic and the other anaerobic. And in eukaryotes (aerobes), ubiquinone is synthesized from tyrosine, presumably via an O_2-dependent pathway. An equally interesting example is the oxidative decarboxylation of coproporphyrinogen III to protoporphyrinogen during the synthesis of protoporphyrin, the precursor of the chlorophylls. In aerobes this is an O_2-requiring step (Bogorad, 1979) with the decarboxylation of the propionic acid side chain to form a vinyl group proceeding through an O_2-derived β-OH derivative. However, in anaerobes (and in amphiaerobes metabolizing anaerobically) this is not the case; the decarboxylation proceeds independently of O_2, involving, among other cofactors, Mg^{2+}, ATP, and the amino acid methionine (Tait, 1972). Moreover, the amphiaerobe *Rhodopseudomonas spheroides* exhibits both aerobic and anaerobic coproporphyrinogenase activity. Thus, one is tempted to speculate that the apparent ability of many obligate anaerobes to synthesize uroporphyrinogen III and compounds derived from it (and hence presumably coproporphyrinogen III), but apparently not protoporphyrin, may have been related to the difficulty of forming a vinyl group (an oxidative decarboxylation step) in the presence of the strongly anoxic conditions under which Archean anaerobes first evolved. Many, if not most hydroxylations in aerobes, involve molecular oxygen (e.g. via Cytochrome P-450); anaerobes, however, carry out hydroxylation by anaerobic processes, with either similar or differing specificity. Anaerobic and amphiaerobic photosynthetic bacteria, for example, hydroxylate hydrocarbon carotenes (but not in the C-3 or C-3′ positions, steps that apparently require O_2). And amphiaerobes of the Rhodospirillaceae can apparently hydroxylate cyclic triterpene hopanes at the C-3 position by a non-oxygen requiring process (Howard, 1980) in contrast to the O_2-requiring (via squalene epoxidase) hydroxylation occurring in the formation of the cyclic triterpene sterols in aerobes.

Finally, as a general rule one finds that the oxygen-dependent metabolites of aerobes are often used simply as a means of refining, in a chemical structure-function relationship, pre-existing anaerobic systems. Sterols and polyunsaturated fatty acids are essential components of eukaryotic membranes. The earlier-evolved anaerobes, however, appear to have used less "sophisticated" analogues, in particular the hopanes, carotenoids, and the saturated and monounsaturated fatty acids. This particular example of "development of sophistication," which was first proposed by Ourisson's group (Rohmer et al., 1979), has received support from recent hopane studies (Howard, 1980) and agrees well with numerous other aspects of metabolic evolution (Cairns-Smith and Walker, 1974; Chapman and Ragan, 1980).

13-4. Quantitative Aspects of Oxygen Biochemistry

13-4-1. O_2 Concentrations and Metabolism

Very little definitive work has as yet been done on the effect of the partial pressure of available oxygen on either O_2-requiring biochemical systems, or on organisms in general. This lack of activity may stem, in part, from a widespread failure to realize that the present atmosphere level of O_2 (1 PAL O_2 = 20.9% by volume, or ~0.2 atm) is far in excess of the minimum (or even the optimum) amount required for the O_2-based systemic biochemistry or organismic biology of extant microorganisms. The oxygen physiology of mammalian metazoans has undoubtedly influenced much of the thinking and experimental design in this area. It is well known, for example, that among the various tissues of higher mammals, probably the most complex—the specialized nerve tissue that forms most of the human brain—inevitably dies if it is deprived of molecular oxygen for more than about four minutes. The concept of the "Pasteur point" (0.01 PAL O_2), the level at which amphiaerobes switch between aerobic respiration and fermentation, has also been a contributor to popular misconceptions. Certainly this is a useful colloquial concept (provided that the value is specified for each organism in terms of culture conditions, pH, E_h, etc), but the Pasteur point is not a universally fixed value; indeed, it is far more variable than has generally been recognized, and for most amphiaerobes it has in fact yet to be determined.

With regard to the quantitative aspects of O_2-sensitive systemic biochemistry, as is summarized in Table 13-9, a number of papers between 1969 and 1975 clearly established the importance of O_2 concentrations in determining enzymatic activity (whether inhibitory or stimulatory in effect). Few of these experiments, however, considered very low pO_2 (e.g. 0.01 PAL O_2 = 0.2% O_2 = 2×10^{-3} atm), and in general, only two or three levels of oxygen partial pressure were investigated. Earlier studies by Amoore (1961) demonstrated a very interesting relationship between the completion of mitotic cell division and low O_2 concentrations (see Table 13-10). More recently, Rogers and Stewart

TABLE 13-9. Oxygen relations and limits on physiological activity.

(1) Effect of E_h changes on anaerobic bacteria	
Bacteroides fragilis, Clostridium perfringens, Peptococcus magnus:	
Growth unchanged	−50 mV to +325 mV (Anaerobic)
Bacteriostatic	at +300 mV (Aerobic)
(Onderdonk et al., 1976; Walden and Hentges, 1975)	
(2) Effect of O_2 on Nitrogen fixation and nitrogenase activity in a bacterium (*Spirillum*) and cyanobacteria (*Anabaena, Nostoc*)	
Spirillum lipoferum:	Optimum at pO_2 0.005 atm; inhibited above pO_2 0.01 atm
Anabaena flos-aquae and Nostoc muscorum:	Maximum reduction in heterocysts at pO_2 0.1 atm; inhibited above pO_2 0.01 atm in cell-free extracts
(Okon et al., 1977; Stewart and Pearson, 1975)	
(3) Oxygen inhibition of growth of anaerobic bacteria	
Selenomonas ruminantium:	Inhibited above pO_2 0.1 atm
Campylobacter sputorum:	Inhibited above pO_2 0.15 atm
Clostridium acetobutylicum:	Inhibited above pO_2 0.04 atm (E_h − 50 mV)
(Wimpenny and Samah, 1978; Niekus et al., 1977; O'Brien and Morris, 1971)	
(4) Oxygen inhibition of enzymatic activity*	
L-Lactate dehydrogenase:	Maximum inhibition pO_2 0.025–0.05 atm
D-Lactate dehydrogenase:	Maximum inhibition pO_2 0.025–0.05 atm
Fumarase	Maximum inhibition pO_2 0.025–0.05 atm
Succinate dehydrogenase:	Maximum inhibition pO_2 0.025–0.05 atm
2-oxoglutarate dehydrogenase:	Inhibited in absence of O_2 and CO_2
(Pritchard et al., 1977; Keevil et al., 1979)	
(5) Effect of O_2 on anaerobic-aerobic metabolic switch in amphiaerobic bacteria (threshold pO_2 or "Pasteur Point")	
Beneckea natriegens:	Fermentation below pO_2 0.01 atm
Propionibacterium shermanii:	Anaerobic growth below pO_2 0.025 atm
(Linton et al., 1975; Pritchard et al., 1977)	

* Reduced inhibition both above and below these "oxygen trap" O_2 values.

TABLE 13-10. Relationship of mitosis to oxygen concentration in *Pisum sativum* (Pea).*

O_2 *level (PAL* O_2*)*	*State of mitosis*
2×10^{-5}	Mitosis completely arrested
4×10^{-5} to 1×10^{-3}	Cells in mitosis complete division; initiation of division inhibited
2×10^{-3} to 1×10^{-2}	Cells in interphase begin, but do not finish, division
$>2 \times 10^{-2}$	Mitosis not inhibited

* After Amoore (1961).

(1973), Whittaker and Klein (1977), and Jahnke and Klein (1979) have provided the most useful analysis of O_2 concentrations and systemic biochemistry (summarized in Table 13-11).

With regard to the organismic aspects of the problem, and despite the importance of O_2 and its role in aerobic metabolism, there is a woeful shortage of knowledge. The work of Rogers and Stewart (1973) and Klein's group (Whittaker and Klein, 1977; Jahnke and Klein, 1979) was performed on amphiaerobic eukaryotic yeasts. The quantitative oxygen relations of amphiaerobic prokaryotes are as yet largely unstudied.

Although decidedly limited in scope, the data that are available point to two very important conclusions: (*i*) systemic aerobic biochemistry can be operative at values of 0.002 PAL O_2; and (*ii*) organismic aerobiosis can be operative at oxygen concentrations as low as 0.01 or 0.02 PAL O_2. These values are very low in comparison with the present atmospheric level of O_2, but because of the limited amount of data available they should not be regarded as quantitative definitions of the minimum pO_2 required for aerobiosis. They do, however, illustrate well an important problem for studies in the field; namely, to be useful, laboratory experiments involving the "anaerobic" must measure and specify the pO_2 being used and must be designed to go below a pO_2 of 10^{-4} atm.

TABLE 13-11. Half maximum values for aerobic biochemistry in *Saccharomyces cerevisiae* (yeast).*

Types of Compounds or Functions	*Values*
COMPOUNDS	*Half-maximal* O_2 *concentration for full complement*:
Cytochrome oxidase	0.002 PAL O_2
Sterols	0.002 PAL O_2
Unsaturated fatty acids	0.002 PAL O_2
Cytochromes *b*, c_1	<0.001 PAL O_2
COMPOUNDS/FUNCTIONS	O_2 *concentration for half maximal activity*:
Cytochrome oxidase	0.002 PAL O_2
Cytochrome *c* peroxidase	0.001 PAL O_2
Succinate-cytochrome *c* reductase	0.001 PAL O_2
Succinate dehydrogenase	0.002 PAL O_2
Palmitoyl CoA-desaturase	0.002 PAL O_2
Respiratory rate	0.01 to 0.08 PAL O_2
Growth rate	0.01 to 0.08 PAL O_2

* Data from Rogers and Stewart (1973); Whitaker and Klein (1977), and Jahnke and Klein (1979).

13-4-2. AT WHAT pO_2 DID AEROBES FIRST EVOLVE?

In Glossary II included in this volume are provided nonquantitative definitions of the terms "aerobe" and "anaerobe." One can, however, pose the question "At what pO_2 in the evolution of the Earth's environment could aerobic biochemistry and aerobic microorganisms have become established?" A conservative estimate based on the data at hand suggests that systemic biochemistry must have been totally anaerobic prior to the occurrence of at least 0.0005 PAL O_2 (i.e., 0.01% $O_2 = 10^{-4}$ atm), while organismic aerobiosis (and thus aerobic organisms) would not have been possible before the pO_2 had reached at least an order of magnitude higher. It should be noted that the Pasteur point, the level of O_2 at which amphiaerobes switch from anaerobic to aerobic metabolism, is about 0.01 PAL O_2, a value similar to the probable minimum O_2 level capable of supporting organismic aerobiosis. What is perhaps most significant, however, is that obligate aerobiosis could not have evolved and been sustained on a widespread scale until a stable oxygenic environment had become established, perhaps about 1.7 Ga ago (see discussion in Chapter 11), once the rate of photosynthetic oxygen production exceeded the rate of supply of reduced materials to the atmosphere and oceans. Amphiaerobes probably evolved earlier, and transient levels of pO_2 high enough to have supported aerobiosis may have also occurred at an earlier time, but the diversification of strict aerobes was dependent not only on pO_2 but on the continued presence of a relatively high level of molecular oxygen.

13-5. The First Oxygen Producers

It has been widely assumed that the first oxygen producing microorganisms were cyanobacteria ("blue-green algae"). Certainly this would appear to be a reasonable assumption. Such prokaryotes are the simplest oxygenic photosynthesizers now known, and the fossil evidence indicates that cyanobacteria were well-established during the Proterozoic and that the group has exhibited a rather remarkable degree of morphological (and presumably biochemical) evolutionary conservatism (see, for example, Chapter 14 and Schopf and Walter, 1981). Modern cyanobacteria are probably little changed from their Precambrian precursors.

Nevertheless, extant cyanobacteria are themselves products of the evolutionary effects of oxygen. Indeed, such organisms require molecular oxygen to produce biliproteins (via heme oxygenase), sterols (via squalene epoxidase), polyunsaturated fatty acids (via O_2-dependent desaturase), plastoquinone (probably via an oxygenase), plastocyanin (a Cu-protein), cytochrome oxidase (a Cu + heme protein), and C-3 hydroxylated carotenoids (via O_2-dependent hydroxylation). The first O_2-producer—that is, a "proto-cyanobacterium" containing a chlorophyll-*a* reaction center with E° at $\simeq 0.82$V (corresponding to current Photosystem II)—would probably not have been able to synthesize the above compounds prior to

TABLE 13-12. Comparison of the biochemical characteristics (including oxygen-related biochemistry) of cyanobacteria, Purple non-sulfur photosynthetic bacteria, and their postulated evolutionary precursors, hypothetical "proto-cyanobacteria."

Oxygen-related Metabolites	*Extant cyanobacteria*	*Hypothetical "proto-cyanobacteria"*	*Extant purple non-sulfur photosynthetic bacteria*
	Lost	Succinate-glycine: ALA	+
	+	Glutamate: ALA	(+)
	Lost	Bacteriochlorophyll *a*	+
	+	Chlorophyll *a*	Lost
	Photosystems I and II Connected (PS II disconnected anaerobically)	Disconnected photosystems I and II	Photosystem I only (PS II lost)
Bilins and biliproteins	+	−	−
	+	Squalene cyclase (hopanes)	+
Sterols	+	−	−
	Lost	Ubiquinones	+
	Lost	Menaquinones	+
Plastoquinones	+	−	−
	+	Cyclic carotenes	Lost*
	+	Acyclic xanthophylls	+
O_2-xanthophylls	+	−	−
	+	Saturated and monounsaturated fatty acids	+
Polyunsaturated fatty acids	+	−	−
	+	Cytochrome oxidase	+
	+	RuBP carboxylase	+
	+	Mn-SOD and Fe-SOD	Mn-SOD only
	+	Nitrogenase	+
	+	TCA Cycle enzymes	+

+ = Feature present.
(+) = Feature probably present.
− = Feature never present.
Lost = Feature lost during derivation of group.
Lost* = Cyclic carotenes present only in *Rhodomicrobium vannielii* and *Rhodopseudomonas acidophila.*
ALA = δ-aminolevulinic acid (see Figure 12-3).
Fe = Iron.
Mn = Manganese.
RuBP carboxylase = Ribulose 1,5 bisphosphate carboxylase.
(PS II disconnected anaerobically) = In some cyanobacteria, Photosystem II is disconnected under anaerobic conditions (Padan, 1979).
SOD = Superoxide dismutase.
TCA = Tricarboxylic acid.

the buildup of environmental oxygen. As noted above, it is of course possible that such oxygen may have been provided by non-biologic sources (e.g., photolysis of water vapor); nevertheless, there is no reason to assume a priori the presence of a "fully aerobic" biochemistry in the earliest oxygen-producing photosynthesizer.

What was the nature of the first O_2-producer? A primitive precursor of the purple non-sulfur bacteria (like members of the modern Rhodospirillaceae) that had developed a "chlorophyll-*a*-like" reaction center with redox potential capable of mediating H_2O photolysis would be a logical possibility (see also discussion in Chapter 6). In terms of biochemistry, such an organism would be much "closer" to a photosynthetic bacterium than to a modern cyanobacterium and would presumably have the morphology of a single spheroidal cell (like extant chroococcalean cyanobacteria). Subsequent evolution of this hypothetical organism could have given rise to the Rhodospirillaceae, the Cyanobacteria, and even to non-photosynthetic aerobic bacteria (via loss of photosynthetic capability). Similar relationships have previously been proposed (Dickerson, 1980), based principally on cytochrome sequence data, but not to the extent of suggesting that such an ancestral organism was the first O_2 producer. The relationships suggested are summarized in Table 13-12. It would, of course, be of considerable evolutionary interest to detect such an organism in the modern biota: a purple non-sulfur photosynthetic bacterium containing a chlorophyll-*a*-like reaction center and thus capable of producing O_2.

13-6. Summary and Conclusions

(1) Of the four "biogenetic elements"—carbon, hydrogen, oxygen, and nitrogen—oxygen stands out as being a geochemical and biochemical anomaly. In a combined state it is a highly abundant, fundamental constituent of the inner planets, their satellites, and of all known living systems; yet of all the planets of the solar system molecular O_2 is a major component only of the atmosphere of the Earth, and in this uncombined state it is at the same time both lethal to many organisms and an absolute requirement for many others.

(2) The seemingly paradoxical relationship of O_2 to organisms—the fact that it can be either a requirement or an anathema to living systems—is a reflection of the same factor that has resulted in its high abundance in the crust of the Earth; namely, molecular oxygen and its derivatives are markedly reactive, capable to combining readily with other terrestrial elements to produce common rock-forming minerals or of reacting with the reduced (H-rich) biochemicals of living systems with such oxidation resulting either in the generation of cellularly useable chemical energy (as, for example, during aerobic respiration) or in the loss of biologic function (as, for example, results from the oxidation of numerous enzymes).

(3) During or prior to the Early Archean subera of Earth history, life originated in an essentially anoxic milieu, with molecular oxygen being present only in trace amounts as a result of the photolysis of water vapor by ultraviolet light. The earliest organisms were thus anaerobic, presumably chemoheterotrophic (fermenting) prokaryotes; subsequent evolution resulted in the derivation of numerous other types of anaerobic bacteria (viz., of other chemoheterotrophs, chemoautotrophs, and of both photoheterotrophs and photoautotrophs) and, ultimately—apparently near the beginning of the Early Proterozoic—the appearance of aerotolerant, oxygen-producing photosynthesizers (viz., "proto-cyanobacteria"). The earliest, more primitive, cellular prokaryotes (here referred to as "concept organisms") were aerointolerant; the Archean-Early Proterozoic history of biologic oxygen relations was thus characterized by the development among prokaryotes of the biochemical systems required for intracellular oxygen-protection, oxygen-production, and metabolic and biosynthetic oxygen-use.

(4) Oxygen interactions with organisms can be conveniently divided into two principal categories: (*i*) oxygen metabolism and enzymology; and (*ii*) oxygen tolerance or toxicity. With regard to oxygen metabolism, three subcategories of interactions can be recognized: (*ia*) use of oxygen for biosynthesis of metabolites (e.g., of polyunsaturated fatty acids); (*ib*) use of oxygen for energy capture (e.g., via aerobic respiration); and (*ic*) interactions resulting in production of toxic, reactive, "side products" (viz., the superoxide radical, $O_2^{1-\cdot}$; hydrogen peroxide, H_2O_2; the hydroxide radical, $HO\cdot$; and singlet oxygen, $^1O_2$).

(5) Such interactions are either "systemic," the oxygen interacting with specific, identifiable biochemical reactions or pathways as in (*ia*), (*ib*), and

(*ic*), above, or "organismic," a result of the synergistic or aggregate effect of systemic oxygen interactions on the total organism. And the interactions can either be a result of "direct oxygen intervention," for example in the direct incorporation of oxygen into substrates as in (*ia*), (*ib*), and (*ic*), above, or of "indirect oxygen intervention," for example when changes of environmental E_h effect the availability to organisms of mineral nutrients (e.g., of copper, nitrate, etc.). Both direct and indirect oxygen interventions have played a role during the evolution of biologic oxygen relations.

(6) Over the course of evolution, numerous biochemical mechanisms have been developed to protect intracellular systems from the deleterious effects of the toxic "side products" of molecular oxygen. Among the most important and best known of these are superoxide dismutase, an iron-, manganese-, or copper/zinc-containing enzyme complex that converts the superoxide radical to hydrogen peroxide, and that may also scavenge hydroxide radicals; hemeprotein and other catalases that convert peroxide to water and molecular oxygen; peroxidases that scavenge peroxide by mediating an oxidative reaction with an organic substrate; and carotenoids that quench singlet oxygen. Most, and perhaps all, of these protective mechanisms evolved early in Earth history in response to the biologically (i.e., photosynthetically) produced transformation of the global environment from an essentially anoxic to an aerobic condition.

(7) Subsequent to the development of biochemical mechanisms involved in the production of and protection from molecular oxygen, capability evolved for the metabolic and biosynthetic utilization of this newly abundant, and highly reactive, environmental component. In general, such capability involved the extension and refinement of pathways already extant in previously evolved anaerobes, with particularly important examples including the development of aerobic respiration by "addition" of a complete citric acid cycle to the previously well-established pathway of glycolytic fermentation; see, however, Chapman and Ragan, 1980, for a discussion of the origin of the citric acid cycle, (also Gest, 1981) particularly of α-ketoglutarate dehydrogenase, and the use of molecular oxygen for formation of such structurally important compounds as sterols (by extension of the anaerobic pathway leading to squalene) and polyunsaturated fatty acids (via the introduction of the mechanisms of oxidative desaturation).

(8) Although the quantitative aspects of O_2-related biochemistry have as yet been relatively little investigated (especially at low pO_2), extrapolation from data currently available suggests that systemic aerobic biochemistry could have occurred once the atmospheric pO_2 had attained a level of about 10^{-4} atm, with the threshold for organismic aerobiosis being about an order of magnitude higher. The known fossil record (viz., the oldest evidence of probable eukaryotes) suggests that a pO_2 sufficient to support organismic aerobiosis had become established at least as early as 1.4 Ga ago, and it is likely that such levels may have been attained much earlier. Nevertheless, obligate aerobes could not have evolved and been sustained on a widespread scale until this level was not only attained but had become stabilized, a development that the geologic evidence suggest probably did not occur earlier than about 1.7 Ga ago.

(9) In sum, it is evident that photosynthetically produced molecular oxygen has played a pervasive role in the history of life on Earth; indeed, it was the origin of oxygenic photosynthesis and the evolution of biologic oxygen relations that together have been responsible for the development of the anaerobic-aerobic ecosystem that currently characterizes the planet. The beginnings of this ecosystem can be traced well into the Precambrian, apparently into the Early Proterozoic, but numerous aspects of its history are at present only poorly understood. It is here postulated, for example, that the earliest photosynthesizers ("protocyanobacteria") may have more closely resembled purple non-sulfur photosynthetic bacteria than modern cyanobacteria, but this as yet is only speculation and little is known regarding quantitative aspects of the rate of change of pO_2 in the Early Proterozoic environment. Clearly, much remains to be learned with regard to the development of biologic oxygen relations, a signal aspect of early biospheric evolution.

Acknowledgments

This paper was prepared while D. J. C. was Distinguished Visiting Scientist at the Atlantic Research Laboratory (NRC-Canada); his research relevant to this work has been supported by NSF Grants GB-42461 and PCM 78-25852.

CHAPTER 14

Early Proterozoic Microfossils

By Hans J. Hofmann and J. William Schopf

14-1. Introduction

The beginnings of Precambrian paleontology as a subdiscipline of the natural sciences date back to the 1850s and the discovery of laminated, megascopic structures in Proterozoic marbles of the Grenville Province of southern Quebec and eastern Ontario, Canada (Hofmann, 1971a, p. 6–12), geologic units that were at the time considered to be among the oldest rocks on Earth. The discovery of these objects, later named *Eozoon canadense* and interpreted as (atypically large) foraminifera by J. W. Dawson, occurred at about the same time as the publication of Darwin's *Origin of Species*. The supposed biogenicity of *Eozoon* at once became the subject of acrimonious controversy, a long-term debate that subsided only many years later when the nonbiologic origin of these structures was conclusively demonstrated. In the meantime, however, during the 1870s and the 1880s, the interest thus aroused led to the discovery of bona fide biologic remnants (viz., stromatolites) in Proterozoic carbonates elsewhere in Canada, most notably of *Archeozoon acadiense* described in 1890 by G. F. Matthew from Precambrian limestones of New Brunswick.

As is discussed by P. Cloud in the first chapter of this volume, following these earliest studies Precambrian paleontology experienced a minor, but significant renaissance in the first two decades of the 1900s, stimulated largely by the investigations carried out by C. D. Walcott. The first descriptions and illustrations of microscopic remains from the Early Proterozoic, however, are those by Moore (1918), who reported structures that he interpreted as coccoid and filamentous cyanobacteria from near the base of the Kipalu Iron Formation of the Belcher Islands in Hudson Bay, Canada.

Reports of Early Proterozoic microfossils and possible microfossils appeared sporadically over the following 40 years, based mainly on material from the Great Lakes region of North America (e.g., Gruner, 1922, 1923, 1924a, 1924b; Tyler and Barghoorn, 1954; Barghoorn et al., 1954; Barghoorn and Tyler, 1962; Moorhouse and Beales, 1962; Barghoorn, 1963). But it was not until the mid-1960s that systematic and taxonomic studies of Early Proterozoic microfossils per se began to be published. Papers by Barghoorn and Tyler (1965a, 1965b) and by others (Cloud, 1965; Cloud and Hagen, 1965; Schopf and Barghoorn, 1965; Schopf et al., 1965) based on studies of exceptionally well-preserved material from microfossiliferous cherts of the Gunflint Formation of Ontario, Canada, stimulated research not only on Early Proterozoic life, but also into still older as well as younger Precambrian stratigraphic units. Since then, the number of publications on Early Proterozoic microfossils has risen to more than 160; a large proportion of these deal with the Gunflint microbiota, demonstrating the great importance of this occurrence to the development of the field.

At least 40 occurrences of microfossils and microfossil-like objects are now known from rocks 2.5 to 1.6 Ga in age. The geographic distribution of these units is shown in Figure 14-1; their approximate stratigraphic position is tabulated in Table 14-1; and a cumulative curve summarizing the time of first report of these occurrences, and illustrating well the dramatic recent growth of the field, is shown in Figure 14-2 (see Figure 9-2 for a comparable illustration regarding reported occurrences from the Archean). However, of the 40 occurrences only 24 are composed of unequivocal microfossils; of the remainder, 10 consist of dubiomicrofossils, and 6 are composed of mineralic pseudomicrofossils or of objects that are known not to be of Precambrian age (Table 14-2; microstructures reported from the Hamersley Group of Western Australia, rock units formerly considered to be Early Proterozoic, are not here included because recent radiometric dating has resulted in referral of this group to the Late Archean as indicated in Section 9-4-3). And of the 24 authentic

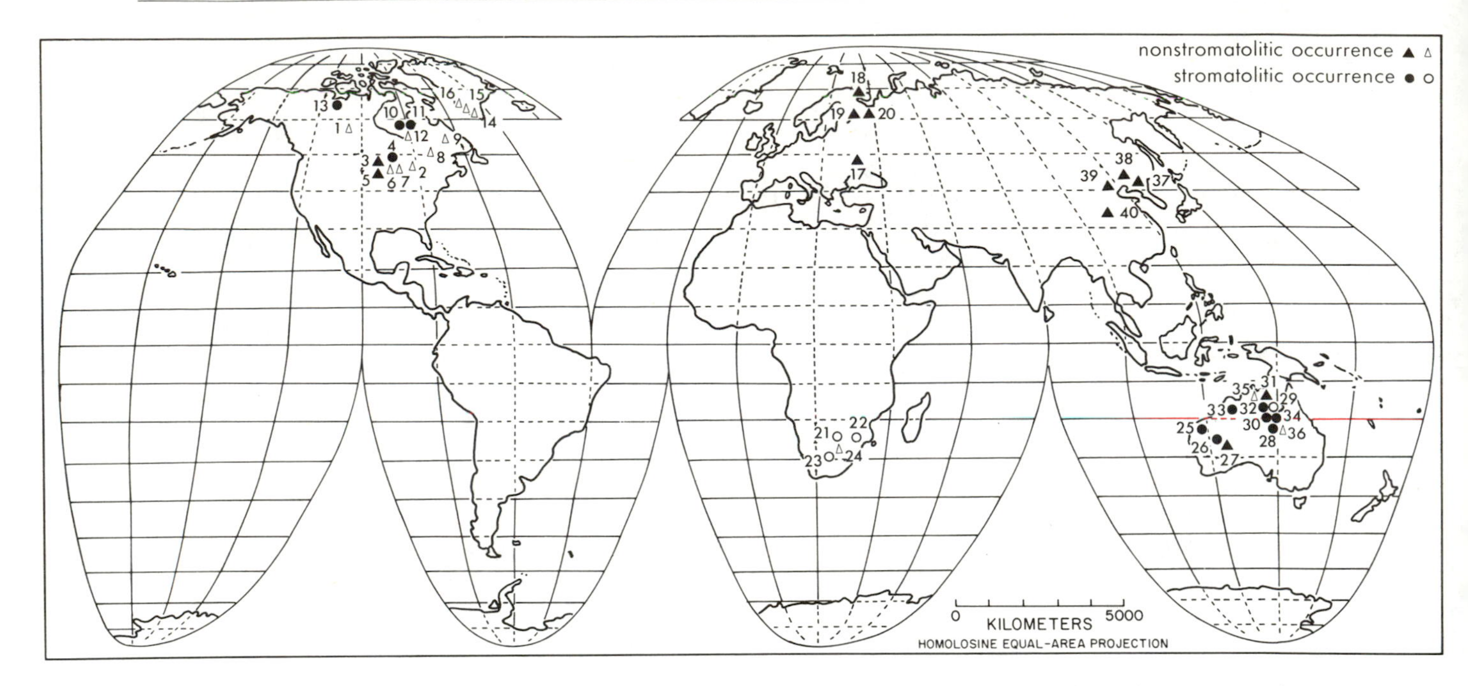

FIGURE 14-1. Georgraphic distribution of Early Proterozoic occurrences of microfossils, dubiomicrofossils, and pseudomicrofossils. ▲ Nonstromatolitic occurrences of microfossils. △ Nonstromatolitic occurrences of dubiomicrofossils (?) and pseudomicrofossils (×). ● Stromatolitic occurrences of microfossils. ○ Stromatolitic occurrences of dubiomicrofossils (?) and pseudomicrofossils (×).

NORTH AMERICA	
1 Archean or Aphebian	× △
2 Gowganda Fm.	? △
3 Pokegama Fm.	▲
4 Gunflint-Biwabik Fms.	●▲
5 Tyler Fm.	▲
6 Michigamme Fm.	? △
7 Negaunee Fm.	× △
8 Temiscamie Fm.	× △
9 Sokoman Fm.	× △▲
10 Kasegalik Fm.	●
11 McLeary Fm.	●
12 Kipalu Fm.	? △
13 Rocknest Fm.	●
14 Zig Zag Land Fm.	? △
15 Graensesø Fm.	? △
16 Foselv Fm.	? △

EUROPE	
17 Krivoirog Ser.	▲
18 Pechenga Ser.	▲
19 Ladoga Ser.	▲
20 Jatulian Ser.	▲

AFRICA	
21 Dolomite Ser.	? ○
22 Malmani Dol.	× ○
23 Malmani Dol.	? ○
24 Pretoria Gp.	? △

AUSTRALIA	
25 Duck Creek Dol.	●▲
26 Frere Fm.	●
27 Windidda Fm.	▲
28 Amelia Dol.	●
29 Emmerugga Dol.	? ○
30 Cooley Dol.	●
31 HYC Pyritic Shale	▲
32 Balbirini Dol.	●
33 Bungle Bungle Dol.	●▲
34 Paradise Creek Fm.	●
35 Vizard Fm.	× △
36 Mount Isa Shale	? △

ASIA	
37 Anshan Gp.	▲
38 Liaohe Gp.	▲
39 Changcheng System	▲
40 Xiyanghe Gp.	▲

assemblages, only about ten are sufficiently well known from studies of the microbiotas in petrographic thin sections to warrant consideration in the discussion that follows. Most of these are from shallow water stromatolitic carbonate facies that have been subject to early diagenetic silicification; such assemblages are thus preserved in cherts in a restricted and relatively unusual environment, and the possibility therefore exists that they may comprise an ecologically biased, "unrepresentative sample" of Early Proterozoic life. Although microfloras are also known from offshore shaly facies, these assemblages have been studied largely or exclusively in acid-resistant residues rather than in thin sections, and are composed of compressed, carbonaceous remains, chiefly of spheroidal acritarchs presumed to have been planktonic. The original relationships of such acritarchs to

TABLE 14-1. Chronostratigraphic summary of reported occurrences of Early Proterozoic microfossils and microfossil-like objects.

Ga	*North America*	*Europe*	*Africa*	*Australia*	*Asia*
				? △ 36 Mount Isa Sh.	
				× △ 35 Vizard Fm.	
1.6–				● 34 Paradise Creek Fm.	
				●▲ 33 Bungle Bungle Dol.	
				● 32 Balbirini Dol.	
				▲ 31 H. Y. C. Pyritic Sh.	
				● 30 Cooley Dol.	
1.7–				? ○ 29 Emmerugga Dol.	
				● 28 Amelia Dol.	
		▲ 20 Jatulian Ser.			
		▲ 19 Ladoga Ser.			
				▲ 27 Windidda Fm.	
		▲ 18 Pechenga Ser.		● 26 Frere Fm.	▲ 40 Xiyanghe Gp. ⎤*
1.8–	? △ 16 Foselv Fm.				▲ 39 Changcheng Syst. ⎥
	? △ 15 Graensesø Fm.				▲ 38 Liaohe Gp. ⎦
	? △ 14 Zig Zag Land Fm.				
	● 13 Rocknest Fm. ⎤*				
	? △ 12 Kipalu Fm. ⎥				
1.9–	● 11 McLeary Fm. ⎥				
	● 10 Kasegalik Fm. ⎥				
	× △▲ 9 Sokoman Fm. ⎥				
	× △ 8 Temiscamie Fm. ⎥				
	× △ 7 Negaunee Fm. ⎥				
	? △ 6 Michigamme Fm. ⎥				
2.0–	▲ 5 Tyler Fm. ⎥				
	●▲ 4 Gunflint-Biwabik Fms. ⎥	▲ 17 Krivoirog Ser.			
	▲ 3 Pokegama Fm. ⎦			●▲ 25 Duck Creek Dol.	

▲ = Nonstromatolitic occurrences of microfossils.
△ = Nonstromatolitic occurrences of dubiomicrofossils (?) and pseudomicrofossils (×).
● = Stromatolitic occurrences of microfossils.
○ = Stromatolitic occurrences of dubiomicrofossils (?) and pseudomicrofossils (×).
* Two or more units within the brackets may be stratigraphically correlative.

Table 14-1. (*continued*).

Ga	*North America*	*Europe*	*Africa*	*Australia*	*Asia*
2.1–					
2.2–					
	? △ 2 Gowganda Fm.		? △ 24 Pretoria Ser.		
			? ○ 23 } Malmani Dol.		
			○ 22 } Malmani Dol.		
			? ○ 21 Dolomite Ser.		
2.3–					
2.4–					
2.5–	× △ 1 Archean or Aphebian				▲ 37 Anshan Gp.

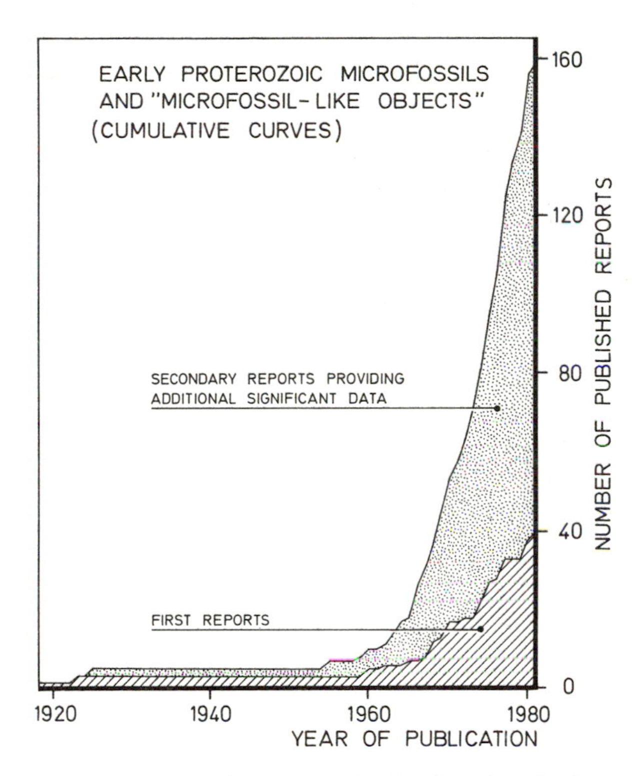

FIGURE 14-2. Cumulative curves showing date of publication and number of published reports describing Early Proterozoic microfossils and "microfossil-like objects."

their encompassing shale matrix are thus in general essentially unknown, and their flattened, compressed nature makes it difficult to meaningfully compare them with the three-dimensionally preserved, morphologically much more diverse microfossils studied in chert thin sections. However, the microflora of the H. Y. C. Pyritic Shale, the best known and most diverse deep water assemblage so far studied in both macerations and thin sections, contains many taxa that are similar or identical to those occurring in coeval shallow water environments (Oehler, 1977c, p. 321), suggesting that the distribution of Early Proterozoic microfossils may not be as markedly facies-dependent as might first have been anticipated. The differences between chert-carbonate and shale assemblages are at least in part due to differences of preservation and the different techniques employed in their study.

In this chapter we concentrate only on the relatively better known assemblages of the 2.5 to 1.6 Ga period, in addition to summarizing and evaluating previously available data regarding the Early Proterozoic fossil record (Section 14-3-1, Table 14-2) and to illustrating representative fossils of this age (Photos 14-1 through 14-11); newly detected fossils are described and illustrated from four Early Proterozoic formations (Section 14-3-2).

14-2. Preservation

Considering the data currently available, the highest fidelity of preservation exhibited by Early Proterozoic microfossils appears to have been effected under conditions of early diagenetic permineralization by silica, particularly in saline, peritidal, stromatolitic carbonate environments. Diagenesis of carbonate minerals tends to destroy such microfossils (Hofmann, 1975a, 1975b); thus, in general, it is only in the silicified (cherty) portions of such sequences that microorganisms are cellularly preserved.

Microorganisms become chemically and structurally altered during progressive diagenesis and catagenesis; secondary morphologic features are thus produced that can be subject to multiple interpretions (Awramik et al., 1972). Moreover, because of differences in their original chemistry and structural integrity, different parts of microscopic organisms have differing potential for preservation. Thus, for example, the physically resilient sheaths and envelopes that commonly encompass cyanobacterial cells appear to have a greater potential for fossilization than the cell contents or cell walls of such organisms. This can lead to divergent interpretations of preserved microfossils, the question being whether the preserved cell-like objects seen in thin sections are in fact cells, delimited by preserved cell walls, or whether they might more properly be interpreted as remnants of originally mucilagenous envelopes that as a result of differential degradation lack internal contents (Golubić and Campbell, 1979, p. 205). Similar difficulties are encountered in the interpretation of atypically enlarged cells that occur in various Early Proterozoic septate filaments and that are regarded by some workers as heterocysts or akinetes. Such features could also be explained as osmotically enlarged vegetative cells, as suggested by the occurrence in some specimens of cells with an "exploded" aspect (e.g., Photo 14-1C, D), or as a result of other types of diagenetic alteration (as discussed in Section 14-4-1).

Diffusion phenomena affect the form of microfossils insofar as they redistribute chemical substances along concentration gradients (e.g., Photos 14-1F, G, and 14-5A), and originally organic filaments may become replaced by one or more minerals (e.g., Photos 14-3G-J, and 14-7A, B) changing the morphology and size of the preserved organisms, sometimes drastically. Experimental studies simulating diagenetic processes have proved valuable in evaluating such effects on ancient microbiotas (Oehler and Schopf, 1971; J. Oehler, 1976; Knoll et al., 1978); many more should be undertaken.

Preservational modifications of cell morphology obviously have direct bearing on taxonomic interpretations of microfossils, namely, whether individual fossils of slightly differing aspect represent different biological species or whether they are in fact degradational variants of a single taxon. Demonstration of gradational relationships with intermediate morphologic stages occurring within the same thin section has proven particularly useful in resolving such questions (Hofmann, 1976, p. 1070).

Microfossils recovered from clastic rocks by acid maceration also exhibit varying effects of diagenetic alteration, such as the occurrence of granulose or reticulate surface textures resulting from impressions of minute mineral grains on the organic walled vesicles. Similarly, as studied in acid-resistant organic residues, the walls of acritarchs are not uncommonly folded or otherwise distorted; although such features seem obviously to have been produced during compaction of the enclosing sediment, they have nevertheless been used by several workers as diagnostic characters by which to establish fossil taxa. The taxonomy of microfossils described from Early Proterozoic shales is thus currently in a rather confused state, one much in need of detailed, careful re-evaluation.

14-3. Occurrences

The compilation of the geographic distribution of the 40 reported occurrences of Early Proterozoic microscopic remains (Figure 14-1) shows a concentration in 5 main regions; however, bona fide fossils occur in only 4 of these: North America, Australia, Europe, and eastern Asia. The remains from South Africa are of dubious paleontological pertinence inasmuch as either their biogenicity or their Precambrian age has not been confirmed. No data are apparently available from South America and Antarctica.

A historical perspective of the first reports of new occurrences is given in Figure 14-2, which also shows the great burst of activity beginning in 1959 after a prolonged period of quiescence that followed the initial spurt between 1918 and 1924. The greatest rate of publication was attained in the mid-1970s, reflecting particularly the concentration of work and discoveries in Australia in the last decade. During the time between 1960 and 1980, the spread in the ratio of primary to secondary reports decreased from 1:1 to approximately 1:4, an indication of the importance and growing interest generated by the discoveries.

14-3-1. Synopsis of Previously Reported Occurrences

Table 14-2 presents an annotated listing and interpretation of all reported occurrences known to us (through 1981) from the Early Proterozoic, as well as material here newly reported. The numbering system in the table follows that employed in Figure 14-1 and Table 14-1, which is in approximate geochronological order, continent by continent. The occurrences are further distinguished on the basis of whether or not they are stromatolitic, and taxa are grouped by morphological types where appropriate. The most important microbiotas are illustrated in Photos 14-1 through 14-11. These figures are prepared from new material (about half of the illustrations) discovered in 1979 and 1980 during the course of the P.P.R.G. research project, as well as from previously published photographs. Table 14-3 summarizes, in alphabetic order, described Early Proterozoic taxa, their basic morphology and possible affinities, and their distribution among the 40 occurrences. The remains include a wide variety of shapes and sizes, ranging from submicrometric dimensions to those visible to the naked eye, and hence "megascopic." Not all the taxa listed are bona fide microfossils, and some of the taxa listed are no doubt synonymous with others in the list.

14-3-2. Newly Detected Fossils

In this section we report and briefly discuss taxa not previously recognized in four of the Australian assemblages, specifically, those of the Duck Creek,

TABLE 14-2. Early Proterozoic microfossils and microfossil-like objects.

Lithology	*Reported objects*	*Intepretation*
	Archean or Aphebian *Uranium City area, Saskatchewan, Canada* *<2.7, >1.7 Ga*	
(1) Chert	(1) "Possibly biogenic" spheroids containing dark elongate mass (Beck, 1969).	(1) NONFOSSIL: chemical or radioactive alteration zones around dark grains (Hofmann, 1971a, p. 64).
	Gowganda Formation *Blind River area, Ontario, Canada* *2.237 ± 0.085 Ga*	
(2) Sulfides in varved argillites	(2A) "Actinomycetes" and other "ultramicrofossils" detected by transmission electron microscopy (Jackson, 1967). (2B) Sulfides, carbonaceous inclusions and "fossil gas bubbles" of presumed biologic origin (Jackson, 1971).	(2A) NONFOSSIL: modern contaminants (Schopf, 1975, p. 235) or goethite crystallites (Hofmann, 1971a, p. 52). (2B) DUBIOFOSSILS: possibly, but not certainly of biologic origin.
	Pokegama Quartzite *northeastern Minnesota, U.S.A.* *<2.1, >1.8 Ga*	
(3) Chert pebbles and nonstromatolitic chert	(3A) *Gunflintia minuta* Barghoorn, "blue-green algae" (Gruner, 1922, 1924; Cloud & Licari, 1972; Darby, 1972). (3B) *Horoniospora*-like spheroids (Cloud & Licari, 1972). (3C) *Palaeomicrocoleus gruneri* Korde (Vologdin & Korde, 1965; Gruner, 1923)	(3A) MICROFOSSILS: filamentous bacteria and/or cyanobacteria. (3B) DUBIOMICROFOSSILS: spheroidal microstructures, possibly biogenic. (3C) NONFOSSIL: trails of ambient pyrite grains (Tyler & Barghoorn, 1963; Darby, 1972).
	Gunflint Formation *northwestern Ontario, Canada* *and Biwabik Formation* *northern Minnesota, U.S.A.* *<2.1, >1.8 Ga*	
(4) Stromatolitic and nonstromatolitic, ferruginous and non-ferruginous chert	(4A) COCCOID MICROFOSSILS: (4Aa) *Chlamydomonopsis primordialis* Edhorn (1973). (4Ab) *Corymbococcus hodgkissii* Awramik & Barghoorn (1977). (4Aa) "Eucaryotic cells" (Edhorn, 1972). (4Ac) *Huroniospora macroreticulata* Barghoorn (Barghoorn & Tyler, 1965d; Cloud & Hagen, 1965; Hofmann, 1971a; Awramik & Barghoorn, 1977). (4Ac) *H. microreticulata* Barghoorn (Barghoorn & Tyler, 1965a; Cloud & Hagen, 1965; LaBerge, 1967; Licari & Cloud, 1968; Cloud & Licari, 1968; Hofmann, 1971a; Darby, 1974; Akiyama & Imoto, 1975; Awramik, 1976; Tappan, 1976; Awramik & Barghoorn, 1977; Knoll et al., 1978; Awramik & Semikhatov, 1979). (4Ac) *H. psilata* Barghoorn (Barghoorn & Tyler, 1965a; Cloud & Hagen, 1965; Hofmann, 1971a; Awramik & Barghoorn, 1977). (4Ad) *Leptoteichos golubicii* Knoll et al. (Barghoorn, 1971; Knoll et al., 1978). (4Ae) *Megalytrum diacenum* Knoll et al. (1978).	(4A) MICROFOSSILS: spheroidal bacteria and/or cyanobacteria. (4Aa) cf. *Huroniospora* spp. (based on restudy by H. J. H.; also see Awramik & Barghoorn, 1977, p. 129). (4Ab) Colonial prokaryote, probably *Aphanocapsa*-like cyanobacterium (Awramik & Barghoorn, 1977, p. 132–134). (4Ac) Unicellular (and colonial?) cyanobacteria and/or bacteria, previously suggested to be cyanobacterial (Awramik & Barghoorn, 1977, p. 140), fungal (Darby, 1974, p. 1595–1596) and red algal (Tappan, 1976, p. 635); see Photo 14-1I-N, and Q. (4Ad) Relatively large form, previously compared with isolated cells of *Cylindrospermum* (Knoll et al., 1978, p. 987–990). (4Ae) Sheath of colonial bacterium or chroococcalean cyanobacterium.

TABLE 14-2. (*continued*).

Lithology	Reported objects	Intepretation
	(4B) SEPTATE FILAMENTOUS MICROFOSSILS:	(4B) MICROFOSSILS: cellular, filamentous, bacteria and/or cyanobacteria.
	(4Ba) *Anabaenidium barghoornii* Edhorn (1973). (4Bb) *Gunflintia grandis* Barghoorn (Barghoorn & Tyler, 1965a; Moorhouse & Beales, 1962; Licari & Cloud, 1968; Hofmann, 1971a; Awramik & Barghoorn, 1977). (4Bb) *G. minuta* Barghoorn (Barghoorn & Tyler, 1965a; Tyler & Barghoorn, 1954; Cloud, 1965; Cloud & Hagen, 1965; LaBerge, 1967; Cloud & Licari, 1968; Licari & Cloud, 1968; Hofmann, 1971a; Awramik, 1976; Awramik & Barghoorn, 1977; Awramik & Semikhatov, 1979). (4Ba) *Palaeoscytonema moorhousei* Edhorn (1973). (4Ba) *Palaeospiralis canadensis* Edhorn (1973). (4Ba) *Palaeospirulina arcuata* Edhorn (1973). (4Ba) *P. minuta* Edhorn (1973). (4Ba) *Primorivularia thunderbayensis* Edhorn (1973). (4Ba) "*Schizothrix*" *atavia* Edhorn (1973).	(4Ba) cf. *Gunflintia* spp. (based on restudy by H. J. H.). (4Bb) Filamentous bacteria and/or cyanobacteria, previously suggested to be nostocacean cyanobacteria (Licari & Cloud, 1968) based on the presence of "heterocysts" and "akinetes" (Licari & Cloud, 1968, p. 1056; Awramik & Barghoorn, 1977, p. 126–127), structures that are here interpreted to be diagenetic artifacts (Section 14-4-1); see Photo 14-1B-F and H.
	(4C) TUBULAR UNBRANCHED MICROFOSSILS:	(4C) MICROFOSSILS: sheaths of filamentous bacteria and/or cyanobacteria.
	(4Ca) *Animikiea septata* Barghoorn (Barghoorn & Tyler, 1965a; Cloud, 1965; Hofmann, 1971a; Awramik & Barghoorn, 1977). (4Cb) *Entosphaeroides amplus* Barghoorn (Barghoorn & Tyler, 1965a; Hofmann, 1971a; Akiyama & Imoto, 1975; Awramik, 1976; Awramik & Barghoorn, 1977).	(4Ca) Relatively broad, thick-walled, tubular sheath with corrugations; see Photo 14-1A. (4Cb) Apparently, tubular sheaths that contain either endospores, unrelated coccoid prokaryotes, or remnants of trichomic cells; specimen illustrated by Akiyama & Imoto (1975, p. 381) is a probable contaminant.
	(4D) UNUSUAL, BIZZARRE MICROFOSSILS:	(4D) MICROFOSSILS: fossil microorganisms of unusual morphology and uncertain systematic position.
	(4Da) *Archaeorestis magna* Awramik & Barghoorn (1977). (4Da) *A. schreiberensis* Barghoorn (Barghoorn & Tyler, 1965a; Barghoorn, 1971; Hofmann, 1971a; Awramik & Barghoorn, 1977). (4Db) "Desmid-like" microorganism (Edhorn, 1973). (4Dc) *Eoastrion bifurcatum* Barghoorn (Barghoorn & Tyler, 1965a; Hofmann, 1971a; Awramik & Barghoorn, 1977). (4Dc) *E. simplex* Barghoorn (Barghoorn & Tyler, 1965a; Cloud, 1965; Hofmann, 1971a; Kline, 1975a, 1975b; Awramik & Barghoorn, 1977). (4Dd) *Eomicrhystridium barghoornii* Deflandre (1968; Hofmann, 1971a; Awramik & Barghoorn, 1977, p. 123–124 [reported as *E. aremoricanum*]). (4De) *Eosphaera tyleri* Barghoorn (Barghoorn & Tyler, 1965a; LaBerge, 1967; Barghoorn, 1971; Hofmann, 1971a; Kazmierczak, 1976, 1979; Tappan, 1976; Awramik & Barghoorn, 1977; Knoll et al., 1978). (4Df) *Exochobrachium triangulum* Awramik & Barghoorn (1977). (4Dg) *Frutexites microstroma* Walter & Awramik (1979; Hofmann, 1969). (4Dh) *Galaxiopsis melanocentra* Awramik & Barghoorn (1977; Barghoorn, 1971). (4Di) *Kakabekia umbellata* Barghoorn (Barghoorn & Tyler, 1965a; Siegel & Giumarro, 1966; Siegel & Siegel, 1968; Hofmann, 1971a;	(4Da) Budding bacterium incertae sedis (Awramik & Barghoorn, 1977, p. 124). (4Db) cf. *Kakabekia* (based on restudy by H. J. H.; compare with Barghoorn & Tyler, 1965, fig. 7, pt. 10). (4Dc) Bacterium incertae sedis, similar to *Metallogenium personatum* Perfil'ev & Gabe, considered by Awramik & Barghoorn (1977, p. 124) to be a budding bacterium; see Photo 14-1R and S. (4Dd) Acanthomorph acritarch? or degraded sphaeromorph?; see Photo 14-1M and O. (4De) Prokaryote? incertae sedis, previously compared with green algae *Volvox* and *Eovolvox* (Kazmierczak, 1976, 1979) and with red alga *Porphyridium* (Tappan, 1976); see Photo 14-1P. (4Df) Prokaryote incertae sedis; see Photo 14-1V; (4Dg) Interpreted as microstromatolite (Hofmann, 1969, p. 15) and as scytonematacean cyanobacterium (Walter & Awramik, 1979, p. 22–23). (4Dh) Prokaryote incertae sedis (degraded sphaeromorph? or cf.? *Eomicrhystridium* or cf.? *Exochobrachium*). (4Di) Budding bacterium? (Awramik & Barghoorn, 1977, p. 124); see Photo 14-1U. (4Dj) cf. *Eoastrion* (based on restudy by H. J. H.) (4Dk) Prokaryote incertae sedis. (4Dl) Polygonomorph acritarch; see Photo 14-1T.

TABLE 14-2. (*continued*).

Lithology	*Reported objects*	*Intepretation*
	Awramik & Barghoorn, 1977). (4Dj) "Radiolarian-like" organism (Edhorn, 1973). (4Dk) *Thymos halis* Awramik & Barghoorn (1977). (4Dl) *Veryhachium*? (Hofmann, 1971b). (4Dk) *Xenothrix inconcreta* Awramik & Barghoorn (1977).	
	(4E) DUBIOMICROFOSSILS:	(4E) DUBIOMICROFOSSILS: microstructures possibly but not certainly biogenic.
	(4Ea) *Glenobotrydion aenigmatis* Schopf (Edhorn, 1973; Awramik & Barghoorn, 1977). (4Eb) *Menneria levis* Lopukhin (1975, 1976). (4Ec) *Palaeoanacystis irregularis* Edhorn (1973). (4Ec) *Sphaerophycus gigas* Edhorn (1973).	(4Ea) Probably biogenic cf. *Corymbococcus* or cf. *Palaeoanacystis irregularis* Edhorn (could not be relocated in original thin sections). (4Eb) Probably a sphaeromorph acritarch cf. *Megalytrum diacenum* Knoll et al. (not available for study). (4Ec) Hematitic, poorly preserved spheroids, possibly cf. *Corymbococcus* (based on restudy by H. J. H.).
	(4F) PSEUDOMICROFOSSILS:	(4F) NONFOSSIL: pseudomicrofossils and mordern contaminants.
	(4Fa) "Coccoid bacteria" (Schopf et al., 1965). (4Fb) *Cumulosphaera lamellosa* Edhorn (1973; Moorhouse & Beales, 1962; Kazmierczak, 1979). (4Fc) *Palaeorivularia ontarica* Korde (Tyler & Barghoorn, 1954; Vologdin & Korde, 1965; Barghoorn, 1971; Hofmann, 1971a; D. Oehler, 1976; Awramik & Barghoorn, 1977). (4Fa) "Rod-shaped bacteria" (Schopf et al., 1965; Barghoorn, 1971; Hofmann, 1971a). (4Fd) "Spherical microstructures" (LaBerge, 1973).	(4Fa) Modern bacterial contaminants (Schopf, 1975, p. 235). (4Fb) Mineralic pseudomicrofossils (based on restudy by H. J. H.). (4Fc) Pseudomicrofossil produced by displacement of organic matter by growth of fibrous chalcedony (J. Oehler, 1976a, p. 123–128; Awramik & Barghoorn, 1977, p. 123) with filament mass at center possibly being remnants of *Gunflintia* or *Eoastrion*. (4Fd) Mineralic pseudomicrofossils (compare with Klein & Fink, 1976, p. 471, 474).
	Tyler Formation *Gogebic County, northern Michigan, U.S.A.* *<2.5, >1.6 Ga*	
(5) Siliceous clasts in pyritic chert	(5A) COCCOID MICROFOSSILS:	(5A) MICROFOSSILS: spheroidal bacteria and/or cyanobacteria.
	(5Aa) cf. *Huroniospora* (Cloud & Morrison, 1979b, 1980).	(5Aa) Unicellular prokaryote (cf. 4Ac, above).
	(5B) SEPTATE FILAMENTOUS MICROFOSSILS:	(5B) MICROFOSSILS: cellular, filamentous, bacteria and/or cyanobacteria.
	(5Ba) *Gunflintia* sp. (Cloud & Morrison, 1980). (5Bb) *Gunflintia* cf. *G. grandis* Barghoorn (Cloud & Morrison, 1979b, 1980).	(5Ba) Filamentous bacteria and/or cyanobacteria (cf. 4Bb, above). (5Bb) Filamentous bacteria and/or cyanobacteria (cf. 4Bb, above).
	(5C) TUBULAR UNBRANCHED MICROFOSSILS:	(5C) MICROFOSSILS: sheaths of filamentous bacteria and/or cyanobacteria.
	(5Ca) cf. *Animikiea* sp. (Cloud & Morrison, 1980).	(5Ca) Relatively broad tubular sheath (see 4Ca, above).
	(5D) UNUSUAL, BIZARRE MICROFOSSILS:	(5D) MICROFOSSILS: fossil microorganisms of unusual morphology and uncertain systematic position.
	(5Da) *Kakabekia* cf. *K. umbellata* Barghoorn (Cloud & Morrison, 1979b, 1980). (5Db) *Metallogenium* (*Eoastrion*) sp. (Cloud & Morrison, 1979b, 1980).	(5Da) Budding bacterium? (cf. 4Di, above). (5Db) Bacterium incertae sedis (cf. 4Dc, above).
	(5E) DUBIOMICROFOSSILS:	(5E) DUBIOMICROFOSSILS: microstructures possibly but not certainly biogenic.
	(5Ea) Filament 20 μm in diameter (Cloud & Morrison, 1980).	(5Ea) Probably biogenic, possibly prokaryotic sheath.

TABLE 14-2. (*continued*).

Lithology	*Reported objects*	*Intepretation*
	Michigamme Formation *Baraga County, northern Michigan, U.S.A.* *<2.5, >1.6 Ga*	
(6) Phosphatic layer,	(6) Microfossil-like pyritic chains (Cloud & Morrison, 1979b).	(6) DUBIOMICROFOSSILS: insufficient data as yet available (reported in abstract only, without illustrations).
	Negaunee Iron Formation *Empire Mine, northern Michigan, U.S.A.* *<2.5, >1.6 Ga*	
(7) Iron formation (jaspilite)	(7A) "Irregularly segmented structures" (Mancuso et al., 1971).	(7A) NONFOSSIL: mineralic pseudomicrofossils.
	(7B) Hematitic framboids and associated carbonaceous material (Lougheed & Mancuso, 1973).	(7B) DUBIOFOSSILS: hematite pseudomorphic after pyrite and thus possibly indicative of sulfate-reducing bacteria.
	(7C) Spheroidal microstructures (LaBerge, 1973).	(7C) NONFOSSIL: mineralic pseudomicrofossils of diagenetic or metamorphic origin.
	Temiscamie Formation *Lake Albanel, central Quebec, Canada* *<2.5, >1.749 ± 0.054 Ga*	
(8) Iron formation (jaspilite)	(8) Ferruginous and carbonaceous(?) spheroidal microstructures (LaBerge, 1967, 1973).	(8) NONFOSSIL: spherulitic crystallites of colloidal precursors (Hofmann, 1971a, p. 53–54).
	Sokoman Iron Formation *Schefferville area, northeastern Quebec, Canada* *<2.5, >1.8 Ga*	
(9) Iron formation	(9A) Pigmented "cell-like" spheroids (LaBerge, 1967, 1973).	(9A) NONFOSSIL: spherulitic crystallites (cf. 8, above; see also Klein & Fink, 1976, p. 471; and Reeves, 1979).
	(9B) "*Eosphaera*-like microfossils" (LaBerge, 1967, 1973).	(9B) NONFOSSIL: mineralic pseudomicrofossil (cf. 9A, above).
	Kasegalik Formation *Belcher Islands, Hudson Bay, Canada* *<2.5, >1.76 ± 0.037 Ga*	
(10) Cherty stromatolitic dolostone	(10) MICROFOSSILS—Taxa noted are reported in: Hofmann & Jackson, 1969; Hofmann, 1974, 1975, 1976; Golubić & Hofmann, 1976; Golubić & Campbell, 1979.	
	(10A) COCCOID MICROFOSSILS:	(10A) MICROFOSSILS: spheroidal bacteria and/or cyanobacteria.
	(10Aa) *Eoentophysalis belcherensis* Hofmann. (10Ab) *Eosynechococcus moorei* Hofmann. (10Ac) *E. medius* Hofmann. (10Ad) *Globophycus* sp. (10Ad) *Leptoteichos golubicii* Knoll et al. (=Type 2 structure of Hofmann & Jackson, 1969). (10Ad) *Melasmatosphaera magna* Hofmann. (10Ad) *M. media* Hofmann. (10Ad) *M. parva* Hofmann. (10Ae) *Myxococcoides minor* Schopf. (10Ae) *Palaeoanacystis vulgaris* Schopf. (10Af) *Pleurocapsa*? sp. (10Ag) *Sphaerophycus parvum* Schopf.	(10Aa) Colonial prokaryote cf. *Entophysalis major* Ercegović; see Photo 14-2G-J. (10Ab) Colonial prokaryote, compared to *Gloeothece coerulea* Geitler (Golubić Campbell, 1979); see Photo 14-2N. (10Ac) Ellipsoidal prokaryote; see Photo 14-2L. (10Ad) Prokaryote incerate sedis. (10Ae) Prokaryote, probably chroococcacean. (10Af) Prokaryote, possibly pleurocapsacean; see Photo 14-2P. (10AG) Chrococcacean-like prokaryote; see Photo 14-2K.
	(10B) SEPTATE FILAMENTOUS MICROFOSSILS:	(10B) MICROFOSSILS: cellular, filamentous bacteria and/or cyanobacteria.
	(10Ba) *Biocatenoides sphaerula* Schopf. (10Bb) *Halythrix* sp. (10Bc) *Rhicnonema antiquum* Hofmann	(10Ba) Prokaryote, probably bacterium; see Photo 14-2A. (10Bb) Degraded bacterium or oscillatoriacean; see Photo 14-2D. (10Bc) Degraded sheath-enclosed bacterium or oscillatoriacean; see Photo 14-2E,F.

TABLE 14-2. (*continued*).

Lithology	*Reported objects*	*Intepretation*
	(10C) TUBULAR UNBRANCHED MICROFOSSILS:	(10C) MICROFOSSILS: sheaths of filamentous bacteria and/or cyanobacteria.
	(10Ca) *Eomycetopsis* sp. [reported as *Archaeotrichion* sp.]. (10Ca) *E. filiformis* Schopf.	(10Ca) Sheath, bacterial or cyanobacterial; see Photo 14-2B,C.
	(10D) UNUSUAL, BIZARRE MICROFOSSILS:	(10D) MICROFOSSILS: fossil microorganisms of unusual morphology and uncertain systematic position.
	(10Da) Acritarch-like structures. (10Db) cf. *Eosphaera* sp. (=Type 4 microstructures of Hofmann & Jackson, 1969).	(10Da) Prokaryote incertae sedis, probably coccoid bacteria and/or cyanobacteria. (10Db) Prokaryote? incertae sedis (cf. 4De, above); see Photo 14-2Q.
	McLeary Formation *Belcher Islands, Hudson Bay, Canada* *<2.5, >1.76 ± 0.037 Ga*	
(11) Cherty stromatolitic dolostone	(11) MICROFOSSILS—Taxa noted are reported in: Hofmann, 1974, 1975, 1976; Golubić & Hofmann, 1976.	
	(11A) COCCOID MICROFOSSILS:	(11A) MICROFOSSILS: spheroidal bacteria and/or cyanobacteria.
	(11Aa) Acritarcha, (11Ab) *Caryosphaeroides* sp. (11Ac) *Eoentophysalis belcherensis* Hofmann. (11Ac) *Eosynechococcus medius* Hofmann. (11Ad) *E. grandis* Hofmann. (11Ae) *Eozygion minutum* Schopf & Blacic. (11Ae) *Glenobotrydion majorinum* Schopf & Blacic. (11Af) *Globophycus* sp. (11Ag) *Leptoteichos golubicii* Knoll et al. (11Ah) *Melasmatosphaera magna* Hofmann. (11Ai) *M. media* Hofmann. (11Aj) *M. parva* Hofmann. (11Ae) *Myxococcoides* sp. (11Ae) *M. inornata* Schopf. (11Ak) *M. minor* Schopf. (11Al) *Palaeoanacystis vulgaris* Schopf. (11Ae) *Sphaerophycus parvum* Schopf. (11Am) *Zosterosphaera* sp.	(11Aa) Prokaryote incertae sedis. (11Ab) Prokaryote incertae sedis; see Photo 14-2T. (11Ac) Colonial prokaryote (cf. 10Aa, above). (11Ad) Ellipsoidal prokaryote; see Photo 14-2M. (11Ae) Prokaryote, probably chroococcacean. (11Af) Prokaryote incertae sedis; see Photo 14-2U. (11Ag) Prokaryote incertae sedis; see Photo 14-2R. (11Ah) Prokaryote incertae sedis; see Photo 14-2W. (11Ai) Prokaryote incertae sedis; see Photo 14-2V. (11Aj) Prokaryote incertae sedis; see Photo 14-2X. (11Ak) Prokaryote, probably chroococcacean; see Photo 14-2S. (11Al) Prokaryote, probably chroococcacean; see Figure 14-4O. (11Am) Prokaryote? incertae sedis; see Photo 14-2Z.
	(11B) SEPTATE FILAMENTOUS MICROFOSSILS:	(11B) MICROFOSSILS: cellular, filamentous prokaryotes.
	(11Ba) *Biocatenoides sphaerula* Schopf.	(11Ba) Prokaryote, probably bacterium.
	(11C) TUBULAR UNBRANCHED MICROFOSSILS:	(11C) MICROFOSSILS: sheaths of filamentous bacteria and/or cyanobacteria.
	(11Ca) *Eomycetopsis filiformis* Schopf.	(11Ca) Sheath, probably cyanobacterial.
	(11D) UNUSUAL, BIZARRE MICROFOSSILS:	(11D) MICROFOSSILS: fossil microorganisms of unusual morphology and uncertain systematic position.
	(11Da) *Kakabekia*? sp.	(11Da) Budding bacterium? (cf. 4Di, above) or degraded sphaeromorph acritarch?; see Photo 14-2Y.
	Kipalu Iron Formation *Belcher Islands, Hudson Bay, Canada* *<2.5, >1.7 Ga*	
(12) Iron formation (jaspilite)	(12A) Spheroidal "algal cells" (Moore, 1918). (12A) "Filamentous algae" (Moore, 1918).	(12A) DUBIOMICROFOSSILS: probably biogenic microstructures, illustrated by line drawings only (original thin section not located; Hofmann, 1971a, p. 53).
	(12B) Mineralic spheroids (LaBerge, 1967, 1973; Hofmann, 1971a).	(12B) NONFOSSIL: spherulitic crystallites (cf. 8, above).

TABLE 14-2. (*continued*).

Lithology	*Reported objects*	*Intepretation*
	Rocknest Formation *northern District of Mackenzie, Canada* *<2.5, >1.7 Ga*	
(13) Stromatolitic carbonate	(13) "Filament molds" (Gebelein, 1974, p. 591).	(13) MICROFOSSILS: Scytonematacean-like, suberect filament molds 8 μm across, cf. *Palaeoleptophycus* (based on restudy of Geol. Surv. Canada field sample HY-ROKU-1 by H. J. H.).
	Zig Zag Formation *Graenseland, southwestern Greenland* *<2.0, >1.7 Ga*	
(14) Black quartzite	(14) Microscopic spheroids and carbonaceous fragments with "cell-like" structure (Bondesen et al., 1967).	(14) DUBIOMICROFOSSILS: illustrated material too ill-defined to permit interpretation.
	Graensesø Formation *Graenseland, southwestern Greenland* *<2.0, >1.7 Ga*	
(15A) Dolostone	(15A) *Vallenia erlingi* Pedersen, "planktonic alga" (Pedersen, 1966; Bondesen et al., 1967; Häntzschel, 1975; Glaessner, 1979).	(15A) DUBIOMICROFOSSIL: "doubtfully biogenic" (Glaessner, 1979, p. A111; possibly are replaced ooids.
(15B) Black quartzite	(15B) Microscopic spheroids and carbonaceous fragments composed of fine threads (Bondesen et al., 1967).	(15B) DUBIOMICROFOSSILS: illustrated material material indistinct, not interpretable (cf. 14, above).
	Foselv Formation *Graenseland, southwestern Greenland* *<2.0, >1.7 Ga*	
(16) Carbonaceous shale	(16A) Microscopic spheroids (Bondesen et al., 1967; Pedersen & Lam, 1970). (16B) "Bacterium-like" filaments (Bondesen et al., 1967).	(16A) DUBIOMICROFOSSILS: possibly sphaeromorph acritarchs. (16B) DUBIOMICROFOSSILS: "bacterioid" but too indistinct for interpretation (cf. 14, above).
	Krivoirog Series *Ukrainian Shield, Ukraine, U.S.S.R.* *<2.1, >1.81 Ga*	
(17) Siliceous rocks, chiefly shales	(17) "Sphaeromorph acritarchs" (Timofeev, 1969, 1973b, 1979): *Favososphaeridium* sp.; *Gloeocapsomorpha priscata* Tim.; *Orygmatosphaeridium* sp.; *Protosphaeridium acis* Tim.; *P. densum* Tim.; *P. flexuosum* Tim.; *P. laccatum* Tim.; *P.? palaceum* Tim; *P. patelliforme* Tim.; *P. rigidulum* Tim.; *Sticto-sphaeridium* sp.; *S. implexum* Tim.; *S. sinapticuliferum* Tim.; *Trachysphaeridium laminaritum* Tim.; *Trematosphaeridium* sp.	(17) MICROFOSSILS: spheroidal, planktonic organic-walled microfossils of uncertain systematic position; several of the taxa are probably synonymous (based on restudy by J. W. S. of material described in 1969 and 1973).
	Pechenga Series *Kola Peninsula, U.S.S.R.* *<1.89, >1.71 Ga*	
(18) Shales	(18) "Sphaeromorph acritarchs" Timofeev, 1979); *Gloeocapsomorpha* sp.; *G. priscata* Tim.; *Orygmatosphaeridium* sp.; *O. distributum* Tim.; *Protosphaeridium acis* Tim.; *P. asaphum* Tim.; *P. densum* Tim.; *P. flexuosum* Tim.; *P. laccatum* Tim.; *P. rigidulum* Tim.; *P. tuberculiferum* Tim.; *Pterospermopsimorpha* sp.; *Stictosphaeridium implexum* Tim.; *S. sinapticuliferum* Tim.; *Symplassosphaeridium* sp.; *Synsphaeridium* sp.; *Trematosphaeridium* sp.	(18) MICROFOSSILS: spheroidal, chiefly planktonic, organic-walled microfossils of uncertain (but apparently largely of cyanobacterial) affinities; numerous of these taxa are probably synonymous (material not available to us for study).

TABLE 14-2. (*continued*).

Lithology	*Reported objects*	*Intepretation*
	Ladoga Series *Lake Ladoga, Karelia, U.S.S.R.* *<2.0, <1.6, Ga*	
(19) Shales (from at least three localities)	(19) "Sphaeromorph acritarchs" (Timofeev, 1959 (1979): *Favososphaeridium* sp.; *Protosphaeridium* sp.; *P. densum* Tim.; *P. flexuosum* Tim.; *P. laccatum* Tim.; *P. parvulum* Tim.; *P. rigidulum* Tim.; *Pterospermopsimorpha* sp.; *Stictosphaeridium implexum* Tim.; *Turuchanica alara* Tim.	(19) MICROFOSSILS: spheroidal, chiefly planktonic, organic-walled microfossils of uncertain (but apparently largely of cyanobacterial) affinities; several of the species of *Protosphaeridium* are apparently synonymous (based on restudy by J. W. S. of material described in 1959).
	Jatulian Series *Lake Onega, Karelia, U.S.S.R.* *<1.81, >1.63 Ga*	
(20) Shales (for at least six localities)	(20) "Sphaeromorph acritarchs" (Timofeev, 1959, 1979); *Favososphaeridium* sp.; *Gloeocapsomorpha* sp.; *Protosphaeridium* sp.; *P. densum* Tim.; *P. flexosum* Tim.; *P. parvulum* Tim; P. *scabridum* Tim.; *P. tuberculiferum* Tim.; *Pterospermopsimorpha* sp.; *Stictosphaeridium implexum* Tim.; *Turuchanica ternata* Tim.	(20) MICROFOSSILS: sphaeromorph acritarchs of uncertain affinities (cf. 19, above; based on restudy by J. W. S. of material described in 1959).
	Dolomite Series *Ootse area, southeastern Botswanaland* *~2.25 Ga*	
(21) Manganese "stromatolite"	(21) Coccoid and filamentous "microfossils" (Nagy, 1980)	(21) DUBIOMICROFOSSILS: microstructures in need of detailed study; manganese "stromatolites" may be nonbiogenic.
	Dolomite Series, *Boetsap (Hol) River area, northern Cape Province, South Africa* *~2.25 Ga*	
(22) Carbonate stromatolite	(22) Filamentous "fossil blue-green algae" detected by scanning electron microscopy (MacGregor et al., 1974).	(22) NONFOSSIL: modern microbial contaminants (Cloud & Morrison, 1979a, p. 84, 86).
	Malmani Dolomite *Abel Erasmus Pass, eastern Transvaal, South Africa* *~2.25 Ga*	
(23) Stromatolitic dolostone	(23) *Petraphera vivescenticula* Nagy, "*Raphidiopsis*-like blue-green alga," and other filamentous and coccoid "microfossils" (Nagy et al., 1973; Nagy, 1974, 1975, 1978). [Nagy, 1975, erroneously attributed this material to the "Wolkberg Group".]	(23) NONFOSSIL: although considered by Cloud & Morrison (1979a, p. 88) to be probably syngenetic with the surrounding carbonate, restudy of the original thin section (by H. J. H. and J. W. S.) shows the filaments to be reddish-brown modern endoliths (probably fungi), occurring in cylindrical voids, that have bored into the carbonate matrix that contains dark gray organic-rich laminae.
	Pretoria Group *northern Cape Province, South Africa* *~2.22 Ga*	
(24) Banded iron formation (jaspilite)	(24) "*Huroniospora*-like" spheroids (Cloud & Licari, 1968)	(24) DUBIOMICROFOSSILS: not sufficiently characterized for taxonomic assignment.

TABLE 14-2. (*continued*).

Lithology	*Reported objects*	*Intepretation*
	Duck Creek Dolomite *Mount Stuart area, Western Australia* *~2.02 Ga*	
(25) Carbonaceous cherts in dolostone	(25A) COCCOID MICROFOSSILS:	(25A) MICROFOSSILS: spheroidal bacteria and/or cyanobacteria.
	(25Aa) "Coccoid bacteria" (Section 14-3-2). (25Ab) *Huroniospora* sp. (Knoll & Barghoorn, 1976). (25Ac) ?*Leptoteichos golubicii* (Knoll & Barghoorn, 1976; Knoll et al., 1978).	(25Aa) Prokaryote incertae sedis; see Photo 14-4C. (25Ab) Unicellular prokaryote (cf. 4Ac, above); see Photo 14-4D, E. (25Ac) Prokaryote incertae sedis.
	(25B) SEPTATE FILAMENTOUS MICROFOSSILS:	(25B) MICROFOSSILS: cellular, filamentous, bacteria and/or cyanobacteria.
	(25Ba) *Ferrimonilis* sp. (Section 14-3-2). (25Bb) *Gunflinta* sp. (Knoll & Barghoorn, 1976). (25Bc) *Rhicnonema* sp. (Section 14-3-2).	(25Ba) Mineralized prokaryote, probably bacterial; see Photo 14-3H. (25Bb) Filamentous bacteria and/or cyanobacteria (cf. 4Bb, above); see Photos 14-3D, E, G, I, J, and 14-4E. (25Bc) Degraded sheath-enclosed prokaryote (cf. 10Bc, above); see Photo 14-3F.
	(25C) TUBULAR UNBRANCHED MICROFOSSILS:	(25C) MICROFOSSILS: sheaths of filamentous bacteria and/or cyanobacteria.
	(25Ca) *Siphonophycus* sp. (Section 14-3-2).	(25Ca) Sheath, probably cyanobacterial; see Photo 14-3A-C.
	(25D) UNUSUAL, BIZARRE MICROFOSSILS:	(25D) MICROFOSSILS: fossil microorganisms of unusual morphology and uncertain systematic portion.
	(25Da) *Kakabekia* sp. (Section 14-3-2), (25Db) *Eoastrion* sp. (Knoll & Barghoorn, 1976, 1978).	(25Da) Budding bacterium? (cf. 4Di, above); see Photo 14-4A. (25Db) Bacterium incertae sedis (cf. 4Dc, above).
	Frere Formation *Camel Well area, Weastern Australia* *~1.7 Ga*	
(26) Silicfied oncolites and stromatolites in iron formation	(26) MICROFOSSILS—Taxa noted are reported in: Walter, 1975; Walter et al., 1976; Photo 14-5F.	
	(26A) COCCOID MICROFOSSILS:	(26A) MICROFOSSILS: spheroidal bacteria and/or cyanobacteria.
	(26Aa) *Huroniospora* spp.	(26Aa) Unicellular prokaryote (cf. 4Ac, above); see Photo 14-5F.
	(26B) SEPTATE FILAMENTOUS MICROFOSSILS:	(26B) MICROFOSSILS: cellular, filamentous, bacteria and/or cyanobacteria.
	(26Ba) *Gunflintia minuta* Barghoorn.	(26Ba) Filamentous prokaryote (cf. 4Bb, above); see Photo 14-5D.
	(26C) TUBULAR UNBRANCHED MICROFOSSILS:	(26C) MICROFOSSILS: sheaths of filamentous bacteria and/or cyanobacteria.
	(26Ca) *Animikiea* sp. (= *Rhicnonema* sp.).	(26Ca) Tubular microbial sheath (cf. 4Ca, above); see Photo 14-5E.
	(26D) UNUSUAL, BIZARRE MICROFOSSILS	(26D) MICROFOSSILS: fossil microorganisms of unusual morphology and uncertain systematic position.
	(26Da) cf. *Archaeorestis schreiberensis* Barghoorn. (26Db) *Eoastrion bifurcatum* Barghoorn. (26Db) *E. simplex* Barghoorn. (26Dc) ?*Eomicrhystridium barghoornii* Deflandre. (26Dd) *Kakabekia umbellata* Barghoorn.	(26Da) Prokaryote? incertae sedis (cf. 4Da, above). (26Db) Bacterium incertae sedis (cf. 4Dc, above); see Photo 14-5G. (26Dc) Acanthomorph acritarch? or degraded sphaeromorph?; see Photo 14-5F. (26Dd) Budding bacterium? (cf. 4Di, above).

TABLE 14-2. (*continued*).

Lithology	*Reported objects*	*Intepretation*
	Windidda Formation *Tooloo Bluff, Western Australia* *~1.7 Ga*	
(27) Apatitic and sulfidic cherts	(27A) Mention of occurrence of "unstudied microfossils" (Walter et al., 1976, p. 221).	(27A) MICROFOSSILS: see 27b and 27c, below.
	(27B) *Archaeotrichion* sp. (Section 14-3-2).	(27B) MICROFOSSILS: filamentous prokaryote, probably bacterial; see Photo 14-5C.
	(27C) *Eomycetopsis* sp. (Section 14-3-2).	(27C) MICROFOSSILS: tubular sheaths of filamentous bacteria and/or cyanobacteria; see Photo 14-5A, B.
	Amelia Dolomite *McArthur River area, Northern Territory, Australia* *~1.7 Ga*	
(28) Carbonaceous stromatolitic chert	(28) MICROFOSSILS—Taxa noted are reported in: Croxford et al., 1973; Muir, 1974, 1976; Oehler et al., 1976; Section 14-3-2.	
	(28) COCCOID MICROFOSSILS:	(28A) MICROFOSSILS: colonial and unicellular, spheroidal bacteria and/or cyanobacteria.
	(28Aa) *Ameliaphycus croxfordii* Muir. (28Ab) *Eoentophysalis belcherensis* Hofmann. (28Ac) *Eosynechococcus moorei* Hofmann. (28Ad) ?*Huroniospora* sp. (28Ad) *H. microreticulata* Barghoorn. (28Ad) *H. psilata* Barghoorn (cf. *Eoentophysalis belcherensis* Hofmann). (28Ad) *H. ornata* Muir. (28Aa) *Sphaerophycus parvum* Schopf. (28Aa) *S. tetragonalis* Muir. (28Ae) *Tetraphycus* sp. (28Ae) cf. *T. major* D. Oehler. (28Af) ?*Myxococcoides* sp. (28Af) *M. reniformis* Muir. (28Af) *M. minutus* Muir. (28Af) *M. kingii* Muir (cf.? *Eoentophysalis belcherensis* Hofmann). (28Af) *M. konzalovae* Muir. (28Af) *M. minor* Schopf. *28Ag. Myxomorpha janecekii* Muir. (28Ah) *Palaeoanacystis* sp. (28Ah) *P. vulgaris* Schopf. (28Ai) *P. plumbii* Muir.	(28Aa) Colonial prokaryote, probably chroococcacean. (28Ab) Colonial prokaryote (cf. 10Aa, above); see Photo 14-6F?, L?, M. (28Ac) Colonial prokaryote (cf. 10Ab, above); see Photo 14-6N. (28Ad) Spheroidal prokaryote; see Photo 14-6B, D. (28Ae) Colonial prokaryote, probably chroococcacean; see Photo 14-6C, G-I. (28Af) Spheroidal prokaryote, probably chroococcacean; see Photo 14-6D. (28Ag) Colonial prokaryote, probably dermocarpacean or pleurocapsacean; see Photo 14-6O. (28Ah) Colonial prokaryote, probably chroococcacean; see Photo 14-6K. (28Ai) Colonial prokaryote, probably chroococcacean; see Photo 14-6J.
	(28B) SEPTATE FILAMENTOUS MICROFOSSILS:	(28B) MICROFOSSILS: cellular, filamentous, bacteria and/or cyanobacteria.
	(28Ba) *Gunflintia minuta* Barghoorn. (28Ba) *G. oehlerae* Muir.	(28Ba) Filamentous bacteria and/or cyanobacteria.
	(28C) TUBULAR UNBRANCHED MICROFOSSILS:	(28C) MICROFOSSILS: sheaths of filamentous bacteria and/or cyanobacteria.
	(28Ca) *Eomycetopsis filiformis* Schopf. (28Ca) *E. robusta* Schopf.	(28Ca) Sheath, bacterial or cyanobacterial (cf. 10Ca, above).
	Emmerugga Dolomite *McArthur Basin, Northern Territory, Australia* *~1.7 Ga*	
(29) Chert? in Dolostone	(29) Mention of unstudied "microfossils" (Muir, 1976, p. 143, fig. 1).	(29) DUBIOFOSSILS: occurrence insufficiently documented.
	Cooley Dolomite *McArthur Basin, Northern Territory, Australia* *~1.6 Ga*	
(30) Chert	(30) "Microfossils," cf. *Metallogenium personatum* Perfil'ev & Gabe and *Huroniospora* sp. (Muir et al., 1974; Oehler, 1977c; Muir, 1978, p. 940; Marshall, 1979, Figure 5.15).	(30) MICROFOSSILS: structurally preserved prokaryotes; for *Huroniospora* sp., see Photo 14-7Z.

TABLE 14-2. (*continued*).

Lithology	*Reported objects*	*Intepretation*
	H. Y. C. Pyritic Shale *McArthur Basin, Northern Territory, Australia* *~1.6 Ga*	
(31) Nonstromatolitic chert	(31) MICROFOSSILS—Taxa noted are reported in: Hamilton & Muir, 1974; Oehler & Croxford, 1976; Oehler, 1977c; Oehler & Logan, 1977).	
	(31A) COCCOID MICROFOSSILS:	(31A) MICROFOSSILS: spheroidal bacteria and/or cyanobacteria.
	(31Aa) *Bigeminococcus mucidus* Schopf & Blacic. (31Ab) *Bisacculoides tabeoviscus* J. Oehler. (31Ac) *B. vacua* J. Oehler. (31Ad) *B. grandis* J. Oehler. (31Ae) *Clonophycus elegans* J. Oehler. (31Af) *Globophycus minor* J. Oehler. (31Ag) *Huronoiospora* sp. (31Ah) *Myxococcoides kingii* Muir. (31Ai) *Nanococcus vulgaris* J. Oehler. (31Aj) *Sphaerophycus parvum* Schopf.	(31Aa) Colonial prokaryote, probably chroococcacean; see Photo 14-7V. (31Ab) Colonial prokaryote, probably chroococcacean; see Photo 14-7X. (31Ac) Spheroidal prokaryote, probably chroococcacean; see Photo 14-7U. (31Ad) Spheroidal prokaryote, probably chroococcacean; see Photo 14-7W. (31Ae) Prokaryote incertae sedis; see Photo 14-7T. (31Af) Prokaryote incertae sedis; see Photo 14-7Q. (31Ag) Spheroidal unicellular prokaryote; see Photo 14-7Y. (31Ah) Colonial prokaryote, probably chroococcacean; see Photo 14-7S. (31Ai) Spheroidal prokaryote, probably chroococcacean; see Photo 14-7R. (31Aj) Spheroidal prokaryote, probably chroococcacean; see Photo 14-7P.
	(31B) SEPTATE FILAMENTOUS MICROFOSSILS:	(31B) MICROFOSSILS: cellular, filamentous, bacteria and/or cyanobacteria.
	(31Ba) *Biocatenoides rhabdos* Schopf. *31Bb. B. incrustata* J. Oehler. (31Bc) *B. pertenuis* J. Oehler. (31Bd) *Coleobacter primus* J. Oehler. (31Be) *Cyanonema inflatum* J. Oehler. (31Be) *C. minor* J. Oehler. (31Bf) *Ferrimonilis variabile* J. Oehler. (31Bg) Gunflintia septata (Schopf) J. Oehler. (31Bh) cf. *Obconicophycus* sp. (31Bi) *Oscillatoriopsis schopfii* J. Oehler. (31Bj) *Ramacia carpentariana* J. Oehler.	(31Ba) Filamentous prokaryote, probably bacterial; see Photo 14-7C. (31Bb) Filamentous prokaryote, probably bacterial; see Photo 14-7D. (31Bc) Filamentous prokaryote, probably bacterial; see Photo 14-7E. (31Bd) Filamentous prokaryote, probably bacterial; see Photo 14-7F. (31Be) Filamentous prokaryote, bacterial or cyanobacterial; see Photo 14-7I, L. (31Bf) Filamentous prokaryote, probably bacterial; see Photo 14-7A, B. (31Bg) Filamentous prokaryote, bacterial or cyanobacterial; see Photo 14-7J. (31Bh) Filamentous prokaryote, probably cyanobacterial; see Photo 14-7N. (31Bi) Filamentous prokaryote, probably oscillatoriacean cyanobacterium; see Photo 14-7K. (31Bi) Filamentous prokaryote, probably bacterial; see Photo 14-7G.
	(31C) TUBULAR UNBRANCHED MICROFOSSILS:	(31C) MICROFOSSILS: sheaths of filamentous bacteria and/or cyanobacteria.
	(31Ca) cf. *Animikiea* sp. (31Cb) cf. *Eomycetopsis* sp.	(31Ca) Sheath, probably oscillatoriacean; see Photo 14-7O. (31Cb) Sheath, bacterial or cyanobacterial; see Photo 14-7M.
	(31D) UNUSUAL, BIZARRE MICROFOSSILS:	(31D) MICROFOSSILS: fossil microorganisms of unusual morphology and uncertain systematic position.
	(31Da) *Eoastrion* sp.	(31Da) Bacterium incertae sedis (cf. 4Dc, above); see Photo 14-7H.

TABLE 14-2. (*continued*).

Lithology	*Reported objects*	*Intepretation*
	Balbirini Dolomite *Top Crossing area, Northern Territory, Australia* *~1.6 Ga*	
(32) Carbonaceous stromatolitic chert	(32) MICROFOSSILS—Taxa noted are reported in: D. Oehler, 1976, 1978.	
	(32A) COCCOID MICROFOSSILS:	(32A) MICROFOSSILS: spheroidal bacteria and/or cyanobacteria.
	(32Aa) *Balbiriniella praestans* D. Oehler, (32Ab) *Clonophycus* sp. (32Ac) *C. biattina* D. Oehler. (32Ad) *C. ostiolum* D. Oehler. (32Ae) *C. refringens* D. Oehler. (32Af) *C. vulgaris* D. Oehler. (32Ag) *Eoentophysalis belcherensis* Hofmann. (32Ah) *Myriasporella pyriformis* D. Oehler. (32Ai) *Myxococcoides* sp. (32Aj) *M. cracens* D. Oehler. (32Ai) *M. kingii* Muir. (32Ak) *M. minuta* Muir. (32Al) *Myxomorpha janecekii* Muir. (32Am) *Nanococcus vulgaris* J. Oehler (32An) *Palaeoanacystis plumbii* Muir. (32Ao) *P. vulgaris* Schopf. (32Ap) *Sphaerophycus parvum* Schopf. (32Aq) *S. reticulatum* Muir. (32Ar) *Tetraphycus acinulus* D. Oehler. (32As) *T. diminutivus* D. Oehler. (32At) *T. gregalis* D. Oehler. (32Au) *T. major* D. Oehler.	(32Aa) Prokaryote incertae sedis. (32Ab) Prokaryote incertae sedis; see Photo 14-8V. (32Ac) Prokaryote incertae sedis; see Photo 14-8T. (32Ad) Prokaryote incertae sedis; see Photo 14-8S. (32Ae) Prokaryote incertae sedis; see Photo 14-8W. (32Af) Prokaryote incertae sedis; see Photo 14-8U. (32Ag) Colonial prokaryote (cf. 10Aa, above); see Photo 14-8C. (32Ah) Colonial prokaryote, probably dermocarpacean or pleurocapsacean; see Photo 14-8Q; (32Ai) Prokaryote, probably chroococcacean. (32Aj) Prokaryote, probably chroococcacean; see Photo 14-8K. (32Ak) Prokaryote, probably chroococcacean; see Photo 14-8I. (32Al) Colonial prokaryote, probably democarpacean or pleurocapsacean; see Photo 14-8R. (32Am) Prokaryote, probably chroococcacean; see Photo 14-8L, M. Prokaryote, probably chroococcacean; see Photo 14-8O. (32Ao) Prokaryote, probably chroococcacean; see Photo 14-8N. (32Ap) Prokaryote, probably chroococcacean; Photo 14-8P. (32Aq) Prokaryote, probably chroococcacean; see Photo 14-8J. (32Ar) Prokaryote, possibly chroococcacean; see Photo 14-8D. (32As) Prokaryote, possibly chroococcacean; see Photo 14-8E. (32At) Prokaryote, possibly chroococcacean; see Photo 14-8G. (32Au) Prokaryote, possibly chroococcacean; see Photo 14-8F.
	(32B) SEPTATE FILAMENTOUS MICROFOSSILS:	(32B) MICROFOSSILS: cellular, filamentous, bacteria and/or cyanobacteria.
	(32Ba) *Palaeolyngbya* sp.	(32Ba) Prokaryote, probably oscillatoriacean.
	(32C) TUBULAR UNBRANCHED MICROFOSSILS:	(32C) MICROFOSSILS: sheaths of filamentous bacteria and/or cyanobacteria.
	(32Ca) cf. *Eomycetopsis* sp. (32Cb) *Siphonophycus* sp.	(32Ca) Sheath, cyanobacterial or bacterial; see Photo 14-8A, B. (32Cb) Sheath, probably oscillatoriacean.
	Bungle Bungle Dolomite *Osmond Range, Western Australia* *~1.6, Ga*	
(33A) Carbonaceous stromatolitic chert	(33A) Unnamed "microfossils" described and illustrated by Diver (1974); taxa noted below are described in Section 14-3-2.	
	(33Aa) COCCOID MICROFOSSILS:	(33Aa) MICROFOSSILS: spheroidal bacteria and/or cyanobacteria.

TABLE 14-2. (*continued*).

Lithology	Reported objects	Intepretation
	(33Aaa) *Eoentophysalis belcherensis* Hofmann. (33Aab) *Leptoteichos golubicii* Knoll et al. (33Aac) *Melasmatosphaera media* Hofmann. (33Aad) *Myxococcoides minor* Schopf. (33Aae) *Sphaerophycus parvum* Schopf.	(33Aaa) Spheroidal prokaryote; see Photo 14-9O-Q. (33Aab) Prokaryote incertae sedis; see Photo 14-9J. (33Aac) Prokaryote incertae sedis; see Photo 14-9K, L. (33Aad) Prokaryote, probably chroococcacean; see Photo 14-9M, N. (33Aae) Prokaryote, probably chroococcacean; see Photo 14-9H, I.
	(33Ab) POSSIBLY SEPTATE FILAMENTS:	(33Ab) MICROFOSSILS: filamentous bacteria and/or cyanobacteria.
	(33Aba) *Archaeotrichion* sp. or *Biocatenoides* sp. (33Abb) *Ricnonema antiquum* Hofmann.	(33Aba) Septate or nonseptate prokaryotic filaments, probably bacterial; see Photo 14-9D. (33Abb) Degraded bacterium or oscillatoriacean; see Photo 14-9C.
	(33Ac) TUBULAR UNBRANCHED MICROFOSSILS:	(33Ac) MICROFOSSILS: sheaths of filamentous bacteria and/or cyanobacteria.
	(33Aca) *Eomycetopsis* sp. (33Acb) cf. *Obconicophycus*.	(33Aca) Sheath, bacterial or cyanobacterial; see Photo 14-9A, B. (33Acb) Sheath, probably oscillatoriacean; see Photo 14-9G.
	(33Ad) UNUSUAL, BIZARRE MICROFOSSILS:	(33Ad) MICROFOSSILS: fossil microorganisms of unusual morphology and uncertain systematic position.
	(33Ada) cf. *Eosphaera* sp.	(33Ada) Prokaryote? incertae sedis (cf. 4De, above); see Photo 14-9F.
(33B) Non-stromatolitic carbonate mudstone	(33B) Megascopic carbonaceous films resembling *Beltina* and carbonaceous discs similar to *Chuaria* (Section 14-3-2).	(33B) DUBIOFOSSILS: *Beltina*-like films (see Photo 14-10A-O) possibly remnants of sheet-like green (e.g., *Monostroma*) or red (e.g., *Porphyra*) eukaryotic algae, or remnants of cyanobacterial mats, or unstructured organic detritus; *Chuaria*-like discs (see Photo 14-10M, P-S) could be compressions of megaplanktonic algae or of colonies of microscopic algae; additional examples of better preserved specimens are needed to resolve the origin of both types of structures.
	Paradise Creek Formation *north of Mount Isa, Queensland, Australia* *~1.6, Ga*	
(34) Stromatolitic chert	(34) MICROFOSSILS—Taxa noted are reported in: Licari et al., 1969; Licari & Cloud, 1972; Kline, 1975a, 1975b; Tappan, 1976; Oehler, 1977C.	(34A) MICROFOSSILS: spheroidal bacteria
	(34A) COCCOID MICROFOSSILS:	(34A) MICROFOSSILS: spheroidal bacteria and/or cyanobacteria.
	(34Aa) *Eucapsis*(?) sp. 34Ab Colonial unicells. (34Ac) Solitary unicells.	(34Aa) Cuboidal colonies of bacteria or cyanobacteria; see discussion in Section 14-4-1, Table 14-4, and Photo 14-11A-C. (34Ab) Colonial prokaryote, compared by Licari & Cloud (1972) to chroococcacean *Anacystis*. (34Ac) Spheroidal prokaryote, possibly cf. *Huroniospora* or cf. *Myxococcoides*.

TABLE 14-2. (*continued*).

Lithology	*Reported objects*	*Intepretation*
	(34B) SEPTATE FILAMENTOUS MICROFOSSILS:	(34B) MICROFOSSILS: cellular, filamentous, bacteria and/or cyanobacteria.
	(34Ba) Filaments cf. *Gunflintia* and cf. *Cephalophytarion*. (34Ba) Filaments cf. *Veteronostocale*.	(34Ba) Filamentous prokaryotes incertae sedis.
	(34D) UNUSUAL, BIZARRE MICROFOSSILS:	(34D) MICROFOSSILS: fossil microorganisms of unusual morphology and uncertain systematic position.
	(34Da) *Eoastrion simplex* Barghoorn. (34Da) *E. bifurcatum* Barghoorn.	(34Da) Bacteria incertae sedis (cf. 4Dc, above).
	Vizard Formation *Roper River area, Northern Territory, Australia* *~1.6 Ga*	
(35) Fine-grained cherty sediments	(35) "Triact sponge spicules" (Dunn, 1964).	(35) NONFOSSIL: pseudofossils, apparently remnants of volcanic shards (Cloud, 1968, p. 56).
	Mount Isa Shale *Mount Isa, northwestern Queensland, Australia* *~1.6 Ga*	
(36) Pyrite in black shale	(36) "Sphaeromorph acritarchs similar to *Leiominmscula minuta* Naumova," "di-sphaeromorph acritarchs," and aggregates of spheroids compared with *Symplassosphaeridium subcoalitum* Tim. (Love & Zimmerman, 1961; Love, 1965).	(36) DUBIOMICROFOSSILS: probably pseudo-microfossils, aggregated particulate organic matter; in need of restudy.
	Anshan Group *Eastern Liaoning Province, China* *<2.5, >2.2 Ga*	
(37) Iron formation (jaspilite)	(37A) "Microfossils" detected by transmission electron microscopy of peels of polished rock sections (Ouyang, 1979): *Huroniospora*? *minima* Ouyang; *H. crassa* Ouyang; *Sphaerocongregus*? *anshanensis* Ouyang; *Marsupicapsa pristina* Ouyang.	(37A) DUBIOMICROFOSSILS: probably nonfossil (viz., nonbiogenic microstructures, artifacts of of preparation, modern contaminants, etc., see Table 9-1); material not available to us for study.
	(37B) "Microfossils" detected by optical microscopy of petrographic thin sections (Ouyang, 1979): *Anshania triangularis* Ouyang; *Lophosphaeridium pyriforme* Ouyang; *Trachysphaeridium granulatum* Ouyang; *T. parapertum* Ouyang; *T.*? *ellipsoidale* Ouyang; *T.* sp. 1 Ouyang; *T.* sp. 2 Ouyang; "algal filaments"; 11 other types of microstructures.	(37B) DUBIOMICROFOSSILS: all of these apparently poorly preserved and rather indistinctly defined specimens are difficult to assess from published illustrations; material not available to us for study.
	(37C) "Fossil bacteria," identified as fossil members of living genera, detected by transmission electron microscopy of peels of polished rock specimens (Xü & Chu, 1979): *Caulobacter* sp.; *Clostridium sporulosum* Xü & Chu; *C. anshanense* Xü & Chu; *Gallionella ferrugineoides* Xü & Chu; *Leptothrix minuta* Xü & Chu; *L. microsiphonia* Xü & Chu; *Siderococcus archilimoniticus* Xü & Chu; *Thiobacillus* sp.; *Vibrio anshanense* Xü & Chu.	(37C) DUBIOMICROFOSSILS: probably nonfossil (viz., modern contaminants?, see Table 9-1); specimens not illustrated in published report; material not available to us for study.

TABLE 14-2. (*continued*).

Lithology	Reported objects	Intepretation
	(37D) "Sphaeromorph acritarchs" and carbonaceous filamentous "microfossils" studied by optical microscopy of acid-resistant residues (Yin, 1979): *Brocholaminaria* cf. *nigrita* Ouyang, Yin & Li; *Entretisphaeridium abnormum* Yin; *Entosphaeroides*? sp.; *Gloeocapsomorpha* cf. *heveica* Tim.; *Gloeocapsomorpha* sp.; *Huroniospora compacta* Yin; *Leiomarginata* cf. *desma* Yin & Li; *L.* cf. *simplex* Naumova; *Leiominuscula* spp.; *Myxococcoides*? sp.; *Oscillatoriopsis anshanensis* Yin; *Palaeoanacystis*? *antiquus* Yin; *Paleamorpha* sp.; *Polyedrosphaeridium* sp.; *Protoleiosphaeridium densum* (Tim.) Yin; *P. parvulum* (Tim.) Yin; *P. sorediforme* Tim.; *Prototracheites* sp.; *Retinarites* sp.; *R.* cf. *irregularis* Ouyang, Yin & Li; *Trachysphaeridium* spp.; *Trematosphaeridium* sp.; *T.* cf. *minutum* Sin & Liu; *Triangumorpha prima* Yin; "2 types of algae"; "5 indeterminable types"; "Bodeibinei" (opaque spheroids).	(37D) MICROFOSSILS and DUBIOMICROFOSSILS: assemblage contains numerous authentic acritarchs (organic-walled microfossils of uncertain systematic position), several apparently bona fide fossil filaments, as well as other microstructures of more doubtful status; in general, the reported taxa are difficult to assess based on published information; among other doubtful forms reported: "*Entosphaeroides*?" appears to us to be misidentified, possibly being a contaminant; "*Paleamorpha*" and the "5 indeterminable types" may be modern contaminants; and the "2 types of algae" are cf. mineral framboids; material not available to us for a study.
	Liaoho Group *Heiniuzhang, northeast of Anshan, northeastern China* *<2.0, >1.5 Ga*	
(38) Ferruginous quartzite	(38A) "Sphaeromorph acritarchs" and carbonaceous filaments in petrographic thin sections (Ouyang, 1979): *Arctacellularia*? *gigantea* Ouyang; *Asperatopsophosphaera* sp.; *Ostiana prisca* Ouyang; *Petraphaera*? *mirable* Ouyang.	(38A) DUBIOMICROFOSSILS: material difficult to interpret from published illustrations (N.B.: filaments identified as "*Petraphaera*?" are not assignable to this nonfossil genus; see 23, above); material not available to us for study.
	(38B) "Sphaeromorph acritarchs" and carbonaceous filaments in acid-resistant residues (Yin, 1979): *Brocholaminaria* sp.; *Dictyosphaera* cf. *macroreticulata* Sin & Liu; *Leiomarginata* sp.; *Leiominuscula compacta* Yin & Li; *L.* cf. *compacta* Yin & Li; *L. minuta* Naumova; *Lophominuscula* sp.; *Macroptycha biplicata* Tim.; *Microconcentrica* cf. *simplex* Naumova; *Orygmatosphaeridium* sp.; *Polyedrosphaeridium* sp.; *P. hidusense* Rudavskaya; *P.* cf. *hidusense* Rudavskaya; *Protoleiosphaeridium* sp.; *P. densum* (Tim.) Yin; *P. sorediforme* Tim.; *Prototracheites* spp.; *Retinarites* sp.; *Taeniatum crassum* Sin & Liu; *Trematosphaeridium holtedahlii* Tim.; *Turuchanica alara* Rudavskaya; "Bodaibinei" (opaque spheroids).	(38B) MICROFOSSILS and DUBIOMICROFOSSILS: assemblage apparently contains both bona fide microfossils and microstructures of doubtful biogenicity (cf. 37D, above).

TABLE 14-2. (*continued*).

Lithology	*Reported objects*	*Intepretation*
	Changcheng System *Jixian (Chihsien) and Ming Tombs areas, northern China* *<1.95, >1.7 Ga*	
(39A) Clastic sediments, mostly shales	(39A) "Sphaeromorph acritarchs" and other carbonaceous microstructures in acid-resistant residues (Sin & Liu, 1973; Chen et al., 1980; Wang et al., 1980, p. 333): *Asperatopsophosphaera bacca*; *A. bavlensis* Schepeleva; *A. concentrica*; *A. partialis* Schepeleva; *A. umishanensis* Sin & Liu; *Dictyosphaera* sp.; *D. macroreticulata* Sin & Liu; *D. rugosa*; *D. sinica* Sin & Liu; *Laminarites* cf. *antiquissimus* Eichwald; *Leiofusa digitata* Sin; *Leiominuscula* sp.; *L. incrassata* Sin & Liu; *L.* aff. *minuta* Naumova; *L. orientalis* Sin & Liu; *L. pellucentis* Sin & Liu; *Leiopsophosphaera* sp.; *L. apertus* Schepeleva; *L.* aff. *effusus* Schepeleva; *L. minor* Schepeleva; *L. pelucidus* Schepeleva; *Lignum nematoideum* Sin; *Margominuscula antiqua* Naumova; *M. rugosa* Naumova; *M.* aff. *tennela* Naumova; *Polyporata microporosa* Sin & Liu; *P. obsoleta* Sin & Liu; *Protoleiosphaeridium* sp.; *P. solidum* Liu & Sin; *Protosphaeridium densum* Tim.; *P. gibberosum*; *Pseudozonosphaera verrucosa* Sin & Liu; *Symplassosphaeridium infriatum*; *Trachysphaeridium adornatum*; *T. conglomeratum*; *T. hyalinum* Sin & Liu; *T. incrassatum* Sin; *T. rugosum* Sin; *T. simplex* Sin ex Sin & Liu; *Trematosphaeridium minutum* Tim.; *Zonosphaeridium* cf. *minutum* Sin.	(39A) MICROFOSSILS: chiefly apparently planktonic sphaeromorphs of uncertain systematic position.
(39B) Shaly carbonaceous dolostone, near Jixian (Chihsien), China	(39B) "Fossils" preserved as carbonaceous compressions: *Chuaria* sp.; *Tyrasotaenia* sp.; and angulate films resembling *Beltina* (Hofmann & Aitken, 1979, p. 165; Hofmann & Chen, 1981).	(39B) FOSSILS: microscopic to megascopic biogenic structures (based on studies by H. J. H.)
	Xiyanghe Group *eastern Qinlin (Tsinling) Shan, western Honan Province, China* *<2.0, >1.5 Ga*	
(40) Clastic rocks	(40) "Sphaeromorph acritarchs" and carbonaceous filamentous "microfossils" (Wang et al., 1980); *Glottimorpha* sp.; *Leiominuscula* aff. *minuta* Naumova; *Polyporata* sp.; *Trematosphaeridium* sp.; *T. holtedahlii* Tim.; *T. minutum* Sin & Liu; *T.* sp.; *Taeniatum crassum* Sin.	(40) MICROFOSSILS and/or DUBIOMICROFOSSILS: material not fully described; not available to us for study.

TABLE 14-3. Synopsis of reported Early Proterozoic fossil occurrences. ● = Stromatolitic microfossils; ○ = Stromatolitic pseudomicrofossils and/or dubiomicrofossils; ▲ = Nonstromatolitic microfossils; △ = Nonstromatolitic pseudomicrofossils and/or dubiomicrofossils. Numbers before geologic units refer to Figure 14-1.

Taxa reported (genera)	*Morphology*						*Postulated Affinities*					*Occurrences*																																								
												NORTH AMERICA																EUROPE				AFRICA				AUSTRALIA												ASIA			
(>0.2 mm)*	Coccoids (spheroids and ellipsoids)	Filaments, septate	Tubular structures nonseptate filaments)	Branching filaments	Unusual, bizarre forms	Dubiofossils, pseudofossils	Bacteria	Cyanobacteria	Chlorophyta	Rhodophyta	Problematica	1 △Archean or Aphebian	2 △Gowganda Fm.	3 ▲Pokegama Fm.	4 ●▲Gunflint-Biwabik Fms.	5 ▲Tyler Fm.	6 △Michigamme Fm.	7 △Negaunee Fm.	8 △Temiscamie Fm.	9 ▲△Sokoman Fm.	10 ●Kasegalik Fm.	11 ●McLeary Fm.	12 △Kipalu Fm.	13 ●Rocknest Fm.	14 △Zig Zag Land Fm.	15 △Graensesø Fm.	16 △Foselv Fm.	17 ▲Krivoirog Ser.	18 ▲Pechenga Ser.	19 ▲Ladoga Ser.	20 ▲Jatulian Ser.	21 ○Dolomite Ser.	22 ○Malmani Dol.	23 ○Malmani Dol.	24 △Pretoria Gp.	25 ●▲Duck Creek Dol.	26 ●Frere Fm.	27 ▲Windidda Fm.	28 ●Amelia Dol.	29 ○Emmerugga Dol.	30 ●Cooley Dol.	31 ▲H. Y. C. Pyritic Sh.	32 ●Balbirini Dol.	33 ●▲Bungle Bungle Dol.	34 ●Paradise Creek Fm.	35 △Vizard Fm.	36 △Mount Isa Sh.	37 ▲Anshan Gp.	38 ▲Liaohe Gp.	39 ▲Changcheng Syst.	40 ▲Xiyanghe Gp.
1 *Ameliaphycus*	●							●																															●												
2 *Anabaenidium* (cf. *47*)		●						●							●																																				
3 *Animikiea*			●					●							●	▲				▲																	●					?									
4 *Anshania*						●																																										▲			
5 *Archaeorestis*				●			?				●				●																						?														
6 *Arctacellularia*						●																																											△		
7 *Asperatopsosphaera*	●										●																																						△	▲	
8 *Archaeotrichion*			●				●	?																														▲						?							
9 *Balbiriniella*	●							?	?	?	●																																●								
10 *Beltina**						●		?	?	?	●																																	▲						▲	
11 *Bigeminocossus*	●							●																																		▲									
12 *Biocatenoides*		●					●														●	●																				▲		?							
13 *Bisacculoides*	●							●																																		▲									
14 *Brocholaminaria*					●						●																																					△	△		
15 *Caryosphaeroides*	●							●														●																													
16 *Caulobacter*						●																																										△			
17 *Cephalophytarion*		●						●																																				?							
18 *Chlamydomonopsis* (cf. *49*)	●							●							●																																				
19 *Chuaria**	●								?	?	●																																	?						▲	
20 *Clonophycus*	●							?	?	?	●																															▲	●								
21 *Clostridium*						●																																										△			
22 *Coleobacter*		●					●																																			▲									
23 *Corymbococcus*	●							●							●																								?												
24 *Cumulosphaera*						●									○																																				
25 *Cyanonema*		●					?	●																																		▲									
26 *Dictyosphaera*	●										●																																						▲	▲	
27 *Enretisphaeridium*	●										●																																					▲			
28 *Entosphaeroides*			●				?	?							●																																	△			
29 *Eoastrion*					●		●								●	▲				▲																▲	●				●	▲			●						
30 *Eoentophysalis*	●							●													●	●																	●				●	●							

31 *Eomicrhystridium*	●	? ●	●						●			
32 *Eomycetopsis*	●	? ●			● ●				▲ ●	▲ ●	●	
33 *Eosphaera*	●	? ? ●	●		?						?	
34 *Eosynechococcus*	●	●			● ●				●			
35 *Eozygion*	●	●			●							
36 *Eucapsis?*	●	? ? ●									●	
37 *Exochobrachium*	●	? ●	●									
38 *Favososphaeridium*	●	? ●					▲ ▲ ▲					
39 *Ferrimonilis*	●	? ? ●							●	▲		
40 *Frutexites*	●	? ●	●									
41 *Galaxiopsis*	●	? ●	▲									
42 *Gallionella*	●											△
43 *Glenobotrydion*	●	●	●		●							
44 *Globophycus*	●	●			● ●					▲		
45 *Glottimorpha*	●	●										▲
46 *Gloeocapsomorpha*	●	? ●					▲ ▲ ▲					▲
47 *Gunflintia*	●	? ●	▲ ●	▲	▲				● ● ●	▲	?	
48 *Halythrix*	●	●			●							
49 *Huroniospora*	●	? ●	▲ ●	▲	▲ ●				● ● ●	● ▲	?	△
50 *Kakabekia*	●	? ●	▲	▲	?				● ●			
51 *Laminarites**	●	? ? ? ●										▲
52 *Leiofusa*	●	●										△
53 *Leiomarginata*	●	●										▲ ▲
54 *Leiominuscula*	●	●										▲ ▲ ▲ ▲
55 *Leiopsophosphaera*	●	? ●										▲
56 *Leptoteichos*	●	●	▲		● ●				▲		●	
57 *Leptothrix*	●											△
58 *Lignum*	●	●										▲
59 *Lophominuscula*	●	? ●										▲
60 *Lophosphaeridium*	●	? ●										▲
61 *Macroptycha*	●	●										△
62 *Margominuscula*	●	●										▲
63 *Marsupicapsa*	●											△
64 *Megalytrum*	●	●	●									
65 *Melasmatosphaera*	●	●			● ●						●	
66 *Menneria*	●	? ●	?									
67 *Microconcentrica*	●	? ●										▲
68 *Myriasporella*	●	●								●		
69 *Myxococcoides*	●	? ●			● ●				●	▲ ●	● ?	▲
70 *Myxomorpha*	●	●							●	●		

TABLE 14-3. (*continued*).

Taxa reported (genera) (* >0.2 mm)	Morphology						Postulated Affinities					Occurrences																																							
												NORTH AMERICA																EUROPE				AFRICA				AUSTRALIA												ASIA			
	Coccoids (spheroids and ellipsoids)	Filaments, septate	Tubular structures nonseptate filaments)	Branching filaments	Unusual, bizarre forms	Dubiofossils, pseudofossils	Bacteria	Cyanobacteria	Chlorophyta	Rhodophyta	Problematica	1 △Archean or Aphebian	2 △Gowganda Fm.	3 ▲Pokegama Fm.	4 ●▲Gunflint-Biwabik Fms.	5 ▲Tyler Fm.	6 △Michigamme Fm.	7 △Negaunee Fm.	8 △Temiscamie Fm.	9 ▲△Sokoman Fm.	10 ●Kasegalik Fm.	11 ●McLeary Fm.	12 △Kipalu Fm.	13 ●Rocknest Fm.	14 △Zig Zag Land Fm.	15 △Graensesø Fm.	16 △Foselv Fm.	17 ▲Krivoirog Ser.	18 ▲Pechenga Ser.	19 ▲Ladoga Ser.	20 ▲Jatulian Ser.	21 ○Dolomite Ser.	22 ○Malmani Dol.	23 ○Malmani Dol.	24 △Pretoria Gp.	25 ●▲Duck Creek Dol.	26 ●Frere Fm.	27 ▲Windidda Fm.	28 ●Amelia Dol.	29 ○Emmerugga Dol.	30 ●Cooley Dol.	31 ▲H. Y. C. Pyritic Sh.	32 ●Balbirini Dol.	33 ●▲Bungle Bungle Dol.	34 ●Paradise Creek Fm.	35 △Vizard Fm.	36 △Mount Isa Sh.	37 ▲Anshan Gp.	38 ▲Liaohe Gp.	39 ▲Changcheng Syst.	40 ▲Xiyanghe Gp.
71 *Nanococcus*	●						?	●																																		▲	●								
72 *Obconicophycus*		●						●																																		?		?							
73 *Orygmatosphaeridium*	●										●																	▲	▲																				△		
74 *Oscillatoriopsis*		●						●																																		▲						△			
75 *Ostiana*						●																																											△		
76 *Palaeoanacystis*	●						?	●							●						●	●																	●				●					▲			
77 *Palaeolyngbya*		●						●																																			●								
78 *Palaeoleptophycus*			●	?				●																●																											
79 *Palaeomicrocoleus*						●								△							○																														
80 *Palaeorivularia* (cf. 47)						●									●																																				
81 *Palaeoscytonema* (cf. 47)		●					?	●							●																																				
82 *Palaeospiralis* (cf. 47)		●					?	●							●																																				
83 *Palaeospirulina* (cf. 47)		●					?	●							●																																				
84 *Paleamorpha*						●																																										△			
85 *Petraphera*						●	modern endolith																											○															△		
86 *Pilavia*	●							?	?	?	●																																●								
87 *Pleurocapsa?*	●							●													●																														
88 *Polyedrosphaeridium*	●										●																																					▲	▲		
89 *Polyporata*					●						●																																							▲	▲
90 *Primorivularia* (cf. 47)		●					?	●							●																																				

91 *Protoleiosphaeridium*	●	? ●										▲ ▲ △
92 *Protosphaeridium* (cf. 91)	●	? ●					▲ ▲ ▲ ▲					▲
93 *Prototracheites*	●											△ △
94 *Pseudozonosphaera*	●	●										△
95 *Pterospermopsimorpha*	●	●					▲ ▲ ▲					
96 *Ramacia*	●	●								▲		
97 *Retinarites*	●	●										▲ △
98 *Rhicnonema*	●	? ●			●				●		●	
99 "*Schizothrix*" (cf. 47)	●	? ●	●									
100 *Siderococcus*	●											△
101 *Siphonophycus*	●	●							●	●		
102 *Sphaerocongregus*	●											△
103 *Sphaerophycus*	●	? ●	●		● ●				●	▲ ●	● ?	
104 *Stictosphaeridium*	●	●					▲ ▲ ▲ ▲					
105 *Symplassosphaeridium*	●	●					▲					▲
106 *Synsphaeridium*	●	●					▲					
107 *Taeniatum*	●	? ? ? ●										▲ ▲
108 *Tetraphycus*	●	? ●							●	●		
109 *Thiobacillus*	●											△
110 *Thymos*	●	●	●									
111 *Trachysphaeridium*	●	? ? ? ●					▲					▲ ▲
112 *Trematosphaeridium*	●	? ? ? ●					▲ ▲					▲ ▲ ▲ ▲
113 *Triangumorpha*	●	●										△
114 *Turuchanica*	●	? ? ? ●					▲ ▲					▲
115 *Tyrasotaenia**	●	? ? ? ●										▲
116 *Vallenia**	●					△						
117 *Veteronostocale*	●	? ●									?	
118 *Veryhachium?*	●	●	●									
119 *Vibrio*	●											△
120 *Xenothrix*	●	? ●	●									
121 *Zonosphaeridium*	●	●										△
122 *Zosterosphaera*	●	● ? ?			●					?		
Others, Without Linnean designations			△ △ ●	▲ △ △ △	△ ● ● △	△ △ △		○ ○ ○ △		○ ▲	▲ ● △ ▲	△

Windidda, Amelia, and Bungle Bungle Formations; we also illustrate representative specimens of the taxa newly detected.

DUCK CREEK DOLOMITE

Microfossils from the Duck Creek Dolomite have been described previously by Knoll and Barghoorn (1975b, 1976) and Knoll et al. (1978). These authors reported *Huroniospora*, *Gunflintia*, *Eoastrion*, and later, *Leptoteichos*? from the Mount Stuart area. To this we can now add the taxa that follow (in addition to the possible coccoid bacteria shown in Photo 14-4C).

Ferrimonilis sp. Photo 14-3H. This genus, first described from the H. Y. C. Pyritic Shale (Oehler, 1977c), comprises clumpy filaments preserved by pyrite crystals. Such filaments were found in P.P.R.G. Sample Number 049-1-A (for locality data, see Appendix I). They are of special interest because some individuals are partially preserved by hematite at one end, suggesting that the structures may have originally been *Gunflintia*-like threads, about 1 μm across, which later became mineralized by pyrite and hematite. The diameter in the portion replaced by pyrite represents an enlargement of the filament by diagenesis, whereas the portion replaced by hematite remained close to its original diameter.

Rhicnonema sp. Photo 14-3F. The genus *Rhicnonema* was established for unbranched filaments composed of degraded uniserial trichomes surrounded by a thick tubular sheath, from the Belcher Islands of Canada (Hofmann, 1976, p. 1053). Organically preserved filaments attributable to this taxon occur in P.P.R.G. Sample No. 049-1-A; these vividly demonstrate diverse degradational and diagenetic modifications of the morphology of fossil microorganisms.

Siphonophycus sp. Photo 14-3A-C. *Siphonophycus*, first reported from the Bitter Springs Formation (Schopf, 1968), includes microscopic tubular structures more than 15 μm in diameter. Such tubes, with diameters in the range of 20–23 μm, were found in P.P.R.G. Samples No. 062-1-A and 062-1-B. These constitute the oldest known tubes with such a large diameter.

Kakabekia sp. Photo 14-4A. A single, unusually large specimen attributable to *Kakabekia* Barghoorn was found in P.P.R.G. Sample No. 062-1-A. It comprises an ellipsoidal bulb, 35 × 20 μm in size, that is connected to a 50 μm wide umbrella via a 10 μm long neck. This is the largest non-filamentous organism now known from 2-Ga-old rocks. Unfortunately, the affinities of this genus are no better known than before, though the size of the specimen makes it an implausible candidate for a bacterium.

As a general remark concerning the Duck Creek fossils illustrated here, one can say that almost all of them are unusually large representatives of the genera to which they belong, including the *Eoastrion* shown in Photo 14-4B. We may refer to this as a possible case of Early Proterozoic "gigantism", as yet unexplained.

WINDIDDA FORMATION

Microfossils were initially reported by Walter et al. (1976, p. 221) from the Tooloo Bluff area of the Nabberu Basin in Western Australia, but this was only an incidental mention of unstudied material. Our material from the same area has yielded tubular filaments from the cherty filling between rounded detrital grains.

Archaeotrichion sp. Photo 14-5C. Narrow, nonseptate threads, less than 1 μm across, and often 100 μm or more long, are attributed to *Archaeotrichion* Schopf. Such filaments were found in P.P.R.G. Sample No. 091-2-B. They are morphologically similar to representatives now known from rocks as old as 3.5 Ga (see Section 9-5-1), and they thus provide a link in the continuum to occurrences in younger rocks.

Eomycetopsis sp. Photo 14-5A, B. Organically preserved, nonseptate tubes, approximately 2–5 μm in width, are assigned to *Eomycetopsis* Schopf. They are a common constituent of Early Proterozoic and younger microbiotas, and probably are of diverse affinities. Filaments of this type occur together with *Archaeotrichion* in P.P.R.G. Sample No. 091-2-B. They are somewhat unusual, however, in that they are to a large extent collapsed, and in that some individuals are exceptionally long and can be traced over distances of 1 to 2 mm as they wind around detrital grains. Moreover, along some of them (Photo 14-5A) one can observe, within the brownish chert matrix, a number of discoloration spheres, (possibly from oxidation) centered on the surface of the collapsed tubes, near wrinkles or constrictions.

AMELIA DOLOMITE

A treatment of the microbiota of the Amelia Dolomite, dominated by coccoid microfossils, is found in several publications: Croxford et al. (1973), Muir (1974, 1976), and Oehler et al. (1976). Many of the taxa identified by these authors occur in the material we examined. In addition to taxa reported earlier, we also have detected those that follow.

Eoentophysalis belcherensis Hofmann. Photo 14-6L, M. This taxon includes palmelloid and globular colonies of coccoids 2.5–9 μm across, with individual and common envelopes. It is comparable to the modern *Entophysalis major* Ercegović (Golubić and Hofmann, 1976), and appears to be of relatively widespread geographic distribution, having now been found in several Proterozoic stromatolite assemblages. We here report it also from the Amelia Dolomite (P.P.R.G. Sample No. 111-1-A), and include in it specimens that Muir (1976) referred to other taxa, such as *Myxococcoides kingii* Muir and *Huroniospora psilata* Barghoorn (see Table 14-2, occurrence 28).

Eosynechococcus moorei Hofmann. Photo 14-6N. This species, occuring as closely packed ellipsoidal and polyhedral cells 1.5–4 μm in size, is found associated with *Eoentophysalis belcherensis* in P.P.R.G. Sample No. 111-1-A, an association previously known from the Kasegalik Formation of the Belcher Islands (Hofmann, 1976); however, it is relatively rare in the Amelia material studied. Golubić and Campbell (1979) compared the taxon with the modern, terrestrial, lithophytic cyanobacterium *Gloeothece coerulea* Geitler.

Tetraphycus sp. Photo 14-6C, G-I. The genus *Tetraphycus* was erected for tetrad clusters in the Balbirini Dolomite (Oehler, 1978a). It comprises colonies of micrometric coccoids with envelopes. The main distinguishing feature is a tendency toward the development of clustered tetrads, rather than to palmelloid colonies. However, such tetrad clusters are also found among *Eoentophysalis belcherensis* in the Kasegalik and McLeary Formations of the Belcher Islands (Hofmann, 1976, pl. 6), and it is thus conceivable that the *Tetraphycus* of the Amelia could in fact be small colonies of *Eoentophysalis*. The form was found in P.P.R.G. Samples No. 109-1-A and 111-1-A. Its significance is limited to an extension of its stratigraphic and geographic ranges.

BUNGLE BUNGLE DOLOMITE

In a preliminary report, Diver (1974) illustrated coccoid and filamentous microfossils from cherts of the Bungle Bungle Dolomite of the Osmond Range of Western Australia. No genus-level taxonomic assignments were made, but the microfossils were interpreted to represent both prokaryotic and eukaryotic organisms. We have studied new material from the Bungle Bungle stromatolitic cherts that permits us to report the occurrence of the taxa noted below; in addition, in nonstromatolitic carbonates of the formation we have also detected megascopic carbonaceous films and discs that resemble, respectively, *Beltina* Walcott and *Chuaria* Walcott.

Eoentophysalis belcherensis Hofmann. Photo 14-9O-Q. In common with other chertified stromtolitic dolostones from the Proterozoic, the Bungle Bungle Formation has yielded organically preserved colonies of *E. belcherensis*. The cell-like units are 3–6 μm in size, and many clumps show multiple envelopes and cell degradation features. Particularly well-preserved colonies were found in P.P.R.G. Sample No. 159-1-B.

Leptoteichos golubicii Knoll et al. Photo 14-9J. Organically preserved spheroids about 15 μm and bearing an internal dark object about 1 μm across, are present in P.P.R.G. Sample No. 153-1-B. Their affinities are obscure; the spheroids resemble *Melasmatosphaera media* Hofmann, with which they occur, but which differ by the great number of micrometric internal dark bodies. The species is now also known from the Gunflint, Kasegalik, McLeary, and Duck Creek Formations.

Melasmatosphaera media Hofmann. Photo 14-9K, L. Large organically preserved spheroids 11–18 μm across, exhibiting scattered multiple micrometric dark bodies, are assigned to *M. media* Hofmann. This species was previously known only from the Kasegalik and McLeary Formations of the Belcher Islands, where its size range is intermediate between a smaller and a larger species of *Melasmatosphaera*. The affinity of these spheroids is problematic; they may represent degradational variants of relatively large-celled taxa, or or of globular, gloeocapsoid colonies of small coccoids.

Myxococcoides minor Schopf. Photo 14-9M, N. Solitary and clustered spheroids of intermediate size (6–12 μm), without internal dark bodies, are referred to this taxon, which was previously known from the Kasegalik and McLeary Formations (Hofmann, 1976), the Amelia Dolomite (Muir, 1976), as well as the Late Proterozoic Bitter Springs Formation (Schopf, 1968). It occurs in P.P.R.G. Samples No. 153-1-B and 159-1-A.

Sphaerophycus parvum Schopf. Photo 14-9H, I. *S. parvum* includes small, loosely packed spheroids (2–4 μm) that are generally clustered and without multiple envelopes. Its association with *Eoentophysalis belcherensis* and *Myxococcoides minor* in the Bungle Bungle Dolomite (in P.P.R.G. Samples No. 153-1-A and 159-1-A) is duplicated in the Kasegalik and McLeary Formations, and also the Amelia Dolomite. The species has been reported from the H. Y. C. Pyritic Shale and the Balbirini Dolomite, as well as the Late Proterozoic Bitter Springs Formation.

Archaeotrichion sp. or *Biocatenoides* sp. Photo 14-9D. Narrow threads less than 1 μm in diameter occur in P.P.R.G. Sample No. 153-1-B. The structures are too degraded to permit a conclusion as to whether they are septate or nonseptate. Similar degraded threads have been illustrated from the Kasegalik and McLeary Formations (Hofmann 1976, pl. 1, figs. 6–8, 10) and the Bitter Springs Formation (Schopf, 1968).

Eomycetopsis sp. Photo 14-9A, B. Sinous, thick-walled, nonseptate tubes 2.5–5 μm in width (in P.P.R.G. Samples No. 154-1-A and 159-1-B) are referred to *Eomycetopsis* sp. They are generally not collapsed. Other Early Proterozoic occurrences of the genus are in the Kasegalik, McLeary, Windidda, Amelia, Balbirini, and H. Y. C. Formations. This genus also occurs in many Middle and Late Proterozoic formations on a worldwide basis, and represents a truly cosmopolitan taxon that appears to maintain constant morphology through time.

Cf. *Obconicophycus*. Photo 14-9G. A single specimen of a large degraded sheath, 12–16 μm in width, and containing faint remnants of disc-shaped cells, is present in P.P.R.G. Sample No. 154-1-A. It likely represents a large oscillatoriacean cyanobacterium. Comparable forms are found in the H. Y. C. Pyritic Shale (Photo 14-7N), and in the much younger Bitter Springs Formation, where they are attributed to *Obconicophycus amadeus* Schopf and Blacic (1971).

Rhicnonema antiquum Hofmann. Photo 14-9C. Under this heading are grouped unbranched filaments composed of degraded mutlicellular trichomes surrounded by a thick tubular sheath 3–4 μm in width. Such structures occur in P.P.R.G. Sample No. 159-1-B. The species is known from the Kasegalik Formation, and the genus also occurs in the Duck Creek Dolomite (Photo 14-3F).

Cf. *Eosphaera*. Photo 14-9F. A single specimen of a small individual resembling *Eosphaera* Barghoorn was observed in P.P.R.G. Sample No. 153-1-B. Two concentric ellipsoids, the outer 8.5 × 11 μm, and the inner 6 × 9 μm in size, are separated by what appear to be small faint spheroids slightly larger than 1 μm in size. The specimen is much smaller than typical *Eosphaera* from the Gunflint (Photo 14-1P), but comparable to a small specimen from the Kasegalik Formation (Photo 14-2Q).

MEGASCOPIC CARBONACEOUS FILMS

Beltina-like films. Photo 14-10A-O. A nonstromatolitic carbonate unit of the Bungle Bungle Dolomite (P.P.R.G. Sample No. 144) has yielded irregular carbonaceous films that resemble the problematic structures *Beltina* Walcott from the Middle Proterozoic Belt Supergroup and Little Dal Formation of North America (e.g. Hofmann and Aitken, 1979). The films are characteristically angulate in outline, and folded, or at least overlapping (Photo 14-10I). Some show delicate crenulations, visible only under low oblique illumination (Photo 14-10D, F, H), suggesting a delicate but coherent nature. Other specimens are intricately deformed (Photo 14-10J, L, O). They are very thin and appear amorphous at high magnification, showing only the disruption caused by impressions of mineral grains. The structures are of undetermined affinities, but could be pieces of larger sheet-like algae (similar in morphology to the modern chlorophyte genus *Monostroma* and the rhodophyte *Porphyra*), or they could be fragmented mats or scum. Their angulate and overlapping nature indicates fragmentation of larger entities of carbonaceous material as well as a detrital origin.

Chuaria-like discs. Photo 14-10M, N, P-S. A second type of carbonaceous film encountered in nonstromatolitic beds of the formation, and intimately associated with the *Beltina*-like films, is represented by submillimetric discs 0.3 to 0.9 mm across. The discs resemble *Chuaria* Walcott, but they do not exhibit the strong concentric wrinkling that would allow them to be confidently placed in this genus (one specimen with possible wrinkling is seen in the middle left of Photo 14-10P). However, the concentric wrinkling, in any case, is a compactional artifact, and spheroids with very thin membranes could deform without production of wrinkles, as in the case in some specimens from the late mid-Proterozoic Little Dal Group (Hofmann and Aitken, 1979, fig. 13 M).

14-4. Paleobiology

14-4-1. Morphologic Categories of Early Proterozoic Microfossils

As is summarized in Table 14-3, 97 genera of authentic microfossils, 2 genera of "megafossils" (forms >0.2 mm), and 23 "genera" of dubiofossils or pseudofossils have thus far been described from sediments of the Early Proterozoic. The majority of the bona fide taxa have been established on the basis of organically preserved (i.e., permineralized, kerogenous) specimens as studied in petrographic thin sections; a number of these forms are also known from mineral-replaced specimens (e.g., hematitic, limonitic, or pyritic microfossils like those shown in Photos 14-3G-I, 14-5D-G, and 14-7A, B, D). In general, Early Proterozoic microfossils are notable both for their minute size, tending to cluster near the "small end" of the microscopic size range, and for their decidedly simple, undifferentiated morphology. Five morphological categories of such fossils can be recognized: (i) coccoid (spheroidal to ellipsoidal) "unicell-like" bodies; (ii) septate, unbranched, filaments; (iii) tubular, unbranched, microstructures; (iv) branched filaments; and (v) "bizarre forms" (fossils exhibiting a type of organization that is of uncommon or unknown occurrence among modern microorganisms).

COCCOIDS

In terms both of their described taxonomic diversity and, apparently, their overall abundance, coccoid unicell-like microfossils are the predominant components of assemblages of this age; representatives of some 58 genera of such forms have been reported from all but three of the 23 fossiliferous Early Proterozoic deposits now known (Table 14-3). In general morphology, they range from simple microscopic spheres (e.g., *Huroniospora microreticulata*, Photo 14-1I; *Leptoteichos golubicii*, Photos 14-2R, and 14-9J; and *Clonophycus* spp., Photos 14-8S, U, V) to elongate ellipsoids (e.g., *Eosynechococcus* spp., Photo 14-2I-N). They occur as solitary bodies (e.g., *Huroniospora* spp., Photos 14-1I, N, 14-4D, E, 14-6B, and 14-7Y, Z; *Leptoteichos golubicii*, Photos 14-2R, and 14-9J; *Melasmatosphaera* spp., Photos 14-2W, and 14-9K, L; *Zosterosphaera* spp., Photos 14-2Z, and 14-8W?; and cf. *Chuaria* spp., Photo 14-10N, P-S) and in small groupings such as cell-pairs (e.g., *Eoentophysalis belcherensis*, Photos 14-2H, and 14-9O; *Sphaerophycus parvum*, Photos 14-2K, and 14-7P; *Eosynechococcus medius*, Photo 14-2L; *Palaeoanacystis plumbii*, Photo 14-6J; and *Myxococcoides minor*, Photo 14-9N), planar tetrads (e.g., *Eoentophysalis belcherensis*, Photos 14-2I-J, and 14-9O, P; and *Tetraphycus* sp., Photo 14-6G, H) and decussate tetrads (e.g., *Melasmatosphaera media*, Photo 14-2V; and *Eoentophysalis belcherensis*, Photo 14-9O). Many of the described taxa also occur as relatively large colonial aggregates that range in form from irregularly organized cell masses (e.g., *Huroniospora* sp., Photo 14-2Q?; *Eoentophysalis belcherensis*, Photos 14-2G, 14-6I?, M, and 14-8C; *Eosynechococcus moorei*, Photos 14-2N, and 14-6N; *Palaeoanacystis* spp., Photos 14-2O, 14-6K, and 14-8N, O; *Myxococcoides* spp., Photos 14-6F, 14-7S, and 14-8K; *Myxomorpha janecekii*, Photos 14-6O, and 14-8R; and *Bisacculoides tabeoviscus*, Photo 14-7X) to distinctive, highly ordered, cuboidal colonies (e.g., *Eucapsis*? sp., Photo 14-11A-C). And in many such colonies, the component cells are embedded in a diffuse, originally mucilagenous, organic matrix Photos 14-2O, 14-6F, K, 14-8K, N, and 14-9H) or are surrounded by prominent, commonly multilamellated, sheaths (Photos 14-2G, J, 14-6H, I, 14-7U, V, and 14-9P).

Although a few Early Proterozoic coccoids have been described as exhibiting "porelike" apertures (Licari and Cloud, 1968, figs. 19, 20), such openings are of exceptionally rare occurrence and are best regarded as preservational or diagenetic artifacts. Indeed, and unlike spheroidal acritarchs and similar fossils of the latest Proterozoic and early Paleozoic, coccoids of this age are simple and unornamented; they lack pylomes, appendages and similar structural elaborations (with the possible exception of *Eomicrhystridium*?, discussed below); their surface texture is simple, varying from psilate to coarsely and irregularly granular (the latter apparently being a result of diagenetic alteration); and although some such coccoids contain particulate or globular organic bodies (e.g., Photos 14-2H-J, T-X, 14-6E, M, 14-7X, 14-8H, M, and 14-9K, L), the bodies are of relatively irregular form, variable occurrence, and seem certain to represent diagenetic artifacts (e.g., condensed protoplasm) rather than preserved intracellular organelles or similar, original, biologic structures. However, the nature of rounded to elongate "bud-like" protuberances that arise from the surface of certain such coccoids (viz., *Huroniospora* spp., Photo 14-1J, K, M) is more difficult to evaluate. Such structures, although of rare occurrence, have been illustrated in a sufficiently large number of examples (e.g., Darby, 1974) as to suggest that they might be biogenic, rather than diagenetic. If so, at least some Early Proterozoic coccoids may have reproduced by "budding."

With regard to cell size, virtually all Early Proterozoic coccoids are of small dimensions; microfossils of this category larger than about 25 μm in diameter are of rare occurrence (e.g., such forms as *Zosterosphaera*, shown in Photo 14-2Z, and 14-8W?, spheroids that range to more than 40 μm in diameter, and—if actually biogenic—the *Chuaria*-like discs shown in Photo 14-10P-S, bodies that are more than 0.5 mm across). Indeed, of the approximately 3,000 Early Proterozoic solitary and colonial coccoid unicells for which size data are available (as based on measurement in petrographic thin sections of fossiliferous cherts from the Gunflint, Kasegalik, McLeary, Frere, and Paradise Creek Formations, and the Amelia, Balbirini, and Bungle Bungle Dolomites), more than 99.5% are less than 20 μm in diameter, about 98% are less than 15 μm in diameter, and the vast majority are even smaller, falling in a modal size grouping that ranges from 2 to about 7 μm (Schopf, 1977a, p. 153–157). Detailed size data of this type are as yet not available for Early Proterozoic assemblages preserved in shales. Nevertheless, the few data that are available (Timofeev, 1966, 1969, 1973a, 1979) suggest that (*i*) the relatively large coccoids of this age (viz., forms > 50 μm in diameter) were predominantly or solely planktonic in habit; that (*ii*) such forms occur in relatively greater abundance in shaly, offshore facies rather than in shallow-water, stromatolitic units of the same age (Schopf, 1978); and that (*iii*) with the exception of possibly biogenic *Chuaria*-like discs that have recently been detected in shales of the Bungle Bungle Dolomite of Australia (Section 14-3-2) and of the Changcheng System of China (Hofmann and Chen, 1981, fig. 3), coccoid microfossils larger than about 100 μm in diameter appear to be unknown in geologic units older than about 1,400 Ma (apparently the oldest such examples being ~100 μm-diameter coccoids reported by Horodyski and Bloeser, 1978, from shales of the ~1,400-Ma-old Newland Limestone of Montana, and 600 μm-diameter spheroids described by Peat et al., 1978, from shales of the 1,300 to 1,400-Ma-old McMinn Formation of northern Australia).

SEPTATE FILAMENTS

Septate, unbranched filaments, now known from about two-thirds of the fossiliferous Early Proterozoic units listed in Table 14-3, similarly tend to be of small diameter and of simple, undifferentiated morphology. The 18 described genera of such forms comprise five principal subgroups: (*i*) fine, sinuous, threads, less than 1 μm in diameter, composed of a uniseriate row of minute cylindrical or spheroidal cells (e.g., *Biocatenoides* spp., Photos 14-2A, 14-7C, and 14-9D?; *Archaeotrichion* spp., Photos 14-5C, and 14-9D?; and *Coleobacter primus*, Photo 14-7F); (*ii*) narrow filaments, ranging from about 1 to 2.5 μm in diameter, composed of cylindrical, uniseriate cells (e.g., *Gunflintia* spp., Photos 14-1B, F?, H?, 14-3D, 14-4E, and 14-7J; and *Cyanonema inflatum*, Photo 14-7I, L); (*iii*) intermediate-diameter forms, 2 to about 4 μm in width, composed of equant to somewhat elongate uniseriate cells (e.g., *Gunflintia grandis*, Photo 14-1C, D; and *Halythrix* sp., Photo 14-2D); (*iv*) somewhat broader filaments, about 5 μm in diameter, made up of a uniseriate row of elongate, cylindrical cells (e.g., *Gunflintia*? sp., Photo 14-3J); and (*v*) broad cellular filaments, less than 5 μm to as much as 15 μm in width, composed of a uniseriate row of short, disc-shaped cells that are commonly capped by a distinctive, hemispherical to obconical terminal cell (e.g., *Oscillatoriopsis schopfii*, Photo

14-7K; and cf. *Obconicophycus* sp., Photos 14-7N, and 14-9G?). Rather commonly there appears to be a rough inverse correlation between the diameter of such preserved filaments and both their abundance and their total length; broad morphotypes tend to occur as rare, relatively short filaments, whereas the more narrow varieties are generally of much greater abundance and of greater length. Thus, and although relatively broad septate taxa are known from the Early Proterozoic, the vast majority of such filaments are of small size, generally between about 1 and 2.5 μm in diameter (Schopf, 1977a, fig. 9).

Like most similar fossils now known from the younger Precambrian (e.g., Schopf, 1968; Schopf and Blacic, 1971), Early Proterozoic septate filaments are most commonly not surrounded by an encompassing sheath (e.g., Photos 14-1B-D, 14-2D, 14-3J, and 14-7K, N). Sheaths, however, do occur, commonly enclosing but a single cellular trichome (e.g., Photos 14-1F?, H, and 14-9E) the diagenetic alteration of which (viz., the collapse and condensation of the internal chain of cells) can produce spiraliform and other types of degradational variants (e.g., Photo 14-2E, F). And at least one example of an Early Proterozoic sheath containing remnants of two cellular trichomes has been reported, viz., *Gunflintia minuta* from the ~2.0-Ga-old Gunflint Formation (Awramik and Barghoorn, 1977, fig. 4d); but the authors of this report are mistaken that this occurrence "strengthens" the interpretation of *G. minuta* as being of "blue-green algal affinity . . . as known bacteria contain only one trichome per sheath" (Awramik and Barghoorn, 1977, p. 140) for, on the contrary, both *Desmanthos*, among extant colorless gliding bacteria, and *Thioploca*, among living colorless sulfur gliding bacteria, exhibit multi-trichomic sheaths (Buchanan and Gibbons, 1974, pp. 115, 127).

Other degradational variants of such filaments (especially of taxa included in subgroups *i* and *ii*, above) include specimens in which one or more intercalary cells have been variously expanded, swollen, or even ruptured during diagenesis and preservation (or in which only one or a few "normal-size" cells are preserved within a diagenetically collapsed trichome), giving the impression of "cellular differentiation" (e.g., Photos 14-1D, and 14-4E). Indeed, relatively large cells of this type were interpreted initially as being "heterocysts" (Licari and Cloud, 1968; Cloud and Licari, 1972), examples of a specialized cell type characteristic of nostocaceans, whose presence would provide a potentially useful criterion by which to distinguish such cyanobacteria from filamentous microbes of otherwise similar morphology. True heterocysts would be expected to be decidedly thick-walled; to exhibit rather commonly a regular pattern of distribution within the trichomes; to have polar pores; to exhibit distinct organic "plugs" at such pores; to be of relatively abundant occurrence within single filaments; and to occur in many or most filaments of a given taxon. However, as has been noted elsewhere (Schopf, 1975, p. 226–227), none of these characteristics is evident in known Early Proterozoic "heterocyst-like" cells. In addition, the "heterocystous" fossil filaments are commonly of smaller diameter than most modern nostocaceans (Schopf, 1977a, p. 162), and individual specimens occur that contain a gradation of "cell types" ranging from collapsed, degraded cells, to "normal" relatively well-preserved cells, to swollen cellular remnants of the type interpreted as "heterocysts" (e.g., Licari and Cloud, 1968, figs. 2, 4, 13). Gradational series of this type seem most reasonably interpreted as evidencing variable preservation rather than "cellular differentiation."

TUBULAR STRUCTURES

The third category of Early Proterozoic microfossils here recognized is composed of tubular, unbranched microstructures, forms now known to occur in about half of the 23 fossiliferous units listed in Table 14-3. The nine described genera of such fossils comprise three principal subgroups: (*i*) narrow tubules, less than 2 to about 5 μm in diameter, of the "*Eomycetopsis*-type" (e.g., Photos 14-2B, C, 14-5A, B, 14-6A, 14-7M, 14-8A, B, and 14-9A, B); (*ii*) somewhat broader tubules, 6 to about 12 μm in diameter, of the "*Animikiea*-type" (e.g., Photos 14-1A, and 14-7O); and (*iii*) broad, ~20 μm-diameter, thick-walled forms of the type commonly referred to the genus *Siphonophycus* (e.g., Photo 14-3A-C, and Awramik and Barghoorn, 1977, fig. 7c, d). Like the septate filaments discussed above, there tends to be an inverse correlation between the diameter of these microfossils and their abundance; indeed, whereas *Eomycetopsis*-like tubules are of common occurrence in several Early Proterozoic assemblages, only two specimens of "*Siphonophycus*" have as yet been reported from deposits of this age and exceptionally broad fossils of this type, tubes that are up to 30 μm or more in diameter (Schopf and Sovietov, 1976), are known only from sediments of Late Proterozoic or younger age (Schopf, 1977a, fig. 7). However, flattened tubular structures (*Tyrasotaenia*) have recently been found in the mid-Proterozoic Changcheng System of northern China (Hofmann and Cheng, 1981).

BRANCHING FILAMENTS

Branched filaments are here grouped together as a fourth category of Early Proterozoic microfossils. As is summarized in Table 14-3, however, forms of this type are of exceedingly rare occurrence, represented by only two or three genera (*Archaeorestis*, *Ramacia* and, possibly, *Palaeoleptophycus*) that are known to occur in but three, or possibly four formations. *Archaeorestis*, a rare but distinctive member of the Gunflint microbiota (Barghoorn and Tyler, 1965; Awramik and Barghoorn, 1977), is of uncertain systematic position; *Ramacia* (Photo 14-7G) bears some resemblance to actinomycetacean bacteria; and *Palaeoleptophycus* (Hofmann, unpublished) is as yet little known and incompletely characterized.

UNUSUAL, BIZARRE FORMS

Finally, in the fifth category are here included bizarre forms, microfossils that have been grouped together because of their unusual morphologies and their uncertain biological affinities. More than a dozen such genera have been described, including such forms as *Eoastrion* Photos 14-1R, S, 14-4B, and 14-7H), actinomorphic microfossils that are similar in gross morphology to *Metallogenium* spp., modern iron- and manganese-oxidizing "budding bacteria"; *Eomicrhystridium*?, (Photo 14-1M, O), forms that are at least roughly comparable to Paleozoic acanthomorphic acritarchs (but that might actually be preservational variants of coccoid unicells or, possibly, of the umbrella-like "crown" of *Kakabekia*); *Eosphaera* (Photo 14-1P, and the "dwarf" specimens shown in Photos 14-2Q?, and 14-9F?), distinctive spheroidal microfossils of uncertain systematic position (for a summary of the suggested affinities of this genus, see Awramik and Barghoorn, 1977, p. 129–130); *Exochobrachium*, a monospecific genus known from but two specimens that exhibit an unusual tri-radiate morphology (Photo 14-1V, and Awramik and Barghoorn, 1977, fig. 5g); *Galaxiopsis*, forms that bear some resemblance to *Eomichrystridium*? (and that similarly might simply be unusually-preserved coccoids rather than members of a separate, bizarre taxon; Awramik and Barghoorn, 1977, p. 137); *Kakabekia*, a decidedly unusual type of microfossil that consists of an umbrella-like "crown" connected via a tubular stipe to an ellipsoidal or spheroidal basal(?) bulb (e.g., Photos 14-1U, 14-2Y?, and the "giant" specimen shown in Photo 14-4A); *Veryhachium*?, a highly distinctive form known from the single specimen shown in Photo 14-1T (and a form that we consider assuredly biogenic despite the reservations noted by Awramik and Barghoorn, 1977, table 1); and *Xenothrix*, a monospecific "dumbbell-shaped" taxon similarly known from but a single specimen (and a form that may in fact be congeneric with *Archaeorestis*; compare *X. inconcreta*, as shown in Awramik and Barghoorn, 1977, fig. 5e, with the unbranched specimens of *A. schreiberensis* shown in Barghoorn and Tyler, 1965, pts. 6 and 7). At present, most such bizarre taxa are best known from the ~2.0-Ga-old Gunflint Formation; a few also occur in approximately coetaneous sediments of the Frere Formation, Duck Creek Dolomite and, possibly, the Kasegalik and McLeary Formations; and all are of uncertain systematic position.

14-4-2. Biological Affinities and Level of Organization

Of the five morphological categories of microfossils thus far recognized in Early Proterozoic sediments, members of two groups—the branching filaments and the bizarre forms—comprise a numerically insignificant fraction of the known total biota. Moreover, as is noted above, the biological affinities of taxa included in these two categories are poorly understood. Thus, the following discussion will be limited to a consideration of the three principal morphological categories, members of which collectively represent well over 99 percent of all known Early Proterozoic fossils: (*i*) coccoid (spherical to ellipsoidal), solitary and colonial, unicell-like forms; (*ii*) septate, unbranched, sheath-lacking or sheath-enclosed, threadlike filaments; and (*iii*) non-septate, unbranched, cylindrical or compressed, tubular microstructures.

COCCOIDS

As is discussed above, coccoid Early Proterozoic microfossils (*i*) range in morphology from simple spheres to elongate ellipsoids; (*ii*) occur as solitary bodies, cell pairs, tetrads, and (in commonly sheath-enclosed) irregular to highly-ordered colonies; and (*iii*) are of small size, generally less than 15 μm, and commonly about 5 μm in diameter. In all of these characteristics they resemble various types of extant cyanobacteria, a possible affinity that has thus been commonly suggested and one that is consistent with, and seems reinforced by, the occurrence of such fossils in laminated stromatolitic communities, microbial biocoenoses that in the modern environment are dominated by various cyanobacterial (i.e., "blue-green algal") taxa. Indeed, in a number of reports dealing with Proterozoic life this possible affinity has in essence been simply assumed, the problem considered being not whether such fossils are in fact cyanobacteria but, rather, what modern cyanobacterial genus do they most resemble (a practice well-illustrated by the selection of such generic names as *Palaeoanacystis*, *Eoentophysalis*, *Eosynchococcus*, and *Gloeocapsomorpha* for the fossil coccoids—*Anacystis*, *Entophysalis*, *Synechococcus*, and *Gloeocapsa* all being genera of extant chroococcaleans—and by the suggestion that at least some such fossils might actually be representatives of living cyanobacterial genera, e.g., of *Pleurocapsa* and *Eucapsis*).

However, concomitant with this emphasis on cyanobacteria it has been commonly overlooked that essentially identical morphologies can also occur (and in fact are rather widely represented) among non-cyanobacterial microbes as well. A more or less typical example in this regard (but one more carefully presented than other examples that might be cited) is provided by the report of fossil "*Eucapsis*?" (Photo 14-11), 2 to 3 μm-diameter conccoids that occur in four-tiered (64-celled) cuboidal colonies preserved in association with digitate stromatolites of the late Early Proterozoic Paradise Creek Formation of northeastern Australia (Licari et al., 1969). According to Licari et al. (1969, p. 60), these well-ordered cuboidal aggregates "are indistinguishable from colonies of the living myxophycean alga [=cyanobacterium] *Eucapsis*," and although these authors consider possible affinities with two non-cyanobacterial genera as well, such possibilities are rejected (viz., with *Lampropedia*, because of its occurrence in planar rather than three-dimensional colonies, and with *Sarcina*, because they regard it as forming only eight-celled, isolated cuboids, rather than larger aggregates of the type detected). Licari et al. (1969, p. 61) thus "designate these fossils simply as *Eucapsis*? . . . photosynthesizing units referable to the chroococcacean blue-green algae." In actuality, however, comparable 64-celled (or larger) cuboidal colonies, similarly composed of spheroidal unicells in the 2 to 3 μm size range and equally "indistinguishable" from the fossil cuboids, also occur in at least five genera of extant non-cyanobacterial prokaryotes (Table 14-4; Buchanan and Gibbons, 1974). And, most importantly, the physiological characteristics of members of these five genera differ markedly from those of cyanobacteria: unlike *Eucapsis*, a typical cyanobacterial genus of aerobic photoautotrophs, the other morphological analogues range from anaerobic photosynthesizers (*Thiosarcina* spp.), to anaerobic, fermentative chemoheterotrophs (*Sarcina* spp.), to anaerobic, methanogenic chemoautotrophs (*Methanosarcina* spp.), to aerobic, respiratory, chemoheterotrophs (*Micrococcus* spp. and *Sporosarcina* spp.). Thus, even if it is assumed that these Early Proterozoic fossil coccoids are not representatives of some now extinct lineage (cube-forming unicells that may have differed significantly in physiology from their extant morphological counterparts), the size and form of these fossils provides little insight regarding either their biological affinities or their physiological capabilities; on the basis of morphology alone it cannot be inferred whether these microorganisms were aerobic or anaerobic, autotrophic or heterotrophic, photosynthetic or chemometabolic.

Difficulties of this sort abound in the study of Early Proterozoic coccoids, for although all of the basic morphologies exhibited by such fossils (viz., spheres, ellipsoids, cell pairs, tetrads, and mucilage-embedded colonies) are well-representated among cyanobacteria, as is summarized in Table 14-4 they occur also among non-cyanobacterial prokaryotes as well, forms that are often of disparate physiological characteristics. (Table 14-4 is intended merely to illustrate the nature of the problem involved; although the taxa there listed have been chosen to highlight the range of metabolic variability exhibited among non-cyanobacterial microbes of the various morphological types, the table is by no means exhaustive in this regard and particular emphasis has been placed on atypically large-celled species rather than on smaller, more common, and thus more representative taxa).

The biochemical components (e.g., chlorophylls, accessory pigments, enzyme complexes, etc.) used in classification of modern microbes are not detectable in fossil microorganisms; there are thus relatively few potentially preservable characters that can be used to distinguish fossil cyanobacteria from morphologically similar non-cyanobacterial

TABLE 14-4. Comparison of morphologically similar coccoid (spheroidal and ellipsoidal) cyanobacteria and non-cyanobacterial prokaryotes (data from Desikachary, 1959; Buchanan and Gibbons, 1974).

Cell type, taxa, and dimensions (μm)	*Physiological characteristics*
SOLITARY SPHEROIDAL UNICELLS	
Cyanobacteria:	
Chroococcus westii, 13.0–27.0	Aerobic photoautotrophy
Aphanocapsa elachista, 1.5–2.0	Aerobic photoautotrophy
Non-cyanobacterial prokaryotes:	
Thiovulum majus, 5.0–25.0	Aerobic chemoautotrophy
Micrococcus spp., 0.5–3.5	Aerobic chemoheterotrophy
Rhodopseudomonas sphaeroides, 0.7–4.0	Anaerobic photoautotrophy
Rhodopseudomonas sphaeroides, 0.7–4.0	Anaerobic photoheterotrophy
Methanococcus vannielii, 0.5–4.0	Anaerobic chemoautotrophy
Peptococcus spp., 0.5–1.6	Anaerobic chemoheterotrophy
SOLITARY ELLIPSOIDAL CELLS	
Cyanobacteria	
Synechococcus aeruginosus, 5.0–16.0 × 25.0–30.0	Aerobic photoautotrophy
Aphanothece saxicola, 1.0–2.0 × 2.0–6.0	Aerobic photoautotrophy
Non-cyanobacterial prokaryotes:	
Achromatium volutans, 5.0–25.0 × 5.0–40.0	Aerobic chemoautotrophy
Azomonas spp., 2.0–6.0 × 8.0–10.0	Aerobic chemoheterotrophy
Chromatium spp., 1.0–6.0 × 2.0–25.0	Anaerobic photoautotrophy
Rhodopseudomonas capsulata, 0.5–1.2 × 2.0–6.0	Anaerobic photoautotrophy
Rhodopseudomonas capsulata, 0.5–1.2 × 2.0–6.0	Anaerobic photoheterotrophy
CELL PAIRS	
Cyanobacteria:	
Chroococcus giganteus, 54.0–58.0	Aerobic photoautotrophy
Gloeocapsa punctata, 0.7–2.8	Aerobic photoautotrophy
Non-cyanobacterial prokaryotes:	
Nitrococcus mobilis, 1.5–3.5	Aerobic chemoautotrophy
Azomonas spp., 2.0–10.0	Aerobic chemoheterotrophy
Lamprocystis roseopersicina, 3.0–3.5	Anaerobic photoautotrophy
Methanococcus vannielii, 0.5–4.0	Anaerobic chemoautotrophy
Megasphaera elsdenii, 2.4–2.6	Anaerobic chemoheterotrophy
CELL TETRADS	
Cyanobacteria:	
Chroococcus macrococcus, 25.0–50.0	Aerobic photoautotrophy
Aphanocapsa koordersi, 2.2–2.8	Aerobic photoautotrophy
Non-cyanobacterial prokaryotes:	
Nitrosococcus oceanus, 1.8–2.2	Aerobic chemoautotrophy
Micrococcus spp., 0.5–3.5	Aerobic chemoheterotrophy
Thiocapsa spp., 1.2–3.0	Anaerobic photoautotrophy
Peptococcus spp., 0.5–1.6	Anaerobic chemoheterotrophy
IRREGULAR, MUCILAGE-EMBEDDED CELL COLONIES	
Cyanobacteria:	
Microcystis robusta, 6.0–9.0	Aerobic photoautotrophy
Aphanocapsa delicatissima, 0.5–0.8	Aerobic photoautotrophy
Non-cyanobacterial prokaryotes:	
Nitrococcus mobilis, 1.5–3.5	Aerobic chemoautotrophy
Azomonas spp., 2.0–10.0	Aerobic chemoheterotrophy
Lamprocystis roseopersicina, 3.0–3.5	Anaerobic photoautotrophy
Methanococcus vannielii, 0.5–4.0	Anaerobic chemoautotrophy
Peptococcus spp., 0.5–1.6	Anaerobic chemoheterotrophy
REGULAR (CUBOIDAL) CELL COLONIES	
Cyanobacteria:	
Eucapsis minuta, 1.0–1.5	Aerobic photoautotrophy
Non-cyanobacterial prokaryotes:	
Micrococcus roseus, 1.0–2.5	Aerobic chemoheterotrophy
Sporosarcina ureae, 1.2–2.5	Aerobic chemoheterotrophy
Thiosarcina rosea, 2.0–3.0	Anaerobic photoautotrophy
Methanosarcina methanica, 2.0–2.5	Anaerobic chemoautotrophy
Sarcina maxima, 2.0–3.0	Anaerobic chemoheterotrophy

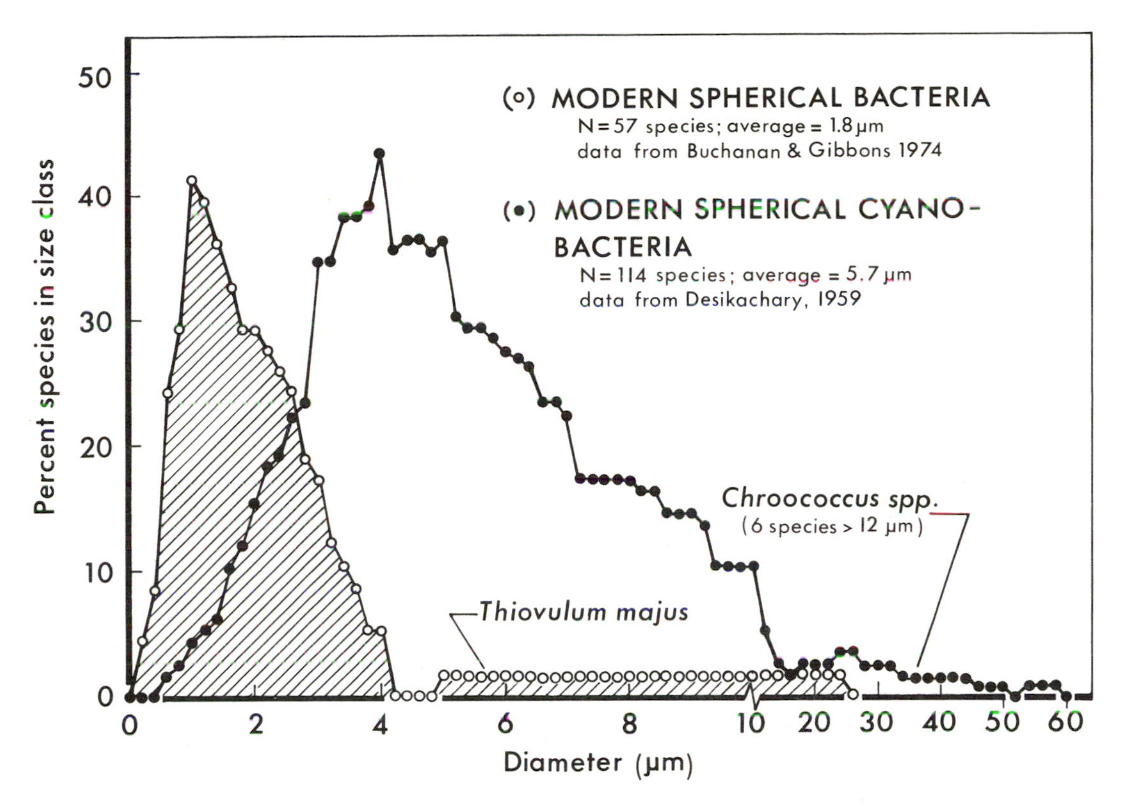

FIGURE 14-3. Size ranges of modern spherical bacteria and cyanobacteria.

analogues. Of these, and perhaps together with the nature of preserved encompassing sheaths (multilamellated sheaths being apparently of limited occurrence among non-cyanobacteria), one of the most promising is cell size. Figure 14-3 summarizes the size distribution of extant species of spherical (solitary and colonial) cyanobacteria, compared with that of morphologically comparable (spherical, non-filament-forming, solitary, and colonial), free-living, non-cyanobacterial prokaryotes. As is there shown, cyanobacterial cells tend to be of relatively large size, ranging from about 0.5 to 58 μm in diameter (with an average diameter of 5.7 μm, N = 114); about 70% of described taxa (80 of the 114 forms included in Figure 14-3) are 4 μm or more in diameter. In contrast, cells of modern free-living spherical bacterial are generally quite small; only one taxon of such forms (*Thiovulum majus*, 5 to 25 μm in diameter) is of typical cyanobacterial dimensions (although prochlorophytes, spherical bacterial symbionts in ascidians, are 10 to 20 μm in diameter), and 98% of such coccoid taxa are 4 μm or less in diameter (with an average diameter of 1.8 μm, N = 57). A similar, if less pronounced distinction can be drawn among species of ellipsoidal prokaryotes (Figure 14-4): although a number of large-celled non-cyanobacterial taxa are known (viz., colorless ovoid gliding bacteria of the genus *Achromatium*, and photosynthetic purple sulfur bacteria of the genera *Chromatium* and *Chromatiopsis*), such forms have average dimensions of about 2.9 × 5.7 μm (N = 28) and thus tend to be of smaller size than known ellipsoidal cyanobacteria (which have average dimensions of 4.4 × 7.5 μm, N = 42). Although there is a good deal of overlap between the size ranges of cyanobacterial and non-cyanobacterial coccoids (especially for spheroids between 0.5 and 4.0 μm in diameter, and for ellipsoids with lengths between about 3.0 and 5.0 μm), relatively large cell size is decidedly more typical of cyanobacteria than of their non-cyanobacterial counterparts.

A number of assumptions (presumably reasonable, but nonetheless unproved) must be made for the criterion of cell size to be applied to the fossil record, among others that the modern microbial biota includes a representative sample of past microbial assemblages, and that the size ranges of prokaryotic taxa have not changed significantly

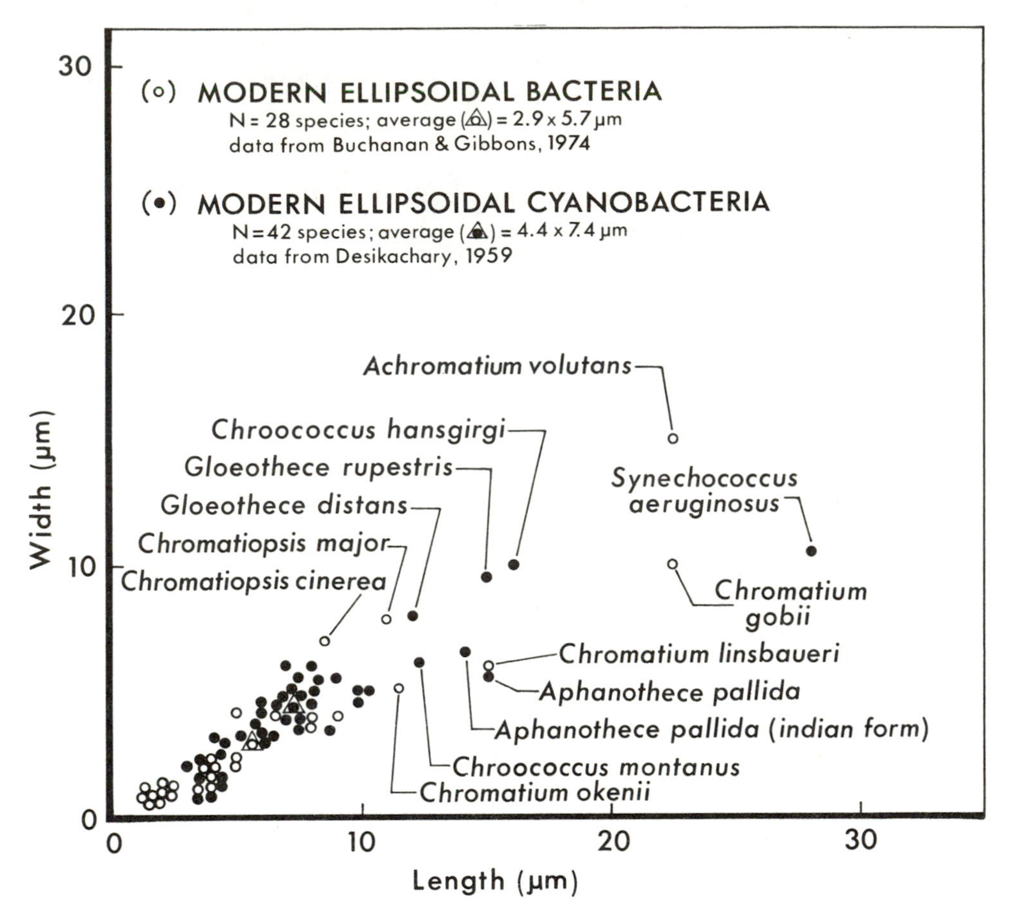

FIGURE 14-4. Size ranges of modern elliposidal bacteria and cyanobacteria.

over the course of evolutionary history. If such assumptions are made, however, it is evident that not only are virtually all known Early Proterozoic coccoid microfossils of a size and morphology consistent with a cyanobacterial affinity, but that many such forms are "typically cyanobacterial" (and unlike all but the most atypical of other coccoid prokaryotes) in their cellular dimensions. Thus, for numerous fossil taxa (especially for such small-celled forms as "*Eucapsis*?") the available data seem equally consistent with either a cyanobacterial or a non-cyanobacterial affinity, it seems reasonable to suppose—although it is by no means firmly established—that the majority of such forms were probably cyanobacteria.

SEPTATE FILAMENTS

For much the same reasons as those discussed above with regard to coccoid Early Proterozoic fossils, the biological affinities of septate filaments of this age are difficult to assess: typically, such filaments are of small size and of simple, undifferentiated morphology and although evidently of microbial affinities, they are morphologically comparable to such a diverse range of modern prokaryotes that neither their phylogenetic relations nor their physiological capabilities can be inferred confidently on the basis of morphology alone.

As outlined above (Section 14-4-1), five principal subgroups of such fossils have been recognized: (*i*) sinuous cellular threads 1 μm or less in diameter; (*ii*) narrow septate filaments, 1 to 2.5 μm in width; (*iii*) intermediate-diameter filaments, 2 to 4 μm in width, composed commonly of essentially equant cells; (*iv*) broader filaments, about 5 μm in diameter composed of elongate cells; and (*v*) relatively broad filaments, a few microns to as much as 15 μm across, composed of short, disc-shaped cells.

As a first approximation, based on analogy with modern septate microbes, it would seem reasonable to suggest that the relatively narrow filaments—those of categories (*i*) and (*ii*), above, less than 2 or perhaps 3 μm in width—may be largely of non-cyanobacterial affinities, and that the relatively broad forms—those of categories (*iii*), (*iv*), and (*v*)—may be principally cyanobacteria. Although such a relationship could well be valid, the size ranges exhibited by extant prokaryotes do not permit such a straight-forward solution to the problem. Most septate, filamentous, non-cyanobacterial prokaryotes are, in fact, of small diameter (e.g., *Flexithrix* spp., 0.3–0.5 μm wide; *Methanospirillum hungatii*, 0.5–0.7 μm; *Peloploca* spp., 0.4–1.0 μm; *Herpetosiphon* spp., 0.5–1.5 μm; *Saprospira* spp., 0.8–1.5 μm; *Beggiatoa minima*, ~1.0 μm, and *B. leptomitiformis*, 1.0–2.0 μm; *Vitreoscilla* spp., 1.2–2.0 μm; *Chloroflexus aurantiacus*, 0.5–1.0 μm, *C. aurantiacus* var. *mesophilus*, 0.7–1.1 μm, *Chloronema spiroidem*, 1.5–2.0 μm, *Chloronema giganteum*, 2.0–2.5 μm; *Pelonema* spp., 0.6–2.2 μm; and 10 spp. of *Acroonema*, 0.3–3.0 μm). But cyanobacteria of similarly small dimensions are also known (viz., in the 0.6–2.0 μm-diameter size range, 19 spp. of *Lyngbya*, 16 spp. of *Oscillatoria*, 12 spp. of *Phormidium*, 7 spp. of *Spirulina*, 5 spp. of *Schizothrix* and 2 spp. of *Symploca*). And although modern nostocacean and oscillatoriacean cyanobacteria are typically greater than 2 or 3 μm in diameter (Schopf, 1977a, fig. 11), separate non-cyanobacterial prokaryotes of comparable size are also known, including forms with equant cells—see Category (*iii*), *Oscillochloris chrysea* and *Thioploca schmidlei*—elongate cells—Category (*iv*), *Acroonema splendens*, *A. macromeres*, *Thioploca ingrica* and *Beggiotoa alba*—and disc-shaped cells, the latter including taxa substantially broader than any yet described from the Early Proterozoic (e.g., *Beggiatoa arachnoidea*, 5.0–14.0 μm wide; *B. mirabilis*, 15.0–21.0 μm; and *B. gigantea*, 26.0–55.0 μm). Thus, although it is plausible, and seems to us likely, that cellular filamentous cyanobacteria were represented in fossil assemblages of this age, available data do not prove the point; further evidence, such as the discovery of unquestionable heterocysts, akinetes, or of other relatively advanced levels of cellular organization (e.g., of branched, heterotrichous stigonemataleans) seems needed before the commonly presumed Early Proterozoic occurrence of such oxygen-producing photoautotrophs can be considered firmly established.

TUBULAR STRUCTURES

Early Proterozoic non-septate tubules occur either as cylinders that are hollow, devoid of organic contents (e.g., Photos 14-1A, and 14-9A), or that encompass narrow septate filaments (e.g., Photos 14-1F.?, H, and 14-9E) or the degraded remnants of such filaments (e.g., Photo 14-2-E, F). Structures of this type—cylindrical, mucilagenous (mucopolysaccharide), trichome-encompassing "sheaths"—are of common occurrence among modern filamentous prokaryotes, especially those forms that exhibit "gliding motility." Hollow, empty sheaths thus occur commonly in both modern and fossil microbial populations, largely as a result of the motile, phototactic or chemotactic internal trichomes having glided away leaving behind their non-motile extracellular encasements. (It should be noted, however, that in moribund populations, empty sheaths can also result from the in situ degradation of internal trichomes and that cylindrical "sheath-like" microstructures can be produced as a result of the differential breakdown of transverse trichomic septa; thus, the presence of preserved sheaths or sheath-like tubules in fossil populations cannot be taken as definitive evidence of the occurrence of gliding microorganisms.) The possibility that such fossils might be remnants of originally hollow, tubular (i.e., coenocytic) microorganisms seems excluded by the fact that all such forms now known are open-ended rather than exhibiting the closed, rounded end-walls that would be expected of coenocytic microbes. Thus, the majority, and probably all such Early Proterozoic tubules can be safely interpreted as representing preserved microbial sheaths. What can be said with regard to their specific microbial affinities?

In terms of tube diameter, three principal categories of such fossils have been recognized: tubules less than 2 to about 5 μm across; those 6 to about 12 μm in width; and broad, thick-walled forms about 20 μm across. Non-tapered, unbranched sheaths of these diameters are well-represented among extant nostocalean cyanobacteria (viz., Oscillatoriaceae and some species of Microchaetaceae; for a summary of sheath diameters, see Schopf, 1977, figs. 8 and 10). However, sheaths of comparable form occur as well in numerous non-cyanobacterial microbes (viz., chloroflexaceans, pelonemataceans, *Toxothrix* spp., gliding bacteria of the Cytophagaceae and Beggiatoaceae, and the "sheathed bacteria"), and although as a general rule such sheaths tend to be of small diameter, broad-sheathed taxa are also known (e.g., *Thioploca minima*, with sheaths up to 30 μm in diameter; *T. ingrica*, having sheaths as much as 80 μm in width; and *T. schmidlei*, exhibiting sheaths 50 to as much as 160 μm across). Thus, like that of the two other major types of Early Proterozoic microfossils, the morphology of these tubular carbonaceous microstructures is consistent with either a cyanobacterial or a non-cyanobacterial affinity.

LEVEL OF ORGANIZATION

In summary, (*i*) the size, organization and simple undifferentiated morphology of members of each of the three major categories of Early Proterozoic microfossil are consistent with a prokaryotic (i.e., non-nucleated, microbial) biologic affinity; and (*ii*) many of the fossil taxa are comparable in detail to specific, living, cyanobacterial, and non-cyanobacterial prokaryotes. Moreover, in terms both of their overall morphological simplicity (e.g., the absence of evidence of complex, differentiated reproductive structures; of pylomic apertures or elaborate surface ornamentation; of true-branching and a resultant polarity of organization; and of multicellularity and tissue formation) and of their consistently small, prokaryote-like size (except for the possibly biogenic *Chuaria*-like discs discussed in Section 14-3-2), they are unlike most known eukaryotes (for a more detailed discussion, see Schopf and Oehler, 1976, and Schopf, 1977a, pp. 153–157). Available data thus indicate that in level of organization, the Early Proterozoic biota consisted largely, and perhaps entirely, of primitive prokaryotic microbes.

However, the physiological characteristics of these Early Proterozoic microorganisms cannot be inferred on the basis of their morphology alone. In general, for all of the major categories, the relatively large-diameter forms seem more suggestive of cyanobacteria than of other prokaryotic microbes, but in no case has this possible affinity been unequivocally established. The argument for the Early Proterozoic presence of such aerobic photoautotrophs thus reduces to three main lines of reasoning: (*i*) the morphology and mode of occurrence of most commonly occurring Early Proterozoic microorganisms are not inconsistent with a cyanobacterial affinity; (*ii*) the presence of such oxygen-producing photoautotrophs seems required by currently preferred models for the Early Proterozoic development of an aerobic environment (as discussed in Chapter 11); and (*iii*) if it is assumed that the sediment-water interface was aerobic during all or part of the Early Proterozoic, consideration of the biochemical characteristics of extant prokaryotes (especially the fact that bacterial photoautotrophy and the biosynthesis of bacteriochlorophyll are obligately anaerobic processes; see Chapter 13) and of the mode of formation of extant, stromatolitic biocoenoses would seem to require that the upper portions of benthonic microbial communities of this age were dominated by aerotolerant (i.e., cyanobacterial) photosynthesizers. Although we recognize that these lines of reasoning are notably indirect and that available data are insufficient to prove that cyanobacteria were in fact extant during the Early Proterozoic, it seems to us likely that microorganisms of this type (viz., aerobic photoautotrophs that if alive today would be referred to the cyanobacteria, although possibly to non-modern "fossil families") must have been major components of the Early Proterozoic biota.

14-4-3. Early Proterozoic Paleobiology

Although much remains to be learned regarding the paleobiology of the Early Proterozoic, a number of generalizations seem warranted by the data currently at hand.

First, in mode of occurrence, cellular organization and in major morphologic types, known Early Proterozoic microorganisms are fundamentally similar to those occurring in assemblages both of much greater (Section 9-5-1) and substantially younger (viz., Late Proterozoic) Precambrian age; with the exception only of megascopic organisms known from the latest Proterozoic, the Precambrian appears to have been dominated by simple, microscopic, chiefly prokaryotic, coccoid, and filamentous microorganisms.

Second, the Early Proterozoic biota included both benthonic (viz., prokaryotic filaments and colonial coccoids) and planktonic microorganisms (e.g., *Eosphaera*, *Kakabekia*, and various of the solitary coccoid unicells); in this respect it is comparable to assemblages of the younger Precambrian but possibly differs from those of the Archean that, to date, are known only to have included benthonic forms (but from which presumably planktonic "dubiomicrofossils" have also been described; Section 9-4).

Third, and despite these general similarities, the biota of the Early Proterozoic may have differed significantly from those both of older and younger Precambrian age: Early Proterozoic microbial communities appear to lack (or to contain only as minor components) non-septate "long, sinuous, narrow threads" of the *Archaeotrichion*-type that are the dominant components of the one Archean assemblage thus far studied in some detail (Section 9-5-1), and they seem substantially less diverse (especially with regard to assured oscillatoriaceans) than comparably preserved communities of the Late Proterozoic.

Fourth, there appears to be a general trend of increasing cell and filament diameter (suggested both by changes in average and "modal" size and in maximum size range), and of increasing biologic diversity (as reflected by numbers of described taxa), throughout the Proterozoic. Such trends may be a result both of evolutionary diversification among prokaryotes and of the origin and early diversification of relatively large-celled eukaryotic microorganisms.

And fifth, the biota of the mid-portion of the Early Proterozoic (viz., about 2 Ga ago) appears to have been characterized by a distinctive component of bizarre microorganisms, microbes of decidedly atypical morphology. The possible paleobiologic significance of these forms has yet to be demonstrated, although it has been suggested that they might evidence a phase of evolutionary experimentation among prokaryotes correlative with the development of an aerobic environment (Awramik and Barghoorn, 1977). Further studies are also needed to demonstrate whether such forms can be used effectively as time-restricted index fossils and, indeed, whether any of the differences now noted between Precambrian assemblages of various ages (e.g., trends in cell and filament size) will ultimately prove useful as a basis for biostratigraphic zonation.

14-5. Summary and Conclusions

(1) During the past 60 years, microfossils and microfossil-like objects have been reported from at least 40 Early Proterozoic geologic units from North America, Europe, Africa, Australia, and Asia (Figure 14-1, Table 14-1); nearly 90% of the more than 160 publications dealing with these occurrences have appeared since 1965 (Figure 14-2).

(2) However, of these 40 occurrences, only 24 include authentic microfossils; the remainder are composed solely of dubiomicrofossils or of mineralic pseudomicrofossils (Table 14-2). Moreover, of the 24 bona fide assemblages now known, fewer than a dozen have been carefully characterized on the basis of microscopic study in petrographic thin sections.

(3) Two types of such assemblages have been detected: (*i*) shallow water stromatolitic communities preserved by silica permineralization in cratonal carbonate sediments; and (*ii*) relatively deeper water assemblages preserved commonly as carbonaceous compressions in offshore black shales. In both of these lithotypes, interpretation of the organic-walled microfossils detected is made difficult by the diagenetic alteration that invariably accompanies their preservation.

(4) Five morphological categories of Early Proterozoic microfossils have been recognized: (*i*) coccoid bodies; (*ii*) septate filaments; (*iii*) tubular microstructures; (*iv*) branched filaments; and (*v*) bizarre forms (fossils of highly atypical morphology). Members of the latter two categories comprise a numerically insignificant fraction of the total Early Proterozoic biota as now known.

(5) Coccoid (spherical to ellipsoidal) Early Proterozoic microfossils are of small size and of simple undifferentiated morphology, occurring singly and in cell pairs, tetrads, and in irregular to highly ordered (and commonly mucilage-embedded) colonies. Representatives of some 58 genera of such forms have been reported from about 90% of the 24 fossiliferous deposits of this age that are now known; in terms both of their taxonomic diversity and their overall abundance, such coccoids are the predominant components of Early Proteroic assemblages (Table 14-3). Virtually all such fossils are morphologically comparable to specific modern prokaryotes (i.e., non-nucleated microbes).

(6) Septate cellular filaments, reported from about two-thirds of the Early Proterozoic fossiliferous units now known, are similarly of small size and simple morphology; most are only a few microns in diameter and all are unbranched and cellularly undifferentiated—morphological characteristics that they share with small-diameter cyanobacteria and with numerous types of non-cyanobacterial prokaryotes.

(7) Tubular Early Proterozoic microfossils, known from about half of the 24 fossiliferous units, are similarly unbranched and of small diameter (<20 μm across); the majority, and probably all such forms represent preserved sheaths that originally encompassed cellular microbial trichromes.

(8) Thus, as now known, the Early Proterozoic biota appears to have been composed largely, and perhaps entirely, of primitive prokaryotic microbes. Many such organisms may have been of non-cyanobacterial affinities, but several lines of evidence suggest that oxygen-producing photosynthetic microorganisms of the cyanobacterial type were also represented.

(9) In mode of occurrence, cellular organization, and in the major morphological types of microorganisms represented, the Early Proterozoic biota seems fundamentally similar to those known both from the Archean and from later portions of the Proterozoic. Nevertheless, in several respects Early Proterozoic assemblages appear to differ, perhaps significantly, from those of both older and younger geologic age, differences that may ultimately prove useful in Precambrian biostratigraphy.

(10) In sum, while the Early Proterozoic fossil record as yet remains relatively poorly known, progress during the past 15 years has been substantial; the task now at hand is to uncover new, more definitive evidence that will better define the physiology and phylogeny of Early Proterozoic microbes, especially that needed to differentiate between preserved remnants of oxygen-producing photoautotrophs (viz., cyanobacteria) and their non-cyanobacterial morphological counterparts.

CHAPTER 15

Evolution of Earth's Earliest Ecosystems: Recent Progress and Unsolved Problems

J. William Schopf, J. M. Hayes, and Malcolm R. Walter

15-1. Introduction

The ultimate goal of Precambrian paleobiology is to decipher and document both the timing and nature of major events in the early history of life. As is evident from the foregoing chapters of this volume, however, myriad questions are as yet unanswered. Despite the progress of recent years—the discovery of numerous bits and pieces of the enormous puzzle—a well-grounded understanding of the antiquity, biologic composition, and sucessional development of the Earth's earliest ecosystems seems beyond our present reach. Nevertheless, it is worthwhile to attempt to order the facts and interpretations that are available into some sort of coherent whole, to bring sharply into focus the present status of the field and the areas of uncertainty that especially need ongoing detailed study. This chapter summarizes results of one such attempt toward construction of a current "best guess scenario" for the early history of life.

15-2. Limitations of the Early Fossil Record

The limitations inherent in any such synthesis are many and varied. Chiefly, however, they derive from three sources.

First, there are limitations imposed by solely geologic aspects of the problem: (*i*) no rocks are yet known that have survived to the present from the earliest (Hadean) stage of the development of the planet, the earliest 600 Ma of Earth history, and a time span that may well have encompassed the beginnings of biological activity; (*ii*) sediments of the Early, Middle, and Late Archean suberas, potential repositories of relevant primary data, are commonly so highly altered or severely metamorphosed that the fossil evidence they may have once contained has been all but obliterated (see Chapters 3, 5, and 9); and (*iii*) as a result of the geologic evolution of the planet (summarized in Chapters 3 and 10), sediments deposited in shallow water, shelf-type environments, the principal source of paleobiologic data in the younger Precambrian (see, for example, Chapter 14), are in the Archean of only limited areal extent and of relatively rare occurrence.

Second, the nature of early biologic evolution itself serves as a major limiting factor. During the Phanerozoic (and latest Precambrian), biotic evolution was typified by the development of new morphologies and innovative body plans among megascopic, multicellular forms of life. In contrast, microbial evolution of the Archean and Early Proterozoic occurred almost exclusively at an intracellular, biochemical level. Organisms of this age were of microscopic size, prokaryotic organization, and of a generalized, simple morphology (see Chapters 9 and 14). Yet prokaryotes of indistinguishable morphology can exhibit disparate physiological characteristics (e.g., Tables 9-2 and 14-4). In fact, the evolution of such microbes appears to have been characterized by a sort of "Volkswagen syndrome," a lack of change in external body form that has served to mask the evolution of internal biochemical machinery. To document such evolution, the paleobiologist must therefore turn to sources of data other than mere morphology, in particular to the evidence sequestered in the known organic geochemical record. But here, too, difficulties abound: (*i*) although in general the extractable organic components (i.e., the molecular "chemical fossils") of ancient sediments can be readily interpreted biochemically, geologic considerations indicate that they nevertheless tend to be an unreliable basis of paleobiologic inference (see Section 5-3-1);

(*ii*) well-established techniques are as yet unavailable for chemical analysis of individual organic-walled microfossils, an approach that may be required for differentiation between physiologically (and biochemically) differing, but morphologically indistinguishable fossil microbes; and (*iii*) the state-of-the-art is such that numerous fundamental questions are far from being resolved—chemical criteria have yet to be defined, for example, that would permit differentation between the kerogenous remnants of aerobes and anaerobes, of oxygen-producing and anoxygenic photosynthesizers, or even between ancient kerogens produced solely via abiotic rather than biotic processes (see Section 4-3-3).

Third, understanding of the early history of life is limited by the very novelty of such studies, and in particular by the resulting absence of an extensive base of data and experience on which valid inferences can be firmly grounded. For example, fewer than a dozen occurrences of stromatolites are now known from Archean terranes, more than half of which were discovered or first described only during the past decade; they are forms that prior to the P.P.R.G. project had not been considered together and compared in detail (see Chapter 8). Similarly, well over three-quarters of all published reports dealing with Archean microfossils (Figure 9-2), and nearly 90% of those concerned with fossil microbes of the Early Proterozoic (Figure 14-2), have appeared since 1965. And of the elemental analyses that have thus far been carried out on kerogens from 45 Precambrian units, all but one were first reported during the past decade; more than half are recorded in the present volume for the first time (Table 5-8). In these circumstances, data are as yet too fragmentary for numerous important questions to be answered—questions, for example, relating the potential biostratigraphic usefulness of Archean microfossils and (or) stromatolites to the precise mechanisms of the early evolutionary process, and to the biologic interrelations and detailed phylogeny of primitive, early evolving lineages. Questions dealing with such quantitative aspects of early paleoecology as the rates of primary productivity, the influx and utilization of limiting nutrients, the biological recycling of carbon, and the rates of change of biomass or of biotic diversity during early Earth history are just beginning to be considered.

15-3. Major Benchmarks in Archean-Early Proterozoic Evolution

Bearing these limitations in mind, what aspects of the early record of life can be profitably addressed at the present time? Because of the particular nature of this problem area, it seems to us that questions regarding such aspects must be phrased biochemically and metabolically (rather than in terms solely of organismal morphology); that their answers—tentative as they are likely to be at present—should be based on consideration of the totality of chemical, biological, and geological evidence now available, not on data from any single discipline alone; and that at present, any fruitful approach to such questions must be largely model-dependent, determined to a substantial degree by theoretical and indirect (e.g., neobiological) considerations rather than being based solely on directly observable features of the fossil record as now known.

Following this strategy, and recognizing that development of the Earth's earliest ecosystems can probably best be described as a continuum punctuated by the sequential occurrence of a series of "benchmark" metabolic and biochemical innovations, it is possible to construct a conceptual model of the early evolutionary progression. For reasons detailed elsewhere in this volume, as referenced below, we thus assume as the organizing model for our discussion that: (*i*) the origin of life occurred early in Earth history as a result of a complex series of abiological organic reactions that took place in the essentially anoxic atmosphere and hydrosphere of the primordial planet (Sections 4-3-2, 11-5-1, 13-1-1 and 13-1-2); (*ii*) these earliest organisms were primitive prokaryotes, among them anaerobic chemoheterotrophs that obtained their carbon and energy via fermentation of simple abiotically produced organic compounds such as hexose sugars (Section 6-2); (*iii*) subsequent evolution led to the appearance of anaerobic chemoautotrophs, microbes such as methanogenic bacteria that were capable of metabolically fixing CO_2 present in the early environment (Chapter 12); (*iv*) via photoheterotrophic intermediates, these in turn led to anaerobic photoautotrophs, evolutionary precursors of extant photosynthetic bacteria, the earliest microorganisms capable of photophosphorylation (Section 6-8); (*v*) these latter anaerobes gave rise to the earliest aerotolerant oxygen-producing (viz., cyanobacterium-like) photosynthesizers (Sections 6-8, 13-2-1 and 13-5); (*vi*) refinement of the biochemical systems necessary for oxygen-protection and for metabolic and biosynthetic oxygen-use resulted in the evolution of amphiaerobic prokaryotes (Sections 11-5-3, 13-2 and 13-3); and (*vii*) as a biologically generated, stable oxygenic environment became established, obligate aerobes diversified to become widespread on a global scale.

The seven principal metabolic benchmarks thus identified are: (*i*) the origin of life and the advent of (*ii*) anaerobic chemoheterotrophy, (*iii*) anaerobic chemoautotrophy, (*iv*) anaerobic photoautotrophy, (*v*) aerobic photoautotrophy, (*vi*) amphiaerobic metabolism, and (*vii*) obligate aerobiosis. Section 15-5, below, is devoted to evaluation of this model, an assessment of available paleobiologic evidence relating to the probable time of occurrence of each of these major developments. Both the availability and reliability of data relevant to these benchmarks, however, are highly varied. In the following section, therefore, we first consider the problems attendant on the assessment of such data.

15-4. Categories of Evidence

Ideally, a synthesis such as that here attempted should be based to the greatest extent possible on observations, rather than on any model, the synthesis being generated to accomodate established facts (rather than the known facts being "fitted" to the model). Unfortunately, however, available relevant data are in short supply; some reliance on a model thus seems required. Such an approach is charged with difficulties, for if the model is fundamentally in error it will result in the wrong questions being asked, in the evidence of the wrong process being sought in materials of the wrong age, or in significant lines of evidence being discarded simply because they fail to mesh with preconceived, model-dependent notions. In the task at hand, such difficulties are especially acute because the available facts are not all of equal weight—some being far better established than others—and because assessment of their apparent reliability is necessarily a subjective matter.

To deal with these difficulties, we have here adopted the approach first introduced to such studies by Schopf (1975) and by Cloud (1976a, 1980) of establishing a multi-tiered hierarchy of categories for evaluation of the evidence and its interpretation. As defined in the following sections of this chapter, these categories, ordered in terms of decreasing persuasiveness, are: (*i*) compelling evidence; (*ii*) presumptive or preponderant evidence; (*iii*) permissive or consistent evidence; and (*iv*) missing evidence, a category including those aspects of the problem for which "no direct interpretable evidence has apparently yet been identified." It is important to note that such categories are necessarily subjective, the compelling evidence of one investigator perhaps being regarded as only presumptive or even permissive by another. Despite this inherent subjectivity, use of this hierarchical system seems useful in the present context since it explicitly identifies the degree of confidence with which the various lines of evidence (and their interpretation) are here regarded, an especially important matter in studies of Precambrian life for which many of the major generalizations have yet to be firmly established.

In the following paragraphs, we illustrate the application of this hierarchical system by considering a series of interpretations relating to the well-known fossil biocoenose preserved in carbonaceous stromatolitic cherts of the Gunflint Iron Formation of northwestern Ontario, Canada, the most diverse and intensively studied microbial community now known from the Early Proterozoic (viz., including at least 31 taxa of bona fide microfossils that have been described over the past two decades in a series of nearly three dozen publications; for a listing of relevant references, see Table 14-2, No. 4).

15-4-1. Compelling Evidence

This category we define as including abundant evidence that permits only one "reasonable" interpretation. As an example, we regard the statement that "microbial life was extant ~2.0 Ga ago," the approximate age of the Gunflint cherts, as being supported by compelling evidence. The geologic source and probable age limits of the Gunflint fossils are well defined (Table 14-2, No. 4). The provenance of several comparably aged microbial fossil assemblages is similarly well established (e.g., those of the Pokegama, Tyler, Kasegalik, McLeary, Krivoirog, and Duck Creek deposits; see Table 14-2). The Gunflint fossils are demonstrably indigenous to, and syngenetic with deposition of, the stromatolitic cherts in which they occur. Finally, the biogenicity, and for most taxa the prokaryotic biological affinities of the Gunflint fossils have been firmly documented (e.g., they are diverse, abundant, and of organic composition; they exhibit ontogenetic growth stages and a "biological" size distribution; and they are morphologically comparable to specific modern microbes; for illustration of representative taxa from the assemblage, see Figure 14-3). Evidence is abundant. Only one interpretation seems permitted.

15-4-2. Presumptive Evidence

This category is defined as including situations in which the preponderance of evidence suggests a "most likely" interpretation but for which "less probable" hypotheses also merit consideration. For example, we regard the statement that "the Gunflint microbiota included CO_2-fixing photoautotrophs" as being supported by presumptive (preponderant) evidence. Certainly, there are numerous lines of evidence consistent with this interpretation, among them the $\delta^{13}C$ values of Gunflint (and virtually all other $\sim$2.0 Ga old) kerogens and carbonates; the morphology and mat-forming mode of occurrence of the major Gunflint taxa; the inferred paleoecologic setting of the deposit; and analogies between the Gunflint stromatolites and modern stromatolitic structures of known photoautotrophic origin. However, the possibility remains that the observed $\delta^{13}C$ values might be a result of other than photoautotrophic processes (e.g., produced by chemoautotrophs or via abiotic syntheses, diagenetic alteration, and/or some as yet undefined biologic or inorganic process), and it is conceivable that the laminated, stromatolitic organization of the biocoenose could be of other than autotrophic origin (a result, for example, of the mat-forming mode of growth of a photoheterotroph-dominated assemblage). In view of the preponderance of available evidence, it certainly seems "most likely" that the Gunflint microbiota included (and was in fact dominated by) cyanobacterium-like photoautotrophs, but current data do not rule out the possibility that this interpretation could be in error.

15-4-3. Permissive Evidence

Here we include evidence that seems consistent with at least two "more-or-less equally tenable" competing interpretations. With regard to the Gunflint biota, for example, by themselves available $\delta^{13}C$ values seem equally consistent with the presence either of oxygen-producing (i.e., cyanobacterial) or of anoxygenic (e.g., chloroflexacean) photosynthetic microorganisms. Similarly, morphology alone cannot be used to determine whether the small spheroidal unicells of the Gunflint assemblage are of cyanobacterial or of non-cyanobacterial prokaryotic affinities (see Section 14-4-2). As a final example, investigated organic geochemical criteria (Section 4-3-3) cannot be used by themselves to distinguish unambiguously between biogenic and any abiogenic organic matter preserved in the Gunflint cherts. In such examples, the permissive (consistent) evidence currently available does not permit differentiation between various, evidently equally plausible, interpretations.

15-4-4. Missing Evidence

Finally, there are numerous important questions for which no direct interpretable evidence has apparently yet been identified. Virtually all aspects of the evolution of intracellular structures (e.g., of thylakoids, fibrillar DNA, organelles, etc.) fall into this category, for with but few exceptions (e.g., Niklas and Brown, 1981; and Schopf and Oehler, 1976, and references cited therein) such structures are either not preservable, or not identifiable, in fossil microorganisms. Similarly, direct evidence is simply unavailable regarding such questions as whether "the earliest forms of life were (or were not) acellular aggregates of nucleic acid and protein," or whether "gene transfer via the processes of bacterial parasexuality did (or did not) play an important role in Archean-Early Proterozoic evolution." Indeed, strictly speaking, statements regarding the time of first appearance of essentially all evolutionary innovations must also be included in this category, for available geologic data can in general serve to suggest a minimum age only, evidence supporting a proposition that one or another innovation occurred "at least as early as" (and in some cases, "evidently later than") a geologically definable date, rather than establishing the actual time of the occurrence itself.

15-5. Assessment of the Evidence

In this section we evaluate in turn available paleobiologic evidence relating to the seven evolutionary benchmarks identified above (Section

15-3), in each case by summarizing what we consider to be the oldest compelling, presumptive, and permissive evidence now known regarding their probable time of occurrence. Questions regarding these metabolic benchmarks have seldom been raised by paleobiologists. It is therefore not surprising that relevant data are less plentiful (and also, less persuasive) than one might wish. Thus, the chief value of this exercise is more a matter of highlighting major areas of uncertainty than of resolving fully the numerous issues at hand.

15-5-1. Origin of Life

In what age sediments should relevant evidence be sought?

Based on data presented elsewhere in this volume, we regard the direct evidence from organic geochemistry (Sections 5-5 and 7-2-5), stromatolites (Section 8-4-1), and microfossils (Section 9-5-1) as establishing beyond reasonable doubt that living systems were extant on Earth at least as early as ~3.5 Ga ago, the age of the Warrawoona Group of Western Australia (for details regarding the radiometric dates herein mentioned, see Appendix I). Evidence consistent with this interpretation is available also from comparably aged sediments of the "Swaziland Sequence" of South Africa (viz., from the Kromberg Formation of the Onverwacht Group; see Tables 5-7 and 5-8 and Sections 8-4-2 and 9-4-2). Thus, documentation of the beginnings of biologic activity on the planet must be sought in some still older geologic terrane, the most obvious candidate being that of the ~3.8-Ga-old Isua Supracrustal Belt of western Greenland, the oldest Early Archean sedimentary sequence that has as yet been investigated paleobiologically in some detail.

Was life extant in Isua time?

In recent years, several lines of evidence have been brought to bear on the problem of the possible existence of life in Isua time. At least three of these, however, have been seriously questioned: (*i*) "Yeast-like microfossils" reported from the Isua sequence by H. D. Pflug and his colleagues (Pflug, 1978a, 1978b, 1978c; Pflug and Jaeschke-Boyer, 1979; Pflug et al., 1979) are now regarded by other workers as being of solely inorganic, nonbiologic origin (Bridgwater et al., 1981, and Table 9-1, No. 1, and Photo 9-1 herein). (*ii*) Pyrolysis of Isua graphite has yielded small organic fragments ($\leq C_2$; B. Nagy et al., 1975; Leventhal et al., 1975; Walters et al., 1981) and, from a single graphite sample, minute quantities (in the subnanogram/gram levels) of four larger hydrocarbon fragments (up to m/e = 175), these latter being initially interpreted as both "indigenous to these rocks" and apparently "syngenetic with deposition" (Walters et al., 1979) and therefore regarded as organic remnants that "may well be the oldest molecular fossils yet found on the Earth" (Walters et al., 1981). However, investigations by B. Nagy et al. (1981) have led them to suggest that these hydrocarbon fragments are probably post-depositional, rather than syngenetic in origin, and Walters and Ponnamperuma (1981) have recently indicated that because it is now "uncertain whether the [graphitic] precursor material was syngenetic or biogenic . . . it is impossible to determine the significance of these fragments." (*iii*) The occurrence of banded iron-formation in the Isua sequence (and in the comparably aged Limpopo Fold Belt) has been suggested by P. Cloud (1980 and earlier references there cited) as possibly requiring the presence of photosynthetically produced free oxygen and, thus, the existence of life. For reasons detailed in Chapter 11 of this volume, however, it is conceivable that these particular ferruginous sediments may have been deposited "slowly using oxygen produced abiotically in the atmosphere"; as currently understood, "the mere presence of [such] banded iron-formation does not necessarily require the existence of oxygenic photosynthesizers (nor, apparently, the necessary occurrence of any type of biological activity)" (Section 11-5-2; see also Schopf, 1978, 1981).

Despite these questionable lines of evidence, it is nevertheless well established that carbonaceous matter—particulate kerogen—is indigenous to (and at least partly syngenetic with deposition of) the Isua sediments (B. Nagy et al., 1975; Oehler and Smith, 1977; Schidlowski et al., 1979; and Table 5-8 herein). The question thus becomes whether or not this kerogenous material is demonstrably of biological origin. The salient characteristics of the Isua kerogen are that (*i*) it occurs as black, opaque, microscopic particles of irregular shape (B. Nagy et al., 1975, and Table 5-8 herein); (*ii*) it exhibits a variable (Walters et al., 1980), but generally marked degree of graphitic crystallinity (B. Nagy et al., 1975; Perry and Ahmad, 1977; Walters and Ponnamperuma, 1981; and Table 5-8 herein); (*iii*) it has very low H/C (0.1 to 0.01) and

N/C ($\sim$0.001) ratios (see Walters and Ponnamperuma, 1981, and Table 5-8 herein); and (*iv*) it exhibits a notably broad range of $\delta^{13}C_{PDB}$ values (viz., -5.9 ‰ to -28.2 ‰ with an average of ~ -13 ‰; see Oehler and Smith, 1977, Schidlowski et al., 1979, Walters and Ponnamperuma, 1981, and Tables 5-7 and 5-8 and Figure 7-3 herein). Similarly, the coexisting carbonate minerals of the Isua sequence exhibit an anomalously broad range of carbon isotopic compositions (viz., $\delta^{13}C_{PDB}$ values ranging from $+5.4$ ‰ to -7.7 ‰ with an average of ~ -2 ‰; see Oehler and Smith, 1977, Perry and Ahmad, 1977, Schidlowski et al., 1979, and Figure 7-3 herein). Geologic and mineralogic evidence establish that the Isua supracrustal units have been subjected to at least six episodes of faulting, shearing, and large-scale deformation, including two phases of alteration at the high temperatures (500° to 600° C) and pressures of amphibolite-grade metamorphism (summarized by Bridgwater et al., 1981).

It is obvious from the data summarized above that the molecular structure of the Isua kerogen has been severely altered (and that the Isua carbonates have been highly metamorphosed). Indeed, careful consideration of the relationship between the H/C ratios and the $\delta^{13}C$ values of this kerogen indicates that due to this alteration "it is impossible ... to estimate [reliably] the initial isotopic composition of the carbonaceous material in samples of the Isua Supracrustal Group" (Section 5-5-2). As discussed in Chapter 7 of this volume (Section 7-2-6), it is conceivable that both they and the isotopic composition of carbonates from the sequence may have initially been consistent with the existence of autotrophic forms of life, but it is equally conceivable that this may not have been the case (e.g., that the Isua kerogen may have been solely of abiotic origin or, if biotic, that it may represent the remnants of a primitive, entirely heterotrophic ecosystem).

CONCLUSIONS

Because there are numerous, plausible, solely abiotic pathways by which organic matter could have been synthesized in the early environment (Section 4-3-2), the mere presence of kerogen in the Isua rocks cannot by itself be considered indicative of the existence of life. Similarly, as preserved and available for current study, the chemical characteristics of the Isua kerogen cannot be regarded as necessarily indicating a biological origin. Neither, however, is this kerogen demonstrably a product of abiotic organic synthesis. A third interpretation, suggested by Perry and Ahmad (1977), is that the Isua kerogen may be a product of the metamorphic alteration of carbonate minerals [viz., produced at 400° to 500° C by the reaction $6\ FeCO_{3(siderite)} \rightarrow 2\ Fe_3O_4 + 5\ CO_2 + C_{(graphite)}$, with an expected $\Delta\delta^{13}C_{(siderite\text{-}graphite)}$ of $\sim$8 ‰ at 400° C and $\sim$6 ‰ at 600° C]. However, while there is no doubt that this kerogen has been severely altered as a result of intense regional metamorphism, it is unlikely that it is solely of inorganic, metamorphic origin. This conclusion is supported by four lines of evidence: (*i*) the kerogen contains traces of hydrogen and nitrogen (Walters and Ponnamperuma, 1981, and Table 5-8 herein), elements that would not be expected to occur in pure, metamorphically produced graphite; (*ii*) it is of relatively high abundance in the Isua metasediments with total organic carbon (T.O.C.) values as great as 22 to 25 mgC/g rock (Table 5-7) and reportedly up to 10% in some units of the sequence (Walters et al., 1981); (*iii*) it exhibits a notably broad range of carbon isotopic compositions; and (*iv*) the coexisting carbonate-graphite pairs from the Isua metasediments exhibit markedly varied $\Delta\delta^{13}C$ values (ranging from 2.1 ‰ to 22.9 ‰ with an average $\Delta\delta^{13}C = 10.9$ ‰, $n = 13$; see Oehler and Smith, 1977, Perry and Ahmad, 1977, and Schidlowski et al., 1979).

Thus, it seems that the available data do not allow a decisive choice among three plausible sources for the Isua kerogen—biologically produced organic matter, abiotically synthesized organic matter, and metamorphically altered carbonate minerals all remain as possible sole or partial contributors. Moreover, as noted above, it seems likely that some portion of this carbonaceous material is post-depositional (and thus, possibly post-metamorphic), rather than syngenetic in origin; a multi-phase origin, from differing sources and possibly at decidedly differing times, must therefore be seriously considered. In this regard, it is perhaps significant that of the 13 values of $\Delta\delta^{13}C$ for the carbonate-graphite pairs cited above, three seem anomalous—viz., $\Delta\delta^{13}C$ values of 22.9 ‰ and 19.8 ‰ (Oehler and Smith, 1977) and of 21.3 ‰ (Schidlowski et al., 1979)—whereas the average of the remaining ten values, $\Delta\delta^{13}C = 7.8$ ‰, is within the range expected to have been produced via amphibolite-grade metamorphism. The low concentrations of organic carbon in the anomalous pairs measured by Oehler and Smith

(1977) led them to suggest that post-depositional contamination may have been the source of the ^{13}C-depleted kerogen in these two samples. Total organic carbon values are not available for the single anomalous carbonate-graphite pair reported by Schidlowski et al. (1979). Thus, based on the fragmentary data now available, it seems reasonable to suggest that the Isua kerogen may be partly, or even chiefly, of inorganic, metamorphic origin; the remainder of this material may be of organic, non-metamorphic origin, but if so its source—whether biotic and/or abiotic, entirely epigenetic or in part syngenetic—is at present difficult, if not impossible to discern.

Clearly, interpretation of the Isua kerogen is complicated. In fact, in our view the statement that "living systems were (or were not) extant in Isua time" falls in the category for which no direct interpretable evidence has apparently yet been identified (see Section 15-4-4). Like many others in the field, it is our hunch that life may well have existed in Isua time, but rigorously considered, the established facts do not seem to us even "permissive" of such a statement. With regard to the timing of the origin of life, we conclude that living systems must have been extant earlier than ~3.5 Ga ago, but how much earlier is at present a matter of speculation.

15-5-2. Anaerobic Chemoheterotrophy

Anaerobic chemoheterotrophs are those anaerobes, such as fermenting bacteria, that obtain their cellular carbon, reducing power, and energy, from the assimilation and catabolism of exogenous organic compounds. Despite the fact that there is abundant circumstantial evidence for the existence of such anaerobes since very early in Earth history, direct evidence of their presence is (like that of several of the other metabolic categories discussed below) notably sparse. With the probable exception of bacterium-like fossils observed in Phanerozoic coprolites and inferred by various workers to be cellular remnants of anaerobic chemoheterotrophs (e.g., Renault, 1900; Moodie, 1920; and other reports reviewed by ZoBell, 1957), to our knowledge such prokaryotes have not been definitely identified in the fossil record. Nevertheless, the existence of invertebrate metazoans since the late Precambrian (Figure 15-1)—organisms for which a microbial, anaerobic, gut flora seems absolutely required—appears to us to be compelling evidence for the occurrence of bacterial chemoheterotrophs at least since ~0.7 Ga ago.

Evidence we regard as presumptive of the existence of anaerobic chemoheterotrophs extends into the Late Archean: (*i*) As is discussed in Chapter 7 (Section 7-3-3), $\delta^{34}S$ values suggestive of the presence of sulfate-reducing (anaerobic chemoheterotrophic) bacteria can be traced to ~2.7 Ga ago. (*ii*) Pyrite-rich laminae occurring in stromatolites of the ~2.8 Ga-old-Fortescue Group of Western Australia similarly may reflect the presence of such bacteria. And (*iii*) as is documented in Chapter 8 (Section 8-5-3), fenestrate sedimentary structures are known earliest in the geologic record from sediments ~2.6 and ~2.8 Ga old; these millimetric to centimetric mineral-infilled voids appear to have been produced originally within cohesive stromatolitic laminae via the microbial decomposition of organic matter, either by anaerobic or by aerobic chemoheterotrophs.

An additional line of evidence, perhaps less firm and certainly less direct than the foregoing, comes from studies of the total organic carbon (T.O.C.) content and the carbon isotopic composition of ancient sediments. As has been recently documented in some detail by Schidlowski (1981a; see also Table 5-7 herein), the average total organic carbon content of the major sedimentary lithotypes (viz., relatively unmetamorphosed shales, carbonates, and sandstones) appears to have varied within relatively narrow limits over geologic time (e.g., analyses of about 2,500 Archean and Proterozoic shales and of some 9,000 Phanerozoic shales indicate that the average T.O.C. content of these sediments has varied between about 0.2% and 1.6% with a mean value of ~0.7%). If it could be assumed that the sedimentation rates of the sampled units have similarly held relatively constant, and that the rates of primary productivity in these facies have not varied appreciably, then it could be inferred that the rates of biologic recycling of carbon must also have not varied greatly over the sampled interval. Since anaerobic chemoheterotrophs can reasonably be expected to have played an important role in such recycling in the primitive, presumably anaerobic environment, their existence since the beginning of the relatively unmetamorphosed sampled record—extending to the ~3.5 Ga old Warrawoona and Onverwacht sediments—might plausibly be inferred. In essence, so the argument goes, if such anaerobic recycling had not occurred then very highly organic-rich

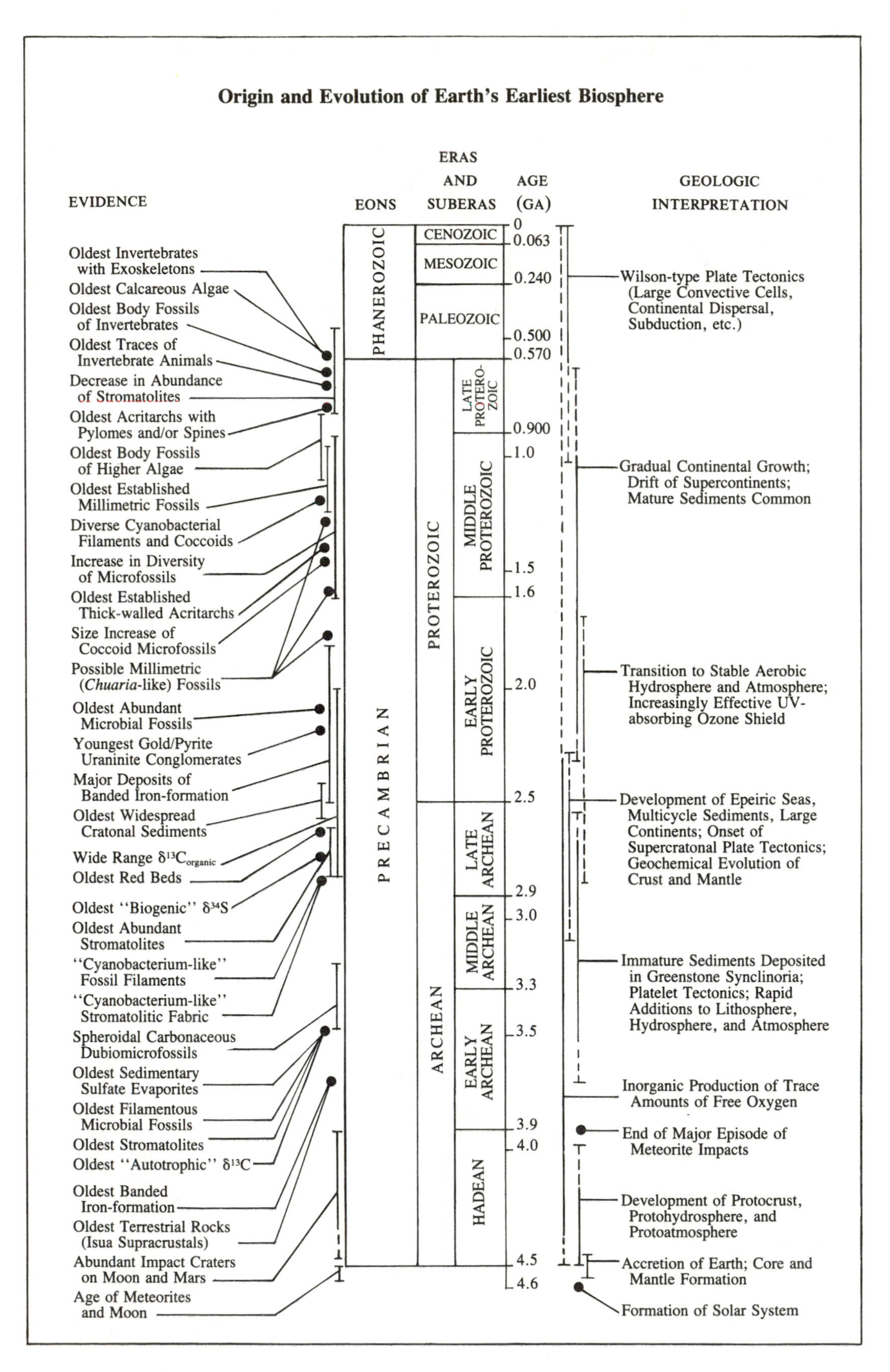

FIGURE 15-1. A geochronologically ordered summary of the major lines of evidence now available relating to the Precambrian history of life and a synopsis of the geologic and biologic interpretations of these data that are here favored.

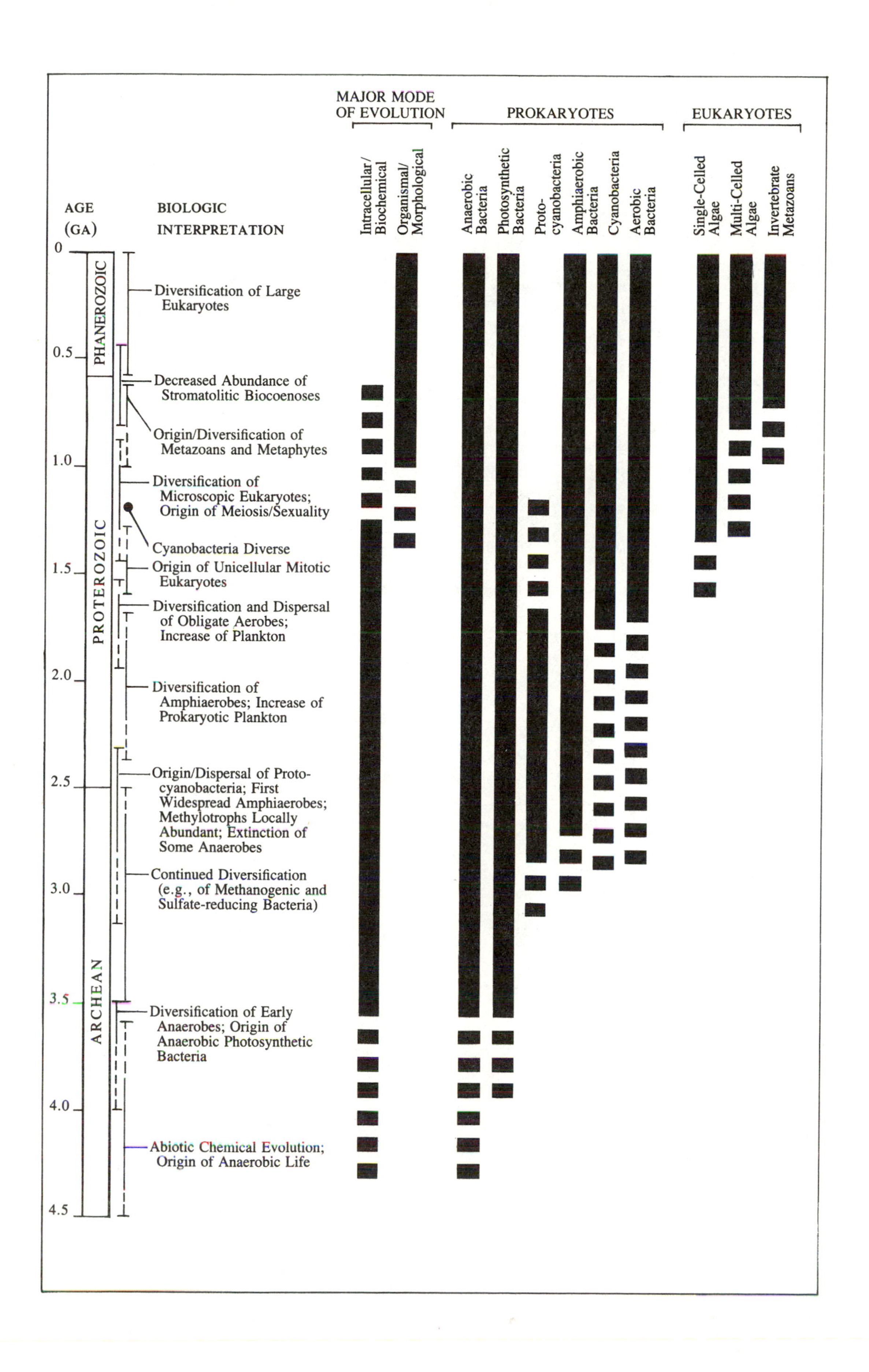

MAJOR MODE OF EVOLUTION
PROKARYOTES
EUKARYOTES
Intracellular/ Biochemical
Organismal/ Morphological
Anaerobic Bacteria
Photosynthetic Bacteria
Proto-cyanobacteria
Amphiaerobic Bacteria
Cyanobacteria
Aerobic Bacteria
Single-Celled Algae
Multi-Celled Algae
Invertebrate Metazoans
AGE (GA)
BIOLOGIC INTERPRETATION
0
0.5
1.0
1.5
2.0
2.5
3.0
3.5
4.0
4.5
PHANEROZOIC
PROTEROZOIC
ARCHEAN
Diversification of Large Eukaryotes
Decreased Abundance of Stromatolitic Biocoenoses
Origin/Diversification of Metazoans and Metaphytes
Diversification of Microscopic Eukaryotes; Origin of Meiosis/Sexuality
Cyanobacteria Diverse
Origin of Unicellular Mitotic Eukaryotes
Diversification and Dispersal of Obligate Aerobes; Increase of Plankton
Diversification of Amphiaerobes; Increase of Prokaryotic Plankton
Origin/Dispersal of Proto-cyanobacteria; First Widespread Amphiaerobes; Methylotrophs Locally Abundant; Extinction of Some Anaerobes
Continued Diversification (e.g., of Methanogenic and Sulfate-reducing Bacteria)
Diversification of Early Anaerobes; Origin of Anaerobic Photosynthetic Bacteria
Abiotic Chemical Evolution; Origin of Anaerobic Life

sediments (e.g., coal seams) would be expected to be markedly abundant in the early geologic record, but this demonstrably is not the case. A postulated early (and continued) existence of such heterotrophs would be consistent also with the relative constancy of the $\delta^{13}C_{PDB}$ values of the bulk of preserved kerogens (viz., -27 ± 7 ‰) and of carbonates (viz., 0.4 ± 2.6 ‰) measured in sediments ~3.5 Ga in age to the present (Figure 7-3), a relationship which can be interpreted as suggesting the occurrence of an autotrophic-heterotrophic biologic carbon cycle since the Early Archean (Van Valen, 1976; Schidlowski et al., 1979).

There are, however, major uncertainties in any such interpretation, either with regard to the global ecosystem (Van Valen, 1976) or to that of a restricted basin (e.g., Reimer et al., 1979), a principal problem being the assumption of relatively uniform rates of primary productivity. As is discussed in Chapter 11 (Section 11-5; see also Garrels et al., 1976, and Holland, 1978), the autotrophic biomass (e.g., of photosynthetic bacteria) on an anaerobic Earth is likely to have been substantially smaller than it is at present. However, this, in turn, would have served to limit the heterotrophic biomass, and aerobic heterotrophs, the major recycling vector in the modern ecosystem, would not have been extant. Thus, on the early Earth it is conceivable that the expected lower rates of primary productivity (due to a smaller autotrophic biomass) may have been offset by correspondingly lower rates of biological recycling, resulting in a more or less constant influx of organic matter into sediments and the observed limited range of average T.O.C. values. Although such arguments seem at least roughly consistent with the evidence now available, data from early sediments are so few and the required assumptions so poorly established that this interpretation can be considered neither compelling nor even presumptive; we thus regard this line of evidence as only permissive.

Nevertheless, it is difficult to conceive of any terrestrial ecosystem, including that of the primitive Earth, that would completely lack anaerobic chemoheterotrophs. The occurrence at any time in the sampled geologic past of a solely autotrophic biota, one in which there was no heterotrophic recycling of biologically fixed carbon whatever, appears to be ruled out by available paleontologic evidence—the fact that the great majority (generally >99%) of kerogen in all known ancient sediments, including those that are profusely microfossiliferous, occurs in a finely divided triturated state (evidently a product of microbial degradation) rather than in structurally preserved organic-walled fossils, and that "perfectly preserved" microorganisms (forms wholly unaltered and demonstrably not biologically decomposed) are unknown even in the earliest deposits (see, for example, Photos 9-7 through 9-10 and Photos 14-1 through 14-11). Further, the possibility that the biological recycling of fixed carbon may have ever been due solely to aerobic chemoheterotrophs (e.g., prior to a postulated later appearance of anaerobes) seems obviated by mineralogic and other geologic evidence apparently indicative of an anaerobic surficial environment during the Archean and early portion of the Proterozoic (Section 11-5-3), a time span from which there is ample evidence that living systems were already extant (Sections 8-4 and 9-5).

These observations, together with numerous biochemical and theoretical considerations apparently indicative of the primitive nature of anaerobic chemoheterotrophic metabolism (e.g., Bernal, 1967; Oparin, 1968; and Sections 6-2, 6-3, 13-1-5 and 13-3-2 herein), seem to us persuasive of the very early appearance of fermenting microbes; in our view, presumptive evidence of their occurrence extends to at least ~3.5 Ga ago.

15-5-3. Anaerobic Chemoautotrophy

Among anaerobic chemoautotrophs are included anaerobes, such as methanogenic bacteria, capable of using CO_2 (fixed by light-independent metabolic processes) as their immediate and major source of cellular carbon. Such forms are evidently quite primitive, members of a very early evolving lineage (see, for example, Woese and Fox, 1977; Fox et al., 1980; and Chapter 12, herein). Moreover, the presence of sedimentary carbonates in the Isua Supracrustal Belt, for example, indicates that the CO_2 metabolically required by such organisms has been readily available since early in Earth history. Nevertheless, fossil evidence of their past existence is notably sparse, morphological remnants of such microbes having to our knowledge never been identified in the fossil record.

As is noted in Chapters 7 and 12, the methane produced by methanogenic anaerobic chemoautotrophs is isotopically distinctive (with $\delta^{13}C_{PDB}$ values generally lower than -55 ‰), a fact suggesting at least one means of recognizing evidence of such organisms in geologic materials. Indeed, apparently the only geologic evidence thus far interpreted in terms of anaerobic chemoautotrophy has been the occurrence of anomalously light (^{13}C-poor) carbon isotopic values recorded for the non-volatile, kerogenous components of one Phanerozoic (Kaplan and Nissenbaum, 1966) and now numerous Late Archean sediments (e.g., one graphite sample from the ~2.75-Ga-old Abitibi-Wawa igneous belt of Canada having a $\delta^{13}C_{PDB}$ value < -40 ‰ reported by Schoell and Wellmer, 1981, and a total of 29 kerogens in this range reported in Tables 5-7 and 5-8 with 3 samples from the ~2.5-Ga-old Hamersley Group of Australia, 5 samples from the ~2.64-Ga-old Ventersdorp Group of South Africa, and 21 samples from the ~2.77-Ga-old Fortescue Group of Australia). The interpretation of these values is based on the observation that the most plausible source of such isotopically distinctive kerogen is the methane produced by methanogenic bacteria (for discussion, see Chapter 12).

Clearly, this interpretation is not "compelling"; relevant data are by no means plentiful and, as emphasized elsewhere in this volume (viz., Chapter 12), other interpretations may develop as a wider range of information becomes available. Based on the discussion presented in Chapter 12, however, we do regard the evidence now available for the Archean presence of such microbes as certainly "permissive" and, we think, "presumptive." Other interpretations need to be investigated, but in view of the known paleobiologic data it seems to us most likely that anaerobic chemoautotrophs (viz., methanogenic bacteria) were extant at least as early as ~2.8 Ga ago, with neobiological considerations suggesting their even earlier origin.

15-5-4. Anaerobic Photoautotrophy

Anaerobic photoautotrophs are those prokaryotes capable of anoxygenic, but not oxygen-producing, photosynthesis—microbes such as extant photosynthetic bacteria that during photosynthetic growth can use H_2, H_2S, or various organic substrates (but not H_2O) as a source of electrons for the light-driven reduction of CO_2 to produce cellular organic compounds and the concomitant release (via photophosphorylation) of cellularly usable chemical energy. While photosynthetic bacteria are ubiquitous components of numerous, widespread, near-surface aqueous environments in the modern ecosystem (see, for example, Pfennig, 1978), their fossil record is essentially unknown; few if any such organisms have been positively identified in geologic materials.

To exist, photosynthetic bacteria need only light, CO_2 (and in some cases an organic carbon source), a source of reducing power (e.g., H_2S), and a suitably habitable aqueous environment containing various trace elements; they are especially successful in ecologic niches that are characterized by the microbial breakdown of organic matter under conditions of oxygen limitation or anaerobiosis (Pfennig, 1978, p. 12). All of these conditions have been abundantly present since the beginning of the Early Archean (and probably well into the Hadean as well). Based on a diverse array of biochemical and physiological considerations (Sections 6-8, 13-2-1 and 13-5), it seems evident that such organisms were the evolutionary precursors of later evolving oxygenic photosynthesizers (viz., oxygen-producing cyanobacteria). Thus, it is a reasonable postulate that photosynthetic bacteria must have been extant prior to the biologically mediated development of an oxygenic environment (i.e., prior to the Early Proterozoic). Indeed, although the evidence in support of this proposition may not be compelling, it clearly is at least permissive: (*i*) the mode of occurrence and cellular morphology of several types of filamentous microfossils reported from Archean terranes are comparable to those of extant photosynthetic bacteria (and to those of various cyanobacteria as well; Section 9-5 and Table 9-2); (*ii*) the form, distribution and microstructural fabric of known Archean stromatolites are consistent with a photosynthetic bacterial origin (although they are similarly permissive of a cyanobacterial origin; Section 8-5); and (*iii*) the arguments summarized above (Section 15-5-2) with regard to the recycling of biologically fixed carbon in early ecosystems are not inconsistent with the occurrence of autotrophs (whether chemoautotrophic or photoautotrophic) throughout much of the Archean.

The relevant paleobiologic data are thus consistent with the occurrence of photosynthetic bacteria, cyanobacteria, or both of these groups, at least as early as ~3.5 Ga ago. But neobiological considerations suggest that if the relatively more advanced cyanobacteria were already extant at that time, then the evolutionary history of photosynthetic bacteria must extend even farther into the geologic past. Although not compelling, we regard these observations as constituting presumptive evidence for the existence of anaerobic photoautotrophy at least as early as ~3.5 Ga ago.

15-5-5. Aerobic Photoautotrophy

Aerobic photoautotrophs are those chlorophyll-*a*-containing photosynthetic prokaryotes (viz., cyanobacteria and prochlorophytes), protists, and plants (algae, bryophytes, and tracheophytes) that are capable of using light as their energy source, CO_2 as their immediate (and major) carbon source, and water as the source of electrons necessary for the reduction of CO_2 to form cellular organic matter. In such organisms, free molecular oxygen is a byproduct of photosynthetic activity; all such organisms must therefore be aerotolerant (i.e., as discussed in Section 13-2, they must possess intracellular biochemical systems capable of effectively scavenging O_2 or its reactive derivatives with appropriate reductants).

As evidenced by the well-documented record of Paleozoic plant fossils, there can be no doubt that aerobic photoautotrophs had become established at least as early as the end of the Silurian, ~0.4 Ga ago. However, as is summarized in Figure 15-1, additional compelling evidence for the existence of this level of organization extends well into the Proterozoic: the oldest known calcareous algae (~0.6 Ga old), the oldest reported acritarchs with pylomes and spines (~0.85 Ga old; Vidal, 1981), the oldest detected multicellular algae (viz., *Tawuia*, 0.8 to 1.1 Ga old, described by Hofmann and Aitken, 1979, and, possibly, *Proterotainia*, *Lanceforma*, *Grypania*, and *Helminthoidichinites*?, ~1.3 Ga old, reported by Walter et al., 1976), and the oldest assured megaplanktonic algae now known (viz., *Chuaria*, ~1.0 Ga old; Hofmann, 1977a), were all evidently oxygen-producing photoautotrophs.

Presumptive evidence of the existence of such organisms extends appreciably farther into the geologic past: (*i*) Numerous lines of evidence (e.g., as shown in Figure 15-1, the occurrence of the oldest known thick-walled acritarchs in sediments ~1.4 Ga old, reported by Horodyski and Bloeser, 1978, and by Vidal and Knoll, 1981; the well-documented size increase of coccoid microfossils beginning at ~1.4 to 1.5 Ga ago reported by Timofeev, 1973a, and Schopf, 1977a; the increased diversity of known microfossils beginning ~1.6 Ga ago presented in Schopf, 1978; together with the several other lines of evidence summarized by Schopf and Oehler, 1976, and by Schopf, 1978) can be interpreted as reflecting the presence of aerobic photoautotrophic eukaryotes as early as the beginning of the Middle Proterozoic. (*ii*) *Chuaria*-like fossils, possibly remnants of megaplanktonic oxygen-producing photosynthesizers, have been reported from sediments as old as 1.75 Ga (Hofmann and Chen, 1981; see Figure 15-1 herein). (*iii*) Microbial biocoenoses containing fossil taxa most reasonably interpreted as oxygen-producing cyanobacteria are well known from the Early Proterozoic (see Sections 8-5-4 and 14-4-2). And (*iv*) as noted in Chapter 11 (Section 11-5-2), the inferred mode of formation and rate of accumulation of certain Early Proterozoic banded iron-formations (viz., those of the Hamersley Basin of Western Australia) seem to require the existence of oxygen-producing photoautotrophs at least as early as ~2.5 Ga ago.

Indeed, as is discussed in the foregoing section of the present chapter, it is conceivable that aerobic photoautotrophs may have been extant as early as ~3.5 Ga ago; several types of evidence—among them, available $\delta^{13}C$ and T.O.C. data and the known Early Archean stromatolitic and micropaleontological fossil records—are not inconsistent with this supposition. Nevertheless, there are a number of lines of reasoning that suggest (but do not prove) that aerobic photosynthesizers may have first appeared somewhat later, possibly on the order of ~2.9 Ga ago.

First, as is discussed in Chapter 8 of this volume (Section 8-5-3), the oldest stromatolite evidence that can be regarded as presumptive of the presence of cyanobacteria occurs in sediments of Late Archean age; these are morphologically and microstructurally diverse stromatolites, some closely comparable to modern stromatolites of known cyanobacterial origin and exhibiting fabrics (such as those in stromatolites of the ~2.8-Ga-old Fortescue Group) that are very similar to those now constructed by oscillatoriacean cyanobacteria.

Second, known microfossils from the Late Archean (viz., from the Fortescue Group; Section 9-5-2 and Photos 9-9 and 9-10) bear considerable resemblance to members of well-known genera of extant, oxygen-producing, oscillatoriaceans.

Third, as noted above (Section 15-5-2), sedimentary fenestrae are known earliest in the geologic record from this same suite of ~2.8-Ga-old sediments (Section 8-5-3). Analogy with examples from the Recent suggests that these mineral-infilled voids may be a result of aerobic (rather than of anaerobic) decay; as such, they could be interpreted as evidence of aerobic respiration, of at least local aerobic conditions, and, indirectly, of the probable existence of oxygenic photosynthesizers.

Fourth, as discussed in Chapter 12, carbon isotopic data from sediments ~2.8 to ~2.5 Ga in age can be reasonably interpreted as requiring the presence of methane-utilizing methylotrophs; since virtually all such modern microbes are aerobes, this interpretation (like that of the fenestrae noted above) would seem to imply the occurrence of local aerobic conditions and, plausibly, at least the local presence of aerobic photoautotrophs.

Fifth, as now known, stromatolites appear to have first become widespread and abundant in geologic units ~2.8 Ga in age. Although it is conceivable that this pattern of distribution might simply be a result of the vagaries of preservation of such structures—perhaps compounded by the geologic evolution of the planet and the onset of widespread cratonal sedimentation at about this time (Sections 3-6 and 10-4-2)—it is also possible that their earliest widespread appearance might be correlative with, and a product of, the origin and early diversification of O_2-producing, stromatolite-building, cyanobacteria (see Schopf, 1975, p. 224).

And sixth, geologic evidence detailed in Chapter 11 (viz., Sections 11-5-3 and Figure 11-12) indicates that a stable aerobic environment probably did not become established on a global scale prior to ~1.7 Ga ago. Certainly, as is there argued, following the origin of oxygenic photosynthesis there must have been a prolonged period of fluctuating imbalance between oxygen sources and sinks prior to the establishment of widespread aerobic conditions. A cardinal question is over how long an interval this transition occurred. Recent calculations, based on consideration of the numerous parameters involved, suggest a probable duration for this transition on the order of "a few, or perhaps several, hundred millions of years" (Schopf, 1980); thus, a Late Archean origin for aerobic photoautotrophy seems decidedly more plausible than a far earlier date during the Early Archean.

In summary, it seems that compelling evidence of oxygenic photosynthesis can be traced to at least ~1.0 Ga ago (e.g., via the fossil record of structurally complex acritarchs, megascopic algae, and megaplanktonic algal protists) and probably earlier (evidenced, for example, by the earliest records of putative eukaryotes and by the evidently cyanobacterium-dominated biocoenoses of the mid-portion of the Early Proterozoic). Presumptive evidence of their existence extends to ~2.8 Ga ago (as based on oxygen fluxes apparently required for deposition of ~2.5-Ga-old banded iron-formations and on stromatolitic, sedimentologic, micropaleontologic and carbon isotopic data from sediments ~2.5 to ~2.8 Ga in age). And some lines of evidence (viz., from microfossils, stromatolites and $\delta^{13}C$ and T.O.C. values) seem permissive of their occurrence as early as ~3.5 Ga ago.

15-5-6. Amphiaerobic Metabolism

Amphiaerobes are those organisms that can use molecular oxygen as the terminal electron acceptor for intracellular energy conversion (viz., via aerobic respiration) but that in the absence of sufficient oxygen concentrations can survive and grow using an alternative, anaerobic, metabolic process. Such organisms thus occupy a "facultative," metabolically intermediate position between strict anaerobes and obligate aerobes. In the present environment they occupy a similarly intermediate position ecologically, being widespread, abundant, and particularly diverse in the transitional (and spatially fluctuating) zone between aerobic and anaerobic conditions that occurs beneath the sediment-water interface in the uppermost few centimeters of the sediment profile. It seems evident that the evolutionary development of this metabolic versatility occurred during the Precambrian, prior to the origin of oxygen-requiring eukaryotic cells (Section 13-1-2), either earlier than or coincident with the transition from essentially anoxic to oxygenic environmental conditions brought about by the origin and dispersal of aerobic photosynthesizers (for discussion, see Schopf, 1978, 1981, and Sections 11-5-3 and 13-1-2 herein).

Presumptive evidence of the occurrence of biologic aerotolerance (viz., of aerobic photoautotrophy; Section 15-5-5) dates back to ~2.8 Ga ago, whereas evidence permissive of the occurrence of amphiaerobiosis (including the presence of conical stromatolites ~3.1 Ga old that may have been constructed by microaerophiles; Section 8-5-2) extends even farther. Unfortunately, however, all such interpretations either are model-dependent, are based on indirect evidence, or are both, for direct evidence that might be used to differentiate unambiguously among anaerobes, amphiaerobes, and aerobes (e.g., differences in chemistry or in cellular morphology) has yet to be discovered in the early fossil record.

In sum, available data seem permissive of the presence of amphiaerobes at, or even before (Section 11-5-3), the time of origin of O_2-producing photoautotrophs, and suggest that such organisms probably first became diverse and abundant during the Late Archean-Early Proterozoic. Based on the data and arguments summarized above (especially in Section 15-5-5), we conclude that amphiaerobic metabolism probably first appeared earlier than ~2.8 Ga ago.

15-5-7. Obligate Aerobiosis

For reasons detailed elsewhere in this volume (viz., Sections 11-5, 13-1-2 and 13-2), it seems unlikely that obligate aerobes—organisms wholly dependent on the use of molecular oxygen as the terminal electron acceptor for energy conversion—could have become abundant and ubiquitous components of the Earth's biota prior to the establishment of a stable, global, oxygenic environment. Such organisms may have evolved far earlier (in principal, nearly as early as their presumed amphiaerobic progenitors), possibly having been protected from asphyxia in mucilage-embedded stromatolitic communities that served as local "oxygen oases" (Fischer, 1965). Yet it is difficult to envision how obligate aerobes could have become dispersed and been sustained on a global scale (e.g., as plankton) were the photic zone of the Earth's oceans to have been anaerobic, even intermittently and for short periods of time. It is thus interesting to note that as currently known, the earliest records of widespread (presumably aerobic) phytoplankton date from the mid-Early Proterozoic (Section 14-4-3), more or less coincident with the earliest geologic evidence of widespread oxygenic conditions (Section 11-5-3).

For reasons similar to those discussed above regarding amphiaerobic metabolism, the time of origin (as opposed to that of the global dispersal) of obligately aerobic microorganisms is uncertain. In general, evidence permissive of amphiaerobiosis is similarly permissive (but because of the sustained oxygen levels required, presumably somewhat less so) of the occurrence of obligate aerobes. At present, perhaps the oldest compelling evidence of such organisms is that cited in Section 15-5-5 for the Proterozoic occurrence of oxygen-producing (and obligately oxygen-using) eukaryotic photoautotrophs. Presumptive evidence from earlier sediments includes the presence of probable cyanobacteria in Early Proterozoic biocoenoses (which, however, might conceivably have been neither oxygen-producing nor oxygen-utilizing, exhibiting a capability for anaerobic growth possessed by many modern strains of cyanobacteria—see, Oren et al., 1977); and Early Proterozoic occurrences of *Eoastrion* spp. (Table 14-3), actinomorphic microfossils that are similar in gross morphology to the enigmatic, but reportedly obligately aerobic iron- and manganese-oxidizing "budding bacteria" of the modern genus *Metallogenium* (Buchanan and Gibbons, 1974, p. 163; Zavarzin, 1981).

15-5-8. Synopsis

From the foregoing analysis—but here focusing on only the relatively more persuasive lines of evidence known to us (viz., those we regard as either "compelling" or "presumptive")—we conclude that (*i*) the origin of life occurred earlier than ~3.5 Ga ago; that (*ii*) anaerobic chemoheterotrophs, (*iii*) anaerobic chemoautotrophs, and (*iv*) anaerobic photoautotrophs were probably extant as early as ~3.5 Ga ago (although as now know, firm evidence of anaerobic chemoautotrophy apparently extends only to ~2.8 Ga ago); that (*v*) aerobic photoautotrophs and (*vi*) amphiaerobes were probably established as early as ~2.8 Ga ago; and that although probably originating earlier than at least ~2.0 Ga ago, (*vii*) obligately aerobic microorganisms first became abundant and globally widespread between ~1.7 and ~1.5 Ga ago.

15-6. A Current "Best Guess Scenario" for the Early History of Life

Figure 15-1 presents a geochronologically ordered summary of the principal lines of geological, organic geochemical, and paleontological evidence now available relevant to the Precambrian history of life and a synopsis of the geologic and biologic interpretations of these data that are here favored. Based on this summary and synopsis, and on assessment of the probable time of occurrence of the seven major metabolic benchmarks discussed above, in this section we present our version of a "best guess scenario" for the sequential, four-stage development of the Earth's earliest ecosystems. In doing so, we claim no omniscience, and certainly no great degree of prescience; numerous aspects of that which is now "known" are likely to be modified and amended, possibly to a major extent, as understanding of these problems continues to advance. Indeed, we claim only that what follows is our current best guess.

15-6-1. Stage I. The Primitive Anaerobic Ecosystem

The Earth has been a habitable planet since at least ~3.8 Ga ago, and was probably habitable much earlier. Just how much earlier can only be speculated, for direct evidence of the preceding (Hadean) development of the planet has yet to be discovered; but the geology, mineralogy, and geochemistry of the oldest sediments now known, those of the ~3.8-Ga-old Isua Supracrustal Belt of western Greenland, establish beyond doubt that at least small granitic (i.e., "continental") land masses, a liquid water ocean, an atmosphere containing CO_2 (possibly at relatively high concentrations, but presumably composed largely of N_2), and the processes involved in erosion, transport, deposition, and formation of both clastic and chemical sediments have been extant at least since that time. Volcanism and related crustal movements (viz., platelet tectonics) were the dominant geologic forces that shaped this early environment, releasing gaseous effluents that fed the Earth's primitive atmosphere (and that partly condensed as rain to erode exposed crustal rocks and to add to the increasing volume of the planet's oceans) and at the same time adding rock material to the growing crustal mass and, thereby, to the forming protocontinents. Land forms were rugged, angular, and relatively immature; the landscape was speckled with volcanic cones, lava flows, and eroded remnants of meteorite craters; sedimentary deposits were generally "first-cycle" and immature, relatively poorly sorted mixes of crustal debris that were subject to geologically rapid recycling; and both the salinity and the pH of the early oceans were not greatly different from those of the present (both being determined largely by inorganic, rather than biologic processes). The Earth's magnetosphere, shading the planet's surface from the effects of the solar wind, had long been established. Day length was evidently shorter than, tidal amplitudes were somewhat higher than, and the average global surface temperature was probably rather comparable to, those of the present and the relatively recent epochs of the geologic past.

The early Earth was by no means a potentially inhospitable planet. Indeed, the only environmental parameter of major biologic import that appears to have differed markedly from its current condition was the availability of free molecular oxygen, the early atmosphere and hydrosphere being anaerobic—in fact, essentially anoxic—with O_2 being produced only via the inorganic (UV-induced) photodissociation of atmospheric water vapor and being consumed virtually as rapidly as it was produced via oxidative reactions with previously unoxidized, volcanically derived, gaseous and mineralic substrates. A biologically protective UV-absorbing ozone layer had yet to be established; life was possible, but probably not in the atmosphere, on exposed land surfaces, nor in the uppermost reaches of the water column.

It is in this primitive environment—a setting distinctive principally because of its anoxia and high UV-flux—that life arose, earlier than 3.5 and quite possibly earlier than 3.8 Ga ago.

The origin of life required the presence of (*i*) potentially interactive precursor molecules, (*ii*) energy to power the numerous reactions involved, and (*iii*) liquid water to serve as a solvent and UV-protective respository for the products formed, as well as the absence of (*iv*) free molecular oxygen that in all but trace, nonequilibrium concentrations would have served to oxidize these products and to thus inhibit the life-generating process. All of these conditions were met on the primitive Earth: (*i*) carbon-, hydrogen-, nitrogen-, and oxygen-containing precursor molecules (e.g., CO_2, CO, and possibly CH_4, H_2O and H_2, N_2, and possibly NH_3), derived principally via the outgassing of the interior of the formative planet but altered by photochemical and other reactions in the early environment, were abundant, both in the atmosphere and hydrosphere of the early Earth; (*ii*) solar-derived energy, important especially in the ultraviolet portion of the spectrum but including also visible light and such energy sources as atmospheric lightning and cavitation (occurring in the high pressure, low temperature microenvironments formed in gas bubbles at the frontal edge of breaking wave crests in the early seas), was in plentiful supply; (*iii*) liquid water, the condensed product of volcanism and geologic outgassing, was similarly abundant having begun to accumulate as a surficial planetary veneer as soon as crustal cooling had permitted ambient temperatures to drop beneath its boiling point; and (*iv*) in the absence of life (viz., of oxygenic photosynthesizers), environmental concentrations of free oxygen were vanishingly low, produced slowly by photochemical reactions and scavenged rapidly by available oxidizable substrates.

In this milieu, a series of relatively simple chemical reactions, involving first the two highly reactive intermediate compounds formaldehyde (an intermediate in the synthesis of sugars and of amino acids) and hydrogen cyanide (involved in the production of amino acids and such nitrogen-containing heterocyclics as the purines and pyrimidines of nucleic acids), gave rise to a vast array of small monomeric organic compounds. From these monomers via dehydration-condensation reactions (perhaps catalyzed by mineralic substrates or via interaction with such abiotically produced organic condensing agents as cyanimide or cyanoacetylene) were derived oligomers and, subsequently, polymers of progressively increasing length. Aggregation of these polymers, a result of physical and chemical interactions and of the inherent hydrophobic nature of many of their constituents, led to formation of phase-separated systems, microscopic droplet-like bodies capable of chemical exchange with their surroundings and within which, over time, increasing physico-chemical interaction gave rise to increasing physico-chemical complexity. Ultimately, via such processes, chemical coupling developed within such bodies between nucleic acid and amino acid polymers (the latter serving as primitive organic catalysts that facilitated the production of the former); the first vestiges of the genetic code became established; inflow and assimilation of exogenous organic compounds, and outflow of the residual byproducts, served as an early form of energy- and carbon-yielding metabolism; and peripheral, membranous, "wall-like" structures were developed, enveloping investments that were of increasing effectiveness in mediating the flow of material into and out of these now decidedly cell-like bodies. At some stage in this long chain of events—events that took place sporadically in various chemically and physically differing microenvironments within the primordial milieu—the threshold was bridged between nonliving and living systems: the Earth's earliest ecosystem became established.

What were the components of this ecosystem and how did it operate? Physiologically the earliest organisms were strict anaerobes, biochemically incapable of coping with greater than trace amounts of free molecular oxygen; metabolically they were chemoheterotrophs, relying on abiotically produced foodstuffs for their source of carbon and energy; ecologically they inhabited an aqueous and presumably benthonic environment, the muds and silts of bottom sediments beneath wave base where organic detritus could accumulate and where the overlying water column served to protect them from the deleterious effects of UV-irradiation; and phylogenetically they were prokaryotes, morphologically simple, undifferentiated, nonmotile, noncolonial and presumably spheroidal microorganisms that if alive today would probably be classed among the anaerobic, fermentative, coccoid bacteria. Such systems may have originated on multiple occasions. There may have been gene flow and/or a primitive form of symbiosis among the early biotic components. There probably developed intense competition for available foodstuffs among the coexisting primitive taxa (leading to rapid refinement of metabolic and reproductive capabilities and a concomitant increase in biologic diversity). And such forms may have been ecologically stratified or otherwise organized at the sediment-water interface, distributed as a result of their metabolic interactions and their ability to metabolize chemically differing abiotic substrates or the byproducts of other members of the biota. Nevertheless, all such forms shared the common traits of anaerobiosis, fermentative chemoheterotrophy, and the ecologic role of being consumers. Indeed, the most striking aspect of this earliest ecosystem was the complete absence of biological primary producers, a role in the food chain played wholly by the abiological fixation of carbon into metabolically usable organic compounds. In this respect, this primitive anaerobic ecosystem thus differed from all ecosystems that were to evolve subsequently in the history of the planet. Biomass was probably "carbon-limited" (i.e., constrained by the availability of potentially usable organic substrates rather than by available nitrogen and phosphorus as it is at present) with primary production via abiotic means likely to have been substantially less efficient than later evolving biologic autotrophy. Yet as long as the conditions required for abiotic organic syntheses continued to be met, a constant "rain" (or, more likely, a continuous drizzle) of foodstuffs would have been provided to the biota and at least a small-scale, steady-state ecosystem (with carbon inflow from abiotic sources equaling carbon outflow due to sedimentation and burial of the remnants of dead organisms) could have been sustained, essentially indefinitely.

Evidence of this ecosystem, not yet identified in the fossil record, can be expected to (*i*) occur in rocks older than ~3.5 Ga and to be characterized by kerogens (*ii*) consisting of a mix of biotic (remnants of chemoheterotrophs) and abiotic (metabolically unutilizable) organic matter, and (*iii*) having elemental and isotopic compositions consistent with their primary abiotic derivation; (*iv*) preserved microorganisms should be morphologically simple and non-planktonic; and (*v*) both biomass (possibly reflected, for example, by globally averaged T.O.C. values) and biotic diversity should be low relative to later evolving ecologic assemblages.

15-6-2. Stage II. The Advanced Anaerobic Ecosystem

From this primary biota evolved a secondary ecosystem, one still strictly anaerobic (and thus decidedly less advanced than later evolving assemblages), but at the same time more diverse and complex than its primitive precursor. This secondary ecosystem was characterized by the presence of the earliest biological primary producers, a stage in the ecological succession evidently attained at least as early as ~3.5 Ga ago.

Environmentally there seems little difference between the world inhabited by this derivative ecosystem and that of its primitive forerunner. But biologically the differences were great. Although anaerobic chemoheterotrophs persisted, an additional layer of ecological organization had been added—autotrophic microbes that were capable of fixing CO_2 into cellularly metabolizable organic compounds, the presence of which thus freed the biota forever from its previous dependence on abiologically ready-made foodstuffs. The earliest such organisms may well have been anaerobic chemoautotrophs, prokaryotes such as methanogenic bacteria that obtained their carbon and energy via reduction of CO_2 to CH_4 (using hydrogen either of volcanic origin or, more likely, generated by associated and similarly primitive, anaerobic hydrogen-producing bacteria). Yet light-utilizing autotrophs, the precursors of modern anaerobic photosynthetic bacteria, were also early represented (presumably derived from metalloporphyrin-containing anaerobic photoheterotrophs, organisms that could use the chemical energy generated by light-driven reactions to photo-assimilate exogenous organic matter). Primary production in this diverse, multicomponent ecosystem was thus carried out by chemoautotrophs, photoautotrophs, and, presumably, by continuing abiotic syntheses, whereas carbon was recycled by various types of chemo- and photoheterotrophic consumers (and by degradation due to photochemical reactions in the early atmosphere).

The earliest evolving photo-responsive microorganisms were no doubt facultative, capable of using light as an energy source when available but capable also of dark metabolism. Because of the relatively greater efficiency of light-driven metabolic processes, there was selective advantage for photosynthetic and photoheterotrophic members of the biota to inhabit the photic zone. However, because of the absence of an effective UV-absorbing ozone layer in the contemporary anaerobic atmosphere, their distribution was probably first restricted to beneath wave base (and they were probably thus best adapted to zones of low light intensity and to use of only the most penetrative portions of the solar spectrum, adaptations that persist in many such microbes to the present). Subsequently, mechanisms for UV-protection evolved permitting habitation of the relatively shallow-water littoral zone, a physically energetic environment where the availability of solar energy had been initially offset by the concomitant high flux of potentially damaging UV-irradiation (and by the susceptibility of microbes in this zone to being swept up by wave action into the upper reaches of the water column where such radiation was particularly intense). Principal among these was the development of effective systems for repair of UV-induced damage to genetic material (for a recent review of these systems, see Webb, 1977), and the development of the biochemical systems required for production of copious amounts of extracellular, polysaccharide mucilage, a viscous, physically resilient material that served to embed and fix in place the photo-responsive microorganisms and to which mineralic and other UV-absorbing, and thus UV-protective material could adhere (for discussion, see Awramik et al., 1976).

Cell division within the early, mucilage-embedded aggregates resulted in the formation of colonies, the earliest of which were probably unordered and irregular in form. Presumably because of intracolony shading effects (lower members of large colonies being thus light-limited by the overlying cell mass in the aggregate), there was selection among these photo-responsive microorganisms for small colony size and for geometrically ordered cell division leading to formation of small ordered aggregates, planar colonies (via cell division in a plane parallel to the habitat surface) and, subsequently, to production of uniseriate, sheath-enclosed, strand-like filaments (via division in one direction only). The benthonic habit of such ensheathed colonies presented an additional problem, namely that they could become buried and thus light-limited by accumulating detritus adhering to the upper surface of the coherent mucilagenous layer that they had constructed. Under these circumstances, phototropism and, subsequently, phototaxis were of selective advantage. And the morphology of the sheath-enclosed strand-like filaments permitted "discovery" of an effective solution to this problem, the development of the capacity for gliding motility (a common mode of locomotion among phototactic filamentous microbes of the modern biota). Thus, this (or some more or less similar) series of developments can be envisioned as having resulted in formation of multicomponent, filament-dominated microbial biocoenoses, biologically stratified as a function of their internal photic gradient—structures preserved in the geologic record as stromatolites and known to have existed as early as $\sim$3.5 Ga ago. Indeed, the geologic setting of these oldest known stromatolites indicates that shallow water environments were being biologically exploited at least this early in the Archean and that by this time mat-forming, apparently filamentous, microorganisms were capable of coping in such environments with the ambient UV-flux and evidently with both high salinity and desiccation.

As a result of the presence of autotrophs, this derivative anaerobic ecosystem would not be expected to have been as carbon-limited as was presumably its precursor, the solely chemoheterotrophic biota. And although initially this ecosystem may have been nitrogen-limited, with metabolically usable nitrogenous compounds being in short supply in the early environment (ammonia being photochemically unstable and short-lived in the presence of a high UV-flux and the photochemical production of nitrate occurring only slowly in the anaerobic atmosphere), this limitation was at least partly offset by biochemical evolution among photosynthetic bacteria, all of which in the modern biota are apparently capable of N_2-fixation under anaerobic conditions. Rather, it seems more likely that although available phosphate may have also played a limiting role, the distribution (and biomass) of this advanced anaerobic ecosystem was perhaps chiefly UV-limited, with exploitation of widespread, and otherwise potentially habitable niches (e.g., exposed land surfaces and, especially, the photic zone of the open ocean) being largely precluded by the intense ambient UV-flux.

In comparison with the chemoheterotrophic ecosystem described in the foregoing section of this chapter, it can be expected that this secondary ecosystem should be reflected in the geologic record by the occurrence of (*i*) kerogens that are largely of biotic (autotrophic) origin or derivation, as evidenced in their isotopic and elemental signatures; (*ii*) multicomponent, stromatolitic biocoenoses dominated by sheath-enclosed, filamentous, mat-forming and phototactic microorganisms; and (*iii*) evidence indicative of a relative, and perhaps rather marked increase in both biomass and biotic diversity. However, because evidence of the postulated earlier, more primitive, anaerobic ecosystem has not been identified in the fossil record, no such comparison can yet be made.

15-6-3. Stage III. The Transitional (Anaerobic-Amphiaerobic) Ecosystem

Geologically, the Late Archean (2.9 to 2.5 Ga ago) appears to have been a time of profound transformation of the surficial planetary environment—a time encompassing the formation of supercontinents by growth and coalescence of their smaller granitic progenitors; the transition from deposition of sequences dominated by volcaniclastics and immature first-cycle sediments to the multicycle, shelf and platform assemblages typical of emergent cratons; of the onset of the large-scale crustal movements of supercratonal plate tectonics; and of the first development of widespread, shallow, epeiric seas. Biologically it was no less significant, and no less transitional, a time characterized by the occurrence of such evolutionary advances as oxygenic photosynthesis, aerobic metabolism, and the development of an amphiaerobic mode of life, critical events in biologic history that set the stage for the subsequent transformation of the Earth's surficial environment from an anaerobic to an aerobic state and for the derivation of a physiologically fully modern anaerobic-aerobic ecosystem. This transitional stage in ecosystem development may have begun earlier than $\sim$2.9 Ga ago, and it seems certain to have extended into the Early Proterozoic, well past the close of the Archean.

During, or perhaps somewhat earlier than the Late Archean, oxygen-producing photoautotrophs—microorganisms that if alive today would be classed among the cyanobacteria—evolved from their more primitive, anaerobic, photoautotrophic progenitors, the ancestors of modern photosynthetic bacteria. Morphologically they were probably indistinguishable from their primitive precursors. But physiologically they differed significantly; their evolutionary derivation involved the development of phycobiliprotein accessory pigments and the establishment of a serial linkage between the two photoacts and electron transport systems of chlorophyll-*a* based photosynthesis, innovations that permitted them to use water as the source of electrons for the reduction of CO_2 and, thus, to produce gaseous molecular oxygen as a byproduct of the photosynthetic process.

Compared to their photoautotrophic bacterial progenitors, the newly evolved proto-cyanobacteria had several advantages. Perhaps most important among these was their capability to biologically, photodissociate H_2O a ubiquitous component of the environment (unlike H_2S and other potential hydrogen-donors), and to thus produce during daylight hours a continuously renewed enshrouding "envelope" of gaseous and dissolved oxygen, a substance toxic to contemporary anaerobes and an inhibitor of anaerobic photosynthesis (with even the biosynthesis of bacteriochlorophyll, the key compound in such anaerobic photoautotrophy, being obligately anaerobic). Like many modern cyanobacteria, these earliest oxygen-producers were probably facultative photosynthesizers, forms that depending on environmental conditions were able to use either water or reduced compounds such as H_2S (like some photosynthetic bacteria) as their source of reducing power. Because (at low light intensities) photosynthetic oxygen production has been shown to have an inhibitory effect on the pathways necessary for aerobiosis in modern cyanobacteria (Carr and Hallaway, 1965; Holm-Hansen, 1968), it is uncertain whether they were capable of aerobic respiration. At night, when respiration might be expected to have occurred, local oxygen concentrations probably would have decreased rapidly to a minimal level as oxygen escaped into the surrounding anaerobic environment and was consumed in the oxidation of various inorganic substrates (e.g., ferrous iron and sulfides). Thus, at night, these primitive cyanobacteria may have reverted to anaerobic heterotrophy, a capability exhibited by certain modern members of the group (Holm-Hansen, 1968; Fay, 1965); they may have generated chemical energy from the breakdown of photosynthetically manufactured cellular components; or they may have respired O_2 produced during daylight via oxygenic photosynthesis and trapped in their mucilagenous, benthonic, mat-like communities.

At least three additional and particularly important aspects of the physiology and ecology of these early cyanobacteria seem well established: (*i*) As oxygen-producers, they must have possessed the intracellular biochemical systems necessary for protection of oxygen-labile enzymes and other biochemicals from the destructive effects of molecular oxygen and its highly reactive derivatives. (*ii*) Like their progenitors and their modern descendants, they almost certainly were relatively resistant to UV-irradiation (Gerloff et al., 1950), possessed effective genetic repair mechanisms (Witkin, 1966), and exhibited structural (e.g., mucilagenous investments) and behavioral adaptations (phototaxis, clumping, mat-building, etc.) that enabled them to cope with the relatively high UV-flux at the sediment-water interface in shallow littoral environments. (*iii*) Finally, because of their metabolic versatility (viz., their facultative photosynthesis), the fact that they had the capability to release uncombined O_2 as a photosynthetic byproduct (a material toxic to contemporary anaerobes) and, possibly, because they and photosynthetic bacteria absorb light in complementary regions of the visible and near infrared portions of the solar spectrum (although this may have been a secondary, later, adaptation), they were presumably highly successful competitors for available photosynthetic space in benthonic habitats. The development of these organisms thus resulted in a more highly stratified biocoenose than that of the forerunning ecosystem, one layered not only in response to a gradient in the availability of light but also, and even more importantly, to an intrabiocoenose gradient in molecular oxygen concentration.

The earliest mechanisms for intracellular oxygen-protection may have developed prior to the advent of aerobic photosynthesis—a pre-adaptation perhaps brought about as a response to free oxygen produced solely via inorganic photodissociation of water vapor in the early environment. If so, aerobiosis may have similarly evolved at some earlier time. Yet it seems evident that concomitant with the development of biological oxygen production there must have been considerable selective advantage for the development of additional such mechanisms, and that aerobiosis, even operating in an amphiaerobic mode, could not have become widespread prior to the origin and dispersal of oxygenic photosynthesizers. Thus, in general, it appears that the sequential order of development of such biochemical processes was probably for O_2-protection, O_2-production, then O_2-use, first in metabolism and, later, in biosyntheses, the latter two processes being prime examples of the addition of aerobic pathways onto previously established, solely anaerobic systems.

The earliest aerobic respirers, like the earliest O_2-producers, were no doubt facultative; that is, they were amphiaerobes, capable of O_2-utilizing metabolism when ambient oxygen concentrations attained a sufficient level (thereby taking advantage of its highly effective ATP-generating potential), but capable also of reverting to more primitive (and less efficient) anaerobic energy-yielding processes when oxygen concentrations dropped beneath some minimum level. Clearly, amphiaerobiosis is an adaptation to fluctuating pO_2, with such fluctuations in the Late Archean environment presumably being a result of a combination of two principal causes: (*i*) day-night diurnal cycles, oxygen being produced and locally accumulating during daylight hours as a result of photosynthesis, but dropping to low levels during the night; and (*ii*) geologically induced fluctuations with the ambient level of pO_2 decreasing as a result of episodic influxes of potentially oxidizable substrates into the environment (due, for example, to sporadic local volcanism or to seasonal upwelling into the photic zone of reduced substrates, such as ferrous iron, from the deep sea).

Interestingly, many extant cyanobacteria are best adapted to relatively low oxygen tensions even today—living, for example, in thick mats, hot springs (where the concentration of dissolved oxygen is lowered at the relatively high temperatures) and in thick, oxygen-deficient blooms—with both nitrogen-fixation and photosynthetic carbon-fixation being inhibited in such organisms by high concentrations of molecular oxygen (Fogg et al., 1973, p. 265). Similarly, modern members of the group are particularly well adapted to habitats of low light intensity (e.g., caves, mats, water blooms, and the deeper portions of the photic zone of lakes and seas), in planktonic forms with population maxima tending "to occur towards the bottom of the photic zone, where light intensity is around one percent of that at the surface," under which conditions photoassimilation of exogenous organic matter may occur, the advantage of this process "in dimly lit situations [being] that by starting with an already reduced source of carbon, the limited assimilatory power from the photochemical reactions may be used to provide a much greater amount of growth than would be possible with carbon dioxide as the carbon source" (Fogg et al., 1973, p. 264). These two interrelated adaptations—to low oxygen tensions and to low light intensities—may both be "evolutionary relics," adaptations that first evolved as means enabling cyanobacteria to survive, and flourish, in habitats where they would be protected from UV-induced genetic damage and from the effects of intracellular photo-oxidation (e.g., of chlorophyll-*a*) prior to, and during, the buildup of environmental oxygen concentrations.

The ecosystem of this transitional stage in biotic development was thus more complex, and decidedly more diverse, than its evolutionary precursors. Primary producers included facultatively oxygen-producing photoautotrophs (cyanobacteria), anaerobic photoautotrophs (photosynthetic bacteria), anaerobic chemoautotrophs (e.g., methanogenic bacteria), and, possibly, anaerobic abiotic syntheses (in times or places where such syntheses were not inhibited by reaction with free molecular oxygen). Biological recycling of fixed carbon was accomplished by the metabolic processes of photoheterotrophy, anaerobic chemoheterotrophy (including that by sulfate-reducing bacteria), and by aerobic chemoheterotrophy (carried out by amphiaerobes and, possibly, by strict aerobes in localized "oxygen oases"). Most such organisms inhabited layered stromatolitic biocoenoses at the sediment-water interface, stratified as a function both of the availability of light and of the internal variation of pO_2 (with aerotolerant microbes, including photo-responsive forms, localized in the upper reaches of such communities; anaerobic light-using prokaryotes occurring at slightly lower levels; and light-independent anaerobes concentrated at progressively greater depths). Indeed, such stromatolitic biocoenoses appear to have first become widespread and abundant during this transitional phase of ecosystem development, perhaps partly because of the ecologic advantages conferred by the newly evolved process of oxygen-producing photosynthesis and perhaps also because of the coincident geological evolution of the planet and the new development of widespread, shallow water, cratonal environments that were potentially habitable by such communities.

This was a time characterized by fluctuating pO_2, both locally, as a result of diurnal variation in biological oxygen production, and regionally, as a result of the episodic influx of oxygen-scavenging mineralic and dissolved gaseous substrates. Globally, this was thus a time of prolonged, fluctuating imbalance between sources and sinks of molecular oxygen. However, as the oxygen-producing biomass gradually increased (permitted by the increasing efficiency of carbon and nutrient recycling via oxygen-using heterotrophs and presumably reflected by an increase during this period in the abundance of stromatolitic deposits), and as the available reservoir of potentially oxidizable substrates became increasingly depleted (as evidenced, for example, by the large-scale deposition of Late Archean-Early Proterozoic banded iron-formations), environmental oxygen levels slowly, sporadically, but inexorably began to rise. Concomitant with this rise was a gradual buildup of the UV-absorbing, biologically protective ozone layer in the upper atmosphere, the development of which had at least two important biological consequences: (*i*) It permitted increased occupation of the lower reaches of the photic zone of open oceanic waters by planktonic prokaryotes (a niche habitable by cyanobacteria because of their adaptation to low light intensity and their ability under such conditions to produce intracellular gas vacuoles, structures that enabled them to control buoyancy and, thus, to float beneath wave base, avoiding UV-exposure within the turbulent upper reaches of the water column). Perhaps even more importantly, (*ii*) it played a significant role in facilitating dispersal of benthonic microorganisms and, thus, the global spread of stromatolitic biocoenoses into numerous previously unoccupied locales. Moreover, coupled with an increase in the efficiency of carbon and nutrient recycling via aerobiosis during this period, both of these consequences would have combined to produce a feedback effect, the increasing dispersal of cyanobacteria resulting in an increase in photosynthetic biomass; this increased biomass resulting in an increase in the sedimentation and burial of carbon derived from oxygen-producing photosynthesizers and a concomitant increase in environmental oxygen levels; the increased oxygen levels leading to an increase in atmospheric ozone concentrations; and the increase in the ozone shield facilitating further biotic dispersal.

It seems probable that like the photosynthetic biomass, the total global biomass would have increased during this period, no longer being principally carbon-limited (as in Stage I), and increasingly becoming less UV-limited (than was the ecosystem of Stage II); rather, as the transition developed, it seems likely that biomass probably became progressively more nutrient-limited (e.g., by available, biologically usable nitrogen, and especially phosphorus), as is the modern biota. Anaerobic fixation of atmospheric N_2 continued to be carried out by photosynthetic bacteria, now joined by various types of cyanobacteria. However, the nitrogenase enzyme system involved in such fixation is highly oxygen-labile; thus, as environmental oxygen levels increased, biological nitrogen-fixation may have initially decreased, a development, however, that was probably soon offset by the evolution of heterocysts (specialized cells that permit N_2-fixation under aerobic conditions) in filamentous cyanobacteria and by the increasing production of biologically usable nitrate by inorganic reactions in the evolving atmosphere. The development of heterocysts in filamentous cyanobacteria indicates that available nitrogen was probably an important ecologically limiting factor, at least locally, but it seems likely that during the development of this transitional ecosystem, the availability of phosphorus became of comparable, and probably of even greater importance.

Throughout this transition the photosynthetic biomass, and primary productivity, apparently increased. This was possible chiefly because the recycling rate of organic carbon would have increased as well (because of the increased efficiency of this process due to the presence of amphiaerobic and, possibly, obligately aerobic consumers). Indeed, the increase in photosynthetic biomass and of recycling rates were probably tightly coupled, the former being permitted by the increasing availability of required nutrients due to the increased efficiency of the latter. Thus, the global, average, T.O.C. values in sediments of this age may have remained more or less constant in comparison with those reflecting the presence of the previously wholly anaerobic ecosystem of Stage II (see discussion in Chapter 12). In comparison with the succeeding aerobe-dominated ecosystem discussed below, however, the $\delta^{13}C$ values of organic debris derived from this transitional anaerobic-amphiaerobic ecosystem may have been distinctive and relatively variable, the prevalence of methanogenic anaerobes and microaerophilic methylotrophs, for example, being reflected by the local incorporation of unusually ^{13}C-depleted kerogens into sediments of this age, kerogens of a type only rarely encountered in younger deposits because as environmental pO_2 levels subsequently increased, the biogenic methane giving rise to such kerogens became more effectively oxidized inorganically precluding its widespread biologic fixation and subsequent burial.

15-6-4. Stage IV. Onset of the "Modern" (Anaerobic-Amphiaerobic-Aerobic) Ecosystem

By the middle to late portion of the Early Proterozoic (i.e., ~2.1 to 1.6 Ga ago), the changes set in motion during the Late Archean led to emergence of the final stage of ecologic development here considered with the diversification and dispersal of strict aerobes and the establishment of a physiologically modern ecosystem.

As potential oxygen sinks became increasingly saturated, the fluctuating imbalance in environmental oxygen levels that characterized the preceding stage of ecologic development gave way to increasingly stable aerobic conditions. Oxygen concentrations, in both the atmosphere and hydrosphere, continued their inexorable rise; a biologically effective UV-protective ozone shield became established; and land surfaces and the photic zone of open oceanic environments became increasingly habitable by photoautotrophs (and accompanying heterotrophs). Strict aerobes diversified to become widespread and abundant, ultimately becoming the dominant components of the global ecosystem.

Numerous other ramifications of this final phase of ecologic development were equally significant: (*i*) Environmental pO_2 came principally under biological control—a result of interaction between biological sources (aerobic photosynthesis) and sinks (aerobic respiration)—rather than being determined largely by inorganic processes. (*ii*) The geochemical redox cycles of numerous elements became increasingly biologically mediated. (*iii*) As atmospheric oxygen levels increased, chemical reactions in the atmosphere added increasing quantities of nitrate to the environment, obviating the need for the metabolically expensive process of nitrogen-fixation in later evolving organisms (e.g., eukaryotes) and consolidating still further the relative importance of available phosphate as the ultimate ecologically limiting factor. And (*iv*) the establishment of a stable, sustained, relatively high level of environmental oxygen precluded the need for anaerobic capability (e.g., amphiaerobiosis) in most subsequently evolving forms of life.

It seems possible that the metabolic (and phylogenetic) diversification brought about by the advent of stable oxygenic conditions could have resulted in an increase in the efficiency of carbon and nutrient recycling and a concomitant increase in global biomass. This, however, need not necessarily have been the case for if phosphate (or some other similar nutrient) were already limiting, then this development may have simply resulted in a redistribution of biomass, for example from planktonic anaerobic environments (beneath the photic zone) or from benthonic anaerobic or amphiaerobic niches, into fully aerobic environments, with the Earth's total biomass remaining relatively unchanged. In any case, there is no doubt that the global establishment of aerobic conditions was a signal event in the evolution of the planet, one that played a major role in permitting the later development (apparently early in the Middle Proterozoic) of strictly aerobic eukaryotic cells and that thus paved the way for the advent of a new phase in biologic evolution, a phase based sequentially on exploitation of the refined reproductive capabilities of eukaryotic mitosis, of meiotic sexuality, and then of multicellularity and the development of tissues, organs, and of innovative morphologies and body plans among megascopic forms of life. The early, metabolically and biochemically based successional development of the Earth's ecosystems had been completed, supplanted during the late Precambrian by a new, successful, morphologically based mode of evolutionary advance.

15-7. Summary and Conclusions

(1) Understanding of the sequential development of the Earth's earliest ecosystems is markedly incomplete, hampered chiefly by (*i*) a paucity of preserved, ancient sedimentary sequences and, thus, of direct, geologically based, relevant data; (*ii*) the predominantly biochemical, intracellular nature of the early evolutionary process and the difficulties inherent in the preservation of evidence of such a process in the fossil record; and (*iii*) the newness of studies in this problem area and the resulting absence of an extensive base of data and experience on which reliable interpretations can be firmly grounded.

(2) Bearing such limitations in mind, it seems evident that questions regarding the earliest phases of ecologic development must be phrased biochemically and metabolically (rather than solely in terms of organismal morphology); that their answers should be based on consideration of the totality of relevant chemical, biological, and geological evidence now available (rather than on data from any single discipline alone); and that at present, any fruitful approach to such questions must, of necessity, be largely model-dependent (determined to a substantial degree by theoretical and indirect considerations, rather than solely by directly observable features of the fossil record as now known).

(3) Using this approach, we can identify seven metabolic benchmarks as having played crucial roles in early ecologic advance. Available data relevant to these events, however, are highly varied in both quantity and quality. To specify the degree of confidence with which these data (and their derivative interpretations) can be regarded, a hierarchy of four necessarily subjective categories of evidence is here recognized, in order of decreasing degree of certainty: (*i*) abundant, compelling evidence, consideration of which permits only one reasonable interpretation; (*ii*) presumptive evidence, the preponderance of which suggests a most likely interpretation but for which less probable hypotheses also merit consideration; (*iii*) permissive evidence, data of a type that are consistent with at least two more or less equally tenable competing interpretations; and (*iv*) missing evidence, a category relating to those aspects of early evolution for which no direct, interpretable evidence has apparently yet been uncovered.

(4) Evaluation of the various lines of evidence now available relating to the time of occurrence of early metabolic benchmarks indicates that (*i*) the origin of life occurred earlier than ~3.5 Ga ago; that (*ii*) anaerobic chemoheterotrophs, (*iii*) anaerobic chemoautotrophs, and (*iv*) anaerobic photoautotrophs (and photoheterotrophs) were probably extant as early as ~3.5 Ga ago (although as now known, firm evidence of anaerobic chemoautotrophy apparently extends only to ~2.8 Ga ago); that (*v*) aerobic photoautotrophs and (*vi*) amphiaerobes were evidently extant as early as ~2.8 Ga ago; and that (*vii*) although probably originating prior to ~2.0 Ga ago, obligately aerobic microorganisms first became abundant and globally widespread between ~1.7 and ~1.5 Ga ago.

(5) Based on this assessment of the time of appearance of these seven metabolic benchmarks, and on consideration of a summary and synopsis of the relevant geological, organic geochemical, and paleontological evidence now known (Figure 15-1), a four-stage "best guess scenario" is here suggested for the sequential evolution of the Earth's earliest ecosystems.

(6) Stage I, that of a primitive anaerobic ecosystem, is envisioned as having followed immediately the origin of living systems and as having been composed of anaerobic, prokaryotic, fermenting heterotrophs with the role of primary producer being performed solely by abiotic organic syntheses; because of the relative inefficiency of such syntheses in producing metabolizable organic substrates, the biomass of this primary ecosystem was probably "carbon-limited" (i.e., limited by the availability of usable organic substrates).

(7) Stage II, that of an advanced anaerobic ecosystem, is regarded as having commenced with the origin of the earliest autotrophs, microbes capable of biologically fixing carbon dioxide, the advent of which forever freed the Earth's biota from its previous dependence on abiotically ready-made foodstuffs. Primary production in this multicomponent, derivative ecosystem is envisioned as having been carried out by chemoautotrophs, anaerobic photoautotrophs, and continuing abiotic syntheses, with carbon being recycled by various types of chemo- and photoheterotrophic consumers. The establishment of this ecosystem was probably coincident with the first appearance of microbially laminated stromatolites, biocoenoses stratified as a function of their internal photic

gradient, and with an increase both in global biomass and in biologic diversity. Unlike its presumably carbon-limited predecessor, the distribution (and biomass) of this ecosystem may have been limited chiefly by the availability of habitable, UV-protected, environmental niches.

(8) Stage III, that of a transitional (anaerobic-amphiaerobic) ecosystem, is viewed as having been initiated toward the beginning of the Late Archean by evolution of the capabilities of facultative (i.e., oxygenic-anoxygenic) photosynthesis and respiration. As a result of their metabolic and ecologic versatility and their capacity to produce molecular oxygen as a photosynthetic byproduct (a substance toxic to contemporary anaerobes), the earliest, cyanobacterium-like, oxygen-producing photosynthesizers were presumably highly successful competitors for available photosynthetic space. They thus probably spread rapidly, serving as the surficial components of increasingly widespread, multicomponent, stromatolitic biocoenoses that were biologically stratified in response to internal gradients both of available light and of free molecular oxygen. Coupled with an increase in the efficiency of carbon and nutrient recycling during this period (resulting from the development of facultative aerobiosis), the increasing dispersal of cyanobacteria led to an increase in photosynthetic biomass; this increased biomass (and the concomitant sedimentation and burial of carbon increasingly derived from oxygen-producing photosynthesizers) led to an increase in environmental pO_2; the increased pO_2 levels resulted in an increase in the concentration of atmospheric ozone; and the gradual buildup of this ozone shield facilitated further biotic dispersal into not only near-shore, benthonic environments but also into the lower portions of the open oceanic photic zone. Thus, the global environment during Stage III was characterized by a fluctuating imbalance between oxygen sources and sinks and by a slow, sporadic, but inexorable rise in environmental oxygen concentrations. Although the relatively high UV-flux on exposed land surfaces and penetrating the upper reaches of the water column probably remained an important ecologically limiting factor (and was presumably particularly so during the early stages of development of this ecosystem), as the transition to fully aerobic conditions continued the limited availability of usable nutrients, such as nitrogen and especially phosphorus, became increasingly important.

(9) State IV, that of a "modern" (anaerobic-amphiaerobic-aerobic) ecosystem, is regarded as having begun with the establishment during the mid- to late-Early Proterozoic of a stable oxygenic environment capable of supporting and sustaining obligately aerobic microorganisms. With this development, a biologically effective UV-absorbing atmospheric ozone layer became established; the photic zone of the open oceanic environment (and the exposed land surface) became wholly habitable; strict aerobes diversified to become widespread, abundant, and ultimately the dominant components of the Earth's biota; environmental pO_2 came chiefly under biologic (rather than largely inorganic) control; and the geochemical redox cycles of numerous elements became increasingly biologically mediated. A physiologically modern ecosystem had become established; the earlier, biochemically and metabolically based phase of ecologic development had been completed to become supplanted by the close of the Precambrian by a new, highly successful mode of evolutionary advance, one based chiefly on the development of new morphologies and innovative body plans among megascopic, multicellular, sexually reproducing eukaryotes.

(10) Although this broadly painted picture of the early ecological succession seems consistent with the evidence now available, it is appropriate to emphasize that the data are few and their interpretation tentative—that this scenario is, in fact, a current best guess. Indeed, in this discussion, as in other chapters of this volume, one overriding theme is evident: while recent progress in the field has been encouraging, there is a myriad of basic questions that are yet outstanding, questions that although currently unanswered seem likely to be resolved only through intensive interdisciplinary analysis like that here attempted, and ultimately only by referral to the single known source of direct evidence now available, the Archean-Early Proterozoic geologic record.

APPENDIX I

Geographic and Geologic Data for Processed Rock Samples

By Malcolm R. Walter, Hans J. Hofmann, and J. William Schopf

I-1. Introduction

This appendix provides background data for all samples included in the P.P.R.G. collections (Section I-2), and gives detailed geologic and geographic information for the 368 rock samples that were processed during the course of the project as outlined in Appendix II (i.e., those samples for which results of various types of organic geochemical analyses are presented elsewhere in this volume) and for 40 additional unprocessed samples.

I-2. Numerical Tabulation of Samples in the P.P.R.G. Collections

Table I-1 presents a listing, ordered by P.P.R.G. sample number, of the 530 samples available for study during this project. For each sample or group of samples the relevant major stratigraphic unit (supergroup, group or formation), tectonic unit, and regional geographic source are indicated. Specific data regarding the stratigraphy, geochronology, geographic locality, etc., of each of the processed samples indicated in Table I-1 are presented in Section I-3.

I-3. Geographic and Geologic Data for Processed Samples

Of the 530 samples listed in Table I-1, 368 were processed following the flow chart and procedures outlined in Appendix II. Tabulated elsewhere in this volume are the $\delta^{13}C_{organic}$ and total organic carbon (T.O.C.) values of 298 of these processed samples (Table 5-7); the $\delta^{13}C_{carbonate}$ and $\delta^{18}O_{carbonate}$ values of 70 P.P.R.G. samples (Table 7-2); the $\delta^{13}C_{PDB}$ values of 39 additional carbonates from the collection (Table 7-2); and the results of detailed elemental, isotopic, and x-ray diffraction analyses of kerogens isolated from 42 of the processed samples (Table 5-8). In addition, the stromatolitic fabrics of 24 P.P.R.G. samples are illustrated in Chapters 8 and 9; microfossils detected in 16 of the samples are figured in Chapters 9 and 14; and dubiomicrofossils or pseudomicrofossils from three other P.P.R.G. samples are illustrated in Chapter 9.

Background information pertaining to the 368 processed samples (and to 40 additional, but unprocessed samples for which only petrographic thin sections were prepared or on which only isotopic analyses of carbonates were performed) is given below, a listing in which the rock specimens from each sampled tectonic unit are tabulated in numerical order according to their P.P.R.G. sample numbers with the various tectonic units of each continent being arranged in approximate geochronologic order, from oldest to youngest.

For the samples listed from each tectonic unit, the tabulation below summarizes available information (with pertinent literature citations) regarding (*i*) their regional geographic source; (*ii*) their stratigraphic setting; (*iii*) the radiometric age of the sampled rock units; (*iv*) the inferred paleoenvironmental setting of these rock units; (*v*) the apparent degree of metamorphic alteration of these units; and (*vi*) the local stratigraphic and geographic source of each sample. The stratigraphic tables for each tectonic unit list only sampled geologic units.

As is discussed in Appendix II, each locality represented in the P.P.R.G. collections has been designated by a three-digit number (e.g., 001). In the numerically ordered listings below, each of the P.P.R.G. locality numbers is followed immediately by one or more of the following series of symbols, the occurrence of which indicates that additional data of the type specified below is presented elsewhere in this volume for at least one rock specimen from that locality:

* $\delta^{13}C_{organic}$ and total organic carbon (T.O.C.) values for this processed sample are presented in Table 5-7.

One or more T.O.C. values (but no $\delta^{13}C_{organic}$ values) for this processed sample is presented in Table 5-7.

Neither a $\delta^{13}C_{organic}$ nor a T.O.C. value (determined via the "survey method"; see Appendix II) is available for this pro-

TABLE I-1. Samples Included in the P.P.R.G. Collections.

P.P.R.G. sample no.	*Processed samples*	*Stratigraphic unit*	*Tectonic unit*	*Geographic source*
001–018	001–018	Warrawoona Group	Pilbara Block	Western Australia
019–023	019, 020, 022, 023	Gorge Creek Group	Pilbara Block	Western Australia
024–041	024–033, 035–041	Fortescue Group	Hamersley Basin	Western Australia
042–043	042, 043	Hamersley Group	Hamersley Basin	Western Australia
044–045	044, 045	Fortescue Group	Hamersley Basin	Western Australia
046	046	Warrawoona Group	Pilbara Block	Western Australia
047–048	047, 048	Hamersley Group	Hamersley Basin	Western Australia
049–062	049–057, 059–062	Wyloo Group	Ashburton Trough	Western Australia
063	063	Turee Greek Group	Hamersley Basin	Western Australia
064–102	066, 073, 076, 078, 080, 082, 086, 089, 091–093, 095–097, 099, 100	Earaheedy Group	Naberru Basin	Western Australia
103	103	McArthur Group	McArthur Basin	Northern Territory, Australia
104	104	Earaheedy Group	Naberru Basin	Western Australia
105–113	105–113	McArthur Group	McArthur Basin	Northern Territory, Australia
114–131	114–131	Roper Group	McArthur Basin	Northern Territory, Australia
132	132	Fortescue Group	Hamersley Basin	Western Australia
133	133	Bitter Springs Fm.	Amadeus Basin	Northern Territory, Australia
134	134	Gunflint Iron Fm.	Port Arthur Homocline	Ontario, Canada
135	135	Uinta Mountain Group	Uinta Uplift	Utah, U.S.A.
136–161	137–141, 143–159	Bungle Bungle Dolomite	Birrindudu Basin	Western Australia
162–178		"Isua Supracrustal Sequence"	Isua Supracrustal Belt	southwestern Greenland
179–201	179, 181, 182, 184–189, 191–201	Onverwacht Group	Barberton Greenstone Belt	northeastern South Africa
202–203	202, 203	Fig Tree Group	Barberton Greenstone Belt	northeastern South Africa
204–205	204, 205	Transvaal Supergroup	Transvaal Basin	northeastern South Africa
206	206	Bulawayan Group	Bulawayan Greenstone Belt	Zimbabwe, Africa
207		Fig Tree Group	Barberton Greenstone Belt	northeastern South Africa
208	208	Onverwacht Group	Barberton Greenstone Belt	northeastern South Africa
209	209	Moodies Group	Barberton Greenstone Belt	northeastern South Africa
210–211		Onverwacht Group	Barberton Greenstone Belt	northeastern South Africa

Table I-1. (*continued*)

P.P.R.G. sample no.	*Processed samples*	*Stratigraphic unit*	*Tectonic unit*	*Geographic source*
212–217	212–216	Fig Tree Group	Barberton Greenstone Belt	northeastern South Africa
218–228	218–228	Manjeri Formation	Belingwe Greenstone Belt	Zimbabwe, Africa
229	229	"Lower Greenstones"	Belingwe Greenstone Belt	Zimbabwe, Africa
230–246	240, 242	Transvaal Supergroup	Transvaal Basin	northeastern South Africa
247–255	253, 254, 255	Bulawayan Group	Bulawayan Greenstone Belt	Zimbabwe, Africa
256–265	256–265	Insuzi Group	Wit Mfolozi Inlier	eastern South Africa
266–270	266, 267, 269, 270	Mozaan Group	Wit Mfolozi Inlier	eastern South Africa
271	271	Archean(?) Granite	Wit Mfolozi Inlier	eastern South Africa
272–294	272–293	Ventersdorp Supergroup	Ventersdorp Basin	central South Africa
295	295	Griqualand Group	Transvaal Basin	northeastern South Africa
296–307	296–300, 302, 304–307	Transvaal Supergroup	Transvaal Basin	northeastern South Africa
308	308	Ventersdorp Supergroup	Ventersdorp Basin	central South Africa
309	309	Insuzi Group	Wit Mfolozi Inlier	eastern South Africa
310–320	310, 312, 314–320	Transvaal Supergroup	Transvaal Basin	northeastern South Africa
321–329	321–326	Steeprock Group	Wabigoon Belt	Ontario, Canada
330	330	"Keewatin" (Archean)	Superior Province	Ontario, Canada
331	331	Michipocoten Iron Fm.	Wawa Belt	Ontario, Canada
332	332	Vermillion Limestone	Sudbury Basin	Ontario, Canada
333	333	Huronian Supergroup	Penokean Fold Belt	Ontario, Canada
334	334	Onwatin Formation	Sudbury Basin	Ontario, Canada
335–336	335, 336	Gunflint Iron Fm.	Port Arthur Homocline	Ontario, Canada
337		Epworth Group	Epworth Basin	central District of Mackenzie, Canada
338–339	338	Pethei Group	East Arm Fold Belt, Athapuscow Aulacogen	southern District of Mackenzie, Canada
340	340	Kahochella Group	East Arm Fold Belt, Athapuscow Aulacogen	southern District of Mackenzie, Canada
341	341	Belcher Group	Belcher Fold Belt, Circum-Ungava Geosyncline	southeastern Hudson Bay, Canada
342–343	342, 343	Sibley Group	Lake Superior Basin	Ontario, Canada
344–345	344, 345	Manitounuk Group	Belcher Fold Belt, Circum-Ungava Geosyncline	southeastern Hudson Bay, Canada
346–347		Miscellaneous Samples		
	346	Dharwar Supergroup	Dharwar Craton	southern India
348		No Sample		
349	349	Warrawoona Group	Pilbara Block	Western Australia
350–355		"Isua Supracruatal Sequence"	Isua Supracrustal Belt	southwestern Greenland

TABLE I-1. (*continued*)

P.P.R.G. sample no.	*Processed samples*	*Stratigraphic unit*	*Tectonic unit*	*Geographic source*
356–374	356–374	Hamersley Group	Hamersley Basin	Western Australia
375–377	375–377	Fortescue Group	Hamersley Basin	Western Australia
378–383	383	Hamersley Group	Hamersley Basin	Western Australia
384–385	384, 385	Fortescue Group	Hamersley Basin	Western Australia
386–387	386, 387	Hamersley Group	Hamersley Basin	Western Australia
388	388	Gorge Creek Group(?)	Pilbara Block	Western Australia
389		Biwabik Iron Fm.	Canadian Southern Province	northeastern Minnesota, U.S.A.
390–399		No samples		
400		Gunflint Iron Fm.	Port Arthur Homocline	Ontario, Canada
401		Miscellaneous Sample		
	401	Krivoi-Rog Iron Fm.	Ukranian Platform	Ukrainian S.S.R.
402	402	Proterozoic anthraxolite	Sudbury Basin	Ontario, Canada
403		Great Slave Supergroup	East Arm Fold Belt, Athapuscow Aulacogen	southern District of Mackenzie, Canada
404		Pethei Group	East Arm Fold Belt, Athapuscow Aulacogen	southern District of Mackenzie, Canada
405–406	406	Great Slave Supergroup	East Arm Fold Belt, Athapuscow Aulacogen	southern District of Mackenzie, Canada
407	407	Kahochella Group	East Arm Fold Belt, Athapuscow Aulacogen	southern District of Mackenzie, Canada
408		Pethei Group	East Arm Fold Belt, Athapuscow Aulacogen	southern District of Mackenzie, Canada
409–411		Great Slave Supergroup	East Arm Fold Belt, Athapuscow Aulacogen	southern District of Mackenzie, Canada
412	412	Pethei Group	East Arm Fold Belt, Athapuscow Aulacogen	southern District of Mackenzie, Canada
413	413	Yellowknife Supergroup	Slave Province	central District of Mackenzie, Canada
414–420	414, 417	Holocene Samples		
421–423	421	McNamara Group	Lawn Hill Platform	Queensland, Australia
424		No sample		
425–440	426, 435, 436, 439, 440	Earaheedy Group	Nabberu Basin	Western Australia
441	441	Lewin Shale	Hamersley Basin	Western Australia
442		Miscellaneous Sample		
	442	Aravalli Group	Rajasthan Craton	northern India
443	443	Upper Albanel Fm.	Mistassini Homocline	Quebec, Canada
444	444	Lower Albanel Fm.	Mistassini Homocline	Quebec, Canada
445–449	445–449	Belcher Group	Belcher Fold Belt, Circum-Ungava Geosyncline	southeastern Hudson Bay, Canada
450	450	Chelmsford Formation	Sudbury Basin	Ontario, Canada

Table I-1. *(continued)*

P.P.R.G. sample no.	*Processed samples*	*Stratigraphic unit*	*Tectonic unit*	*Geographic source*
451	451	Onwatin Formation	Sudbury Basin	Ontario, Canada
452	452	McArthur Group	McArthur Basin	Northern Territory, Australia
453–454		Oronto Group	Lake Superior Basin	northern Michigan, U.S.A.
455	455	Burra Group	Adelaide Geosyncline	South Australia
456	456	Clarno Formation	John Day Basin	north-central Oregon, U.S.A.
457–459	457–459	Roper Group	McArthur Basin	Northern Territory, Australia
460		Miscellaneous Sample		
461–469	461, 462, 464–469	"Isua Supracrustal Sequence"	Isua Supracrustal Belt	southwestern Greenland
470–471	470, 471	Onverwacht Group	Barberton Greenstone Belt	northeastern South Africa
472–479		Holocene Samples		
480–491	480–491	Hamersley Group	Hamersley Basin	Western Australia
492–494	492–494	"Isua Supracrustal Sequence"	Isua Supracrustal Belt	southwestern Greenland
495	495	Belcher Group	Belcher Fold Belt, Circum-Ungava Geosyncline	southeastern Hudson Bay, Canada
496		"Isua Supracrustal Sequence"	Isua Supracrustal Belt	southwestern Greenland
497	497	Oronto Group	Lake Superior Basin	northern Michigan, U.S.A.
498	498	Rove Formation	Port Arthur Homocline	Ontario, Canada
499	499	Campbell Group	Transvaal Basin	northeastern South Africa
500	500	Michigamme Formation	Penokean Fold Belt	northern Michigan, U.S.A.
501–503	501–503	Randville Dolomite	Penokean Fold Belt	northern Michigan, U.S.A.
504–508	504–508	Hamersley Group	Hamersley Basin	Western Australia
509–510	509, 510	Huronian Supergroup	Penokean Fold Belt	Ontario, Canada
511–514		Warrawoona Group	Pilbara Block	Western Australia
515	515	Huronian Supergroup	Penokean Fold Belt	Ontario, Canada
516	516	Woman Lake Marble	Uchi Greenstone Belt	Ontario, Canada
517	517	Warrawoona Group	Pilbara Block	Western Australia
518	518	Allamoore Formation	Van Horn Mobile Belt	western Texas, U.S.A.
519		Holocene Sample		
520	520	Hamersley Group	Hamersley Basin	Western Australia
521	521	Archean shale	Yilgarn Block	Western Australia
522	522	Warrawoona Group	Pilbara Block	Western Australia
523–532		Transvaal Supergroup	Transvaal Basin	northeastern South Africa
533	533	Fortescue Group	Hamersley Basin	Western Australia
534–542		Warrawoona Group	Pilbara Block	Western Australia

cessed sample.

k Results of elemental, isotopic, and x-ray diffraction analyses of kerogen isolated from this processed sample are presented in Table 5-8.

c One or more $\delta^{13}C_{carbonate}$ (and in most cases, $\delta^{18}O_{carbonate}$) values for this processed sample is presented in Table 7-2.

NPc A $\delta^{13}C_{carbonate}$ value for this non-processed sample is presented in Table 7-2.

s The stromatolitic fabric of this sample is illustrated in Chapter 8 or in Chapter 9 (see the photo indicated).

μf Microfossils detected in petrographic thin sections of this sample are illustrated in Chapter 9 or in Chapter 14 (see the photo indicated).

"μf" Dubiomicrofossils or pseudomicrofossils detected in petrographic thin sections of this sample are illustrated in Chapter 9 (see the photo indicated).

NPt Although this sample was not processed (as outlined in Appendix II), petrographic thin sections were prepared for optical microscopic study.

For those listed samples that were not collected during the P.P.R.G. field work of May and June, 1979 (see Preface to this volume), the source of the sample and/or the collector's field number are indicated in parentheses following the above listed symbols. The abbreviations there used to indicate the sources of these samples are as follows:

A. G. Alan D. T. Goode, BHP Company Ltd., Melbourne, Australia;

A. T. Alec Trendall, Geological Survey of Western Australia;

A. V. N. Albert V. Nyberg, University of California, Los Angeles;

B. N. Bartholomew Nagy, University of Arizona;

B. V. T. Boris V. Timofeev, Institute of Precambrian Geology and Geochronology, Akademia Nauk, Leningrad, U.S.S.R.;

D. B. David Bridgwater, Geological Survey of Greenland, Copenhagen, Denmark;

D. E. G. David E. Grandstaff, Temple University, Philadelphia, Pennsylvania;

E. C. P. Eugene C. Perry, Jr., Northern Illinois University, DeKalb;

E. G. N. Euan G. Nisbet, Oxford University, England;

E. S. B. Elso S. Barghoorn, Harvard University, Cambridge, Massachusetts;

F. M. Field Museum of Natural History, Chicago, Illinois;

G.S.W.A. Geological Survey of Western Australia, Perth;

H. D. P. Hans D. Pflug, University of Giessen, Germany;

H. I. P. Hamersley Iron Pty. Ltd.;

J. D. John Dunlop, University of Western Australia, Nedlands;

J. S. Janine Sarfati, Centre Geologique et Geophysique, Montpellier, France;

J. S. N. J. Swami Nath, Geological Survey of India, Mysore Circle;

J. W. S. J. William Schopf, University of California, Los Angeles;

K. A. P. Kenneth A. Plumb, Bureau of Mineral Resources, Geology and Geophysics, Canberra, Australia;

M. D. M. Marjorie D. Muir, CRA Exploration Pty. Ltd., Canberra, Australia;

M. J. J. Michael J. Jackson, Bureau of Mineral Resources, Canberra, Australia;

M. R. W. Malcolm R. Walter, Baas Becking Geobiological Laboratory, Canberra, Australia;

M. S. Manfred Schidlowski, Max-Planck-Institut für Chemie, Mainz, Germany;

N.A.S.A. National Aeronautics and Space Administration, Ames Research Center, Moffett Field, California;

P. G. Paul Grant, Imperial College, London;

R. H. Rudi Horwitz, CSIRO Division of Mineralogy, Perth, Australia; and

S. M. A. Stanley M. Awramik, University of California, Santa Barbara.

Petrographic thin sections and, as available, reserve portions of each processed rock sample (viz., as indicated in Figure II-1, reserve rock specimens, reserve rock fragments, reserve thin section chips, etched rock chips, reserve rock powders, rock powders for Rock-Eval analysis, and total kerogens) are stored in the P.P.R.G. collections (J. William Schopf, curator), housed in the Paleobiological Laboratories of the Department of Earth and Space Sciences at the University of

California, Los Angeles. At or near the end of certain of the entries below, the location of duplicate rock specimens, paleontological type specimens, and/or petrographic thin sections of several of these samples are indicated by the abbreviations for the institutions as follows, followed by catalogue numbers:

C.P.C. The Commonwealth Palaeontological Collections at the Bureau of Mineral Resources, Geology and Geophysics, Canberra, Australia;
F.M. The Field Museum of Natural History, Chicago, Illinois;
G.S.C. The Geological Survey of Canada, Ottawa;
U.W.A. The University of Western Australia, Nedlands.

The abbreviation "DDH" in certain of the entries is for "diamond drill hole" and refers to drill core samples; and "I.G.C." noted in some entries for Western Australian rocks refers to the 25th International Geological Congress held in Sydney, Australia, in 1976. For those samples for which stromatolitic fabric (e.g., 001s), microfossils (e.g., 001μf), or dubiomicrofossils or pseudomicrofossils (e.g., 001"μf") are illustrated elsewhere in this volume, the relevant photos are noted at the end of the entry.

I-3-1. AFRICA

BARBERTON GREENSTONE BELT

Tectonic unit/location. Barberton Greenstone Belt, northeastern South Africa.

Stratigraphy. "Swaziland Sequence":
- Moodies Group
- Fig Tree Group
- Onverwacht Group:
  - Zwartkoppie Formation
  - Kromberg Formation
  - Hooggenoeg Formation
  - Komati Formation
  - Theespruit Formation
  - Sandspruit Formation.

Geologic age. Post-tectonic granites in the sequence are older than 3.3 Ga (see discussion in Eriksson, 1978); volcanics in the Onverwacht Group have yielded a Sm-Nd date of 3.540 ± 0.030 Ga (Hamilton et al., 1979).

Paleoenvironments. Sediments of the Moodies Group were deposited in diverse alluvial and marginal marine environments (Eriksson, 1978). The Fig Tree Group grades from relatively deep-water facies in the north to shallow and locally emergent facies in the south, associated with volcanism (see summary in Lowe, 1980b). The volcanic complexes of the Onverwacht Group were mostly subaqueous, but locally formed subaerial cones around which shallow-water sediments, including evaporites, were deposited (for discussion, see Lowe, 1980b).

Metamorphic grade. Lower greenschist to amphibolite facies (R. P. Viljoen and M. J. Viljoen, 1969). All samples from the Theespruit Formation processed in the present study seem to have come from close to intrusive granites and have a rodded and foliated fabric indicating considerable deformation.

P.P.R.G. samples/localites (38 total samples, 31 processed).

179* (N.A.S.A. no. 5-3). Black chert from the lower third of the Theespruit Fm., 32 km southwest of Barberton, eastern Transvaal.

181*k (N.A.S.A. no. 5-6). Siliceous shale from the upper third of the Theespruit Fm., locality cf. 179.

182*k (N.A.S.A. no. 5-7). Black chert from the lower third of the Hooggenoeg Fm., 27 km southwest of Barberton.

183°NPc (N.A.S.A. no. 5-8). Carbonate and chert, locality cf. 182.

184* (N.A.S.A. no. 5-10). Black chert from the middle third of the Hooggenoeg Fm., 37 km south-southwest of Barberton.

185* (N.A.S.A. no. 5-11). Carbonate(?) and shale from the middle third of the Kromberg Fm., 29 km south-southwest of Barberton.

186* (N.A.S.A. no. 5-12). Black chert, locality cf. 185.

187* (N.A.S.A. no. 5-13). Black chert, locality cf. 185.

188* (N.A.S.A. no. 5-14). Black chert, locality cf. 185.

189*k (N.A.S.A. no. 5-15). Black chert, locality cf. 185.

190°NPc (N.A.S.A. no. 5-16). Carbonate, locality cf. 185.

191* (N.A.S.A. no. 5-18). Black chert from the upper third of the Kromberg Fm., 29 km south of Barberton.

192* (N.A.S.A. no. 5-19). Black chert, locality cf. 191.

193* (N.A.S.A. no. 5-20). Black chert from the middle third of the Kromberg Fm., 27 km south of Barberton.
194* (N.A.S.A. no. 5-21). Black and gray chert, labeled "lower third of Fig Tree Group" (but probably from the Szwartkoppie Fm.), 10 km northeast of Barberton.
195* (N.A.S.A. no. 5-22). Carbonate schist(?) from the Szwartkoppie Fm., 11 km northeast of Barberton.
196* (N.A.S.A. no. 5-23). Black chert, locality cf. 195.
197* (N.A.S.A. no. 5-24). Black chert with quartz veins, locality cf. 195.
198*k (N.A.S.A. no. 5-25). Black and gray chert with quartz veins from the Szwartkoppie Fm., 24 km east-northeast of Barberton.
199*k (N.A.S.A. no. 5-26). Graphitic, pyritic quartz vein(?), locality cf. 198.
200* (N.A.S.A. no. 5-27). Black chert from the Zwartkoppie Fm., 22 km east-northeast of Barberton.
201* (N.A.S.A. no. 5-28). Black chert from the middle third of the Fig Tree Group, 13 km south-southeast of Barberton.
202* (N.A.S.A. no. 5-32). Gray shale from the Fig Tree Group collected at the Sheba Gold Mine dump, near Barberton.
203* (N.A.S.A. no. 5-41). Black shale with iron stains from the Fig Tree Group, Montore Gold Mine, near Barberton.
208* (B. N. no. H2). Black chert from the Hooggenoeg Fm., 36 km south-southwest of Barberton.
209* (B. N. no. M3). Black shale from the Moodies Group, 8 km north-northeast of Barberton.
212* (H. D. P. no. 136). Black shale with pyrite laminae from the Fig Tree Group at the Sheba Gold Mine, near Barberton.
213* (E. S. B. no. SWFT-1). Black chert, locality cf. 212.
214* (H. D. P. no. 156). Pyritic black shale, locality cf. 212.
215*k (E. S. B. no. FT-D). Black chert from the Fig Tree Group at a road cut close to the portal to Daylight Mine, near Louws Creek, eastern Transvaal.
216* (E. S. B. sample "Fig Tree Shale at French Bob Mine"). Black chert with pyrite, veins and slickensides, from the Fig Tree Group at French Bob Mine, eastern Transvaal.
470* (N.A.S.A. no. 5-4). Basic volcanic from the top of the lower third of the Theespruit Fm., 32 km southwest of Barberton.
471° (N.A.S.A. no. 5-17). Tuff from the upper third of the Kromberg Fm., 29 km south of Barberton.

WIT MFOLOZI INLIER

Tectonic unit/location. Wit Mfolozi Inlier, eastern South Africa.

Stratigraphy. Pongola Supergroup:
Mozaan Group
Insuzi Group.

Geologic age. As summarized by Von Brunn and Mason (1977), the Mozaan Group is apparently older than 2.87 Ga and younger than 3.06 Ga. Zircons from lavas of the underlying Insuzi Group have yielded a date of 3.09 $\pm$ 0.09 Ga and a whole rock Rb-Sr date of 3.15 $\pm$ 0.15 has been obtained on Insuzi felsites (see Von Brunn and Mason, 1977).

Paleoenvironments. The paleoenvironmental setting of the Insuzi Group is discussed in Section 8-4-3 of the present volume. The Mozaan Group has been interpreted as a marine peritidal sequence (Von Brunn and Mason, 1977).

Metamorphic grade. According to Mason and Von Brunn (1977), "evidence of regional metamorphism is restricted to incipient recrystallisation of the arenaceous and carbonate rocks; the lavas show evidence of deuteric alteration."

Field photographs. The Insuzi stromatolites are shown in Photo 8-3A and 8-4A, herein.

P.P.R.G. samples/localities (17 total samples, 16 processed).

256*. Volcanics of the Insuzi Group at the Wit Mfolozi River, 50 m upstream from an old bridge; 28°13′55″S, 31°11′12″E.
257*. Dark gray siltstone (above uppermost dolomite) of the Insuzi Group at the Wit Mfolozi River, 100 m downstream from a new bridge; 28°14′12″S, 31°10′40″E.
258*c. Uppermost dolomite of the Insuzi Group, locality cf. 257.
259°. Volcanics of the Insuzi Group, near a road cut about 1 km south of the Wit Mfolozi River; 28°15′39″S, 31°10′55″E.
260*. Argillite from the lower Insuzi Group at the Wit Mfolozi River; 28°15′01″S, 31°10′28″E.
261*c. Stromatolitic carbonate of the Insuzi Group at the Wit Mfolozi River; 28°14′25″S, 31°10′40″E.
262°c. Oncolitic(?) carbonate from the Insuzi Group, locality cf. 261.

263′cs. Oncolitic(?) dolomite from the Insuzi Group at the Wit Mfolozi River; 28°14′13″S, 31°10′21″E; stromatolitic fabric in Photos 8-4B and 8-5A, B.
264*c. Stromatolitic cherty limestone from the Insuzi Group at the Wit Mfolozi River, between locality 263 and a new bridge; 28°14′13″S, 31°10′22″E.
265*c (K. A. P.) Stromatolitic dolomite, locality cf. 263.
266*k. Black shale from the Mozaan Group at the Wit Mfolozi River; 28°13′53″S, 31°12′07″E.
267*. Black shale, locality cf. 266.
269*. Jaspilite from the Mozaan Group at the Wit Mfolozi River; 28°13′52″S, 31°11′53″E.
270*. Jaspilite, locality cf. 269.
271*. Archean(?) granite boulder in Holocene river gravels from the Wit Mfolozi River at the south end of an old bridge, near locality 256; 28°13′55″S, 31°11′12″E.
309*s. Stromatolitic carbonate from the Insuzi Group at the Wit Mfolozi River, collected in a talus slope upstream from the new bridge noted in locality 257; stromatolitic fabric in Photo 8-3A-C.

BELINGWE GREENSTONE BELT

Tectonic unit/location. Belingwe Greenstone Belt, Zimbabwe, Africa.

Stratigraphy. "Upper Greenstones"
Manjeri Formation
"Lower Greenstones"

Geologic age. Hickman (1976) has obtained a Rb-Sr date of 2.625 ± 0.025 Ga on adamellite intrusive into the sequence; a Sm-Nd date of 2.64 ± 0.14 Ga has been reported by Hamilton et al. (1977) for volcanics from the "Upper Greenstones"; and the "Lower Greenstones" are probably about 3 Ga old (as summarized by Martin et al., 1980).

Paleoenvironments. All but one of the P.P.R.G. samples processed are black cherts from the Manjeri Formation; according to Martin et al. (1980), these cherts are interbedded with stromatolitic limestones, but these authors do not offer a paleoenvironmental interpretation.

Metamorphic grade. Greenschist facies, at least locally, for the "Lower Greenstones" (Hawkesworth et al., 1979); the "Upper Greenstones" are also reported to be within the greenschist facies (Nisbet et al., 1977).

Field photographs. The Belingwe stromatolites are shown in Photos 8-7A-C and 8-8, herein.

P.P.R.G. samples/localities (12 total samples, 12 processed).
218* (E. G. N. no. B1). Black chert from the Manjeri Fm. near the Ngesi River, north of Belingwe (see map in Bickle et al., 1975).
219* (E. G. N. no. B2). Black chert, locality cf. 218.
220° (E. G. N. no. B4). Black chert, locality cf. 218.
221* (E. G. N. no B5). Black chert, locality cf. 218.
222° (E. G. N. no. B6). Black chert, locality cf. 218.
223* (E. G. N. no B8). Black chert, locality cf. 218.
224*c (E. G. N. no. U1). Black chert from the type section of the Manjeri Fm., from the eastern portion of the Belingwe Greenstone Belt, just above the unconformity figured in Bickle et al. (1975).
225*kc (E. G. N. no. U3). Black chert, locality cf. 224.
226* (E. G. N. no. U4). Dolomitic black chert, locality cf. 224.
227* (E. G. N. no. U5). Black chert, locality cf. 224.
228*c (E. G. N. no. CM/14). Black chert, locality cf. 224.
229* (E. G. N. no. SH 15). Black shale from the "Lower Greenstones," from the greenstone terrane beneath the unconformity noted in 224.

BULAWAYAN GREENSTONE BELT

Tectonic unit/location. Bulawayan Greenstone Belt, Zimbabwe, Africa.

Stratigraphy. Bulawayan Group:
Lower Formation:
Zwankendaba Member (known as "Huntsman Limestone").

Geologic age. A Sm-Nd date of 2.64 ± 0.14 Ga has been obtained for volcanics of the closely associated Belingwe Greenstone Belt (Hamilton et al., 1977); for details, see the discussion in Wilson et al. (1978).

Paleoenvironments. The paleoenvironment inferred for this sequence is discussed Section 8-4-6 of the present volume.

Metamorphic grade. According to Bond et al. (1973), organic carbon in the Bulawayan limestones has been crystallized to graphite, indicating a metamorphic grade within (or greater than) greenschist facies.

P.P.R.G. samples/localities (10 total samples, 4 processed).

206°c (N.A.S.A. no. 7-3). Gray stromatolitic carbonate from the Zwankendaba Member at the Huntsman Limestone Quarries, near Turk Mine, about 55 km north-northeast of Bulawayo (for locality, see text fig. 1 in Schopf et al., 1971).
247°NPc (N.A.S.A. no. Bul-1). Black chert with minor carbonate, locality cf. 206.
248°NPc (N.A.S.A. no. Bul-2). Black and gray chert with carbonate, locality cf. 206.
249°NPc (N.A.S.A. no. Bul-3). Black chert with carbonate, locality cf. 206.
250°NPts (N.A.S.A. no. Bul-4). Black stromatolitic carbonate, locality cf. 206; stromatolitic fabric in Photo 8-9A.
251°NPc (N.A.S.A. no. Bul-5). Gray stromatolitic carbonate, locality cf. 206.
253*s (N.A.S.A. no. Bul-7). Black stromatolitic carbonate, locality cf. 206; stromatolitic fabric in Photo 8-9B.
254*kcs (N.A.S.A. no. Bul-9). Black stromatolitic carbonate, locality cf. 206; stromatolitic fabric in Photo 8-9C.
255* (J. W. S. "Zwankendaba Limestone"). Black stromatolitic carbonate, locality cf. 206.

VENTERSDORP BASIN

Tectonic unit/location. Ventersdorp Basin, central South Africa.

Stratigraphy. Ventersdorp Supergroup:
Platberg Group:
Bothaville Formation
Klippan Formation
Rietgat Formation.

Geologic age. Zircons in volcanics from the sequence have yielded a U-Pb date of 2.643 $\pm$ 0.080 Ga (Van Niekerk and Burger, 1978).

Paleoenvironments. According to Buck (1980), the sequence is composed of lacustrine sediments deposited in intermontane basins; Grobler and Emslie (1976a), however, consider that the units in the T'Kuip Hills area are marine.

Metamorphic grade. According to S. G. Buck (pers. comm. to M. R. W., 1980), in the Welkom Goldfield Area at least 7 km of sediments originally must have overlain the Ventersdorp Supergroup; he thus judges that the grade of metamorphism is in the greenschist facies.

Field photographs. Stromatolites from the Ventersdorp Supergroup are shown in Photos 8-14A, B, and 8-15A, B, herein.

P.P.R.G. samples/localities (24 total samples, 23 processed).
272*. Calcareous stromatolitic chert from the basal black chert of the Rietgat Fm. at the Omdraii Vlei 93 Farm, in the T'Kuip Hills area; 30°02′43″S, 23°10′13″E (for details, see Grobler and Emslie, 1976, p. 50).
273*. Cherty stromatolitic limestone, locality cf. 272.
274*s. Calcitic stromatolitic chert, locality cf. 272; stromatolitic fabric in Photo 8-16A.
275*. Andesite, locality cf. 272.
276°. Chert, locality cf. 272.
277°. Stromatolitic chert, locality cf. 272.
278*s. Stromatolitic chert, locality cf. 272; stromatolitic fabric in Photo 8-16B.
279*. Chert, locality cf. 272.
280*c. Carbonate and siltstone from the Ventersdorp Supergroup at the Welkom Goldfield, Geduld Gold Mine, DDH UT1 at 1715′, locality given in Figure I-1.

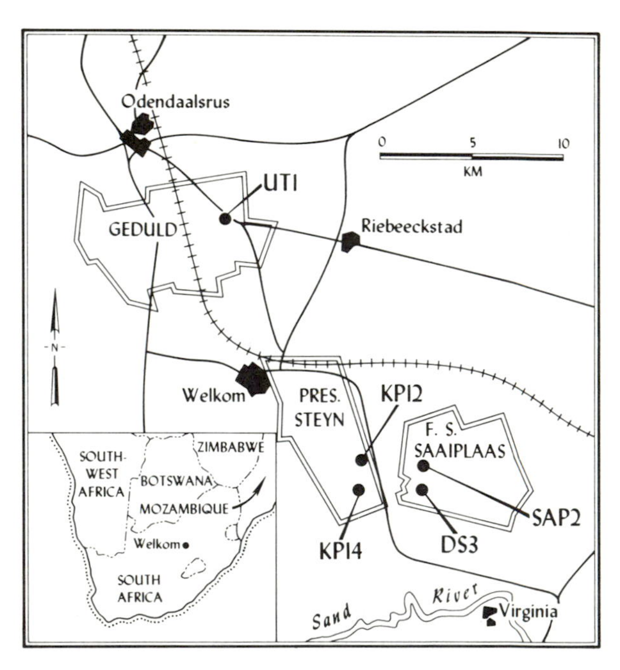

FIGURE I-1. Sketch map showing localities of processed drill core samples from the Ventersdorp Supergroup in goldfields administered by the Anglo-American Corporation in the Welkom area, South Africa; P.P.R.G. Sample Nos. 280, 281, 282 are from the Geduld Gold Mine, DDH UT1; Nos. 283, 284, 285, 286 are from the F. S. Saaiplaas Gold Mine, DDH DS3; Nos. 287, 288, 289, 290 are from the F. S. Saaiplaas Gold Mine, DDH SAP2; Nos. 291, 292 are from the Pres. Steyn Gold Mine, DDH KP12; No. 293 is from the Pres. Steyn Gold Mine, DDH KP14.

281*cs. Dolomitic stromatolitic limestone, locality cf. 280 except at 1721′; stromatolitic fabric in Photo 8-18B.

282*c. Dolomitic limestone and shale, locality cf. 280 except at 1726′.

283*c. Dolomitic limestone, locality cf. 280 except from F. S. Saaiplaas Gold Mine, DDH DS3 at 612.6 m (see Figure I-1).

284*. Limestone, locality cf. 283.

285*. Carbonate, locality cf. 283.

286*c. Limestone, locality cf. 283 except at 610.7 m.

287*cs. Stromatolitic limestone, locality cf. 283 except from DDH SAP2 at 655.3 m (see Figure I-1); stromatolitic fabric in Photo 8-18A.

288*c. Stromatolitic limestone, locality cf. 287 except at 701.0 m.

289°. Carbonate, locality cf. 288.

290°. Stromatolitic limestone, locality cf. 288.

291*cs. Stromatolitic limestone, locality cf. 280 except from Pres. Steyn Gold Mine, DDH KP12 at 1380′ (see Figure I-1); stromatolitic fabric in Photo 8-17B, C.

292*s. Stromatolitic limestone, locality cf. 291 except 1383′; stromatolitic fabric in Photo 8-17A, D.

293*kc. Pyritic argillaceous limestone, locality cf. 291 except from DDH KP14 at 3908′.

294°NPts (no. S535) Stromatolitic carbonate from the Ventersdorp Supergroup; stromatolitic fabric in Photo 8-19A-C.

308′. Cherty stromatolitic carbonate, locality cf. 272.

TRANSVAAL BASIN

Tectonic unit/location. Transvaal Basin, northeastern South Africa.

Stratigraphy.

"Griqualand West Sequence":	Transvaal Supergroup:
Campbell Group	Chuniespoort Group
Griqualand Group	Malmani Subgroup
Asbestos Hills Banded Iron Fm.	Crocodile River Formation
	Wolkberg Group.

For details regarding the stratigraphy of the Transvaal Basin, see Button (1976).

Geologic age. As summarized by Button (1976a), the Transvaal Supergroup is older than between 2.224 ± 0.021 Ga and 2.263 ± 0.085 Ga (the age of the Pretoria Group, the uppermost unit of the Supergroup), and is younger than the Ventersdorp Supergroup (viz., 2.643 ± 0.080 Ga; Van Niekerk and Burger, 1978).

Paleoenvironments. Results of the extensive paleoenvironmental studies of the Transvaal Basin sediments carried out by A. Button, K. A. Eriksson, J. F. Truswell and others are summarized by Button (1976a).

Metamorphic grade. Apart from contact metamorphism associated with the intrusive Bushveld Complex, the sequence appears to be essentially unmetamorphosed.

P.P.R.G. samples/localities (54 total samples, 25 processed).

204* (N.A.S.A. no. 6-1). Chert from the "Transvaal System" in the Elands River Valley.

205*c (N.A.S.A. no. 6-2). Dolomite and chert, locality cf. 204.

240* (M. S. no. 64). Dolomite from the Crocodile River Fm. at Farms Tweefontein and Uitkoms, northwest of Johannesburg (Swartkops).

242* (M. S. no. 93). Dolomite of the Malmani Subgroup (transition beds) from the geologic section between Schmidtsdrift and Ghaap Plateau, west of Kimberley.

295*. Banded iron-formation of the Asbestos Hills Banded Iron Fm. along the main road near Prieska.

296*c. Dolomitic oncolitic limestone from the Malmani Subgroup at the Groot Boetsap River (Hol River), 50 km northwest of Warrenton, at "Fall 2" of Truswell and Eriksson (1973, their fig. 2).

297*c. Stromatolitic limestone, locality cf. 296, a 10 to 20 cm-thick tabular biostrome near the top of "Fall 2".

298*c. Stromatolitic limestone, locality cf. 296, from the top of the waterfall.

299*kc. Stromatolitic dolomite, locality cf. 296, 150 m upstream from the waterfall.

300*c. Stromatolitic carbonate, locality cf. 296 but at "Fall 1" of Truswell and Eriksson (1973).

302*c (M. R. W. no. 6.3/2/73). Stromatolitic carbonate, locality cf. 300.

304*kc (M. R. W. no. 7.4/2/73). Calcitic stromatolitic dolomite, locality cf. 296.

305° (M. R. W. no. 7.2/2/73). Stromatolitic dolomite, locality cf. 296.

306* (M. R. W. no. 8.1/2/73). Stromatolitic dolomite, locality cf. 296 but just above "Fall 2".

307* (M. R. W. no. 8.4/2/73). Dolomite, locality cf. 296.

310* (M. R. W. no. 9.2/2/73). Chert from Zone 9 of the Malmani Subgroup collected at the road cut for the Hartbeesport Dam Road, northeast of Johannesburg.
311°NPt (M. R. W. no. 13.1/2/73). Stromatolitic carbonate from 250 to 300 m above the base of the Malmani Subgroup on the road to Wolkberg, eastern Transvaal.
312*c (M. R. W. no. 14.3/2/73). Stromatolitic limestone from the Malmani Subgroup, 25 to 30 m above the base of the Mixed Zone at the road cut at Abel Erasmus Pass.
314*c (M. R. W. no. 8.5a/2/73). Calcitic stromatolitic dolomite, locality cf. 296 except between "Fall 2" and "Fall 3".
315*c (M. R. W. no. 8.5b/2/73). Stromatolitic limestone, locality cf. 314.
316*c (M. R. W. no. 7.2/2/73). Stromatolitic dolomite, locality cf. 296.
317*c (M. R. W. no. 8.4/2/73). Stromatolitic limestone, locality cf. 296.
318*c (M. R. W. no. 8.6/2/73). Stromatolitic limestone, locality cf. 296 except at "Fall 3".
319° (M. R. W. no. 14.2/2/73). Stromatolitic carbonate from the Wolkberg Group at the mouth of Blyde River Gorge, Transvaal.
320*c (M. R. W. no. 14.4/2/73). Stromatolitic carbonate, locality cf. 312.
499° (H. D. P. no. 104). "Carbon" from the "Campbell Series" at Limeacres via Danielskuit.
523°NPt (J. S.). Stromatolitic dolomite, locality cf. 296 except almost at the top of "Fall 3".
524°NPt (J. S.). Stromatolitic limestone, locality cf. 296 except at flat between "Fall 2" and "Fall 3".
525°NPt (J. S.). Stromatolitic carbonate, locality cf. 296 except near the top of "Fall 2".
526°NPt (J. S.). Stromatolitic dolomite, locality cf. 296 except at river bed near "Fall 4".
527°NPt (J. S.). Stromatolitic limestone, locality cf. 296 except at base of "Fall 3".
528°NPt (J. S.). Dolomitic stromatolitic limestone, locality cf. 524.
529°NPt (J. S.). Stromatolitic dolomite, locality cf. 296 except at upper part of "Fall 1".
530°NPt (J. S.). Stromatolitic dolomite of the Malmani Subgroup from the "Schmidtschrift area" in the "upper part (Visser)".
531°NPt (J. S.). Stromatolitic limestone, locality cf. 296 except at the upper part of "Fall 3".
532°NPt (J. S.). Stromatolitic dolomite, locality cf. 296 except at the base of "Fall 4".

I-3-2. Australia

PILBARA BLOCK

Tectonic unit/location. Pilbara Block, Western Australia.

Stratigraphy. Gorge Creek Group
Warrawoona Group:
Towers Formation
McPhee Formation.

Geologic age. All available relevant dates are listed by Hickman (1980). Of these, three of the more significant are a date of 3.050 ± 0.180 obtained on granites intrusive into the sequence (Compston and Arriens, 1968); a U-Pb date of 3.453 ± 0.016 obtained on felsic volcanics from near the middle of the Warrawoona Group (Pidgeon, 1978a); and a Sm-Nd date of 3.556 ± 0.032 obtained on basalt from near the base of the Warrawoona Group (Hamilton et al., 1980).

Paleoenvironments. The paleoenvironmental setting of the Towers Formation is discussed in some detail in Section 8-4-1 of the present volume. Although such information is not available for the banded iron-formations and shales here studied from the overlying Gorge Creek Group, they appear probably to be relatively deep-water marine sediments comparable to the "basin plain suspension and chemical sediments" known elsewhere in the Group (Eriksson, 1980a).

Metamorphic grade. Sediments of the Warrawoona Group (from the North Pole Dome area) are generally within the greenschist facies, although locally they are of lower metamorphic grade, within the prehnite-pumpellyite facies (Hickman and Lipple, 1978). X-ray diffraction analyses of Warrawoona kerogens are consistent with metamorphism to the prehnite-pumpellyite and lower greenschist facies (Tables 5-8 and 5-10, herein). Comparable analyses of kerogens from the Gorge Creek Group are consistent with greenschist facies metamorphism (Tables 5-8 and 5-10).

P.P.R.G. samples/localities (41 total samples, 27 processed).
001*. Chert from a chert dike in the Warrawoona Group collected in a small gorge at a waterfall and rockhole 150 m from the entrance to the gorge; Marble Bar 1:250,000 map sheet area grid reference no. 223364.

022*ks"μf". Chert of the Towers Fm. at the entrance to a small gorge, from an outcrop between the creek and a minor unpaved road (track), 1 m below the top of the outcrop; North Shaw 1:100,000 topographic sheet grid reference no. 536648 (C.P.C. 20205, 20206, 20209); stromatolitic fabric in Photos 8-2A, B, and 9-5E; possible microfossils in Photo 9-6A-G.

003*. Chert from a chert dike, locality cf. 001.

004*k. Chert from the Towers Fm., locality cf. 002 except at the top of the outcrop.

005°. Chert from the "lower chert" horizon of the Towers Fm. at a bluff below which the track crosses a ridge, on a steep hillslope; Marble Bar 1:250,000 map sheet area grid reference no. 224362.

006*. Chert from the "lower chert" of the Towers Fm. at North Pole B Deposit Mine, from the upper lip of the open cut on the west side; Marble Bar 1:250,000 map sheet grid reference no. 223357.

007*. Chert from the "lower chert" of the Towers Fm. at North Pole B Deposit Mine, from the upper bench on the west side; Marble Bar 1:250,000 map sheet grid reference no. 223357.

008*. Chert from the "lower chert" of the Towers Fm., locality cf. 007 except 5 m lower.

009*. Chert from a chert dike at the North Pole B Deposit Mine, from the base of an open cut on the east wall; Marble Bar 1:250,000 map sheet grid reference no. 223357.

010*. Barite from the "lower chert" of the Towers Fm. at the North Pole B Deposit Mine, from the west wall of the open cut; Marble Bar 1:250,000 map sheet grid reference no. 223357.

011°. Chert and barite from the "lower chert" of the Towers Fm. at the North Pole B Deposit Mine; Marble Bar 1:250,000 map sheet grid reference no. 223357.

012°. Chert from the "lower chert" of the Towers Fm. at the North Pole H Deposit; Marble Bar 1:250,000 map sheet grid reference no. 222356.

013*c. Chert from the "lower chert" of the Towers Fm., 0.5 km west of the old Dresser Minerals camp site, in a creek bed, about 30 m northeast of a creek junction; North Shaw 1:100,000 topographic sheet grid reference no. 533657.

014° (J. D. no. 10). Oncolitic(?) chert from the Towers Formation.

015*. Basalt from a creek-crossing on the ore haulage road; Marble Bar 1:250,000 map sheet grid reference no. 217369.

016*kc. Chert from Getty DDH NP3 at 87 m, North Pole area.

017°. Chert from Getty DDH NP5 at 114 m, North Pole area.

018*. Basalt from Getty DDH NP5 at 118 m, North Pole area.

019°. Chert from the Gorge Creek Group at the Shay Gap No. 4 Orebody pit; Yarrie 1:250,000 map sheet area.

020°. Chert from the Gorge Creek Group at the Shay Gap No. 3 Orebody pit; Yarrie 1:250,000 map sheet area.

022*k. Shale from the Gorge Creek Group at the Mt. Goldsworthy pit from a wall beside the ore haulage road about one-third the distance below the pit rim, from a shale bed about 3 m thick; Pt. Hedland 1:250,000 map sheet area.

023*. Shale and chert from the Gorge Creek Group at the Mt. Goldsworthy pit, from a wall beside the ore haulage road about two-thirds the distance below the pit rim (total pit depth, 730′); Pt. Hedland 1:250,000 map sheet area.

046°. Archean granite (map unit Agm, Archean) from a low isolated outcrop on the south side of the track from Mt. Edgard Homestead to Meentheena (Mt. Edgar hill bears 10° magnetic); Nullagine 1:250,000 map sheet, grid reference no. 304338.

349°s. Stromatolitic chert from the Towers Fm. from the locality specified in Walter et al. (1980) (U.W.A. 91338 and C.P.C. 21450); stromatolitic fabric in Frontispiece (top and middle) and Photo 8-1A-D.

388° (J. W. S. no. II-1 of 1976). Archean chert, from the locality listed in the 25th I.G.C. Guide Book 43A as Stop 2-1.

517°sμf (S. M. A. locality A4 of 9/10/77-2, S. M. A. nos. 298 and 299). Chert of the Towers Fm. collected on the west side of a 50 m-high east-west trending bedded chert ridge 400 m due east of the intersection of the Dresser Minerals haulage road (from the occupied mine site of June 1980) and the unpaved road to Marble Bar; North Shaw 1:100,000 topographic sheet grid reference no. 504683.5 (S. M. A. no. 298 or C.P.C. 20210, 20211 and 20212; S. M. A. no. 299 or C.P.C. 20201, 20202, 20203, 20204 and 20207); stromatolitic fabric in Photo 9-5F, G; microfossils (S. M. A. 298) in Figures 9-5D and 9-7E; microfossils (S. M. A. 299) in Photos 9-4A-I, 9-5A-C, 9-7A-D, F-I, and 9-8A-C.

522° (M. R. W. no. 18.2/5/80). Black chert from the

"lower chert" horizon of the Towers Fm., Warrawoona Group, at North Pole Dome; North Shaw 1:100,000 map sheet area grid reference no. 536669.

534°NPt (S. M. A. no. Fl of 6/9/80). Gray chert (float) from the "upper chert" of the Warrawoona Group at North Pole Dome; North Shaw 1:100,000 topographic sheet map coordinates 504668.35.

535°NPt (S. M. A. no. F2 of 6/9/80). Black chert (float), locality cf. 534.

536°NPt (S. M. A. no. 3 of 6/9/80). Chert and carbonate(?), locality cf. 534 except map coordinates 508668.3.

537°NPt (S. M. A. no. 4 of 6/9/80). Black and gray chert, locality cf. 534 except map coordinates 511668.3.

538°NPt (S. M. A. no. F5 of 6/9/80). Gray chert, locality cf. 537.

539°NPt (S. M. A. no. 5 of 6/9/80). Black chert, locality cf. 537.

540°NPt (S. M. A. no. 1 of 6/10/80). Black chert and carbonate(?), locality cf. 536.

541°NPt (S. M. A. no. 2 of 6/10/80). Black chert and brown carbonate(?), locality cf. 536.

542°NPt (S. M. A. no. 3 of 6/10/80). Red and gray chert, locality cf. 536.

YILGARN BLOCK

Tectonic unit/location. Yilgarn Block, Warriedar Fold Belt (Murchison Province), Western Australia.

Stratigraphy. Shale in Association 4 of Baxter and Lipple (1980).

Geologic age. Granites in this region of the Yilgarn Block have been dated at 2.62 Ga, an age that may be approximately equivalent to that of the Woodley's Find No. 1 shale; a maximum age for the sequence is unknown.

Paleoenvironments. No published paleoenvironmental interpretations are currently available.

Metamorphic grade. Undetermined.

P.P.R.G. sample/locality (1 total sample, 1 processed).

521°. Black shale from DDH Woodley's Find No. 1 of the Newport Mining Co.; Perenjory 1:250,000 map sheet area.

HAMERSLEY BASIN

Tectonic unit/location. Hamersley Basin, Western Australia.

Stratigraphy. Turee Creek Group
- Hamersley Group:
  - Brockman Iron Formation
    - Whaleback Shale Member
    - Dales Gorge Member
  - Mount McRae Shale
  - Mount Sylvia Formation
  - Wittenoom Dolomite/Carawine Dolomite
  - Marra Mamba Iron Formation
- Fortescue Group:
  - Jeerinah Formation/Lewin Shale
  - Maddina Basalt
  - Tumbiana Formation:
    - Meentheena Carbonate Member
  - Pillingini Tuff
  - Hardey Sandstone

Geologic age. Relevant information available as of 1976 has been summarized by Trendall (1976). More recently, W. Compston (pers. comm. to M. R. W., 1980) has obtained a U-Pb date of 2.470 $\pm$ 0.050 Ga for the Woongarra Volcanics of the Hamersley Group and a U-Pb date of 2.490 $\pm$ 0.030 Ga for the Brockman Iron Formation, and studies by R. T. Pidgeon (pers. comm. to M. R. W., 1979) have yielded a U-Pb date of 2.768 $\pm$ 0.024 Ga for the Spinaway Porphyry, an intrusive sill considered to have extrusive equivalents in the Fortescue Group.

Paleoenvironments. The sedimentary environments both of the Fortescue Group (Section 8-4-9) and the Hamersley Group (Section 8-4-10) are summarized elsewhere in this volume. According to Trendall (1976), the banded iron-formations of the Hamersley Group were deposited as chemical precipitates in an "exceptionally still and quiet subaqueous environment," probably within a barred basin in an arid environment, whereas the interbedded shales of the sequence mark influxes of volcanic debris. With regard to units of the Turee Creek Group, no paleoenvironmental interpretations have been published except for the suggestion that a diamictite in the group may be glaciogenic (Trendall, 1979).

Metamorphic grade. Relevant metamorphic maps have been published by Smith et al. (1978). On the basis of further studies in the region, R. E. Smith has kindly provided the detailed information regarding the localities sampled during the course of the P.P.R.G. project (Table I-2):

TABLE I-2. Burial metamorphic grade of units sampled from the Hamersley Basin.

P.P.R.G. sample no.	*Geologic formation*	*Burial metamorphic grade*	*Estimated burial depth (km)*
366–371	Wittenoom Dolomite and Marra Mamba Iron Fm.	Prehnite-Pumpellyite Facies (at the boundary between the prehnite-pumpellyite and the pumpellyite-epidote zones)	2.6 ± 0.4
041	Tumbiana Formation	Prehnite-Pumpellyite Facies (pumpellyite-epidote zone)	3.1 ± 0.25
025–033	Tumbiana Formation	Prehnite-Pumpellyite Facies (middle portion of the pumpellyite-epidote zone)	3.2 ± 0.4
383, 483–491	Brockman Iron Formation	Prehnite-Pumpellyite Facies (middle portion of the pumpellyite-epidote zone)	3.5 ± 0.25
024	Tumbiana Formation	Prehnite-Pumpellyite Facies (middle portion of the pumpellyite-epidote zone)	3.5 ± 0.8
059–062	Duck Creek Dolomite	Prehnite-Pumpellyite Facies (pumpellyite-epidote zone)	~ 4
036–040	Jeerinah Formation	At the boundary between the Prehnite-Pumpellyite Facies (pumpellyite-epidote zone) and Pumpellyite-Actinolite Facies (pumpellyite-actinolite zone)	4.5 ± 0.25
372–377	Marra Mamba Iron Fm. and Jeerinah Formation	Grade apparently between those of 036–040 and 049–057	
049–057	Duck Creek Dolomite	Pumpellyite-Actinolite Facies (middle portion of the pumpellyite-actinolite zone)	5.2 ± 0.5
356–361	Mt. McRae Shale and Mt. Sylvia Formation	Pumpellyite-Actinolite Facies (pumpellyite-actinolite zone)	5.5 ± 0.5
480–482	Brockman Iron Formation	Grade apparently between those of 356–361 and 047–048	
047–048	Mt. McRae Shale	At the boundary between the Pumpellyite-Actinolite Facies (pumpellyite-actinolite zone) and the Greenschist Facies (actinolite zone)	6.0 ± 0.25

Field photographs. Illustrated elsewhere in this volume are stromatolites from the Carawine Dolomite (Photo 8-26A) and from the Fortescue Group (Photo 8-20A-C).

P.P.R.G. samples/localities (77 total samples, 71 processed).

024*cs. Cherty stromatolitic carbonate from the Meentheena Carbonate Member of the Tumbiana Fm. at Nullagine River; Nullagine 1:250,000 map sheet area; 21°18′S, 120°23′49″E (C.P.C. 21433); stromatolitic fabric in Photo 8-21A-C.

025*csμf. Cherty stromatolitic limestone of the Tumbiana Fm. from a small quarry in the cliff to the east of the maintenance road at km 211.3 on the Newman-Port Hedland railroad; Roy Hill 1:250,000 map sheet area; this is Locality RHL 3 of Grey (1979) and the "Knossos" locality discussed in Chapter 8 of the present volume (Section 8-4-9) and referred to in the entries below in which the stratigraphic locations of the sampled strata are given in terms of their stratigraphic distance above an unconformity with the underlying Archean gneiss (C.P.C. 21434, 21435, 21436, 21437); stromatolitic fabric in Photo 8-24A, B, D; microfossils in Photo 9-9A-C.

026*cs. Cherty stromatolitic limestone from the "Knossos" locality (cf. 025), about 37 m above the unconformity with the underlying gneiss (C.P.C. 21438); stromatolitic fabric in Photo 8-25A, B.

027*kc. Calcitic chert from the "Knossos" locality, from float 4 m below the top of the cliff.

028*c. Cherty stromatolitic limestone from the "Knossos" locality, about 35 m above the unconformity.

029*csμf. Limestone from the "Knossos" locality, about 30 m above the unconformity and 4.2 m above the base of the dolomite unit (C.P.C. 21439, 21440); stromatolitic fabric in Photo 8-23B, C, and 8-24B; microfossils in Photo 9-10A-D.

030*c. Cherty limestone from the "Knossos" locality, about 27 m above the unconformity and 2 m above the base of the carbonate unit.

031*. Cherty limestone, locality cf. 028.

032*c. Cherty stromatolitic limestone from the "Knossos" locality, about 39 m above the unconformity, from the lower part of a large domal stromatolite.

033*cs. Cherty stromatolitic limestone from the "Knossos" locality, about 40 m above the unconformity, from the upper part of a large domal stromatolite (C.P.C. 21441); stromatolitic fabric in Photo 8-22A-E.

035*. Basalt (possibly from the Maddina Basalt) collected at the side of the maintenance road at km 210.0 on the Newman-Port Hedland railroad, about 2 km north of the radio mast; Roy Hill 1:250,000 map sheet area.

036*. Pyritic chert of the Jeerinah Fm. from the Hamersley Exploration Co. Drill Hole 182 at 182.9 m; Mount Bruce 1:250,000 map sheet grid reference no. 616219.

037*. Pyritic chert and shale, locality cf. 036 but at 185.3 m.

038*k. Pyritic shale, locality cf. 036 but at 194.4 m.

039*. Pyritic shale, locality cf. 036 but at 194.7 m.

040*. Pyritic shale, locality cf. 036 but at 198.1 m.

041*c (M. D. M. no. MDT 65). Cherty stromatolitic limestone of the Tumbiana Fm. from a long road and rail cutting on the Tom Price-King Bay railroad; Pyramid 1:250,000 map sheet grid reference no. 523334.

042*cs. Stromatolitic chert of the Carawine Dolomite from the lip on the north side of a gorge, about 5 m stratigraphically below the top of a hill; Nullagine 1:250,000 map sheet grid reference no. 363369 (C.P.C. 21442); stromatolitic fabric in Photo 8-26B, C.

043°. Carbonate ("oncolites"-calcrete), possibly of the Carawine Dolomite, at locality cf. 042 but on the west side of the tributary creek near the top of a hill.

044°. Porphyry (map unit Pp, an intrusive rock equivalent to volcanics of the Fortescue Group), beside the track north of Meentheena; Nullagine 1:250,000 map sheet grid reference no. 341352.

045°. Basalt (map unit Pb of the Fortescue Group) from the south side of the Mt. Edgar-Meentheena track; Nullagine 1:250,000 map sheet grid reference no. 333340.

047*. Shale, from near the top of the Mt. McRae Shale, at the open cut of the Tom Price Mine.

048*k. Shale, locality cf. 047.

063* (M. D. M.). Stromatolitic carbonate of the Turee Creek Group, 1 km east of the junction of Turee Creek and O'Brien Creek; Turee Creek 1:250,000 map sheet; 23°12′30″S, 118°5′30″E.

132* (J. W. S. no. II-6 of 1976). Shale from the Jeerinah Fm. collected at a road cut on the Wittenoom-Roebourne road about 110 km southeast of Karratha during Stop 2-6 of the

25th I.G.C. Excursion 43A; Pyramid 1:250,000 map sheet area.

356* (G.S.W.A. no. 52031). Black pyritic shale of the Mt. McRae Shale from Rhodes Ridge core RD1 from the interval between 430′9″ and 432′6″; 23°05′S, 119°03′E.

357*k (G.S.W.A. nos. 52043 and 52044). Cherty black shale, locality cf. 356 except between 532′9″ and 534′6″.

358* (G.S.W.A. no. 52056). Sideritic black shale, locality cf. 356 except between 615′3″ and 616′9″.

359* (G.S.W.A. no. 52065). Sideritic black shale, locality cf. 356 except between 707′ and 709′3″.

360* (G.S.W.A. no. 52067). Sideritic black and red shale, locality cf. 356 except between 725′ and 727′.

361*c. Banded iron-formation from Bruno's Band of the Mt. Sylvia Fm., locality cf. 356 except between 758′ and 760′6″.

362*. Dolomite of the Carawine Dolomite from G.S.W.A. Bore Hole Woodie Woodie 2, from the interval between 70′ and 72′6″.

363*. Dolomite of the Carawine Dolomite, locality cf. 362 except between 101′ and 104′.

364*. Dolomite of the Carawine Dolomite from G.S.W.A. Bore Hole Woodie Woodie 1, from the interval between 60′1″ and 63′; 21°38′S, 121°14′E.

365*. Dolomite, locality cf. 364 except between 168′ and 170′.

366*c (G.S.W.A. no. 42316). Dolomite with shale bands from the Wittenoom Dolomite from Bore Hole Millstream 9, from the interval between 205.4 m and 205.9 m; 20°37′S, 117°01′E.

367*c (G.S.W.A. nos. 42306 and 42307). Dolomite, cf. locality 366 except between 190.3 m and 190.8 m.

368*kc (G.S.W.A. nos. 42333 and 42334). Dolomite, cf. locality 366 except between 221.3 m and 221.8 m.

369*c (G.S.W.A. nos. 42348L and 42348D). Banded iron-formation from the Marra Mamba Iron Fm., locality cf. 366 except between 228.3 m and 228.7 m.

370*c. Dolomite, locality cf. 366 except between 93.0 m and 93.5 m.

371*c. Dolomite, locality cf. 366 except between 143.9 m and 144.4 m.

372*. Banded iron-formation of the Marra Mamba Iron Fm. from Hamersley Exploration Co. DDH 186, from the interval between 93.3 m and 93.7 m; AMA coordinates 594600 E, 7522600 N.

373*. Banded iron-formation, locality cf. 372 except between 116.3 m and 116.9 m.

374*. Banded iron-formation, locality cf. 372 except between 139.0 m and 139.3 m.

375*. Pyritic shale from the Jeerinah Fm. locality cf. 372 except between 143.1 m and 145.0 m.

376*. Pyritic shale, locality cf. 375 except between 158.5 m and 158.8 m.

377*. Pyritic shale, locality cf. 375 except between 194.1 m and 194.4 m.

383° (J. W. S. no. III-2 of 1976). Dales Gorge Member of the Brockman Iron Fm. collected at Stop 3-2 during 25th I.G.C. Excursion 43A.

384° (J. W. S. no. II-4 of 1976). Pillingini Tuff of the Fortescue Group collected at Stop 2-4 during 25th I.G.C. Excursion 43A.

385° (J. W. S. no. II-5 of 1976). Maddina Basalt of the Fortescue Group collected at Stop 2-5 during 25th I.G.C. Excursion 43A.

386° (J. W. S. no. V-5 of 1976). Black chert from the Wittenoom Dolomite collected at Stop 5-5 during 25th I.G.C. Excursion 43A.

387* (J. W. S. no. V-2 of 1976). Black shale from the Wittenoom Dolomite collected at Stop 5-2 during 25th I.G.C. Excursion 43A.

441* (A. G.). Shale cuttings from the Lewin Shale (stratigraphic equivalent of the Jeerinah Fm.) at Sunday Hill, east Pilbara, Western Australia.

480*. Pyritic shale of the Whaleback Shale Member of the Brockman Iron Fm., from the North Deposit at Mt. Tom Price, DDH 85 at 163 m.

481*. Shale, locality cf. 480 except at 158 m.

482*. Black shale, locality cf. 480 except at 159 m.

483*c. Banded iron-formation from the Dales Gorge Member of the Brockman Iron Fm., "BIF13" of Hole 51, Wittenoom, at 312′.

484°. Shale from the Dales Gorge Member, "S12" of Hole 51, Wittenoom, at 350′.

485*. Shale, locality cf. 484 except "S11" at 371′.

486*. Shale, locality cf. 484 except "S11" at 390′.

487*. Shale and chert, locality cf. 484 except "S10" at 407′.

488*. Shale, locality cf. 484 except "S9" at 430′.

489*c. Banded iron-formation, locality cf. 484 except "BIF7" at 457′.

490*. Shale, locality cf. 484 except "S6" at 494′.

491*c. Banded iron-formation, locality cf. 484 except "BIF4" at 522′.

504° (H. I. P. no. 101). Black shale of the Brockman Iron Fm. from DDH Paraburdoo.

505° (H. I. P. no. 224). Shale, locality cf. 504.

506° (H. I. P. no. 286). Shale, locality cf. 504.

507°(H. I. P. no. 287). Shale, locality cf. 504.
508° (H. I. P. no. 347). Shale, locality cf. 504.
520°"μf" (A. T. no. 40126B). Carbonaceous chert and banded iron-formation from the Marra Mamba Iron Fm. at Kungarra Gorge; 22°47′S, 116°58′E (C.P.C. 21443); dubiomicrofossils in Photo 9-3A-K.
533° (G.S.W.A. nos. 7433, 7435, 7440, 7451, 7454, and 7461). Black shale (powdered) from the Hardey Sandstone, east-southeast of Marble Bar (see Hickman and de Laeter, 1977).

ASHBURTON TROUGH

Tectonic unit/location. Ashburton Trough (Gee, 1979b), Western Australia.

Stratigraphy. Wyloo Group:
Duck Creek Dolomite.

Geologic age. As summarized by Trendall (1976) and Gee (1979b), the group may be between 1.85 and 2.2 Ga in age. However, recent data from the Hamersley Basin (see discussion in the preceding section of this appendix) indicate that the age of the Wyloo Group could approach 2.5 Ga, that of the underlying Hamersley Group. In addition, the only firm minimum age limit for the Wyloo units is set by the age of intrusive granodiorites dated at about 1.7 Ga. "Layered acid igneous rocks" interbedded in the Wyloo Group have yielded a date of 2.020 ± 0.165 Ga "but more samples are needed" (Compston and Arriens, 1968).

Paleoenvironments. No sedimentological analyses of the Group have been published. The abundantly stromatolitic Duck Creek Dolomite most probably is a shallow-water marine deposit including facies that range from upper intertidal-supratidal (for the microfossiliferous cherty dolomite) to shallow subtidal (for those units containing columnar stromatolites). The Wyloo carbonates have $\delta^{13}C_{PDB}$ values near zero (Figure 7-3, herein) supporting their interpretation as being of marine origin.

Metamorphic grade. As tabulated above (in the discussion of the metamorphic grade of samples from the Hamersley Basin), samples of the Duck Creek Dolomite here studied range in metamorphic grade from the prehnite-pumpellyite to the pumpellyite-actinolite facies.

P.P.R.G. samples/localities (14 total samples, 13 processed).
049*μf. Chert and dolomite of the Duck Creek Dolomite from a stratigraphic level at the top of a thick-bedded dolomite and below a ferruginous dolomite, 20 m north of where the Mt. Stuart-Duck Creek Homestead track crosses Duck Creek, at the edge of the brown cliff, just above the base of the tree-line; Wyloo 1:250,000 map sheet area grid reference no. 436195 (C.P.C. 21421,21422); microfossils in Photo 14-5E, G-I.
050*. Chert, locality cf. 049 except 2 m stratigraphically higher in the section.
051*. Chert, locality cf. 049, 050, and 052, except from a ferruginous dolomite unit above 050 and below 052.
052*. Chert and carbonate, locality cf. 049 but at the west end of the outcrop and 1 m below the bulbous stromatolite unit above the ferruginous dolomite.
053*. Stromatolitic carbonate and chert, locality cf. 049 but from the unit containing bulbous stromatolites at the west end of the outcrop.
054*kc. Stromatolitic dolomite from the Duck Creek Dolomite on the west bank of Duck Creek, from a north-facing cliff, 30 m above the base of the cliff; Wyloo 1:250,000 map sheet area grid reference no. 436196.
055*c. Stromatolitic dolomite, locality cf. 054.
056*c. Stromatolitic dolomite, locality cf. 054.
057*c. Stromatolitic dolomite, locality cf. 054 except 30 m stratigraphically lower on the cliff.
059*μf. Chert of the Duck Creek Dolomite from a 2 m-thick unit of ferruginous dolomite on a southwest-facing hillside at the foot of a cliff; Wyloo 1:250,000 map sheet area grid reference no. 459195 (C.P.C. 21423); microfossils in Photo 14-3F.
060*kcμf. Chert, locality cf. 059 but from a unit of light-brown weathered dolomite with chert lenses above 059 (C.P.C. 21424); microfossils in Photo 14-4C.
061*. Carbonate, locality cf. 059 and 060, but above 060 from a unit containing nodular and bulbous stromatolites.
062*μf. Chert, locality cf. 059 but from talus (C.P.C. 21425, 21426); microfossils in Photo 14-3A-D, J, and 14-4A, B, D, E.

NABBERU BASIN

Tectonic unit/location. Nabberu Basin, Western Australia.

Stratigraphy. Earaheedy Group:
Kulele Creek Limestone
Windidda Formation
Frere Formation
Yelma Formation.

Geologic age. At present, the age of the Earaheedy Group is poorly defined. Hall and Goode (1978) consider the Nabberu Basin sediments to be Early Proterozoic, possibly a shallow-water time-equivalent of the Hamersley Basin units (and thus ~2.5 Ga old). Gee (1979a), however, infers that the Earaheedy Group may be younger than the Wyloo Group of the Ashburton Trough, and summarizes geochronologic data that suggest an age of about 1.6 to 1.7 Ga for the Nabberu sediments.

Paleoenvironments. The Earaheedy Group was deposited within the photic zone on a shallow platform, with reworking by wave action (see Hall and Goode, 1978, for a detailed discussion).

Metamorphic grade. The samples here studied come from the "unmetamorphosed zone" of Hall and Goode (1978) and have a metamorphic grade lower than the lower greenschist facies.

P.P.R.G. samples/localities (56 total samples, 22 processed).

066* (M. R. W. no. 23.1c/9/77). Stromatolitic chert from the Frere Fm.; Wiluna 1:250,000 map sheet area grid reference no. 436751.

073* (M. R. W. no. 23.4/9/77). Chert from the Frere Fm., 0.5 to 1.0 km stratigraphically above 066 (locality on a fence line); Wiluna 1:250,000 map sheet area grid reference no. 435762.

075°NPt (M. R. W. no. 22.2/9/77). Oncolitic chert from the Frere Fm., 2.5 km northeast of Camel Well; Wiluna 1:250,000 map sheet area grid reference no. 433753.

076*μf (M. R. W. no. 22.3/9/77). Chert from the Frere Fm., a float specimen collected 2.5 km northeast of Camel Well; Wiluna 1:250,000 map sheet area grid reference no. 433753 (C.P.C. 21446); microfossils in Photo 14-5D, E.

078* (M. R. W. no. 24.3/9/77). Chert from the Frere Fm.; Kingston 1:250,000 map sheet area grid reference no. 491727.

079°NPt (M. R. W. no. 27.1/9/77). Carbonate and chert from the Windidda Fm., 50 to 100 m above the base of the formation; Kingston 1:250,000 map sheet area grid reference no. 53157170.

080*cμf (M. R. W. no. 27.2/9/77). Stromatolitic limestone and chert from the transition beds between the Frere and Windidda Fms.; Kingston 1:250,000 map sheet area grid reference no. 53157170 (C.P.C. 21447, 21448); microfossils in Photo 14-5F, G.

082* (M. R. W. no. 26.3/9/77). Dolomitic limestone and chert of the Windidda Fm. from immediately below the unconformity with the Wandiwarra Fm., on a roadside; Kingston 1:250,000 map sheet area grid reference no. 502742.

086*c (M. R. W. no. 27.4/9/77). Dolomitic limestone and chert from the basal part of the transition beds between the Frere and Windidda Fms.; Kingston 1:250,000 map sheet area grid reference no. 53157160.

089* (M. R. W. no. 27.7/9/77). Chert, locality cf. 086.

091*μf (M. R. W. no. 28.1/9/77). Chert and carbonate from the Windidda Fm. east of Tooloo Bluff; Kingston 1:250,000 map sheet area grid reference no. 534714 (C.P.C. 21449); microfossils in Photo 14-5A-C.

092* (M. R. W. no. 28.2/9/77). Chert of uncertain stratigraphic position from Mt. Russel; Glengarry 1:250,000 map sheet area.

093* (M. R. W. no. 29.1/9/77). Chert of uncertain stratigraphic position from a hilltop in Meekatharra; Glengarry 1:250,000 map sheet area.

094°NPt (R. H. no. 9251). Brown stromatolitic carbonate from the Frere Fm. near Lake Gregory, between Combine Well and Midway Well; Peak Hill 1:250,000 map sheet area; 25°43′S, 119°48′E.

095*c (R. H. no. 9252). Stromatolitic dolomite from the Frere Fm. near Lake Gregory, between Combine Well and Midway Well; Peak Hill 1:250,000 map sheet area; 25°43′S, 119°48′E.

096*c (R. H. no. 9263). Stromatolitic dolomite from the Frere Fm. near Sweetwaters Well; Nabberu 1:250,000 map sheet area; 25°37′S, 120°22′E.

097*c (R. H. no. 9250). Stromatolitic dolomite, locality cf. 095.

098°NPt (M. J. J. no. 72880233). Stromatolitic carbonate, possibly from the Frere Fm.; Duketon 1:250,000 map sheet area grid reference no. 609632.

099* (no. 318A). Stromatolitic dolomite, possibly from the Frere Fm. (locality unknown).

100* (no. 318B). Stromatolitic carbonate, locality cf. 099.

104* (no. 237). Stromatolitic dolomite, locality and stratigraphic position unknown.

426* (M. R. W. no. 19.5/9/77). Calcareous chert, possibly from the Yelma Fm.; Nabberu 1:250,000 map sheet area grid reference no. 369795.

435*c (M. R. W. no. 25.4/9/77). Stromatolitic carbonate from the Kulele Creek Limestone; Stanley 1:250,000 map sheet area grid reference no. 514778.

436* (M. R. W. no. 25.5/9/77). Stromatolitic carbonate, locality cf. 435 except grid reference no. 514777.

439*c (M. R. W. no. 27.3/9/77). Limestone from the transition beds between the Frere and Windidda Fms.; Kingston 1:250,000 map sheet area grid reference no. 53157170.

440*kc (M. R. W. no. 27.5/9/77). Cherty carbonate, locality cf. 086.

LAWN HILL PLATFORM

Tectonic unit/location. Lawn Hill Platform, Queensland, Australia.

Stratigraphy. McNamara Group:
Gunpowder Creek Formation.

Geologic age. On the basis of correlation with the nearby, radiometrically dated (U-Pb ages of zircons in a tuff) Mount Isa Group, the McNamara Group is considered to be about 1.65 to 1.67 Ga old, an age consistent with a wealth of other available geochronological data (Plumb et al., 1981).

Paleoenvironments. According to Derrick and Sweet (1978), the McNamara units were deposited in a low-energy lagoonal environment.

Metamorphic grade. At the locality here sampled, the Gunpowder Creek Formation occurs in a fault crush zone (Edwards, 1953).

PPRG sample/locality (3 total samples, 1 processed).

421° (M. R. W. no. MO13.59.1). Carbonaceous siltstone of the Gunpowder Creek Fm. from the pit of Mt. Oxide Mine.

BIRRINDUDU BASIN

Tectonic unit/location. Birrindudu Basin, northwestern Australia.

Stratigraphy. Bungle Bungle Dolomite.

Geologic age. Sediments of the Birrindudu Basin unconformably overlie granites dated at 1.765 to 1.685 Ga and contain glauconites giving minimum radiometric dates of about 1.56 Ga. Regional correlations suggest an age of about 1.6 Ga for the Bungle Bungle Dolomite (Plumb et al., 1981).

Paleoenvironments. There are no published paleoenvironmental interpretations of this unit. However, in an unpublished report, M. D. Muir records the occurrence of pseudomorphs after gypsum, intraclast grainstone, and *Conophyton* stromatolites; the sequence is likely to be marine peritidal.

Metamorphic grade. Available data on kerogens from the sequence indicate that they have been only moderately altered (Table 5-10, herein); evidently the sequence is only little metamorphosed.

P.P.R.G. samples/localities (26 total samples, 22 processed).

137* (M. D. M. no. WBB3). Gray shale of the Bungle Bungle Dolomite from an unnamed tributary to Osmond Creek; Dixon Range 1:250,000 map sheet area; 17°12′S, 128°22′E.

138°kc (M. D. M. no. WBB4). Calcareous dolomite and chert, locality cf. 137.

139*c (M. D. M. no. WBB6). Carbonate and chert, locality cf. 137.

140* (M. D. M. no. WBB7). Carbonate, locality cf. 137.

141* (M. D. M. no. WBB8). Carbonate, locality cf. 137.

143*c (M. D. M. no. WBB10). Carbonate, locality cf. 137.

144*μf (M. D. M. no. WBB11). Green shale, locality cf. 137 (C.P.C. 21427); fossils in Photo 14-10A-S.

145* (M. D. M. no. WBB12). Gray shale, locality cf. 137.

146* (M. D. M. no. WBB13). Gray shale, locality cf. 137.

147* (M. D. M. no. WBB15). Carbonate, locality cf. 137.

148* (M. D. M. no. WBB16). Carbonate, locality cf. 137.

149* (M. D. M. no. WBB17). Green shale, locality cf. 137.

150* (M. D. M. no. WBB18). Black shale, locality cf. 137.

151* (M. D. M. no. WBB19). Carbonate and chert, locality cf. 137.

152*kc (M. D. M. no. WBB20). Dolomite and chert, locality cf. 137.

153°cμf (M. D. M. no. WBB21). Stromatolitic, dolomitic chert, locality cf. 137 (C.P.C. 21428, 21429); microfossils in Photo 14-9D, F, I-L, N.

154*kcμf (M. D. M. no. WBB22). Cherty stromatolitic dolomite from the Bungle Bungle Dolomite, locality unknown (C.P.C. 21430); microfossils in Photo 14-9 A, G.

155*k (M. D. M. no. WBB23). Black shale from the Bungle Bungle Dolomite on an unnamed tributary to Osmund Creek; Dixon Range 1:250,000 map sheet area; 17°11′S, 128°27′E.

156* (M. D. M. no. WBB24). Black shale, locality cf. 155.
157* (M. D. M. no. WBB25). Black shale, locality cf. 155.
158* (M. D. M. no. WBB27). Carbonate from the Bungle Bungle Dolomite, locality unknown.
159*kcμf (M. D. M. no. WBB29). Carbonate and chert, locality cf. 137 (C.P.C. 21431, 21432); microfossils in Photo 14-9B, C, E, H, M, O-Q.

MCARTHUR BASIN

Tectonic unit/location. McArthur Basin, Northern Territory, Australia.

Stratigraphy. Roper Group:
McMinn Formation
Sherwin Ironstone Member
McArthur Group:
Emmerugga Dolomite
Tooganinie Formation
Amelia Dolomite
Barney Creek Formation
HYC Pyritic Shale Member
Mallapunyah Formation.

Geologic age. The geologic succession in the McArthur Basin is younger than 1.688 Ga. On the bases of dates available from within the basin and on correlations with the better defined sequences of the nearby Mount Isa Orogen, Plumb et al. (1981) infer an age of between 1.65 and 1.67 Ga for the McArthur Group. Dolerites intrusive into the overlying Roper Group are at least 1.28 Ga old, with preliminary data from the dating of Roper Group micas suggesting an age of about 1.4 Ga for this sequence (Plumb et al., 1981).

Paleoenvironments. The HYC Pyritic Shale Member of the Barney Creek Formation (McArthur Group) has been variously interpreted as a subaqueous deposit from below the photic zone (Oehler, 1977b) or as a coastal sabkha or saline lake deposit (Williams, 1980). The Amelia Dolomite and the Mallapunyah Formations (McArthur Group) were both deposited in marine lagoons and sabkhas in a peritidal, evaporitic, arid setting (Muir, 1979). Similar settings are probable for the Tooganinie Formation and the Emmerugga Dolomite of this group. Little is known about the depositional environment of the Sherwin Ironstone Member of the McMinn Formation (Roper Group), but it may have been deposited in a restricted, shallow-water, marine environment (Peat et al., 1978).

Metamorphic grade. The Sherwin Ironstone Member (McMinn Formation) is "unmetamorphosed" except close to intrusive dolerite sills (with results of a study of metamorphism adjacent to these sills being given by Peat et. al., 1978). A study of the carbon, sulfur, and oxygen isotope geochemistry of the HYC Pyritic Shale Member (Barney Creek Formation) suggests that the sulfide minerals in this unit may have been deposited from waters with temperatures of 150° to 260°C, but these suggested temperatures depend on a number of assumptions and can be regarded only as tentative (Williams, 1979). No other data on the metamorphic grade of the units here studied are currently available.

Photomicrographs of fossils. In addition to the photomicrographs of fossils from the Amelia Dolomite noted in certain of the entries below, microfossils from the HYC Pyritic Shale Member of the Barney Creek Formation are shown in Photo 14-7A-Y, herein.

P.P.R.G. samples/localities (32 total samples, 32 processed).
103*c (M. R. W. no. 11.1a/9/73). Stromatolitic dolomite and chert from the Amelia Dolomite, from a roadside; Bauhinia Downs 1:250,000 map sheet area grid reference no. 363916.
105*c (M. R. W. no. 7095). Carbonate nodule from the Mallapunyah Formation, collected about 1 km east of Archie Springs; Walhallow 1:250,000 map sheet area.
106*c (M. R. W. no. 12.1/9/73). Stromatolitic carbonate and chert from near the middle of the Tooganinie Fm.; Mallapunyah 1:100,000 map sheet grid reference no. 772378.
107*c (M. R. W. no. 11.4/9/73). Stromatolitic dolomite and chert from the Tooganinie Fm., collected at a roadside; Bauhinia Downs map sheet area grid reference no. 368879.
108*c. Carbonate and chert of the Amelia Dolomite from DDH Tawallah Pocket Number 1 at 25 m; 16°02′00″S, 135°38′45″E.
109*cμf. Carbonate and chert, locality cf. 108 except at 42.1 m (C.P.C. 21444); microfossils in Photo 14-6 H, J. O.
110*. Carbonate and chert, locality cf. 108 except at 25.6 m.
111*μf. Carbonate and chert, locality cf. 108 except 41.9 m (C.P.C. 21445); microfossils in Photo 14-6A-G, I, K.N.
112*. Carbonate and chert, locality cf. 108 except at 42.8 m.

113*k (M. R. W. no. 19.1/8/77). Calcareous shale and chert from the HYC Pyritic Shale Member; Bauhinia Downs map sheet area grid reference no. 408928.
114*k. Shale of the McMinn Fm. from DDH B7 at 39′6″, 20 km west of the Roper River Police Station; 14°42′S, 134°20′E.
115*. Shale, locality cf. 114 except at 56′4″.
116′. Shale, locality cf. 114 except from the interval between 74′2″ and 74′6″.
117*. Shale, locality cf. 114 except at 95′6″.
118*. Shale, locality cf. 114 except at 111′0″.
119*. Shale, locality cf. 114 except at 124′6″.
120*k. Shale, locality cf. 114 except at 131′6″.
121*. Shale, locality cf. 114 except at 163′4″.
122*. Shale, locality cf. 114 except at 171′0″.
123*k. Shale, locality cf. 114 except at 190′0″.
124*. Shale, locality cf. 114 except at 200′8″.
125′. Shale, locality cf. 114 except at 204′0″.
126*. Shale, locality cf. 114 except at 217′5″.
127′. Shale, locality cf. 114 except at 376′6″.
128*. Shale, locality cf. 114 except at 385′2″.
129*. Shale, locality cf. 114 except at 400′9″.
130*. Shale, locality cf. 114 except at 413′2″.
131*. Shale, locality cf. 114 except at 428′4″.
452* (M. R. W. no. 14.6/9/73). Stromatolitic dolomite and chert from the Amelia Dolomite; Mallapunyah 1:100,000 map sheet area grid reference no. 722665.
457*. Shale, locality cf. 114 except at 452′2″.
458*. Shale, locality cf. 114 except at 462′0″.
459*. Shale, locality cf. 114 except at 490′6″.

AMADEUS BASIN

Tectonic unit/location. Amadeus Basin, Northern Territory, Australia.

Stratigraphy. Bitter Springs Formation:
Loves Creek Member.

Geologic age. Dolerite dikes unconformably underlying the sedimentary sequence have yielded a maximum Rb-Sr age for the Bitter Springs Formation of 0.897 ± 0.009 Ga (Black et al., 1980). A minimum age for the formation is provided by unconformably overlying tillites which are probably about 0.750 Ga old (Preiss et al., 1978).

Paleoenvironments. Sediments of the Bitter Springs Formation were deposited in a marine peritidal environment (Walter, 1972a, and unpublished observations by M. R. W.).

Metamorphic grade. No prior published information is available (however, see Table 5-10).

P.P.R.G. sample/locality (1 total sample, 1 processed).
133*kc (J. W. S. no. 1-9/11/73). Carbonaceous, microfossiliferous chert from the Loves Creek Member at Ellery Creek, 80 km west of Alice Springs, N.T.

ADELAIDE GEOSYNCLINE

Tectonic unit/location. Adelaide Geosyncline, South Australia.

Stratigraphy. Burra Group:
Kalachalta Formation
"Transition Unit"
Skillogalee Dolomite.

Geologic age. The sequence is older than 0.750 ± 0.053 Ga, the Rb-Sr whole rock shale age of the unconformably overlying Tapley Hill Formation; it is possibly younger than 0.80 to 0.85 Ga; and it is probably younger than 1.076 ± 0.034 Ga, the Rb-Sr age of volcanics that almost certainly underlie the Burra Group (Preiss and Forbes, 1980).

Paleoenvironments. According to Ambrose et al. (1981), the sequence was deposited in a paralic environment.

Metamorphic grade. Most probably from the lower greenschist facies (but the only published data are from studies of areas 20 km and more from the locality here sampled; Ambrose et al., 1981).

P.P.R.G. sample/locality (1 total sample, 1 processed).
455*kc (J. W. S. no. 1-8/22/73). Stromatolitic, carbonaceous, microfossiliferous dolomitic chert from the "Transition Unit" between the Skillogalee Dolomite and the Kalachalta Fm. at Boorthanna in the Peake and Denison Ranges; 28°40′S, 135°53′E.

I-3-3. Greenland

ISUA SUPRACRUSTAL BELT

Tectonic unit/location. Isua Supracrustal Belt, southwestern Greenland.

Stratigraphy. "Isua Supracrustal Sequence."

Geologic age. Among the numerous age determinations available for this highly metamorphosed sequence are a Pb-Pb isochron date of 3.71 ± 0.07 Ga (Moorbath et al., 1973; recalculated by Michard-Vitrac et al., 1977); a Rb-Sr whole rock date of 3.71 (+0.00, −0.09) Ga maximum (Moorbath et al., 1975); $^{207}Pb-^{206}Pb$ dates on zircons of 3.67 Ga, 3.77 Ga (Baadsgaard, 1976) and 3.769 (+0.011, −0.008) Ga (Michard-Vitrac et al., 1977); and a Sm-Nd date of 3.770 ± 0.042 Ga (Hamilton et al., 1978).

Paleoenvironments. The Isua Supracrustals are a subaqueous, volcanic-exhalative sequence (Appel, 1980; see also Chapters 3 and 10 in the present volume).

Metamorphic grade. As summarized by Bridgwater et al. (1981), the Isua units were subjected to at least five regional metamorphic events between 3.7 and 1.6 Ga ago, including two episodes of metamorphism in the amphibolite facies.

P.P.R.G. samples/localities (36 total samples, 11 processed).

171°NPt"μf" (M. S. no. 2377). Greenish white metaquartzite, locality given in Figure I-2; pseudomicrofossils in Photo 9-1A-I.

461* (E. C. P. no. EP5.1-75). Quartz-magnetite-amphibole-graphite gneiss (see Figure I-2).

462* (E. C. P. no. EP6.1-75). Quartz-grunerite-magnetite gneiss (see Figure I-2).

464* (E. C. P. no. EP35-75). Micaceous quartzite (see Figure I-2).

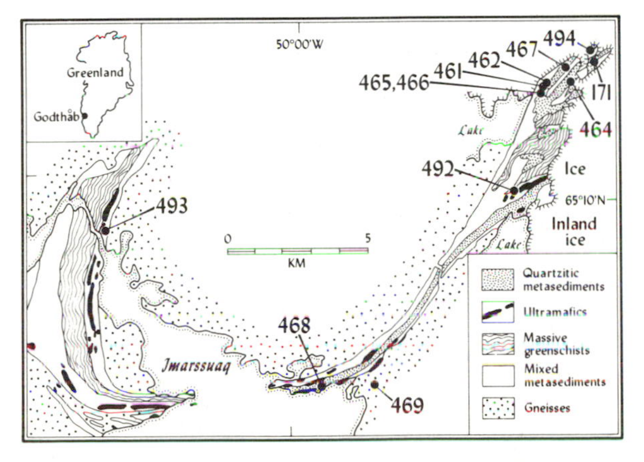

FIGURE I-2. Generalized geologic sketch map showing the location of pseudomicrofossil-containing (P.P.R.G. Sample No. 171) and processed samples (Nos. 461, 462, 464–469, 492–494) from the "Isua Supracrustal Sequence" in the Godthåb area of southwestern Greenland (modified after fig. 1 in Oskvarek and Perry, 1976).

465*kc (E. C. P. no. EP51.1-75). Magnetite-graphite-carbonate rock (see Figure I-2).

466*k (E. C. P. no. 51.2-75). Garnet schist (see Figure I-2).

467° (E. C. P. no. 87-75). Quartzite with grunerite (see Figure I-2).

468* (E. C. P. no. 134-75). Gneiss (see Figure I-2).

469* (E. C. P. no. 185-75). Gneiss (see Figure I-2).

492*c (M. S. no. PR17). Silicate iron-formation (see Figure I-2).

493° (M. S. no. 2914). Silicate iron-formation (see Figure I-2).

494* (M. S. no. PA4). Oxide iron-formation (see Figure I-2).

I-3-4. North America

UCHI GREENSTONE BELT

Tectonic unit/location. Uchi Greenstone Belt (Superior Province), Ontario, Canada.

Stratigraphy. Woman Lake Marble.

Geologic age. Zircons from rhyolite in the succession have yielded a U-Pb date of 2.787 Ga (Thurston, 1980).

Paleoenvironments. The Woman Lake carbonates were deposited in a volcanic sequence of a greenstone belt (Thurston, 1980).

Metamorphic grade. Greenschist facies (Thurston and Jackson, 1978).

P.P.R.G. sample/locality (1 total sample, 1 processed).

516*. Limestone marble from the top of Cycle II volcanism (map unit 5d of Thurston and Jackson, 1978; see also Thurston, 1980) at Woman Lake Narrows; 51°08′29″N, 92°44′42″W.

WAWA BELT

Tectonic unit/location. Wawa Belt (Superior Province), Ontario, Canada.

Stratigraphy. Michipocoten Iron Formation.

Geologic age. Zircons from associated metavolcanics have yielded a U-Pb date (Krough and Davis, 1971) of 2.710 Ga (recalculated, using new decay constants of Steiger and Jäger, 1978). A whole rock "errorchron" age of 2.680 ± 0.055 Ga ($\lambda^{87}Rb = 1.42 \times 10^{-11} a^{-1}$) from analyses of 31 lavas has been obtained by C. Brooks (pers. comm. to H. J. H., 1981).

Paleoenvironments. This banded iron-formation has been ascribed to the occurrence of large-scale hot spring and fumarolic activity in the vicinity of expanding volcanic piles (see Goodwin, 1962).

Metamorphic grade. Subgreenschist facies.

P.P.R.G. sample/locality (1 total sample, 1 processed).

331*. Loose blocks of iron-formation from a waste dump at the Sir James Dunn Iron Mine, at the northeast end of Wawa Lake; G.S.C. locality no. 78420.

WABIGOON BELT

Tectonic unit/location. Wabigoon Belt (Superior Province), Ontario, Canada.

Stratigraphy. Steeprock Group.

Geologic age. The radiometric age of this Archean sequence is poorly defined; according to Lepp and Goldich (1964), a sample of "phyllite" from a drill hole at Strawhat Lake has yielded an age of 2.7 Ga.

Paleoenvironments. Carbonates and cherts deposited in a localized basin within a volcanic belt (see Shklanka, 1972).

Metamorphic grade. Greenschist facies.

Field photographs. The Steeprock Group stromatolites are shown in Photo 8-10, herein.

P.P.R.G. samples/localities (9 total samples, 6 processed).

321*. Limestone with "*Atikokania*" from a hill on the northeastern side of the dam on the Southeast Arm of Steep Rock Lake; 48°48′36″N, 91°35′36″W; G.S.C. locality no. 79200.

322*c. Limestone from the footwall bench near a hilltop at Elbow Point, Steep Rock Lake; 48°47′43″N, 91°37′58″W.

323*. Pyritic chert from the Errington Pit at Steep Rock Lake; 48°47′37″N, 91°37′57″W.

324*cs. Large domal carbonate stromatolite from the Hogarth Pit at Steep Rock Lake; 48°48′59″N, 91°38′26″W; stromatolitic fabric in Photo 8-11A, B.

325*kc. Chert from the South Roberts Pit at Steep Rock Late; 48°47′54″N, 91°38′23″W.

326*. Carbonaceous sediment from the Hogarth Pit at Steep Rock Lake, locality cf. 324.

327°NPts (M. R. W. no. 13.1/7/71). Stromatolitic carbonate in a forest east of the open cut at Steep Rock Late; stromatolitic fabric in Photo 8-13A, B.

328°NPts (M. R. W. no. 13.2/7/71). Stromatolitic carbonate, locality cf. 327; stromatolitic fabric in Photo 8-12A, B.

329°NPt. Carbonate from a large domal stromatolite at Steep Rock Lake; 48°48′55″N, 91°38′29″W; G.S.C. locality no. 79199.

SUPERIOR PROVINCE

Tectonic unit/location. Superior Province, Ontario, Canada.

Stratigraphy. "Keewatin" iron-formation.

Geologic age. Radiometric data for this unit are not available, but the iron-formation is part of a basement that is older than 2.5 Ga.

Paleoenvironments. Metasediments occurring within a sequence of metavolcanic units.

Metamorphic grade. Greenschist facies (at the locality sampled, the unit is heavily sheared).

P.P.R.G. sample/locality (1 total sample, 1 processed).

330*. "Black chert" (viz., a fine-grained quartzite) from a road cut on Ontario Highway 17, 0.8 km northwest of Schreiber, Ontario; 48°49′18″N, 87°16′43″W; G.S.C. locality no. 78419 (it is from this locality that Madison, 1958, has reported Archean "protozoans"; see Table 9-1, No. 36).

SLAVE PROVINCE

Tectonic unit/location. Slave Province, central District of Mackenzie, Canada.

Stratigraphy. Yellowknife Supergroup.

Geologic age. As summarized by Henderson (1975a), the age of metamorphism of the Yellowknife Supergroup is between 2.6 and 2.3 Ga. A ^{207}Pb–^{206}Pb discordant age of 2.510 Ga has been obtained on zircons from felsic volcanic breccias occurring 3.5 km southwest of the stromatolitic locality here sampled (Wanless *in* Henderson, 1975a, p. 1628).

Paleoenvironments. Carbonate stromatolites occur in (faulted?) lenses that overlie graywacke-mudstone turbidites and that underlie felsic volcanics, including tuffs; the volcanics were probably emergent, providing narrow shallow-water environments in which the stromatolites were formed (Henderson, 1975a).

Metamorphic grade. Greenschist facies (the unit collected is located near a shear zone).

P.P.R.G. sample/locality (1 total sample, 1 processed).

413*cs. Ferruginous carbonate from near the base of the exposure of "Carbonate Unit 6" of Padgham (1973) at Snofield Lake; 67°17.5′N, 110°45′W; Locality 1 of Henderson (1975a); stromatolitic fabric in Photos 8-5C and 8-6.

PENOKEAN FOLD BELT

Tectonic unit/location. Penokean Fold Belt (Canadian Southern Province), Ontario, Canada, and northern Michigan, U.S.A.

Stratigraphy. Huronian Supergroup (Canada):
Gowganda Formation
Espanola Formation
Matinenda Formation
Pronto Paleosol/Denison Paleosol
Lower Proterozoic (Michigan):
Michigamme Formation
Randville Dolomite.

Geologic age. In the Huronian sequence in Canada, Fairbairn et al. (1969) have obtained a whole rock date of 2.238 ± 0.085 Ga (recalculated, using λ^{87}Rb = 1.42×10^{-11} a^{-1}) on the Gowganda Formation; this unit is intruded by the Nipissing Diabase, dated at 2.109 ± 0.078 Ga (recalculated, using λ^{87}Rb = 1.42×10^{-11} a^{-1}) by van Schmus (1965). The sequence in northern Michigan is between 2.1 and 1.9 Ga in age (van Schmus, 1976).

Paleoenvironments. The Canadian sediments include paleosols, varved mudstones with dropstones, and basin-margin to fluvial deposits. In northern Michigan, the Randville Dolomite is a platform carbonate, and the Michigamme Formation is a highly carbonaceous, basinal black shale.

Metamorphic grade. Subgreenschist to greenschist facies.

P.P.R.G. samples/localities (8 total samples, 8 processed).

333°. Varved argillite, with dropstones, of the Gowganda Fm. from a cliff on the northwestern side of Ontario Highway 129 (in the northeastern corner of Wells Township), 32 km northeast of Thessalon, Ontario; 46°26′18″N, 83°20′05″W; G.S.C. Locality no. 78421.

500° (J. W. S. "Morrison Creek Exploration Site" Locality of 1964). Carbonaceous, coaly shale of the Michigamme Fm. from an exploration pit 30 km northwest of Iron River, northern peninsular Michigan.

501° (F. M.). Stromatolitic dolomite of the Randville Fm., 9.6 km north of Felch, Dickinson County, Michigan; 5½SW¼ Sec. 3, T42N, R28W (F. M. PP2141).

502° (F. M.). Stromatolitic dolomite, locality cf. 501 (F. M. PP2142).

503° (F. M.). Stromatolitic dolomite of the Randville Dolomite from a road metal quarry 1.6 km east of Iron Mountain, Michigan, on the eastern shore of Lake Antoine (F. M. PP2144).

509° (D. E. G.). Denison Paleosol, 2 m below an unconformity with the overlying Matinenda Fm., from the Consolidated Denison Mine at Elliot Lake, Ontario (see Gay and Grandstaff, 1980).

510° (D. E. G.). Pronto Paleosol, 0.5 m below an unconformity with the overlying Matinenda Fm., west of Pronto Mine at Elliot Lake, Ontario (see Gay and Grandstaff, 1980).

515°. Stromatolitic carbonate of the Espanola Fm., near the No. 2 Shaft of Denison Mine, on the northwest shore of Quirke Lake, Ontario; 46°29′30″N, 82°35′50″W (see Hofmann et al., 1980).

EAST ARM FOLD BELT, ATHAPUSCOW AULACOGEN

Tectonic unit/location. East Arm Fold Belt (Athapuscow Aulacogen), southern District of Mackenzie, Canada.

Stratigraphy. Great Slave Supergroup:
Pethei Group:
Hearne Formation
Taltheilei Formation
Kahochella Group:
Gilbralter Formation.

See Hoffman (1968) for a detailed discussion of the stratigraphy.

Geologic age. The minimum age of the supergroup is 1.790 Ga, based on analyses of intrusive quartz diorite laccoliths; the maximum age of the sequence is 2.170 Ga, based on studies of dikes that cut basement rocks but that do not cut units of the Great Slave Supergroup (Hoffman et al., 1974, p. 41).

Paleoenvironments. Carbonate sediments deposited in platform and basinal environments (for details, see Hoffman et al., 1974).

Metamorphic grade. Essentially unmetamorphosed.

P.P.R.G. samples/localities (13 total samples, 5 processed).

338*c. Stromatolite biostrome (*Stratifera hearnica*) of the Hearne Fm., from the south shore of Blanchet Island at Great Slave Lake, N.W.T.; 62°04′30″N, 112°13′00″W; G.S.C. Locality no. 75130.

340*c. Stromatolitic ferruginous carbonate with *Stratifera kahochella* (caliche?) of the Kahochella Group from the southwestern side of Shelter Bay, Great Slave Lake, N.W.T.; 62°48′45″N, 110°31′10″W; G.S.C. Locality no. 75133.

406° (M. R. W. no. 14.2/8/72). Carbonate from the Great Slave Supergroup from the northeastern end of Blanchet Island, Great Slave Lake, N.W.T., Canada.

407°c (M. R. W. no. 12.3/8/72). Calcrete(?) from the Gibralter Fm. from the East Arm of Great Slave Lake, N.W.T., Canada.

408°NPt (M. R. W. no. 14.4/8/72). Carbonate from the Hearne Fm. from the south-central part of Blanchet Island, Great Slave Lake, N.W.T., Canada.

411°NPt (M. R. W. no. 15.5/8/72). Stromatolitic carbonate from the Great Slave Supergroup.

412*c. Stromatolitic carbonate (*Conophyton*) of the Taltheilei Fm., from the East Arm of Great Slave Lake, N.W.T., Canada.

PORT ARTHUR HOMOCLINE

Tectonic unit/location. Port Arthur Homocline (Canadian Southern Province), Ontario, Canada.

Stratigraphy. Rove Formation
Gunflint Iron Formation.

Geologic age. An isochron date of 1.600 Ga (recalculated, using λ^{87}Rb = 1.42×10^{-11} a^{-1}) has been obtained on whole rock samples from the Gunflint and Rove Formations (Faure and Kovach, 1969); the basement underlying the Gunflint Formation is older than 2.5 Ga.

Paleoenvironments. The sequence includes both shallow-water, nearshore facies, with stromatolites encrusting boulders of an underlying basal conglomerate, and deeper water shale and cherty facies, in part containing planktonic microfossils; some cherty ferruginous stromatolites of the Gunflint Iron Formation may have been deposited in localized hot spring environments (Walter, 1972b).

Metamorphic grade. Essentially unmetamorphosed to subgreenschist facies.

Photomicrographs of fossils. Microfossils from the Gunflint Iron Formation are shown in Photo 14-1A-J, herein.

P.P.R.G. samples/localities (5 total samples, 4 processed).

134*k (J. W. S. "Schreiber Beach Locality" of 1968). Chert cobbles of the Gunflint Iron Fm. ("lower algal chert") from a beach opposite Flint Island, 6 km west of Schreiber, Ontario, on the northern shore of Lake Superior.

335*. Anthraxolitic silicified grainstone from the basal part of the Gunflint Iron Fm. at a creek bed at Hillside, Whitefish River, 2.5 km south-southwest of Nolalu, Ontario; 48°17′08″N, 89°50′30″W; G.S.C. Locality no. 78410.

336*. Hematitic, stromatolitic chert of the Gunflint Fm. collected along an unpaved road on the west side of Mink Mountain, 7.3 km west of Mackies, Ontario; 48°14′25″N, 90°09′30″W; G.S.C. Locality no. 82436.

400°NPt. Hematitic, stromatolitic chert, locality cf. 336.

498*k (J. W. S. "Rove Quarry Locality" of 1968). Calcareous concretion from the Rove Fm. collected at Rove Quarry, near Pass Lake, 45 km northeast of Thunder Bay, Ontario.

BELCHER FOLD BELT, CIRCUM-UNGAVA GEOSYNCLINE

Tectonic unit/location. Belcher Fold Belt, Circum-Ungava Geosyncline (Churchill Province), southeastern Hudson Bay, N.W.T., Canada (Belcher Group) and Quebec, Canada (Manitounuk Group).

Stratigraphy.

Belcher Group	Manitounuk Group.
Mavor Formation	
McLeary Formation:	
Upper Member	
Lower Member	
Kasegalik Formation.	

For a summary of the stratigraphy of these groups and of suggested stratigraphic correlations, see Dimroth et al. (1970).

Geologic age. Volcanics and shales overlying these units have been dated by use of a composite isochron at 1.760 ± 0.037 Ga (recalculated, using $\lambda^{87}Rb = 1.42 \times 10^{-11}\ a^{-1}$) by Fryer (1972); the underlying basement is older than 2.5 Ga.

Paleoenvironments. Both peritidal (Kasegalik and McLeary Formations) and carbonate platform facies (Mavor Formation and Manitounuk Group) are represented in these sequences.

Metamorphic grade. Essentially unmetamorphosed to subgreenschist facies.

Photomicrographs of fossils. Elsewhere in this volume are illustrated microfossils from the Kasegalik (Photo 14-2A-1, N, P, Q) and McLeary Formations (Photo 14-2M, O, R-Z).

P.P.R.G. samples/localities (9 total samples, 9 processed).

341*. Stromatolitic, evaporitic dolomite from the Lower Member (and 105.8 m above the base) of the McLeary Fm. at Sanikiluaq, Belcher Islands, N.W.T.; 56°32.6′N, 79°14.1′W; see Locality no. 42774 in Hofmann (1976, text-fig. 2).

344*. Stromatolitic dolomite of the Manitounuk Group from a small island near to the shore (exposed at low tide) of the Maver Islands of Hudson Bay, 5.6 km north-northeast of Great Whale River, Quebec; 55°19′38″N, 77°43′35″W; G.S.C. Locality no. 79201.

345*. Stromatolitic dolomite of the Manitounuk Group, collected 5.9 km northeast of Manitounuk Sound; 55°45′10″N, 76°59′42″W; G.S.C. Locality no. 79204.

445*. Black chert from the Lower Member of the McLeary Fm. at Sanikiluaq, Belcher Islands, N.W.T., 20 cm above locality 341.

446*. Black chert from the Upper Member of the McLeary Fm. at Sanikiluaq, Belcher Islands, N.W.T.; see Locality no. 42771 in Hofmann (1976, text-fig. 2).

447*kc. Black chert from the Upper Member of the McLeary Fm. at Sanikiluaq, Belcher Islands, N.W.T.; see Locality no. 42778 in Hofmann (1976, text-fig. 2).

448°. Dolomite with large stromatolites from the Mavor Fm. at Sanikiluaq, Belcher Islands N.W.T.; 56°32.7′N, 79°14.6′W; for details of the locality, see Hofmann (1977, fig. 4, near top of Unit D).

449*c. Black chert and dolomite of the Kasegalik Fm. from an unnamed island of the Belcher Islands located at 56°29.5′N, 70°30.2′W; Locality A of Hofmann (1976, text-fig. 1).

495°. Stromatolitic dolomite with *Tungussia* stromatolites from the McLeary Formation at Sanikiluaq, Belcher Islands, N.W.T. (viz., from Marker Bed T, 12 m northeast of Locality no. 42776 in Hofmann, 1976, text-fig. 2).

MISTASSINI HOMOCLINE

Tectonic unit/location. Mistassini Homocline (Superior Province), Quebec, Canada.

Stratigraphy. Upper Albanel Formation
Lower Albanel Formation.

Geologic age. Fryer (1972) has obtained a whole rock isochron date on units of the immediately overlying Temiscamie Formation of 1.749 ± 0.054 Ga (recalculated, using $\lambda^{87}Rb = 1.42 \times 10^{-11}\ a^{-1}$); the basement underlying the Lower Albanel Formation is older than 2.5 Ga.

Paleoenvironments. Stromatolitic carbonates deposited on a subtidal platform.

Metamorphic grade. Essentially unmetamorphosed to subgreenschist facies.

P.P.R.G. samples/localities (2 total samples, 2 processed).

443*kc. Stromatolitic dolomite with anthraxolite cavity fillings from the Upper Albanel Fm. at Lake Albanel, Quebec; 51°01.5′N, 73°06.5′W; for details, see Hofmann (1977, fig. 2; and 1978, locality 3).

444*c. Stromatolitic dolomite containing the stromatolite *Mistassinia wabassinon* of the Lower Albanel Fm. from the north shore of Lake Mistassini, Quebec; 51°12′42″N, 73°25′06″W; locality 1 of Hofmann (1978).

SUDBURY BASIN

Tectonic unit/location. Sudbury Basin, Ontario, Canada.

Stratigraphy. Chelmsford Formation
Onwatin Formation
Vermillion Limestone.

Geologic age. The sequence is older than the Sudbury Irruptive which has been dated by Fairbairn et al. (1968) at 1.683 ± 0.029 Ga (recalculated using $\lambda^{87}Rb = 1.42 \times 10^{-11}\ a^{-1}$).

Paleoenvironments. Tuffaceous sediments and turbidites accumulated in a local basin (for discussion of the turbidites of the Chelmsford Formation, see Rousell, 1972).

Metamorphic grade. Subgreenschist facies.

P.P.R.G. samples/localities (4 total samples, 4 processed).

332*. Carbonate fracture fillings in a carbonate block of the Vermillion Limestone from the mine dump of Errington No. 2 Mine, 6.5 km southwest of Chelmsford, Ontario; G.S.C. Locality no. 78423 (for description, see Thomson, 1960).

334*. Black, carbonaceous, tuffaceous shale from the Onwatin Fm., collected at a road cut on the south side of Ontario Highway 144 (544), 10.6 km west of Chelmsford, Ontario; 46°35′05″N, 81°19′12″W.

402° (J. W. S.). Proterozoic anthraxolite, Ontario, Canada.

450*. Graywacke turbidite of the Chelmsford Fm. from the north side of Ontario Highway 144 (544), 3 km west-southwest of Chelmsford, Ontario; 46°34′15″N, 81°14′20″W.

451*. Black shale from the Onwatin Fm., collected on the east side of Gordon Lake Road, 0.6 km north of Vermillion Lake Road, in Dowling Township, Ontario, 10 km west-southwest of Chelmsford; 46°32′49″N, 81°18′48″W.

LAKE SUPERIOR BASIN

Tectonic unit/location. Lake Superior Basin (Canadian Southern Province), Ontario, Canada (Sibley Group), and northern Peninsular Michigan, U.S.A. (Oronto Group).

Stratigraphy.

Oronto Group (Upper Keweenawan):
- Nonesuch Shale:
  - Upper Shale Member

Sibley Group:
- Rossport Formation.

For a discussion of the stratigraphy of the Lake Superior Basin, see Franklin et al. (1980).

Geologic age. The Nonesuch Shale has yielded a Rb-Sr whole rock date of 1.046 ± 0.046 Ga (see Barghoorn et al., 1965). As determined by Franklin et al. (1980), the Sibley Group has a radiometric age of 1.339 ± 0.033 Ga (λ^{87}Rb = 1.42×10^{-11} a^{-1}).

Paleoenvironments. The Nonesuch Shale accumulated in a shallow, quiet, reducing environment (Barghoorn et al., 1965; Brown, 1971). The Rossport Formation was deposited in a shallow saline basin (see Franklin et al., 1980).

Metamorphic grade. Essentially unmetamorphosed.

P.P.R.G. samples/localities (5 total samples, 3 processed).

342*. Pink dolomite containing the stromatolite *Conophyton* cf. *garganicum* from the Rossport Fm. near the northwest corner of Disraeli Lake, 55 km west-northwest of Nipigon, Ontario; G.S.C. Locality no. 79193 (see fig. 8 of Hofmann, 1969).

343*. Pink dolomite containing the stromatolite *Conophyton* from the Rossport Fm. collected from a cliff on the northwest side of Stewart Lake; G.S.C. Locality no. 79198 (see fig. 8 of Hofmann, 1969).

497* (J. W. S. "White Pine Mine Locality" of 1964). Black shale from the Upper Shale Member of the Nonesuch Shale, collected at the White Pine Copper Mine, northern peninsular Michigan, U.S.A. (see Barghoorn et al., 1965).

VAN HORN MOBILE BELT

Tectonic unit/location. Van Horn Mobile Belt (Flawn, 1956), western Texas, U.S.A.

Stratigraphy. Allamoore Formation.

Geologic age. The Allamoore Formation is older than overlying rhyolite flows that have yielded K-Ar dates of about 1.0 Ga, and is younger than the underlying Carrizo Mountain Formation which is regarded as at least 1.2 Ga in age (Denison, 1980).

Paleoenvironments. Studies by Nyberg and Schopf (1981) indicate that the microfossiliferous unit here sampled was deposited in a shallow, marine, subtidal, intertidal or supratidal setting (see also King, 1965).

Metamorphic grade. Essentially unmetamorphosed to subgreenschist facies.

P.P.R.G. sample/locality (1 total sample, 1 processed).

518* (A. V. N.). Carbonaceous, microfossiliferous, stromatolitic chert from the Allamoore Fm. at Garren Ranch, 10 km north of the village of Allamoore and 25 km northwest of Van Horn, Texas.

UINTA UPLIFT

Tectonic unit/location. Uinta Uplift, northeastern Utah, U.S.A.

Stratigraphy. Uinta Mountain Group:
Red Pine Shale.

Geologic age. A preliminary Rb-Sr whole rock age of 0.932 ± 0.005 Ga (recalculated, using $\lambda^{87}Rb = 1.42 \times 10^{-11} a^{-1}$) has been obtained from this unit by Crittenden and Peterman (1975).

Paleoenvironments. Carbonaceous shales deposited in a shallow, basinal, marine environment.

Metamorphic grade. Essentially unmetamorphosed.

P.P.R.G. sample/locality (1 total sample, 1 processed).

135* (J. W. S. no. i of 8/13/77). Carbonaceous, gray to black shale collected in the Ashley National Forest, 50 km north of Duchesne, Utah, in T3N R4W Sec. 32 of Duchesne County; 40°36′49″N, 110°22′04″W (see Hofmann, 1977, p. 2).

JOHN DAY BASIN

Tectonic unit/location. John Day Basin, north-central Oregon, U.S.A.

Stratigraphy. Clarno Formation.

Geologic age. Eocene.

Paleoenvironments. Megafossiliferous, carbonaceous cherts (a silicified peat bog) accumulated in a terrigenous basin associated with volcaniclastics.

Metamorphic grade. Unmetamorphosed.

P.P.R.G. sample/locality (1 total sample, 1 processed).

456° (J. W. S. "Perret Ranch Locality" of 1972). Black, carbonaceous chert, containing numerous axes of fossil vascular plants, collected on the H. J. Perret Ranch about 10 km northeast of Redmond, central Oregon.

I-3-5. Holocene Samples

P.P.R.G. samples/localities (16 total samples, 2 processed).

414° (M. R. W. no. 24.3/6/72). Holocene high temperature opaline geyserite from the Shoshone Geyser Basin at Yellowstone National Park, Wyoming, U.S.A.

417° (M. R. W.). Holocene pedogenic calcrete from western Victoria, Australia.

519°NPt (M. R. W.). Tufted cyanobacterial mat collected at Fisherman's Bay, Spencer Gulf, South Australia.

I-3-6. Miscellaneous Samples

P.P.R.G. samples/localities (7 total samples, 5 processed).

346° (J. S. N. no. 125/MR/73). Kyanite-garnet-graphite-feldspar gneiss from the Archean Dharwar Supergroup, Sargur Group, at Manchahalli, Mysore State, southern India.

401° (B. V. T.). Early Proterozoic banded iron-formation from the Krivoi-Rog Iron Fm., Ukrainian S. S. R.

442° (P. G.). Phosphatic stromatolite from the Proterozoic Aravalli Group at Jhamar Kotra near Udaipur, Rajasthan, northern India.

APPENDIX II

Flow Chart and Processing Procedure for Rock Samples

By J. William Schopf, J. M. Hayes, Udo Matzigkeit, Malcolm R. Walter, and Kim W. Wedeking

The several hundred rock samples investigated during the course of the P.P.R.G. project were processed via the series of steps outlined in the flow chart shown in Figure II-1. One or more intact, homogeneous, large rock samples (generally ≥ 295 g) from each locality was selected for analysis, with each such sample being designated by a two-part number, the first part referring to the sample locality (e.g., P.P.R.G. No. 001; see Appendix I) and the second designating the specific sample selected from that locality (e.g., P.P.R.G. No. 001-1).

Each of the rock samples was fragmented and the weathered, surficially contaminated exterior portions were discarded. Fragments were then selected for the preparation of thin sections, generally two such sections for each sample: one principally for petrographic study (A series, 30 μm thick; e.g., Thin Section 001-1-A) and a second for micropaleontologic study (B series, 150 μm thick; e.g., Thin Section 001-1-B); a third section (C series, 100 μm thick; e.g., Thin Section 001-1-C) was prepared of each stromatolitic sample.

The interior fragments from each sample were then further broken to pieces about 2 cm in size; the residuum was sieved and chips <2 mm in size were discarded. Fragmentation and chipping of each sample were carried out in a clean, aluminum foil-lined box using a geologic hammer; clean, disposable plastic gloves were worn throughout these and subsequent rock-processing procedures. To remove contaminants that may have penetrated the samples along fractures or cleavage or bedding planes, the remaining interior rock chips from each sample (e.g., P.P.R.G. No. 001-1-RC) were then etched in hydrofluoric and hydrochloric acids, rinsed in distilled water, and dried in a forced-air oven at 105° C until a 10 to 15% loss in dry weight had been obtained (in general, yielding a total of ≥ 250 g of dry, etched rock chips). A 25 g portion of these etched rock chips (e.g., P.P.R.G. No. 001-1-ERC) was then crushed in a shatterbox (using a thoroughly cleaned ring and puck mill) to yield a reserve rock powder (e.g., P.P.R.G. No. 001-1-RRP) that was then analyzed both for its total organic carbon content (T.O.C., mg $C_{organic}$/g rock powder) and its organic carbon isotopic composition ($\delta^{13}C_{organic}$ vs. PDB) using the procedures for "survey analyses" specified in Appendix IV; for those samples containing ≥ 1 mg $C_{organic}$/g rock powder, a portion of this powder (e.g., P.P.R.G. 001-1-RXE) was dispatched to the U.S. Geological Survey in Denver for rock-eval analysis.

About 50 kerogeneous samples of particular interest were selected for subsequent detailed study; the kerogens isolated (e.g., P.P.R.G. No. 001-1-RK), generally amounting to ≥ 200 mg kerogen from each sample studied, were evaluated for their color, x-ray diffraction characteristics, and their carbon, hydrogen and nitrogen elemental and isotopic compositions using the analytical procedures specified in Appendix IV. As is indicated in Figure II-1, at numerous steps in the rock processing procedure, unused aliquots were retained for possible future studies (including reserve specimens of each rock sample, e.g., P.P.R.G. No. 001-1; reserve thin section chips, e.g., P.P.R.G. No. 001-1-RTS; reserve rock powder, e.g., P.P.R.G. No. 001-1-RRP; etched rock chips from most samples, e.g., P.P.R.G. No. 001-1-ERC; and reserve fragments, e.g., P.P.R.G. No. 001-1-RF, of which a 10 g portion of each chert sample processed was specifically set aside for D/H and $^{18}O/^{16}O$ analyses).

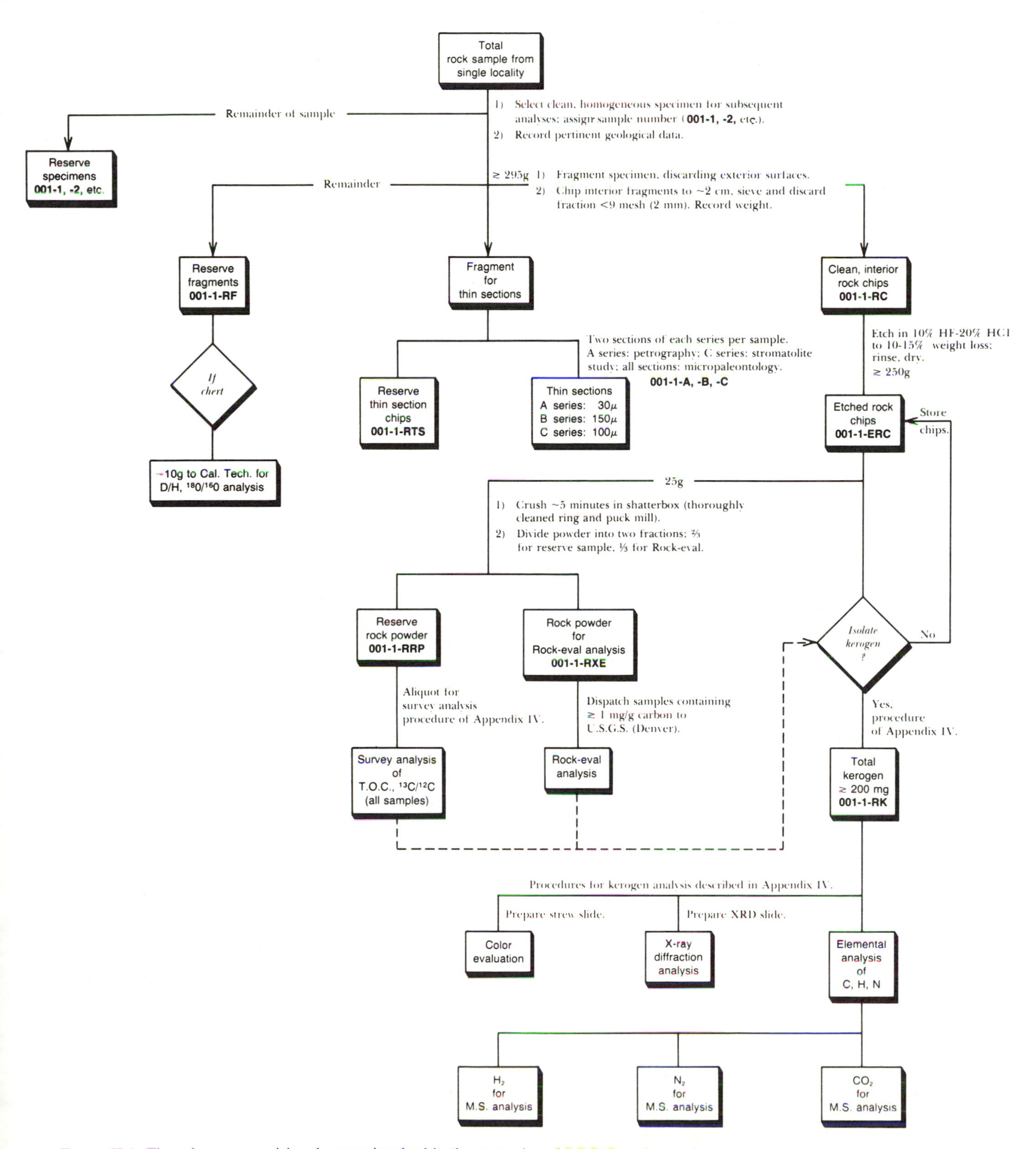

FIGURE II-1. Flow chart summarizing the steps involved in the processing of P.P.R.G. rock samples.

APPENDIX III

Processing Procedure for Abiotic Samples and Calculation of Model Atmospheric Compositions

By David DesMarais and Sherwood Chang

III-1. Experimental Design of Studies of Abiotic Syntheses

III-1-1. Electric Discharge Experiments

EQUIPMENT

The reaction vessel for all spark discharge experiments was a modified 5-liter Pyrex round-bottom flask with two ports for removable tungsten electrodes, an adaptor that allowed sampling into an evacuated tube, and a Teflon-coated stirring bar. Glassware was cleaned by heating in air to 800° K. A model BD-50 tesla coil was used for most experiments: in quenched spark discharges, one electrode was positioned with a gap of about 1 cm to the water surface; in gas spark discharges, two electrodes were positioned with a gap of about 3 cm between them.

MATERIALS

The following reagents were used: methane, CP grade; N_2, research grade; NH_3, anhydrous grade; NH_4OH and HCl, reagent grade; water, freshly distilled from glass apparatus commercial gas mixture, CO_2 15%, H_2 85% (Matheson Gas Products).

PROCEDURE

The reaction vessel, generally containing 100 ml water (exceptions noted below), was attached to the vacuum manifold and evacuated until release of dissolved gases appeared complete, then frozen. Gases were added to the vessel from glass reservoirs.

Ammonium chloride present. In experiments where NH_4Cl at pH 8.5 to 9 was specified (Table III-1), 95 ml water plus 5 ml of 1 M HCl (5 mmoles) was outgassed and frozen. Then a previously prepared tube containing 0.4 ml concentrated NH_4OH under vacuum (about 6 mmoles) was opened to the flask and ammonia distilled in. This resulted in 0.05 M ammonium chloride, or 267 mg NH_4Cl plus 1 mmole NH_4OH. Contamination with atmospheric CO_2 was avoided by this procedure.

Methane-ammonia-water series. Both gaseous components were added to the reaction vessel containing frozen H_2O at a pressure of 300 mm with the use of gas reservoirs (Table III-2). Ammonia rapidly condensed onto the water-ice surface, resulting, when melted, in 100 ml of 0.9 *M* NH_4OH.

Carbon dioxide-hydrogen series. No water was present initially in these experiments (Table III-3). The flask was filled to 200 mm with a commercially prepared mixture of CO_2 and H_2. Analysis of an aliquot from the reaction vessel before sparking was: CO_2 15%, H_2 85%. A gas-phase spark was generated across a 3 cm gap by a BD-50 tesla coil.

III-1-2. Ultraviolet Experiments

PHOTOLYSIS AT 254 NANOMETERS

A 12.7 liter quartz flask containing approximately 183 mm CH_4, 210 mm C_2H_6, 186 mm NH_3, and 98 mm H_2S was irradiated for 29 hours with a quartz high-pressure mercury vapor lamp (Hanovia, 450 W). Radiation at 254 nm was 5.8 W. The gases were sampled, then 50 ml degassed H_2O and additional H_2S were added and the vessel was irradiated an additional 374 hours (Table III-3). The sum of reactants, corrected for sampling losses

TABLE III-1. Methane-nitrogen-water spark discharges: Reaction conditions and compositions of reactants and products.

Gas composition		*Reaction conditions*				*Non-volatile product*		
CH_4 *(mm)*	N_2 *(mm)*	NH_4Cl[a]	*N/C*	*Electrodes*[b]	*% Reaction*[c]	*Wt. (mg)*[d]	*N/C*	$\delta^{13}C_{PDB}$ (‰)[e]
50	300	+	12.4	1	64 ± 10	284	0.17	−47.6
50	300	−	12.0	1	≥95	58	0.27	−43.7
187	198	−	2.12	2	48 ± 3	260	0.05	−44.6
200	200	+	2.1	2	39 ± 6	408	0.08	−46.8
200	100	−	1.0	1	39 ± 6	57	0.02	−48.8
200	100	+	1.05	1	55 ± 5	2640	0.10	−42.5
340	10	+	0.12	1	49 ± 3	286	0.07	−46.2
370	10	−	0.05	1	≤5	40	0.003	−50.4

[a] When NH_4Cl present (+), see procedural details in Sections III-1-1 ("ammonium chloride present") and III-2-2 ("removal of residual ammonia").

[b] 1 electrode: quenched spark; 2 electrodes: gas spark.

[c] % CH_4 consumed, determined by GC peak heights and combustion data.

[d] Includes approximately 267 mg NH_4Cl in experiments where NH_4Cl was present.

[e] Standard deviation = ±0.2 ‰. Isotopic composition of initial CH_4: −38.6 ‰. Products analyzed as CO_2 (see Section III-2-3).

TABLE III-2. Composition of products obtained from methane-ammonia-water quenched spark discharges.[a]

	Total non-volatile product				*Acid-insoluble residue*			
% Rxn[b]	*Wt. (mg)*	*N/C*[c]	$\delta^{13}C_{PDB}$ (‰)[d]	*% of total C*	*Wt. (mg)*	*N/C*[c]	$\delta^{13}C_{PDB}$ (‰)[d]	*% of total C*
44 ± 5	112	0.132	−47.4	4.7	46[e]	0.039	−47.0	2.1
51 ± 3					39	0.030	−45.1	1.1
70 ± 10					43	0.026	−45.8	2.6
70 ± 10	110	0.097	−43.2	6.8	68[e]	0.028	−42.7	4.4
70 ± 10	507	0.057	−41.1	33.2				
≥95	583	0.48	−39.5	23.7	150[e]	0.039	−36.9	7.8

[a] Experimental and analytical procedures described in Sections III-1-1 ("equipment" and "methane-ammonia-water series") and III-2-2 ("electric discharge experiments" and "acid hydrolysis and isolation of glycine").

[b] Methane consumption derived from GC peak heights and combustion data.

[c] Commercial CHN analysis performed by Microanalytical Laboratories, Mountain View, California.

[d] Standard deviation = ±0.2 ‰. Isotopic composition of initial CH_4: −38.6 ‰. Products analyzed as CO_2.

[e] Extrapolated from hydrolysis of fraction of non-volatiles.

but not adjusted for photoproducts of previous irradiation was: 168 mm CH_4, 193 mm C_2H_6, 171 mm NH_3, 159 mm H_2S, and 50 ml H_2O (see Khare and Sagan, 1973).

PHOTOLYSIS AT 124 NANOMETERS

A 356 ml vessel equipped with MgF_2 windows was filled with 2.0 mm CH_4 and 153 mm H_2 and was irradiated for 80 minutes with a low-pressure Krypton-Argon lamp powered by a KIVA microwave generator at 20 W (Table III-4). Radiation at 124 nm was 4×10^{15} photons cm^{-1} sec^{-1}.

III-1-3. AMMONIUM CYANIDE POLYMERIZATION EXPERIMENT

In a glass flask, 22.16 mmoles each of NH_3 and HCN were condensed in 20 ml H_2O under vacuum. The flask was sealed, and stored at room temperature for approximately 3 months.

TABLE III-3. Carbon isotopic compositions and N/C ratios of products of abiotic syntheses.

Type of abiotic syntheses	% Rxn[a]	$\delta^{13}C_{PDB}$(‰)[b]									N/C atomic ratio in bulk solids			Procedures
		CH_4	C_2H_6	C_2H_x to C_6H_y	CO_2	CO	CN^-	Bulk amino acids	Glycine	Bulk solid products	Wt. % C[c]	Wt. % N[c]	N/C	
254 nm Irradiation	0	−39.8	−39.8										0.33	III-1-2
	15 ± 10	−37.2	−38.6	−42.2										III-2-1
	50 ± 20									−49.2	23.9	16.4	0.41	III-2-2, -2-3
CH_4–NH_3–H_2O Spark	0	−38.6											1.0	III-1-1
	51 ± 3	−34.4		−43.7	−39.3	−42.0		−47.9	−55.7	−45.1	29.0	1.0	0.03	III-2-1, -2-2, -2-3
CO_2–H_2 Spark	0				−41.5	−52.7								III-1-1
	14				−39.6	−44.5								III-2-1
	86				−25.8									III-2-1
	≥95		−64.2											III-2-1
NH_4CN Polymerization	0						−28.2						2.0	III-1-3, -2-3
	80						−23.1			−29.8	35.0	36.9	0.90	III-2-2, -2-3

[a] Measured as % consumption of carbon source.
[b] Standard deviation = ±0.2 ‰. Products analyzed as CO_2.
[c] Commercial C and N analyses supplied by Microanalytical Laboratories, Mountain View, California.

TABLE III-4. Carbon isotopic composition of hydrocarbons generated from methane by spark discharge and by irradiation at 124 nm.[a]

	Spark discharge[b]				124 nm irradiation[c]	
Time (hr)	*0*	*0.25*	*1*	*19*	*0*	*1.33*
Methane	−38.6	−39.3	−37.6	−22.3	−30.1	
Ethane		−53.2	−49.3	−35.2		−30.9
Ethylene		−48.5	−49.4	−38.9		
Acetylene		−48.6	−49.8	−40.1		
Propane			−52.3	−41.3		−34.5
Saturated C_4's + Propylene			−51.7			−36.8
Saturated C_5's + Unsat. C_4's			−48.0	−39.7		−40.6
Benzene			−62.5	−44.3		

[a] Products analyzed as CO_2 generated on GC-combustion line (see Section III-2-1). $\delta^{13}C_{PDB}$ values corrected for combustion blank of 0.02 to 0.05 μmole; standard deviation = ±0.2 ‰.
[b] CH_4, 200 mm; no water or nitrogen.
[c] See Section III-1-2.

III-2. Methods of Analysis of Abiotic Products

III-2-1. Analysis of Volatile Products

GAS CHROMATOGRAPHY

Volatile products were analyzed by gas chromatography (GC) on a column consisting of 6′ × 1/8″ Porapak Q followed by 6′ × 1/8″ Porapak R (50 p.s.i. helium, programmed from 293° K or 353° K to 523° K). The effluent was split for simultaneous detection by flame ionization and nitrogen-selective detectors. This system provided separation of hydrocarbons to C_6 and enhanced response to HCN and nitriles.

GAS CHROMATOGRAPHY-COMBUSTION

The mixture of light hydrocarbons was injected onto a 2 m × 2 mm (internal diameter) column packed with Porasil B (operated at 15 cc/min helium flow, isothermal for 10 min. at 253° K, then programmed at 6 degrees per minute to 423° K). The eluting hydrocarbons were combusted to CO_2 and trapped immediately using methods described by Matthews and Hayes (1978). The CO_2 was purified using a cold trap described by Des Marais (1978) and transferred to a Nuclide 6-60 RMS mass spectrometer for isotopic analysis.

SEPARATION OF COMPONENTS AND COMBUSTION

The separation of mixtures of CO, CO_2 CH_4, and low molecular weight hydrocarbons and conversion of the components to CO_2 suitable for isotopic determinations were accomplished using the methods of Sakai et al. (1974).

III-2-2. Preparation of Non-Volatile Products

ELECTRIC DISCHARGE EXPERIMENTS

When the spark was terminated and gaseous products had been sampled, the vessel was briefly evacuated. The aqueous phase and solids were poured out and the flask was rinsed with a little water, followed by methanol and chloroform if necessary to remove the residue. The products were then dried.

Removal of residual ammonia. In order to obtain a N/C value on samples that initially contained NH_4Cl, residual NH_3 was removed as follows: 2 to 3 ml of 1 molar KOH solution was added to an aqueous suspension/solution of non-volatile product, which was next evaporated to a damp residue on a rotary evaporator. The residue was then suspended in water and the pH was measured using pH paper. KOH addition was repeated until the pH of the re-suspended residue was greater than 9. Then HCl was added until a pH of less than 4 and was obtained, and the residue was evaporated to dryness in order to remove adsorbed CO_2.

Acid hydrolysis and isolation of glycine. The reaction vessel was briefly evacuated to remove undissolved gases and part of the free ammonia. Then HCl was added under vacuum to acidify the reaction mixture, and pentane-ice-volatile gases were removed. The suspension was heated at 378° K overnight in 3 M HCl, then cooled and centrifuged. The insoluble product was hydrolyzed overnight in 6 M HCl; the acid-insoluble residue was recovered by centrifugation. The residue was washed with water, alcohol, and chloroform, then dried. The acid-soluble fraction was desalted first on a Dowex 50, then on a Dowex 2-OH column. The product, designated "bulk amino acids," contained some brown material that accompanied the amino acids on adsorption and desorption from both columns. An aliquot of "bulk amino acids" was removed and set aside for future study. The remainder was chromatographed on Whatman 3 MM filter paper in a vented tank in the solvent-mixture chloroform: methanol: concentrated ammonium hydroxide (5:5:1). This technique optimized the separation of amino acids from the bulk of very slow-moving brown material. Glycine was eluted with 2 mM HCl and freed of cellulose residues on a short column of Dowex 50. The 2 M NH_4OH eluate was evaporated to dryness, and then acidified and re-evaporated in order to remove adsorbed CO_2.

PHOTOLYSIS AT 254 NANOMETERS

Gases not condensable at 77° K were pumped away and the condensable gases were sampled. After additional pumping, the non-volatile product was scraped from the wall of the vessel and extracted with benzene, leaving a benzene-insoluble residue (see Khare and Sagan, 1973).

AMMONIUM CYANIDE POLYMERIZATION

The sample was opened and transferred to a flask that was then attached to a vacuum manifold, quickly frozen and evacuated to minimize adsorption of atmospheric CO_2. Water and volatiles were distilled into a flask cooled with liquid nitrogen leaving the non-volatile residue. Aliquots of the distillate (having a pH about 10.5) were titrated with $AgNO_3$ by the Liebig method (Day and Underwood, 1958, pp. 186 and 194). Additional $AgNO_3$ was added to a two-fold excess of the volume at which AgCN precipitation was observed, and AgCN was recovered by centrifugation after the samples had stood overnight at room temperature. An aliquot of the original HCN was condensed in a tube containing 0.2 M NH_4OH and precipitated as AgCN. Carbon isotopic compositions obtained by bomb combustion of AgCN derived from the original HCN, and by direct combustion of original HCN on the gas combustion line, were identical within the precision of the measurements.

III-2-3. Combustion Analysis of Solid Residues for Carbon and Nitrogen Abundances

In this section we indicate only the procedures in analysis of the products of abiotic syntheses. Methods used for determination of H, C, and N abundances and δD and $\delta^{15}N$ values of the heated samples (Table III-5) are essentially the same as those described in Appendix IV for the corresponding analyses of kerogens. Some H, C, and N analyses were obtained commercially from Microanalytical Laboratories, Mountain View, California.

Two methods were used to combust solid samples to CO_2 for isotope analysis. For boat combustion, the sample was placed in a porcelain boat within a quartz tube and heated in the presence of O_2, using the methods of Sakai et al. (1974). For bomb combustion, the sample was sealed in vacuum in a quartz tube containing CuO and silver wire, and heated to 1123° K. Both CO_2 and N_2 were recoverable by this method. Results obtained can be compared with previously published measurements of the N/C ratios in acid-insoluble abiogenic organic matter (Table III-6) and with the abundance and isotopic composition of abiotic carbon in various types of meteorites (Table III-7).

Table III-5. Effects of heating experiments on the elemental and isotopic compositions of products of abiotic syntheses.

Type of abiotic synthesis	*Temperature* (°K)	*Time* (*wk*)	*Wt. loss*[a] (*wt.* %)	*H/C*	*N/C*	$\delta^{13}C_{PDB}$ (‰)	δD_{SMOW}[b] (‰)	$\delta^{15}N_{AIR}$[c] (‰)
NH_4CN Poly-merization	unheated	—	0	—	0.90	—	—	—
	unheated	—	0	0.43	0.75	−30.0	—	—
	373	2.0	35	1.11	0.94	−29.4	−25.9	—
	373	2.0	5	0.96	0.89	−29.0	−33.3	—
	473	2.0	16	0.39	0.39	−30.0	—	+4.3
	473	2.0	—	0.54	0.69	−29.7	−13.2	—
	673	0.5	12	0.59	0.75	−29.4	−40.2	+1.5
	673	0.5	39	0.42	0.76	—	−4.7	—
	673	1.0	—	—	—	−30.0	—	—
	673	1.0	16	0.37	0.75	−29.8	−17.2	—
	673	2.0	45	0.36	0.78	−29.7	−15.7	—
	673	2.0	41	0.18	0.95	−30.0	−56.1	+2.0
	673	4.0	50	—	—	—	—	—
	673	4.0	51	0.45	0.74	−29.8	—	—
Electric Discharge	unheated	—	0	1.10	0.092	−41.7	−219	—
	unheated	—	0	—	0.057	—	—	—
	373	2.0	47	1.05	2.1	—	−227	—
	373	2.0	—	0.84	0.42	−41.8	—	+2.6
	473	2.0	—	0.65	2.53	−41.5	−239	+2.1
	673	0.5	—	—	—	—	—	—
	673	1.0	—	0.56	1.31	−41.8	−206	—
	673	2.0	—	0.50	1.84	−41.8	−221	—
	673	4.0	—	0.44	0.060	−41.8	−195	—

[a] Control experiments showed that of the solid material remaining after heating ≥ 90% could be recovered for analysis.
[b] Hydrogen/deuterium ratios relative to Standard Mean Ocean Water.
[c] $^{15}N/^{14}N$ ratios relative to air.

Table III-6. Atomic ratios of N/C in acid-insoluble abiogenic organic matter.

Source	*Reactants* *N/C*	*Product* *C* (*wt.* %)	*N* (*wt.* %)	*N/C*	*Comments*	*References*
METEORITES:						
Orgueil (CI)	—	70.4	1.59	0.019	Also contained S, Cl, F.	1
Murchison (CM)	—	73.1	2.68	0.031		2
	—	5.00	0.15	0.026	Phyllosilicate separates from which extractable organics were removed. *No acid used.*	3
	—	3.10	0.15	0.042		
Allende (CV3)	—	3.4 to 85	0.02 to 0.37	0.0053	Excellent C–N correlation among 5 samples. Variations in C and N abundances due to variable amounts of inorganic matter. Fe, Cr, O, and S also present.	4

References: 1. Hayes (1967). 2. Hayatsu et al. (1977). 3. Chang et al. (1978). 4. Ott et al. (1981). 5. Miller (1955). 6. Matthews and Moser (1967). 7. Voelker (1960). 8. Fox and Harada (1960). 9. Henley (1968). 10. Kenyon and Blois (1965).

TABLE III-6. (*continued*)

Meteorite (type)[a]	Bulk %C	Bulk $\delta^{13}C_{PDB}$	Acid insoluble fraction %C	Acid insoluble fraction $\delta^{13}C_{PDB}$	Carbonate %C	Carbonate $\delta^{13}C_{PDB}$	Extract[b] $\delta^{13}C_{PDB}$	References
ABIOTIC SYNTHESES:								
FTT Reaction	0.1	76.3	1.23	0.014	$CO/H_2/NH_3$ = 1/1/0.1, P = 2 atm, catalyst: Al- and Fe-oxides, T = 673°K (1 day), 573°K (2 days).			2
Electric Discharge	1.0	37.4	7.54	0.173	Contains 42.2% Ash. CH_4–NH_3–H_2–H_2O; CH_4/NH_3 = 1/1; but most of NH_3 dissolved in H_2O			5
HCN-Polymerization	1.0	44.5	52.5	1.01	No H_2O; HCN/NH_3 = 64/1			6
	1.0	40.7	43.5	0.92	Sub-fractions of above after chromatography on Sephadex G-15.			6
	1.0	39.1	38.4	0.84				
	1.23	41.4	42.3	0.88	HCN/NH_3 = 4.28. H_2O present.			7
Fox Microstructures	~0.2	51.0	13.2	0.22	Reactants were 10 g Glu, 10 g Asp, 5 g of a 16-amino acid mixture.			8
	~0.2	52.7	10.4	0.17	Same as above.			9
Phenylalanine Photo-degradation	0.13	46.2	12.4	0.23	Aqueous phenylalanine solution irradiated with 258 nm light.			10

TABLE III-7. Abundance (wt. %) and isotopic composition (‰) of carbon in meteorites.

Meteorite (type)[a]	Bulk %C	Bulk $\delta^{13}C_{PDB}$	Acid insoluble fraction %C	Acid insoluble fraction $\delta^{13}C_{PDB}$	Carbonate %C	Carbonate $\delta^{13}C_{PDB}$	Extract[b] $\delta^{13}C_{PDB}$	References
Ivuna (CI)	3.3	−6.6						1
	4.03	−7.5	1.57	−17.1	0.20	+65.8	−24.1	5
Orgueil (CI)	2.8	−11.4						1
							−10.1	2
					0.064 to 0.23	+58.6 to +64.3		3
		−15.6						4
		−51.9[c]		−24.8		+52.3	−12.9	4
		−9.7		−25.0		+55.1	−12.9	4
		−10.9		−23.9			−12.5	4
				−18.0		+61.8		4
	3.75	−11.6	2.15	−16.9	0.13	+70.2	−18.0	5
			2.50	−15.2				6

[a] Classification scheme of Wasson (1974). Carbonaceous meteorite types indicated by "C".

[b] Unless otherwise indicated, extracts were obtained with organic solvents (e.g., benzene-methanol).

[c] Disparate values of questionable significance.

[d] Average of two measurements.

[e] Average of three measurements.

[f] Range of values obtained from a water extract after separation into neutral, acidic and amino acid fractions.

[g] Pantar was separated into portions that contained light primordial noble gases (Pantar-I) and no primordial noble gases (Pantar-II). Lt. and Dk. refer to visually light and dark sub-portions.

References: 1. Boato (1954). 2. Briggs (1963). 3. Clayton (1963). 4. Krouse and Modzeleski (1970). 5. Smith and Kaplan (1970). 6. Belsky and Kaplan (1970). 7. Kvenvolden et al. (1970). 8. Chang et al. (1978). 9. Begemann and Heinzinger (1969).

Table III-7. *(continued)*

Meteorite (type)[a]	Bulk		Acid insoluble fraction		Carbonate		Extract[b]	References
	% C	$\delta^{13}C_{PDB}$	% C	$\delta^{13}C_{PDB}$	% C	$\delta^{13}C_{PDB}$	$\delta^{13}C_{PDB}$	
Revelstroke (CI)		−14.8		−26.4		+4.0		4
		−14.5		−28.0				4
Cold Bokkeveld (CM)	1.6	−9.4						1
	2.35	−7.2	1.27	−16.4	0.07	+50.7	−17.8	5
			1.41	−13.8				6
Erakot (CM)	2.30	−7.6	0.89	−15.1	0.05	+44.4	−19.1	5
			1.29	−13.8				6
Haripura (CM)	1.6	−3.7						1
							−3.1	2
				−17.8		+59.8		4
				−23.2		+60.6		4
	0.83	−21.3[c]	0.6	−16.6				6
Mighei (CM)	2.6	−9.9						1
	2.85	−10.3	0.72	−16.8	0.21	+41.6	−17.8	5
			1.56	−14.2				6
Murchison (CM)	2.5	−7.2 ± 0.1[d]		−10.6	0.114	+45.4	+5.0 ± 0.6[e]	7
	2.15	−0.5	1.45	−15.3	0.23	+44.4	+13 to +44[f]	8
				−16.3				
Murray (CM)	1.9	−3.9						1
							−4.0	2
		−16.4		−22.7		+43.8		4
		−14.0		−22.9		+37.7		4
		−16.4				+49.3		4
	2.24	−5.6	1.06	−14.8	0.13	+42.3	−5.3	5
		−6.4	1.20	−11.7				6
Nawapali (CM)	1.9	−10.0						1
		−8.3		−23.8		+25.0		4
Santa Cruz (CM)	2.2	−4.3						1
Felix (CO3)	0.47	−16.4						1
Lancé (CO3)	0.34	−15.7						1
		0.0		−26.4		+61		4
Mokoia (CV2)	0.84	−17.4						1
	0.65	−18.8						1
							−17.9	2
		−18.0						4
	0.74	−18.3	0.47	−15.8			−27.2	5
			0.51	−16.6				6
Allende (CV3)	0.30	−17.3						6
	0.27	−16.4						8
Karoonda (CO4)	0.10	−25.2	0.03	−21.4				6
Abee (E4)	0.37	−8.1		−6.1				6
Indarch (E4)	0.40	−12.9						1
Hvitas (E6)	0.25	−4.2						6
Norton County	0.14	−23.9	−23.9					6

TABLE III-7. (*continued*)

	Bulk		Acid insoluble fraction		Carbonate		Extract[b]	
Meteorite (type)[a]	% C	$\delta^{13}C_{PDB}$	% C	$\delta^{13}C_{PDB}$	% C	$\delta^{13}C_{PDB}$	$\delta^{13}C_{PDB}$	*References*
Forest County (H5)	0.06	−30.2						6
	0.08	−24.3						1
Richardton (H5)	0.08	−25.8						6
	0.02	−24.6						1
Weston (H4)	0.22	−24.0						6
Breitscheid (H)	0.32	−25.3						9
	0.60	−21.5						9
Pantar-I-Lt. (H5)[g]	0.56	−23.2						9
	0.07	−22.5						9
-I-Dk.	2.0	−19.9						9
	0.72	−16.0						9
	1.2	−16.2						9
	0.4	−15.9						9
	0.24	−16.6						9
-II-Lt.	1.2	−21.5						9
	0.2	−19.9						9
-II-Dk.	1.9	−21.9						9
	0.17	−21.7						9
Bjurbole (L4)	0.10	−26.3		−24.4				6
Modoc (L6)	0.06	−26.0						6
Saratove (L4)	0.06	−25.6						6
Bruderheim (L6)	0.03	−28.4						6

III-3. Calculation of Model Atmospheric Compositions

In the calculation of model compositions for outgassed atmospheres, the following equilibria are used (after Holland, 1962):

$$H_2 + \tfrac{1}{2}O_2 \rightleftarrows H_2O, \quad K_1 = \frac{P_{H_2O}}{P_{H_2}(P_{O_2})^{0.5}},$$

$$CO + \tfrac{1}{2}O_2 \rightleftarrows CO_2, \quad K_2 = \frac{P_{CO_2}}{P_{CO}(P_{O_2})^{0.5}},$$

$$CO_2 + 4\,H_2 \rightleftarrows 2\,H_2O + CH_4, \quad K_3 = \frac{P_{CH_4}(P_{H_2O})^2}{P_{CO_2}(P_{H_2})^4},$$

$$\tfrac{1}{2}N_2 + \tfrac{3}{2}H_2 \rightleftarrows NH_3, \quad K_4 = \frac{P_{NH_3}}{(P_{N_2})^{0.5}(P_{H_2})^{1.5}}.$$

Values for the four equilibrium constants at various temperatures are given in Table III-8.

In addition to thermodynamic data, values are needed for the abundances of H, C, and N in the

TABLE III-8. Equilibrium constants used in calculation of model atmospheric compositions.[a]

$T(K)$	$\log K_1$	$\log K_2$	$\log K_3$	$\log K_4$
298	39.58	44.74	19.72	2.87
400	29.24	32.43	12.44	0.78
500	22.89	25.02	7.94	−0.50
600	18.63	20.08	4.86	−1.38
700	15.58	16.57	2.61	−2.03
800	13.28	13.91	0.89	−2.52
873	11.96	12.39	−0.13	−2.81
900	11.50	11.86	−0.47	−2.91
973	10.42	10.63	−1.29	−3.23
1000	10.06	10.22	−1.56	−3.50
1100	8.88	8.88	−2.48	−3.66
1173	8.15	8.05	−3.05	−3.72
1200	7.90	7.76	−3.24	−3.90
1300	7.06	8.62	−3.90	−4.08
1400	6.35	6.01	−4.46	−4.17
1473	5.88	5.50	−4.83	−4.20
1500	5.73	5.32	−4.95	−4.24

[a] Calculated from data compiled in the JANAF *Thermochemical Tables* by Stull and Prophet (1971).

TABLE III-9. Abundances of H, C, and N in earth expressed as atmospheres of H_2O, CO_2, and N_2.

Source	H_2O	CO_2	N_2	H/C
Hydrosphere +				
Crust + Atmosphere[a]	515	43	1.8	24
Excess volatiles[b]	494	35	1.0	28
Used in this work:				
Nominal H/C	500	40	1.5	25
Maximum H/C	2000	40	1.5	100

[a] From Turekian and Clark (1975).
[b] From Walker (1977, p. 200–201).

primitive atmosphere. These are here based on estimates of "excess volatiles" (Rubey, 1951; Walker, 1977, p. 200–201) and of the sum of these elements in the atmosphere, hydrosphere, and crust (Turekian and Clark, 1975). The values used in our calculations are given in Table III-9, where they are expressed arbitrarily as atmospheres of H_2O, CO_2, and N_2. Two sets of initial abundances are used, one with a nominal H/C ratio of 25:1 and the other with a hypothetical maximum H/C ratio of 100:1. For both sets, the total inventories of C and N are the same. For our purposes we neglect the abundances of H, C, and N residing in the mantle, which are highly uncertain. We assume that volatiles now in the mantle did not contribute to the hypothetical initially outgassed atmosphere and, therefore, were not involved subsequently in geochemical cycles at the Earth's surface. Total abundances of H, C, and N, respectively, can be expressed in atmospheres as

$$P_{(H_2)T},\ P_{(C)T},\ \text{and } P_{(N_2)T};$$

the first term represents the total abundance of H_2 derivable from all hydrogen-containing gases, the second corresponds to the abundance of C from all carbon-containing gases, and the last to that of N_2 from nitrogen-containing gases.

For our first model atmosphere we consider the gases released at 1473° K in equilibrium with magmas containing metallic iron. The equilibrium oxygen partial pressure is taken to be $10^{-12.5}$ atm after Holland (1962). Since $K_1 = 10^{5.88}$, $K_2 = 10^{5.5}$, and $K_3 = 10^{-4.83}$ (from Table III-8),

$$\frac{P_{H_2O}}{P_{H_2}} = 0.43, \qquad \text{(III-1)}$$

$$\frac{P_{CO_2}}{P_{CO}} = 0.18, \qquad \text{(III-2)}$$

and

$$\frac{P_{CH_4}(P_{H_2O})^2}{P_{CO_2}(P_{H_2})^4} = \frac{P_{CH_4}}{P_{CO_2}} \cdot \left(\frac{P_{H_2O}}{P_{H_2}}\right)^4 \cdot \frac{1}{(P_{H_2O})^2} = 10^{-4.83}. \qquad \text{(III-3)}$$

Substitution of Equation (III-1) in (III-3) and rearrangement yields

$$P_{CH_4} = 4.33 \times 10^{-4}(P_{H_2O})^2(P_{CO_2}). \qquad \text{(III-4)}$$

For a nominal-H/C value (Table III-9),

$$P_{(C)T} = P_{CO_2} + P_{CO} + P_{CH_4} = 40 \text{ atm}, \qquad \text{(III-5)}$$

and

$$P_{(H_2)T} = P_{H_2O} + P_{H_2} + 2\,P_{CH_4} = 500 \text{ atm}. \qquad \text{(III-6)}$$

We neglect contributions of NH_3 and H_2S to Equation (III-6) because $C \gg N > S$ (see Table III-9 and Walker, 1977, p. 200–201). Substitution of P_{CO} from Equation (III-2) and P_{CH_4} from Equation (III-4) into Equation (III-5) gives after simplification:

$$P_{CO_2}(6.56 + 4.33 \times 10^{-4}\,P_{H_2O}^2) = 40 \text{ atm}. \qquad \text{(III-7)}$$

Substitution of P_{H_2} from Equation (III-1) and P_{CH_4} from Equation (III-4) into Equation (III-6) affords

$$P_{H_2O} = 500 - 2.33\,P_{H_2O} - 2[4.33 \times 10^{-4}(P_{H_2O})^2(P_{CO_2})],$$

which rearranges to

$$P_{CO_2} = \frac{(3.33\,P_{H_2O} - 500)}{-8.66 \times 10^{-4}(P_{H_2O})^2}. \qquad \text{(III-8)}$$

Substitution of P_{CO_2} from Equation (III-8) into Equation (III-7) gives

$$\frac{(3.33\,P_{H_2O} - 500)(4.33 \times 10^{-4}P_{H_2O}^2 + 6.56)}{-8.66 \times 10^{-4}(P_{H_2O})^2} = 40$$

which on simplification becomes Equation (III-9),

$$0.00144(P_{H_2O})^3 - 0.182(P_{H_2O})^2 + 21.8\, P_{H_2O} - 3280 = 0, \quad \text{(III-9)}$$

the solution of which by successive approximations yields $P_{H_2O} = 137$ atm. Thus, it can be calculated from Equations (III-1) and (III-6) that $P_{H_2} = 319$ atm and $P_{CH_4} = 22$ atm. Equation (III-8) can be used to calculate $P_{CO_2} = 2.7$ atm, which when substituted into Equation (III-2) affords $P_{CO} = 15$ atm.

To calculate the abundances of N_2 and NH_3 for this first model atmosphere, we substitute $P_{H_2} = 319$ into the equation for K_4 to obtain

$$\frac{P_{NH_3}}{(P_{N_2})^{0.5}} = 0.385. \quad \text{(III-10)}$$

Squaring both sides of Equation (III-10) and rearranging terms yields

$$P_{N_2} = 6.74(P_{NH_3})^2.$$

Using the value of $P_{(N_2)T} = 1.5$ atm (Table III-9), we find

$$P_{(N_2)T} = P_{N_2} + 0.5\, P_{NH_3} = 1.5 \text{ atm}. \quad \text{(III-11)}$$

Substitution of

$$P_{N_2} = 6.74(P_{NH_3})^2$$

into Equation (III-11) produces a quadratic equation

$$6.74(P_{NH_3})^2 + 0.5\, P_{NH_3} - 1.5 = 0,$$

which when solved leads to $P_{NH_3} = 0.44$ atm and $P_{N_2} = 1.28$ atm. Thus, starting with a nominal $H/C = 25$, we find the model atmosphere outgassed at 1473° K in the presence of metallic iron is composed of 319 atm H_2, 137 atm H_2O, 22 atm CH_4, 15 atm CO, 3 atm CO_2, 1.3 atm N_2, and 0.4 atm NH_3.

The procedure described above is also used to calculate the model compositions of atmospheres outgassed from or equilibrated with metallic iron-free source regions at 873° K using appropriate values for K_1, K_2, K_3, and K_4. The difference between such calculations and those outlined above is that equations analogous to (3), (4), (7), (8) and (9) are expressed in terms of P_{H_2} rather than P_{H_2O}. A value of $P_{O_2} = 10^{-20.5}$ atm is used based on the synthetic fayalite-magnetite-quartz buffer system at 873° K (see Buddington and Lindsley, 1964; and Haggerty, 1978).

For the next illustrative calculation we consider an "unequilibrated" atmosphere outgassed at 1473° K with a maximum H/C ratio. In this case the first 10% of the atmosphere (the "precore" portion) is presumed to have outgassed in equilibrium with metallic iron, to have remained "unequilibrated" with a growing mantle depleted in metallic iron, and to have simply mixed with the bulk of the atmosphere having been released subsequently in the "post-core" stage.

For the pre-core portion of this atmosphere,

$$[P_{(C)T}]_{\text{pre-core}} = P_{CO_2} + P_{CO} + P_{CH_4} = 4.0 \text{ atm}, \quad \text{(III-12)}$$

$$[P_{(H_2)T}]_{\text{pre-core}} = P_{H_2} + P_{H_2O} + 2\, P_{CH_4} = 200 \text{ atm}, \quad \text{(III-13)}$$

$$[P_{(N_2)T}]_{\text{pre-core}} = P_{N_2} + 0.5\, P_{NH_3} = 0.15 \text{ atm}. \quad \text{(III-14)}$$

Using the procedure described for the preceding example, we can use Equations (III-1), (III-2), (III-4), (III-12), and (III-13) to generate an equation analogous to Equation (III-9), from which P_{H_2O} can be obtained. The "pre-core" atmosphere that results is composed of 60 atm H_2O, 137 atm H_2, 1.5 atm CH_4, 2.2 atm CO, and 0.4 atm CO_2. Following the procedure in the first example, we can use K_4, $P_{H_2} = 137$ atm, and Equation (III-14) to calculate $P_{N_2} = 0.13$ atm and $P_{NH_3} = 0.04$ atm.

For the "post-core" portion of this atmosphere (outgassed in the absence of metallic Fe) we take the equilibrium partial pressure of O_2 to be $10^{-8.4}$ atm based on the synthetic fayalite-magnetite-quartz buffer system at 1473° K (see Buddington and Lindsley, 1964; and Haggerty, 1978). Using values of K_1, K_2, and K_3 from Table III-8, we obtain the following relationships:

$$\frac{P_{H_2O}}{P_{H_2}} = 48, \quad \text{(III-15)}$$

$$\frac{P_{CO_2}}{P_{CO}} = 20, \quad \text{(III-16)}$$

and

$$\frac{P_{CH_4}}{P_{CO_2}}\left(\frac{P_{H_2O}}{P_{H_2}}\right)^2 \cdot \frac{1}{(P_{H_2})^2} = 1.48 \times 10^{-5}. \quad \text{(III-17)}$$

Substitution of Equation (III-15) into Equation (III-17) affords

$$\frac{P_{CH_4}}{P_{CO_2}} = 6.46 \times 10^{-9}(P_{H_2})^2. \quad \text{(III-18)}$$

The mass balance relationships for H and C are

$$[P_{(C)_T}]_{\text{post-core}} = P_{CO_2} + P_{CO} + P_{CH_4} = 36 \text{ atm}, \quad \text{(III-19)}$$

and

$$[P_{(H_2)_T}]_{\text{post-core}} = P_{H_2} + P_{H_2O} + 2\, P_{CH_4} = 1800 \text{ atm}. \quad \text{(III-20)}$$

Since

$$P_{CH_4} \leq 36 \text{ atm and } P_{H_2O} = 48\, P_{H_2},$$

inequalities based on Equation (III-20) can be written as 1800 atm $\geq 49\, P_{H_2} \geq$ 1728 atm; thus, 37 atm $\geq P_{H_2} \geq$ 35 atm. When $P_{H_2} = 37$ atm is chosen so as to maximize the value of P_{CH_4} calculated from Equation (III-18), it follows that $P_{CH_4} \leq 8.7 \times 10^{-6}\, P_{CO_2}$; and P_{CH_4} in Equation (III-19) can be neglected. Since $P_{CO} = 0.05\, P_{CO_2}$ from Equation (III-16), use of Equation (III-19) leads to $P_{CO_2} = 34.3$ atm and $P_{CO} = 1.7$ atm. Thus, P_{CH_4} amounts to less than 3×10^{-4} atm and can therefore also be neglected in Equation (III-20). Substituting $P_{H_2O} = 48\, P_{H_2}$ from Equation (III-15) into Equation (III-20) and solving for P_{H_2} affords 37 atm H_2 and, thus, 1763 atm H_2O.

For this post-core portion of the atmosphere,

$$[P_{(N_2)_T}]_{\text{post-core}} = P_{N_2} + 0.5\, P_{NH_3} = 1.35 \text{ atm}. \quad \text{(III-21)}$$

Using K_4, $P_{H_2} = 37$ atm and Equation (III-21), we calculate $P_{NH_3} = 0.02$ atm and $P_{N_2} = 1.34$ atm.

Combining the pre-core and post-core portion of this second type of model atmosphere yields a final composition containing 1823 atm H_2O, 174 atm H_2, 1.5 atm CH_4, 3 atm CO, 35 atm CO_2, 1.47 atm N_2, and 0.06 atm NH_3.

The composition of an "unequilibrated" atmosphere evolved at 1473° K and characterized by a nominal H/C value of 25 is calculated in analogous fashion. In addition, the model compositions of those atmospheres discussed in Chapter 4 equilibrated at 1473° K with metallic iron-free source magmas are obtained by the same method used for the post-core portion of the "unequilibrated" atmospheres.

It is noteworthy that the ratios in Equations (III-15) and (III-16) are about a factor or two lower than those calculated originally by Holland (1962). This discrepancy occurs because our value of $P_{O_2} = 10^{-8.4}$ atm differs from the value of $10^{-7.1}$ atm used by Holland (1962). In addition, our use of more recent thermochemical data (from the 1971 JANAF Tables; Stull and Prophet, 1971) yields lower values for log K_3 (above 298° K) than those supplied by Holland (1962) [e.g., at 1473° K, log $K_3 = -4.83$ (Table III-8) versus log $K_3 = -3.15$ (Holland, 1962)]. Consequently, values of the numerical constants in Equations (III-4) and (III-18), which are based in part on data for log K_3 in Table III-8, also differ from the analogous constants calculated from data in Holland (1962). None of these differences, however, affects the inferences and conclusions drawn from the calculations.

APPENDIX IV

Procedures of Organic Geochemical Analysis

By Kim W. Wedeking, J. M. Hayes, and Udo Matzigkeit

IV-1. Survey Analyses: Determinations of Total Organic Carbon (T.O.C.) Content and Organic Carbon Isotopic Composition ($\delta^{13}C_{PDB}$)

All rock powder samples produced (refer to Appendix II) were analyzed both for total organic carbon content (T.O.C., mg C_{org}/g rock powder) and for isotopic composition ($\delta^{13}C_{PDB}$). Use of an automated commercial induction furnace/CO_2 analyzer system (Laboratory Equipment Co., "LECO analyzer," model 589-600), modified to allow collection of the CO_2 produced after sample combustion, facilitated analyses of T.O.C. with simultaneous production of gas for carbon isotope ratio determination. As indicated in Figure II-1, this approach led to results with precision adequate for selection of samples for kerogen isolation and further analyses, as well as providing a large set of data worthy of geochemical interpretation.

The analytical procedure involved the following major steps: (*i*) weighing of samples, followed by acid treatment to dissolve carbonates; (*ii*) combustion of samples for T.O.C. determination using the LECO analyzer; and (*iii*) collection, purification, and packaging of CO_2 produced in (*ii*) for isotopic analyses. Samples were carried through these steps in batches of 24. To each batch of samples were added at least two isotopic standards and two glass powder blanks (in order to monitor the accuracy of $\delta^{13}C$ determinations and any contamination associated with the rock crushing procedure). Thus, each batch involved 28 or more analyses.

In step (*i*), rock powder was weighed into pre-combusted, porous (filter-type), ceramic crucibles. The amount of powder used was such that the total organic carbon in the sample did not exceed the upper limit of the LECO analyzer system ($\sim$4.0 mg C) and also such that the total residue of material in the crucible after carbonate removal was $\leq$1 gram (since greater amounts of such residue led to incomplete combustion in the induction furnace). Estimation of T.O.C. by powder color and knowledge of sample lithology allowed selection of appropriate sample quantities. After weighing of samples, standards, and blanks, the crucibles were placed in a 28-port suction rack for acid treatment. All samples (except the water-soluble dextrose isotopic standard; see following discussion of standards and calibration) were first treated with 6 N HCl until effervescence was no longer noted in any of the crucibles, the samples then being treated three times with boiling 12 N HCl, a 10 to 15 minute reaction period being allowed after each treatment. Finally, samples were washed to neutrality with distilled water and dried overnight at 105°C. When the carbonate-free samples were dry, 1 g both of copper and iron flakes (having a low carbon content) were added as "accelerators" to facilitate inductive energy transfer to samples within the analyzer furnance. Replicate analyses with more vigorous HCl treatment (viz., hot 12 N HCl, overnight) showed that the above treatment could not be improved by increasing reaction time.

Before analysis of each batch of samples, the LECO analyzer was calibrated by combustion of four steel ring carbon standards containing 0.201, 0.610, 1.58, and 3.79 mg carbon, respectively. A linear regression of the data obtained on standard input provided a response function that unambiguously related instrument readout to sample-derived carbon, contributions by crucible and furnace blanks being absorbed in the intercept term of the response function.

Samples to be analyzed were placed in the furnace chamber, which was then purged with O_2 to remove atmospheric gases. The automated analysis cycle was initiated, resulting in combustion of the sample in the O_2 stream at 1600°C,

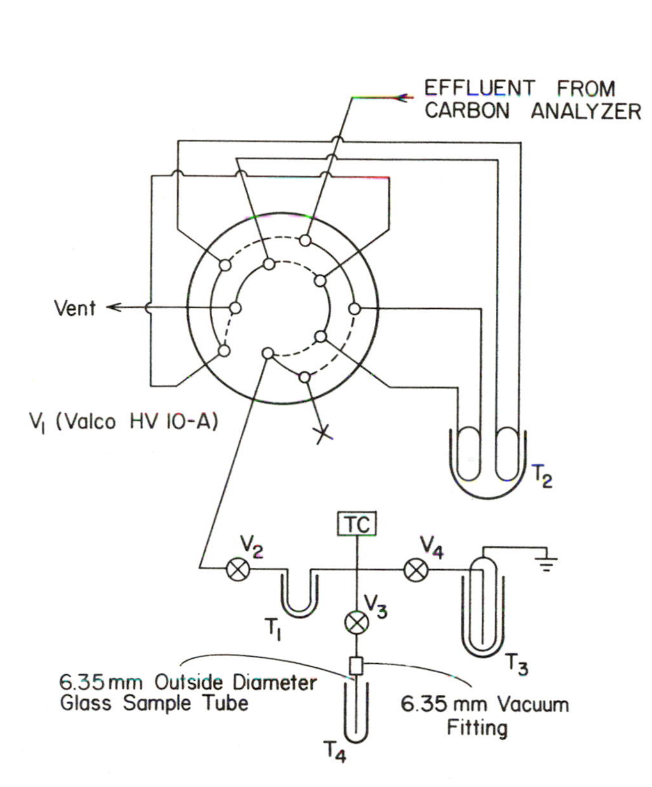

FIGURE IV-1. (A) System used for collection, purification, and packaging of carbon dioxide for isotopic analysis: schematic view in which T designates traps; V designates values; and TC indicates a thermal conductivity pressure transducer.

	Operation				
Valve status	Collect peak	Remove He		Transfer and package CO_2	Standby
V_1[a]	——	-----	----	----	----
V_2[b]	O	C	O	O	O
V_3[b]	O	C	C	O	O
V_4[b]	O	O	O	C	O
Temperatures of traps, °C					
T_1	−120 ⟶				
T_2	−196 ⟶			25	−196
T_3	−196 ⟶				
T_4	25 ⟶			−196	25

a. Solid and broken lines correspond to flow paths shown at left.

b. O, open; C, closed.

Solid lines indicate tubing, which was 3.18-mm outside diameter stainless steel upstream from V_2 and 9-mm outside diameter Pyrex downstream from V_2. (B) Summary of system operation.

followed by CO_2 "cleanup" (i.e., removal of metal oxide dust, SO_2 and H_2O), conversion of CO to CO_2 over CuO at 300°C, and selective adsorption of the CO_2 from the O_2 stream on a molecular sieve (Linde 5A) trap held at 25° C. When combustion was completed and all CO_2 had been purged from the furnace, automated analysis continued with desorption (at 230° C) of the trapped CO_2, a helium carrier gas being utilized to sweep the CO_2 through a silica gel column and to a gas chromatographic detector (of the thermal conductivity type). The CO_2 peak was electronically integrated in order to provide a digital readout of peak area. After passing through the detector, the CO_2 was swept into the collection system.

Figure IV-1 shows the CO_2-collection system schematically. A ten-port valve (Valco model HV-10A) joined the carbon analyzer and the collection apparatus. During each analytical cycle, the valve was placed in "stand-by" position (Figure IV-1) during the combustion/trapping phase. The ten-port valve was toggled to the "collect peak" position during CO_2 desorption/chromatographic purification but before peak integration began. When collection was complete, residual helium was removed with a second toggling of the valve (see Figure IV-1 for status of other valves) after which the CO_2 was condensed in a length of 6 mm Pyrex tubing, passing through trap T_1 (at −130° C) to remove any residual H_2O. The Pyrex tubing was sealed with a torch in order to package the sample (Des Marais and Hayes, 1976) for transfer to the mass spectrometer.

Calibration of the CO_2 analyzer by combustion of carbon standards gave calculated standard deviations of the regression estimates (i.e., the uncertainties due to scatter in the calibration points) of typically one percent, with a representative analysis providing a result of, for example, 0.451 ± 0.005 mg C. Since relative uncertainties

TABLE IV-1. Replicate T.O.C. analyses of P.P.R.G. Sample No. 229–1.

Date of analysis	*Organic carbon content, mg/g*
18 February 1980	1.76
14 March 1980	1.77
	1.66
	1.55
	1.92
	1.65
	1.63
	1.63
19 March 1980	1.70
	1.40
	1.68
	1.70
25 March 1980	1.71
	1.74
26 March 1980	1.63
	1.55
1 April 1980	1.68
	1.63
	1.68
	1.61

Note: The mean is 1.66; the standard deviation of population is 0.10; and the 95% confidence interval for a single observation is 0.21.

of sample weights were negligible in comparison to uncertainties associated with carbon analysis, precision of reported T.O.C. results (in mg/g) was primarily limited by the reproducibility of sample work-up and sample analysis, as well as the instrument calibration. Observed precision of replicate analyses of actual rock powder samples showed that these additional (uncharacterized) sources of error significantly affected analytical reproducibility; hence, reported confidence limits for T.O.C. results were empirically derived by replicate analysis of representative samples. Table IV-1 shows typical results obtained from repeated analysis of powder Sample No. 229-1. It should be noted that contamination of samples during the work-up procedure could have had no significant effect upon T.O.C. precision, considering results obtained from analyses of glass blanks, which were treated exactly as actual rock samples throughout the processing steps. The mean of 58 individual blank samples was 0.03 mg/g, with a standard deviation of 0.02 mg/g (0.04 mg/g, 95% confidence limits).

In the initial testing of the LECO system for precision and accuracy of $\delta^{13}C$ results, two samples previously analyzed by quartz bomb combustion were used as isotopic standards. The first was a reagent grade dextrose sample ($\delta^{13}C_{PDB} = -9.18 \pm 0.05$ ‰); the second, a rock powder sample (P.P.R.G. Sample No. 229-1, $\delta^{13}C_{PDB} = -24.10 \pm 0.05$ ‰). Both standards, when analyzed by the LECO procedure, gave isotopic compositions that showed a dependence upon the amount of sample combusted. This dependence was attributed to the influence of a carbon blank that was present in the combustion of crucibles containing only the iron and copper accelerators. Since attempts to eliminate this carbon blank were not successful, it was necessary to determine quantitatively both the magnitude and isotopic composition of the blank under actual sample combustion conditions. Consider the exact equation:

$$(n_x + n_b)F_{obs} = F_x n_x + F_b n_b$$

in which F_{obs} represents the observed fractional isotopic abundance $[^{13}C/(^{12}C + ^{13}C)]$ in the carbon derived from the sample and the blank, F_x represents the true abundance of ^{13}C in the sample alone, and F_b represents the abundance of ^{13}C in the blank. The terms n_x and n_b represent the amounts of sample- and blank-derived carbon, respectively. Combustion and isotopic analysis (to give F_{obs}) of two standards (F_x known) provided two equations, allowing calculation of amount and isotopic composition of the blank. Since the characteristics of the blank were found to be essentially constant both within and between sample batches, a mean was taken of calculated blank quantities and isotopic compositions after all samples had been analyzed. The values $\bar{n}_b = 0.07 \pm 0.02$ mg C and $\bar{F}_B = (1.069 \pm 0.002) \times 10^{-2}$ (equivalent to -38.3 ± 2 ‰ vs. PDB) were accepted as best estimates of the true values of n_b and F_b and were used to correct observed isotopic compositions for the influence of the blank by use of the equation above. Uncertainties reported for $\delta^{13}C$ values are calculated 95% confidence intervals based on propagation of error in the correction formula, since uncertainties of F_{obs}, n_x, n_b, and F_b were well characterized. Table IV-2 shows typical results of determinations of $\delta^{13}C$ for both isotopic standards. Given in the table are uncorrected and corrected compositions, along with calculated

TABLE IV-2. Accuracy and precision of carbon isotopic analyses (survey method).

	P.P.R.G. sample 229–1[a]			Dextrose[b]		
Date of analysis	*mgC*	$\delta^{13}C_{obs}$	$\delta^{13}C_{corr}$[c]	*mgC*	$\delta^{13}C_{obs}$	$\delta^{13}C_{corr}$[c]
18 February 1980	0.55	−26.37	−24.8 ± 0.2	0.93	−11.20	−9.2 ± 0.3
14 March 1980	0.22	−27.66	−24.3 ± 0.6	1.17	−11.00	−9.4 ± 0.3
	0.40	−26.29	−24.2 ± 0.4			
	0.66	−25.50	−24.1 ± 0.2			
	0.94	−24.92	−23.9 ± 0.2			
	1.16	−24.87	−24.1 ± 0.1			
	1.39	−24.65	−24.0 ± 0.1			
	1.55	−24.67	−24.0 ± 0.1			
19 March 1980	1.70	−24.81	−24.2 ± 0.1	1.15	−11.01	−9.4 ± 0.2
	0.42	−26.27	−24.3 ± 0.3			
	1.84	−24.59	−24.1 ± 0.1			
	0.42	−26.39	−24.4 ± 0.3			
25 March 1980	0.41	−26.15	−24.1 ± 0.4	0.92	−11.44	−9.4 ± 0.3
	0.58	−25.94	−24.4 ± 0.3			
26 March 1980	0.52	−25.73	−24.0 ± 0.3	0.93	−11.34	−9.3 ± 0.3
	0.41	−26.44	−24.4 ± 0.3			
1 April 1980	0.47	−26.13	−24.3 ± 0.3	0.77	−11.35	−8.9 ± 0.4
	0.70	−25.41	−24.1 ± 0.2	1.32	−10.63	−9.2 ± 0.2
	0.88	−25.28	−24.2 ± 0.2	1.02	−10.92	−9.0 ± 0.3
	1.11	−24.90	−24.0 ± 0.1	0.69	−11.76	−9.1 ± 0.4

[a] Known isotopic composition from quartz-bomb combustion, $\delta^{13}C_{PDB} = -24.10 \pm 0.05$ ‰.
[b] Known isotopic composition from quartz-bomb combustion, $\delta^{13}C_{PDB} = -9.18 \pm 0.05$ ‰.
[c] Corrected for influence of blank as described in text: uncertainties are calculated 95% confidence intervals.

uncertainties and sample quantities (as mg carbon). These data show that the blank correction calculation is accurate within the stated uncertainty over a wide range of sample sizes, and that observed precision does not differ greatly from that calculated by propagation of error. Table IV-3 gives similar results for actual samples that were randomly chosen for duplicate analysis. In addition to sample quantities (as mg C) and uncorrected and blank-corrected $\delta^{13}C$ results, the table gives T.O.C. results (as mg C/g rock) for these samples.

IV-2. Kerogen Isolation

Following the completion of survey analyses, rock samples were selected from the P.P.R.G. collection for further study involving isolation and analysis of kerogen from representative specimens. Whenever carbonates, cherts, and shales were available from a particular unit, an attempt was made to select at least one sample representing each lithology. To avoid processing of impractically large (> 300 g) amounts of rock in order to obtain the desired quantity of kerogen (~200 mg), samples containing less than 0.7 mg/g C were usually not selected. Samples that were known to be stromatolitic and/or to contain microfossils were favored, as were samples that were isotopically representative of the geologic source sequence (based upon organic carbon $\delta^{13}C_{PDB}$ from survey analyses).

The selected samples were treated to give kerogen isolates according to the following steps: (*i*) weighing of sample fragments in quantities to yield approximately 200 mg of organic carbon in the final residue; (*ii*) crushing of samples to <100 μm in a steel mill; (*iii*) treatment with HF and/or HCl to dissolve acid-soluble minerals; (*iv*) treatment to remove neoformed fluoride residues such as fluorite or ralstonite often formed when demineralization involved the use of hydrofluoric acid; and (*v*) treatment to remove pyrite, if present

TABLE IV-3. Duplicate T.O.C. and $\delta^{13}C$ analyses (survey method).

P.P.R.G. sample no.	*Date*	*mgC*	*T.O.C. (mg/g)*	$\delta^{13}C_{obs}$	$\delta^{13}C_{corr}$ [a]
040–1	14 December 1979	0.65	70.3 ± 13	−41.33	−41.7 ± 0.2
	14 December 1979	0.82	75.6 ± 14	−40.85	−41.1 ± 0.2
066–1	6 December 1979	0.05	0.07 ± 0.04	−31.50	−22.0 ± 4
	14 February 1980	0.07	0.08 ± 0.04	−28.21	−18.1 ± 4
133–1	10 December 1979	0.48	0.40 ± 0.1	−23.80	−21.7 ± 0.6
	19 December 1979	0.63	0.57 ± 0.1	−23.74	−22.1 ± 0.6
134–1	10 December 1979	0.55	0.43 ± 0.1	−31.61	−30.8 ± 0.4
	14 December 1979	0.38	0.33 ± 0.1	−31.59	−30.4 ± 0.6
139–1	10 December 1979	0.67	0.59 ± 0.1	−29.58	−28.7 ± 0.4
	4 January 1980	0.79	0.67 ± 0.1	−29.72	−29.0 ± 0.2
196–1	10 December 1979	1.88	1.87 ± 0.3	−24.03	−23.5 ± 0.2
	14 December 1979	0.84	2.02 ± 0.4	−24.35	−23.2 ± 0.4
198–1	11 December 1979	1.33	1.60 ± 0.3	−32.80	−32.5 ± 0.1
	14 December 1979	1.49	1.56 ± 0.3	−32.55	−32.3 ± 0.1
200–1	12 December 1979	0.62	1.52 ± 0.3	−31.59	−30.8 ± 0.3
	17 March 1980	0.68	1.61 ± 0.3	−30.97	−30.2 ± 0.3
204–1	12 December 1979	1.37	3.32 ± 0.6	−29.46	−29.0 ± 0.2
	17 March 1980	1.24	3.54 ± 0.6	−29.23	−28.7 ± 0.2
224–1	19 February 1980	0.77	0.60 ± 0.1	−10.94	−8.4 ± 0.8
	17 March 1980	0.50	0.56 ± 0.1	−12.66	−9.1 ± 1
261–1	17 March 1980	0.55	0.54 ± 0.1	−16.97	−14.2 ± 0.8
	27 March 1980	0.62	0.52 ± 0.1	−16.26	−13.8 ± 0.8
322–1	10 March 1980	0.70	0.49 ± 0.1	−27.39	−26.3 ± 0.4
	20 March 1980	0.53	0.51 ± 0.1	−27.95	−26.6 ± 0.5
368–1	10 March 1980	2.07	2.05 ± 0.4	−42.76	−42.9 ± 0.1
	17 March 1980	1.12	2.33 ± 0.5	−42.71	−43.0 ± 0.2
370–1	10 March 1980	0.29	0.20 ± 0.1	−30.86	−29.1 ± 0.8
	17 March 1980	0.19	0.19 ± 0.1	−31.66	−29.2 ± 1
435–1	29 February 1980	0.18	0.12 ± 0.05	−24.66	−19.4 ± 2
	8 March 1980	0.12	0.08 ± 0.04	−27.09	−20.6 ± 2
440–1	10 March 1980	3.11	3.29 ± 0.6	−33.42	−33.3 ± 0.1
	25 March 1980	2.67	2.89 ± 0.5	−32.39	−32.2 ± 0.1
489–1	10 March 1980	1.27	2.76 ± 0.5	−21.87	−21.0 ± 0.3
	25 March 1980	0.76	2.46 ± 0.5	−23.16	−21.8 ± 0.4
494–1	25 March 1980	0.02	0.11 ± 0.05	−36.14	−28.6 ± 7
	2 April 1980	0.11	0.11 ± 0.05	−31.95	−27.9 ± 2

[a] Corrected for influence of blank as described in text; uncertainties are calculated 95% confidence intervals.

in excess of 50% by weight in the kerogen residue. Figure IV-2 summarizes the kerogen isolation procedure here used.

Appropriate quantities of samples (previously chipped and etched as described in Appendix II) were each crushed for two minutes using the procedure of Appendix II, but with a shorter crushing period—the steel mill being thoroughly cleaned between samples, as noted in that appendix. Weighed powder samples were carried through an acid dissolution procedure. In the first step, an excess of 60% HF (for carbonate-poor cherts or shales), or 37% HCl (for carbonates), or a 5:2 (V:V) mixture of the two acids (for cherts or shales bearing carbonate) was employed. Rock powder was slowly added to the appropriate acid solution with stirring,

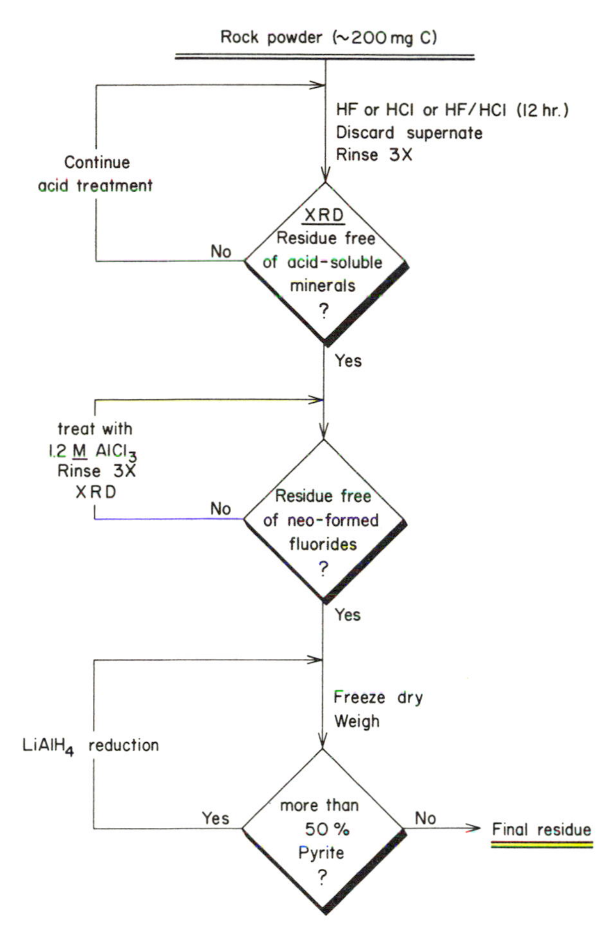

FIGURE IV-2. Flow chart showing the scheme employed for isolation of kerogen from Precambrian rock samples.

while solution temperature was held below 60° C (with all operations being carried out in acid-resistant, polypropylene labware). The resulting slurry was stirred intermittently over the course of 12 hours (at 25° C), after which the liquid phase was removed and discarded, the solid residue being washed with ~1 liter of distilled water. The acid treatment was carried out a second time using an HF–HCl mixture, and the washed residue was subjected to an x-ray diffraction analysis in order to determine if quartz, silicates, carbonates, or other minerals susceptible to acid dissolution (oxides, hydroxides, etc.) had survived the above treatments. If such minerals were present, acid digestion was continued until these species were no longer detectable by x-ray diffraction. The effects of mineral acid treatment upon the elemental composition of kerogen have been studied (Durand, 1980, p. 42; Saxby 1976) and are generally negligible, especially for geochemically mature samples.

Over the course of acid digestions when HF was employed, the dissolution of quartz and silicate minerals was (in most cases) accompanied by the appearance of hydrous, neoformed fluoride phases, a commonly encountered problem (Durand, 1980, p. 37; Forsman and Hunt, 1958) known to interfere with elemental analysis of kerogen (Durand, 1980, p. 117). The use of aluminum chloride for removal of such mineral species from kerogen residues has been mentioned (Grew, 1974) and was employed in our procedure. Flouride-containing samples, previously rinsed to neutrality, were shaken with approximately 200 ml of 1.2 *M* $AlCl_3$ (prepared by dissolving 250 g of $AlCl_3 \cdot 6\ H_2O$ in 750 g H_2O) for 24 hours at 25° C, after which the solution was removed and discarded. The resulting residue was washed and analyzed by x-ray diffraction, the above steps being repeated (as many as five times) until no detectable fluoride minerals remained. Usually, three such treatments were effective in complete removal of the typically encountered quantities of fluoride complexes.

Aluminum halides such as $AlCl_3$ have been reported to solubilize kerogen from shale (Aarna and Lippmaa, 1957; Jones and Dickert, 1965), probably by cleavage of ether functions. However, in these studies the reactions were carried out under rigorously anhydrous conditions and at elevated temperatures (80 to 100° C) in order to maximize the Lewis acid reactivity of aluminum halides. Ether cleavage reactions catalyzed by $AlCl_3$ are carried out in the absence of H_2O (Fieser and Fieser, 1967); the Lewis acid catalytic activity of hydrated $AlCl_3$ in aqueous solution is very low (Olah, 1963, p. 307). Therefore, solubilization of kerogen under the conditions here used for fluoride removal is extremely unlikely. No evidence of solubilization was noted during the treatment described, and material recovery was never inexplicably low.

After acid digestion and fluoride removal were completed, samples were again washed and then freeze-dried. After drying, the samples were weighed and, based on the known organic carbon content of the starting material (±20% from survey T.O.C. analysis) and the weight of rock powder initially digested, an estimate of carbon content (and therefore of maximal ash content) of each kerogen residue was obtained. In almost all cases, if the ash content estimated in this way was greater than 10%, x-ray diffraction and visual analysis (by transmitted light microscopy) indicated that the most abundant mineral impurity was pyrite. Those samples that contained more than 50% pyrite (by weight) were further processed, with the goal of reducing the pyrite abundance to or below this amount, a level that can generally be tolerated in CHN elemental analysis of kerogen (Durand, 1980, p. 120; see also CHN analysis, Section IV-3).

Removal of pyrite was accomplished by $LiAlH_4$ reduction, using a procedure similar to that of Lawlor et al. (1963). Pyrite-containing samples were weighed in amounts such that at least 10 mg of carbon in the form of pyrite-free kerogen could be recovered; this involved the use of as much as 700 mg of pyrite-containing starting material. To each sample was added a two-fold excess of $LiAlH_4$ (0.3 mg $LiAlH_4$ per mg FeS_2) for reduction of all pyrite estimated to be present. Reductions were carried out in 30 to 40 ml anhydrous tetrahydrofuran, the mixtures being refluxed overnight, after which solution volumes were reduced to 5 ml by evaporation of solvent at 70 to 80° C. In order to liberate H_2S and destroy excess $LiAlH_4$, 40 ml of 6 N HCl were slowly added to the reaction flasks and the samples were refluxed for an additional 2 to 3 hours. The resulting residues were washed repeatedly and then freeze-dried and weighed. Calculation of remaining pyrite within samples treated in this manner revealed that pyrite removal was not quantitative. However, in all but one case (P.P.R.G. Sample No. 113-1, the HYC shale—initially 98% pyrite), one such treatment was adequate to reduce pyrite content to the desired level of 50% or less.

Treatment of kerogen with $LiAlH_4$ results in minor chemical modifications, primarily involving reduction of carboxyl groups to alcohols (Lawlor et al., 1963), reaction with mature kerogen being less problematic (Saxby, 1976). No substantial alteration of bulk elemental composition of kerogen results from $LiAlH_4$ treatment (Lawlor et al., 1963; Saxby, 1976), nor is kerogen solubilized by this reagent (Jones and Dickert, 1965). Elemental analysis of P.P.R.G. Sample No. 225-1, before and after $LiAlH_4$ treatment (as described above), gave H/C, N/C, $\delta^{13}C_{PDB}$, $\delta^{15}N_{Air}$ and δD_{SMOW} results that can be found in Table IV-7.

Forty comparisons of $\delta^{13}C_{PDB}$ determined by CHN analysis of kerogen samples versus $\delta^{13}C_{PDB}$ of organic carbon in corresponding rock powder samples were obtained, thus allowing evaluation of the extent to which contamination and/or fractionation of kerogen may have occurred during the isolation procedure itself. Figure IV-3 shows the $\delta^{13}C_{PDB}$ values determined for the kerogen samples obtained versus the corresponding values from T.O.C. analysis of whole rock samples. In statistical treatment of these data, the point corresponding to P.P.R.G. Sample No. 113-1 has been omitted, since the observed $\Delta\delta^{13}C_{PDB}$ (T.O.C. vs. kerogen) of of 7.21 ‰ can be considered atypically large and an assignable cause for this unusual disagreement may be the extreme pyrite abundance of both the 113-1 rock powder sample and the corresponding kerogen sample (rock sample 113-1 was a sulfide ore, the only such sample in the P.P.R.G. collection). The mean value of $\Delta\delta^{13}C_{PDB}$ (T.O.C. vs. kerogen) for the 39 remaining comparisons was found to be 0.61 ‰ with a standard deviation of 1.42 ‰. This mean difference is not significantly different from zero, as a two-tailed *t*-test (Davies, 1967, p. 65), indicates that the null hypothesis: $\Delta\delta^{13}C_{PDB}$ (T.O.C. vs. kerogen) = 0 should be accepted at the 99% confidence level. Thus the available data do not suggest that any systematic shift in the $^{13}C/^{12}C$ ratio of kerogen results from the isolation procedure. Such a shift, if observed, could be caused by organic contamination during kerogen isolation, or by a systematic carbon isotopic fractionation of kerogen samples during processing (for example, by partial dissolution of the kerogen or by mechanical loss of some isotopically distinct fraction).

If no single mechanism, such as one of those mentioned above, played a dominant role during kerogen isolation, do available data suggest that a random combination of such processes has affected the carbon isotopic composition of the samples obtained? An equivalent question is: Is the scatter of points about the line in Figure IV-3 greater than the scatter expected due only to the combined imprecision of the two analytical methods? This question is amenable to statistical treatment. It can be shown, by standard methods of error analysis and the additive properties of variance (Davies, 1967), that the observed standard deviation of $\Delta\delta^{13}C_{PDB}$ (T.O.C. vs. kerogen), designated by $s_{\Delta,obs}$, is given by:

$$s^2_{\Delta,obs} = s^2_{CHN} + s^2_{T.O.C.} + s^2_{isolation}$$

where s^2_{CHN}, $s^2_{T.O.C.}$, and $s^2_{isolation}$ are estimates of the individual variances that determine the magnitude

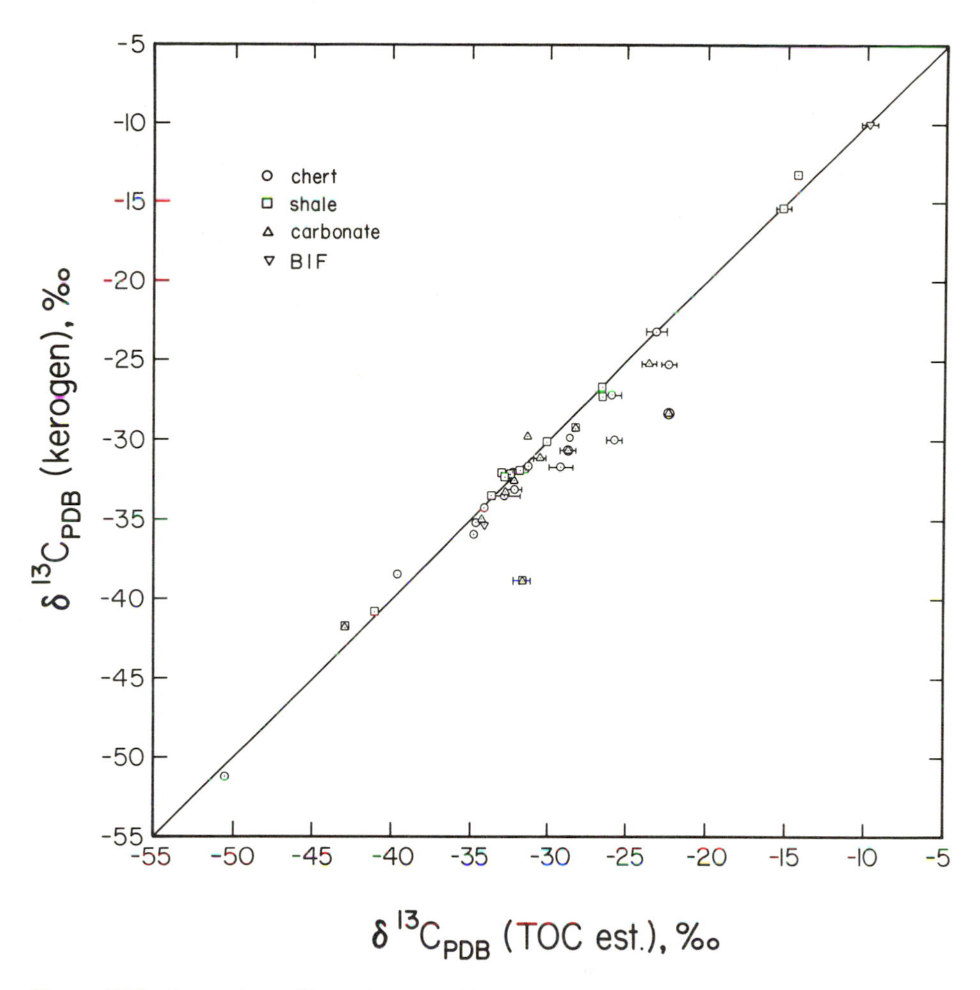

FIGURE IV-3. Comparison of isotopic compositions of carbon determined by CHN analysis of isolated kerogen and survey analysis of total organic carbon in corresponding rock samples.

of $s^2_{\Delta,obs}$ and are variances in $\delta^{13}C_{PDB}$ associated with the analytical procedures for determination of kerogen $\delta^{13}C_{PDB}$ ("CHN" method), determination of $\delta^{13}C_{PDB}$ of organic carbon in rock powder (survey T.O.C. method), and isolation of kerogen, respectively. Values of $s_{T.O.C.}$ and s_{CHN} calculated from replicate analyses of actual samples (Tables IV-3 and IV-7) are found to be 0.8 ‰ (18 degrees of freedom) and 0.2 ‰ (12 degrees of freedom), respectively.

One way to approach the questions above concerning the possible role of contamination and fractionation of kerogen carbon during the isolation procedure is to test the null hypothesis:

$$s^2_{\Delta,obs} = s^2_{CHN} + s^2_{T.O.C.},$$

using the F-test (Davies, 1967, p. 66), with the alternative hypothesis being

$$s^2_{\Delta,obs} > s^2_{CHN} + s^2_{T.O.C.}.$$

Results of this test (99% confidence level) are somewhat ambiguous. With all 39 points ($s_{\Delta,obs} = 1.42$) included in such a test, the null hypothesis must be rejected. However, elimination of a single point comparison with an unusually large discrepancy [viz., P.P.R.G. Sample No. 447-1; $\Delta\delta^{13}C_{PDB}$(T.O.C. vs. kerogen) = 5.96] gives $s_{\Delta,obs} = 1.14$ and in this case the null hypothesis would be accepted. More data would be required for a conclusive test, but available data suggest that $s_{isolation}$ is not greater than 1 ‰, and that if shifts in kerogen $^{13}C/^{12}C$

ratios do occur by several mechanisms during kerogen isolation, these effects are not severe. Generally, it is apparent that the product of the isolation procedure described here is closely representative of material present within the host rock specimen.

IV-3. Carbon, Hydrogen, Nitrogen (CHN) Analysis

All kerogen isolates were analyzed to give elemental abundances of hydrogen and nitrogen relative to carbon (reported as H/C and N/C atomic ratios) and the isotopic compositions of these elements (reported as $\delta^{13}C_{PDB}$, δD_{SMOW}, and $\delta^{15}N_{Air}$) in the samples. The method employed in these analyses was an adaptation of a technique (Frazer, 1962; Frazer and Crawford, 1963; Frazer and Crawford, 1964) involving combustion of samples in quartz bombs (with CuO as oxidant), followed by separation and measurement of the N_2, CO_2, and H_2O produced (H_2O being converted to H_2 over uranium metal at 800° C before measurement). In the method described here, each of the three gases was packaged after abundance measurement in order to provide gas samples for isotope ratio determination.

The analytical procedure consisted of two steps: (*i*) preparation of bombs and sample combustion; and (*ii*) separation, measurement, and packaging of the gaseous combustion products. In step (*i*), quartz tubes (9 mm o.d., 7 mm i.d. × 25 cm; sealed at one end), cupric oxide (wire form), and silver sample boats (15 × 3 × 3 mm) were cleaned by heating in a muffle furnace, open to air at 860° C for one hour, prior to use in subsequent steps. Dry kerogen samples (previously treated to remove pyrite and/or other residual minerals; see Section IV-2) were weighed into the silver boats and carefully transferred into combustion tubes, 500 mg of cupric oxide having been previously added to each tube. Selection of amount of kerogen for combustion was based upon estimated ash content, with total kerogen weights chosen to yield up to 10 mg C (since use of greater quantities of material would have generated excessive pressures during combustion). When six tubes had been loaded, they were placed on a vacuum line and slowly evacuated to 10^{-3} torr. After samples had been subject to evacuation for at least twelve hours, and showed no detectable pressure increase over a ten-minute period when isolated from the pumps, the tubes were sealed under vacuum and transferred to the furnace for combustion. Combustion was usually carried out in lots of 12, with sealed tubes being placed in the furnace after it had reached 900° C. After two hours at that temperature, the furnace was cooled slowly to 600° C, was allowed to remain at this temperature for another two hours, and was then cooled to 500° C and held at that temperature overnight prior to cooling to room temperature. This cooling cycle was employed to facilitate conversion of SO_2 to $CuSO_4$ and of oxides of nitrogen to N_2 (Frazer and Crawford, 1963).

The gaseous contents of the combustion tubes were removed, separated, and measured manometrically, employing the vacuum apparatus shown schematically in Figure IV-4. In routine operation, a sample tube was attached at the sample port P and the head space of the tube cracking device C (Des Marais and Hayes, 1976) was evacuated. When the vacuum system reached its minimal pressure (as indicated by thermal conductivity vacuum gauges TC_1 and TC_2; typically less than 5×10^{-3} torr), the sample tube S and trap T_1 were cooled to −196° C in preparation for removal of N_2 from the tube. The sample tube was cracked with valves V_1, V_3, V_5, V_6, and V_7 closed; after a brief delay (to insure trapping of all H_2O and CO_2 either at S or T_1), valve V_3 was opened and the N_2 was quantitatively transferred (by use of a Toepler pump) to the calibrated gas buret. The gas quantity was measured (with a precision of $\pm 1\%$ or better) and then displaced to the length of Pyrex tubing T, in which it was sealed for transfer to the mass spectrometer for isotopic analysis.

When the N_2-gas sample tube at T had been replaced and the new tube evacuated, the sample tube S and trap T_1 were warmed to −125° C, and T_2 was cooled to −196° C, resulting in retention of water at S and T_1 and transfer of CO_2 to trap T_2 (with progress of the CO_2 distillation being monitored by vacuum gauge TC_2). With valve V_3 closed, trap T_2 was warmed to room temperature and Toepler pumping was initiated to transfer CO_2 into the buret for measurement and packaging (as described above for N_2), followed by replacement and evacuation of the gas sample tube, T.

With trap T_1 at −125° C and valve V_3 closed, the sample tube S was heated gently with a torch (to ~600° C) to facilitate complete transfer of water to the trap (during this transfer, the metal

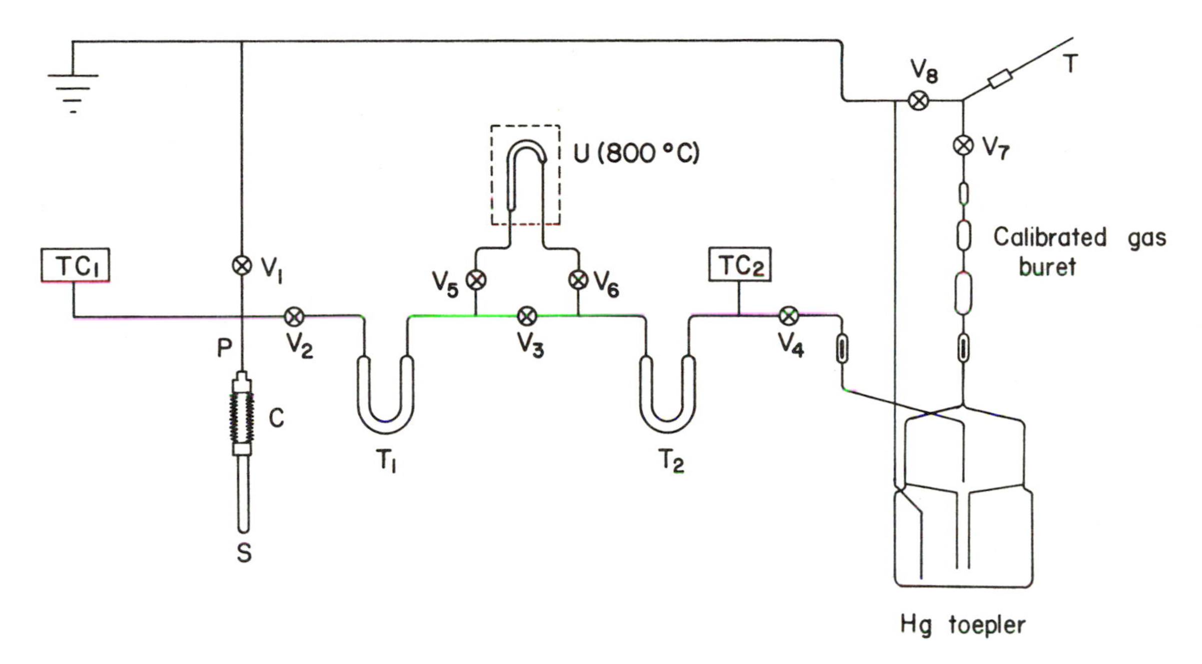

FIGURE IV-4. Schematic representation of the vacuum system employed in the separation, measurement, and packaging of gases for CHN analyses (see text for details of operation). Symbols used in the figure are as follows: V_1–V_8 represent valves; T_1 and T_2, 20-mm outside diameter Pyrex traps; TC_1 and TC_2, thermal conductivity pressure transducers; P, sample attachment port; S, 9-mm outside diameter quartz combustion tube; C, sample-tube cracking device (Des Marais and Hayes, 1976); U, 9-mm outside diameter quartz, uranium-packed reactor (dashed line indicates heating jacket for operation at 800° C); T, 6-mm outside diameter, 15-cm length pyrex tube for packaging of gas samples. Solid lines indicate 9-mm outside diameter pyrex tubing.

bellows of the sample cracking apparatus was also warmed to ~80° C with a laboratory heat gun). When gauge TC_1 indicated complete transfer of water to trap T_1, valves V_2, and V_4 were closed, V_5 and V_6 were opened, and T_1 was warmed slowly to room temperature, allowing the H_2O to be reduced to H_2 in the uranium reactor U. Before opening valve V_4 in order to collect the H_2, T_1 was heated to 100° C and T_2 was cooled to −196° C (in order to minimize loss of H_2O due to adsorption on the walls of trap T_1 and to trap any unreacted H_2O at T_2). Cycling of the Hg Toepler was begun, and continued until gauge TC_2 indicated baseline pressure (usually requiring at least 20 minutes) after which the H_2 was measured and packaged as indicated above (but only after the absence of water at trap T_2 had been verified by closing V_6 and V_4 and warming T_2, with gauge TC_2 being used to indicate the presence of any significant quantity of unreacted H_2O).

In order to characterize the analytical scheme outlined above, several "control" experiments were conducted prior to, and between, actual kerogen analyses. Blank samples of two types were run to allow estimation of C, H, and N elemental blanks under conditions of sample analysis. Table IV-4 gives the results of these analyses. All blanks were combusted and processed as described above. The runs of 30 May were made prior to actual sample processing as a test for the effect of pyrite (present in most kerogen samples) upon the magnitude of

TABLE IV-4. Determinations of CHN analytical blanks.

		Blank (*μmole*)		
Date[a]	*Tube contents*[b]	*C*	*H*	*N*
30 May 1980	CuO, pyrite	0.40	1.98	<0.001
30 May 1980	Ag, CuO, pyrite	0.55	1.40	<0.001
24 June 1980	Ag, CuO	0.16	1.66	<0.001
1 July 1980	Ag, CuO	0.06	0.16	<0.001

[a] Analyses of 30 May were conducted prior to kerogen analyses; runs of 24 June and 1 July were preceeded by 10 and 52 kerogen analyses, respectively.

[b] See text for explanation of sample preparation.

the hydrogen and carbon blanks. The silver sample boat was omitted from one of the two runs and both tubes contained 1 mg of pyrite. The results from these pyrite-containing blanks show that no analytically significant amount of SO_2 was recovered from the tubes after combustion, either in the H_2O or CO_2 fractions, and that the most probable fate of any SO_2 formed during combustion was conversion to $CuSO_4$. Results of analyses of a pure organic compound in the presence of added pyrite are presented and discussed below. The blank analyses of 24 June and 1 July (Table IV-4) included only empty sample boats and cupric oxide, and were conducted after analysis of 10 and 52 kerogen samples, respectively (including a total of 63 kerogen analyses).

Taken collectively, the data presented in Table IV-4 clearly indicate that the blanks of carbon and nitrogen are insignificant in the procedure, both being at least two orders of magnitude less than the smallest recovered quantity of the corresponding elements in actual analyses. Replicate analyses of kerogen samples (Table IV-7) showed no dependence of N/C, $\delta^{13}C$, or $\delta^{15}N$ upon the amount of sample combusted, a result which is consistent with analytically unimportant carbon and nitrogen blanks.

The significance of the hydrogen blank is more difficult to assess. Apparently, the hydrogen blank either (*i*) decreased steadily over the course of the analyses (a not unreasonable possibility, since the uranium furnace employed contained freshly prepared material and may have become "conditioned" over the period of analyses) or (*ii*) was highly variable (viz., varying from 2 to 0.2 μmole H) during kerogen analysis. Regardless of whether (*i*) or (*ii*) is correct, the temporal variations in the hydrogen blank cannot be resolved, and this limits the precision of the H/C and δD_{SMOW} determinations. It is not useful to deal quantitatively with the effect of blank variation upon the precision of H/C or δD_{SMOW} measurements, because there is a large systematic variation in these quantities associated with the presence of pyrite (see discussion below).

Analyses of pure organic compounds provided data for evaluation of the performance of the analytical procedure, specifically with regard to the precision of measured isotopic compositions and atomic ratios, both in the presence and absence of added pyrite. Table IV-5 gives data obtained from analyses of various organic compounds and Table IV-6, gives the results of analyses of 1,10-phenanthroline (run prior to or between kerogen analyses, with 1 mg pyrite added to the sample tube as indicated in the table). Phenanthroline analytical results with pyrite added show distinct shifts of H/C and δD_{SMOW} relative to analyses without pyrite added, whereas carbon and nitrogen results appear unaffected by pyrite addition. The cause of H/C shifts to higher values (actually resulting in improved absolute accuracy in both cases) and of the lower observed deuterium abundances in analyses including pyrite is not clear.

It cannot be determined which of the two populations of phenanthroline deuterium abundances (pyrite present vs. no pyrite) is the more accurate, since the δD_{SMOW} of this sample was unknown and no alternate method was available for its determination. Unfortunately, observed variations in δD_{SMOW} for phenanthroline samples are extreme under conditions like those of kerogen analysis, and reported δD_{SMOW} values for kerogen samples must be considered suspect (i.e., with a possibility

TABLE IV-5. CHN analyses of pure organic compounds.

		Relative elemental abundances			
		H/C		*N/C*	
Date	*Compound and formula*	*Found*	*Calculated*	*Found*	*Calculated*
2 June 1980	valine, $C_5H_{11}N$	2.200	2.200	0.205	0.200
9 June 1980	chrysene, $C_{18}H_{12}$	0.650	0.667	0.003	0.0
9 June 1980	triphenylamine, $C_{18}H_{15}N$	0.860	0.833	0.058	0.056
9 June 1980	benzoic acid, $C_7H_6O_2$	0.853	0.857	0.003	0.0

TABLE IV-6. Elemental and isotopic analyses of 1,10-phenanthroline, $C_{12}H_8N_2$.

Date	*Relative elemental abundances*		*Isotopic abundances, ‰*		
	H/C	*N/C*	δD_{SMOW}	$\delta^{15}N_{AIR}$	$\delta^{13}C_{PDB}$
(calculated values)	0.667	0.167			
5 June 1980[a]	0.646	0.168	90.0	0.0	−28.28
10 June 1980[a]	0.657	0.168	92.7	—[b]	−28.39
24 June 1980[c]	0.592	0.167	175.4	−0.87	−28.37
25 June 1980	0.601	0.170	151.6	0.08	−28.30
26 June 1980	0.586	0.169	136.4	−0.87	−28.25
27 June 1980	—[b]	0.174	—[b]	−0.96	−28.41
1 July 1980	0.593	0.169	152.1	−0.64	−28.27
2 July 1980	0.594	0.168	139.6	−0.49	−28.27

[a] 1 mg pyrite added to combustion tube to give ~50% FeS_2.
[b] Value not determined.
[c] Analyses of 24 June through 2 July were preceeded by 10, 16, 23, 31, 45, and 63 kerogen analyses, respectively (analysis of kerogen samples began on 20 June 1980).

of analytical error large enough to preclude any useful interpretations of these data).

Some assessment of the uncertainties that should be associated with kerogen H/C ratios should be made. Considering the analytical problems mentioned above (viz., those of the hydrogen blank and of pyrite influence), both the accuracy and the the precision of reported H/C results should be evaluated. The data of Table IV-5 and IV-6 show that all H/C determinations of "control" samples are within 12% of the theoretical values, with significantly better agreement in six of the eleven cases there reported. It should be noted that the presence of pyrite does not result in anomalous H/C results, and that pyrite thus cannot be regarded as a source of "analytical interference" on the basis of the available data. It should also be noted that H/C values obtained for actual kerogen samples were always consistent with other pertinent information available (e.g., kerogen color, degree of graphitization, metamorphic grade, sample age, etc.). Therefore, the absolute accuracies of H/C values reported are adequate for geochemical interpretation, being better than ±15% (three times the average deviation here observed) for any reported result. The precision of the H/C determinations is estimated to be ±15% (95% confidence limits) for any single result, based on a pooled relative standard deviation of 0.0679 and eleven degrees of freedom (as calculated from the data of Table IV-7, omitting the outlying comparison furnished by the pair of results for P.P.R.G. Sample No. 225-1).

Taken collectively, the data of Tables IV-5, IV-6, and IV-7 indicate that N/C, $\delta^{15}N$, and $\delta^{13}C$ determinations do not reflect the analytical problems associated with the hydrogen analyses, and provide a basis for realistic estimates of uncertainties in these values reported for kerogen samples. A calculation based on the data of Table IV-7 (carried out as described above for estimation of precision in H/C results) show that the 95% confidence limits for a single determination of N/C is ±12%. Pooled standard deviations calculated from the data of Table IV-7 show that the uncertainties in $\delta^{15}N_{air}$ and $\delta^{13}C_{PDB}$ determinations are ±2‰ and ±0.4‰, respectively, at the 95% confidence level.

IV-4. X-Ray Diffraction Analysis of Kerogen

Kerogen specimens, obtained by the procedure described in Section IV-2, were subjected to x-ray diffraction analysis in order to yield information regarding the extent of graphitization of these

TABLE IV-7. Replicate CHN analyses.

P.P.R.G. sample no.	Date	Elemental yield (μmole) CO_2	H_2	N_2	Atomic ratios H/C	N/C	Isotopic compositions, ‰ $\delta^{13}C_{PDB}$	δD_{SMOW}	$\delta^{15}N_{Air}$
004–1	24 June 1980	174	14.7	0.535	0.169	0.006	−35.89	−116.2	—[a]
004–1	1 July 1980	408	33.6	1.34	0.165	0.007	−36.09	−110.8	17.87
022–1	24 June 1980	347	16.5	1.89	0.095	0.011	−29.88	−118.1	1.56
022–1	27 June 1980	217	8.50	1.14	0.078	0.010	−30.64	−94.6	2.98
022–1	29 June 1980	545	23.3	7.35	0.085	—[a]	−29.92	−99.6	—[a]
054–1	24 June 1980	255	18.7	1.48	0.147	0.012	−25.20	−110.2	—[a]
054–1	24 June 1980	97.1	8.32	0.562	0.171	0.012	−25.25	−118.0	—[a]
054–1	28 June 1980	263	18.9	1.51	0.144	0.011	−25.27	−94.5	4.77
054–1	29 June 1980	297	20.8	1.72	0.140	0.012	−25.16	−100.2	3.30
054–1	1 July 1980	391	28.1	2.23	0.144	0.011	−25.00	−99.6	2.85
120–1	26 June 1980	266	—[a]	3.23	—[a]	0.024	−32.10	—[a]	3.07
120–1	27 June 1980	306	61.6	3.79	0.402	0.025	−32.13	−94.6	2.98
133–2	28 June 1980	214	35.9	2.69	0.336	0.025	—[a]	−120.6	7.43
133–2	2 July 1980	172	30.3	2.17	0.353	0.025	−25.32	−122.2	8.67
182–1	26 June 1980	253	14.7	0.473	0.116	0.004	−31.65	−102.5	—[a]
182–1	2 July 1980	569	34.5	1.04	0.121	0.004	−31.66	−99.3	2.86
215–1	27 June 1980	142	11.3	0.460	0.159	0.006	−31.88	−111.5	—[a]
215–1	1 July 1980	411	33.8	1.28	0.165	0.006	−31.86	−95.6	3.20
225–1[b]	30 June 1980	116	8.55	0.262	0.147	0.005	−33.02	−101.7	—[a]
225–1	2 July 1980	227	12.0	0.463	0.106	0.004	−33.17	−106.4	—[a]
325–1	28 June 1980	60.8	5.18	0.262	0.170	0.009	−27.20	−99.2	—[a]
325–1	3 July 1980	205	17.7	0.832	0.173	0.008	−27.16	−94.8	—[a]

[a] Value not determined.
[b] Sample analyzed on 2 July was $LiAlH_4$ treated prior to analysis; sample analyzed on 30 June was not $LiAlH_4$ treated.

samples, thus providing a measure of metamorphic "maturity" (as discussed in Chapter 5).

The technique employed was similar to that of Griffin (1967). A small amount of dry sample was evenly dispersed (using a few drops of water) within a 1 × 2 × 0.05 cm depression in an aluminum slide. Samples were allowed to dry and were then scanned with nickel-filtered Cu$K\alpha$ radiation at a rate of 1° (2θ) per minute in the forward direction, beginning at 10° and continuing to at least 40°. All samples were analyzed under identical instrumental operating conditions. The use of glass slides was avoided in these analyses due to the broad, diffuse reflection, from 15° to 40° (2θ), produced in scans without a sample present. Aluminum showed a featureless diffraction pattern in the region of interest (10° to 35°); its use thus precluded the production of misleading patterns that may have resulted from inadequate coverage of sample upon a glass substrate. The aluminum (111) reflection at 38.47° provided a convenient internal standard for peak position.

Strip chart recordings of the x-ray diffraction patterns were used to measure both the position (measured at peak centroid in degrees 2θ; converted to d_{002} in Å) and peak width at half height (measured in °2θ at half the centroid peak height) of the graphite (002) reflection. Replicate slide preparation and analysis of P.P.R.G. Samples No. 038, 159, 198, and 199 showed that the uncertainty (95% confidence limits) in peak position (converted to d spacing by the Bragg equation) was ±0.01 Å and was ±0.4° (2θ) for width at half height.

IV-5. Evaluation of Kerogen Color

Kerogen samples were examined by transmitted light microscopy to allow evaluation of their color. Strew mounts of samples were prepared and studied under oil immersion at approximately 1000 × magnification under conditions of critical illumination. Observed kerogen colors were compared directly with the Inter-Society Color Council (I.S.C.C.)–U.S. National Bureau of Standards (N.B.S.) Centroid Color Charts under "recommended conditions of lighting and viewing" as described in detail by Kelly and Judd (1976). Kerogen particles were generally amorphous, varying in optical thickness such that colors other than black were observable only in the more diaphanous regions of larger particles. Some degree of subjectivity was involved in choosing appropriate particles or areas of particles for direct comparison to color standards; reported color descriptions correspond to the "lightest" (tending toward yellow) repeatedly observable color for each specimen. Descriptive terms used in Chapter 5 (Table 5-8) correspond to names of I.S.C.C.–N.B.S. Centroid Color Chart plates (Kelly and Judd, 1976).

GLOSSARY I

Technical Terms

By J. William Schopf

Abiotic. A term pertaining to substances or objects that are of nonbiologic origin; used especially in reference to organic matter produced via chemical reactions in the absence of living systems.

Acetate. Any of various organic salts or esters of acetic acid.

Acetic acid. An organic compound, the two-carbon carboxylic acid CH_3COOH.

Acetyl-coenzyme A (*acetyl-coA*). An organic complex involved in enzymatic acetyl transfer reactions, $CH_3COSCoA$.

Acritarch. An "alga-like" microfossil of unknown or uncertain biologic relationship, unicellular or apparently unicellular, organic-walled, commonly spheroidal, smooth or spiny; "Acritarcha" is an artificial group composed of such microfossils.

Acrolein. An organic compound, acrylic aldehyde, $CH_2{=}CHCHO$.

Actinolite. A mineral of the amphibole group, $Ca_2(Mg, Fe)_5Si_8O_{22}(OH)_2$.

Adenosine triphosphate (*ATP*). An organic compound, adenosine 5′-triphosphoric acid, $C_{10}H_{16}N_5O_{13}P_3$, a coenzyme involved in the transfer of phosphate bond energy.

Adiabatic. In thermodynamics, a term pertaining to the relationship of pressure and volume when a gas or fluid is compressed or expanded without either giving or receiving heat.

Aerobe. See Glossary II.

Aerotaxis. The locomotory movement of an organism in response to an environmental gradient of molecular oxygen, either toward a higher oxygen concentration ("positive aerotaxis") or away from a higher oxygen concentration ("negative aerotaxis").

Akinete. Specialized reproductive cells (resting spores) occurring singly or in uniseriate groups, commonly adjacent to heterocysts, in various types of (predominantly filamentous) cyanobacteria.

Albite. A mineral of the feldspar group, $NaAlSi_3O_8$.

Albite-epidote-amphibolite facies. Metamorphic rocks formed under intermediate pressures (3 to 7 kb) and temperatures (250° to 450° C) with conditions intermediate between those of the greenschist and amphibolite facies; also referred to as the "epidote-amphibolite facies" or the "quartz-albite-epidote-almadine subfacies."

Alcohol. Any of various organic compounds analogous to ethyl alcohol, hydroxyl (OH-containing) derivatives of hydrocarbons.

Aldehyde. Any of various organic compounds characterized by the group CHO.

Algae. Photosynthetic, eukaryotic, nonvascular (thallophytic) plants, unicellular or multicellular, commonly aquatic; "seaweeds" and their freshwater equivalents.

Aliphatic hydrocarbon. Any of various straight or branched, open-chain (non-cyclic) organic compounds having the empirical formula C_nH_{2n+2}, such as methane, CH_4.

Alkali. Any strongly basic substance such as a hydroxide or carbonate of an alkali metal (e.g., sodium or potassium).

Allochthonous. Formed or occurring elsewhere than in place, of foreign origin; opposite of autochthonous.

Alluvium. A general term for unconsolidated detrital sediment (clay, silt, sand, etc.) deposited during relatively recent geologic time by running water, especially as a result of flood.

Amide. Any of various organic compounds resulting from replacement of an atom of hydrogen in ammonia by an element or radical or of one or more atoms of hydrogen in ammonia by univalent acid radicals.

Amine. Any of various basic organic compounds derived from ammonia by replacement of hydrogen by one or more univalent hydrocarbon radicals, or a compound containing one or more halogen atoms attached to nitrogen.

Aminolevulinic acid. An organic compound, the five-carbon monocarboxylic acid $HOOCCH_2 \cdot CH_2COCH_2NH_2$.

Amphiaerobe. See Glossary II.

Amphibolite facies. Metamorphic rocks formed under moderate to high pressures (3 to 8 kb) and temperatures (450° to 700° C).

Anaerobe. See Glossary II.

Andalusite. A mineral, Al_2SiO_5, trimorphous with kyanite and sillimanite.

Andesine. A mineral of the plagioclase feldspar group common in andesite and diorite.

Andesite. Igneous, extrusive rock, composed chiefly of acid plagioclase (especially andesine), one or more mafic minerals (e.g., biotite, hornblende, pyroxene), and some quartz; extrusive equivalent of diorite.

Anhydrite. A mineral, anhydrous calcium sulfate, $CaSO_4$ (see *Gypsum*).

Ankerite. A mineral, $Ca(Fe,Mg,Mn)(CO_3)_2$.

Anorthosite. Igneous, plutonic rock, composed almost entirely of calcic plagioclase feldspar (usually labradorite).

Anoxic. A term pertaining to the absence of uncombined molecular oxygen.

Anthracite. Clastic sedimentary rock, a coal of the highest metamorphic rank; metamorphosed derivative of bituminous coal.

Anticline. A geologic structure, a fold convex upward whose core contains the stratigraphically older rocks.

Antiform. An anticline-like geologic structure in which the stratigraphic sequence is unknown.

Apatite. A calcium phosphatic mineral with the general formula $Ca_5(PO_4,CO_3)_3(F,OH,Cl)$.

Aragonite. A mineral, $CaCO_3$, a metastable form of calcite.

Archaebacteria. Members of the Archaebacteriae, a kingdom proposed by Woese and Fox (1977) to include methanogenic, extremely halophilic and some thermoacidophilic bacteria, taxa regarded by them as being separable from all other prokaryotes (viz., the Eubacteriae) based on the chemistry of their cell walls, membranes, transfer RNA's, and RNA polymerase subunits.

Archean. An era of Earth history, as used here composed of four suberas: the Hadean (4.5 to 3.9 Ga ago), the Early Archean (3.9 to 3.3 Ga ago), the Middle Archean (3.3 to 2.9 Ga ago), and the Late Archean (2.9 to 2.5 Ga ago); the Archean and the subsequent Proterozoic Era (2.5 to 0.57 Ga ago) together comprise the Precambrian Eon.

Argillite. Clastic sedimentary rock, derived from, but more indurated than, either a mudstone or a shale.

Aromatic hydrocarbon. Any of various cyclic organic compounds composed of carbon and hydrogen and characterized by the presence of at least one benzene ring.

Aspartic acid. An organic compound, the amino acid $HOOCCH_2CH(NH_2)COOH$.

Asteroid. Any of the small, celestial, "planetoid" bodies in orbit about the Sun.

Asthenosphere. The structurally weak shell of the Earth below the lithosphere; essentially equivalent to the upper mantle.

Augite. A mineral, the clinopyroxene $(Ca,Na)(Mg,Fe^{2+},Al)(Si,Al)_2O_6$.

Aulocogen. A geologic structure, a fault-bounded intracratonal trough.

Autochthonous. Formed or occurring in the place where found; opposite of allochthonous.

Autotroph. See Glossary II.

Bacteria. Prokaryotic organisms (including archaebacteria and prochlorophytes) other than cyanobacteria.

Banded iron-formation. Chemical sedimentary rock, typically thin bedded and/or finely laminated, containing at least 15% iron and commonly containing layers of chert; see also *Iron-formation*, *Jaspilite*; following Western usage (e.g., Brandt et al., 1972), "iron-formation" is here hyphenated when the term is used in a lithologic sense, but is capitalized and non-hyphenated when used in a stratigraphic sense as in "Marra Mamba Iron Formation."

Barite. A mineral, $BaSO_4$.

Basalt. Igneous, extrusive rock, composed chiefly of calcic plagioclase feldspar (usually labradorite) and clinopyroxene; extrusive equivalent of gabbro.

Batholith. A mass of plutonic igneous rocks that has more than 100 km^2 in surface exposure and is composed predominantly of granodiorite and quartz monzonite.

Benthos. Subaqueous (usually marine) bottom-dwelling organisms.

Benzene. An organic compound, the aromatic (cyclic) hydrocarbon C_6H_6.

Biocoenose. A community of organisms that live closely together and that form a natural ecologic unit.

Biogenetic element. A general term for the principal chemical elements of living systems, C, H, O, N (S and P).

Biogenic. A term pertaining to substances or objects of biological origin; formed by the activity of organisms.

Biostrome. A laterally extensive or broadly lenticular commonly "reef-like" rock mass built by and composed mainly of the remains of sedentary organisms.

Biota. All of the organisms of a particular area or time.

Biotite. A mineral of the mica group, $K(Mg,Fe^{2+})_3(Al,Fe^{3+})Si_3O_{10}(OH)_2$.

Bituminous coal. Clastic sedimentary rock, a coal of metamorphic rank between lignite and anthracite, common in Carboniferous "coal measures" strata.

Blue-green algae. Cyanobacteria.

Blueschist facies. Metamorphic rocks formed under high pressures (>5 kb) and moderate temperatures (300° to 400° C); also referred to as the "glaucophane schist facies."

Breccia. Sedimentary, igneous, or tectonic rock, clastic and coarse-grained, composed of large (>2 mm), angular, broken rock fragments cemented together in a finer-grained matrix; consolidated equivalent of rubble.

Butyric acid. An organic compound, butyl monocarboxylic acid, $CH_3CH_2CH_2COOH$.

Bytownite. A mineral of the plagioclase feldspar group common in mafic and ultramafic igneous rocks.

C3 metabolism. A metabolic pathway exhibited by autotrophs in which the three-carbon compound phosphoglyceric acid is the first identified product of carbon fixation.

C4 metabolism. A metabolic pathway exhibited by autotrophs in which the four-carbon compound oxalacetic acid is the first identified product of carbon fixation.

Calcite. A mineral, $CaCO_3$, trimorphic with aragonite and vaterite.

Calvin cycle. Ribulose bis-phosphate-utilizing biosynthetic cycle of carbon fixation exhibited by C3 autotrophs.

Cambrian. The earliest geologic period of the Paleozoic Era (and the Phanerozoic Eon) of Earth history.

CAM plant. An autotroph exhibiting "crassulacean acid metabolism," a specialized pathway of carbon fixation typical especially of plants of the Crassulaceae.

Carbohydrate. Any of various organic compounds consisting of a chain of carbon atoms in which hydrogen and oxygen are attached in a 2:1 ratio as in cellulose, $(C_6H_{10}O_5)_n$.

Carbonaceous. A term pertaining to an object or substance that contains or is composed of organic matter.

Carbonate. A mineral, characterized by CO_3^{2-} (e.g., calcite, $CaCO_3$), or a rock consisting chiefly of carbonate minerals (e.g., limestone, dolomite, or carbonatite).

Carbonatite. Igneous rock of carbonate composition.

Carbonyl. The functional group $-C{=}O$ of many organic compounds.

Carboxylic acid. Any of various organic acids containing one or more carboxylic (–COOH) groups.

Carnot cycle. A sequence of operations forming the working cycle of an ideal heat engine of maximum thermal efficiency (a Carnot engine) consisting of isothermal expansion, adiabatic expansion, isothermal compression, and adiabatic compression to the initial state.

Catagenesis. The pre-metamorphic stage of alteration of sedimentary organic matter intermediate between diagenesis and metagenesis occurring between about 50° and 150° C.

Catalase. Any of a group of protein complexes with hematin groups that catalyzes the decomposition of hydrogen peroxide into water and oxygen.

Cenozoic. Most recent era of the Phanerozoic Eon of Earth history.

Chalcedony. A mineral, crytocrystalline quartz, SiO_2, commonly microscopically fibrous.

Chamosite. A mineral of the chlorite group, $(Mg,Fe)_3Fe_3^{3+}(Al,Si_3)O_{10}(OH)_8$.

Charnockite. Igneous or metamorphic rock, a hypersthene-bearing granite, commonly found only in granulite facies terranes.

Chasmolith. Any of various microorganisms growing in the crevices of a rock.

Chemical fossil. Broadly, any direct chemical evidence (geochemical compound, isotope effect, etc.) indicative of pre-existent life; in a more restricted sense, pertaining to organic molecular fossils containing specific structures of identifiable biosynthetic origin or derivation.

Chemical sediment. Sediment or sedimentary rock composed primarily of material formed by precipitation from solution or colloidal suspension, as by evaporation, that typically exhibits a crystalline texture (e.g., gypsum, halite, and many cherts and limestones).

Chemotroph. See Glossary II.

Chert. Chemical sedimentary rock (or a product of secondary replacement), consisting chiefly of microcrystalline or cryptocrystalline quartz and lesser amounts of chalcedony.

Chlorite. Minerals of a group having the general formula $(Mg,Fe^{2+},Fe^{3+})_6AlSi_3O_{10}(OH)_8$.

Chloroplast. Chlorophyll-containing, membrane-bound organelle of plants and some protists; the site of photosynthesis in eukaryotic cells.

Chondrite. A stony meteorite characterized by the presence of chondrules, spheroidal granules usually about 1 mm in diameter that consist chiefly of olivine and/or orthopyroxene.

Citric acid cycle. The energy-yielding, molecular oxygen-requiring metabolic cycle of aerobic respiration.

Clastic. A term pertaining to a rock or sediment composed principally of broken fragments derived from pre-existing rocks that have been transported individually from their places of origin; sandstones and shales are commonly occurring clastic rocks.

Clinopyroxene. A mineral group composed of pyroxenes of the monoclinic crystal system and sometimes containing considerable calcium (see *Orthopyroxene*).

Coal ball. A concretionary sedimentary structure, predominantly calcareous and pyritic, containing permineralized plant fragments and typical especially of Carboniferous-age coal deposits.

Coenocytic. A term pertaining to a filamentous organism that is tubular, lacking transverse walls to separate protoplasts into a series of cells; in eukaryotes (especially thallophytes) the term is used commonly in reference to a multinucleate cell.

Conglomerate. Clastic sedimentary, coarse-grained rock, composed of rounded fragments larger than 2 mm in diameter cemented in a finer-grained matrix; consolidated equivalent of gravel.

Continental shelf. That part of the continental margin between the shoreline and the continental slope.

Convection cell. In plate tectonics, a pattern of mass movement of mantle material in which the central area is uprising and the outer area is downflowing, due to heat variations.

Cordierite. A mineral, $(Mg,Fe)_2Al_4Si_5O_{18}$, occurring in granites and typical of metamorphic rocks formed under low pressure.

Core of the Earth. The spherical central zone of the Earth's interior, below the Gutenberg Discontinuity at a depth of 2,900 km to the center of the planet (at a depth of 6,371 km), divided into two zones, the liquid outer core (2,900 to 5,080 km deep) and the solid inner core (5,080 km to the center of the Earth).

Coriolis effect. The apparent deflective component in the centrifugal force produced by the rotation of the Earth.

Craton. The extensive, central, tectonically stable region of a continent composed of shields and platforms.

Critical point. In a system of one component, the temperature and pressure at which a liquid and its vapor become identical in all properties.

Crossbedding. A sedimentary structure with an internal arrangement of the layers of a stratified rock characterized by the laminae being inclined in concave forms by changing currents of air (as in dune sandstones) or water (e.g., in stream, channel or delta deposits).

Crust of the Earth. The outermost shell of the Earth (representing less than 0.1% of the Earth's total volume) that lies above the Mohorovičić Discontinuity (i.e., extending to a depth of 30 to 50 km beneath most continents and about 10 to 12 km beneath most oceans).

Cryptozoic. Precambrian.

Cumulate. A term pertaining to an igneous rock formed by the accumulation of crystals that due to gravity settled out from a magma.

Curie temperature. The temperature of transition from ferromagnetism to paramagnetism above which thermal agitation prevents spontaneous magnetic ordering.

Cyanamide. A caustic, acidic, organic compound, $H_2NC{\equiv}N$.

Cyanobacteria. Prokaryotic, bacterium-like microorganisms containing phycocyanin and/or phycoerythrin, chlorophyll *a* (but not chlorophyll *b*), and capable of aerobic (oxygen-producing) photosynthesis; numerous strains are also capable of anaerobic (non-oxygen-producing) photosynthesis; also referred to as blue-green algae, cyanophytes, and myxophytes (see *Prochlorophyte* and *Proalgae*).

Cyanogen. The organic compound ethane dinitrile, $N{\equiv}C{-}C{\equiv}N$.

Cyanophytes. Cyanobacteria.

Cysteine. An organic compound, the sulfur-containing amino acid $HSCH_2CH(NH_2)COOH$.

Cystine. An organic compound, the sulfur-containing amino acid $HOOCCH(NH_2)CH_2 \cdot SSCH_2CH(NH_2)COOH$.

Cytochrome. Any of a group of iron-containing

hydrogen or electron carriers involved in cellular metabolism.

Decollement. A type of geologic structure produced by the detachment of strata during folding and overthusting resulting in independent styles of deformation in the rocks above and below.

Dehydrogenase. Any of various enzymes capable of mediating the biochemical removal of hydrogen from an appropriate substrate.

Dendrite. A mineralic object, occurring either as a surficial deposit or an inclusion, that has crystallized in a branching pattern.

Diabase. Igneous, intrusive rock, composed chiefly of labradorite and pyroxene and exhibiting ophitic texture.

Diagenesis. Broadly, all of the chemical, physical, and biological changes undergone by a sediment and its component minerals, fossils, etc., after its initial deposition and during and after its lithification exclusive of subsequent weathering or metamorphism; in organic geochemistry, the low temperature ($<50°$ C), in part microbially mediated, earliest stage of alteration of organic matter in sediments preceding catagenesis, metagenesis, and metamorphism.

Dike. A tabular intrusion of igneous rock that cuts across the planar structure of the surrounding rock (see *Sill*).

Diorite. Igneous, intrusive rock, composed chiefly of acid plagioclase feldspars (oligoclase and andesine), amphiboles (especially hornblende), pyroxene, and sometimes small amounts of quartz; intrusive equivalent of andesite.

DNP. An organic compound, nicotinamide adenine dinucleotide.

Dolomite. A mineral, $CaMg(CO_3)_2$, or sedimentary rock (also referred to as "dolostone") consisting chiefly of the mineral dolomite.

Dolostone. A sedimentary rock consisting chiefly of the mineral dolomite.

Dubiofossil. A megascopic or microscopic fossil-like object or structure (or chemical component) of possible but unestablished biogenicity; structures (or chemical components) are assigned to this category temporarily, pending availability of additional information that would permit them to be classed either as bona fide fossils or as assuredly nonfossil.

Dubiomicrofossil. A mineralic or organic microfossil-like object or structure of possible but unestablished biogenicity; a morphological dubiofossil bearing resemblance to a fossil microscopic organism or other microfossil.

Dunite. Igneous, plutonic rock, a peridotite composed almost entirely of olivine.

Duricrust. A general term for a hard soil crust formed in semiarid environments due to evaporation of vadose waters and resultant precipitation of calcareous material (calcrete), siliceous material (silcrete), phosphatic material (phoscrete), and the like.

Eclogite facies. Metamorphic rocks formed under high pressures (7.5 to >10 kb) and high temperatures (generally, 600° to 700° C).

Ediacarian System. A proposed name for rocks of the latest Precambrian, including and stratigraphically equivalent to those containing megascopic soft-bodied invertebrate fossils of the Ediacaran Fauna of South Australia; also referred to as Vendian and Eocambrian.

Electrophosphorylation. See Glossary II.

Embden-Meyerhof-Parnas (*E-M-P*) *Pathway*. The energy-yielding, anaerobic, metabolic pathway resulting in formation of lactic acid and adenosine triphosphate from the catabolism of glucose; also referred to as glycolysis.

Enantiomer. Either of the two forms of various organic compounds (e.g., many amino acids, sugars, etc.) or of inorganic compounds (e.g., quartz crystals) that have the property of being mirror images of each other.

Endogenic. A term meaning derived from within; pertaining to a geologic process or its resultant features that originates within the Earth (e.g., volcanism, extrusive igneous rocks, etc.); opposite of exogenic.

Endolith. Any of various microorganisms growing within the pore spaces of a rock or lithified soil crust.

Ensialic. A term pertaining to material, usually sedimentary, accumulating on a sialic crust.

Eolian. A term pertaining to a geologic process or its resultant features caused by wind (e.g., loess, dunes, and similar wind-deposited or wind-eroded materials).

Epeiric sea. A sea on the continental shelf or within a continent; also referred to as epicontinental sea.

Epidote. A mineral, $Ca_2(Al,Fe)_3Si_3O_{12}(OH)$.

Ester. Any of various organic compounds characterized by a $-C(O)O-$ linkage.

Ethanol. An organic compound, ethyl alcohol, CH_3CH_2OH.

Ether. Any of various organic compounds characterized by a –C–O–C– linkage.

Eucaryotes. Eukaryotes.

Eugeosyncline. A geosyncline in which volcanism is associated with clastic sedimentation; the portion of an orthogeosyncline located away from the craton; see *Miogeosyncline*.

Eukaryotes. Unicellular or multicellular organisms (viz., protists, fungi, plants and animals) characterized by nucleus-, mitochondrion-, and (in plants and some protists) chloroplast-containing cells that are capable typically of mitotic cell division.

Evaporite. Chemical sedimentary rock, composed principally of minerals (e.g., gypsum, anhydrite, rock salt, etc.) precipitated as a result of evaporation of an aqueous saline solution.

Exogenic. Derived from without; pertaining to a geologic process or its resultant features that originates externally to the Earth (e.g., weathering, erosion, clastic rocks, etc.); opposite of edogenic.

Exosphere. The upper region of an atmosphere where some particles achieve the velocity necessary for their escape from the planet's gravitational field.

Fabric. In sedimentology, the distribution and orientation of elements (particles, cement, etc.) of which a sedimentary rock is composed; with regard to stromatolites, the microstructure, laminar shape, and related microscopic details of a stromatolitic sediment.

Facies. The sum of all primary lithologic and paleontologic characteristics exhibited by a sedimentary rock and from which its origin and environment of formation may be inferred (see *Metamorphic facies*).

Fault. A surface or zone of rock fracture along which there has been displacement.

Fayalite. A mineral of the olivine group, Fe_2SiO_4.

Feldspar. A mineral group composed of compounds of the general formula $(K,Na,Ca,Ba,Rb,Sr,Fe)Al(Al,Si)_3O_8$.

Feldspathic. A term pertaining to a rock or mineral aggregate containing feldspars.

Felsic. A general term for light-colored igneous rocks consisting largely of feldspar and silica (see *Mafic*).

Fenestra. In sedimentology, a shrinkage pore or open space structure in a rock that may be completely or partly filled by secondarily introduced sediment and cement.

Fermentation. See Glossary II.

Ferredoxins. Any of a group of non-heme iron-containing proteins functioning as cellular electron carriers in photosynthesis, nitrogen fixation, and other biological oxidation-reduction reactions.

Filament. In microbiology, a collective term referring to the cylindrical external sheath and the internal cellular trichome of a filamentous prokaryotic microorganism.

Fischer-Tropsch Reaction. The chemical synthesis of hydrocarbons, aliphatic alcohols, aldehydes, and ketones by the catalytic hydrogenation of carbon monoxide using enriched synthesis gas from the passage of steam over heated coke.

Flavin. Any of various multiple ring organic compounds that form a component of riboflavin and such hydrogen carriers as flavin adenine dinucleotide (FAD) and flavin mononucleotide (FMA).

Flysch. A marine sedimentary facies characterized by a thick sequence of poorly fossiliferous, thinly bedded, graded deposits (chiefly of calcareous shales and muds) rhythmically interbedded with conglomerates, coarse-grained sandstones, and graywackes; the term is commonly applied to turbidites.

Fold. A bend of rock strata or bedding planes resulting from tectonic deformation.

Formaldehyde. An organic compound, the simplest aldehyde, HCHO.

Formic acid. An organic compound, the simplest monocarboxylic acid, HCOOH.

Fossil. Any direct physical (viz., morphological fossil) or chemical (viz., chemical fossil) remains, object, structure, trace, imprint, etc., indicative of pre-existent life.

Fumarate reductase (*FR*). An enzyme capable of mediating the reduction of fumarate to succinate.

Fumaric acid. An organic compound, the dicarboxylic acid HOOCCH=CHCOOH.

Ga. Giga anna, 10^9 years.

Gabbro. Igneous, intrusive rock, composed chiefly of basic plagioclase (commonly labradorite or bytownite) and clinopyroxene; intrusive equivalent of basalt.

Galena. A mineral, PbS.

Garnet. A mineral group composed of compounds with the general formula $(Ca,Mg,Fe^{2+},Mn^{2+})_3(Al,Fe^{3+},Mn^{3+},Cr)_2(SiO_4)_3$.

Geosyncline. A mobile, geographically extensive, elongate or basin-like downwarping of Earth's crust where thick sequences (thousands of meters) of sedimentary and volcanic rocks accumulate.

Geothermal gradient. The increase in temperature within the Earth's crust occurring with increasing depth, approximately 25° C/km.

Glaucophane. A mineral of the ampibole group, $Na_2(Mg,Fe^{2+})_3Al_2Si_8O_{22}(OH)_2$.

Glaucophane schist facies. Blueschist facies.

Gliding motility. A type of biologic locomotion, slow, smooth to jerky, not involving flagella, pseudopodia or similar structures, typical especially of the sheath-enclosed trichomes of filamentous prokaryotes.

Glucose. An organic compound, the six-carbon (hexose) sugar $C_6H_{12}O_6$.

Glutamic acid. An organic compound, the amino acid $HOOCCH_2CH_2CH(NH_2)COOH$.

Glycerol. An organic compound, the three-carbon alcohol $CH_2OHCHOHCH_2OH$.

Glycine. An organic compound, the simplest amino acid, H_2NCH_2COOH.

Glycolysis. Anaerobic fermentation of sugars via the Embden-Meyerhof-Parnas pathway to yield lactic acid and energy.

Gneiss. Metamorphic rock foliated and banded as a result of regional deformation.

Goethite. A mineral, α-FeO(OH).

Gondwana. A supercontinent of the Paleozoic and early Mesozoic composed of India and the present-day southern hemisphere continents; derived from the splitting of Pangea.

Grainstone. Sedimentary, clastic rock, essentially mud-free and composed of grain-supported carbonate particles.

Granite. Igneous, plutonic rock with quartz and feldspar as principal constituents; intrusive equivalent of rhyolite.

Granodiorite. Igneous, plutonic rock, intermediate in composition between quartz diorite and quartz monzonite.

Granulite facies. Metamorphic gneissic rocks formed at low to high pressures (3 to 12 kb) and high temperatures (>650° C).

Graphite. A mineral, crystalline carbon, dimorphous with diamond.

Graywacke. Clastic sedimentary rock, a firmly indurated coarse-grained sandstone consisting of poorly sorted, angular to subangular grains of quartz, feldspar, and abundant rock and mineral fragments.

Greenalite. A mineral, $(Fe^{2+}, Fe^{3+})_{5-6} \cdot Si_4O_{10}(OH)_8$.

Greenschist facies. Metamorphic schistose rocks formed under low to moderate pressures (3 to 8 kb) and low to moderate temperatures (250° to 450° C).

Greenstone belt. A folded, structurally distinct, generally elongate region containing abundant, dark-green, altered mafic to ultramafic igneous rocks.

Green sulfur bacteria. Anaerobic photosynthetic prokaryotes of the Chlorobiaceae.

Gypsum. A mineral, hydrous calcium sulfate, $CaSO_4 \cdot 2H_2O$ (see *Anhydrite*).

Hadean. Earliest subera of the Archean Era of Earth history.

Halite. A mineral, rock salt, NaCl.

Halophile. An organism that is well adapted to a high salinity (including hypersaline) environment.

Hematite. A mineral, ferric oxide, α-Fe_2O_3.

Heterocyclic. A term pertaining to ring-forming organic compounds composed of atoms of more than one kind.

Heterocyst. A specialized thick-walled cell, the site of nitrogen fixation, occurring in members of several families of filamentous cyanobacteria.

Heterotrichous. A term pertaining to organisms, chiefly algae and one group of cyanobacteria, composed of two or more structurally distinct types of trichomes or filaments.

Heterotroph. See Glossary II.

Hexose. Any of various six-carbon sugars such as glucose.

Hopanoid. Any of various members of a subgroup of organic pentacyclic triterpanes.

Hornblende. A mineral, the commonest member of the amphibole group, $Ca_2Na(Mg, Fe^{2+})_4 \cdot (Al, Fe^{3+}, Ti)(Al, Si)_8O_{22}(O, OH)_2$.

Humic acid. A general term for dark-colored, acidic organic matter extractable by alkali from soils, low rank coals, and decayed plant material that is insoluble in acids and organic solvents.

Hydrazine. An organic compound, H_2NNH_2.

Hydrocarbon. Any of various members of a class of organic compounds composed only of hydrogen and carbon.

Hydrogenase. Any of various enzymes capable of mediating the biochemical production of hydrogen from an appropriate substrate.

Hydrogen cyanide. A highly reactive organic compound, H−C≡N.

Hypersthene. A mineral of the orthopyroxene group, (Mg, Fe)SiO_3.

Igneous. A term pertaining to a rock or mineral that has solidified from molten or partly molten material (i.e., from a magma).

Immature sediment. A clastic sediment that has been differentiated from its parent rock by processes acting over a short time and/or at low intensity such that it is characterized by relatively unstable minerals and poorly sorted and angular grains (see *Mature sediment*).

Index fossil. Morphologically distinctive, abundant, and widespread fossil taxon restricted to, and thus characteristic of, a defined stratigraphic range.

Indole. An organic compound, 2,3-benzopyrrole, C_8H_7N.

Intraclast. A component of a sedimentary rock, usually of a limestone, that represents a torn-up and reworked fragment of a penecontemporaneous sediment.

Iron-formation. Banded iron-formation.

Island arc. A curved chain of islands, chiefly volcanic, rising from the deep sea floor near to a continent.

Isocline. A geologic structure, a fold with limbs so compressed that they have the same dip; characteristic of strong regional deformation.

Isoprenoid hydrocarbon. Any of various branched organic compounds composed of one or more isoprene (C_5H_8) subunits.

Jaspilite. Banded iron-formation.

Kaolinite. A mineral of the kaolin clay group, $Al_2Si_2O_5(OH)_4$.

Karst. A type of topography formed over limestone, dolomite, gypsum, or other soluble rock formed by dissolution and that is characterized by closed depressions ("sinkholes"), caves, and underground drainage.

Kepler's three laws of planetary motion. The statements that (*i*) each planet of the solar system moves in an elliptical orbit with the Sun at one focus of the ellipse; (*ii*) the line from the Sun to any planet sweeps out equal areas of space in equal intervals of time; and (*iii*) the squares of the sideral periods of the several planets are proportional to the cubes of their mean distances from the Sun.

Kerogen. Particulate, geochemically altered, macromolecular organic matter lacking regular chemical structure, insoluble in organic solvents and mineral acids, present in sedimentary rocks.

α-ketoglutarate dehydrogenase. An enzyme involved in removal of hydrogen from α-ketoglutaric acid, the five-carbon dicarboxylic acid $HOOCCH_2CH_2COCOOH$.

Ketone. Any of various organic compounds, such as acetone, that contain a carbonyl (−C=O) group attached to two carbon atoms.

Kimberlite. Igneous, plutonic rock, a porphyritic alkalic peridotite containing abundant phenocrysts of olivine and phlogopite.

Komatiite. Igneous extrusive rock, picritic (i.e., olivine- and pyroxene-rich) and of ultramafic composition.

Kyanite. A mineral, Al_2SiO_5, trimorphous with andalusite and sillimanite.

Labradorite. A mineral, the plagioclase feldspar with the approximate composition $(NaAlSi_3O_8) \cdot (CaAl_2Si_2O_8)$.

Lactic acid. An organic compound, the three-carbon monocarboxylic acid $CH_3CHOHCOOH$.

Lahar. A mudflow (or similar mass move) composed chiefly of vocaniclastic materials on the flank of a volcano.

Large ion lithophile (*LIL*). Any of various silicate-forming chemical elements of relatively high atomic number.

Lawsonite. A mineral, $CaAl_2(Si_2O_7)(OH)_2 \cdot (H_2O)$.

Lawsonite-albite facies. Metamorphic rocks of the glaucophane schist (blueschist) facies formed at high pressures (6 to 7.5 kb) and moderate temperatures (250° to 400° C).

Lherzolite. Igneous, plutonic rock, a periodotite composed chiefly of olivine, orthopyroxene, and clinopyroxene.

Lignin. A general term for a group of polyphenolic, aromatic organic compounds that occur with cellulose in the woody portions of vascular plants.

Lignite. Clastic sedimentary rock, a minimally metamorphosed coal, the altered product of peat intermediate in rank between peat and subbituminous coal.

Limestone. Clastic or chemical sedimentary rock composed chiefly of calcium carbonate minerals.

Limonite. A mineral group composed of naturally occurring hydrous ferric oxides and hydroxides.

Lipid. Any of various saponifiable oxygenated fats or fatty acid-containing substances such as waxes, exclusive of hydrocarbons, that generally are soluble in organic solvents.

Lithification. The conversion of unconsolidated sediment into a coherent, solid rock, involving such processes as cementation, compaction, desiccation, and the like.

Lithosphere. The solid portion of the Earth as compared with the hydrosphere and atmosphere; also, the crust and the upper portion of the upper mantle of the Earth.

Ma. Mega anna, 10^6 years.

Maceration. The act or process of disintegrating sedimentary rocks (such as shale, chert, or coal) by various chemical and physical techniques in order to extract and concentrate acid-resistant kerogen and kerogenous and other microfossils from them.

Mafic. A general term for dark-colored igneous rocks consisting largely of magnesium and ferric (i.e., ferromagnesian), basic minerals (see *Felsic*).

Magma. Molten or partly molten rock material.

Magnetite. A mineral, ferro-ferric oxide, $(Fe, Mg)Fe_2O_4$.

Maillard reaction. The chemical reaction of amino groups of amino acids, peptides, or proteins with the glycosidic hydroxyl groups of sugars resulting ultimately in the formation of brown insoluble products; known also as the "browning reaction."

Malic acid. An organic compound, the four-carbon dicarboxylic acid $HOOCCHOHCH_2COOH$.

Mantle of the Earth. The shell of the Earth below the crust and above the core, divided into three zones: the upper mantle (extending from the Mohorovičić Discontinuity to a depth of about 250 to 400 km), probably peridotite, eclogite, and pyrolite in composition; the transition zone (extending from a depth of about 300 to 900 km); and the lower mantle (900 to 2,900 km deep), composed probably of material of pyrolite composition.

Mature sediment. A clastic sediment that has been differentiated from its parent rock by processes acting over a long time and/or at high intensity such that it is characterized by stable minerals (e.g., quartz) and well-sorted but subangular to angular grains.

Meiosis. The type of division of a cell nucleus leading to formation of four daughter cells ("spores" or their equivalents) each containing a copy of one-half of the chromosomes of the parent cell; meiosis and syngamy (fusion of gametes) are fundamental aspects of eukaryotic sexual reproduction.

Megafossil. A morphological fossil large enough to be studied without the aid of a microscope (generally >0.2 mm); also referred to as macrofossil.

Mesozoic. An era of the Phanerozoic Eon of Earth history intermediate between the Paleozoic and Cenozoic Eras.

Meta-. In igneous or sedimentary petrology, a prefix indicating that the rock type has been metamorphosed as in "metabasalt" or "metaquartzite."

Metagenesis. The pre-metamorphic stage of alteration of sedimentary organic matter intermediate between catagenesis and greenschist facies metamorphism and occurring at about 150° to 250° C.

Metamorphic facies. All the rocks (or any chemical composition and mineralogy) that have reached chemical equilibrium within the limits of a certain pressure-temperature range defined by the stability of specific index minerals (see *Amphibolite facies*, *Blueschist facies*, *Ecologite facies*, *Granulite facies*, *Greenschist facies*, *Prehnite-pumpellyite facies*, and *Zeolite facies*).

Metamorphic rock. Rock derived as a result of mineralogical, chemical, and structural changes in pre-existing rocks in response to marked changes in temperature, pressure, and chemical environment at depth in the Earth's crust.

Metamorphism. The mineralogical and structural adjustment of rocks to physical and chemical conditions that have been imposed at depth in the Earth's crust and that differ from the conditions under which the rocks in question originated.

Metaphyte. Eukaryotic, multicellular, usually megascopic plant, whether vascular or nonvascular (e.g., "seaweeds").

Metasediment. Metamorphosed sedimentary rock, as in metaconglomerate, metashale, metachert, and metaquartzite.

Metazoan. Eukaryotic, multicellular, usually megascopic animal.

Methanogen. Bacterium capable of producing methane (CH_4) as a metabolic byproduct of the reduction of carbon dioxide.

Methanol. An organic compound, methyl alcohol, CH_3OH.

Methionine. An organic compound, the amino acid $CH_3SCH_2CH_2CH(NH_2)COOH$.

Methyl cyanide. An organic compound, CH_3CN.

Methylotroph. Bacterium capable of metabolically oxidizing methane; extant methylotrophs are predominantly (but not exclusively) aerobes.

Mica. A mineral group composed of phyllosilicates (i.e., silicates with a sheet-like substructure) with the general formula $(K, Na, Ca)\cdot(Mg, Fe, Li, Al)_{2-3}(Al, Si)_4O_{10}(OH, F)_2$.

Microbiota. A localized group of microscopic

organisms that comprise a biocoenose, used especially in reference to communities of fossil microorganisms that occur within a stromatolite or a particular stromatolitic horizon.

Microfossil. A morphological fossil too small to be studied without the aid of a microscope (generally, <0.2 mm), either the remains of a microscopic organism of or a larger organism (e.g., spores and pollen of higher plants).

Migmatite. Rock, composed of igneous or igneous-looking and/or metamorphic materials that are megascopically distinguishable.

Minnesotaite. A mineral, $(Fe, Mg)_3Si_4O_{10}(OH)_2$.

Miogeosyncline. A geosyncline in which volcanism is not associated with sedimentation; the portion of an orthogeosyncline located near the craton (see *Eugeosyncline*).

Mitochondrion. Organelle of eukaryotic cells, the site of aerobic respiration.

Mitosis. The type of division of a cell nucleus leading to formation of two daughter cells each of which is an exact copy of the parent cell; in unicellular eukaryotic organisms, a type of asexual reproduction.

Mobile belt. A long, relatively narrow crustal region of tectonic activity having the potential to develop into an "orogenic belt," a folded and deformed linear region giving rise to a "mountain belt" via post-orogenic processes.

Mohorovičić Discontinuity. The boundary surface (or seismic-velocity discontinuity) that separates the Earth's crust from the subjacent mantle; its depth varies from about 5 to 15 km beneath the ocean floor to about 30 to 50 km below the continents (up to 70 km beneath some mountain ranges).

Monzonite. Igneous, plutonic rock containing little or no quartz, composed chiefly of orthoclase and plagioclase feldspars and augite.

Morphological fossil. Any direct morphological evidence (remains, trace, imprint, etc.) indicative of pre-existent life.

Myxophytes. Cyanobacteria.

Nappe. Sheet-like, allochthonous geologic unit introduced into an area via thrust faulting and/or recumbent folding.

Nebula. Interstellar cloud of gas and dust.

Nicotinamide adenine dinucleotide (*NAD*). An electron carrier involved in glycolysis; also abbreviated as DPN.

Nitrile. Any of various organic compounds containing the group $C{\equiv}N$ that on hydrolysis yields an acid with the elimination of ammonia.

Nitrogenase. An enzyme complex present in some prokaryotes involved in the fixation and reduction of N_2 to yield NH_3 or similarly reduced nitrogen compounds.

Nonfossil. A structure or object (or organic geochemical or biochemical component) that may be mistaken for a morphological (or a chemical) fossil but that is not a true fossil (e.g., nonbiogenic organic or inorganic artifacts of sample preparation; mineralic or organic pseudofossils; and modern biogenic contaminants, both morphological and chemical).

Nucleic acid. Any of various organic acids (such as DNA, deoxyribonucleic acid, or RNA, ribonucleic acid) composed of joined nucleotides, made up of a five-carbon sugar (deoxyribose or ribose), a phosphate, and a nitrogenous organic base (e.g., adenine, guanine, uracil, thymine, or cytosine).

Olivine. A mineral, especially common in ultramafic igneous rocks, $(Mg, Fe)_2SiO_4$.

Oncolite. An accretionary organosedimentary structure, viz., an unattached stromatolite, usually small (<10 cm) and more or less spheroidal, with encapsulating, concentric, or overlapping laminae.

Oolite. Small, spheroidal accretionary granules, 0.25 to 2 mm in diameter, commonly products of inorganic precipitation, occurring in sedimentary rocks; also referred to as ooliths (in which case "oolite" refers to the rock or mineral made up of ooliths); small-size pisolites (pisoliths).

Opal. A mineral, amorphous hydrous silica, $SiO_2 \cdot nH_2O$.

Open space structure. A small-scale sedimentary structure, commonly lenticular, usually in chert or carbonate rock, produced by the partial or complete mineralic infilling of an originally gas-filled pocket or void.

Ophiolite. A general term for a group of mafic and ultramafic igneous rocks, ranging from basalt to gabbro and peridotite, the origin of which is associated with an early phase in the development of a geosyncline.

Ophitic texture. A term pertaining to the holocrystalline texture of an igneous rock, especially diabase, in which lath-shaped plagioclase crystals are partially or completely included in pyroxene crystals (typically augite).

Organelle. An intracellular structure or body having a specific function (e.g., mitochondria, nuclei, and chloroplasts).

Organic. A term pertaining or relating to a compound, structure, or substance containing carbon, and usually hydrogen, oxygen and/or nitrogen, of the type characteristic of, but not limited to (e.g., non-biogenic organic matter), biologic systems.

Organic preservation. A term pertaining to morphological fossils preserving some portion of the original (but geochemically altered) organic matter, as in permineralization.

"*Organized elements*". A term coined by Claus and Nagy (1961) to refer to spheroidal, microscopic, kerogenous particles occurring in carbonaceous chondritic meteorites.

Orogeny. The process of the formation of mountains.

Ortho-. In sedimentary petrology, a prefix that indicates the primary origin of a crystalline sedimentary rock as in "orthoquartzite" as distinguished from "metaquartzite," or in "orthoconglomerate" as distinguished from "metaconglomerate."

Orthopyroxene. A mineral group composed of pyroxenes of the orthorhombic crystal system and usually containing no calcium and little aluminum (see *Clinopyroxene*).

Orthoquartzite. Clastic sedimentary rock composed almost entirely of quartz sand; a "pure quartz sandstone."

Overthrust. A geologic structure, a low-angle thrust fault of large scale.

Oxalic acid. An organic compound, the simplest dicarboxylic acid HOOCCOOH.

Oxaloacetic acid (*OAA*). An organic compound, the four-carbon dicarboxylic acid $HOOCCOCH_2COOH$.

Oxic. A term pertaining to the presence of uncombined molecular oxygen.

Oxidase. Any of various enzymes that catalyze oxidations, especially one capable of interacting directly with molecular oxygen.

Oxidation-reduction (*O/R*) *reaction*. A chemical reaction in which one or more electrons are transferred from one atom or molecule to another.

Oxidizing. A term pertaining to a process (or environment) in which a chemical element or ion is oxidized via the loss of electrons (i.e., changed from a lower to a higher positive valence); an "oxidizing environment" need not necessarily be "aerobic" (i.e., capable of supporting aerobiosis).

Oxygenic. A term pertaining to the presence of uncombined molecular oxygen (also referred to as oxic).

Paleozoic. Earliest era of the Phanerozoic Eon.

Palimpsest stromatolitic microstructure. Microstructure in a stromatolitic sediment in which the distribution of kerogen, iron oxide, pyrite, or some other pigmenting material indicates the former distribution of microbial remains.

Palynology. A subdiscipline of botany concerned chiefly with the study of the pollen and spores of higher plants, whether living or fossil.

Pangea. Pre-Mesozoic supercontinent, the precursor of Gondwana and Laurasia.

Paraformaldehyde. An organic compound, polymerized formaldehyde, $(CH_2O)_n$.

Pegmatite. Igneous, plutonic rock, exceptionally coarse-grained, commonly having the composition of granite.

Pelagic. A term pertaining to the open ocean environment or to the organisms inhabiting that environment.

Pelite. Clastic sedimentary rock composed of very fine-grained detritus (viz., of clay- or mud-size particles).

Pentose. Any of various five-carbon sugars.

Peptide. Any of various amides that are derived from two or more amino acids by combination of the amino group of one acid and the carboxylic group of another, thus characterized by the peptide linkage, the bivalent group HN–OC.

Peraluminous. A term pertaining to an igneous rock in which the molecular proportion of aluminum oxide is greater than that of sodium and potassium oxides combined.

Peridotite. Igneous, plutonic, coarse-grained rock composed chiefly of olivine with or without other mafic minerals.

Permineralization. A process of fossilization whereby the original hard parts of an animal, the test of a protist, or the physically resiliant organic structures (especially, the cell walls) of a plant, fungus, protist or prokaryote have mineral material (commonly silica or calcium carbonate) deposited in pore spaces, cell lumina, intramicellar spaces, etc.; cf. petrifaction in which original (but geochemically altered) organic matter is preserved.

Peroxidase. Any of a series of enzymes that catalyzes the oxidation of various substrates by peroxides.

Petrographic thin section. A slice of rock or mineral mechanically ground to a thickness of approximately 30 μm (for petrologic study) or

150 μm (for micropaleontolgic study of translucent chert), mounted on a glass microscope slide, for study by optical microscopy.

Phanerozoic. The more recent eon of Earth history, composed of the Paleozoic (570 to 230 Ma ago), Mesozoic (230 to 62 Ma ago), and the Cenozoic Eras (62 Ma ago to the present); the Phanerozoic and the preceding Precambrian Eon (4500 to 570 Ma ago) together comprise all of Earth history.

Phase diagram. A graph showing the boundaries of the fields of stability, usually in terms of pressure and temperature, of the various phases of a chemical system.

Phenocrysts. Relatively large, conspicuous crystals in a porphyritic rock.

Phenylacetylene. An organic compound, ethynylbenzene, $C_6H_5C{\equiv}CH$.

Phenylalanine. An organic compound, the aromatic amino acid $C_6H_5CH_2CH(NH_2)COOH$.

Phlogopite. A mineral of the mica group, $K(Mg, Fe)_3AlSi_3O_{10}(OH, F)_2$.

Phosphoenolpyruvate (*PEP*). An organic compound, a phosphorylated reactive intermediate in glycolysis, $CH_2{=}COPO_3^{2-}COO^-$.

3-*phosphoglycerate.* An organic compound, a phosphorylated reactive intermediate in glycolysis, $CH_2OPO_3^{2-}CHOHCOO^-$.

Photic zone. That part of an aqueous body where there is sufficient light penetration to support biological photosynthesis (of variable depth to about 200 m, but commonly about 50 m).

Photophosphorylation. In biologic systems, the light-driven conversion of adenosine diphosphate (ADP) to adenosine triphosphate (ATP).

Photorespiration. In biologic systems, the light-driven oxidation of organic compounds resulting in generation of carbon dioxide.

Photosynthetic bacteria. Prokaryotic microorganisms capable of anaerobic, but not oxygen-producing, photosynthesis.

Phototaxis. Locomotory movement of an organism (e.g., via gliding motility) in response to an environmental gradient of light, either toward a higher light concentration ("positive phototaxis") or away from a higher light concentration ("negative phototaxis").

Phototroph. See Glossary II.

Phylogeny. The lineal evolutionary relationships among a particular group of organisms.

Pillow basalt. Igneous, extrusive volcanic rock, solidified in a subaqueous environment to form discontinuous pillow-shaped masses generally 30 to 60 cm across.

Pisolite. Small, spheroidal, accretionary granules, 2 to 10 mm in diameter, commonly products of inorganic precipitation, occurring in sedimentary rocks; also termed pisoliths (in which case "pisolite" refers to the rock of mineral made up of pisoliths); pea-size oolites (ooliths).

Placer. A surficial mineral deposit (e.g., of a heavy metal such as gold) formed by mechanical concentration of mineral particles from weathered debris, commonly of alluvial origin.

Plagioclase. A mineral group composed of feldspars having the general formula $(Na, Ca)Al \cdot (Si, Al)Si_2O_8$.

Planetesmal. Small (1 to 10 km), rocky, "meteorite-like" bodies, the accretion of which leads to the formation of protoplanets.

Plankton. Aquatic, floating (or weakly swimming) pelagic organisms.

Plate tectonics. Global tectonics based on an Earth model characterized by several to many thick plates (i.e., blocks composed of both continental and oceanic crust and upper mantle) each of which "floats" on a viscous underlayer within the mantle and which moves slowly across the global surface, propelled via convection cells within the mantle, commonly with centers of sea-floor spreading at the rear of the plate and a subduction zone at its leading edge.

Platform. That part of a continent covered by flat-lying or gently tilted, mainly sedimentary, strata; together with shields, platforms are components of cratons.

Pluton. A deep-seated, igneous intrusive rock mass.

Polythionate. A sulfur-containing organic compound, $S_nO_6^{2-}$.

Porphyrin. Any of various metal-free organic compounds derived from pyrrole-containing (C_4H_5N-containing) compounds, especially from chlorophyll or hemoglobin.

Porphyry. An igneous rock of any composition that contains phenocrysts in a fine-grained groundmass.

Potash. Potassium carbonate, K_2CO_3; also, a term used loosely for potassium oxide, potassium hydroxide, or potassium.

Precambrian. The earlier eon of Earth history, composed of the Archean (4.5 to 2.5 Ga ago) and the Proterozoic Eras (2.5 to 0.57 Ga ago); the Precambrian and the subsequent Phanerozoic Eon (0.57 Ga ago to the present) together comprise all of Earth history.

Prehnite. A mineral, $Ca_2Al_2Si_3O_{10}(OH)_2$.

Prehnite-pumpellyite facies. Metamorphic rocks formed under conditions of very low grade metamorphism; see *Zeolite facies.*

Proalgae. A term proposed by Van Valen and Maiorana (1980) to include prokaryotes capable of oxygen-producing photosynthesis (viz., cyanobacteria and prochlorophytes).

Procaryotes. Prokaryotes.

Prochlorophyte. Members of the Prochlorophyta, a systematic division proposed by Lewin (1976) to include prokaryotic, bacterium-like microorganisms containing chlorophylls *a* and *b*, lacking phycocyanin, phycoerythrin, and other bilin pigments, and capable of aerobic (oxygen-producing) photosynthesis; all of the few species of known prochlorophytes are symbionts in tropical ascidians (see *Cyanobacteria* and *Proalgae*).

Prokaryotes. Microbial microorganisms (viz., bacteria, cyanobacteria, archaebacteria, and prochlorophytes) characterized by cells that lack membrane-bound nuclei, mitochondria, chloroplasts, and similar organelles and that reproduce by non-mitotic and non-meiotic division.

Propionaldehyde. An organic compound, propanol aldehyde, CH_3CH_2CHO.

Propionic acid. An organic compound, the three-carbon monocarboxylic acid CH_3CH_2COOH.

Protein. An organic compound, any of numerous polymers composed of amino acids.

Proterozoic. An era of Earth history, as here used composed of three suberas: the Early Proterozoic (2.5 to 1.6 Ga ago), the Middle Proterozoic (1.6 to 0.9 Ga ago), and the Late Proterozoic (0.9 to 0.57 Ga ago); the Proterozoic and the preceding Archean Era (4.5 to 2.5 Ga ago) together comprise the Precambrian Eon.

Pseudofossil. A naturally occurring, non-biogenic, mineralic or organic object or structure (viz., morphological pseudofossil) or chemical component (viz., chemical pseudofossil) that bears resemblance to, and may be mistaken for, a morphological or chemical fossil.

Pseudomicrofossil. A naturally occurring, non-biogenic, mineralic or organic, microfossil-like object or structure; a morphological pseudofossil bearing resemblance to a fossil microscopic organism or other microfossil.

Ptygmatic fold. Geologic structure in pegmatitic material within migmatite or gneiss in which the form of the fold is modified by the varying materials through which it passes.

Purine. Organic compounds with the general formula $C_5H_4N_4$, nitrogen-containing bases such as adenine or guanine that are components of nucleotides and of nucleic acids.

Purple sulfur bacteria. Anaerobic photosynthetic bacteria of the Chromatiaceae.

Pycnocline. A density gradient, especially a vertical gradient making a sharp change as in a layer of oceanic water that is characterized by a rapid change of density with depth.

Pylome. A pore-like opening, commonly circular, in the wall of an acritarch, probably functioning in excystment.

Pyrimidine. Organic compounds with the general formula $C_4H_4N_2$, nitrogen-containing bases such as thymine, cytosine, and uracil that are components of nucleotides and of nucleic acids.

Pyrite. A mineral, FeS_2.

Pyrobitumen. Dark-colored, solid, fairly hard, nonvolatile substances composed of geochemically altered, largely insoluble, hydrocarbon complexes.

Pyroclastic. A term pertaining to clastic rock material formed by a volcanic explosion.

Pyrolite. The proposed composition of mantle material, consisting of one part basalt and three parts dunite.

Pyrolysis. Chemical breakdown brought about by heating.

Pyroxene. A mineral group composed of compounds with the general formula $(Ca, Na, Mg, Fe^{2+})(Mg, Fe^{3+}, Al)Si_2O_6$; see *Clinopyroxene* and *Orthopyroxene.*

Pyrrhotite. A mineral, $Fe_{1-x}S$ (having a defect lattice in which some of the ferrous ions are lacking).

Pyruvic acid. An organic compound, the three-carbon carboxylic acid $CH_3COCOOH$, the precursor of lactic acid in fermentation.

Quartz. A mineral, crystalline silica, SiO_2 (see *Opal* and *Chalcedony*).

Quartzite. Clastic sedimentary rock, sandstone consisting chiefly of cemented quartz grains.

Rare earth element (*REE*). Any of a group of trivalent metalic elements with atomic numbers 57 to 71, inclusive.

Recumbent fold. Geologic structure, an overturned fold with a nearly horizontal axial surface.

Red bed. Clastic sedimentary rock, chiefly sandstone, siltstone, and shale, red or reddish-brown in color due to the presence of ferric oxide minerals (mainly hematite) usually coating individual grains.

Reducing. A term pertaining to a process (or environment) in which a chemical element or ion is reduced via the gain of electrons (i.e., changed from a higher to a lower positive valence); opposite of oxidizing.

Refractory. An element or substance resistant to a particular treatment and/or process.

Respiration. See Glossary II.

Rheology. The study of the deformation and flow of matter.

Rhyolite. Igneous, extrusive rock generally porphyritic with phenocrysts of quartz and feldspar and exhibiting flow structure; extrusive equivalent of granite.

Ribulose 1.5 bis-phosphate (RuBP) carboxylase/oxygenase. Enzyme of the Calvin cycle catalyzing the carboxylation (with CO_2) and cleavage of ribulose 1,5-diphosphate to yield two molecules of 3-phosphoglycerate.

Ribulose monophosphate (RuMP) cycle. Biosynthetic pathway leading from formaldehyde to sugars exhibited by many methylotrophic bacteria.

Riebeckite. A mineral of the amphibole group, $Na_2(Fe, Mg)_5Si_8O_{22}(OH)_2$.

Ripidolite. A mineral of the chlorite group, $(Mg, Fe^{2+})_9Al_6Si_5O_{20}(OH)_{16}$.

Ripple mark. A sedimentary structure with an undulatory surface formed at the interface between a fluid (e.g., wind or water currents) and incoherent sedimentary material (e.g., snow particles or loose sand).

Saccharolytic. A term pertaining to the breakdown of sugars.

Schist. Metamorphic rock, highly foliated as a result of regional deformation.

Schlieren. In some igneous rocks, irregular streaks or masses that contrast with the surrounding matrix.

Second law of thermodynamics (Kelvin-Planck statement). The statement that it is impossible to construct an engine that, operating in a cycle, will produce no effect other than the extraction of heat from a reservoir and the performance of an equivalent amount of work (Clausius statement of the second law: that it is impossible to construct a device that, operating in a cycle, will produce no effect other than the transfer of heat from a cooler to a hotter body).

Secular. A term pertaining to a process or event persisting for an indefinitely (e.g., geologically) long period of time.

Serine. An organic compound, the amino acid $HOCH_2CH(NH_2)COOH$.

Serpentine. A mineral group composed of compounds having the general formula $(Mg, Fe)_3Si_2O_5(OH)_4$.

Serpentinization. The process of hydrothermal alteration by which magnesium-rich silicate minerals (olivine, pyroxene, etc.) are converted into minerals of the serpentine group.

Sexual reproduction. In eukaryotes, a type of biological reproduction involving meiosis and syngamy (fusion of gametes) and most commonly a regularized alternation of haploid and diploid phases in the life cycle; analogous (but not evolutionarily homologous), non-meiotic, parasexual processes of reproduction occur in some prokaryotes.

Shale. Clastic sedimentary rock, fine-grained and finely layered; a thinly laminated claystone, siltstone or mudstone.

Sheath. In prokaryotic microorganisms, an extracellular, generally mucopolysaccharide, mucilagenous investment surrounding individual cells or colonies of cells; in filamentous prokaryotes, a cylindrical, hollow, mucilagenous organic tube that encompasses the cellular trichome.

Shield. A large area of exposed continental basement rock in a craton, generally of Precambrian age, surrounded by sediment-covered platforms.

Sial. A petrologic term for the granitic, silica- and alumina-rich, upper layer of the Earth's crust characteristic of continental masses (see *Sima*).

Siderite. A mineral, $FeCO_3$.

Silcrete. Clastic sedimentary rock, a silica-cemented duricrust.

Sill. A tabular intrusion of igneous rock that parallels the planar structure of the surrounding rock (see *Dike*).

Sillimanite. A mineral, Al_2SiO_5, trimorphous with kyanite and andalusite.

Sima. A petrologic term for the basaltic, silica- and magnesium-rich, oceanic crust and the lower portion of the Earth's continental crust (see *Sial*).

Solar wind. The stream of interplanetary plasma or ionized particles emitted from the Sun.

Solidus. On a temperature-composition diagram, the locus of points in a chemical system above which solid and liquid are in equilibrium and below which the system is completely solid.

Sparry calcite. Coarse-grained crystalline calcite.

Spinel. A mineral, $MgAl_2O_4$, or members of a mineral group having similar general composition.

Spreading center. In plate tectonics, a linear, principally submarine ridge system (For example, the mid-Atlantic ridge) at which new oceanic crust is added, and from which adjacent tectonic plates diverge (at a rate of about 1 to 10 cm per year) as a result of the convective upwelling of magma.

Staurolite. A mineral, $(Fe, Mg)_2Al_9Si_4O_{23}(OH)$.

Stefan-Boltzmann constant. In the Stefan-Boltzmann Law showing that the radiant emittance of a black body at any temperature θ is equal to $R_B(\theta) = \sigma\theta^4$, the term "$\sigma$" is the Stefan-Boltzmann constant.

Sterol. Any of various cyclic organic compounds such as cholesterol, containing a molecular skeleton of four fused carbon rings.

Stilpnomelane. A mineral, approximately of the composition $K(Fe, Mg, Al)_3Si_4O_{10}(OH)_2 \cdot nH_2O$.

Stiriolite. A subaerial evaporitic deposit such as a coniatolite, geyserite, or stalactite (speleothem).

Stochastic. A term pertaining to a process in which the dependent variable is random and the outcome at any instant cannot be predicted with certainty.

Stratigraphy. The arrangement (or study of the arrangement) of rock strata, especially with regard to the geographic position and chronologic order of the sequence.

Stromatolite. An accretionary organosedimentary structure, commonly laminated, megascopic, and calcareous, produced as a result of the growth and metabolic activities of (and usually due to the attendant trapping, binding, and/or precipitation of mineralic material by) benthonic, mat-building communities of mucilage-secreting microorganisms, principally filamentous photoautotrophic prokaryotes; broadly, a lithified or unlithified, commonly laminated, accretionary sedimentary structure produced by a microbial biocoenose; stromatolites can be stratiform (with flat-lying, continuous laminae), columnar (with the dimension in the direction of accretion greater than at least one of the transverse dimensions), or spheroidal (viz., oncolites); a "thrombolite" is an unlaminated stromatolite characterized by a megascopic clotted fabric.

Stromatoporid. Any of a group of invertebrate animal fossils of uncertain systematic position currently placed in the hydrozoan (coelenterate) order Stromatoporoidea characterized by a calcareous exoskeleton and a colonial, massive, sheet-like or dendroid growth pattern.

Stylolite. A thin seam usually occurring in more or less homogeneous carbonate rocks produced via pressure solution in which the insoluble constituents of the rock (e.g., clay, kerogen, etc.) form an irregular, interlocking layer of mutually interpenetrating projections.

Subduction zone. In plate tectonics, an elongate region along which a crustal block descends relative to an adjacent crustal block.

Subera. A subdivision of a geologic era (e.g., as here used, the Hadean, Early Archean, Middle Archean, and Late Archean are suberas of the Archean Era).

Subsolidus. A chemical system below its melting point in which reactions may occur in the solid state.

Substrate-level phosphorylation. See Glossary II.

Succinic acid. An organic compound, the four-carbon dicarboxylic acid $HOOCCH_2CH_2 \cdot COOH$.

Supracrustal. A term pertaining to rocks that overlie a complex of basement rocks.

Synclinorium. A composite geologic structure of regional extent composed of a series of synclinal folds the cores of which contain the stratigraphically younger rocks of the sequence.

Syngenetic. A term pertaining to structures (e.g., ripple marks, microfossils) or substances (e.g., organic compounds, minerals) of primary origin, formed or deposited contemporaneously with the deposition of the surrounding sediment.

Szent-Györgyi pathway. A metabolic pathway from hexose sugar via a three-carbon intermediate compound to oxaloacetate and, ultimately, to succinate.

Talc. A mineral, $Mg_3Si_4O_{10}(OH)_2$.

Taxon. In taxonomy, a unit of any rank (e.g., a particular species, genus, family, or class) or the formal name applied to that unit.

Tectonic. A term pertaining to the forces involved in, or the resulting structures or features produced by, diastrophism, orogeny, etc. (i.e., "tectonism").

Terrestrial planets. The inner planets of the solar system, similar to the Earth in terms of size, mean density, and rocky composition (viz., Mercury, Venus, Earth, Moon, and Mars).

Thallophyte. A eukaryotic nonvascular plant (viz., alga) or fungus composed of a "thallus," a multicellular or coenocytic body without differentiation into true roots, leaves or tracheid-containing stems; although similarly nonvascular, bryophytes (mosses and liverworts) exhibit a greater degree of cellular differentiation than do thallophytes.

Tholeiite. Igneous, extrusive rock, a member of a major subgroup of alkali basalts.

Thrust fault. A fault with a dip of 45° or less whose overlying side (the "hanging wall") appears to have moved upward relative to the other wall.

Thucholite. Sedimentary, organic-rich rock, a brittle, jet-black mixture of solidified hydrocarbons and uraninite, with some sulfide minerals, occurring especially in Early Proterozoic gold-bearing conglomerates.

Tillite. Clastic sedimentary rock, lithified glacial till (unsorted and unstratified glacial debris).

Tommotian. A time-stratigraphic subdivision of the Cambrian System of rocks, the earliest stage of the Lower Cambrian Series.

Tonalite. Igneous, intrusive rock, essentially identical to quartz diorite.

Trichome. In filamentous prokaryotic microorganisms, the threadlike, usually many-celled strand that is encompassed commonly by a tubular sheath to form a filament.

Trimethylamine-N-oxide (*TMAO*). An organic compound, $(CH_3)_3NO$.

Tryptophan. An organic compound, the cyclic amino acid α-aminoindole-3-propionic acid, $C_{11}H_{12}N_2O_2$.

Tuff. Clastic sedimentary rock, a compacted pyroclastic deposit of volcanic ash and dust.

Turbidite. Clastic sedimentary rock characterized by graded bedding, moderate sorting and well-developed primary structures and inferred to have been deposited from a turbidity current.

Tyrosine. Organic compound, the aromatic amino acid β-(*p*-hydroxyphenyl) alanine, $C_9H_{11}NO_3$.

Ultramafic. A term pertaining to an igneous rock composed chiefly of mafic (basic) minerals.

Uraninite. A mineral approximately of the composition UO_2.

Urea. An organic compound, H_2NCONH_2.

Vitrain. A coal lithotype (i.e., a macroscopically visible band), especially in bituminous coal, characterized by brilliant vitreous luster, black color, and cubical cleavage with conchoidal fracture.

Vitrinite. An oxygen-rich coal maceral (i.e., a micropetrological unit) characteristic of vitrain and composed of humic material associated with peat formation.

Volcaniclastic. A term pertaining to a clastic rock containing volcanic material.

Volcanogenic. A term pertaining to a volcanic origin.

Wackestone. Clastic sedimentary rock, a mud-supported carbonate containing more than 10% of grains (i.e., particles $>20\ \mu m$).

Wilson cycle (*Wilson-type*) *plate tectonics*. The plate tectonic regime characteristic of the Phanerozoic Eon, as described by J. T. Wilson, involving large convection cells, rifting, and dispersal of large continental fragments, formation of paired metamorphic belts, mountain building at plate margins, and related features.

Wüstite. A mineral, FeO.

Xenotopic. A term pertaining to the fabric of a crystalline sedimentary rock where the majority of the constituent crystals are anhedral (i.e., lacking well-defined crystal faces).

Zeolite facies. Metamorphic rocks formed in the zone of transition from diagenesis to metamorphism, under low pressures (2 to 3 kb) and temperatures (200° to 300° C).

Zircon. A mineral, $ZrSiO_4$.

Terms Used to Describe Biological Energy Conversions, Electron Transport Processes, Interactions of Cellular Systems with Molecular Oxygen, and Carbon Nutrition*

By David J. Chapman and Howard Gest

Introduction

With the increased understanding of biological energy conversion systems, cellular interactions with O_2, and carbon nutrition, a number of older definitions relating to these aspects of biology have gradually become inaccurate or absolete. Along with great progress, nomenclature has become more cumbersome, and different investigators have sometimes applied different meanings to the same term. Accordingly, we are suggesting revised and new definitions for a number of basic terms useful for describing phenomena in the three general categories noted here. The definitions given are short and simple; sub-definitions are avoided in order to minimize the need to record exceptions.

Energy Conversion and Electron Transport Processes

Fermentation. Any anaerobic process in which energy derived from metabolism or catabolism of organic substrates is used for generation of ATP by substrate-level phosphorylation. Maintenance of oxidation-reduction balance is obligatory, and is achieved in either of two ways:

(*i*) Internally balanced oxidation-reduction, with redox balance being maintained by production of a mixture of end products of different oxidation states.

(*ii*) Accessory oxidant-dependent fermentation, with maintenance of the oxidation-reduction balance requiring provision of an exogenous oxidant.

Fermentation of glucose to lactic acid (glycolysis) is a classical example of (*i*). In fermentations of this category, NADH produced during sugar catabolism is reoxidized by reduction of organic intermediates, usually pyruvate or its derivatives. In some instances, the final oxidation-reduction balance involves utilization of fermentatively generated CO_2 as an electron sink, or the reduction of H^+ to H_2. *Clostridium lacto-acetophilum*, which will not ferment lactate unless exogenous acetate is provided to serve as an electron acceptor for reoxidation of NADH (Bhat and Barker, 1947), is an excellent example of (*ii*). Similarly, *Rhodopseudomonas capsulata* can use fructose or glucose as its sole carbon and energy source for dark anaerobic, fermentative growth, but only if an accessory oxidant such as trimethylamine-N-oxide is supplied (Madigan et al., 1980).

Substrate-level phosphorylation. Generation of ATP by transfer of a phosphoryl group from an organic compound to ADP. As far as is known, substrate-level phosphorylation is catalyzed only by soluble (cytoplasmic) enzyme systems. The most frequently encountered phosphoryl donors for such ATP synthesis are acetyl phosphate, phosphoenolpyruvate, and 1,3-bisphosphoglycerate; butyrylphosphate and carbamyl phosphate have also been identified as phosphoryl donors in certain microorganisms (Gottschalk and Andreesen, 1979; Thauer et al., 1977). Acetyl phosphate is particularly important in the bioenergetics of

* National Research Council of Canada contribution number 18252.

anaerobic bacteria, being produced from a variety of substrates (including carbohydrates, organic acids, alcohols, amino acids, etc.), usually via acetyl-coenzyme A. The organic phosphoryl donor may have a transient existence, as in the type of reaction catalyzed by succinate thiokinase:

$$RCOSCoA + Pi + ADP\ (\text{or } GDP) \longrightarrow RCOO^- + CoASH + ATP\ (\text{or } GTP).$$

Electrophosphorylation. Any process in which phosphorylation of ADP (yielding ATP) is dependent upon electron flow via membrane-bound carriers (see Gest, 1980).

(*i*) *Aerobic electrophosphorylation.* Electrophosphorylation with O_2 serving as the terminal electron acceptor.

(*ii*) *Anaerobic electrophosphorylation.* Electrophosphorylation with an oxidant other than O_2 serving as the terminal electron acceptor.

The most common alternatives to O_2 are CO_2, NO_3^-, NO_2^-, or fumarate. Methanogens are here regarded as anaerobic electrophosphorylating organisms, even though the energy-yielding electron flow pathways in these bacteria have not yet been fully elucidated (Wolfe, 1979).

In energy-yielding oxidation-reduction processes the electrochemical potential difference between redox couples is a measure of the maximum energy available for ATP synthesis (and membrane energization). It should also be noted that the term "oxidative phosphorylation" is used by many workers only in the context of ATP synthesis associated with aerobic electron transport and is consequently ambiguous.

(*iii*) *Anaerobic photophosphorylation.* A type of anaerobic electrophosphorylation employed by photosynthetic bacteria,* in which exogenous reductants or oxidants are unnecessary. Absorption of light by a membrane-bound photopigment system simultaneously generates reducing equivalents (electrons) and oxidized bacteriochlorophyll ($BChl^+$) which serves as the terminal oxidant. The flow of electrons is said to be "cyclic" because the electrons originate from light-activated BChl and eventually return to $BChl^+$. Similar systems involving other forms of chlorophyll apparently operate in cyanobacteria that can catalyze light-dependent (anaerobic) reduction of CO_2 with H_2S as the electron donor.

(*iv*) *Oxygenic photophosphorylation.* Light-dependent phosphorylation (catalyzed by cyanobacteria, algae, photosynthetic protists, and higher plants) associated with electron flow from water to the oxidized form of pyridine nucleotide. Since water is the electron donor, O_2 is produced as a photosynthetic byproduct.

Respirations. Electron transport processes (independent of light) in which electrons flow from inorganic or organic substrates to molecular oxygen or some other terminal oxidant.

(*i*) *Aerobic respiration.* Respiration in which O_2 serves as the terminal electron acceptor.

(*ii*) *Anaerobic respiration.* Respiration in which an oxidant other than O_2 serves as the terminal electron acceptor.

If a respiratory system is membrane-bound and catalyzes phosphorylation of ADP dependent upon electron flow, it is by definition an electrophosphorylation system. Respirations, however, need not be associated with phosphorylation, and they may have such other functions such as the removal of an inhibitory oxidant or of "excess" electrons, or the generation of heat. Processes in which an electron acceptor is reduced and then used primarily or exclusively for biosynthesis or incorporation into cell material are here not considered to be respirations. An example of an "assimilatory reduction" is the reductive conversion of sulfate to the –SH group of an amino acid.

Some ambiguities in use of the term "respiration" can be traced to the original definition or understanding of this word in the context of the pulmonary physiology of higher animals. Many biochemists mistakenly consider "respiration" to be synonymous with "oxidative phosphorylation."

* Bacteria that catalyze only non-oxygenic photosynthetic processes are designated here as *photosynthetic bacteria*; as far as is known, all cyanobacteria ("blue-green algae") can perform oxygenic photosynthesis.

Relations of Cells to Molecular Oxygen

Aerobe. An organism or cell that must use O_2 as the terminal electron acceptor for energy conversion.

Anaerobe. An organism or cell that cannot use O_2 for energy conversion and, consequently, employs some alternative energy-yielding process that is independent of O_2.

Amphiaerobe. An organism or cell that can either use O_2 (like an aerobe), or as an alternative, some other energy conversion process that is independent of O_2.

It is evident that aerobes and amphiaerobes must possess O_2-reducing oxidases that are linked to electron transport systems capable of energy conversion, whereas anaerobes do not. These definitions follow, to some extent, those proposed by Decker et al. (1970). In contrast to these authors, however, we have deleted qualification regarding the presence or absence of oxygenases. This avoids complicating definitions by implying relations between two quite different kinds of catalysts (viz., oxidases and oxygenases). Note that some organisms that are clearly anaerobes, as defined above, may require O_2 for a specialized process or function. An interesting example is given by *Schistosoma mansoni*, which obtains energy only from anaerobic glycolysis (Bueding and Fisher, 1982); the female of the species, however, requires O_2 for egg formation (quinone produced from tyrosine and O_2 by a phenol oxidase is needed for forming cross-linked constituents of the egg shell). We eschew use of terms such as "obligate," "facultative aerobe," "facultative anaerobe," "aerophile," "microaerophile," etc., because they often have a quantitative connotation that cannot be satisfactorily specified. More significantly, however, such terms are usually applied to organisms on the basis of observations made on growing cultures, and are thus more "physiological" than biochemical. The definitions given above are strictly systemic (i.e., based upon discrete, identifiable biochemical processes), and if used in conjunction with the terms *aerotolerant* and *aerointolerant* (see below) should define all responses of organisms to the presence of O_2 and its use in biological energy conversion.

It has long been common practice to use the adjective "aerobic" to describe a process that requires O_2 or a condition in which O_2 is present, with the opposite meaning for "anaerobic" (e.g., as in anaerobic metabolism or anaerobic atmosphere). We encourage continued use of these terms in connection such processes and conditions as, for example, in reference to aerobic or anaerobic electrophosphorylation, aerobic or anaerobic respiration, aerobic or anaerobic organism, and aerobic or anaerobic biosynthesis (e.g., sterol biosynthesis requires O_2 and would thus be described as aerobic, whereas bacteriochlorophyll synthesis can be termed an anaerobic process since it can occur in the absence of O_2).

Use of the term "anoxic" in connection with biologic systems is not encouraged because its meaning is the same as anaerobic, and there seems to be no corresponding antonym.

We caution about the indiscriminate use of terms such as "anaerobic atmosphere" or "anaerobic conditions" since such terms have different meanings for different investigators. In such usage, the partial pressure of oxygen should be specified. For example, many aerobic processes are still operative (Chapman and Ragan 1980) at oxygen concentrations as low as 0.02% (a pO_2 of 2×10^{-4} atm), conditions that have been rather commonly referred to as "anaerobic."

Aerotolerant/Aerointolerant. These terms are usually applied to various species of anaerobes to designate whether or not the presence of O_2 inhibits growth and are difficult to define with precision. Aerobes, by definition, are aero tolerant. Aerotolerance or resistance to oxygen toxicity (Cole, 1976) in an anaerobe is usually correlated with the presence of systems that are capable of effectively scavenging O_2 or its immediate derivatives with appropriate reductants.

Superoxide dismutase (S.O.D.) that dismutates O_2^- (superoxide anion) to H_2O_2 is much discussed in this connection, and the presence or absence of S.O.D. in an organism has been used by some investigators (McCord et al., 1971) as the basis for distinguishing between aerobes and anaerobes. However, this notion cannot be accepted because S.O.D. has been found in anaerobes (Hewitt and Morris, 1975) that do not show any kind of metabolism involving O_2 and, in fact, are aerointolerant (Morris, 1975, 1976). We also believe it is unnecessary to recognize "microaerophilic" formally as an intermediate stage, between aerotolerant and aerointolerant. The term "microaerophilic" (Stouthamer et al., 1979), however, is colloquially useful; such organisms apparently can use or may require O_2 for some metabolic functions, but cannot tolerate O_2 at its normal atmospheric partial pressure (the extent of O_2 tolerance being variable depending upon environmental or culture conditions, physiological state of the organism, pO_2 etc.).

Aerointolerant organisms, which are invariably strict anaerobes, usually require low E_h (e.g., for methanogens, -0.30 v) as well as the absence of molecular oxygen. Incorporation of specific E_h values or ranges into definitions, however, is not feasible.

Carbon Nutrition: Autotrophy and Heterotrophy

Autotroph. An organism that uses CO_2, present in the environment or generated from some other compound, as the immediate (and major) source of cellular carbon.

Heterotroph. An organism that uses organic carbon compounds (defined for most purposes as compounds of carbon containing covalent C–H bonds) as sources of cellular carbon.

The key feature of these definitions is the immediate form of carbon fixed by the cell. The recommended definition of autotrophy is more restrictive than that used by some investigators (Whittenbury and Kelly, 1977; Quayle, 1980) who suggest that the use of any C_1 compound as the source of cellular carbon be considered as the basis of autotrophy. We regard "methylotrophs" (Higgins, 1979) as heterotrophs. These organisms can use CH_4 or related C_1 compounds as sole carbon sources for growth; formaldehyde or methylene folate, generated from the C_1 compound, serves as the immediate precursor of cellular carbon. Bacteria that grow on CO, however, are here regarded as autotrophs since they oxidize CO to CO_2, which then is directly assimilated (Hegeman, 1980). A number of organisms are known that have both autotrophic and heterotrophic capabilities.

It should be noted that all organisms "fix" CO_2. The extent of CO_2-fixation, however, varies over a wide range. During heterotrophic metabolism, CO_2 is ordinarily required in small quantity and is generated from the catabolism of organic substrates; in some instances additional exogenous CO_2 is required or is stimulatory for growth. The essence of the definition of heterotrophy is that an exogenous organic compound is mandatory as the principal source of cellular carbon. Conversely, an organism that uses CO_2 as the immediate precursor of virtually all cell material, but which requires trace amounts of an organic vitamin (as an enzymic cofactor) is considered to be an autotroph.

Some complexity of nomenclature has arisen from combining terms relating to carbon nutrition with terms concerned with energy metabolism. When the energy source is light, the situation is reasonably straightforward. Thus,

Phototroph. An organism that can use light as the energy source.

(*i*) *Photoautotroph.* An organism that can use light as the energy source and CO_2 as the immediate (and major) source of cellular carbon.

(*ii*) *Photoheterotroph.* An organism that can use light as the energy source and organic carbon compounds as sources of cellular carbon.

Organisms that cannot use light as the source of energy are called chemotrophs. Accordingly,

Chemotroph. An organism that uses inorganic or organic substances as energy sources (it is implied that these substances also provide reducing power for biosynthesis).

(*i*) *Chemoautotroph.* An autotroph that uses one or more inorganic compounds as the source of electrons (reducing power) and energy.

(*ii*) *Chemoheterotroph.* An organism that uses organic compounds as the source of cellular carbon, reducing power, and energy.

Using the system here proposed, it is unnecessary to use -litho- and -organo- for subdivisions of chemotrophy. So-called "chemolithotrophs" would simply be "chemoautotrophs." Although of considerable biochemical interest, metabolic patterns in which heterotrophic carbon nutrition is combined with energy conversion, based on the oxidation of inorganic compounds seem to be quite uncommon. Use of a special term for organisms with this capacity seems unwarranted at present. Most heterotrophs are, in fact, chemoheterotrophs.

The term "mixotrophy" has been discussed by Rittenberg (1972). He uses this term to imply "concomitant generation of energy by supposedly alternative mechanisms—that is, chemolithotrophically and chemoheterophically—or the concomitant use of autotrophic and heterotrophic mechanisms for biosynthesis". Thus, the term describes organisms (see Matin, 1978) that are bioenergetic and/or carbon nutrition hybrids. Since the energy conversion and carbon nutrition categories are already specified, "mixotrophy" does not represent a fundamental definition.

Recognizing the multiplicity of metabolic combinations that are possible, even in a single organism, we believe it is wise to minimize the number of definitions used to describe "physiological patterns" in so far as possible. As a specific case in point we cite *Rhodopseudomonas capsulata,* one of the most versatile prokaryotes known. At least five growth modes are available to this remarkable bacterium (Madigan and Gest, 1979): (*i*) anaerobic growth as a photoautotroph on H_2 plus CO_2, with light as the energy source; (*ii*) anaerobic growth as a photoheterotroph on various organic carbon sources, with light as the energy source; (*iii*) growth as a fermentative (heterotrophic) anaerobe, in darkness, on sugars as sole carbon and energy sources; (*iv*) aerobic growth as an ordinary chemoheterotroph in darkness; and (*v*) aerobic growth as a chemoautotroph, in darkness, with H_2 as the source of electrons and energy. It is likely that certain combinations of these modes can be expressed simultaneously, to greater or lesser extents, depending on environmental conditions.

Acknowledgments

This paper was prepared while D. J. C. was Distinguished Visiting Scientist at the Atlantic Research Laboratory (NRC-Canada) and while H. G. was recipient of a fellowship from the John Simon Guggenheim Foundation; research relevant to this work has been supported by N. S. F. Grants GB-42461 and PCM 78-25852 (to D. J. C.) and PCM 79-10747 (to H. G.).

REFERENCES CITED

Material in brackets indicates the Chapters (1–15), Appendixes (I–IV), Glossaries (I–II) and/or other location (e.g., Introduction) where each paper is cited in this book.

Aarna, A. Ya. and Lippmaa, E. T. 1957. Cleavage of ether oxygen bonds in the kerogen of esthonian bituminous shale. *J. Appl. Chem. (USSR) 30*: 323–326. [Appendix IV.]

Abelson, P. H. 1959. Geochemistry of organic substances. In P. H. Abelson, ed., *Researches in Geochemistry*, New York, Wiley, 79–103. [Chapter 1.]

———. 1966. Chemical events on the primitive Earth. *Proc. Natl. Acad. Sci. USA 55*: 1365–1372. [Chapter 4.]

———. 1978. Organic matter in the Earth's crust. *Ann. Rev. Earth Planet. Sci. 6*: 325–351. [Chapter 5.]

Abelson, P. H. and Hoering, T. C. 1961. Carbon isotope fractionation in formation of amino acids by photosynthetic organisms. *Proc. Natl. Acad. Sci. USA 47*: 623–632. [Chapters 5, 7.]

Abrahamson, J. 1977. Saturated platelets are new intermediates in hydrocarbon pyrolysis and carbon formation. *Nature 266*: 323–327. [Chapter 4.]

Abt, H. A. 1977. The companions of sunlike stars. *Scient. Am. 256*: 96–104. [Chapter 2.]

Aitken, J. D. 1977. New data on correlation of the Little Dal Formation and a revision of Proterozoic map-unit H-5. In Report of Activities, Part A, *Geol. Surv. Can. Pap. 77-1A*, 131–135. [Chapter 11.]

Aizenshtat, Z.; Baedecker, M. J.; and Kaplan, I. R. 1973. Distribution and diagenesis of organic compounds in JOIDES sediment from Gulf of Mexico and Western Atlantic. *Geochim. Cosmochim. Acta 37*: 1881–1898. [Chapter 5.]

Ajtay, G. L.; Ketner, P.; and Duvigneaud, P. 1979. Terrestrial primary production and phytomass. In B. Bolin, F. T. Degens, S. Kempe and P. Ketner, eds., *The Global Carbon Cycle*, Chichester, Wiley, 29–101. [Chapter 5.]

Akhtar, K. 1976. Ganugarh shale of Southeastern Rajasthan, India: A Precambrian regressive sequence of lagoon-tidal flat origin. *J. Sed. Petrol. 46*: 14–21. [Chapter 11.]

Akiyama, M. and Imoto, N. 1975. Scanning electron microscopic examination of some microfossils from the Gunflint chert. *Chikyu Kagaku (Earth Science), J. Assoc. Geol. Collab. Japan 29* (6): 280–282. [Chapter 14.]

Aldrich, L. T.; Hart, S. R.; Tilton, G. R.; Davis G. L.; Rama, S. N. I.; Steiger, R.; Richards, J. R.; and Gerken, J. S. 1964. Isotope Geology. *Carnegie Inst. Wash. Yearbook 63*: 331–340. [Chapter 11.]

Alexander, M. 1961. *Introduction to Soil Microbiology*, New York, Wiley, 432 pp. [Chapter 7.]

Allaart, J. H. 1976. The pre-3760 Myr old supracrustal rocks of the Isua area, central West Greenland and the associated occurrence of quartz banded ironstone. In B. F. Windley, ed., *The Early History of the Earth*, London, Wiley, 177–189. [Chapter 3.]

Allard, P. 1979. $^{13}C/^{12}C$ and $^{34}S/^{32}S$ ratios in magmatic gases from ridge volcanism in Afar. *Nature 282*: 56–58. [Chapter 4.]

Allard, P.; Tazieff, H.; and Dajlevic, D. 1979. Observations of seafloor spreading in Afar during the November 1978 fissure eruption. *Nature 279*: 30–33. [Chapter 4.]

Allègre, C. J. 1980. La géodynamique chimique. *Mém. Sér. Soc. Géol. France 10*: 84–104. [Chapter 10.]

———. 1982. Chemical geodynamics. *Tectonophys. 81*: 109–132. [Chapter 10.]

Allègre, C. J. and Ben Othman, D. 1980. Nd-Sr isotopic relationship in granitoid rocks and continental crust development: A chemical approach to orogenesis. *Nature 286*: 335–342. [Chapter 10.]

Allen, W. V. and Ponnamperuma, C. 1967. A possible prebiotic synthesis of monocarboxylic acids. *Curr. Mod. Biol. 1*: 24–28. [Chapter 4.]

Allsopp, H. L. 1964. Rubidium/stronium ages from the western Transvaal. *Nature 204*: 361–363. [Chapter 11.]

Ambrose, G. J.; Flint, R. V.; and Webb, A. W. 1981. Precambrian and Palaeozoic geology of the Peake and Denison Ranges. *Geol. Surv. S. Austr. Bull. 50*, 71 pp. [Appendix I.]

Amoore, J. E. 1961. Dependence of mitosis and respiration in roots upon oxygen tension. *Proc. Roy. Soc. London Ser. B 154*:109–129. [Chapter 13.]

Anders, E. 1978. Most stony meteorites come from the asteroid belt. In D. Morrison and W. C. Wells, eds., *Asteroids: An Exploration Assessment*, Washington, D.C., NASA Conf. Pub. 2053, 57–76. [Chapter 4.]

Anders, E.; Hayatsu R.; and Studier, M. H. 1973. Organic compounds in meteorites. *Science 182*: 781–790. [Chapter 4.]

Anders, E. and Owen, T. 1977. Mars and Earth: Origin and abundance of volatiles. *Science 198*: 453–465. [Chapter 4.]

Anderson, A. T. 1975. Some basaltic and andesitic gases. *Rev. Geophys. Space Phys. 13*:37–55. [Chapter 4.]

Anderson, C. A.; Blacet, P. M.; Silver, L. T.; and Stern, T. W. 1971. Revision of Precambrian stratigraphy in the Prescott-Jerome area, Yavapai County, Arizona. *U.S. Geol. Surv. Bull. 1324C*:C1–C16. [Chapter 11.]

Anderson, D. L. 1979. Chemical stratification of the mantle. *J. Geophys. Res. 84*:6297–6298. [Chapter 2.]

Anderson, R. L. and Ordal, E. J. 1961. CO_2-dependent fermentation of glucose by *Cytophaga succinicans*. *J. Bacteriol. 81*:139–146. [Chapter 6.]

Anderton, R. 1975. Tidal flat and shallow marine sediments from the Craignish Phyllites, middle Dalradian Argyll, Scotland. *Geol. Mag. 112*: 337–348. [Chapter 11.]

Anhaeusser, C. R. 1971. The Barberton Mountain Land, South Africa—a guide to the understanding of the Archaean geology of western Australia. In J. E. Glover, ed., *Symposium on Archaean Rocks*, *Geol. Soc. Austr. Spec. Pub. 3*. [Chapter 3.]

Anhaeusser, C. R.; Mason, R.; Viljoen, M. J.; and Viljoen, R. P. 1969. A reappraisal of some aspects of Precambrian shield geology. *Geol. Soc. Am. Bull. 80*:2175–2200. [Chapter 3.]

Anonymous. 1972. Ophiolites. *Geotimes 17* (12): 24–25. [Chapter 3.]

Appel, P. W. U. 1980. On the early Archaen Isua iron-formation, West Greenland. *Precambrian Res. 11*:73–87. [Appendixes I, II.]

Arculus, R. J. 1981. Island arc magmatism in relation to the evolution of the crust and mantle. *Tectonophys. 75*:113–133. [Chapter 3.]

Arculus, R. J. and Delano, J. W. 1980. Implications for the primitive atmosphere of the oxidation state of Earth's upper mantle. *Nature 288*:72–74. [Chapter 4.]

Armstrong, D. W.; Seguin, R.; and Fendler, J. H. 1977. Partitioning of amino acids and nucleotides between water and micellar hexadecyltrimethylammonium halides. The prebiotic significance of cationic surfaces. *J. Mol. Evol. 10*:241–250. [Chapter 4.]

Armstrong, H. S. 1960. Marbles in the Archean of the southern Canadian Shield. In *Pre-Cambrian Stratigraphy and Correlations*, *Rept. 21st Internat. Geol. Congr.* (Paris), *Pt. 9*, *Sect. 9*:7–20. [Chapter 8.]

Armstrong, R. L. 1981. Radiogenic isotopes: The case for crustal recycling on a near-steady-state no-continental-growth Earth. *Phil. Trans. Roy. Soc. London Sect. A 301*:443–472. [Chapter 10.]

Arriens, P. A. 1971. Archaean geochronology of Australia. In J. E. Glover, ed., *Symposium on Archaean Rocks*, *Geol. Soc. Austr. Spec. Pub. 3*: 1–23. [Chapter 11.]

Asada, K. and Kanematsu, S. 1978. Distribution of cuprozinc, manganic, and ferric superoxide dismutase in plants and fungi: An evolutionary aspect. In K. Matsubara and T. Yamanaka, eds. *Evolution of Protein Molecules*, Tokyo, Japan Scient. Soc. Press, 361–372. [Chapter 13.]

Ashley, B. E. 1937. Fossil algae from the Kundelungu Series of Northern Rhodesia. *J. Geol. 45*: 332–335. [Chapter 1.]

Awramik, S. M. 1971. Precambrian columnar stromatolite diversity: Reflection of metazoan appearance. *Science 174*:825–827. [Chapters 8, 9.]

———. 1976. Gunflint stromatolites: Microfossil distribution in relation to stromatolite morphology. In M. R. Walter, ed., *Stromatolites*, Amsterdam, Elsevier, 311–320. [Chapters 8, 14.]

———. 1981. The pre-Phanerozoic biosphere—three billion years of crises and opportunities. In M. Nitecki and D. Raup, eds., *Biotic Crises in Ecological and Evolutionary Time*, New York, Academic Press, 83–102. [Chapter 9.]

Awramik, S. M. and Barghoorn, E. S. 1977. The Gunflint microbiota. *Precambrian Res. 5*:121–142. [Chapters 1, 9, 14.]

———. 1978. Bibliography of Precambrian paleontology and paleobiology. *Harvard Univ. Bot. Mus. Leaflets 26* (2–4):65–175. [Chapter 1.]

Awramik, S. M. and Semikhatov, M. A. 1979. The relationship between morphology, microstructure, and microbiota in three vertically intergrading stromatolites from the Gunflint Iron Formation. *Can. J. Earth Sci. 16*:484–495. [Chapters 8, 14.]

Awramik, S. M.; Golubić, S.; and Barghoorn, E. S. 1972. Blue-green algal cell degradation and its implication for the fossil record. *Geol. Soc. Am. Abstr. Progr. 4* (7):438 (Abstr.). [Chapter 14.]

Awramik, S. M.; Margulis, L.; and Barghoorn, E. S. 1976. Evolutionary processes in the formation of stromatolites. In M. R. Walter, ed., *Stromatolites*, Amsterdam, Elsevier, 149–162. [Chapters 8, 15.]

Awramik, S. M.; Schopf, J. W.; and Walter, M. R. 1983. Filamentous fossil bacteria 3.5×10^9 years old from the Archean of Western Australia. *Precambrian Res.*, in press. [Introduction; Chapters 7, 8, 9.]

Ayres, L. D. 1978. A transition from subaqueous to subaerial eruptive environments in the middle Precambrian Amisk Group at Amisk Lake, Saskatchewan—a progress report. *Centre for Precambrian Studies, Univ. Manitoba, 1977 Ann. Rept.*:36–51. [Chapter 10.]

Baadsgaard, H. 1976. Further U-Pb dates on zircons from the early Precambrian rocks of the Godthaabsfjord area, West Greenland. *Earth Planet. Sci. Lett. 33*:261–267. [Appendix I.]

Bada, J. L. and Miller, S. L. 1968. Ammonium ion concentration in the primitive ocean. *Science 159*:423–425. [Chapter 4.]

Bada, J. L. and Schroeder, R. A. 1975. Amino acid racemization reactions and their geochemical implications. *Naturwiss. 62*:71–79. [Chapter 5.]

Badham, J. P. N. and Stanworth, C. W. 1977. Evaporites from the lower Proterozoic of the East Arm, Great Slave Lake. *Nature 268*:516–517. [Chapter 11.]

Baer, A. J. 1981. Geotherms, evolution of the lithosphere and plate tectonics. *Tectonophys. 72*: 203–227. [Chapter 3.]

Bahadur, K.; Ranganayaki, S.; and Santamaria, L. 1958. Photosynthesis of amino acids from paraformaldehyde involving the fixation of nitrogen in the presence of colloidal molybdenum oxide as catalyst. *Nature 182*:1668. [Chapter 4.]

Bailes, A. H. 1980. Origin of Early Proterozoic volcaniclastic turbidites, south margin of the Kisseynew sedimentary gneiss belt, File Lake, Manitoba. *Precambrian Res. 12*:197–225. [Chapter 10.]

Baker, D. R. and Claypool, G. E. 1970. Effects of incipient metamorphism on organic matter in mudrock. *Am. Assoc. Petrol. Geol. Bull. 54*: 456–468. [Chapter 5.]

Barghoorn, E. S. 1957. Origin of life. In H. S. Ladd, ed., *Treatise on Marine Ecology and Paleoecology, Vol. 2, Geol. Soc. Am. Mem. 67*:75–85. [Chapter 5.]

———. 1963. Precambrian flora. *McGraw-Hill Yearbook Science and Technology*:453–455. [Chapter 14.]

———. 1971. The oldest fossils. *Scient. Am. 224* (5):30–42. [Chapters 9, 14.]

———.1974. Two billion years of prokaryotes and the emergence of eukaryotes. *Taxon 23*:259. [Chapter 14.]

Barghoorn, E. S. and Schopf, J. W. 1965. Microorganisms from the Late Precambrian of central Australia. *Science 150*:337–339. [Chapter 9.]

———. 1966. Microorganisms three billion years old from the Precambrian of South Africa. *Science 152*:758–763. [Chapter 9.]

Barghoorn, E. S. and Tyler, S. A. 1962. Microfossils from the Middle Precambrian of Canada. *Pollen et Spores 4*:331 (Abstr.). [Chapter 14.]

———. 1965a. Microorganisms from the Gunflint chert. *Science 147*:563–577. [Chapters 1, 9, 14.]

———. 1965b. Microorganisms of Middle Precambrian age from the Animikie Series, Ontario, Canada. In *Current Aspects of Exobiology*, Pasadena, Calif., Jet Propulsion Lab., 93–118. [Chapter 14.]

Barghoorn, E. S.; Knoll, A. H.; Dembicki, H.; and Meinschein, W. G. 1977. Variation in stable carbon isotopes in organic matter from the Gunflint Iron Formation. *Geochim Cosmochim. Acta 41*:425–430. [Chapter 7.]

Barghoorn, E. S.; Meinschein, W. G.; and Schopf, J. W. 1965. Paleobiology of a Precambrian shale. *Science 148*:461–472. [Appendix I.]

Barghoorn, E. S.; Tyler, S. A.; and Hurley, P. 1954. Oldest plant discovered. *Traveler*, February 16, 1954. [Chapter 14.]

Barker, F. and Friedmann, I. 1969. Carbon isotopes in pelites of the Precambrian Uncompahgre formation, Needle Mountains, Colorado. *Geol. Soc. Am. Bull. 80*:1403–1408. [Chapter 5.]

Barker, F.; Arth, J. G.; and Hudson, D. 1981. Tonalites in crustal evolution. *Phil. Trans. Roy.*

Soc. London Sect. A 301:293–303. [Chapter 10.]

Barker, J. F. and Fritz, P. 1981. Carbon isotope fractionation during microbial methane oxidation. *Nature 293*:289–291. [Chapter 12.]

Barley, M. E. 1978. Shallow-water sedimentation during deposition of the Archaean Warrawoona Group, east Pilbara Block, Western Australia. In J. E. Glover and D. I. Groves, eds., *Archaean Cherty Metasediments: Their Sedimentology, Micropalaeontology, Biogeochemistry, and Significance to Mineralization*, Perth, *Univ. West. Austr. Spec. Pub. 2*:22–29. [Chapter 8.]

Barley, M. E.; Dunlop, J. S. R.; Glover, J. E.; and Groves, D. E. 1979. Sedimentary evidence for an Archean shallow-water volcanic-sedimentary facies, eastern Pilbara Block, Western Australia. *Earth Planet. Sci. Lett. 43*:74–84. [Chapters 3, 8, 9, 11.]

Bar-Nun, A. 1979. Acetylene formation on Jupiter: Photolysis or thunderstorms? *Icarus 38*:180–191. [Chapter 4.]

Bar-Nun, A. and Hartman, H. 1978. Synthesis of organic compounds from carbon monoxide and water by UV photolysis. *Origins of Life 9*:93–101. [Chapter 4.]

Bar-Nun, A. and Shaviv, A. 1975. Dynamics of the chemical evolution of Earth's primitive atmosphere. *Icarus 24*:197–210. [Chapter 4.]

Barrett, E. C. and Curtis, L. F. 1976. *Introduction to Environmental Remote Sensing*, London, Chapman and Hall, 336 pp. [Chapter 11.]

Barrois, C. 1892. Sur la présence de fossiles dans les terrains azoïques de Bretagne. *Compte Rend. Acad. Sci. Paris 115*:327. [Chapter 1.]

Baskov, Y. A.; Vetshteyn, V. Y.; Surikov, S. N.; Tolstikhin, I. N.; Malyuk, G. A.; and Mishina, T. A. 1973. Isotope composition of H, O, C, Ar, and He in hot springs and gases in the Kuril-Kamchatka volcanic region as indicators of formation conditions. *Geochem. Internat. 1973*: 130–138. [Chapter 4.]

Basu, A. R.; Ray, S. L.; Saha, A. K.; and Sarkar, S. N. 1981. 3.8 B.Y.-old tonalities from eastern India: Evidence for early mantle differentiation. *Am. Geophys. Union Trans. 62*:420 (Abstr.). [Chapter 3.]

Bathurst, R. G. C. 1975. *Carbonate Sediments and Their Diagenesis*, Amsterdam, Elsevier, 658 pp. [Chapter 10.]

Baur, M. E. 1978. Thermodynamics of heterogeneous iron-carbon systems: Implications for the terrestrial primitive reducing atmosphere. *Chem. Geol. 22*:189–206. [Chapter 4.]

Baxter, J. and Lipple, S. 1980. Explanatory notes on the Perenjory 1:250,000 Geological Sheet, Perth, W. Austr. Geol. Surv., 12 pp. [Appendix I.]

Baxter, M. S.; Stenhouse, M. J.; and Drndarski, N. 1980. Fossil carbon in coastal sediments. *Nature 287*:35–36. [Chapter 5.]

Bayley, R. W. and James, H. L. 1973. Precambrian iron-formations of the United States. *Econ. Geol. 68*:934–960. [Chapter 11.]

Beaty, C. B. 1978. The causes of glaciation. *Am. Scient. 66*:452–459. [Chapter 11.]

Beck, L. S. 1969. Uranium deposits of the Athabasca region, Saskatchewan. *Sask. Dept. Mineral Resources Geol. Sci. Branch Rept. 126*, 139 pp. [Chapter 14.]

Becker, R. H. and Clayton, R. N. 1972. Carbon siotopic evidence for the origin of a banded iron formation in Western Australia. *Geochim. Cosmochim. Acta 36*:577–595. [Chapters 7, 12.]

Becker, R. S.; Hong, K.; and Hong, J. H. 1974. Hot hydrogen-atom reactions of interest in molecular evolution and interstellar chemistry. *J. Mol. Evol. 4*:157–172. [Chapter 4.]

Begemann, F. and Heinzinger, L. 1969. Content and isotopic composition of carbon in the light and dark portions of gas-rich chondrites. In P. M. Millman, ed., *Meteorite Research*, New York, Springer, 87–92. [Appendix III.]

Behrens, E. W. and Frishman, S. A. 1971. Stable carbon isotopes in blue-green algal mats. *J. Geol. 79*:94–100. [Chapter 7.]

Bell, K.; Blenkinsop, J.; Cole, T. J. S.; and Menagh, D. P. 1974. Sr isotope composition of the Bulawayan Limestone. *Geol. Soc. Am. Abstr. Progr. 6* (7):650–651 (Abstr.). [Chapter 8.]

Bell, P. M. and Mao, H. K. 1975. Preliminary evidence of disproportionation of ferrous iron in silicates at high pressures and temperatures. *Carnegie Inst. Wash. Yearbook 74*:557–559. [Chapter 4.]

Bell, R. T. 1970. The Hurwitz Group, a prototype for deposition on metastable cratons. *Geol. Surv. Can. Pap. 70-40*:159–170. [Chapter 11.]

Bell, R. T. and Jackson, D. E. 1974. Aphebian halite and sulfate indications in the Belcher Group, Northwest Territories. *Can. J. Earth Sci. 11*:722–728. [Chapter 11.]

Belsky, T. and Kaplan, I. R. 1970. Light hydrocarbon gases, C^{13}, and origin of organic matter in carbonaceous chondrites. *Geochim. Cosmochim. Acta 34*:257–278. [Chapter 4; Appendix III.]

Bence, A. E.; Grove, R. L.; and Papike, J. J. 1980. Basalts as probes of planetary interiors: Constraints on the chemistry and mineralogy of their source regions. *Precambrian Res. 10*: 249–279. [Chapter 4.]

Benedict, C. R. 1978. The fractionation of stable carbon isotopes in photosynthesis. *What's New in Plant Physiology 9*:13–16. [Appendix I.]

Benlow, A. and Meadows, A. J. 1977. The formation of the atmospheres of the terrestrial planets by impact. *Astrophys. Space Sci. 46*: 293–300. [Chapter 4.]

Berkner, L. V. and Marshall, L. C. 1964a. The history of oxygenic concentration in the earth's atmosphere. *Disc. Faraday Soc. 37*:122–141. [Chapter 1.]

Berkner, L. V. and Marshall, L. C. 1964b. The history of growth of oxygen in the Earth's atmosphere. In P. J. Brancazio and A. G. W. Cameron, eds., *The Origin and Evolution of Oceans and Atmospheres*, New York, Wiley, 102–126. [Chapter 11.]

Berkner, L. V. and Marshall, L. C. 1965. On the origin and rise of oxygen concentration in the Earth's atmosphere. *J. Atmos. Sci. 22*:225–261. [Chapter 11.]

Berkner, L. V. and Marshall, L. C. 1966. Limitation on oxygen concentration in a primitive planetary atmosphere. *J. Atmos. Sci. 23*:133–143. Chapter 11.]

Berkner, L. V. and Marshall, L. C. 1967. The rise of oxygen in the Earth's atmosphere with notes on the martian atmosphere. *Adv. Geophys. 12*:309–331. [Chapter 11.]

Bernal, J. D. 1951. *The Physical Basis of Life*, London, Routledge and Kegan Paul, 80 pp. [Chapter 4.]

———. 1967. *The Origin of Life*, New York, World, 345 pp. [Chapter 15.]

Bernatowicz, T. J. and Podosek, F. A. 1978. Nuclear components in the atmosphere. In E. C. Alexander, Jr. and M. Ozima, eds., *Terrestrial Rare Gases*, Tokyo, Japan Scient. Soc. Press, 99–135. [Chapter 4.]

Bertrand-Sarfati, J. 1976a. An attempt to classify late Precambrian stromatolite microstructures. In M. R. Walter, ed., *Stromatolites*, Amsterdam, Elsevier, 251–259. [Chapter 8.]

———. 1976b. Pseudomorphoses de gypse en rosettes dans un calcaire cryptalgo-laminaire du Precambrian inferieur (Système du Transvaal, Afrique du Sud). *Bull. Soc. Géol. France Suppl. 3*:99–102. [Chapter 11.]

———. 1980. Columnar stromatolites from early Proterozoic (S. Africa). Part II. Microstructures and environments of growth. *Abstracts, 26th Internat. Geol. Congr.*, (Paris), *2*:435 (Abstr.). [Chapter 8.]

Bertrand-Sarfati, J. and Eriksson, K. A. 1977. Columnar stromatolites from the Early Proterozoic Schmidtsdrift Formation, Northern Cape Province, South Africa—Part I: Systematic and diagnostic features. *Palaeontol. Afr. 20*:1–26. [Chapter 8.]

Bertrand-Sarfati, J. and Walter, M. R. 1981. Stromatolite biostratigraphy. *Precambrian Res., 15*:353–371. [Chapter 8.]

Beutner, R. 1938. *Life's Beginning on the Earth*, Baltimore, Williams and Wilkins, 99 pp. [Chapter 4.]

Bhat, J. V. and Barker, H. A. 1947. *Clostridium lacto-acetophilum* nov. spec. and the role of acetic acid in the butyric acid fermentation of lactate. *J. Bacteriol. 54*:381–391. [Chapter 6; Glossary II.]

Bickle, M. J.; Martin, A.; and Nisbet, E. G. 1975. Basaltic and periodotitic komatiites and stromatolites above a basal unconformity in the Belingwe Greenstone Belt, Rhodesia. *Earth Planet. Sci. Lett. 27*:155–162. [Chapter 8; Appendix I.]

Biemann, K.; Oró, J.; Toulmin, P. III; Orgel, L. E.; Nier, A. O.; Anderson, D. M.; Simmonds, P. G.; Flory, D.; Diaz, A. V.; Rushneck, D. R.; Biller, J. E.; and Lafleur, A. L. 1977. The search for organic substances and inorganic volatile compounds in the surface of Mars. *J. Geophys. Res. 82*:4641–4658. [Chapter 4.]

Binda, P. L. 1976. Preliminary observations on the palynology of the Precambrian Katanga sequence, Zambia. *Geol. Mijnbouw 51* (3):315–319. [Chapter 1.]

Black, L. P.; Shaw, R. D.; and Offe, L. A. 1980. The age of the Stuart Dyke Swarm and its bearing on the onset of late Precambrian sedimentation in central Australia. *J. Geol. Soc. Austr. 27*:151–155. [Appendix I.]

Black, M. 1933. The algal sediments of Andros Island, Bahamas. *Phil. Trans. Roy. Soc. London Sect. B 222*:165–192.

Blair, N. E. and Bonner, W. A. 1980. The radiolysis of tryptophan and leucine with ^{32}P β-radiation. *J. Mol. Evol. 15*:21–28. [Chapter 4.]

Blake, A. J. and Carver, J. H. 1977. The evolutionary role of atmospheric ozone. *J. Atmos. Sci. 34*: 720–728. [Chapter 11.]

Bliss, N. W. and Stidolph, P. A. 1969. The Rhodesian Basement Complex: A review. *Geol. Soc. S. Afr. Spec. Pub. 2*:305–333. [Chapter 11.]

Bloch, K. E. 1979. Speculations on the evolution of sterol structure and function. *CRC Crit. Rev. Biochem. 7*:1–5. [Chapter 6.]

Boato, G. 1954. The isotopic composition of hydrogen and carbon in carbonaceous chondrites. *Geochim. Cosmochim. Acta 6*:209–220. [Appendix III.]

Bodine, M. W.; Holland, H. D.; and Borcsik, M. 1965. Coprecipitation of manganese and strontium with calcite. In M. Štemprok and M. Rieder, eds., *Symposium on Problems of Postmagmatic Ore Deposition, Vol. 2.*, Prague, Czech Acad. Soc., 401–406. [Chapter 10.]

Boettcher, A. L. 1973. Volcanism and orogenic belts—the origin of andesites. *Tectonophys. 17*: 223–240. [Chapter 3.]

Boettcher, A. L.; Mysen, B. O.; and Modreski, P. J. 1975. Melting in the mantle: Phase relationships in natural and synthetic periodotite-H_2O and peridotite-H_2O-&O_2 systems at high pressures. *Phys. Chem. Earth 9*:855–867. [Chapter 4.]

Bogard, D. D. and Gibson, E. K., Jr. 1978. The origin and relative abundances of C, N and the noble gases on the terrestrial planets and in meteorites. *Nature 271*:150–153. [Chapter 4.]

Bogorad, L. 1979. Biosynthesis of porphyrins. In D. Dolphin, ed., *Porphyrins, Vol. 6, Biochemistry Part A*, New York, Academic Press, 125–178. [Chapter 13.]

Bolin, B.; Degens, E. T.; Duvigneaud, P.; and Kempe, S. 1979. The global biogeochemical carbon cycle. In B. Bolin, E. T. Degens, S. Kempe, and P. Ketner, eds., *The Global Carbon Cycle*, Chichester, Wiley, 1–53. [Chapter 5.]

Bonatti, B. 1978. The origin of metal deposits in the oceanic lithosphere. *Scient. Am. 238* (2): 54–61. [Chapter 10.]

Bond, G. and Falcon, R. 1973. The palaeontology of Rhodesia, with a section of the palynology of the middle Zambezi Basin. *Rhodesia Geol. Surv. Bull. 70*:121. [Chapter 8.]

Bond, G.; Wilson, J. F.; and Winnall, N. J. 1973. Age of the Huntsman Limestone (Bulawayan) stomatolites. *Nature 244*:275–276. [Chapter 8, Appendix I.]

Bondesen, E.; Raunsgaard Pedersen, K.; and Jørgensen, 0. 1967. Precambrian organisms and the isotopic composition of organic remains in the Ketilidian of southwest Greenland. *Meddel. Grønl. 164* (4):1–41. [Chapter 14.]

Boström, K.; Peterson, M. N. A.; Joensuu, O.; and Fisher, D. E. 1969. Aluminum-poor ferromanganoan sediments on active oceanic ridges. *J. Geophys. Res. 74*:3261–3270. [Chapter 10.]

Bottinga, Y. 1969. Calculated fractionation factors for carbon and hydrogen isotope exchange in the system calcite-carbon dioxide-graphite-methane-hydrogen-water vapor. *Geochim. Cosmochim. Acta 33*:49–64. [Chapters 5, 7.]

Boureau, E. 1977. Characteres et évolution de certains organismes du Précambrien terminal de l'ouest Africain. *Compte Rend. 102d Congr. Nat. des Soc. Savantes* (Limoges), *1*:11–33. [Chapter 1.]

Boureau, E. 1978. Sur l'origine et la formation des grands sphéroides pré-cellulaires du Précambrien de l'ouest Africain. *Compte Rend. 103d Congr. nat. des soc. savantes* (Nancy), *2*:7–25. [Chapter 1.]

Bowen, H. J. M. 1979. *Environmental Chemistry of the Elements*, London, Academic Press, 333 pp. [Chapter 12.]

Bowen, N. L. 1928. *The Evolution of the Igneous Rocks*, New York, Dover, 334 pp. [Chapter 3.]

Boyer, P. D., ed. 1975. *The Enzymes. Vol. 12, Oxidation-Reduction, Part B*, 3rd ed., New York, Academic Press, 647 pp. [Chapter 13.]

———. 1976. *The Enzymes. Vol. 13, Oxidation-Reduction, Part C*, 3rd ed., New York, Academic Press, 542 pp. [Chapter 13.]

Bradley, W. H. 1929. Algae reefs and oolites of the Green River Formation. *U. S. Geol. Surv. Prof. Pap. 154-G*:203–223. [Chapter 8.]

Brand, S. W.; Chung, H. M.; Grizzle, P. L.; Kaplan, I. R.; and Sweeney, R. E. 1980. Geochemical characterization of petroleum using carbon, sulfur, hydrogen and nitrogen stable isotopes. *Ann. Rept. to Dept. of Energy, Contract No. EY-77-C-19-0034*, 20 pp. [Chapter 7.]

Brand, U. and Viezer, J. 1980. Chemical diagenesis of a multicomponent carbonate system—I: Trace elements. *J. Sed. Petrol. 50*:1219–1236. [Chapter 10.]

Brand, U. and Veizer, J. 1981. Chemical diagenesis of a multicomponent carbonate system—II: Stable isotopes. *J. Sed. Petrol. 51*: 987–997. [Chapter 10.]

Brandt, R. T.; Dorr, J. V. N.; Gross, G. A.; Gruss, H.; and Semenenko, N. P. 1972. Problems of nomenclature for banded ferruginous-cherty sedimentary rocks and their metamorphic equivalents. *Econ. Geol. 67*:682–684. [Chapter 11; Glossary II.]

Breger, I. A.; Chandler, J. C.; Daniels, G. J.; and Simons, G. 1962. Biochemical implications from a study of Precambrian organic matter. Abstract of Informal Communication Before Geological Society of Washington, 28 March 1962 (Abstr.). [Chapter 5.]

Bremner, J. W. and Tabatabai, M. A. 1973. Nitrogen-15 enrichment of soils and soil derived nitrate. *J. Environ. Qual. 2*:363–365. [Chapter 7.]

Bremner, J. M.; Cheng, H. H.; and Edwards, A. P. 1966. Assumptions and errors in nitrogen-15 tracer research. In *Report FAO/IAEA Tech. Meeting* (Braunschweig, Germany, 1963), New York, Pergamon, 429–442. [Chapter 7.]

Brett, R. 1976. The current status of speculations on the composition of the core of the Earth. *Rev. Geophys. Space Phys. 14*:375–383. [Chapter 4.]

Briden, J. C. 1973. Applicability of plate tectonics to pre-Mesozoic time. *Nature 244*:400–405. [Chapter 3.]

Bridgwater, D.; McGregor, V. R.; and Myers, J. S. 1974. A horizontal tectonic regime in the Archaean of Greenland and its implications for early crustal thickening. *Precambrian Res. 1*: 179–197. [Chapter 11.]

Bridgwater, D.; Allaart, J. H.; Schopf, J. W.; Klein, C.; Walter, M. R.; Barghoorn, E. S.; Strother, P.; Knoll, A. H.; and Gorman, B. E. 1981. Microfossil-like objects from the Archaean of Greenland: A cautionary note. *Nature 289*: 51–53. [Introduction; Chapters 3, 9, 15; Appendix I.]

Briggs, M. H. 1963. Evidence of an extraterrestrial origin for some organic constituents of meteorites. *Nature 197*:1290. [Appendix III.]

Brinkmann, R. T. 1969. Dissociation of water vapor and evolution of oxygen in the terrestrial atmosphere. *J. Geophys. Res. 74*:5355–5368. [Chapter 11.]

Brock, T. D. 1978. *Thermophilic Microorganisms and Life at High Temperatures*, New York, Springer, 465 pp. [Chapter 11.]

Broda, E. 1975a. *The Evolution of the Bioenergetic Processes*, Oxford, Pergamon, 220 pp. [Chapters 6, 7.]

———. 1975b. The history of inorganic nitrogen in the biosphere. *J. Mol. Evol. 7*:87–100. [Chapter 13.]

Broda, E. and Peschek, G. A. 1979. Did respiration or photosynthesis come first? *J. Theoret. Biol. 81*:201–212. [Chapter 6.]

———. 1980. Evolutionary considerations on the thermodynamics of nitrogen fixation. *Biosystems 13*:47–56. [Chapter 6.]

Broecker, W. S. 1963. Radioisotopes and large scale ocean mixing. In M. N. Hill, ed., *The Sea, Vol. 3.*, New York, Wiley, 88–108. [Chapter 10.]

———. 1970. A boundary condition on the evolution of atmospheric oxygen. *J. Geophys. Res. 75*:3553–3557. [Chapter 7.]

———. 1971. A kinetic model for the chemical composition of sea water. *Quaternary Res. 1*: 188–207. [Chapter 11.]

———. 1974. *Chemical Oceanography*, New York, Harcourt Brace, 214 pp. [Chapter 11.]

Bronner, G. and Chauvel, J. J. 1979. Precambrian banded iron-formations of the Ijil Group (Kediat Ijil, Reguibat Shield, Mauritania). *Econ. Geol. 74*:77–94. [Chapter 11.]

Brooks, J. 1970. In the beginning. *BP Shield* (Nov., 1970):6–10. [Chapter 9.]

Brooks, J. and Muir, M. D. 1971. Morphology and chemistry of the organic insoluble matter from the Onverwacht Series Precambrian chert and the Orgueil and Murray carbonaceous meteorites. *Grana 11*:9–14. [Chapter 9.]

———. 1974. Early Precambrian microorganisms in the Onverwacht Group (3.4–3.7 × 10^9 years old) from the Swaziland Sequence of southern Africa. In S. N. Naumova, ed., *Palinologiya Proterofita i Paleofita*, Trudy III Mezhdunarodnoy Palinoligischeskoy Konferentsii (*Palynology of the Proterophytic and Paleophytic*, Proc. 3rd Internat. Palynol. Conf.), Moscow, Nauka, 15–19. [Chapter 9.]

Brooks, J. and Shaw, G. 1971. Evidence for life in the oldest known sedimentary rocks—the Onverwacht Series chert, Swaziland system of southern Africa. *Grana 11*:1–8. [Chapter 5.]

Brooks, J. and Shaw, G. 1973. *Origin and Development of Living Systems*, New York, Academic Press, 412 pp. [Chapter 9.]

Brooks, J.; Muir, M. D.; and Shaw, G. 1973. Chemistry and morphology of Precambrian micro-organisms. *Nature 244*:215–217. [Chapter 9.]

Brooks, J. D. 1971. Organic matter in Archaean

rocks. In J. E. Glover, ed., *Symposium on Archaean Rocks*, *Geol. Soc. Austr. Spec. Pub. 3*: 413–418. [Chapter 5.]

Broshe, P. and Sünderman, J., eds., 1978. *Tidal Friction and the Earth's Rotation*, New York, Springer, 242 pp. [Chapter 11.]

Brown, A. C. 1971. Zoning in the White Pine copper deposit, Ontonagon County, Michigan. *Econ. Geol. 66*:543–573. [Appendix I.]

Brown, F. S.; Baedecker, M. J.; Nissenbaum, A.; and Kaplan, I. R. 1972. Early diagenesis in a reducing fjord, Saanich Inlet, British Columbia—III. Changes in organic constituents of sediment. *Geochim. Cosmochim. Acta 36*:1185–1203. [Chapter 5.]

Brown, H. 1949. Rare gases and the formation of the Earth's atmosphere. In G. P. Kupier, ed., *The Atmospheres of the Earth and Planets*, Chicago, Univ. Chicago Press, 258–266. [Chapter 4.]

Brown, J. C. 1981. Space and time in granite plutonism. *Phil. Trans. Roy. Soc. London Sect. A 301*: 321–336. [Chapter 10.]

Brown, J. S. 1973. Sulfur isotopes of Precambrian sulfates and sulfides in the Grenville of New York and Ontario. *Econ. Geol. 68*:362–369. [Chapters 7, 11.]

Buchanan, B. B. 1979. Ferredoxin-linked carbon dioxide fixation in photosynthetic bacteria. In M. Gibbs and E. Latzko eds., *Encyclopedia of Plant Physiology*, new series, *Vol. 6*, Berlin, Springer, 416–424. [Chapter 7.]

Buchanan, R. E. and Gibbons, N. E., eds. 1974. *Bergey's Manual of Determinative Bacteriology*, Baltimore, Williams and Wilkins, 1246 pp. [Chapters 9, 14, 15.]

Buck, S. G. 1980. Stromatolite and ooid deposits within the fluvial and lacustrine sediments of the Precambrian Ventersdorp Supergroup of South Africa. *Precambrian Res. 12*:311–330. [Chapters 8, 10, 11; Appendix I.]

Buddington, A. F.; Jensen, M. L.; and Mauger, R. L. 1969. Sulfur isotopes and origin of northwest Adirondack sulfide deposits. *Geol. Soc. Am. Mem. 115*:423–451. [Chapter 7.]

Buddington, A. F. and Lindsley, D. H. 1964. Iron-titanium oxide minerals and synthetic equivalents. *J. Petrol. 5*:310–357. [Chapter 4; Appendix III.]

Budyko, M. I. 1977. *Climatic Changes*, Washington, D.C., Am. Geophys. Union, 261 pp. [Chapter 11.]

Bueding, E. and Fisher, J. 1982. Metabolic requirements of schistosomes. *J. Parasitol. 68*:208–212. [Chapter 6, Glossary II.]

Buick, R.; Dunlop, J. S. R.; and Groves, D. I. 1981. Stromatolite recognition in ancient rocks: An appraisal of irregularly laminated structures in an Early Archaean chert-barite unit from North Pole, Western Australia. *Alcheringa 5*:161–181. [Chapter 9.]

Bunch, T. E. and Chang, S. 1980. Carbonaceous chondrites—II. Carbonaceous chondrite phyllosilicates and light element geochemistry as indicators of parent body processes and surface conditions. *Geochim. Cosmochim. Acta 44*:1543–1577. [Chapter 4.]

Bunch, T. E.; Chang, S.; Frick, U.; Neil, J.; and Moreland, G. 1979. Carbonaceous chondrites—I. Characterization and significance of carbonaceous chondrite (CM) xenoliths in the Jodzie howardite. *Geochim. Cosmochim. Acta 43*:1727–1742. [Chapter 4.]

Burchard, R. P. 1980. Gliding motility of bacteria. *Bioscience 30*:157–162. [Chapter 8.]

Burnett, D. S. 1975. Lunar science: The Apollo legacy. *Rev. Geophys. Space Phys. 13*:13–34. [Chapter 3.]

Burnie, S. W.; Schwarcz, H. P.; and Crocket, J. H. 1972. A sulfur isotopic study of the White Pine Mine, Michigan. *Econ. Geol. 67*:895–914. [Chapter 7.]

Button, A. 1976a. Transvaal and Hamersley Basins—review of basin development and mineral deposits. *Minerals Sci. Engng. 8*:262–293. [Chapters 8, 11; Appendix I.]

———. 1976b. Halite casts in the Umkondo System, southeastern Rhodesia. *Geol. Soc. S. Afr. Trans. 79*:177–178. [Chapter 7.]

———. 1976c. Iron-formation as an end member in carbonate sedimentary cycles in the Transvaal Supergroup, South Africa. *Econ. Geol. 71*:193–201. [Chapter 10.]

———. 1979. Early Proterozoic weathering profile on the 2200 m.y. old Hekpoort Basalt, Pretoria Group, South Africa: Preliminary results. *Univ. Witwatersrand Econ. Geol. Res. Unit Info. Circ. 133* (Johannesburg), 21 pp. [Chapters 1, 11.]

Button, A. and Tyler, N. 1979. Precambrian palaeoweathering and erosion surfaces in Southern Africa: Review of their character and economic significance. *Univ. Witwatersrand Econ. Geol. Res. Unit Info. Circ. 135* (Johnanesburg), 89 pp. [Chapter 11.]

Button, A.; Brock, T. D.; Cook, P. J.; Eugster, H. P.; Goodwin, A. M.; James, H. L.; Margulis, L.; Nealson, K. H.; Nriagu, J. O.; Trendall, A. F.; and Walter, R. M. 1982. Sedimentary iron deposits, evaporites and phosphorites. In H. D. Holland and M. Schidlowski, eds., *Mineral Deposits and the Evolution of the Biosphere*, Dahlem Konferenzen-Dahlem Workshop Rept. 16, Berlin, Springer, 259–273. [Chapter 10.]

Cairns-Smith, A. G. 1965. The origin of life and the nature of the primitive gene. *J. Theoret. Biol. 10*:53–88. [Chapter 4.]

———. 1971. *The Life Puzzle*, Edinburgh, Oliver and Boyd, 165 pp. [Chapter 4.]

———. 1978. Precambrian solution photochemistry, inverse segregation, and banded iron formations. *Nature 276*:807–808. [Chapter 11.]

Cairns-Smith, A. G. and Walker, G. L. 1974. Primitive metabolism. *Biosystems 5*:173–180. [Chapter 13.]

Calder, J. A. and Parker, P. L. 1973. Geochemical implications of induced changes in ^{13}C fractionation by blue-green algae. *Geochim. Cosmochim. Acta 37*:133–140. [Chapter 7.]

Calvin, M. 1956. Chemical evolution and the origin of life. *Am Scient. 44*:248–263. [Chapter 4.]

Cameron, A. G. W. 1975. Abundances of the elements in the solar system. *Space Sci. Rev. 15*: 121–146. [Chapter 4.]

———. 1978a. Physics of the primitive solar nebula and of giant gaseous protoplanets. In T. Gehrels, ed., *Protostars and Planets*, Tucson, Univ. Arizona Press, 453–487. [Chapter 4.]

———. 1978b. Physics of the primitive solar accretion disk. *Moon and Planets 18*:5–40. [Chapters 2, 4.]

Cameron, A. G. W. and Truran, J. W. 1977. The supernova trigger for the formation of the solar system. *Icarus 30*:447–461. [Chapter 2.]

Cameron, E. M. and Baumann, A. 1972. Carbonate sedimentation during the Archean. *Chem. Geol. 10*:17–30. [Chapter 10.]

Cameron, E. M. and Garrels, R. M. 1980. Geochemical compositions of some Precambrian shales from the Canadian Shield. *Chem. Geol. 28*:181–197. [Chapters 5, 10.]

Canavan, F. and Edwards, A. B. 1938. The iron ores of Yanpi Sound, Western Australia. *Australas. Inst. Min. Metall. Proc. 110*:59–101. [Chapter 11.]

Cannon, R. T. 1965. Age of the transition in the Precambrian atmosphere. *Nature 205*:586. [Chapter 11.]

Cao, R. and Zhao, W. 1978. Manicosiphoniaceae, a new family of fossil algae from the Sinian System of S. W. China with reference to its systematic position. *Acta Paleontologica Sinica 17* (1):29–42 (in Chinese with English abstract). [Chapter 1.]

Carmichael, I. S. E. and Nicholls, J. 1967. Iron-titanium oxides and oxygen fugacities in volcanic rocks. *J. Geophys. Res. 72*:4665–4687. [Chapter 4.]

Carr, N. G. and Hallaway, M. 1965. The presence of α-tocopherolquinone in blue-green algae. *Biochem. J. 97*:9c–10c. [Chapter 15.]

Carver, J. H. 1974. The origin of atmospheric oxygen. *Search 5*:130–135. [Chapter 11.]

Cayeux, M. L. 1894a. Les preuves de l'existence d'organismes dans le terrains précambrien: Première note sur les radiolaires précambriens. *Bull. Soc. Géol. France 22*:197–228. [Chapter 1.]

———. 1894b. Sur la présence des restes de foraminiferes dans les terrains précambriens de Bretagne. *Annals Soc. Géol. du Nord 22*:116–119. [Chapter 1.]

———. 1895. De l'existence de nombreux débris de spongiaires dans le Précambrien de Bretagne. *Annals Soc. Géol. du Nord 23*:52–65. [Chapter 1.]

Chambers, L. A. 1979. Sulphur isotope fractionation in intertidal sediments, Spencer Gulf, South Australia. *4th Internat. Symp. Envir. Biogeochem. Progr. Abstr.* (Stockholm):25 (Abstr.). [Chapter 7.]

Chambers, L. A. and Trudinger, P. A. 1979. Microbiological fractionation of stable sulfur isotopes: A review and critique. *Geomicrobiol. J. 1*:249–293. [Chapter 7.]

Chameides, W. L. and Walker, J. C. G. 1981. Rates of fixation by lightning of carbon and nitrogen in possible primitive atmospheres. *Origins of Life 11*:291–302. [Introduction; Chapters 4, 11.]

Chandler, F. W. 1980. Proterozoic redbed sequences of Canada. *Geol. Surv. Can. Bull. 311*:1–53. [Chapter 1.]

Chang, S. 1979. Comets: Cosmic connections with carbonaceous meteorites, interstellar molecules and the origin of life. In M. Neugebauer, D. K. Yeomans, and J. C. Brandt, eds., *Space Missions to Comets*, Washington, D.C., NASA Conf. Pub. 2089, 59–111. [Chapter 4.]

Chang, S.; Mack, R.; and Lennon, K. 1978. Carbon chemistry of separated phases of Murchison and Allende meteorites. In *Lunar and Planetary Science, Vol. 9*, Houston, Lunar Planet. Sci. Inst., 157–159. [Appendix III.]

Chang, S.; Scattergood, T.; Aronowitz, S.; and Flores, J. 1979. Organic chemistry on Titan. *Rev. Geophys. Space Phys. 17*:1923–1933. [Chapter 4.]

Chapman, D. J. and Ragan, M. A. 1980. Evolution of biochemical pathways: Evidence from comparative biochemistry. *Ann. Rev. Pl. Physiol. 31*:639–678. [Chapter 13; Glossary II.]

Charlot, R. 1976. The Precambrian of the Anti-Atlas (Morocco): A geochronological synthesis. *Precambrian Res. 3*:273–299. [Chapter 11.]

Chatton, E. 1938. *Titres et travaux scientifiques de Edouard Chatton 1906–1937*, Sete, Paris, Imprimerie E. Sottano. [Chapter 1.]

Chauvel, J. J. and Schopf, J. W. 1978. Late Precambrian microfossils from Brioverian cherts and limestones of Brittany and Normandy, France. *Nature 275*:640–642. [Chapter 1.]

Chen, J.; Zhang, H.; Wang, C.; Lu, S.; Li, Q.; and Li, B. 1979. *The Sinian Suberathem of China*, Tianjin, Chinese Acad. Geol. Sci. and Tianjin Inst. Geol. Min. Res., 32 pp. [Chapter 11.]

Chen, J.; Zhang, H.; Zhu, S.; Zhao, Z.; and Wang, Z. 1980. Research on Sinian Suberathem of Jixian, Tianjin. In Y. Wang, ed., *Research on Precambrian Geology, Sinian Suberathem in China*, Tianjin, Tianjin Sci. Tech. Press, 56–114 (in Chinese with English abstract). [Chapter 14.]

Cheng, H. H.; Bremner, J. M.; and Edwards, A. P. 1964. Variations of nitrogen-15 abundance in soils. *Science 146*:1574–1575. [Chapter 7.]

Cheshire, P. 1978. Archaean volcano-sedimentary basins in the Bulawayan Group in the Midlands area. *Annals Rhodesia Geol. Surv. 4*:46–54. [Chapter 8.]

Chown, E. V. and Caty, J. L. 1973. Stratigraphy, petrography and paleocurrent analysis of the Aphebian clastic formations of the Mistassini-Otish Basins. *Geol. Assoc. Can. Spec. Pap. 12*:49–79. [Chapter 11.]

Chumakov, N. M. 1978. Precambrian tillites and tilloids. *Akad. Nauk. SSSR Geol. Inst. Trudy 308*, 202 pp. (in Russian). [Chapter 11.]

Chung, H. M. and Sackett, W. M. 1979. Use of stable carbon isotope compositions of pyrolytically derived methane as maturity indices for carbonaceous materials. *Geochim. Cosmochim. Acta 43*:1979–1988. [Chapter 5.]

Clark, S. P. and Ringwood, A. E. 1964. Density distribution and constitution of the mantle. *Rev. Geophys. 2*:35–88. [Chapter 3.]

Claus, G. and Nagy, B. 1961. A microbiological examination of some carbonaceous chondrites. *Nature 192*:594–596. [Glossary I.]

Clay, A. N. 1978. An Archaean limestone and volcanic association from a Rhodesian greenstone belt. Evidence of volcanogenic control in a sedimentary system. *Annals Rhodesia Geol. Surv. 4*:55–63. [Chapter 8.]

Claypool, G. A.; Holser, W. T.; Kaplan, I. R.; Sakai, H.; and Zak, I. 1980. The age curves of sulfur and oxygen isotopes in marine sulfate and their mutual interpretation. *Chem. Geol. 28*:199–260. [Chapters 7, 10.]

Clayton, R. N. 1963. Carbon isotope abundance in meteoritic carbonates. *Science 140*:192–193. [Appendix III.]

Clemmey, H. 1978. A Proterozoic lacustrine interlude from the Zambian Copperbelt. *Internat. Assoc. Sedimentol. Spec. Pub. 2*:259–278. [Chapter 11.]

Cline, J. D. and Kaplan, I. R. 1975. Isotopic fractionation of dissolved nitrate during denitrification in the Eastern Tropical North Pacific Ocean. *Marine Chem. 3*:271–299. [Chapter 7.]

Cloud, P. 1942. Notes on stromatolites. *Am. J. Sci. 240*:363–379. [Appendix I.]

———. 1948. Some problems and patterns of evolution exemplified by fossil invertebrates. *Evol. 2*:322–350. [Appendix I.]

———. 1965. Significance of the Gunflint (Precambrian) microflora. *Science 148*:27–45. [Chapters 1, 9, 14.]

———. 1968a. Atmospheric and hydrospheric evolution on the primitive earth. *Science 160*: 729–736. [Chapters 1, 11.]

———. 1968b. Pre-Metazoan evolution and the origins of the Metazoa. In E. T. Drake, ed., *Evolution and Environment*, New Haven, Conn., Yale Univ. Press, 1–72. [Chapters 1, 9, 14.]

———. 1972. A working model of the primitive earth. *Am. J. Sci. 272*:537–548. [Chapters 1, 11.]

———. 1973a. Paleoecological significance of the banded iron-formation. *Econ. Geol. 68*:1135–1143. [Chapters 1, 10, 11.]

———. 1973b. Pseudofossils: A plea for caution. *Geology 1*:123–127. [Chapter 1.]

———. 1973c. Some early microbiotas and their bearing on the evolution of the primitive earth. In M. J. Neishtadt, ed., *Problems of Palynology, Proc. 3rd Internat. Palynol. Congr.*, (Novosibirsk, U.S.S.R.), Moscow, Nauka, 91–94. [Chapter 1.]

———. 1974. Evolution of ecosystems. *Am. Scient. 62*:54–66. [Chapters 1, 11.]

———. 1976a. Beginnings of biospheric evolution and their biogeochemical consequences. *Paleobiol. 2*:351–387. [Chapters 1, 9, 11, 15.]

———. 1976b. Major features of crustal evolution. *Geol. Soc. S. Afr. Annexure 79*:1–33. [Chapters 1, 3, 10.]

———. 1980. Early biogeochemical systems. In P. A. Trudinger, M. R. Walter, and B. J. Ralph, eds., *Biogeochemistry of Ancient and Modern Environments*, Canberra, Austr. Acad. Sci., 7–21. [Chapters 1, 15.]

Cloud, P. and Abelson, P. H. 1961. Woodring conference on major biologic innovations and the geologic record. *Proc. Natl. Acad. Sci. USA 47*:1705–1712. [Appendix I.]

Cloud, P. and Hagen, H. 1965. Electron microscopy of the Gunflint microflora: Preliminary results. *Proc. Natl. Acad. Sci. USA 54*:1–8. [Chapters 1, 14.]

Cloud, P. and Licari, G. R. 1968. Microbiotas of the banded iron formations. *Proc. Natl. Acad. Sci. USA 61*:779–786. [Chapters 9, 14.]

———. 1972. Ultrastructure and geologic relations of some two-aeon old nostocacean algae from northeastern Minnesota. *Am. J. Sci. 272*:138–149. [Chapter 14.]

Cloud, P. and Morrison, K. 1979a. On microbial contaminants, micropseudofossils, and the oldest records of life. *Precambrian Res. 9*:81–91. [Chapters 1, 9, 14.]

———. 1979b. New microbiotas from the Animikie and Baraga Groups (~2 Gyr old), north Michigan. *Geol. Soc. Am. Abstr. Progr. 11* (5): 227 (Abstr.). [Chapter 14.]

———. 1980. New microbial fossils from 2 Gyr old rocks in northern Michigan. *Geomicrobiol. J. 2*:161–178. [Chapter 14.]

Cloud, P. and Semikhatov, M. A. 1969. Proterozoic stromatolite zonation. *Am. J. Sci. 267*:1017–1061. [Chapter 8.]

Cloud, P.; Gruner, J. W.; and Hagen, H. 1965. Carbonaceous rocks of the Soudan Iron Formation (Early Precambrian). *Science 148*:1713–1716. [Chapters 1, 9.]

Cloud, P.; Gustafson, L. B.; and Watson, J. A. L. 1980. The works of living social insects as pseudofossils and the age of the oldest known Metazoa. *Science 210*:1013–1015. [Chapter 1.]

Cloud, P.; Licari, G. R.; Wright, L. A.; and Troxel, B. W. 1969. Proterozoic eucaryotes from eastern California. *Proc. Natl. Acad. Sci. USA 62*:623–631. [Chapter 1.]

Cloud, P.; Moorman, M.; and Pierce, D. 1975. Sporulation and ultrastructure in a late Proterozoic cyanophyte: Some implications for taxonomy and plant phylogeny. *Quart. Rev. Biol. 50*:131–150. [Chapter 1.]

Cochrane, G. W. and Edwards, A. B. 1960. The Roper River oolitic ironstone formations. *Austr. Commonwealth Sci. Industr. Res. Org. Min. Inv. Tech. Pap. 1*, 28 pp. [Chapter 11.]

Cohen, S. S. 1963. On biochemical varibility and innovation. *Science 139*:1017–1026. [Chapter 6.]

Cohen, Y. and Gordon, L. I. 1979. Nitrous oxide production in the ocean. *J. Geophys. Res. 84*:347–353. [Chapter 7.]

Cohen, Y.; Padan, E.; and Shilo, M. 1975. Facultative anoxygenic photosynthesis in the cyanobacterium *Oscillatoria limnetica*. *J. Bacteriol. 123*:855–861. [Chapter 7.]

Colby, J., Dalton, H. and Whittenbury, R. 1979. Biological and biochemical aspects of microbial growth on C_1 compounds. *Ann. Rev. Microbiol. 33*:481–517. [Chapter 7.]

Cole, J. A. 1976. Microbial gas metabolism. *Adv. Microbial Physiol. 14*:1–92. [Chapter 13; Glossary II.]

Cole, J. A. and Brown, C. M. 1980. Nitrite reduction to ammonia by fermentative bacteria: A short circuit in the biological nitrogen cycle. *FEMS Microbiol. Lett. 7*:65–72. [Chapter 6.]

Coleman, D. D.; Risatti, J. B.; and Schoell, M. 1981. Fractionation of carbon and hydrogen isotopes by methane-oxidizing bacteria. *Geochim. Cosmochim. Acta 45*:1033–1038. [Chapter 12.]

Colombo, U.; Gazzarini, F.; Gonfiantini, R.; Kneuper, G.; Teichmüller, M.; and Tecihmüller, R. 1970. Das $^{12}C/^{13}C$-verhältnis von Kohlen und kohlenbürtigem Methan. *Compte Rend. 6e Congr. Internat. Strat. Geol. Carbonif.* (Paris), *2*:557–574. [Chapter 7.]

Compston, W. and Arriens, P. A. 1968. The Precambrian geochronology of Australia. *Can. J. Earth Sci. 5*:561–583. [Chapter 11; Appendix I.]

Compston, W.; Trendall, A. F.; Williams, I. S.; Arriens, P. A.; and Foster, J. 1981. A revised age for the Hamersley Group. Unpublished manuscript. [Chapter 11.]

Condie, K. C. 1973. Archean magmatism and crustal thickening. *Geol. Soc. Am. Bull. 84*: 2981–2992. [Chapter 10.]

Condie, K. C. 1980. Origin and early development of the Earth's crust. *Precambrian Res. 11*: 183–197. [Chapter 10.]

Condie, K. C.; Macke, J. E.; and Reimer, T. O. 1970. Petrology and geochemistry of Early Precambrian graywackes from the Fig Tree Group, South Africa. *Geol. Soc. Am. Bull. 81*:2759–2776. [Chapter 10.]

Connan, J. 1974. Time-temperature relation in oil genesis. *Am. Assoc. Petrol. Geol. Bull. 58*: 2516–2521. [Chapter 5.]

Cook, F. A.; Albaugh, D. S.; Brown, L. D.; Kaufman, S.; Oliver, J. E.; and Hatcher, J. D., Jr. 1979. Thin-skinned tectonics in the crystalline southern Appalachians; COCORP seismic-reflection profiling of the Blue Ridge and Piedmont. *Geology 7*:563–567. [Chapter 3.]

Cook, F. A. and Turcotte, D. L. 1981. Parameterized convection and the thermal evolution of the Earth. *Tectonophys. 75*:1–17. [Chapter 3.]

Cooper, T. G.; Tchen, T. T.; Wood, H. G.; and Benedict, C. R. 1968. The carboxylation of phosphoenolpyruvate and pyruvate. I. The active species of "CO_2" utilized by phosphoenolpyruvate carboxykinase, carboxytransphosphorylase, and pyruvate carboxylase. *J. Biol. Chem. 243*:3857–3863. [Chapter 6.]

Cornell, D. H. 1978. Petrologic studies at T'Kuip: Evidence for metamorphism and metasomatic alteration of volcanic formations beneath the Transvaal volcanosedimentary pile. *Geol. Soc. S. Afr. Trans. 81*:261–270. [Chapter 11.]

Cowie, J. W. 1967. Life in Pre-Cambrian and early Cambrian time. In *The Fossil Record, a Symposium with Documentation*, London, Geol. Soc. London, 17–35. [Chapter 1.]

———. 1978a. Symposium on the Precambrian-Cambrian boundary. *Geol. Mag. 115*:81–82. [Chapter 1.]

———. 1978b. U.U.G.S./I.G.C.P. Project 29 Precambrian-Cambrian boundary working group in Cambridge, 1978. *Geol. Mag. 115*:151–152. [Chapter 1.]

Cox, D. P. 1967. Regional environment of the Jacobina auriferous conglomerate, Brazil. *Econ. Geol. 62*:773–780. [Chapter 11.]

Cox, J. C.; Madigan, M. T.; Favinger, J. L.; and Gest, H. 1980. Redox mechanisms in "oxidant-dependent" hexose fermentation by *Rhodopseudomonas capsulata. Arch. Biochem. Biophys. 204*:10–17. [Chapter 6.]

Craig, H. 1953. The geochemistry of the stable carbon isotopes. *Geochim. Cosmochim. Acta 3*: 53–92. [Chapters 4, 5.]

———. 1957. Isotopic standards for carbon and oxygen and correction fractors for mass spectrometric analysis of carbon dioxide. *Geochim. Cosmochim. Acta 12*:133–149. [Chapter 4.]

———. 1961. Isotopic variations in meteoric waters. *Science 133*:1702–1703. [Chapter 7.]

Craig, H. and Lupton, J. E. 1976. Primordial neon, helium and hydrogen in oceanic basalts. *Earth Planet. Sci. Lett. 31*:369–385. [Chapter 4.]

Craig, H.; Welhan, J. A.; Kim, K.; Poreda, R.; and Lupton, J. E. 1980. Geochemical studies of of the 21°N EPR hydrothermal fluids. *Am. Geophys. Union. Trans. 61*:992 (Abstr.). [Chapter 4.]

Crick, F. H. C. 1968. The origin of the genetic code. *J. Mol. Biol. 38*:367–379. [Chapter 4.]

Crick, I. H. and Muir, M. D. 1979. Evaporites and uranium mineralization in the Pine Creek Geosyncline. In *International Uranium Symposium on the Pine Creek Geosyncline, Australia, 1979, Proc. Internat. Atomic Energy Agency* (Vienna), 30–33. [Chapter 11.]

Crick, I. H. and Muir, M. D. 1980. Evaporites and uranium mineralization in the Pine Creek Geosyncline. In J. Ferguson and A. B. Goleby, eds., *Uranium in the Pine Creek Geosyncline*, Vienna, Internat. Atomic Energy Agency, 30–33. [Chapter 7.]

Crittenden, M. D. and Peterman, Z. E. 1975. Provisional Rb/Sr age of the Precambrian Uinta Mountain Group, northeastern Utah. *Utah Geol. 2*:75–77. [Appendix I.]

Crittenden, M. D., Jr.; Stewart, J. H.; and Wallace, C. A. 1972. Regional correlation of Upper Precambrian strata in western North America. *Proc. 24*th *Internat. Geol. Congr.* (Montreal, Can), *Sec. 1*:334–341. [Chapter 11.]

Crockett, R. N. and Jones, M. T. 1975. Some aspects of the geology of the Waterberg System in eastern Botswana. *Geol. Soc. S. Afr. Trans. 78*:1–10. [Chapter 11.]

Crowell, J. C. and Frakes, L. A. 1970. Phanerozoic glaciations and the causes of ice ages. *Am. J. Sci.*:193–224. [Chapter 3.]

Croxford, N. J. W.; Janecek, J.; Muir, M. D.; and Plumb, K. A. 1973. Microorganisms of Carpentarian (Precambrian) age from the Amelia

Dolomite, McArthur Group, Northern Territory, Australia. *Nature 245*:28–30. [Chapter 14.]

Crutzen, P. J. 1976. Upper limits on atmospheric ozone reductions following increased application of fixed nitrogen to the soil. *Geophys. Res. Lett. 3*:169–172. [Chapter 7.]

Dansgaard, W. 1964. Stable isotopes in precipitation. *Tellus 16*:436–468. [Chapter 7.]

Dansgaard, W.; Johnsen, S. J.; Moller, J.; and Langway, C. C. 1969. One thousand centuries of climatic record from Camp Century on the Greenland ice sheet. *Science 166*:377–381. [Chapter 7.]

Darby, D. G. 1972. Evidences of Precambrian life in Minnesota. In P. K. Sims and G. B. Morey, eds., *Geology of Minnesota: A Centennial Volume*, St. Paul, Minn., Minnesota Geol. Surv., 264–271. [Chapter 14.]

———. 1974. Reproductive modes of *Huroniospora microreticulata* from cherts of the Precambrian Gunflint Formation. *Geol. Soc. Am. Bull. 85*:1595–1596. [Chapter 14.]

Davidson, C. F. 1957. On the occurrence of uranium in ancient conglomerates. *Econ. Geol. 52*:668–693. [Chapter 11.]

———. 1965. Geochemical aspects of atmospheric evolution. *Proc. Natl. Acad. Sci. USA 53*:1194–1204. [Chapter 11.]

Davidson, C. F. and Cosgrove, M. E. 1955. On the impersistence of uraninite as a detrital mineral. *Geol. Surv. Great Brit. Bull. 10*:74–80. [Chapter 11.]

Davies, F. G. 1979. Thickness and thermal history of continental crust and root zones. *Earth Planet Sci. Lett. 44*:231–238. [Chapter 10.]

———. 1980. Thermal histories of convective Earth models and constraints on radiogenic heat production in the Earth. *J. Geophys. Res. 85*:2517–2530. [Chapters 4, 10.]

Davies, G. R. 1970. Algal-laminated sediments, Gladstone Embayment, Shark Bay, Western Australia. *Am. Assoc. Petrol. Geol. Mem. 13*:169–205. [Chapter 8.]

Davies, O. L., ed. 1967. *Statistical Methods in Research and Production*, New York, Hafner, 396 pp. [Appendix IV.]

Davies, R. D.; Allsopp, H. L.; Erlank, A. J.; and Manton, W. I. 1970. Sr-isotopic studies on various layered mafic intrusions in southern Africa. *Geol. Soc. S. Afr. Spec. Pub. 1*:576–593. [Chapter 11.]

Day, R. A., Jr. and Underwood, A. L. 1958. *Quantitative Analysis*, Englewood-Cliffs, N.J., Prentice-Hall, 465 pp. [Appendix III.]

Day, W. 1979. *Genesis on Planet Earth: The Search for Life's Beginnings*, East Lansing, Mich., House of Talos, 408 pp. [Chapter 4.]

Deamer, D. W. and Oró, J. 1980. Role of lipids in prebiotic structures. *Biosystems 12*:167–175. [Chapter 4.]

Debrenne, F. and La Fuste, J. G. 1979. *Buschmannia roeringi* (Kaever and Richter, 1976) a so-called archaeocyatha, and the problem of the Precambrian or Cambrian age of the Nama System (S. W. Africa). *Geol. Mag. 116*:143–144. [Chapters 8, 11.]

DeCampli, W. M. and Cameron, A. G. W. 1979. Structure and evolution of isolated giant gaseous protoplanets. *Icarus 38*:367–391. [Chapter 4.]

Decker, K.; Jungermann, K.; and Thauer, R. K. 1970. Energy production in anaerobic organisms. *Angew. Chem. Internat. Edit. 9*:138–158. [Glossary II.]

Deflandre, G. 1949. Les soi-disant radiolaires du précambrien de Bretagne et la question de l'existence de radiolaires embryonnaires fossiles. *Soc. Zool. France Bull. 74*:351–352. [Chapter 1.]

———. 1968. Sur l'existence, des le Précambrien, d'Acritarches du type Acanthomorphitae: *Eomicrhystridium* nov. gen. Typification du genre *Palaeocryptidium* Defl. 1955. *Compte Rend. Acad. Sci. Paris 266 Ser. D 26*:2385–2389. [Chapter 14.]

Degens, E. T. 1969. Biogeochemistry of stable carbon isotopes. In G. Eglinton and M. T. J. Murphy, eds., *Organic Geochemistry*, New York, Springer, 304–328. [Chapters 4, 5, 7.]

Degens, E. T. and Ross, D. A., eds. 1969. *Hot Brines and Recent Heavy Metal Deposits in the Red Sea*, New York, Springer, 600 pp. [Chapter. 10.]

Deines, P. 1980a. The isotopic composition of reduced organic carbon. In P. Friz and J. Ch. Fontes, eds., *Handbook of Environmental Isotope Geochemistry*, Amsterdam, Elsevier, 329–406. [Chapters 5, 12.]

———. 1980b. The carbon isotopic composition of diamonds: Relationship to diamond shape, color, occurrence and vapor composition. *Geochim. Cosmochim. Acta 44*:943–961. [Chapter 4.]

de Laeter, J. R. and Blockley, J. G. 1972. Granite ages within the Pilbara Block, Western Australia. *J. Geol. Soc. Austr. 19*:363–370. [Chapter 9.]

Delsemme, A. H. 1975. The volatile fraction of the cometary nucleus. *Icarus 24*:95–110. [Chapter 4.]

———. 1977. The pristine nature of comets. In A. H. Delsemme, ed., *Comets-Asteroids-Meteorites*, Toledo, Ohio, Univ. of Toledo, 3–12. [Chapter 4.]

Delwiche, C. C. 1970. The nitrogen cycle. *Scient. Am. 223*:136–147. [Chapter 7.]

Delwiche, C. C. and Steyn, P. L. 1970. Nitrogen isotope fractionation in soils and microbial reactions. *Environ. Sci. Tech. 4*:929–935. [Chapter 7.]

DeNiro, M. J. and Epstein, S. 1977. Mechanism of carbon isotope fractionation associated with lipid synthesis. *Science 197*:261–263. [Chapter 7.]

DeNiro, M. J.; Epstein, S.; Kenealy, W. R.; Belayev, S. S.; Wolkin, R.; and Zeikus, G. J. 1981. Isotopic fractionation during carbon dioxide reduction by three methanogens. Unpublished manuscript. [Chapter 12.]

Denison, R. E. 1980. Pre-Bliss (Precambrian) rocks in the Van Horn region, Trans-Pecos Taxas. In P. W. Dickerson and J. M. Hoffer, eds., *New Mexico Geological Society, 31*st *Field Conf., Trans-Pecos Region*, Socorro, N. M., New Mexico Geol. Soc., 155–158. [Appendix I.]

DePaolo, D. J. 1979. Implications of correlated Nd and Sr isotopic variations for the chemical evolution of the crust and mantle. *Earth Planet. Sci. Lett. 43*:201–211. [Chapters 4, 10.]

———. 1980a. The sources of continental crust: Nd isotope evidence from the Sierra Nevada and Peninsular Ranges. *Science 209*:684–687. [Chapter 10.]

———. 1980b. Crustal growth and mantle evolution: Inferences from models of element transport and Nd-Sr isotopes. *Geochim. Cosmochim. Acta 44*:1185–1196. [Chapter 10.]

———. 1981. Nd isotopic studies: Some new perspectives on Earth structure and evolution. *Am. Geophys. Union Trans. 62*:137–140. [Chapter 3.]

Dermott, S. F., ed. 1978. *The Origin of the Solar System*, New York, Wiley, 668 pp. [Chapter 2.]

Derrick, G. M. and Sweet, I. P. 1978. Mount Isa-Lawn Hill project. *Bur. Min. Resour. Geol. Geophys. Austr. Rec. 1978/89*:119–127. [Appendix I.]

Desikachary, T. V. 1959. *Cyanophyta*, New York, Academic Press and New Delhi, Indian Coun. Agricul. Res., 686 pp. [Chapter 14.]

DesMarais, D. J. 1978. Variable temperature cryogenic trap for the separation of gas mixtures. *Analyt. Chem. 50*:1405–1406. [Appendix III.]

DesMarais, D. J. and Hayes, J. M. 1976. A tube-cracker for opening glass-sealed ampoules under vacuum. *Analyt. Chem. 48*:1651–1652. [Appendix IV.]

DesMarais, D. J.; Donchin, J. H.; Nehring, N. L.; and Truesdell, A. H. 1980. Carbon isotopes in individual hydrocarbon gases from geothermal systems. *Geol. Soc. Am. Abstrs. Progr.*, Atlanta, Georgia (Abstr.). [Appendix III.]

DesMarais; D. J.; Donchin, J. H.; Nehring, N. L. and Truesdell, A. H. 1981. Molecular carbon isotopic evidence for the origin of geothermal hydrocarbons. *Nature 292*:826–828. [Appendix III.]

DeVooys, C. G. N. 1979. Primary production in aquatic environments. In B. Bolin, E. T. Degens, S. Kempe, and P. Ketner, eds., *The Global Carbon Cycle*, Chichester, Wiley, 259–292. [Chapter 5.]

Dewey, J. F. and Bird, J. M. 1971. Origin and emplacement of the ophiolote suite: Appalachian ophiolites in Newfoundland. *J. Geophys. Res. 76*:3179–3206. [Chapter 3.]

Dickerson, R. E. 1978. Chemical evolution and the origin of life. *Scient. Am. 239*:70–86. [Chapter 1.]

———. 1980. Cytochrome *c* and the evolution of energy metabolism. *Scient. Am. 242*:137–153. [Chapter 13.]

———. 1981. The cytochromes *c*: An exercise in scientific serendipity. In D. S. Sigma and M. A. B. Brizier, eds., *Evolution of Protein Structure and Function*, New York, Academic Press, in press. [Chapter 13.]

Diessel, C. F. K. and Offler, R. 1975. Changes in physical properties of coalified and graphitised phytoclasts with grade of metamorphism. *Neues Jahrb. Miner. Mh. 1975*:11–26. [Chapter 5.]

Diessel, C. F. K.; Brothers, R. N.; and Black, P. M. 1978. Coalification and graphitization in high-pressure schists in New Caledonia. *Contrib. Mineral. Petrol. 68*:63–78. [Chapter 5.]

Dimroth, E. 1968. The evolution of the central segment of the Labrador geosyncline, Part 1: Stratigraphy, facies and paleogeography. *Neues*

Jahrb. Geol. Palaeont. Abh. 132:22–54. [Chapter 1.]

———. 1970. Evolution of the Labrador geosyncline. *Geol. Soc. Am. Bull. 81*:2717–2742. [Chapter 1.]

———. 1977. Facies models 6. Diagenetic facies of iron-formation. *Geosci. Can. 4*:83–88. [Chapter 11.]

Dimroth, E. and Kimberley, M. M. 1976. Precambrian atmospheric oxygen: Evidence in the sedimentary distributions of carbon, sulfur, uranium, and iron. *Can. J. Earth Sci. 13*:1161–1185. [Chapter 11.]

Dimroth, E.; Baragar, W. R. A.; Bergeron, R.; and Jackson, G. D. 1970. The filling of the Circum-Ungava Geosyncline. In A. J. Baer, ed., *Basins and Geosynclines of the Canadian Shield, Geol. Surv. Can. Dept. Energy, Mines and Resources, Pap. 70–40*:45–142. [Chapter 11; Appendix I.]

Dimroth, E.; Donaldson, J. A.; and Veizer, J., eds. 1980. Early Precambrian volcanology and sedimentology in the light of the recent. *Precambrian Res. 12*:1–470. [Chapter 10.]

Diver, W. L. 1974. Precambrian microfossils of Carpentarian age from Bungle Bungle Dolomite of Western Australia. *Nature 247*:361–363. [Chapter 14.]

Dodge, F. C. W.; Fleck, R. J.; Hadley, D. G.; and Millard, H. T., Jr. 1978. Geochemistry and $^{87}Sr/^{86}Sr$ ratios of Halaban rocks of the Central Arabian Shield. *Precambrian Res. 6*:A13 (Abstr.). [Chapter 11.]

Doemel, W. N. and Brock, T. D. 1974. Bacterial stromatolites: Origin of laminations. *Science 184*:1083–1085. [Chapter 8.]

———. 1977. Structure, growth, and decomposition of laminated algal-bacterial mats in alkaline hot springs. *Appl. Environ. Microbiol. 34*:433–452. [Chapter 8.]

Dole, M.; Lane, G. A.; Rudd, D. P.; and Zaukelies, D. A. 1954. Isotopic composition of atmospheric oxygen and nitrogen. *Geochim. Cosmochim. Acta 6*:65–78. [Chapter 7.]

Donaldson, J. A. 1976. Paleoecology of *Conophyton* and associated stromatolites in the Precambrian Dismal Lake and Rae Groups, Canada. In M. R. Walter, ed., *Stromatolites*, Amsterdam, Elsevier, 523–534. [Chapter 11.]

Donaldson, J. A. and Platt, J. P. 1975. Structure and sedimentology of Archean metasediments near Lawlers, Yilgarn Block, Western Australia. *Geol. Soc. Austr. 1st Conf.* (Adelaide) *Abstr. 106* (Abstr.). [Chapter 11.]

Donnelly, T. H. and Ferguson, J. 1980. A stable isotope study of three deposits in the Alligator Rivers uranium field, N. T. In J. Ferguson and A. B. Goleby, eds., *Uranium in the Pine Creek Geosyncline*, Vienna, Internat. Atomic Energy Agency, 397–407. [Chapter 7.]

Donnelly, T. H.; Lambert, I. B.; Oehler, D. Z.; Hallberg, J. A.; Hudson, D. R.; Smith, J. W.; Bavinton, O. A.; and Golding, L. 1977. A reconnaissance study of stable isotope ratios in Archaean rocks from the Yilgarn block, Western Australia. *J. Geol. Soc. Austr. 24*:409–420. [Chapters 5, 7.]

Dorr, J. V. N., Jr. 1969. Physiographic, stratigraphic and structural development of the Quadrilatero Ferrifero Minas Gerais, Brazil. *U.S. Geol. Surv. Prof. Pap. 641-A*:A1–A110. [Chapter 11.]

———. 1973. Iron-formation in South America. *Econ. Geol. 68*:1005–1022. [Chapter 11.]

Dort, W. and Dort, D. S. 1970. Low temperature origin of sodium sulfate deposits, particularly in Antarctica. In J. L. Rau and L. R. Dellwig, eds., *Third Symposium on Salt*, Cleveland, Ohio, Northern Ohio Geol. Soc., 181–203. [Chapter 11.]

Drever, J. J. 1974. Geochemical model for the origin of Precambrian banded iron formations. *Geol. Soc. Am. Bull. 85*:1099–1106. [Chapters 10, 11.]

Dungworth, G. and Schwartz, A. W. 1972. Kerogen isolates from the Precambrian of South Africa and Australia: Analysis for carbonised micro-organisms and pyrolysis gas liquid chromatography. In H. R. van Gaertner and H. Wehner, eds., *Advances in Organic Geochemistry 1971*, New York, Pergamon, 699–706. [Chapters 5, 9.]

———. 1974. Organic matter and trace elements in Precambrian rocks from South Africa. *Chem. Geol. 14*:167–172. [Chapter 5.]

Dunham, R. J. 1969. Vadose pisolite in the Capitan Reef (Permian), New Mexico and Texas. In G. M. Friedman, ed., *Depositional Environments in Carbonate Rocks, Soc. Econ. Paleontol. Mineral. Spec. Pub. 14*:182–191. [Chapter 8.]

Dunlop, D. J. 1981. Paleomagnetic evidence for Proterozoic continental development. *Phil. Trans. Roy. Soc. London Sect. A 301*:265–277. [Chapter 11.]

Dunlop, J. S. R. 1978. Shallow water sedimentation at North Pole, Pilbara, Western Australia. In J. E. Glover and D. I. Groves, eds., *Archaean Cherty Metasediments: Their Sedimentology, Micropalaeontology, Biochemistry, and Significance to Mineralization* Perth, *Univ. W. Austr., Spec. Pub. 2*:30–38. [Chapters 7, 8, 10, 11.]

Dunlop, J. S. R. and Buick, R. 1980. The sedimentary environment of an Archaean chert-barite deposit, North Pole, Pilbara Block, Western Australia. In J. E. Glover and D. I. Groves, eds., *2nd Internat. Archaean Symposium* (Perth), Canberra, Austr. Acad. Sci., 20–21 (Abstr.). [Chapters 8, 11.]

Dunlop, J. S. R. and Groves, D. I. 1978. Sedimentary barite of the Barberton Mountain Land: A brief review. In J. E. Glover and D. I. Groves, eds., *Archaean Cherty Metasediments: Their Sedimentology, Micropalaeontology, Biochemistry, and Significance to Mineralization*, Perth, *Univ. W. Austr. Spec. Pub. 2*:39–44. [Chapter 11.]

Dunlop, J. S. R.; Muir, M. D.; Milne, V. A.; and Groves, D. I. 1978. A new microfossil assemblage from the Archaean of Western Australia. *Nature 274*:676–678. [Chapter 8, 9.]

Dunn, P. R. 1964. Triact spicules in Proterozoic rocks of the Northern Territory of Australia. *J. Geol. Soc. Austr. 11*:195–197. [Chapter 14.]

Durand, B., ed., 1980. *Kerogen*, Paris, Editions Technip, 519 pp. [Chapter 5; Appendix IV.]

Durand, B. and Monin, J. C. 1980. Elemental analysis of kerogens (C, H, O, N, S, Fe). In B. Durand, ed., *Kerogen*, Paris, Editions Technip, 113–142. [Chapter 5.]

Durand, B. and Nicaise, G. 1980. Procedures for kerogen isolation. In B. Durand, ed., *Kerogen*, Paris, Editions Technip, 35–54. [Chapter 5.]

Edgell, H. S. 1964. Precambrian fossils from the Hamersley Range, Western Australia, and their use in stratigraphic correlation. *J. Geol. Soc. Austr. 11*:235–262. [Chapter 8.]

Edhorn, A. S. 1973. Further investigations of fossils from the Animikie, Thunder Bay, Ontario. *Geol. Assoc. Can. Proc. 25*:37–66. [Chapter 14.]

Edmond, J. M.; Measures, C.; McDuff, R. E.; Chan, L. H.; Collier, R.; Grant, B.; Gordon, L. I.; and Corliss, J. B. 1979a. Ridge crest hydrothermal activity and balance of major and minor elements in the ocean: The Galapagos data. *Earth Planet. Sci. Lett. 46*:1–18. [Chapter 10.]

Edmond, J. M.; Measures, C.; Magnam, B.; Grant, B.; Sclatter, F. R.; Collier, R.; Hudson, A.; Gordon, L. I.; and Corliss, J. B. 1979b. On the formation of metal-rich deposits at ridge crest. *Earth Planet Sci. Lett. 46*:19–30. [Chapter 10.]

Edwards, A. B. 1953. The Mount Oxide copper mine. In A. B. Edwards, ed., *Geology of Australian Ore Deposits*, Canberra, Austr. Acad. Sci., 391–395. [Appendix I.]

———. 1958. Oolitic iron-formation in northern Australia. *Geol. Rundsch. 47*:668–682. [Chapter 11.]

Edwards, A. P. 1973. Isotopic tracer techniques for identification of sources of nitrate pollution. *J. Environ. Qual. 2*:382–387. [Chapter 7.]

Egami, F. 1973. A comment to the concept on the role of nitrate fermentation and nitrate respiration in an evolutionary pathway of energy metabolism. *Z. Allg. Microbiol. 13*:177–181. [Chapter 13.]

———. 1975. Origin and early evolution of transition element enzymes. *J. Biochem. 77*:1165–1169. [Chapter 13.]

———. 1976. Comment on the position nitrate respiration in metabolic evolution. *Origins of Life 7*:71–72. [Chapter 13.]

Egami, F.; Ohmachi, K.; Iida, K.; and Taniguchi, S. 1957. Nitrate reducing systems in cotyledons and seedlings of bean seed embryos, *Vigna sesquipedalis*, during the germinating stage. *Biokhim. 22*:122–134. [Chapter 6.]

Eglinton, G. and Murphy, M. T. J., eds. 1969. *Organic Geochemistry*, New York, Springer, 828 pp. [Chapter 1.]

Egorov, E. W. and Timofeieva, M. W. 1973. Effusive iron-silica formations and iron deposits of the Maly Khingan. In *Genesis of Precambrian Iron and Manganese Deposits* (Proc. Kiev Symp. 1970), Paris, UNESCO, 181–186. [Chapter 11.]

Ehhalt, D. H. and Schmidt, U. 1978. Sources and sinks of atmospheric methane. *Pure and Appl. Geophys. 116*:452–464. [Chapter 5.]

Eichmann, R. and Schidlowski, M. 1975. Isotopic fractionation between coexisting organic carbon-carbonate pairs in Precambrian sediments. *Geochim. Cosmochim. Acta 39*:585–595. [Chapters 7, 8.]

Eigen, M. 1971. Self-organization of matter and the evolution of biological macromolecules. *Naturwiss. 58*:465–523. [Chapter 4.]

Elsasser, W. M. 1963. Early history of the Earth. In J. Geiss and E. Goldberg, eds., *Earth Science and Meteorites*, Amsterdam, North Holland, 1–30. [Chapters 2, 3.]

Embleton, B. J. J. 1978. The paleomagnetism of 2400 m.y. old rocks from the Australian Pilbara craton and its relation to Archean-Proterozoic tectonics. *Precambrian Res. 6*:275–291. [Chapter 3.]

Emery, K. O. and Rittenberg, S. C. 1952. Early diagenesis of California Basin sediments in relation to the origin of oil. *Am. Assoc. Petrol. Geol. Bull. 36*:735–806. [Chapter 5.]

Emery, K. O.; Orr, W. L.; and Rittenberg, S. C. 1955. Nutrient budget in the ocean. In K. O. Emery, ed., *Essays in the Natural Sciences in Honor of Captain A. Hancock*, Berkeley, Univ. Calif. Press, 299–310. [Chapter 7.]

Engel, A. E. J. and Kelm, D. L. 1972. Pre-Permian global tectonics: A tectonic test. *Geol. Soc. Am. Bull. 83*:2325–2340. [Chapter 3.]

Engel, A. E. J.; Itson, S. P.; Engel, C. G.; Stickney, D. M.; and Cray, E. J. 1974. Crustal evolution and global tectonics: A petrogenic view. *Geol. Soc. Am. Bull. 85*:843–858. [Chapter 10.]

Engel, A. E. J.; Nagy, B.; Nagy, L. A.; Engel, C. G.; Kremp, G. O. W.; and Drew, C. M. 1968. Algalike forms in Onverwacht Series, South Africa: Oldest recognized lifelike forms on earth. *Science 161*:1005–1008. [Chapter 9.]

Epstein, S.; Sharp, R. P.; and Gow, A. J. 1970. Antarctic ice sheets stable isotope analysis of Byrd Station cores and interhemispheric climatic implications. *Science 168*:1570–1572. [Chapter 7.]

Epstein, S.; Thompson, P.; and Yapp, B. J. 1977. Oxygen and hydrogen isotopic ratios in plant cellulose. *Science 198*:1209–1214. [Chapter 7.]

Epstein, S.; Yapp, C. J.; and Hall, J. H. 1976. The determination of the D/H ratio of non-exchangeable hydrogen in cellulose extracted from aquatic and land plants. *Earth Planet. Sci. Lett. 30*:241–251. [Chapter 7.]

Eriksson, K. A. 1977. Tidal deposits from the Archean Moodies Group, Barberton Mountain Land, South Africa. *Sediment. Geol. 18*:257–281. [Chapter 11.]

———. 1978. Alluvial and destructive beach facies from the Archean Moodies Group, Barberton Mountain Land, South Africa and Swaziland. *Can. Soc. Petrol. Geol. Mem. 5*:287–311. [Appendix I.]

———. 1979. Marginal marine depositional processes from the Archean Moodies Group, Barberton Mountain Land, South Africa: Evidence and significance. *Precambrian Res. 8*:153–182. [Chapter 11.]

———. 1980a. Platform to trough sedimentation in the East Pilbara Block, Australia. In J. E. Glover and D. I. Groves, eds., *2nd Internat. Archaean Symposium* (Perth), Canberra, Austr. Acad. Sci., 21–22 (Abstr.). [Chapter 11; Appendix I.]

———. 1980b. Hydrodynamic and paleogeographic interpretation of turbidite deposits from the Archean Fig Tree Group of the Barberton Mountain Land, South Africa. *Geol. Soc. Am. Bull. 91*:21–26. [Chapter 11.]

———. 1980c. Transitional sedimentation styles in the Moodies and Fig Tree Groups, Barberton Mountain Land, South Africa: Evidence favouring an Archean continental margin. *Precambrian Res. 12*:141–160. [Chapter 10.]

Eriksson, K. A. and Vos, R. G. 1979. A fluvial fan depositional model for Middle Proterozoic red beds from the Waterberg Group, South Africa. *Precambrian Res. 9*:169–188. [Chapter 11.]

Eriksson, K. A.; Truswell, J. F.; and Button, A. 1976. Palaeoenvironmental and geochemical models from an Early Proterozoic carbonate succession in South Africa. In M. R. Walter, ed., *Stromatolites*, Amsterdam, Elsevier, 635–643. [Chapter 8.]

Ernst, W. G. 1972. Occurrence and mineralogic evolution of blueschist belts with time. *Am. J. Sci. 272*:657–668. [Chapter 3.]

Espitalié, J.; Laporte, J. L.; Madec, M.; Marquis, F.; Leplat, P.; Poulet, J.; and Boutefeu, A. 1977. Méthode rapide de caractérisation des roches mères, de leur potential pétrolier et de leur degré d'évolution. *Rev. Inst. Fr. Pét. 32*:23–42. [Chapter 5.]

Esteban, M. 1976. Vadose pisolite and caliche. *Am. Assoc. Petrol. Geol. Bull. 60*:2048–2057. [Chapter 8.]

Estep, M. E. and Hoering, T. C. 1980. Biogeochemistry of the stable hydrogen isotopes. *Geochim. Cosmochim. Acta 44*:1197–1206. [Chapter 7.]

Eucken, A. 1944. Physikalisch-chemische Betrachtungen über der früeste Entwicklungs-gesichte der Erde. *Nach. Akad. Wiss. Göttingen, Math.-Phys. Kl* (1):1–25. [Chapter 4.]

Eugster, H. P. and Chou, I-Ming. 1973. The depositional environments of Precambrian banded iron-formations. *Econ. Geol. 68*:1144–1169. [Chapter 11.]

Evans, C. R.; Rogers, M. A.; and Bailey, N. J. L. 1971. Evolution and alteration of petroleum in Western Canada. *Chem. Geol. 8*:147–170. [Chapter 5.]

Ewers, W. E. 1981. Chemical conditions for the precipitation of banded iron-formations. In P. A Trudinger and M. R. Walter, eds., *Fourth International Symposium on Environmental Biogeochemistry*, New York, Springer, in press. [Chapter 11.]

Fairbairn, H. W.; Faure, G.; Pinson, W. H.; and Hurley, P. M. 1978. Rb-Sr whole-rock age of the Sudbury lopolith and basin sediments. *Can. J. Earth Sci. 5*:707–714. [Appendix I.]

Fairbairn, H. W.; Hurley, P. M.; Card, K. D.; and Knight, C. J. 1979. Correlation of radiometric ages of Nipissing diabase and Huronian metasediments with Proterozoic orogenic events in Ontario. *Can. J. Earth Sci. 6*:489–497. [Appendix I.]

Fairchild, T. R. and Dardenne, M. A. 1978. First report of well-preserved Precambrian microfossils in Brazil (Paraopeba Formation, Bambui Goup, near Brasilia). *Bol. IG. Inst. Geociências, Univ. São Paulo 9*:62–68. [Chapter 1.]

Fanale, F. 1971. A case for catastrophic early degassing of the Earth. *Chem. Geol. 8*:79–105. [Chapter 4.]

Farley, J. 1977. *The Spontaneous Generation Controversy from Descartes to Oparin*, Baltimore, Johns Hopkins Univ. Press, 225 pp. [Chapter 4.]

Farquhar, G. D. 1980. Carbon isotope discrimination by plants: Effects of carbon dioxide concentration and temperature via the ratio of intracellular and atmospheric CO_2 concentrations. In G. I. Pearman, ed., *Carbon Dioxide and Climate: Australian Research*, Canberra, Austr. Acad. Sci., 105–110. [Chapter 12.]

Faure, G. and Kovach, J. 1969. The age of the Gunflint Iron Formation of the Animikie Series in Ontario, Canada. *Geol. Soc. Am. Bull. 80*: 1725–1736. [Appendix I.]

Faure, G. and Powell, J. L. 1972. *Strontium Isotope Geology*, Heidelberg, Springer, 188 pp. [Chapter 10.]

Fay, P. 1965. Heterotrophy and nitrogen fixation in *Chlorogloea fritschii*. *J. Gen. Microbiol. 39*: 11–20. [Chapter 15.]

Feather, C. E. and Koen, G. M. 1975. The mineralogy of the Witwatersrand reef. *Minerals Sci. Engng. 7*:189–224. [Chapter 11.]

Fedonkin, M. A. 1977. Precambrian-Cambrian ichnocoenoses of the east European platform. In T. P. Crimes and J. C. Harper, eds., *Trace Fossils, Vol. 2*, Liverpool, Seel House Press, 183–194. [Chapter 1.]

Fee, J. A. and Valentene, J. S. 1977. Chemical and physical properties of superoxide. In A. M. Michelson, J. M. McCord, and I. Fridovich, eds., *Superoxide and Superoxide Dismutases*, London, Academic Press, 19–60. [Chapter 13.]

Fenchel, T. and Blackburn, T. H. 1979. *Bacteria and Mineral Cycling*, London, Academic Press, 225 pp. [Chapter 12.]

Fenton, C. L. and Fenton, M. A. 1931. Algae and algal beds in the Belt Series of Glacier National Park. *J. Geol. 39*:670–686. [Chapter 1.]

———. 1937. Belt Series of the north: Stratigraphy, sedimentation, paleontology. *Geol. Soc. Am. Bull. 48*:1873–1970. [Chapter 1.]

———. 1939. Pre-Cambrian and Paleozoic algae. *Geol. Soc. Am. Bull. 50*:89–126. [Chapter 1.]

Ferris, J. P. and Chen, C. T. 1975. Chemical evolution XXVI. Photochemistry of methane, nitrogen and water mixtures as a model for the atmosphere of the primitive Earth. *J. Am. Chem. Soc. 97*:2962–2967. [Chapter 4.]

Feux, A. N. 1977. The use of stable carbon isotopes in hydrocarbon exploration. *J. Geochem. Explor. 7*:155–188. [Chapter 4.]

Fieser, L. F. and Fieser, M. 1967. *Reagents for Organic Synthesis, Vol. 1*, New York, Wiley, 1457 pp. [Appendix IV.]

Fischer, A. G. 1965. Fossils, early life and atmospheric history. *Proc. Natl. Acad. Sci. USA 53*: 1205–1215. [Chapter 15.]

Flawn, P. T. 1956. Basement rocks of Texas and southeast New Mexico. *Texas Bur. Econ. Geol. Pub. 5605*, 261 pp. [Appendix I.]

Fogg, G. E.; Stewart, W. D. P.; Fay, P.; and Walsby, A. E. 1973. *The Blue-Green Algae*, New York, Academic Press, 459 pp. [Chapter 15.]

Folk, R. L. 1976. Reddening of desert sands: Simpson Desert, N. T., Australia. *J. Sediment. Petrol. 46*:604–615. [Chapter 11.]

Folsome, C. E.; Allen, R. D.; and Ichinose, N. K. 1975. Organic microstructures as products of Miller-Urey electric discharges. *Precambrian Res. 2*:263–275. [Chapter 9.]

Ford, T. D. and Breed, W. J. 1973. The problematical Precambrian fossil *Chuaria. Palaeontol. 16*:535–550. [Chapter 1.]

Forsman, J. P. and Hunt, J. M. 1958. In L. G. Weeks, ed., *Habitat of Oil*, Tulsa, Ok., Am. Assoc. Petrol. Geol. [Appendix IV.]

Fox, G. E.; Stackebrandt, E.; Hespell, R. B.; Gibson, J.; Maniloff, J.; Dyer, T. A.; Wolfe, R. S.; Balch, W. E.; Tanner, R. S.; Magrum, L. J.; Zablen, L. B.; Blakemore, R.; Gupta, R.; Bonen, L.; Lewis, B. J.; Stahl, D. A.; Luehrsen, K. R.; Chen, K. N.; and Woese, C. R. 1980. The phylogeny of prokaryotes. *Science 209*: 457–463. [Chapters 1, 4, 13, 15.]

Fox, S. W. and Dose, K. 1977. *Molecular Evolution and the Origin of Life*, New York, M. Dekker, 370 pp. [Chapter 4.]

Fox, S. W. and Harada, K. 1960. The thermal copolymerization of amino acids common to proteins. *J. Am. Chem. Soc. 82*:3745–3751. [Appendix III.]

Fox, S. W. and Yuyuma, S. 1963. Abiotic production of primitive protein and formed microparticles, *Annals N. Y. Acad. Sci. 108*:487–494. [Chapter 9.]

Frakes, L. A. 1979. *Climates Throughout Geologic Time*, Amsterdam, Elsevier, 310 pp. [Chapter 11.]

Franklin, J. M.; McIlwaine, W. H.; Poulsen, K. H.; and Wanless, R. K. 1980. Stratigraphy and depositional setting of the Sibley Group, Thunder Bay District, Ontario, Canada. *Can. J. Earth Sci. 17*:633–651. [Appendix I.]

Frarey, M. J. and Roscoe, S. M. 1970. The Huronian Supergroup north of Lake Huron. In A. J. Baer, ed., *Symposium on Basins and Geosynclines of the Canadian Shield, Geol. Surv. Can. Pap. 70–40*:143–158. [Chapter 11.]

Frazer, J. W. 1962. Simultaneous determination of carbon, hydrogen, and nitrogen. Part II. An improved method for solid organic compounds. *Mikrochim. Acta 1962*:993–999. [Appendix IV.]

Frazer, J. W. and Crawford, R. W. 1963. Modifications in the simultaneous determination of carbon, hydrogen, and nitrogen. *Mikrochim. Acta 1963*:561–566. [Appendix IV.]

———. 1964. The handling of volatile compounds for the simultaneous determination of carbon, hydrogen, and nitrogen. *Mikrochim. Acta 1964*: 676–678. [Appendix IV.]

French, B. M. 1964. Graphitization of organic material in a progressively metamorphosed Precambrian iron formation. *Science 146*:917–918. [Chapter 5.]

———. 1966. Some geological implications of equilibrium between graphite and a C—H—O gas phase at high temperatures and pressures. *Rev. Geophys. 4*:223–253. [Chapter 4.]

Freund, F.; Kathrein, H.; Wengeler, H.; Knobel, R.; and Heinen, H. J. 1980. Carbon in solid solution in forsterite—a key to the untractable nature of reduced carbon in terrestrial and cosmogenic rocks. *Geochim. Cosmochim. Acta 44*:1319–1333. [Chapter 4.]

Frey, F. and Green, D. H. 1974. The mineralogy, geochemistry, and origin of lherzolite inclusions in Victorian basanites. *Geochim. Cosmochim. Acta 38*:1023–1059. [Chapter 4.]

Frick, U. and Chang, S. 1977. Ancient carbon and noble gas fractionation. *Proc. 8th Lunar Sci. Conf.*:263–272. [Chapter 4.]

———. 1978. Elimination of chromite and novel sulfides as important carriers of noble gases in carbonaceous meteorites. *Meteoritics 13*:465–470. [Chapter 4.]

Fridovich, I. 1975. Superoxide dismutases. *Ann. Rev. Biochem. 44*:147–159. [Chapter 13.]

———. 1976. Oxygen radicals, hydrogen feroxide and oxygen toxicity. In W. G. Pryor, ed., *Free Radicals in Biology, Vol. 1*, New York, Academic Press, 239–277. [Chapter 13.]

———. 1978. Biology of oxygen radicals. *Science 201*:875–880. [Chapter 13.]

Friedman, I. and O'Neil, J. R. 1972. Hydrogen. In K. H. Wedepohl, ed., *Handbook of Geochemistry, Vol. 1.*, Berlin, Springer, chap. 1. [Chapter 7.]

Friedman, I.; Schoen, B.; and Harris, J. 1961. The deuterium content of Arctic ice. *J. Geophys. Res. 66*:1861–1864. [Chapter 7.]

Friedman, I. and Smith, G. I. 1970. Deuterium content of snow cores from the Sierra Nevada area. *Science 169*:467–470. [Chapter 7.]

Friedmann, N. and Miller, S. L. 1969. Phenylalanine and tyrosine synthesis under primitive Earth conditions. *Science 166*:766–767. [Chapter 4.]

Frietch, R. 1973. Precambrian iron ore of sedimentary origin in Sweden. In *Genesis of Pre-*

cambrian Iron and Manganese Deposits (Proc. Kiev Symp. 1970), Paris, UNESCO, 77–84. [Chapter 11.]

Frietsch, R. 1976. The iron ore deposits in Sweden. *Iron Ore Deposits of Europe 1*:279–293. [Chapter 11.]

Fripp, R. E. P.; Donnelly, T. H.; and Lambert, I. B. 1979. Sulphur isotope results for Archaean banded iron formation, Rhodesia. *Geol. Soc. S. Afr. Spec. Pub. 5*:205–208. [Chapter 7.]

Frost, B. R. 1978. Some aspects of the sedimentary and diagenetic environment of Proterozoic banded iron-formations—a discussion. *Econ. Geol. 73*:1369–1371. [Chapter 11.]

Fryer, B. J. 1972. Age determinations in the Circum-Ungava Geosyncline and the evolution of Precambrian iron-formations. *Can. J. Earth Sci. 9*:652–663. [Chapter 11; Appendix I.]

Fryer, B. J.; Fyfe, W. S.; and Kerrich, A. 1979. Archean volcanogenic oceans. *Chem. Geol. 24*: 25–33. [Chapter 10.]

Fuchs, G.; Thauer, R.; Ziegler, H.; and Stichler, W. 1979. Carbon isotope fractionation by *Methanobacterium thermoautotrophicum*. *Arch. Microbiol. 120*:135–139. [Chapter 7.]

Fudali, R. F. 1965. Oxygen fugacities of basaltic and andesitic magmas. *Geochim. Cosmochim. Acta 29*:1063–1075. [Chapter 4.]

Fuller, R. C.; Smillie, R. M.; Sisler, E. C.; and Kornberg, H. L. 1971. Carbon metabolism in *Chromatium*. *J. Biol. Chem. 236*:2140–2149. [Chapter 7.]

Gáal, G.; Mikkola, A.; and Söderholm, B. 1978. Evolution of the Archean Crust in Finland. *Percambrian Res. 6*:199–215. [Chapter 11.]

Gabel, N. 1977. Chemical evolution: A terrestrial reassessment. In F. E. Hahn, ed., *Progress in Molecular and Subcellular Biology, Vol. 5*, Heidelberg, Springer, 145–172. [Chapter 4.]

Gaffron, H. 1965. The role of light in evolution: The transition from a one quantum to a two quanta mechanism. In S. W. Fox, ed., *The Origins of Prebiological Systems and of Their Molecular Matrices*, New York, Academic Press, 437–455. [Chapter 6.]

Galimov, E. M. 1973. *Izotopy ugleroda v neftgazovoy geologii*, Moscow, Nedra, 384 pp. 1974 transl. *Carbon isotopes in oil-gas geology*, NASA Tech. Transl. TTF-682, Washington, D.C. [Chapters 4, 5.]

———. 1980. $^{13}C/^{12}C$ in kerogen. In B. Durand, ed., *Kerogen*, Paris, Editions Technip, 271–299. [Chapter 7.]

Galimov, E. M.; Migdisov, A. A.; and Ronov, A. B. 1975. Variation in the isotopic composition of carbonate and organic carbon in sedimentary rocks during the Earth's history. *Geochem. Internat. 12*:1–19. [Chapter 7.]

Games, L. M. and Hayes, J. M. 1976. On the mechanisms of CO_2 and CH_4 production in natural anaerobic environments. In J. O. Nriagu, ed., *Environmental Biogeochemistry*, Ann Arbor, Mich., Ann Arbor Sci. Pub., 51–73. [Chapter 12.]

Games, L. M.; Hayes, J. M.; and Gunsalus, R. P. 1978. Methane-producing bacteria: Natural fractionation of the stable carbon isotopes. *Geochim. Cosmochim. Acta 42*:1295–1297. [Chapter 7.]

Ganapathy, R.; Keays, R. R.; Laul, J. C.; and Anders, E. 1970. Trace elements in Apollo 11 lunar rocks: Implications for meteoritic influx and origin of Moon. *Proc. 2nd Lunar Sci. Conf.*: 1117–1142. [Chapter 4.]

Ganapathy, R. and Anders, E. 1974. Bulk compositions of the Moon and Earth, estimated from meteorites. *Proc. 5th Lunar Sci. Conf.*: 1181–1206. [Chapter 4.]

Gancarz, A. J. and Wasserburg, G. J. 1977. Initial Pb of the Amitsoq gneiss, West Greenland, and implications for the age of the Earth. *Geochim. Cosmochim. Acta 41*:1283–1301. [Chapter 4.]

Garlick, S.; Oren, A.; and Padan, E. 1977. Occurrence of facultative anoxygenic photosynthesis among filamentous and unicellular cyanobacteria. *J. Bacteriol. 129*:623–629. [Chapter 8.]

Garrels, R. M. and Christ, C. L. 1965. *Solutions, Minerals, and Equilibria*, San Francisco, Freeman Cooper, 765 pp. [Chapter 11.]

Garrels, R. M. and Mackenzie, F. T. 1971. *Evolution of Sedimentary Rocks*, New York, Norton, 397 pp. [Chapters 4, 5, 7, 10, 11, 12.]

———. 1972. A quantitative model for the sedimentary rock cycle. *Marine Chem. 1*:27–41. [Chapter 5.]

Garrels, R. M. and Perry, E. A., Jr. 1974. Cycling of carbon, sulfur, and oxygen through geologic time. In E. D. Goldberg, ed., *The Sea, Vol. 5*, New York, Wiley, 303–336. [Chapters 5, 10.]

Garrels, R. M.; Lerman, A.; and Mackenzie, F. T. 1976. Controls of atmospheric O_2 and CO_2:

Past, present and future. *Am. Scient. 64*:306–315. [Chapters 5, 15.]

Garrels, R. M.; Mackenzie, F. T.; and Hunt, C. 1975. *Chemical Cycles and the Global Environment*, Los Altos, Calif., Kaufmann, 206 pp. [Chapter 7.]

Garrett, P. 1970. Phanerozoic stromatolites: Noncompetitive ecologic restriction by grazing and burrowing animals. *Science 169*:171–173. [Chapter 9.]

Garrison, W. M.; Morrison, D. C.; Hamilton, J. G.; Benson, A. A.; and Calvin, M. 1951. Reduction of carbon dioxide in aqueous solutions by ionizing radiation. *Science 114*:416–418. [Chapters 1, 4.]

Gast, P. W.; Kulp, J. L.; and Long, L. E. 1958. Absolute age of Early Precambrian rocks in the Bighorn Basin of Wyoming and Montana, and southeastern Manitoba. *Am. Geophys. Union Trans. 39*:322–334. [Chapter 11.]

Gaudette, H. E.; Hurley, P. M.; and Naylor, R. S. 1972. Geochronological anomaly in the Imataca Complex, Venezuela. *Geol. Soc. Am. Abstr. Progr. 4*:515 (Abstr.). [Chapter 11.]

Gavelin, S. 1957. Variations in isotopic composition of carbon from metamorphic rocks in northern Sweden and their geological significance. *Geochim. Cosmochim. Acta 12*:297–314. [Chapter 5.]

Gay, A. L. and Grandstaff, D. E. 1980. Chemistry and mineralogy of Precambrian paleosols at Elliot Lake, Ontario, Canada. *Precambrian Res. 12*:349–373. [Chapters 1, 11; Appendix I.]

Gebelein, C. D. 1969. Distribution, morphology, and accretion rate of recent subtidal algal stomatolites, Bermuda. *J. Sed. Petrol. 39*:49–69. [Chapters 8, 9.]

———. 1974. Biologic control of stromatolite microstructure: Implications for Precambrian time stratigraphy. *Am. J. Sci. 274*:575–598. [Chapters 8, 14.]

———. 1976. The effects of the physical, chemical and biological evolution of the Earth. In M. R. Walter, ed., *Stromatolites*, Amsterdam, Elsevier, 499–515. [Chapter 8.]

Gebelein, C. D. and Hoffman, P. F. 1973. Algal origin of dolomite laminations in stromatolitic limestone. *J. Sed. Petrol. 43*:603–613. [Chapter 8]

Gee, R. D. 1979a. The geology of the Peak Hill area. *Geol. Surv. W. Austr. Ann. Rept. 1978*: 55–62. [Appendix I.]

———. 1979b. Structure and tectonic style of the Western Australian Shield. *Tectonophys. 58*: 327–369. [Chapter 11; Appendix I.]

Geitler, L. 1979. Einige kritische Bermerkungen zu neuen zusammenfassenden Darstellungen der Morphologie und Systematik der Cyanophyceea. *Pl. Syst. Evol. 132*:153–160. [Chapter 13.]

Gerling, E.; Kratz, K.; and Lobach-Zhuchenko, S. 1968. Precambrian Geochronology of the Baltic Shield. *23rd Internat. Geol. Congr.*, (Prague), *Sec. 4*:265–273. [Chapter 11.]

Gerloff, G. C.; Fitzgerald, G. P.; and Skoog, F. 1950. The isolation, purification and culture of blue-green algae. *Am. J. Bot.* 37:216–218. [Chapter 15.]

Gest, H. 1951. Metabolic patterns in photosynthetic bacteria. *Bacteriol. Rev. 15*:183–210. [Chapter 6.]

Gest, H. 1954. Oxidation and evolution of molecular hydrogen by microorganisms. *Bacteriol. Rev. 18*:43–73. [Chapter 6.]

Gest, H. 1980. The evolution of biological energy-transducing systems. *FEMS Microbiol. Lett. 7*: 73–77. [Chapter 6; Glossary II.]

Gest, H. 1981. Evolution of the citric acid cycle and respiratory energy conversion in prokaryotes. *FEMS Microbiol. Lett. 12*:209–215. [Chapter 6, Glossary II.]

Getoff, N. G.; Scholes, G.; and Weiss, J. 1960. Reduction of carbon dioxide in aqueous solution under the influence of radiation. *Tetrahedron Lett. 1960*:17–23. [Chapter 4.]

Gieskes, J. M. and Lawrence, J. R. 1981. Geochemical significance of diagenetic reactions in ocean sediments: An evaluation of interstitial water data. *Oceanologica Acta 66*:111–113. [Chapter 10.]

Gimmel'farb, G. B. 1975. Karbonatnoeosadkonakoplenie v dokembrii Aldanskogo i Baltijskogo schitov (Carbonate sedimentation in the Precambrian of the Aldan and Baltic Shields). In A. V. Sidorenko, ed., *Problems of Precambrian Sedimentary Geology*, *Vol. 2*, Moscow, Nedra, 167–180 (in Russian). [Chapter 10.]

Glaessner, M. F. 1958a. New fossils from the base of the Cambrian in South Australia. *Roy. Soc. S. Austr. Trans. 81*:185–188. [Chapter 1.]

———. 1958b. The oldest fossil faunas of South Australia. *Geol. Rundsch. 47* (2):522–531. [Chapter 1.]

———. 1966. Precambrian paleontology. *Earth Sci. Rev. 1*:29–50. [Chapter 1.]

———. 1979. Precambrian. In R. A. Robison and C. Teichert, eds., *Treatise on Invertebrate Paleontology*, *Part A*, Boulder, Colorado, Geol. Soc. Am. and Lawrence, Kansas, Univ. Kansas, A79–A118. [Chapter 14.]

———. 1980. Pseudofossils from the Precambrian, including "Buschmannia" and "Praesolenopora". *Geol. Mag. 117*:199–200. [Chapters 8, 11.]

Glikson, A. Y. 1976. Stratigraphy and evolution of primary and secondary greenstones: Significance of data from shields of the Southern Hemisphere. In B. F. Windley, ed., *The Early History of the Earth*, London, Wiley, 257–277. [Chapter 3.]

———. 1979. Early Precambrian tonalite—trodhjemite sialic nuclei. *Earth Sci. Rev. 15*: 1–73. [Chapter 10.]

Goldhaber, M. B. and Kaplan, I. R. 1974. The sulfur cycle. In E. D. Goldberg, ed., *The Sea*, *Vol. 5*, New York, Wiley, 569–655. [Chapter 7.]

———. 1980. Mechanisms of sulfur incorporation and isotope fractionation during early diagenesis in sediments of the Gulf of California. *Marine Chem. 9*:95–143. [Chapter 7.]

Goldich, S. S. 1973. Ages of Precambrian iron-formations. *Econ. Geol. 68*:1126–1135. [Chapter 11.]

Goldich, S. S. and Hedge, C. E. 1974. 3800 Myr granitic gneiss in southwestern Minnesota. *Nature 252*:467–468. [Chapter 3.]

Golding, L. Y. and Walter, M. R. 1979. Evidence of evaporite minerals in the Archaean Black Flag Beds, Kalgoorlie, Western Australia. *BMR J. Austr. Geol. Geophys. 4*:67–71. [Chapter 11.]

Goldreich, P. 1966. History of the lunar orbit. *Rev. Geophys. 4*:411–439. [Chapter 11.]

Goldreich, P. and Ward, W. R. 1973. The formation of planetesimals. *Astrophys. J. 183*:1051–1061. [Chapter 2.]

Gole, M. J. and Klein, C. 1981. Banded iron-formations through much of Precambrian time. *J. Geol. 89*:169–183. [Introduction; Chapter 11.]

Goubić, S. 1973. The relationship between blue-green algae and carbonate deposits. In N. G. Carr and B. A. Whitton, eds., *The Biology of Blue-green Algae*, Oxford, Blackwell, 434–472. [Chapter 8.]

———. 1976a. Organisms that build stromatolites. In M. R. Walter, ed., *Stromatolites*, Amsterdam, Elsevier, 113–126. [Chapter 8.]

———. 1976b. Taxonomy of extant stromatolite-building cyanophytes. In M. R. Walter, ed., *Stomatolites*, Amsterdam, Elsevier, 127–140. [Chapter 8.]

Golubić, S. and Barghoorn, E. S. 1977. Interpretation of microbial fossils with special reference to the Precambrian. In E. Flügel, ed., *Fossil Algae*, New York. Springer, 1–14. [Chapter 1.]

Golubić, S. and Campbell, S. E. 1979. Analogous microbial forms in recent subaerial hatitats and in Precambrian cherts: *Gloeothece coerulea* Geitler and *Eosynechococcus moorei* Hofmann. *Precambrian Res. 8*:201–217. [Chapter 14.]

Golubić, S. and Focke, J. W. 1978. *Phormidium hendersonii* Howe: Identity and significance of a modern stromatolite-building microorganism. *J. Sed. Petrol. 48*:751–764. [Chapter 8.]

Golubić, S. and Hofmann, H. J. 1976. Comparison of Holocene and mid-Precambrian Entophysalidaceae (Cyanophyta) in stomatolitic algal mats: Cell division and degradation. *J. Paleontol. 50*:1074–1082. [Chapters 1, 14.]

Goode, A. D. T. 1981. The Proterozoic geology of Western Australia. In D. R. Hunter, ed., *The Precambrian Geology of the Southern Hemisphere*, Amsterdam, Elsevier, 882. [Chapters 8, 11.]

Gooding, J. L. 1978. Chemical weathering on Mars. Thermodynamic stabilities of primary minerals (and their alteration products) from mafic igneous rocks. *Icarus 33*:483–513. [Chapter 4.]

Goodwin, A. M. 1962. Structure, stratigraphy, and origin of iron formations, Michipicoten area, Algoma District, Ontario, Canada. *Geol. Soc. Am. Bull. 73*:561–585. [Appendix I.]

———. 1977. Archean basin-craton complexes and the growth of Precambrian shields. *Can. J. Earth Sci. 14*:2737–2759. [Chapter 11.]

Goodwin, A. M. and Ridler, R. H. 1970. The Abitibi orogenic belt. *Geol. Surv. Can. Pap. 70-40*:1–29. [Chapter 3.]

Goodwin, A. M.; Monster, J.; and Thode, H. G. 1976. Carbon and sulfur isotope abundances in Archaean iron-formations and early Precambrian life. *Econ. Geol. 71*:870–891. [Chapter 7.]

Goody, R. M. and Walker, J. C. G. 1972. *Atmospheres*, Englewood Cliffs, N.J., Prentice-Hall, 150 pp. [Chapter 11.]

Gormly, J. R. and Sackett, W. M. 1977. Carbon isotope evidence for the maturation of marine lipids. In R. Compos and J. Goni, eds., *Advances in Organic Geochemistry 1975*, Madrid, Enadisma, 321–339. [Chapter 5.]

Gorrell, T. E. and Uffen, R. L. 1977. Fermentative metabolism of pyruvate by *Rhodospirillum rubrum* after anaerobic growth in darkness. *J. Bacteriol. 131*:533–543. [Chapter 6.]

Gottschalk, G. and Andreesen, J. 1979. Energy metabolism in anaerobes. In J. R. Quayle, ed., *Microbial Biochemistry*, *Internat. Rev. Biochem. 21*:85–116. [Glossary II.]

Gottwald, M.; Andreesen, J. R.; LeGall, J.; and Ljungdahl, L. G. 1975. Presence of cytochrome and menaquinone in *Clostridium formicoaceticum* and *Clostridium thermoaceticum*. *J. Bacteriol. 122*:325–328. [Chapter 6.]

Goudie, A. 1973. *Duricrusts in Tropical and Subtropical Landscapes*, Oxford, Claredon, 174 pp. [Chapter 8.]

Govett, G. J. S. 1966. Origin of banded iron formations. *Geol. Soc. Am. Bull. 77*:1191–1212. [Chapter 11.]

Gowda, S. S. and Sreenivasa, T. N. 1969. Microfossils from the Archean Complex of Mysore. *J. Geol. Soc. India 10*:201–208. [Chapters 9, 11.]

Grabert, H. 1969. Präkambrian Südamerikas. *Zentralb. Geol. Pälaontol. 3*:523–540. [Chapter 11.]

Grambling, J. A. 1981. Pressures and temperatures in Precambrian metamorphic rocks. *Earth Planet Sci. Lett.*, in press. [Chapter 3.]

Grandstaff, D. E. 1976. A kinetic study of the dissolution of uraninite. *Econ. Geol. 71*:1493–1506. [Chapter 11.]

———. 1980. The origin of uraniferous conglomerates at Elliot Lake, Canada, and Witwatersrand, South Africa: Implications for the Precambrian atmosphere. *Precambrian Res. 13*:1–26. [Chapters 1, 10, 11.]

Gray, C. T. and Gest, H. 1965. Biological formation of molecular hydrogen. *Science 148*:186–192. [Chapter 6.]

Green, D. H. 1975. Genesis of Archaean periodotitic magmas and constraints on Archaean geothermal gradients and tectonics. *Geology 3*: 15–18. [Chapter 3.]

Green, D. H. and Ringwood, A. E. 1967a. The stability fields of aluminous pyroxene peridotite and garnet peridotite and their relevance in upper mantle structure. *Earth Planet. Sci. Lett. 3*:151–160. [Chapter 3.]

———. 1967b. An experimental investigation of the gabbro to eclogite transformation and its petrological applications. *Geochim. Cosmochim. Acta 31*:767–833. [Chapter 3.]

Green, D. H.; Nicholls, I.; Viljoen, M.; and Viljoen, R. 1975. Experimental demonstration of the existence of peridotitic liquids in earliest Archaean magmatism. *Geology 3*:11–14. [Chapter 4.]

Greenberg, R. 1979. Growth of large, late-stage planetesimals. *Icarus 39*:141–150. [Chapter 4.]

———. 1980. Numerical simulation of planet growth: Early runaway growth. In *Lunar and Planet. Sci., Vol. 9*, Houston, Lunar Planet. Sci. Inst., 365–367. [Chapter 4.]

Grew, E. S. 1974. Carbonaceous material in some metamorphic rocks of New England and other areas. *J. Geol. 82*:50–73. [Chapter 5; Appendix IV.]

Grey, K. 1979. Preliminary results of biostratigraphic studies of Proterozoic stromatolites in Western Australia. *Geol. Surv. W. Austr. Record 1979/2*, 26 pp. [Chapter 8; Appendix I.]

———. 1981. Small conical stromatolites from the Archaean near Kanowna, Western Australia. *Ann. Rept. Geol. Surv. W. Austr. 1980*:90–94. [Chapter 8.]

Grieve, R. A. F. 1980. Impact bombardment and its role in proto-continental growth on the early Earth. *Precambrian Res. 10*:217–247. [Chapter 4.]

Griffith, E. J.; Ponnamperuma, C.; and Gabel, N. W. 1977. Phosphorus, a key to life on the primitive Earth. *Origins of Life 8*:71–85. [Chapter 4.]

Griffin, G. M. 1967. X-ray diffraction techniques applicable to studies of diagenesis and low rank metamorphism in humic sediments. *J. Sed. Petrol. 37*:1006–1011. [Chapter 5; Appendix IV.]

Grizzle, P. I.; Coleman, J. H.; Sweeney, R. E.; and Kaplan, I. R. 1979. Correlation of crude oil source with nitrogen, sulfur and carbon isotope ratios. In *Symp. Preprints, ACS/CSJ Div. of Petroleum Chem.*, Honolulu, Hawaii, 39–57. [Chapter 7.]

Grobler, N. J. and Emslie, D. P. 1976a. Stromatolitic limestone and chert in the Ventersdorp Supergroup at the T'Kuip Hills area and surroundings, Britstown District, South Africa. *Geol. Soc. S. Afr. Trans. 79*:49–52. [Chapter 8; Appendix I.]

Grobler, N. J. and Emslie, D. P. 1976b. A re-examination of the Zoetlief-Ventersdorp relationship at the T'Kuip Hills, Britstown District. *Annals Geol. Surv. S. Afr. 11*:99–113. [Chapter 8.]

Gross, G. A. 1973. The depositional environment of principal types of Precambrian iron-formations. In *Genesis of Precambrian Iron and Manganese Deposits* (Proc. Kiev. Symp. 1970), Paris, UNESCO, 15–21. [Chapter 11.]

Groth, W. and Von Weyssenhoff, H. 1960. Photochemical formation of organic compounds from mixtures of simple gases. *Planet. Space Sci. 2*: 79–85. [Chapter 4.]

Groves, D. I.; Dunlop, J. S. R.; and Buick, R. 1981. An early habitat of life. *Scient. Am. 245* (4):64–73. [Chapter 9.]

Gruner, J. W. 1922. Organic matter and the origin of the Biwabik iron-bearing formation of the Mesabi Range. *Econ. Geol. 17*:407–460. [Chapter 14.]

———. 1923. Algae, believed to be Archean. *J. Geol. 31*:146–148. [Chapters 1, 9, 14.]

———. 1924a. Discovery of life in the Archean. *J. Geol. 33*:151–152. [Chapter 9.]

———. 1924b. Contributions to the geology of the Mesabi range with special reference to the magnetites of the iron-bearing formation west of Mesaba. *Minn. Geol. Surv. Bull. 19*:1–71. [Chapter 14.]

———. 1946. *The mineralogy and geology of the taconites and iron ores of the Mesabi range, Minnesota*. Iron Range Resources and Rehabilitation Comm. and Minn. Geol. Surv., Minneapolis, 127 pp. [Chapter 11.]

Guan, B.; Pan, Z.; Geng, W.; Rong, Z.; and Du, H. 1980. Sinian Suberathem in the northern slope of eastern Quinling Ranges. In Y. Wang, ed., *Research on Precambrian Geology, Sinian Suberathem in China*, Tianjin, Tianjin Sci. Technol. Press, 288–313 (in Chinese, with English abstract). [Chapter 14.]

Gunn, B. M. 1976. A comparison of modern and Archean oceanic crust and island-arc petrochemistry. In B. F. Windley, ed., *The Early History of the Earth*, New York, Wiley, 389–403. [Chapter 4.]

Gunsalus, I. C. and Sligar, S. G. 1978. Oxidation reduction by the P-450 monoxygenase system. *Adv. Enzymol. 47*:1–44. [Chapter 13.]

Gunter, B. D. and Musgrave, B. C. 1971. New evidence on the origin of methane in hydrothermal gases. *Geochim. Cosmochim. Acta 35*:113–118. [Chapters 4, 7.]

Gutstadt, A. M. and Schopf, J. W. 1969. Possible algal microfossils from the Late Pre-Cambrian of California. *Nature 233*:165–167. [Chapter 8.]

Haggerty, S. E. 1978. The redox state of planetary basalts. *Geophys. Res. Lett. 5*:443–446. [Chapter 4; Appendix III.]

Haldane, J. B. S. 1929. The origin of life. *Rationalist Annual 148*:3–10. [Chapter 4.]

———. 1954. The origins of life. *New Biology 16*: 12–27. [Chapter 4.]

Halfen, L. N. and Castenholz, R. W. 1971. Energy expenditure for gliding motility in a blue-green alga. *J. Phycol. 7*:258–260. [Chapter 8.]

Hall, J. 1884. Cryptozoön n.g.; *Cryptozoön proliferum* n. sp. *Ann. Rept. New York State Mus. Nat. Hist. 36*: Description of plate 6. [Appendix 1.]

Hall, W. D. M. and Goode, A. D. T. 1978. The Early Proterozoic Nabberu Basin and associated iron ore formations of Western Australia. *Precambrian Res. 7*:129–184. [Chapter 10; Appendix I.]

Hallbauer, D. K. 1975. The plant origin of the Witwatersrand "carbon". *Min. Sci. Engng. 7*: 111–131. [Chapter 9.]

———. 1978. Witwatersrand gold deposits. Their genesis in light of morphological studies. *Gold Bulletin 11* (1):18–23. [Chapter 9.]

Hallbauer, D. K. and van Warmelo, K. T. 1974. Fossilized plants in thucholite from Precambrian rocks of the Witwatersrand, South Africa. *Precambrian Res. 1*:199–212. [Chapter 9.]

Hallbauer, D. K.; Jahns, H. M.; and Beltmann, H. A. 1977. Morphological and anatomical observations on some Precambrian plants from the Witwatersrand, South Africa. *Geol. Rundsch. 66*:477–491. [Chapter 9.]

Halliwell, B. 1974. Superoxide dismutase, catalase and glutathione peroxidase: Solutions to the problems of living with oxygen. *New Phytol. 73*: 1075–1086. [Chapter 13.]

———. 1978. Biochemical mechanisms accounting for the toxic action of oxygen on living organisms: The key role of superoxide dismutase. *Cell. Biol. Internat. Rept. 2*:113–129. [Chapter 13.]

Halmann, M.; Aurian-Blajeni, B.; and Bloch, S. 1980. Photoassisted carbon dioxide reduction and formation of two- and three-carbon compounds. In Y. Wolman, ed., *Origin of Life*, Dordrecht, Reidel, 143–150. [Chapter 4.]

Hamano, Y. and Ozima, M. 1978. Earth-atmosphere evolution model based on Ar isotopic data. In E. C. Alexander, Jr. and M. Ozima, eds., *Terrestrial Rare Gases*, Tokyo, Japan Scient. Soc. Press, 155–171. [Chapter 4.]

Hamilton, L. H. 1976. Biogenic aspects of Precambrian iron formations from Africa and elsewhere. *Abstracts, 25th Internat. Geol. Congr.* (Sydney, Austr.), *1*, *Sect. 1* (1A): 30 (Abstr.). [Chapter 9.]

Hamilton, L. J. and Muir, M. D. 1974. Precambrian microfossils from the McArthur River lead-zinc-silver deposit Northern Territory, Australia. *Mineral. Deposita 9*: 83–86. [Chapter 14.]

Hamilton, P. J.; Evensen, V. M.; O'Nions, R. K.; Smith, H. S.; and Erlank, A. J. 1979. Sm-Nd dating of Onverwacht Group volcanics, southern Africa. *Nature 279*: 298–300. [Appendix I.]

Hamilton, P. J.; Evensen, N. M.; O'Nions, R. K.; Glikson, A. Y.; and Hickman, A. H. 1980. Sm-Nd dating of the Talga-Talga Subgroup, Warrawoona Group, Pilbara Block, Western Australia. In J. E. Glover and D. I. Groves, eds., *2nd Internat. Archaean Symp.* (Perth), Canberra, Austr. Acad. Sci., 11–12 (Abstr.). [Chapter 9; Appendix I.]

Hamilton, P. J.; O'Nions, R. K.; and Evensen, N. M. 1977. Sm-Nd dating of Archaean basic and ultrabasic volcanic rocks. *Earth Planet. Sci. Lett. 36*: 263–268. [Appendix I.]

Hamilton, P. J.; O'Nions, R. K.; Evensen, N. M.; Bridgwater, D.; and Allaart, J. H. 1978. Sm-Nd isotopic investigations of Isua supracrustals and implications for mantle evolution. *Nature 272*: 41–43. [Appendix I.]

Hamilton, P. J.; O'Nions, R. K.; and Pankhurst, R. J. 1980. Isotopic evidence for the provenance of some Caledonian granites. *Nature 287*: 279–284. [Chapter 10.]

Hampicke, W. 1979. Net transfer of carbon between the land biota and the atmosphere, induced by man. In B. Bolin, E. T. Degens, S. Kempe, and P. Ketner, eds., *The Global Carbon Cycle*, Chichester, Wiley, 219–236. [Chapter 5.]

Hanks, T. C. and Anderson, D. L. 1969. The early thermal history of the Earth. *Phys. Earth Planet. Interiors 2*: 19–29. [Chapter 4.]

Hanson, R. S. 1980. Ecology and diversity of methylotrophic organisms. *Adv. Appl. Microbiol. 26*: 3–39. [Chapter 12.]

Häntzschel, W. 1975. Trace fossils and problematica. In C. Teichert, ed., *Treatise on Invertebrate Paleontology*, *Part W*, *Suppl. 1*, Boulder, Colorado, Geol. Soc. Am. and Lawrence, Kansas, Univ. Kansas, 269 pp. [Chapter 14.]

Harder, W. and van Dijken, J. P. 1976. Theoretical considerations on the relation between energy production and growth of methane-utilizing bacteria. In H. G. Schlegel, G. Gottschalk, and N. Pfenning, eds., *Symposium on Microbial Production and Utilization of Gases* (H_2, CH_4, *CO*), Gottingen, E. Goltze KG, 403–418. [Chapter 12.]

Hardie, L. A. 1967. The gypsum-anhydrite equilibrium at one atmosphere pressure. *Am. Mineral. 52*: 171–200. [Chapter 11.]

Hardie, L. A. and Garret, P. 1977. Some miscellaneous implications and speculations. In L. A. Hardie, ed., *Sedimentation on the Modern Carbonate Tidal Flats of Northwest Andros Island, Bahamas*, Baltimore, Johns Hopkins Univ. Press, 184–187. [Chapter 8.]

Hardie, L. A. and Ginsburg, R. N. 1977. Layering: The origin and environmental significance of lamination and thin bedding. In L. A. Hardie, ed., *Sedimentation on the Modern Carbonate Tidal Flats of Northwest Andros Island, Bahamas*, Baltimore, Johns Hopkins Univ. Press, 50–123. [Chapter 8.]

Hargraves, R. B. 1976. Precambrian geologic history. *Science 193*: 363–371. [Chapters 3, 4.]

Harold, F. M. 1978. Vectorial metabolism. In L. N. Ornston and J. R. Sokatch, eds., *The Bacteria, Vol. 6*: *Bacterial Diversity*, New York, Academic Press, 463–521. [Chapter 6.]

Harper, C. T., ed. 1973. *Geochronology*: *Radiometric Dating of Rocks and Minerals*, Stroundsburg, Penna., Dowden, Hutchison and Ross, 469 pp. [Chapter 1.]

Harris, P. G.; Hutchinson, R.; and Paul, D. K. 1972. Plutonic xenoliths and their relation to the upper mantle. *Phil. Trans. Roy. Soc. London Sect. A 271*: 313–323. [Chapter 4.]

Harrison, A. G. and Thode, H. G. 1958. Mechanism of bacterial reduction of sulphate from isotope fractionation studies. *Trans. Faraday Soc. 54*: 84–92. [Chapter 7.]

Harrison, J. E. and Peterman, Z. E. 1980. North American Commission on Stratigraphic Nomenclature Note 52—a preliminary proposal for a chronometric time scale for the Precambrian of the United States and Mexico. *Geol. Soc. Am. Bull. Part 1, 91*:377–380. [Chapter 3.]

Hart, M. 1978. The evolution of the atmosphere of the Earth. *Icarus 33*:23–39. [Chapter 4.]

Hart, M. H. 1979. Was the prebiotic atmosphere of the Earth heavily reducing? *Origins of Life 9*: 261–266. [Chapter 4.]

Hart, S. R. and Allègre, C. J. 1980. Trace element constraints on magma genesis. In R. B. Hargraves, ed., *Physics of Magmatic Processes*, Princeton, N. J., Princeton Univ. Press, 121–159. [Chapter 10.]

Hartman, H. 1975. Speculations on the origin and evolution of metabolism. *J. Mol. Evol. 4*:359–370. [Chapter 4.]

Hartmann, M. and Nielsen, H. 1969. δ^{34}S-Werte in rezenten Meeressedimenten and ihre Deutung am Beispiel einiger Sedimentprofile aus der westlichen Ostsee. *Geol. Rundsch. 58*:621–655. [Chapter 7.]

Hartmann, W. K. 1977. Relative crater production rates on planets. *Icarus 31*:260–276. [Chapter 2.]

Harvey, G. R.; Degens, E. T.; and Mopper, K. 1975. Synthesis of nitrogen heterocycles on kaolinite from carbon dioxide and ammonia. *Naturwiss. 58*:624–625. [Chapter 4.]

Hasan, M. and Hall, J. B. 1977. Dissimilatory nitrate reduction in *Clostridium tertium*. *Z. Allg. Mikrobiol. 17*:501–506. [Chapter 6.]

Hatchikian, C. E. and Henry, Y. A. 1977. An iron-containing superoxide dismutase from the strict anaerobe *Desulfovibrio desulfuricans*. *Biochimie 59*:153–161. [Chapter 13.]

Haughton, S. H. 1969. *Geological History of Southern Africa*, Cape Town, Geol. Soc. S. Afr., 535 pp. [Chapters 10, 11.]

Hawkesworth, C. J.; Bickle, M. J.; Gledhill, A. R.; Wilson, J. F.; and Orpen, J. L. 1979. A 2.9 b.y. event in the Rhodesian Archaean. *Earth Planet. Sci. Lett. 45*:285–287. [Appendix I.]

Hawley, J. E. 1926. An evaluation of the evidence of life in the Archaean. *J. Geol. 34*:441–461. [Chapter 9.]

Hay, R. L. and Reeder, R. J. 1978. Calcretes of Olduvai Gorge and the Ndolanya Beds of northern Tanzania. *Sedimentol. 25*:649–673. [Chapter 8.]

Hayaishi, O. and Asada, K. 1977. *Biochemical and Medical Aspects of Active Oxygen*, Baltimore, University Park Press, 313 pp. [Chapter 13.]

Hayashi, C.; Nakazawa, K.; and Mizuno, H. 1979. Earth's melting due to the blanketing effect of the primordial dense atmosphere. *Earth Planet. Sci. Lett. 43*:22–28 [Chapter 4.]

Hayatsu, R.; Studier, M. H.; Matsuoka, S.; and Anders, E. 1972. Origin of organic matter in early solar system VI. Catalytic synthesis of nitrites, nitrogen bases and porphyrin-like pigments. *Geochim Cosmochim. Acta 36*:555–571. [Chapter 4.]

Hayatsu, R.; Matsuoka, S.; Scott, R. G.; Studier, M. H.; and Anders, E. 1977. Origin of organic matter in the early solar system. VII. The organic polymer in carbonaceous chondrites. *Geochim. Cosmochim. Acta 41*:1325–1339. [Appendix III.]

Hayatsu, R.; Scott, R. G.; Studier, M. H.; Lewis, R. S.; and Anders, E. 1980. Carbynes in meteorites: Detection, low-temperature origin, and implications for interstellar molecules. *Science 209*:1515–1518. [Chapter 4.]

Hayes, J. M. 1967. Organic constituents of meteorites—a review. *Geochim. Cosmochim. Acta 31*:1395–1440. [Chapter 4; Appendix III.]

Heald, E. F.; Naughton, J. J.; and Barnes, I. L., Jr. 1963. The chemistry of volcanic gases. 2. Use of equilibrium calculations in the interpretation of volcanic gas samples. *J. Geophys. Res. 68*:545–557. [Chapter 4.]

Hegeman, G. 1980. Oxidation of carbon monoxide by bacteria. *Trends in Biochem. Sci. 5*:256–259. [Glossary II.]

Heinrichs, T. 1980. Lithostratigraphische Untersuchungen in der Fig Tree Gruppe des Barberton greenstone belt zwischen Umsoli und Lomati (Südafrika). *Göttinger Arbeit. Geol. Pälaontol. 22*:1–113. [Chapter 10.]

Heinrichs, T. K. and Reimer, T. O. 1977. A sedimentary barite deposit from the Archaean Fig Tree Group of the Barberton Mountain Land (South Africa). *Econ. Geol. 72*:1426–1441. [Chapters 7, 10, 11.]

Henderson, J. B. 1975a. Archean stromatolites in the northern Slave Province, Northwest Territories, Canada. *Can. J. Earth Sci. 12*:1619–1630. [Chapter 8; Appendix I.]

———. 1975b. Sedimentological studies of the Yellowknife Supergroup in the Slave Structural Province. *Geol. Surv. Can. Pap. 75-1* (A):325–330. [Chapter 8.]

———. 1977. Archean geology and evidence of ancient life in the Slave Structural Province, Canada. In C. Ponnamperuma, ed., *Chemical Evolution of the Early Precambrian*, New York, Academic Press, 41–54. [Chapter 8.]

Henderson-Sellers, A. and Schwartz, A. W. 1980. Chemical evolution and ammonia in the early Earth's atmosphere. *Nature 287*:526–528. [Chapters 4, 11.]

Henley, R. R. 1968. Degradation of Insoluble Natural Polymers Under Simulated Primitive Earth Conditions and the Resultant Geochemical Cycle. Ph.D. Thesis, Stanford Univ. (Dept. Biophysics), 114 pp. [Chapter 4; Appendix III.]

Hennecke, E. W. and Manual, O. K. 1975. Noble gases in an Hawaiian xenolith. *Nature 257*: 778–780. [Chapter 4.]

Henry, L. and Hall, D. O. 1977. Superoxide dismutases in green algae: An evolutionary survey. In S. Miyachi, S. Katoh, Y. Fujita, and K. Shibata, eds., *Photosynthetic Organelles*, Tokyo, Center Acad. Pub., 377–382. [Chapter 13.]

Héroux, Y.; Chagnon, A.; and Bertrand, R. 1979. Compilation and correlation of major thermal maturation indicators. *Am. Assoc. Petrol. Geol. Bull. 63*:2128–2144. [Chapter 5.]

Hesstvedt, E.; Henriksen, S. E.; and Hjartarson, H. 1974. On the development of an aerobic atmosphere. A model experiment. *Geophys. Norvegica 31*:1–8. [Chapter 11.]

Hewitt, J. and Morris, J. G. 1975. Superoxide dismutase in some obligately anaerobic bacteria. *FEBS Microbiol. Lett. 50*: 315–317. [Chapter 13; Glossary II.]

Hickman, A. H. 1973. The North Pole barite deposits, Pilbara Goldfield. *Ann. Rept. Geol. Surv. W. Austr. 1972*:57–60. [Chapters 8, 11.]

———. 1980. Archaean geology of the Pilbara Block. *2nd Internat. Archaean Symposium* (Perth), *Excursion Guide*, Canberra, Geol. Soc. Austr., 55 pp. [Chapter 8; Appendix I.]

Hickman, A. H. and de Laeter, J. R. 1977. The depositional environment and age of a shale within the Hardey Sandstone of the Fortescue Group. *Ann. Rept. Geol. Surv. W. Austr. 1976*: 62–68. [Chapter 8; Appendix I.]

Hickman, A. H. and Lipple, S. L. 1978. Explanatory notes on the Marble Bar 1:250,000 Geological Sheet, Perth, W. Austr. Geol. Surv., 24 pp. [Chapter 9; Appendix I.]

Hickman, M. H. 1976. Isotopic evidence for crustal reworking in the Rhodesian Archaean craton, southern Africa. *Geology 6*:215–216. [Appendix I.]

Higgins, I. J. 1979. Methanotrophy. In J. R. Quayle, ed., *Microbial Biochemistry, Internat. Rev. Biochem. 21*:300–354. [Glossary II.]

Higgins, I. J.; Best, D. J.; and Hammond, R. C. 1980. New findings in methane-utilizing bacteria highlight their importance in the biosphere and their commercial potential. *Nature 286*: 561–564. [Chapter 6.]

Hill, H. A. O. 1978. The superoxide iron and the toxicity of molecular oxygen. In R. J. P. Williams and J. R. R. F. Da Silva, eds., *New Trends in Bio-inorganic Chemistry*, New York, Academic Press, 173–208. [Chapter 13.]

Hillmer, P. and Gest, H. 1977. H_2 metabolism in the photosynthetic bacterium *Rhodopseudomonas capsulata*: H_2 production by growing cultures. *J. Bacteriol. 129*:724–731. [Chapter 6.]

Hirsch, P. 1978. Microbial mats in a hypersaline solar lake: Types, composition and distribution. In W. E. Krumbein, ed., *Environmental Biogeochemistry and Geomicrobiology, Vol. 1*, Ann Arbor, Mich., Ann Arbor Sci. Pub., 189–201. [Chapter 8.]

Hochacka, P. W. 1976. Design of metabolic and enzymic machinery to fit lifestyle and environment. In R. M. S. Smellie and J. F. Pennock, eds., *Biochemical Adaptation to Environmental Change, Biochem. Soc. Symp. 41*:3–31. [Chapter 6.]

Hochachka, P. W. and Mustafa, T. 1972. Invertebrate facultative anaerobiosis. *Science 178*: 1056–1060. [Chapter 6.]

Hoefs, J. 1965. Ein Beitrag zur Geochemie des Kohlenstoffs in magmatischen und metamorphen Gesteinen. *Geochim. Cosmochim. Acta 29*: 399–428. [Chapter 4.]

Hoefs, J. and Frey, M. 1976. Isotopic composition of carbonaceous matter in a metamorphic profile from the Swiss Alps. *Geochim. Cosmochim. Acta 40*:945–951. [Chapters 5, 7.]

Hoefs, J. and Schidlowski, M. 1967. Carbon isotope composition of carbonaceous matter from the Precambrian of the Witwatersrand System. *Science 155*:1096–1097. [Chapter 7.]

Hoering, T. C. 1955. Variations of nitrogen-15 abundance in naturally occurring substances. *Science 122*:1233–1234. [Chapter 7.]

———. 1961. The stable isotopes of carbon in the carbonate and reduced carbon of Precambrian sediments. *Carnegie Inst. Wash. Yearbook 61*: 190–191. [Chapter 1.]

———. 1964. The hydrogenation of kerogen from sedimentary rocks with phosphorous and anhydrous hydrogen iodide. *Carnegie Inst. Wash. Yearbook 63*:258–262. [Chapter 5.]

———. 1967a. The organic geochemistry of Precambrian rocks. In P. H. Abelson, ed., *Researches in Geochemistry*, New York, Wiley, 89–111. [Chapter 7.]

———. 1967b. Criteria for suitable rocks in Precambrian organic geochemistry. *Carnegie Inst. Wash Yearbook 65*:365–372. [Chapter 5.]

———. 1977. The stable isotopes of hydrogen in Precambrian organic matter. In C. Pannamperuma, ed., *Chemical Evolution of the Early Precambrian*, New York, Academic Press, 81–86. [Chapter 7.]

Hoering, T. C. and Abelson, P. H. 1964. Hydrocarbons from the low-temperature heating of kerogen. *Carnegie Inst. Wash. Yearbook 63*: 256–258. [Chapter 5.]

Hoering, T. C. and Ford, H. T. 1960. The isotope effect in the fixation of nitrogen by *Azotobacter*. *Am. Chem. J. 82*:376–378. [Chapter 7.]

Hoering, T. C. and Moore, H. E. 1958. The isotopic composition of the nitrogen in natural gases and associated crude oils. *Geochim. Cosmochim. Acta 13*:225–232. [Chapter 7.]

Hoffman, P. F. 1968. Stratigraphy of the lower Proterozoic (Aphebian), Great Slave Supergroup, East Arm of Great Slave Lake, District of Mackenzie. *Geol. Surv. Can. Pap. 68-42*, 93 pp. [Chapter 11; Appendix I.]

———. 1976. Environmental diversity of middle Precambrian stromatolites. In M. R. Walter, ed., *Stromatolites*, Amsterdam, Elsevier, 599–611. [Chapter 8.]

———. 1978. Speleothems and evaporite solution collapse in Athapuscow Aulacogen (Middle Precambrian), Great Slave Lake, Northwest Territories. *Am. Assoc. Petrol. Geol. Bull. 62*: 523–537. [Chapter 11.]

Hoffman, P. F.; Dewey, J. F.; and Burke, K. 1974. Aulacogens and their genetic relation to geosynclines, with a Proterozoic example from Great Slave Lake, Canada. *Soc. Econ. Paleontol. Mineral. Spec. Pub. 19*:38–55. [Appendix I.]

Hoffman, P. F.; Fraser, J. A.; and McGlynn, J. C. 1970. The Coronation Geosyncline of Aphebian age, District of Mackenzie. *Geol. Surv. Can. Pap. 70–40*:201–212. [Chapter 11.]

Hofmann, H. J. 1969. Stromatolites from the Proterozoic Animikie and Sibley Groups, Ontario. *Geol. Surv. Can. Pap. 68–69*, 77 pp. [Chapter 14; Appendix I.]

———. 1971a. Precambrian fossils, pseudofossils, and problematica in Canada. *Geol. Surv. Can. Bull. 189*, 146 pp. [Chapters 8, 9, 11, 14.]

———. 1971b. Polygonomorph acritarch from the Gunflint Formation (Precambrian), Ontario. *J. Paleontol. 45*:522–524. [Chapter 14.]

———. 1974. Mid-Precambrian procaryotes (?) from the Belcher Islands. *Nature 249*:87–88. [Chapter 14.]

———. 1975. Stratiform Precambrian stromatolites, Belcher Islands, Canada: Relations between silicified microfossils and microstructure. *Am. J. Sci. 275*:1121–1132. [Chapters 8, 14.]

———. 1976. Precambrian microflora, Belcher Islands, Canada: Significance and systematics. *J. Paleontol. 50*:1040–1073. [Chapters 1, 14; Appendix I.]

———. 1977a. The problematic fossil *Chuaria* from the Late Precambrian Uinta Mountain Group, Utah. *Precambrian Res. 4*:1–11. [Chapters 1, 15.]

———. 1977b. On Aphebian stromatolites and Riphean stromatolite stratigraphy. *Precambrian Res. 5*:175–205. [Appendix I.]

———. 1978. New stromatolites from the Aphebian Mistassini Group, Quebec. *Can. J. Earth Sci. 15*:571–585. [Appendix I.]

Hofmann, H. J. and Aitken, J. D. 1979. Precambrian biota from the Little Dal Group, Mackenzie Mountains, northwestern Canada. *Can. J. Earth Sci. 16*:150–166. [Chapters 14, 15.]

Hofmann, H. J. and Chen, J. 1981. Carbonaceous megafossils from the Precambrian (1800 Ma) near Jixian, northern China. *Can. J. Earth Sci. 18*:443–447. [Chapters 14, 15.]

Hofmann, H. J. and Jackson, G. D. 1969. Precambrian (Aphebian) microfossils from Belcher Islands, Hudson Bay. *Can. J. Earth Sci. 6*: 1137–1144. [Chapter 14.]

Hofmann, H. J.; Pearson, D. A. B.; and Wilson, B. H. 1980. Stromatolites and fenestral fabric in Early Proterozoic Huronian Supergroup, Ontario. *Can. J. Earth Sci. 17*:1351–1357. [Appendix I.]

Holland, H. D. 1962. Model for the evolution of the Earth's atmosphere. In A. E. J. Engel, H. L. James, and B. F. Leonard, eds., *Petrologic*

Studies: A Volume to Honor A. F. Buddington, New York, Geol. Soc. Am., 447–477. [Chapter 4; Appendix III.]

———. 1970. Ocean water, nutrients and atmospheric oxygen. In *Proc. Internat. Symp. Hydrogeochemistry and Biogeochemistry*, (Tokyo), 68–81. [Chapter 7.]

———. 1972. The geologic history of sea water—an attempt to solve the problem. *Geochim. Cosmochim. Acta 36*:637–651. [Chapter 11.]

———. 1973. Ocean water, nutrients, and atmospheric oxygen. In *Proceedings of Symposium on Hydrogeochemistry and Biogeochemistry*, Washington, D.C., Clarke, 66–81. [Chapter 11.]

———. 1973. The oceans: A possible source of iron in iron-formations. *Econ. Geol. 68*:1169–1172. [Chapters 1, 10, 11.]

———. 1974. Marine evaporites and the composition of sea water during the Phanerozoic. In W. W. Hay, ed., *Studies in Paleo-Oceanography, Soc. Econ. Paleontol. Mineral. Spec. Pub. 20*:187–192. [Chapter 11.]

———. 1976. The evolution of sea water. In B. F. Windley, ed., *The Early History of the Earth*, London, Wiley, 559–567. [Chapters 10, 11.]

———. 1978. *The Chemistry of the Atmosphere and Oceans*, New York, Wiley, 351 pp. [Chapters 1, 7, 10, 11, 13, 15.]

Holloway, J. R. and Burnham, C. W. 1972. Melting relations of basalt with equilibrium water pressure less than total pressure. *J. Petrol. 13*:1–29. [Chapter 3.]

Holloway, J. R. and Eggler, D. H. 1976. Fluid-absent melting of peridotite containing phlogopite and dolomite. *Carnegie Inst. Wash. Yearbook 75*:636–639. [Chapter 4.]

Holm-Hansen, O. 1968. Ecology, physiology, and biochemistry of blue-green algae. *Ann. Rev. Microbiol. 22*:47–70. [Chapter 15.]

Holser, W. T. and Kaplan, I. R. 1966. Isotope geochemistry of sedimentary sulfate. *Chem. Geol. 1*:93–135. [Chapter 7.]

Horodyski, R. J. 1976. Stromatolites from the Middle Proterozoic, Altyn Limestone Belt Supergroup, Glacier National Park, Montana. In M. R. Walter, ed., *Stromatolites*, Amsterdam, Elsevier, 585–597. [Chapter 11.]

Horodyski, R. J. 1977. *Lyngbya* mats at Laguna Mormona, Baja California, Mexico: Comparison with Proterozoic stromatolites. *J. Sed. Petrol. 47*:1305–1320. [Chapter 8.]

———. 1980. Middle Proterozoic shale-facies microbiota from the lower Belt Supergroup, Little Belt Mountains, Montana. *J. Paleontol. 54*:649–663. [Chapter 1.]

Horodyski, R. J. and Bloeser, B. 1978. 1400-million-year-old shale-facies microbiota from the Lower Belt Supergroup, Montana. *Science 199*:682–684. [Chapters 14, 15.]

Horodyski, R. J.; Donaldson, J. A.; and Kerans, C. 1980. A new shale-facies microbiota from the Middle Proterozoic Dismal Lake Group, District of Mackenzie, Northwest Territories, Canada. *Can. J. Earth Sci. 17*:1166–1173. [Chapter 1.]

Horowitz, N. H. 1945. On the evolution of biochemical syntheses. *Proc. Natl. Acad. Sci. USA 31*:153–157. [Chapters 1, 4.]

Horowitz, N. H. and Miller, S. L. 1962. Current theories on the origin of life. *Fortschr. Chem. Org. Naturst. 20*:423–459. [Chapter 4.]

Horowitz, N. H.; Hobby, G. L.; and Hubbard, J. S. 1977. Viking on Mars: The carbon assimilation experiments. *J. Geophys. Res. 82*:4659–4662. [Chapter 4.]

Horwitz, R. C. and Smith, R. E. 1978. Bridging the Yilgarn and Pilbara Blocks, Western Australia. *Precambrian Res. 6*:293–322. [Chapter 8.]

Howard, D. 1980. Polycyclic Triterpenes of the Anaerobic Photosynthetic Bacterium *Rhodomicrobium vanielii*. Ph.D. Thesis, Univ. Calif., Los Angeles (Dept. Biology), 272 pp. [Chapter 13.]

Hoyle, F. 1960. Formation of the planets. *Quart. J. Roy. Astron. Soc. 1*:28–42. [Chapter 2.]

Hubbard, J. 1979. Laboratory simulations of the pyrolytic release experiments: An interim report. *J. Mol. Evol. 14*:211–221. [Chapter 4.]

Hubbard, J. S.; Hardy, J. P.; Voecks, G. E.; and Golub, E. E. 1973. Photocatalytic synthesis of organic compounds from CO and water: Involvement of surfaces in the formation and stabilization of products. *J. Mol. Evol. 2*:149–166. [Chapter 13.]

Hubbard, J. S.; Voecks, G. E.; Hobby, G. L.; Ferris, J. P.; Williams, E. A.; and Nicodem, D. E. 1975. Ultraviolet-gas phase and -photocatalytic synthesis from CO and NH_3. *J. Mol. Evol. 5*:223–241. [Chapter 13.]

Huc, A. Y. 1980. Origin and formation of organic matter in recent sediments and its relation to kerogen. In B. Durand, ed., *Kerogen*, Paris, Editions Technip, 445–474. [Chapter 5.]

Huc, A. Y. and Durand, B. M. 1977. Occurrence and significance of humic acids in ancient sediments. *Fuel 56*:73–80. [Chapter 5.]

Hulston, J. R. and McCabe, W. J. 1962. Mass spectrometer measurements in the thermal areas of New Zealand. Part 2. Carbon isotopic ratios. *Geochim. Cosmochim. Acta 26*:399–410. [Chapter 4.]

Hunt, B. G. 1979. The effects of past variations of the Earth's rotation rate on climate. *Nature 281*:188–191. [Chapter 11.]

Hunt, J. M. 1972. Distribution of carbon in crust of Earth. *Am. Assoc. Petrol. Geol. Bull. 56*: 2273–2277. [Chapter 7.]

———. 1979. *Petroleum Geochemistry and Geology*, San Francisco, Freeman, 617 pp. [Chapter 5.]

Hunter, D. R. 1974. Crustal development in the Kaapvaal Craton, II. The Proterozoic. *Precambrian Res. 1*:295–326. [Chapter 11.]

Hurlbut, C. S., Jr. and Klein, C. 1977. *Manual of Mineralogy*, New York, Wiley, 532 pp. [Chapter 5.]

Hurley, P. M. 1968. The confirmation of continental drift. *Scient. Am. 218* (4):52–64. [Chapter 3.]

Hurley, P. M. and Rand, J. R. 1969. Pre-drift continental nuclei. *Science 164*:1229–1242. [Chapter 10.]

Hurley, P. M.; Leo, G. W.; White, R. W.; and Fairbairn, H. W. 1971. Liberian Age Province (about 2,700 m.y.) and adjacent provinces in Liberia and Sierra Leona. *Geol. Soc. Am. Bull. 82*:3483–3490. [Chapter 11.]

Hurley, P. M.; Melcher, G. C.; Pinson, W. H.; and Fairbairn, H. W. 1968. Some orogenic episodes in South America by K-Ar and whole rock Rb-Sr dating. *Can. J. Earth Sci. 5*:633–638. [Chapter 11.]

Hutchinson, E. 1944. Nitrogen in the biogeochemistry of the atmosphere. *Am. Scient. 32*: 178–195. [Chapter 7.]

Hyde, R. S. 1980. Sedimentary facies in the Archean Timiskaming Group, and their tectonic implications, Abitibi greenstone belt, northeastern Ontario, Canada. *Precambrian Res. 12*: 161–195. [Chapter 10.]

Iltchenko, L. N. 1972. Late Precambrian acritarchs of Antarctica. In R. J. Adie, ed., *Antarctic Geology and Geophysics*, Internat. Union Geol. Sci., B-1, Oslo, Universitetsforlaget, 599–602. [Chapter 1.]

Ingmanson, D. E. and Dowler, M. J. 1977. Chemical evolution and the evolution of the Earth's crust. *Origins of Life 8*:221–224. [Chapter 4.]

———. 1980. Unique amino acid composition of Red Sea brine. *Nature 286*:51–52. [Chapter 4.]

Irving, A. J. and Wyllie, P. J. 1973. Melting relationships in CaO-CO_2 and Mg-CO_2 to 36 kilobars with comments on CO_2 in the mantle. *Earth Planet. Sci. Lett. 20*:220–225. [Chapter 4.]

———. 1975. Subsolidus and melting relationships for calcite, magnesite and the join $CaCO_3$-$MgCO_3$ to 36 kb. *Geochim. Cosmochim. Acta 39*:35–53. [Chapter 4.]

Irwin, W. P. and Barnes, I. 1980. Tectonic relations of carbon dioxide discharges and earthquakes. *J. Geophys. Res. 85*:3115–3121. [Chapter 4.]

Ito, K. and Kennedy, G. C. 1967. Melting and phase relations in a natural peridotite to 40 kilobars. *Am. J. Sci. 265*:519–538. [Chapter 3.]

Ivanov, M. V.; Gogotova, G. I.; Matrosov, A. G.; and Zyakun, A. M. 1976. Fractionation of sulfur isotopes by phototrophic sulfur bacteria, *Ectothiorhodospira shaposhnikovii*. *Microbiol 45*: 655–659 (English transl.). [Chapter 7.]

Izawa, E. 1968. Carbonaceous matter in some metamorphic rocks in Japan. *J. Geol. Soc. Japan 74*:427–432. [Chapter 5.]

Jackson, T. A. 1967. Fossil actinomycetes in Middle Precambrian glacial varves. *Science 155*:1003–1005. [Chapter 14.]

———. 1971. Carbonaceous inclusions, sulfides, and "fossil gas bubbles" of presumably biologic origin associated with rafted erratics in Huronian (Precambrian) glacial-like argillites. *J. Sed. Petrol. 41*:313–315. [Chapter 14.]

Jackson, T. A.; Fritz, P.; and Drimmie, R. 1978. Stable carbon isotope ratios and chemical properties of kerogen and extractable organic matter in pre-Phanerozoic and Phanerozoic sediments: Their interrelations and possible paleobiological significance. *Chem. Geol. 21*: 335–350. [Chapter 7.]

Jacobsen, S. B. and Wasserburg, G. J. 1981. Transport models for crust and mantle evolution. *Tectonophys. 75*:163–179. [Chapter 3.]

Jahnke, L. and Klein, H. P. 1979. Oxygen as a factor in eukaryote evolution: Some effects of low levels of oxygen on *Saccharomyces cerevisiae*. *Origins of Life 9*:329–334. [Chapter 13.]

Jakosky, B. M. and Ahrens, T. J. 1979. The history of an atmosphere of impact origin. *Proc.*

10th Lunar Sci. Conf.: 2727–2739. [Chapters 2, 4.]

James, H. L. 1954. Sedimentary facies of iron-formation. *Econ. Geol. 49*:235–293. [Chapter 11.]

———. 1966. Chemistry of the iron-rich sedimentary rocks. In *Data of Geochemistry, U.S. Geol. Surv. Prof. Pap. 440-W*, chapt. W. [Chapter 11.]

James, H. L. and Hedge, C. E. 1980. Age of the basement rocks of southwest Montana. *Geol. Soc. Am. Bull. 91*:11–15. [Chapter 11.]

James, H. L. and Sims, P. K., eds., 1973. Precambrian iron-formations of the world. *Econ. Geol. 68*:913–1179. [Chapter 1.]

Jenkins, W. J.; Edmond, J. M.; and Corliss, J. B. 1978. Excess ^{3}He and ^{4}He in Galapagos submarine hydrothermal waters. *Nature 272*:156–158. [Chapter 4.]

Joliffe, A. W. 1955. Geology and iron ores of Steep Rock Lake. *Econ. Geol. 50*:373–398. [Chapter 8.]

Jolly, W. T. 1980. Development and degradation of Archean lavas, Abitibi area, Canada, in light of major element geochemistry. *J. Petrol. 21*: 323–363. [Chapter 3.]

Jones, D. G. and Dickert, J. J., Jr. 1965. Composition and reactions of oil shale of the Green River Formation. *Chem. Eng. Progr. Symp. Ser. 54*: 33–41. [Appendix IV.]

Junge, C. E.; Schidlowski, M.; Eichmann, R.; and Pietrek, H. 1975. Model calculations for the terrestrial carbon cycle: Carbon isotope geochemistry and evolution of photosynthetic oxygen. *J. Geophys. Res. 80*:4542–4552. [Chapter 7.]

Junk, G. and Svec, H. J. 1958. The absolute abundance of the nitrogen isotopes in the atmosphere and compressed gas from various sources. *Geochim. Cosmochim. Acta. 14*:234–243. [Chapter 7.]

Kahn, P. G. K. and Pompea, S. M. 1978. Nautiloid growth rhythms and dynamical evolution of the Earth-Moon system. *Nature 275*:606–611. [Chapter 11.]

Kalkowsky, E. 1908. Oölith and stromatolith in norddeutschen Buntsandstein. *Z. Deutsch. Geol. Gesell. 60*:68–125. [Chapter 1.]

Kalliokoski, J. 1975. Chemistry and mineralogy of Precambrian paleosols in northern Michigan. *Geol. Soc. Am. Bull. 86*:371–376. [Chapter 11.]

Kaneoka, I. and Takaoka, N. 1980. Rare gas isotopes in Hawaiian ultramafic nodules and volcanic rocks: Constraints on genetic relationships. *Science 208*:1366–1368. [Chapter 4.]

Kaneoka, I.; Takaoka, N.; and Aoki, K. I. 1977. Rare gases in a phlogopite nodule and phlogopite-bearing peridotite in South African kimberlites. *Earth Planet. Sci. Lett. 36*:181–186. [Chapter 4.]

Kaplan, I. R. 1975. Stable isotopes as a guide to biogeochemical processes. *Proc. Roy. Soc. London Ser. B 189*:183–211. [Chapter 7.]

Kaplan, I. R. and Nissenbaum, A. 1966. Anomalous carbon isotope ratios in nonvolatile organic material. *Science 153*:744–745. [Chapters 7, 15.]

Kaplan, I. R. and Rittenberg, S. C. 1964. Microbiological fractionation of sulfur isotopes. *J. Gen. Microbiol. 34*:195–212. [Chapter 7.]

Kaplan, I. R.; Emery, K. O.; and Rittenberg, S. C. 1963. The distribution and isotopic abundance of sulfur in recent marine sediments off southern California. *Geochim. Cosmochim. Acta 27*:297–331. [Chapter 7.]

Kaplan, I. R.; Sweeney, R. E.; and Nissenbaum, A. 1969. Sulfur isotope studies on Red Sea geothermal brines and sediments. In E. T. Degens and D. A. Ross, eds., *Hot Brines and Recent Heavy Metal Deposits in the Red Sea*, New York, Springer, 474–498. [Chapter 10.]

Karkhanis, S. N. 1976. Fossil iron bacteria may be preserved in Precambrian ferroan carbonate. *Nature 261*:406–407. [Chapter 9.]

Karl, D. M.; Wirsen, C. O.; and Jannasch, H. W. 1980. Deep-sea primary production at the Galapagos hydrothermal vents. *Science 207*: 1345–1347. [Chapter 5.]

Kasting, J. F. and Donahue, T. M. 1980. The evolution of atmospheric ozone. Unpublished manuscript. [Chapter 11.]

Kasting, J. F.; Liu, S. C.; and Donahue, T. M. 1979. Oxygen levels in the prebiological atmosphere. *J. Geophys. Res. 84*:3097–3107. [Chapters 4, 11.]

Kasting, J. F. and Walker, J. C. G. 1981. Limits on oxygen concentration in the prebiological atmosphere and the rate of abiotic fixation of nitrogen. *J. Geophys. Res. 86*:1147–1158. [Chapters 4, 11.]

Katsumori, M. 1979. Photochemical-radiative equilibrium of the Earth's paleo-atmospheres with various amounts of oxygen. *J. Meteorol. Soc. Japan 57*:243–253. [Chapter 11.]

Kaula, W. M. 1979. Thermal evolution of Earth and Moon growing by planetesimal impacts. *J. Geophys. Res. 84*:999–1008. [Chapters 2, 3, 4.]

———. 1980. The beginnings of the Earth's thermal evolution. In D. W. Strangway, ed., *The Continental Crust and Its Mineral Deposits*, Toronto, Geol. Assoc. Can., 25–34. [Chapters 2, 4.]

Kazmierczak, J. 1976. Devonian and modern relatives of the Precambrian *Eosphaera*: Possible significance for early eukaryotes. *Lethaia 9*: 39–50. [Chapter 14.]

———. 1979. The eukaryotic nature of *Eosphaera*-like ferriferous structures from the Precambrian Gunflint Iron Formation, Canada: A comparative study. *Precambrian Res. 9*:1–22. [Chapter 14.]

Keevil, T. and Mason, H. S. 1978. Molecular oxygen in biological oxidations—an overview. In S. Fleischer and L. Paker, eds., *Biomembranes: Methods in Enzymology, Vol. 3*, New York, Academic Press, pp. 3–40. [Chapter 13.]

Keevil, C. W.; Hough, J. S.; and Cole, J. A. 1979. Regulation of 2-oxoglutarate dehydrogenase synthesis in *Citrobacter freundii* by traces of oxygen in commercial nitrogen gas and by glutamate. *J. Gen. Microbiol. 114*:355–359. [Chapter 13.]

Keller, B. M., ed. 1963. *Stratigraphiya SSSR: Verkhnii Kokembrii*. (*Stratigraphy of the USSR: Upper Precambrian*), Akad. Nauk SSSR, Gosudarstvennyi Geol. Kom. SSSR, Ministerstvo Visshevo i Srednevo Spetsial nevo Obrazovovaniya SSSR, Moscow, Gosudarstvennoye Nauchno-Techniskoye Izdatelstova Literaturi po Geologii i Okhrane Nedr, 716 pp. (in Russian). [Chapter 1.]

Kelly, K. L. and Judd, D. B. 1976. *Color: Universal Language and Dictionary of Names*, U.S. Natl. Bur. Stand. Spec. Pub. 440, 184 pp. [Appendix IV.]

Kemp. A. L. W. and Thode, H. G. 1968. The mechanism of bacterial reduction of sulphate and sulphite from isotope fractionation studies. *Geochim. Cosmochim. Acta 32*:71–91. [Chapter 7.]

Kempe, S. 1979. Carbon in the rock cycle. In B. Bolin, E. T. Degens, S. Kempe, and P. Ketner, eds., *The Global Carbon Cycle*, Chichester, Wiley, 343–377. [Chapter 5.]

Kennedy, C. S. and Kennedy, G. C. 1976. The equilibrium boundary between graphite and diamond. *J. Geophys. Res. 81*: 2467–2470. [Chapter 4.]

Kenyon, D. H. and Blois, M. S. 1965. π-electron photochemistry of DL-phenylalanine in oxygen saturated and oxygen free solutions. *Photochem. Photobiol. 4*:335–349. [Appendix III.]

Kenyon, D. H. and Steinman, G. 1969. *Biochemical Predestination*, New York, McGraw-Hill, 301 pp. [Chapter 4.]

Key, R. M. 1977. The chronology of Botswana. *Geol. Soc. S. Afr. Trans. 80*:31–42. [Chapter 11.]

Khare, B. N. and Sagan, C. 1973. Red clouds in reducing atmospheres. *Icarus 20*:311–321. [Appendix III.]

Kimberley, M. M. 1979. Origin of oolitic iron formations. *J. Sed. Petrol. 49*:111–132. [Chapter 11.]

King, P. B. 1965. Geology of the Sierra Diablo region, Texas. *U.S. Geol. Surv. Prof. Pap. 480*, 185 pp. [Appendix I.]

Kinsman, D. J. J. and Holland, H. D. 1969. The co-precipitation of cations with $CaCO_3$—IV. The co-precipitation of Sr^{2+} with aragonite between 16 and 96°C. *Geochim. Cosmochim. Acta 33*:1–18. [Chapter 10.]

Kirschenbaum, I.; Smith, J. S.; Crowell, T.; Graff, J.; and McKee, R. 1947. Separation of isotopes by the exchange reaction between ammonia and solution of ammonium nitrate. *J. Chem. Phys. 15*:440–446. [Chapter 7.]

Klappa, C. F. 1979. Lichen stromatolites: Criterion for subaerial exposure and a mechanism for the formation of laminar calcretes (caliche). *J. Sed. Petrol. 49*:387–400. [Chapter 8.]

Klein, C. 1974. Greenalite, stilpnomelane, minnesotaite, crocidolite and carbonates in a very low-grade metamorphic Precambrian iron-formation. *Can. Mineral. 12*:475–498. [Chapter 11.]

———. 1978. Regional metamorphism of Proterozoic iron-formation, Labrador Trough, Canada. *Am. Mineral. 63*:898–912. [Chapter 11.]

Klein, C. and Bircker, O. P. 1977. Some aspects of the sedimentary and diagenetic environment of Proterozoic banded iron-formation. *Econ. Geol. 72*:1457–1470. [Chapter 11.]

———. 1978. Some aspects of the sedimentary and diagenetic environment of Proterozoic banded iron-formations—a reply. *Econ. Geol. 73*:1371–1373. [Chapter 11.]

Klein, C. and Fink, R. P. 1976. Petrology of the Sokoman Iron Formation in the Howells River area, at the western edge of the Labrador Trough. *Econ. Geol. 71*:453–487. [Chapter 9.]

Kline, G. L. 1975a. Proterozoic Budding Bacteria From Australia and Canada. M. Sc. Thesis, Univ. Calif., Santa Barbara (Dept. Geol.) 92 pp. [Chapter 14.]

———. 1975b. *Metallogenium*-like microorganisms from the Paradise Creek Formation. *Geol. Soc. Am. Abstr. Progr. 7* (3):336 (Abstr.). [Chapter 14.]

Klein, H. P. 1977. The Viking biological investigation: General aspects. *J. Geophys. Res. 82*:4677–4680. [Chapter 4.]

Klots, C. E. and Benson, B. B. 1963. Isotope effect in the solution of oxygen and nitrogen in distilled water. *J. Chem. Phys. 38*:890–892. [Chapter 7.]

Knauth, L. P. and Epstein, S. 1976. Hydrogen and oxygen isotope ratios in nodular and bedded cherts. *Geochim. Cosmochim. Acta 40*: 1095–1108. [Chapters 10, 11.]

Knauth, L. P. and Lowe, D. R. 1978. Oxygen isotope geochemistry of cherts from the Onverwacht Group (3.4 billion years), Transvaal, South Africa, with implications for secular variations in the isotopic composition of cherts. *Earth Planet. Sci. Lett. 41*:209–222. [Chapters 10, 11.]

Knoll, A. H. 1979. Archean photoautotrophy: Some alternatives and limits. *Origins of Life 9*: 313–327. [Chapter 11.]

Knoll, A. H. and Barghoorn, E. S. 1975a. Precambrian eucaryotic organisms: A reassessment of the evidence. *Science 190*:52–54. [Chapter 1.]

———. 1975b. A Gunflint-type flora from the Duck Creek Dolomite, Western Australia. *Abstracts, College Park Symposium on Chemical Evolution of the Early Precambrian* (Univ. of Maryland): 61 (Abstr.). [Chapter 14.]

———. 1976. A Gunflint-type microbiota from the Duck Creek Dolomite, Western Australia. *Origins of Life 7*:417–423. [Chapter 14.]

———. 1977. Archean microfossils showing cell division from the Swailand System of South Africa. *Science 198*:396–398. [Chapter 9.]

Knoll, A. H. and Simonson, B. 1981. Early Proterozoic microfossils and penecontemporaneous quartz cementation in the Sokoman Iron Formation, Canada. *Science 211*:478–480. [Chapter 14.]

Knoll, A. H.; Barghoorn, E. S.; and Awramik, S. M. 1978. New microorganisms from the Aphebian Gunflint Iron Formation, Ontario. *J. Paleontol. 52*:976–992. [Chapters 1, 14.]

Knoll, A. H.; Barghoorn, E. S.; and Golubić, S. 1975. *Paleopleurocapsa wopfneri* gen. et sp. nov.: A late Precambrian alga and its modern counterpart. *Proc. Natl. Acad. Sci. USA 72*:2488–2492. [Chapter 1.]

Kohlmiller, E. F., Jr. and Gest, H. 1951. A comparative study of the light and dark fermentations of organic acids by *Rhodospirillum rubrum*. *J. Bacteriol. 61*:269–282. [Chapter 6.]

Kondratieva, E. N. and Mekhtieva, V. L. 1966. Fractionation of stable isotopes of sulfur by green photosynthesizing bacteria. *Microbiol. 35*:481–483 (English transl.). [Chapter 7.]

Korolyuk, I. K. 1958. Znachenie stromatolitoz dlya stratigrafii kembriya i dokembriya na primere yuga sibirskoi platformy (Significance of stromatolites for the Cambrian and Precambrian stratigraphy in the case of the south Siberian platform). *Trudy Mezhdu. Sovesch. Raz. Unifitsirovan, Stratig. Skhem Sib., Akad. Nauk SSSR*, 103–109 (in Russian). [Chapter 1.]

Kouvo, O. and Tilton, G. R. 1966. Mineral ages from the Finnish Precambrian. *J. Geol. 74*:421–442. [Chapter 11.]

Krasnovskii, A. A. 1959. Development of the mode of action of the photocatalytic system in organisms. In A. I. Oparin, A. G. Pasynskii, A. E. Braunschtein, and T. E. Pavlovskaya, eds., *Proc. 1st Internat. Symp. on the Origin of Life on the Earth*, New York, Pergamon, 606–618. [Chapter 6.]

Krebs, H. A. and Kornberg, H. L. 1957. Energy transformations in living matter. *Ergeb. Physiol. Biol. Chem. Exp. Pharmakol. 49*:212–298. [Chapter 6.]

Krinsky, N. I. 1978. Non-photosynthetic function of carotenoids. *Phil. Trans. Roy. Soc. London Sect. B 284*:581–590. [Chapter 13.]

———. 1979. Carotenoid protection against oxidation. *Pure and Appl. Chem. 51*:649–660. [Chapter 13.]

Kröger, A. 1977. Phosphorylative electron transport with fumarate and nitrate as terminal hydrogen acceptors. In B. A. Haddock and W. A. Hamilton, eds., *Microbial Energetics*

(27th Symp. Gen. Microbiol.), Cambridge, Cambridge Univ. Press, 61–93. [Chapter 6.]
———. 1978. Fumarate as terminal electron acceptor of phosphorylative electron transport. *Biochim. Biophys. Acta 505*: 129–145. [Chapter 6.]
Krogh, T. E. and Davis, G. L. 1971. Zircon U-Pb ages of Archean metavolcanic rocks in the Canadian Shield. *Carnegie Inst. Wash. Yearbook 70*:241–242. [Appendix I.]
Kröner, A. 1981. Precambrian plate tectonics. In A. Kröner, ed., *Precambrian Plate Tectonics*, Amsterdam, Elsevier, 57–90. [Chapter 10.]
Krouse, H. R. and Modzeleski, V. E. 1970. $^{13}C/^{12}C$ abundances in components of carbonaceous chondrites and terrestrial samples. *Geochim. Cosmochim. Acta 34*:459–474. [Appendix III.]
Krumbein, W. E. 1979a. Calcification by bacteria algae. In P. A. Trudinger and D. J. Swaine, eds., *Biogeochemical Cycling of Mineral-forming Elements*, Amsterdam, Elsevier, 47–68. [Chapter 8.]
———. 1979b. Photolithotrophic and chemoorganotrophic activity of bacteria and algae as related to beachrock formation and degradation (Gulf of Aqaba, Sinai). *Geomicrobiol. J. 1*:139–203. [Chapter 8.]
Krumbein, W. E. and Cohen, Y. 1977. Primary production, mat formation and lithification: Contribution of oxygenic and facultative anoxygenic cyanobacteria. In E. Flügel, ed., *Fossil Algae*, New York, Springer, 37–56. [Chapter 8.]
Krumbein, W. E. and Giele, C. 1979. Calcification in a coccoid cyanobacterium associated with the formation of desert stromatolites. *Sedimentol. 26*:593–604. [Chapter 8.]
Krumbein, W. E. and Potts, M. 1979. Girvanella-like structures formed by *Plectonema gloeophilum* (Cyanophyta) from the Borrego Desert in southern California. *Geomicrobiol. J. 1*:211–217. [Chapter 8.]
Krumbein, W. E.; Cohen, Y.; and Shilo, M. 1977. Solar Lake (Sinai). 4. Stromatolitic cyanobacterial mats. *Limnol. Oceanogr. 22*:635–656. [Chapter 8.]
Krylov, I. N. 1959a. Stromatolites from the Riphean of the Urals. *Dokl. Akad. Nauk SSSR 126*:1312–1315. 1960 transl. American Geophysical Institute, Washington, D.C., 512–514. [Chapter 1.]
———. 1959b. Riphean stromatolites of Kil'din Island. *Dokl. Akad. Nauk SSSR 127*:888–891. 1960 transl. American Geophysical Institute, Washington, D.C., 797–799. [Chapter 1.]
———. 1968. Drevneishiye sledi zhizni na zemlye (The earliest traces of life on Earth). *Priroda 11*: 41–54 (in Russian). Chapter 1.]
Kuhn, T. S. 1970. *The Structure of Scientific Revolutions*, 2nd ed., Chicago, Univ. Chicago Press, 210 pp. [Chapter 1.]
Kuhn, W. R. and Atreya, S. K. 1979. Ammonia photolysis and the greenhouse effect in the primordial atmosphere of the Earth. *Icarus 37*: 207–213. [Chapters 4, 11.]
Kvenvolden, K. A. 1975. Advances in the geochemistry of amino acids. *Ann. Rev. Earth Planet. Sci. 3*:183–212. [Chapter 1.]
Kvenvolden, K. A.; Lawless, J.; Pering, K.; Peterson, E.; Flores, J.; Ponnamperuma, C.; Kaplan, I. R.; and Moore, C. B. 1970. Evidence for extraterrestrial amino acids and hydrocarbons in the Murchison meteorite. *Nature 228*:923–926. [Chapter 4; Appendix III.]
Kvenvolden, K. A.; Peterson, E.; and Pollock, G. E. 1969. Optical configuration of amino acids in Precambrian Fig Tree chert. *Nature 221*: 141–143. [Chapter 5.]
LaBerge, G. L. 1967. Microfossils in Precambrian iron-formations. *Geol. Soc. Am. Bull. 78*:331–342. [Chapters 9, 14.]
———. 1973. Possible biological origin of Precambrian iron-formations. *Econ. Geol. 68*:1098–1109. [Chapters 9, 14.]
Lahav, N. and Chang, S. 1976. The possible role of solid surface area in condensation reactions during chemical evolution: Reevaluation. *J. Mol. Evol. 8*:357–380. [Chapter 4.]
Lahave, N. and White, D. H. 1980. A possible role of fluctuating clay-water systems in the production of ordered prebiotic oligomers. *J. Mol. Evol. 16*:11–21. [Chapter 4.]
Lahav, N.; White, D.; and Chang, S. 1978. Peptide formation in the prebiotic era: Thermal condensation of glycine in fluctuating clay environments. *Science 201*:67–69. [Chapter 4.]
Lambeck, K. 1977. Tidal dissipation in the oceans: Astronomical, geophysical and oceanographic consequences. *Phil. Trans. Roy. Soc. London Sect. B 287*:545–594. [Chapter 11.]
Lambeck, K. 1978. The Earth's paleorotation. In P. Brosche and J. Sünderman, eds., *Tidal Friction and the Earth's Rotation*, New York, Springer, 145–153. [Chapter 11.]

———. 1979. The history of the Earth's rotation. In M. W. McElhinny, ed., *The Earth: Its Origin, Structure and Evolution*, New York, Academic Press, 59–81. [Chapter 11.]

Lambert, I. B. 1978. Sulfur isotope investigations of Archaean mineralization and some implications concerning geobiochemical evolution. *Pub. Geol. Dept. and Extension Service, Univ. W. Austr. 2*:45–56. [Chapter 7.]

Lambert, I. B.; Donnelly, T. H.; Dunlop, J. S. R.; and Groves, D. I. 1978. Stable isotope studies of early Archaean sulfate deposits of probable evaporitic and volcanogenic origins. *Nature 276*: 808–811. [Chapters 7, 8.]

Lambert, R. St. J. 1976. Archean thermal regimes, crustal and upper mantle temperatures, and a progressive evolutionary model for the Earth. In B. F. Windley, ed., *The Early History of the Earth*, New York, Wiley, 363–376. [Chapters 3, 4.]

Lambert, R. St. J.; Chamberlain, V. E.; and Holland, J. G. 1976. The geochemistry of Archean rocks. In B. F. Windley, ed., *The Early History of the Earth*, New York, Wiley, 377–403. [Chapter 4.]

Lancet, M. S. and Anders, E. 1970. Carbon isotope fractionation in Fischer-Tropsch synthesis and in meteorites. *Science 170*:980–982. [Chapters 4, 7.]

Landis, C. R. 1971. Graphitization of dispersed carbonaceous material in metamorphic rocks. *Contrib. Mineral. Petrol. 30*:34–45. [Chapters 5, 9.]

Lange, M. A. and Ahrens, T. J. 1980. The evolution of an impact generated atmosphere. In *Lunar and Planetary Science, Vol. 9*, Houston, Lunar Planet. Sci. Inst., 596–598. [Chapter 4.]

Lasaga, A. C.; Holland, H. D.; and Dwyer, M. J. 1971. Primordial oil slick. *Science 174*:53–55. [Chapter 4.]

Lawless, J. G. and Boynton, C. D. 1973. Thermal synthesis of amino acids from a simulated primitive atmosphere. *Nature 243*:405–407. [Chapter 4.]

Lawless, J. G. and Edelson, E. H. 1980. The possible role of metal ions and clays in prebiotic chemistry. In R. Holmquist, ed., *Life Sciences and Space Research, Vol. 8*, New York, Pergamon, 83–88. [Chapter 4.]

Lawless, J. G. and Levi, N. 1979. The role of metal ions in chemical evolution: Polymerization of alanine and glycine in a cation-exchanged clay environment. *J. Mol. Evol. 13*:281–286. [Chapter 4.]

Lawlor, D. L.; Fester, J. I.; and Robinson, W. E. 1963. Pyrite removal from oil-shale concentrates using lithium aluminum hydride. *Fuel 42*:239–244. [Chapter 5; Appendix IV.]

Lawrence, J. R. and Gieskes, J. M. 1981. Constraints on water transport and alteration in the oceanic crust from the isotopic composition of pore water. *J. Geophys. Res. 86*:7924–7934. [Chapter 10.]

Lebedev, A. P. 1957. The geochemistry of carbon in Siberian Trap rocks and some other basic rocks of the USSR. *Geochem. 1975*:193–197. [Chapter 4.]

Lee, T. 1979. New isotopic clues to solar system formation. *Rev. Geophys. Space Phys. 17*:1591–1611. [Chapter 4.]

Lee, T.; Pappanastassiou, D.; and Wasserburg, G. J. 1977. Aluminum-26 in early solar system: Fossil or fuel? *Astrophys. J. 211*:L107–L110. [Chapter 2.]

Lemmon, R. 1970. Chemical evolution. *Chem. Rev. 70*:95–109. Chapter 4.

Lepp, H. S. and Goldich, S. S. 1964. Origin of the Precambrian iron formations. *Econ. Geol. 59*: 1025–1060. [Chapters 1, 11.]

Lerman, A. 1979. *Geochemical Processes Water and Sediment Environments*, New York, Wiley, 481 pp. [Chapters 5, 12.]

———. 1982. Sedimentary balance through geological time. In H. D. Holland and M. Schidlowski, eds., *Mineral Deposits and the Evolution of the Biosphere*, Dahlem Konferenzen-Dahlem Workshop Rept. 16, Berlin, Springer, 237–256. [Chapter 10.]

Leventhal, J. 1976. Nitrogen in gas wells as an indicator of atmospheric evolution and the nitrogen cycle. Unpublished manuscript. [Chapter 7.]

Leventhal, J.; Suess, S. E.; and Cloud, P. 1975. Nonprevalence of biochemical fossils in kerogen from pre-Phanerozoic sediments. *Proc. Natl. Acad. Sci. USA 72*:4706–4710. [Chapters 5, 7, 15.]

Levine, J. S.; Hays, P. B.; and Walker, J. C. G. 1979. The evolution and variability of atmospheric ozone over geological time. *Icarus 39*: 295–309. [Chapter 11.]

Lewin, R. A. 1976. Prochlorophyta as a new division of algae. *Nature 261*:697–698. [Glossary I.]

———. 1979. Formal taxonomic treatment of Cyanophytes. *Internat. J. Syst. Bacteriol. 29*: 411–412. [Chapter 13.]

Lewis, J. and Prinn, R. G. 1980. Kinetic inhibition of CO and N_2 reduction in the solar nebula. *Astrophys. J. 238*:357–364. [Chapter 4.]

Lex, M.; Silverster, W. B.; and Stewart, W. D. P. 1972. Photorespiration and nitrogenase activity in the blue-green alga *Anabaena cylindrica*. *Proc. Roy. Soc. London Ser. B 180*:87–102. [Chapter 11.]

Libby, L. M. and Pandolfi, L. J. 1974. Temperature dependence of isotope ratios in tree rings. *Proc. Natl. Acad. Sci. USA 71*:2482–2486. [Chapter 7.]

Licari, G. R. 1978. Biogeology of the late pre-Phanerozoic Beck Spring Dolomite of eastern California. *J. Paleontol. 52*:767–792. [Chapter 1.]

Licari, G. R. and Cloud, P. 1968. Reproductive structures and taxonomic affinities of some nannofossils from the Gunflint Iron Formation. *Proc. Natl. Acad. Sci. USA 59*:1053–1060. [Chapter 14.]

Licari, G. R. and Cloud, P. 1972. Prokaryotic algae associated with Australian Proterozoic stromatolites. *Proc. Natl. Acad. Sci. USA 69*:2500–2504. [Chapter 14.]

Licari, G. R.; Cloud, P.; and Smith, W. D. 1969. A new chroococcacean alga from the Proterozoic of Queensland. *Proc. Natl. Acad. Sci. USA 62*: 52–62. [Chapter 14.]

Liebenberg, W. R. 1955. The occurrence and origin of gold and radioactive minerals in the Witwatersrand System, the Dominion Reef, the Ventersdorp Contact Reef, and the Black Reef. *Geol. Soc. S. Afr. Trans. 58*:101–227. [Chapter 11.]

Linton, J. D.; Harrison, D. E. F.; and Bull, A. T. 1975. Molar growth yields, respiration and cytochrome patterns of *Beneckea natriegens* when grown at different medium dissolved-oxygen tensions. *J. Gen. Microbiol. 90*:237–246. [Chapter 13.]

Lipmann, F. 1946. Metabolic process patterns. In D. E. Green, ed., *Currents in Biochemical Research*, New York, Interscience, 137–148. [Chapter 6.]

Liu, C. L. and Peck, H. D., Jr. 1980. Comparative bioenergetics of sulfate reduction in *Desulfovibrio* and *Desulfotomaculum*. *Abstracts, Ann. Meeting Am. Soc. Microbiol.* (Boston): 157 (Abstr.). [Chapter 6.]

Liu, K. K. 1979. Geochemistry of Inorganic Nitrogen Compounds in Two Marine Environments: The Santa Barbara Basin and the Ocean off Peru. Ph.D. Thesis, Univ. Calif., Los Angeles (Dept. Earth and Space Sci.), 354 pp. [Chapter 7.]

Ljungdahl, L. G. and Andreesen, J. R. 1976. Reduction of CO_2 to acetate in homoacetate fermenting clostridia and the involvement of tungsten in formate dehydrogenase. In H. G. Schlegel, G. Gottschalk, and N. Pfennig, eds., *Microbial Production and Utilization of Gases*, Gottingen, E. Goltze KG, 163–172. [Chapter 6.]

Logan, B. W. 1974. Inventory of diagenesis in Holocene-Recent carbonate sediments, Shark Bay, Western Australia. *Am. Assoc. Petrol. Geol. Mem. 22*:195–249. [Chapter 8.]

Logan, B. W. and Semeniuk, V. 1976. Dynamic metamorphism; processes and products in Devonian carbonate rocks, Canning Basin, Western Australia. *Geol. Soc. Austr. Spec. Pub. 6*, 138 pp. [Chapter 8.]

Logan, B. W.; Hoffman, P.; and Gebelein, C. D. 1974. Algal mats, cryptalgal fabrics and structures, Hamelin Pool, Western Australia. *Amer. Assoc. Petrol. Geol. Mem. 13*:38–84. [Chapter 8.]

Logan, J. A.; Prather, M. J.; Wofsy, S. C.; and McElroy, M. B. 1978. Atmospheric chemistry: Response to human influence. *Phil. Trans. Roy. Soc. London Sect. A 290*:187–234. [Chapter 4.]

Lohrmann, R.; Bridson, P. K.; and Orgel, L. E. 1980. Efficient metal-ion catalyzed template-directed oligonucleotide synthesis. *Science 208*: 1464–1465. [Chapter 4.]

Long, G.; Neglia, S.; and Favretto, L. 1968. The metamorphism of the kerogen from Triassic black shales, southeast Sicily. *Geochim. Cosmochim. Acta 32*:647–656. [Chapter 5.]

Lopuchin, A. S. 1975. Structures of biogenic origin from early Precambrian rocks of Euro-Asia. *Origins of Life 6*:45–57. [Chapters 9, 14.]

———. 1976. Probable ancestors of Cyanophyta in sedimentary rocks of the Precambrian and Paleozoic. *Geol. Fören. Stockholm Förhandl. 98*: 297–315. [Chapters 9, 14.]

Lopuchin, A. S. and Moralev, V. M. 1973. Vodorslepodobnye mikroiskopaemye v arkheyskikh porodakj Yuzhnoy Indii (Algae-like microstructures in Archean rocks of South India). *Izv. Vysshikh Ucheb. Zaved., Geol. Raz. 1973* (7): 185–187 (in Russian). [Chapter 9.]

Lougheed, M. S. and Mancuso, J. J. 1973. Hematite framboids in the Negaunee Formation, Michigan, evidence for their biologic origin. *Econ. Geol. 68*:202–209. [Chapter 14.]

Love, L. G. 1965. Micro-organic material with diagenetic pyrite from the Lower Proterozoic Mount Isa shale and a Carboniferous shale. *Proc. Yorkshire Soc. 35*:187–202. [Chapter 14.]

Love, L. G. and Zimmerman 1961. Bedded pyrite and microorganisms from the Mount Isa Shale. *Econ. Geol. 56*:873–896. [Chapter 14.]

Lowe, D. R. 1980a. Stromatolites 3,400-Myr old from the Archaean of Western Australia. *Nature 284*:441–443. [Chapters 1, 7, 8, 9.]

———. 1980b. Archean sedimentation. *Ann. Rev. Earth Planet. Sci. 8*:145–167. [Chapters 3, 8, 10; Appendix I.]

Lowe, D. R. and Knauth, L. P. 1977. Sedimentology of the Onverwacht Group (3.4 billion years), Transvaal, South Africa, and its bearing on the characteristics and evolution of the early Earth. *J. Geol. 85*:699–723. [Chapters 3, 7, 8, 10, 11.]

Lumsden, J.; Cammack, R.; and Hall, D. O. 1976. Purification and physicochemical properties of superoxide dismutase from two photosynthetic micro-organisms. *Biochem. Biophys. Acta 438*: 380–392. [Chapter 13.]

Lumsden, J.; Henry, L.; and Hall, D. O. 1977. Superoxide desmutase in photosynthetic organisms. In A. M. Michelson, J. M. McCord, and I. Fridovich, eds., *Superoxide and Superoxide Dismutases*, London, Academic Press, 437–450. [Chapter 13.]

Luth, W. C.; Jahns, R. H.; and Tuttle, O. F. 1964. The granite system at pressures of 4 to 10 kilobars. *J. Geophys. Res. 69*:759–773. [Chapter 3.]

Lyon, G. L. 1974. Geothermal gases. In I. R. Kaplan, ed., *Natural Gases in Marine Sediments*, New York, Plenum, 141–150. [Chapter 7.]

MacGreehan, P. J. and MacLean, W. H. 1980. An Archean sub-seafloor geothermal system, "calc-alkali" trends, massive sulphide genesis. *Nature 286*:767–771. [Chapter 11.]

MacGregor, A. M. 1927. The problem of the Precambrian atmosphere. *S. Afr. J. Sci. 24*:155–172. [Chapters 1, 11.]

———. 1941. A Pre-Cambrian algal limestone in Southern Rhodesia. *Geol. Soc. S. Afr. Trans. 43*:9–15. [Chapter 8.]

MacGregor, I. M.; Truswell, J. F.; and Eriksson, K. A. 1974. Filamentous algae from the 2,300 m.y. old Transvaal Dolomite. *Nature 247*:538–540. [Chapter 14.]

Madigan, M. T. and Gest, H. 1978. Growth of a photosynthetic bacterium anaerobically in darkness, supported by "oxidant-dependent" sugar fermentation. *Arch. Microbiol. 117*:119–122. [Chapter 6.]

———. 1979. Growth of the photosynthetic bacterium *Rhodopseudomonas Capsulata* chemoautotrophically in darkness with H_2 as the energy source. *J. Bacteriol. 137*:524–530. [Glossary II.]

Madigan, M. T.; Cox, J. C.; and Gest, H. 1980. Physiology of dark fermentative growth of *Rhodopseudomonas capsulata. J. Bacteriol. 142*:908–915. [Chapter 6; Glossary II.]

Madison, K. M. 1958. Fossil protozoans from the Keewatin sediments. *Illinois Acad. Sci. Trans. 50*:287–290 (dated 1957, issued 1958). [Chapter 9; Appendix I.]

Mahler, H. R. and Raff, R. A. 1975. The evolutionary origin of the mitochondrion: A nonsymbiotic origin. *Internat. Rev. Cytol. 43*:1–124. [Chapter 1.]

Maignien, R. 1966. *Review of Research on Laterites*, Leige, UNESCO, 148 pp. [Chapter 8.]

Maithy, P. K. 1975. Micro-organisms from the Bushimay System (late Precambrian) of Kanshi, Zaire. *The Paleobotanist 22*:133–149 (dated 1973, issued 1975). [Chapter 1.]

Maithy, P. K. and Shukla, M. 1977. Microbiota from the Suket shales, Ramapura, Vindhyan System (Late Pre-Cambrian), Madhya Pradesh. *The Paleobotanist 23*:176–188 (dated 1974, issued 1977). [Chapter 1.]

Mäkelä, M. 1974. A study of sulfur isotopes in the Outokumpu ore deposits, Finland. *Geol. Surv. Finl. Bull. 267*, 45 pp. [Chapter 7.]

Malmstrom, B. G.; Andreasson, L. A.; and Reinhammar, B. 1975. Copper containing oxidases and superoxide dismutase. In Boyer, P. D. ed., *The Enzymes*, 3rd ed., *Vol. 12*, New York, Academic Press, 507–579. [Chapter 13.]

Mancusco, J. J.; Lougheed, M. S.; and Wygant, T. 1971. Possible biogenic structures from the Precambrian Negaunee (Iron) Formation, Marquette Range, Michigan. *Am. J. Sci. 271*:181–186. [Chapter 14.]

Mann, J. 1978. *Secondary Metabolism*, Oxford, Oxford Univ. Press, 316 pp. [Chapter 13.]

Margulis, L. 1970. *Origin of Eucaryotic Cells*, New Haven, Conn., Yale Univ. Press, 349 pp. [Chapter 1.]

Margulis, L.; Walker, J. C. G.; and Rambler, M. B. 1976. Reassessment of roles of oxygen and ultraviolet light in Precambrian evolution. *Nature 264*:620–624. [Chapters 8, 11.]

Markhinin, E. K. and Podkletnov, N. E. 1977. The phenomenon of formation of prebiological compounds in volcanic processes. *Origins of Life 8*:225–235. [Chapter 4.]

Marowsky, G. 1969. Schwefel-, Kohlenstoff- und Sauerstoff-Isotopenuntersuchungen am Kupferschiefer als Beitrag zur genetischen Deutung. *Contr. Mineral. Petrol. 22*:290–334. [Chapter 7.]

Marshall, C. G. A.; May, J. W.; and Perret, C. J. 1964. Fossil microorganisms: Possible presence in Precambrian Shield of Western Australia. *Science 144*:290–292. [Chapter 9.]

Marshall, K. C. 1979. Biogeochemistry of manganese minerals. In P. A. Trudinger and D. J. Swaine, eds., *Biogeochemical Cycling of Mineral-forming Elements, Studies in Environmental Science 3*, Amsterdam, Elsevier, 253–292. [Chapter 14.]

Marston, R. J. and Travis, G. A. 1976. Stratigraphic implications of heterogeneous deformation in the Jones Creek conglomerate (Archaean), Kathleen Valley, Western Australia. *J. Geol. Soc. Austr. 23*:141–156. [Chapter 11.]

Martens, C. S. and Berner, R. A. 1977. Interstitial water chemistry of anoxic Long Island Sound sediments. I. Dissolved gases. *Limnol. Oceanogr. 22*:10–25. [Chapter 12.]

Martens, C. S. and Klump, J. V. 1980. Biogeochemical cycling in an organic-rich coastal marine basin—I. Methane sediment-water exchange processes. *Geochim. Cosmochim. Acta 44*:471–490. [Chapter 5.]

Martin, A.; Nisbet, E. G.; and Bickle, M. J. 1980. Archaean stromatolites of the Belingwe Greenstone Belt, Zimbabwe (Rhodesia). *Precambrian Res. 13*:337–362. [Chapters 8, 11; Appendix I.]

Maslov, V. P. 1939. Nizhnepaleozoiskiye porodoobrazurjuskchie vodorosli vostochnoi Sibiri (Lower Paleozoic rock-building algae of East Siberia). *Problemi Paleontol. 2-3*:249–325 (in Russian with English summary). [Chapter 1.]

———. 1960. *Stromatoliti (Stromatolites), Trudy Geol. Inst. Akad. Nauk. SSSR 41*, 188 pp. (in Russian). [Chapter 1.]

Mason, B. 1975. Mineralogy and geochemistry of two Amîtsoq gneisses from the Godthaab region, West Greenland. *Rapp. Grønlands Geol. Unders. 71*:1–11. [Chapter 3.]

Mason, T. R. and Von Brunn, V. 1977. 3-Gyr-old stromatolites from South Africa. *Nature 266*:47–49. [Chapters 8, 10; Appendix I.]

Matson, D. L.; Veeder, G. L.; and Lebofsky, L. A. 1978. Infrared observations of asteroids from Earth and space. In D. Morrison and W. C. Wells, eds., *Asteroids: An Exploration Assessment*, Washington, D.C., NASA Conf. Pub. 2053, 127–144. [Chapter 4.]

Matin, A. 1978. Organic nutrition of chemolithotrophic bacteria. *Ann. Rev. Microbiol. 32*:433–468. [Glossary II.]

Matthew, D. E. and Hayes, J. M. 1978. Isotope ratio monitoring gas chromatography-mass spectrometry. *Analyt. Chem. 50*:1465–1473. [Appendix III.]

Matthew, G. F. 1980. On the existence of organisms in the Pre-Cambrian rocks. *Bull. Nat. Hist. Soc. New Brunswick 2* (9):28–33. [Chapters 1, 14.]

Matthews, C. N. and Moser, R. E. 1967. Peptide synthesis from hydrogen cyanide and water. *Nature 215*:1230–1234. [Appendix III.]

McCartney, J. T. and Teichmüller, M. 1972. Classification of coals according to degree of coalification by reflectance of the vitrinite component. *Fuel 51*:64–68. [Chapter 5.]

McClendon, J. H. 1976. Elemental abundance as a factor in the origins of mineral nutrient requirements. *J. Mol. Evol. 8*:175–195. [Chapter 13.]

McCord, J. M.; Keele, B. B., Jr.; and Fridovich, I. 1971. An enzyme based theory of obligate anaerobiosis: The physiological function of superoxide dismutase. *Proc. Natl. Acad. Sci. USA 68*:1024–1027. [Glossary II.]

McCready, R. G. L. and Krouse, H. R. 1980. Sulfur isotope fractionation by *Desulfovibrio vulgaris* during metabolism of $BaSO_4$. *Geomicrobiol. J. 2*:55–62. [Chapter 7.]

McCulloch, M. T. and Wasserburg, G. J. 1978. Sm-Nd and Rb-Sr chronology of continental crust formation. *Science 200*:1003–1011. [Chapter 10.]

McElhinney, M. W. 1966. Rb-Sr and K-Ar age measurements on the Modipe Gabbro and Gaberones Granite, Bechuanaland and South Africa. *Earth Planet. Sci. Lett. 1*:439–442. [Chapter 11.]

———. 1973. *Palaeomagnetism and Plate Tectonics*, Cambridge, Cambridge Univ. Press, 358 pp. [Chapter 3.]

McElhinny, M. W. and Senanayake, W. E. 1980. Paleomagnetic evidence for the existence of the geomagnetic field 3.5 Ga ago. *J. Geophys. Res. 85*:3523–3528. [Chapter 2.]

McElhinny, M. W.; Giddings, J. W.; and Embleton, B. J. J. 1974. Paleomagnetic results and late Precambrian glaciations. *Nature 248*:557–561. [Chapter 3.]

McElhinny, M. W.; Taylor, S. R.; and Stevenson, D. J. 1978. Limits to the expansion of the Earth, Moon, Mars and Mercury and to changes in the gravitational constant. *Nature 271*:316–321. [Chapter 11.]

McElroy, M. B.; Elkins, J. W.; Wolfsy, S. C.; and Yung, T. L. 1976. Sources and sinks for atmospheric N_2O. *Rev. Geophys. Space Phys. 14*:143–150. [Chapter 7.]

McGovern, W. E. 1969. The primitive Earth: Thermal models of the upper atmosphere for a methane-dominated environment. *J. Atmos. Sci. 26*:623–635. [Chapter 4.]

McIver, R. D. 1967. Composition of kerogen: Clue to its role in the origin of petroleum. *Proc. 7th World Petrol. Congr. 2*:25–36. [Chapter 5.]

McKirdy, D. M. 1974. Organic geochemistry in Precambrian research. *Precambrian Res. 1*:75–137. [Chapter 5.]

———. 1976. Biochemical markers in stromatolites. In M. R. Walter, ed., *Stromatolites*, Amsterdam, Elsevier, 163–191. [Chapter 5.]

McKirdy, D. M. and Hahn, J. H. 1982. Composition of kerogen and hydrocarbons in Precambrian rocks. In H. D. Holland and M. Schidlowski, eds., *Mineral Deposits and the Evolution of the Biosphere*, Dahlem Konferenzen-Dahlem Workshop Rept. 16, Berlin, Springer, 123–154. [Chapter 5.]

McKirdy, D. M. and Kantsler, A. J. 1980. Oil geochemistry and potential source rocks of the Officer Basin, South Australia. *Austr. Petrol. Explor. Assoc. J. 20*:68–86. [Chapter 5.]

McKirdy, D. M. and Powell, T. G. 1974. Metamorphic alteration of carbon isotopic composition in ancient sedimentary organic matter: New evidence from Australia and South Africa. *Geology 2*:591–595. [Chapters 5, 7.]

McKirdy, D. M.; McHugh, D. J.; and Tardif, J. W. 1980. Comparative analysis of stromatolitic and other microbial kerogens by pyrolysis-hydrogenation-gas chromatography (PHGC). In P. A. Trudinger, M. R. Walter, and B. J. Ralph, eds., *Biogeochemistry of Ancient and Modern Environments*, Canberra, Austr. Acad. Sci., 187–200. [Chapter 5.]

McKirdy, D. M.; Sumartojo, J.; Tucker, D. H.; and Gostin, V. 1975. Organic, mineralogic and magnetic indications of metamorphism in the Tapley Hill Formation, Adelaide Geosyncline. *Precambrian Res. 2*:345–373. [Chapter 5.]

McLennan, S. M. 1982. Timing and relationships among Precambrian crustal and atmospheric evolution and banded iron-formations. In P. A. Brudinger and M. R. Walter, eds., *Proceedings 4th International Symposium on Environmental Biogeochemistry*, New York, Springer, in press. [Chapter 11.]

McWilliams, M. O. and McElhinny, M. W. 1980. Late Precambrian paleomagnetism of Australia: The Adelaide Geosyncline. *J. Geol. 88*:1–26. [Chapter 11.]

Meinschein, W. G.; Rinaldi, G. G. L.; Hayes, J. M.; and Schoeller, D. A. 1974. Intramolecular isotopic order in biologically produced acetic acid. *Biomed. Mass Spec. 1*:172–174. [Chapter 5.]

Mekhtieva, V. L. 1971. Isotope composition of sulfur and plants and animals from reservoirs of different salinity. *Geokhimiya 6*:725–730. [Chapter 7.]

Mekhtieva, V. L. and Kondratieva, E. N. 1966. Fractionation of stable isotopes of sulfur by photosyntheiszing purple sulfur bacteria, *Rhodopseudomonas* sp. *Dokl. Biol. Sci. 166*:80–83 (English transl.). [Chapter 7.]

Melander, L. and Saunders, W. H. 1980. *Reaction Rates of Isotopic Molecules*, New York, Wiley, 331 pp. [Chapter 4.]

Mel'nik, Yu. P. 1973. *Physicochemical Conditions of Formation of the Precambrian Ferruginous Quartzites*, Kiev, 272 pp. 1975 transl. by D. Vitalinao, U.S. Geol. Surv. Washington, D.C. [Chapter 11.]

Melton, C. E. and Giardini, A. A. 1976. Experimental evidence that oxygen is the principal impurity in natural diamonds. *Nature 263*:309–310. [Chapter 4.]

Melton, C. E.; Salotti, C. A.; and Girdini, A. A. 1972. The observation of nitrogen, water, carbon dioxide, methane and argon as impurities in natural diamonds. *Am. Mineral. 57*:1518–1523. [Chapter 4.]

Mendelson, C. V. and Schopf, J. W. 1982. Proterozoic microfossils from the Sukhaya Tunguska, Shorikha, and Yudoma Formations of the Siberian Platform, USSR. *J. Paleontol. 56*:42–83. [Chapter 9.]

Merek, E. L. 1973. Imaging and life detection. *Bioscience 23*:153–159. [Chapter 9.]

Michaelis, W. and Albrecht, P. 1979. Molecular fossils of archaebacteria in kerogen. *Naturwiss. 66*:420–422. [Chapter 5.]

Michard-Vitrac, A.; Lancelot, J.; and Allègre, C. J. 1977. U-Pb ages on single zircons from the Early Precambrian rocks of west Greenland and the Minnesota River Valley. *Earth Planet. Sci. Lett. 35*:449–453. [Appendix I.]

Michelson, A. M.; McCord, J. M.; and Fridovich, I. eds. 1977. *Superoxide and Superoxide Dismutases*, London, Academic Press. 508 pp. [Chapter 13.]

Miller, S. L. 1953. A production of amino acids under possible primitive Earth conditions. *Science 117*:528–529. [Chapters 1, 4.]

———. 1955. Production of some organic compounds under possible primitive earth conditions. *J. Am. Chem. Soc. 77*:2351–2361. [Chapter 4; Appendix III.]

———. 1957. The mechanism of synthesis of amino acids by electric discharges. *Biochim. Biophys. Acta 23*:480–489. [Chapter 4.]

Miller, S. L. and Urey, H. C. 1959. Organic compound synthesis on the primitive Earth. *Science 130*:245–251. [Chapter 4.]

Miller, S. L. and Orgel, L. E. 1974. *The Origins of Life on the Earth*, Englewood Cliffs, N.J., Prentice-Hall, 299 pp. [Chapters 1, 4, 11, 13.]

Miller, S. L. and Schlesinger, G. 1983. The atmosphere of the primitive earth and the prebiotic synthesis of organic compounds. In R. Holmquist, ed., *Life Sciences and Space Research*, New York Pergamon, in press. [Chapter 4; Appendix III.]

Miller, S. L.; Urey, H. C.; and Oró, J. 1976. Origin of organic compounds on the primitive Earth and in meteorites. *J. Mol. Evol. 9*:59–72. [Chapter 4.]

Miller, S. L.; Schlesinger, G.; and Ring, D. 1981. Prebiotic syntheses in atmospheres containing CO, CO_2 and H_2. Unpublished manuscript. [Chapter 4.]

Mitchell, P. 1979. Keilin's respiratory chain concept and its chemiosmotic consequences. *Science 206*:1148–1159. [Chapter 6.]

Mittlefehldt, D. W. 1979. The nature of asteroidal differentiation processes: Implications for primordial heat sources. *Proc. 10th Lunar Planet. Sci. Conf.*:1975–1994. [Chapter 4.]

Miyake, Y. and Wada, E. 1967. The abundance ratio of $^{15}N/^{14}N$ in marine environments. *Records Oceanogr. Works, Japan 9*:37–53. [Chapter 7.]

———. 1971. The isotope effect on the nitrogen in biochemical oxidation-reduction reactions. *Records Oceanogr. Works, Japan 11*:1–6. [Chapter 7.]

Miyashiro, A. 1961. Evolution of metamorphic belts. *J. Petrol. 2*:277–311. [Chapter 3.]

———. 1967. Orogeny, regional metamorphism and magmatism in the Japanese islands. *Medd. Dansk Geol. Foren. København 17*:390–446. [Chapter 3.]

Moldowan, J. M. and Siefert, W. K. 1979. Head-to-head linked isoprenoid hydrocarbons in petroleum. *Science 204*:169–171. [Chapter 5.]

Monster, J.; Appel, P. W. U.; Thode, H. G.; Schidlowski, M.; Carmichael, C. M.; and Birdgwater, D. 1979. Sulfur isotope studies on Early Archaean sediments from Isua, West Greenland: Implications for the antiquity of bacterial sulfate reduction. *Geochim. Cosmochim. Acta 43*:405–413. [Chapter 7.]

Monty, C. L. V. 1967. Distribution and structure of Recent stromatolitic algal mats, eastern Andros Island, Bahamas. *Annals Soc. Géol. Belg. 90*:55–100. [Chapters 8, 9.]

———. 1973. Precambrian background and Phanerozoic history of stromatolite communities, an overview. *Annals Soc. Géol. Belg. 96*:585–624. [Chapter 9.]

———. 1976. The origin and development of cryptalgal fabrics. In M. R. Walter, ed., *Stromatolites*, Amsterdam, Elsevier, 193–249. [Chapter 8.]

———. 1977. Evolving concepts on the nature and ecological significance of stromatolites: A review. In E. Flügel, ed., *Fossil Algae*, Berlin, Springer, 15–35. [Chapter 1.]

Moodie, R. L. 1920. Thread moulds and bacteria in the Devonian. *Science 51*:14–15. [Chapter 15.]

Moorbath, S. 1975. Evolution of Precambrian crust from strontium isotopic evidence. *Nature 254*:395–399. [Chapter 3.]

Moorbath, S. 1976. Age and isotope constraints for the evolution of Archean crust. In B. F. Windley, ed., *The Early History of the Earth*, New York, Wiley, 351–360. [Chapter 4.]

———. 1977. The oldest rocks and the growth of the continents. *Scient. Am. 236* (3):92–104. [Chapter 10.]

Moorbath, S.; O'Nions, R. K.; and Pankhurst, R. J. 1973. Early Archaean age for the Isua iron-formation, West Greenland. *Nature 245*:138–139. [Chapter 11; Appendix I.]

———. 1975. The evolution of Early Precambrian crustal rocks at Isua, West Greenland: Geochemical and isotopic evidence. *Earth Planet. Sci. Lett. 27*:229–239. [Chapter 3; Appendix I.]

Moorbath, S.; O'Nions, R. K.; Pankhurst, J. R.; Gale, N. H.; and McGregor, V. R. 1972. Further Rb-Sr age determinations on the Early Precambrian rocks of the Godthaab district, West Greenland. *Nature 240*:78–82. [Chapter 3.]

Moore, E. S. 1918. The iron-formation on Belcher Islands, Hudson Bay, with special reference to its origin and its associated algal limestones. *J. Geol. 26*:412–438. [Chapters 1, 14.]

Moore, J. G.; Batchelder, J. N.; and Cunningham, C. G. 1977. CO_2-filled vesicles in mid-ocean basalts. *J. Volcanol. Geotherm. Res. 2*:309–327. [Chapter 4.]

Moorhouse, W. W. and Beales, F. W. 1962. Fossils from the Animikie, Port Arthur, Ontario. *Trans. Roy. Soc. Can. Sect. 4* (3):97–110. [Chapter 14.]

Mopper, K. and Degens, E. T. 1979. Organic carbon in the ocean: Nature and cycling. In B. Bolin, E. T. Degens, S. Kempe, and P. Ketner, eds., *The Global Carbon Cycle*, Chichester, Wiley, 293–316. [Chapter 5.]

Moraghan, J. T. and Buresh, R. J. 1977. Chemical reduction of nitrite and nitrous oxide by ferrous iron. *Soil Sci. Soc. Am. J. 41*:47–50. [Chapter 4.]

Morris, J. G. 1975. Physiology of obligate anaerobiosis. *Adv. Microbiol. Physiol. 12*:169–245. [Chapter 13; Glossary II.]

———. 1976. Oxygen and the obligate anaerobe. *J. Appl. Bacteriol. 40*:229–244. [Chapter 13; Glossary II.]

Morrison, D. 1978. Physical observations and taxonomy of asteroids. In D. Morrison and W. C. Wells, eds., *Asteroids: An Exploration Assessment*, Washington, D.C., NASA Conf. Pub. 2053, 81–98. [Chapter 4.]

Morss, D. A. and Kuhn, W. R. 1978. Paleoatmospheric temperature structure. *Icarus 33*:40–49. [Chapter 4.]

Muehlenbachs, K. and Clayton, R. N. 1976. Oxygen isotope composition of the oceanic crust and its bearing on sea water. *J. Geophys. Res. 81*:4365–4369. [Chapter 10.]

Muller, G. 1972. Organic microspheres from the Precambrian of South-West Africa. *Nature 235*:90–95. [Chapter 9.]

Muir, M. D. 1974. Microfossils from the Middle Precambrian McArthur Group, Northern Territory, Australia. *Origins of Life 5*:105–118. [Chapters 1, 14.]

———. 1976. Proterozoic microfossils from the Amelia Dolomite, McArthur Basin, Northern Territory. *Alcheringa 1*:143–158. [Chapters 1, 9, 14.]

———. 1978a. Microenvironments of some modern and fossil iron- and manganese-oxidizing bacteria. In W. E. Krumbein, ed., *Environmental Biogeochemistry and Geomicrobiology, Vol. 3*, Ann. Arbor, Mich., Ann Arbor Sci. Pub., 937–944. [Chapter 14.]

———. 1978b. Occurrence and potential uses of Archaean microfossils and organic matter. In J. E. Glover and D. I. Groves, eds., *Archaean Cherty Metasediments: Their Sedimentology, Micropalaeontology, Biogeochemistry, and Significance to Mineralization*, Perth., *Univ. W. Austr. Spec. Pub. 2*:11–21. [Chapter 9.]

———. 1979. A sabkha model for the deposition of part of the Proterozoic McArthur Group of the Northern Territory and its implications for mineralization. *BMR J. Austr. Geol. Geophys. 4*:149–162. [Chapter 11; Appendix I.]

Muir, M. D. and Grant, P. R. 1976. Micropalaeontological evidence from the Onverwacht Group, South Africa. In B. F. Windley, ed., *The Early History of the Earth*, London, Wiley, 595–604. [Chapters 8, 9.]

Muir, M. D. and Hall, D. O. 1974. Diverse microfossils in Precambrian Onverwacht Group rocks of South Africa. *Nature 252*:376–378. [Chapter 9.]

Muir, M. D. and Plumb, K. A. 1976. Precambrian microfossils in Australia—distribution, significance and problems. *Abstracts, 25th Internat. Geol. Congr.* (Sydney, Austr.), *1, Sect. 1* (1A): 32–33 (Abstr.). [Chapter 9.]

Muir, M. D.; Grant, P. R.; Bliss, G. M.; Diver, W. L.; and Hall, D. O. 1977. A discussion of biogenicity criteria in a geological context with examples from a very old greenstone belt, a Late Precambrian deformed zone, and tectonized Phanerozoic rocks. In C. Ponnamperuma, ed., *Chemical Evolution of the Early Precambrian*, New York, Academic Press, 155–170. [Chapter 9.]

Muir, M. D.; Spicer, R. A.; Grant, P. R.; and Giddens, R. 1974. X-ray microanalysis in the SEM for the determination of elements in modern and fossil micro-organisms. In J. V. Sanders and D. J. Goodchild, eds., *Electron Microscopy 1974, Vol. 2, Biological* (8th Internat. Congr. Electron Microscopy), Canberra, Austr. Acad. Sci., 104–105. [Chapter 14.]

Mukhin, L. 1974. Evolution of organic compounds in volcanic regions. *Nature 251*:50–51. [Chapter 4.]

Mukhin, L. M. 1976. Volcanic processes and synthesis of simple organic compounds on primitive Earth. *Origins of Life 7*:355–368. [Chapter 4.]

Mysen, B. O. 1977. The solubility of H_2O and CO_2 under predicted magma genesis conditions and some petrological and geophysical implications. *Rev. Geophys. Space Phys. 15*:351–361. [Chapter 4.]

Nagy, B. 1970. Porosity and permeability of the early Precambrian Onverwacht chert. *Geochim. Cosmochim. Acta 34*:525–527. [Chapter 5.]

———. 1975. *Carbonaceous Meteorites*, New York, Elsevier. 747 pp. [Chapter 4.]

———. 1976. Organic chemistry on the young Earth. *Naturwiss. 63*:499–505. [Chapter 5.]

Nagy, B. and Nagy, L. A. 1969. Early Precambrian Onverwacht microstructures: Possibly the oldest fossils on earth? *Nature 223*:1226–1229. [Chapters 5, 9.]

———. 1976. Interdisciplinary search for early life forms and for the beginning of life on earth. *Interdisciplinary Sci. Rev. 1*:291–310. [Chapter 9.]

Nagy, B.; Engel, M. H.; Zumberge, J. E.; Ogino, H.; and Chang, S. Y. 1981. Amino acids and hydrocarbons 3,800-Myr old in the Isua rocks, southwestern Greenland. *Nature 289*:53–56. [Chapter 15.]

Nagy, B.; Nagy, L. A.; Zumberge, J. E.; Sklarew, D. S.; and Anderson, P. 1976. Biological evolutionary trends and related aspects of carbon chemistry during the Precambrian, between 3,800 m.y. and 2,300 m.y. *Abstracts, 25th Internat. Geol. Congr.* (Sydney, Austr.), *1, Sect. 1* (1A): 33–34 (Abstr.). [Chapter 9.]

Nagy, B.; Zumberge, J. E.; and Nagy, L. A. 1975. Abiotic graphitic microstructures in micaeous metaquartzite about 3700 million years old from southwestern Greenland: Implications for Early Precambrian microfossils. *Proc. Natl. Acad. Sci. USA 72*:1206–1209. [Chapters 9, 15.]

Nagy, L. A. 1971. Ellipsoidal microstructures of narrow size range in the oldest known sediments on earth. *Grana 11*:91–94. [Chapter 9.]

———. 1972. Microstructures from the Onverwacht and Fig. Tree Precambrian sediments of the Barberton Mountain Land of South Africa. *Geosci. Man 4*:133–134 (Abstr.). [Chapter 9.]

———. 1974. Transvaal stromatolite: First evidence for the diversification of cells about 2.2×10^9 years ago. *Science 183*:514–516. [Chapters 9, 14.]

———. 1975. Comparative micropaleontology of a Transvaal stromatolite ($\sim 2.3 \times 10^9$ y. old) and in Witwatersrand carbon seam ($\sim 2.6 \times 10^9$ y. old). *Geol. Soc. Am. Abstr. Progr. 7* (7): 1209–1210 (Abstr.). [Chapter 14.]

———. 1978. New filamentous and cystous microfossils, 2,300 M.Y. old, from the Transvaal sequence. *J. Paleontol. 52*:141–154. [Chapter 14.]

Nagy, L. A. and Zumberge, J. E. 1976. Fossil microorganisms from the approximately 2800 to 2500 million-year-old Bulawayan stromatolites: Application of ultramicrochemical analyses. *Proc. Natl. Acad. Sci. USA 73*:2973–2976. [Chapter 9.]

Nagy, L. A.; Zumberge, J. E.; and Nagy, B. 1973. Early Precambrian life: Problems and significance. *Prog. 4th Internat. Conf. Origin Life*

(Barcelona, Spain), *Session 2, Pap. 65* (Abstr.). [Chapter 14.]

Naqvi, S. M. 1976. Physico-chemical conditions during the Archean as indicated by Dharwar geochemistry. In B. F. Windley, ed., *The Early History of the Earth*, London, Wiley, 289–298. [Chapter 10.]

Naqvi, S. M.; Divakara Rao, V.; and Narain, H. 1978. The primitive crust: Evidence from the Indian Shield. *Precambrian Res. 6*:323–345. [Chapter 10.]

Nash, W. P. and Wilkinson, J. F. G. 1970. Shonkin Sag Laccolith, Montana. *Contrib. Mineral. Petrol. 25*:241–269. [Chapter 4.]

Naumova, S. N. 1968. Zonal'nye kompleksy rastityel'nykh mikrofossilii Dokemriya i Nizhnevo Kembrigy Yevrazii i ikh stratigrafichyeskoye znachyeniye (Zonal assemblages of Precambrian and Lower Cambrian plant microfossils of Eurasia and their stratigraphic importance). *Mezhdu. Geol. Kong. 23rd Session, Dokl. Soveyetskikh Geologov, Problema 9*:30–39 (in Russian). [Chapter 1.]

Naumova, S. N. and Pavlovsky, Ye. V. 1961. Nakhodka rastiyelnykh ostatkov (spor) v slantsakh Torridona Schotlandi (Discovery of plant remains [spores] in the Torridonian shales of Scotland). *Dokl. Akad. Nauk SSSR 141*:181–182 (in Russian). [Chapter 1.]

Nelsestuen, G. L. 1978. Amino acid-directed nucleic acid synthesis. *J. Mol. Evol. 11*:109–120. [Chapter 4.]

———. 1980. Origin of life: Consideration of alternatives to proteins and nucleic acids. *J. Mol. Evol. 15*:59–72. [Chapter 4.]

Nesbitt, R. W. 1971. Skeletal crystal forms in the ultramafic rocks of the Yilgarn Block, Western Australia: Evidence for an Archean ultramafic liquid. *Geol. Soc. Austr. Spec. Pub. 3*:331–350. [Chapter 4.]

Nesbitt, R. W.; Sun, S. S.; and Purvis, A. C. 1979. Komatiites: Geochemistry and genesis. *Can. Mineral. 17*:165–186. [Chapter 3.]

Neukem, G. and Wise, D. U. 1976. Mars: A standard crater curve and possible new timescale. *Science 194*:1381–1387. [Chapter 2.]

Neumann-Redlin, C. and Zitzmann, A. 1976. The iron ore deposits of Denmark and Greenland. *Iron Ore Deposits of Europe 1*:125–127. [Chapter 11.]

Newman, M. J. and Rood, R. T. 1977. Implications of solar evolution for the Earth's early atmosphere. *Science 198*:1035–1037. [Chapter 11.]

Newton, R. C. 1978. Experimental and thermodynamic evidence for the operation of high pressures in Archaean metamorphism. In B. F. Windley and S. M. Naqvi, eds., *Archaean Geochemistry*, Amsterdam, Elsevier, 221–240. [Chapter 3.]

Nicholls, I. A. 1971. Petrology of Santorini Volcano, Cyclades, Greece. *J. Petrol. 12*:67–119. [Chapter 4.]

Neikus, H. G. D.; deVries, W.; and Stouthamer, A. H. 1977. The effect of different dissolved oxygen tensions on growth and enzyme activities of *Campylobacter sputorum* subspecies *bubulus*. *J. Gen. Microbiol. 103*:215–222. [Chapter 13.]

Nielsen, H. 1965. Schwefeliosotope im marinen Kreislauf und das δ^{34}S der früheren Meere. *Geol. Rundsch. 55*:160–172. [Chapter 7.]

Nielsen, H. 1978. Sulfur: Isotopes in Nature. In K. H. Wedepohl, ed., *Handbook of Geochemistry, Vol. 2*, Berlin, Springer, chap. 16B, 1–40. [Chapter 7.]

Niklas, K. H. and Brown, R. M., Jr. 1981. Ultrastructural and paleobiochemistry correlations among fossil leaf tissues from the St. Maries River (Clarkia) area, northern Idaho, USA. *Am. J. Bot. 68*:332–341. [Chapter 15.]

Nisbet, E. G.; Bickle, M. J.; and Martin, A. 1977. The mafic and ultramafic lavas of the Belingwe Greenstone Belt, Rhodesia. *J. Petrol. 18*:521–566. [Chapter 3; Appendix I.]

Nissenbaum, A. 1974a. The organic chemistry of marine and terrestrial humic substances: Implications of carbon and hydrogen isotope studies. In B. Tissot and F. Bienner, eds., *Advances in Organic Geochemistry 1973*, Paris, Editions Technip, 39–52. [Chapter 7.]

———. 1974b. Deuterium content of humic acids from marine and nonmarine environments. *Marine Chem. 2*:59–63. [Chapter 7.]

———. 1976. Scavenging of soluble organic matter from the prebiotic oceans. *Origins of Life 7*:413–416. [Chapter 4.]

Nissenbaum, A. and Kaplan, I. R. 1972. Chemical and isotopic evidence for the *in situ* origin of marine humic substances. *Limnol. Oceanogr. 17*:570–582. [Chapters 5, 7.]

Nissenbaum, A.; Kenyon, D. H.; and Oró, J. 1975. On the possible role of organic melanoidin polymers as matrices for prebiotic activity. *J. Mol. Evol. 6*:253–270. [Chapter 4.]

Nordlie, B. E. 1971. The composition of the magmatic gas of Kilauea and its behavior in the near surface environment. *Am. J. Sci. 271*:417–463. [Chapter 4.]

Norris, T. L. 1980. Kinetic model of ammonia synthesis in the solar nebula. *Earth Planet Sci. Lett. 47*:43–50. [Chapter 4.]

Novokhatsky, I. P. 1973. Precambrian ferruginous-siliceous formations of Kazakhstan. In *Genesis of Precambrian Iron and Manganese Deposits* (Proc. Kiev Symp. 1970), Paris, UNESCO, 153–158. [Chapter 11.]

Nozaki, M. 1979. Oxygenases and dioxygenases. *Topics Curr. Chem. 79*:146–183. [Chapter 13.]

Nunes, P. D. and Tilton, G. R. 1971. Uranium-lead ages of minerals from the Stillwater Igneous Complex and associated rocks. Montana. *Geol. Soc. Am. Bull. 82*:2231–2249. [Chapter 11.]

Nursall, J. R. 1959. Oxygen as a prerequisite to the origin of the Metazoa. *Nature 183*:1170–1172. [Chapter 1.]

Nussinov, M. D. and Vekhov, A. A. 1978. Formation of the early Earth regolith. *Nature 275*: 19–21. [Chapter 4.]

Nyberg, A. V. and Schopf, J. W. 1981. Microfossils in stromatolitic cherts from the Proterozoic Allamoore Formation of west Texas. *Precambrian Res. 16*:129–141. [Chapter 9; Appendix I.]

Oberlies, F. and Prashnowsky, A. A. 1968. Biogeochemische und elektronenmikroskopische Untersuchung präkambrischer Gestein. *Naturwiss. 55*:25–28. [Chapter 9.]

O'Brien, R. W. and Morris, J. G. 1971. Oxygen and the growth and metabolism of *Clostridium acetobutylicum*. *J. Gen. Microbiol. 68*:307–318. [Chapter 13.]

Ochiai, E. I. 1978. The evolution of the environment and its influence on the evolution of life. *Origins of Life 9*:81–91. [Chapters 11, 13.]

Oehler, D. Z. 1976. Biology, mineralization, and biostratigraphic utility of microfossils from the mid-Proterozoic Balbirini Dolomite, McArthur Group, N. T., Australia. *Abstracts, 25th Internat. Geol. Congr.* (Sydney, Austr.), *1*, *Sect. 1*(1A): 34–35 (Abstr.). [Chapter 14.]

———. 1978. Microflora of the middle Proterozoic Balbirini Dolomite (McArthur Group) of Australia. *Alcheringa 2*:269–309. [Chapters 1, 8, 14.]

Oehler, D. Z. and Smith, J. W. 1977. Isotopic composition of reduced and oxidized carbon in Early Archaean rocks from Isua, Greenland. *Precambrian Res. 5*:221–228. [Chapters 5, 7, 15.]

Oehler, D. Z.; Schopf, J. W.; and Kvenvolden, K. A. 1972. Carbon isotopic studies of organic matter in Precambrian rocks. *Science 175*: 1246–1248. [Chapter 7.]

Oehler, J. H. 1976a. Experimental studies in Precambrian paleontology: Structural and chemical changes in blue-green algae during simulated fossilization in synthetic chert. *Geol. Soc. Am. Bull. 87*:117–129. [Chapter 14.]

———. 1976b. Hydrothermal crystallization of silica gel. *Geol. Soc. Am. Bull. 87*:1143–1152. [Chapter 9.]

———. 1977a. Irreversible contamination of Precambrian kerogen by ^{14}C-labelled organic compounds. *Precambrian Res. 4*:221–227. [Chapter 5.]

———. 1977b. Precambrian microfossils and associated mineralisation in the McArthur deposit, Northern Territory, Australia. *Alcheringa 1*:315–349. [Appendix I.]

———. 1977c. Microflora of the H. Y. C. Pyritic Shale Member of the Barney Creek Formation (McArthur Group), middle Proterozoic of northern Australia. *Alcheringa 2*:72–97. [Chapters 1, 14.]

Oehler, J. H. and Croxford, N. J. W. 1976. Precambrian microfossils and associated mineralization in the McArthur deposit, N. T., Australia. *Abtracts, 25th Internat. Geol. Congr.* (Sydney, Austr.), *1*, *Sect. 1*(1A):35–36 (Abstr.). [Chapter 14.]

Oehler, J. H. and Logan, R. G. 1977. Microfossils, cherts, and associated mineralization in the Proterozoic McArthur (H.Y.C.) lead-zinc-silver deposit. *Econ. Geol. 72*:1393–1409. [Chapter 14.]

Oehler, J. H. and Schopf, J. W. 1971. Artificial microfossils: Experimental studies of permineralization of blue-green algae in silica. *Science 174*:1229–1231. [Chapter 14.]

Oehler, J. H.; Oehler, D. Z.; and Muir, M. D. 1976. On the significance of tetrahedral tetrads of Precambrian algal cells. *Origins of Life 7*: 259–267. [Chapter 14.]

O'Hara, M. J. 1968. The bearing of phase equilibria studies in synthetic and natural systems on the origin and evolution of basic and ultrabasic rocks. *Earth Sci. Rev. 4*:69–133. [Chapter 3.]

———. 1977. Thermal history of excavation of Archaean gneisses from the base of the continental crust. *J. Geol. Soc. London 134*:185–200. [Chapter 3.]

Ohomoto, H. 1972. Systematics of sulfur and carbon isotopes in hydrothermal ore deposits. *Econ. Geol. 67*:551–578. [Chapter 7.]

Ohomoto, H. and Rye, R. O. 1979. Isotopes of sulfur and carbon. In H. L. Barnes, ed., *Geochemistry of Hydrothermal Ore Deposits*, 2nd ed., New York, Wiley, 509–567. [Chapter 7.]

Okon, Y.; Houchins, J. P.; Albrecht, S. L.; and Burris, R. H. 1977. Growth of *Spirillum lipoferum* at constant partial pressures of oxygen, and the properties of its nitrogenase in cell-free extracts. *J. Gen. Microbiol. 98*:87–93. [Chapter 13.]

Olah, G. A., ed., 1963. *Friedel-Crafts and Related Reactions, Vol. 1*, New York, Interscience, 1031 pp. [Appendix IV.]

O'Leary, M. H. 1981. Carbon isotope fractionation in plants. *Phytochem. 20*:553–567. [Chapters 7, 12.]

Onderdonk, A. B.; Johnston, J. J.; Mayhew, J. W.; and Gorbach, S. L. 1976. Effect of dissolved oxygen and E_h on *Bacteroides fragilis* during continuous culture. *Appl. Environ. Microbiol. 31*:108–172. [Chapter 13.]

O'Nions, R. K.; Carter, S. R.; Evensen, N. M.; and Hamilton, P. J. 1979. Geochemical and cosmochemical application of Nd-isotopes. *Ann. Rev. Earth Planet. Sci. 7*:11–38. [Chapter 10.]

O'Nions, R. K.; Hamilton, P. J.; and Evensen, N. M. 1980. The chemical evolution of the Earth's mantle. *Scient. Am. 242* (5):120–133. [Chapter 10.]

Oparin, A. I. 1924. *Proiskhozhdenie Zhizni* (*The Origin of Life*), Moscow, Izd. Moskovskiĭ Rabochiĭ (in Russian). [Chapter 4.]

———. 1938. *The Origin of Life*, New York, Macmillan. Republished in 1953, New York, Dover, 270 pp. [Chapters 1, 4, 6.]

———. 1957. *The Origin of Life On the Earth*, New York, Academic Press, 495 pp. [Chapter 4.]

———. 1968. *Genesis and Evolutionary Development of Life*, New York, Academic Press, 203 pp. [Chapter 15.]

Oren, A.; Padan, E.; and Avron, M. 1977. Quantum yields for oxygenic and anoxygenic photosynthesis in the cyanobacterium *Oscillatoria limnetica*. *Proc. Natl. Acad. Sci. USA 74*:2152–2156. [Chapter 15.]

Orgel, L. E. 1968. Evolution of the genetic apparatus. *J. Mol. Biol. 38*:381–393. [Chapter 4.]

Orgel, L. E. and Lohrmann, R. 1974. Prebiotic chemistry and nucleic acid replication. *Accounts Chem. Res. 7*:368–377. [Chapter 4.]

Oró, J. 1961. Comets and the formation of biochemical compounds on the primitive Earth. *Nature 190*:389–390. [Chapter 4.]

O'Rourke, J. E. 1961. Paleozoic banded iron-formations. *Econ. Geol. 56*:331–361. [Chapter 11.]

Orpen, J. L. and Wilson, J. F. 1981. Stromatolites at ~3,500 Myr and a greenstone-granite unconformity in the Zimbabwean Archean. *Nature 291*:218–220. [Chapter 8.]

Oskvarek, J. D. and Perry, E. C., Jr. 1976. Temperature limits on the early Archaean ocean from oxygen isotope variations in the Isua supracrustal sequence, West Greenland. *Nature 259*:192–194. [Appendix I.]

Osterberg, R. 1974. Origins of metal ions in biology. *Nature 249*:382–383. [Chapter 13.]

Ott, U.; Mack, R.; and Chang, S. 1981. Noble-gas-rich separates from the Allende meteorite. *Geochim. Cosmochim. Acta, 45*:1751–1788. [Appendix III.]

Ottaway, J. H. and Mowbray, J. 1977. The role of compartmentation in the control of glycolysis. In B. L. Horecker and E. R. Stadtman, eds., *Current Topics in Cellular Regulation, Vol. 12*, New York, Academic Press, 107–208. [Chapter 6.]

Otroschenko, V. A. and Vasilyeva, N. V. 1977. The role of mineral surfaces in the origin of life. *Origins of Life 8*:25–31. [Chapter 4.]

Ourisson, G.; Albrecht, P.; and Rohmer, M. 1979. The hopanoids, paleochemistry and biochemistry of a group of natural products. *Pure and Appl. Chem. 51*:709–729. [Chapter 5.]

Ouyang, S. 1979. Ultramicro- and micro-fossils from the Anshan Group and the Liaohe Group in eastern Liaoning, northeastern China. In *Selected Works for a Symposium on Iron-geology of China, Sponsored by Academia Sinica, 1977*, Beijing, Science Press, 1–30 (in Chinese with English abstract). [Chapter 14.]

Oversby, V. M. 1976. Isotopic ages and geochemistry of Archaean acid igneous rocks from the Pilbara, Western Australia. *Geochim. Cosmochim. Acta 40*:817–829. [Chapter 9.]

Owen, T.; Cess, R. D.; and Ramanathan, V. 1979. Enhanced CO_2 greenhouse to compensate for reduced solar luminosity on early Earth. *Nature 277*:640–642. [Chapter 11.]

Padan, E. 1979. Facultative anoxygenic photosynthesis in cyanobacteria. *Ann. Rev. Pl. Physiol. 30*:27–40. [Chapter 13.]

Padgham, W. A. 1973. Geology High Lake 76 M/7, District of Mackenzie. *Dept. of Indian and Northern Affairs, Can., Open File 208.* [Appendix I.]

Paecht-Horowitz, M. 1978. The influence of various cations on the catalytic properties of clays. *J. Mol. Evol. 11*:101–107. [Chapter 4.]

Paecht-Horowitz, M.; Berger, J.; and Katchalsky, A. 1970. Prebiotic synthesis of polypeptides by heterogeneous polycondensation of amino acid adenylates. *Nature 228*:636–639. [Chapter 4.]

Panganiban, A. T.; Patt, T. E.; Hart, W.; and Hanson, R. S. 1979. Oxidation of methane in the absence of oxygen in lake water samples. *Appl. Environ. Microbiol. 37*:303–309. [Chapter 7.]

Pannella, G. 1976. Geophysical inferences from stromatolite lamination. In Walter, M. R., ed., *Stromatolites*, Amsterdam, Elsevier, 673–685. [Chapter 11.]

Papanastassiou, D. A. and Wasserburg, G. J. 1971. Lunar chronology and evolution for Rb-Sr studies of Apollo 11 and 12 samples. *Earth Planet. Sci. Lett. 11*:37–62. [Chapter 3.]

———. 1975. Rb-Sr study of a lunar dunite and evidence for early lunar differentiates. *Proc. 6th Lunar Sci. Conf.*:1467–1489. [Chapter 4.]

Papenfuss, G. T. 1955. Classification of the algae. In R. C. Miller, ed., *A Century of Progress in the Natural Sciences*, San Francisco, Calif. Acad. Sciences, 575–582. [Chapters 1, 13.]

Papike, J. J. and Bence, A. E. 1979. Planetary basalts: Chemistry and petrology. *Rev. Geophys. Space Phys. 17*:1612–1617. [Chapter 4.]

Pappenheimer, A. M., Jr. and Shaskan, E. 1944. Effect of iron on carbohydrate metabolism of *Clostriidium welchii. J. Biol. Chem. 155*:265–275. [Chapter 6.]

Pardue, J. W.; Scalan, R. S.; Van Baalen, C.; and Parker, P. L. 1976. Maximum carbon isotope fractionation in photosynthesis by blue-green algae and a green alga. *Geochim. Cosmochim. Acta 40*:309–312. [Chapter 7.]

Park, R. and Epstein, S. 1960. Carbon isotope fractionation during photosynthesis. *Geochim. Cosmochim. Acta 21*:110–126. [Chapter 7.]

———. 1961. Metabolic fractionation of C^{13} and C^{12} in plants. *Pl. Physiol. 35*:133–138. [Chapter 5.]

Parsons, B. and Sclater, J. G. 1977. An analysis of the variation of ocean floor bathymetry and heat flow with age. *J. Geophys. Res. 82*:803–827. [Chapter 10.]

Patterson, C. C. 1956. Age of meteorites and the earth. *Geochim. Cosmochim. Acta 10*:230–237. [Chapter 3.]

Peale, S. J. and Cassen, P. 1978. Contribution of tidal dissipation to lunar thermal history. *Icarus 36*:245–269. [Chapter 11.]

Peat, C. J.; Muir, M. D.; Plumb, K. A.; McKirdy, D. M.; and Norvick, M. S. 1978. Proterozoic microfossils from the Roper Group, Northern Territory, Australia. *BMR J. Austr. Geol. Geophys. 3*:1–17. [Chapter 14; Appendix I.]

Peck, H. D., Jr.; Smith, O. H.; and Gest, H. 1957. Comparative biochemistry of the biological reduction of fumaric acid. *Biochim. Biophys. Acta 25*:142–147. [Chapter 6.]

Pedersen, R. K. 1966. Precambrian fossils from the Ketilidian of southwest Greenland. *Grønl. Geol. Unders. Rept. 11*:40–41. [Chapter 14.]

Pedersen, R. K. and Lam, J. 1970. Precambrian organic compounds from the Ketilidian of south-west Greenland, Part III. *Meddel. Grønl. 185*:1–142. [Chapter 14.]

Perry, E. C., Jr. and Ahmad, S. N. 1977. Carbon isotope composition of graphite and carbonate minerals from 3.8-AE metamorphosed sediments, Isukasia, Greenland. *Earth Planet. Sci. Lett. 36*:280–284. [Chapters 7, 15.]

Perry, E. C., Jr. and Tan, F. C. 1972. Significance of oxygen and carbon isotope variations in Early Precambrian cherts and carbonate rocks of southern Africa. *Geol. Soc. Am. Bull. 83*:647–664. [Chapter 8.]

Perry, E. C.; Ahmad, S. N.; and Swulius, T. M. 1978. The oxygen isotopic composition of 3,800 m.y. old metamorphosed chert from Isukasia, west Greenland. *J. Geol. 86*:223–239. [Chapters 3, 10.]

Perry, E. C.; Monster, J.; and Reimer, T. 1971. Sulfur isotopes in Swaziland System barites and the evolution of the Earth's atmosphere. *Science 171*:1015–1016. [Chapter 7.]

Perry, E. C.; Tan, F. C.; and Morey, G. B. 1973. Geology and stable isotope geochemistry of the

Biwabik iron formation, northern Minnesota. *Econ. Geol. 68*:1110–1125. [Chapter 7.]

Peterman, Z. E. 1979. Geochronology and the Archean of the United States. *Econ. Geol. 74*: 1544–1562. [Chapter 11.]

Peterman, Z. E.; Hedge, C. E.; and Tourtelot, H. A. 1970. Isotopic composition of Sr in seawater throughout Phanerozoic time. *Geochim. Cosmochim. Acta 34*:105–120. [Chapter 10.]

Peters, K. E.; Ishiwatari, R.; and Kaplan, I. R. 1977. Color of kerogen as index of organic maturity. *Am. Assoc. Petrol. Geol. Bull. 61*:504–510. [Chapter 5.]

Peters, K. E.; Rohrback, B. G.; and Kaplan, I. R. 1981. Carbon and hydrogen stable isotope variations in kerogen during laboratory-simulated thermal maturation. *Am. Assoc. Petrol. Geol. Bull. 65*:501–508. [Chapters 5, 7.]

Peters, K. E.; Simoneit, B. R. T.; Brenner, S.; and Kaplan, I. R. 1978. Vitrinite reflectance-temperature determination for intruded Cretaceous black shale in the eastern Atlantic. In D. F. Oltz, ed., *Low Temperature Metamorphism of Kerogen and Clay Minerals*, Los Angeles, Pacific Sect. Soc. Econ. Paleontol. Mineral., 53–58. [Chapters 5, 7.]

Peters, K. E.; Sweeney, R. E.; and Kaplan, I. R. 1978. Correlation of carbon and nitrogen stable isotopes and sedimentary organic matter. *Limnol. Oceanogr. 23*:687–692. [Chapter 7.]

Petersilje, I. A. and Pripachkin, W. A. 1977. Hydrogen, carbon, nitrogen and helium in gases of igneous rocks. In L. H. Ahrens, ed., *Orgin and Distribution of the Elements*, New York, Pergamon, 541–545. [Chapter 4.]

Pfenning, N. 1978. General physiology and ecology of photosynthetic bacteria. In R. K. Clayton and W. R. Sistrom, eds., *The Photosynthetic Bacteria*, New York, Plenum, 3–18. [Chapter 15.]

Pflug, H. D. 1966. Structured organic remains from the Fig Tree Series of the Barberton Mountain Land, *Univ. Witwatersrand Econ. Geol. Res. Unit Info. Circ. 28* (Johannesburg), 14 pp. [Chapter 9.]

———. 1967. Structured organic remains from the Fig Tree Series (Precambrian) of the Barberton Mountain Land (South Africa). *Rev. Palaeobot. Palynol. 5*:9–29. [Chapter 9.]

———. 1976. Strukturiert erhaltene Fossilien aus dem Archaikum von Sudafrika. *Paläont. Z. 50*: 15–26. [Chapter 9.]

———. 1978a. Yeast-like microfossils detected in the oldest sediments of the earth. *Naturwiss. 65*: 611–615. [Chapters 9, 15.]

———. 1978b. Darwinism in earliest evolution? In *Proc. Symp. Natural Selection* (Liblice), Prague, ČSAV, 219–230. [Chapters 9, 15.]

———. 1978c. Früheste, bisher bekannte Lebewesen: *Isuasphaera isua* n. gen. n. spec. aus der Isua-Serie von Grönland (ca. 3800 Mio. J.). *Oberhess. Naturwiss. Z. 44*:131–145. [Chapters 9, 15.]

———. 1979. Archean fossil finds resembling yeasts. *Geol. Palaeontol. 13*:1–8. [Chapter 9.]

Pflug, H. D. and Jaeschke-Boyer, H. 1979. Combined structural and chemical analysis of 3,800-Myr-old microfossils. *Nature 280*:483–486. [Chapters 9, 15.]

Pflug, H. D. and von Klopotek, A. 1978. Eucaryonten im Archaikum? *Oberhess. Naturwiss. Z. 44*: 19–28. [Chapter 9.]

Pflug, H. D.; Jaeschke-Boyer, H.; and Sattler, E. L. 1979. Analysis of Archaean microfossils by the laser molecular microprobe. *Microscopica Acta 82*:255–266. [Chapters 9, 15.]

Pflug, H. D.; Neumann, K. H.; and Meinel, M. 1969. Entwicklungstendenzen des frühen Lebens auf der Erde. *Naturwiss. 56*:10–14. [Chapter 9.]

Phillips, R. J.; Kaula, W. M.; McGill, G. E.; and Malin, M. C. 1981. Tectonics and evolution of Venus. *Science 212*:879–887. [Chapter 3.]

Philp, R. P. and Calvin, M. 1976. Possible origin for insoluble organic (kerogen) debris in sediments from insoluble cell-wall materials of algae and bacteria. *Nature 262*:134–136. [Chapter 5.]

Philp, R. P. and Calvin, M. 1977. Kerogenous material in recent algal mats at Laguna Mormona, Baja California. In R. Campos and J. Goni, eds., *Advances in Organic Geochemistry 1975*, Madrid, Enadisma, 735–752. [Chapter 5.]

Philp, R. P.; Calvin, M.; Brown, S.; and Yang, E. 1978. Organic geochemical studies on kerogen precursors in recently deposited algal mats and oozes. *Chem. Geol. 22*:207–231. [Chapter 5.]

Pidgeon, R. T. 1978a. 3450 m.y.-old volcanics in the Archaean layered greenstone succession of the Pilbara Block, Western Australia. *Earth Planet. Sci. Lett. 37*:421–428. [Chapter 9; Appendix I.]

———. 1978b. Big Stubby and the early history of the Earth. In R. Zartman, ed., *4th Internat. Conf. Geochronology, Cosmochronology and Isotope Geology, U. S. Geol. Surv. Open File Rept. 78–701*:334–335. [Chapter 4.]

Pineau, F.; Javoy, M.; and Bottinga, Y. 1976. $^{13}C/^{12}C$ ratios of rocks and inclusions in popping rocks of the mid-Atlantic ridge and their bearing on the problem of isotopic composition of deep-seated carbon. *Earth Planet. Sci. Lett. 29*:413–421. [Chapter 4.]

Pinto, J. P.; Gladstone, G. R.; and Yung, Y. L. 1980. Photochemical production of formaldehyde in Earth's primitive atmosphere. *Science 210*:183–185. [Chapters 4, 11.]

Piper, J. D. A.; Briden, J. C.; and Lomax, K. 1973. Precambrian Africa and South America as a single continent. *Nature 245*:244–248. [Chapter 3.]

Pittendrigh, C. S.; Vishniac, W.; and Pearman, I. P. T., eds. 1966. *Biology and the Exploration of Mars*, Washington, D. C., Natl. Acad. Sci.-Natl. Res. Coun. Pub. 1296, 516 pp. [Chapter 1.]

Plumb, K. A.; Derrick, G. M.; Needham, R. S.; and Shaw, R. D. 1981. The Proterozoic of northern Australia. In D. R. Hunter, ed., *Precambrian of the Southern Hemisphere*, Amsterdam, Elsevier, 205–307. [Appendix I.]

Pollack, J. B. and Yung, Y. L. 1980. Origin and evolution of planetary atmospheres. *Ann. Rev. Earth Planet. Sci. 8*:425–487. [Chapters 2, 4.]

Ponomarev, V. V.; Burkova, V. N.; Serebrennikova, O. V.; and Titov, V. I. 1979. A method of extracting volcanogenic organic matter. *Geochem. Internat. 16*:76–78. [Chapter 4.]

Powell, T. G.; Cook, P. J.; and McKirdy, D. M. 1975. Organic geochemistry of phosphorites: Relevance to petroleum genesis. *Am. Assoc. Petrol. Geol. Bull. 59*:618–632. [Chapter 5.]

Prashnowsky, A. A. and Oberlies, F. 1972. Über Lebenszeugnisse im Präkambrium Afrikas und Südamerikas. In H. R. von Gaertner and H. Wehner, eds., *Advances in Organic Geochemistry 1971*, New York, Pergamon, 683–698. [Chapter 9.]

Preiss, W. V. 1976. Intercontinental correlations. In M. R. Walter, ed., *Stromatolites*, Amsterdam, Elsevier, 359–370. [Chapter 1.]

Preiss, W. V. and Forbes, B. G. 1980. Stratigraphy, correlation and sedimentary history of Adelaidean (late Proterozoic) basins in Australia. *S. Austr. Dept. Mines and Energy Rept. Bk. 80/7*, 60 pp. [Chapter 11; Appendix I.]

Preiss, W. V.; Walter, M. R.; Coats, R. P.; and Wells, A. T. 1978. Lithological correlations of Adelaidean glaciogenic rocks in parts of the Amadeus, Ngalia, and Georgina Basins. *BMR J. Austr. Geol. Geophys. 3*:43–53. [Appendix I.]

Presti, D. and Delbrück, M. 1978. Photo-receptors for biosynthesis, energy storage and vision. *Plant, Cell and Environment 1*:81–100. [Chapter 8.]

Pretorius, D. A. 1976. The nature of the Witwatersrand gold-uranium deposits. In K. H. Wolfe, ed., *Handbook of Stratabound and Stratiform Ore Deposits, Vol. 2, Regional Studies and Specific Deposits*, Amsterdam, Elsevier, 29–88. [Chapter 11.]

Prinn, R. G. and Owen, T. 1976. Chemistry and spectroscopy of the Jovian atmosphere. In T. Gehrels, ed., *Jupiter*, Tucson, Univ. Arizona Press, 319–371. [Chapter 4.]

Pritchard, G. G.; Wimpenny, J. W. T.; Morris, H. A.; Lewis, M. W. A.; and Hughes, D. E. 1977. Effects of oxygen on *Propionibacterium shermanii* grown in continuous culture. *J. Gen. Microbiol. 102*:223–233. [Chapter 13.]

Puchelt, H. 1972. Barium: Abundance in common sediments and sedimentary rock types. In K. H. Wedepohl, ed., *Handbook of Geochemistry, Vol. 2, Pt. 3*, Berlin, Springer, K1–K8. [Chapter 10.]

Purser, B. H. and Loreau, J. P. 1973. Aragonitic, supratidal encrustations on the Trucial Coast, Persian Gulf. In B. H. Purser, ed., *The Persian Gulf*, Berlin, Springer, 343–376. [Chapter 8.]

Quastel, J. H.; Stephenson, M.; and Whetham, M. D. 1925. Some reactions of resting bacteria in relaion to anaerobic growth. *Biochem. J. 19*: 304–317. [Chapter 6.]

Quayle, J. R. 1980. Microbial assimilation of C_1 compounds. *Biochem. Soc. Trans. 8*:1–10. [Chapter 6; Glossary II.]

Quayle, J. R. and Ferenci, T. 1978. Evolutionary aspects of autotrophy. *Microbiol. Rev. 42*:251–273. [Chapters 6, 7.]

Rabinowitch, E. I. 1945. *Photosynthesis and Related Processes, Vol. 1*, New York, Interscience, 61–98. [Chapter 4.]

Radke, B. M. 1980. Epeiric carbonate sedimentation of the Ninmaroo Formation (Upper Cambrian-Lower Ordovician), Georgina Basin. *BMR J. Austr. Geol. Geophys. 5*:183–200. [Chapter 8.]

Rambler, M. and Margulis, L. 1976. Comment on Egami's concept of the evolution of nitrate respiration. *Origins of Life 7*:73–74. [Chapter 13.]

———. 1980. Bacterial resistance to ultraviolet irradiation under anaerobiosis: Implications for pre-Phanerozoic evolution. *Science 210*:638–640. [Chapters 8, 11.]

Rambler, M.; Margulis, L.; and Barghoorn, E. S. 1977. Natural mechanisms of protection of a blue-green alga against ultraviolet light. In C. Ponnamperuma, ed., *Chemical Evolution of the Early Precambrian*, New York, Academic Press, 133–141. [Chapters 8, 11.]

Ramdohr, P. 1955. Neues Beobachtungen an Erzen des Witwatersrandes in Südafrika und ihre genetische Bedeutung. *Abh. dt. Akad. Wiss. Berlin, Kl. Chem. Geol. Biol. 1954* (5), 43 pp. [Chapter 11.]

———. 1958. Die Uran- und Goldlagerstätten Witwatersrand, Blind River District, Dominion Reef, Serra de Jacobina: Erzmikroskopische Untersuchungen und ein geologischer Vergleich. *Abh. dt. Akad. Wiss. Berlin, Kl. Chem. Geol. Biol. 1958* (3), 35 pp. [Chapter 11.]

Ramsay, J. G. 1980. The crack-seal mechanism of rock deformation. *Nature 284*:135–139. [Chapter 9.]

Rankama, K. 1955. Geologic evidence of chemical composition of the Precambrian atmosphere. In A. Poldervaart, ed., *Crust of the Earth*, *Geol. Soc. Am. Spec. Pap. 62*:651–664. [Chapter 11.]

Rao, M.; Odom, D. G.; and Oró, J. 1980. Clays in prebiological chemistry. *J. Mol. Evol. 15*:317–331. [Chapter 4.]

Ratner, M. I. and Walker, J. C. G. 1972. Atmospheric ozone and the history of life. *J. Atmos. Sci. 29*:803–808. [Chapter 11.]

Raymond, P. E. 1935. Pre-Cambrian life. *Geol. Soc. Am. Bull. 46*:375–392. [Chapter 1.]

Read, J. F. 1976. Calcretes and their distinction from stromatolites. In M. R. Walter, ed., *Stromatolites*, Amsterdam, Elsevier, 55–71. [Chapter 8.]

Redding, C. 1978. Hydrogen and carbon isotopes in coals and kerogens. In *4th Internat. Conf. Geochronology, Cosmochronology and Isotope Geology*, *U. S. Geol. Surv. Open File Rept. 78–701*. [Chapter 7.]

Redding, C. E.; Schoell, M.; Monin, J. C.; and Durand, B. 1980. Hydrogen and carbon isotopic composition of coals and kerogens. In A. G. Douglas and J. R. Maxwell, eds., *Advances in Organic Geochemistry 1979*, London, Pergamon, 711–723. [Chapter 5.]

Redfield, A. C. 1958. The biological control of chemical factors in the environment. *Am. Scient. 46*:205–221. [Chapter 12.]

Reeburgh, W. S. 1980. Anaerobic methane oxidation: Rate depth distributions in Skan Bay sediments. *Earth Planet. Sci. Lett. 47*:345–352. [Chapters 5, 12.]

Rees, C. E. 1970. The sulfur isotope balance of the ocean: An improved model. *Earth Planet Sci. Lett. 7*:366–370. [Chapter 7.]

———. 1973. A steady state model for sulphur isotope fractionation in bacterial reduction processes. *Geochim. Cosmochim. Acta 37*:1141–1162. [Chapter 7.]

Reeves, P. L. 1979. Precambrian microfossils or inorganic microspheres in the Sokoman Iron Formation, Schefferville, Quebec. *Geol. Assoc. Can. Mineral. Assoc. Can. Progr. Abstr. 4*:73 (Abstr.). [Chapter 14.]

Reibach, P. H. and Benedict, C. R. 1977. Fractionation of stable carbon isotopes by phosphoenolpyruvate carboxylase from C_4 plants. *Pl. Physiol. 59*:564–568. [Chapter 7.]

Reiche, H. and Bard, A. J. 1979. Heterogeneous photosynthetic production of amino acids from methane-ammonia-water at Pt/TiO_2. Implications in chemical evolution. *J. Am. Chem. Soc. 101*:3127–3128. [Chapter 4.]

Reimer, T. O.; Barghoorn, E. S.; and Margulis, L. 1979. Primary productivity in an early Archean microbial ecosystem. *Precambrian Res. 9*:93–104. [Chapters 5, 15.]

Renault, B. 1900. Sur quelques microorganisms des combustibles fossiles. *Soc. de l'Industrie Minerale Bull. 14*:6–159. [Chapter 15.]

Richey, J. E.; Brock, J. T.; Maiman, R. J.; Wissmar, R. C.; and Stallard, R. F. 1980. Organic carbon: Oxidation and transport in the Amazon River. *Science 207*: 1348–1351. [Chapter 5.]

Rickard, D. T.; Zweifel, H.; and Donnelly, T. H. 1979. Sulfur isotope systematics in the Åsen pyrite-barite deposits, Skellefte District, Sweden. *Econ. Geol. 74*:1060–1068. [Chapter 7.]

Riding, R. 1977. Skeletal stromatolites. In E. Flügel, ed., *Fossil Algae*, New York, Springer, 57–60.

Ridler, R. H. 1976. Stratigraphic keys to the gold metallogeny of the Abitibi Belt. *Can. Mining J. 97* (6):81–88. [Chapter 10.]

Riga, A.; VanPraag, H. J.; and Birgode, N. 1971. Rapport isotopique naturel de l'azote dans quelques sols forestiers et agricoles de Belgique doumis a divers traitements culturaux. *Geoderma 6*: 213–222. [Chapter 7.]

Rinaldi, G.; Meinschein, W. G.; and Hayes, J. H. 1974. Intramolecular carbon isotopic distribution in biologically produced acetoin. *Biomed. Mass Spec. 1*:415–417. [Chapter 5.]

Ring, D.; Wolman, Y.; Friedmann, N.; and Miller, S. L. 1972. Prebiotic synthesis of hydrophobic and protein amino acids. *Proc. Natl. Acad. Sci. USA 69*:765–768. [Chapter 4.]

Ringwood, A. E. 1966. Chemical evolution of the terrestrial planets. *Geochim. Cosmochim. Acta 30*:41–104. [Chapter 4.]

———. 1975. *Composition and Petrology of the Earth's Mantle*, New York, McGraw-Hill, 618 pp. [Chapters 2, 3.]

———. 1979. *Origin of the Earth and Moon*, New York, Springer, 295 pp. [Chapters 2, 4, 11.]

Ringwood, A. E. and Green, D. H. 1966. An experimental investigation of the gabbro-eclogite transformation and some geophysical implications. *Tectonophys. 3*:383–427. [Chapter 2.]

Rippka, R.; Dervelles, J.; Waterbury, J. B.; Herdman, M.; and Stanier, R. Y. 1979. Generic assignments, strain histories and properties of pure cultures of cyanobacteria. *J. Gen. Microbiol. 111*:1–61. [Chapter 8.]

Rittenberg, S. C. 1972. The obligate autotroph—the demise of a concept. *Ant. van Leeuwenhoek 45*:5–12. [Glossary II.]

Robin, P. L.; Rouxhet, P. G.; and Durand, B. 1977. Caracterisation des kerogenes et de leu evolution par spectroscopie infrarouge: Fonctions hydrocarbonees. In R. Campos and J. Goni, eds., *Advances in Organic Geochemistry 1975*, Madrid, Enadisma, 693–716. [Chapter 5.]

Roblot, M. M. 1964. Sporomorphes du Précambrien Armoricain. *Annals Paléontol. (Invertebres) 50*:105–110. [Chapter 1.]

Rogers, M. A.; McAlary, J. D.; and Bailey, N. J. L. 1974. Significance of reservoir bitumens to thermal-maturation studies, Western Canada Basin. *Am. Assoc. Petrol. Bull. 58*:1806–1824. [Chapter 5.]

Rogers, P. J. and Stewart, P. R. 1973. Respiratory development in *Saccharomyces cerevisiae* grown at controlled oxygen tension. *J. Bacteriol. 115*: 88–97. [Chapter 13.]

Rohmer, M.; Bouvier, P.; and Ourisson, G. 1979. Molecular evolution of biomembranes: Structural equivalents and phylogenetic precursors of sterols. *Proc. Natl. Acad. Sci. USA 76*:847–851. [Chapter 13.]

Ronov, A. B. 1964. Common tendencies in the chemical evolution of the Earth's crust, ocean and atmosphere. *Geochem. Internat. 1*:713–737. [Chapter 10.]

———. 1968. Probable changes in the composition of sea water during the course of geological time. *Sedimentol. 10*:25–43. [Chapters 7, 10.]

———. 1980. *Osadochnaja Obolochka Zemli* (*Sedimentary Layer of the Earth*), Moscow, Nauka, 78 pp. (in Russian). [Chapters 5, 10.]

Ronov, A. B. and Midgisov, A. A. 1971. Evolution of the chemical composition of the rocks in the shields and sediment cover of the Russian and North American platforms. *Sedimentol. 16*:137–185. [Chapter 10.]

Ronov, A. B. and Yaroshevsky, A. A. 1969. Chemical composition of the Earth's crust. In P. J. Hart, ed., *The Earth's Crust and Upper Mantle*, Richmond, William Byrd, 37–57. [Chapter 4.]

Roscoe, S. M. 1969. Huronian rocks and uraniferous conglomerates in the Canadian Shield. *Geol. Surv. Can. Pap. 68–40*, 205 pp. [Chapter 11.]

———. 1973. The Huronian Supergroup, a Paleoaphebian succession showing evidence of atmospheric evolution. *Geol. Assoc. Can. Spec. Pap. 12*:31–48. [Chapter 11.]

Rosenfeld, W. D. and Silverman, S. R. 1959. Carbon isotope fractionation in bacterial production of methane. *Science 130*:1658–1659. [Chapter 7.]

Rosenhauer, M.; Woermann, E.; Knecht, B.; and Ulmer, C. G. 1977. The stability of graphite and diamond as a function of the oxygen fugacity in the mantle. In *Extended Abstracts, Contributed to the Second International Kimberlite Conference*, Washington, D.C., Am. Geophys. Union. [Chapter 4.]

Rossignol-Strick, M. and Barghoorn, E. S. 1971. Extraterrestrial abiogenic organization of organic matter: The hollow spheres of the Orgueil meteorite. *Space Life Sci. 3*:89–107. [Chapter 9.]

Rothpletz, A. 1961. Über die systematische Deutung und die stratigraphische Stellung der ältesten Versteinerungen Europas und Nordamerikas mit besonderer Berücksichtigung der Cryptozoon und Oolithe. Teil II. Über *Cryptozoon*, *Eozoon* und *Atikokania*. *Abh. Bayer. Akad. Wiss. 28* (4), 91 pp. [Chapter 8.]

Rousell, D. H. 1972. The Chelmsford Formation of the Sudbury Basin—a Precambrian turbidite.

Geol. Assoc. Can. Spec. Pap. 10:79–91. [Appendix I.]

Rouxhet, P. G.; Robin, P. L.; and Nicaise, G. 1980. Characterization of kerogens and of their evolution by infrared spectroscopy. In B. Durand, ed., *Kerogen*, Paris, Editions Technip, 163–190. [Chapter 5.]

Roy, A. B. and Trudinger, P. A. 1970. *The Biochemistry of Inorganic Compounds of Sulphur*, London, Cambridge Univ. Press, 400 pp. [Chapter 7.]

Rubey, W. W. 1951. Geological history of sea water. An attempt to state the problem. *Geol. Soc. Am. Bull. 62*:1111–1148. [Chapters 1, 4, 10; Appendix III.]

———. 1955. Development of the hydrosphere and atmosphere, with special reference to probable composition of the early atmosphere. In A. Poldervaart, ed., *Crust of the Earth*, New York, Geol. Soc. Am., 631–650. [Chapters 1, 4.]

Runcorn, S. K. 1979. Nautiloid growth rhythms and lunar dynamics. *Nature 279*:452–453. [Chapter 11.]

Saager, R. 1970. Structures in pyrite from the Basal Reef in the Orange Free State gold-field. *Geol. Soc. S. Afr. Trans. 73*:29–46. [Chapter 9.]

Saager, R. and Muff, R. 1980. A new discovery of possible primitive life-forms in conglomerates of the Archean Pietersburg Greenstone Belt, South Africa. *Geol. Rundsch. 69*:179–185. [Chapter 9.]

Sackett, W. M. and Chung, H. M. 1979. Experimental confirmation of the lack of carbon isotope exchange between methane and carbon oxides at high temperatures. *Geochim. Cosmochim. Acta 43*:273–276. [Chapter 4.]

Sackett, W. M.; Eadie, B. J.; and Exner, M. E. 1974. Stable isotope composition of organic carbon in recent Antarctic sediments. In B. Tissot and F. Bienner, eds., *Advances in Organic Geochemistry 1973*, Paris, Editions Technip, 661–671. [Chapter 5.]

Sackett, W. M.; Nakaparksin, S.; and Dalrymple, D. 1968. Carbon isotope effects in methane production by thermal cracking. In G. D. Hobson and G. C. Speers, eds., *Advances in Organic Geochemistry 1966*, Oxford, Pergamon, 37–53. [Chapter 7.]

Sackett, W. M.; Poag, C. W.; and Eadie, B. J. 1974. Kerogen recycling in the Ross Sea, Antarctica. *Science 185*:1045–1047. [Chapter 5.]

Safronov, V. S. 1972. *Evolution of the Protoplanetary Cloud and Formation of the Earth and Planets*, NASA Tech. Transl., Washington, D.C., 206 pp. [Chapter 2.]

———. 1978. The heating of the Earth during its formation. *Icarus 33*:3–12. [Chapter 4.]

Sagan, C. 1973. Ultraviolet selection pressure on the earliest organisms. *J. Theoret. Biol. 39*:195–200. [Chapter 11.]

Sagan, C. and Khare, B. N. 1971. Long wavelength ultraviolet photoproduction of amino acids in the primitive Earth. *Science 173*:417–520. Also in *Nature 232*:577–578. [Chapter 4.]

Sagan, C. and Mullen, G. 1972. Earth and Mars: Evolution of atmospheres and surface temperatures. *Science 177*:52–56. [Chapter 11.]

Saito, K.; Basu, A. R.; and Alexander, C. A., Jr. 1978. Planetary-type rare gases in an upper mantle-derived amphibole. *Earth Planet. Sci. Lett. 39*:274–280. [Chapter 4.]

Sakai, H. 1957. Fractionation of sulfur isotopes in Nature. *Geochim. Cosmochim. Acta 12*:150–169. [Chapter 7.]

Sakai, H.; Smith, J. W.; and Kaplan, I. R. 1974. Microdeterminations of C, N, S, H, He, metallic Fe, δ^{13}C, δ^{15}N and δ^{34}S in geologic samples. *Geochem. J. 10*:85–96. [Appendix III.]

Salop, L. I. 1968. Pre-Cambrian of the U.S.S.R. *Proc. 23rd Internat. Geol. Congr.* (Prague.), *Sec. 4*:61–73. [Chapter 11.]

———. 1977. *Precambrian of the Northern Hemisphere*, Amsterdam, Elsevier, 378 pp. [Chapter 11.]

Sangster, D. F. 1980. Distribution and origin of Precambrian massive sulphide deposits of North America. *Geol. Assoc. Can. Spec. Pap. 20*:723–739. [Chapter 10.]

Sangster, D. F. and Brooks, W. A. 1977. Primitive lead in an Australian Zn-Pb-Ba deposit. *Nature 270*:423. [Chapter 9.]

Sanyal, S. K.; Kvenvolden, K. A.; and Marsden, S. S., Jr. 1971. Permeabilities of Precambrian Onverwacht cherts and other low permeability rocks. *Nature 232*:325–327. [Chapter 5.]

Saratovkin, D. D. 1959. *Dendritic Crystallization*, 2nd ed., trans. J. E. S. Bradley, New York, Consultants Bureau, 126 pp. [Chapter 9.]

Sarkar, S. N.; Saha, A. K.; and Miller, J. A. 1969. Geochronology of the Precambrian of Singhbhum and adjacent regions, eastern India. *Geol. Mag. 106*:15–45. [Chapter 11.]

Sato, M. 1978. Oxygen fagacity of basaltic magmas and the role of gas-forming elements. *Geophys. Res. Lett. 5*:447–449. [Chapter 4.]

Saxby, J. D. 1970. Isolation of kerogen in sediments by chemical methods. *Chem. Geol. 6*:173–184. [Chapter 5; Appendix IV.]

———. 1976. Chemical separation and characterization of kerogen from oil shales. In T. F. Yen and G. V. Chilingarin, eds., *Oil Shale*, Amsterdam, Elsevier, 103–128. [Chapter 5; Appendix IV.]

Scalan, R. S. 1959. The Isotopic Composition, Concentration and Chemical State of Nitrogen in Igneous Rocks. Ph.D. Thesis, Univ. Arkansas (Dept. Geology). University Microfilm No. 59-1379, Ann Arbor, Mich. [Chapter 7.]

Schidlowski, M. 1965. Probable life-forms from the Precambrian of the Witwatersrand System (South Africa). *Nature 205*:895–896. [Chapter 9.]

———. 1966. Zellular strukturierte Elemente aus dem Präkambrium des Witwatersrand-Systems (Südafrika). *Z. Deutsch. Geol. Ges. 115*:783–786. [Chapter 9.]

———. 1969. Critical remarks on a postulated relationship between Precambrian thucolite and boghead coal. In P. A. Schenck and I. Havenaar, eds., *Advances in Organic Geochemistry 1968*, New York, Pergamon, 579–592. [Chapter 9.]

———. 1970a. Elektronenoptische Identifizierung zellartiger Mikrostrukturen aus dem Präkambrium des Witwatersrand System. *Paleontol. Z. 44*:128–133. [Chapter 9.]

———. 1970b. Unterzuchungen zur Metallogenese in südwestlichen Witwatersrand-Becken (Oranje-Freistaat-Goldfeld, Südafrika). *Geol. Jahrb. Beih. 85*, 80 pp. [Chapter 11.]

———. 1973. Sulfur in the Precambrian metallogeny. *Geol. Rundsch. 62*:840–863. [Chapter 7.]

———. 1979. Antiquity and evolutionary status of bacterial sulfate reduction: Sulfur isotope evidence. *Origins of Life 9*:299–311. [Chapter 7.]

———. 1980a. Antiquity of photosynthesis: Possible contraints from Archaean carbon isotope record. In P. A. Trudinger, M. R. Walter, and B. J. Ralph, eds., *Biogeochemistry of Ancient and Modern Environments*, Canberra, Austr. Acad. Sci., 47–54. [Introduction.]

———. 1980b. Origin and early evolution of life on Earth. *Abstracts, 26th Internat. Geol. Congr.* (Paris), *2*:794 (Abstr.). [Introduction.]

———. 1981a. Content and isotopic composition of reduced carbon in sediments. In H. D. Holland and M. Schidlowski, eds., *Mineral Deposits and the Evolution of the Biosphere*, Dahlem Konferenzen-Dahlem Workshop Rept. 16, Berlin, Springer, 103–122. [Introduction; Chapter 15.]

———. 1981b. Uraniferous constituents of the Witwatersrand conglomerates: Ore-microscopic observations and implications for the Witwatersrand metallogeny. In F. C. Armstrong, ed., *Genesis of Uranium- and Gold-Bearing Precambrian Quartz-Pebble Conglomerates, U.S. Geol. Surv. Prof. Pap. 1161*:N1–N29. [Chapter 11.]

Schidlowski, M. and Junge, C. E. 1981. Coupling among the terrestrial sulfur, carbon and oxygen cycles: Numerical modeling based on revised Phanerozoic carbon isotope record. *Geochim. Cosmochim. Acta 45*:589–594. [Introduction; Chapter 7.]

Schidlowski, M.; Appel, P. W. U.; Eichmann, R.; and Junge, C. E. 1979. Carbon isotope geochemistry of the 3.7×10^9-yr-old Isua sediments, West Greenland: Implications for the Archaean carbon and oxygen cycles. *Geochim. Cosmochim. Acta 43*:189–199. [Chapters 1, 5, 7, 10, 15.]

Schidlowski, M.; Eichmann, R.; and Fiebiger, W. 1976a. Isotopic fractionation between organic carbon and carbonate carbon in Precambrian banded ironstone series from Brazil. *Neues Jahrb. Miner. Mh. 1976*:344–353. [Chapter 7.]

Schidlowski, M.; Eichmann, R.; and Junge, C. E. 1975. Precambrian sedimentary carbonates: Carbon and oxygen isotope geochemistry and implication for the terrestrial oxygen budget. *Precambrian Res. 2*:1–69. [Chapters 7, 8, 10.]

———. 1976b. Carbon isotope geochemistry of the Precambrian Lomagundi carbonate province, Rhodesia. *Geochim. Cosmochim. Acta 40*:449–455. [Chapters 7, 12.]

Schidlowski, M.; Junge, C. E.; and Pietrek, H. 1977. Sulfur isotope variations in marine sulfate evaporites and the Phanerozoic oxygen budget. *J. Geophys. Res. 82*:2557–2565. [Chapter 7.]

Schiegl, W. E. 1972. Deuterium content of peat as a paleoclimatic recorder. *Science 175*:512–513. [Chapter 7.]

Schiegl, W. E. and Vogel, J. C. 1970. Deuterium content of organic matter. *Earth Planet. Sci. Lett. 7*:307–313. [Chapter 7.]

Schneider, A. 1970. The sulfur isotope composition of basaltic rocks. *Contrib. Mineral. Petrol. 25*: 95–124. [Chapter 7.]

Schneour, E. A. and Ottesen, E. A., eds. 1966. *Extraterrestrial Life: An Anthology and Bibliography*, Washington, D.C., Natl. Acad. Sci.-Natl. Res. Coun. Pub. 1296A, 478 pp. [Chapter 1.]

Schoell, M. 1980. The hydrogen and carbon isotopic composition of methane from natural gases of various origins. *Geochim. Cosmochim. Acta 44*:649–661. [Chapter 7.]

Schoell, M. and Wellmer, F. W. 1981. Anomalous ^{13}C depletion in early Precambrian graphites from Superior Province, Canada. *Nature 290*: 696–699. [Chapters 12, 15.]

Schopf, J. W. 1968. Microflora of the Bitter Springs Formation, late Precambrian, central Australia. *J. Paleontol. 42*:651–688. [Chapters 1, 9, 14.]

———. 1970. Precambrian micro-organisms and evolutionary events prior to the origin of vascular plants. *Biol. Rev. Cambridge Phil. Soc. 45*:319–352. [Chapters 1, 9.]

———. 1974. Paleobiology of the Precambrian: The age of blue-green algae. In T. Dobzhansky, M. K., Hecht, and W. C. Steere, eds., *Evolutionary Biology, Vol. 7*, New York, Plenum, 1–43. [Chapters 1, 9.]

———. 1975. Precambrian paleobiology: Problems and Perspectives. *Ann. Rev. Earth Planet. Sci. 3*:213–250. [Chapters 1, 9, 14.]

———. 1976. Are the oldest "fossils," fossils? *Origins of Life 7*:19–36. [Chapters 1, 9.]

———. 1977a. Biostratigraphic usefulness of stromatolitic Precambrian microbiotas: A preliminary analysis. *Precambrian Res. 5*:143–173. [Chapters 1, 8, 14, 15.]

———. 1977b. Evidences of Archean life. In C. Ponnamperuma, ed., *Chemical Evolution of the Early Precambrian*, New York, Academic Press, 101–105. [Chapter 9.]

———. 1978. The evolution of the earliest cells. *Scient. Am. 239* (3):110–134. [Chapters 1, 13, 14, 15.]

———. 1980a. Archean "microfossils": The oldest evidences of life? *Abstracts, 6th Internat. Conf. Origins of Life*, (Jerusalem, Israel): 85 (Abstr.). [Introduction; Chapter 9.]

———. 1980b. Evidences of Early Precambrian (Archean) life. *Abstracts, 5th Internat. Palynol. Conf.* (Cambridge, England): 356 (Abstr.). [Introduction; Chapter 9.]

———. 1980c. The origin and Archean evolution of life. In M. H. Hickman, ed., *The Primitive Earth Revisited*, Oxford, Ohio, Dept. Geol., Miami Univ., 94–103. [Introduction; Chapter 15.]

———. 1981. The Precambrian development of an oxygenic atmosphere. In F. C. Armstrong, ed., *Genesis of Uranium- and Gold-Bearing Precambrian Quartz-Pebble Conglomerates, U.S. Geol. Surv. Prof. Pap. 1161*:B1–B11. [Chapter 15.]

Schopf, J. W. and Barghoorn, E. S. 1965. Electron microscopy of Precambrian microfossils. *Geol. Soc. Am. Progr. Abstr. 1965 Ann. Mtg.*:147 (Abstr.). [Chapter 14.]

———. 1967. Alga-like fossils from the Early Precambrian of South Africa. *Science 156*:508–512. [Chapter 9.]

Schopf, J. W. and Blacic, J. M. 1971. New microorganisms from the Bitter Springs Formation (late Precambrian) of the north-central Amadeus Basin, Australia. *J. Paleontol. 45*:925–960. [Chapters 9, 14.]

Schopf, J. W. and Fairchild, T. R. 1973. Late Precambrian microfossils: A new stromatolitic biota from Boorthanna, South Australia. *Nature 242*:537–538. [Chapter 1.]

Schopf, J. W. and Oehler, D. Z. 1976. How old are the eukaryotes? *Science 193*:47–49. [Chapters 1, 9, 13, 14, 15.]

Schopf, J. W. and Prasad, K. N. 1978. Microfossils in Collenia-like stromatolites from the Proterozoic Vempalle Formation of the Cuddapah Basin, India. *Precambrian Res. 6*:347–366. [Chapters 1, 9.]

Schopf, J. W. and Sovietov, Yu. K. 1976. Microfossils in *Conophyton* from the Soviet Union and their bearing on Precambrian biostratigraphy. *Science 193*:143–146. [Chapters 8, 9, 14.]

Schopf, J. W. and Walter, M. R. 1980. Archean microfossils and "microfossil-like" objects—a critical appraisal. In J. E. Glover and D. I. Groves, eds., *2nd Internat. Archaean Symposium* (Perth), Canberra, Austr. Acad. Sci., 23–24 (Abstr.). [Introduction; Chapter 9.]

———. 1982. Origin and early evolution of cyanobacteria: the geological evidence. In N. G. Carr and B. A. Whitton, eds., *The Biology of Cyanobacteria*, London, Blackwell, pp. 543–564. [Introduction; Chapters 9, 13.]

Schopf, J. W.; Barghoorn, E. S.; Maser, M. D.; and Gordon, R. O. 1965. Electron microscopy of fossil bacteria two billion years old. *Science 149*: 1365–1367. [Chapter 14.]

Schopf, J. W.; Kvenvolden, K. A.; and Barghoorn, E. S. 1968. Amino acids in Precambrian sediments: An assay. *Proc. Natl. Acad. Sci. USA 59*:639–646. [Chapter 5.]

Schopf, J. W.; Oehler, D. Z.; Horodyski, R. J.; and Kvenvolden, K. A. 1971. Biogenicity and significance of the oldest known stromatolites. *J. Paleontol. 45*:477–485. [Chapters 7, 8.]

Schrauzer, G. N. 1975. Nonenzymatic simulation of nitrogenase reactions and the mechanism of biological nitrogen fixation. *Angew. Chem. Internat. 14*:514–522. [Chapter 4.]

Schrauzer, G. N. and Guth, T. D. 1977. Photolysis of water and photo-reduction of nitrogen on titanium dioxide. *J. Am. Chem. Soc. 99*:7189–7193. [Chapter 4.]

Schrauzer, G. N.; Guth, T. D.; and Palmer, M. R. 1979. Nitrogen reducing solar cells. In R. R. Hautala, R. B. King, and C. Kutal, eds., *Solar Energy: Chemical Conversion and Storage*, Clifton, N. J., Humana Press, 261–269. [Chapter 4.]

Schubert, G. 1979. Subsolidus convection in the mantles of terrestrial planets. *Ann. Rev. Earth Planet. Sci. 7*:289–342. [Chapter 2.]

Schubert, G.; Stevenson, D.; and Cassen, P. 1980. Whole planet cooling and the radiogenic heat source contents of the Earth and Moon. *J. Geophys. Res. 85*:2531–2538. [Chapter 4.]

Schwab, F. L. 1978. Secular trends in composition of sedimentary rock assemblages—Archean through Phanerozoic time. *Geology 6*:532–536. [Chapter 10.]

Schwartz, A. W. 1972. Prebiotic phosphorylation-nucleotide synthesis with apatite. *Biochim. Biophys. Acta 281*:477–480. [Chapter 4.]

Schwartz, A. W. and Chittenden, G. J. F. 1977. Synthesis of uracil and thymine under simulated prebiotic conditions. *Biosystems 9*:87–92. [Chapter 4.]

Schweigart, H. 1965. Genesis of the iron ore of the Pretoria Series, South Africa. *Econ. Geol. 60*: 269–299. [Chapter 11.]

Scott, E. R. D. 1979. Origin of iron meteorites. In T. Gehrels, ed., *Asteroids*, Tucson, Univ. Arizona Press, 892–925. [Chapter 4.]

Scott, W. M.; Modzeleski, V. E.; and Nagy, B. 1970. Pryolysis of early Precambrian organic matter (> 3×10^9 year-old). *Nature 225*:1129–1130. [Chapter 5.]

Scrutton, C. T. 1978. Periodic growth features in fossil organisms and the length of the day and month. In P. Brosche and J. Sünderman, eds., *Tidal Friction and the Earth's Rotation*, New York, Springer, 154–196. [Chapter 11.]

Searcy, D. G.; D. B. Stein; and G. R. Green. 1978. Phylogenetic affinities between eukaryotic cells and a thermophilic mycoplasma. *Biosystems 10*: 19–28. [Chapter 1.]

Seccombe, P. K. 1977. Sulphur isotope and trace metal composition of stratiform sulphides as an ore guide in the Canadian Shield. *J. Geochem. Explor. 8*:117–137. [Chapter 7.]

Seckbach, J. and Kaplan, I. R. 1973. Growth pattern and $^{13}C/^{12}C$ isotope fractionation of *Cyanidium caldarium* and hot spring algal mats. *Chem. Geol. 12*:161–169. [Chapter 7.]

Seifert, W. K. and Moldowan, J. M. 1979. The effect of biodegradation on steranes and terpanes in crude oils. *Geochim. Cosmochim. Acta 43*: 111–126. [Chapter 5.]

Sekiya, M.; Nakazawa, K.; and Hayashi, C. 1980. Dissipation of the rare gases contained in the primordial Earth's atmosphere. *Earth Planet. Sci. Lett. 50*:197–201. [Chapters 2, 4.]

Sellers, W. D. 1965. *Physical Climatology*, Chicago, Univ. Chicago Press, 272 pp. [Chapter 11.]

———. 1969. A global climatic model based on the energy balance of the Earth atmosphere system. *J. Appl. Meteorol. 8*:392–400. [Chapter 11.]

———. 1973. A new global climatic model. *J. Appl. Meteorol. 12*:241–254. [Chapter 11.]

Semenenko, N. 1973. The iron-chert formations of the Ukrainian Shield. In *Genesis of Precambrian Iron and Manganese Deposits*. (Proc. Kiev. Symp. 1970), Paris, UNESCO, 135–142. [Chapter 11.]

Semenenko, N. P.; Schcherbak, N. P.; and Bartnitsky, E. N. 1972. Geochronology, stratigraphy and tectonic structure of the Ukrainian Shield. *Proc. 24th Internat. Geol. Congr.* (Montreal), *Sec. 1*:363–370. [Chapter 11.]

Semikhatov, M. A. 1976. Experience in stromatolite studies in the U.S.S.R. In M. R. Walter, ed., *Stromatolites*, Amsterdam, Elsevier, 337–357. [Chapter 1.]

———. 1978. Nekotorye karbonatnye stromatolity afebiya kanadskogo shchita (Some carbonate stromatolites of the Aphebian of the Canadian Shield). In *Nizhnyaya granitza rifeya: stromatolity afebiya* (*Lower Boundary of the Riphean and Stromatolites of the Aphebian*). *Akad. Nauk*

SSSR, Trudy 312:111–147 (in Russian). [Chapter 8.]

Semikhatov, M. A.; Gebelein, C. D.; Cloud, P.; Awramik, S. M.; and Benmore, W. C. 1979. Stromatolite morphogenesis—progress and problems. *Can. J. Earth Sci. 16*:992–1015. [Chapter 1.]

Seward, A. C. 1933. *Plant Life Through the Ages*, Cambridge, Cambridge Univ. Press, 603 pp. [Chapter 9.]

Shanks, W. C. and Bischoff, J. L. 1980. Geochemistry, sulfur isotope composition, and accumulation rates of Red Sea geothermal deposits. *Econ. Geol. 75*:445–459. [Chapter 10.]

Sharp, R. P. 1940. Ep-Archean and Ep-Algonkian erosion surfaces, Grand Canyon, Arizona. *Geol. Soc. Am. Bull. 51*:1235–1270. [Chapter 11.]

Shaw, G. H. 1978. Effects of core formation. *Phys. Earth Planet. Interiors 16*:361–369. [Chapters 2, 4.]

Shegelski, R. J. 1978. Stratigraphy and Geochemistry of Archean Iron Formations in the Sturgeon Lake-Savant Lake Greenstone Terrain, Northwestern Ontario. Ph.D. Thesis, Univ. Toronto (Dept. Geol.), 251 pp. [Chapter 10.]

Shegelski, R. J. 1980. Archean cratonization, emergence and red bed development, Lake Shebandowan area, Canada. *Precambrian Res. 12*:331–347. [Chapter 1.]

Sheppard, S. M. F. and Schwarcz, H. P. 1970. Fractionation of carbon and oxygen isotopes and magnesium between coexisting metamorphic calcite and dolomite. *Contrib. Mineral. Petrol. 26*:161–198. [Chapter 7.]

Shieh, Y. H. and Taylor, H. P. 1969. Oxygen and carbon isotope studies of contact metamorphism of carbonate rocks. *J. Petrol. 10*:307–331. [Chapter 7.]

Shima, M.; Gross, W. H.; and Thode, H. G. 1963. Sulfur isotope abundances in basic sills, differentiated granites, and meteorites. *J. Geophys. Res. 68*:2835–2847. [Chapter 7.]

Shimizu, M. 1976. Instability of a highly reducing atmosphere on the primitive Earth. *Precambrian Res. 3*:463–470. [Chapter 4.]

———. 1979. An evolutional model of the terrestrial atmosphere from a comparative planetological view. *Precambrian Res. 9*:311–324. [Chapter 4.]

Shinn, E. A. 1968. Practical significance of bridseye structures in carbonate rocks. *J. Sed. Petrol. 38*:215–223. [Chapter 8.]

Shklanka, R. 1972. Geology of the Steep Rock Lake area, District of Rainy River. *Ontario Dept. Mines Northern Affairs Geol. Rept. 93*, 114 pp. [Appendix I.]

Sibley, D. F. and Wilband, J. T. 1977. Chemical balance of the Earth's crust. *Geochim. Cosmochim. Acta 41*:545–554. [Chapter 10.]

Siedlecka, A. 1976. Silicified Precambrian evaporite nodules from northern Northway: A preliminary report. *Sed. Geol. 16*:161–175. [Chapter 11.]

Siegel, S. M. and Guimarro, C. 1966. On the culture of a microorganism similar to the Precambrian microfossil *Kakabekia umbellata* Barghoorn in NH_3-rich atmospheres. *Proc. Natl. Acad. Sci. USA 55*:343–353. [Chapter 14.]

Siegel, S. M. and Siegel, B. Z. 1968. A living organism morphologically comparable to the Precambrian genus *Kakabekia*. *Am. J. Bot. 55*: 684–687. [Chapter 14.]

Siever, R. 1968. Sedimentological consequences of steady-state ocean-atmosphere. *Sedimentol. 11*: 5–29. [Chapter 11.]

Siever, R. and Woodford, N. 1979. Dissolution kinetics and the weathering of mafic minerals. *Geochim. Cosmochim. Acta 43*:717–724. [Chapter 1.]

Sill, G. I. and Wilkening, L. L. 1978. Ice clathrate as a possible source of the atmospheres of the terrestrial planets. *Icarus 33*:13–22. [Chapter 4.]

Sillén, L. G. 1961. The physical chemistry of sea water. In M. Sears, ed., *Oceanography*, Washington, D.C., Am. Assoc. Advance. Sci., 549–581. [Chapter 11.]

Simoneit, B. R.; Christiansen, P. C.; and Burlingame, A. L. 1973. Volatile element chemistry of selected lunar, meteoritic, and terrestrial samples. *Proc. 4th Lunar Sci. Conf.*:1635–1650. [Chapter 4.]

Simpson, P. R. and Bowles, J. F. W. 1977. Uranium mineralization of the Witwatersrand and Dominion Reef systems. *Phil. Trans. Roy. Soc. London Sect. A 286*:527–548. [Chapter 11.]

Sims, P. K. and Morey, G. B. 1972. Resume of geology of Minnesota. In P. K. Sims and G. B. Morey, eds., *Geology of Minnesota: A Centennial Volume*, Minneapolis, Minn. Geol. Surv., 3–17. [Chapter 11.]

Sin, Y. and Liu, K. 1973. Sinian micro-flora in the Yenliao region of China and its geological significance. *Acta Geologica Sinica 1*:1–64 (in

Chinese with English summary). [Chapters 1, 14.]

Sin, Y. and Liu, K. 1976. *Micropaleoflora from the Sinian Subera of W. Hupeh and its Stratigraphic Significance*, Peking, Inst. Geol. Mineral Resources Chinese Acad. Geol. Sci., 23 pp. [Chapter 1.]

Singh, A. P. and Bragg, P. D. 1975. Reduced nicotinamide adenine dinucleotide dependent reduction of fumarate coupled to membrane energization in a cytochrome deficient mutant of *Escherichia coli* K12. *Biochim. Biophys. Acta 396*:229–241. [Chapter 6.]

———. 1976. Anaerobic transport of amino acids coupled to the glycerol-3-phosphate-fumarate oxidoreductase system in a cytochrome-deficient mutant of *Escherichia coli. Biochim. Biophys. Acta 423*:450–461. [Chapter 6.]

Sirevåg, R.; Buchanan, B. B.; Berry, J. A.; and Troughton, J. H. 1977. Mechanisms of CO_2 fixation in bacterial photosynthesis studied by the carbon isotope fractionation technique. *Arch. Microbiol. 112*:35–38. [Chapter 7.]

Sklarew, D. S. and Nagy, B. 1979. 2,5-Dimethylfuran from the $\sim 2.7 \times 10^9$-yr-old Rupemba-Belingwe stromatolite, Rhodesia: Potential evidence for remnants of carbohydrates. *Proc. Natl. Acad. Sci. USA 76*:10–14. [Chapter 5.]

Skyring, G. W. and Donnelly, T. H. 1982. Precambrian sulfur isotopes and a possible role for sulfite in the evolution of biological sulfate reduction. *Precambrian Res. 17*:41–61. [Chapter 1.]

Smale, D. 1973. Silcretes and associated diagenesis in southern Africa and Australia. *J. Sed. Petrol. 43*:1077–1089. [Chapter 8.]

Smith, B. N. and Epstein, S. 1970. Biogeochemistry of stable isotopes of hydrogen and carbon in salt marsh biota. *Pl. Physiol. 46*:738–742. [Chapter 7.]

———. 1971. Two categories of $^{13}C/^{12}C$ ratios for higher plants. *Pl. Physiol. 47*:380–384. [Chapter 7.]

Smith, J. V. 1979. Mineralogy of the planets: A voyage in space and time. *Mineral. Mag. 43*:1–89. [Chapters 3, 4.]

Smith, J. W. and Batts, B. D. 1974. The distribution and isotopic composition of sulfur in coal. *Geochim. Cosmochim. Acta 38*:121–133. [Chapter 7.]

Smith, J. W. and Croxford, N. J. W. 1973. Sulphur isotope ratios in the McArthur lead-zinc-silver deposit. *Nature 245*:10–12. [Chapter 7.]

———. 1975. An isotopic investigation of the environment of deposition of the McArthur mineralization. *Mineral. Deposita 10*:269–276. [Chapter 7.]

Smith, J. W. and Kaplan, I. R. 1970. Endogenous carbon in carbonaceous meteorites. *Science 167*:1367–1370. [Chapter 4; Appendix III.]

Smith, J. W.; Gould, K. W.; and Rigby, D. 1981. The stable isotope geochemistry of Australian coals. In Austr. Commonwealth Sci. Industr. Res. Org. Fuel Geosci. Unit, Unpublished report. [Chapter 7.]

Smith, J. W.; Schopf, J. W.; and Kaplan, I. R. 1970. Extractable organic matter in Precambrian cherts. *Geochim. Cosmochim. Acta 34*:659–675. [Chapters 5, 7.]

Smith, R. E. 1979. Interpretation of volcanic relations and low-grade metamorphic alteration, Maddina Volcanics, Western Australia. *Austr. Commonwealth Sci. Industr. Res. Org. Div. Mineral. Rept. FP21*, 14 pp. [Chapter 8.]

Smith, R. E.; Green, A. A.; Roberts, G.; and Honey, F. R. 1978. Use of Landsat-1 imagery in exploration for Keweenawan type copper deposits. *Remote Sensing of Environ. 7*:129–144. [Appendix I.]

Snyman, C. P. 1965. Possible biogenic structures in Witwatersrand thucolite. *Geol. Soc. S. Afr. Trans. 68*:225–235. [Chapter 9.]

Sobolev, N. V. 1977. *Deep-Seated Inclusions in Kimberlites and the Problem of the Composition of the Upper Mantle*, Washington D.C., Am. Geophys. Union, 279 pp. [Chapter 4.]

Soederlund, R. and Svensson, B. H. 1975. The global nitrogen cycle. In B. H. Svensson and R. Soederlund, eds., *Nitrogen, Phosphorus and Sulphur Global Cycles*, SCOPE Workshop, Oresundsbro, Sweden. [Chapter 7.]

Sokolov, B. S. 1972. The Vendian Stage in Earth history. *Proc. 24th Internat. Geol. Congr.* (Montreal), *Sec. 1*:78–84. [Chapter 1.]

Solomon, S. C. 1980. Differentiation of crusts and cores of the terrestrial planets: Lessons for the early Earth. *Precambrian Res. 10*:177–194. [Chapter 4.]

Sonett, C. P. and Reynolds, R. T. 1979. Primordial heating of asteroid parent bodies. In T. Gehrels, ed., *Asteroids*, Tucson, Univ. Arizona Press, 822–848. [Chapter 4.]

Southward, A. J.; Southward, E. C.; Dando, P. R.; Rau, G. H.; Felbeck, H.; and Flugel, H. 1981. Bacterical symbionts and low $^{13}C/^{12}C$ ratios in tissues of Pogonophora indicate unusual nutrinition and metabolism. *Nature 293*:616–620. [Chapters 7, 12.]

Spooner, C. M.; Berrangé, J. P.; and Fairbairn, H. W. 1971. Rb-Sr whole-rock age of the Kanuku Complex, Guyana. *Geol. Soc. Am. Bull. 82*: 207–210. [Chapter 11.]

Sprigg, R. C. 1947. Early Cambrian (?) jellyfishes from the Flinders Ranges, South Australia. *Trans. Roy. Soc. S. Austr. 71* (2):212–224. [Chapter 1.]

———. 1949. Early Cambrian "jellyfishes" of Ediacara, South Australia and Mount John, Kimberley District, Western Australia. *Trans. Roy. Soc. S. Austr. 73* (1):72–99. [Chapter 1.]

Srinavasan, R. and Sreenivas, B. L. 1976. Greenstone-granite pluton and gneiss-granulite belts of the type Dharwar craton, Karnataka, India. In *Abstracts, 25th Internat. Geol. Congr.* (Sydney, Austr.), *1*, Sect. 1 (1B):19 (Abstr.). [Chapter 11.]

Stach, E.; Mackowsky, M.-Th.; Teichmüller, M.; Taylor, G. H.; Chandra, D.; and Teichmüller, R. 1975. *Stach's Textbook of Coal Petrology*, Berlin, Gebruder Borntraeger, 428 pp. [Chapter 5.]

Stanier, R. Y.; Sistrom, W. R.; Hansen, T A.; Whitton, B. A.; Castenholz, R. W.; Pfennig, N.; Gorlenko, V. N.; Kondratieva, E. M.; Eimhjellen, K. E.; Whittenbury, R.; Gherna, R. L.; and Trüper, H. G. 1978. Proposal to place the nomenclature of the cyanobacteria (blue-green algae) under the rules of the international code of nomenclature of bacteria. *Internat. J. Syst. Bacteriol. 28*:335–336. [Chapter 13.]

Stanley, S. M. 1976. Fossil data and the Precambrian-Cambrian evolutionary transition. *Am. J. Sci. 276*:56–76. [Chapter 1.]

Steacy, H. R. 1953. An occurrence of uraninite in a black sand. *Am. Mineral. 38*:549–550. [Chapter 11.

Steiger, R. H. and Jäger, E. 1978. Subcommission on Geochronology: Convention on the use of decay constants in geochronology and cosmochronology. *Am. Assoc. Petrol. Geol. Studies in Geol. 6*:67–71. [Appendix I.]

Stevenson, D. J. 1980a. Lunar asymmetry and paleomagnetism. *Nature 287*:520–521. [Chapter 2.]

———. 1980b. Core formation dynamics and primordial planetary dynamos. *Proc. 11th Lunar Sci. Conf.*:1091–1093. [Chapters 2, 3, 4.]

Stevenson, D. J. and Turner, J. S. 1979. Fluid models in mantle convection. In M. W. McElhinny, ed., *The Earth: Its Origin, Structure and Evolution*, New York, Academic Press, 227–264. [Chapters 2, 4.]

Stewart, A. D. 1975. "Torridonian" rocks of western Scotland. In A. L. Harris, ed., *A Correlation of Precambrian Rocks in the British Isles, Geol. Soc. London Spec. Rept. 6*:43–51. [Chapter 11.]

Stewart, A. J. 1979. A barred-basin marine evaporite in the Upper Proterozoic of the Amadeus Basin, Central Australia. *Sedimentol. 26*:33–62. [Chapter 11.]

Stewart, W. D. P. and Pearson, H. W. 1975. Effects of aerobic and anaerobic conditions on growth and metabolism of blue-green algae. *Proc. Roy. Soc. London Ser. B 175*:293–311. [Chapter 13.]

Stillwell, W. 1980. Facilitated diffusions as a method for selective accumulation of materials from primordial oceans by 2 lipid-vesicle protocell. *Origins of Life 10*:277–292. [Chapter 4.]

Stockwell, C. H.; McClynn, J. C.; Emslie, R. F.; Sanford, B. V.; Norris, A. W.; Donaldson, J. A.; Fahrig, W. F.; and Currie, K. L. 1970. Geology of the Canadian Shield. In R. J. W. Douglas, ed., *Geology and Economic Minerals of Canada.* Economic Geology Report 1, Ottawa, Queen's Printer, 44–150. [Chapter 10.]

Stolper, E. 1977. Experimental petrology of aucritic meteorites. *Geochim. Cosmochim. Acta 41*:587–611. [Chapter 4.]

Stolper, E.; McSween, H. Y.; and Hays, J. F. 1979. A petrogenetic model of the relationships among achondritic meteorites. *Geochim. Cosmochim. Acta 43*:589–602. [Chapter 4.]

Stouthamer, A. H.; de Vries, W.; and Niekus, H. G. D. 1979. Microaerophily. *Ant. van Leeuwenhoek 45*:5–12. [Glossary II.]

Strobel, D. F. 1973. The photochemistry of hydrocarbons in the Jovian atmosphere. *J. Atmos. Sci. 30*:489–498. [Chapter 4.]

Strother, P. K. and Barghoorn, E. S. 1980. Microspheres from Swartkoppie Formation: A review. In *The Origins of Life and Evolution*, New York, A. R. Liss, 1–18. [Chapter 9.]

Studier, M. H.; Hayatsu, R.; and Anders, E. 1968. Origin of organic matter in early solar system. I. Hydrocarbons. *Geochim. Cosmochim. Acta 32*: 151–173. [Chapter 4.]

Stuermer, D. H.; Peters, K. E.; and Kaplan, I. R. 1978. Source indicators of humic substances and proto-kerogen. Table isotopic ratios, elemental compositions and electron spin resonance spectra. *Geochim. Cosmochim. Acta 42*:989–997. [Chapter 5.]

Stull, D. R. and Prophet, H. 1971. *JANAF Thermochemical Tables*, Washington, D.C., U.S. Government Printing Office. [Chapter 4; Appendix III.]

Suess, H. 1949. Die Häufigkeit der Edelgase auf der Erde und im Kosmos. *J. Geol. 57*:600–607. [Chapter 4.]

Sünderman, J. and Brosche, P. 1978. The numerical computation of tidal friction for present and ancient oceans. In P. Brosche and J. Sünderman, J., eds., *Tidal Friction and the Earth's Rotation*, New York, Springer, 125–144. [Chapter 11.]

Sweeney, R. E. and Kaplan, I. R. 1980a. Stable isotope composition of dissolved sulfate and hydrogen sulfide in the Black Sea. *Marine Chem. 9*:145–152. [Chapter 7.]

———. 1980b. Isotope fractionation of sulfur during sulfate reduction in marine sediments. Unpublished manuscript. [Chapter 7.]

Sweeney, R. E.; Liu, K. K.; and Kaplan, I. R. 1978. Oceanic nitrogen isotopes and their uses in determining the source of sedimentary nitrogen. In *Stable Isotopes in the Earth Sciences, New Zealand Dept. Scient. Industr. Res. Bull. 220*:9–26. [Chapter 7.]

Swift, W. H., ed. 1961. An outline of the geology of Southern Rhodesia. *Geol. Surv. S. Rhodesia Bull. 50*. [Chapter 11.]

Tait, G. H. 1972. Coproporphyrinogenase activities in extracts of *Rhodopseudomonas spheroides* and *Chromatium* Strain D. *Biochem. J. 128*: 1159–1169. [Chapter 13.]

Tappan, H. 1976. Possible eucaryotic algae (Bangiophycidae) among early Proterozoic microfossils. *Geol. Soc. Am. Bull. 87*:633–639. [Chapter 14.]

Tarney, J.; Dalziel, I. W. D.; and deWit, M. J. 1976. Marginal basin "Rocas Verdes" complex from S. Chile: A model for Archaean greenstone belt formation. In B. F. Windley, ed., *The Early History of the Earth*, London, Wiley, 131–146. [Chapter 3.]

Taylor, H. P. 1978. Oxygen and hydrogen isotope studies of plutonic granitic rocks. *Earth Planet. Sci. Lett. 38*:177–210. [Chapter 10.]

Taylor, R. S. 1979. Chemical composition and evolution of continental crust: The rare earth element evidence. In M. W. McElhinny, ed., *The Earth, Its Origin, Structure and Evolution*, New York, Academic Press, 353–376. [Chapter 10.]

Taylor, R. S. and McLennan, S. M. 1982. Composition of the Archean crust. *Geol. Soc. Austr. Spec. Pub. 7*, in press. [Chapter 7.]

Tera, F.; Papanastassiou, D. A.; and Wasserburg, G. J. 1974. Isotopic evidence for a terminal lunar cataclysm. *Earth Planet. Sci. Lett. 22*:1–21. [Chapter 4.]

Termier, H. and Termier, G. 1960. L'Ediacarien, premier étage paléontologique. *Rev. Gén. Sci. Bull. Assoc. Français Avan. Sci. 67*(3–4):79–87. [Chapter 1.]

Thauer, R. K. and Fuchs, G. 1979. Methanogene Bakterien. *Naturwiss. 66*:89–94. [Chapter 7.]

Thauer, R. K.; Jungermann, K.; and Decker, K. 1977. Energy conservation in chemotrophic anaerobic bacteria. *Bacteriol. Rev. 41*:100–180. [Chapter 6; Glossary II.]

Thode, H. G.; Dunford, H. B.; and Shima, M. 1962. Sulfur isotope abundances in rocks of the Sudbury District and their geological significance. *Econ. Geol. 57*:565–578. [Chapter 7.]

Thode, H. G. and Monster, J. 1965. Sulfur isotope geochemistry of petroleum, evaporites, and ancient seas. *Am. Assoc. Petrol. Geol. Mem. 4*: 367–377. [Chapter 7.]

Thompson, R. I.; Struttmatter, P.; Erikson, E.; Witteborn, F.; and Strecker, D. 1977. Observation of preplanetary disks around MWC 349 and LkHα 101. *Astron. J. 218*:170–180. [Chapter 2.]

Thomsen, L. 1980. ^{129}Xe on the outgassing of the atmosphere. *J. Geophys. Res. 85*:4374–4378. [Chapter 4.]

Thorsteinsson, R. and Tozer, E. T. 1962. Banks, Victoria and Stefansson Islands, Arctic Archipelago. *Geol. Surv. Can. Mem. 330*, 85 pp. [Chapter 11.]

Thrailkill, J. 1976. Speleothems. In M. R. Walter, ed., *Stromatolites*, Amsterdam, Elsevier, 73–86. [Chapter 8.]

Thurston, P. C. 1980. Subaerial volcanism in the Archean Uchi-Confederation volcanic belt. *Precambrian Res. 12*:79–98. [Appendix I.]

Thurston, P. C. and Jackson, M. C. 1978. Confederation Lake Area, District of Kenora (Patricia portion). *Ontario Geol. Surv. Geol. Ser. Prelim. Map P. 1975*. [Chapter 8; Appendix I.]

Timofeev, B. V. 1958. Spori Proterozoishikh i rannego Paleozoishikh otlozhenni vostochnoi Sibiri i ikh stratigraphicheskoe znachenie (Spores of the Proterozoic and early Paleozoic deposits of East Siberia and their stratigraphic significance). *Trudy Mezhdu. Sovesc. Raz. Unific. Strat. Sibiu*:226–230 (in Russian). [Chapter 1.]

———. 1959. Drevneishaya flora Pribaltiki i ee stratigraficheskoe znachenie (The oldest flora of Prebaltica and its stratigraphic significance). *Trudy Vsesoy. Neft. Nauchno-Issledov. Geolog. Inst.* (*VNIGRI*) *129*, 320 pp. (in Russian). [Chapters 1, 14.]

———. 1966. *Mikropaleofitologicheskoe issledovanie drevnikh svit*, Moscow, Nauka, Lab. Geol. Dokembriya Akad. Nauk SSSR, 147 pp. 1974, English transl., *Micropalaeophytological Research Into Ancient Strata*, British Library-Lending Div., London, 214 pp. [Chapters 1, 14.]

———. 1969. *Sferomorfidy proterozoya* (*Proterozoic Sphaeromorphida*), Leningrad, Nauka, Inst. Geol. Geokhron. Dokembriya, Akad. Nauk SSSR, 146 pp. (in Russian). [Chapter 14.]

———. 1973a. Mikrofitofossilii proterozoya i rannego paleozoya (Proterozoic and early Paleozoic microfossils). In T. F. Vozzhennikova and B. V. Timofeev, eds., *Mikrofossilii drevneishikh otlozhenii* (*Microfossils of the Oldest Deposits*, Proc. 3rd Internat. Palynol. Conf.), Moscow, Nauka, 7–12 (in Russian with English abstract). [Chapters 1, 14.]

———. 1973b. *Microfitofossilii dokembriya Ukrainy* (*Precambrian Microfossils from the Ukraine*), Leningrad, Nauka, 58 pp. (in Russian). [Chapter 14.]

———. 1974. *Mikrofitofossilii Proterozoya i Rannego Paleozoya S.S.S.R.* (*Plant Microfossils in the Proterozoic and Lower Paleozoic of the U.S.S.R.*), Moscow, Nauka, Inst. Geol. Geokhron. Dokembriya, Akad. Nauk SSSR, 80 pp. (in Russian). [Chapter 1.]

———. 1979. Mikrofitossilii Pechengskoi serii (Microfossils from the Pechenga Series). In B. V. Sokolov, ed., *Paleontologiya dokembriya i rannego kembriya* (*Paleontology of the Precambrian and Early Cambrian*), Leningrad, Nauka, 119–120 (in Russian). [Chapter 14.]

Timofeev, B. V.; German, T. N.; and Mikhailova, N. 1976. *Mikrofitofossilii Dokembriya, Kembriya i Ordovika* (*Plant Microfossils of the Precambrian, Cambrian and Ordovician*), Leningrad, Nauka, Inst. Geol. Geokhron. Dokembriya, Akad. Nauk SSSR, 87 pp. (in Russian). [Chapter 9.]

Tissot, B. P. and Welte, D. H. 1978. *Petroleum Formation and Occurrence*, New York, Springer, 538 pp. [Chapters 4, 5.]

Toulmin, P. III; Baird, A. K.; Clark, B. C.; Keil, K.; Rose, H. J. Jr.; Christian, R. P.; Evans, P. H.; and Kelliher, W. C. 1977. Geochemical and mineralogical interpretation of the Viking inorganic chemical results. *J. Geophys. Res. 82*: 4625–4634. [Chapter 4.]

Toupance, G.; Raulin, F.; and Buvet, R. 1975. Formation of prebiochemical compounds in models of the primitive Earth's atmosphere: 1. CH_4-NH_3 and CH_4-N_2 atmospheres. *Origins of Life 6*:83–90. [Chapter 4.]

Towe, K. M. 1970. Oxygen-collagen priority and the early Metazoan fossil record. *Proc. Natl. Acad. Sci. USA 65*:781–788. [Chapter 1.]

Trendall, A. F. 1973. Time-distribution and type-distribution of Precambrian iron-formations in Australia. In *Genesis of Precambrian Iron and Manganese Deposits* (Proc. Kiev Symp. 1970), Paris, UNESCO, 49–57. [Chapter 11.]

———. 1975. Hamersley Basin. In *Geology of Western Australia, Geol. Surv. W. Austr. Mem. 2*:118–141. [Chapters 8, 9.]

———. 1976. Geology of the Hamersley Basin. *25th Internat. Geol. Congr.* (Sydney) *Excursion Guide 43A*, 44 pp. [Chapters 8, 9; Appendix I.]

———. 1979. A revision of the Mount Bruce Supergroup. *Geol. Surv. W. Austr. Ann. Rept. 1978*:63–71. [Chapter 11; Appendix I.]

Trendall, A. F. and Blockley, J. G. 1970. The iron formations of the Precambrian Hamersley Group, Western Australia. *Geol. Surv. W. Austr. Bull. 119*:336 pp. [Chapters 8, 11.]

Trudinger, P. A. 1969. Assimilatory and dissimilatory metabolism of inorganic sulfur compounds by microorganisms. *Adv. Microbiol. Physiol. 3*: 111–158. [Chapter 7.]

Trüper, H. G. 1982. Microbial processes in the sulfuretum through time. In H. D. Holland and M. Schidlowski, eds., *Mineral Deposits and the Evolution of the Biosphere*, Dahlem Konferenzen-Dahlem Workshop Rept. 16, Berlin, Springer, 5–30. [Chapter 7.]

Truswell, J. F. and Eriksson, K. A. 1972. The morphology of stromatolites from the Transvaal

Dolomite north-west of the Johannesburg, South Africa. *Geol. Soc. S. Afr. Trans. 75*:99–100. [Chapter 8.]

———. 1973. Stromatolitic associations and their palaeo-environmental significance: A re-appraisal of a Lower Proterozoic locality from the northern Cape Province, South Africa. *Sed. Geol. 10*:1–23. [Chapters 8, 11; Appendix I.]

———. 1975a. A palaeo-environmental interpretation of the Early Proterozoic Malmani Dolomite from Zwartkops, South Africa. *Precambrian Res. 2*:277–303. [Chapter 8.]

———. 1975b. Facies and laminations in the Lower Proterozoic Transvaal Dolomite, South Africa. In G. D. Rosenberg and S. K. Runcorn, eds., *Growth Rhythms and the History of the Earth's Rotation*, New York, Wiley, 57–73. [Chapter 8.]

Tseng, S. S. and Chang, S. 1974. Photo-induced free radicals on a simulated Martian surface. *Nature 248*:42–43. [Chapter 4.]

Tucker, M. E. 1976. Replaced evaporites from the late Precambrian of Finnmark, Arctic Norway. *Sed. Geol. 16*:193–204. [Chapter 11.]

Tudge, A. P. and Thode, H. G. 1950. Thermodynamic properties of isotopic compounds of sulfur. *Can. J. Res. 28B*:567–578. [Chapter 7.]

Turcotte, D. L. 1980. On the thermal evolution of the Earth. *Earth Planet. Sci. Lett. 48*:53–58. [Chapter 10.]

Turcotte, D. L. and Oxburgh, E. R. 1972. Mantle convection and the new global tectonics. *Ann. Rev. Fluid Mech. 4*:33–68. [Chapter 2.]

Turekian, K. K. and Clark, S. P. Jr. 1969. Inhomogeneous accumulation of the Earth from the primitive solar nebula. *Earth Planet. Sci. Lett. 6*:346–348. [Chapter 4.]

———. 1975. The non-homogeneous accumulation model for terrestrial planet formation and the consequences for the atmosphere of Venus. *J. Atmos. Sci. 32*:1257–1261. [Chapter 4; Appendix III.]

Turner, B. 1980. Interstellar molecules. *J. Mol. Evol. 15*:79–101. [Chapter 4.]

Turner, C. G. and Walker, R. G. 1973. Sedimentology, stratigraphy and the crustal evolution of the Archaean greenstone belt near Sioux Lookout, Ontario. *Can. J. Earth Sci. 10*:817–845. [Chapters 10, 11.]

Turner, F. J. 1968. *Metamorphic Petrology*, New York, McGraw-Hill, 403 pp. [Chapter 5.]

Tyler, S. A. and Barghoorn, E. S. 1954. Occurrence of structurally preserved plants in pre-Cambrian rocks of the Canadian shield. *Science 119*:606–608. [Chapters 1, 9, 14.]

———. 1963. Ambient pyrite grains in Precambrian cherts. *Am. J. Sci. 261*:424–432. [Chapters 9, 14.]

Uffen, R. L. and Wolfe, R. S. 1970. Anaerobic growth of purple nonsulfur bacteria under dark conditions. *J. Bacteriol. 104*:462–472. [Chapter 6.]

Urey, H. C. 1952a. *The Planets: Their Origin and Development*, New Haven, Conn., Yale Univ. Press, 245 pp. [Chapter 4.]

———. 1952b. On the early chemical history of the Earth and the origin of life. *Proc. Natl. Acad. Sci. USA 38*:351–363. [Chapter 4.]

Usher, D. A. 1977. Early chemical evolution of nuclei acids. *Science 196*:311–313. [Chapter 4.]

Valley, J. N. and O'Neil, J. R. 1981. $^{13}C/^{12}C$ exchange between calcite and graphite: A possible thermometer in Grenville marbles. *Geochim. Cosmochim. Acta 45*:411–419. [Chapters 5, 7.]

Vander Wood, T. B. and Thiemens, M. H. 1980. The fate of the hydroxyl radical in the Earth's primitive atmosphere and implications for the production of molecular oxygen. *J. Geophys. Res. 85*:1605–1610. [Chapters 4, 11, 13.]

Van Hoeven, W.; Maxwell, J. R.; and Calvin, M. 1969. Fatty acids and hydrocarbons as evidence of life processes in ancient sediments and crude oils. *Geochim. Cosmochim. Acta 33*:877–881. [Chapter 14.]

Van Houten, F. B. 1973. Origin of red beds. A review, 1961–1972. In *Ann. Rev. Earth Planet. Sci.*, Palo Alto, Calif., Annual Reviews, Inc., 39–61. [Chapter 11; Appendix I.]

Van Niekerk, C. B. and Burger, A. J. 1978. A new age for the Ventersdorp acidic lavas. *Geol. Soc. S. Afr. Trans. 81*:155–163. [Appendix I.]

Van Schmus, W. R. 1965. The geochronology of the Blind River-Bruce Mines area, Ontario, Canada. *J. Geol. 73*:755–780. [Appendix I.]

———. 1976. Early and Middle Proterozoic history of the Great Lakes area, North America. *Phil. Trans. Roy. Soc. London Sect. A 280*:605–628. [Appendix I.]

Van Trump, J. E. and Miller, S. L. 1973. Carbon monoxide on the primitive Earth. *Earth Planet. Sci. Lett. 20*:145–150. [Chapter 4.]

———. 1980. The Strecker synthesis in the primitive ocean. In Y. Wolman, ed., *Proc. 3rd Meeting*

Internat. Soc. Study Origin of Life (Dordrecht, Reidel), 135–141.

Van Valen, L. 1971. The history and stability of atmospheric oxygen. *Science 171*:439–443. [Chapter 11.]

———. 1976. Energy and evolution. *Evolutionary Theory 1*:179–229. [Chapter 15.]

Van Valen, L. M. and Moriarana, V. C. 1980. The Archaebacteria and eukaryotic origins. *Nature 287*:248–250. [Chapter 1; Glossary I.]

Vassoevich, N. B.; Korchagina, Yu. I.; Lopatin, N. V.; and Chernyshev, V. V. 1969. Principal phase of oil formation. *Moscow Univ. Vestnik 6*:3–37. 1970 English transl., *Internat. Geol. Rev. 12*:1276–1296. [Chapter 5.]

Vaughan, F.; Eby, D.; and Meyers, W. J. 1977. Calcitization of evaporites: Do fabrics reflect diagenetic environment? *Geol. Soc. Am. Abstr. Progr. 9*:1209 (Abstr.). [Chapter 11.]

Veizer, J. 1973. Sedimentation in geologic history: Recycling *vs.* evolution or recycling with evolution. *Contrib. Mineral. Petrol. 38*:261–278. [Chapter 10.]

———. 1974. Chemical diagenesis of belemnite shells and possible consequences for paleotemperature determinations. *Neues Jahrb. Geol. Pal. Abh. 147*:91–111. [Chapter 10.]

———. 1976. Evolution of ores of sedimentary affiliation through geologic history, relation to the general tendencies in evolution of the crust hydrosphere, atmosphere and biosphere. In K. H. Wolf, ed., *Handbook of Strata-bound and Stratiform Ore Deposits, Vol. 3*, Amsterdam, Elsevier, 1–41. [Chapter 10.]

———. 1977. Diagenesis of pre-Quaternary carbonates as indicated by tracer studies. *J. Sed. Petrol. 47*:565–581. [Chapter 10.]

———. 1978a. Strontium: Abundance in common sediments and sedimentary rock types. In K. H. Wedepohl, ed., *Handbook of Geochemistry, Vol. 2, Pt. 5*, Berlin, Springer, K1–K13. [Chapter 10.]

———. 1978b. Secular variations in the composition of sedimentary carbonate rocks—II. Fe, Mn, Ca, Mg, Si and minor constituents. *Precambrian Res. 6*:381–413. [Chapter 10.]

———. 1979a. Secular variations in chemical composition of sediments: A review. In L. H. Ahrens, ed., *The Origin and Evolution of the Elements*, London, Pergamon, 269–278. [Chapter 10.]

———. 1979b. Chemistry of the early oceans: Implications for crustal development. *RSES, Austr. Natl. Univ. Yearbook 1979*:170–172. [Chapter 10.]

———. 1982. The evolving terrestrial exogenic cycle. In R. M. Garrels, C. B. Gregor, and F. T. Mackenzie, eds., *Chemical Cycles in the Evolution of the Earth*, New York, Wiley, in press. [Chapter 10.]

———. 1983. Chemical diagenesis of carbonates: Theory and application of trace element technique. In M. A. Arthur, T. F. Anderson, I. R. Kaplan, L. S. Land and J. Veizer, *Stable Isotopes in Sedimentary Geology, Short Course Notes 10*, Soc. Econ. Paleontol. Mineral., Tulsa, Oklahoma, III-1–III-112. [Chapter 10.]

Veizer, J. and Compston, W. 1976. $^{87}Sr/^{86}Sr$ in Precambrian carbonates as an index of crustal evolution. *Geochim. Cosmochim. Acta 40*:905–914. [Chapters 8, 10.]

Veizer, J. and Hoefs, J. 1976. The nature of O^{18}/O^{16} and C^{13}/C^{12} secular trends in sedimentary carbonate rocks. *Geochim. Cosmochim. Acta 40*: 1387–1395. [Chapters 8, 10.]

Veizer, J. and Jansen, S. 1979. Basement and sedimentary recycling and continental evolution. *J. Geol. 87*:341–370. [Chapters 3, 10.]

Veizer, J; Compston, W.; Hoefs, J.; and Nielsen, H. 1982. Mantle buffering of the early oceans. *Naturwiss. 69*:173–180. [Chapter 10.]

Veizer, J.; Holser, W. T.; and Wilgus, C. K. 1980. Correlation of $^{13}C/^{12}C$ and $^{34}S/^{32}S$ secular variations. *Geochim. Cosmochim. Acta 44*:579–587. [Chapters 7, 10.]

Veizer, J.; Lemieux, J.; Jones, B.; Gibling, R. M.; and Savelle, J. 1978. Paleosalinity and dolomitization of a Lower Paleozoic carbonate sequence, Somerset and Prince of Wales Islands, Arctic Canada. *Can. J. Earth Sci. 15*:1448–1461. [Chapter 10.]

Vidal, G. 1976. Late Precambrian microfossils from the Visingö Beds in southern Sweden. *Fossils and Strata 9*, 57 pp. [Chapter 1.]

———. 1979. Acritarchs from the Upper Proterozoic and Lower Cambrian of East Greenland. *Grønlands Geol. Undersøg. Bull. 134*, 40 pp. [Chapter 1.]

———. 1981. Aspects of problematic acid-resistant, organic-walled microfossils (acritarchs) in the Upper Proterozoic of the North Atlantic region. *Precambrian Res. 15*:9–23. [Chapter 15.]

Vidal, G. and Knoll, A. H. 1981. Proterozoic

plankton. *The Proterozoic Eon*, Madison, Wis., In press [Chapter 15.]

Vidal, P. and Dosso, L. 1978. Core formation: Catastrophic or continuous? Sr and Pb isotope geochemistry constraints. *Geophys. Res. Lett.* *5*:169–172. [Chapter 4.]

Viljoen, M. J. and Viljoen, R. P. 1969. A reappraisal of the granite-greenstone terrains of shield areas based on the Barberton model. *Geol. Soc. S. Afr. Spec. Pub.* *2*:245–274. [Chapter 3.]

Viljoen, R. P. and Viljoen, M. J. 1969. The effects of metamorphism and serpentinization on the volcanic and associated rocks of the Barberton region. *Geol. Soc. S. Afr. Spec. Pub.* *2*:29–53. [Appendix I.]

———. 1971. The geological and geochemical evolution of the Onverwacht volcanic group of the Barberton Mountain Land, South Africa. In J. E. Glover, ed., *Symposium on Archean Rocks*, *Geol. Soc. Austr. Spec. Pub.* *3*:133–149. [Chapter 3.]

Vinogradov, A. P.; Grinenko, V. A.; and Ustinov, U. S. 1962. Izotopny sostav soedinenii sery v chornom more (Isotopic composition of sulfur compounds in the Black Sea). *Geokhim.* *10*: 851–873. [Chapter 7.]

Vinogradov, V. I.; Reimer, T. O.; Leites, A. M.; and Smelov, S. B. 1976. The oldest sulfates in Archaean formations of the South African and Aldan Shields and the evolution of the Earth's oxygenic atmosphere. *Lithol. Min. Resources* *11*: 407–420. [Chapters 7, 11.]

Visser, J. N. J. 1971. The deposition of the Griquatown glacial member in the Transvaal Supergroup. *Geol. Soc. S. Afr. Trans.* *74*:187–199. [Chapter 11.]

Visser, J. N. J.; Grobler, N. J.; Joubert, C. W.; Potgieter, C. D.; Potgieter, G. J. A.; McLaren, C. H.; and Liebenberg, J. 1976. The Ventersdorp Group between Taung and Britstown, Northern Cape Province. *Annals. Geol. Surv. S. Afr.* *11*: 15–28. [Chapter 8.]

Vitorovic, D. 1980. Structure elucidation of kerogen by chemical methods. In B. Durand, ed., *Kerogen*, Paris, Editions Technip, 301–338. [Chapter 5.]

Vityazev, A. V. and Mayeva, S. V. 1976. Model of the Earth. *Izv. Phys. Solid Earth* *12*:79–85. [Chapter 2.]

Völker, T. 1960. Polymere Blausäure. *Angew. Chemie* *72*:370–384. [Appendix III.]

Vogler, E. A. and Hayes, J. M. 1980. Carbon isotopic compositions of carboxyl groups of biosynthesized fatty acids. In A. G. Douglas and J. R. Maxwell, eds., *Advances in Organic Geochemistry 1979*, London, Pergamon, 697–704. [Chapter 5.]

Vologdin, A. G. and Drozdova, N. A. 1969. Noviye sinezeleniye vodorosli Dokembriiskovo vozrasta iz Batenevskovo Kryazha (New blue-green algae of Precambrian age from Batenovski Ridge). *Dokl. Akad. Nauk SSSR* *187* (2):440–442 (in Russian). [Chapter 1.]

Vologdin, A. G. and Korde, K. B. 1965. Neskol'ko vidov drevnikh Cyanophyta i ikh tsenozy (Several species of ancient Cyanophyta and their coenoses). *Dokl. Akad. Nauk SSSR* *164* (2): 429–432(in Russian). [Chapters 9, 14.]

Von Brunn, V. and Mason, T. R. 1977. Siliciclastic-carbonate tidal deposits from the 3000 m.y. Pongola Supergroup, South Africa. *Sed. Geol.* *18*:245–255. [Chapter 8; Appendix I.]

Vorona, I. D.; Kravchenko, V. M.; Pervago, V. A.; and Frumkin, I. M. 1973. Precambrian ferruginous formations of the Aldan Shield. In *Genesis of Precambrian Iron and Manganese Deposits* (Proc. Kiev Symp. 1970), Paris, UNESCO, 243–248. [Chapter 11.]

Wada, E. and Hattori, A. 1975. Natural abundance of ^{15}N in particulate organic matter in the North Pacific Ocean. *Geochim. Cosmochim. Acta* *40*:249–251. [Chapter 7.]

Wada, E.; Kadonaga, T.; and Matsuo, S. 1975. ^{15}N abundance in nitrogen of naturally-occurring substances and global assessment of denitrification from an isotopic viewpoint. *Geochem. J.* *9*:139–148. [Chapter 7.]

Walcott, C. D. 1883. Pre-Carboniferous strata of the Grand Canyon of Colorado. *Am. J. Sci.* *26*: 437–442. [Chapter 1.]

———. 1895. Algonkian rocks of the Grand Canyon of the Colorado. *J. Geol.* *3*:312–330. [Chapter 1.]

———. 1899. Pre-Cambrian fossiliferous formations. *Geol. Soc. Am. Bull.* *10*:199–244. [Chapter 1.]

———. 1912. Notes on fossils from limestone of Steeprock Series, Ontario, Canada. *Geol. Surv. Can. Mem.* *28*:16–23. [Chapter 8.]

———. 1914. Cambrian geology and paleontology. III. No. 2, Pre-Cambrian Algonkian algal flora. *Smithsonian Misc. Coll.* *64*:77–156. [Chapter 1.]

———. 1915. Discovery of Algonkian bacteria.

Proc. Natl. Acad. Sci. USA 1:256–257. [Chapter 1.]

———. 1919. Cambrian Geology and Paleontology IV: Middle Cambrian algae. *Smithsonian Misc. Coll. 67*:217–260. [Chapter 1.]

Walden, W. C. and Hentges, D. J. 1975. Differential effects of oxygen and oxidation-reduction potential on the multiplication of three species of anaerobic intestinal bacteria. *Appl. Microbiol. 30*:781–785. [Chapter 13.]

Walker, D. 1979. *Energy, Plants and Man*, Chichester, Packard, 31 pp. [Chapter 7.]

Walker, D. C., ed., 1979. *Origins of Optical Activity in Nature*, New York, Elsevier, 261 pp. [Chapter 4.]

Walker, J. C. G. 1974. Stability of atmospheric oxygen. *Am. J. Sci. 274*:193–214. [Chapter 11.]

———. 1976. Implications for atmospheric evolution of the inhomogeneous accretion model of the origin of the Earth. In B. F. Windley, ed., *The Early History of the Earth*, New York, Wiley, 535–546. [Chapter 4.]

———. 1977. *Evolution of the Atmosphere*, New York, Macmillan, 318 pp. [Chapters 1, 4, 11; Appendix III.]

———. 1978a. Oxygen and hydrogen in the primitive atmosphere. *Pure and Appl. Geophys. 116*: 222–231. [Chapters 4, 11.]

———. 1979. The early history of oxygen and ozone in the atmosphere. *Pure and Appl. Geophys. 117*:498–512. [Chapters 11, 13.]

———. 1980a. Biogeochemical cycles: Oxygen. In O. Hutzinger, ed., *Handbook of Environmental Chemistry*, Heidelberg, Springer, 87–104. [Chapter 11.]

———. 1980b. Atmospheric constraints on the evolution of metabolism. *Origins of Life 10*:93–104. [Chapter 11.]

———. 1980c. The influence of life on the evolution of the atmosphere. *Life Sci. Space Res. 8*: 89–100. [Chapters 4, 11.]

———. 1981. The earliest atmosphere of the Earth. *Precambrian Res.*, in press. [Introduction.]

Walker, J. C. G.; Hays, P. B.; and Kasting, J. F. 1981. A negative feedback mechanism for the long-term stabilization of Earth's surface temperature. *J. Geophys. Res.*, in press. [Introduction; Chapter 11.]

Walker, R. N.; Muir, M. D.; Diver, W. L.; Williams, N.; and Wilkins, N. 1977. Evidence of major sulphate evaporite deposits in the Proterozoic McArthur Group, Northern Territory, Australia. *Nature 265*:526–529. [Chapters 8, 11.]

Walls, R. A.; Burleigh, H. W.; and Nunan, W. E. 1975. Calcareous crust (caliche) profiles and early subaerial exposure of Carboniferous carbonates, northeastern Kentucky. *Sedimentol. 22*: 417–440. [Chapter 8.]

Walpole, B. P.; Dunn, R. P.; and Randal, M. A. 1968. Geology of the Katherine-Darwin region, Northern Territory. *Austr. Bur. Min. Resources Bull. 82*, 169 pp. [Chapter 11.]

Walter, M. 1979. Precambrian glaciation. *Am. Scient. 67*:142. [Chapter 11.]

Walter, M. R. 1970a. Stromatolites used to determine the time of nearest approach of Earth and Moon. *Science 170*:1331–1332. [Chapter 11.]

———. 1970b. Stromatolites and the Biostratigraphy of the Australian Precambrian. Ph.D. Thesis, Univ. Adelaide (Dept. Geol.), 466 pp. [Chapter 8.]

———. 1972a. Stromatolites and the biostratigraphy of the Australian Precambrian and Cambrian. *Spec. Pap. Palaeont. 11* (Palaeont. Assoc. London), 190 pp. [Chapters 8, 9; Appendix I.]

———. 1972b. A hot spring analog for the depositional environment of Precambrian iron formations of the Lake Superior region. *Econ. Geol. 67*: 965–972. [Appendix I.]

———. 1976a. Introduction. In M. R. Walter, ed., *Stromatolites*, Amsterdam, Elsevier, 1–3. [Chapters 1, 8.]

———. 1976b. Geyserites of Yellowstone National Park: An example of abiogenic "stromatolites". In M. R. Walter, ed., *Stromatolites*, Amsterdam, Elsevier, 87–112. [Chapter 8.]

———. 1976c. Hot spring sediments in Yellowstone National Park. In M. R. Walter, ed., *Stromatolites*, Amsterdam, Elsevier, 489–498. [Chapter 8.]

———. 1977. Interpreting stromatolites. *Am. Scient. 65*:563–571. [Chapter 8.]

———. 1978. Recognition and significance of Archaean stromatolites. In *Archaean Cherty Metasediments: Their Sedimentology, Micropalaeontology, Biogeochemistry, and Significance to Mineralization*, Perth, *Univ. W. Austr. Spec. Pub. 2*:1–10. [Chapter 8.]

———. 1980. Palaeobiology of Archaean stromatolites. In J. E. Glover and D. I. Groves, eds., *2nd Internat. Archean Symposium* (Perth), Canberra, Austr. Acad. Sci., 22–23 (Abstr.). [Introduction.]

———. 1981. Adelaidean and Early Cambrian stratigraphy of the southwestern Georgina Basin: Correlation chart and explanatory notes. *Austr. Bur. Min. Resources Geol. Geophys. Rept. 214*, in press. [Chapter 11.]

Walter, M. R. and Awramik, S. M. 1979. *Frutexites* from stromatolites of the Gunflint Iron Formation of Canada, and its biological affinities. *Precambrian Res. 9*:23–33. [Chapters 8, 14.]

Walter, M. R.; Bauld, J.; and Brock, T. D. 1972. Siliceous algal and bacterial stromatolites in hot spring and geyser effluents of Yellowstone National Park. *Science 178*:402–405. [Chapter 8.]

———. 1976. Microbiology and morphogenesis of columnar stromatolites (*Conophyton*, *Vacerrilla*) from hot springs in Yellowstone National Park. In M. R. Walter, ed., *Stromatolites*, Amsterdam, Elsevier, 273–310. [Chapters 8, 9.]

Walter, M. R.; Buick, R.; and Dunlop, J. S. R. 1980. Stromatolites 3,400–3,500 Myr old from the North Pole area, Western Australia. *Nature 284*:443–445. [Introduction; Chapters 1, 7, 8, 9.]

Walter, M. R.; Golubić, S.; and Preiss, W. V. 1973. Recent stromatolites from hydromagnesite and aragonite depositing lakes near the Coorong Lagoon, South Australia. *J. Sed. Petrol. 43*: 1021–1030. [Chapter 8.]

Walter, M. R.; Goode, A. D. T.; and Hall, W. D. M. 1976. Microfossils from a newly discovered Precambrian stromatolitic iron formation in Western Australia. *Nature 261*:221–223. [Chapters 1, 14.]

Walter, M. R.; Oehler, J. H.; and Oehler, D. Z. 1976. Megascopic algae 1300 million years old from the Belt Supergroup, Montana: A reinterpretation of Walcott's Helminthoidichnites. *J. Paleontol. 50*:872–881. [Chapters 1, 15.]

Walters, C. and Ponnamperuma, C. 1981. Charracterization of graphite isolated from ~3,800 m.y. old metasediments of Isua, Greenland: Implications for organic geochemical studies. *Nature*, in press. [Chapter 15.]

Walters, C.; Shimoyama, A.; and Ponnamperuma, C. 1980. Organic geochemistry of the Isua Supracrustals. *Abstracts, 3rd Internat. Meeting Soc. Study Origin Life* (Jerusalem, Israel): 84 (Abstr.). [Chapter 15.]

Walters, C.; Shimoyama, A.; and Ponnamperuma, C. 1981. Organic geochemistry of the Isua Supracrustals. In Y. Wolman, ed., *Origin of Life*, Amsterdam, Reidel, 473–479. [Chapter 15.]

Walters, C.; Shimoyama, A.; Ponnamperuma, C.; Barghoorn, E. S.; Moorbath, S.; and Schidlowski, M. 1979. Organic geochemistry of 3.8 × 10^9 year-old metasediments from Isua, Greenland. *Abstracts, 178th Natl. Meeting, Am. Chem. Soc.* (Washington D.C.), (Abstr.). [Chapter 15.]

Wang, C.; Xiao, Z.; Shi. F.; Xu, H.; and Li, Z. 1980. Sinian Suberathem in the Ming Tombs, Beijing. In Y. Wang, ed., *Research on Precambrian Geology, Sinian Suberathem in China*, Tianjin, Tianjin Sci. Technol. Press, 332–340 (in Chinese with English abstract). [Chapter 14.]

Wasserburg, G. J.; Papanastassiou, D. A.; and Lee, T. 1979. Isotopic heterogeneities in the solar system. In *Les Elements et Leurs Isotopes dans l'Universe* (Proc. 22nd Internat. Astrophys. Colloquium, Liege, Belgium), Liege, Belgium Acad. Sci., 203–255. [Chapter 1.]

Wasserburg, G. J.; Papanastassiou, D. A.; Tera, F.; and Huneke, J. G. 1977. The accumulation and bulk composition of the Moon. Outline of a lunar chronology. *Phil. Trans. Roy. Soc. London Sect. A 285*:7–22. [Chapter 4.]

Wasson, J. T. 1974. *Meteorites*, New York, Springer, 316 pp. [Chapter 4; Appendix III.]

Webb, R. B. 1977. Lethal and mutagenic effects of near-ultraviolet radiation. In K. C. Smith, ed., *Photochemical and Photobiological Reviews, Vol. 2*, New York, Plenum, 169–261. [Chapter 15.]

Weber, A. and Miller, S. L. 1981. Reasons for the occurrence of the twenty coded portein amino acids. *J. Mol. Evol. 17*:273–284. [Chapter 4.]

Wedeking, K. W. and Hayes, J. M. 1982. Graphitization of Precambrian kerogens. In H. R. v. Gaertner and H. Wehner, eds., *Advances in Organic Geochemistry 1981*, New York, Pergamon, in press. [Introduction.]

Wehhan, J. A. 1981. Carbon and Hydrogen Gases in Hydrothermal Systems: The Search for a Mantle Source. Ph.D. Thesis, Univ. Calif., San Diego (Dept. Geophys.). [Chapter 7.]

Welhan, J. 1980. Gas concentrations and isotope ratios at the 21° N EPR hydrothermal site. *Am. Geophys. Union Trans. 61*:996 (Abstr.) [Chapter 4.]

Welhan, J. A. and Craig, H. 1979. Methane and hydrogen in East Pacific Rise hydrothermal fluids. *Geophys. Res. Lett. 6*:829–831. [Chapters 2, 4, 5.]

Weller, D.; Doemel, W. N.; and Brock, T. D. 1975. Requirement of low oxidation-reduction potential for photosynthesis in a mat-forming blue-green alga, *Phormidium* sp. *Arch. Mikrobiol. 104*:7–13. [Chapter 8.]

Wellman, R. P.; Cook, E. D.; and Krouse, H. R. 1968. Nitrogen-15 microbiological alteration of abundance. *Science 161*:269–270. [Chapter 7.]

Wells, J. W. 1963. Coral growth and geochronometry. *Nature 197*:948–950. [Chapter 11.]

Wells, P. R. A. 1979. Chemical and thermal evolution of Archaean sialic crust, south West Greenland. *J. Petrol. 20*:187–226. [Chapter 3.]

Welte, D. H.; Kalreuth, W.; and Hoefs, J. 1975. Age trend in carbon isotopic composition in Paleozoic sediments. *Naturwiss. 62*:482–483. [Chapter 7.]

Wendt, I. 1968. Fractionation of carbon isotopes and its temperature dependence in the system CO_2 (gas) – CO_2 (in solution) and HCO_3-CO_2 in solution. *Earth Planet. Sci. Lett. 4*:64–68. [Chapter 7.]

Wetherill, G. W. 1972. The beginning of continental evolution. *Tectonophys. 13*:31–45. [Chapter 3.]

———. 1975a. Radiometric chronology of the early solar system. *Ann. Rev. Nuc. Sci. 25*:283–328. [Chapter 4.]

———. 1975b. Late heavy bombardment of the moon and terrestrial planets. *Proc. 6th Lunar Sci. Conf.*: 1539–1561. [Chapter 4.]

———. 1976. The role of large bodies in the formation of the Earth and Moon. *Proc. 7th Lunar Sci. Conf.*:3245–3257. [Chapters 3, 4.]

———. 1977. Evolution of the Earth planetisimal swarm subsequent to the formation of the Earth and Moon. *Proc. 8th Lunar Sci. Conf.*:1–16. [Chapter 4.]

———. 1978. Dynamical evidence regarding the relationship between asteroids and meteorites. In D. Morrison and W. C. Wells, eds., *Asteroids: An Exploration Assessment*, Washington, D.C., NASA Conf. Pub. 2053, 17–36. [Chapter 4.]

———. 1980. Formation of the terrestrial planets. *Ann. Rev. Astron. Astrophys. 18*:77–113. [Chapter 4.]

Wetzel, Otto. 1933. Die in organischen Substanz erhaltenen Mikrofossilien des baltischen Kreide-Feuersteins mit einem sediment petrographischen und stratigraphischen Anhang. *Palaeontographica 77*:141–188 and 78:1–110. [Chapter 1.]

Whipple, R. 1976. A speculation about comets and the Earth. *Mem. Soc. Roy. Sci. Liege Ser. 6, Vol. 9*:101–111. [Chapter 4.]

Whistance, G. R. and Threlfall, D. R. 1970. Biosynthesis of phytoquinones. Homogentisic acid. *J. Biochem. 117*:593–600. [Chapter 13.]

Whitaker, N. S. and Klein, H. P. 1977. Low oxygen levels and the palmitoyl CoA desaturase of yeast. Relation to primitive biological evolution. In C. Ponnamperuma, ed., *Chemical Evolution of the Early Precambrian*, New York, Academic Press, 211–214. [Chapter 13.]

White, B. 1977. Silicified evaporites from the middle Proterozoic Altyn Formation, lower Belt Supergroup of Montana. *Geol. Soc. Am. Abstr. Progr. 9*:1223–1224 (Abstr.). [Chapter 11.]

White, D. 1980. A theory for the origin of a self-replicating chemical system. I: Natural selection of the autogen from short random oligomers. *J. Mol. Evol. 16*:121–147. [Chapter 4.]

White, D. and Waring, G. A. 1963. Volcanic emanations. In *Data of Geochemistry, U.S. Geol. Surv. Prof. Pap. 440-K*:1–29. [Chapters 4, 11.]

White, D. C.; Bryant, M. P.; and Caldwell, D. R. 1962. Cytochrome-linked fermentation in *Bacteriodes ruminicola. J. Bacteriol. 84*:822–828. [Chapter 6.]

White, R. E. and Coon, M. J. 1980. Oxygen activation by cytochrome P-450. *Ann. Rev. Biochem. 49*:315–356. [Chapter 13.]

Whittaker, A. G.; Lewis, R. S.; and Anders, E. 1980. Carbynes: Carriers of primordial noble gases in meteorites. *Science 209*:1512–1514. [Chapter 4.]

Whittaker, R. H. and Likens, G. E. 1973. Carbon in the biota. In C. M. Woodwell and E. V. Pecan, eds., *Carbon in the Biosphere*, Washington, D.C., U.S. Atomic Energy Commission, 281–302. [Chapter 7.]

Whittenbury, R. and Kelly, D. P. 1977. Autotrophy—a conceptual phoenix. *Symp. Soc. Gen. Microbiol. 27*:121–149. [Glossary II.]

Wilde, P. and Berry, B. N. 1980. Progressive ventilation of the oceans—potential for return to anoxic conditions in the post-Paleozoic. Unpublished manuscript, 32 pp. [Chapter 10.]

Williams, G. E. 1968. Torridonian weathering and

its bearing on Torridonian palaeoclimate and source. *Scot. J. Geol. 4*:164–184. [Chapter 11.]
———. 1975. Late Precambrian glacial climate and the Earth's obliquity. *Geol. Mag. 112*:441–544. [Chapter 11.]
Williams, N. 1979. The timing and mechanisms of formation of the Proterozoic stratiform Pb-Zn and related Mississippi Valley-type deposits at McArthur River, N. T., Australia. *Soc. Econ. Geol. Am. Inst. Min. Eng., Joint Meeting* (New Orleans), *Preprint 79–51*, 15 pp. [Appendix I.]
———. 1980. Precambrian mineralisation in the McArthur-Cloncurry region, with special reference to stratiform lead-zinc deposits. In R. A. Henderson and P. J. Stephenson, eds., *The Geology of North-Eastern Australia*, Canberra, Austr. Acad. Sci., 89–107. [Appendix I.]
Williams, N.; Cohen, Y.; von Gehlen, K.; Haack, U.; Hallberg, R. O.; Kaplan, I. R.; Nielsen, H.; Sangster, D.; Trudinger, P. A.; and Trüper, H. G. 1982. Stratified sulfide deposits. In H. D. Holland and M. Schidlowski, eds., *Mineral Deposits and the Evolution of the Biosphere*, Dahlem Konferenzen-Dahlem Workshop Rept. 16, Berlin, Springer, 275–286. [Chapter 10.]
Williams, R. J. P. and Da Silva, J. J. R. F. 1978. High redox potential chemicals in biological systems. In R. J. P. Williams and J. J. R. F. Da Silva, eds., *New Trends in Bio-inorganic Chemistry*, London, Academic Press, 121–171. [Chapter 13.]
Wilson, J. R.; Bickle, M. J.; Hawkesworth, C. J.; Martin, A.; Nisbet, E. G.; and Orpen, J. L. 1978. Granite-greenstone terrains of the Rhodesian Archaean craton. *Nature 271*:23–27. [Chapter 8; Appendix I.]
Wilson, L. and Head, J. W. 1980. Ascent and eruption of basaltic magma on the Earth and Moon. *J. Geophys. Res. 85*, in press. [Chapter 4.]
Wiman, C. 1894. Ein präkambrisches Fossil. *Bull. Geol. Inst. Upsala 2* (*Paleont. Notiz 1*):109–113, 116, pl. 5. [Chapter 1.]
Wimpenny, J. W. T. 1969. Oxygen and carbon dioxide regulation of microbial growth and metabolism. *Symp. Soc. Gen. Microbiol. 19*: 169–197. [Chapter 13.]
Wimpenny, J. W. T. and Samah, O. A. 1978. Some effects of oxygen on the growth and physiology of *Selenomonas ruminantium*. *J. Gen. Microbiol. 108*:329–332. [Chapter 13.]
Windley, B. F. 1977. *The Evolving Continents*, New York, Wiley, 385 pp. [Chapters 3, 10.]
———. 1979. Tectonic evolution of continents in the Precambrian. *Episodes, Internat. Union Geol. Sci. Newsletter 4*:12–16. [Chapter 3.]
Winter, H. de la R. 1963. Algal structures in the sediments of the Ventersdorp System. *Geol. Soc. S. Afr. Trans. 66*:115–121. [Chapter 8.]
———. 1976. A lithostratigraphic classification of the Ventersdorp succession. *Geol. Soc. S. Afr. Trans. 79*:31–48. [Chapter 8.]
Wise, D. U. 1974. Continental margins, freeboard and the volumes of continents and oceans through time. In C. A. Burk and C. L. Drake, eds., *The Geology of Continental Margins*, New York, Springer, 45–48. [Chapter 10.]
Witkin, E. M. 1966. Radiation-induced mutations and their repair. *Science 152*:1345–1353. [Chapter 15.]
Wlotzka, F. 1972 Nitrogen. In K. H. Wedepohl, ed., *Handbook of Geochemistry, Vol. 2*, Berlin, Springer, chap. 7. [Chapter 7.]
Woese, C. R. 1977. A comment on methanogenic bacteria and the primitive ecology. *J. Mol. Evol. 9*:369–371. [Chapter 4.]
Woese, C. R. and Fox, G. E. 1977. Phylogenetic structure of the prokaryotic domain: The primary kingdoms. *Proc. Natl. Acad. Sci. USA 74*:5088–5090 [Chapter 15; Glossary I.]
Wolfe, R. S. 1979. Methanogenesis. In J. R. Quayle, ed., *Microbial Biochemistry, Internat. Rev. Biochem. 21*:270–300. [Glossary II.]
Wolery, T. J. and Sleep, N. H. 1976. Hydrothermal circulation and geochemical flux at mid-ocean ridges. *J. Geol. 64*:249–275. [Chapter 10.]
Wolman, Y.; Haverland, W. J.; and Miller, S. L. 1972. Non-protein amino acids from spark discharges and their comparison with Murchison meteorite amino acids. *Proc. Natl. Acad. Sci. USA 69*:809–811. [Chapter 4.]
Wood, W. A. 1961. Fermentation of carbohydrates and related compounds. In I. C. Gunsalus and R. Y. Stanier, eds., *The Bacteria, Vol. 2, Metabolism*, New York, Academic Press, 59–149. [Chapter 6.]
Woodwell, G. M.; Whittaker, R. H.; Reiners, W. A. Likens, G. E.; Delwiche, C. C.; and Botkin, D. D. 1978. The biota and the world carbon budget. *Science 199*:141–146. [Chapter 7.]
Wong. J. R. F. and Bronskill, P. M. 1979. Inadequacy of prebiotic synthesis as origin of proteinous amino acids. *J. Mol. Evol. 13*:115–125. [Chapter 4.]

Wong, W. W.; Benedict, C. R.; and Kohel, R. J. 1979. Enzymic fractionation of the stable carbon isotopes of carbon dioxide by ribulose-1.5-bisphosphate carboxylase. *Pl. Physiol. 63*:852–856. [Chapter 7.]

Wong, W. W. and Sackett, W. M. 1978. Fractionation of stable carbon isotopes by marine phytoplankton. *Geochim. Cosmochim. Acta 42*:1809–1815. [Chapter 7.]

Wong, W. W.; Sackett, W. M.; and Benedict, C. R. 1975. Isotope fractionation in photosynthetic bacteria during carbon dioxide assimilation. *Pl. Physiol. 55*:475–479. [Chapter 7.]

Wszolek, P. C.; Simoneit, B. R.; and Burlingame, A. L. 1973. Studies of magnetic fines and volatile rich soils: Possible meteoritic and volcanic contributions to lunar carbon and light element chemistry. *Proc. 4th Lunar Sci. Conf.*:1693–1706. [Chapter 4.]

Wyllie, P. J. 1977. Mantle fluid compositions buffered by carbonates in periodotite-CO_2-H_2O. *J. Geol. 85*:187–207. [Chapter 4.]

———. 1978. The effect of H_2O and CO_2 on planetary mantles. *Geophys. Res. Lett. 5*:440–442. [Chapter 4.]

Xü, R. and Chu, W. 1979. Fossil microbiota of the Anshan Group and preliminary investigation on the genesis of the Anshan iron ore. In *Selected Works for a Symposium on Iron-Geology of China, Sponsored by Academia Sinica, 1977*, Beijing, Science Press, 33–38 (in Chinese with English abstract). [Chapter 14.]

Yapp, C. J. and Epstein, S. 1977. Climatic implications of D/H ratios of meteoric water over North America (9,500–22,000 B.P.) as inferred from ancient wood cellulose C-H hydrogen. *Earth Planet. Sci. Lett. 34*:333–350. [Chapter 7.]

Yeh, H. W. and Epstein, S. 1978. D/H and O^{18}/O^{16} ratios of Precambrian cherts of Swaziland sequence and others. *Am. Geophys. Union Trans. 59*:386–387 (Abstr.). [Chapter 3.]

Yen, H.-C. and Marrs, B. 1977. Growth of *Rhodopseudomonas capsulata* under anaerobic dark conditions with dimethyl sulfoxide. *Arch. Biochem. Biophys. 181*:411–418. [Chapter 6.]

Yin, L. 1979. Microflora from the Anshan Group and the Liaohe Group in E. Liaoning with its stratigraphic significance. In *Selected Works for a Symposium on Iron-geology of China, Sponsored by Academia Sinica*, 1977, Beijing, Science Press, 39–58 (in Chinese with English abstract). [Chapter 14.]

Yin, L. and Li, Z. 1978. Pre-Cambrian microfloras of southwest China, with reference to their stratigraphical significance. *Mem. Nanjing Inst. Geol. Paleont. Acad. Sinica 10*:41–118 (in Chinese with English abstract). [Chapter 1.]

Young, G. M. 1973a. Tillites and aluminous quartzites as possible time markers for middle Precambrian (Aphebian) rocks of North America. In G. M. Young, ed., *Huronian stratigraphy and sedimentation, Geol. Assoc. Can. Spec. Pap. 12*:97–127. [Chapter 11.]

———. 1976. Iron-formation and glaciogenic rocks of the Rapitan Group, Northwest Territories, Canada. *Precambrian Res. 3*:137–158. [Chapter 11.]

Young, G. M., ed. 1973b. *Huronian stratigraphy and sedimentation, Geol. Assoc. Can. Spec. Pap. 12*, 271 pp. [Chapter 11.]

Young, G. M.; Jefferson, C. W.; Delaney, G. D.; and Yeo, G. M. 1979. Middle and late Proterozoic evolution of the northern Canadian Cordillera and Shield. *Geology 7*:125–128. [Chapter 11.]

Yung, Y. L. and McElroy, M. B. 1979. Fixation of nitrogen in the prebiotic atmosphere. *Science 203*:1002–1004. [Chapters 4, 11.]

Yung, Y. L. and Pinto, J. P. 1978. Primitive atmosphere and implications for the formation of channels on Mars. *Nature 273*:730–732. [Chapter 4.]

Zartman, R. E. and Stern, T. W. 1967. Isotopic age and geologic relationship of the Little Elk Granite, northern Black Hills, South Dakota. In *Geological Survey Research 1967, U.S. Geol. Surv. Prof. Pap. 575-D*:157–163. [Chapter 11.]

Zelitch, I. 1971. *Photosynthesis, Photorespiration, and Plant Productivity*, New York, Academic Press, 347 pp. [Chapter 11.]

Zeschke, G. 1961. Transportation of uraninite in the Indus River, Pakistan. *Geol. Soc. S. Afr. Trans. 63*:87–94. [Chapter 11.]

Zimen, K. E. P.; Offerman, P.; and Hartmann, G. 1977. Source functions of CO_2 and future CO_2 burden in the atmosphere. *Z. Naturforsch. 32A*:1544–1554. [Chapter 5.]

ZoBell, C. E. 1957. Bacteria. In H. S. Ladd, ed., *Treatise on Marine Ecology and Paleoecology, Vol. 2, Geol. Soc. Am. Mem. 67*:693–698. [Chapter 15.]

Zumberge, J. E. and Nagy, B. 1975. Alkyl substituted cyclic ethers in 2,300 M year-old Transvaal algal stromatolite. *Nature 255*:695–696. [Chapter 5.]

Zumberge, J. E.; Sigleo, A. C.; and Nagy, B. 1978. Molecular and elemental analyses of the carbonaceous matter in the gold and uranium bearing Vaal Reef carbon seams, Witwatersrand Sequence. *Minerals Sci. Engl. 10*:223–246. [Chapter 5.]

SUBJECT INDEX

Library of Congress Cataloging in Publication Data

Main entry under title:

Earth's earliest biosphere.

Bibliography: p.
Includes index.
1. Paleontology—Pre-Cambrian.
I. Schopf, J. William,
1941–
QE724.E27 1983 560'.1'71 82-61383
ISBN 0-691-08323-1
ISBN 0-691-02375-1 (pbk.)